International Deep-Drawing Research Group Conference (IDDRG 2022)

IOP Conference Series: Materials Science and Engineering
Volume 1238

Lorient, France
6 - 10 June 2022

Editors:

Sandrine Thuillier
Vincent Grolleau
Herve Laurent

ISBN: 978-1-7138-5439-5
ISSN: 1757-8981

Printed from e-media with permission by:

Curran Associates, Inc.
57 Morehouse Lane
Red Hook, NY 12571

Some format issues inherent in the e-media version may also appear in this print version.

Institute of Physics
Dirac House, Temple Back
Bristol BS1 6BE UK

Phone: 44 1 17 929 7481
Fax: 44 1 17 920 0979

techtracking@iop.org

Additional copies of this publication are available from:

Curran Associates, Inc.
57 Morehouse Lane
Red Hook, NY 12571 USA
Phone: 845-758-0400
Fax: 845-758-2633
Email: curran@proceedings.com
Web: www.proceedings.com

TABLE OF CONTENTS

IOP Conf. Series: Materials Science and Engineering 1238 (2022) 011001 doi:10.1088/1757-899X/1238/1/011001

41st International Deep Drawing Research Group (IDDRG) conference

International Deep Drawing Research Group (IDDRG) annual conference was held in Lorient, France from 6 to 10 June 2022. The conference was co-organised by Université de Bretagne Sud and ArcelorMittal France and was chaired by Prof Sandrine Thuillier. This was a fantastic opportunity to gather the IDDRG community after the pandemics, though some colleagues were still not able to travel due to COVID-19.

Conference topics

The topics covered by IDDRG 2022 are the core topics related to forming and mechanical joining of metallic sheets and organized in the following topics:

- material constitutive modelling
- warm and hot forming
- damage and fracture in sheet metal forming
- shear cutting, blanking and sheared edged formability (MS). *Coordinator Prof. Dr.-Ing. Wolfram Volk (Technical University of Münich) - Co-organisers: Dr. Christoph Hartmann, Florian Steinlehner*
- necking, instability and formability limit prediction
- non-conventional testing methods for appropriate material characterization (MS). *Coordinators: Dr Lionel Léotoing (INSA Rennes) and Prof. Hervé Laurent (Université Bretagne Sud)*
- tribology, friction and wear in forming
- AI in sheet metal forming (MS). *Coordinators: Prof. Dr.-Ing. Mathias Liewald and Dr. sc. techn. Celalettin Karadogan (Stuttgart Universität)*
- intelligent tools for sheet metal forming
- advanced simulation methods for sheet metal forming

Moreover, a special session was organized in the honor of Prof Dr Pavel Hora, coordinated by Prof. Dirk Mohr and Prof Vincent Grolleau, ETH Zürich.

Three topics were organized as thematic mini-symposia (MS), cf. the list above, leading to a semi-plenary talk in parallel sessions.

Scientific presentations were possible either with a paper published in the conference proceedings, with only an abstract and an oral presentation. Moreover, a special issue in relation to IDDRG in Mechanics & Industry was organized, on the topic "What's up ... in forming and mechanical joining of sheet metals?". Finally, eleven plenary talks were presented by internationally recognized researchers in academy and industry, listed below in alphabetical order:

- Prof. Stefania Bruschi, University of Padova, Italy, *Sheet metal forming at elevated temperatures*
- Dr Bart Carleer, Autoform Engineering, *The Autoform Pareto Principle effectively defines the Digital Process Model*
- Dr Lander Galdos, Mondragon University, Spain, *Friction and contact modelling for sheet and tube forming processes. Review of experimental characterization techniques and industrial examples*
- Dr Lars Greve, Volkswagen AG, *AI-based surrogate models for crash simulation applications*
- Prof. em. Dr Pavel Hora, INSPIRE AG, *Impact of CO_2 constraints on industrial sheet metal forming*
- Prof Kaan Inal, University of Waterloo, Canada, *An advanced artificial intelligence framework to simulate metal forming processes*
- Prof Myoung-Gyu Lee, Seoul National University, *Significance of microstructure effect in constitutive modeling for sheet metal forming simulations and applications*
- Prof Dr Ing Marion Merklein, FAU, Germany, *Mechanical joining of high-strength multi-materials systems – trends and innovation*
- Prof Dirk Mohr, ETH Zürich, *Creating big data on sheet metal plasticity and failure*
- Prof Fabrice Pierron, University of Southampton & MatchID, *The virtual fields method for anisotropic plasticity: a Material Testing 2.0 perspective*
- Mr Stéphane Tondo, ArcelorMittal, *Steel as a carbon neutral material of choice: towards local decarbonized steel production in Europe*

Patronage
IDDRG 2202 was held under the patronage of IDDRG (with a new website, have a look!), and also the French Association of Mechanics (AFM) and the French Society for Metallurgy and Materials (SF2M) and especially the committee TOMMI. The year 2022 was "special" for mechanics in France, with the promotion of this scientific field I, see the nice picture below.

IOP Conf. Series: Materials Science and Engineering 1238 (2022) 011001 doi:10.1088/1757-899X/1238/1/011001

Committees
The composition of the local organizing committee and of the scientific committee are given below.

Local organizing committee

Prof Sandrine Thuillier, Université Bretagne Sud

Prof Hervé Laurent, Université Bretagne Sud
Prof Vincent Grolleau (Université Bretagne Sud/on leave at ETH Zürich, Switzerland)
Prof Laurent Mahéo, Académie Militaire Saint-Cyr Coëtquidan
Dr Bertrand Galpin, Académie Militaire Saint-Cyr Coëtquidan
Dr Lionel Léotoing, INSA Rennes
Dr Bernardete Coelho, Université Bretagne Sud
Dr Ahmed Kacem, Université Bretagne Sud
Dr Nathan Demazel, Université Bretagne Sud

International scientific committee

Asnafi N. (Sweden)	Korkolis Y. (USA)	Rolfe B. (Australia)
Banabic D. (Romania)	Kräusel V. (Germany)	Saenz de Argandoña E. (Spain)
Barlat F. (South Korea)	Kuwabara T. (Japan)	Santos A. (Portugal)
Behrens B.A. (Germany)	Langerak N. (The Netherlands)	Stoughton T. (USA)
van den Boogaard T. (The Netherlands)	Lee M.G. (Korea)	Thibaud S. (France)
Clarke K. (USA)	Léotoing L. (France)	Thuillier S. (France)
Duroux P. (France)	Liewald M. (Germany)	van Tyne C. (USA)
Gantar G. (Slovenia)	Lukács Z. (Hungary)	Volk W. (Germany)
Ghiotti A. (Italy)	Merklein M. (Germany)	Wagner L. (Austria)
Green D. (Canada)	Mohr D. (Switzerland)	Worswick M. (Canada)
Hora P. (Switzerland)	Narasimhan K. (India)	Yoon J.W. (South Korea/Australia)
Izuka E. (Japan)	Oliveira M. (Portugal)	Yoshida F. (Japan)
Karadogan C. (Germany)	Penazzi L. (France)	Zhang S.H. (China)
Kim H. (USA)	Peura P. (Finland)	

Acknowledgements

The organizing committee wishes to express their most sincere gratitude to all who supported IDDRG 2022, including reviewers, mini-symposium coordinators and co-organisers, plenary speakers, scientific committee and local organizing committee members, conference staff, conference sponsors (cf. page 5) and exhibitors.

Special thanks to Prof Frederic Barlat (Postech) and Prof Jeong Whan Yoon (KAIST), from South Korea, for supporting actively the project of having IDDRG hosted in Brittany.

Editor

Prof Sandrine Thuillier
IRDL – Université Bretagne Sud

Co-editors

Prof Vincent Grolleau
IRDL – Université Bretagne Sud, on leave at ETH Zürich

Prof Hervé Laurent
IRDL – Université Bretagne Sud

IOP Conf. Series: Materials Science and Engineering 1238 (2022) 011001 doi:10.1088/1757-899X/1238/1/011001

Sponsors

IOP Conf. Series: Materials Science and Engineering 1238 (2022) 011002 doi:10.1088/1757-899X/1238/1/011002

Peer Review Statement

All papers published in this volume have been reviewed through processes administered by the Editors. Reviews were conducted by expert referees to the professional and scientific standards expected of a proceedings journal published by IOP Publishing.

- **Type of peer review:** Single Anonymous
- **Conference submission management system:** Morressier
- **Number of submissions received:** 94
- **Number of submissions sent for review:** 94
- **Number of submissions accepted:** 88
- **Acceptance Rate (Submissions Accepted / Submissions Received × 100):** 94.7
- **Average number of reviews per paper:** 2.02
- **Total number of reviewers involved:** 37
- **Contact person for queries:**
  Name: Prof Sandrine Thuillier
  Email: sandrine.thuillier@univ-ubs.fr
  Affiliation: Université Bretagne Sud

IOP Conf. Series: Materials Science and Engineering 1238 (2022) 012001 doi:10.1088/1757-899X/1238/1/012001

Identification Protocol of Yoshida-Uemori Hardening Model

X Lemoine[1*], Q Chichery[2], J M Devin[2] and P Duroux[2]

[1]Global R&D ArcelorMittal Maizières, voie Romaine, BP30320, F-57283 Maizières-lès-Metz, France
[2]Global R&D ArcelorMittal Montataire, 1 Route de Saint-Leu, F-60160 Montataire, France

xavier.lemoine@arcelormittal.com

Abstract. The Yoshida-Uemori model is a combined hardening model coupled with the elastic modulus evolution which depends on the equivalent plastic strain. A lot of publications show that it allows to improve the springback prediction compared to a classical isotropic hardening model. It is implemented in the most used FEA stamping codes (Autoform®, Ls-Dyna® and Pamstamp®) with some variations and it is more and more used in automotive industry. ArcelorMittal decided to provide material cards to its customers. To achieve this goal, ArcelorMittal has at one's disposal the equipment to perform reverse shear tests and hysteresis loops for all steel grades. An in-house methodology has been developed to identify the Yoshida-Uemori adjusting parameters in a very precise and robust manner. The material cards are produced for each FEA stamping code, considering the specificities of the implementations.

1. Introduction

A lot of authors recommend the use of combined hardening models to improve prediction of the springback [1-5]. The combined hardening Yoshida-Uemori model [6-8] is implemented in the most used FEA stamping codes (Autoform®, Pamstamp® and LS-Dyna®) with occasionally some variations [9, 10]. This model describes the specificities of stress-strain response to cyclic hardening (Figure 1):

- The plastic strain dependence of the elastic modulus.
- The Bauschinger effect characterized by early re-yielding strength, transient effect and work-hardening stagnation. The work-hardening stagnation appears later when the Ultimate Tensile Strength increases, and the length of the plateau increases with the pre-strain level.
- The permanent softening, characterized by a stress offset observed after the transient stage, which increases with the pre-strain level.

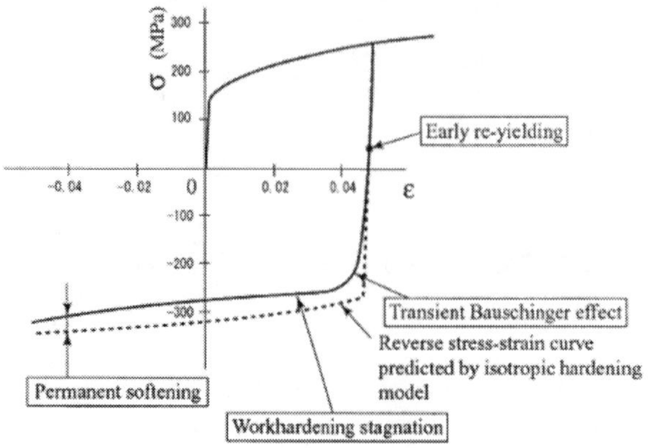

Figure 1. Description of the different stages of Bauschinger effect on a Stress-strain curve in tension-compression [8]

2. Yoshida-Uemori models

2.1. *Combined hardening model.*

The Yoshida-Uemori hardening model, presented originally in 2002 [6-8], involves three different surfaces in the plane-stress space. A pure kinematic hardening model is imposed for the Yield surface with a combined hardening model for the Bounding surface. As the evolutions of Yield and Bounding surfaces are not enough to model permanent softening and work-hardening stagnation, an additional surface has been introduced. In 2008, Shi et al. [9] suggested to replace the Voce law by Swift law to improve the prediction of formability. In 2015, Yoshida et al. [10] added the possibility to use a linear combination between Swift and Voce isotropic hardening laws.

2.2. *Elastic modulus evolution model.*

The Yoshida-Uemori combined hardening model is coupled with a non-linear elastic modulus model. The evolution of the elastic modulus is a function of the plastic pre-strain level (equation (1)).

$$E(p) = E_0 - (E_0 - E_a)(1 - exp(-\xi p)) \quad (1)$$

where E_0 is initial Young modulus; E_a and ξ are two material parameters, p is equivalent plastic strain.

3. ArcelorMittal experimental tests needed for the identification of material parameters

3.1. *Reverse shear tests needed for the identification of combined hardening model*

Many reverse experimental tests (tension-compression, compression-tension, reverse shear or bending-unbending tests) allow to evaluate the material combined hardening by imposing a displacement in one direction and then imposing another displacement in the opposite direction. The advantages and disadvantages of each reverse test are detailed in [12].

We advise the reverse shear test for several reasons. First because there is no limitation due to diffuse necking like in compression-tension or tension-compression tests (Figure 2). Next, the buckling limit of very thin sheets is observed later compared to a compression loading. Last, very large plastic deformation (0.3~0.5 in equivalent plastic strain) and pre-strain levels can be reached, which are representative of strain levels in stamping. And finally, our reserve shear equipment allows to determine the behavior of all steel grades from 250 to 2000 MPa and from 0.5 to 4 mm sheet thickness.

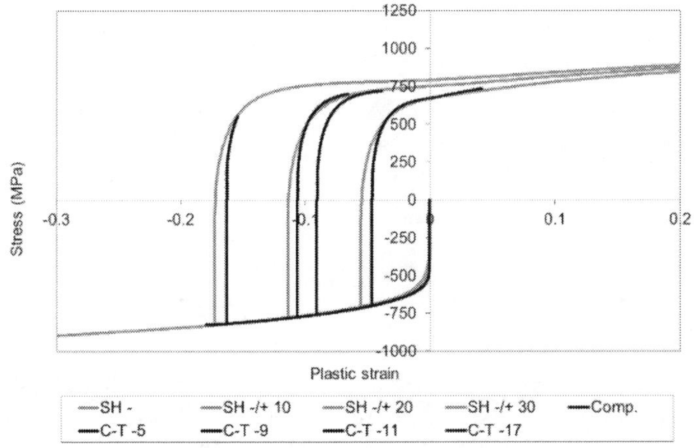

Figure 2. Comparison flow curves for a reverse shear test (SH +/-) (orange curves) vs compression-tension test (C-T) (black curves) for DP600 [12]

A shear test is performed by imposing a displacement on a piston, while maintaining constant the gap between grip wedges. Since the area of interest in the simple shear specimen is very small, a high-resolution optical technique must be used to determine the shear plastic strain γ in the middle area of the specimen. This approach is a reliable way to avoid boundary effects creating normal stresses at the vicinity of specimen corners. To ensure accurate measurements, it is not recommended to consider pre-strain levels below 0.04.

The identification tools are programmed with tension-compression data inputs. Then, it is necessary to generate an equivalence between tension-compression and the shear flow curve (red curve on figure 3). A coefficient χ is determined by an inverse method to match the equivalent shear curve (green curve) with the uniaxial tensile curve (blue curve).

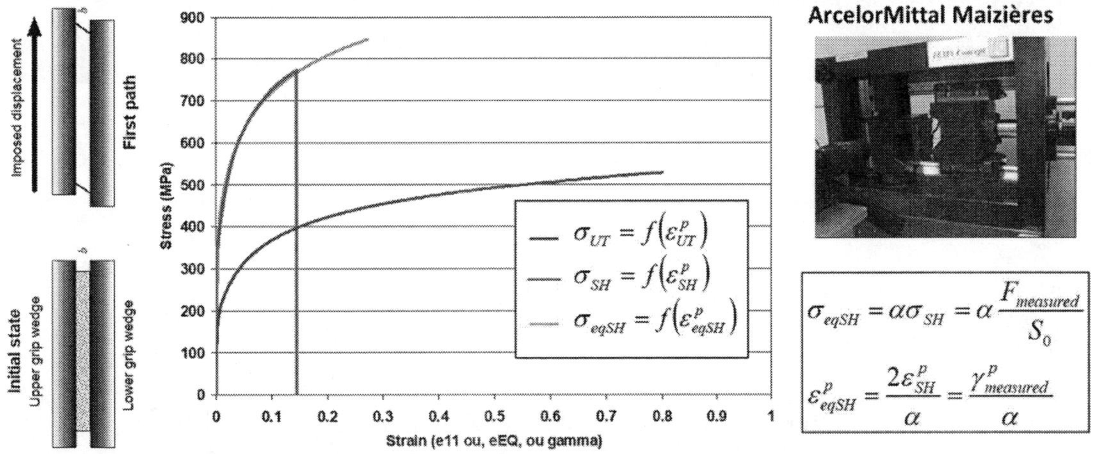

Figure 3. Equivalent tension-compression stress-strain determination.

3.2. Hysteresis loop used for the identification of elastic modulus evolution model

The adjusting parameters governing the evolution of the elastic modulus are identified thanks to hysteresis loops conducted between 0.2% of plastic deformation and uniform elongation (Uel) every 1% on ISO 12.5x50 uniaxial tensile specimens (Figure 4a). The strain is measured with a high-resolution Zwick® digiClip extensometer as per Annex G of ISO 6892-1 standard. For each unloading/reloading loop, the elastic modulus is considered as the slope crossing points A (maximal stress at the begin of the loop) and B (minimal stress of loop) (figure 4b). The slope is deducted from the linear regression on unloading path between the points A and B.

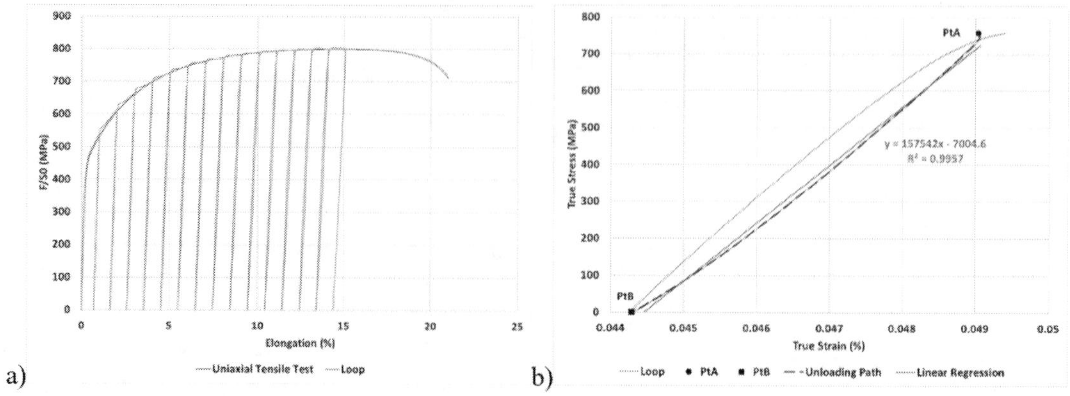

Figure 4. Determination of the pre-strain dependence of the elastic modulus for DP780 a) Hysteresis loop (orange curve) conducted every 1% until the end of the uniform elongation b) the elastic modulus is the slope deducted from the linear regression (blue curve) on unloading path (black curve) between points A and B.

4. Methodology to produce material cards according Yoshida-Uemori model for ArcelorMittal steel grades

The identification of adjusting parameters governing the evolution of the elastic modulus is treated separately from the identification of combined hardening parameters defined in Yoshida-Uemori model.

4.1. Identification of the evolution of the elastic modulus

The two adjusting parameters E_a and ξ are fitted according to a gradient method (red curve on figure 5) in order to get the better correspondence between equation (1) and experimental results derived from hysteresis loops tests (blue points).

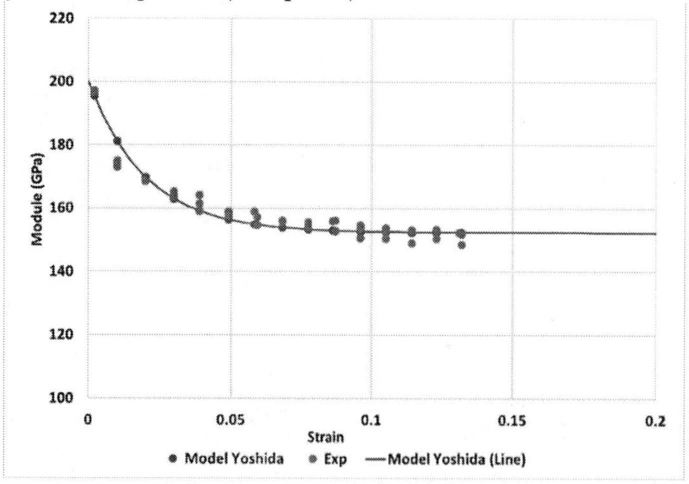

Figure 5. Experimental points (blue dots) versus Elastic modulus evolution model (red line) for a DP780 steel ($E_0 = 200GPa$, $E_a = 152GPa$ and $\xi = 50.03$)

4.2. Identification of combined hardening parameters defined in Yoshida-Uemori model

We have developed an in-house methodology to identify Yoshida-Uemori material parameters, so that the model fits at best: 1) the experimental curves (one uniaxial tensile load and three uniaxial compressive loads are considered at the same time), 2) the following mechanical characteristics of the uniaxial tensile test : Yield Stress (YS), Ultimate Tensile Strength (UTS), Uniform Elongation (UEL%) and hardening coefficient (n).

The central component of the methodology is the direct solving of Yoshida-Uemori equations for the case of uniaxial loads. Then, fitting of the experimental curves and mechanical characteristics are performed by optimization technics available in LS-OPT [13] using an inverse approach.

When compared to MatPara® software developed by Prof. Yoshida for the identification of the Yoshida-Uemori material parameters, the present methodology considers: 1) several experimental curves at the same time, 2) all the combined hardening parameters at the same time (especially parameters for Swift and Voce laws and the mixing coefficient between these two laws all together).

4.2.1. Direct solving of the equations for the case of uniaxial loads

The solving of the Yoshida-Uemori model equations for the "one dimension" case has been developed to calculate the model variables (α, β, R and σ – see table 1) for a given set of material parameters (Y, B, C_1, C_2, m, R_{sat}, b, K, ε_0, n, μ – see table 1).

Table 1. Yoshida-Uemori model equations for the "one dimension" case.

Equations	$R_{voce}(p) = R_{Sat}\{1 - exp(-mp)\}$	(2)
	$R_{swift}(p) = K\{(\varepsilon_0 + p)^n - \varepsilon_0^n\}$	(3)
	$R(p) = \mu R_{swift}(p) + (1 - \mu)R_{voce}(p)$	(4)
	$\dot{\theta}(p) = C\left[a - \sqrt{a\lvert\theta(p)\rvert}\right]\dot{p}$ with $a = B + R - Y$	(5)
	where $C = C_1$ when $Max(\theta(p)) < B - Y$ and $C = C_2$ otherwise	(6)
	$\dot{\beta}(p) = m(b - \beta(p))\dot{p}$	(7)
	$\alpha(p) = \theta(p) + \beta(p)$	(8)
	$Y^2 = [\sigma(p) - \alpha(p)]^2$	(9)
Variables	$\alpha(p)$[MPa]	Translation of the Yield Surface (blue cross in figure 6)
	$\beta(p)$[MPa]	Translation of the Bounding surface (orange circle in figure 6)
	$R(p)$[MPa]	Expansion of the Bounding surface (red square in figure 6)
	$\sigma(p)$[MPa]	True stress (large black dot in figure 6)
Parameters	Y [MPa]	Initial size of Yield surface
	B [MPa]	Initial size of Bounding surface
	C_1 [MPa]	Rate of translation of the Yield surface, with respect to the centre of the Bounding surface, until plastic strain is below a threshold value calculated in the model using parameters Y and B
	C_2 [MPa]	Rate of translation of the Yield surface, with respect to the center of the Bounding surface when the threshold value of plastic strain has been exceeded $(C_1 > C_2)$
	EPREF	Additional parameter in PamStamp. The equation (6) of the Yoshida-Uemori model allows to calculate the threshold value of plastic strain to switch from C_1 to C_2. But ESI asks the user to choose a value of plastic strain for the threshold value to switch from C_1 to C_2, called EPREF in the PamStamp material card
	m	Parameter of Voce law (used for expansion of the Bounding surface) and rate of translation of the Bounding surface
	R_{Sat} [MPa]	Saturated value for expansion of the Bounding surface, parameter of Voce law
	b [MPa]	Saturated amount of translation of the Bounding surface
	K [MPa]	Parameter for Swift hardening law (used for expansion of the Bounding surface)
	ε_0	Parameter for Swift hardening law (used for expansion of the Bounding surface)

n	Parameter for Swift hardening law (used for expansion of the Bounding surface)
μ	Mixing factor between Swift and Voce (combined laws for expansion of the Bounding surface)

The solving time is very quick. Results are illustrated at two states; at the end of a tensile load in figures 6.a and during a compressive load in figures 6.b. In both cases, the position of the Yield surface, Bounding surface and symbols for the model variables α, β, R and σ are illustrated (left figures) with the corresponding evolution of those variables (right figures).

a)
b)

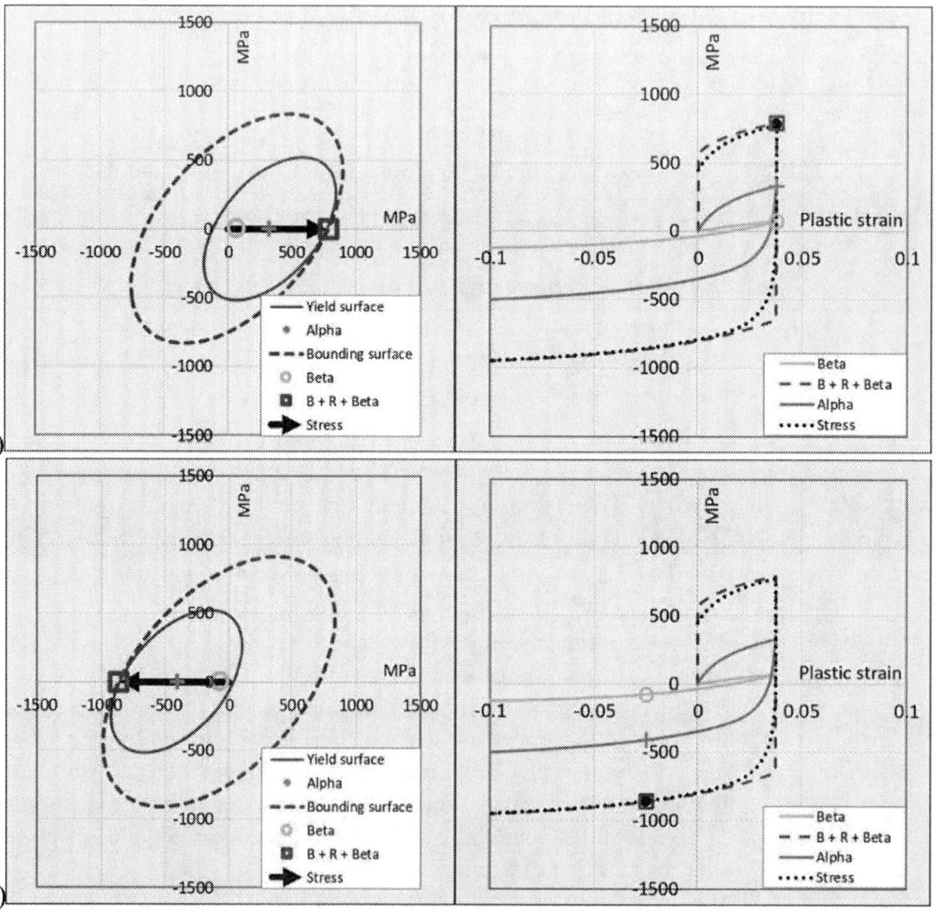

Figure 6. Yoshida-Uemori combined hardening model at two states: a) end of a tensile load and b) during a compressive load.

Note that the two parameters of the Yoshida-Uemori model for work hardening stagnation (h and r0) have not been included in the solving for two reasons. First, the work hardening stagnation is negligible for AHSS steels. Second, there are some discrepancies in the work hardening stagnation implementation: a) between MatPara and PamStamp® (different results when r0=0), b) between MatPara and LS-DYNA® (r0 is not required in this stamping software). The ways of implementation of work hardening stagnation in these two FEA codes must be clarified with the support of the stamping editors before introducing these two work-hardening stagnation parameters in material cards.

At the opposite, the direct solving includes the parameter EPREF, which is required in the PamStamp® material card. In PamStamp®, it is the threshold value of plastic strain to switch from C_1 to C_2. Direct solving can be performed for two options: a) calculation of EPREF being the value of p for which Max(θ(p)) exceeds B-Y for the first time (according equation (6) in table 1), b) the threshold value EPREF is chosen by the user as an input value and equation (6) is replaced by $C = C_1$ when p < EPREF, $C=C_2$ otherwise.

The direct solving results have been validated for several sets of material parameters: same results as with MatPara software and as with LS-Dyna® and Pamstamp® simulations of tensile-compressive uniaxial loads.

4.2.2. Inverse approach developed to fit the experimental curves and the mechanical characteristics of tensile test

An inverse approach has been developed with LS-OPT software which proposes a lot of optimization methods and algorithms for material parameters identification.

A multi-objective function to be minimized has been defined, which includes 8 objectives. Four objectives deal with the tensile test material properties (YS, UTS, UEL% and *n*): the calculated material properties must be as close as possible to experimental values. Four other objectives deal with the minimization of gaps between the four experimental and predicted stress-strain curves.

Several optimization technics available in LS-OPT (Genetic Algorithm, Adaptive Simulated Annealing, ...) and several metamodels (Polynomial, Radial Basis Function Network, Feedforward Neural Network, ...) have been tested. The "Adaptive Simulated Annealing" (ASA) optimization algorithm and the "quadratic polynomial" metamodel were selected, as the most suitable for this problem. At each optimization process iteration, direct solving approach is performed for several hundred sets of Yoshida-Uemori material parameters from a DOE (design of experiences) defined by LS-OPT. The metamodel is calculated and the optimization algorithm is run. The design space (minimum and maximum values for the material parameters) of the DOE is automatically updated at each iteration using the "Sequential with Domain Reduction" (SRSM) strategy.

The proposed methodology allows to create material cards for PamStamp® and LS-DYNA® stamping softwares. For Autoform®, the parameters of the simplified kinematic hardening model developed by Autoform (called « AKH » model) [2,3,11] are identified directly from experimental tensile-compressive curves. It is possible to input parameters of Yoshida-Uemori model since version R8 of Autoform®, but the model is not really implemented. It is automatically converted and replaced by the « AKH » model for the FEA simulations. We don't recommend to input the parameters of Yoshida-Uemori model found with our methodology in Autoform®.

5. Very accurate results of our methodology

Two applications are presented for DP780 and Fortiform®1180 steel grades. Very good fitting of experimental curves (figure 7) and tensile test material properties (table 2) are found.

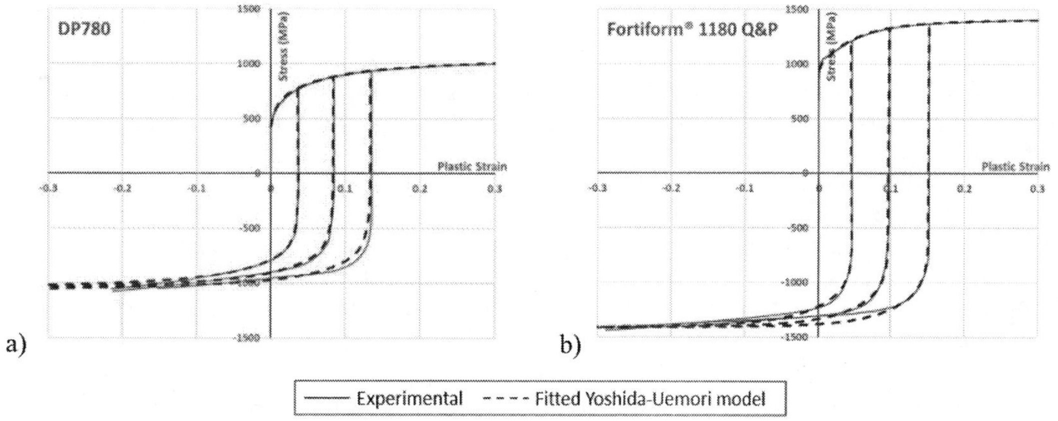

Figure 7. Identification results for a) DP780 and for b) Fortiform®1180 steel grades.

Table 2. Tensile test material properties of DP780 and for Fortiform®1180 steel grades.

	YS [MPa]	UTS [MPa]	UEL [%]	Hardening coefficient n
DP780 - Experimental	495	815	12.7	0.148
DP780 - Fitted	495	816	12.7	0.144
Fortiform®1180 - Experimental	976	1197	9.7	0.118
Fortiform®1180 - Fitted	971	1199	9.7	0.113

There is not a unique solution for the Yoshida-Uemori material parameters. But prediction of springback for an industrial part is very weakly modified by the choice of the solution. We have checked this for 7 industrial parts whose springback has been calculated for 2 solutions that fit both very well the experimental data (table 3). The differences of springback prediction for 6 out of 7 parts are negligible: the cost and the delay for the management of springback in the stamping tools and process setup are the same for the 2 solutions.

Table 3. Comparison of springback prediction for 7 parts with the 2 solutions

Material parameter (PamStamp® name)	Gap between the 2 solutions
B-Y (XSATI)	10%
b (BSAT)	~0%
m (AM)	~0%
C_1 (CX1)	20%
C_2 (CX2)	~0%
ε_0 (YOSHIDA-EPSILON0)	100%
K (YOSHIDA-K)	20%
n (YOSHIDA-N)	20%

Springback prediction for 7 parts	Maximal gap on the part between the 2 solutions
Roof arch reinforcement	0.3 mm
Siderail part	1.8 mm
Side sill closing part	0.7 mm
Door lateral beam	1.1 mm
Rear siderail part	1 mm
Upper A-pillar reinforcement part	0.2 mm
Floor reinforcement part	0.3 mm

The development of the proposed methodology has allowed to get a very good knowledge of the various implementations of Yoshida-Uemori model in the stamping softwares.

First example: results of our direct solving for the 'one dimension' case and exchanges with ESI technical support have helped to confirm recommendations for the values of some material parameters in PamStamp®, and to clarify the user's documentation of this stamping software.

Second example: the writing of the correspondence table between the material parameters of Yoshida-Uemori model for PamStamp® and for LS-DYNA®. This table allows an easy conversion of a LS-DYNA® material card '125' in a PamStamp® material card. Also, this table shows the conditions to make possible the conversion of a PamStamp® material card in LS-DYNA® material card and the cases when this conversion is not possible.

6. Conclusions and prospects

By using the Yoshida-Uemori model in FEA stamping softwares, a better prediction of springback is observed. Then geometrical conformity of stamped parts is got quicker and at a lower cost at our customers.

ArcelorMittal decided to provide its customers with material cards according this model. To achieve this goal, we have at our disposal the equipment to perform reverse shear tests and hysteresis loops for all our steel grades. In 2021, we have developed our own methodology to fit in a very precise manner both the experimental tension-compression curves (equivalent to reverse shear curves) and the mechanical characteristics of tensile test. The central component of the methodology is the direct solving of the equations of Yoshida-Uemori model for the case of uniaxial loads. Fitting is performed by using optimization technics.

ArcelorMittal is now ready to produce accurate Yoshida-Uemori material cards for its products, considering the various implementations of this model in the most used stamping softwares PamStamp®, LS-DYNA® and Autoform®.

References

[1] Firat M, Kaftanoglu B and Eser O 2008 Sheet metal forming analyses with an emphasis on the springback deformation *J. Mater. Process. Technol.* **196** 135

[2] Kubli W, Krasovskyy A and Sester M 2008 Modeling of reverse loading effects including workhardening stagnation and early re-plastification *Proc. of ESAFORM 2008, Int. J. Mater. Form.* ed P Boisse (Lyon France: Springer) **4**

[3] Kubli W, Krasovskyy A and Sester M 2008 Advanced modeling of reverse loading effects for sheet metal forming processes *Proc. of NUMISHEET 2008* ed P Hora (Interlaken Switzerland) 479

[4] Eggersten P A and Mattiasson K 2009 Material Modelling for Accurate Springback Prediction *Proc. of ESAFORM 2009, Int. J. Mater. Form.* (Twente Netherland: Springer) **2** 793

[5] Eggersten P A and Mattiasson K 2009 On the modelling of the bending–unbending behaviour for accurate springback predictions *Int. J. Mech. Sci.* **51** 547

[6] Yoshida F and Uemori T 2002 A model of large-strain cyclic plasticity describing the Bauschinger effect and workhardening stagnation *Int. J. Plast.* **18** 661

[7] Yoshida F, Uemori T and Fujiwara K 2002 Elastic–plastic behavior of steel sheets under in-plane cyclic tension–compression at large strain *Int. J. Plast.* **18** 633

[8] Yoshida F and Uemori T 2003 A model of large-strain cyclic plasticity and its application to springback simulation *Int. J. Mech. Sci.* **45** 1687

[9] Shi M F, Zhu X Z, Xia C and Stoughton T 2008 Determination of nonlinear isotropic/kinematic hardening constitutive parameters for AHSS using tension and compression tests *Proc. of NUMISHEET 2008* ed P Hora (Interlaken Switzerland) 137

[10] Yoshida F, Hamasaki H and Uemori T 2015 Modeling of anisotropic hardening of sheet metals including description of the Bauschinger effect *Int. J. Plast.* **75** 170

[11] Tisza M and Lukàcs M 2014 Theoretical and and Experimental investigation of springback in high strength DP-steels *Prod. Process. Syst.* **7-1** 87

[12] Lemoine X, Durrenberger L, Zhu H and Kergen R 2011 Mixed hardening models: parameters identification on AHSS steels *Proc. of IDDRG 2011* (Bilbao Spain)

[13] Livermore Software Technology Corporation, LS OPT® 2019 A design optimization and probabilistic analysis tool for the engineering analyst *User's Manual*

IOP Conf. Series: Materials Science and Engineering 1238 (2022) 012002 doi:10.1088/1757-899X/1238/1/012002

Transient Hardening and R-value Behavior in Two-step Tension and Loading Reversal for DP980 Sheet

M Kim[1], J Ha[2], J McNally[2] and Y P Korkolis[1]

[1] The Ohio State University, Columbus, Ohio, United States
[2] University of New Hampshire, Durham, New Hampshire, United States

E-mail: korkolis.1@osu.edu

Abstract. The present work deals with transient hardening and R-value behavior in two-step tension and loading reversal for an advanced high strength steel DP980. Mechanical properties under the loading paths were obtained through continuous tension-compression-tension, two-step tension and loading reversal experiments. The Yld2000-2d was employed to describe the initial yielding of DP980 1.0t through conducting monotonic tests. To characterize the transient behavior, a combined isotropic/non-linear kinematic hardening model, based on the 4-term Chaboche model, was selected. The Chaboche model was calibrated with the hardening curve in continuous tension–compression-tension loading. The transient behavior from the two-step tension and the loading reversal tests was then predicted by the model. The model performance was evaluated with the comparison of transient hardening and R-value from experiments and model predictions.

1. Introduction

As design of structures has been increasingly complicated, material points experience complex loading paths during sheet metal forming processes. To understand material behavior under complex loading paths, much research has been conducted regarding changes of yielding, work-hardening rate and R-value (aka plastic strain ratio) under proportional or non-proportional loading conditions. Under the reverse-loading condition, an early-reyielding behavior, so called the Bauschinger effect [1], is observed in subsequent loading. The yield stress in the second loading is generally less than that in the first loading. With work-hardening changes in a low strain, the flow in further deformation tends to be saturated to a smaller value than that in the first loading, i.e. permanent softening [2]. For non-proportional cases, two-step tension, combination of tension in two different orientations, is generally adopted to probe the change in material properties. The yielding, work-hardening rate and R-value change during further strain in the second loading. The amount of change depends on the pre-strain and the angle difference between the loading directions of the first and the second loadings [3, 4].

To take account of changes in hardening behavior in complex loading paths, kinematic hardening models have been developed. The kinematic hardening model adopts the concept of the translation of the yield surface, the center of which is called the back-stress, to describe the Bauschinger effect and the subsequent hardening [5, 6]. For expression of complex behavior in subsequent loading, non-linear kinematic hardening was developed with summation of few non-linear terms for the back-stress [7, 8]. Most researches in modeling of the kinematic hardening under different loading paths were focused on simulating flow stresses in subsequent loading rather than the transient of the R-value.

This study investigates transient hardening and R-value behavior of DP980 sheet in two-step tension and loading reversal. A representative Chaboche kinematic hardening model [7, 8], combined with isotropic hardening, is used to confirm the model performance for simulating the transient behavior.

2. Determination of initial yield locus

The material of this study is a dual-phase advanced high strength steel (AHSS) of DP980 (1 mm thickness). To observe the mechanical responses in monotonic tension, tensile tests were conducted along three different orientations of 0°, 45° and 90° from the rolling direction. Figure 1 shows the obtained stress and R-value behavior with respect to the plastic work. The plastic work (W_p) was calculated by Equation (1). The DP980 sheet of this study shows anisotropic behavior as shown in figure 1(a). The R-value in figure 1(b) was determined by the plastic strain ratio of strain in width and thickness directions as described in Equation (2).

$$W_p = \int \bar{\sigma} \cdot d\bar{\varepsilon}_p = \int \boldsymbol{\sigma} \cdot d\boldsymbol{\varepsilon}_p \tag{1}$$

$$R = \frac{d\varepsilon_p^w}{d\varepsilon_p^t} \tag{2}$$

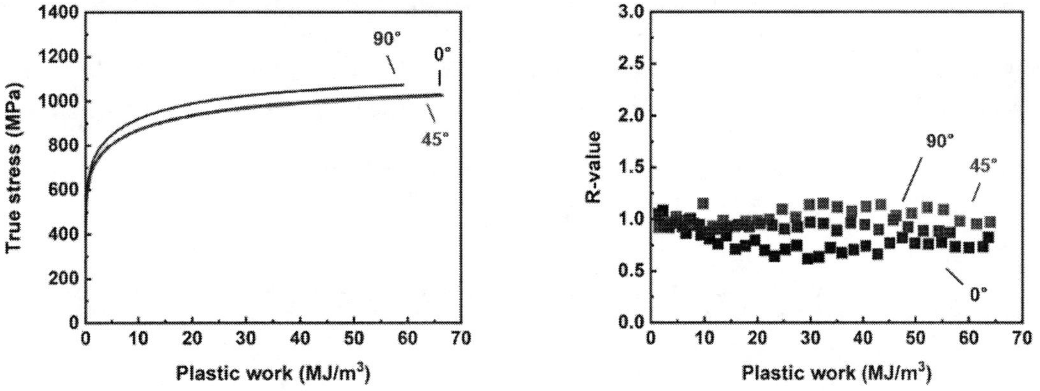

(a) (b)

Figure 1. Mechanical responses in monotonic tension: (a) flow stress; (b) R-value.

To characterize the anisotropic yielding of the DP980 sheet, plane-strain tension (PST) and disk compression tests [9-13] were additionally conducted to calibrate the anisotropic yield function Yld2000-2D [14] as shown in figure 2. These experiments were post-processed as in our earlier work, e.g., [9, 11]. The stress and the R-value data set at the plastic work of 35 MJ/m^3 (which corresponds to a plastic strain of about 3.5%) were used for the calibration. The calibrated parameters are listed in Table 1. The exponent used for Yld2000-2D is 6, as typical for BCC metals.

Table 1. Anisotropic parameters of Yld2000-2D.

α_1	α_2	α_3	α_4	α_5	α_6	α_7	α_8
0.921	0.965	1.016	0.966	0.993	0.828	1.007	1.046

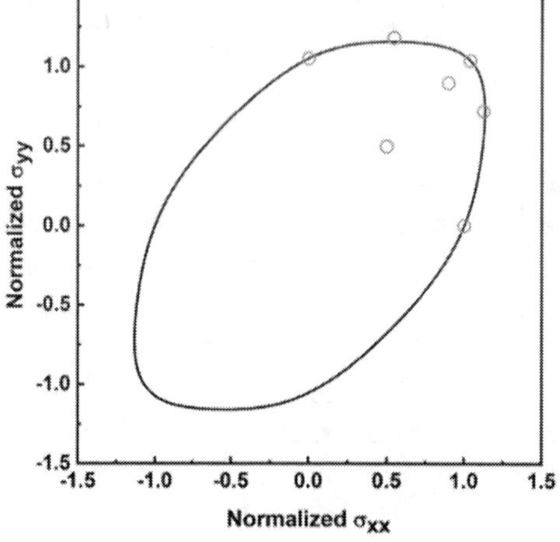

Figure 2. Initial yield locus (solid line) calibrated with experimental data (symbols).

3. Calibration of Chaboche model

For simulating the transient hardening and R-value behavior, a combined isotropic/non-linear kinematic hardening model based on the Chaboche model, was adopted and implemented through the following equations [7, 8, 15, 16]. The constitutive law is defined by Equation (3) with the isotropic hardening term of $R(\bar{\varepsilon}_p)$, where R is the size of the yield surface and $\bar{\varepsilon}_p$ the equivalent plastic strain.

$$f(\sigma - \alpha) = \sigma_0 + R(\bar{\varepsilon}_p) \tag{3}$$

The isotropic hardening indicates the size change of the yield locus with an exponential-decaying function of Equation (4). This term is used for yield surface shrinkage [16] in the present work.

$$R(\bar{\varepsilon}_p) = Q(1 - e^{-b\bar{\varepsilon}_p}) \tag{4}$$

where Q and b are fitting parameters.

For kinematic hardening, the evolution of the back-stress (α) is described by the Chaboche model, using 4 terms to address the non-linear kinematic hardening, as expressed in Equation (5).

$$d\alpha = \sum_{i=1}^{4} d\alpha_i \tag{5}$$

The Chaboche model is based on the Armstrong-Frederic model, which in turn is a non-linear extension of Prager's linear hardening rule [5]. Prager's rule associates the evolution of the back-stress to the plastic strain increment. To avoid erroneous back-stress evolution under plane-stress (i.e., the accumulation of an out-of-plane back-stress, computed from the thickness strain), Ziegler's modification [6] of the Prager rule is adopted in this work. In the end, non-linear hardening is accomplished as represented in Equation (6).

$$d\alpha_i = \frac{C_i}{\bar{\sigma}}\left(\sigma - \sum_{i=1}^{4}\alpha_i\right)d\bar{\varepsilon}_p - \gamma_i\alpha_i d\bar{\varepsilon}_p \tag{6}$$

As an experimental observation, a continuous tension-compression-tension test was performed to obtain the hardening behavior. This experiment was performed on a custom tension-compression testing machine of 50 kN force capacity, using an antibuckling device using comb-shaped teeth, to offer continuous support to the specimen throughout loading (see [17], based on the design described in [16]). The specimen was elongated up to 5% engineering strain and compressed back to 0%. It was then elongated again up to the fracture. Using this experimental result, the Chaboche model was calibrated as shown in figure 3. The fitted curve (solid-line) shows very good agreement with the experiment (symbol). The calibrated parameters are listed in Table 2.

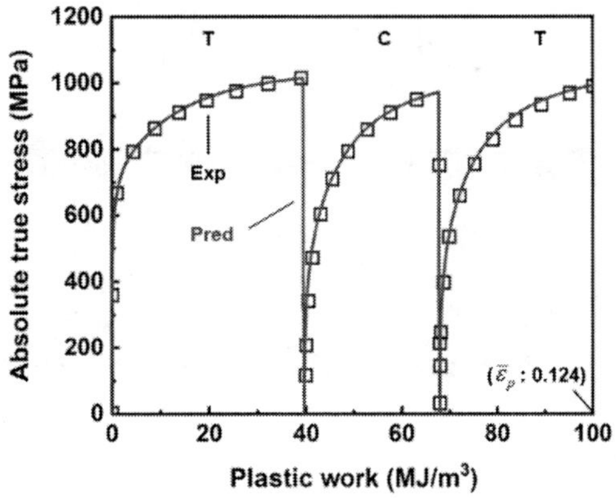

Figure 3. Hardening behavior in continuous tension-compression-tension and model calibration. Shown in parenthesis is also the equivalent plastic strain at the end of the loading.

Table 2. Parameters of Chaboche model.

i	C_i	γ_i	b	Q (MPa)	σ_0 (MPa)
1	300000	3000			
2	110000	500	200	-140	430
3	1500	22.5			
4	30000	80			

4. Model performance

Two-step tension and a loading reversal test were additionally designed to confirm the performance of calibrated Chaboche model for simulating transient hardening and R-value behavior. For the two-step tension test, a specimen was elongated up to 3.2% along the rolling direction (0°) first. Another tensile specimen was cut from the elongated specimen in the first step and was elongated up to fracture along

90° in the second step. Regarding the loading reversal test, a specimen was firstly compressed by 5 % along 0° direction. The compressed specimen was elongated up to fracture along 0° direction in the second step. Figure 4 shows the results of the two-step tension and the loading reversal with monotonic tension and compression curves experimentally obtained. In figure 4(a), the transient hardening behavior is observed in the early part of the second loading while the permanent softening behavior is slightly observed regarding further deformation. On the other hand, the loading reversal case in figure 4(b), the early re-yielding and the permanent softening are more remarkable compared to the two-step tension.

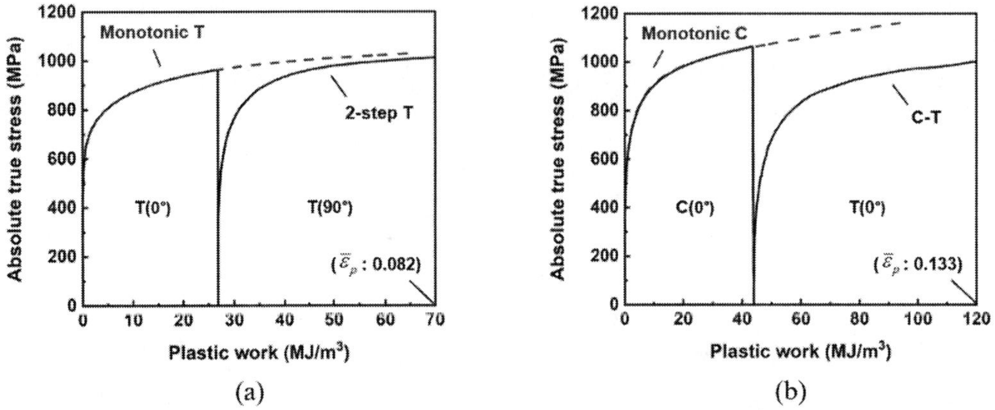

Figure 4. Experimental observation of hardening behavior of DP980 in: (a) Two-step tension; (b) Loading reversal. Shown in parentheses are also the values of the equivalent plastic strain at the end of each loading.

With the calibrated Chaboche model, the hardening and R-value behavior in the two-step tension and the loading reversal were predicted as shown in figure 5 and figure 6, respectively. For both cases of the two-step tension and the loading reversal, the transient R-value behavior were well-predicted by the model. Related to the hardening behavior, however, the predicted stresses are little bit different from the experimental results although 4 non-linear terms were adopted. It seems that the loading history in compression must be considered more carefully to enhance the model performance since the particular DP980 sheet shows asymmetric behavior as shown in figure 6(a). The constitutive framework adopted in this work does not allow for asymmetric yielding.

The fact that the predicted R-value behavior in both two-step tension and loading reversal is showing very good agreement with the experimental results can lead to the conclusion that the shape of yield locus in these regions is represented quite well. According to the loading direction, the back-stress evolves resulting in the translation of the yield locus so that complex transient behavior in R-value and hardening can be simulated by the model.

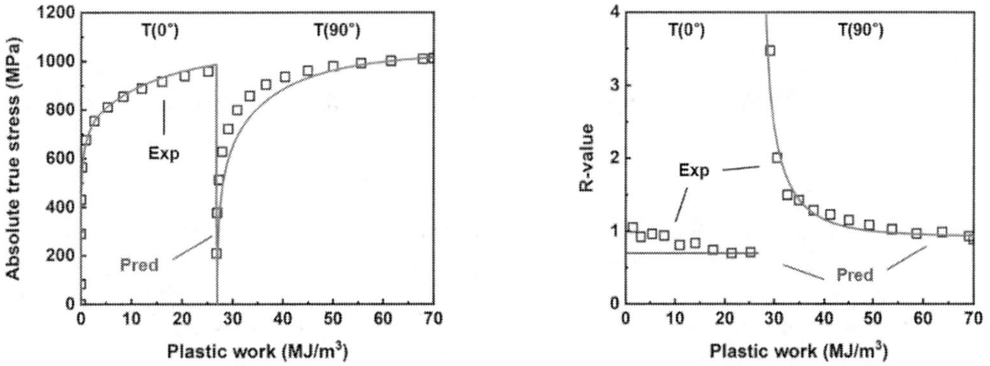

(a) (b)

Figure 5. Prediction of transient behavior by calibrated model in two-step tension: (a) hardening; (b) R-value. The values of the equivalent plastic strain at the end of the loading are given in Fig. 4.

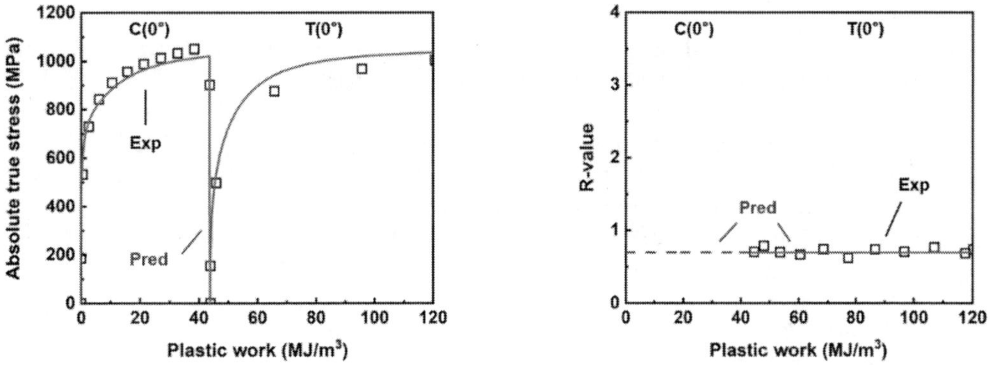

(a) (b)

Figure 6. Prediction of transient behavior by calibrated model in loading reversal: (a) hardening; (b) R-value. The values of the equivalent plastic strain at the end of the loading are given in Fig. 4.

5. Summary and future work

This paper investigates transient hardening and R-value behavior in two-step tension and loading reversals for a DP980 sheet. The anisotropic yield function Yld2000-2D is selected and calibrated to represent initial yielding. To construct a combined isotropic/non-linear kinematic hardening, 4-terms Chaboche model was adopted and calibrated with various monotonic experiments and a continuous tension-compression-tension test. For confirming the model performance, two-step tension (tension in 0° and tension in 90°) and loading reversal (compression in 0° and tension in 0°) tests were conducted. The calibrated model was then used for simulating the transient behavior. Overall predictions for hardening and R-value are showing good agreements with the experimental results. To enhance the performance to describe the hardening and the R-value simultaneously, it would be necessary to consider the loading history in compression more carefully. In future work, the authors will adopt other advanced models, e.g. distortional hardening model, to capture transient behavior of both hardening and R-value under various loading paths.

Acknowledgements
The support of the US National Science Foundation under award CMMI-1929873 is acknowledged with thanks.

References

[1] Bauschinger J 1881 Changes of the elastic limit and the modulus of elasticity on various metals *Civilingenieur* **27** 289-348

[2] Sun L and Wagoner R H 2015 Proportional and non-proportional hardening behavior of dual-phase steels *International Journal of Plasticity* **45** 174-187

[3] Mánik T, Holmedal B, Hopperstad O S 2015 Strain-path change induced transients in flow stress, work hardening and r-values in aluminum *International Journal of Plasticity* **69** 1-20

[4] Ha J, Lee J, Kim J H, Lee M-G, Barlat F 2017 Investigation of plastic strain rate under strain path changes in dual-phase steel using microstructure-based modeling *International Journal of Plasticity* **93** 89-111

[5] Prager W 1956 A new method of analyzing stresses and strains in work-hardening plastic solids *Journal of Applied Mechanics* **23** 493-496

[6] Ziegler H 1959 A modification of Parger's hardening rule *Quartely of Applied Mathematics* **17** 55-65

[7] Chaboche J L 1986 Time-independent constitutive theories for cyclic plasticity *International Journal of Plasticity* **2** 149-188

[8] Lemaitre J and Chaboche J L 1990 Mechanics of Solid Materials *Cambridge Univeristy Press*, Cambridge

[9] Tian H, Brownell B, Baral M and Korkolis Y P 2017 Earing in cup-drawing of anisotropic Al6022-T4 sheets *International Journal of Material Forming* **10** 329-343

[10] Tardif N and Kyriakides S 2012 Determination of anisotropy and material hardening for aluminum sheet metal *International Journal of Solids and Structures* **49** 3496-3506

[11] Ha J, Baral M and Korkolis Y P 2018 Plastic anisotropy and ductile fracture of bake-hardened AA6013 aluminum sheet *International Journal of Solids and Structures* **155** 123-139

[12] Ha J, Fones J, Kinsey B L and Korkolis Y P 2020 Plasticity and formability of annealed, commercially-pure aluminum: Experiments and modeling *Materials* **13**, 4285

[13] Ha J, Coppieters S and Korkolis Y P 2020 On the Expansion of a Circular Hole in an Orthotropic Elastoplastic Thin Sheet *International Journal of Mechanical Sciences* **182** 105706

[14] Barlat F, Brem J C, Yoon J W, Chung K, Dick R E, Lege D J, Pourboghrat F, Choi S-H and Chu E 2003 Plane stress yield function for aluminum alloy sheets-part 1: theory *International Journal of Plasticity* **19** 1297-1319

[15] Bari S and Hassan T 2000 Anatomy of coupled constitutive models for ratcheting simulation *International Journal of Plasticity* **16** 381-409

[16] Deng N, Kuwabara T and Korkolis Y P 2018 On the non-linear unloading behavior of a biaxially loaded dual-phase steel sheet *International Journal of Mechanical Sciences* **138** 383-397

[17] McNally J 2022 Design of a testing machine for continuous-tension-compression of thin sheets and cyclic experiments on steel and aluminum alloys *MS Thesis, Univ. New Hampshire* (in progress).

New approach for the identification of anisotropy material parameters using hydraulic bulge tests

A Khalfallah[123]*, SA Zoueghi[2] and JV Fernandes[1]

[1] CEMMPRE, Department of Mechanical Engineering, University of Coimbra, Rua Luis Reis Santos, Pinhal de Marrocos, 3030-788 Coimbra, Portugal
[2] Laboratoire de Génie Mécanique, Ecole Nationale d'Ingénieurs de Monastir, Université de Monastir, Av.Ibn El Jazzar, 5019 Monastir, Tunisia
[3] DGM, Institut Supérieur des Sciences Appliquées et de Technologie de Sousse, Cite Ibn Khaldoun 4003- Sousse, Université de Sousse, Tunisia
* ali.khalfallah@dem.uc.pt

Abstract. This paper put forwards a new method for the identification of the flow curves of anisotropic sheet metals using hydraulic bulge tests throughout circular and elliptical dies. This method is based on the membrane equilibrium equation, numerical simulation to derive the pole thickness and experimental data involving the measurement of only the polar deflection versus the applied hydraulic pressure curves. The experimental curves of four hydraulic bulge tests were obtained: one bulge test through circular die, and three bulge tests through elliptical dies, where the major axe of the elliptical die was oriented in three directions with respect the rolling direction of the sheet metal. These results were used in the identification of flow parameters and anisotropy coefficients of Hill48' yield criterion. Then, FEA was utilized to perform sensitivity analysis of material parameters on the different hydraulic bulge test responses. The case study was a sheet metal of low carbon steel DC04. The identified material parameters were used in the numerical simulation of hydraulic bulge tests for the accuracy verification of the proposed method. It was shown a good agreement between model's prediction and experimental results.

1. Introduction

The hydraulic bulge test is biaxial tensile test, which can be used as a good complementary or as an alternative to the simple tensile test for material behaviour characterization. Indeed, the homogeneous plastic deformation in simple tension is limited by the appearance of necking which manifests relatively "early" compared to that observed in the hydraulic bulge test. Extrapolation of the hardening curve obtained from simple tensile test for large deformations does not correctly predict the formability and springback in the sheet metal forming process. Therefore, the hydraulic bulge test has attracted the attention of many researchers to be used for the determination of stress-strain curves of sheet metals with the isotropic material assumption [1, 2]. The analysis of the hydraulic bulge test through circular and elliptical dies was reported in some published works in the literature [3, 4]. Several methods were reported on the determination of flow curves using hydraulic bulge tests based on analytical formulation to express the polar thickness and the curvature radius at the pole bulge. A simple model was proposed by Hill [5] to calculate the polar thickness; later improved by Chakrabarty [6] by introducing the hardening effects. Kruglov proposed a simple formula to compute the polar strain [7]. Later, Lăzărescu developed a modified version of Kruglov's model by considering non-uniform strains on the dome surface [2]. Panknin proposed a formula for the determination of polar radius considering the die fillet radius [8]. Some works reported on the application of digital image correlation (DIC) to measure the polar radii [9, 10] and to detect necking and fracture in bulge tests [11]. However, few researches were dedicated for the parameter identification of the anisotropic behaviour of sheet metals using simultaneously hydraulic bulge tests through circular and elliptical dies.

The main goal of this paper is to present a new methodology for an accurate determination of the biaxial stress-strain curve and anisotropy coefficients using hydraulic bulge tests through circular and elliptical die cavity. This methodology combines an analytical model with numerical simulation of the bulge tests to determine the polar thickness as a function of bulge height instead of using approximated formulas. To capture the anisotropy effect, the hydraulic bulge tests through elliptical die are carried out by turning the major axis of the ellipse with a given angle relative to the rolling direction (RD). Numerical simulation of hydraulic bulge tests was compared to experimental results to validate the proposed identification methodology.

2. Description of the analytical model

The hydraulic bulge test consists of applying a pressure using fluid to deform a tightly clamped blank at its extremities through die cavity. Figure 1 shows the schematic drawing used for stress analysis at the pole of the bulging zone for an hydraulic bulge test through elliptical die cavity.

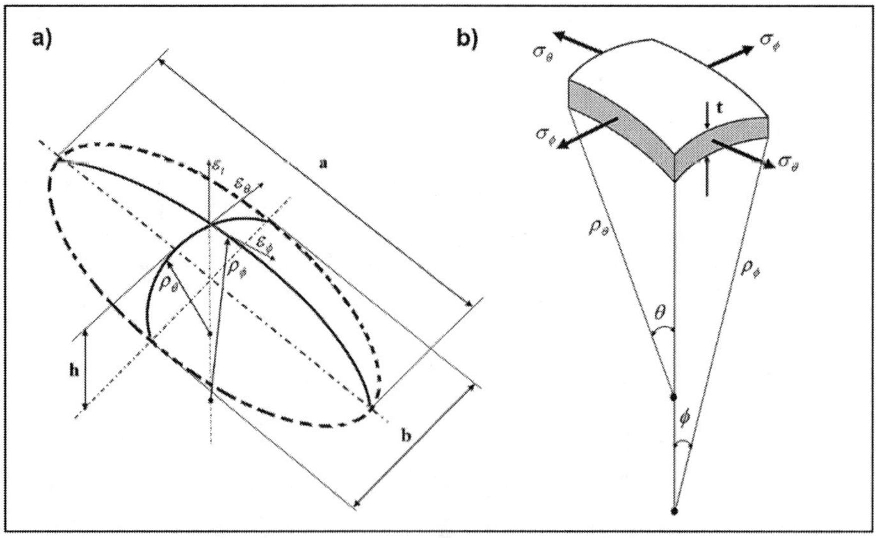

Figure 1. Schematic drawing for stress analysis: a) geometry of hydraulic bulge test through an elliptical die cavity; b) equilibrium of infinitesimal element at the pole of the bulged zone.

The analysis of the hydraulic bulge test is commonly based on the equilibrium of infinitesimal element at the pole of the bulge (see, Figure 1 b). It is assumed that the great axis of the elliptical die is along the rolling direction of the metal sheet. The plastic strains along the three principal axes were determined using the blank thickness at the pole of the bulged area.

$$\varepsilon_t = \ln\frac{t}{t_0}; \quad \varepsilon_\theta = -\frac{\beta}{1+\beta}\varepsilon_t; \quad \varepsilon_\phi = -\frac{1}{1+\beta}\varepsilon_t; \quad \beta = \frac{\varepsilon_\theta}{\varepsilon_\phi} \tag{1}$$

where t, t_0 define the instant and the initial pole thickness. β represents the plastic strain ratio. The biaxial stresses σ_ϕ and σ_θ along ϕ and θ directions define the stress components along the two directions of the major and the minor axes of the elliptical die cavity, respectively. They are given as follows.

$$\sigma_\theta = \frac{P}{t}\left(\frac{1}{\rho_\theta}+\frac{\alpha}{\rho_\phi}\right)^{-1};\ \sigma_\phi = \frac{\alpha P}{t}\left(\frac{1}{\rho_\theta}+\frac{\alpha}{\rho_\phi}\right)^{-1};\ \alpha = \frac{\sigma_\phi}{\sigma_\theta} \tag{2}$$

where P is the applied pressure, ρ_ϕ,ρ_θ define the curvature radii along the major and minor axes, respectively. α represents the stress ratio. The curvature radii along the major axis a and minor axis b of the elliptical die were obtained by the Panknin expressions [8], as given in equation (3). In these relations and for better accuracy, the fillet radius of the die was taken into account. Koç has demonstrated that using the Panknin relations, the measured polar radii were closer to the ones obtained by stepwise or digital image correlation technique [10]. Therefore, Panknin relations are considered in this work to accurately evaluate the polar radii.

$$\rho_\phi = \frac{(a+r_f)^2+h^2-2r_f h}{2h};\ \rho_\theta = \frac{(b+r_f)^2+h^2-2r_f h}{2h} \tag{3}$$

To compute the polar thickness as a function of the bulge height, Hill [5], Chakrabarty [6] and Kruglov [7] have proposed analytical relationships. However, different prediction was obtained for the pole thickness. This latter has a significant effect on the accuracy of the determination of the biaxial stress-strain curves. Therefore, the numerical simulation of polar thickness *vs.* bulge height was carried out and compared to those obtained using analytical formulas. A noticeable difference was revealed. Therefore, the cornerstone on which is based this identification methodology is that the polar thickness was obtained using numerical simulation of the hydraulic bulge test using circular die cavity. The equivalent stress -strain curves were calculated using Hill48 yield criterion for the different hydraulic bulge tests. Equations (4) and (5) express only the equivalent stress and strain obtained for the hydraulic bulge test though the circular die cavity. For each hydraulic bulge test, the equivalent stress- strain curve was determined and not presented here. Figure 2 displays the flow-chart of the computation of the equivalent stress-strain curve using the experimental pressure *vs.* bulge height curve, as a continuous measurement data.

$$\bar{\sigma} = \left(\frac{3}{2}\frac{r_0\sigma_\theta^2 + r_{90}\sigma_\phi^2 + r_0 r_{90}(\sigma_\theta-\sigma_\phi)^2}{r_0 r_{90}(1+\frac{1}{r_0}+\frac{1}{r_{90}})}\right)^{1/2} \tag{4}$$

$$\bar{\varepsilon} = \left\{\frac{2}{3}\frac{r_0+r_{90}+r_0 r_{90}}{1+r_0+r_{90}}\left(\frac{1}{r_{90}}\left(\frac{1}{r_0}\varepsilon_\theta-\varepsilon_t\right)^2+\frac{1}{r_0}\left(\varepsilon_t-\frac{1}{r_{90}}\varepsilon_\phi\right)^2+\left(\frac{\varepsilon_\phi}{r_{90}}-\frac{\varepsilon_\theta}{r_0}\right)^2\right)\right\}^{\frac{1}{2}} \tag{5}$$

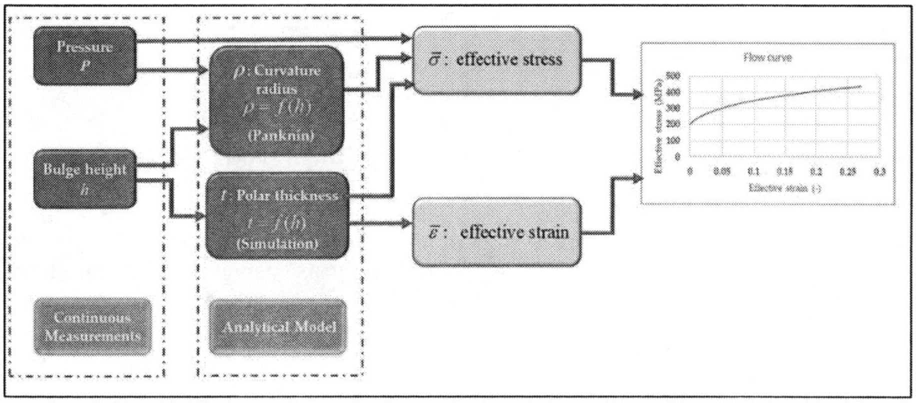

Figure 2. Flow-chart for the computation algorithm of the equivalent stress-strain curves using hydraulic bulge test throughout circular and elliptical die cavity.

The curvature radii were calculated using Panknin formula and the polar thickness was obtained using numerical simulation of the hydraulic bulge test. As it will be shown later, it was demonstrated that the polar thickness as a function of bulge height, was insensitive to anisotropy coefficients and to the die geometry. Therefore, the simple tensile test result along rolling direction was used as input data to run the numerical simulation of the hydraulic bulge thought circular die to derive the polar thickness.

3. Experiments

Experimental hydraulic bulge tests were carried out using a simple stand-alone apparatus which has been manufactured for this purpose. Circular and elliptical die cavities were used to apply throughout them a fluid pressure in order to bulge sheet metal blanks. These latter were firmly constrained at their borders using draw bead implemented at the flange contour of the blanks to ensure their tightness and insure a pure stretching. DC04 metal sheet blank of 1 mm thickness was used as an application for the proposed parameter identification methodology. Figure 3 shows the four samples deformed until blanks burst and they are labelled Test-0, test-1, Test-2 and Test-3. The experimental burst pressure measured for the bulged sample through the circular die cavity was around of 12.4 MPa and the fracture pressure for those bulged using elliptical die cavity was around an average value of 15.1 MPa. The diameter of the circular die cavity is $d_c = 91$ mm, and the die fillet radius is $r_f = 6$ mm. The samples 1, 2 and 3 correspond to the hydraulic bulge tests performed using elliptical die cavity for which the length of the major axis a =110 mm, the length of the minor axis b = 74 mm and the die fillet radius is $r_f = 6$ mm. In addition, uniaxial tensile tests according to ASTM-E8 standards were performed to obtain the uniaxial stress-strain curves and the Lankford coefficients $r_{0°}$ $r_{45°}$ and $r_{90°}$ along three directions: 0°, 45° and 90° with respect to the rolling direction (RD). Figure 4 displays the experimental bulge pressure *vs.* bulge height at the pole obtained for the four samples. The measuring device was composed of pressure gage to monitor the applied pressure and a displacement transducer probe was placed at the centre of the blank to measure the bulge height during the bulging process. It is worth noting that for the test repeatability, these curves represent the average results of three hydraulic bulge replication tests performed for each test case. Only curve plots for a bulge height higher than almost 5 mm were shown.

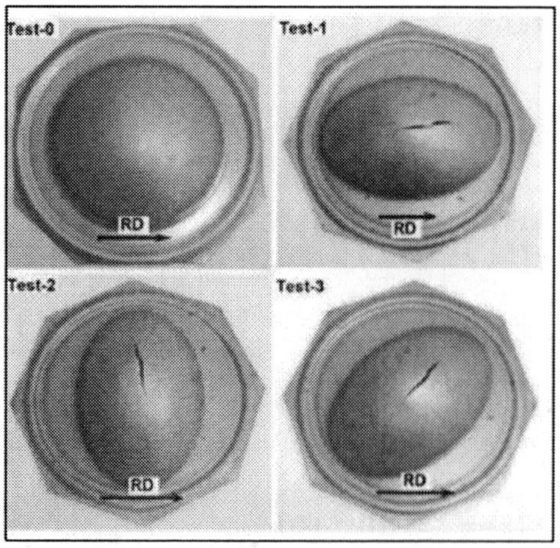

Figure 3. Four hydraulic bulge samples deformed until fracture: the sample labelled Test-0 was bulged using circular die cavity and the samples labelled Test-1, Test-2 and Test-3 were hydroformed using elliptical die.

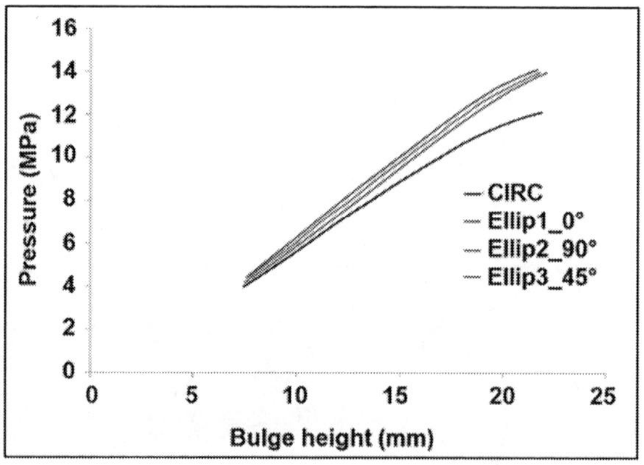

Figure 4. Experimental hydraulic bulge pressure *vs.* bulge height curves obtained at the pole of the four bulge tests, using circular and elliptical die cavity.

4. Numerical models

Numerical simulation of hydraulic bulge tests was performed using ABAQUS® Standard commercial software. The blank was discretised using shell element (S4R) and the die was modelled as analytical rigid body. The draw bead was modelled by a firm clamp at the flange of the blank. It is worth to note that three-dimensional (3D) finite element model was also built using eight note solid elements with reduced integration (C3D8R) and 2 elements were stacked across the blank thickness. Equal simulation results were obtained, but with very short computation time for the 2D finite element model compared to 3D model. Thus, the model using shell elements was definitely considered to run numerical simulation of the hydraulic bulge tests. Taking account of the material, geometry and boundary condition symmetries, one quarter of the FE model was adopted to run numerical simulation for the bulge tests: Test-0, Test-1 and Test-2. The total number of elements was 462. However, for the Test-3, which represents the bulge test with elliptical die cavity in the 45° direction, the whole finite element model composed of 1848 shell elements was used. For the material behaviour, it was assumed as an isotropic elastic behaviour defined by the generalized Hooke's law. Besides, an anisotropic plastic material behaviour, which is described by Hill'48 yield criterion and obeying to the isotropic hardening law, being defined by Swift equation. The Hill'48 yield criterion is expressed as follows:

$$F(\sigma_{22}-\sigma_{33})^2+G(\sigma_{33}-\sigma_{11})^2+H\left(\sigma_{11}-\sigma_{22}\right)^2+2L\sigma_{23}^2+2M\sigma_{13}^2+2N\sigma_{12}^2=\bar{\sigma}^2 \quad (6)$$

where σ_{ij} $(i,j=1..3)$ are the Cauchy stress tensor components and defined in the material's orthotropic coordinate system. The parameters *F, G, H, L, M, N* are the Hill48's anisotropy coefficients and $\bar{\sigma}^2$ is the yielding stress. In the case of the plane stress conditions, and assuming that the yield stress is identified to the uniaxial tensile stress along the RD of the sheet metal, thus the relation ($G+H=1$) is deduced. *L* and *M* parameters are set to the isotropic value, which is equal to 1.5. The Hill48's anisotropy parameters can be expressed as a function of the Lankford anisotropy coefficients, $r_{0°}$, $r_{45°}$ and $r_{90°}$, using the following equations:

$$F=\frac{r_{0°}}{r_{90°}\left(1+r_{0°}\right)};\quad G=\frac{1}{1+r_{0°}};\quad H=\frac{r_{0°}}{1+r_{0°}};\quad N=\frac{\left(r_{0°}+r_{90°}\right)\left(2r_{45°}+1\right)}{2r_{90°}\left(1+r_{0°}\right)} \quad (7)$$

The Swift empirical relation was used to model isotropic strain-hardening behaviour.

$$\bar{\sigma}=K(\varepsilon_0+\bar{\varepsilon})^n \quad (8)$$

where *K*, ε_0 and *n* are the Swift's law constants. Figure 5 shows the numerical simulations of the hydraulic bulge tests through circular and elliptical die cavity, when the major axis coincided with 0°, 90° and 45° with respect to the RD. For a balance between better accuracy and the computing time, the mesh was refined close to the centre of the blank, and a coarse mesh was considered elsewhere.

5. Sensitivity of material parameters

Sensitivity analysis was carried out to investigate the effect of the material parameters (e.g., the hardening exponent *n* and the anisotropy coefficients $r_{0°}$, $r_{45°}$, $r_{90°}$) variation on the pressure and polar thickness *vs.* the bulge height. These responses were determined using numerical simulation of the four hydraulic bulge tests. The aim of this parametric sensitivity analysis is to assess the relevance of responses obtained from hydraulic bulge tests through circular and elliptical die cavity. Figure 6 shows the effect of anisotropy coefficient on the simulated bulge test responses. It is revealed that the anisotropy coefficients $r_{0°}$ and $r_{90°}$ have an effect on the pressure – bulge height curves, noticeably for the $r_{90°}$ coefficient. However, the $r_{45°}$ anisotropy coefficient was insensitive to the bulge responses obtained for all the bulge tests. As consequence, the accuracy of the identification of the $r_{45°}$ anisotropy coefficient using bulge test performed through the elliptical die cavity will be less evident. Figure 7 shows that the anisotropy coefficient has no significant effect on the pole thickness. It is worth noting that anisotropy coefficients were varied and no effect was revealed on the pole thickness for all bulge tests. Thus, they are not shown here. Figure 7a shows only one case studied, where it depicts the thickness at the pole of the bulged area obtained using the numerical simulation of the hydraulic bulge tests; Test-0, Test-1, Test-2 and Test-3.

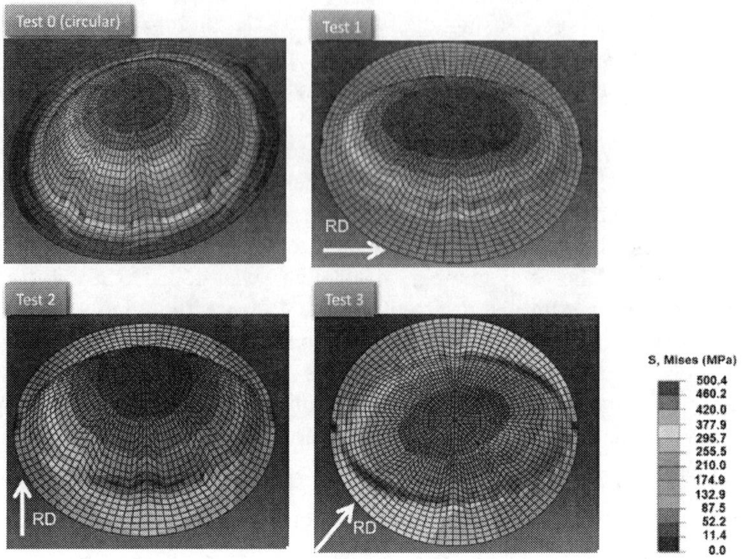

Figure 5. Numerical simulation of the hydraulic bulge tests using circular die cavity for Test-0 and elliptical die cavity for Tests 1, 2 and 3 (von Mises stress contours).

One may conclude that the predicted thickness calculated using three different anisotropy coefficients was insensitive to these coefficients. However, it is clear that the work-hardening exponent has a significant effect on the polar thickness, as exhibited in Figure 7b. It is also noticed that this influence is observable for a bulge height higher than 10 mm. This exponent has also an influence on the pressure *vs.* bulge height for hydraulic bulge tests through the circular and the elliptical die cavity (see, Figure 7c). Therefore, the thickness *vs.* bulge height curve, which is essential to compute the equivalent stress-strain curves for the four hydraulic bulge tests (i.e., Test-0, Test-1, Test-2 and Test-3) was generated using numerical simulation of the hydraulic bulge test through circular die cavity.

6. Parameter identification

The mild steel DC04 is used as an application of the proposed identification methodology to determine the anisotropy coefficients and flow curve of this material. This method was decoupled in two steps to identify the material parameters. In the first step, the anisotropy coefficients of the Hill'48 yield criterion were identified. In the second step, Swift law parameters were determined, as the anisotropy coefficients were already identified in the first step.

6.1 Identification of anisotropy parameters

Hill48's anisotropy coefficients were identified using the equivalent stress-strain curves obtained from the hydraulic bulge tests and the computation algorithm given in Figure 2. The parameter identification was converted as an optimization problem, for which an objective function was defined. The objective function is given in equation (9) and it defines the gap between four equivalent stress-strain curves calculated for each bulge test. The minimization of this function f leads to the convergent solution of the parameter identification problem. This solution represents the set of the identified anisotropy parameters.

$$f = \sum_{i=1}^{N} \left(\bar{\sigma}_{cir} - \bar{\sigma}_{ell_1}\right)^2 + \left(\bar{\sigma}_{cir} - \bar{\sigma}_{ell_2}\right)^2 + \left(\bar{\sigma}_{cir} - \bar{\sigma}_{ell_3}\right)^2 \quad (9)$$

The optimization method used to solve this non-linear problem and to avoid trapping into local minimums, the simulated annealing (SA) algorithm, as a heuristic optimization method was used to reach a global minimum. SA algorithm adopts an iterative movement according to the variable temperature parameter which imitates the annealing transaction of the metals. The base of the annealing process is to generate random points in the neighbourhood of the current best point and evaluate the problem functions there. This strategy evolves iteratively until the value of objective function is less than a given tolerance. Figure 8 displays the flow-chart of the algorithm used in the parameter identification of the anisotropy coefficients based on the four responses of the hydraulic bulge tests.

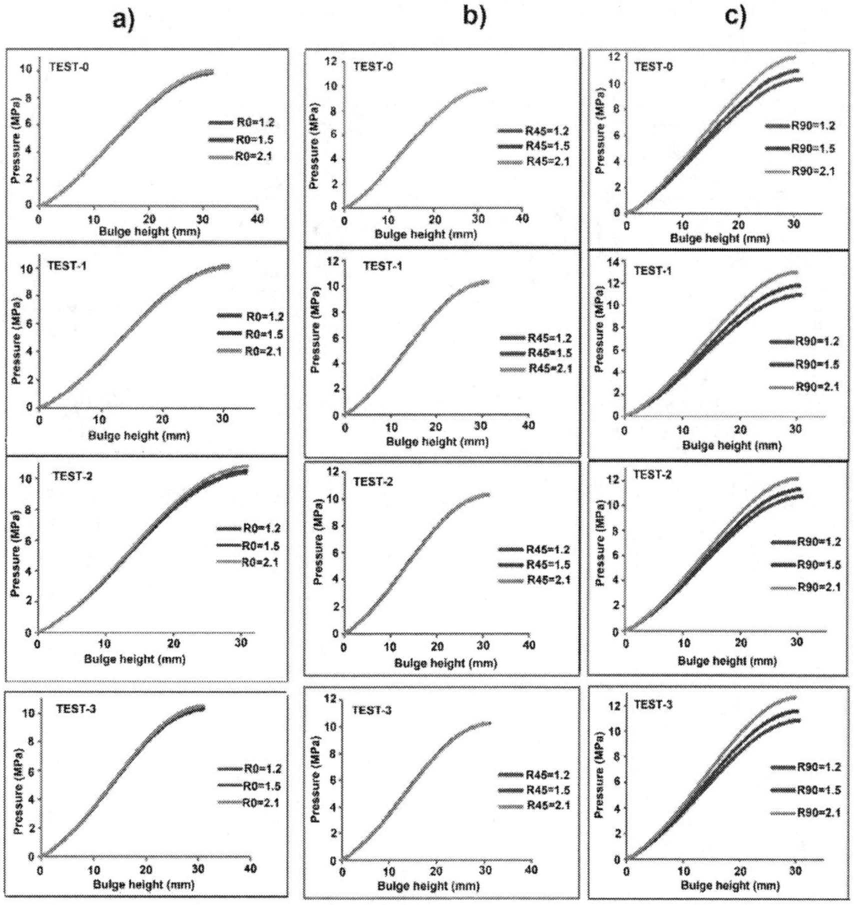

Figure 6. Parametric sensitivity analysis of the anisotropy coefficients on the response of the bulge tests through the circular and the elliptical die cavity: a) effect of the $r_{0°}$ anisotropy coefficient; b) effect of the $r_{45°}$ anisotropy coefficient; c) effect of the $r_{90°}$ anisotropy coefficient.

6.2 Identification of work-hardening parameters

The identification of Swift's parameters was determined by curve fitting procedure. These parameters were determined using the equivalent stress-strain curve. This curve was derived from the hydraulic bulge test which is performed through circular die cavity. Table 1 lists the identified Hill'48 anisotropy coefficients along with the Swift work-hardening parameters using the proposed identification methodology. Moreover, for comparison purposes, the Hill'48 anisotropy coefficients calculated using Lankford anisotropy coefficients measured using uniaxial tensile tests with respect to three directions ($r_{0°}$, $r_{45°}$ and $r_{90°}$) to RD were also given in this table. It is clearly seen that the set of the identified

IOP Conf. Series: Materials Science and Engineering 1238 (2022) 012003 doi:10.1088/1757-899X/1238/1/012003

parameters: anisotropy coefficients and work-hardening parameters are in good agreement with those obtained using uniaxial tensile tests. However, the attained equivalent strains using hydraulic bulge tests are higher than those obtained for tensile test. Figure 9 shows the equivalent stress-strain curves obtained for the four hydraulic bulge tests using circular and elliptical die cavity. It shows that at algorithm convergence, the four responses are overlapping and in good agreement, even a small discrepancy was noticed. Figure 10 displays the comparison between *r*-values calculated using the Hill'48 anisotropy parameters and the Lankford coefficients measured using uniaxial tensile tests along the three directions relative to RD. It is worth noting that despite the observed difference, these results are generally in good agreement with those obtained using uniaxial tensile tests. This could be likely explained by the weak ability of Hill'48 yield criterion for accurate description of the DC04 material behaviour. Moreover, it is interesting to mention that the Test-3 has a shallow sensitivity with respect to the $r_{45°}$ anisotropy coefficient, as probably was not expected in the foremost step of this study.

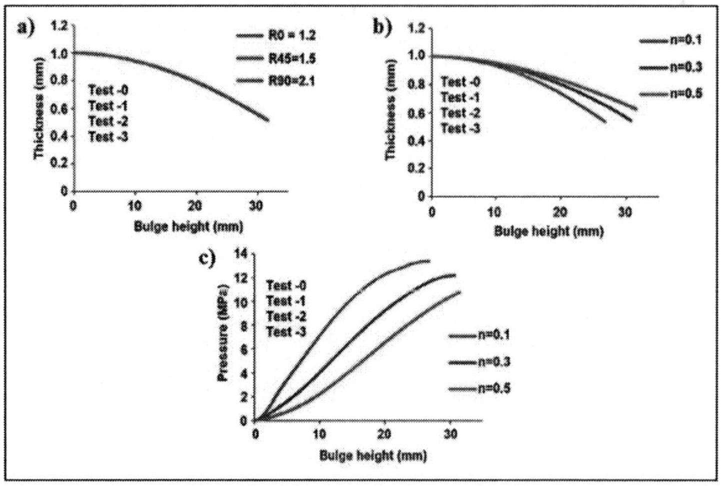

Figure 7. sensitivity analysis obtained for the four bulge tests: a) of the anisotropy coefficients on the pole thickness; b) of the hardening exponent on the thickness at the pole; c) of the hardening exponent on the bulge response: pressure *vs.* bulge height.

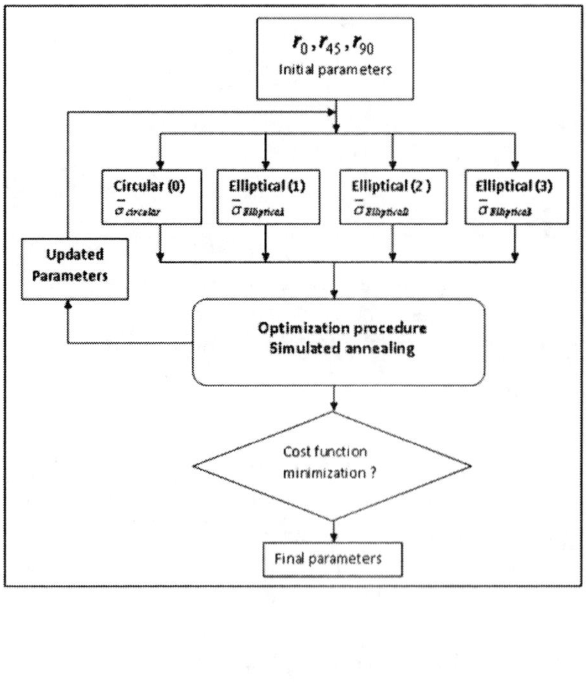

Figure 8. Flow chart algorithm used in the identification of anisotropy parameters.

Table 1. Comparison between the identified parameters using hydraulic bulge tests and those determined by simple tensile tests.

DC04	Swift parameters			Hill's anisotropy coefficients			
	K(MPa)	Y_0 (MPa)	n	F	G	H	N
Hydraulic Bulge tests	692.3	209.3	0.271	0.344	0.425	0.575	1.231
Tensile test	601.8	181.2	0.209	0.361	0.371	0.629	1.363

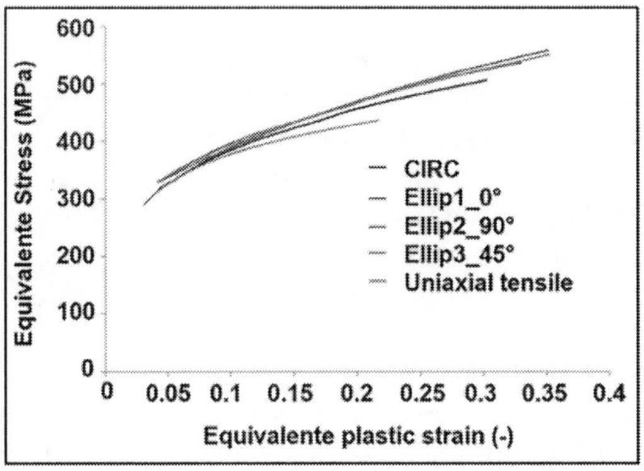

Figure 9. Flow curves obtained using hydraulic bulge test through circular and elliptical die cavity along with stress-strain curve measured from uniaxial tensile test along the rolling direction.

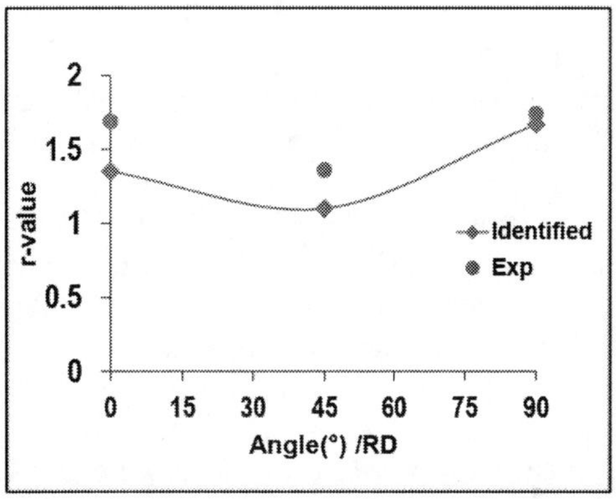

Figure 10. Comparison between r-values identified using hydraulic bulge tests through circular and elliptical die cavity and the Lankford coefficients (i.e., experimental values).

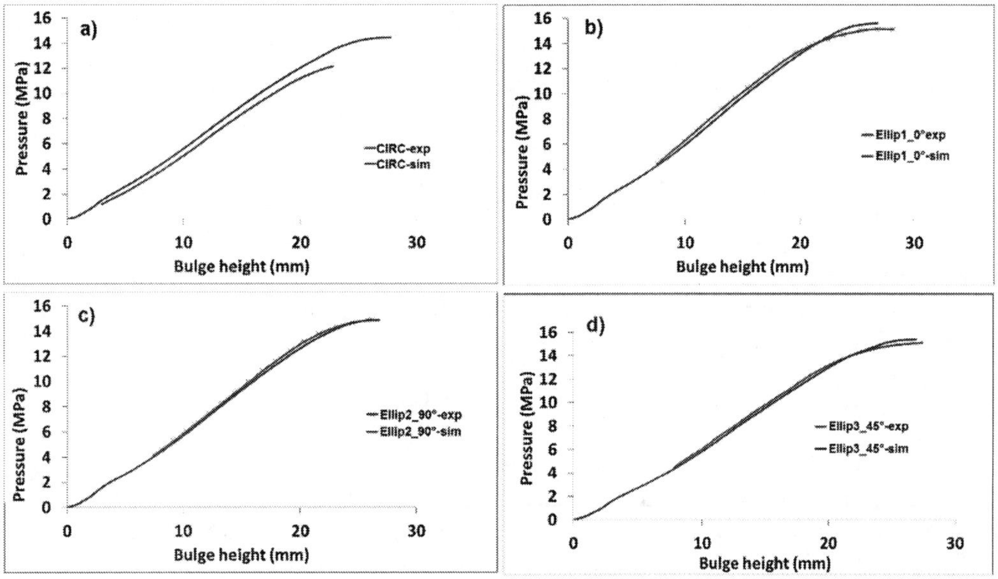

Figure 11. Comparison between experimental and identified pressure *vs.* bulge height curves obtained using circular and elliptical die cavity.

7. Experimental verification

The identified parameters: anisotropy coefficients and work-hardening parameters were used as inputs into finite element model to simulate the hydraulic bulge tests through circular and elliptical die cavity to assess the accuracy of the identification procedure. This could be considered as a first attempt in the validation and verification of the proposed identification method along with the chosen plasticity yield criterion to describe the mechanical behaviour of the sheet metal DC04. Figure 11 shows the comparison between experimental and numerical responses obtained for the hydraulic bulge tests (Test-0, Test-1, Test-2 and Test-3) carried out using circular and elliptical die cavity. A good agreement was revealed for all the considered tests, despite a small discrepancy between numerical and experimental results. However, these findings globally demonstrate an accurate prediction of pressure *vs.* pole height owing to this simple and robust identification procedure.

8. Conclusions

New approach was proposed for material parameter identification of anisotropy and flow curves using hydraulic bulge tests through circular and elliptical die cavity. The use of the experimental continuous measurement of the applied pressure *vs.* the bulge height is the chief advantage of this parameter identification methodology. The instantaneous thickness was obtained by numerical simulation, in contrast to previous works, where the thickness was either determined using analytical formulas or using DIC technique. It was observed that the thickness evolution at the pole as a function of the bulge height, which was calculated for the four bulge tests, was found insensitive to the anisotropy coefficients. However, the curvature radii were calculated using Panknin relation and taking account of the die fillet radius. The experimental validation shows that the predicted results of the numerical simulation using the identified material parameters were in good agreement with the experimental data. Nevertheless, further work is needed to systematically investigate the sensitivity analysis of the input parameters on the different outputs of bulge tests. Furthermore, the application of more metal sheets diversity and using non- quadratic yield criteria is the purpose of the forthcoming research work.

Acknowledgements

This research is sponsored by FEDER funds through the program COMPETE – Programa Operacional Factores de Competitividade and by national funds through FCT – Fundação para a Ciência e a Tecnologia, under the project UIDB/00285/2020. It was also supported by the project RDFORMING (reference PTDC/EME-EME/31243/2017), co-funded by Portuguese Foundation for Science and Technology, by FEDER, through the program Portugal-2020 (PT2020), and by POCI, with reference POCI-01-0145-FEDER-031243. Both supports are gratefully acknowledged.

References

[1] Gutschur G, Wu HC, Nagile G, Altan T 2004 J. Mater. Process. Technol. **146** 1-7
[2] Lăzărescu L, Comşa DS, Banabic D 2011 AIP Conf. Proc. **1353** 1429-1434
[3] Keller S, Hotz W, Friebe H 2009 International deep drawing research group IDDRG 2009 International conference, Golden, CO, USA
[4] Rees DW 1995 Int. J. Mech. Sci, **37** 373-389
[5] Hill R 1950 Philos Mag Ser **7** 41 01133–42
[6] Chakrabraty J and Alexander JM 1970 J. Strain Anal, **5** 155–61
[7] Kruglov AA, Enikeev FU and Lutfullin RY 2002 Mater Sci Eng A **323** 416–26
[8] Panknin W 1959 Ph.D Dissertation, The hydraulic bulge test and the determination of the flow stress curves, University Stuttgart, Germany
[9] Chenk S, Scales M, Kyriakides S and Corona E 2016 Int. J. Sol. & Struct. **82** 70-84
[10] Koç M, Billur E and Cora ÖN 2011 Mater. & Design **32** 272–281
[11] Stoughton TB, Min JY and Carsley JE 2017 J. Phys. Conf. Ser. **896** 012108

IOP Conf. Series: Materials Science and Engineering 1238 (2022) 012004 doi:10.1088/1757-899X/1238/1/012004

Non-Quadratic Pseudo Dual Potentials for Plastic Flow Modeling

Seung-Yong Yang[1], **Jie Sheng**[2] **and Wei Tong**[2]
[1]Korea University of Technology and Education, Byeongchon-myon, Cheonan-si, Chungnam 31253, Korea
[2]Research Center for Advanced Manufacturing, Southern Methodist University, Dallas, Texas 75275, USA

E-mail: ysy@koreatech.ac.kr, wtong@smu.edu

Abstract. The existence of dual flow potentials is well established in mathematical theory of plasticity since the seminal work by Hill in 1987. For a metal undergoing plastic flow, a flow stress potential is used to compute its plastic strain increments when the applied yield stress is known. On the other hand, the corresponding dual flow strain-rate potential is used to compute the stress on the flow surface when the plastic strain increments are given. This work examines some issues associated with plasticity modeling using non-quadratic dual flow potentials. Unlike the quadratic case where flow stress and strain-rate potentials are the exact dual to each other, it is often difficult if not impossible to obtain analytically the dual of a non-quadratic flow stress or strain-rate potential. The study instead focuses on formulating and assessing various non-quadratic pseudo dual flow potentials that approximate the actual flow surfaces in either stress or strain-rate space. The difference and connection between the yield surface and flow surface in non-associated plasticity are also investigated. Although only one of the dual flow potentials is actually needed for their applications in associated and non-associated plasticity modeling, the unique advantage of having both dual flow potentials on hand even in their pseudo forms is pointed out for new computational analyses.

1. Introduction

In classical mathematical theory of plasticity [1, 2, 3], two *positive* scalars called *equivalent* flow stress $\bar{\sigma}$ and plastic strain-rate $\dot{\bar{\varepsilon}}$ are introduced to define the plastic work-rate $\dot{w}^p$ of a plastically deforming material element with a plastic strain-rate $\dot{\boldsymbol{\varepsilon}}^p$ under the applied Cauchy stress $\boldsymbol{\sigma}$

$$\dot{w}^p(\boldsymbol{\sigma};\dot{\boldsymbol{\varepsilon}}^p) = \bar{\sigma}\dot{\bar{\varepsilon}} = f(\boldsymbol{\sigma})\dot{\bar{\varepsilon}}^p = \sigma_f q(\dot{\boldsymbol{\varepsilon}}^p) = \boldsymbol{\sigma} : \dot{\boldsymbol{\varepsilon}}^p > 0, \quad \boldsymbol{\sigma} \neq \mathbf{0}, \quad \dot{\boldsymbol{\varepsilon}}^p \neq \mathbf{0}, \tag{1}$$

where $\bar{\sigma} = f(\boldsymbol{\sigma})$ and $\dot{\bar{\varepsilon}} = q(\dot{\boldsymbol{\varepsilon}}^p)$ are the *flow stress function* and *plastic strain-rate function* respectively[1]. They are positive convex homogeneous functions of degree one with respect to positive multipliers and are what Hill calls dual plastic potentials or polars to each others [4].

Consider a flat thin sheet modeled by Hill's 1948 associated quadratic anisotropic plasticity [5] with either its yield/flow stress function (YLD) given in plane stress $\boldsymbol{\sigma} = (\sigma_x, \sigma_y, \tau_{xy})$

$$\Phi_h(\sigma_x, \sigma_y, \tau_{xy}) = \bar{\sigma}^2 = f^2(\boldsymbol{\sigma}) = A_1\sigma_x^2 + A_2\sigma_x\sigma_y + A_3\sigma_y^2 + A_4\tau_{xy}^2, \tag{2}$$

[1] It is understood here that the use of a set of dual variables $(\boldsymbol{\sigma};\dot{\boldsymbol{\varepsilon}}^p)$ for a function $\dot{w}^p(\boldsymbol{\sigma};\dot{\boldsymbol{\varepsilon}}^p)$ emphasizes only either $\boldsymbol{\sigma}$ or $\dot{\boldsymbol{\varepsilon}}^p$ but not both may be independent variables. In the following, either $\dot{w}_p(\boldsymbol{\sigma})$ or $\dot{w}_p^*(\dot{\boldsymbol{\varepsilon}}^p)$ will be used as two specific examples of $\dot{w}^p(\boldsymbol{\sigma};\dot{\boldsymbol{\varepsilon}}^p)$.

or its (plastic work rate conjugate) plastic strain-rate function (SRP) with the imposed planar plastic strain increments $\dot{\boldsymbol{\varepsilon}}^p = (\dot{\varepsilon}_x^p, \dot{\varepsilon}_y^p, \dot{\gamma}_{xy}^p)$

$$\Psi_h(\dot{\varepsilon}_x^p, \dot{\varepsilon}_y^p, \dot{\gamma}_{xy}^p) = q^2(\dot{\boldsymbol{\varepsilon}}^p) = B_1(\dot{\varepsilon}_x^p)^2 + B_2\dot{\varepsilon}_x^p\dot{\varepsilon}_y^p + B_3(\dot{\varepsilon}_y^p)^2 + B_4(\dot{\gamma}_{xy}^p)^2. \tag{3}$$

A flow strain-rate or stress rule is then used to compute the plastic strain increments $\dot{\boldsymbol{\varepsilon}}^p$ or Cauchy stress $\boldsymbol{\sigma}$ respectively as ($\dot{\bar{\varepsilon}} = q(\dot{\boldsymbol{\varepsilon}}^p)$)

$$\dot{\boldsymbol{\varepsilon}}^p = \dot{\bar{\varepsilon}}^p \frac{\partial f(\boldsymbol{\sigma})}{\partial \boldsymbol{\sigma}} = \frac{\dot{\bar{\varepsilon}}^p}{2\bar{\sigma}} \frac{\partial \Phi_h(\boldsymbol{\sigma})}{\partial \boldsymbol{\sigma}}, \quad \boldsymbol{\sigma} = \sigma_f \frac{\partial q(\dot{\boldsymbol{\varepsilon}}^p)}{\partial \dot{\boldsymbol{\varepsilon}}^p} = \frac{\sigma_f}{2\dot{\bar{\varepsilon}}} \frac{\partial \Psi_h(\dot{\boldsymbol{\varepsilon}}^p)}{\partial \dot{\boldsymbol{\varepsilon}}^p}. \tag{4}$$

By using the above associated flow rules and plastic work rate equivalency, one can show that the polynomial coefficients (A_1, A_2, A_3, A_4) of the quadratic yield function and (B_1, B_2, B_3, B_4) of the quadratic strain-rate function are directly related *algebraically* via [4, 6, 7]

$$A_1 = \frac{4B_3}{4B_1B_3 - B_2^2}, \; A_2 = -\frac{4B_2}{4B_1B_3 - B_2^2}, \; A_3 = \frac{4B_1}{4B_1B_3 - B_2^2}, \; A_4 = \frac{1}{B_4}, \tag{5}$$

$$B_1 = \frac{4A_3}{4A_1A_3 - A_2^2}, \; B_2 = -\frac{4A_2}{4A_1A_3 - A_2^2}, \; B_3 = \frac{4A_1}{4A_1A_3 - A_2^2}, \; B_4 = \frac{1}{A_4}. \tag{6}$$

It is noted that both $4A_1A_3 - A_2^2 > 0$ and $4B_1B_3 - B_2^2 > 0$ per strict convexity requirements [8, 9, 10].

As pointed out by Hill [4], when the yield stress function $f(\boldsymbol{\sigma})$ and the plastic strain-rate function $q(\dot{\boldsymbol{\varepsilon}}^p)$ are *entirely smooth and strictly convex* (obviously they should be positive too as $\sigma_f > 0$ and $\dot{\bar{\varepsilon}}^p > 0$), there is a one-to-one correspondence between the Cauchy stress $\boldsymbol{\sigma}$ and plastic strain-rate $\dot{\boldsymbol{\varepsilon}}^p$. Consequently, only the yield stress (or plastic strain-rate) function is needed to be specified in an associated plasticity theory as the corresponding plastic strain-rate (or yield stress) function can be derived in principle from the flow or normality rules Eq.(4).

2. Non-Quadratic Isotropic Flow Potentials

Unlike the quadratic case above where flow stress and strain-rate potentials are the exact dual to each other, it is often difficult if not impossible to obtain analytically the dual of a non-quadratic flow stress or strain-rate potential even for isotropic materials. We will examine several commonly used ones to identify their pseudo dual potentials instead.

2.1. Hershey-Hosford Flow Potentials

The non-quadratic Hershey-Hosford yield function generalizes the quadratic von Mises yield function [11, 12]. In terms of three principal Cauchy stresses $(\sigma_1, \sigma_2, \sigma_3)$, it has the following form with the exponent $a > 1$

$$f_H^a(\boldsymbol{\sigma}_d) = \frac{|\sigma_1 - \sigma_2|^a + |\sigma_2 - \sigma_3|^a + |\sigma_3 - \sigma_1|^a}{2}. \tag{7}$$

Flow stress surfaces $f_H(\boldsymbol{\sigma}_d) = \sigma_f(\bar{\varepsilon}^p)$ with $a = 6$, 8, and 20 are shown in Figure 1(a) in terms of biaxial deviatoric stresses $\boldsymbol{\sigma}_d = (\sigma_1 - \sigma_3, \sigma_2 - \sigma_3)$.

The plastic principal strain increments are readily obtained per the associated flow rule of Eq.(4)$_1$. Numerically generated flow surfaces with $a = 6$, 8, and 20 are shown in terms of biaxial plastic strain increments $\dot{\boldsymbol{\varepsilon}}_d^p = (\dot{\varepsilon}_1^p, \dot{\varepsilon}_2^p)$ in Figure 1(b). It is noted that the flow surface is given in the parametric form in Cauchy stress on the yield surface per the flow rule $\dot{\boldsymbol{\varepsilon}}_d^p = \dot{\boldsymbol{\varepsilon}}_d^p(\sigma)$. Except for $a = 2$ or $a = 4$, an analytical strain-rate function $q_H(\dot{\boldsymbol{\varepsilon}}_d^p)$ as the *exact* dual plastic potential of Hershey-Hosford yield stress function cannot be obtained in general. Instead, the

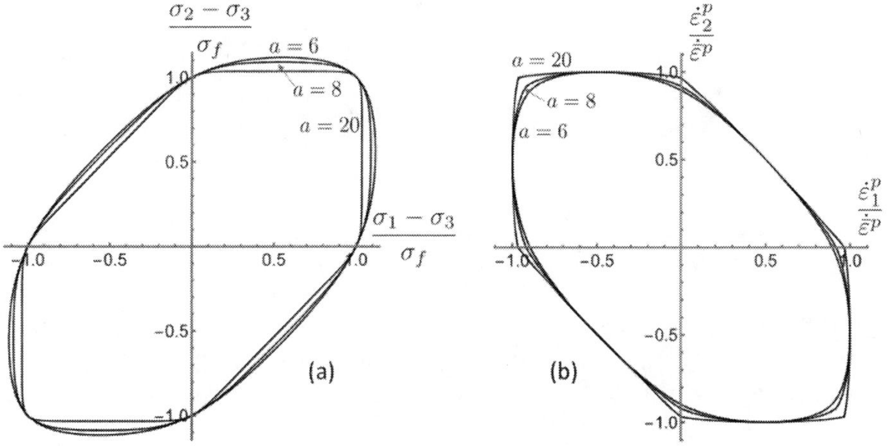

Figure 1. (a) Flow surfaces in stress space given by Hershey-Hosford yield function with $a =$ 6, 8, and 20; (b) the corresponding flow surfaces in strain-rate space given *numerically* by the associated flow rule.

following strain-rate function of the Hershey-Hosford form is used as an approximate or *pseudo* dual potential (noting $\dot{\varepsilon}_3^p = -\dot{\varepsilon}_1^p - \dot{\varepsilon}_2^p$ due to plastic incompressibility)

$$\widetilde{q}_H^b(\dot{\boldsymbol{\varepsilon}}_d^p) = \frac{|\dot{\varepsilon}_1^p|^b + |\dot{\varepsilon}_2^p|^b + |\dot{\varepsilon}_3^p|^b}{1 + 2^{1-b}} = \frac{|\dot{\varepsilon}_1^p|^b + |\dot{\varepsilon}_2^p|^b + |\dot{\varepsilon}_1^p + \dot{\varepsilon}_2^p|^b}{1 + 2^{1-b}}. \tag{8}$$

where the exponent $b = a$ is assumed first for simplicity. As shown in Figs.2(a)-2(c), the approximate flow surface $\widetilde{q}_H(\dot{\boldsymbol{\varepsilon}}_d^p) = \dot{\bar{\varepsilon}}^p$ with $b = a$ gives quite close matching of actual flow surfaces.

For a better approximation, one may find the least-square best-fit b numerically for a given value of stress exponent a by minimizing the sum-squared errors

$$SSD_H = \frac{1}{K}\sum_{k=1}^{K}\delta_H^2(k) = \frac{1}{K}\sum_{k=1}^{K}(\widetilde{q}_H(\dot{\varepsilon}_1^p(k)/\dot{\bar{\varepsilon}}^p, \dot{\varepsilon}_2^p(k)/\dot{\bar{\varepsilon}}^p) - 1)^2, \tag{9}$$

where $(\dot{\varepsilon}_1^p(k)/\dot{\bar{\varepsilon}}^p, \dot{\varepsilon}_2^p(k)/\dot{\bar{\varepsilon}}^p)$ are the numerical values of plastic strain increments computed from the flow rule of Hershey-Hosford yield function given in the parametric form in Cauchy stress on the yield surface with a total of $K = 1440$ discrete points. As an example (for the case of $a = 6$), the best-fit b is found to be 1.52234 with the resulting deviation $\delta_H(k)$ being shown in Figure 3 (where the largest absolute deviation is about 0.00218319). More detailed best-fit results are summarized in Table 1 for $a = 6, 8, ..., 100$. Besides the corresponding best-fit b values and resulting sum-squared errors SSD_H, the cases of $b = a$ and $b = a/(a-2)$ and their sum-squared errors SSD_H are also listed in the table. All three choices of the strain-rate exponent b of the pseudo dual potential Eq.(8) give rather small sum-of-square errors.

2.2. *Drucker's 1949 Sixth-Order Yield Function*

We consider next the flow stress surface given by Drucker's 1949 sixth-order yield function for isotropic materials [13]

$$f_D^6(\boldsymbol{\sigma}_d) = 27\alpha[J_2^3(\sigma_1, \sigma_2, \sigma_3) - cJ_3^2(\sigma_1, \sigma_2, \sigma_3)] = \sigma_f^6, \tag{10}$$

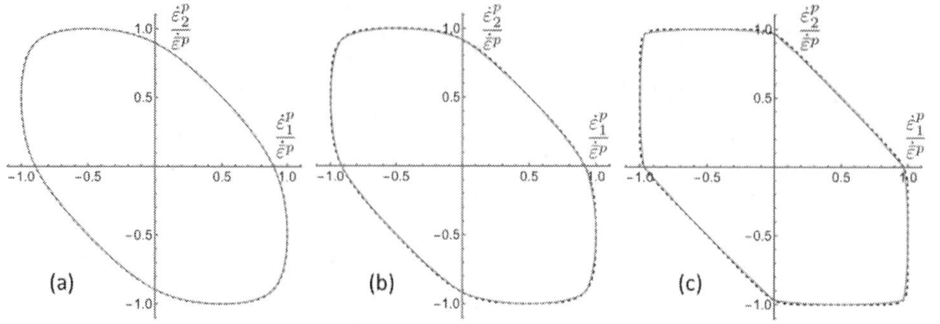

Figure 2. Comparison of the approximate flow surface based on the pseudo dual strain-rate function $\widetilde{q}_H(\dot{\boldsymbol{\varepsilon}}) = \dot{\bar{\varepsilon}}^p$ in solid lines with the actual flow surface per flow rule in dashed lines: (a) $b = 6$; (b) $b = 8$; (c) $b = 20$.

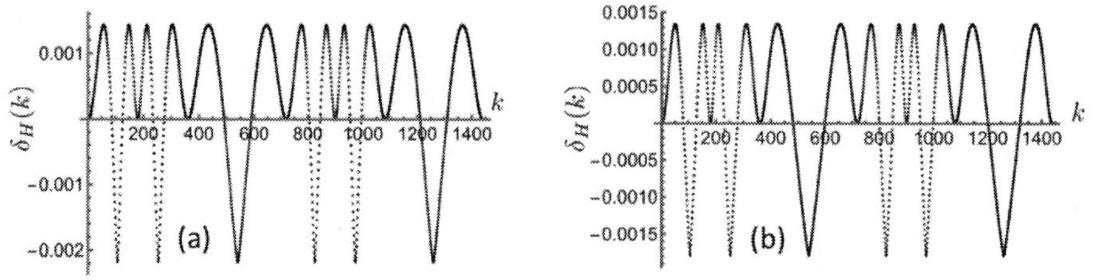

Figure 3. Deviations of the approximate from actual flow surfaces $\delta_H = \widetilde{q}_H(\dot{\varepsilon})/\dot{\bar{\varepsilon}}^p$-1 along the stress states of the yield surface: (a) $b = 6$; (b) $b = 8$.

Table 1. A summary of exponents a and b values plus deviations in flow surfaces.

a	6	8	12	20	50	100
best-fit b	1.52234	1.3467	1.2059	1.11314	1.04196	1.02048
SSD_H	1.1×10^{-6}	9.0×10^{-7}	2.6×10^{-7}	2.9×10^{-8}	3.4×10^{-10}	1.1×10^{-11}
$b = a$	6	8	12	20	50	100
SSD_H	3.8×10^{-6}	2.1×10^{-5}	3.7×10^{-5}	2.6×10^{-5}	5.0×10^{-6}	1.0×10^{-6}
$b = a/(a-2)$	1.5	1.3333	1.2	1.11111	1.04167	1.02041
SSD_H	3.8×10^{-6}	3.3×10^{-6}	1.3×10^{-6}	2.5×10^{-7}	8.3×10^{-9}	5.5×10^{-10}

where

$$\begin{aligned} 6J_2(\sigma_1,\sigma_2,\sigma_3) &= (\sigma_1-\sigma_2)^2+(\sigma_2-\sigma_3)^2+(\sigma_3-\sigma_1)^2, \\ 27J_3(\sigma_1,\sigma_2,\sigma_3) &= (2\sigma_1-\sigma_2-\sigma_3)(2\sigma_2-\sigma_3-\sigma_1)(2\sigma_3-\sigma_1-\sigma_2), \end{aligned} \tag{11}$$

c is an adjustable material constant, $\alpha = 27/(27-4c)$ and σ_f is the current yield strength of the material in uniaxial tension. Dodd and Naruse [14] subsequently showed that $c < 27/4$ and $-27/8 \leq c \leq 9/4$ are the *necessary and sufficient* conditions that are required to ensure Drucker's yield function be positive and convex respectively. Flow stress surfaces $f_D(\boldsymbol{\sigma}_d) = \sigma_f(\bar{\varepsilon}^p)$ with c = -27/8 and 9/4 are shown in Figure 4(a) in terms of biaxial deviatoric stresses $\boldsymbol{\sigma}_d = (\sigma_1-\sigma_3, \sigma_2-\sigma_3)$ and the corresponding flow strain-rate surfaces numerically generated per the associated flow rule are shown in Figure 4(b).

Guided by the results obtained for the Hershey-Hosford yield function above, the following

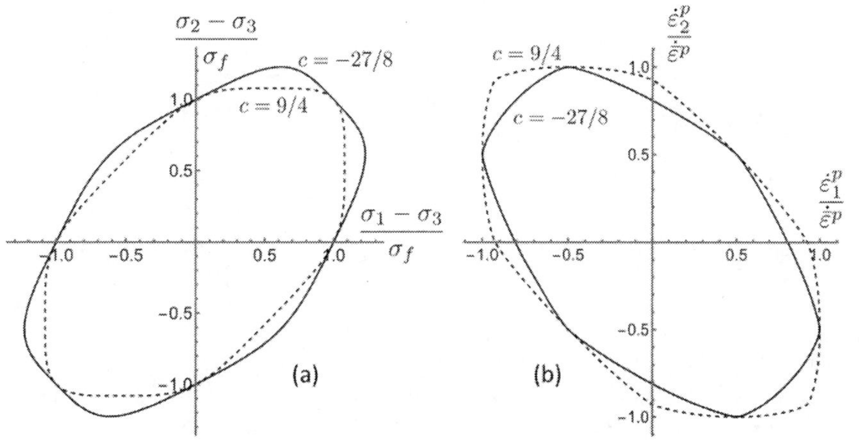

Figure 4. (a) Flow surfaces given by Drucker's 1949 yield function with c = -27/8 (solid lines) and 9/4 (dashed lines): (a) in stress space; (b) in strain-rate space.

strain-rate function of the Drucker's 1949 form is proposed as an approximate or pseudo dual potential

$$\tilde{q}_D^6(\dot{\boldsymbol{\varepsilon}}_d^p) = 27\beta[I_2^3(\dot{\varepsilon}_1^p, \dot{\varepsilon}_2^p, \dot{\varepsilon}_3^p) - dI_3^2(\dot{\varepsilon}_1^p, \dot{\varepsilon}_2^p, \dot{\varepsilon}_3^p)], \tag{12}$$

where

$$\begin{aligned} 6I_2(\dot{\varepsilon}_1^p, \dot{\varepsilon}_2^p, \dot{\varepsilon}_3^p) &= (\dot{\varepsilon}_1^p - \dot{\varepsilon}_2^p)^2 + (\dot{\varepsilon}_2^p - \dot{\varepsilon}_3^p)^2 + (\dot{\varepsilon}_3^p - \dot{\varepsilon}_1^p)^2, \\ 27I_3(\dot{\varepsilon}_1^p, \dot{\varepsilon}_2^p, \dot{\varepsilon}_3^p) &= (2\dot{\varepsilon}_1^p - \dot{\varepsilon}_2^p - \dot{\varepsilon}_3^p)(2\dot{\varepsilon}_2^p - \dot{\varepsilon}_3^p - \dot{\varepsilon}_1^p)(2\dot{\varepsilon}_3^p - \dot{\varepsilon}_1^p - \dot{\varepsilon}_2^p), \end{aligned} \tag{13}$$

d is an adjustable material constant between -27/8 and 9/4 per the convexity requirement and $\beta = 64/(27(27-4d))$.

Clearly, the value of constant d is dependent on the value of constant c. One may find the least-square best-fit d numerically for a given value of c by minimizing the sum-squared errors

$$SSD_D = \frac{1}{K}\sum_{k=1}^{K}\delta_D^2(k) = \frac{1}{K}\sum_{k=1}^{K}(\tilde{q}_D(\dot{\varepsilon}_1^p(k)/\dot{\bar{\varepsilon}}^p, \dot{\varepsilon}_2^p(k)/\dot{\bar{\varepsilon}}^p) - 1)^2, \quad -27/8 \le d \le 9/4, \tag{14}$$

where $(\dot{\varepsilon}_1^p(k)/\dot{\bar{\varepsilon}}^p, \dot{\varepsilon}_2^p(k)/\dot{\bar{\varepsilon}}^p)$ are the numerical values of plastic strain increments computed from the flow rule of Drucker's 1949 yield function given in the parametric form in Cauchy stress on the yield surface with a total of K = 1440 discrete points. Table 2 summarized some typical results for selected c values between -27/8 and 9/4 and the corresponding best-fit d values and resulting sum-squared errors SSD_D. As shown in Figs.5(a) and 5(b), the approximate flow strain-rate surfaces $\tilde{q}_D(\dot{\boldsymbol{\varepsilon}}_d^p) = \dot{\bar{\varepsilon}}^p$ with d = 9/4 and -27/8 are compared with the numerically generated ones based on the flow rule applied to $f_D(\boldsymbol{\sigma}_d) = \sigma_f(\bar{\varepsilon}^p)$ with c = -27/8 and 9/4.

Table 2. A summary of constants c and d values plus deviations in flow surfaces.

c	-27/8	-2	-1	0	1	9/4
best-fit d	9/4	1.65891	0.909222	0	-1.11579	-3.0723
SSD_D	4.8×10^{-5}	6.7×10^{-6}	5.5×10^{-7}	0	8.7×10^{-7}	2.9×10^{-5}

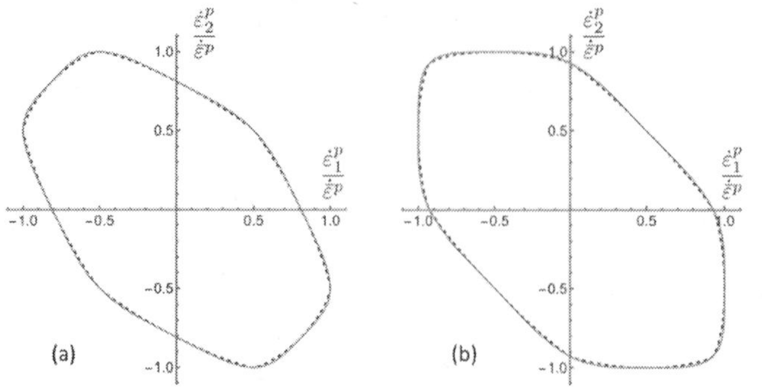

Figure 5. Comparison of the approximate flow surface based on the pseudo dual strain-rate function $\widetilde{q}_D(\dot{\boldsymbol{\varepsilon}}) = \dot{\bar{\varepsilon}}^p$ in solid lines with the actual flow surface per flow rule in dashed lines: (a) c=-27/8, d =9/4; (b) c = 9/4, d = -27/8.

3. Non-Quadratic Orthotropic Flow Potentials

For modeling plastic flow in sheet metals, orthotropic yield functions and plastic potentials are often used. The earliest and perhaps also simplest non-quadratic orthotropic yield function that has been fully applied to sheet metal plasticity modeling is Gotoh's 1977 quartic yield function [15] in the following form in plane stress

$$f_G^4(\boldsymbol{\sigma}) = A_1\sigma_x^4 + A_2\sigma_x^3\sigma_y + A_3\sigma_x^2\sigma_y^2 + A_4\sigma_x\sigma_y^3 + A_5\sigma_y^4 + A_6\sigma_x^2\tau_{xy}^2 + A_7\sigma_x\sigma_y\tau_{xy}^2 + A_8\sigma_y^2\tau_{xy}^2 + A_9\tau_{xy}^4, \tag{15}$$

where $(A_1, A_2, ..., A_9)$ are nine polynomial coefficients (material constants). Gotoh considered also a fourth-order strain-rate potential

$$\widetilde{q}_G^4(\dot{\boldsymbol{\varepsilon}}) = B_1\dot{\varepsilon}_x^4 + B_2\dot{\varepsilon}_x^3\dot{\varepsilon}_y + B_3\dot{\varepsilon}_x^2\dot{\varepsilon}_y^2 + B_4\dot{\varepsilon}_x\dot{\varepsilon}_y^3 + B_5\dot{\varepsilon}_y^4 + B_6\dot{\varepsilon}_x^2\dot{\gamma}_{xy}^2 + B_7\dot{\varepsilon}_x\dot{\varepsilon}_y\dot{\gamma}_{xy}^2 + B_8\dot{\varepsilon}_y^2\dot{\gamma}_{xy}^2 + B_9\dot{\gamma}_{xy}^4, \tag{16}$$

where $(B_1, B_2, ..., B_9)^2$ are its nine polynomial coefficients. He showed that the fourth-order stress-based yield function and the fourth-order strain-rate-based plastic potential are not equivalent in general. That is, only under some special occasions (e.g., square of Hill's 1948 yield function), they are exact dual plastic potentials. Nevertheless, a complete 3D fourth-order plastic strain-rate potential with orthotropic symmetry and its analytical calculations from texture functions was subsequently applied to effectively model polycrystalline sheet metals [16, 17, 18].

Here we examine the degree of approximation of a fourth-order strain-rate function as a pseudo dual to Gotoh's fourth-order yield stress function for some selected sheet metals. As the initial effort, we limit our attention to the special case of biaxial loading only without any shear stress or strain rate components. It is noted that simple algebraic necessary and sufficient convex conditions have recently been given in [19] for biaxial loading on their five polynomial coefficients as

$$\begin{pmatrix} 12A_1 & 3A_2 & 2A_3 \\ 3A_2 & 2A_3 & 3A_4 \\ 2A_3 & 3A_4 & 12A_5 \end{pmatrix} \geq 0, \quad \begin{pmatrix} 12B_1 & 3B_2 & 2B_3 \\ 3B_2 & 2B_3 & 3B_4 \\ 2B_3 & 3B_4 & 12B_5 \end{pmatrix} \geq 0. \tag{17}$$

For a calibrated Gotoh's yield function (i.e., with a set of known $A_1, A_2, ..., A_5$), one may find the least-square best-fit set of $(B_1, B_2, ..., B_5)$ by minimizing the sum-squared errors

$$SSD_G = \frac{1}{K}\sum_{k=1}^{K}\delta_G^2(k) = \frac{1}{K}\sum_{k=1}^{K}(\widetilde{q}_G(\dot{\varepsilon}_1^p(k)/\dot{\bar{\varepsilon}}^p, \dot{\varepsilon}_2^p(k)/\dot{\bar{\varepsilon}}^p) - 1)^2, \tag{18}$$

[2] Coefficients $(E_1, ..., E_9)$ were used instead in Eq.(19) of [15].

IOP Conf. Series: Materials Science and Engineering 1238 (2022) 012004 doi:10.1088/1757-899X/1238/1/012004

where $(\dot{\varepsilon}_1^p(k)/\dot{\bar{\varepsilon}}^p, \dot{\varepsilon}_2^p(k)/\dot{\bar{\varepsilon}}^p)$ are the numerical values of plastic strain increments computed from the flow rule of Gotoh's 1977 yield function given in the parametric form in Cauchy stress on the yield surface with a total of $K = 1440$ discrete points.

Biaxial flow stress surfaces and the corresponding flow strain-rate surfaces are shown in Fig.6(a) and (b) for an aluminum sheet and a steel sheet respectively. Table 3 summarized the best-fit results for the two selected sheet metals and the resulting sum-squared errors SSD_G. The convexity conditions (positive-semidefinite on those two 3-by-3 matrices in Eq.(17)) were verified numerically. As shown in Figs.7(a) and 7(b), the approximate flow strain-rate surfaces $\widetilde{q}_G(\dot{\boldsymbol{\varepsilon}}_d^p) = \dot{\bar{\varepsilon}}^p$ are compared very well with the numerically generated ones based on the flow rule applied to $f_G(\boldsymbol{\sigma}_d) = \sigma_f(\bar{\varepsilon}^p)$ for the aluminum and steel sheets.

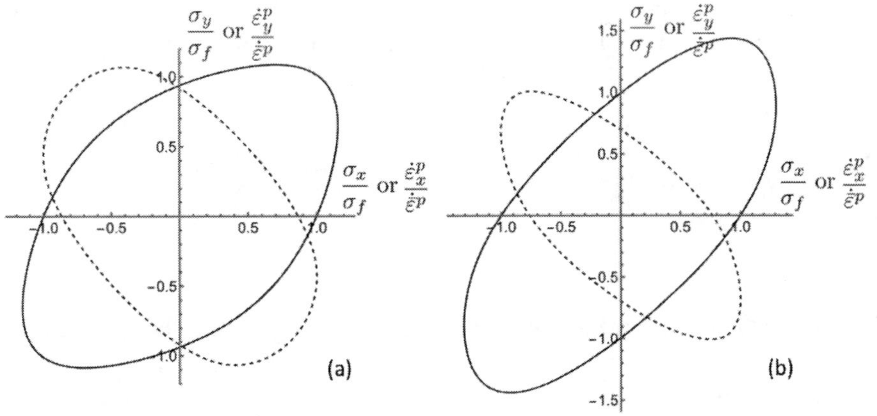

Figure 6. (a) Flow stress surfaces (solid lines) and flow strain-rate surfaces (dashed lines) given by Gotoh's 1977 yield function: (a) an aluminum 60616-T4 sheet [10]; (b) an IF steel sheet [10].

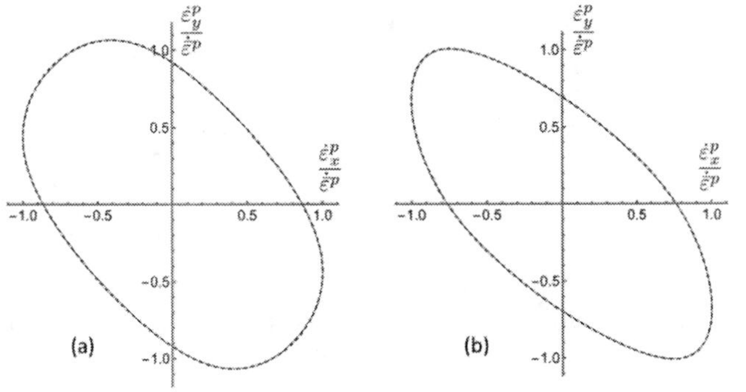

Figure 7. Comparison of the approximate flow surface based on the pseudo dual strain-rate function $\widetilde{q}_G(\dot{\boldsymbol{\varepsilon}}) = \dot{\bar{\varepsilon}}^p$ in solid lines with the actual flow surface per flow rule in dashed lines: (a) an aluminum 60616-T4 sheet [10]; (b) an IF steel sheet [10].

4. Discussion and Conclusions

Historically, two forms of the same plastic work rate in plasticity are the dual (the function and its conjugate) as noted much earlier by Hill in 1956 [20]

$$\frac{1}{2}\dot{w}_p(\boldsymbol{\sigma}) + \frac{1}{2}\dot{w}_p^*(\dot{\boldsymbol{\varepsilon}}^p) = \frac{1}{2}\dot{\bar{\varepsilon}}^p f(\boldsymbol{\sigma}) + \frac{1}{2}\sigma_f q(\dot{\boldsymbol{\varepsilon}}^p) = \boldsymbol{\sigma} : \dot{\boldsymbol{\varepsilon}}^p, \tag{19}$$

Table 3. A summary of polynomial coefficient values plus deviations in flow surfaces.

	A_1	A_2	A_3	A_4	A_5	
AA6016-T4	1	-1.7273	2.3675	-1.9619	1.2945	
IF steel	1	-2.7179	4.1770	-3.0688	1.0327	
best-fit	B_1	B_2	B_3	B_4	B_5	SSD_G
AA6016-T4	1.78004	3.9593	6.22603	3.4712	1.37219	2.0×10^{-6}
IF steel	2.95682	8.11137	13.2123	11.006	4.28155	9.4×10^{-7}

where the first and second terms on the left-hand side of the above equation have been called by Hill as work and complementary functions respectively. In fact, dual plastic potentials may also be regarded as examples of Legendre dual transformation on convex differentiable functions or their conjugates [21, 22]. One function may be obtained from the other function via the Legendre transform [23]

$$\frac{1}{2}\sigma_f q(\dot{\boldsymbol{\varepsilon}}^p) = \boldsymbol{\sigma}(\dot{\boldsymbol{\varepsilon}}^p) : \dot{\boldsymbol{\varepsilon}}^p - \frac{1}{2}\dot{\bar{\varepsilon}}^p f(\boldsymbol{\sigma}(\dot{\boldsymbol{\varepsilon}}^p)), \quad \dot{\boldsymbol{\varepsilon}}^p = \dot{\bar{\varepsilon}}^p \frac{\partial f(\boldsymbol{\sigma})}{\partial \boldsymbol{\sigma}} \quad \Rightarrow \quad \boldsymbol{\sigma}(\dot{\boldsymbol{\varepsilon}}^p), \tag{20}$$

$$\frac{1}{2}\dot{\bar{\varepsilon}}^p f(\boldsymbol{\sigma}) = \boldsymbol{\sigma} : \dot{\boldsymbol{\varepsilon}}^p(\boldsymbol{\sigma}) - \frac{1}{2}\sigma_f q(\dot{\boldsymbol{\varepsilon}}^p(\boldsymbol{\sigma})), \quad \boldsymbol{\sigma} = \sigma_f \frac{\partial q(\dot{\boldsymbol{\varepsilon}}^p)}{\partial \dot{\boldsymbol{\varepsilon}}^p} \quad \Rightarrow \quad \dot{\boldsymbol{\varepsilon}}^p(\boldsymbol{\sigma}). \tag{21}$$

Although these two Legendre transforms appear at the first glance to be different from the early introduction on the dual plastic potentials given by Eq.(1)-Eq.(6), they are in fact equivalent as $\sigma_f q(\dot{\boldsymbol{\varepsilon}}^p) = \dot{\bar{\varepsilon}}^p f(\boldsymbol{\sigma}) = \boldsymbol{\sigma} : \dot{\boldsymbol{\varepsilon}}^p$ due to the two particular positive scalars σ_f and $\dot{\bar{\varepsilon}}^p$ used here.

As a convex even-order homogeneous polynomial [5, 1, 15, 10] is often used to define the yield stress function, one may obtain directly the strain-rate function $q(\dot{\boldsymbol{\varepsilon}}^p)$ using the normalized Cauchy stress $\boldsymbol{\sigma}/\sigma_f$ in terms of the normalized plastic strain-rate $\dot{\boldsymbol{\varepsilon}}^p/\dot{\bar{\varepsilon}}^p$ from the flow strain-rate rule

$$\frac{\dot{\boldsymbol{\varepsilon}}^p}{\dot{\bar{\varepsilon}}^p} = \frac{\partial f(\boldsymbol{\sigma})}{\partial \boldsymbol{\sigma}} = r(\frac{\boldsymbol{\sigma}}{\sigma_f}) \Rightarrow \frac{\boldsymbol{\sigma}}{\sigma_f} = r^{-1}(\frac{\dot{\boldsymbol{\varepsilon}}^p}{\dot{\bar{\varepsilon}}^p}) \Rightarrow \frac{q(\dot{\boldsymbol{\varepsilon}}^p)}{\dot{\bar{\varepsilon}}^p} = f(r^{-1}(\frac{\dot{\boldsymbol{\varepsilon}}^p}{\dot{\bar{\varepsilon}}^p})) = 1. \tag{22}$$

An analytical solution of $\boldsymbol{\sigma}$ in terms of $\dot{\boldsymbol{\varepsilon}}^p$ from inverting the given flow rule result of Eq.$(4)_1$ may not always be feasible in general even when the yield stress function $f(\boldsymbol{\sigma})$ is both smooth and convex. The resulting plastic strain-rate surface may thus only strictly be in the discrete numerical form for non-quadratic yield stress function for either isotropic or anisotropic materials. Nevertheless, results from our current study as presented here indicate that a plastic *strain-rate* function formulated simply in the same form as the non-quadratic yields *stress* function may be used as its pseudo dual with good approximation of the actual flow surface.

So far in this study the dual plastic potentials have been presented and discussed in the framework of associated plasticity where yield and flow surfaces are treated to be identical and exchangeable. In non-associated anisotropic plasticity modeling of sheet metals, the yield surface and flow surface in the stress space are distinct as they are defined by a yield stress function $f(\boldsymbol{\sigma})$ and a flow stress function $g(\boldsymbol{\sigma})$ respectively where $g(\boldsymbol{\sigma}) \neq f(\boldsymbol{\sigma})$. The notion of a *dual* plastic strain-rate potential is applied to the flow stress function only with the flow strain-rate and stress rules as

$$\dot{\boldsymbol{\varepsilon}}^p = \dot{\lambda}\frac{\partial g(\boldsymbol{\sigma})}{\partial \boldsymbol{\sigma}}, \quad \boldsymbol{\sigma} = \tilde{\sigma}\frac{\partial q(\dot{\boldsymbol{\varepsilon}}^p)}{\partial \dot{\boldsymbol{\varepsilon}}^p}, \quad \dot{\lambda}(\dot{\boldsymbol{\varepsilon}}^p) > 0 \text{ if } \dot{\boldsymbol{\varepsilon}}^p \neq 0, \quad \tilde{\sigma}(\boldsymbol{\sigma}) > 0 \text{ if } \boldsymbol{\sigma} \neq 0, \tag{23}$$

where $\dot{\lambda}(\dot{\boldsymbol{\varepsilon}}^p)$ is not necessarily the equivalent plastic strain rate $\dot{\bar{\varepsilon}}$ and $\tilde{\sigma}(\boldsymbol{\sigma})$ is not necessarily the equivalent yield/flow stress $\bar{\sigma}(\boldsymbol{\sigma})$ (which has been set to be $f(\boldsymbol{\sigma}) = \sigma_f$ with $\dot{\bar{\varepsilon}} = \dot{\bar{\varepsilon}}^p$ for associated plasticity in this study per Eq.(1)). In fact, one has per the plastic work rate equivalency

$$\boldsymbol{\sigma} : \dot{\boldsymbol{\varepsilon}}^p = \bar{\sigma}\dot{\bar{\varepsilon}} = g(\boldsymbol{\sigma})\dot{\lambda} = \widetilde{\sigma} q(\dot{\boldsymbol{\varepsilon}}^p) \quad \Rightarrow \quad \dot{\lambda} = \frac{\bar{\sigma}}{g(\boldsymbol{\sigma})}\dot{\bar{\varepsilon}}, \quad \widetilde{\sigma} = \frac{\dot{\bar{\varepsilon}}}{q(\dot{\boldsymbol{\varepsilon}}^p)}\bar{\sigma}. \tag{24}$$

For simplicity, one may choose $\widetilde{\sigma} = \bar{\sigma} = g(\boldsymbol{\sigma})$ so $\dot{\lambda} = \dot{\bar{\varepsilon}} = q(\dot{\boldsymbol{\varepsilon}}^p)$ for non-associated plasticity. It is noted here that the flow surface in the strain-rate space $q(\dot{\boldsymbol{\varepsilon}}^p) = \dot{\bar{\varepsilon}}^p$ is no longer dual to the yield surface $f(\boldsymbol{\sigma}) = \sigma_f$. As illustrated in Figure 8 in biaxial stress space, the yield surface $f(\boldsymbol{\sigma}) = \sigma_f$ is defined in terms of the actual Cauchy stress $\boldsymbol{\sigma}$ while the nominal flow surface $g(\boldsymbol{\sigma}^*) = \sigma_f$ is defined in terms of the nominal Cauchy stress $\boldsymbol{\sigma}^*$ instead. It is this nominal flow surface that is associated with the dual plastic strain-rate potential $q(\dot{\boldsymbol{\varepsilon}}^p)$. This can be seen from the flow stress rule of Eq.$(23)_2$

$$\boldsymbol{\sigma} = g(\boldsymbol{\sigma})\frac{\partial q(\dot{\boldsymbol{\varepsilon}}^p)}{\partial \dot{\boldsymbol{\varepsilon}}^p} \quad \Rightarrow \quad \boldsymbol{\sigma}^* = \sigma_f \frac{\partial q(\dot{\boldsymbol{\varepsilon}}^p)}{\partial \dot{\boldsymbol{\varepsilon}}^p}, \quad \boldsymbol{\sigma}^* = \eta\boldsymbol{\sigma} \text{ and } g(\boldsymbol{\sigma}^*) = g(\eta\boldsymbol{\sigma}) = \eta g(\boldsymbol{\sigma}) = \sigma_f, \tag{25}$$

where the positive scalar η can be obtained from the flow stress rule and yield condition as

$$f(\boldsymbol{\sigma}) = f\Big(g(\boldsymbol{\sigma})\frac{\partial q(\dot{\boldsymbol{\varepsilon}}^p)}{\partial \dot{\boldsymbol{\varepsilon}}^p}\Big) = g(\boldsymbol{\sigma}) f\Big(\frac{\partial q(\dot{\boldsymbol{\varepsilon}}^p)}{\partial \dot{\boldsymbol{\varepsilon}}^p}\Big) = \sigma_f \quad \Rightarrow \quad \eta = f\Big(\frac{\partial q(\dot{\boldsymbol{\varepsilon}}^p)}{\partial \dot{\boldsymbol{\varepsilon}}^p}\Big). \tag{26}$$

Although the development of a particular plasticity model and its implementation in a finite element code need only either flow stress or plastic strain-rate function, there are certain advantages to have both functions on hand. For example, when using a texture-adjusted plastic strain-rate potential in sheet metal forming modeling applications, the principle of maximal work has been used to test whether a given stress state lies within the yield surface instead of invoking directly the yield condition [17, 18]. When one obtains in very good approximation the pseudo dual yield stress function from its plastic strain-rate potential, one may then use the existing finite element code commonly based on the yield stress function where the yield condition can be used directly to check a given stress state in the numerical simulation.

In summary, the existence of a dual plastic strain-rate potential has been well established for any flow stress function in plasticity [4] and its applications appear increasingly in anisotropic plasticity modeling of crystalline metals [16, 17, 18, 24, 25, 26, 27]. For example, a yield polyhedron and its dual were derived for a single crystal [24] and a textured polycrystal [25]. The strain rate potential Srp2004-18p analogous to Yld2004-18p was constructed based on experimental measurements [26]. Rabahallah et al. [27] considered several advanced strain-rate potentials (quartus, Srp93, Srp2004-18p) identified with crystallographic texture data and mechanical testing inputs. Analytical derivation of the exact dual potential may be only feasible in general for quadratic yield functions. Numerical results presented in this study show that one

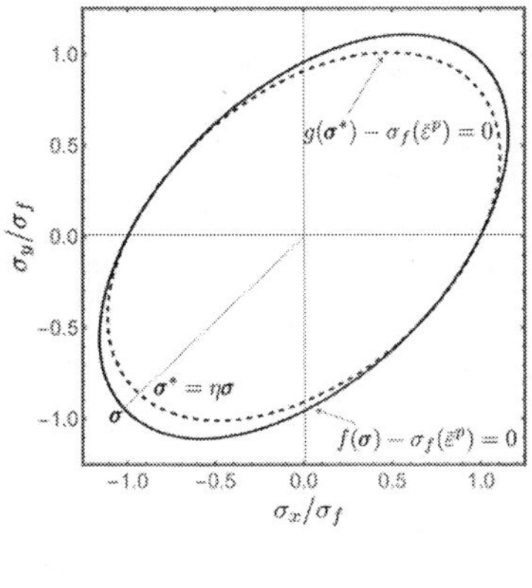

Figure 8. An illustration of yield surface $f(\boldsymbol{\sigma}) = \sigma_f$ (solid line) and the corresponding flow surface $g(\boldsymbol{\sigma}^*) = \sigma_f$ (dashed line) for non-associated plasticity $f \neq g$: where the nominal stress is given as $\boldsymbol{\sigma}^* = \eta\boldsymbol{\sigma}$, $\eta = \sigma_f/g(\boldsymbol{\sigma}) > 0$.

may nevertheless obtain for both isotropic and anisotropic materials a very good pseudo-dual non-quadratic strain-rate potential of the same mathematical form as the given non-quadratic flow stress function. As our work so far has only studied three simple non-quadratic yield stress functions, our future effort will focus on pseudo-dual strain-rate potentials of more advanced yiedl stress functions such as Yld2000-2d, Yld2004-18p and CB2001 and its variants [28, 29, 30].

References

[1] Hill R 1950 *The Mathematical Theory of Plasticity.* Oxford University Press, London
[2] Lubliner J 1990 *Plastic Theory.* Macmillan, New York
[3] Maugin G A 1992 *The Thermomechanics of Plasticity and Fracture.* Cambridge University Press, Cambridge
[4] Hill R 1987 Constitutive dual potentials in classical plasticity *J. Mech. Phys. Solids* **35**(1) 23–33
[5] Hill R 1948 A theory of yielding and plastic flow of anisotropic metals *Proc. Royal Soc.* **193A** 281–97
[6] Arminjon M and Bacroix B 1990 On plastic potentials for anisotropic metals and their derivation from texture function *Acta Mechanica* **88** 219–43
[7] Darrieulat M and Montheillet F 2003 A texture based continuum approach for predicting the plastic behaviour of rolled sheet *Int. J. Plasticity* **19** 517–46
[8] Hill R 1990 Constitutive modeling of orthotropic plasticity in sheet metals *J Mech. Phys. Solids* **38**(2) 405–17
[9] Stoughton T B 2002 A non-associated flow rule for sheet metal forming *Int. J. Plasticity* **18** 687–714
[10] Tong W and Alharbi M 2017 Comparative evaluation of non-associated quadratic and associated quartic plasticity models for orthotropic sheet metals *Int. J. Solids & Struct.* **128** 133–48
[11] Hershey A V 1954 The plasticity of an isotropic aggregate of anisotropic face centered cubic crystals *J. Appl. Mech.* **21** 241–9
[12] Hosford W F 1972 A generalized isotropic yield criterion *J. Appl. Mech.* **39** 607–9
[13] Drucker D C 1949 Relations of experiments to mathematical theories of plasticity *J. Appl Mech.* **16** 349–57
[14] Dodd B and Naruse K 1989 Limitation on isotropic yield criterion *Int. J. Mech. Sci.* **31** 511–19
[15] Gotoh M 1977 A theory of plastic anisotropy based on a yield function of fourth order (plane stress state) *Int. J. Mech. Sci.* **19** 505–20
[16] Arminjon M, Bacroix B, Imbault D and Raphanel J L 1994 A fourth-order plastic potential for anisotropic metals and its analytical calculation from the texture function *Acta Mechanica* **107** 31–51
[17] Bacroix B and Gilormini P 1995 Finite-element simulations of earing in polycrystalline materials using a texture-adjusted strain-rate potential *Modelling Simul. Mater. Sci. Eng.* **3** 1–21
[18] Zhou Y, Jonas J J, Szabo I L, Makinde A, Jain M and MacEwen S R 1997 Incorporation of an anisotropic strain-rate potential into three dimensional finite element simulations *International Journal of Plasticity* **13** 165–81
[19] Tong W 2018 Algebraic convexity conditions for Gotoh's non-quadratic yield function *ASME Journal of Applied Mechanics* **85** 074501-1
[20] Hill R 1956 New horizons in the mechanics of solids *J. Mech. Phys. Solids* **5** 66–74
[21] Hiriart-Urruty J-B and Lemaréchal C 1993 *Convex Analysis and Minimization Algorithms. I. Fundamentals; II.Advanced Theory and Bundle Methods.* The Grundlehren der mathematischen Wissenschaften book series (GL, volume 305), Springer
[22] Lasserre J B 1998 Homogeneous functions and conjugacy *J. Convex Anal.* **5** 397–403
[23] Zia R K P, Redish E F and McKay S R 2009 Making sense of the Legendre transform *Am. J. Phys.* **77**(7) 614–22
[24] Fortunier R 1989 Dual potentials and extremum work principles in single crystal plasticity *J. Mech. Phys. Solids* **37** 779–90
[25] Van Houtte P, Van Bael A and Winters J 1995 The incorporation of texture-based yield loci into elasto-plastic finite element programs *Texture Microstructures* **24** 225–72
[26] Kim D, Barlat F, Bouvier S, Rabahallah M, Balan T and Chung K 2007 Nonquadratic anisotropic potentials based on linear transformation of plastic strain rate *Int. J. Plasticity* **23** 1380–99
[27] Rabahallah M, Balan T, Bouvier S, Bacroix B, Barlat F and Chung K 2009 Parameter identification of advanced plastic strain rate potentials and impact on plastic anisotropy prediction *Int. J. Plasticity* **25** 491–512.
[28] Cazacu O and Barlat F 2001 Generalization of Drucker's yield criterion to orthotropy *Mathematics and Mechanics of Solids* **6**(6) 613–30.
[29] Cazacu O and Barlat F 2003 Application of the theory of representation to describe yielding of anisotropic aluminum alloys *Int. J. Engineering Sci.* **41** 1367–85
[30] Banabic D, Barlat F, Cazacu O and Kuwabara T 2020 Advances in anisotropy of plastic behavior and formability of sheet metals *International Journal of Forming Materials* **13**(5) 749–87

IOP Conf. Series: Materials Science and Engineering 1238 (2022) 012005 doi:10.1088/1757-899X/1238/1/012005

Numerical analysis of AISI 321 alloy material parameters on rubber pad diaphragm forming

P S Barros[1,2*], N Otegi[1], L Galdós[1] and E Sáenz de Argandoña[1]

[1] Mondragon Unibertsitatea, Loramendi Kalea, 4, 20500 Mondragon, Spain
[2] ITP-Aero, Parke Tecnológico de Bizkaia, Edificio 902, 48160 Derio, Spain

* psanchez@mondragon.edu

Abstract. Rubber pad diaphragm forming is a technology that uses a rigid die in conjunction with a flexible polyurethane tool and fluid pressure action in order to shape metal sheets. In this process, the action of the fluid forms a polyurethane tool, which forces the sheet to acquire the geometry of the die, being this process quite different from the classical stamping processes. The rubber tool is compatible with different die geometries, which allows a quick reconfiguration of the system, making it adequate for the production of small batches in the aeronautical sector. In this study, an advanced numerical model of the process is developed, in order to obtain a simulation that allows comprehensive understanding of the technology. An advanced characterization of the AISI 321 stainless steel is carried out, evaluating different yielding criteria. On the other side, the hyperelastic behaviour of the polyurethane rubber is taken into account. The main goal of the study has been to perform a sensitivity analysis of the sheet material parameters and to understand their influence on the final numerical results.

1. Introduction

Nowadays the use of numerical methods and finite elements is more widespread in the daily use of the industry. The numerical simulations, allows the companies to reduce the designing time, the try outs and consequently the costs, increasing the competitiveness of the industries. In the field of sheet metal forming processes, simulation plays a fundamental role, allowing us to detect the defects, before forming the blanks. This detection of defects is important in industries such as the aeronautical industry, where despite small size of the batches, the presence of defects in the parts, like earing, splits, wrinkles and springback or a high number of try outs, supposes a loss of competitiveness. In order to avoid this problem, it is important to understand which ones are the main control variables in the process and in the simulation.

This work will focus on the rubber pad diaphragm forming process, a rubber forming process introduced in the industry in the last years, and based on the Verson-Wheelon process [1]. This process was designed with the aeronautical industry in mind, since it allows the production of small batches and a relatively fast die exchange. As its name indicates, the process is framed within the rubber forming processes, being the simulation field of this type of technology less developed. Different elements are used in the rubber pad diaphragm forming process. On the one hand, it is a die with the geometry that it is desired to obtain. The sheet metal is placed on this die, and on the blank, a rigid rubber tool known as throw pad is placed. The machine contains a volume of oil, which is supported by a soft rubber diaphragm. To perform the forming, the pressure is increased in the oil, which deforms the diaphragm like a balloon, and pushes the other elements in a similar way to the hydroforming process.

In order to better understand the influence of the characterization of the sheet material on the process, different yield criteria will be evaluated by finite element simulation. In this work, the yield constitutive models that will be fitted and simulated are, the Hill48's criterion, proposed by R.Hill in 1948 [2], being one of the simplest and more used in sheet metal forming simulation, the Yld-2000's criterion, proposed by F.Barlat et al. in 2003 [3] and the 6th order Yoshida's criterion, proposed by Yoshida et al. in 2013 [4].

To carry out this study, the metallic stainless steel alloy AISI 321 will be used. This kind of alloy was selected for its use in the aeronautical industry, as well as its good mechanical properties and formability [5]. In this work, the throw pad will also be introduced, and its hyperelastic mechanical behavior will be described [6]. The selected material for the rubber will be a polyurethane rubber with ShoreA 90 hardness and 10 mm thickness. In order to evaluate the influence of the mechanical properties of the AISI 321 in the process, a battery of simulations will be carried out in which different frictions and yield criteria will be mixed, in order to see which one of the variables has a greater influence on the process.

2. Materials and methods

2.1. *Constitutive material laws*

The AISI 321 blank mechanical behavior during the forming process is described in the numerical simulation by the constitutive models. The elastic behavior of the material is defined by its Young's modulus E and the Poisson's coefficient υ. The material hardening defines the stress-strain relation when the material is in the plastic regime and the yield criterions, sets the boundaries between the elastic and plastic regimes, determined by the yield stress σ_y. The yield criteria compared in this article, are given by:

- Hill's criterion (1948) [2]: Extended the Von Mises criterion, introducing stress weighting coefficients in each direction, in order to take into account the anisotropy of the material, where F, G, H, L, M, N are constants to be defined by experimental tests. It is a symmetric and quadratic criterion in shape. These coefficients are generally calculated starting from the anisotropy coefficients (r) at 0°, 45° and 90° from the rolling direction. This criterion is widely used and provide good results in the simulation of steel materials.

$$F(\sigma_{22}-\sigma_{33})^2+G(\sigma_{33}-\sigma_{11})^2+H(\sigma_{11}-\sigma_{22})^2+2L\sigma_{23}{}^2+2M\sigma_{31}{}^2+2N\sigma_{12}{}^2=2\sigma_y{}^2 \tag{1}$$

The previous equation is used for describe the three-dimensional yield behaviour; however, as in metal forming the work material are sheets, we can remove the thickness components in order to work with a bi-dimensional criteria.

$$(H+F)(\sigma_{22})^2+(G+H)(\sigma_{11})^2-2H\sigma_{11}\sigma_{22}+2N\sigma_{12}{}^2=2\sigma_y{}^2 \tag{2}$$

- Yld-2000's criterion (2003) [3]: In order to improve their 1991 and 1994 criteria, Barlat et al. propose this new criterion, which aims to reduce the complexity of their previous models, but without losing flexibility. This yield criterion, uses eight coefficients, in order to describe the yield surface [7].

$$X=X'+X''=2\sigma_y{}^M \tag{3}$$

$$\begin{Bmatrix} X'=\left|X'_1-X'_2\right|^M \\ X''=\left|2X''_2+X''_1\right|^M+\left|2X''_1+X''_2\right|^M \end{Bmatrix};\begin{Bmatrix} X'_i=\frac{1}{2}\left(X'_{11}+X'_{22}\pm\sqrt{(X'_{11}-X'_{22})^2+4X'^2_{12}}\right) \\ X''_j=\frac{1}{2}\left(X''_{11}+X''_{22}\pm\sqrt{(X''_{11}-X''_{22})^2+4X''^2_{12}}\right) \end{Bmatrix} \quad (4)(5)$$

$$\begin{Bmatrix} X' = L'\sigma \\ X'' = L''\sigma \end{Bmatrix} \tag{6}$$

$$\begin{bmatrix} L'_{11} \\ L'_{12} \\ L'_{21} \\ L'_{22} \\ L'_{66} \end{bmatrix} = \begin{bmatrix} 2/3 & 0 & 0 \\ -1/3 & 0 & 0 \\ 0 & 0 & 0 \\ 0 & -1/3 & 0 \\ 0 & 2/3 & 1 \end{bmatrix} \begin{bmatrix} \alpha_1 \\ \alpha_2 \\ \alpha_7 \end{bmatrix} ; \begin{bmatrix} L''_{11} \\ L''_{12} \\ L''_{21} \\ L''_{22} \\ L''_{66} \end{bmatrix} = \begin{bmatrix} -2 & 2 & 8 & -2 & 0 \\ 1 & -4 & -4 & 1 & 0 \\ 4 & -4 & -4 & 4 & 0 \\ -2 & 8 & 2 & -2 & 0 \\ 0 & 0 & 0 & 1 & 9 \end{bmatrix} \begin{bmatrix} \alpha_3 \\ \alpha_4 \\ \alpha_5 \\ \alpha_6 \\ \alpha_8 \end{bmatrix} \tag{7)(8}$$

- Yoshida's criterion (2013) [4]: This criterion, developed by Yoshida et al. uses a sixth-order polynomial yield function, in order to describe the yielding surface of the material. The Yoshida's criterion, uses sixteen anisotropy coefficients. Thanks to the amount of coefficients, this criteria proves good flexibility to adjust the experimental points. On the other side, the drawback of the Yoshida's criterion is the number of experimental points that should be obtained in order to fit the model. The model needs at least sixteen experimental points to be fitted.

$$C_1\sigma_{11}{}^6 - 3C_2\sigma_{11}{}^5\sigma_{22} + 6C_3\sigma_{11}{}^4\sigma_{22}{}^2 - 7C_4\sigma_{11}{}^3\sigma_{22}{}^3 + 6C_5\sigma_{11}{}^2\sigma_{22}{}^4 - 3C_6\sigma_{11}\sigma_{22}{}^5 + C_7\sigma_{22}{}^6 + 9\left(C_8\sigma_{11}{}^4 - 2C_9\sigma_{11}{}^3\sigma_{22} + 3C_{10}\sigma_{11}{}^2\sigma_{22}{}^2 - 2C_{11}\sigma_{11}\sigma_{22}{}^3 + C_{12}\sigma_{22}{}^4\right)\sigma_{12}{}^2 + 27\left(C_{13}\sigma_{11}{}^2 - C_{14}\sigma_{11}\sigma_{22} + C_{15}\sigma_{22}{}^2\right)\sigma_{12}{}^4 + 27C_{16}\sigma_{12}{}^6 = \sigma_y{}^6 \tag{9}$$

In the case of the material hardening, the tensile test provides experimental data of the stress-strain relation between the yield stress and the ultimate strength of the material. In order to extrapolate this behaviour to bigger deformations, different hardening models exists. In this study only one model will be used, the Hollomon's hardening law, a simple isotropic hardening model, widely use in the practice. This model is given by:

- Hollomon's hardening law (1945) [8]: The Hollomon's hardening equation is a power law that defines a relation between the true stress and the true strain. In this equation K is the strength coefficient, and n is the strain hardening exponent.

$$\sigma(\varepsilon) = K\,\varepsilon^n \tag{10}$$

The ShoreA 90 polyurethane rubber it is characterized by its hyperelastic behaviour. The hyperelasic material behaviour is defined by the Green's theory. Unlike linear elasticity, where the relationship between stress and strain is obtained directly, in this case the relationship is derived from an energy potential function [9]. One of the most used hyperelastic models was set by R.S. Rivlin and proposes the following formulation for the potential energy equation:

- Mooney-Rivlin (1948) [10]: The Mooney-Rivlin model is typically used in the modelling of rubber tools in rubber forming processes numerical simulation. In this model, the values of C_{10} and C_{01} are material dependent constants and I_1 and I_2 the first and second invariants of the strain tensor.

$$\sigma = \partial W/\partial \boldsymbol{\varepsilon} \quad ; \quad W(I_1, I_2) = C_{10}(I_1 - 3) + C_{01}(I_2 - 3) \tag{11)(12}$$

2.2. Parameter identification

The characterized materials will be, on the one hand, the austenitic stainless steel AISI 321, in a 0.76 mm thick sheet format, and on the other hand, the thermoplastic polyurethane with a Shore A 90 hardness with a thickness of 10 mm. To carry out all the tests, the equipment used was a universal testing machine. In the case of the AISI 321 alloy, the tests used have been tensile tests at different angles of the rolling direction, as well as plane strain tests, simple shear tests and stack compression tests. In the case of polyurethane, only uniaxial tensile tests have been performed. The strain rate obtained for the samples in all the test were 0.016 s^{-1} approximately. A summary of the experimental results can be seen in the Table 1.

Table 1. Experimental points for the yielding fitting

Test	*Sample angle [°]*	*Normalize yield stress [-]*	*Lankford Coefficient [-]*
Uniaxial	0	1.000	0.505
Uniaxial	15	0.978	0.637
Uniaxial	30	0.970	0.813
Uniaxial	45	0.935	0.951
Uniaxial	60	0.968	0.917
Uniaxial	75	1.005	0.721
Uniaxial	90	1.002	0.637
Plane strain	0	1.058	-
Plane strain	90	1.085	-
Simple shear	45	0.573	-
Stack compression	-	-	0.744

2.2.1. Uniaxial tensile test. To carry out the tensile tests, tensile specimens have been cut every 15° with respect to the rolling direction of the original steel blank. For the measurement of the deformations and strains, a digital image correlation (DIC) system has been used.

2.2.2. Plane strain test. For the design of the geometries used in this study, the starting point for the geometries have been the designs used by other authors, such as G. Gruben et al. [11] and K. Zang et al. [12]. The final geometry is described in Figure 2. Starting from the previously mentioned geometries, a modification has been done to them, in order to adapt the test to the conditions that exist in the case of the equipment and material. These variations in geometry have been validated by finite element simulation, with the material properties previously obtained from the uniaxial test. In this way, the desired state of charge has been obtained. The test of the plane strain specimens was carried out on specimens cut at 0° and at 90° with respect to the rolling direction. In order to ensure the deformation state during the test, the strain was measured by DIC.

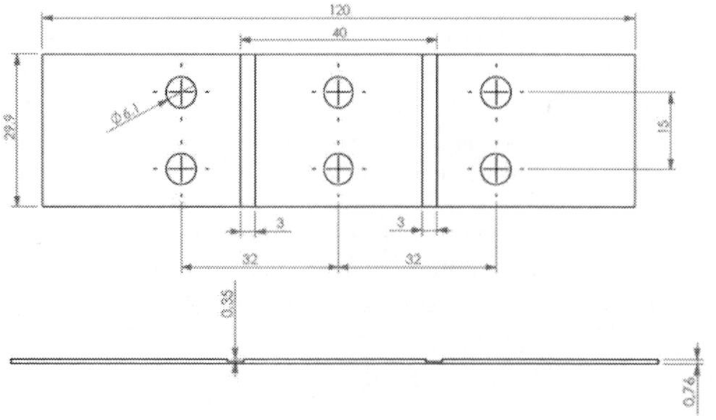

Figure 1. Sample geometry used for simple shear tests

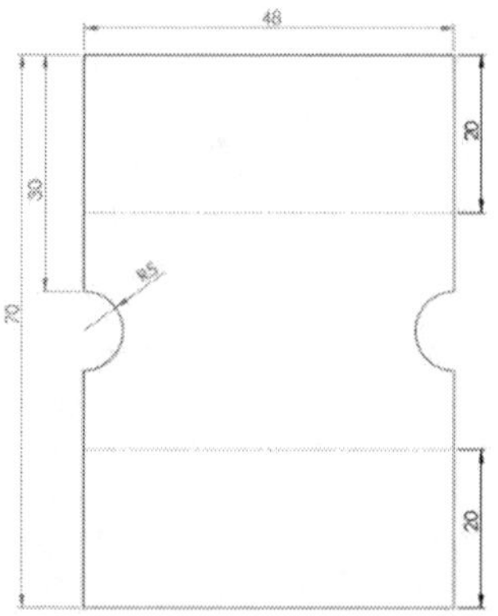

Figure 2. Sample geometry used for plane strain test

2.2.3. Stack compression test. In order to determine the r_{biaxial}, the stack compression test will be used. The process followed is described by Barlat et al. [3] in their work that describes the characterization of an aluminium alloy. In order to obtain a sample with the necessary high, fourteen blanks have been stacked, with a total thickness of 10.64 mm and then a diameter of 7.5 mm was cut using EDM. The experimental points were obtained under reductions in thickness of 4 mm, 5 mm and 6 mm.

2.2.4. Simple shear test. The shear test had been performed following the methodology described by M.Abspoel et al. [13], but measuring the deformations by DIC. The test was carried out with the sample at 45° respect to the rolling direction, in order to obtain the yield stress point with $\sigma_{xx}=\sigma_{yy}$ and $\tau_{xy}=0$. In the tooling, the central part of the sample, described in the Figure 1, has vertical movement and the side parts are fixed, reaching the simple shear state in the regions in between.

2.2.5. Material parameter fitting. In order to determine the parameters of the different yield criteria, the strategy followed was to minimize the error between the experimental parameters from the Table 1 and the values provided by the different models. The results of the fitting process of the yield criteria are exposed in the Table 2, and the comparison between the different models can be appreciated in the Figure 4. The error formulation used, was proposed by the members of the CERTETA research group [7], being d the distance between the experimental points, and its projection in the predicted yield locus :

$$\beta = \varphi + \delta + \gamma \tag{13}$$

$$\varphi = \sqrt{\sum_{i=1}^{n}(d^2)\frac{1}{Y}}\ 100 \quad ; \quad \delta = \sqrt{\sum_{i=1}^{n}\left(\frac{\sigma_{\theta i}^{exp}-\sigma_{\theta i}^{t}}{\sigma_{\theta i}^{exp}}\right)}\ 100 \quad ; \quad \gamma = \sqrt{\sum_{i=1}^{n}\left(\frac{r_{\theta i}^{exp}-r_{\theta i}^{t}}{r_{\theta i}^{exp}}\right)}\ 100 \qquad (14)(15)(16)$$

Table 2. AISI 321 yield parameters

Yoshida	C_1	C_2	C_3	C_4	C_5	C_6	C_7	C_8	C_9	C_{10}
	1	0.591	0.846	1.129	1	0.737	0.930	1.055	0.966	1.464
	C_{11}	C_{12}	C_{13}	C_{14}	C_{15}	C_{16}	-	-	-	-
	1.024	1.030	1.186	1.780	1.225	1.698	-	-	-	-
Yld 2000	α_1	α_2	α_3	α_4	α_5	α_6	α_7	α_8	M	-
	1	0.938	1.2	1.041	1.087	1.292	1.066	0.985	8	-
Hill 48	F	G	H	N	-	-	-	-	-	-
	0.527	0.664	0.336	1.734	-	-	-	-	-	-

On the other side, in order to fit the Hollomon hardening model, the procedure used was the minimization of the sum of existing relative errors, between the points obtained experimentally and the points provided by the Hollomon model. The obtained results are detailed in Table 3 and Figure 3.

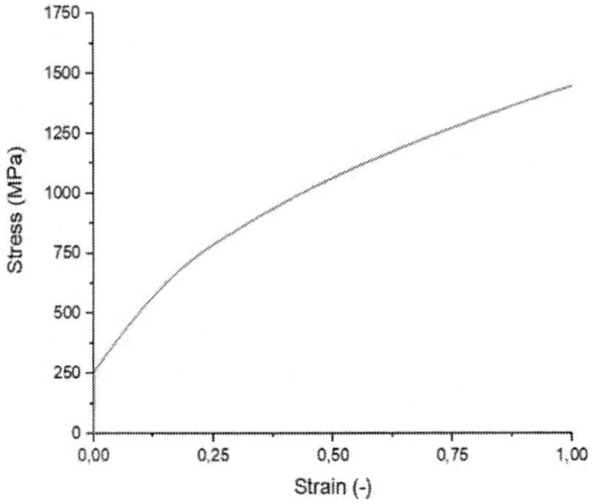

Figure 3. AISI 321 hardening curve

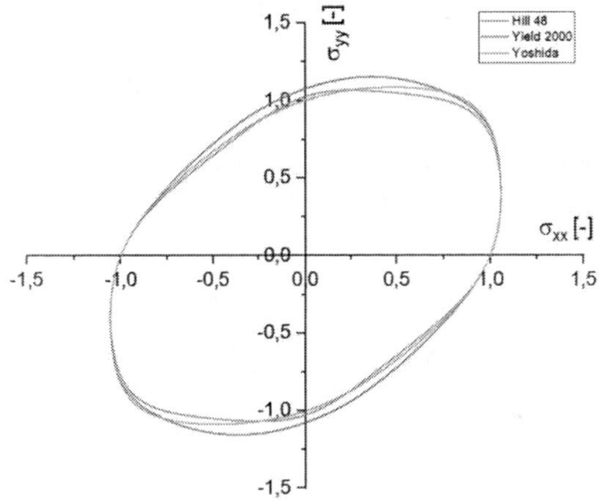

Figure 4. Yield surfaces comparison

IOP Conf. Series: Materials Science and Engineering 1238 (2022) 012005 doi:10.1088/1757-899X/1238/1/012005

Table 3. AISI 321 hardening parameters

	K [MPa]	*n [-]*
Hollomon	1446.64	0.443

2.2.6. Rubber uniaxial test. For the rubber characterization, uniaxial tension and compression tests were performed. The rubber samples were cyclically loaded and unloaded five times, due to its hysteresis softening. In the fifth cycle, the behaviour of the rubber stabilizes. This process is repeated under different deformations, in order to obtain a set of points that were used to fit the Mooney-Rivlin model. The coefficients obtained were C_{10}=1.06 MPa and C_{01}=0.87 MPa, assuming the incompressibility of the rubber.

3. Finite elements model

In order to carry out the analysis, the software "Pam-Stamp Professional 2021.0" has been used. The model that is used can be appreciated in the Figure 5. In this model, the die is defined as a rigid shell, on which the metal sheet is placed, with the material properties obtained from the previously described tests and constitutive models, and also described as a shell. In the case of rubber, it is placed on the sheet, and it is considered that the pressure from the machine is applied uniformly over the entire upper surface of the rubber pad. The rubber is defined using solid elements. Finally a double symmetry is applied in the die in order to reduce the computation time.

The friction is described as constant and between each pair of objects. In the case of the coefficient of friction (COF) between the blank and the die, the values would be 0.05 and 0.1, as it seems to be the limiting values of the friction, obtained after the strip drawing test. In the case of the rubber friction, the values of the coefficient of friction is obtained from bibliography, from the work of R.Elleuch et al [14]. The values of the rubber COF will be 0.5 and 0.4, in order to understand the influence of this variable in the process. The geometry of the sample is based in the geometries that are produced for the aeronautic industry, and the forming pressure used is 80 MPa.

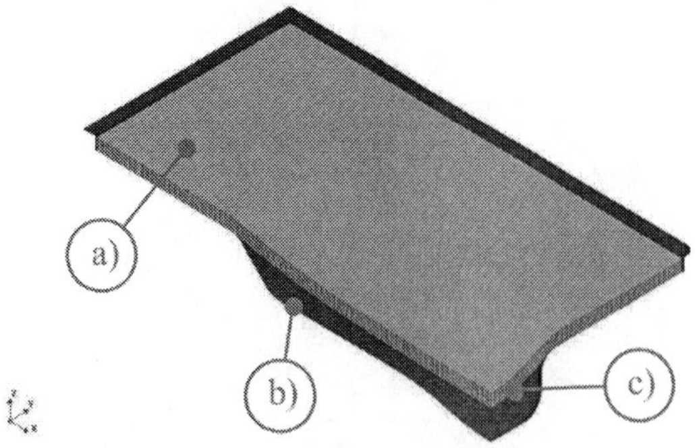

Figure 5. Rubber pad diaphragm forming simulation model a) Blank ; b)Rubber ; c) Die

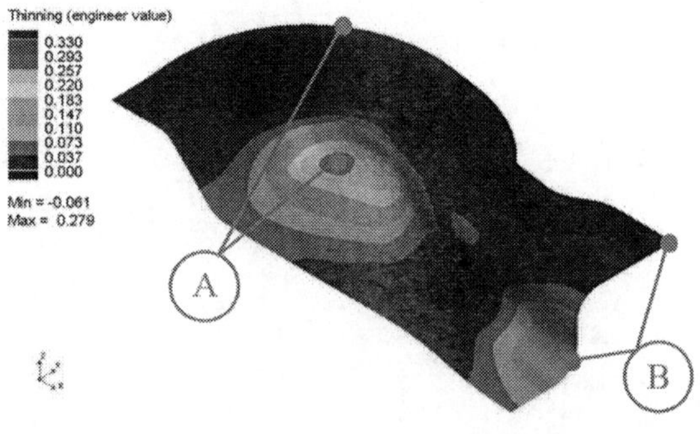

Figure 6. Points used to evaluate thinning, draw-in and springback. Thinning results using Hill 48 model, with blank COF 0.05 and rubber COF 0.4.

Two regions were selected in order to carry out the study, these regions can be seen in the Figure 6. The points used to evaluate the thinning are the two locations that present the maximum value of this parameter. In the case of the springback and the draw-in, the selected points are located in the flange points that are associated with the maximum thinning areas.

4. Results and discussion

The results obtained in the different numerical models, can be appreciated in the Figure 7. In a first sight, the differences in the results provided by all the models can be clearly seen, being the differences smaller between the Yld-2000 and Yoshida's model. The influence of each of the study variables on the results is described below.

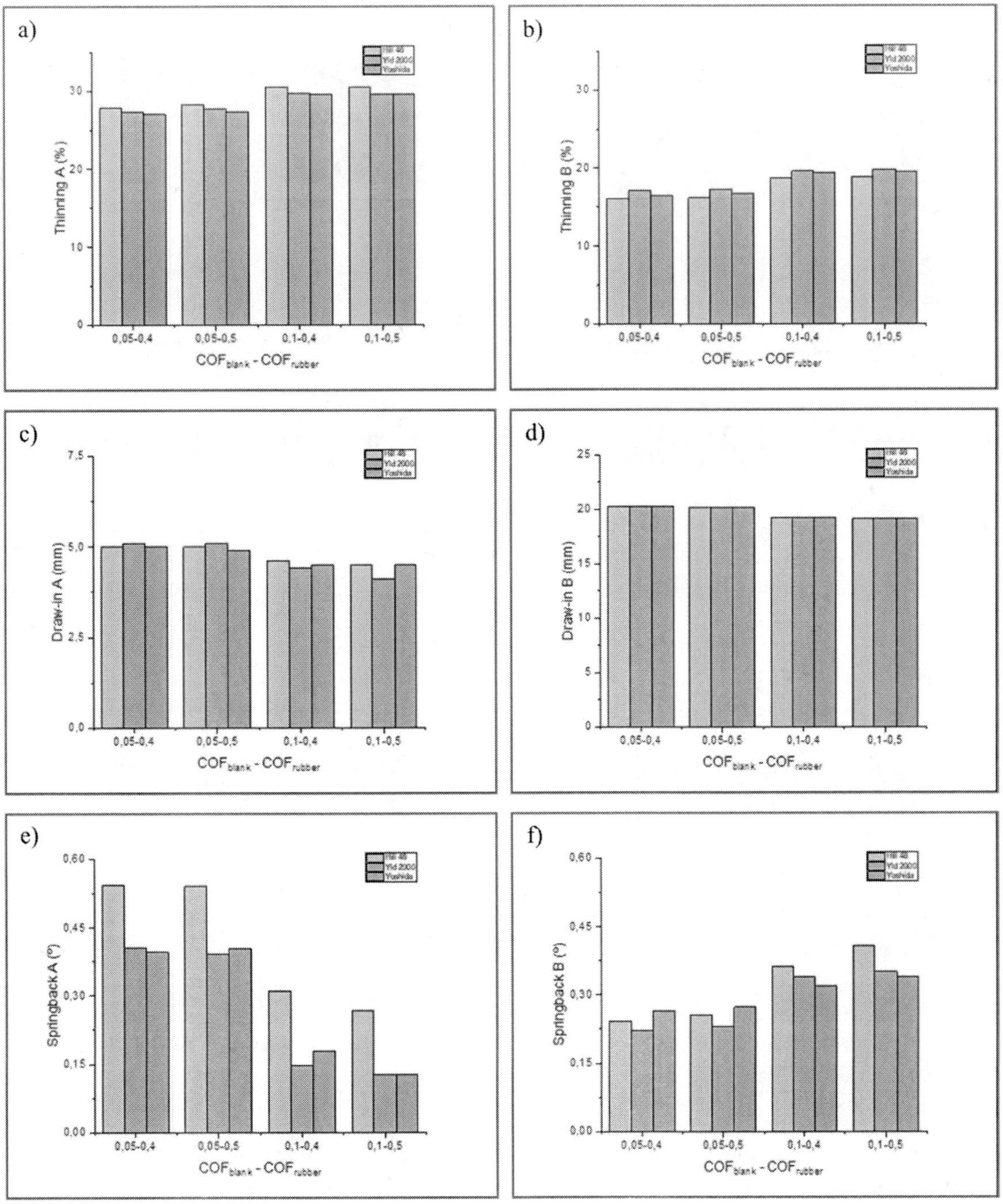

Figure 7. Results of the simulated geometry a) Maximum thinning A ; b) Maximum thinning B ; c) Draw-in A ; d) Draw-in B ; e) Springback A ; f) Springback B

In view of the results, in order to analyse the influence on the one hand of friction, as well as on the other hand of the material model, a factorial analysis has been carried out with the available data. In order to perform the analysis, the material model and the COF in the blank has been taken as control variables, and the COF in the rubber, due to its low influence, and the difficulty to control it in the process, has been taken as noise variable. The analysis is done at each point for each study parameter.

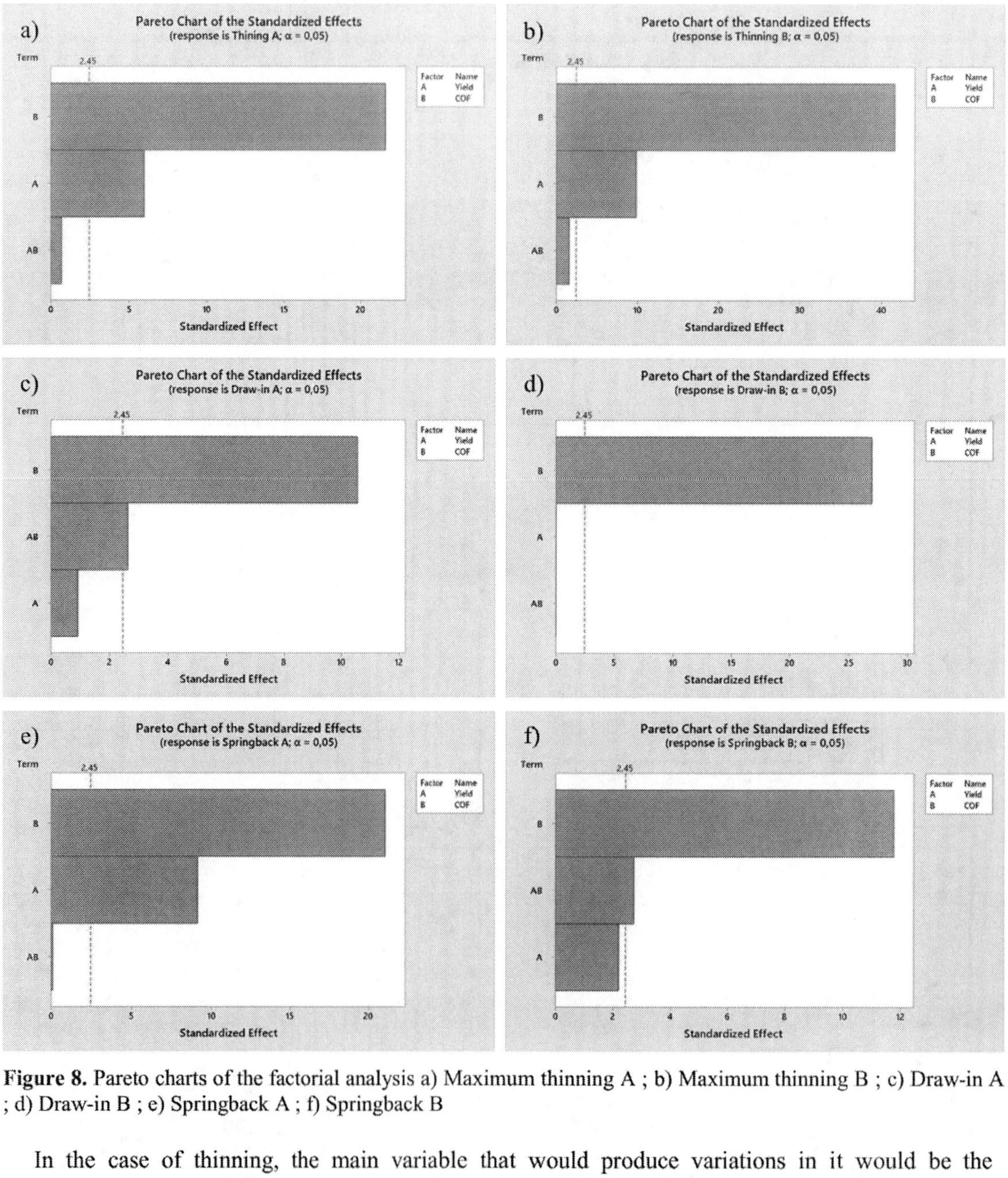

Figure 8. Pareto charts of the factorial analysis a) Maximum thinning A ; b) Maximum thinning B ; c) Draw-in A ; d) Draw-in B ; e) Springback A ; f) Springback B

In the case of thinning, the main variable that would produce variations in it would be the coefficient of friction, followed by the model of the material, and both having an influence that would not be negligible and as can be seen, there would be hardly any influence of the interaction between both.

IOP Conf. Series: Materials Science and Engineering 1238 (2022) 012005 doi:10.1088/1757-899X/1238/1/012005

On the other hand, looking at the draw-in of the blank at the end of the process, this could be considered to be controlled only by the friction between the sheet and the die, this can be seen in the results of the Figure 7 b).

Finally, in the case of springback, this is mostly influenced by the coefficient of friction, however, to a lesser extent, an additional variability arises, conditioned by the yield model used in the material. As can be seen, the main variable that controls the variability of the process is the coefficient of friction, however, the differences that arise in the results due to the use of a yield criterion or another cannot be ignored.

5. Conclusions

In the range of variables studied, the most relevant element is the coefficient of friction then the material model, and finally the friction coefficient of the rubber. The main conclusions that can be drawn from this analysis are, on the one hand, the importance of the control of the friction coefficient and the lubrication in production, in order to reduce the variability of the process. In addition, for modelling, having an adequate characterization of the friction in the process, as well as an adequate characterization of the material.

As future works, a correlation should be carried out with real manufactured parts, in order to determine which material model fits better the material AISI 321 in this process, and perform a complete sensitivity analysis with all the process variables.

6. Acknowledgements

It has been possible to carry out this study, thanks to the collaboration of ITP Aero. This work was supported in part by the Basque Government's R&D program Hazitek under Grant ZE-2020/0000.

References

[1] Z M Ramezani 2012 *Rubber-pad forming processes: Technology and applications* 1st ed. Woodhead Publishing

[2] R Hill 1948 *Proceedings of the Royal Society of London. Series A. Mathematical and Physical Sciences* **193** 281–97

[3] F Barlat, J C Brem, J W Yoon, K Chung, R E Dick, D J Lege, F Pourboghrat, S-H Choi, and E Chu 2003 *International Journal of Plasticity* **19** 1297–1319

[4] F Yoshida, H Hamasaki, and T Uemori 2013 *International Journal of Plasticity* **45** 119–39

[5] H J Batista Alves, T Reis de Oliveira and C G Schön 2017 *Conference paper ESSC & Duplex*

[6] L R Treloar 2005 *The Physics of Rubber Elasticity*, 3rd ed. OUP Oxford

[7] D Banabic 2018 *Sheet Metal Forming Processes: Constitutive Modelling and Numerical Simulation* Springer

[8] J H Hollomon 1945 *Transactions of the Metallurgical Society of AIME* **162** 268–90

[9] R M Hackett, 2018 *Hyperelasticity Primer*. Springer International Publishing

[10] R S Rivlin 1948 *Philosophical Transactions of the Royal Society of London. Series A, Mathematical and Physical Sciences* **240** 459–90

[11] G Gruben, D Morin, M Langseth, and O S Hopperstad 2017 *European Journal of Mechanics A/Solids* **61** 315–29

[12] K Zhang, B Holmedal, O S Hopperstad, and S Dumoulin, 2014 *Materials Science Forum* **794** 596–601

[13] M Abspoel, M E Scholting, M Lansbergen, Y An, and H Vegte 2017 *Journal of Materials Processing Technology* **248** 161–177

[14] R Elleuch, K Elleuch, B Salah, and H Zahouani 2007 *Materials & Design* **28** 824–30

IOP Conf. Series: Materials Science and Engineering 1238 (2022) 012006 doi:10.1088/1757-899X/1238/1/012006

Plasticity and fracture of AA7075 at elevated strain rates and temperatures

X Li, C C Roth, K Pandya, N Karathanasopoulos and D Mohr

Department of Mechanical and Process Engineering, Swiss Federal Institute of Technology (ETH), Zurich, Switzerland

lixue@ethz.ch

Abstract. The accurate description of the strain rate and temperature dependent response of Aluminium alloys is a perpetual quest in the hot forming industry. In the present study, uniaxial tension, and notched tension experiments are conducted for an aluminium AA7075-T6 sheet metal at various temperatures and strain rates. The experimental campaign covers strain rates ranging from 0.001/s to 100/s, and temperatures ranging from 20°C to 360°C. We observe low strain rate sensitivity at room temperature, with an increase in strain rate sensitivity as temperature is increased up to 360°C. An YLD2000-3D model is employed to describe the anisotropy of the material. A machine learning based hardening model is employed to capture the complex strain rate and temperature effect on the observed hardening response. Counter-example regularization is utilized to guarantee a convergence in the numeric return-mapping algorithm. Comparing the experimental force-displacement curves with the numerical predictions, the neural network model accurately describes the large deformation response of the material in the post-necking range.

1. Introduction

The accurate description of the large deformation response of metals at high strain rates and temperatures is critical in solving industrial problems of hot forming, crash-worthiness, machining, etc. In crash-loading scenarios, aluminum 7xxx series has seen many applications in lightweight structural design, as an alternative to advanced high strength steels [1]. These alloys feature negligible strain rate sensitivity at room temperature, but a high strain rate sensitivity at elevated temperatures [2][3]. Existing models, such as the Johnson-Cook [4] and the Zerilli-Armstrong model [5], prescribe a specific analytical formulation to describe the rate- and temperature dependency. These mechanism-inspired models have been widely adopted over the past decades. Data-driven models based on machine learning algorithms are reported to accurately describe mechanical response of materials at various strain rates and temperatures [6][7][8][9][10]. In the present work, notched tensile experiments are carried out at strain rates from 0.001/s to 100/s in the temperature range between 20°C and 360°C. A neural network based hardening model is proposed, which is trained using a hybrid numerical-experimental approach assisted by counter-example regularization [11]. The model showcases accurate prediction on the force-displacement response, at large deformation in the post-necking regime.

2. Experiments at high strain rates and elevated temperatures

2.1. Material and Specimen
The AA7075 material is delivered in 2 mm thick sheets and all specimens are extracted using waterjet cutting. Figure 1 shows the two notched tension (NT) geometries considered. The notched tension specimens feature 20 mm wide shoulder section with symmetric circular cut-outs. In this study, a notch radius of $R = 20$ mm is employed. This reduces the gauge section to a minimum width of 10 mm at the center. The same gauge section is shared by the slow, intermediate and high strain rate tests, with the clamping area adjusted for attaching to different loading systems.

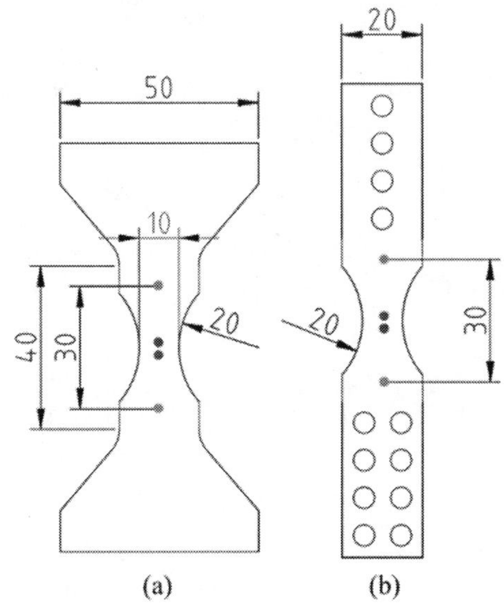

Figure 1. Specimen geometries for (a) low, intermediate and (b) high strain rate testing. The blue dots highlight the position of global (30mm) displacement measurements through DIC. The red dots mark the corresponding locations where surface axial strains (1mm) are extracted.

2.2. Experimental procedure

2.2.1. Low strain rate experiments. The low strain rate experiments are performed on a hydraulic universal testing machine (Instron 8801). Through the build-in control software an actuator speed of 0.4 mm/min is prescribed. This corresponds to an approximate strain rate of 0.001/s. Prior to testing, the specimens are painted with a random black and white speckle pattern for digital image correlation (DIC). The specimens are heated up to the target temperature at a rate of 30K/s using a custom made induction system. This means that for highest temperature of 360°C, the target temperature is reached within 12 seconds. The emissivity of the paint is calibrated to achieve accurate close-loop temperature control using a pyrometer. We verify the homogeneity of the temperature field using an infrared camera (FLIR x6801). A 5MP digital camera (Point Grey GS3-U33-51S5M-C) acquires images for DIC at 1Hz. This achieves a spatial resolution of 25 μm/pixel.

2.2.2. Intermediate strain rate experiments. Experiments at intermediate strain rates are conducted on the same hydraulic testing machine at loading speeds of 40 mm/min and 800 mm/min, corresponding to 100 and 2000 times the speed of the slow tests, respectively. An optical high-speed camera (Photron SA-Z) acquires DIC images at a frequency of 1000 Hz and 8000 Hz. A spatial resolution of 130 μm/pixel is obtained with this setup. To track the temperature rise caused by plastic dissipation, the infrared camera (FLIR X6801SC) is set up to operate at 2000 Hz. The data and image acquisitions are triggered simultaneously by a rise in the force signal. This allows for a full synchronization between the force and DIC measurements.

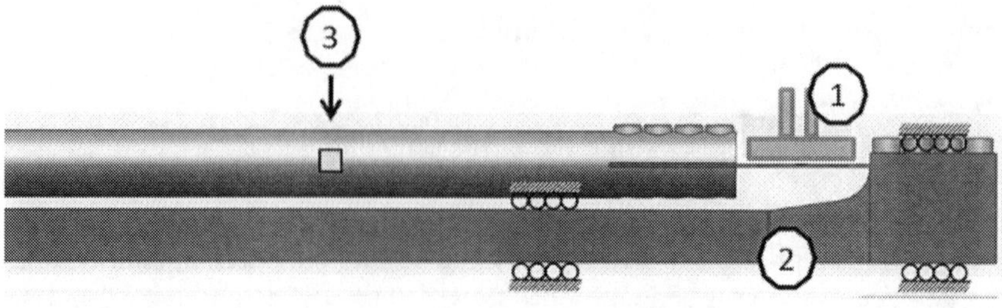

Figure 2. Schematic drawing of the high speed – high temperature experiment setup: ① induction heating coil, ② load-inversion device, ③ strain gauge on the output bar.

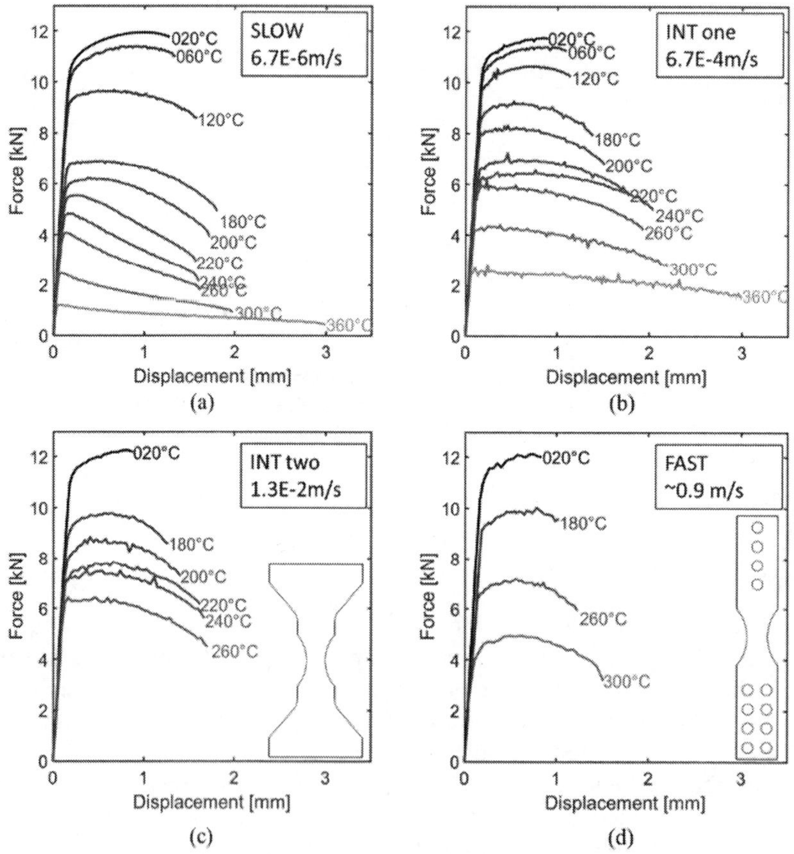

Figure 3. Experimentally measured force-displacement curves for NT20 specimens at loading speeds of (a) 6.7×10^{-6} m/s, (b) 6.7×10^{-4} m/s, (c) 1.3×10^{-2} m/s, and (d) 0.9 m/s. All curves are truncated once a crack is visible on the specimen surface.

2.2.3. High strain rate experiments. Experiments at high strain rates are carried out on a split Hopkinson pressure bar system. The system includes a 4990 mm long steel striker bar, a 5927 mm long input bar, and a 5951 mm long output bar. All three bars feature an outer diameter of 20 mm. A load inversion device connects the specimen to the input side using eight M5 screws [12]. As illustrated in Figure 2, the custom inductor coil heats up the sheet metal specimen from the top side. The axial force histories are measured using strain gauges attached on the output bar, located 430 mm from the specimen/output bar interface. To reduce electro-magnetic interference by the inductor upon the strain gauges, the inductor is turned off using a laser signal prior to the striker/input bar impact. The image and data acquisitions are triggered using the rise in the input bar signal. Details of the high speed – high temperature experimental setup are described in [11].

2.3. *Experimental results*

Figure 3 shows the experimentally measured force-displacement curves from the notched tension specimens at various loading speeds and temperatures. The corresponding force maximum extracted from each curve is plotted in figure 4. For all loading speeds, a monotonic and negative effect of temperature on the force-displacement curve is obtained. At room temperature, the effect of strain rate is negligible. Starting from 120°C we observe a significant effect of strain rate on the hardening response of the material. At 300°C, the reaction force increases by 81% when increasing the loading speed by 100 times. This percentage becomes 133% at 360°C, when the hardening response is governed by the viscosity of the material.

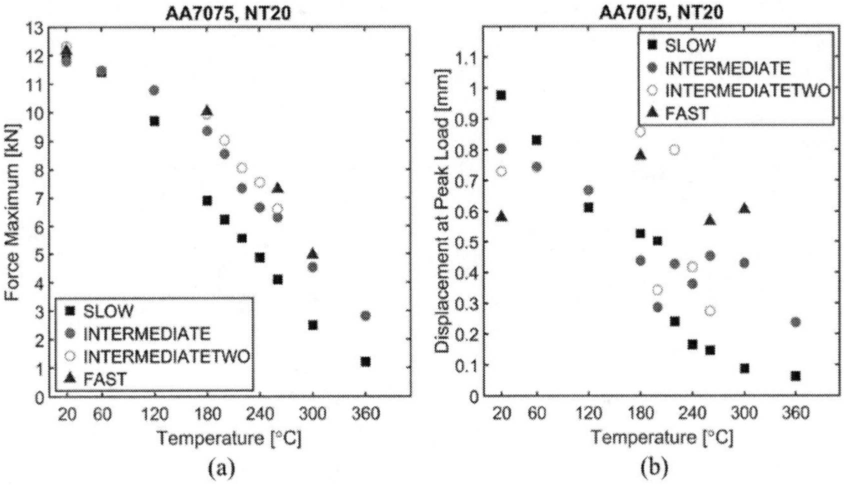

Figure 4. (a) The extracted force-maximum as a function of initial temperature for four loading speeds. (b) Displacement at peak load as a function of initial temperature.

3. Plasticity models

3.1. *Model formulation and parameter identification*

The material is modeled using an elasto-plastic framework, assuming a constant Young's modulus and Poisson ratio. The yield function consists of an anisotropic equivalent stress $\bar{\sigma}_{yld2000}$ and the deformation resistance k,

$$f\left[\boldsymbol{\sigma},k\right]=\bar{\sigma}_{yld2000}-k=0. \tag{1}$$

The anisotropy of the material is described using a Yld2000-3D constitutive framework [13], which is an extension to Barlat's Yld2000-2D model [14]. The evolution of equivalent plastic strain is calculated using an associated flow rule. The eight parameters $\alpha_1 \dots \alpha_8$ describing the yield surface are calibrated from room temperature uniaxial tensile tests along seven different loading orientations from the rolling direction. The formulation of the deformation resistance closely follows [7][11], in which the deformation resistance is decomposed into a reference mixed Swift-Voce term, multiplied by a neural network term dealing with the effect of strain rate and temperature. Equation 3 details the formulation of the seven parameter hardening function. In the present study, a feed-forward neural network architecture featuring three hidden layers, with ten neurons per layer is selected. Its core concept is subsequent activation of the hidden layers $\boldsymbol{H}^i$ using the hyperbolic tangent transfer function [15].

$$k\left[\bar{\varepsilon}_p, \dot{\bar{\varepsilon}}_p, T\right] = k_{SV}\left[\bar{\varepsilon}_p\right] k_{NN}\left[\bar{\varepsilon}_p, \dot{\bar{\varepsilon}}_p, T\right] \tag{2}$$

$$k_{SV}\left[\bar{\varepsilon}_p\right] = \alpha A\left(\varepsilon_0 + \bar{\varepsilon}_p\right)^n + (1-\alpha)\left(k_0 + Q\left(1 - e^{-\beta\bar{\varepsilon}_p}\right)\right) \tag{3}$$

$$\boldsymbol{H}^0 = \left[\bar{\varepsilon}_p, \dot{\bar{\varepsilon}}_p, T\right], \boldsymbol{H}^{n+1} = f(\boldsymbol{W}^n \boldsymbol{H}^n), k_{NN} = f\left(\boldsymbol{W}^3 \boldsymbol{H}^3\right) \tag{4}$$

$$f(x) = \tanh(x) \tag{5}$$

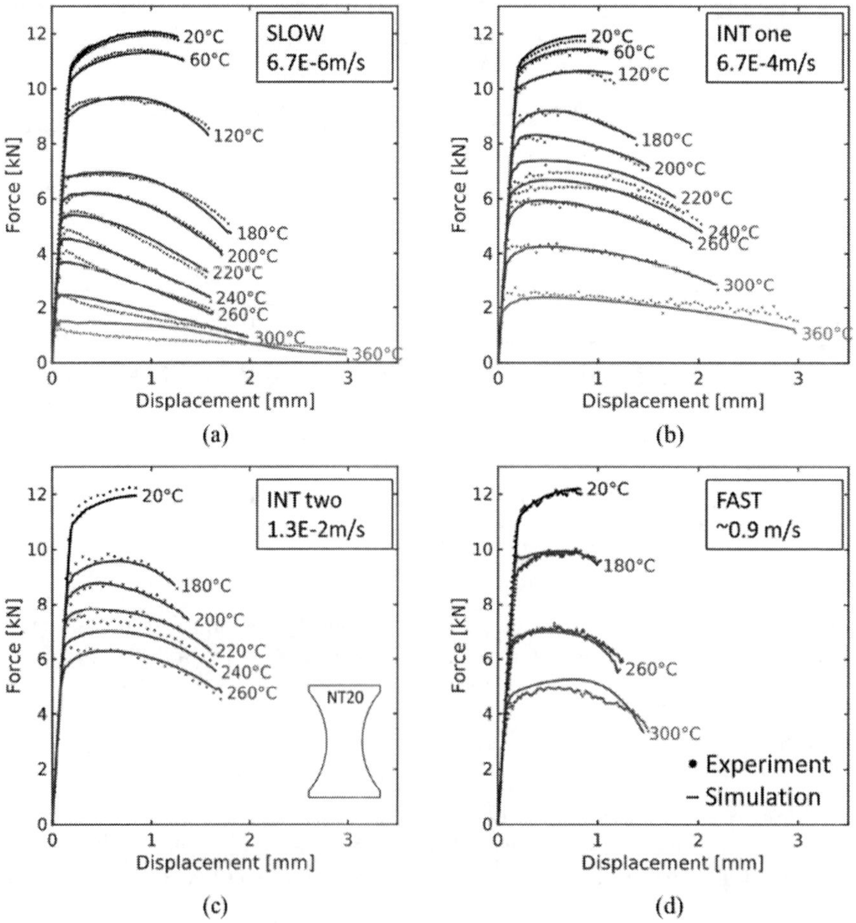

Figure 5. Numerically predicted and experimentally measured force-displacement curves at loading speeds of (a) 6.7×10^{-6} m/s, (b) 6.7×10^{-4} m/s, (c) 1.3×10^{-2} m/s, and (d) 0.9 m/s.

The neural network is trained through an inverse-analysis approach in conjunction with back-propagation. To achieve this, the NN is implemented into a user-defined material subroutine in Abaqus/Explicit. To avoid any convergence problems in the numeric return-mapping algorithm due to negative strain rate sensitivity, counter-example regularization is employed. Details of the parameter identification and the network training process can be found in [11].

3.2. *Results of numerical modelling*

Figure 5 compares the experimentally measured force-displacement curves (solid dots) with the numerical predictions (solid lines) for the NT20 experiments. Overall a good agreement is obtained with the neural network based model, with the largest difference witnessed at room temperature and 0.013m/s. This is due to the slight negative strain rate sensitivity at room temperature from the experiments (Figure 4). At temperatures above 200°C, diffuse necking occurs at very small displacements, followed by a profound post-necking regime before fracture. For all four loading speeds, the neural network model showcases excellent agreement in this range.

4. Conclusion

A comprehensive numerical and experimental investigation is performed from quasi-static (0.001/s) to dynamic (~100/s) strain rates on AA7075-T6 sheet metal at temperatures ranging from 20°C to 360°C. The fracture experiments reveal a minimum strain rate sensitivity for AA7075 at room temperature, with an increased strain rate effect as the temperature rises. A machine learning based plasticity model is employed to capture this phenomenon, which decomposes the deformation resistance into a reference mixed Swift-Voce strain hardening term, multiplied with a neural network function that deals with the rate- and temperature dependency. The model is trained using a hybrid numerical-experimental approach, by iteratively running full 3D finite element simulations on NT20 experiments. Counter-example training is employed to circumvent any potential numerical convergence problems in the return-mapping algorithm. The model exhibits good agreement with the experimental force-displacement curves, particularly in the large deformation post-necking regime.

References

[1] Fourmeau M, Børvik T, Benallal A and Hopperstad O S 2013 Anisotropic failure modes of high-strength aluminium alloy under various stress states *Int. J. Plast.* **48** 34–53
[2] Roth C C, Fras T and Mohr D 2020 Dynamic perforation of lightweight armor: Temperature-dependent plasticity and fracture of aluminum 7020-T6 *Mech. Mater.* **149** 103537
[3] Kabirian F, Khan A S and Pandey A 2014 Negative to positive strain rate sensitivity in 5xxx series aluminum alloys: Experiment and constitutive modeling *Int. J. Plast.* **55** 232–46
[4] Johnson G R and Cook W H 1985 Fracture characteristics of three metals subjected to various strains, strain rates, temperatures and pressures *Eng. Fract. Mech.* **21** 31–48
[5] Zerilli F J and Armstrong R W 1987 Dislocation-mechanics-based constitutive relations for material dynamics calculations *J. Appl. Phys.* **61** 1816–25
[6] Pandya K S, Roth C C and Mohr D 2020 Strain rate and temperature dependent fracture of aluminum alloy 7075: Experiments and neural network modeling *Int. J. Plast.* **135** 102788
[7] Li X, Roth C C and Mohr D 2019 Machine-learning based temperature- and rate-dependent plasticity model: Application to analysis of fracture experiments on DP steel *Int. J. Plast.* **118** 320–44
[8] Abueidda D W, Koric S, Sobh N A and Sehitoglu H 2021 Deep learning for plasticity and thermo-viscoplasticity *Int. J. Plast.* **136** 102852
[9] Jordan B, Gorji M B and Mohr D 2020 Neural network model describing the temperature- And rate-dependent stress-strain response of polypropylene *Int. J. Plast.* **135** 102811
[10] Shang H, Wu P, Lou Y, Wang J and Chen Q 2022 Machine learning-based modeling of the coupling effect of strain rate and temperature on strain hardening for 5182-O aluminum alloy *J. Mater. Process. Technol.* **302** 117501
[11] Li X, Roth C C, Bonatti C and Mohr D 2022 Counterexample-trained Neural Network Model of Rate and Temperature Dependent Hardening with Dynamic Strain Aging *Int. J. Plast.* **151** 103218

IOP Conf. Series: Materials Science and Engineering 1238 (2022) 012006 doi:10.1088/1757-899X/1238/1/012006

[12] Roth C C, Gary G and Mohr D 2015 Compact SHPB System for Intermediate and High Strain Rate Plasticity and Fracture Testing of Sheet Metal *Exp. Mech.* **55** 1803–11

[13] Dunand M, Maertens A P, Luo M and Mohr D 2012 Experiments and modeling of anisotropic aluminum extrusions under multi-axial loading - Part I: Plasticity *Int. J. Plast.* **36** 34–49

[14] Barlat F, Brem J C, Yoon J W, Chung K, Dick R E, Lege D J, Pourboghrat F, Choi S-H and Chu E 2003 Plane stress yield function for aluminum alloy sheets — part 1 : theory *Int. J. Plast.* **19** 1297–319

[15] Bishop 2006 *Pattern Recognition and Machine Learning* Springer

IOP Conf. Series: Materials Science and Engineering 1238 (2022) 012007 doi:10.1088/1757-899X/1238/1/012007

Implicitization of the Vegter Yield Criterion

J Sheng[1], S-Y Yang[2] and W Tong[1]

[1]Research Center for Advanced Manufacturing, Southern Methodist University, Dallas, Texas 75275, USA

[2]Korea University of Technology and Education, Byeongchon-myon, Cheonan-si, Chungnam 31253, Korea

E-mail: ysy@koreatech.ac.kr, wtong@smu.edu

Abstract. An advanced yield criterion plays a central role in rate-independent plasticity modeling of sheet metals in automotive and other industrial applications. Different from many non-quadratic yield criteria based on implicit polynomials, Vegter yield criterion is piecewise and quadratic as it is originally constructed by interpolation of several biaxial stress states using multiple second-order Bézier curves. Additional efforts in recent years have made Vetger yield criterion more user-friendly, flexible and robust for sheet metal forming applications. Nevertheless, Vetger yield criterion has only been presented in its parametric formulation in the literature. This work describes a method of reformulating the existing Vetger Lite yield criterion into the implicit form that is more commonly used in the conventional sheet metal plasticity modeling. The potential benefits of using both the original parametric and the new implicit forms of Vegter yield criterion for material parameter calibration, convexity certification, and computational simulations are discussed.

1. Introduction

A yield stress function $f(\boldsymbol{\sigma})$ plays an essential role in mathematical theory of anisotropic plasticity for sheet metal modeling since Hill's seminal work [1, 2]. The yield stress function is used to establish the plastic yielding condition of a sheet metal subjected to a given Cauchy stress $\boldsymbol{\sigma}$ and obtain the corresponding plastic strain increments via an associated flow rule. For example, a flat thin sheet may be modeled by Hill's 1948 associated quadratic anisotropic plasticity [1] with its yield stress function given in plane stress $\boldsymbol{\sigma} = (\sigma_x, \sigma_y, \tau_{xy})$ as

$$\Phi_h(\sigma_x, \sigma_y, \tau_{xy}) = \bar{\sigma}^2 = f^2(\boldsymbol{\sigma}) = A_1\sigma_x^2 + A_2\sigma_x\sigma_y + A_3\sigma_y^2 + A_4\tau_{xy}^2, \tag{1}$$

where polynomial coefficients (A_1, A_2, A_3, A_4) are material constants calibrated from yield stress and/or plastic strain ratio data. The yield condition and plastic strain increments from the associated flow rule are given respectively as

$$f(\boldsymbol{\sigma}) = \sigma_f(\bar{\varepsilon}^p), \quad \dot{\boldsymbol{\varepsilon}}^p = \dot{\bar{\varepsilon}}^p \frac{\partial f(\boldsymbol{\sigma})}{\partial \boldsymbol{\sigma}} = \frac{\dot{\bar{\varepsilon}}^p}{2\bar{\sigma}} \frac{\partial \Phi_h(\boldsymbol{\sigma})}{\partial \boldsymbol{\sigma}}, \tag{2}$$

where $\sigma_f(\bar{\varepsilon}^p)$ is the equivalent flow strength of the sheet metal, $\dot{\bar{\varepsilon}}^p$ is the plastic-work equivalent strain increment and $\bar{\varepsilon}^p = \int \dot{\bar{\varepsilon}}^p \mathrm{d}t$.

Graphically, the yield condition represents a closed surface and the plastic strain increments form an outer vector normal to the convex yield surface (the flow rule is thus also called normality

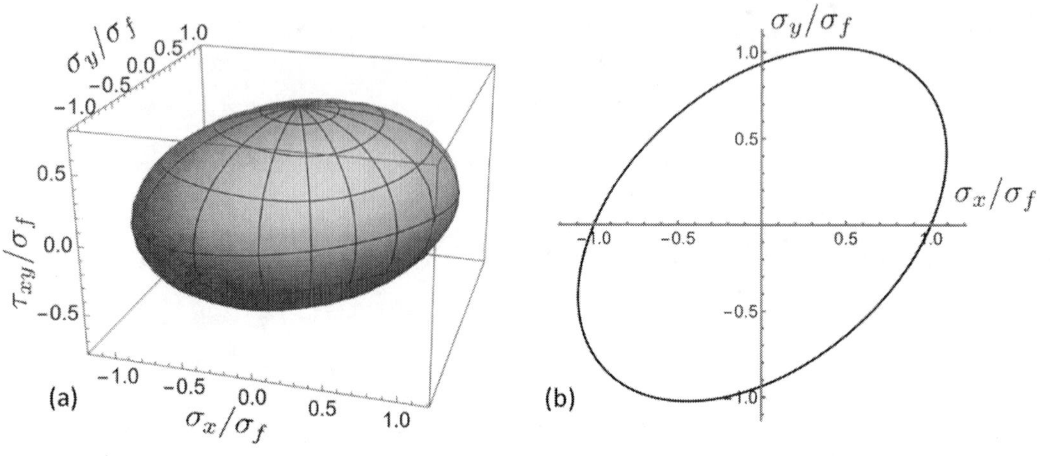

Figure 1. (a) The 3D yield surface given by a yield condition $\sigma_x^2 - 0.8636\sigma_x\sigma_y + 1.1397\sigma_y^2 + 1.9396\tau_{xy}^2 = \sigma_f^2$; (b) the 2D yield surface (a plane curve) given by a yield condition $\sigma_x^2 - 0.8636\sigma_x\sigma_y + 1.1397\sigma_y^2 = \sigma_f^2$.

rule). For example, the yield surface in tri-component plane stress $(\sigma_x, \sigma_y, \tau_{xy})$ and in biaxial stress (σ_x, σ_y) are shown in Figures 1(a) and 1(b) respectively for an aluminum sheet metal modeled by the yield condition $F(\boldsymbol{\sigma}) = \Phi_h(\boldsymbol{\sigma}) - \sigma_f^2 = 0$. In analytical geometry, the yield condition as presented is regarded as an *implicit* representation of a curve or surface [3]. A *parametric* curve or surface is better suited to generate points and thus plots in computer graphics. For example, the 2D yield surface in Figure 1(b) is created in fact by connecting many discrete points $(\sigma_x(\alpha), \sigma_y(\alpha))$ which are computed from the following parametric equations using polar coordinates for $\alpha \in [0, 2\pi]$

$$\sigma_x(\alpha) = r(\alpha)\,\cos\alpha, \quad \sigma_y(\alpha) = r(\alpha)\,\sin\alpha, \tag{3}$$

where

$$r(\alpha) = \frac{\sigma_f}{\sqrt{\cos^2\alpha - 0.8636\,\sin\alpha\cos\alpha + 1.1397\,\sin^2\alpha}}. \tag{4}$$

It is noted that the 2D yield surface (a plane curve or more precisely the boundary contour of a convex plane figure) in Figure 1(b) may also be written in the explicit form as $\sigma_y(\sigma_x) = 0.378872\sigma_x \pm \sqrt{0.877425\sigma_f^2 - 0.733881\sigma_x^2}$.

Although the subsequent development in advanced or non-quadratic yield criteria for sheet metal plasticity modeling has been primarily in the implicit formulation similar to Hill's 1948 quadratic yield stress function [4, 5, 6, 7], a parametric formulation of advanced yield criteria has also been proposed in the literature [8, 9, 10]. In particular, the Vegter yield criterion has a more direct connection to computer graphics and uses multiple second-order Bézier curves to construct the entire closed 2D yield surface. Its methodology has also been called interpolation of biaxial stress states to emphasize its piecewise nature and its flexibility [11, 12, 13]. In terms of the commonly used terminology in analytical geometry and computer graphics, it belongs to the parametric instead of implicit formulation or representation of yield criteria. While there have been works in the field of computer graphics and computer-aided geometric design that address the conversion or transformation of parametric curves or surfaces into the implicit form [14, 15, 16], it appears that implicitization of the Vegter yield criterion has not been carried out at all for sheet metal plasticity modeling. It is the aim of the current study to remedy this situation.

2. Implicitization of Quadratic Bézier Curves by Sylvester's Matrix Method

As a preparation for describing implicitization of the Vegter yield criterion, a single second-order or quadratic parametric Bézier curve is first introduced in terms of its three control points. Then its implicitization by Sylvester's matrix elimination method is presented which has its origin from the classical work in analytical geometry [17, 18].

2.1. Quadratic Bézier Plane Curves

Parametric curves are used predominately in computer graphics and Bézier curves are often adapted in computer-aided auto body design and many other applications [19]. A quadratic Bézier curve in a plane (x, y) is specified by three control points (x_1^r, y_1^r),(x^h, y^h) and (x_2^r, y_2^r) (where the first and third points are called *reference* (end) points and the second or middle point is called a *hinge* point by Vegter and co-workers [10, 11, 12, 13]). It may be treated as a linear interpolation of two linear Bézier curves

$$\begin{aligned} x(\lambda) &= (1-\lambda)[(1-\lambda)x_1^r + \lambda x^h] + \lambda[(1-\lambda)x^h + \lambda x_2^r] = (1-\lambda)^2 x_1^r + 2(1-\lambda)\lambda x^h + \lambda^2 x_2^r, \\ y(\lambda) &= (1-\lambda)[(1-\lambda)y_1^r + \lambda y^h] + \lambda[(1-\lambda)y^h + \lambda y_2^r] = (1-\lambda)^2 y_1^r + 2(1-\lambda)\lambda y^h + \lambda^2 y_2^r, \end{aligned} \tag{5}$$

where the parameter $0 \le \lambda \le 1$.

As illustrated in Figure 2(a), the two reference (first and last control) points are two endpoints of the quadratic Bézier curve shown in solid line while the hinge (the middle control) point does not lie on the curve at all. In fact, the hinge point is the intersection of two tangent lines (shown as dashed lines) passing through the two reference points, namely

$$y^h = y_1^r + k_1(x^h - x_1^r), \quad y^h = y_2^r + k_2(x^h - x_2^r), \tag{6}$$

where (k_1, k_2) are tangents of the plane curve at these two reference points.

Although two quadratic Bézier curves with four control points can form a closed plane curve, at least three quadratic Bézier curves with six control points are needed to form a *smooth* closed plane curve. Using four quadratic Bézier curves with eight control points, Figure 2(b) gives an example similar to a planarly isotropic yield surface shown as Figure 1(a) in [8] . The continuity and smoothness are ensured as two adjacent curves share a common reference point with the same tangent. One can also observe in this example that the closed plane curve is enclosed by a convex polygon (in fact an isosceles trapezoid) in dashed lines formed by the hinge points.

2.2. Implicitization by the Matrix Elimination Method

The parametric Bézier curve given in Eq.(5) consists of two quadratic polynomials in the parameter λ and can be implicitized by Sylvester's matrix elimination method [14, 15]. The two equations may be rewritten as

$$\alpha_2\lambda^2 + \alpha_1\lambda + (\alpha_0 - x) = 0, \quad \beta_2\lambda^2 + \beta_1\lambda + (\beta_0 - y) = 0 \tag{7}$$

where

$$\begin{aligned} \alpha_0 &= x_1^r, \quad \alpha_1 = 2(x^h - x_1^r), \quad \alpha_2 = (x_1^r + x_2^r - 2x^h), \\ \beta_0 &= y_1^r, \quad \beta_1 = 2(y^h - y_1^r), \quad \beta_2 = (y_1^r + y_2^r - 2y^h). \end{aligned} \tag{8}$$

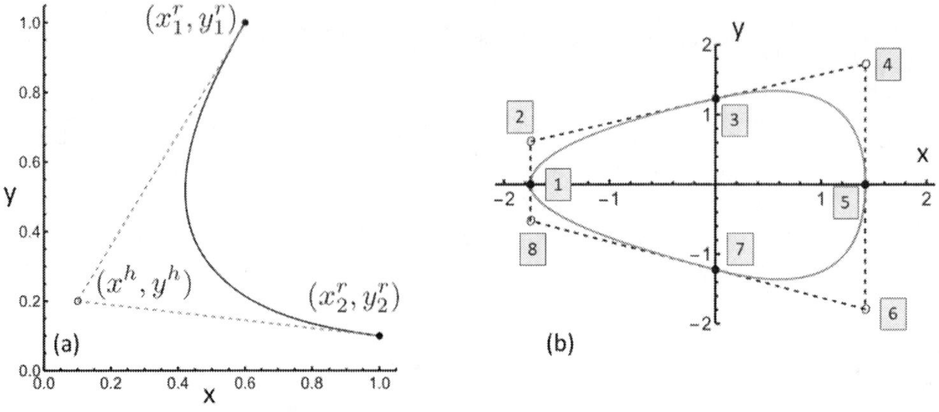

Figure 2. (a) A single quadratic Bézier curve (solid line) defined by two reference (end) points and one hinge point; (b) a closed plane curve consisting of four quadratic Bézier curves (solid lines). Filled and open circles are respectively reference and hinge points.

The implicit expression of two polynomials of Eq.(7) is given by the determinant of Sylvester's matrix

$$\begin{vmatrix} \alpha_2 & \alpha_1 & \alpha_0 - x & 0 \\ 0 & \alpha_2 & \alpha_1 & \alpha_0 - x \\ \beta_2 & \beta_1 & \beta_0 - y & 0 \\ 0 & \beta_2 & \beta_1 & \beta_0 - y \end{vmatrix} = 0, \tag{9}$$

that is,

$$A_1 x^2 + A_2 xy + A_3 y^2 + B_1 x + B_2 y + C = 0, \tag{10}$$

where

$$\begin{aligned} &A_1 = \beta_2^2, \quad A_2 = -2\alpha_2\beta_2, \quad A_3 = \alpha_2^2, \quad B_1 = 2\alpha_2\beta_0\beta_2 + \alpha_1\beta_1\beta_2 - \alpha_2\beta_1^2 - 2\alpha_0\beta_2^2, \\ &B_2 = \alpha_2\beta_1\alpha_1 + 2\alpha_2\beta_2\alpha_0 - 2\alpha_2^2\beta_0 - \alpha_1^2\beta_2, \\ &C = (\alpha_2\beta_0 - \alpha_0\beta_2)^2 + (\alpha_1\beta_2 - \alpha_2\beta_1)(\alpha_1\beta_0 - \alpha_0\beta_1). \end{aligned} \tag{11}$$

The implicit form of a parametric quadratic Bézier curve is simply an equation of a conic curve in terms of a non-homogeneous quadratic polynomial with six coefficients $(A_1, A_2, A_3, B_1, B_2, C)$. It depicts actually a parabola as $4A_1A_3 - A_2^2 = 0$.

In particular, the quadratic Bézier curve shown in Figure 2(a) are defined by the three control points (x_1^r, y_1^r)=(0.6,1.0),(x^h, y^h)=(0.1,0.2) and (x_2^r, y_2^r)=(1.0,0.1). Its parametric and implicit representations are given respectively as

$$\begin{aligned} &x(\lambda) = 1.4\lambda^2 - \lambda + 0.6, \quad y(\lambda) = 0.7\lambda^2 - 1.6\lambda + 1, \quad 0 \le \lambda \le 1, \\ &0.49x^2 - 1.96xy + 1.96y^2 - 1.092x - 1.204y + 0.8988 = 0. \end{aligned} \tag{12}$$

The four quadratic Bézier curves that form the closed plane curve shown in Figure 2(b) are specified clockwise by the following eight control points (x_1, y_1)=(-1.75,0), (x_2, y_2)=(-1.75,0.619), (x_3, y_3)=(0,1.2281), (x_4, y_4)=(1.419,1.722), (x_5, y_5)=(1.419,0), (x_6, y_6)=(1.419,-1.79), (x_7, y_7)=(0,-1.22177) and (x_8, y_8)=(-1.75,-0.521). Their parametric and implicit

IOP Conf. Series: Materials Science and Engineering 1238 (2022) 012007 doi:10.1088/1757-899X/1238/1/012007

representations are $(0 \leq \lambda \leq 1)$

$$\begin{aligned} &x(\lambda) = 1.75\lambda^2 - 1.75, \quad y(\lambda) = -0.00989618\lambda^2 + 1.238\lambda, \\ &x(\lambda) = -1.419\lambda^2 + 2.838\lambda, \quad y(\lambda) = -2.2159\lambda^2 + 0.987792\lambda + 1.2281, \\ &x(\lambda) = -1.419\lambda^2 + 1.419, \quad y(\lambda) = 2.35823\lambda^2 - 3.58\lambda, \\ &x(\lambda) = 1.75\lambda^2 - 3.5\lambda, \quad y(\lambda) = -0.179773\lambda^2 + 1.40155\lambda - 1.22177, \end{aligned} \tag{13}$$

and

$$\begin{aligned} &0.0000979344x^2 + 0.0346366xy + 3.0625y^2 - 2.68178x + 0.0606141y - 4.69342 = 0, \\ &4.9102x^2 - 6.28871xy + 2.01356y^2 + 2.89582x + 8.92368y - 13.9961 = 0, \\ &5.56123x^2 + 6.69265xy + 2.01356y^2 + 2.40369x - 9.49687y - 14.6087 = 0, \\ &0.0323184x^2 + 0.629206xy + 3.0625y^2 - 1.78697x + 1.10111y - 3.22618 = 0. \end{aligned} \tag{14}$$

It is noted that the numerical values of coefficients for both parametric and implicit functions are given only up to six significant digits here. For example, β_2 for the first quadratic Bézier curve of Eq.(13) is -0.009896181760807865 and the corresponding $A_1 = \beta_2^2$ of Eq.(14) is 0.00009793441344294626 in terms of 16 significant digits. Such approximation or slight inconsistency in precision is thus unavoidable due to numerical round-offs.

3. The Vegter Yield Criterion and Its Implicitization

We are now ready to present results of the implicit representation of the parametric Vegter yield criterion for applications used in sheet metal plasticity modeling. We consider here a variant of the original Vegter Lite yield criterion in biaxial loading [12]. The so-called Vegter Lite model uses a smaller number of reference points and thus reduces the total number of material constants from seventeen to eight for modeling a sheet metal under plane stress [11, 12].

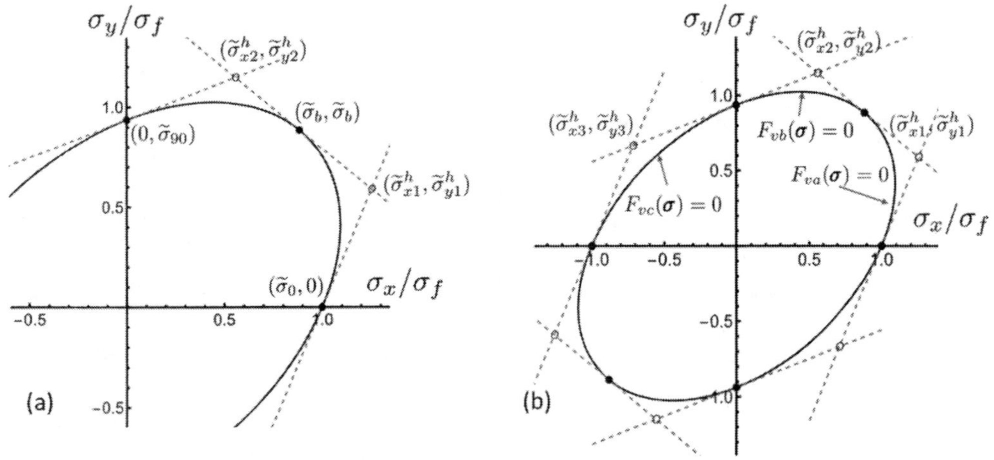

Figure 3. The Vegter yield loci in biaxial loading ($\tilde{\sigma} = \sigma/\sigma_f$): (a) two quadratic Bézier curves for biaxial tension; (b) a set of six quadratic Bézier curves for the entire closed yield surface with tension-compression symmetry.

3.1. The Vegter Lite yield criterion

As shown in Figure 3(a), only three reference points plus two hinge points are needed for the Vegter Lite model in biaxial tension. Assuming the weight $w = 1$ for the Vegter Lite yield criterion for simplicity, its second-order NURB interpolation is reverted back to the second-order Bézier interpolation. The two quadratic parametric Bézier curves for the yield criterion are counterclockwise from uniaxial tension $(\sigma_0, 0)$ along the rolling direction to equal biaxial tension (σ_b, σ_b) and then to uniaxial tension $(0, \sigma_{90})$ along the transverse direction

$$\begin{aligned} &\sigma_x = (1-\lambda)^2\sigma_0 + 2(1-\lambda)\lambda\sigma^h_{x1} + \lambda^2\sigma_b, \quad \sigma_y = 2(1-\lambda)\lambda\sigma^h_{y1} + \lambda^2\sigma_b, \quad \sigma_x \geq \sigma_y \geq 0, \\ &\sigma_x = (1-\lambda)^2\sigma_b + 2(1-\lambda)\lambda\sigma^h_{x2}, \quad \sigma_y = (1-\lambda)^2\sigma_b + 2(1-\lambda)\lambda\sigma^h_{y2} + \lambda^2\sigma_{90}, \; 0 \leq \sigma_x \leq \sigma_y, \end{aligned} \tag{15}$$

where $0 \leq \lambda \leq 1$ and the biaxial stresses at the two hinge points are obtained from the yield stresses and plastic strain ratios per Eq.(6)

$$\begin{aligned} \sigma^h_{x1} &= \frac{k_0\sigma_0 + (1-k_b)\sigma_b}{k_0 - k_b}, \quad \sigma^h_{y1} = \frac{k_0 k_b(\sigma_0 - \sigma_b) + k_0\sigma_b}{k_0 - k_b}, \\ \sigma^h_{x2} &= \frac{(1-k_b)\sigma_b - \sigma_{90}}{k_{90} - k_b}, \quad \sigma^h_{y2} = \frac{k_{90}\sigma_b(1-k_b) - k_b\sigma_{90}}{k_{90} - k_b}. \end{aligned} \tag{16}$$

It is noted that the slope of a tangent line to the yield surface, the vector normal to the yield surface and the plastic strain ratios are related as follows

$$(1, k)\cdot(\dot\varepsilon^p_x, \dot\varepsilon^p_y) = 0 \quad \text{or} \quad k = -\frac{\dot\varepsilon^p_x}{\dot\varepsilon^p_y}: \quad k_0 = 1 + \frac{1}{R_0}, \quad k_{90} = \frac{R_{90}}{1+R_{90}}, \quad k_b = -\frac{1}{R_b}. \tag{17}$$

For the entire closed yield surface with tension-compression symmetry as shown in Figure 3(b), there is one additional unique quadratic parametric Bézier curve from uniaxial tension $(0, \sigma_{90})$ along the transverse direction to uniaxial compression $(-\sigma_0, 0)$ along the rolling direction as

$$\sigma_x = 2(1-\lambda)\lambda\sigma^h_{x3} - \lambda^2\sigma_0, \quad \sigma_y = (1-\lambda)^2\sigma_{90} + 2(1-\lambda)\lambda\sigma^h_{y3}, \quad \sigma_x \leq 0, \; \sigma_y \geq 0, \tag{18}$$

where

$$\sigma^h_{x3} = \frac{-k_0\sigma_0 + \sigma_{90}}{k_0 - k_{90}}, \quad \sigma^h_{y3} = \frac{-k_0 k_{90}\sigma_0 + k_0\sigma_{90}}{k_0 - k_{90}}. \tag{19}$$

Due to the tension-compression symmetry of the entire closed yield surface shown in Figure 3(b), the reference and hinge points with $\sigma_y < 0$ will just have negative values of coordinates of the reference and hinge points with $\sigma_y > 0$. So the other three quadratic parametric Bézier curves for the yield surface are readily obtained but not presented here for brevity.

3.2. Implicit Form of the Vegter Lite yield criterion for selected sheet metals

To illustrate its implicit representation, the simplified Vegter Lite yield criterion is applied to model an IF-steel and an Al-Mg-Si alloy as studied in [11]. The six experimental inputs in terms of three yield stresses $(\sigma_0, \sigma_{90}, \sigma_b)$ normalized by the equivalent yield stress σ_f and corresponding three plastic strain ratios (R_0, R_{90}, R_b) for the two sheet metals are listed in Table 1. The corresponding biaxial stresses normalized by σ_f at the three hinge points are computed from Eqs.(16)-(19) and the results are given in Table 2.

Table 1. Yield stresses and plastic strain ratios for two sheet metals [11]

	σ_0/σ_f	σ_{90}/σ_f	σ_b/σ_f	R_0	R_{90}	R_b
IF steel	1.004	0.997	1.157	1.85	2.51	0.777
Al alloy	1.021	1.009	1.004	0.64	0.76	0.889

Table 2. Normalized stress values of three hinge points for two sheet metals

$(\widetilde{\sigma} = \sigma/\sigma_f)$	$\widetilde{\sigma}^h_{x1}$	$\widetilde{\sigma}^h_{y1}$	$\widetilde{\sigma}^h_{x2}$	$\widetilde{\sigma}^h_{y2}$	$\widetilde{\sigma}^h_{x3}$	$\widetilde{\sigma}^h_{y3}$
IF steel	1.48283	0.737657	0.823665	1.58600	-0.665950	0.520779
Al alloy	1.28810	0.684432	0.722281	1.32089	-0.754365	0.683251

The yield loci of the simplified Vegter Lite yield criterion are shown in Figures 4(a) and 4(b) in solid lines for IF steel and Al-Mg-Si aluminum alloy sheets respectively. Due to the center symmetry (see Figure 3(b)), only half of the entire closed yield surface with $\sigma_y \geq 0$ is shown and is defined by three quadratic parametric Bézier curves. The dashed lines form a convex polygon (again only half of the polygon is shown) that encloses the convex yield surface for each sheet metal. For IF steel, its three quadratic parametric Bézier curves are given in terms of the normalized stresses as:

$$\begin{aligned}
&\widetilde{\sigma}_x = 1.004 + 0.95766\lambda - 0.80466\lambda^2,\ \widetilde{\sigma}_y = 1.47531\lambda - 0.31831\lambda^2, \quad \widetilde{\sigma}_x \geq \widetilde{\sigma}_y \geq 0,\\
&\widetilde{\sigma}_x = 1.157 - 0.66667\lambda - 0.49033\lambda^2,\ \widetilde{\sigma}_y = 1.157 + 0.85801\lambda - 1.01801\lambda^2, \quad 0 \leq \widetilde{\sigma}_x \leq \widetilde{\sigma}_y,\\
&\widetilde{\sigma}_x = -1.3319\lambda + 0.3279\lambda^2,\ \widetilde{\sigma}_y = 0.997 - 0.95244\lambda - 0.04456\lambda^2, \quad \widetilde{\sigma}_x \leq 0,\ \widetilde{\sigma}_y \geq 0.
\end{aligned} \tag{20}$$

For the Al-Mg-Si alloy, its three quadratic parametric Bézier curves are given as

$$\begin{aligned}
&\widetilde{\sigma}_x = 1.021 + 0.53419\lambda - 0.55119\lambda^2,\ \widetilde{\sigma}_y = 1.36886\lambda - 0.36487\lambda^2, \quad \widetilde{\sigma}_x \geq \widetilde{\sigma}_y \geq 0,\\
&\widetilde{\sigma}_x = 1.004 - 0.56344\lambda - 0.44056\lambda^2,\ \widetilde{\sigma}_y = 1.004 + 0.63379\lambda - 0.62879\lambda^2, \quad 0 \leq \widetilde{\sigma}_x \leq \widetilde{\sigma}_y,\\
&\widetilde{\sigma}_x = -1.50873\lambda + 0.48773\lambda^2,\ \widetilde{\sigma}_y = 1.009 - 0.6515\lambda - 0.3575\lambda^2, \quad \widetilde{\sigma}_x \leq 0,\ \widetilde{\sigma}_y \geq 0.
\end{aligned} \tag{21}$$

Using Sylvester's matrix elimination method detailed in Section 2.2, one can readily obtain the implicit form of the above Vegter yield loci in terms of three piecewise non-homogeneous quadratic polynomials. For IF steel, its three implicit yield functions are obtained as

$$\begin{aligned}
F_{va}(\widetilde{\boldsymbol{\sigma}}) &= -1.20472 + 1.0982\widetilde{\sigma}_x + 0.10132\widetilde{\sigma}_x^2 - 0.33062\widetilde{\sigma}_y - 0.51227\widetilde{\sigma}_x\widetilde{\sigma}_y + 0.64748\widetilde{\sigma}_y^2 = 0,\\
F_{vb}(\widetilde{\boldsymbol{\sigma}}) &= -1.56662 - 0.29975\widetilde{\sigma}_x + 1.03633\widetilde{\sigma}_x^2 + 1.33164\widetilde{\sigma}_y - 0.99832\widetilde{\sigma}_x\widetilde{\sigma}_y + 0.24042\widetilde{\sigma}_y^2 = 0,\\
F_{vc}(\widetilde{\boldsymbol{\sigma}}) &= -0.38665 - 0.38311\widetilde{\sigma}_x + 0.00199\widetilde{\sigma}_x^2 + 0.28061\widetilde{\sigma}_y + 0.02922\widetilde{\sigma}_x\widetilde{\sigma}_y + 0.10752\widetilde{\sigma}_y^2 = 0.
\end{aligned} \tag{22}$$

For the Al-Mg-Si alloy, its three implicit yield functions are obtained as

$$\begin{aligned}
F_{va}(\widetilde{\boldsymbol{\sigma}}) &= -0.64333 + 0.49417\widetilde{\sigma}_x + 0.13313\widetilde{\sigma}_x^2 + 0.11174\widetilde{\sigma}_y - 0.40222\widetilde{\sigma}_x\widetilde{\sigma}_y + 0.30381\widetilde{\sigma}_y^2 = 0,\\
F_{vb}(\widetilde{\boldsymbol{\sigma}}) &= -0.72577 + 0.16385\widetilde{\sigma}_x + 0.39538\widetilde{\sigma}_x^2 + 0.52346\widetilde{\sigma}_y - 0.55404\widetilde{\sigma}_x\widetilde{\sigma}_y + 0.1941\widetilde{\sigma}_y^2 = 0,\\
F_{vc}(\widetilde{\boldsymbol{\sigma}}) &= -1.06264 - 0.91029\widetilde{\sigma}_x + 0.12781\widetilde{\sigma}_x^2 + 0.81314\widetilde{\sigma}_y + 0.34873\widetilde{\sigma}_x\widetilde{\sigma}_y + 0.23788\widetilde{\sigma}_y^2 = 0.
\end{aligned} \tag{23}$$

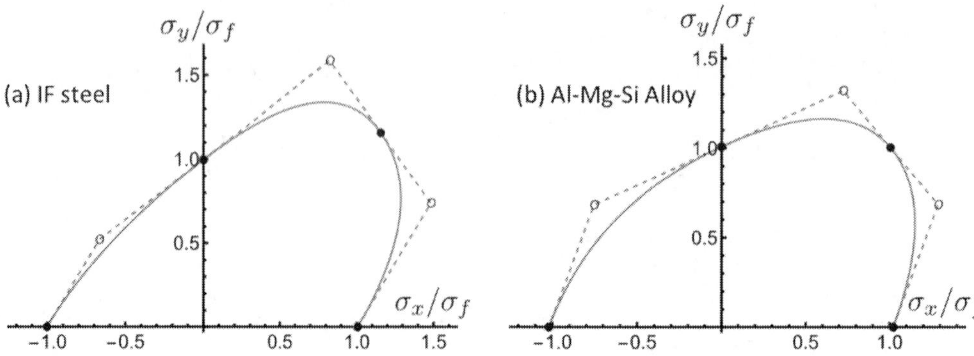

Figure 4. The Vegter yield loci in biaxial loading with $\sigma_y \geq 0$ defined by three quadratic parametric Bézier curves: (a) IF steel sheet (Eq.20); (b) Al-Mg-Si alloy sheet (Eq.21).

Similar to each parametric curve that is limited to $0 \leq \lambda \leq 1$, those implicit yield functions are applicable only within certain biaxial stress states, namely

$$\begin{aligned}
\sigma_x \geq \sigma_y \geq 0 \quad \text{for} \quad & F_{va}(\boldsymbol{\sigma}) = \sigma_f^2 F_{va}(\widetilde{\boldsymbol{\sigma}}) = \Phi_{2a}(\boldsymbol{\sigma}) - \sigma_f^2 = f_{2a}^2(\boldsymbol{\sigma}) - \sigma_f^2 = 0, \\
\sigma_y \geq \sigma_x \geq 0 \quad \text{for} \quad & F_{vb}(\boldsymbol{\sigma}) = \sigma_f^2 F_{vb}(\widetilde{\boldsymbol{\sigma}}) = \Phi_{2b}(\boldsymbol{\sigma}) - \sigma_f^2 = f_{2b}^2(\boldsymbol{\sigma}) - \sigma_f^2 = 0, \\
\sigma_y \geq 0 \text{ and } \sigma_x \leq 0 \quad \text{for} \quad & F_{vc}(\boldsymbol{\sigma}) = \sigma_f^2 F_{vc}(\widetilde{\boldsymbol{\sigma}}) = \Phi_{2c}(\boldsymbol{\sigma}) - \sigma_f^2 = f_{2c}^2(\boldsymbol{\sigma}) - \sigma_f^2 = 0.
\end{aligned} \tag{24}$$

The yield loci specified by those implicit yield functions $F_{va}(\boldsymbol{\sigma}) = 0$, $F_{vb}(\boldsymbol{\sigma}) = 0$ and $F_{vc}(\boldsymbol{\sigma}) = 0$ are shown as dashed, dotted and dotted-and-dashed lines in Figures 5(a) and 5(b) for these two sheet metals respectively. As expected, within their applicability ranges, the yield loci by the implicit form of the Vegter yield criterion match completely those by its parametric form (solid lines). Due to the tension-compression central summery of the yield surface under consideration, three implicit yield conditions $F_{va}(-\boldsymbol{\sigma}) = 0$, $F_{vb}(-\boldsymbol{\sigma}) = 0$ and $F_{vc}(-\boldsymbol{\sigma}) = 0$ completes the description of the remaining half of the entirely closed 2D yield surface for each sheet metal.

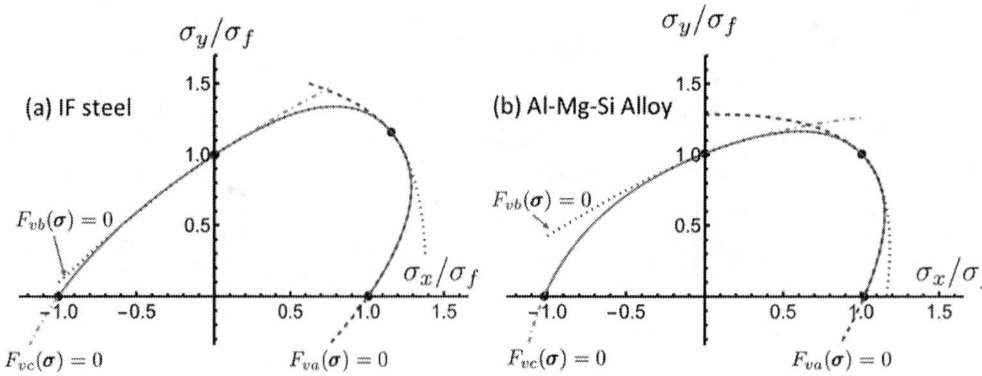

Figure 5. The Vegter yield loci in biaxial loading with $\sigma_y \geq 0$ defined by three implicit quadratic yield stress functions: (a) IF steel sheet (Eq.22); (b) Al-Mg-Si alloy sheet (Eq.23).

4. Discussion and Conclusions

The Vegter yield criterion is flexible and can be made to accept various numbers of experimental inputs for its calibration. The Lite version of the Vegter yield criterion uses only eight inputs ($\sigma_0, \sigma_{45}, \sigma_{90}, \sigma_b, R_0, R_{45}, R_{90}, R_b$), identical to the requirements of YLD2000-2D [5, 6]. Just like YLD2000-2D, the Vegter Lite yield criterion needs only six out of those eight experimental inputs for modeling sheet metals with tension-compression symmetry under biaxial loading. In this study, it is shown that the parametric form of the biaxial Vegter Lite yield criterion with six quadratic Bézier curves can be converted into the more conventional implicit form with six piecewise non-homogeneous quadratic polynomials (only three of them are given due to tension-compression symmetry). As each quadratic polynomial stress function has six coefficients, the total number of polynomial coefficients for the entire closed 2D yield surface is 36 (identical to six times of the six total number of stress values from three control points per a quadratic Bézier curve shown in Figure 3(b)). Nevertheless, due to continuity, smoothness and symmetry conditions imposed on the yield surface, there are only 6 out of 36 coefficients are independent. That is, they can all be determined from six inputs ($\sigma_0, \sigma_{90}, \sigma_b, R_0, R_{90}, R_b$) as given in this study.

Although the parametric form of Vegter yield criterion has been used over the years in various sheet metal forming analyses and numerical simulations without much problems at all [10, 11, 12, 13], there are certain advantages when its implicit form is also known. For example, the determination about the onset or switch-off of plastic yielding is crucial in almost all iterative algorithms for analyzing elastic-plastic deformation of a sheet metal. For a given (trial) stress state $\boldsymbol{\sigma}^* = (\sigma_x^*, \sigma_y^*)$ in a simulation, it is much easier to use the yield stress function $f(\boldsymbol{\sigma}) = \sqrt{\Phi_2(\boldsymbol{\sigma})}$ in the *implicit* form of the yield condition $F(\boldsymbol{\sigma}) = \Phi_2(\boldsymbol{\sigma}) - \sigma_f^2(\bar{\varepsilon}^p) = 0$ (Eq.24). That is, it is very simple to check if $F(\boldsymbol{\sigma}^*) < 0$ (elastic) or $F(\boldsymbol{\sigma}^*) \geq 0$ (plastic). Completion of the same task will be a little bit more complicated by using the parameter form of the yield loci. That is, one needs to solve first (λ^*, η^*) with $0 \leq \lambda^* \leq 1$ and $\eta^* > 0$ from two equations $\sigma_x(\lambda^*) = \eta^* \sigma_x^*$ and $\sigma_y(\lambda^*) = \eta^* \sigma_y^*$ to find out if $\eta^* > 1$ (elastic) and $\eta^* \leq 1$ (plastic). The complexity of this approach will increase if one deals with a parametric yield loci in tri-component plane stress or full 3D stresses.

Given an applied stress $\boldsymbol{\sigma} = (\sigma_x, \sigma_y)$ on the yield surface and its increment $\dot{\boldsymbol{\sigma}} = (\dot{\sigma}_x, \dot{\sigma}_y)$, plastic strain increments are more readily computed directly from the flow rule based on the implicit yield stress function $f(\sigma_x, \sigma_y)$ as (assuming the equivalent stress-strain relation $\sigma_f(\bar{\varepsilon}^p)$ is known)

$$\dot{\varepsilon}_x^p = \dot{\bar{\varepsilon}}^p \frac{\partial f}{\partial \sigma_x}, \quad \dot{\varepsilon}_y^p = \dot{\bar{\varepsilon}}^p \frac{\partial f}{\partial \sigma_y}, \quad \text{where} \quad \dot{\bar{\varepsilon}}^p(\boldsymbol{\sigma}, \dot{\boldsymbol{\sigma}}) = \frac{1}{\sigma_f'(\bar{\varepsilon}^p)}(\dot{\sigma}_x \frac{\partial f}{\partial \sigma_x} + \dot{\sigma}_y \frac{\partial f}{\partial \sigma_y}). \tag{25}$$

On the other hand, the computation of plastic strain increments from the parametric Vegter yield criterion is indirect by invoking the relation between tangent and normal vectors of the yield surface (Eq.17) and the plastic work rate equivalence [11], namely

$$k = \frac{\mathrm{d}\sigma_y(\lambda)}{\mathrm{d}\sigma_x(\lambda)} = \frac{\sigma_y'(\lambda)}{\sigma_x'(\lambda)} = -\frac{\dot{\varepsilon}_x^p}{\dot{\varepsilon}_y^p} \quad \Rightarrow \quad \frac{\dot{\varepsilon}_x^p}{\sigma_y'(\lambda)} = \frac{\dot{\varepsilon}_y^p}{-\sigma_x'(\lambda)}, \quad \text{and} \quad \sigma_f \dot{\bar{\varepsilon}}^p = \sigma_x \dot{\varepsilon}_x^p + \sigma_y \dot{\varepsilon}_y^p. \tag{26}$$

This leads to the final parametric results on plastic strain increments as

$$\dot{\varepsilon}_x^p = \dot{\bar{\varepsilon}}^p \frac{\sigma_f \sigma_y'(\lambda)}{\sigma_x(\lambda)\sigma_y'(\lambda) - \sigma_y(\lambda)\sigma_x'(\lambda)}, \quad \dot{\varepsilon}_y^p = \dot{\bar{\varepsilon}}^p \frac{-\sigma_f \sigma_x'(\lambda)}{\sigma_x(\lambda)\sigma_y'(\lambda) - \sigma_y(\lambda)\sigma_x'(\lambda)}, \tag{27}$$

where the equivalent plastic strain increment $\dot{\bar{\varepsilon}}^p$ and the parameter increment $\dot{\lambda}$ are obtained from stress increments $\dot{\boldsymbol{\sigma}} = (\dot{\sigma}_x, \dot{\sigma}_y)$

$$\sigma_x = \sigma_f \widetilde{\sigma}_x(\lambda), \sigma_y = \sigma_f \widetilde{\sigma}_y(\lambda): \quad \dot{\sigma}_x = \sigma_f' \widetilde{\sigma}_x(\lambda)\dot{\bar{\varepsilon}}^p + \sigma_f \widetilde{\sigma}_x'(\lambda)\dot{\lambda}, \quad \dot{\sigma}_y = \sigma_f' \widetilde{\sigma}_y(\lambda)\dot{\bar{\varepsilon}}^p + \sigma_f \widetilde{\sigma}_y'(\lambda)\dot{\lambda} \tag{28}$$

as

$$\dot{\bar{\varepsilon}}^p = \frac{1}{\sigma'_f(\bar{\varepsilon}^p)} \frac{\widetilde{\sigma}'_y(\lambda)\dot{\sigma}_x - \widetilde{\sigma}'_x(\lambda)\dot{\sigma}_y}{\widetilde{\sigma}_x(\lambda)\widetilde{\sigma}'_y(\lambda) - \widetilde{\sigma}_y(\lambda)\widetilde{\sigma}'_x(\lambda)}, \quad \dot{\lambda} = \frac{1}{\sigma_f(\bar{\varepsilon}^p)} \frac{\widetilde{\sigma}_x(\lambda)\dot{\sigma}_y - \widetilde{\sigma}_y(\lambda)\dot{\sigma}_x}{\widetilde{\sigma}_x(\lambda)\widetilde{\sigma}'_y(\lambda) - \widetilde{\sigma}_y(\lambda)\widetilde{\sigma}'_x(\lambda)}. \tag{29}$$

In summary, it has been shown in this study that implicitization of the parametric Vegter Lite yield criterion in biaxial loading [12] lead to the conventional (implicit) representation of the yield criterion as piecewise non-homogeneous quadratic polynomials in Cauchy stress components (σ_x, σ_y). The methodology presented here can be readily extended to implicitization of the full parametric Vegter yield criterion [11] with six independent quadratic Bézier curves for accommodating additional inputs from both plane strain tension and pure shear tests. One may thus implement both Vegter yield criteria using more widely available iterative algorithms in computational anisotropic plasticity literature and finite element codes based on initial Hill's 1948 (implicit) formulation of yield conditions. In addition, simple algebraic inequalities and numerical methods [20, 21] may be used to certify the convexity of a calibrated Vegter yield criterion in both biaxial loading and tri-component plane stress. Implicitization of the parametric Vegter yield criterion in tri-component plane stress $(\sigma_x, \sigma_y, \tau_{xy})$ and its convexity certification are currently under investigation and its outcomes will be reported in the near future.

References

[1] Hill R 1948 A theory of yielding and plastic flow of anisotropic metals *Proc. Royal Soc.* **193A** 281–97

[2] Hill R 1950 The *Mathematical Theory of Plasticity* Oxford University Press, London

[3] Hoffmann C 1993 Implicit curves and surfaces in CAGD *IEEE Computer Graphics and Applications* **13**(1) 79–88

[4] Gotoh M 1977 A theory of plastic anisotropy based on a yield function of fourth order (plane stress state) *Int. J. Mech. Sci.* **19** 505–20

[5] Barlat F, Brem J, Yoon J W, Chung K, Dick R E, Lege D, Pourboghrat F, Choi S H, Chu E 2003 Plane stress yield function for aluminum alloy sheets Part 1: theory *International Journal of Plasticity* **19** 1297–319

[6] Aretz H 2005 A non-quadratic plane stress yield function for orthotropic sheet metals *Journal of Materials Processing Technology* **168** 1–9

[7] Yoshida F, Hamasaki H, Uemori T 2013 A user-friendly 3D yield function to describe anisotropy of steel sheets *International Journal of Plasticity* **45** 119–39

[8] Budiansky B 1984 Anisotropic plasticity of plane-isotropic sheets *Studies in Applied Mechanics* **6** 15–29

[9] Ferron G, Makkouk R, and Morreale J 1994 A parametric description of orthotropic plasticity in metal sheets *International Journal of Plasticity* **10** 431–49

[10] Vegter H, Drent P, Huetink J 1995 A planar isotropic yield criterion based on mechanical testing at multi-axial stress states. In: S-F Shen and P R Dawson (Eds) *Simulation of Materials Processing: Theory, Methods and Applications.* Balkema, Rotterdam pp 345–50

[11] Vegter H and van den Boogaard A H 2006 A plane stress yield function for anisotropic sheet material by interpolation of biaxial stress states *International Journal of Plasticity* **22** 557–80

[12] Vegter H, ten Horn C and Abspoel M 2011 The Vegter Lite material model: simplifying advanced material modelling *Int. J. Mater. Form.* **4** 85–92

[13] Vegter H, Abspoel M and Mulder J 2019 A plane stress yield surface using Bézier curve interpolation in two directions *IOP Conf. Series: Materials Science and Engineering* **651** (2019) 012052

[14] Sederberg T W and Goldman R N 1986 Algebraic geometry for computer-aided geometric design *IEEE Comput. Graph. Appl.* **6**(6) 52–9

[15] Hoffmann C 1989 *Geometric and Solid Modeling: An Introduction* Morgan Kaufmann Publishers Inc

[16] Floater M S 1995 Rational cubic implicitization. In: M Daehlen, T Lyche, and L L Schumaker (Eds) *Mathematical Methods for Curves and Surfaces* Vanderbilt University Press pp 151–9

[17] Dixon A L (1908) The eliminant of three quantics in two independent variables *Proc. London Mathematical Society* (Series 2) **7** 49–69

[18] Salmon G 1915 *A Treatise on the Analytic Geometry of Three Dimensions* Longmans, Green and Co., London p 264

[19] Faux I D and Pratt M J 1980 *Computational Geometry for Design and Manufacture* Ellis Horwood Ltd

[20] Hill R 1990 Constitutive modelling of orthotropic plasticity in sheet metals *J. Mech. Phys. Solids* **38**(3) 405–17

[21] Tong W and Alharbi M 2017 Comparative evaluation of non-associated quadratic and associated quartic plasticity models for orthotropic sheet metals *Int. J. Solids & Struct.* **128** 133–48

Die Concepts and formability for simultaneous forming of sheet metals and FRPs

X F Fang, T Kloska and A Hajdarevic

Institute of automotive lightweight design, University of Siegen, Germany

E-mail: Xiangfan.fang@uni-siegen.de

Abstract. Fiber-reinforced plastics (FRPs) are increasingly being used to reinforce steel or aluminum (Al) automotive structure components. Currently, these multi-material applications are realized via three-stage manufacturing, which require the corresponding costs. Thus, a new hybrid forming technology has been developed and will be presented here. It is based on the principle of the hydro-mechanical sheet metal- forming technique and uses a heated thermoplastic FRP material (up to 280 °C) in a half melting state as the pressure media to form steel or aluminum sheets. Since the sheet metals are pre-coated with an adhesion promoter, the FRPs are also joined with the sheet metal during this hybrid forming process. In this work, the die and especially the sealing concepts for hybrid forming will be introduced based on different part geometries. The formability of the sheet metals (Nakajima test) at different temperatures up to 280 °C, for each two steel and Al alloys, will be presented. The hybrid formed parts were analyzed using optical method and section analysis. Good forming results were confirmed for both sheet metal and FRPs. A vehicle chassis component was successfully hybrid formed using CP-W800 steel and FRP, and a body component was produced using an Al5182 alloy and an FRP.

1. Introduction

Lightweight design is one of the most important measures to reduce fuel consumption of vehicles using internal combustion engines (ICEs) or extending the driving range of pure battery electric vehicles (BEVs) [1–3]. For ICE vehicles, a fuel consumption reduction of 0.3 L/100 km was determined for a primary weight reduction of 100 kg. For BEVs, a 100kg weight reduction may increase the driving range by 4% in urban traffic [3]. To realize economically acceptable lightweight designs, vehicle body-in-white (BIW) structures are increasingly being built using multi-material designs consisting of steel, aluminum (Al), magnesium, and fiber-reinforced plastic (FRP) [4,5].

The multi-material designs in [4, 5] were both realized using the so-called "post molding assembly" (PMA) and "in-mold assembly" (IMA) approaches. In the PMA process, metal and FRP components are initially manufactured separately and then joined together in a third assembly step. Here, the matching of the two parts is a problem due to different part tolerances, in addition to the increased costs due to the usage of three separate manufacturing steps. In contrast, in the IMA process, a preformed sheet metal part is over-molded by FRPs using, for example, the injection molding process. Alternatively, either a heated thermoplastic or thermoset prepreg can be placed in a preformed and preheated sheet metal part and formed by a well-sealed press die [6]. After the FRPs are press-formed, an adhesive bonding can be created as well when the sheet metal is pre-coated using an adhesion promoter such as VESTAMELT (company Evonik) [7] or adhesion film from company Nolax [8]. Here, the step of making a pure FRP part is eliminated.

To further reduce the processing cost of IMAs, the forming of FRPs and sheet metals or organo-sheets was carried out in the same die but at a separate time step for the injection molding process [9,10]. However, these types of processes were only realized for very simple geometries, such as rotational symmetrical parts made from low-strength mild steels [9]. For high strength steels or even advanced high strength steels and complex geometries, this process has not been realized.

In addition, due to the extruder of the injection molding system and the very small section size of the injection nozzle, the long fibers of the thermoplastics are cut or broken into short fibers with fiber length much lower than 1 mm. Compared to injection molding, compression molding of FRPs results in much longer fiber lengths in the final parts. In addition to plastics with short fibers prepared by the extruder, additional long fibers are placed at the very end of the extruding process (figure 2). These thermoplastics with longer fibers (up to 25 mm) are then press-formed in a well-sealed die to form the desired part. Moreover, since the fibers are only subjected to shear stress when filling the die cavity (no nuzzles), the fiber length degradation is significantly reduced compared to injection molding. Fiber lengths of up to 10 mm could be achieved for a PP glass fiber (GF) material [11], which is thus referred to as long fiber thermoplastic (LFT). According to [12], the material properties significantly depend on fiber length. As shown in figure 1 (a), the stiffness of the materials increases significantly up to ca. 1-2 mm while the strength increases considerably to more than 10 mm. The impact resistance may even increase further with longer fibers of up to ca. 50 mm length. Therefore, in addition to the LFTs with fiber lengths restricted to a maximum of 10 mm in the final part, glass mat thermoplastic (GMT) materials were also investigated in this work. GMT material contains glass mats with very long fibers in the order of 50 mm in addition to the short fibers, and may significantly further enhance impact resistance.

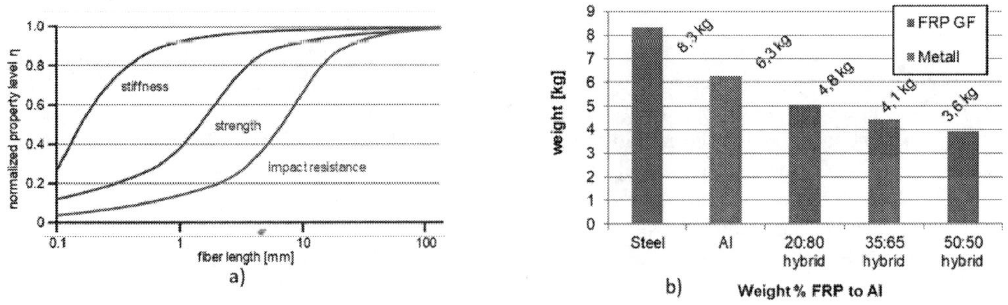

Figure 1. a) Influence of fiber length on mechanical properties of FRPs [11], b) estimated weight of rectangular profiles for energy absorption (for axial crashing) using metal and metal-FRP (glass fiber) composite (100 KJ energy input)

Although FRPs may have high stiffness and strength at a lower weight, they fail suddenly without warning when cyclically loaded. Moreover, if they are overloaded, the car driver may not directly recognize the damage. In contrast, metals show crack formations and stable crack growth during cyclic loading and plastic deformation at overloading; these can be easily recognized due to changes in vehicle driving behaviors. These kinds of properties of metals are called fail-safe.

2. Principle of the hybrid forming method

Due to the lack of fail-safe behavior of FRPs, the idea of hybrid forming was developed to produce a metal-FRP multi-material component with material-based large-area joining between metal sheets and FRPs [11]. The FRP may have different thicknesses and rib structures to effectively reinforce the sheet metal panels. In this way, the lightweight potential of FRPs and the fail-safe behavior of metals can be combined. In addition to the fail-safe properties, the FRP-metal hybrid may absorb higher energy during a crash since FRPs generally possess much higher energy absorption capabilities than metals (figure 1 (b)). Therefore, these kinds of hybrid materials are highly suitable both for vehicle chassis and body applications. Furthermore, the well-known hydro-mechanical sheet metal forming and FRP compression

IOP Conf. Series: Materials Science and Engineering 1238 (2022) 012008 doi:10.1088/1757-899X/1238/1/012008

molding techniques can be combined to make such kind of multi-material parts in a one-step process and, thus, be very cost-efficient.

During the well-known hydro-mechanical forming process, the sheet metals are deformed by the hydrostatic pressure of the pressure media, such as water or oil, in a well-sealed die cavity. Since the thermoplastic composite materials for FRP compression molding, such as PP or PA (with fibers like GF), are in a half-liquid state before compression molding, they can also be used as pressure media. This combined method is called the hybrid forming of sheet metal and FRPs [11].

In figure 2, the entire process chain of hybrid forming is illustrated. In the first step, the long fibers with ca. 50 mm length are mixed with the extruded short fiber materials at the end of the extrusion process. This material, which was in a half-liquid state, is then placed onto the sheet metal blanks (e.g., steel or Al) that have already been positioned in a forming die. The sheet metal blanks are pre-coated with a bonding agent, for example, VESTAMELT (Evonik), in this work [7]. The sheet metal blanks are preheated in an oven between 180 and 220 °C to activate the bonding agent, while a PA GF 40 should have a temperature of approx. 280 °C, and a PP of 220 °C. Instead of direct feeding from an extruder, also pre-extruded LFT or GMT materials can also be reheated in an IR-heater and subsequently placed onto the sheet metal blanks. Therefore, the process can also be applied by all metal forming companies having a hydraulic press and an oven or IR heater.

In the third step, a special sealed hybrid forming die forms the LFTs or GMTs with the sheet metals simultaneously. At first, the punch presses and forms the LFT/GMT material, and the LFT/GMT on the other side forms the sheet metal. Since the LFT/GMT materials are enclosed by die sealing, high pressures are created, which finally form the LFT/GMT over the entire surface of the sheet metal with a pre-defined thickness, and the sheet metal is formed to the final shape as well. In the same time, ribs for reinforcement and functional integration are formed. In the final step, after ca. 20 s die closing time (when the LFT/GMT is cooled to ca. 80 °C), the adhesive bonding of the materials is completed, and the part can be removed from the press.

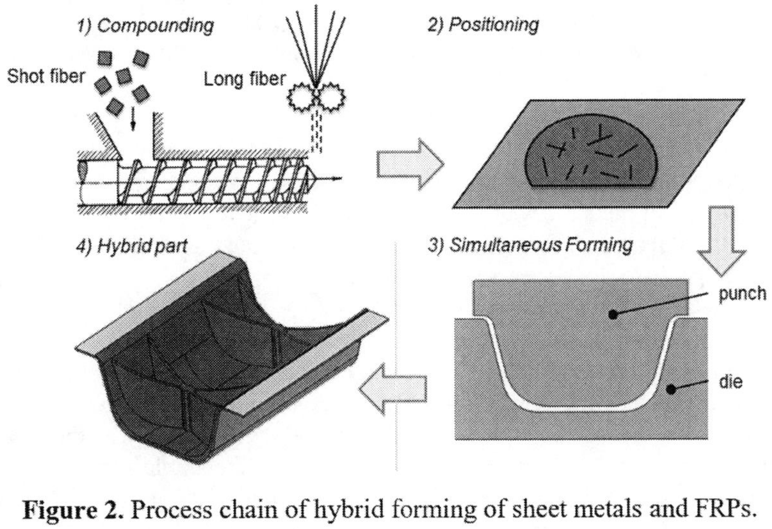

Figure 2. Process chain of hybrid forming of sheet metals and FRPs.

3. Forming and sealing concept of the hybrid forming die

Real vehicle parts in BIW have very complex 3D geometries. In order to develop generic forming and sealing concepts, which are the most challenging aspects of a hybrid forming die, a tube geometry and a U-profile with two open ends were studied. The second one serves as the basis for complex 3D curved open ended profiles with different widths and draw depths, which is currently being developed. Furthermore, for the open end geometry with a uniform draw depth and width, a simplified sealing concept has been already developed. The general requirements for die sealing is that it should prevent LFT/GMT press-outs, while the formability of sheet metals should not be impaired.

3.1. Tube geometry

In figure 3 (a), a tube geometry with different draw depths, radii, and closed ends is illustrated. The target is to evaluate the formability of sheet metals during hybrid forming, especially with regard to whether the geometry transitions can be formed by LFT pressure and if the LFTs may be pressed out when wrinkles appear in the four corners of the part. Similar to a deep drawing die, draw beads were introduced, which may control both sheet metal forming and wrinkle formation. They should also prevent LTF or GMT press-out in addition to the blank holder. The designed die can be seen in figure 3 (b). The blank holder was supported by a die cushion with adjustable forces. Detailed descriptions can be found in [11].

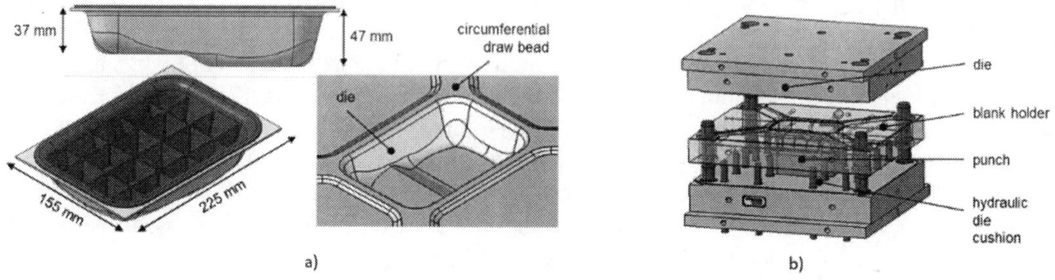

Figure 3. a) Part characteristic of tube geometry and b) the corresponding hybrid forming die

The first hybrid forming experiments with this die showed leakage of LFT in the corner areas due to wrinkles, which are typical when sheet metals deform in a strong deep draw mode. This LFT leakage prevented a complete hybrid forming process for two reasons: 1) there was not enough LFT left in the cavity to complete the metal forming; 2) the leaked LFT prevented the complete closing of the upper and lower die. By analyzing the force curves for the pressing and the blank holder for the first trials (figure 4 (a)), as well as the forming results, it can be determined that an LFT leakage happened when the pressing force slightly decreased (marked by a blue circle). Or in other words, due to the LFT leakage, the hydrostatic pressure could not be further maintained until the gape was closed after 3-4 mm further stroke.

The solution to avoid this leakage was using a travel-dependent hydraulic die cushion force. The principles (i.e., the pre-settings of the press parameters and results after optimization) are depicted in figure 4 (b). At the process beginning, the blank holder force was set low enough to enable a smooth material flow of the sheet metal. Towards the end of the process, especially between -20 and -10 mm distance to the bottom dead end, the blank holder force was increased stepwise to 100 KN so that no distinct decrease of press forces appears in the blue circle. In this way, a constant sealing between the blank holder and metal sheet was ensured, and LFT leakage was prevented. The reduction of the blank holder force from 300 kN to 200 kN (figure 4 (a) compared to (b)) at the process end (-3 mm) is due to the minimization of the necessary blank holder force and has no further influence on the tightness.

Since the punch velocity of the press can only be driven but not be controlled during the pressing process, which is standard for many hydraulic presses, the actual velocity of the upper die is different from the preset values. The pressing force is a reaction force of the whole system, which depends on the preset velocity and the die cushion force. It increases continuously until the whole cavity is filled up with LFT and the sheet metal is almost completely formed – without any leakage. At the end of the process, the LFT is used as a forming medium, and the forming force increases rapidly to the set value of 1800 kN, which is not shown in figure 4 for reasons of clarity. All curves end up at around 2-3 mm above the bottom dead end, which is the thickness of the constant LFT layer in the tub geometry. The thickness of this LFT layer depends on the amount of inserted LFT.

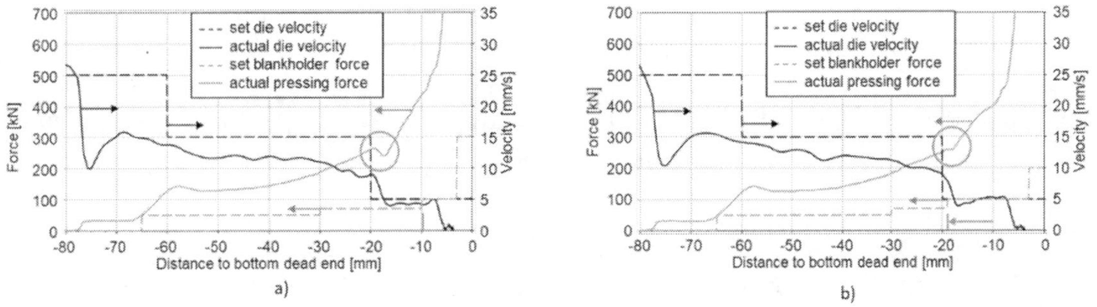

Figure 4. Process forces and velocity of the die with leakage (a) and without leakage (b) for hybrid-forming of a DC04 (2.0 mm) steel together with 300 g LFT PA GF 40

3.2. Open end U-profile – concept 1

Since many of the real car body parts show open ended geometry with various section sizes (width, depth, etc.) with transitions from different sections, the first die concept has been developed using a homogeneous U-profile shape. A generic die sealing concept principle should be developed in this way. Currently, this simplified generic sealing concept is being further developed for arbitrary shapes with open end profiles. For the homogeneous U-profile, it is obvious that the sheet metal forming is very easy (figure 5 (a)). However, the LFT/GMT materials can be very easily pressed out due to the open ends. In order to guarantee the sealing, two measures were introduced. One is the introduction of a small geometry characteristic in the profile, and the other is the separation of the die into several segments.

The first measure is to introduce a small (<5 mm) step at the open end of the U-profile (figure 5 (a)), with which the LFT/GMT flow can be stopped or delayed. The punch was then divided into three segments: the mean punch in the middle and two sealing elements at each end (figure 5 (b)). The sealing elements are supported by gas pressure springs and can deflect by an offset to the major punch, which is responsible for the hybrid forming of sheet metal and LFT/GMT. During the forming process, the sealing elements move ahead. In combination with the counter holders of the lower die (figure 5 (a)), these sealing elements clamp and deform the sheet metal on both lateral ends before the main forming by the major punch and LFT/GMT pressure increases. It was assumed that if both ends of the profile were preformed, these undercuts could work as a mechanical sealing.

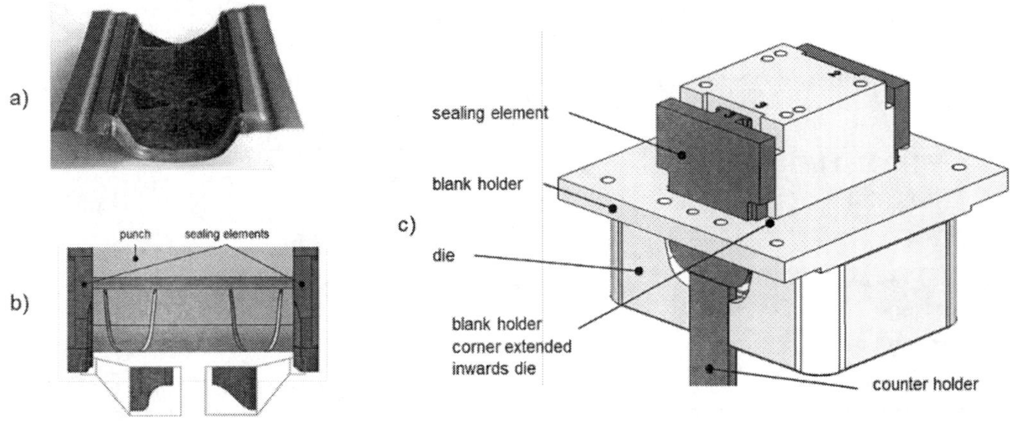

Figure 5. a) Hybrid formed part with homogeneous open ended U-profile, b) and the sealing concept.

However, the LFTs could still be pressed out at the upper end of the profile since the gap in the triangle sealing element - die - BH can only be fully closed at the dead end. To avoid this situation, the blank holder was extended inwards of the die (figure 5 (c)). In this way, the press out of the LFT could be fully avoided, as shown by the pressed part in figure 5 (a).

3.3. Open end U-profile – concept 2

However, the concept in figure 5 is quite complex. Therefore, for an easy homogeneous U-profile, a simplified sealing concept was developed, as shown in figure 6. In contrast to figures 5 (b) and (c), no counter holder for the sealing element is needed, and the sealing elements also do not move separately from the punch. It is mounted together with the punch, and is thus a part of the punch. Through the small step between the sealing element and the main punch, the FLT/GMT press out could be prevented. The height of the step was equal to or slightly larger than the required thickness of LFT/GMT (2-3 mm usually), and its width was between 5-10 mm. If different sheet metal thicknesses are to be tested, the width of the sealing element must be changed accordingly.

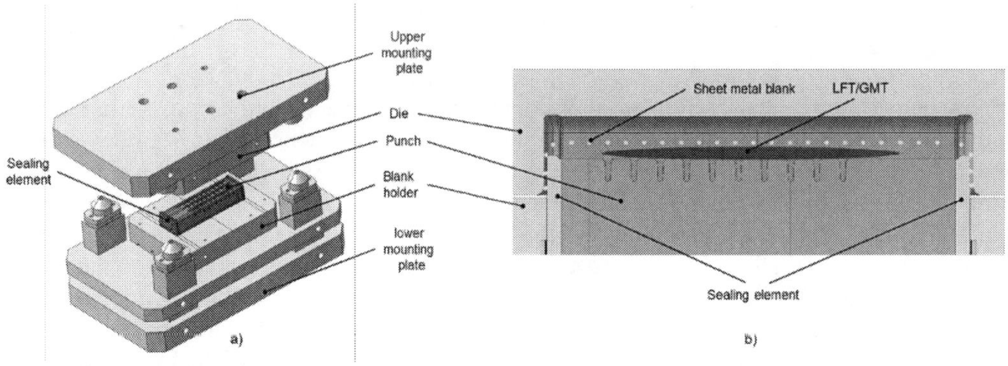

Figure 6. A simplified die sealing concept for homogeneous open ended U-profile.

4. Results and analysis

4.1. Materials and sheet metal formability

In this work, the following materials (table 1) were used for the hybrid forming investigations. The properties of steels were taken from EN10346. Both Al and steels with different strengths were investigated. A PA6 GF 40 with a fiber length of up to 25 mm when placed on the sheet metal was chosen for LFT. In addition, a PP GMT (40% GF) was also included. For bonding, the adhesion promoter VESTAMELT (Evonik) was applied using a powder coating method [7].

Table 1. Properties of sheet metals used for hybrid forming developments

	$R_{p0,2}$ [MPa]	$R_{m,min}$ [MPa]	$A_{80,min}$ [%]
Al 5182 H111/O	135	280	24
Al 6451 T4	130	280	24
DC04	140-210	270-350	38
HC340 LAD	340-420	410-510	21
DP 600	330-430	590-700	20
DP 800	440-550	780	14
CP-W800	660-820	760-960	10

Since the sheet metals have to be preheated to temperatures between 180 and 220 °C to activate the adhesion promoter VESTAMELT [7] and afterward hybrid formed together with 220 °C PP GMT or 280 °C PA LFT, the formability of the metals were, firstly, determined using the Nakajima tests according to the DIN EN ISO 12004 [13]. Five samples with different widths and, thus, stress states were utilized. For each stress state, three samples were tested, and an average value was built. The press velocity was 2mm/s. For Al sheets, the cut-outs in the samples were parallel to the rolling direction, and for steel, perpendicular, as defined by [13].

Prior to the Nakajima tests, the sheet metal blanks were spray-painted with a stochastic black and white speckle pattern that may resist the forming at temperature up to 300 °C so that after the tests, the forming could be measured using the GOM Aramis system, and the failure strain could be determined according to the method defined in [13].

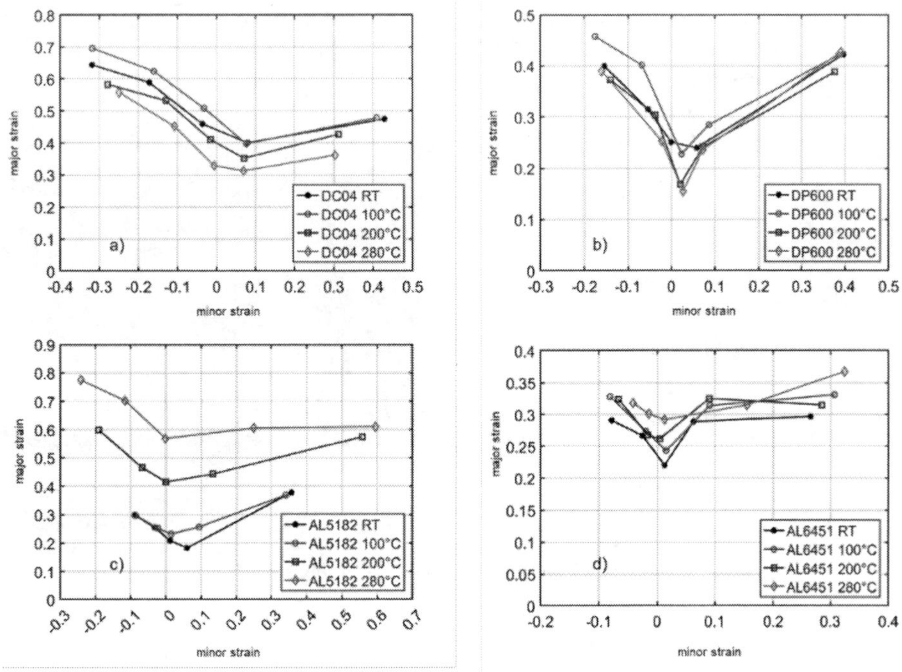

Figure 7. FLCs at four different temperatures for a) 2 mm steel DC04, b) 2mm steel DP 600, c) 1.5 mm Al A5182, and d) 1.5 mm Al A6451

Figures 7 (a-d) show the FLCs determined at RT, 100, 200 and 280°C. The temperature dependencies of the four materials are very different. For the mild steel DC04, the FLC shifts to higher values when testes at 100 °C compared to RT. Further increases of the temperature to 200 °C and 280 °C reduce the level of FLCs. For the AHSS DP 600, the FLC shifts up-wards at 100 °C in both deep drawing and stretch forming area, whereas, in the small minor strain range, the forming limit remains unchanged. At 200 and 280 °C, the forming limit in the deep drawing and stretch forming area is very similar to RT, and in the small minor strain area much lower than RT. The reduction of formability of steels in the temperature range from 200 to 400 °C was often determined and should be a result of blue brittleness or the aging effect of steels [14]. Therefore, the temperature of steel blanks should be kept below 200°C for hybrid forming.

For Al, the FLC movement strongly depends on the type of Al alloy. For Al 5182 alloy, when no aging effect may occur, the formability is significantly improved (figure 7 (c)). For the Al6451 T4 alloy, because of the nucleation and growth of the precipitations in the alloy matrix (aging effect), the FLCs change irregularly (figure 7 (d)). For a first approximation, the RT FLCs may be used for hybrid forming analysis.

4.2. Analysis of hybrid forming results on tube geometry

All sheet metals in table 1 have been investigated, and here, some examples are presented. In order to determine the hybrid forming results, the sheet metal blanks were etched prior to forming with grids so that after the forming the deformation of the sheet metal can be determined using the GOM ARGUS system. In figure 8 (a), the equivalent strain on the tube geometry is shown directly on the part itself with different colors. The corresponding minor and major strains are plotted against the FLC with 10% and 20% margins of the steel used (DP600) at RT, since the steel blanks were preheated to ca. 200°C and the die temperature was 80°C. For more precise selection of FLC, the temperature of sheet metal blanks must be measured in the future. The color of the measurement points in the FLC was chosen to be the same as those indicated in the part. Since the pre-etched grids on the steel sheets were destroyed on the radii of the parts, the strains could not be determined there, as can be seen by the white areas of the tube geometry in figure 8 (a).

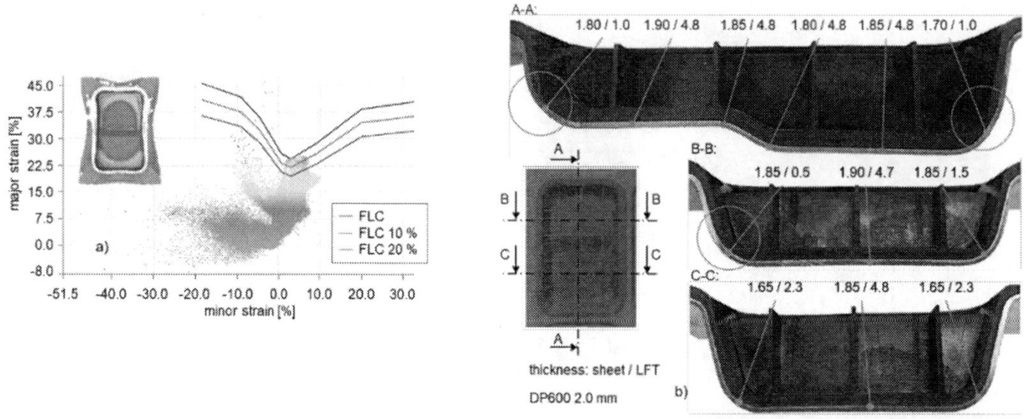

Figure 8. a) Forming analysis on DP 600 steel, b) thickness of DP 600 steel and LFT in tube geometry

In order to determine the forming of the geometrical details and the distribution of LFTs as well as the sheet steel thickness, the parts were cut into different sections. Moreover, the quality of bonding between LFT and steel could be visually inspected. These results are shown in figure 8 (b). The same was done for DC04 steel but not shown here. Due to the higher strength of DP 600, LFT thicknesses at the radii were significantly reduced and often less than 1 mm, whereas it was much thicker in the flat bottom area. For steels with lower strength, i.g. DC04, the thickness distribution was much more homogeneous [11]. In addition, due to the differences of strength of DC04 and DP600 the radii of the part made from DC04 formed more sharply in comparison to DP600 when using the same die and same press force. For forming DP600, higher forces must be applied. This fact must be considered at the CAD design stage of the part and/or when designing the die.

5. Application in vehicle chassis and body parts

Due to the different material properties of steel (higher strength) and Al (higher energy absorption rate), the steel hybrid forming was applied first on chassis (Steel) and BIW parts (Al). For chassis parts, two wheel-guiding components of a compact class vehicle (VW MQB Golf 7) [15] were investigated. The originally manufactured control arms using advanced high strength steels CP-W800 (table 1) were redesigned using steel-LFT hybrid, which should be formed by hybrid forming developed by the authors. The development results are shown on the example of the longitudinal control arm of the rear axle (figure 9 (a)). At first, all the load cases of VW for this control arm were calculated using FEM [15]. The resulting stress values and their distribution then served as target values for the new hybrid design. In the second step, the original sheet metal thickness of 3.5 mm was reduced to only 2 mm by applying LFT materials which is bonded to the entire surface of the steel base part, with various additional reinforcement ribs. The steel thickness must have a minimum of 2mm due to corrosion issues of chassis parts. This hybrid formed part weighed 20% less than the pure steel reference.

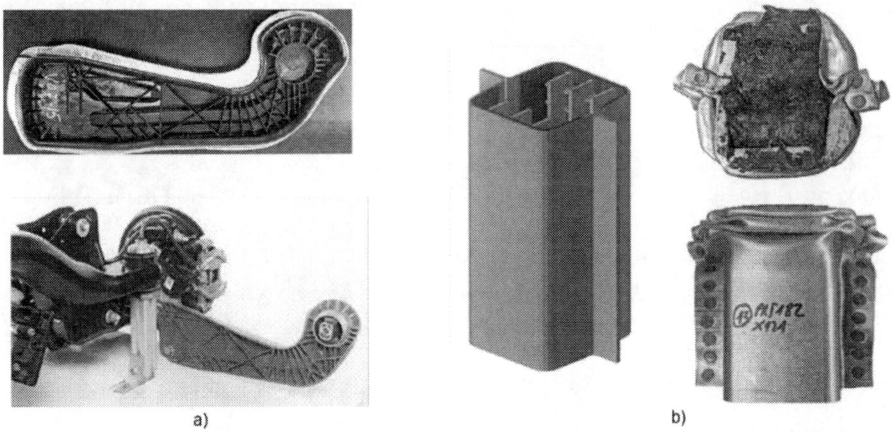

Figure 9. a) A longitudinal control arm of VW Golf rear axle made by hybrid forming of a CP-W800 steel and PA6 GF 40, b) A crash box made from Al 5182 and PP GMT (40% GF) after the axial crash test.

Due to the very good energy absorption rate of Al and Al-FRP hybrid (figure 1 (b)) and its perspective to be used in car body's front rails or crash box, a generic crash box was designed using Al 5182 alloy and a PP GMT (40% GF) (very long fiber and thus higher energy absorption according to figure 1 (a)). It consists of two U-profiles with flanges for joining. Using the die in figure 6, the hybrid forming was performed. The two U-profiles were then assembled using the self-piercing riveting technique. The crash box was finally impact-loaded using a drop tower test (V=6.2 m/s, m= 375 Kg) with 7 KJ kinetic energy. As can be seen in figure 9 (b), through the folding process of Al sheet metal shell and the breakage as well as the delamination of GMTs, high amount of energy could be absorbed.

6. Summary and conclusions

Multi-material design has increasingly been used in lightweight design for vehicle body and chassis structures. Up to now, a multi-stage production process has been applied, in which metal and FRP parts are produced separately and joined together in a subsequent process step. In this work, the idea of hybrid forming, a simultaneous sheet metal forming and LFT/GMT compression molding process, has been successfully realized. Three die sealing concepts have been proposed and developed, covering tube geometry and open ended profiles. For open profiles, a generic sealing concept has been proposed using a segmented die and punch with individually adjustable kinematics. This concept has already worked for U-profiles with homogeneous geometry. It could also be simplified without segmentation of the die and punch and using simple kinematics. Currently, the generic sealing concept is being further developed for complex-shaped open ended profiles. For all concepts, the corresponding process parameters were identified, which are all applicable for industrial companies without difficulty. The hybrid forming results were analyzed using forming, thickness, and part geometry measurements and rated positively.

Using the developed process and die concepts, a real wheel guiding chassis component, the trailing arm of a multi-link suspension axle, was developed and produced in an industrial press line (Company Weber Fibertech, Germany) using high strength steel CP-W 800 and PA6 GF 40 LFT material. A weight reduction of more than 20 % was realized. In addition, a crash box for crash energy absorption was developed using Al 5182 alloy and PP-GMT hybrid. This hybrid part could be successfully crash deformed with high energy absorption.

Acknowledgments
This research was supported by the Federal Ministry of Education and Research (BMBF) with the project "MultiForm" (02PN2081) and the European regional development funds with the project "AKTIV" (EFRE-0801129). We would like to express our gratitude to BMBF and EFRE for their financial support. We also thank our industrial project partners for providing technology consulting and technical support.

References

[1] Rohde-Brandenburger K, 2013 Was bringen 100 kg Gewichtsreduzierung im Verbrauch? Eine physikalische Berechnung, *ATZ,* **07**, 584
[2] Trautwein T, Henn S and Rother K, 2011, Gewichtsspirale – Stellhebel in der Fahrzeugauslegung, *ATZ* **05**, 390
[3] Hoffmann M, Hillebrecht M and Schäfer M, 2018, Leichtbaupotential in urbanen Elektrofahrzeugen, *Lightweight Design*, **6**, 48
[4] Ahlers M, 2016, Carbon Core – die neue BMW 7er Karosserie, *Karosseriebautage Hamburg 2016 Proceedings*, Springer Vieweg, Wiesbaden
[5] Audi AG, Der neue Audi A8, https://www.audi-mediacenter.com/de/pressemappen/audi-a8-9072
[6] Lauter C, Frantz M, Tröster T, 2011, Großserientaugliche Herstellung von Hybridwerkstoffen durch Prepreg-pressen, *Lightweight Design*, **04**, 48
[7] Risthaus M, 2017, VESTAMELT® Hylink Adhesion Promoter (Technical Information). http://adhesivessealants.evonik.com/sites/lists/RE/DocumentsAC/FlyerVESTAMELTHylink.pdf, accessed 30 June 2017
[8] Frey T, 2021, Klebefilme für belastbare Metall-Kunststoff, Leichtbauteil, ***Adhäsion***, **11**
[9] Hussain M, Rauscher B, Tekkaya A, 2008, Wirkmedienbasierte Herstellung hybrider Metall-Kunststoff-Verbundbauteile mit Kunststoffschmelzen als Druckmedium, *Materialwissenschaft und Werkstofftechnik*, **39/9**, 585
[10] Landgrebe D, Kraeusel V, Rautenstrauch A, Albert A, Wertheim R, 2016, Energy-efficiency in a hybrid process of sheet metal forming and polymer injection moulding, *Procedia CIRP*, **40** 109, https://doi.org/10.1016/j.procir.2016.01.068
[11] Fang X F and Kloska T, 2020, Hybrid forming of sheet metals with long Fiber-reinforced thermoplastics (LFT) by a combined deep drawing and compression molding process, *International Journal of Material Forming*, **13(4)**, 561
[12] Buerkle E, Sieverding M and Mitzler J, 2003, Spritzgiessverarbeitung von langfaser-verstärktem PP. *Kunststoffe* **3**, 47
[13] DIN EN ISO 12004, Metallic materials Sheet and strip - Determination of forming-limit curves – Part 2: Determination of forming-limit curves in the laboratory
[14] Saxena K K, Drotleff K and Mukhopadhyay J, 2016, Elevated temperature forming limit strain diagrams of automotive alloy Al6014-T4 and DP600: A case study, *Journal Strain analysis*, DOI:10.1177/0309324716651028
[15] Kloska T and Fang X F, 2021, Lightweight Chassis Components – The Development of a Hybrid Automotive Control Arm from Design to Manufacture, *International Journal of Automotive Technology*, **22/5**, 1245, DOI 10.1007/s12239-021-0109-0

IOP Conf. Series: Materials Science and Engineering 1238 (2022) 012009 doi:10.1088/1757-899X/1238/1/012009

Investigation of the tensile response and the microstructural evolution during hot drawing of multi-layered, ultrafine-grained AA6014 sheets by thermomechanical analyzes

B Zettl[1*], M Merklein[1]

[1] Institute of Manufacturing Technology (LFT), Friedrich-Alexander-Universität Erlangen-Nürnberg, Egerlandstr. 13, 91058 Erlangen, Germany

[*] Author to whom any correspondence should be addressed: bastian.zettl@fau.de

Abstract. Mechanical strengthening of sheet material can be realized by accumulative roll bonding, which belongs to the severe plastic deformation processes. Beside fine-grain hardening also work and precipitation hardening is responsible for the rise of material strength. However, this is also accompanied by a significant decrease in ductility. Thus, the need for an enhancement of ductility is essential for a sufficient formability in later applications and was already introduced by tailored heat treatments. The challenge, however, is to realize a process temperature that leads to a degradation of dislocations and dissolution of MgSi-precipitations, without a recrystallization of the fine-grained microstructure to coarse grain sizes. In order to identify a suitable temperature range to avoid recrystallization, hot forming experiments are carried out at successive temperatures from 20 to 300 °C. Tensile specimens are drawn with the thermomechanical simulator Gleeble 3500 (Dynamic Systems Inc.) aided by the strain measurement system Aramis (GOM GmbH). The aim is to investigate the dependence of the mechanical properties from the forming temperature in order to identify a temperature range, in which high strength with simultaneous enhanced ductility is maintained.

1. Introduction

Beside alloying and heat treatment strategies, inducing high degrees of plastic deformation can lead to strengthening of aluminum alloys. These methods are summarized under the term of severe plastic deformation (SPD) [1]. Belonging to the SPD processes, accumulative roll bonding (ARB) is suitable for producing large-scaled sheets with an increased level of strength compared to the initial material condition [2]. For the investigated wrought aluminum alloy AA6014, beside the hardening effect due to the MgSi-precipitations, also fine-grain as well as work hardening are of an important role in the context of ARB. This is due to the accumulation of the deformation degree by repeated rolling which leads to a high concentration of dislocations and subsequently to the formation of subgrain boundaries. With an increasing misorientation angle, the subgrain boundaries evolve to regular grain boundaries which results in an ultrafine-grained (UFG) microstructure [3]. The fine-grain hardening effect itself can be explained by the Hall-Petch relation in Equation 1 [4].

$$YS = \sigma_0 + k_y/d^{1/2} \tag{1}$$

YS is the yield strength, σ_0 and k_y are material constants and *d* is the average grain diameter. In the term of lightweight applications, it is favorable to establish UFG as the main hardening factor because the high density of grain boundaries results in more possibilities of dislocation gliding mechanisms and thus, to an overall better material formability [5]. Nevertheless, the combination of the hardening effects resulting from ARB lead to a significant decrease of ductility and thus, formability of the multi-layered sheets. In previous investigations, a short-time local heat treatment strategy was already proven to be beneficial for a better local formability of multi-layered AA6016 sheets [6]. The observed softening of material is attributed to the dissolution of the MgSi-clusters. It was shown that the mechanical properties depend from both the maximum temperature as wells as the temperature holding time.

In general, the temperature induced strength decrease for precipitation- and UFG-hardened aluminum alloys can be explained by three basic mechanisms. Firstly, the degradation of grain boundaries by recrystallization (reduction of fine-grain hardening effect). Secondly, the annihilation of dislocations by annealing (reduction of work-hardening effect) and lastly, the already mentioned dissolving of MgSi-clusters (reduction of precipitation-hardening effect) [7]. Thus, the aim of this investigation is to determine the possibility of an enhancement of ductility by introducing elevated temperatures during forming by thermomechanical analyzes. It is objected to identify a temperature range in which the UFG microstructure is maintained but softening mechanisms, like the dissolving of precipitations and the annihilation of dislocations, are already triggered.

2. Experimental setup

2.1. Material

The investigated material is a precipitation-hardening wrought alloy of the Al-Mg-Si system and is used under the numerical designation AA6014. Aluminum alloys of the 6000 series are already used in structural and outer skin lightweight applications for example in the automotive sector, due to their very good formability, excellent corrosion resistance and high hardening potential [8]. One reason for this are the finely distributed MgSi-precipitations which lead to higher resistance against dislocation movement. In Table 1 the chemical composition of the AA6014 aluminum Ac-PX by Novelis is shown. The sheets have a thickness of 1 mm and are solution annealed at 520 °C for 1 h in an air circulation chamber oven in order to accomplish a finely and homogeneously distribution of the MgSi-phase.

Table 1: Chemical composition in weight percentage of Novelis Ac-170 PX [9].

Element	Si	Fe	Cu	Mn	Mg	Cr	Zn	Ti	Others.	Unit
Concentration	0.3 – 0.6	0.35	0.25	0.05 – 0.2	0.4 – 0.8	0.2	0.1	0.1	0.15	Wt. %

2.2. Accumulative roll bonding

The ARB process for the production of multi-layered sheets consists of the process steps of a surface treatment, the stacking of sheets, the cold rolling and a cutting operation. In Figure 1 the cyclic ARB process chain is depicted schematically.

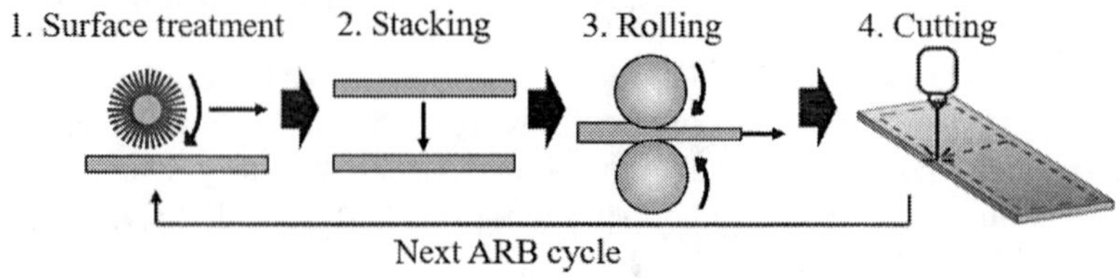

Figure 1: Process steps for cyclic accumulative roll bonding according to [2].

The surface treatment is done by manual wire brushing with the handheld PolyPTX 800 (Eisenblaetter, Germany) with a rotating brush roller in rolling direction of the sheets. The brush has a diameter of 100 mm, a width of 70 mm and curled stainless steel filament. In past investigations it was shown that it provides a high level of roughness and a removal of the oxide layer which both are necessary for the bonding by cold welding in the later rolling step [10]. After the brushing treatment of two sheets, they are stacked together with their roughened side in contact. In the third step of ARB the actual bond formation by cold rolling is realized. For this, the two-layered sheet packages are transferred to an industrial sized rolling stand (Gerd Wolff GmbH, Germany). The targeted relative pass reduction is 50 % which enables a constant sheet thickness of the multi-layered sheets compared to the initial material. The width of the rolled sheets also remains the same whereas its length doubles due to the halving of the sheet thickness. The rolling velocity was set to 3 m/min at a rolling pressure starting at 80 bar for the first rolling cycle. Finally, a laser cutting operation is done to cut the twice as long sheets in order to be able to stack them in the following ARB cycle. In this investigation up to four ARB cycles are carried out in order to produce 16-layered sheet material with a thickness of 1 mm. The number of rolling cycles is further referred as *N*. Thus, N0 stands for the initial sheet material and N4 for the 16-layered sheets after four ARB cycles.

2.3. *Tensile tests at elevated temperatures*

In order to characterize the hot forming properties of multi-layered ARB sheets, tensile tests from 20 °C up to 300 °C were performed on the thermomechanical simulator Gleeble 3500 (Dynamic Systems Inc., USA) parallel to the rolling direction (RD). The aim is to identify a certain temperature range in which a good compromise between the strength and the ductility of the multi-layered sheets is apparent. The tensile tests at elevated temperatures are performed in accordance to DIN EN ISO 6892-2. For the investigation, tensile specimens were cut out of the 16-layered sheets by fine EDM in order to reduce effects on the mechanical properties through heat input (e.g. by laser cutting) or work hardening (e.g. by regular milling). Through the combination of a statistical pattern on the specimens surface and the optical strain measurement system Aramis (GOM GmbH, Germany), the actual strain during forming can be measured via the digital image correlation (DIC) method. The elevation of temperature is done by resistance heating and the controlling of the actual temperature of the specimen is realized via type K thermocouples which are attached in the horizontal and vertical centre of the measuring area in a prior welding step. In Figure 2 the experimental setup for thermomechanical characterization is shown.

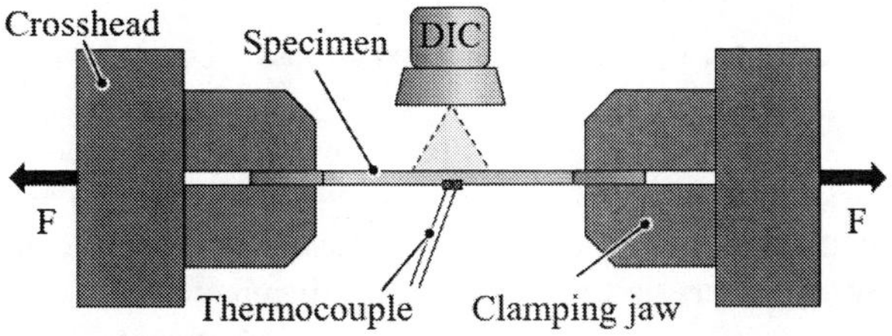

Figure 2: Experimental setup for the thermomechanical analyses.

The temperature window for the hot tensile tests ranges from 80 to 300 °C with an heating rate of 10 K/s respectively and incremental temperature steps of 20 °C. The dwell time for the targeted temperature is 5 s. Thus, twelve different tensile tests are performed. The temperatures and the corresponding strain rates as well as recording frequencies for the DIC-method are depicted in Table 2. The frequency determines the number of images which are recorded during testing and thus, the temporal resolution of the image correlation. To eliminate the influence of the strain rate on the mechanical properties as far as possible, the traverse speed is adjusted. Thus, the otherwise increasing strain rate because of the elevation of temperature can be maintained constant at a value of 0.0067 1/s over the entire temperature range.

Table 2: Parameter values for tensile tests at elevated temperatures.

Temperature in °C	Crosshead speed in mm/s	Targeted strain rate in 1/s	Frequency in Hz
80, 100, 120	2.14	0.0067 ± 0.01	50
140, 160	1.50	0.0067 ± 0.01	50
180, 200	1.15	0.0067 ± 0.01	50, 20
220, 240	1.00	0.0067 ± 0.01	20
260, 280, 300	0.75	0.0067 ± 0.01	20

3. Results and discussion

For the metallurgic analysis, images of the grain-structure were recorded by electron microscopy. In Figure 3 the derived grain sizes are illustrated. An ARB characteristic grain-structure is formed, which is defined by severely elongated grains in rolling direction. Already after two ARB repetitions (N2) the grain length is over 200 µm, starting from mean grain sizes around 20 µm in the initial condition (N0). The elongation is even more pronounced at higher number of ARB cycles. A different grain size distribution is observed considering in direction perpendicular to rolling where the ARB characteristic grain refinement is evident. Starting from mean grain sizes of 15 µm in the initial state (N0), grain sizes decrease to a mean size of only 600 nm after four rolling cycles (N4). This represents a decrease in grain size by 96 % from N0 to N4 in total. The number of samples tested is depicted as small letter *n* in the upcoming diagrams.

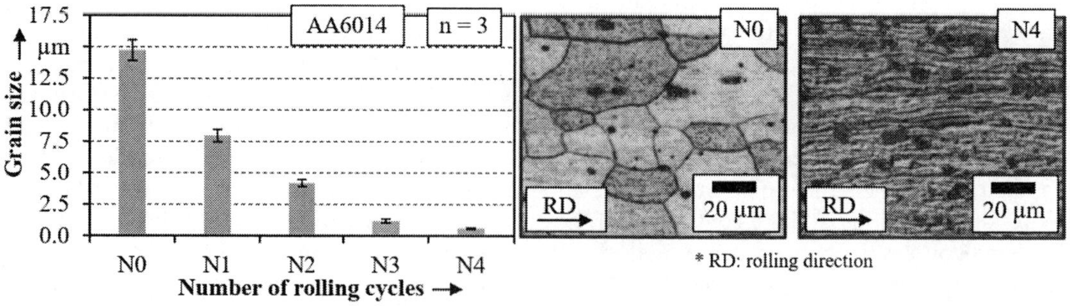

Figure 3: Evolution of grain size for AA6014 in dependence on the number of rolling cycles with exemplary images for the initial material (N0) and the 16-layered sheets (N4).

As already mentioned in the state of art, this leads to a fine-grain hardening of the AA6014 alloy which results in an increase of the overall material strength according to equation 1 (Hall-Petch). One possibility to vary the actual ductility of metals is the elevation of temperature during forming. It can lead to dynamic recovery or to recrystallization which both enhances the materials formability.

In order to analyze this possibility for roll bonded AA6014, the mechanical properties of the yield strength *YS*, the ultimate tensile strength *UTS*, the uniform elongation *UE* and the total elongation *TE* are determined in correlation to the actual testing temperature. In Figure 4, the stress-strain curves, which are derived from the tensile tests at elevated temperatures as well as the associated mechanical properties of the yield strength *YS*, the ultimate tensile strength *UTS*, the uniform elongation *UE* and the total elongation *TE* are depicted in relation to the testing temperature for the 16-layered AA6014 specimens.

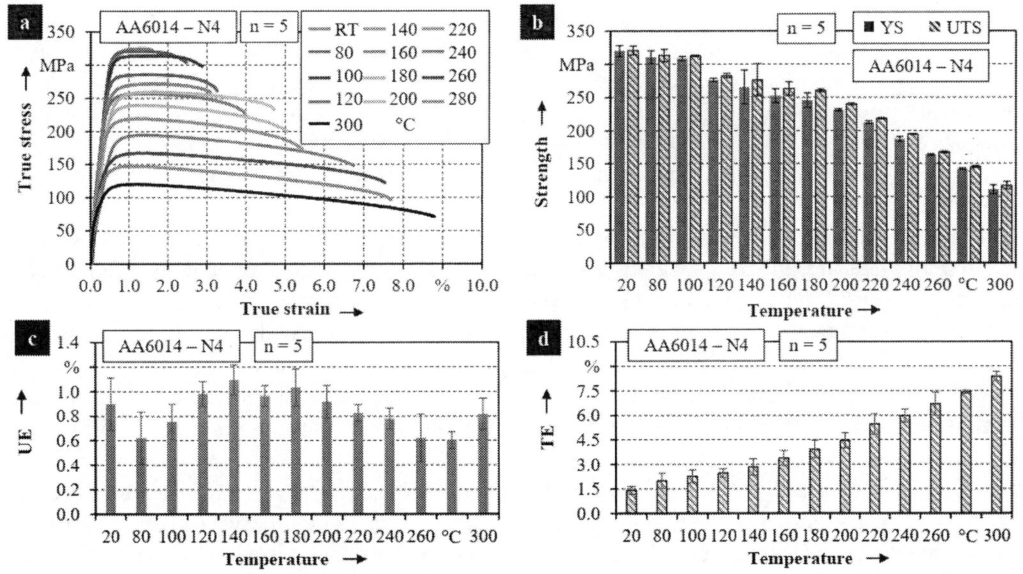

Figure 4: Temperature dependent properties of AA6014-N4 with a) stress-strain curves and b) yield and ultimate tensile strength *YS* and *UTS*, c) the uniform elongation *UE* and d) the total elongation *TE*.

Regarding the exemplary stress-strain curves depicted in Figure 4 a), it is evident that there is a general tendency to a decreasing strength but increasing ductility level with higher temperatures during tensile tests as expected for hot forming of aluminum. Special characteristics show up when looking at the temperatures between 120 °C and 180 °C where the stress strain curves are rather similar regarding the general level of material strength (*YS* and *UTS*) as well as the total elongation *TE*.
Looking at the mean values for both the yield strength *YS* and ultimate tensile strength *UTS* in Figure 4 b), it is apparent that the yield to tensile strength ratio *YS*/*UTS* maintains a high value of over 0.94 in the entire temperature range. This is an indicator for the low plastic formability of the investigated 16-layered AA6014 sheets after four rolling cycles even at higher temperatures. Overall, there is a maximum in strength at 20 °C and in comparison only a minor decrease in strength up to 100 °C. Analyzing the mean value for *UTS* in the temperature range from 120 °C to 180 °C, it is evident that *UTS* declines in comparison to the average level in the temperature field from 80 to 100 °C by a relative value of 14 %. Focusing on the temperature range from 120 °C to 180 °C itself, a tendency to a strength plateau is apparent. Thus, only a minor decline in *UTS* from 120 °C to 180 °C with a relative value of 8 % is present. Regarding the strength properties at temperatures from 200 °C to 300 °C, a nearly linear decline of *UTS* and *YS* is evident. More concretely, with every temperature increment of 20 °C a drop in both yield as well as tensile strength of 24 MPa in average occurs.

Thus, it is assumed, that the thermal stability of the UFG microstructure decreases starting at 200 °C which is in accordance to previous investigations by Sharma et al. [11]. A distinct contribution in the strength increase with higher temperatures is attributed to the already identified dissolving of MgSi-clusters [6]. Regarding the rather short temperature holding time of 5 s in combination with forming times of 2 to 3 s it is suspected, that no pronounced recrystallization is possible. Geiger et al. [7] observed that even for a 256-layered ARB sheet of the relatively similar aluminum alloy AA6016, only a minor increase in grain size is apparent at holding times between 5 to 10 s at 300 °C. Thus, it is assumed that the major contribution for strength decline is initiated by the precipitation solution as well as the relaxation of internal stresses by dislocation annihilation and rearrangement at elevated temperatures. Overall, *UE* lies on a very low level. Taking the standard deviations into account, *UE* only surpasses the 1 % mark in the temperature range from 120 to 200 °C. Thus, it is noticeable that this local plateau of ductility lies in the same temperature window in which the strength values show their slowest rate of decline. Considering the total elongation *TE*, an almost linear progression with increasing temperature is apparent. Pronounced ductile necking occurs only at temperatures above 140 °C where *TE* exceeds the 3 % mark and thus, shows a relative increase in ductility of 42 % in comparison to *TE* at 80 °C. All strain values were corrected in consideration of the coefficient of thermal expansion for aluminum.

4. Conclusion and outlook

In the scope of this work, UFG sheet material processed by accumulative roll bonding the wrought aluminum alloy AA6014 was characterized by thermomechanical analyzes. The aim was to identify a favorable temperature range in which the materials ductility can be enhanced without lowering the strength level significantly. It was observed that in a temperature range from 120 to 180 °C a slower decline in both yield strength *YS* as well as ultimate tensile strength *UTS* is evident and a relatively high strength level of over 250 MPa can be maintained until 180 °C. Additionally, the uniform elongation *UE* shows the highest values in this range of temperature. In combination with a total elongation *TE* over 3 % for temperatures equal or higher than 140 °C the proposed temperature range for forming 16-layered AA6014 lies between 140 and 180 °C. It is assumed that in the range of this temperatures no distinct recrystallization occurs, which is in accordance to previous investigations regarding UFG aluminum alloys of the 6000 series. In future investigational work, the assumptions based on the microstructural mechanisms need to be analyzed more specifically. For this, the method of laser-ultrasonic studies can be applied. It allows in-situ measurements of metallurgical processes like recrystallization, grain growth and phase transitions and thus, a more detailed insight on the temperature driven microstructural changes in UFG materials which could be accompanied by metallographic analysis to validate the method.

5. Acknowledgements

The authors gratefully acknowledge the German Research Foundation (DFG) for funding the research project 392174229 with the title "Improvement of the application characteristics of multi-layered sheet material for forming technology produced via accumulative roll bonding" at the Friedrich-Alexander-Universität Erlangen-Nürnberg.

6. References

[1] Azushima A, Kopp R, Korhonen A, *et al.* Severe plastic deformation (SPD) processes for metals. *CIRP Annals* 2008; **57**(2): 716–35.

[2] Tsuji N, Saito Y, Utsunomiya H, Tanigawa S. Ultra-fine grained bulk steel produced by accumulative roll-bonding (ARB) process. *Scripta Materialia* 1999; **40**(7): 795–800.

[3] Hansen N, Mehl RF, Medalist A. New discoveries in deformed metals. *Metallurgical and Materials Transactions A* 2001; **32**(12): 2917–35.

[4] Hall EO. The Deformation and Ageing of Mild Steel: III Discussion of Results. *Proc. Phys. Soc. B* 1951; **64**(9): 747–53.

[5] Borodachenkova M, Wen W 2017 High-Pressure Torsion: Experiments and Modeling. In: Cabibbo M, editor. *Severe Plastic Deformation Techniques* (London: InTech).

[6] Merklein M, Vogt U. Enhanced Formability of Ultrafine-Grained Aluminum Blanks by Local Heat Treatments. *KEM* 2009; **410-411**: 169–76.

[7] Geiger M, Merklein M, Vogt U. Aluminum tailored heat treated blanks. *Prod. Eng. Res. Devel.* 2009; **3**(4-5): 401–10.

[8] Bloeck M 2012 Aluminium sheet for automotive applications. In: Rowe J, editor. *Advanced Materials in Automotive Engineering* (Sawston: Woodhead Publishing) pp 85–108.

[9] Herrmann J 2020 Kumulatives Walzplattieren: Bewertung der Umformeigenschaften mehrlagiger Blechwerkstoffe der ausscheidungshaertbaren Legierung AA6014. Erlangen: FAU Univ. Press.

[10] Jamaati R, Toroghinejad MR. The Role of Surface Preparation Parameters on Cold Roll Bonding of Aluminum Strips. *J. of Materi Eng and Perform* 2011; **20**(2): 191–7.

[11] Sharma SK, Hassan MS, Kumar BS. Annealing Response of Aluminum Alloy AA6014 Processed By Severe Plastic Deformation. *International Refereed Journal of Engineering and Science (IRJES)* 2017; **6**(7): 59–65.

IOP Conf. Series: Materials Science and Engineering 1238 (2022) 012010 doi:10.1088/1757-899X/1238/1/012010

Investigation of the impact of process parameters on the layer formation of AlSi coated boron-manganese steel

F He[1*] and M Merklein[1]

[1]Institute of Manufacturing Technology, Friedrich-Alexander-Universität Erlangen-Nürnberg, Egerlandstr. 13, 91058 Erlangen, Germany

*Franz.He@fau.de

Abstract. Hot stamping has established itself as a technology for manufacturing safety-relevant car body components. Its usage has found increased importance due to rising political and social pressure e.g. for improved fuel conservation. Due to reduced sheet thickness, hot stamping parts offer lightweight solutions while at the same time providing improved crash performance due to their high strength. Boron-manganese steel is commonly used in hot stamping processes. Due to hot scale formation during heat treatment and the subsequent forming process, additional coatings, typically AlSi-coatings, are applied. Since hot stamped parts are formed at temperatures of 600 °C to 800 °C no suitable lubricants have yet been found. This leads to increased friction and severe wear in the process, which negatively affect tool lifetime and overall part quality. Additionally, process temperatures influence the layer formation and surface roughness due to diffusion processes. A deeper process understanding is needed to acquire detailed knowledge of the resulting impact on tribological behavior. Within this study, the effect of coating thickness on the resulting layer is analyzed by comparing the resulting material composition and surface roughness. To this end, the layer formation and surface roughness are analyzed for two different layer thicknesses (AS150 and AS80) in relation to furnace temperature, holding time and heat up rate. Both are characterized by means of heating experiments conducted on an annealing furnace. The change in layer composition and surface roughness are determined via metallographic investigation and tactile measurement. The results of this study help to improve the process understanding of the tribological conditions within the hot stamping process and to develop future measures for reducing tool wear.

1. Introduction

Current developments in the automotive industry show the reduction of energy consumption as one of the most important challenges. At the same time the specifications for safety-relevant components are increasing which in turn leads to more necessary components and weight [1]. These contrary requirements for modern components can be fulfilled with the usage of ultra-high strength steels like the boron manganese steel 22MnB5 [2]. The hot stamping process combines forming at temperatures of 650 °C to 850 °C and quenching in one process step and is used for the production of high strength components made out of boron manganese steel [3]. This allows for an improved formability alongside a reduction in necessary forming forces and springback behavior [1]. Additionally, the quenching process allows for an improvement of the mechanical properties resulting from the martensitic transformation at cooling rates above 27 K/s [4]. The tensile strength increases from around 500 - 700 MPa [5] to approximately 1500 MPa [4]. The improved mechanical properties of the material allow for a reduction in sheet thickness [2]. Due to furnace temperatures of 900 °C to 950 °C and varying holding times dependent on sheet thickness, the workpiece material undergoes austenitization [6]. Because of the high process temperature, uncoated boron manganese steel will decarburize and experience scale formation on the surface [7]. The oxide layer leads to a reduction in heat transfer between contact partners and damages the tool surface due to hard particles [8]. To prevent this effect from occurring hot stamping parts are coated with aluminium-silicon (AlSi) based coatings [9]. Previous investigations by Sherepenko et al. [10] have found that heat treatment parameters like furnace temperature and holding time in the furnace to have a significant influence on the formation of the coating layer for AlSi-coated 22MnB5 steels. Previous studies came to the same conclusion for the aluminum-silicon based coating AS150 [11]. Due to high process temperatures above 800 °C hot stamping is considered a lubricant free process, which in turn lead to high friction and wear occurences. While the layer coating formation has been investigated regarding weldability because of the coating layer [12], the impact on tribological properties due to layer coating and heat treatment have not yet been analyzed. Detailed knowledge between the cause and effect relation of material properties and the resulting tribological behavior is needed for the calculated design of properties of the final part. In addition to the heat treatment parameters, the thickness of the initial coating was shown to exhibit a significant influence on the layer formation during the austenitisation process [12]. The goal within the scope of this paper is to analyze the influence of heat treatment parameters on the coating layer formation for a reduced coating thickness with AS80 and compare the results to the previously investigated AS150 coating [11]. The study will further help create a baseline ongoing studies on the impact of heat treatment parameters on tribological mechanisms and wear in the industrial hot stamping process.

2. Methodology and experimental setup

2.1. Material and experimental setup

The workpiece material used in the study is the industrial standard boron manganese steel 22MnB5. While the previous investigation utilized an AS150 (150 g/m²) coating, the current experiments used a different coating strength with AS80 (80 g/m²). The sample size consisted of 50 mm x 70 mm. The aluminum-silicon based coating exhibits resistance to both corrosion and elevated temperature oxidation [12]. Due to its excellent properties under hot stamping conditions it has found use in sheet steels in automotive manufacturing and is regarded as the current industrial standard. The usage of the coating is limited to the hot stamping since it exhibits poor forming qualities at room temperature with the coating breaking and therefore losing its barrier effect [10]. The aluminum-silicon coating experiences continuous diffusion of iron particles from the base material during the heat treatment process, creating an AlSiFe layer with a higher melting point, which prevents the melting or evaporation of the AlSi coating [13]. The heat treatment experiments were conducted on an annealing and hardening furnace ME 87/13 by RHODE. The furnace can achieve temperatures of up to 1300 °C with KANTHAL heating elements mounted on support rods on three sides of the furnace walls, providing an even temperature distribution. Double layer insulation prevents heat escaping from the furnace surface while guaranteeing a homogenous furnace atmosphere for the heat treatment of the samples. The samples were not quenched after the heat treatment process and allowed to cool to room temperature. This was done to prevent further impact from the quenching step on the layer formation. Martensite transformation was therefore not achieved.

2.2. Parameters and methodology

For the heat treatment, parameters specific and relevant to the industrial hot stamping process were selected. The parameter levels are furthermore determined by the industrial process windows for the austenitization of the material and were also used in previous studies on the coating layer [11]. The parameters furnace temperature and holding time in the furnace had shown to exhibit a significant influence on the layer formation of AS150 coatings and were selected to be analyzed for the AS80 coating in this study. The parameter of heating rate was found to exhibit no significant influence on the layer formation; therefore, it is excluded from further experiments. The order of experiments was randomized to assure the reproducibility of the findings. The parameter settings were furthermore varied at the same three industrial related level settings (low, mid, high) to ensure the comparability of the findings. The parameters and selected levels are listed below in Table 1.

Table 1. Parameters and level variations selected for heat treatment

Furnace temperature in °C	880; 910; 930
Furnace duration in s	60; 270; 600

Using tactile measurement, the roughness Rz of the samples was determined after heat treatment. The samples were subsequently cold mounted and polished to perform micrograph analysis under a Keyence VK-X200 microscope to identify the effects on layer formation resulting from the heat treatment. Additionally, the resulting micrograph images were separated into different phases according to phase color and the area of each phase given as a percentage in relation to the overall image size to compare the changes in phase area dependent on the heating parameters. The open source software ImageJ was used to apply the classification image sections into the different phases. A detailed explanation of the process can be found in [11].

3. Results and discussion

3.1. Furnace temperature and holding time

The samples were placed in the annealing furnace at 880, 910 and 930 °C to investigate the influence of the austenitisation temperature. Furthermore, the samples were held at different holding times of 60, 270 and 600 seconds after they had reached the desired temperature after approximately 180 seconds. Each parameter combination was repeated 3 times to ensure repeatability, while a quenching process was not performed on the samples after heat treatment. The roughness measurements of the AS80 samples, as displayed in Figure 1 in orange, were performed on a perthometer and show no significant change in surface roughness due to furnace temperature. The AS80 coating exhibits similar behavior compared to the AS150 coating at lower roughness values, shown in blue.

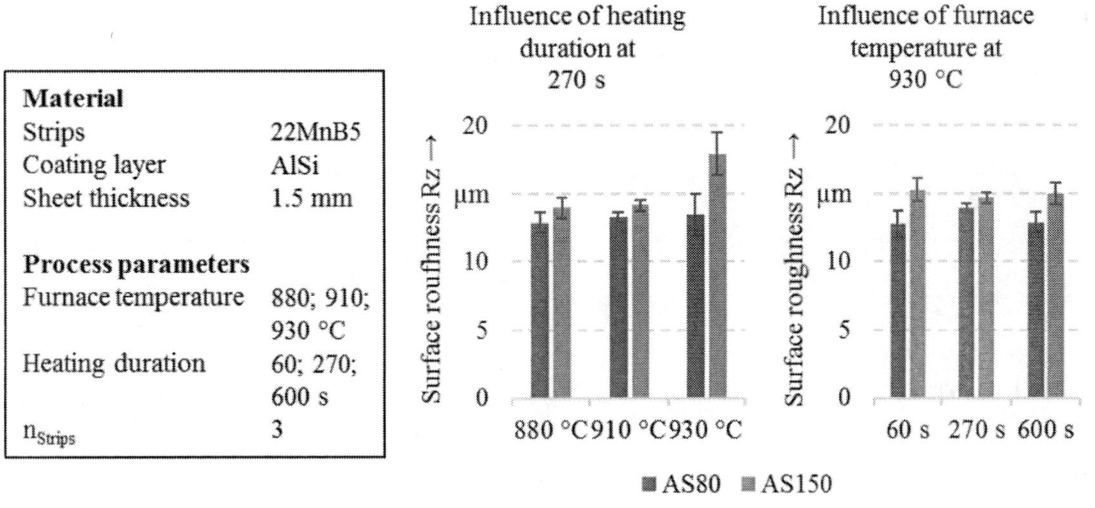

Figure 1. Influence of heat treatment parameters heating duration and furnace temperature on surface roughness Rz for AS80 and AS150 coatings

While the AS150 coating shows a significant non-linear effect of the heating duration with an increase from 910 to 930 °C, the AS80 coating shows no significant change at all parameter settings. The previous study [11] ascribed the jump in surface roughness dependent on the time in the furnace to ongoing diffusion processes in the layer which was ongoing at 60 and 270 seconds. This can be explained by the more active diffusion of iron elements to the coating from the basematerial due to the higher diffusion coefficient of iron into aluminum compared to aluminum into iron [14]. After the coating layer has reached a saturation point and completely transformed into intermetallic AlSiFe, aluminum and silicon elements continue diffusing into the basematerial [15]. Based on this, it can be concluded that the diffusion process on the surface near areas is finished for the thinner AS80 coating. The micrograph images show a porous surface for AS80 at all holding times similar to the surface AS150 exhibits at 600 s. Which supports the theory and also leads to the conclusion that the diffusion of AlSi particles into the base material draws from the surface near area of the coating. Due to the inhomogeneous appearance of the uppermost layer, no viable height measurements could be performed. The AS150 coating layer split into two different phases in previous investigation with two additional phases visible going into the basematerial. Using EDX measurements the same phases could be encountered for the AS80 coating. As seen in Figure 2, the coating again splits into four main phases. While the Phase 1 layer consisted of separate spots initially which gradually formed into a thin line for AS150, the fully formed line is already clearly visible at all parameter settings for AS80.

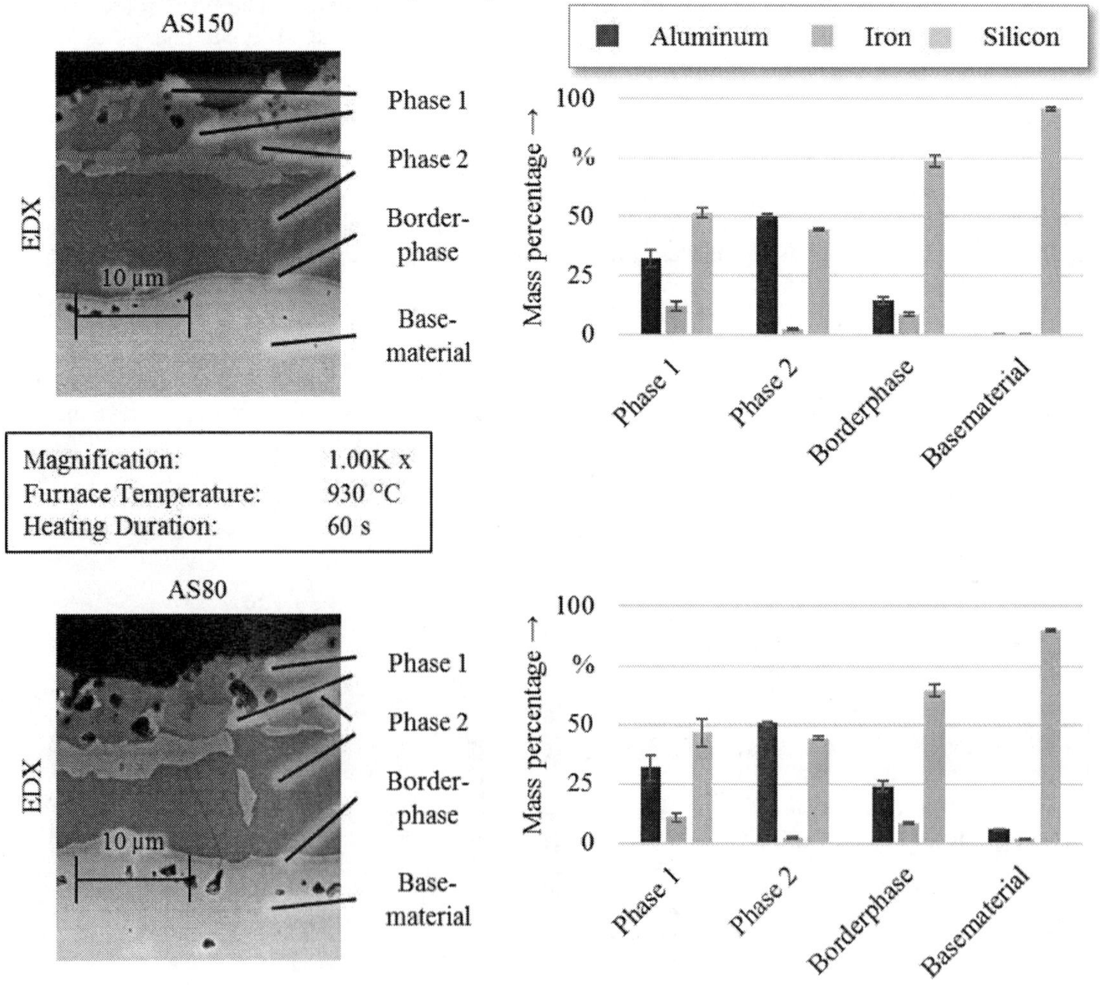

Figure 2. EDX measurement for AS150 (top) and AS80 (bottom)

The intermetallic Phase 1, consisting of AlSiFe, and Phase 2, consisting of AlFe and AlFe2, exhibit similar values for the content of all three elements Al, Si and Fe with no change in general composition visible with ongoing heat treatment. Deviations can be observed in the borderphase with an increase in aluminum, as well as the basematerial, which sees a rise in both aluminum and silicon values, suggesting that the diffusion of aluminum and silicon into the basematerial continues even after the diffusion of iron into the coating layer is finished for AS80. The analysis and comparison of the relative area for the phases in the micrograph images captured via Keyence measurements can be found in Figure 3. For AS150 a significant increase could be identified for Phase 1 with an increase in the holding time from 60 to 600 seconds. The furnace temperature exhibits no significant effect but an interdependency with the heating duration, with a diminishing effect on Phase 1 area at the highest parameter setting.

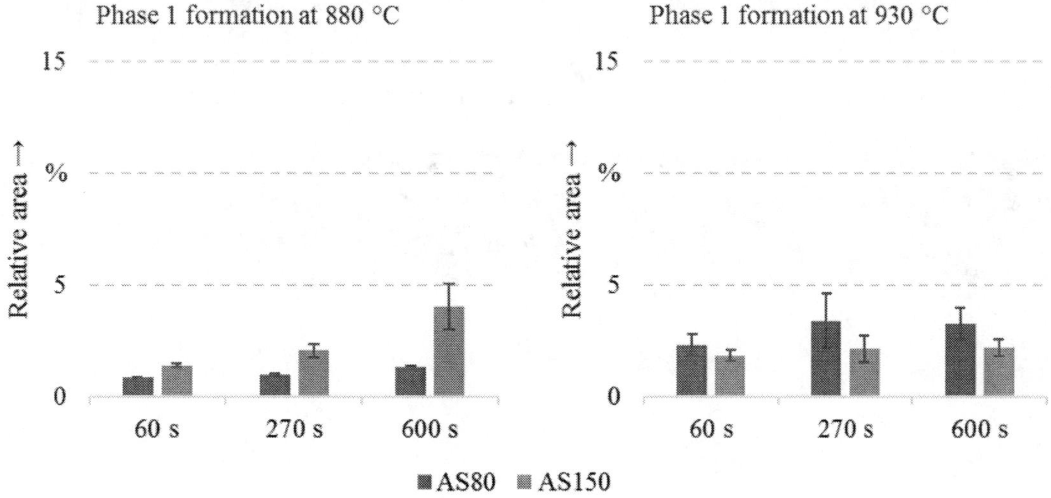

Figure 3. Influence of the holding time on the relative area of Phase 1

Like AS150, the thinner coating layer AS80 sees a significant increase in size due to increasing holding times at a furnace temperature of 880 °C, although the relative area is smaller in size compared to AS150. A trend in increase due to the holding time can also be observed at a furnace temperature of 930 °C similar to AS150, along with a noticeable increase in the phase area compared to AS150. Unlike the AS150 coating, a rise in furnace temperature also exhibits a significant increasing effect for Phase 1. Phase 2 sees a decrease in the total area size with an increase in all parameter settings as seen in Figure 4.

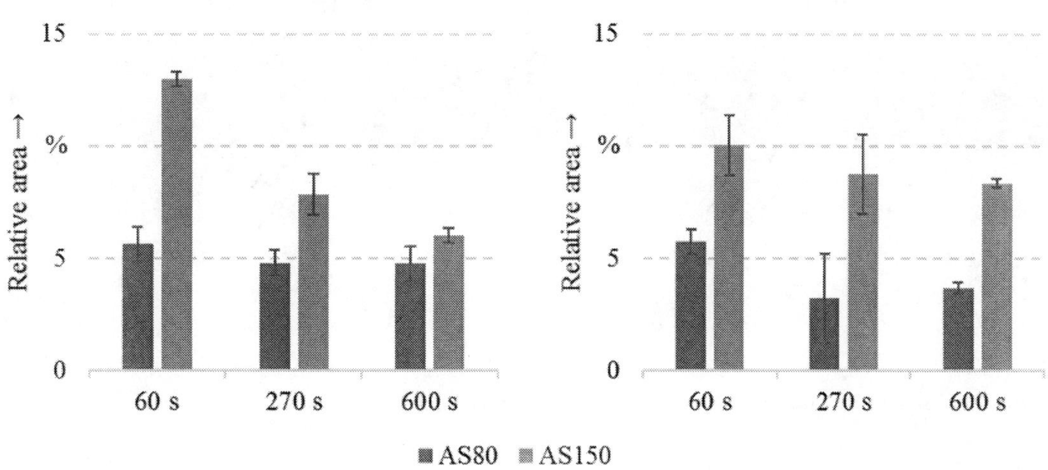

Figure 4. Influence of the holding time on the relative area of Phase 2

An interdependency between both heat treatment parameters can be observed for AS150, with the significant decrease in area size due to increased holding times at 880 °C diminished at 930 °C, correlating with the reduction in total layer height at higher holding times. While only a decreasing trend due to increasing holding times is visible at 880 °C, the decrease becomes significant at 930 °C. Furthermore, the size of the Phase 2 area is reduced compared to AS150. The analysis of the formation of both phases lead to the conclusion, that the diffusion process is influenced by both the austenitization time and the furnace temperature. This can be partially confirmed with the findings from the previous study [11]. The effects of the heating parameters on the layer formation of AS80 appears to be more pronounced, which is likely due to the reduced layer thickness. It can therefore be concluded that the thinner coating can reach a saturation point faster due to there being less aluminum and silicon elements available for diffusion reactions. Further analysis and comparison of the micrograph images of samples held for 60 seconds, as seen in Figure 5, show a porous surface near area which partially explains reduction in Phase 2 in addition to ongoing diffusion processes from the coating to the Base material.

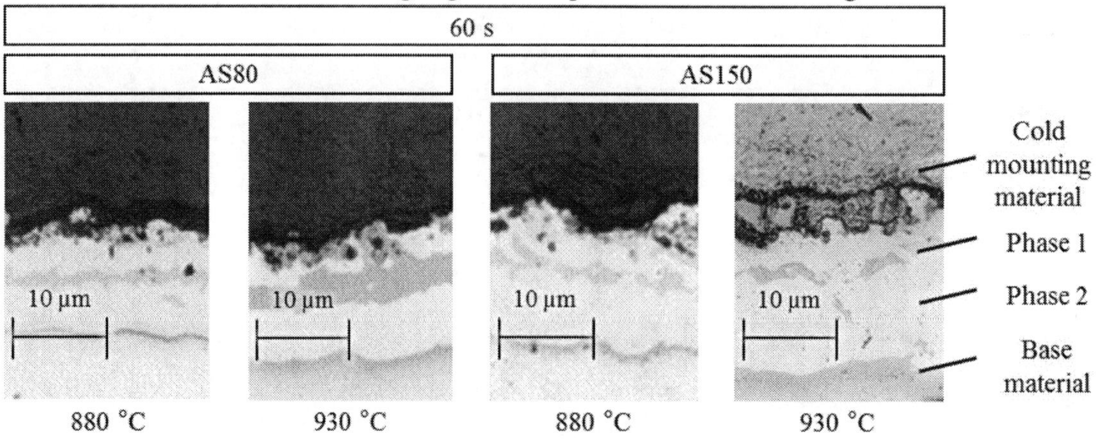

Figure 5. Micrograph images of samples held at 880 and 930 °C for 60 seconds

The micrograph images of samples held for 600 seconds, as seen in Figure 6, confirm the findings of the image analysis with a visible and significant increase of Phase 1 from 60 to 600 seconds. At the highest furnace temperature of 930 °C the relative area of Phase 1 grows further in size and at both 60 and 600 seconds holding time. The size of AS80 Phase 1 grows beyond that of the Phase 1 of AS150 as found in previous studies [12].

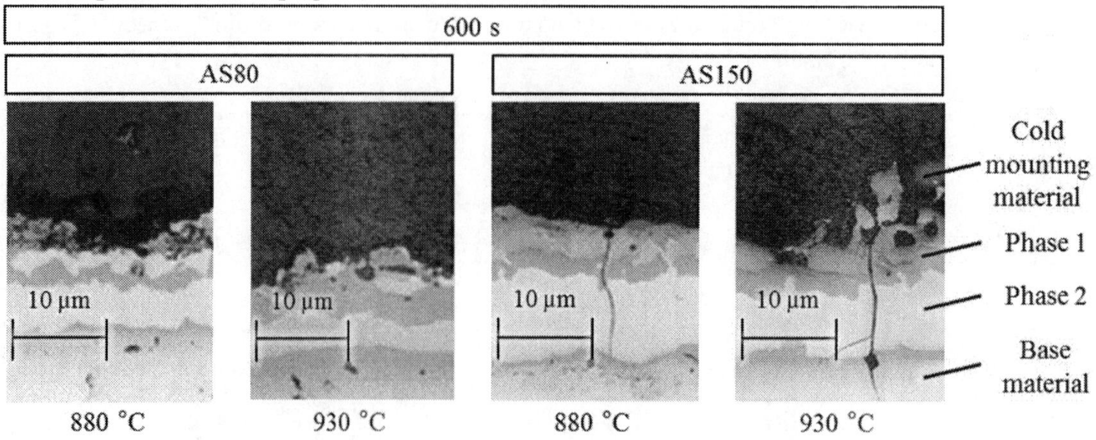

Figure 6. Micrograph images of samples held at 880 and 930 °C for 600 seconds

While the Phase 1 initially consists of single specks which form a discontinuous layer for the AS150 coating, the continuous layer is already formed for AS80. This along with the reduced coating strength and more pronounced porosity of Phase 2 lead to the conclusion that the surface near Phase 2 area is diminished due to diffusion of the elements into the basematerial as described in previous studies [11]. This confirms the theory that the thinner coating can reach a saturation point faster. The reduced coating thickness can negatively impact friction and wear behavior in the hot stamping process due to the increased possibility of the tool material breaking through the coating layer and disrupting the protective barrier effect which prevents decarburization and scale formation. Furthermore findings of Jüttner et al. [12] ascribe a negative impact from the increase of the Phase1 AlSiFe layer on further process steps like welding. It can also be concluded that the thinner AS80 coating is more susceptible to the in- and decreasing influence of heat treatment. It is likely that only lower parameter settings than the ones investigated in this study are needed to achieve similar desired results as with the AS150 coating. This in turn hints at a potential for reducing both the necessary process temperature and cycle time when using AS80 to reach the desired layer composition for hot stamping processes.

4. Summary and outlook

The formation of the layer coating of AlSi coated 22MnB5 steel under hot stamping conditions has been investigated in this study. The thinner AS80 coating has been compared to the previously investigated industrial standard AS150. Neither furnace temperatures nor holding times were found to exhibit a significant influence on the surface roughness Rz. Similar to the surface roughness both parameters were shown to have a significant influence on the phase formation, with the holding time being more significant. While the same phases formed for both AS150 and AS80 coatings, traces of coating elements can be found in the basematerial of AS80. This, along with the micrograph images, leads to the conclusion that although the AS80 coating exhibits a similar behavior compared to the AS150 coating, the influence of the heat treatment parameters is far more pronounced. This can be led back to the reduced layer thickness, which allows the coating to reach a saturation point for the diffusion of iron from the basematerial faster. This opens up the potential of reduced cycle times to reach desired coating properties for the hot stamping process. For further investigations intermediate settings between the investigated ones have to be analyzed, due to the shortened process window to reach the coating saturation. Additionally, future research has to focus on the effect of the AS80 coating on friction and wear in strip drawing test to investigate possible adverse effects due to the reduced coating thickness and interdependencies with other significant process parameters like contact pressure and workpiece temperature. The findings of this study lay the groundwork for further analysis on the impact of the layer formation and heat treatment parameters on friction and wear mechanisms in hot stamping and will help deepen the understanding of cause and effect relations in the process.

Acknowledgement

This study is part of the project P1291 "Workpiece-side Impact on the Tribological Conditions within Hot Stamping" funded by the Research Association for Steel Application (Forschungsvereinigung Stahlanwendung e. V. FOSTA). The authors would like to thank the Research Association for Steel Application for the financial support. Further thanks go to all cooperating industrial partners for their support and cooperation in this project.

References

[1] Geiger M; Merklein M 2007 Sheet Metal Forming - A New Kind of Forge for the Future *KEM* 9–20
[2] Schießl G, Possehn T, Heller T and Sikora S 2004 Manufacturing a roof frame from ultra high strength steel materials by hot stamping *Proceedings of the IDDRG International Deep Drawing Research Group Conference*
[3] Karbasian H; Tekkaya A E 2010 A review on hot stamping *Journal of Materials Processing Technology* **15** 2103–18

[4] Smallman R E; Ngan AHW 2014 Steel Transformations *Modern Physical Metallurgy*: Elsevier) pp 473–498

[5] Wilsius J, Hein P and Kefferstein R 2006 Status and future trends of hot stamping of USIBOR 1500 P *1. Erlanger Workshop Warmblechumformung* 82–101

[6] Schwingenschlögl P, Hildering S and Merklein M 2016 Investigation of Friction Conditions at Elevated Temperatures for Hot Stamping *ICTMP 2016: 7th International Conference on Tribology in Manufacturing Processes* 217–28

[7] Borsetto F, Ghiotti A and Bruschi S 2009 Investigation of the High Strength Steel Al-Si Coating during Hot Stamping Operations *KEM* 289–96

[8] Buckley D H 2010 *Surface effects in adhesion, friction, wear, and lubrication* (Amsterdam, New York, New York: Elsevier Scientific Pub. Co)

[9] Dick P, Scheiker T and Mattes U 2008 Warmumformung bei Daimler - Ein heiß diskutiertes Verfahren *3. Erlanger Workshop Warmblechumformung* 41–55

[10] Sherepenko O, Schreiber V, Wohner M and Luo Y 2019 Influence of Press-Hardening Process on Resistance Spot Weldability of 22MnB5 with Aluminum-Silicon Coating in a Three-Sheet Stack-Up for Automotive Applications *7th International Conference Hot Sheet Metal Forming of High-Performance Steel, June 2-5, 2019, Luleå, Sweden*

[11] He F; Merklein M 2021 Investigation of the impact of heat treatment on the layer formation of AlSi-coated boron-manganese steel *IOP Conf. Ser.: Mater. Sci. Eng.* **1** 12009

[12] Jüttner S *Metallschutzgasschweißen von pressgehärteten höchstfesten Stählen mit unterschiedlichen Beschichtungskonzepten : Schlussbericht zu dem IGF-Vorhaben der Forschungsstelle(n) Otto-von-Guericke Universität Magdeburg/Institut für Werkstoff- und Fügetechnik (IWF)*: Otto von Guericke University Library, Magdeburg, Germany)

[13] Wang K, Gui Z, Liu P, Wang Y and Zhang Y 2014 Cracking Behavior of Al-Si Coating on Hot Stamping Boron Steel Sheet *Procedia Engineering* 1713–18

[14] Kobayashi S; Yakou T 2002 Control of intermetallic compound layers at interface between steel and aluminum by diffusion-treatment *Materials Science and Engineering: A* **1-2** 44–53

[15] Yakubtsov I; Sohmshetty R 2018 Evolution of Al-Si Coating Microstructure during Heat-Treatment of Usibor ® 1500 *IOP Conf. Ser.: Mater. Sci. Eng.* 12015

IOP Conf. Series: Materials Science and Engineering 1238 (2022) 012011 doi:10.1088/1757-899X/1238/1/012011

U-channel forming of an 7075-T6 in warm conditions

S Royne[1], **H Laurent**[1] **and A Maillard**[2]

[1] Univ. Bretagne Sud, UMR CNRS 6027, IRDL, F-56100 Lorient, France

[2] Cetim, F-60300 Senlis, France

E-mail: sylvain.royne@univ-ubs.fr

Abstract. The use of high mechanical strength aluminum coupled with the warm forming process makes it possible to replace certain steel parts in vehicles, thus reducing its mass and consequently greenhouse gas emissions. Warm forming of aluminum alloys requires heating the part between 150 °C and 250 °C to increase formability and reduce springback. In this study, a 7075-T6 alloy U-channel part was warm formed at 200 °C in a short enough time to maintain the T6 state of the material. The process consisted of successively performing the heating by contact and stamping steps on the same mechanical press. From a thermomechanical characterization of the alloy in the temperature range between RT and 200 °C and several strain rates, numerical simulations of the forming of the U-channel part were carried out using Abaqus. The experimental springbacks reduced with temperature were in agreement with the numerical simulations. An analysis of the conductivity and hardness of the material showed also that the 7075 keeps its T6 state until 10 s of warm forming.

1. Introduction

Aluminum alloys with high mechanical strength and low density, such as the alloys of the 7xxx series, are an interesting alternative to conventional steels for reducing the mass of vehicles and therefore greenhouse gas emissions. Despite these good characteristics, these alloys have reduced formability and significant springback at room temperature (RT). Warm forming process, around 200 °C, makes it possible to reduce these two defects [1, 2]. However, this process can lead to metallurgical transformations for temperature above 150 °C. For example, the precipitation of the η' in η appears at a temperature of 200 °C and leads to a decrease in the mechanical characteristics if the temperature holding time is too long [3].

This paper aims to study the warm forming of the U-channel of an AA7075-T6 alloy and to analyse the influence of temperature on its springback. Forming tests are carried out at room temperature and at 200 °C using a multi-step stamping press. Springback results are compared to numerical simulations performed with Abaqus.

Anymore, the influence of holding time on the metallurgical state at the end of process is rarely studied in the literature [4, 5] and determine the optimum holding time is important in order to find the best duration of the warm forming process. Thus, the other objective of this paper is to find the best holding time in temperature to maintain the T6 state of the alloy which offers the highest ultimate and yield strengths. For that, conductivity and hardness tests are carried out for different holding time conditions at 200 °C in order to analyze the influence of this holding time on the metallurgical and mechanical characteristics.

2. Material and methods

2.1. Material

AA7075-T6 sheet aluminum alloy with 0.8 mm thickness is used in this work. Commercially purchased aluminum was directly received in T6 state. The chemical composition of this alloy is presented in table 1. Its mechanical, conductivity and hardness properties measured in the standard and received conditions are given in table 2.

Zn	Mg	Cu	Si	Cr	Fe	Mn	Ti	Al
5.1-6.1	2.1-2.9	1.2-2.0	0.4	0.18-0.28	0.5	0.3	0.2	Balance

Table 1. Chemical composition of AA7075-T6 sheet in %wt.

Condition	$Rp_{0.2\%}$ [MPa]	R_m [MPa]	$A\%$	Conductivity [MS/m]	Hardness [$HV10$]
Standard	≥ 460	≥ 525	≥ 6	17-21	-
Receiving	500	557	14.6	18.5	188

Table 2. Mechanical and electrical properties at RT of the AA7075-T6 measured at receiving condition and standard limit values given for the T6 temper.

2.2. Warm forming tests of a U-channel

U-channel parts are formed under warm forming condition using a multi-step device manufactured by CETIM [6]. This device is presented in figure 1-a. It consists of two blanking stages to switch from a sheet metal strip to blanks of 250 mm×60 mm attached between them. Then, a heating stage makes it possible to reach a uniform temperature of the blank of 200 °C in 0.8 s via clamping contact. Finally, a forming step is performed with tools heated to the target temperature with the help of heating cartridges. An industrial rate of around 30 strokes/min is used. The theoretical dimensions of U-channel part are presented in figure 1-b.

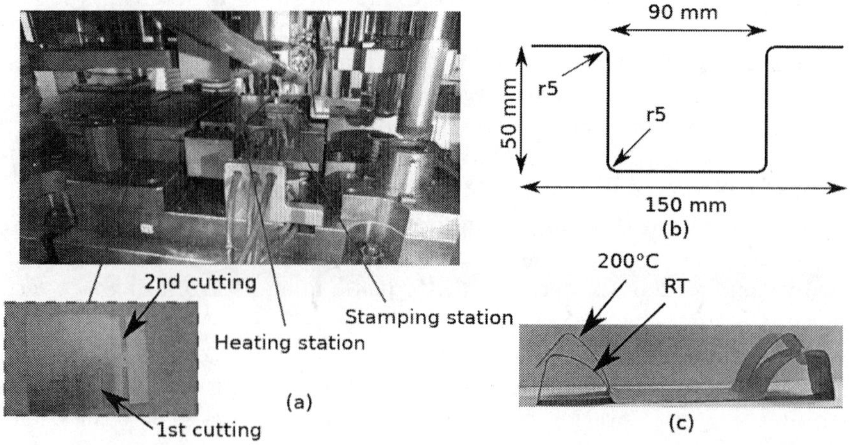

Figure 1. a) Three stages of the multi-step device in the industrial press with blank before and after cutting stages; b) Theoretical dimensions of the U-channel part in mm (width=60 mm); c) Final U-channel part at 20 °C (RT) and at 200 °C after springback.

The dimensions of the tools are presented in figure 2. A clamping force of 5000 N is applied to the blank-holder. Forming tests are performed at RT and at 200 °C. The parts obtained after the springback at RT and at 200 °C are shown in figure 1-c. Three tests per condition are performed to check the reproducibility. Measurements of the part profile after springback are obtained by using a FARO Quantum S 2.5m 7-Axis 3D machine with an integrated Probe-Laser scan.

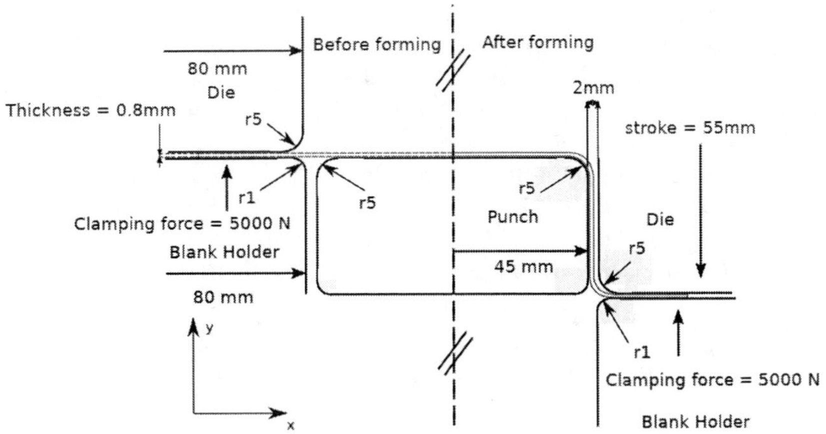

Figure 2. Dimensions of the forming tools in mm

2.3. Numerical simulations of the warm forming process and springback

Isothermal numerical simulations of the forming process and springback are conducted in using Abaqus/Standard. A 2D plane stress model is used with tools modeled as rigid surfaces. A section of the blank is meshed with 2D quadratic quadrilateral elements CPS8. A fine mesh size of approximately 0.15 mm is used in the radius zone of the die whereas the size is increased to approximately 1.5 mm at sections away from this zone. Moreover, 13 elements are used through the thickness. The blank is assumed to have elastic-plastic behaviour described with a Swift law (1) identified through tensile tests at RT and 200 °C performed at a strain rate of $2 \times 10^{-2}\text{s}^{-1}$.

$$\bar{\sigma} = K.(\varepsilon_0 + \bar{\varepsilon})^n \quad (1)$$

K, ε_0, n are material parameters and $\bar{\varepsilon}$ is the equivalent plastic strain.

The parameters of the hardening law at RT and 200 °C are summarized in table 3.

	K [MPa]	ε_0	n
20°C	851.8	0.0518	0.15
200°C	485.7	0.026	0.195

Table 3. Swift's law parameters at RT and 200 °C identified through tensile tests performed at $2 \times 10^{-2}\text{s}^{-1}$.

During the springback simulation, tools movements are simulated as during the experimental process. The purpose of the numerical simulations is to analyze the influence of the friction coefficient on the prediction of the springback.

3. Results and discussion

3.1. Experimental and numerical springbacks

The experimental springback results at RT and 200 °C are presented in figure 3. They are compared to the numerical results. The outer surface of the experimental and numerical profile is taken. The influence of friction coefficient is crucial in the numerical prediction of the forming and springback. After several numerical trials, a friction coefficient making it possible to get as close as possible to the experimental results is fixed at 0.2 between the blank and the die and the blank-holder, whereas a friction coefficient of 0.15 is applied between the punch and the blank. These parameters allow a good agreement between the experimental and numerical in the springback results. As expected, the increase in temperature strongly reduces the springback. Nevertheless, the comparison between the theoretical and numerical/experimental profiles indicates that despite this increase in temperature, the springback remains significant on this type of part with significant free strands.

Finally, it is important to notice that at the die radius, there is a change of curvature in the profile of the blank which could be due to a phenomenon of attachment of the material during the opening of the tools. Other numerical simulations by modifying the friction coefficient only in the zone of the die radius are still in progress to verify this hypothesis.

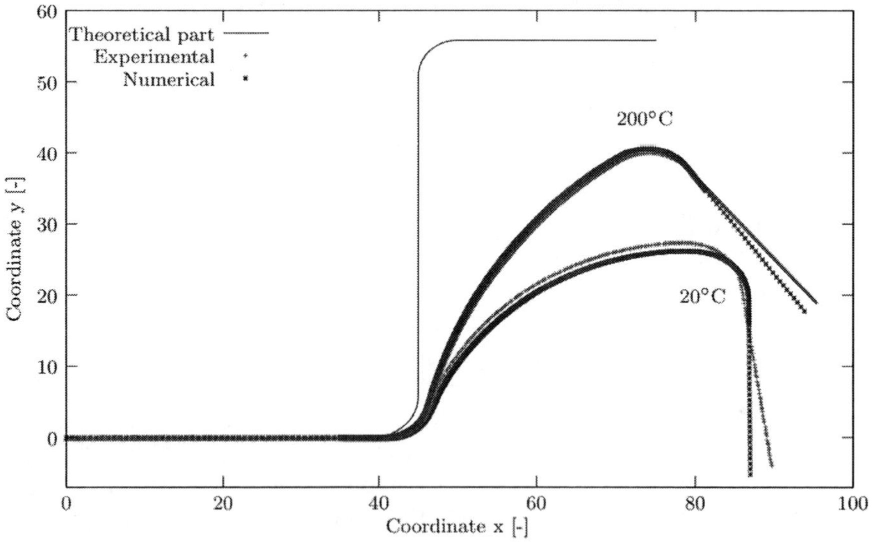

Figure 3. Numerical/experimental comparison of springback of U-channel part at RT and at 200 °C. The clamping force is 5000 N and the coefficient of friction is 0.2 between the blank and the die/blank-holder and 0.15 between the punch and the blank. The outdoor profile of the U-channel part is shown.

3.2. Verification of the T6 state at the end of the process

Measurements of mechanical characteristics, electrical conductivity and hardness after different temperature holding times are easy methods to determine the metallurgical changes of alloys occurring during thermal cycles [7, 8]. These measurements make it possible to define the optimum time for the warm forming process to guarantee the T6 condition after forming.

Experimental thermal cycling tests were performed on a Gleeble 3500 machine to reproduce the warm forming process by varying the holding time at 200 °C. During these tests, Joule heating of a dog-bone specimen was controlled by a thermocouple welded in the center of the specimen. Two thermocouples placed at ±5 mm from the central thermocouple verified also the homogeneity of the temperature over the length of the sample. The thermal cycle used is presented in Figure 4. Its corresponds first to heating with a rate of 250 °C/s (200 °C in 0.8 s) similar to the heating rate used in the experimental warm forming process. Then, after different holding time of 2, 10 and 30 s at 200 °C, a cooling rate of about 2 °C/s which corresponds to cooling in the open air with cooled jaws was imposed on the specimen. After this thermal cycle, tensile tests at RT were performed on an Instron 5969 tensile machine using a DIC system to measure the deformation in the center of the specimen. Finally, at this stage, conductivity and hardness tests were carried out in the homogeneous temperature zone (rectangle of 3 mm× 6 mm in the center of the sample).

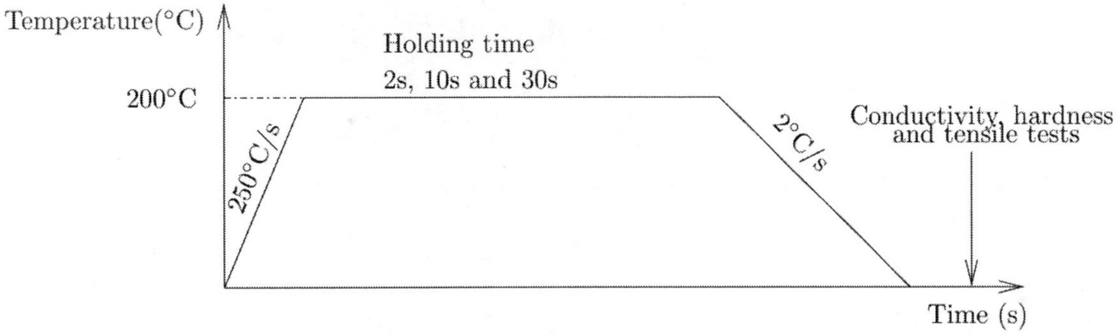

Figure 4. Experimental thermal cycle performed on a Gleeble 3500 machine with the different cycle parameters. Conductivity and hardness tests are carried out after the thermal cycle at RT. Tensile tests are performed on a Instron 5969 tensile machine.

Table 4 presents the results of the electrical conductivity, hardness and mechanical characteristic obtained after different holding times at 200 °C. Results obtained in receiving state T6 are also given in this table. In the case of a holding time of 2 and 10 s, the electrical conductivity remains stable with respect to the receiving condition, whereas after 30 s of holding time, the conductivity increases slightly indicating the occurrence of metallurgical changes.

	Receiving condition	2s	10s	30s
Conductivity [MS/m]	18.5	18.48	18.55	18.71
Hardness [HV10]	188	180	175	172
$Rp_{0.2\%}$ [MPa]	500	495	460	410
R_m [MPa]	557	565	553	539
$A\%$	14.6	12.8	11.6	10.9

Table 4. Mechanical, hardness and electrical characteristics according to the different holding times at 200 °C. The conductivity, hardness and tensile tests were carried out at RT.

The conductivity value obtained after 10 s of holding time indicates that the 7075 alloy still remains in the T6 state since its conductivity complies with the standard for this state (see table 2). As for the hardness, it decreases directly after 2 s of holding time down to a value of 172 HV10 for 30 s. This decrease also indicates metallurgical changes which is in contradiction with the results of conductivity between 2 s and 10 s, i.e. decrease in hardness while conductivity does not change. This reduction in hardness can be associated with the phenomenon of coalescence of precipitates and can therefore imply a reduction in the mechanical characteristics.

Finally, with regard to the mechanical characteristics, in the case of 2 s of holding time, the mechanical characteristics are similar to the reception state, whereas from 10 s, they decrease. By comparing these data with the values of the standard (table 2), this indicates that after 10 s of holding time the alloy is no longer in the T6 state since the elastic limit is then equal to the limit of the standard of 460 MPa.

To validate that 10 s seems an optimum time to keep the material in a T6 state, electrical conductivity and hardness measurements are performed on the U-channel part as presented in Figure 5. Three measurements of the conductivity and hardness are taken in the undeformed area to check reproducibility. Table 5 presents the average of the three values of electrical conductivity and hardness results obtained on parts after forming at RT and 200 °C.

Figure 5. Positions (red dots) of hardness and conductivity measurements after warm forming of the U-channel part.

Temperature	20°C	200 °C
Conductivity [MS/m]	18.5	18.5
Hardness [HV10]	188	184

Table 5. Hardness and electrical conductivity of U-channel parts measured after warm forming. The measurements were made at RT

These results show that the conductivity is maintained at the receiving state (18.5 MS/m) whereas the hardness for 20 °C is equal to the hardness of the receiving state (188 HV10) and for 200 °C, it decreases of -4 HV10 compared to that at 20 °C.

Concerning the time of the warm forming process, the holding time of the part at 200 °C is estimated at 2.6 s. This time corresponds to the transfer of the blank from the heating station to the stamping station under isothermal conditions plus the time of the stamping operation. In fact, the total time of the warm forming process is less than 10 s which confirm the values of hardness and conductivity obtained with the forming parts. It is also important to note that the hardness obtained at 200 °C for an estimated holding time of 2.6 s is slightly greater than that obtained for a holding time of 2 s.

All of these data indicate that the maximum temperature holding time at 200 °C can be estimated at less 10 s in order to preserve the mechanical characteristics of the T6 temper. Nevertheless, the data was obtained on an AA7075-T6 alloy of a certain batch and as indicated in Table 1, the chemical composition may vary according to the batch and therefore also the mechanical characteristics. For the moment, nothing tells us that this value of 10 s is valid whatever the 7075-T6 alloy received. This value of 10 s may therefore change slightly depending on the status of the batch received.

4. Conclusion

Warm forming of a U-channel part of 7075-T6 alloy was performed at RT and 200 °C with a dedicated multi-step process. Numerical springback was compared with experimental results.

As classically observed in the literature, the increase of temperature decreases the springback. Numerical simulations showed that the friction coefficient is an important parameter to obtain good correlation with experimental springback. Electrical conductivity, hardness and tensile tests have made it possible to obtain the maximum temperature holding time at 200 °C to preserve the T6 state after warm forming. A total time of warm process of less 10 s allows to maintain this state. However, a modification of this time is still possible depending on the batch characteristics of the received alloy.

Acknowledgments
The authors gratefully acknowledge the financial support of the Brittany Region (France) via the program RB EMBHOTAL.

References
[1] Kumar M and Ross N G 2016 *Journal of Materials Processing Technology* **231** 189–198
[2] Wang H, Luo Y b, Friedman P, Chen M h and Gao L 2012 *Transactions of Nonferrous Metals Society of China* **22** 1–7
[3] Deschamps A, Texier G, Ringeval S and Delfaut-Durut L 2009 *Materials Science and Engineering: A* **501** 133–139
[4] Lee E W, Oppenheim T, Robinson K, Aridkahari B, Neylan N, Gebreyesus D, Richardson M, Arzate M, Bove C, Iskandar M, Sanchez C, Toss E, Martinez I, Arenas D, Ogren J, Mclennan J, Clark R, Frazier W E and Es-Said O S 2007 *Engineering Failure Analysis* **14** 1538–1549
[5] Zang J, Dai P, Yang Y, Liu S, Huang B, Ru J and Luo X 2021 *Metals* **11** 1483
[6] Maillard A and Piat C 2014 *IDDRG 2014 Conference* 300–306
[7] Miyake J and Fine M E 1992 *Acta Metallurgica et Materialia* **40** 733–741
[8] Mauduit A and Gransac H 2020 *Annales de Chimie - Science des Maériaux* **44** 141–149

Superplastic behaviour of AA5083 sheets in the presence of an oscillating load

M S Dastgiri[1], Z Fuerth[2], L Kiawi[2], E Ryzer[3] and D E Green[4]

[1]PhD Candidate, University of Windsor, Windsor, ON, N9B 3P4, CA
[2]MASc Student, University of Windsor, Windsor, ON, N9B 3P4, CA
[3]Chief Executive Officer, AEM Power Systems, Windsor, ON, N9A 4R8, CA
[4]Professor, University of Windsor, Windsor, ON, N9B 3P4, CA

E-mail: shirinz@uwindsor.ca

Abstract. The superplastic flow behaviour of AA5083 sheets was investigated at 450°C in the presence of a minor oscillating load. Uniaxial tensile specimens were tested to fracture at constant strain rates with a minor oscillating load superimposed onto the monotonically increasing tensile load. In each test, the oscillating load was applied as a sine wave having a constant amplitude of 0.5 N and a constant frequency. A series of tests was conducted in which the strain rate ranged from 0.001 to 0.4 s^{-1} and the frequency of the oscillating load ranged from 10 to 40 Hz. And other tensile specimens were subject to a constant mean load with a superimposed oscillating load until a predetermined elongation was reached. The flow behaviour of AA5083 showed a significant sensitivity to the oscillations although the tensile strengths remained practically unchanged. The addition of a minor oscillating load resulted in a 20% - 50% relative increase in fracture strain for the strain rates considered compared to the same tensile tests without oscillation. Finite element simulation of the tensile tests conducted at constant mean load with superimposed oscillations were carried out with LS-Dyna using material data obtained from the experiments. The numerical simulation results were compared with experimental results obtained under the same loading conditions to confirm their validity. The addition of a minor oscillating load led to greater uniformity in thickness in the specimen gauge. It can be concluded that superimposing a minor oscillating load during superplastic forming results in significantly greater deformation prior to fracture.

1. Introduction

Aluminum AA5083 is a non-heat-treatable aluminum alloy with excellent cold formability and can reach moderate levels of superplasticity [1]. This alloy is affordable and has reasonable corrosion resistance and good mechanical properties, which make it ideal for aerospace, marine, and automotive applications [2].

1.1. Introduction to Superplastic Forming

A material is said to display superplastic behavior when it exhibits significant plastic deformation (an elongation >200%) at elevated temperature without necking prior to fracture [3]. Both the mechanical properties (elongation, UTS, optimum forming temperatures, strain rate dependence, etc.) and the microstructural characteristics of superplastic aluminum alloys have been investigated [4–9]. Of particular interest, published research has shown the effectiveness and results of superplastically deforming AA5083 sheets [10, 11]. The severe deformations occurring in superplastic forming are primarily achieved through Grain Boundary Sliding (GBS). Furthermore, the high levels of grain boundary sliding are accompanied by an additional accommodating deformation mechanism, and traditional GBS models are separated into two categories: diffusional accommodated and dislocation accommodated GBS. Despite the very large deformations that can be achieved, the main drawback to the widespread use of superplastic forming is the significant time that is required to form an industrial part, spanning from two to ten minutes. This limits the numbers of parts that can be formed [12].

Blow forming processes can take advantage of superplastic materials to plastically deform sheets without necking. This has allowed the automobile industry to superplastically form complex parts having a relatively uniform thickness distribution. In this process, a metallic blank may be pre-heated and then placed in a die that is heated to the prescribed forming temperature. A pressure vs. time curve is used to control the pressure of air or argon gas, as the blank is gradually formed against the surface of the die. This process has been successfully used to reduce the weight of parts used in the automotive and aerospace industries.

1.2. Improvement of Superplastic Capabilities and Processes

Industrial forming times could be reduced if superplasticity could be carried out at increased strain rates compared to the conventionally slow superplastic forming strain rates. Much effort has been made to achieve increased superplastic forming times described as High Strain Rate (HSR) superplasticity. HSR research is largely focused on the refinement of the grains in the microstructure due to the relationship between decreasing grain size and increasing maximum elongation. HSR studies include the investigation of severe plastic deformation processes as a pre-treatment to refine grains and theoretically exhibit superplasticity at higher than conventional strain rates.

The effectiveness of an oscillating load on material behavior has been shown in ultrasonic welding [13, 14]. Ultrasonic welding uses high-frequency vibrations to weld two clamped components, and the result is a high strength and highly repeatable weld. Moreover, superimposing ultrasonic oscillations in tensile tests to improve deformation has also been investigated [15, 16, 17, 18]. Superimposed oscillations hypothetically improve deformation effectiveness due to the oscillatory stress generating a mean stress lower than that of the stress strain curve for a static tensile loading.

1.3. Constitutive Modelling of Superplastic Forming Processes

Due to the use of variable pressure vs. time curves, it is of great value to develop constitutive models of the flow stress that can be used to simulate the superplastic forming process. Thus, the objective of this paper is to carry out experimental testing and obtain data that can be used to develop a constitutive model of the superplastic deformation with, and without, oscillations. Additionally, finite element (FE) simulations are very useful to optimize the superplastic forming of specific parts.

Superplastic deformation is regarded as a creep process lying within region II of the sigmodal stress vs strain rate curve [19]. This region is often defined by a strain rate sensitivity index ($m=\Delta(\log(\sigma)/\Delta(\log(\epsilon)))$) surpassing 0.4. Additionally, the activation energy in Region II is close to that required for grain boundary diffusion to take place. As a result, maximum ductility occurs within Region II, so it is often characterized as the Superplastic region. The importance of the strain rate sensitivity index is further seen with its utilization in all the most cited visco-plastic constitutive equations. These constitutive models include the power law, sinh law, the Bird-Mukherjee-Dorn equation, and the unified constitutive model [20]. Despite previous models utilizing a constant *m*-value, recent studies have indicated that this practice is inaccurate and can negatively skew the predictions. Additionally, despite the power law being frequently utilized in FE simulations due to its simplicity it has limitations due to its inability to predict the softening and damage behaviors during superplastic forming.

This present study aims to improve the current framework of superplastic forming by incorporating the material response in the presence of a superimposed oscillating load.

2. Experimental procedures

2.1. *Specimen preparation*

Specimens were made from a 1.4-mm thick AA5083 sheets, whose chemical composition is shown in Table 1. Sub-size specimens were cut to a specified geometry adopted from the literature [10], which consisted of an 18-mm gauge length and an 8-mm gauge width, as recommended in previous studies [21]. Specimens were prepared using wire EDM such that the major stress axis was always parallel to the rolling direction of the sheet. This sub-sized specimen geometry is advantageous because it ensures that the specimen will fracture before the gauge length stretches outside the limits of the available furnace. This specimen geometry also facilitated the comparison of results with data published by other researchers.

Table 1. Chemical composition of aluminum AA5083.

Alloy	Fe	Mg	Mn	Cu	Si	Cr	Ti	Zn	Others
AA5083	0.20	4.75	0.8	0.04	0.15	0.15	0.05	0.05	< 0.15

2.2. *Testing equipment*

The custom-built equipment used to carry out the experiments includes a well-insulated furnace with a double-layer quartz window, a pneumatic opening mechanism, titanium rods and fixture for holding the specimen, an electromagnetic mechanical waveform generator, an electric waveform generator, a National Instruments data acquisition and control device, a secondary load cell, and tests were conducted on an MTS tensile testing machine. Figure 1 shows a schematic of the test setup.

The furnace contains electric heaters that surround its two semi-circular chambers for uniform temperature distribution in accordance with the ASTM E2448-18 standard for determining superplastic properties of aluminum. The pneumatic lifter is used to raise and lower the furnace when mounting the specimen. Two steel rods enter through the openings in the top and bottom of the furnace to hold the specimen as well as transfer the load from the MTS testing machine. The openings in the furnace are sealed with silica cotton when it is operating. One of the rods is attached to a pair of electromagnets that induce oscillations in the loading direction. Sine wave signals are generated by a waveform generator, and the electromagnets convert the signals into an oscillating load which is mechanically transferred to the specimen through the steel rods. The applied oscillating load was calibrated for each strain rate and frequency using the MTS load cell. Figure 2 shows a recorded load oscillation wave at 0.5 N amplitude of 10 Hz frequency with compensated pre-loading.

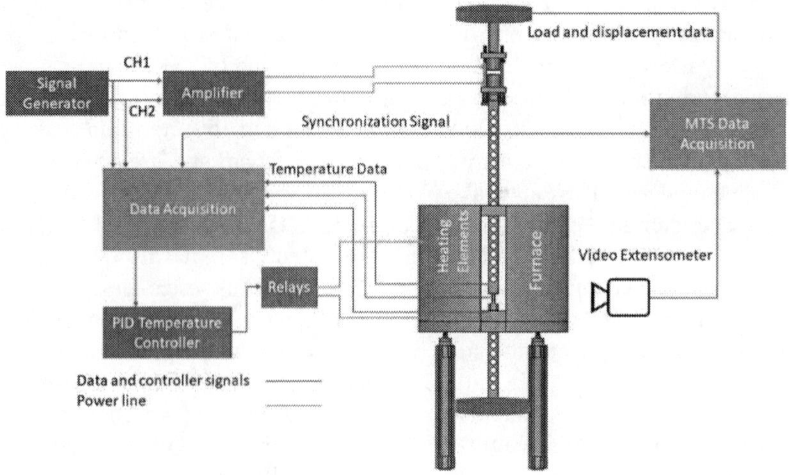

Figure 1. Schematic of the experimental setup.

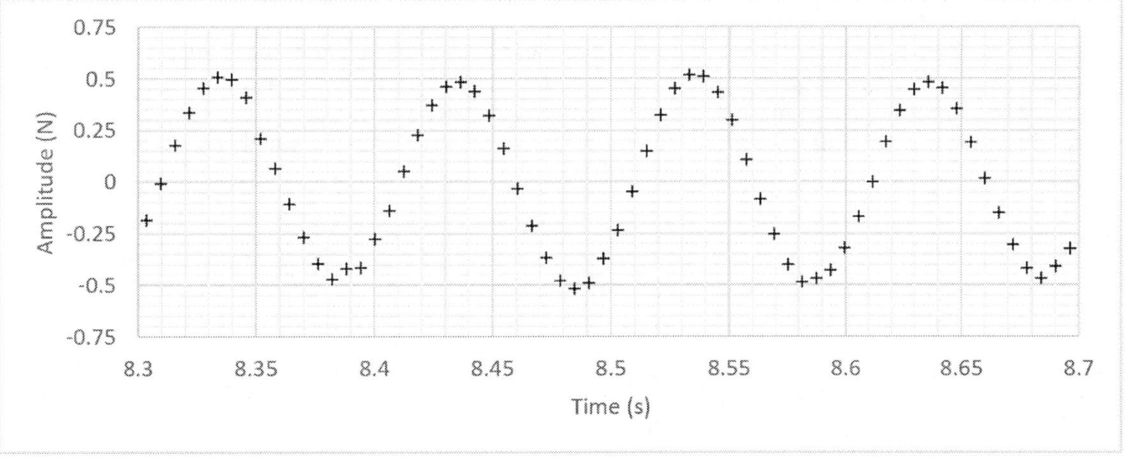

Figure 2. Recorded oscillation wave for a 0.5 N amplitude and 10 Hz frequency oscillating wave.

2.3. Testing procedures
The tensile tests were conducted in accordance with the ASTM E2448-18 standard except for the specimen geometry. For each condition, 3 repeat tests were performed. Before testing, the gauge length and width of each specimen was measured using Mitutoyo calipers for accuracy. Once the oscillating load was calibrated, the furnace was closed and preheated to 450°C. And once the temperature reached 450°C, the furnace was raised up and the specimen was mounted into the gripping fixtures. The top and bottom of the specimen were placed in contact with the temperature probes that are fastened in place with steel wire. The furnace was then lowered and sealed. The start time and temperature were recorded. Once the furnace temperature stabilized at 450°C, the specimen was left to soak for five minutes, in accordance with industrial practice. During this process the crosshead position was adjusted to minimize built-up stress due to thermal expansion. Then the oscillating load was activated, and the tensile test was started. After the test was completed, the specimen was removed, and the chamber was left to preheat for the next test. An MTS Advantage Video Extensometer (AVX) was used to record displacement data in the gauge length. These data were used to confirm the elastic modulus of AA5083 at 450°C, and by comparing the data recorded by the AVX and the crosshead data, a relation between crosshead displacement and gauge length displacement was established. This equation was used to process load-displacement data and calculate the true stress-effective plastic strain for each test.

3. Results
Two types of tensile tests were performed: the first type consisted of conventional tensile testing where a monotonically increasing load was applied at a constant engineering strain rate ranging from 0.001 to 0.4 s^{-1} that leads to a flow curve obtained with and without a superimposed oscillation. The second type consisted of applying a constant load to the specimen and letting the specimen deform until it reached a certain predetermined strain. In this type of test, the time required to reach the target strain was recorded and the minimum thickness in the gauge length was measured after the test. This type of loading leads to a range of strain rates during a single test which provides an opportunity to validate the material model as it is defined by constant strain rate flow curves.

3.1. Conventional tensile testing
To be able to compare results obtained with and without a superimposed oscillating load, a series of benchmark tests were first performed without oscillations. Then the same test conditions were repeated with superimposed oscillation loads. Figure 3 shows the tensile strength versus strain rate, with and without oscillation.

It was observed that for both the benchmark tests and those with oscillations, the variability in the measurement of the tensile stress remained less than 1%, which makes it impractical to show any error bars on the graph. In addition, the variability observed for different frequencies (10, 20 and 40Hz) also remained less than 1%. Figure 3 shows the mean values of each series of tests.

Figure 4 compares the recorded value of the total percent elongation (TPE), with and without oscillations. The variability for these measurements was found to be 7%, which was greater than that of the tensile stress. In Figure 4 the presented value is the mean of each series of tests.

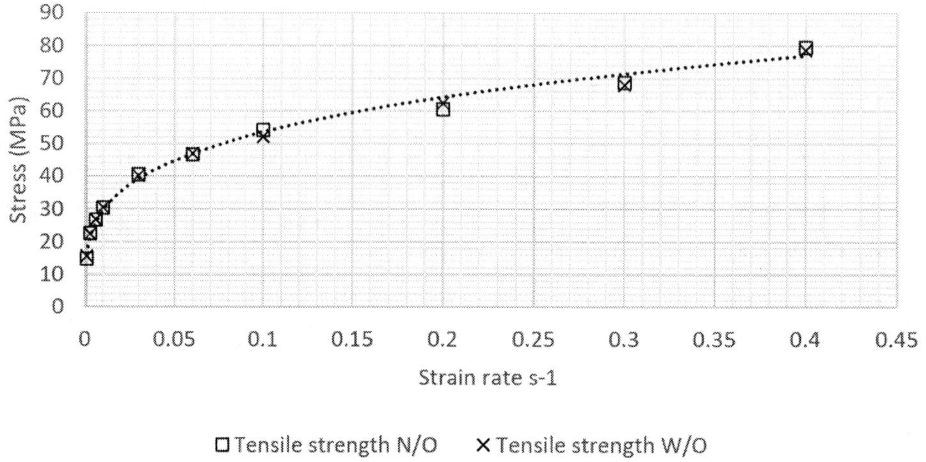

Figure 3. Maximum true tensile stress of AA5083 versus strain rate at 450°C with and without oscillations.

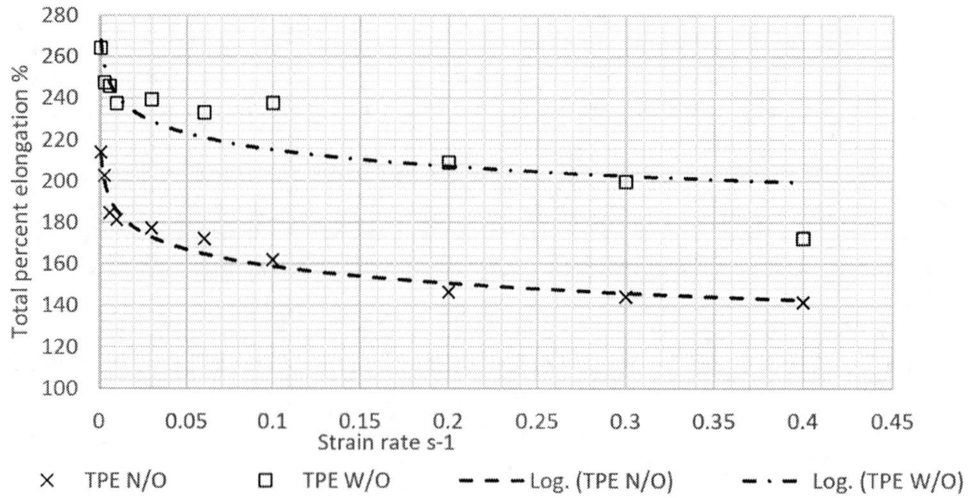

Figure 4. Total percent engineering elongation of AA5083 at 450°C, without and with superimposed oscillations at 0.5N amplitude and 10Hz frequency.

3.2. Constant-load tensile testing

In this second series of tests, specimens were subject to a constant load ranging from 147.1 to 269.7 N, either with or without a superimposed oscillating load. Since the flow behavior in the first series of tests did not show any sensitivity to the frequency of the oscillations, this second series of tests was carried out at a single frequency of 10 Hz. Figure 5 presents the minimum measured thickness across the gauge length for different levels of applied tensile force, both with and without oscillation. The maximum variability for the repeat tests was 1.7% and the data points presented in Figure 5 are the averages.

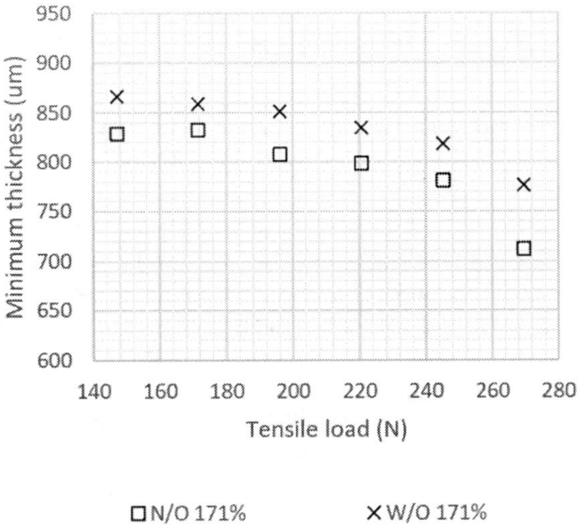

Figure 5. Minimum thickness measured across the gauge length of AA5083 specimens subject to a constant load at 450°C until a strain of 171% was reached.

4. Discussion

According to Figure 3, AA5083 sheet specimens show a positive strain rate sensitivity: indeed, the tensile stress at a strain rate of 0.001 s^{-1} is below 20 MPa, whereas it increases fourfold to reach 80 MPa at a strain rate of 0.4 s^{-1}. And the same trend is observed whether a minor oscillating load is applied or not. Therefore, superimposing an oscillating load during tensile testing did not make any measurable difference on the tensile stress for the entire range of strain rates covered by this study.

In contrast, Figure 4 shows that the total percent elongation decreases with increasing strain rates. At a strain rate of 0.001 s^{-1} the specimen can deform over 200%, but when the strain rate reaches 0.4 s^{-1} the total elongation decreases to about 140%. The effect of a superimposed minor oscillating load on the mechanical behavior of AA5083 can be described by comparing the percent improvement at different strain rate. This information is plotted in Figure 6. The difference in tensile stress at various strain rates is less than 5 percent, however, the improvement in total elongation due to the oscillations varies from 20 to 50%.

The promising results of superimposing an oscillating load on the flow behavior of AA5083 sheets suggests that the production rate of superplastically formed parts can be significantly improved.

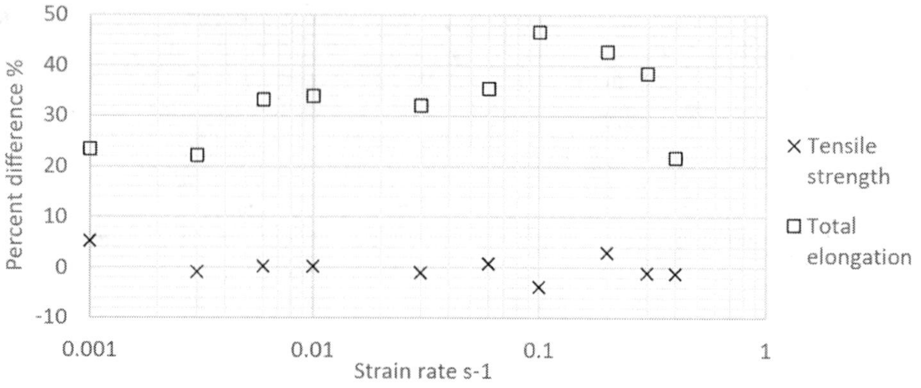

Figure 6. Effect of superimposing an oscillating load on the tensile stress and total elongation of AA5083 at 450°C.

5. Constitutive model

One of the most accurate constitutive models to describe the behavior of AA5083 was developed by Majidi-Jahazi-Bombardier and is known as the VmV model [20]. This model can be calibrated to make accurate predictions at lower strain rates. However, the accuracy of the model decreases at higher strain rates. The absolute values of the residuals predicted by the VmV model compared to AA5083 experimental data remain below 5 MPa for strain rates up to 10^{-2} s^{-1}. However, these values rise to about 20 MPa for a strain rate of 10^{-1} s^{-1}. This experimental study considered the effect of oscillations for strain rates up to 0.4 s^{-1} and, for numerical predictions to be accurate up to this strain rate, a constitutive model should exhibit an acceptable accuracy for this full range of strain rates.

5.1. Proposed constitutive model

As it has been pointed out, the behavior of AA5083 at 450°C is sensitive to strain rate. The trend of this strain rate sensitivity can be determined by observing the changes that occur as a function of the controlling variable. In Figure 3, the fitted line is a power law function with R-squared value of 0.992. This confirms that the strain rate sensitivity can be described by a power law function. On the other hand, the true stress-effective plastic strain curves of AA5083, both with and without oscillation, show a strain softening trend. This trend cannot be described by a power law function. However, strain softening behavior can be adequately described by a second-degree polynomial function. Equation (1) is the proposed function that describes the material behavior in terms of both strain and strain rate. Equation (2) is a second-degree polynomial function that is suggested to describe the flow behavior. And Equation (3) is the suggested function to describe strain rate sensitivity of the stress, where *d* and *e* are simply scalar material parameters. By implementing Equations (2) and (3) into Equation (1) and considering the $g(\dot{\varepsilon})$ as a linear function with slope of 1, Equation (4) will be readily obtained.

$$\sigma = f(\varepsilon) * g(\dot{\varepsilon}) + h(\dot{\varepsilon}) \tag{1}$$

$$f(\varepsilon) = (a\varepsilon^2 + b\varepsilon + c) \tag{2}$$

$$h(\dot{\varepsilon}) = d\dot{\varepsilon}^e \tag{3}$$

$$\sigma = (a\varepsilon^2 + b\varepsilon + c) * \dot{\varepsilon} + d\dot{\varepsilon}^e \tag{4}$$

where ε is the plastic strain, $\dot{\varepsilon}$ is the strain rate, *a*, *b*, *c*, and *d* are material constants and *e* is the strain rate coefficient.

5.2. *Quality of prediction*

Using the MATLAB curve fitting tool, the material parameters in the proposed model were optimized and fitted to the experimental data for AA5083 tested at 450°C and are shown in Table 2.

Table 2. Material parameters in the proposed material model for AA5083 at 450°C.

	a	b	c	d	e
No oscillation	-261.5	85.61	18.27	88.86	0.2758
With Oscillation	-210.3	88.98	12.87	84.11	0.2446

Figure 7 compares the residuals of the predictions for the corresponding flow stress compared to the experimental data. The accuracy of the predictions remains acceptable for the entire range of strain rates considered. At the lower end of strain rates, the absolute value of residuals does not exceed 5 MPa. At the higher end of strain rates this value remains below 7 MPa.

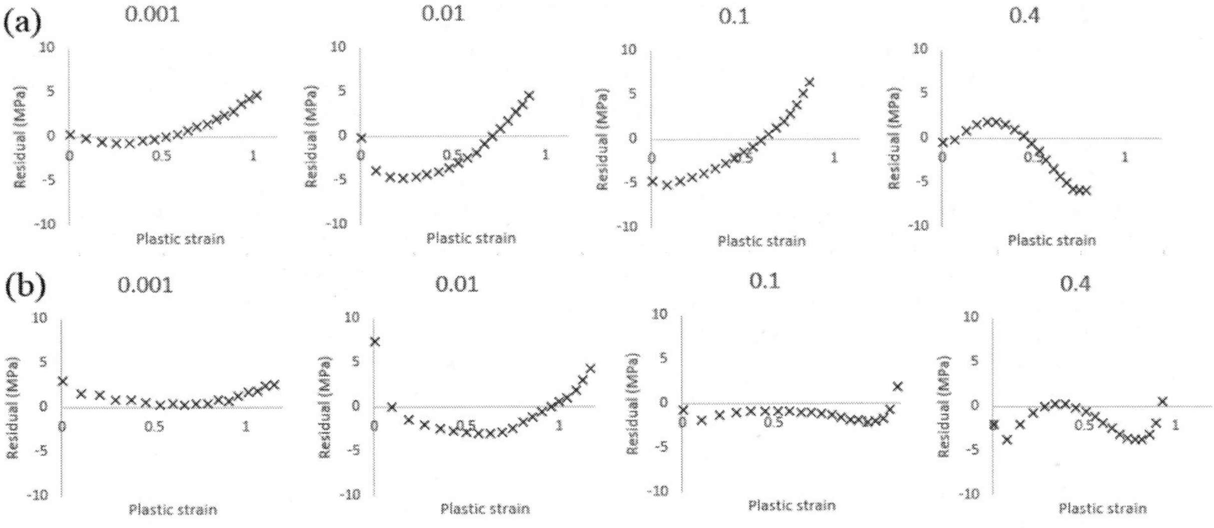

Figure 7. Differences between predicted and experimental data (a) without oscillations and (b) with oscillations.

5.3. *Simulations*

The above-mentioned material model was implemented as a user-defined material model in LS-Dyna. The tensile specimen was modelled with Belytschko-Tsay shell elements having 5 integration points through the thickness. This series of simulations covered the conditions shown in Figure 4. The deformation was applied by allowing contact between the tensile specimen and the solid elements of the test fixture rather than applying constraints as boundary conditions. This strategy considers the flow of material throughout the specimen, from the gripping area through to the gauge area that is observed in experiments. LS-Dyna requires as input the true stress versus effective plastic strain in uniaxial tension. In this case, since the experimental flow curves were not obtained in uniaxial tension due to the specimen geometry, a series of iterative simulations was performed to scale the input flow curve until the experimental force displacement data were accurately predicted. In the second series of simulations the constant load conditions that are described in Figure 5 were predicted. A constant load was applied to the top fixture while keeping the bottom fixture stationary. The simulation was set to terminate upon reaching the 171% strain, and the minimum thickness predicted in the gauge was exported and compared to the experimental data. This comparison is presented in Figure 8 showing a good degree of accuracy under all loading conditions.

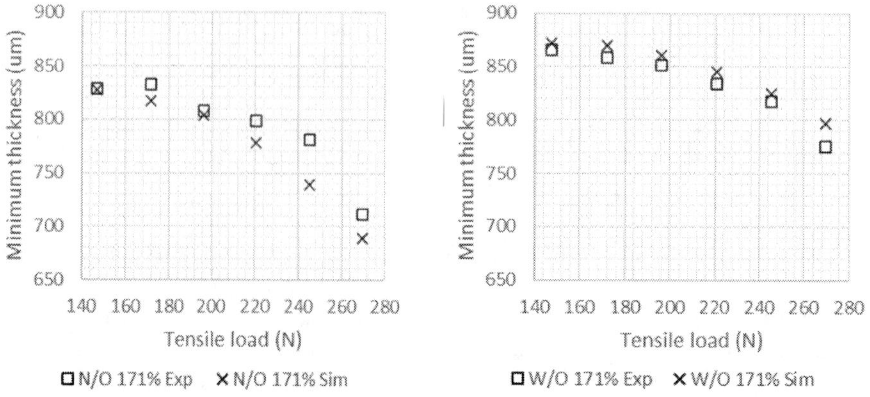

Figure 8. Comparison of minimum thickness in the specimen gauge measured from experiments (Exp) and predicted by simulations (Sim) both with oscillations (W/O) and without oscillations (N/O).

6. Conclusion

Superimposing a minor oscillating load having a sine wave pattern, an amplitude of 0.5 N and a frequency ranging from 10-40 Hz has been shown to have a significant effect on the flow behavior of AA5083 during superplastic forming at 450°C. Although the tensile stress is not sensitive to superimposed oscillations, the total elongation is very sensitive to the oscillating load. Indeed, the relative improvement in total elongation varies from 20 to 50% depending on the applied strain rate. This allows the sheet material to achieve significantly higher levels of plastic deformation before the onset of fracture. Minimum thickness measurements in the gauge of deformed test specimens also confirmed that the presence of oscillating loads during forming result in greater value of the minimum thickness which indicates a slower rate of thinning in the specimen. These results suggest that the addition of a minor oscillating load will allow industrial parts to be safely formed superplastically at great strain rates, and therefore at higher rates of production. Finally, a simple empirical material model has been proposed that is able to accurately describe the superplastic behavior of AA5083 at 450°C with and without the presence of minor oscillations.

References

[1] Giuliano G 2011 Superplastic forming of advanced metallic materials *Superplastic forming of advanced metallic materials* (Woodhead Publishing Limited) p 15
[2] Davoodi B, Payganeh G H and Eslami M R 2012 Cutting forces in dry machining of aluminum alloy 5083 with carbide tools *Advanced Materials Research* vol 445 pp 259–62
[3] Zhang K F and Jiang S S 2014 Superplastic Forming *Comprehensive Materials Processing* vol 5 (Elsevier) pp 371–92
[4] Zelin M G 1997 On micro-superplasticity *Acta Mater.* **45** 3533–42

[5] Mikhaylovskaya A V., Yakovtseva O A, Sitkina M N, Kotov A D, Irzhak A V., Krymskiy S V. and Portnoy V K 2018 Comparison between superplastic deformation mechanisms at primary and steady stages of the fine grain AA7475 aluminium alloy *Mater. Sci. Eng. A, Struct. Mater. Prop. Microstruct. Process.* **718** 277–86
[6] Yakovtseva O A, Sitkina M N, Kotov A D, Rofman O V. and Mikhaylovskaya A V. 2020 Experimental study of the superplastic deformation mechanisms of high-strength aluminum-based alloy *Mater. Sci. Eng. A* **788** 139639
[7] Wang X, Li Q, Wu R, Zhang X and Ma L 2018 A Review on Superplastic Formation Behavior of Al Alloys ed G P Dinda *Adv. Mater. Sci. Eng.* **2018** 7606140
[8] Liu X, Ye L, Tang J, Dong Y and Ke B 2021 Superplastic deformation mechanism of an Al-Mg-Li alloy by high resolution surface studies *Mater. Lett.* **301** 130251
[9] Masuda H and Sato E 2020 Diffusional and dislocation accommodation mechanisms in superplastic materials *Acta Mater.* **197** 235–52
[10] Hosseinipour S J 2009 An investigation into hot deformation of aluminum alloy 5083 *Mater. Des.* **30** 319–22
[11] Das S, Riahi A R, Meng-Burany X, Morales A T and Alpas A T 2012 High temperature deformation and fracture of tribo-layers on the surface of AA5083 sheet aluminum–magnesium alloy *Mater. Sci. Eng. A* **531** 76–83
[12] Kridli G T, Friedman P A and Boileau J M 2021 Manufacturing processes for light alloys *Mater. Des. Manuf. Light. Veh.* 267–320
[13] Singh V and Agrawal P 2008 *Analysis of Bonding Strength of Ultrasonic Welding Process* vol 9001
[14] Drozdov A D and Shtemler Y 1996 *Evaluation of bond strength at ultrasonic welding*
[15] Winsper C E and Sansome D H 1968 A review of the application of oscillatory energy to metals deforming plastically *Adv. Mach. Tool Des. Res. 1967* 1349–60
[16] Ye C 2016 *The effect of ultrasonic waves on tensile behavior of metal The effect of ultrasonic waves on tensile behavior of metal* (Windsor: University of Windsor)
[17] Daud Y, Lucas M and Huang Z 2007 Modelling the effects of superimposed ultrasonic vibrations on tension and compression tests of aluminium *J. Mater. Process. Technol.* **186** 179–90
[18] Kirchner H O K, Kromp W K, Prinz F B and Trimmel P 1985 Plastic deformation under simultaneous cyclic and unidirectional loading at low and ultrasonic frequencies *Mater. Sci. Eng.* **68** 197–206
[19] Mohamed F A 2011 Micrograin superplasticity: Characteristics and utilization *Materials (Basel).* **4** 1194–223
[20] Majidi O, Jahazi M and Bombardier N 2018 A viscoplastic model based on a variable strain rate sensitivity index for superplastic sheet metals *Int. J. Mater. Form. 2018 124* **12** 693–702
[21] Dastgiri M S, Kiawi L, Sarraf I S, Ryzer E and Green D E 2021 Influence of Specimen Preparation Methods on the Mechanical Properties and Superplastic Behavior of AA5083 Sheets *Miner. Met. Mater. Ser.* 1573–83

IOP Conf. Series: Materials Science and Engineering 1238 (2022) 012013 doi:10.1088/1757-899X/1238/1/012013

Process Stability and Application of 1900 MPa Grade Press Hardening Steel with reduced Hydrogen Susceptibility

D Rosenstock[1], T Gerber[1], C Castro Müller[2], S Stille[1] and J Banik[1]

[1] thyssenkrupp Steel Europe AG, Dortmund, Germany
[2] thyssenkrupp Steel Europe AG, Duisburg, Germany

dirk.rosenstock@thyssenkrupp.com

Abstract. While 22MnB5 with aluminium silicon (AS) coating was established as the quasi-standard in press hardening many years ago, the need for steels that offer even higher strength is growing. These needs include corrosion protection of the automotive component and scale prevention during processing in hot forming lines. However, the processing of AS coated ultra-high strength steels involves demands for the furnace atmosphere. Reducing the hydrogen susceptibility and increasing the material strength at the same time was the driving force for the development of MBW 1900 + AS Pro. The paper focuses on the process stability of this steel concerning furnace and stamping parameters as well as the hydrogen induced cracking resistance under different dew points in the furnace atmosphere. In addition, the potential by means of different partial press hardening processes is presented. With tailored tempering, locally heated tools can be used to achieve soft areas in the part. Furthermore, the application potential of MBW 1900 + AS Pro in tailor welded blanks is shown.

1. Introduction

Press hardening (PH) steel is approved since about 20 years in automotive manufacturing. To comply with more demanding crash tests and the increasing vehicle weight of battery electric vehicle (BEV) press hardening steels with strength level above 1500 MPa entered the marked recently. In their first generation, these steels with about 1900 MPa strength level do not have any coating [1, 2]. This reduces the formability in hot forming due to the higher friction coefficient, an additional blast cleaning process of the parts is required and it limits the application to inner automotive parts only. Applying a typical aluminium silicon (AS) coating (e.g. with 11 % Si, 2.5 % Fe and rest Al) to a 1900 MPa grade steel like a 34MnB5 will unfortunately not lead to the desired results. The reason on one hand is a slight decarburization in the furnace of uncoated PH steels, which leads to better folding behaviour during crash. On the other hand, the AS coating is a barrier to the hydrogen effusion, while simultaneously the higher strength grade is more susceptible to hydrogen induced cracking (HIC) [3]. Although there are AS coated concepts that are only based on an adapted chemical composition of the steel [4], an optimization of the coating offers a further increase in the resistance to HIC. Thus, for the development of a coated 1900 MPa press hardening steel grade an adaptation of both the steel and the coating were performed. Although still in development state, the alloying concept of the steel is completed. The applied coating is AS Pro, an improved AS based coating which includes additionally Magnesium. AS Pro is available on the market for 22MnB5 since 2020 and supports a lower hydrogen diffusion into the steel [5].

2. Processing parameters during press hardening of MBW 1900 + AS Pro

2.1. Chemical composition and delivery state

As for most press hardening steels the main alloying elements of MBW 1900 + AS Pro are carbon, manganese and boron. An addition of further micro alloying elements helps to reduce grain growth during furnace heating. A smaller prior austenite grain size improves the tensile strength of press hardening steel [6]. Furthermore, the micro alloying elements decrease the delayed fracture susceptibility and increase the ductility. Concerning niobium this is discussed in [7]. The chemical composition of MBW 1900 + AS Pro is shown in Table 1.

Table 1. Maximum content of selected chemical elements in manganese boron steels.

Grade	C in %	Si in %	Mn in %	Nb + Ti in %	Cr + Mo in %	B in %
34MnB5 uncoated	0.40	0.40	1.50	0.05	0.50	0.005
MBW 1900 + AS Pro	0.35	0.40	1.50	0.05	0.50	0.005

In the as-delivery state the microstructure is ferritic-perlitic and optionally contains traces of harder microconstituents, such as bainite and martensite. The 0.2%-offset yield strength (*YS*) is between 400 and 650 MPa and the tensile strength (*TS*) ranges from 600 to 800 MPa. In the as-delivery state the material can be handled easily such as 22MnB5 + AS e.g. for roller levelling, blanking with an elongation at break A_{80} above 12%. The microstructure and coating layer in delivery state are shown in Figure 1. The appearance and handling of the AS Pro coating layer is the same as for typical AS coatings. In the delivery state, the total thickness is approximately 25 μm when a coating mass of 150 g/m² is applied.

a) 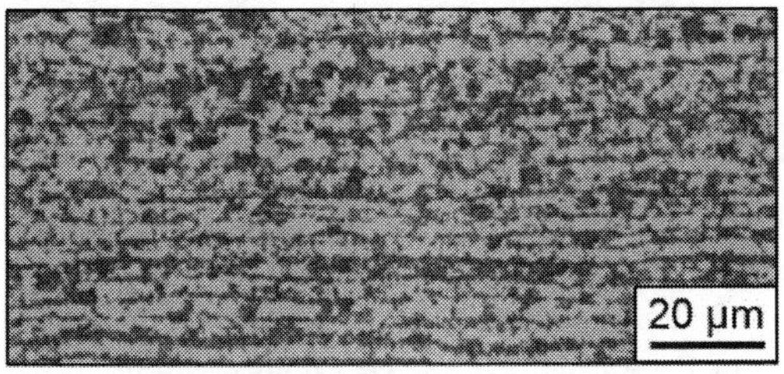 b) 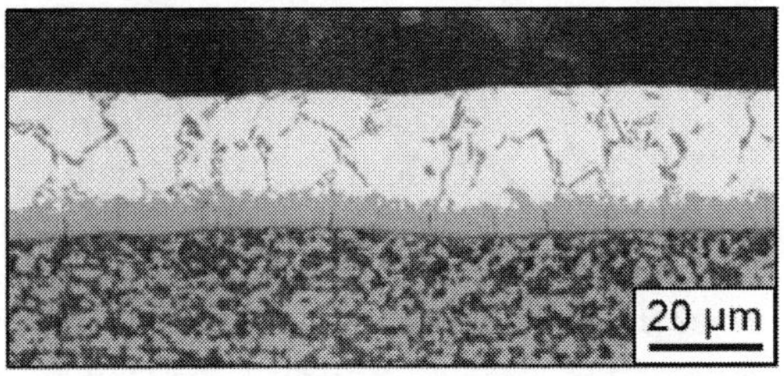

Figure 1. Microstructure in delivery state etched with HNO_3 in longitudinal direction, representing **a)** steel microstructure and **b)** AS Pro layer formation for 150 g/m² coating weight on both sides.

With its critical cooling rate of about 30 K/s and A_{c3} temperature about 820 °C, MBW 1900 + AS Pro is designed for manufacturing automotive parts by the typical direct press hardening process with austenitisation in a furnace at about 900 °C. During this process the AS Pro coating completes its characteristic 5-layered composition (Figure 4f) and the steel microstructure is transformed to mainly martensite with traces of retained austenite (cf. Figure 4a-e). After press hardening the roughness R_a is typically above 2.0 μm.

2.2. Process stability of MBW 1900 + AS Pro in press hardening

MBW 1900 + AS Pro was developed as a process robust hot forming steel to deliver steady properties after press hardening. This is required because of the preferences of different hot forming line designs and for a tolerance regarding the combination of batches with other blanks in the furnace. The process stability was tested under a very wide range of parameters. However, this paper includes only a part of it.

To evaluate the parameters after press hardening, blanks were press hardened under different conditions. If not stated differently, the standard processing parameters were: sheet thickness: 1.5 mm, holding time in a roller hearth furnace: 300 s at a temperature of 920 °C, transfer time from the furnace to the press in 5 s, closing duration 15 s in a water-cooled tool with 3 MPa surface pressure. The forming start temperature was above 760 °C and the cooling rate was partly >300 K/s. A part of the press hardened parts (condition PH) was further heat treated at 170 °C for 20 min to simulate the paint baking process (condition PB) after electrophoretic coating. All testing samples were EDM wire-cut from flat part areas with homogeneous thickness and tested according to international standards. Results of the tensile testing according to DIN EN ISO 6892-1 (sample size 2) have been performed after press hardening with different temperatures and holding times, see Figure 2. The temperature 'Rising' corresponds to a rising furnace temperature inside the furnace over different zones: 800 °C for 60 s, 850 °C / 70 s and 920 °C / 170 s and leads to a lower heating rate in the front area of the furnace as it is performed in a similar way in series roller furnaces.

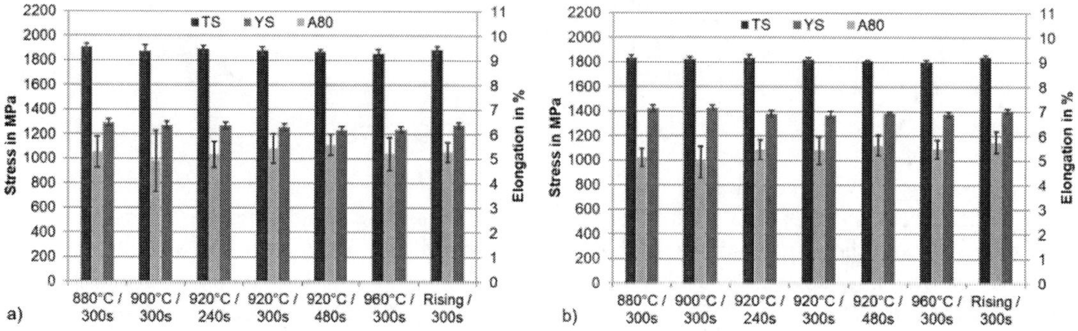

Figure 2. Tensile testing results longitudinal to rolling direction, n ≥ 9; **a)** PH; **b)** PH and PB

For all conditions directly after press hardening the tensile strength is between 1850 and 1910 MPa and the yield strength between 1230 and 1300 MPa. The elongation at break A_{80} is above 4.5 %. After paint baking the bake hardening effect leads to the typical increase of the *YS* of about 130 to 150 MPa to about 1400 MPa, see Figure 2b. The resulting decrease of the *TS* is only about 50 MPa and leads to a *TS* still above 1800 MPa. A_{80} stays on the same level, but the standard deviation is lower. In general, the higher energy input either by higher temperature or longer holding time leads to a slight decrease both in tensile strength and yield strength. Because of its lower A_{c3} temperature of ca. 820 °C compared to 22MnB5 with ca. 845 °C also a lowered furnace temperature could be possible to austenitise the blanks. However, part complexity and formability have to be considered.

The ductility of MBW 1900 + AS Pro can be evaluated by the plate bending test according to VDA 238-100 (07/2020). The results after press hardening and paint bake simulation are shown in Figure 3a according to the optical measurement. In all conditions a bending angle >45° is achieved. Typically for continuous cast and rolled steel the results transversal to the rolling direction are lower than in longitudinal direction. Still for this steel the anisotropy effect can be limited to only about 2° in plate bending test. The highest bending angles are achieved with low temperatures or low holding times, which are favored by most stampers and enable energy savings.

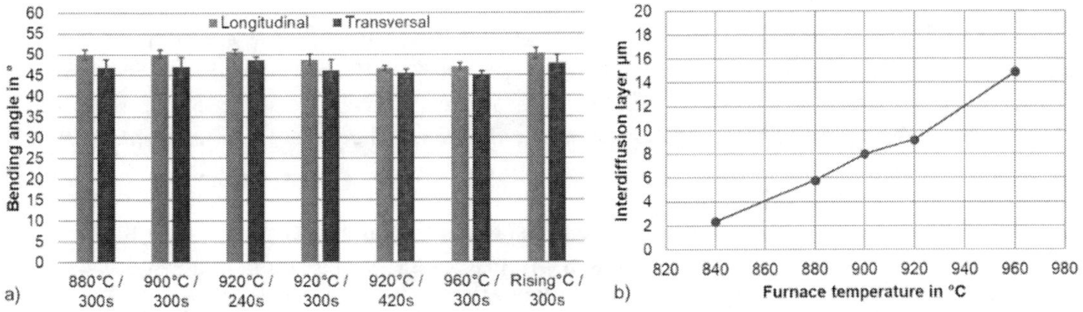

Figure 3. a) Bending angle according to VDA 238-100 plate bending test longitudinal and transversal to rolling direction for different furnace parameters, n ≥ 6; **b)** Interdiffusion layer thickness for 1.5 mm MBW 1900 + AS Pro at different furnace temperatures with dwell time of 300 s.

The microstructure after press hardening is influenced by the amount of thermal energy introduced into the steel in the furnace. As an example, the microstructure for 1.5 mm thick MBW 1900 + AS Pro is shown in Figure 4a-e, representing an increase in the furnace temperature from 840 to 960 °C. The growing heat input increases the grain growth of the austenite microstructure and also affects the microstructure of the martensite after quenching in the same way. As a result, the martensite shows a coarser microstructure with rising temperature, especially at 960 °C. The coarser microstructure is mainly responsible for the slight decrease in the mechanical properties, see Figure 2 and Figure 3a.

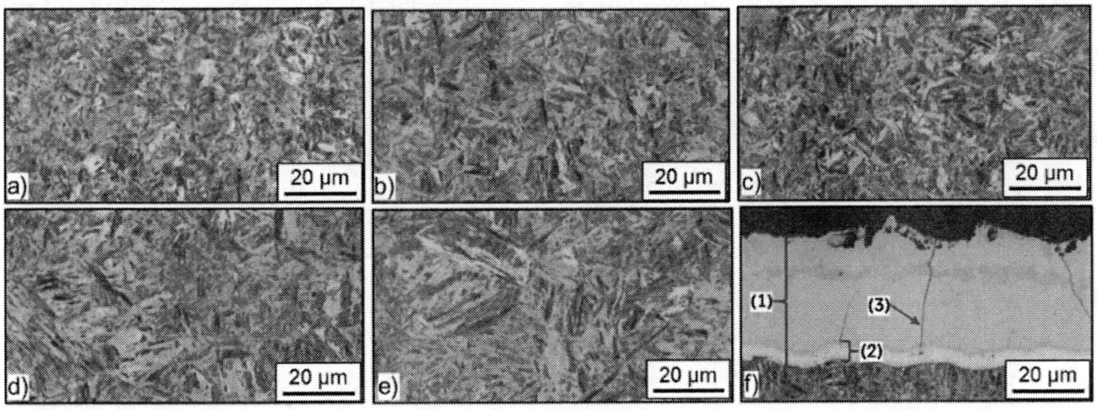

Figure 4. Substrate microstructure: **a)** 840 °C / 300 s; **b)** 880 °C / 300 s; **c)** 900 °C / 300 s; **d)** 920 °C / 300 s; e) 960 °C / 300 s; Coating microstructure: **f)** 880 °C / 300 s

Figure 4f shows the typical 5-layered microstructure (1) of the AS Pro coating after press hardening, with typical total thickness above 30 µm. Cracks, which occur due to different expansion coefficients of substrate and coating during quenching and forming operations (3) stop in the soft ferritic interdiffusion layer (2). The interdiffusion layer grows as a function of energy input due to diffusion of aluminium into the steel substrate. This can be observed in Figure 3b, which represents the interdiffusion layer as a function of the furnace temperature. After heating at 840 °C / 300 s, although the steel is fully martensitic after PH, the coating is not sufficiently alloyed and remains of pure AS Pro coating can be found. These could pollute the forming tools during the stamping operation. Sufficient alloying is achieved even at 880 °C / 300 s, as shown in Figure 4f.

Steels have to have a high resistance against unintentional variation of production parameters or failures, such as problems with the tool cooling system or the transfer unit as this can result in discard of parts. To provide a good process stability against such problems, MBW 1900 + AS Pro has been tested in different conditions. The results for tensile testing under different circumstances are shown in **Figure 5a**. A too long transfer time is especially critical for small sheet thicknesses. With increasing transfer time or tool temperature MBW 1900 + AS Pro shows only minor reduction in the ultimate tensile strength. For 15 s transfer time still a *TS* of above 1730 MPa is achieved even at relatively low 3 MPa of surface pressure. The yield strength shows above 1330 MPa even for 15 s of transfer time or 100 °C tool temperature. The elongation at break A_{80} is not significantly influenced by both tool temperature or transfer time. The surface pressure in the tool also has an influence on the mechanical properties by altering the cooling rate. These results were achieved with relatively low 3 MPa surface pressure. Therefore, by increasing the surface pressure and by this the cooling rate, a further increase of the mechanical properties to compensate longer transfer time or higher tool temperature is possible.

Also the bending angle in Figure 5b is almost not affected by both transfer time or tool temperature changes. In total this makes MBW 1900 + AS Pro a very process robust press hardening steel under different furnace temperatures and dwell times as well as for transfer times and tool temperatures.

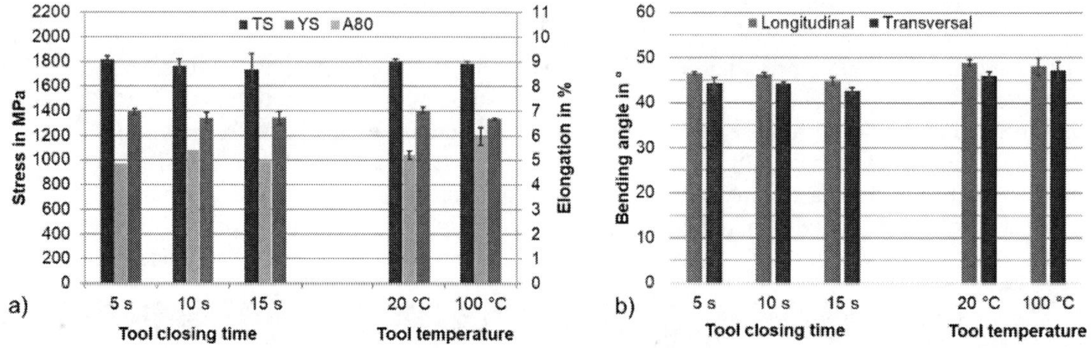

Figure 5. Influence of tool closing time and tool temperature for 1.5 mm sheet thickness; **a)** tensile test, longitudinal, after paint baking and **b)** Plate bending test, paint baking condition

2.3. *Reduced hydrogen susceptibility of MBW 1900 + AS Pro*

During the automotive body assembly of ultra-high strength steel mechanical and thermal stress may occur, which could lead to hydrogen induced cracking if no extra measures are taken into account. A source of hydrogen in press hardening is the surface oxidation and reaction to atmospheric humidity in the furnace during austenitisation [5]. Thus, the dew point of the furnace is generally lowered by injecting dried air. Besides this general processing influence, additional material related measures can be taken. Thus the chemical composition of MBW 1900 + AS Pro with increased micro alloying elements and finer grain make it less susceptible for diffusive hydrogen. Furthermore the recently developed AS Pro coating with the addition of magnesium is one more important step towards a coated press hardening steel with 1900 MPa tensile strength. The driving effect of AS Pro is the lower hydrogen formation and absorption at the surface during heating because of the magnesium-oxide surface layer [5]. The diffusive hydrogen content in steels can be measured by Thermal Desorption Mass-Spectrometric (TDMS) and is compared for different steels and coatings in Figure 6a according to the dew point in the furnace. Already at a dew point of -5 °C MBW 1900 + AS Pro shows a diffusive hydrogen value close to the detection limit of the TDMS. Depending on the situation, about 40% less hydrogen diffuses into the steel as a result of the AS Pro coating [5].

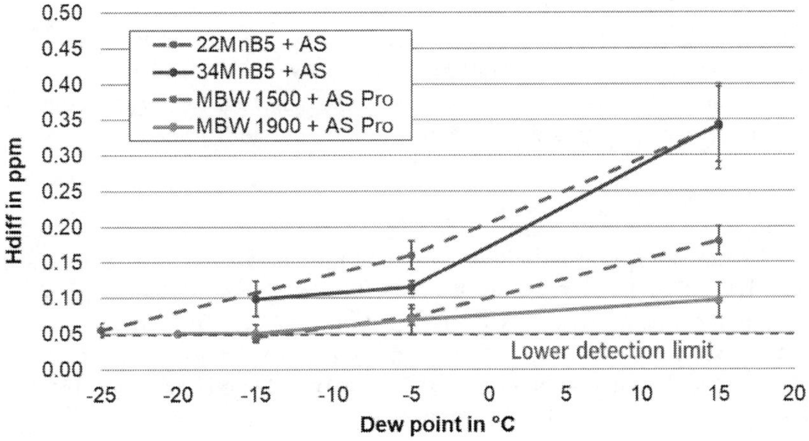

Figure 6. Diffusive hydrogen in dependency of furnace dew point after press hardening in different steels

Besides directly measuring it, the influence of diffused hydrogen can be visualized, by the 4-point bending test under constant load, adapted from DIN EN ISO 7539-2, see Figure 7a. In this test a press hardened part is produced under a controlled dew point in the furnace atmosphere and subsequently set under a specific bending stress for 96 hours or until failure occurs. The bending stress is applied to the sample by turning the screw upwards and measuring the resulting displacement at the center of the sample. According to the mentioned standard the bending stress can be calculated based on the previously measured Young's modulus and normalized to the respective tensile strength. After reaching yield strength, plastic deformation occurs and the linear correlation of displacement and bending stress is no more valid. As a result, values above the *YS* have to be regarded as surrogate values for the stepwise increase of the bending stress.

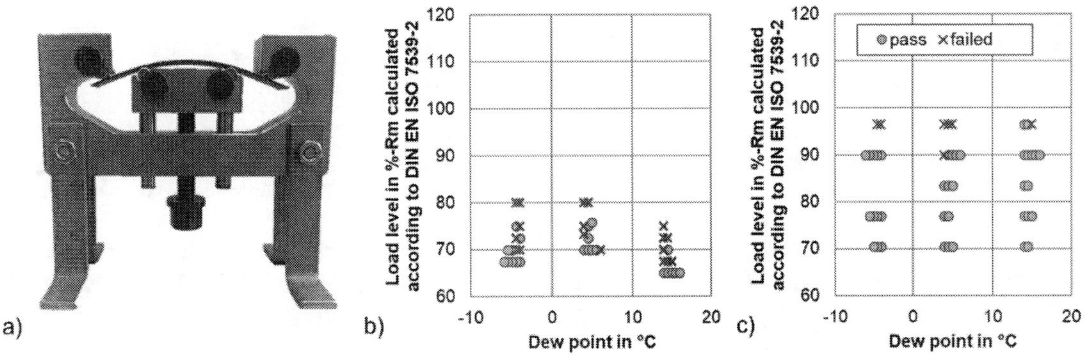

Figure 7. Delayed fracture test; **a)** 4 point bending device; **b)** passed and failed load levels for 34MnB5 + AS and c) MBW 1900 + AS Pro

Figure 7b and c compare the successfully passed and the failed stress levels for a conventional 34MnB5 and MBW 1900 + AS Pro. Both steels reach the level of yield strength. The final successful level for 34MnB5 + AS is 65 to 67.5 % depending on the dew point, while MBW 1900 + AS Pro passes 83 and 90 %, depending on the dew point. This shows the positive effect of the deployed measures to increase the resistance against delayed fracture for MBW 1900 + AS Pro.

3. Application of MBW 1900 + AS Pro

Press hardening steels allow the integration of ductile sections to combine both superior part integrity and high energy absorption within one component. Thus, newly developed grades have to be applicable for such technologies as e.g. tail hang out, schwartz thermal printing, tailored tempering and tailor welded blanks. MBW 1900 + AS Pro is well applicable for these technologies, which will be discussed using the examples tailored tempering and tailor welded blanks.

3.1. *Tailored tempering*

The tailored tempering technology uses partially heated tools of up to 550 °C to prevent full martensite formation during press hardening in certain areas of the part. With a martensite start temperature of Ms = 340 °C, MBW 1900 + AS Pro is well applicable for tailored tempering. With increasing tool temperature, the microstructure changes from full martensite for 20 and 100 °C, to first traces of tempering of full martensite (200 and 250 °C), to tempered martensite with probably lower bainite (300 and 400 °C), to predominantly bainite (450 °C), to Bainite and Perlite (500 °C) finally to Ferrite, Perlite and Bainite at 550 °C. The according mechanical properties are shown in Figure 8a and b. Until about 250 °C tool temperature there is only a slight decrease in the mechanical properties, as already discussed in section 2.2. At 300 °C tool temperature when getting close to the Ms, there is a significant decrease in tensile strength and hardness. Above 300 °C the yield strength starts to decrease as well. At 500 °C and above the properties stabilise again and tensile strength is below 800 MPa, yield strength below 600 MPa and hardness below 250 HV5. As shown in Figure 8b the ductility starts to increase at 350 °C. At 550 °C a bending angle close to the physical maximum of the test >120° is achieved. The elongation at break A_{80} only starts to increase after martensite is mainly vanished at 450 °C. With tool temperatures at 500 °C and above A_{80} of about 11 % are achieved. The presented results show that press hardened under tailored tempering conditions, MBW 1900 + AS Pro shows the desired low strength and high ductility to provide sufficiently ductile integrated areas in the produced parts.

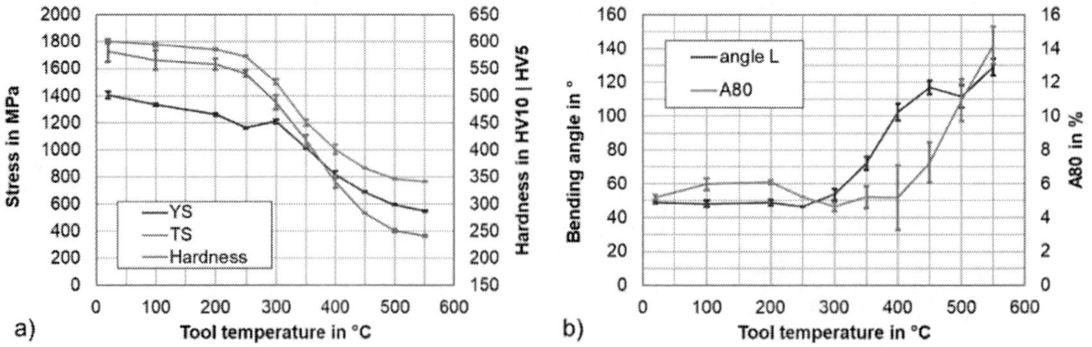

Figure 8. Mechanical properties at different tool temperatures, after paint baking; **a)** Stresses from tensile testing (longitudinal) and hardness results HV10 (500 and 550 °C HV5); **b)** Bending angle longitudinal and elongation at break A_{80}

3.2. *Tailor welded blanks*

The use of tailor welded blanks is one of the earliest technologies to combine areas with different tensile properties in fully press hardened components. Blanks of different steels are (e.g. laser) welded under laboratory conditions and press hardened with the same parameters as used for monolithic blanks. From a larger test series, some exemplary results are displayed in Figure 9. In the study samples of a very ductile 500 MPa grade and a ductile 1200 MPa grade [8] have been combined with MBW 1900 + AS Pro by laser welding. Parts have been press hardened to hat-shaped profiles with a length of 400 mm, which have been spot welded with counterpart sheets of MHZ 340. After paint baking simulation, the samples have been tested on a drop tower with impact energy of about 2.8 kJ.

In all cases the laser welded seam withstood even the direct impact of the impactor and did not show failure. As expected, all deformation is absorbed by the plastic deformation of the softer grade, when the material partition is 37.5% MBW 1900 + AS Pro and 62.5 % of MBW 1200 or MBW 500 (Figure 9a). In axial crash testing (Figure 9b) the same hat-shaped profiles with total length of 350 mm have been used, of which MBW 1900 + AS Pro was applied about 36 % length at the lower end. After the crash with impact energy of about 8.9 kJ, no deformation occurred in MBW 1900. The weld seam was fully intact, also for the combination with MBW 500, where the folding occurred until the weld seam with the MBW 1900.

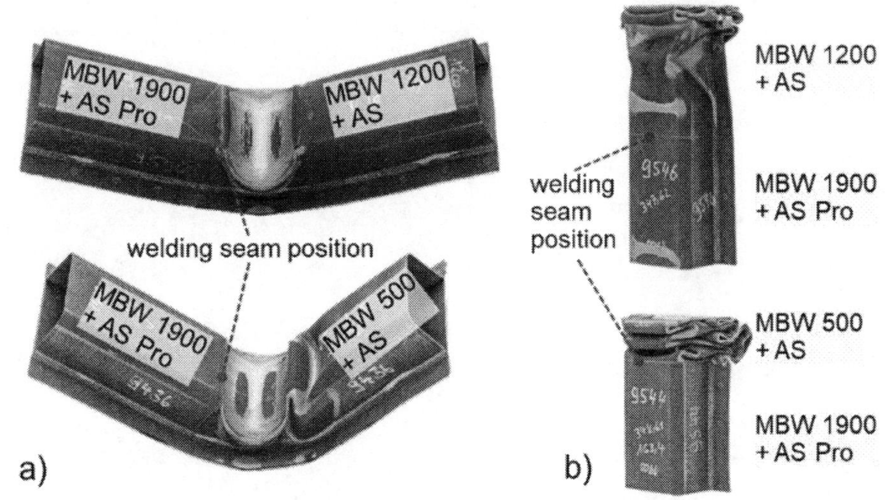

Figure 9. Hat-shaped tailor welded samples after testing with different material combinations; **a)** material partition 37.5% MBW 1900 + AS Pro in longitudinal crash; **b)** material partition 50% MBW 1900 + AS Pro in longitudinal crash; **c)** axial crash test

4. Conclusions

MBW 1900 + AS Pro is developed as a process robust and widely applicable 1900 MPa strength level press hardening steel grade. For typical processing parameters after paint baking, tensile strength ≥1800 MPa and yield strength ≥ 1400 MPa with bending angle >45° are achieved. These parameters stay largely constant despite varying e.g. tool temperature, transfer time, dwell time or furnace temperature. The special AS Pro coating and the micro alloying make it possible to reduce the absorption of diffusive hydrogen in the furnace and the delayed fracture under constant load condition. Different applications of MBW 1900 + AS Pro for providing local soft areas are possible, of which the application of the tailored tempering process with heated tools and the tailor welding process have been discussed and show excellent applicability. Further possible application options for MBW 1900 + AS Pro are tail hang out, schwartz thermal printing technology and the use in patchwork blanks.

5. References

[1] Holzweissig M J, Frost G, Bake K and Frehn A 2019 BTR2000 - A New Uncoated Ultra-High Strength Hot Forming Steel *7th International Conference Hot Sheet Metal Forming of High-Performance Steel, June 2-5, 2019, Luleå, Sweden: Proceedings (CHS²-series* no. 7*)* ed M Oldenburg *et al* 1st edn (Auerbach/Vogtl.: [Verlag Wissenschaftliche Scripten])
[2] Taylor T and McCulloch J 2018 *steel research int.* **89** 1700495
[3] Ramamurthy S and Atrens A 2013 *Corrosion Reviews* **31** 1–31

[4] Ramisetti N, Wang J, Enloe C M, Lu Q and Reini L 2019 Loading Rate Effects on Apparent Bendability of 1.8 GPa PHS *7th International Conference Hot Sheet Metal Forming of High-Performance Steel, June 2-5, 2019, Luleå, Sweden: Proceedings (CHS²-series* no. 7*)* ed M Oldenburg *et al* 1st edn (Auerbach/Vogtl.: [Verlag Wissenschaftliche Scripten])
[5] Ruthenberg M, Banik J, Köyer M and Stille S 2019 Small Change, Substantial Effect - A New Aluminum-Silicon Coating for a Reliable Hot Forming Process *7th International Conference Hot Sheet Metal Forming of High-Performance Steel, June 2-5, 2019, Luleå, Sweden: Proceedings (CHS²-series* no. 7*)* ed M Oldenburg *et al* 1st edn (Auerbach/Vogtl.: [Verlag Wissenschaftliche Scripten])
[6] Okayasu M and Yang L 2019 *J Mater Sci* **54** 5061–73
[7] Zhang S, Huang Y, Sun B, Liao Q, Lu H, Jian B, Mohrbacher H, Zhang W, Guo A and Zhang Y 2015 *Materials Science and Engineering: A* **626** 136–43
[8] Rosenstock D, Banik J, Röttger R P, Graff S and Gerber T 2021 *IOP Conf. Ser.: Mater. Sci. Eng.* **1157** 12015

IOP Conf. Series: Materials Science and Engineering 1238 (2022) 012014 doi:10.1088/1757-899X/1238/1/012014

Development of a process chain for multi-stage sheet metal forming of high-strength aluminium alloys

J Günzel[1,2], J Hauß[2], C Gaedigk[1], J Bergmann[1] and P Groche[1]

[1] Institut für Produktionstechnik und Umformmaschinen, Technische Universität Darmstadt, Otto-Berndt-Straße 2, 64287 Darmstadt, Germany
[2] Werner Schmid GmbH, Weichselstraße 21, 36043 Fulda, Germany

E-Mail: guenzel@ptu.tu-darmstadt.de

Abstract. The high-strength aluminium alloys EN AW-6082 and -7075 are characterized by low density and high strength but also limited cold formability and pronounced springback behaviour in the ultra-high-strength T6 state. In order to exploit their lightweight design potential, temperature-supported process routes such as warm or hot forming are applied. Alternatively, there is the possibility of cold forming preconditioned semi-finished products at the expense of the initial material properties. Common to all variants are complex interrelationships due to linked plant periphery resulting from up- and downstream heat treatments. In addition, occurring heat transfers in temperature-supported process routes or strain hardening effects during cold forming lead to reduced formability. Especially for multi-stage forming processes, as they are required for complex components, the above-mentioned process routes reach their limits. The different requirements of the four single-stages (deep drawing, blanking, collar drawing and upsetting) for the production of a demonstrator geometry with adapted wall thicknesses make a new type of temperature control necessary. This paper shows that the combination of temperature-supported and multi-stage forming contributes to a significant increase in formability. The temperature-controlled forming tool used for this purpose enables an inline heating of the components during the process, so that an industrially feasible and economical overall process chain for the fabrication of the demonstrator geometry out of those alloys is convertible.

1. Introduction

The low density and tensile strengths of at least 540 MPa [1] qualify high-strength aluminium alloys for various lightweight applications in the mobility sector [2]. A broad industrial application is limited by the low cold formability [3] and the high springback during cold forming [4]. To extend the process limits, complex and expensive temperature-supported process routes are required [5]. The most common method is hot forming, in which the sheets are first solution annealed and then quenched parallel to forming in the tool [6]. Alternatively, warm forming at 200 - 250 °C in an isothermal tool [7] or cold forming of preconditioned semi-finished products in the W-Temper-state (solution heat treated and water quenched) [8] or soft-annealed O-state are used [9]. Apart from warm forming, all routes require subsequent heat treatment to return to the high-strength T6-state [10].

If even these processes reach their limits, multi-stage forming is unavoidable. In this way, complicated components can be produced by a combination of various processes or drawing ratios [11].

Studies by Lai [12], Takalkar [13] and Pourkamali Anaraki [14] show the possibilities and challenges of multi-stage forming processes. Depending on the operation and process route, different stress conditions and thus forming requirements arise for the individual tool stages. In addition, there are complex interrelationships due to linked plant peripherals, heat transfers in the case of warm and hot forming [9] or pronounced strain hardening when using preconditioned semi-finished products [15, 16].

The combination of temperature-supported and multi-stage forming is a major challenge and will be illustrated based on the sequence of operations shown in Figure 1 (a). It contains the processes deep drawing, blanking, collar drawing and upsetting and enables the production of a demonstrator geometry with adapted wall thicknesses, which can be seen in Figure 1 (b). The need for targeted, novel temperature control is evident from the cold-formed specimen of EN AW-7075 in the high-strength T6 condition shown in Figure 1 (c). It already breaks down into several parts in stage 1 and illustrates the limited cold formability. For this reason, the development of a process chain for multi-stage sheet metal forming of high-strength aluminium alloys is of particular importance.

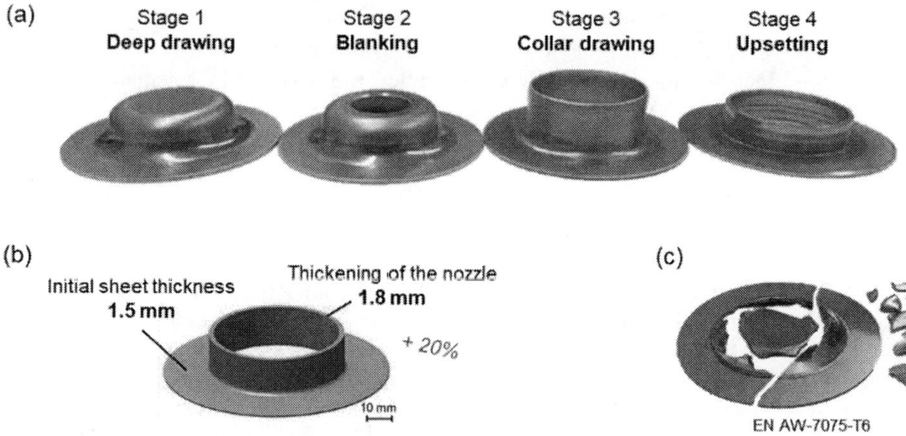

Figure 1. Multi-stage forming process: stadium sequence (a), demonstrator geometry with adapted wall thicknesses (b) and result of cold forming of EN AW-7075-T6 in stage 1 (c).

2. Materials and experimental setup

The high-strength aluminium alloys EN AW-6082 and -7075 and the experimental setup used to study their temperature-dependent process limits and to develop a feasible overall process route are described below.

2.1. High-strength aluminium alloys EN AW-6082 and -7075

The two materials investigated in this study are the high-strength aluminium alloys EN AW-6082 and -7075. Both belong to the heat-treatable alloy classes and are present in the high-strength T6 state (solution heat treated, quenched and artificially aged) with a thickness of 1.5 mm. Characteristic of the 6xxx alloy is the magnesium and silicon content, while zinc and copper are the main alloying elements in 7075. The exact chemical composition of the alloys can be found in Table 1.

Table 1. Chemical composition of EN AW-6082 and -7075.

Chemical elements [wt%]	Si	Fe	Cu	Mn	Mg	Cr	Zn	Ti	Others
EN AW-6082 – T6	0.98	0.46	0.06	0.55	0.94	0.03	0.06	0.02	0.01
EN AW-7075 – T6	0.08	0.12	1.60	0.04	2.70	0.19	5.90	0.05	0.14

Depending on the existing heat treatment condition or temperature, the sheets have very different mechanical properties. To quantify these in more detail, tensile tests are carried out with tensile specimens of shape H longitudinally to the rolling direction at a tensile speed of 0.2 mm/s. Table 2 shows that ultimate tensile strength of 316 MPa and 588 MPa can be achieved in the high-strength T6 state, for the 6082 and 7075, respectively. However, the elongation at break shows only low values of around 12 % each. Preconditioning to the unstable W-Temper- (solution heat treated and quenched) or soft annealed O-state can significantly reduce the strength while increasing the elongation at break, thus extending the formability.

Table 2. Mechanical properties of EN AW-6082 and -7075 in different heat treatment conditions.

Alloy – condition	R_m [MPa]		$R_{p0.2}$ [MPa]		A [%]	
EN AW-6082 – T6	316.1		289.1		12.0	
EN AW-6082 – W	183.7	*-42 %*	59.5	*-79 %*	24.5	*+103 %*
EN AW-6082 – O	115.9	*-63 %*	38.8	*-87 %*	32.4	*+169 %*
EN AW-7075 – T6	587.6		531.3		11.8	
EN AW-7075 – W	374.3	*-36 %*	164.9	*-69 %*	21.8	*+85 %*
EN AW-7075 – O	218.4	*-63 %*	104.8	*-80 %*	15.0	*+27 %*

To set these conditions, defined heat treatment routes are necessary. In the W-Temper, a short solution annealing of 3 minutes is carried out at alloy-dependent temperatures of 530 °C (6082) or 480 °C (7075). Afterwards, the blanks are rapidly quenched in a 15 % polymer solution to maintain the supersaturation state. The polymer quenchant Serviscol 98-AL from Burgdorf leads to a reduction in distortion [17]. To produce the soft-annealed O-state, rapid heating to 410 °C is carried out in accordance with the recommendation of DIN 29 850, including a 2-hour holding time. Subsequently, the sheets are cooled down to 260 °C (6082) or 230 °C (7075) at a controlled rate of 30 °C/h and held for further 2 hours before cooling in air. [18]

2.2. *Experimental setup*

The forming tests are carried out on the specially developed four-stage forming tool which can be seen in Figure 2 (a). Special features are the heatable active parts (die, blankholder and punch) which allow precise temperature-control to ensure the necessary forming temperature or maintain temperatures in the individual stages. The heated active parts made of Uddeholm's hot-work tool steel Unimax® are realized by means of inserted heating cartridges or tubular heaters as well as thermocouples type K for the extraction of the controlled variable. To reduce the heat flow into the adjacent tool components, isolation plates made of AGK's K-Therm® AS 600M and a water cooling are integrated. To ensure high reproducibility with regard to the cycle time and the accuracy of the component alignment, an electro-pneumatic transfer system (not shown in Figure 2) was also implemented.

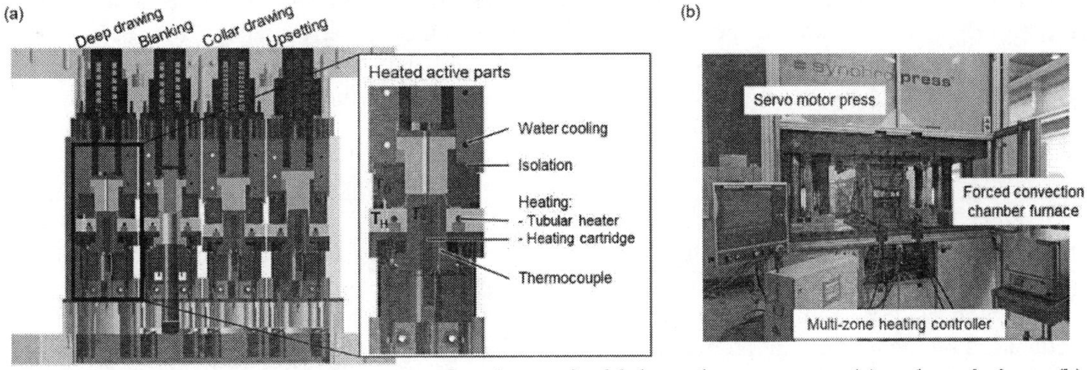

Figure 2. Experimental setup: four-stage forming tool with heated components (a) and used plants (b).

The entire experimental setup with all further equipment is shown in Figure 2 (b). The tool is mounted on the servo motor press Synchropress SWP 2500 with a maximum press force of 2 500 kN. Furthermore, a multi-zone heating controller with a maximum power of 20 kW and 16 individually regulated heating zones is used to control the 12 active parts, as well as the forced convection chamber furnace N 15/65 from Nabertherm for heat treatment or heating the components.

3. Experimental determination of the process limits in the individual tool stages

In order to determine an industrially feasible and economical overall process chain, it is necessary to investigate the process boundaries in the single stages. Accordingly, the four individual stages are examined in more detail below and their process limits and influences are outlined. This also includes factors such as springback, thermal expansion/shrinkage or accuracy.

3.1. Stage 1: Deep drawing

In the first stage, a cup with an internal diameter of 50 mm and a remaining flange is deep drawn from a circular blank with a diameter of 102 mm. The process window shown in Figure 3 demonstrates that EN AW-6082 can be formed in all three initial states, whereas EN AW-7075 necessarily requires preconditioning, because it breaks into several pieces during cold forming. With increasing forming temperatures, the sheet begins to tear in the radius of the bottom before breaking off completely during hot forming. The process limits can be extended by using a distance ring at higher temperatures, but this leads to the formation of wrinkles in the flange, which also remain in the subsequent stages and thus represent rejects.

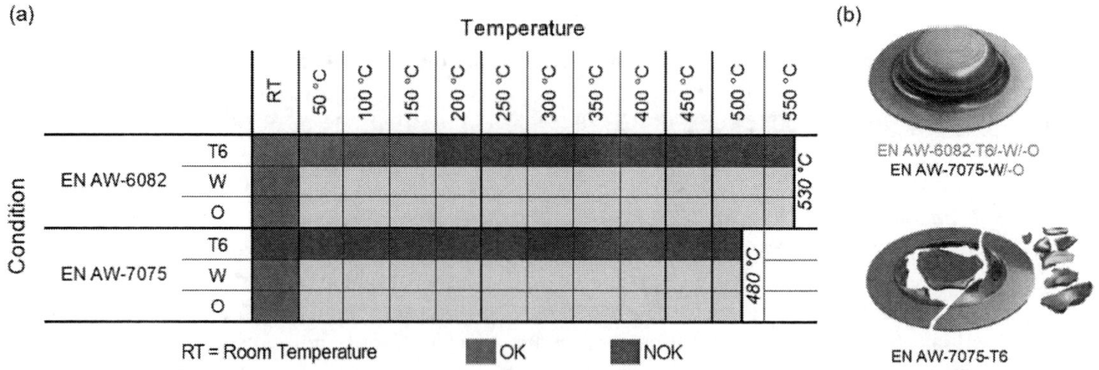

Figure 3. Process window of the 1st stage for EN AW-6082 and -7075 in different heat treatment conditions and temperatures (a) and exemplary presentation of a good and bad part (b).

Forming in the five material-condition combinations (EN AW-6082-T6, -W and -O as well as EN AW-7075-W and -O) shown in green provides good and robust process conditions due to the cold forming. An essential point for the further process route is the wall thickness distribution shown in Figure 4 (b). Therefore, the components were cut through in the respective stages using wire cutting and then measured at significant points using an optical measuring system. In all conditions, thickening occurs in the area of the flange, while the cup wall and, in particular, the radius of the cup bottom become significantly thinner. For the example shown in Figure 4, made of EN AW-6082 in the high-strength T6 condition, the maximum thinning amounts to 10.1 % on the initial sheet thickness of 1.5 mm.

3.2. Stage 2: Blanking

In the second stage, a hole with a diameter of 32 mm is blanked in the bottom of the cup. Blanking proves to be uncritical, as no component failure occurs for any combination, which is why this process step is not considered in more detail below. However, the shape of the cut edge shown in Figure 4 (a) is influenced by the mechanical properties of the material. At low temperatures, strain hardening occurs near the cut surface [19], while at elevated temperatures the microstructure of the materials is influenced by the temperature input.

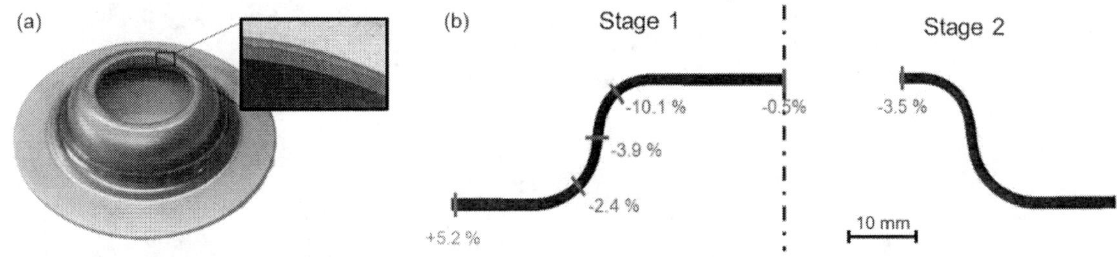

Figure 4. Blanked cup from the 2nd stage incl. enlarged cut edge view (a) and wall thickness profiles of the stages 1 and 2 with the percentage deviation from the initial sheet thickness of 1.5 mm, made of EN AW-6082-T6 (b).

3.3. *Stage 3: Collar drawing*

In the third stage, the remaining bottom is set up by means of collar drawing. In contrast to the first stage, the process window (cf. Figure 5 (a)) clearly shows that only the soft annealed state of EN AW-6082 can be formed successfully at room temperature. All other conditions need forming temperatures above 350 °C. The exact influence of the temperature can be seen in Figure 5 (b). It shows a temperature variation from 200 to 350 °C for the W-Temper state of EN AW-7075. The temperature increase leads to an extended formability and thus also larger expansion ratios. The exact shape of the cut edge and its influences on the later collar geometry under variation of the materials, temperatures and heat treatment conditions are part of current investigations.

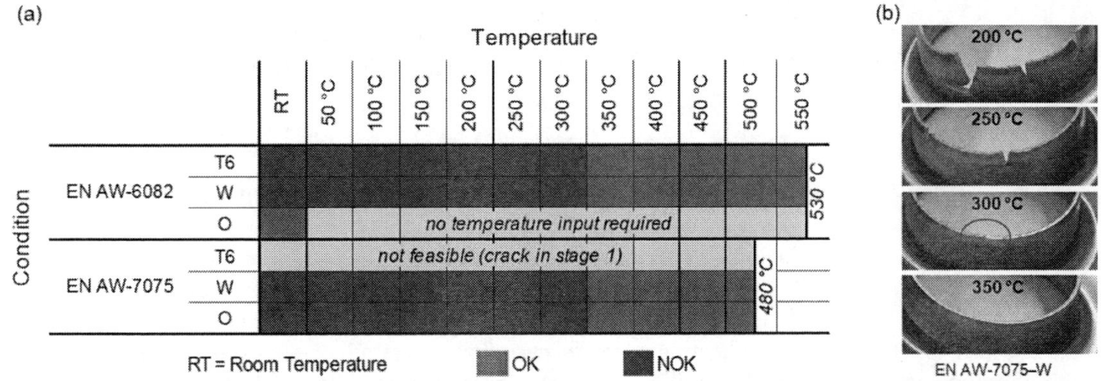

Figure 5. Process window of the 3rd stage for EN AW-6082 and -7075 in different heat treatment conditions and temperatures (a) and temperature influence during collar drawing of EN AW-7075 in the W-Temper condition (b).

During collar drawing and the associated widening of the 32 mm hole to the punch diameter of 50 mm, there is a strong thinning of the collar. The initial sheet consequently becomes up to 31.6 % thinner at the tip, as Figure 6 shows. This leaves a wall thickness of 1.026 mm in the zone with the highest strain hardening through blanking and forming.

3.4. *Stage 4: Upsetting*

And finally, in the fourth stage, an upsetting takes place to increase the initial wall thickness from 1.5 mm to 1.8 mm. The cross-section of the component geometry in Figure 6 shows that the preceding forming stages and the associated strain hardening and thinning effects lead to buckling of the collar. In these notches, the desired sheet thickness is not achieved, but there is a thickening of the previously heavily thinned material so that it is stronger than the initial sheet. In the areas in between, the wall thickness of 1.8 mm is achieved. Consequently, a material thickening of well over 20 % is achieved.

Similar to stage 2, any temperature control can also be implemented for upsetting, but the final geometries differ from each other. A comparison between a cold- and a hot-formed sample at 400 °C shows an improved accuracy in the wall thickness and the springback due to the temperature input, but the overall component shrinks so that the diameter d, shown in Figure 6, falls below the target dimension of 50.0 mm.

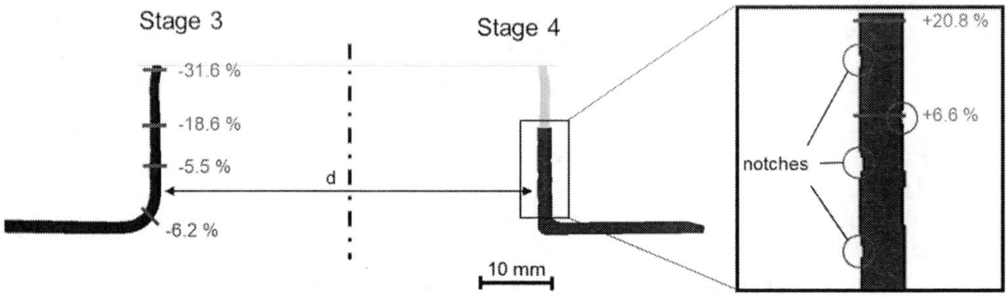

Figure 6. Wall thickness profiles of the stages 3 and 4 with the percentage deviation from the initial sheet thickness of 1.5 mm, made of EN AW-6082-T6 with intermediate heating.

In the example of EN AW-6082-T6 shown above, the diameter d shrinks from 50.09 to 49.80 mm at a measurement height of 7 mm, whereas the wall thickness increases at the same point from 1.77 to 1.79 mm due to the previous temperature input.

4. Possible process routes and industrial implementation

Based on the investigations of the individual forming stages, it becomes apparent that apart from the alloy EN AW-6082 in the soft annealed O-state, no overall process chain can be realized without intermediate heating. For this reason, a new temperature control to realize a coherent overall process chain is developed before its industrial feasibility is demonstrated.

4.1. Possible process routes

The desired demonstrator geometry made of EN AW-6082 can be produced by cold forming in the soft annealed O-state. This has the advantage of not requiring temperature-controlled tools, but subsequent heat treatment to achieve the high-strength material properties.

For the highest-strength alloy EN AW-7075, preliminary investigations indicate that, regardless of preconditioning, heating of the component before stage 3 is indispensable. There are two basic ways to realize this. The first is external heating in the furnace, but this requires extensive peripherals to handle the components. The second possibility is internal heating in the second stage of the temperature-controlled forming tool. This allows the temperature control shown in Figure 7 (a) to be realized.

Due to the heating during the punching process, inline heating within seconds is possible in the case of a two-sided contact under pressure [20]. The influence of holding time and tool temperature on the sheet temperature is shown in Figure 7 (b). The graph illustrates that at a tool temperature of 400 °C, a heating up to 350 °C is possible in about 3 seconds [9]. To determine the heating behaviour, the blanks are prepared with longitudinal notches of 1 mm deep, into which thermocouples type K are inserted. The prepared sheet is then placed on the blank holder and the press is moved so that it is heated under the pressure of the blank holder springs.

However, the parallel stroke of all tool stages must be taken into account. A dwell time at bottom dead centre will inevitably also occur within the other stages, so that shrinkage can occur with hot components and cooler tools. This could be avoided or optimized with a specially developed heating stage.

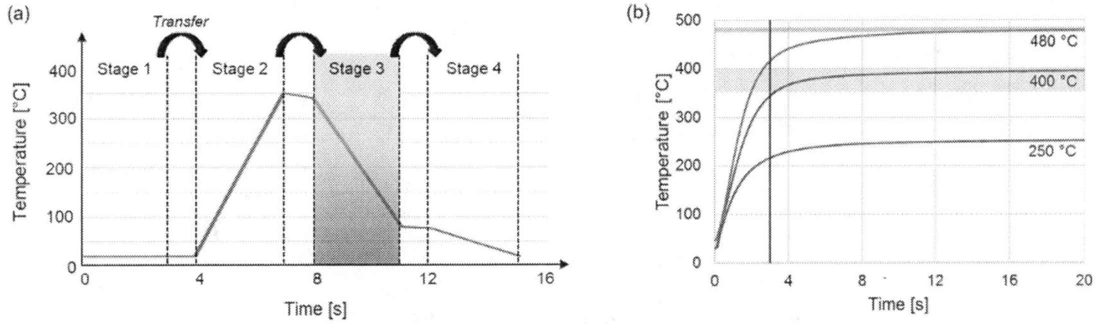

Figure 7. Developed temperature control for multi-stage forming of high-strength aluminium alloys (a) and heating behaviour of the sheet at conductive heating in the forming tool at tool temperatures of 250, 400 and 480 °C (b).

4.2. *Industrial implementation*

For the implementation of the multi-stage process chain in an industrial environment, a precisely designed process sequence is necessary. In addition to robustness, the economic aspects are of course also of high relevance, so that a high degree of automation is required. Only one employee is required for the provision of materials and process monitoring. An industrial process chain for the use case shown above could therefore look as shown in Figure 8.

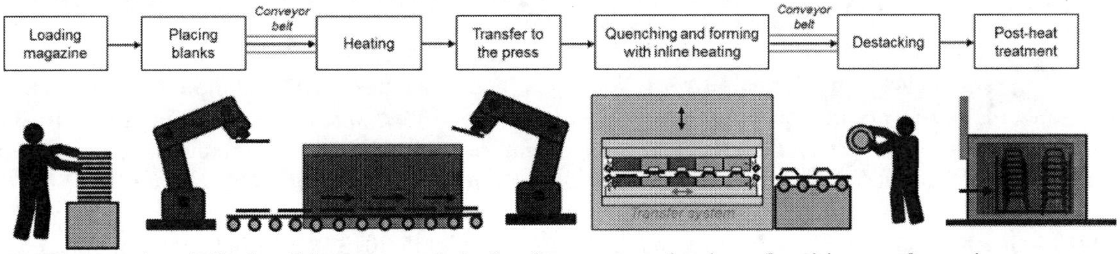

Figure 8. Industrially feasible process chain for the mass production of multi-stage formed components made of high-strength aluminium.

Starting with cut blanks, these are first loaded into a magazine and placed on a conveyer belt by a robot. This passes through a continuous furnace. The tempered sheets can then be removed and placed in a cooled tool within the transfer press. With inline quenching the distortion can be kept low and no drying is required as with external quenching. All subsequent stages are linked by means of the transfer system. The process chain includes a tempered stage for inline heating of the components. After the forming process, a post-heat treatment is required to achieve the high-strength material properties. For both, the soft-annealed alloy EN AW-6082 and the combined process route for the alloy EN AW-7075, the original T6 properties can be reached by means of a subsequent heat treatment (solution heat treated, quenched and artificially aged).

5. Summary and outlook

The paper shows that multi-stage forming operations of the high-strength aluminium alloys EN AW-6082 and -7075 are feasible, although not with the conventional process routes of warm or hot forming, which already fail in stage 1. Due to the different processes and their respective requirements, a precise temperature design and inline control is needed. But the temperature input affects the process and the final geometry of the component in many ways.

Related challenges that continue to be worked on are the tribology under continuous temperature load, the economic and ecological subsequent heat treatment to achieve the high-strength properties while maintaining dimensional accuracy and minimized cycle time through inline heating.

6. Acknowledgement
The project 'Productivity assurance of temperature-supported aluminium forming processes through dual research and training' (HA project no. 1269/21-170) is financed with funds of LOEWE – Landes-Offensive zur Entwicklung Wissenschaftlich-ökonomischer Exzellenz, Förderlinie 3: KMU-Verbundvorhaben (State Offensive for the Development of Scientific and Economic Excellence).

References
[1] Deutsches Institut für Normung e.V. 2016 DIN EN 485-2 - Aluminium und Aluminiumlegierungen Bänder, Bleche und Platten, Teil 2: Mechanische Eigenschaften
[2] Hirsch J 2011 Aluminium in Innovative Light-Weight Car Design *Materials Transactions* Vol. **52** No. 5 pp 818-824
[3] Liu Y, Zhu Z, Wang Z, Zhu B, Wang Y and Zhang Y 2017 Formability and lubrication of a B-pillar in hot stamping with 6061 and 7075 aluminum alloy sheets *Procedia Engineering* **207** (2017) pp 723-728
[4] Cai Z, Batthyany P, Dhawan S, Zhang Q, Sun Y, Luan X, Wang L and Gharbi M M 2019 Study of Springback for High Strength Aluminium Alloys Under Hot Stamping *ICHSU 2018*
[5] Rigas N, Schmid H and Merklein M 2021 Comparison of different forming methods on deep drawing and springback behavior of high-strength aluminum alloys *Materials Science and Engineering* **1157** (2021) 012048
[6] Grohmann T 2016 Forming of AMAG 7xxx Series Aluminium Sheet Alloys *New Developments in Sheet Metal Forming*
[7] Behrens B-A, Nürnberger F, Bonk C, Hübner S, Behrens S and Vogt H 2017 Influences on the formability and mechanical properties of 7000-aluminum alloys in hot and warm forming *Journal of Physics: Conf. Series* **896** 012004
[8] Choi Y, Lee J, Panicker S S, Jin H-K, Kumar Panda S and Lee M-G 2020 Mechanical properties, springback, and formability of W-temper and peak aged 7075 aluminium alloy sheets: Experiments and modeling *International Journal of Mechanical Sciences* **170** (2020) 105344
[9] Günzel J, Hauß J and Groche P 2021 Temperature-controlled tools for multi-stage sheet metal forming of high-strength aluminium alloys *Materials Science and Engineering* **1157** (2021) 012086
[10] Oberhauser P, Sotirov N, Grohmann T and Schulz P 2013 Performance of high strength AlZnMg(Cu) aluminium alloys after W-temper and warm forming *Tools and Technologies for Processing Ultra High Strength Materials* 2013
[11] Chen W, Liu Z J, Hou B and Du R X 2007 Study on multi-stage sheet metal forming for automobile structure-pieces *Journal of Materials Processing Technology* **187-188** (2007) pp 113-117
[12] Lai C P, Chan L C, Chow C L and Yu K M 2010 Warm-Forming of Light-Weight Alloys under Multi-stage Forming Process *Advanced Materials Research* Vols. **83-86** pp 88-93
[13] Takalkar A S and Chinnapandi L B M 2020 Multi-stage deep drawing process of axis-symmetric extra deep drawing steel cylindrical cup *Engineering Research Express* **2** (2020) 025008
[14] Pourkamali Anaraki A, Shahabizadeh M and Babaee B 2012 Finite Element Simulation of Multi-Stage Deep Drawing Processes & Comparison with Experimental Results *World Academy of Science, Engineering and Technology* **61** 2012
[15] Behrens B-A, Hübner S, Vogt H, Golovko O, Behrens S and Nürnberger F 2020 Mechanical properties and formability of EN AW-7075 in cold forming processes *Materials Science and Engineering* **967** (2020) 012017
[16] Günzel J, Bergmann J, Hauß J and Groche P 2021 Verfestigung vorkonditionierter Aluminiumbleche *WT Werkstattstechnik* Bd. **111**-10 pp 677-683
[17] Günzel J, Suckow T, Veitenheimer C, Hauß J and Groche P 2020 Robuste W-Temper-Umformung von hochfestem Aluminium *WT Werkstattstechnik* Bd. **110**-10 pp 697-703
[18] Deutsches Institut für Normung e.V. 1989 DIN 29850 – Wärmebehandlung von Aluminium-

Knetlegierungen
[19] Ostermann F 2014 Anwendungstechnologie Aluminium Berlin Heidelberg Springer Vieweg 3., neu bearbeitete Auflage
[20] Schell L and Groche P 2021 Challenges in Tribometry for Warm and Hot Sheet Metal Forming of High Strength Aluminum with Tool Lubrication, 9th ICTMP – International Conference on Tribology in Manufacturing Processes & Joining by Plastic Deformation, India (online)

IOP Conf. Series: Materials Science and Engineering 1238 (2022) 012015 doi:10.1088/1757-899X/1238/1/012015

Warping and springback reduction in bending of U-profiles through partial heating over the cross-section

E. Hoffmann[1], F. Kolpak[1] and A. E. Tekkaya[1]

[1] Institue of Forming Technology and Lightweight Components, TU Dortmund University, Dortmund, Germany

E-Mail: Eike.hoffmann@iul.tu-dortmund.de

Abstract. Bending of profiles is challenging due to the high stiffness and possible deformations in the process. Especially bending of profiles with asymmetric cross-section regarding the force initiation axis leads to unwanted warping of the bent profile. This warping results from the position difference between the force initiation axis and the shear center, implying torsion moments on the profile. To prevent profile warping the use of shape-bound tools or a change of the force initiation axis position are common practices, though these methods reduce the flexibility of the process. A new method to prevent profile warping during bending is the use of partial, cross-sectional heating. Due to thermal softening in one profile area a quasi-symmetric bending case is achieved, which changes the position of the stress-free fiber and thus reduces the torsion moment. In this work, the warping and springback behavior of partially heated U-profiles consisting of S500MC steel after a three-roll push bending process is investigated using experimental methods. Through partial heating of one profile area to up to 600 °C, a warping reduction of 83 % and a springback reduction of 69 % were achieved.

1. Introduction

The production of lightweight components is focused for automotive body construction due to the increasing requirements regarding CO_2 emissions. Additionally, the range of produced parts is increasing as more model variety is desired. In that sense, the use of kinematic bending processes for profiles is favorable as these are flexible and offer a broad product spectrum. Compared to form-bound bending processes, however, the geometric product accuracy for kinematic processes is reduced.

Especially the bending of asymmetric profiles regarding the force initiation axis is a problematic case due to possible warping. When applying a force central on the profile area torsion moments are initiated in the profile which lead to warping of the cross-section (Figure 1 (a). Different methods exist to eliminate profile warping. Welded blanks can be used to move the force initiation axis in direction of the shear center (Figure 1 (b) [1]. Through the position of the force a counter torsion moment to the geometrical torsion moment is initialized, eliminating the torsion moments in the profile. Another method is to use support elements to constrain the forming zone (Figure 1 (c) [2]. Additionally, special processes use counteracting torsional forces to reduce this effect, like the Torque Superposed Spatial-bending (TSS) process [3].While such methods are able to successfully negate profile warping either additional tool elements or the change of the profile geometry are necessary. Thus the flexibility of the processes is reduced. To maximize the flexibility of these bending processes while maintaining a high geometrical accuracy a new method is required.

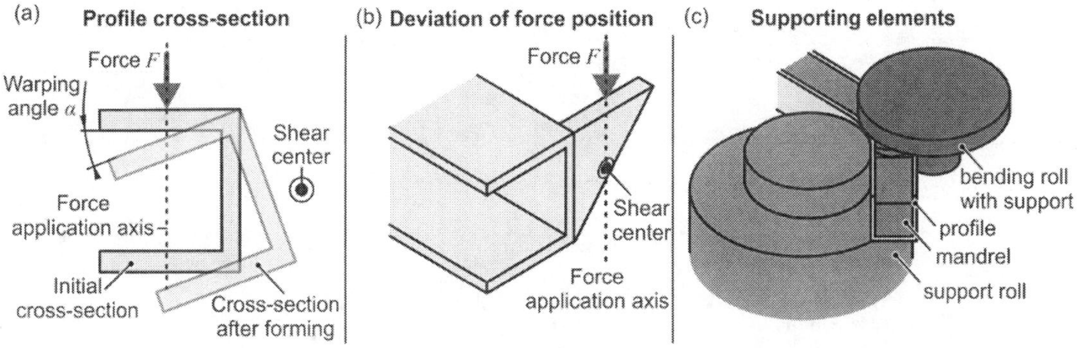

Figure 1. (a) Profile warping due to asymmetry, (b) warping prevention through force positioning, (c) warping prevention through support elements.

The state of the art describes several methods to prevent unfavourable deformation in metal forming through partial heating. Flame straightening for example is a process that is used in ship body production [4]. After welding, large metal sheets are heated to reduce the thermal deformation implied by the bending process. The process of laser bending evolved from this method. In this process, the thermal energy of a laser is used to bend metal sheets and profiles without springback [5]. In the case of springback, Yanagimoto et al. [6] prevented springback in the v-bending of high-strength steel sheets through the use of high temperatures. The prevention of springback is explainable through the reduction of the bending moment due to thermal softening of the steel sheets.

Based on these ideas, this work aims to investigate the suitability of partial heating to reduce springback and warping in asymmetric bending of U-profiles through three-roll-push bending. The use of partial heating for the bending of L-profiles has been proven to reduce warping up to 76 % and springback up to 44 % compared to the room temperature case for the material S500MC and partial heating temperatures of up to 600 °C [7]. The resulting warping and springback data for bending with partial heating of U-profiles is compared with the data for L-profiles.

2. Bending of U-profiles with partially heated cross-section

2.1. Process principle

Warping in bending of profiles will occur if the position of the force application axis differs from the position of the shear center (Figure 2 (a) [8]. In this case, shear stresses are initiated in the profile which imply a torsion moment on the profile cross-section due to its geometry. This torsion moment implies a distortion on the profile by an angle of α (Figure 2 (b).

Partial heating leads to thermal softening in parts of the profile. As the thermal softening reduces the flow stress necessary to bend the profile to the same radius as in the room temperature case, the stresses, especially the shear stresses, in the cross-section are reduced. As thermal softening is not uniform in the cross-section, the position of the shear center changes. In the depicted heating case, the shear center shifts in direction of the force initiation axis, effectively reducing the torsion lever arm length for the bending process. Consequently, the torsion moment, as well as the warping angle, is decreased. This effect can be used to reduce warping in profile bending processes.

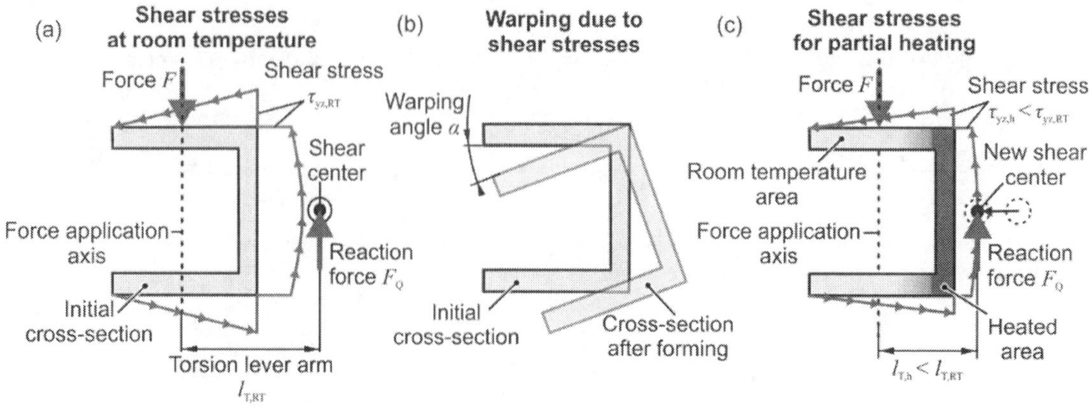

Figure 2. (a) Shear stresses due to force application, (b) resulting warping in the profile cross-section, (c) reduced shear stresses through partial, cross-sectional heating

In the case of springback, the partial heating of the profile will reduce the flow stress locally through thermal softening. As the softening reduces the global stiffness of the profile a smaller bending moment is necessary to bend the desired bending radius, which leads to reduced springback [6].

2.2. Experimental procedure

To analyze the effect of partial, cross-sectional heating on profile springback and warping during kinematic bending, a three-roll push-bending process realized on a rotatory draw bending machine DB 2060-CNC-SE-F (transfluid, Schmallenberg, Germany) is used. The process is divided into three phases: Prebending, kinematic push-bending, and unloading. In prebending, the inner bending radius of the profile is set to 600 mm by rotation of the bending roll (Figure 3 (a). In the kinematic push-bending phase the profile is fed with a constant feed velocity v_f of 8 mm/s. In this phase, the profile is partially heated (as in Figure 2 (c) at 70 mm after the counter roll to temperatures between 200 °C and 600 °C in 100 °C steps, if the profile is not at room temperature. This temperature range is chosen to have minimal influence on the microstructure. The heating is achieved through induction by a Trumpf TruHeat 7040 induction generator with a power of 40 kW. At the position of the induction coil, the temperature is measured by a Sensortherm pyrometer M318. After passing the length of the heating zone of 95 mm the profile is cooled by a water-jet cooling system to localize the heating zone. When the desired bend length of the profile is reached, the profile is unloaded and springback will occur. The geometry after unloading is then digitalized using the 3D scanning system GOM ATOS to evaluate warping and springback.

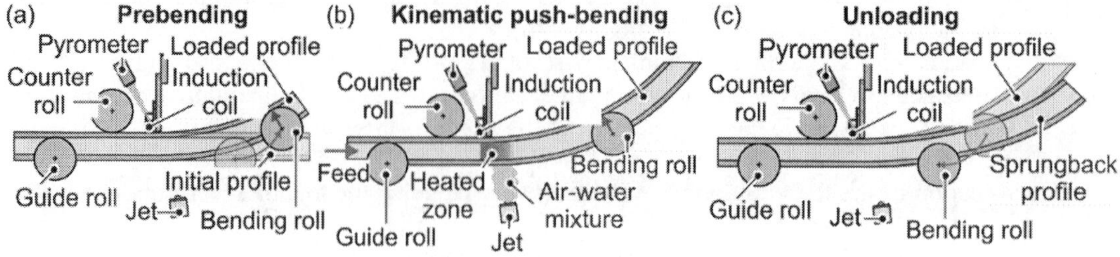

Figure 3. Kinematic profile bending with partial, cross-sectional heating. (a) Prebending phase, (b) kinematic push-bending phase, (c) profile unloading

The heating zone and the cooling zone have been divided by a plate consisting of the material Dotherm, which does not influence the inductive heating but prevents the water spray to reach the heating zone (Figure 4 (a). The tooling in the forming zone consists of the guide, counter, and bending roll, the pyrometer, the induction coil, the cooling jet, and the divider sheet (Figure 4 (b)

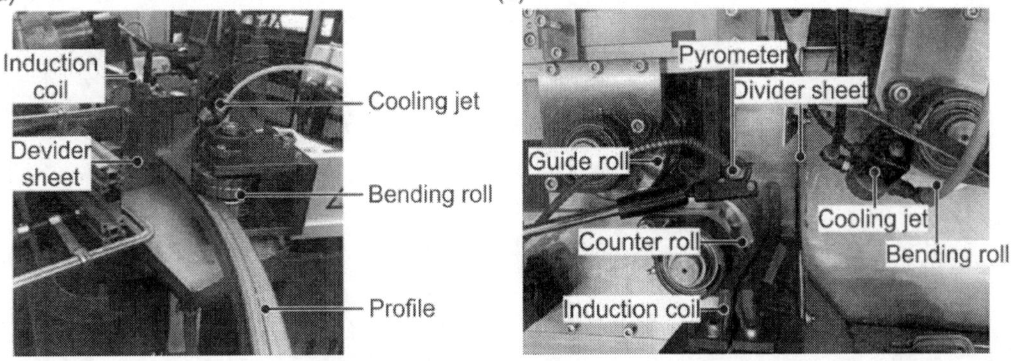

Figure 4. Process set up. (a) process frontal direction, (b) close up of the tools in the forming zone

2.3. Material

The investigated U-profiles are 2000 mm long with a length of the outer edges of 40 mm with a thickness of 1.5 mm and 2.5 mm. These profiles consist of S500MC steel. Delivery conditions of the material are according to EN 10140-2 (Table 1)

Table 1. Delivery conditions of analyzed S500MC steel.

Yield strength $R_{p0,2}$ in MPa	Tensile strength R_m in MPa	Chemical composition in wt%								
		C	Si	Mn	P	S	Al	Nb	Ti	V
585	642	0.045	0.02	0.812	0.013	0.007	0.032	0.013	0.001	0.137

Additionally, temperature-dependent Young's moduli (Figure 5 (a) and temperature and strain rate dependent flow curves (Figure 5 (b) were obtained through isothermal tensile tests on 2.5 mm thick sheet specimens. The tests were carried out on a Zwick Z250 tensile testing machine and the sheets were heated through induction. The investigated temperatures range from 25 °C to 600 °C in 100 °C steps and the analyzed strain rates are 0.0003 1/s, 0.003 1/s, 0.03 1/s and 0.1 1/s.

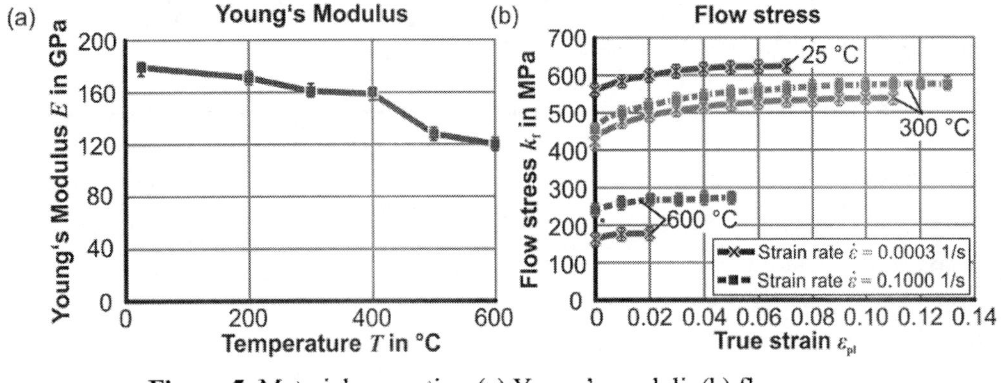

Figure 5. Material properties. (a) Young's moduli, (b) flow curves

3. Results and discussion

3.1. Springback and warping of U-profiles

In Figure 6 the ratio of unloaded inner bending radius r_{iR} to loaded inner bending radius r_i is displayed as a measure of springback dependent on the partial heating temperature for 2.5 mm and 1.5 mm thick specimens. Springback stays constant till 200 °C heating temperature for 1.5 mm thickness and till 400 °C for 2.5 mm thick profiles. At higher temperatures, springback decreases approximately linearly with the lowest obtained springback at 600 °C, with a higher decrease for profiles of lower thickness. The maximum springback reduction amounts to 69 % for 1.5 mm thick profiles and 48 % for 2.5 mm thick profiles.

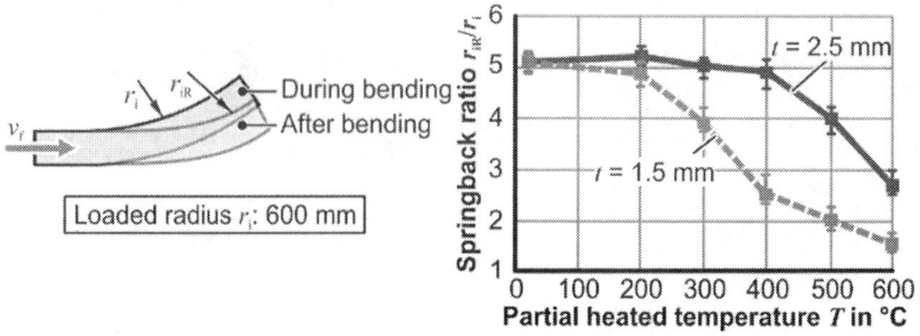

Figure 6. Profile springback ratio for 1.5 mm and 2.5 mm thickness

The reduction of springback occurs due to the thermal softening of the material (Figure 5 (b). As the flow stress in the heated area is reduced, the necessary bending moment to produce the desired bending radius is decreased. Lower bending moments generally lead to a reduced springback.

The springback reduction only occurs if a threshold temperature is reached. The reason for this is the positioning of the heated zone. The maximum bending moment occurs at the x-position of the counter-roll. When the profile is at room temperature, bending would start at this position. By partially heating the profile the position of the bending zone is shifted to the heated area. If the profile strength is decreased enough by thermal softening the profile will bend in the heated zone, while for lower temperatures the profile will still bend at the position of the counter-roll. As the bending zone is still at the position of the counter-roll, at which the profile is at room temperature, for lower partial heating temperatures the bending moment will not decrease. This concludes that springback is only reduced if the forming zone for the process is located in the heated region. For profiles with lower thickness the stiffness is lower. This means less heating suffices to change the position of the heated zone for profiles of lower thickness.

The profile warping is evaluated dependent on the profile arc length starting at the position of the counter roll (Figure 7). Warping develops approximately linearly over the arc length. For both profile variants warping decreases with increasing partial heating temperature. For the lower thickness (Figure 7 (a), maximum warping at room temperature is 60 % less for the 2.5 mm profiles than for the thinner counterparts. Compared to room temperature, the maximum warping reduction is 68 % for 300 °C and 83 % for 600 °C partial heating temperature for the 1.5 mm thick profiles. The warping reduction in the 2.5 mm thick case (Figure 7 (b) is 40 % for 300 °C and 69 % for 600 °C partial heating temperature.

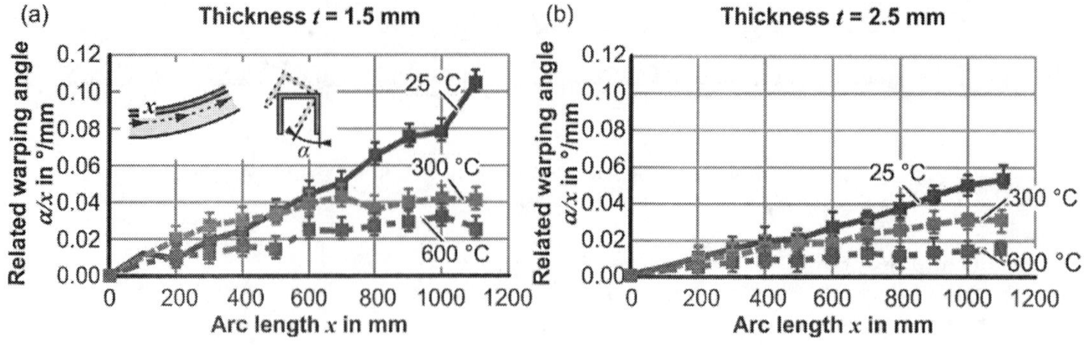

Figure 7. Related profile warping angle for 1.5 mm (a) and 2.5 mm (b) thickness

As the stiffness of the profile with higher thickness is larger the initial warping is lower than for profiles of lower thickness. In fact, the torsion stiffness for 2.5 mm thick profiles is 78 % higher than for 1.5 mm thick profiles (room temperature). As the flow stress decreases in parts of the profiles, the stiffness of the whole profiles is decreased as well. The partial heating temperatures are the same for both profile types and geometrically only thickness changes. This concludes that the warping for higher partial heating temperatures and higher profile thicknesses is lower.

Warping, just as springback, depends on the bending moment of the profile. Still, the decrease in warping is higher than the decrease in profile springback for the same geometry and partial heating strategy. This means, that the warping reduction is not fully achieved through a global stiffness reduction of the profile, but a combined effect of the stiffness reduction and the shift of the shear center position (see section 2.1) which has been proven for L-profiles [7] but needs to be analysed more thoroughly for the U-profiles.

3.2. Comparison of warping and springback between U- and L-profiles

To confirm the transferability of the results in this work with other profiles the results for springback and warping are compared to the behaviour of L-profiles with the same heating strategy and otherwise same geometry [7] (Figure 8). Both L- and U-profiles consist of S500MC material with 2.5 mm thickness and a width of 40 mm.

As for the 2.5 mm and 1.5 mm U-profiles, the springback for the L-profiles is constant until a threshold temperature is reached (Figure 8 (a). This threshold temperature for the L-profiles is 300 °C and is in between the values for the U-profiles (200 °C for 1.5 mm thick U-profiles and 400 °C for the 2.5 mm thick variant). After this temperature the springback also decreases linearly for L-profiles. The springback for the L-profiles a room temperature is 24 % lower than for the U-profiles. At 600 °C partial heating temperature this difference is 33 %.

Profile warping for L-profiles behaves different than for U-profiles (Figure 8 (b). While initially the warping for L-profiles is increasing linearly like the U-profile data, the warping then remains constant after a threshold is reached. This threshold depends on the partial heating temperature. For L-profiles warping at a maximum arc length for the room temperature case is 50 % lower than for U-profiles. At 600 °C the difference amounts to 39 %. The maximum warping reduction for the L-profiles is 76 % while it is 69 % for the U-profiles.

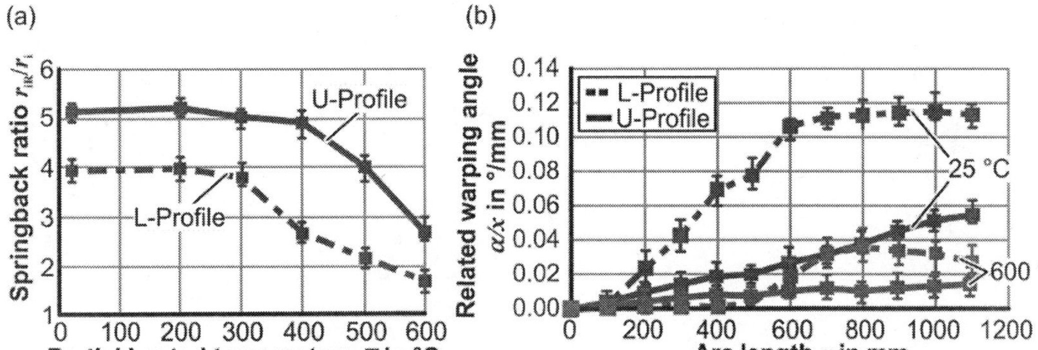

Figure 8. Comparison between results for L-profiles and U-profiles, (a) springback ration, (b) related warping angle

Compared to U-profiles, the bending stiffness is 62 % lower and torsion stiffness 40 % lower for L-profiles at room temperature. Additionally, a larger fraction of the total cross-section is heated for the L-profiles (50 % for L-profiles and 33 % for U-profiles) the stiffness reduction is higher. As the bending stiffness for the L-profiles is lower, less springback occurs. The warping reduction is nearly the same as for the U-profiles. This could mean that for this heating strategy and this combination of profile cross-sections the change of the shear center position is the same, which will be investigated further in the future.

4. Conclusion

It has been proven that partial heating can successfully reduce warping and springback in bending of profiles with an asymmetric cross-section in regards to the force initiation axis. A warping reduction of 83 % and a springback reduction of 69 % have been achieved in the case of bending U-profiles. The springback reduction can be attributed to the global profile stiffness due to the thermal softening of parts of the profile resulting from partial heating. As the bending moment is reduced for partial heated profiles, so is the springback. Consequently, cross-sections of lower stiffness (e. g. L-profiles), lower thicknesses, and thermal softened profiles have lower springback.

Profile warping for U-profiles can be reduced by the use of a partial heating strategy. The reduction of warping can be attributed to the shift of shear center position through the partial heating strategy. The change of shear center position reduces the lever arm for torsion.

The new investigations show that warping reduction for bending of U-profiles through partial, cross-sectional heating is possible. Combining this knowledge with the warping behavior of L-profiles with partial heating [7], it can be expected that warping and springback reduction for other profile types is possible using this heating strategy. As the heating is necessary for the geometric accuracy of the profiles, future work will explore the possibility to use the heat for tailoring of the mechanical profile properties.

References

[1] Seely F B, Putnam W J and Schwalbe W L 1930 *The Torsional Effect of Transverse Bending Loads on Channel Beams* (Bullentin Univ. Illinois 211)
[2] Kreye B 1984 *Verfahren und Vorrichtung zum Biegen von hohlen Metallprofilen, insbesondere Rahmenprofilen für Fenster und Türen* EP0115840
[3] Chatti S, Hermes M, Tekkaya A E and Kleiner M 2010 The new TSS bending process: 3D bending of profiles with arbitrary cross-sections *CIRP Ann.* **59** 315–8
[4] Rothman R L 1975 *Flame Straightening Quenched-and-Tempered Steels in Ship Construction* (US Sh. Struct Com Rep)
[5] Kraus J 1997 *Laserstrahlumformen von Profilen,* Ph.D. Thesis (Friedrich-Alexander-Universität Erlangen-Nürnberg)
[6] Yanagimoto J and Oyamada K 2007 Mechanism of springback-free bending of high-strength steel sheets under warm forming conditions *CIRP Ann. - Manuf. Technol.* **56** 265–8
[7] Hoffmann E, Meya R and Tekkaya A E 2021 Reduction of Warping in Kinematic L-Profile Bending Using Local Heating *Metals.* **11** 1146
[8] Singh D K 2021 *Strength of Materials* (Cham: Springer International Publishing) pp 717–719

IOP Conf. Series: Materials Science and Engineering 1238 (2022) 012016 doi:10.1088/1757-899X/1238/1/012016

Upset bulging as a preforming operation for hot metal gas forming of 22MnB5 tubes

A Alimov[1], R Haase[2] and A Sviridov[1]

[1] Brandenburg University of Technology Cottbus-Senftenberg, Cottbus, Germany

[2] Fraunhofer Institute for Machine Tools and Forming Technology, Chemnitz, Germany

E-mail: alimov@b-tu.de

Abstract. Hot Metal Gas Forming (HMGF) is a coupled process of gas forming and quenching of tubes. This process allows to obtain complex and accurate geometry due to the elimination of springback and high mechanical properties due to the formation of martensite, as a result, the weight of the parts can be reduced. HMGF is widely used in the aerospace and automotive industries to make critical parts from high hardenability steels such as 22MnB5. A typical hot stamped component has 1000 MPa yield stress and 1500 MPa ultimate tensile strength. The main challenge of HMGF process is a significant material thinning and cracking due to the biaxial tension stress state. The paper proposes a preforming method for increasing the forming limits of HMGF by the cold upset bulging of tubes by means of the additional volume of material in the deformation zone. This method allows to obtain one or several waves in the cross section of the tube, which helps to increase the minimum workpiece wall thickness after the forming process by more than 40% and to reduce a probability of the crack formation.

1. Introduction

One of the most important challenges in the modern automotive industry is an improvement of a vehicle efficiency, which is generally defined by the distance traveled per use of a quantified energy [1]. It can be improved by enhancing of the engine efficiency and its hybridization, improving the fuel energy efficiency, decreasing the air resistance due to a more aerodynamic bodywork and a vehicle weight reduction. Reduction of the vehicle weight is one of the important avenue to improve vehicle efficiency and could be achieved both with engineering solutions, such as an architecture and design optimization of a vehicle, and with technological improvements, such as using lightweight materials and innovative manufacturing processes [2]. Substitution of a mild steel or a cast iron with lightweight materials, such as magnesium or aluminum alloys can lead to a vehicle weight loss up to 75%. Using of an advanced high strength steel in critical applications can help to reduce a vehicle weight up to 25%, while increasing the stiffness and safety of the vehicle [3].

Further reduction of a vehicle weight can be reached using the hollow parts and tubes instead of sheet metal welded structures. For example, hydroforming is used today for many structural applications, including frame rails, engine cradles, radiator supports and cross-car beams [4]. The process is well established [5] and usually carried out by the room temperature and widely used for ductile alloys, such as carbon steel and aluminum alloys [6], as well as stainless steel [7]. Hydroforming by the elevated temperature is also possible [8], but the temperature for fluids is still limited to the range below 350 °C [9] and using hot gas or an alternate media is required [10].

High strength steels and other alloys with low cold formability could be hot deformed by a gas forming [11]. This process can be coupled with a simultaneous heat treatment, how it is done in a press hardening (or hot stamping) [12]. The process of tube gas forming coupled with a quenching of a hardenable steel or alloy is known as hot metal gas forming (HMGF) [13]. A tube or a hollow profile made of high hardenability steel, such as 22MnB5 [14], deforms in a fully austenitized state by means of internal gas pressure in order to fill the die cavity [15]. When the workpiece touches the surface of the cooled dies, it is cooled with a cooling rate above the critical one and martensite is formed. Since the phase transformation of a workpiece takes place in a constricted state, there is almost no springback [12], meanwhile the mechanical properties are very high, typically yield stress is up to 1000 MPa and ultimate tensile strength is up to 1500 MPa (even 2000 MPa).

The stress state of a tube during HMGF is a biaxial tension, which significantly limits the potential of the process. The formability could be predicted by flow limit diagrams, which for 22MnB5 were investigated extensively by Li et. al. [16], Rubešová et. al. [17] and Venturato et. al. [18]. However, only relatively small lateral expansion without additional measures is possible. The formability could be enhanced in the following ways:

- an increase of the material formability by providing a favorable microstructure or using rational process parameters;
- an addition of extra amount of metal to the deformation zone;
- a change of a stress state in a workpiece.

The most common methods of the formability enhancement are a heating of a workpiece and an axial feeding of the tube by application of the inner pressure. The success of a tube hydroforming process is highly dependent on the loading paths (axial feed versus pressure) used, which was investigated by Aue-u-Lan et. al. [19] and Jansson et. al. [20]. The combination of these parameters with considering of a material formability and friction condition helps to avoid a crack formation, as well as a buckling or a wrinkling. Yuan et. al. [21] proposed to utilize useful wrinkles and developed an extended process window of tube hydroforming. Lang et. al. [22] expanded this method and proposed a multi-stage tube hydroforming. Nevertheless, axial feeding during a media-based forming has certain restrictions:

- the inner pressure by the axial feeding should be relatively small due to a seal design;
- applicable only for relatively slow processes;
- used only for cold processes (or a limited range of alloys), when the tube stability is high enough and it wrinkles predictably;
- only for a relatively small friction coefficient.

During HMGF complex friction conditions take place and a friction coefficient can rise up to 0.6 combined with a relative low flow stress of a tube material [23], which makes a tube very unstable and makes eventually an axial feeding challenging. In this case it is possible to preform a workpiece in a cold state. The upset bulging process shows a high potential and can be used for this purpose [24]. The figure 1 shows the possible strategies for increasing the volume of metal in the deformation zone during upset bulging.

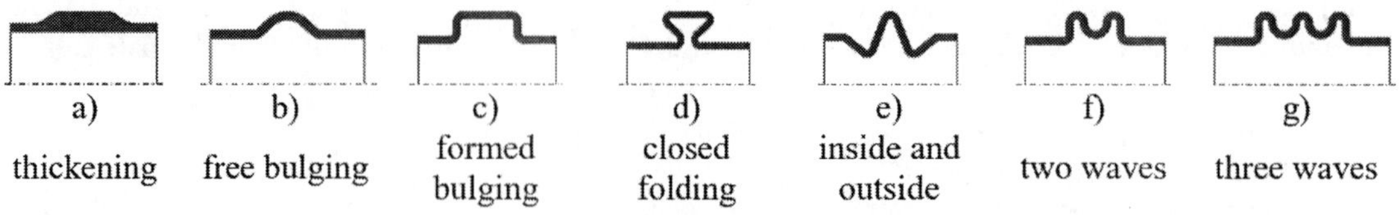

Figure 1. Strategies of the wave formation by the upset bulging.

The aim of the study is to develop and analyze a preforming process by the upset bulging to prevent a tube failure, as well as to increase a tube wall thickness and its uniformity after HMGF.

2. Materials and methods

2.1. Materials

The samples used in the study were made Al-Si coated 22MnB5 welded tubes from ArcelorMittal with 6 m length, outer diameter 40 mm and wall thickness 2 mm. The samples were cut-off with a band-saw and then the ends were turned in order to achieve the desired tube length and perpendicularity of the tube end faces to the tube axis. The chemical composition of as-received tubes is given in table 1.

Table 1. Nominal chemical composition of the 22MnB5 steel (wt.%).

	C	*Si*	Mn	*P*	S	*Al*	N	*Cr*	Ti	*B*
min	0.21	0.15	1.10	-	-	-	-	0.10	0.015	0.0015
max	0.25	0.40	1.35	0.023	0.010	0.080	0.010	0.25	0.045	0.0040

2.2. Experimental procedure

The investigated process chain consists of two steps: preforming of the tube by the upset bulging and subsequent hot metal gas forming. Upset bulging and HMGF experimental setup as well as process parameters are described below. After each step the samples were cut along the axis and the wall thickness distribution was examined.

2.2.1. Measurement of a tube thickness distribution

The samples were cut along the tube axis with a Presi Mecatom T330 cutting machine. A diamond cutting disc with a rotation speed of 3500 RPM and feeding rate of 0.5 mm/sec was used to achieve a flat cut surface. Following the cutting a grinding and polishing operations with a Presi Mecatech 300 SPC were performed. The tube flat cut-sections were scanned with a scanner Canon CanoScan LiDe 1100 with a resolution of 2400 dpi. Then the images were digitized as a .dxf file and a tube thickness distribution was measured by means of Autodesk Inventor. The region with a length of 56 mm along the tube axis was measured (cylindrical part with Ø56 mm, transition zone and short part of cylindrical element with Ø44 mm), which corresponds to the measured profile length of 60 mm. The outer surface of a tube was approximated by a polyline with a segment length of 0.5 mm. The normal distance from this points to inner surface of the deformed tube was measured and taken as a tube thickness.

2.2.2. Upset bulging experimental setup

Upset bulging was perform by the hydraulic press SMG HZPU 100 with a nominal force of 1000 kN and a nominal ram speed of 120 mm/s. The experimental die set is shown in figure 2 and consists of two changeable inserts *a*, mandrel *b* and holders *c*, which are installed on the press ram *d* and the press bed *e*. The tube *e* inserts in the gap between a mandrel *b* and inserts *a* and deforms with a press ram movement. The length of the deforming part of a tube l_k was controlled by a length of a tube and inserts, the final position *h* was set by stopping blocks (are not shown).

Figure 2. Upset bulging experimental setup: in the press (a), the view on the bottom tool (b) and the scheme of the upset bulging process (c).

Special die inserts were used for the upset bulging with a wave support (figure 3). The construction of the inserts is able to prevent a folding of a bulge and to obtain a desired multiwave geometry of the preform.

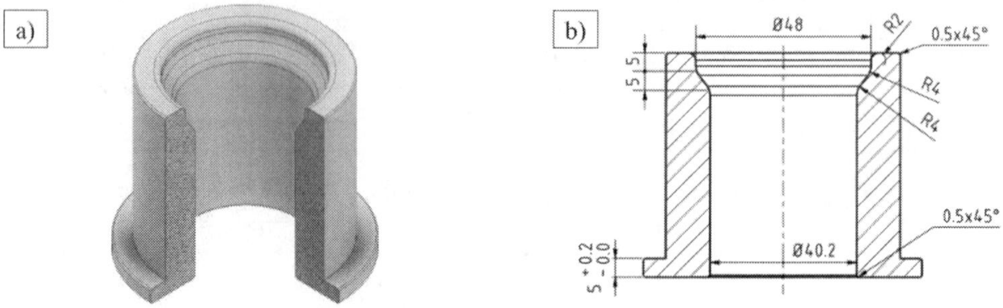

Figure 3. 3D-model (a) and the drawing (b) of the die inserts for the upset bulging with a wave support.

2.2.3. *Hot metal gas forming experimental setup*

Hot metal gas forming was performed by a 3-column hydraulic press hardening machine Dunkes/AP&T HS3-1500 with a 15 MN nominal force and a maximal gas pressure of 700 bar. The geometry of the final part and a bottom tool of a modular tool set for hot metal gas forming are shown in figure 4.

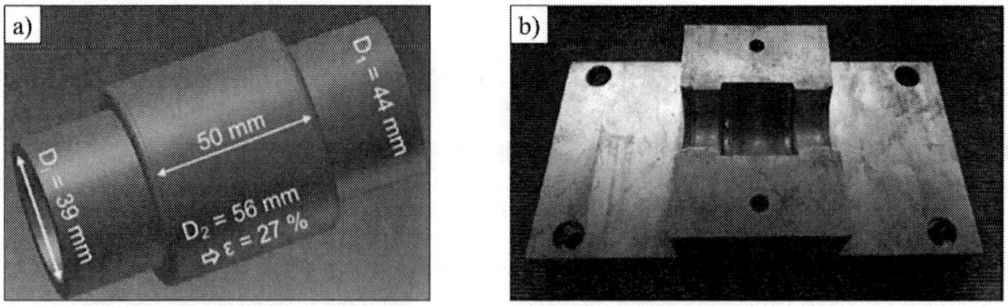

Figure 4. The geometry of the final part (a) and bottom tool for HMGF (b).

Hot metal gas forming experimental setup is shown in figure 5a. A workpiece *a* is heated in a furnace up to 980 °C within 5 minutes. After that the heated tube *a* is placed manually on a bottom die *b* and closed with a top die (not shown). Leak tightness of the workpiece is ensured by closing of side punches *c* with a sealing cone. The handling time before the start of a deformation is about 5 seconds. The gas pressure of 700 bar is supplied through the holes in the side punches in 0.5 seconds with a hold of 3 seconds followed by a release in 2 seconds. It was found that it is not possible to obtain the final part without preforming because of the significant thinning resulting in tearing (figure 5b).

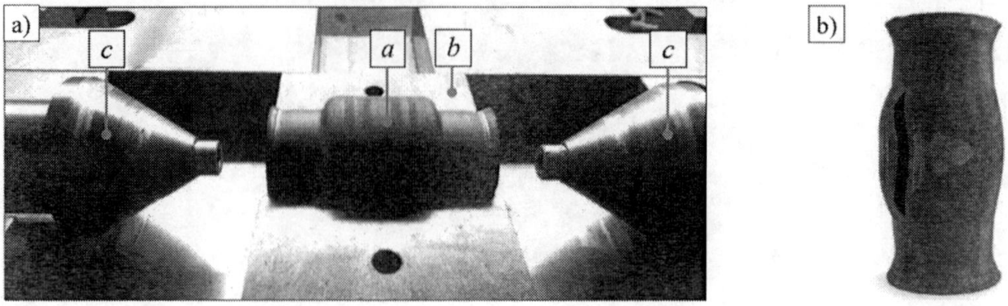

Figure 5. HMGF experimental setup (a) and a teared tube after HMGF without preforming (b).

IOP Conf. Series: Materials Science and Engineering 1238 (2022) 012016 doi:10.1088/1757-899X/1238/1/012016

2.3. *FEM Simulation*
Simulations have been performed by the commercial FEM software QForm (www.qform3d.com) in general forming module.

2.3.1. *Process model*
The workpiece was set as an elastic-plastic body, the tools were set as rigid bodies. The workpiece mesh was created by the QForm mesh generator for parametric geometry. A remeshing was not used. The hexahedral 8-noded volumetric mesh with a mesh density of 4 elements along the radial direction, 75 elements along the tangential direction and 200 elements along the tube axis, totally are approximately 120 000 elements. Thermal processes were taken into account both in cold and hot processes. The initial temperature of the workpiece in HMGF was set as a 980 °C with a subsequent air cooling in 4 s and cooling in dies in 1 s. According to the FEM-simulation the temperature of the tube just before deformation is approximately 850° C and the cooling rate during the HMGF is approximately 150 °C/s, which is significantly higher that critical cooling rate of 30 °C and leads to the formation of fully martensitic structure [25]. Friction was set by a Coulomb law with a $\mu = 0.12$ in the upset bulging and by a Levanov model with $m = 0.8$; $\beta = 1.25$ in the hot metal gas forming. The pressure during HMGF was applied to the inner tube surface of the tube as a time-dependent boundary condition according to the experimental parameters described in 2.2.3.

2.3.2. *Material model*
Flow curves at 20 °C were obtained by the uniaxial tension tests by the authors and approximated as a following function:

$$\sigma_S = 453.393 \cdot \varepsilon^{0.454} + 613.651 \text{ MPa} \quad (1)$$

Thermophysical parameters of the workpiece and the dies were set according the 21Mn4 and 1.2343 (H11) steels from QForm standard database respectively. The environment was set as an "Air" from QForm standard database. Flow curves of 22MnB5 by the hot metal gas forming were set in a tabular form according to Naderi et. al. [26].

3. Results and discussion

3.1. *Analysis of the free upset bulging*
For the investigation of the tube bulging and folding behavior during the free upset bulging a simulative study was performed. Length of the deforming part of a tube, which consists of an initial distance between dies and two radii of a matrix, as well as an initial tube thickness, were used as variable parameters and varied in a range of 20 mm to 60 mm with a step of 5 mm and 2 to 4 mm with a step of 1 mm respectively. Design of experiment of the free upset bulging analysis is presented in the table 3. The final distance between dies was set so that no obvious wrinkles were left with a high probability after the hot metal gas forming.

Table 2. Design of experiment of the free upset bulging analysis.

Variable	Levels								
Length of the deforming part of a tube, mm	20	25	30	35	40	45	50	55	60
Tube wall thickness, mm	2	3	4						

The results of the simulation study are summarized in figure 6. As it can be seen, starting with a certain length of the deforming part of the tube, stability of the tube during the deformation is not enough to form an open wave, and it becomes a closed fold, that cannot be successfully deformed during the hot metal gas forming. For the 2 mm thick tube critical length of the deforming part of the tube is approximately 35-40 mm, for the 3 mm tube – 45-50 mm, and for the 4 mm tube – 50-55 mm.

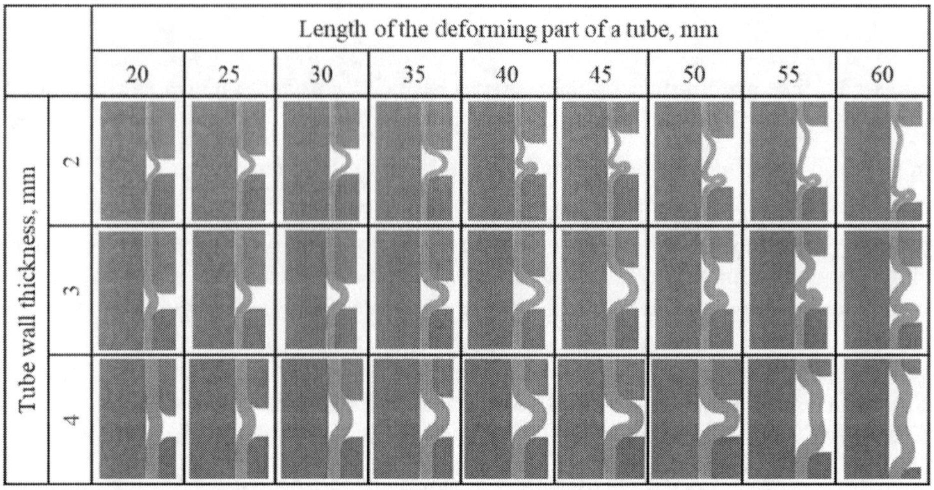

Figure 6. Effect of the process parameters on the wave formation in the free upset bulging.

Folding started with a formation of a second wave, which closes further without significant deformation of the first wave. This mechanism of the tube folding during a free upset bulging is shown in figure 7a. It is possible to avoid a formation of a closed folds by means of a ring-shaped supporting die (figure 7b).

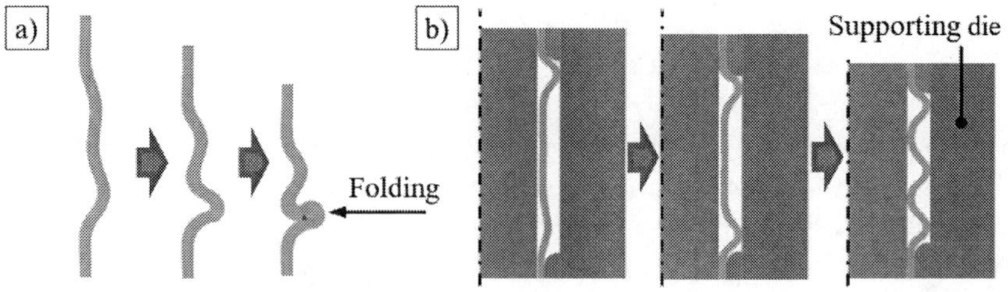

Figure 7. Folding of a tube during the free upset bulging (a) and a principle of the upset bulging with a supporting die (b).

3.2. *Upset bulging with a supporting die*

A supporting die constrains the metal flow outwards and stops a formation of an unstable wave, which helps to deform another wave. A principle of the upset bulging with a supporting die is shown in figure 7b.

A behavior of a tube during an upset bulging with a supporting die was investigated depending on the length of the deforming part of a tube and the diameter of the supporting die in the ranges of 20-80 mm and 42-50 mm respectively. Design of experiment of the upset bulging study with a supporting die is given in table 3.

When the gap between a mandrel and a supporting ring is only 2 mm, by all lengths of the deforming part of the tube in the studied range thickening occurs. Increasing the gap to the length of the deforming part of the tube up to 40 mm also causes thickening, and for larger length of the deforming part of the tube, first two and then three waves are formed. In the range of supporting die diameter of 46-50 mm and length of the deformed part of the tube of 30-40 mm, formation of a formed bulge with an explicit cylindrical area is possible. With an increasing of the diameter of the supporting die a formed bulge is no longer possible because the gap is too large. With a supporting die diameter of 52 mm the formation of three waves is no longer possible due to the loss of tube stability and the formation of a closed fold.

The results of a simulative study of the upset bulging with a supporting die are summarized in figure 8. When the gap between a workpiece and a supporting die is only 2 mm, by all lengths of the deforming part of the tube in the studied range a thickening occurs. Increasing the gap in case of the length of the deforming part of the tube up to 40 mm also causes thickening, and for larger length of the deforming part of the tube, at first two and then three waves are formed. In the range of supporting die diameter of 46-50 mm and length of the deformed part of the tube of 30-40 mm, formation of a formed bulge with an explicit cylindrical area is possible. With an increasing of the diameter of the supporting die a formed bulging is no longer possible because the gap is too large. With a supporting die diameter of 52 mm the formation of three waves is no longer possible due to the loss of tube stability and the formation of a closed fold.

Table 3. Design of experiment of the upset bulging study with a supporting die.

Variable	Levels						
Length of the deforming part of a tube, mm	20	30	40	50	60	70	80
Diameter of the supporting die, mm	42	44	46	48	50	52	-

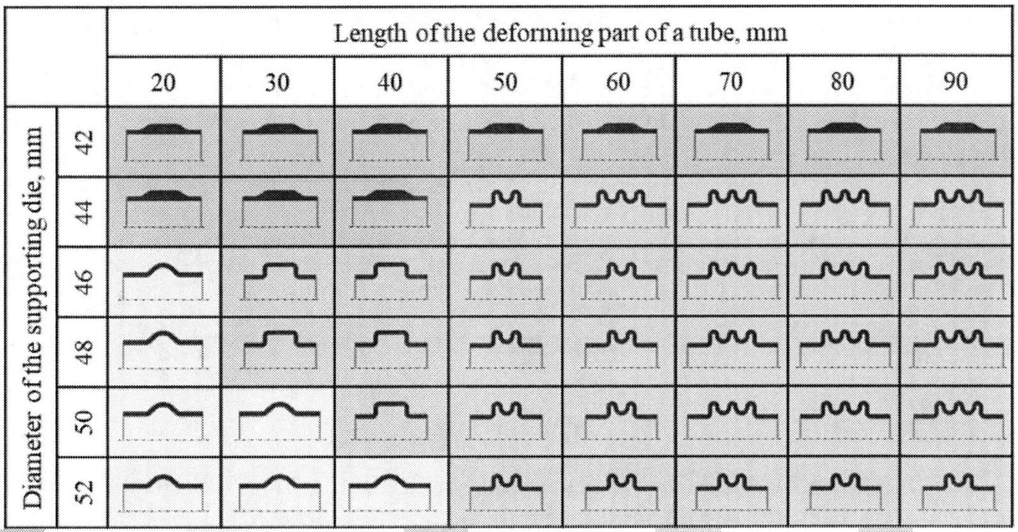

Blue – thickening, yellow – free bulging, purple – formed bulging, orange – two waves, green – tree waves

Figure 8. Effect of the process parameters on the wave formation in the upset bulging with a supporting die.

The formation of one, two or three waves is accompanied by the addition of extra metal volume in increasing order. Thus, for parts with a short-expanded element along the tube axis, one wave can be recommended, and for long parts, preforming with the formation of 3 waves is preferable, as this provides the maximum accumulation of metal in the deformation zone.

3.3. Experimental results of upset bulging

According to the numerical study in 3.2., preformed workpieces with one, two and three waves were produced. The workpiece with one wave was obtained by free upset bulging, while workpieces with two and three waves were obtained by upset bulging with a supporting die with a 48 mm diameter (figure 3). The length of the deforming part of a tube was 32 mm for obtaining of one wave and 68 mm for obtaining of two and three waves. The final distance between dies was 10.8 mm, 32.5 mm and 28.3 in the case of one, two and three waves respectively. Appearance and a cross-section of preformed tubes with one, two and three waves are shown in figure 9. The plastic strain is localized in the wave formation zone and the maximum value is 0.84, 0.50 and 0.55 in the case of one, two and three waves respectively.

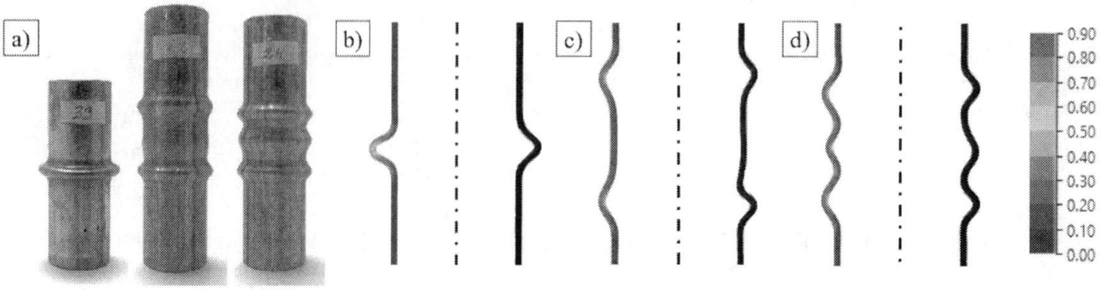

Figure 9. Appearance (a) and a cross-section of preformed tubes with one (b), two (c) and three (d) waves. The plastic strain distribution and a cross-section are shown by the color and in black respectively.

3.4. Hot metal gas forming of preformed tubes

The preformed workpieces shown in figure 9 were deformed by HMGF. Appearance and cross-section after HMGF of preformed tubes with one, two and three waves are shown in figure 10. In the most cases of the hot metal gas forming of workpieces preformed with one wave tearing occurs. Workpieces preformed with two waves in all cases have been cracked. The most stable results with almost no failures were obtained by preforming with 3 waves. In addition, this leads to a maximum accumulation of metal in the deformation zone. It can also be seen that there is a significant thinning of the tube wall in samples preformed with one and two waves after HMGF, while the sample with three waves shows significantly less thinning.

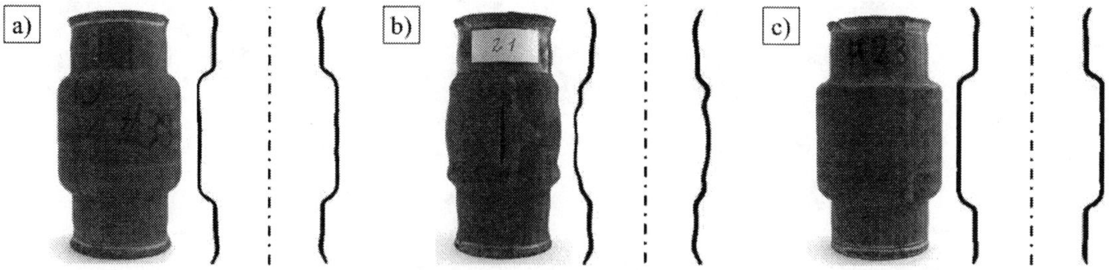

Figure 10. Appearance and cross-section after HMGF of preformed tubes with one (a), two (b) and three (c) waves.

3.5. Comparison of experimental and simulated tube thickness distribution

As mentioned in the section 2.2.3. and can be seen in figure 10, the production of the final part using non-preformed tube as well as preformed tube with 2 waves was unsuccessful. Therefore, for these conditions only the FEM-simulated data are presented, while for the preformed samples with one and three waves the experimental distribution of the wall thickness after HMGF is also presented (figure 11).

HMGF of tubes without preforming leads to a significant thinning up to 1.13 mm. Preforming with one wave significantly reduces a thinning in central part of the workpiece, but thinning up to 1.2 mm in the rounding area still remain. Preforming with two waves could lead to a significant reduction in thinning, but this method was not successful in experimental studies.

The best result was obtained with preforming with three waves, which led to a thinning only up to 1.65 mm, which is approximately 45% thicker than minimal wall thickness of a workpiece without preforming.

Analysis of the results shows, that the simulation is able to predict a final form of the workpiece after the upset bulging with a satisfactory accuracy (figure 9) as well as a tube thickness distribution after HMGF (figure 11).

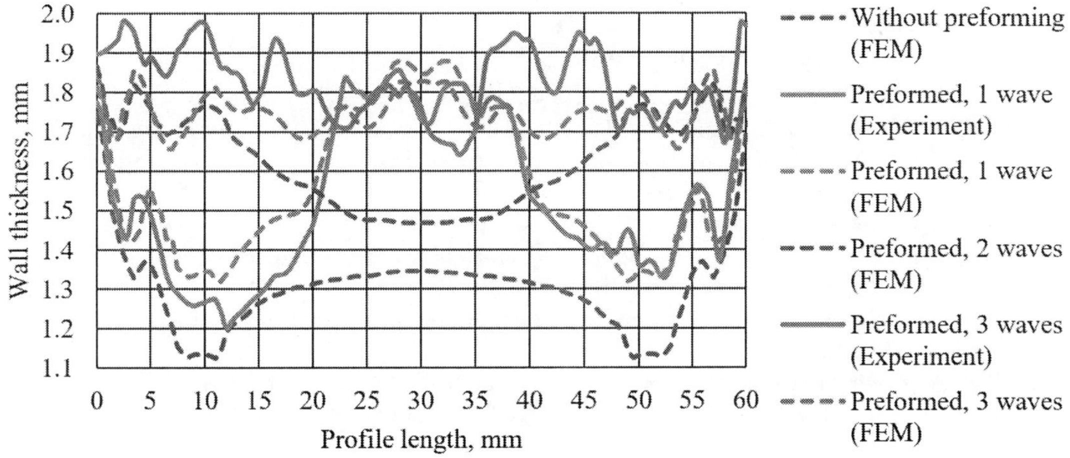

Figure 11. Comparison of experimental and simulated tube thickness distribution after HMGF.

4. Conclusions

The upset bulging process as a preforming operation for hot metal gas forming of 22MnB5 tubes was implemented and investigated.

1. It was shown, that the upset bulging is suitable as a preforming operation for HMGF of 22MnB5 tubes and is an efficient method to prevent a tube failure, as well as to increase a tube wall thickness and its uniformity after HMGF.
2. The best result was obtained with preforming with three waves, which allows to produce a part and led to a thinning only up to 1.65 mm, which is approximately 45% thicker, that minimal wall thickness of a workpiece without preforming, obtained by the simulation.
3. The simulated and experimental geometry of the workpieces after the upset bulging as well as a tube thickness distribution after HMGF are consistent, so the creditability and rationality of simulation are verified.

5. Acknowledgments

The IGF project "Extension of the process limits of internal high-pressure forming – press hardening through material pre-distribution with axial tube upsetting operations" #21359BR of the European Research Association for Sheet Metal Working (EFB) was funded by the AiF as a part of the program for the promotion of industrial community research (IGF) by the Federal Ministry of Economics and Technology based on a resolution of the German Bundestag.

References

[1] Mohanadass A. Making the Most of the Energy We Have: Vehicle Efficiency. In: Quang Dinh T, editor. Intelligent and Efficient Transport Systems - Design, Modelling, Control and Simulation [Internet]. IntechOpen; 2020 [cited 2022 Feb 1]. Available from: https://www.intechopen.com/books/intelligent-and-efficient-transport-systems-design-modelling-control-and-simulation/making-the-most-of-the-energy-we-have-vehicle-efficiency

[2] Taub AI, Krajewski PE, Luo AA, Owens JN. The evolution of technology for materials processing over the last 50 years: The automotive example. JOM. 2007 Feb;**59**(2):48–57.

[3] Luo AA. Recent advances in light metals and manufacturing for automotive applications. CIM J. 2021 Jul 3;**12**(3):79–87.

[4] Taub AI, Luo AA. Advanced lightweight materials and manufacturing processes for automotive applications. MRS Bull. 2015 Dec;**40**(12):1045–54.

[5] Alaswad A, Benyounis KY, Olabi AG. Tube hydroforming process: A reference guide. Mater Des. 2012 Jan;**33**:328–39.

[6] Bell C, Corney J, Zuelli N, Savings D. A state of the art review of hydroforming technology: Its applications, research areas, history, and future in manufacturing. Int J Mater Form. 2020 Sep;**13**(5):789–828.
[7] Yuan SJ, Liu G, Huang XR, Wang XS, Xie WC, Wang ZR. Hydroforming of typical hollow components. J Mater Process Technol. 2004 Sep;**151**(1–3):203–7.
[8] Aginagalde A, Orus A, Esnaola JA, Torca I, Galdos L, García C. Warm Hydroforming of Lightweight Metal Sheets. In: AIP Conference Proceedings [Internet]. Porto (Portugal): AIP; 2007 [cited 2022 Feb 1]. p. 1175–80. Available from: http://aip.scitation.org/doi/abs/10.1063/1.2740969
[9] Grüner M, Gnibl T, Merklein M. Blank Hydroforming Using Granular Material as Medium-investigations on Leakage. Procedia Eng. 2014;**81**:1035–42.
[10] Elsenheimer D, Groche P. Determination of material properties for hot hydroforming. Prod Eng. 2009 Jun;**3**(2):165–74.
[11] Paul A, Reuther F, Neumann S, Albert A, Landgrebe D. Process simulation and experimental validation of Hot Metal Gas Forming with new press hardening steels. J Phys Conf Ser. 2017 Sep;**896**:012051.
[12] Karbasian H, Tekkaya AE. A review on hot stamping. J Mater Process Technol. 2010 Nov;**210**(15):2103–18.
[13] Bach M, Degenkolb L, Reuther F, Psyk V, Demuth R, Werner M. Conductive Heating during Press Hardening by Hot Metal Gas Forming for Curved Complex Part Geometries. Metals. 2020 Aug 17;**10**(8):1104.
[14] Cheng P, Ge Y, Xia Y, Zhou Q. Numerical and Experimental Investigation on Tube Hot Gas Forming Process for UHSS. J Phys Conf Ser. 2018 Jul;**1063**:012172.
[15] Paul A, Strano M. The influence of process variables on the gas forming and press hardening of steel tubes. J Mater Process Technol. 2016 Feb;**228**:160–9.
[16] Li H, Wu X, Li G. Prediction of Forming Limit Diagrams for 22MnB5 in Hot Stamping Process. J Mater Eng Perform. 2013 Aug;**22**(8):2131–40.
[17] Rubešová K, Rund M, Rzepa S, Jirková H, Jeníček Š, Urbánek M, et al. Determining Forming Limit Diagrams Using Sub-Sized Specimen Geometry and Comparing FLD Evaluation Methods. Metals. 2021 Mar 14;**11**(3):484.
[18] Venturato G, Bruschi S, Ghiotti A, Chen X. Numerical modeling of the 22MnB5 formability at high temperature. Procedia Manuf. 2019;**29**:428–34.
[19] Aue-U-Lan Y, Ngaile G, Altan T. Optimizing tube hydroforming using process simulation and experimental verification. J Mater Process Technol. 2004 Feb;**146**(1):137–43.
[20] Jansson M, Nilsson L, Simonsson K. Tube hydroforming of aluminium extrusions using a conical die and extensive feeding. J Mater Process Technol. 2008 Mar;**198**(1–3):14–21.
[21] Yuan S, Wang X, Liu G, Wang ZR. Control and use of wrinkles in tube hydroforming. J Mater Process Technol. 2007 Feb;**182**(1–3):6–11.
[22] Lang L, Li H, Yuan S, Danckert J, Nielsen KB. Investigation into the pre-forming's effect during multi-stages of tube hydroforming of aluminum alloy tube by using useful wrinkles. J Mater Process Technol. 2009 Mar;**209**(5):2553–63.
[23] Merklein M, Lechler J, Stoehr T. Characterization of tribological and thermal properties of metallic coatings for hot stamping boron-manganese steels. In: Proceedings of the Seventh International Conference on Coatings in Manufacturing Engineering. Citeseer; 2008. p. 1–3.
[24] Sizova I, Sviridov A, Bambach M. Avoiding crack nucleation and propagation during upset bulging of tubes. Int J Mater Form. 2017 Jun;**10**(3):443–51.
[25] Stache R. Untersuchungen zur Herstellung von Werkzeugelementen für das Presshärten mittels Laser Beam Melting [PhD]. [Chemnitz]: Technische Universität Chemnitz; 2020.
[26] Naderi M, Durrenberger L, Molinari A, Bleck W. Constitutive relationships for 22MnB5 boron steel deformed isothermally at high temperatures. Mater Sci Eng A. 2008 Apr;**478**(1–2):130–9.

IOP Conf. Series: Materials Science and Engineering 1238 (2022) 012017 doi:10.1088/1757-899X/1238/1/012017

Phenomenological-based constitutive modelling of warm deformation behavior of high-Strength lightweight AL-Li alloy sheets

A Abd El-Aty[1]*, Y Xu[2], S H Zhang[2], X Guo[3], J Tao[3], M G Lee[1]*

[1]Department of Materials Science and Engineering & Research Institute of Advanced Materials (RIAM), Seoul National University, Seoul 08826, Republic of Korea.

[2]Institute of Metal Research, Chinese Academy of Sciences, Shenyang, 110016, P.R. China

[3]College of Material Science and Technology, Nanjing University of Aeronautics and Astronautics, Nanjing 211100, P. R. China.

E-mail: aabdelaty@nuaa.edu.cn
myounglee@snu.ac.kr

Abstract. The flow behavior and formability of Al-Li alloys under warm forming conditions are complicated because they depend on several factors, such as the deformation mode, strain, and strain rates. Therefore, characterizing the mechanical response, and deformation behavior of AA2060-T8 sheets under a wide range of temperatures and strain rates is crucial to develop a new thermo-mechanical processing (TMP) route for their wide industrial applications. Furthermore, determining the activation energy (Q) and predicting the flow behaviour of AA2060-T8 sheets under warm forming temperatures is meaningful for characterizing the mechanical response of AA2060-T8 sheets at warm deformation conditions. Thus, in this study, the Arrhenius constitutive model is developed to investigate the influence of strain rate and temperature on the warm deformation behaviour of AA2060-T8 and determine the activation energy (Q) of AA2060-T8, which is a crucial physical parameter to estimate the difficulties of deforming AA2060-T8 sheets under warm forming conditions.

1. Introduction

Presently, the third generation of Al-Li alloys has demonstrated promise as materials for the components used in military, aircraft, and aerospace applications due to their outstanding physical and mechanical properties [1]. One of the newest candidates in the family of the third generation Al-Li alloys is AA2060-T8 alloy which launched by Alcoa Inc. in 2011 to take the place of AA7075-T6 and AA2024-T3 for fuselage structures and lower, and upper wings [2]. Although AA2060-T8 possesses superior physical and mechanical properties, its board applications are restricted notably at room temperature because of poor formability; anisotropic behavior which caused serious issues during deformation; and springback and wrinkling which are in turn add to the cost of the die and the final components due to the try-out time [3, 4]. Forming at elevated temperatures [5] are considered as significant methods to address the above-mentioned shortcomings of AA2060-T8 sheets and minimize the drawbacks of forming this alloy using traditional cold forming technologies. Hence, it is vitally important to reveal the deformation behavior and the constitutive relations which describe the material flow of AA2060-T8 sheets under warm forming conditions for manufacturing sound thin-walled complex shaped components from AA2060-T8 sheets.

Warm forming technologies are considered as significant techniques to address the drawbacks of cold forming and improve the formability AA2060-T8 sheets [6]. During warm forming, the sheets are subjected to different types of strain under a wide range of temperatures and strain rates. Once the strain reached a threshold value, various types of failures (e.g., necking, wrinkling, fracture, and buckling) may occur. This phenomenon is very interesting and important in sheet metal forming. The flow behavior and formability of Al-Li and Al alloys under warm forming conditions are complicated because they depend on several factors, such as the deformation mode, strain, and strain rates [7-10]. These factors control the strain hardening, which in turn affects the flow behavior and formability of Al-Li and Al alloys [10, 11]. Therefore, characterizing the mechanical response, fracture mechanisms, and deformation behavior of AA2060-T8 sheets under a wide range of temperatures and strain rates is crucial to develop a new thermo-mechanical processing (TMP) route for their wide industrial applications. Furthermore, determining the activation energy (Q), and predicting the flow behavior of AA2060-T8 sheets under warm forming temperatures is meaningful for characterizing the mechanical response of AA2060-T8 sheets at warm deformation conditions. Within this realm, constitutive modelling is usually used to predict the flow behavior of materials in a form that can be used in finite element (FE) codes to simulate the mechanical response of materials under different forming conditions [11].

The motivation of this study is to understand the deformation behavior of AA2060-T8 sheets under warm forming conditions. Thus, this paper starts with investigating the adiabatic heating influence on the stress-strain behaviors of AA2060-T8 sheets. Then, the details of Arrhenius constitutive model to determine the activation energy (Q) of AA2060-T8 is introduced. Q is a crucial physical parameter to estimate the difficulties of plastically deforming the metallic materials.

2. Materials Description and Experimental Procedures

The material used in this investigation is AA2060-T8 alloy, which is a new member of the third-generation Al-Li alloy. AA2060-T8 was produced in 2011 by Alcoa Company for use instead of Al Alloys such as AA7075 and AA2024 in aerospace structures [1]. The microstructure and chemical composition of the as-received sheet of 2060-T8 Al-Li alloy are presented in Fig. 1 and Table.

In this investigation, a Gleeble-3800 material simulator was used to perform the isothermal warm tensile testing at 100℃, 150℃, 200℃, 250℃, and 300℃ and strain rate of 0.001 to 10 s-1strain rates. The details of the test are introduced in Fig. 2. Each experiment was performed three times at least to make sure from repeatability and consistency. The mean values of the three repetitions are considered. Besides, all the experiments contain equal amounts of data and thus are weighted equally. The deformation temperatures were characterized by the welded thermocouples fixed in the middle of the gauge lengths of the tensile specimens and provided signals for accurate feedback, as depicted in Fig. 2. The strain rates, working temperatures, and strains were automatically controlled and reported. Before the warm tensile testing, the specimens heated until the required working temperature.

IOP Conf. Series: Materials Science and Engineering 1238 (2022) 012017 doi:10.1088/1757-899X/1238/1/012017

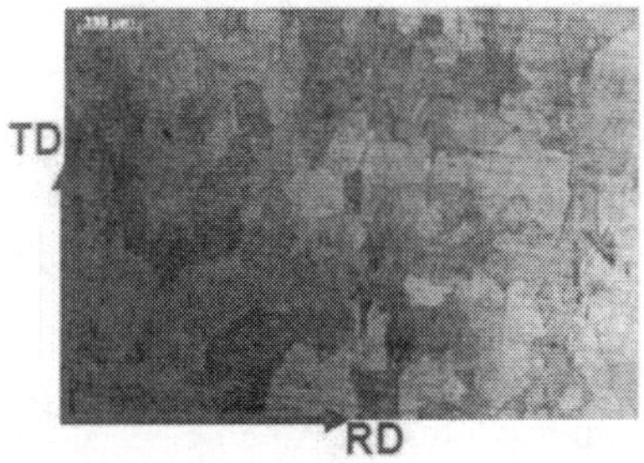

Fig. 1. The microstructure of the as-received AA2060-T8 sheet

Table 1. Uniaxial tensile experiments matrix under warm forming conditions.

Strain rate (s^{-1}) \ Temperature range (℃):	100	150	200	250	300
0.001, 0.01, and 0.1 s^{-1}	√	√	√	√	√

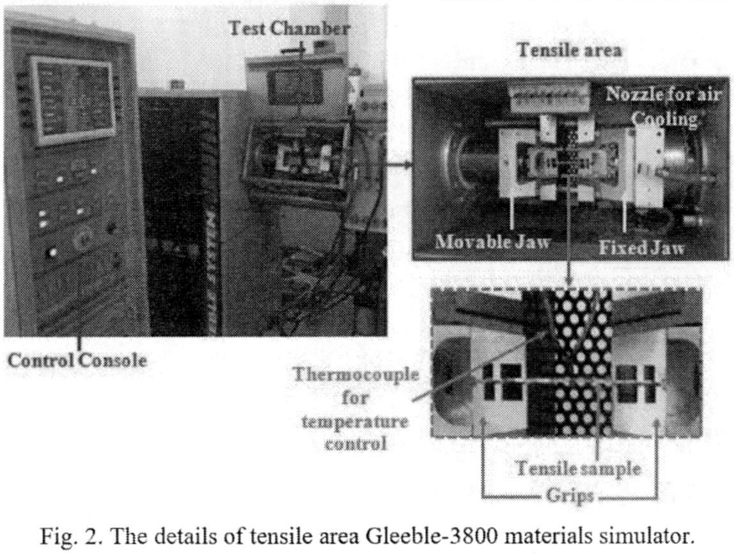

Fig. 2. The details of tensile area Gleeble-3800 materials simulator.

3. Results and Discussions

3.1. Stress-strain behavior

The experimental true stress-strain ($\sigma_t - \varepsilon$) curves of AA2060-T8 sheets obtained from the isothermal warm tensile tests using Gleeble material simulators under warm forming conditions and using the samples machined in rolling direction are depicted in Figs. 3 and 4. These ($\sigma_t - \varepsilon$) curves possess common scale to visualize the direct influences of strain rates and temperature on the deformation behavior of this alloy effectively. As shown in these ($\sigma_t - \varepsilon$) curves, the overall shapes of the flow curves were dependent on temperature and strain rates. Thus, the influences of temperature and strain rate on the deformation behavior and fracture mechanisms are discussed in the next sections. Furthermore, the shapes of these ($\sigma_t - \varepsilon$) curves pointed out some vital characteristics which are helpful to identify the deformation mechanism of AA2060-T8 sheets.

When AA2060-T8 tensile sample was subjected to tensile loading, the mobile dislocations were interacted with the dislocation networks exist in this alloy. Since the structures of the dislocation strongly hinder the dislocations movements, the strength of this alloy tends to increase with increasing strain rate and decreasing the forming temperature. As shown in Figs. 3 and 4, the $(\sigma_t - \varepsilon)$ curves of AA2060-T8 sheets are composed of four stages: stage I (elastic stage); stage II (work hardening stage); stage III (softening stage); and finally, stage IV (steady stage. In the elastic stage, the stresses are linearly proportional to the strains, and by increasing the stresses, the strain exceed the elastic region. At the beginning of deformation (i.e., strain hardening stage), the rapid increase in stresses is associated with increasing in dislocation density, thus, the strain hardening exceeds the dynamic softening. As the plastic deformation progresses (softening stage), the activation of the dynamic softening can partially or completely or offset the influence of strain hardening. At the point of ultimate tensile stress, the strain hardening as well as dynamic softening become balance, therefore, the flow stress decreases slowly or remains unchanged. In general, Strain hardening was attributed to many pinned dislocations of little mobility which led to increase the density and mutual intersection of dislocation under the external stresses. On the other hand, the dynamic softening may result in the microstructures with lower energies induced by reorganization of dislocations and the decreasing the dislocations densities [11].

3.2. Corrections of the adiabatic heating influence.

All the $(\sigma_t - \varepsilon)$ curves of AA2060-T8 sheets in Figs.3, and 4 are depended on the initial temperatures of the tensile specimens. Nevertheless, the actual temperature of the tensile specimen is different from the initial one. This is attributed to the conversion of the plastic work into heat and the losses by radiation, convection, and conduction from the specimen to its surroundings. At low strain rates (i.e., $\dot{\varepsilon} \leq 1$ s^{-1}), the deformation process is assumed to be isothermal since the rise in temperature during deformation process is ignored. In contrast, at high strain rates (i.e., $\dot{\varepsilon} \geq 1$ s^{-1}), the deformation process is adiabatic, thus, the rise in temperature cannot be neglected during deformation and the temperature needs to be corrected to consider the adiabatic effect. Ou et al. [10] mentioned in their investigation that adiabatic heating correction should be applied if the strain rate exceeding 1 s^{-1}. Thus, in this study, the incremental procedure proposed by [12] was used for calculating the temperature increase during tensile testing. They assumed that the rise in temperature is homogeneous throughout the tensile specimen and the variation in specific heat $(\boldsymbol{C})$ and density $(\boldsymbol{\rho})$ during the temperature interval $(\boldsymbol{\delta T})$ is neglectable. Thus, the rise in temperature was calculated using Eq. 1.

$$\int_{T_0}^{T_0+\delta T} \rho C dT = \int_{\varepsilon_0}^{\varepsilon_0+\delta\varepsilon} \sigma d\varepsilon \Rightarrow \delta T = \frac{\bar{\sigma}\delta\varepsilon}{\rho C} \quad (1)$$

where $\bar{\sigma}$ is the mean stress which calculated from the $(\sigma_t - \varepsilon)$ curve of each forming temperature (i.e., 100℃, 150℃, 200℃, 250℃, 300℃) and strain rate of 10 s^{-1} over the strain interval $\boldsymbol{\delta\varepsilon}$ by using trapezoid equation:

$$\bar{\sigma} = \frac{1}{\delta\varepsilon}\int_{\varepsilon_0}^{\varepsilon_0+\delta\varepsilon} \sigma d\varepsilon \quad (2)$$

Thus, the influence of the rise in temperature on the flow stress was assessed using the following equation:

$$\delta\sigma_i = \left[\frac{\delta\sigma}{\delta(1/T)}\right]_{\delta\varepsilon_i,\dot{\varepsilon}} \left[\frac{1}{T_{iso}+\delta T} - \frac{1}{T_{iso}}\right] \quad (3)$$

Fig. 4 shows the corrections of the adiabatic heating influence on $(\sigma_t - \varepsilon)$ curve of AA2060-T8 sheets at different forming temperatures, and strain rate of 10 s^{-1}. At higher strain rates, and lower forming temperatures, the raise in normalized temperature is increased because of the higher flow stresses. Hence, as shown in Fig. 4, the difference between uncorrected and corrected and curves is reduced with increasing the forming temperature because of the decrease of the alloy strength.

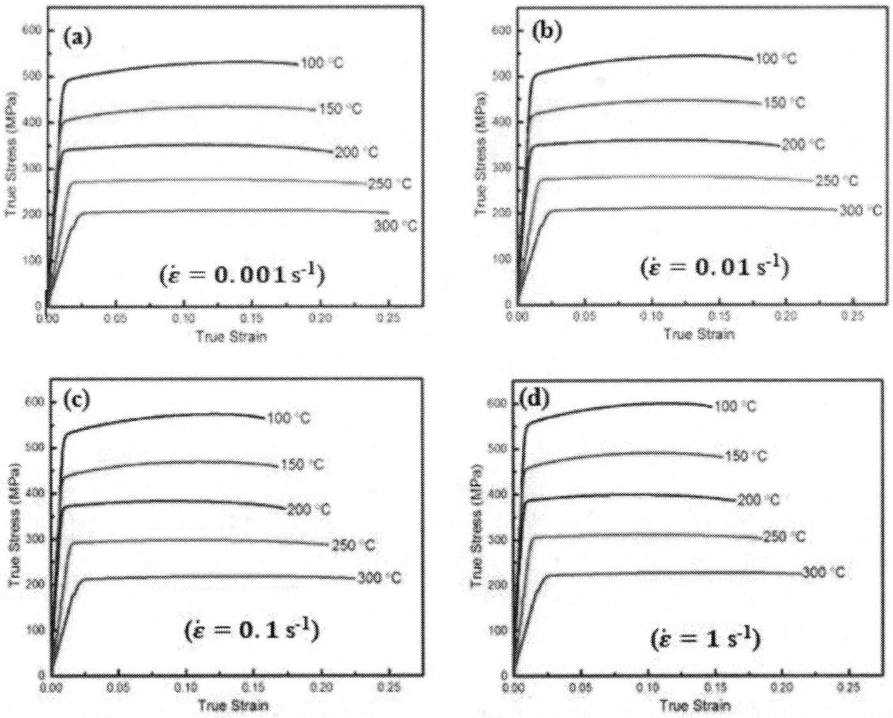

Fig. 3. True stress-strain curves of AA2060-T8 sheets at RD, different forming temperatures (i.e. 100℃, 150℃, 200℃, 250℃, 300℃), and strain rates of (a) 0.001 s^{-1}, (b) 0.01 s^{-1}, and (c) 0.1 s^{-1}.

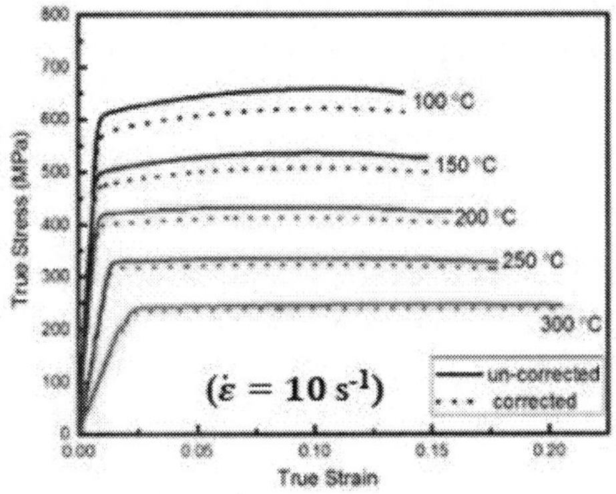

Fig. 4. Correction of adiabatic heating influence on the true stress-strain curves of AA2060-T8 at forming temperatures of 100℃, 150℃, 200℃, 250℃, and 300℃, and strain rate of 10 s^{-1}.

4. Phenomenological -based Constitutive modelling

It is indispensable to propose constitutive equations to reveal and predict the deformation behavior of AA2060-T8 sheets under warm forming conditions. Within this realm, Arrhenius constitutive model is effective phenomenological -based constitutive models to describe the correlation between flow behavior, strain rate and temperature. Thus, in this study, Arrhenius constitutive model is used as the constitutive model of AA2060-T8 sheets to determine the activation energy (*Q*) of AA2060-T8. ***Q*** is crucial physical parameter to estimate the difficulties of plastically deforming the metallic materials.

The combined influence of strain rate and temperature on warm/hot deformation behavior of metals can be characterized by Zener–Hollomon parameter (Z) which is defined by exponent type equation [12-13] as:

$$\mathbf{Z} = \dot{\varepsilon}\, \mathbf{exp}\left(\frac{Q}{RT}\right) \tag{4}$$

where, $\dot{\varepsilon}$ is the strain rate, Q is the activation energy of warm/hot deformation, T is the absolute temperature, and R (8.31 J mol^{-1} k^{-1}) is the universal gas constant.

The relationship between deformation temperature, strain rate, and flow stress, notably at elevated temperature, can be represented by the Arrhenius constitutive model [13] as:

$$\dot{\varepsilon} = AF(\sigma)\, \mathbf{exp}\left(-\frac{Q}{RT}\right) \tag{5}$$

where $F(\sigma)$ can be expressed as power law equation (Eq. 6), or as exponent type equation (Eq. 7) or hyperbolic sine type equation (Eq. 8) as follow:

$$F(\sigma) = \sigma^{n_1} \;, \qquad if\; \alpha\sigma < 0.8 \tag{6}$$

or

$$F(\sigma) = \exp(\beta\sigma) \;, \qquad if, \alpha\sigma > 1.2 \tag{7}$$

or

$$F(\sigma) = [\sinh(\alpha\sigma)]^{n} \;, \qquad for\; all\; \sigma \tag{8}$$

A, n, n_1, and β, are material constants, as well as σ is the flow stress. For Al alloys, σ can be practically replaced in in the constitutive equations by maximum flow stress (σ_p). This assumption is compatible with previous investigations on titanium alloys, magnesium alloys, and steel [10-12]. α is the stress multiplier which is essential to establish the constitutive model of AA2060-T8 under warm forming conditions. For Al alloys, α is varying from 0.01 to 0.08 MPa^{-1} [12]. α is usually calculated by:

$$\alpha = (\beta/n_1) \tag{9}$$

β , and n_1 are calculated from the slopes of the relations between $(\log\dot{\varepsilon} - \sigma_p)$, and $(\log\dot{\varepsilon} - \log\sigma_p)$, respectively. Thereafter, by substituting the values of $= \mathbf{0.113568}$, and $n_1 = \mathbf{6.24}$ in Eq. (9), α was calculated as **0.01392** MPa^{-1}.

In this study Eq. (8) is used to describe the relationship between flow stress, true strain, forming temperature, and strain rates, since it can be used for all stress level. In contrast, Eqs. (6), and (7) are preferred in case of low stress ($\alpha\sigma < 0.8$) such as creep, and high stress ($\alpha\sigma > 1.2$), respectively [12]. Thus, the constitutive equation of AA2060-T8 sheets under warm forming conditions can be expressed as:

$$\dot{\varepsilon} = A[\sinh(\alpha\sigma)]^{n}\, \mathbf{exp}\left(-\frac{Q}{RT}\right) \tag{10}$$

To quantitatively investigate the warm deformation behavior of AA2060-T8 sheets,A, α, n, and Q are calculated by differentiating Eq. (10) as follows:

$$Q = R\, n \left|\frac{\partial \ln[\sinh(\alpha\sigma)]}{\partial(1/T)}\right|_{\dot{\varepsilon}} \tag{11}$$

$$n = \left|\frac{\partial \ln\dot{\varepsilon}}{\partial \ln[\sinh(\alpha\sigma)]}\right|_{T} \tag{12}$$

By combining Eqs. (11) and (12), Q can be calculated by:

$$Q = R \left|\frac{\partial \ln\dot{\varepsilon}}{\partial \ln[\sinh(\alpha\sigma)]}\right|_{T} \left|\frac{\partial \ln[\sinh(\alpha\sigma)]}{\partial(1/T)}\right|_{\dot{\varepsilon}} \tag{13}$$

A, n, and Q are calculated from the linear fitting of the of the relations $(\ln[\sinh\alpha\sigma]) - (1/T)$ and $(\ln[\sinh\alpha\sigma]) - (\ln\dot{\varepsilon})$. Hence, $A = 4.13 \times 10^{10}$s^{-1}, $n = 6.24$, and $Q = 242\; kJ\; mol^{-1}$ are listed in Table 2. Thus, the final constitutive equation of AA2060-T8 sheets can be expressed as:

$$\dot{\varepsilon} = 4.13 \times 10^{10}[\sinh(0.01392\sigma)]^{6.24}\, \mathbf{exp}\left(-\frac{242}{RT}\right) \tag{14}$$

IOP Conf. Series: Materials Science and Engineering 1238 (2022) 012017 doi:10.1088/1757-899X/1238/1/012017

Table 2. The material constants and activation energy (Q) of AA2060-T8

A (s^{-1})	α (MPa^{-1})	n	Q ($kJ\ mol^{-1}$)
4.13×10^{10}	0.01392	6.24	242

5. CONCLUSIONS

The main conclusions can be deduced as follows:

(1) Below forming temperature of 250℃, strain hardening is dominated over the dynamic softening mechanisms. However, at 250℃ and above, the DRV is dominated, and the strain hardening is gradually disappeared. Thus, DRV is the primary reason of causing softening effect and reducing the effect of strain hardening in AA2060-T8 sheets at the temperature of 250℃ and above.

(2) The flow stress can be represented by a Zener-Hollomon parameter in the hyperbolic-sine equation with the warm deformation activation energy of 242 kJ/mol, and the final constitutive equation of AA2060-T8 sheets can be expressed as: $\dot{\varepsilon} = 4.13 \times 10^{10}[\sinh(0.01392\sigma)]^{6.24} \exp\left(-\frac{242}{RT}\right)$.

References

1. Abd El-Aty A, Xu Y, Guo X, Zhang S, Ma Y, Chen D 2018. *J Adv Res.* ***10:*** 49-67.
2. Abd El-Aty A, Xu Y, Zhang S, Ma Y, Chen D 2017. *Procedia Eng.* **207**:13-18.
3. Abd El-Aty A, Xu Y, Zhang S, Ha S, Yan Y, Chen D 2019. *J Adv. Res.* **18**:19-37.
4. Abd El-Aty A, Xu Y, Ha S, Zhang S H 2018. *Mater Sci Eng A.* **731**:583-594
5. Abd El-Aty A, Zhang S, Xu Y, Ha S 2019. *J Mater Res Technol.* **8**:1235-1249.
6. Abd El-Aty A, Zhang S, Guo X, Xu Y, Yan M, Chen D, Tao J. 2021. T*he Minerals, Metals & Materials Series. Springer, Cham*
7. Xu Y, Abd El-Aty A, Zhang S, Ma Y, Chen D 2019. *IOP Conf. Ser.: Mater. Sci. Eng.* **651** (2019).
8. Lin Y, Chen X. 2011. *Mater Des.* **32**:1733-1759.
9. Zheng K, Politis D, Wang L, Lin J 2018. *Int. J. Light. Mater. Manuf.* **1**:55-80
10. Ou L, Zheng Z, Nie Y, Jian H 2015. J *Alloys Compd.* **648**:681-689.
11. Gao H, Weng T, Liu J, Li C, Li Z, Wang L 2016. *Manufacturing Rev.* **9**:9-15.
12. Liang H, Nan Y, Ning Y, Li H, Zhang L, Shi Z, Guo H 2013. *J. Alloy. Compd.* **632(25)**:478-485.

Acknowledgements

The authors greatly acknowledge the financial support from National Natural Science Foundation of China - International (Regional) Cooperation and Exchange Program (No. 5201101342), and Jiangsu Province Science and Technology Project (No. BK20200453), and the National Natural Science Foundation of China (Grand No. 51875548).

Improved formability of HMGF components by preforming in an upset bulging process

R Haase[1], M Werner[1], V Kräusel[1], A Alimov[2], A Sviridov[2] and S Härtel[2]

[1] Fraunhofer Institute for Machine Tools and Forming Technology, Chemnitz, Germany

[2] Brandenburg University of Technology Cottbus-Senftenberg, Cottbus, Germany

E-mail: rico.haase@iwu.fraunhofer.de

Abstract. Hot Metal Gas Forming (HMGF) is a well-established combination of a profile hydroforming operation and an in situ hardening process. Roof rails and camshafts are typical applications. A significant limitation of the component geometry is the strain-related material thinning. In the polar coordinate system of the tube-shaped parts, the strain can be divided into circumferential and longitudinal portions. To achieve higher expansion ratios in circumferential direction, a longitudinal compression is necessary. In a conventional, non-tempered process setup this is achieved by an axial feeding movement during the forming operation. Under tempered conditions, in particular by friction coefficients well above 0.3, this is no longer possible in a controlled process. Anyhow, a beneficial material pre-distribution is required to enlarge the circumferential strain before material failure occurs. In the proposed preforming operation, the longitudinal compression is achieved by offset bulging. Based on aspect ratios between initial diameter, wall thickness, initial length and compression stroke, a specific, controlled wave pattern is achieved on the specimen. Subsequently, the semi-finished parts undergo the HMGF process in an austenitisation, transfer, forming and quenching procedure. By means of continuous, finite element based simulation and related model calibration in experimental trials, a significant increase in geometric complexity is achieved.

1. Hot Metal Gas Forming

The Hot Metal Gas Forming technology does combine the forming operation of predominantly profile- or pipe-typed work pieces with an inline heat treatment by a die quenching process. The process route starts with the initial austenitisation of the material by elevation of its temperature above Ac3 followed by a swift transfer into the forming device, application of the media pressure and rapid cooling due to the pressurized contact of the work piece towards the lower tempered forming die. In a large scale series application, the dies are equipped with an active cooling system to derive the repeated heat input by the processed parts. The forming operation of the tempered specimen followed by the quenching and thus hardening procedure does provide a unique combination of formability and mechanical properties of the finished product.

2. General strain conditions

Due to the pipe-typed characteristic shape of the work pieces, their geometric appearance can be described in a polar coordinate system rather relevant than in the Cartesian tripod. The geometry is then defined by the shape of the cross section along the components characteristic centre line. Starting from a pipe with circular initial section, geometric features can be formed by the application of the media

IOP Conf. Series: Materials Science and Engineering 1238 (2022) 012018 doi:10.1088/1757-899X/1238/1/012018

pressure from the enclosed volume. A minor decrease of the section length is also possible, but heavily restricted by the wrinkling tendency of the sheet metal and therefore is no further regarded in this paper. By the applied media pressure, the metal membrane of the work piece starts to expand freely into the surrounding die cavity. Under those initial conditions, a biaxial stretch strain is observed. Utilizing the aforementioned Cartesian coordinate system, the strain condition can be divided into a longitudinal and circumferential portion. Taking the expansion of a single segment as an elementary forming feature, the strain values can be described as

$$\varepsilon_{circ} = \ln\left(\frac{dU_1}{dU_0}\right) \tag{1}$$

For the free, circular expansion with constant radii along the Cartesian angle it can be reduced to

$$\varepsilon_{circ} = ln\left(\frac{r_1 d\varphi}{r_0 d\varphi}\right) = ln\left(\frac{r_1}{r_0}\right) \tag{2}$$

Following the same consideration in the longitudinal direction of the elementary forming feature the strain can be described through the length ratio in undeformed versus bulged state, as illustrated in Figure 1.

$$\varepsilon_{long,bulge} = ln\left(\frac{l_1}{l_0}\right) = \left(\frac{2r*\sin\left(\frac{\alpha}{2}\right)}{l_0}\right) \tag{3}$$

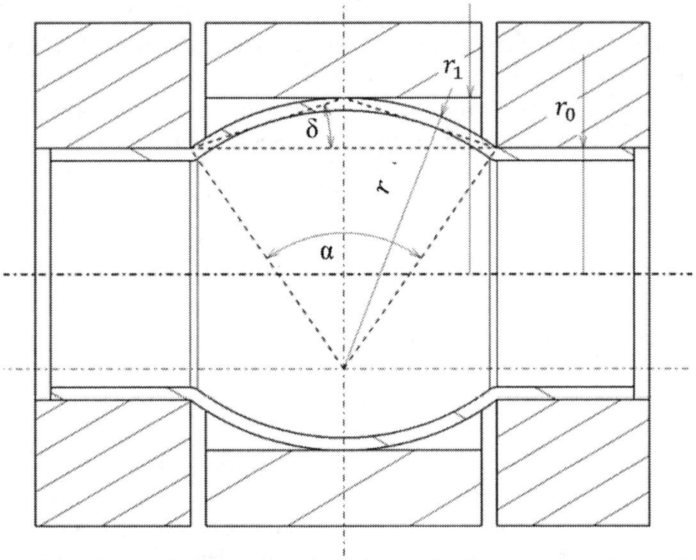

Figure 1. Bulged preform of a hydro formed pipe with dimensions relevant for calculation.

$$\delta = \frac{\alpha}{4} = \arctan\left(2 * \frac{r_1 - r_0}{l_0}\right) \quad ; \quad r = \frac{l_0}{2*\sin\left(\frac{\alpha}{2}\right)} \tag{4}$$

For industrial relevant parameter combinations, this mean longitudinal strain value during the bulged state $\varepsilon_{long,bulge}$ is several factors smaller than the circumferential strain ε_{circ}. As soon as the work piece gets in contact in the middle area of the form feature, the movement is restrained by the severe friction between the wall of the cavity and the pressurized specimen.

Assuming a cylindrical shape of the form feature with conical transitions towards the surrounding geometry as illustrated in Figure 2, the mean longitudinal elongation of the final shaped component can be estimated as:

$$\varepsilon_{long,cyl} = ln\left(\frac{l_0 - 2*\frac{r_1 - r_0}{\tan\varphi} + 2*\frac{r_1 - r_0}{\sin\varphi}}{l_0}\right) = ln\left(\frac{l_0 + 2(r_1 - r_0)*\left(\frac{1}{\sin\varphi} - \frac{1}{\tan\varphi}\right)}{l_0}\right) \quad (5)$$

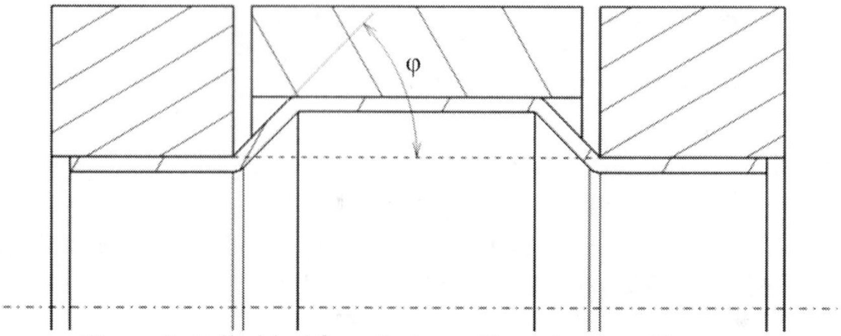

Figure 2. Cylindrical form feature with conical transition area.

As the contact area is growing from the centre towards the conical areas, the width of the formed membrane decreases and therefore longitudinal strain increases progressively. This effect is amplified, the more the pre-shaped bulge and final part geometry diverge. Due to the fact, that the cylindrical areas usually represent functional faces and short transition zones are intended, the localised longitudinal strain plays a crucial role for functional density of hydro formed parts. By this biaxial strain state, the elongations in longitudinal and transversal direction superimpose towards an excessive thickness reduction. Derived from volumetric consistency, thickness strain equals to:

$$\varepsilon_{th} = -(\varepsilon_{circ} + \varepsilon_{long}) \quad (6)$$

For practical applications, not only a compensation of the longitudinal strain during bulging, but even an overcompensation with a resulting, negative value of ε_{long} is aimed for. On the other hand, an excessive compression might result in irreversible wrinkles along the specimen and therefore acts as a boundary condition for the process window.

3. Necessity of a pre-forming operation

In order to avoid excessive thinning or even material failure during the forming operation, a longitudinal feeding and thus compressive strain along the axis of the part is intended. Under non-tempered conditions and thus low friction impact, a feed movement of the sealing punches is performed parallel to the ramp up of the media pressure. Within hot forming conditions, there are two major obstacles which prevent this change in strain state. The friction of a heated austenite blank towards the forming die ranges between 0.3 and 0.45 [1] and therefore predominantly leads to a sticky contact behaviour. In parallel, the axial compression of the profile tends to get unstable due to the lowered strength and potentially inhomogeneous temperature distributions. Due to this lack of process repeatability, axial feeding is not applied to HMGF processes on industrial scale. To provide an alternative solution, the preceding material distribution by a cold preforming operation is proposed. Due to the significantly lowered friction conditions at room temperature and homogeneous stress distribution, a controlled compression is achievable. Geometrically, a macroscopic material movement into the subsequent forming zone is intended. For the considered profiles and pipe-typed components, this can be achieved by an upset bulging process.

The ends of the specimen are restricted in radial movement by outer and potential inner active parts, which are approaching each other along the compression stroke Δl as shown in Figure 3.

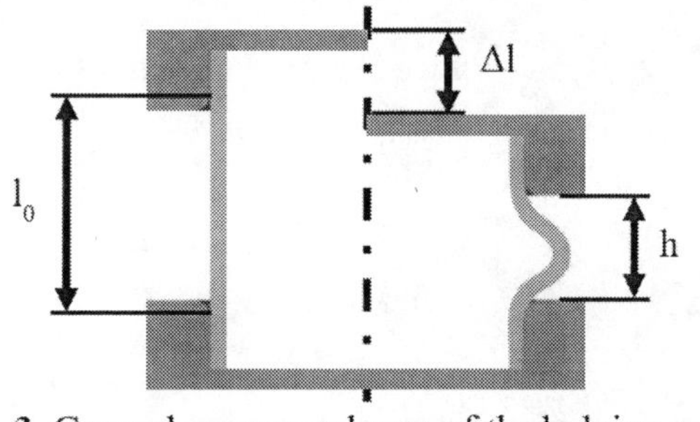

Figure 3. General process scheme of the bulging process.

To maximize the compression effect and prevent remaining humps on the component afterwards, the compression should be equivalent to the expected longitudinal strain as described in the previous chapter. The description of this intention in longitudinal strain delivers:

$$\varepsilon_{pref} + \varepsilon_{long,cyl} = \varepsilon_{pref} + ln\left(\frac{l_0+2(r_1-r_0)*\left(\frac{1}{\sin\varphi}-\frac{1}{\tan\varphi}\right)}{l_0}\right) \cong 0 \tag{7}$$

$$\varepsilon_{pref} = \ln\left(\frac{l_0}{l_0+\Delta l}\right) \cong -\varepsilon_{long,cyl} \tag{8}$$

$$\Delta l \cong 2(r_1 - r_0) * \left(\frac{1}{\sin\varphi} - \frac{1}{\tan\varphi}\right) \tag{9}$$

Aiming for even higher expansion ratios and therefore higher circumferential strains, the longitudinal strain is not only compensated by the preceding longitudinal compression, but even exceeded by higher initial compression strokes.

4. Upset bulging of specimen

The upsetting process is utilized to provide a preform containing a longitudinal compression and radial expansion according to the desired final shape. Following this approach, the material distribution along the longitudinal axis is crucial for the further HMGF process in terms of macroscopic material flow and thus friction effects on the wall of the forming cavity. The parameters defined by the selected pipe material are inner and outer diameter and thus wall thickness. The initial length of the specimen and compression stroke Δl are varied within the trials in order to observe the geometric forming result. The process layout is shown in Figure 3. The restraining forming tools (Figure 4) are equipped with a radius depending on the material thickness $R = 2 * s_0 = 4.0\ mm$ and therefore the effective buckling length is the sum of the initial distance between those dies plus two times that radius. A repeatable compression stroke can be ensured by distance blocks of height $h = l_0 - 2R - \Delta l$.

Figure 4. preforming tool mounted in press (closed state) a)
and detailed review of tool components (opened state) b)

For the cylinder-shaped forming feature, a material distribution with two wave-shaped material depots close to the conical transition zones will minimize the longitudinal material flow during the HMGF process and does therefore provide the optimum result with regards to a uniform wall thickness distribution. Unfortunately, the axial force during the bulging process does not monotonically increase with ongoing compression stroke. As soon as the wave-shaped material depot is formed towards a half circle section, axial force decreases and the wave tends to close in a wrinkle. Due to the small bending radii, this cannot be flattened again during the HMGF process and needs to be avoided. The solution to this concern is a supporting die which limits the radial expansion of the wave-shaped material depot and therefore prevents it from flipping over and collapse. As a result, a pattern of two or even more symmetrical material depots could be achieved. The bulging process is described in [2] more in detail.

5. Numerically assisted process design and prototyping

The overall process chain was elaborated by means of a 3D finite element model undergoing the preforming operation, austenitization in the furnace and subsequent tempering in the handling process and finally the combined forming and quenching operation. The bulging process itself was calculated utilizing the finite element software QForm available at B-TU Cottbus. Based on an application-related selection of input parameters, the following result matrix could be achieved.

Diameter of supporting die in mm	Initial length l_0 in mm: 20	30	40	50	60	70	80	90
42	f) Thickening	f) Thickening	f) Thickening	f) Thickening	f) Thickening	f) Thickening	f) Thickening	f) Thickening
44	f) Thickening	f) Thickening	f) Thickening	b) inverted loop	triple loop	triple loop	triple loop	triple loop
46	a) free bulging	d) cylindric shape	d) cylindric shape	b) inverted loop	b) inverted loop	triple loop	triple loop	triple loop
48	a) free bulging	d) cylindric shape	d) cylindric shape	b) inverted loop	b) inverted loop	triple loop	triple loop	triple loop
50	a) free bulging	a) free bulging	d) cylindric shape	b) inverted loop	b) inverted loop	triple loop	triple loop	triple loop
52	a) free bulging	a) free bulging	a) free bulging	b) inverted loop	b) inverted loop	b) inverted loop	b) inverted loop	b) inverted loop

Figure 5. Parameter matrix of initial length and diameter of supporting tool leading to different shapes of the material depot in the subsequent forming zone.

The geometric result was exported utilizing a Nastran mesh description. By the subsequent heating up to austenite state well above Ac3 temperature, previous strain values and related hardening effects are reset and can be neglected. The wall thickness distribution was transferred by the volumetric elements. In an initial configuration as shown in Figure 6, a single material depot was generated by the combination of a short buckling length and compression stroke. The material thickness distribution was utilized for validation of the FE model and corresponds well to measurements from section cuts.

Parameter	Preform result	Final result - longitudinal material movement in mm
no preforming		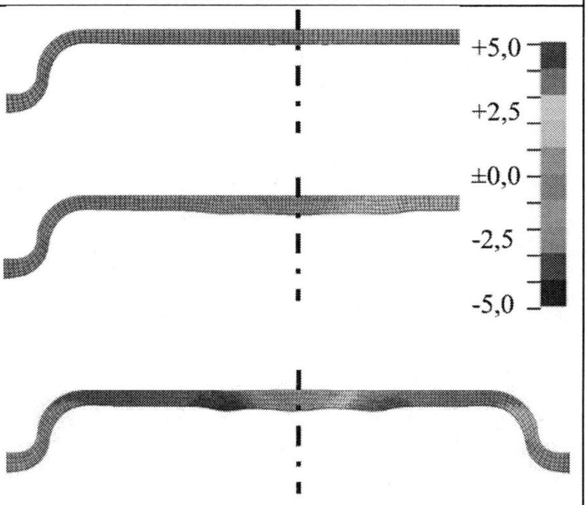
$\Delta l = 13\ mm$ $h_0 = 30\ mm;$ $R = 2 * s_0$ $= 4\ mm$	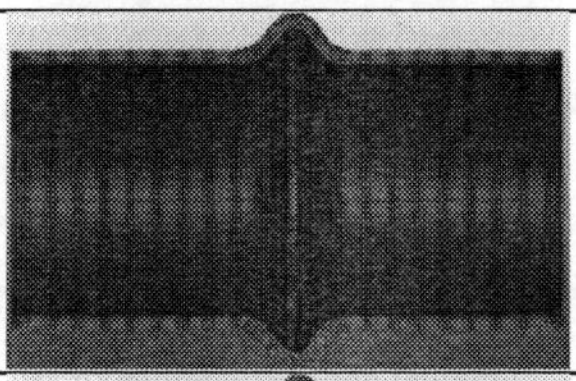	
$\Delta l = 16\ mm$ $h_0 = 30\ mm;$ $R = 2 * s_0$ $= 4\ mm$	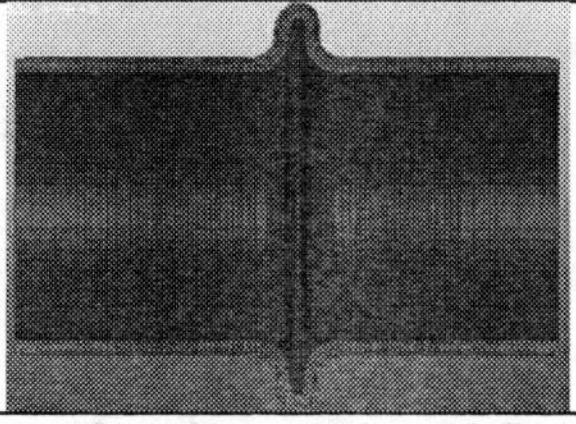	

Figure 6. Preforming result and final material thickness distribution.

Even if the developed length in longitudinal direction is already contained in the preformed specimen and therefore no significant longitudinal strain was expected, the material distribution along the axis of the part does not correspond to the final shape. Under the aforementioned friction conditions in the HMGF process, only a minor longitudinal material flow was observed. As a result, the thickness distribution along the part axis still shows strong peaks with a maximum in the preformed middle section and still considerable thinning towards the conical transition areas. Heading towards a preform closer to the final shape, the buckling length was significantly increased towards the form feature length plus the desired compression stroke. By additional limitation of radial movement by the support die, two material depots close to the conical transition areas could be generated. As Figure 7 illustrates, the preform represents the final shape more precise. Consequently, a homogeneous wall thickness distribution within the cylindrical area of the form feature could be observed.

Optimized preforming parameters:

$\Delta l = \mathbf{17.7}\ \boldsymbol{mm}$
$h_0 = 68\ mm;$
$R = 2 * s_0 = 4\ mm$

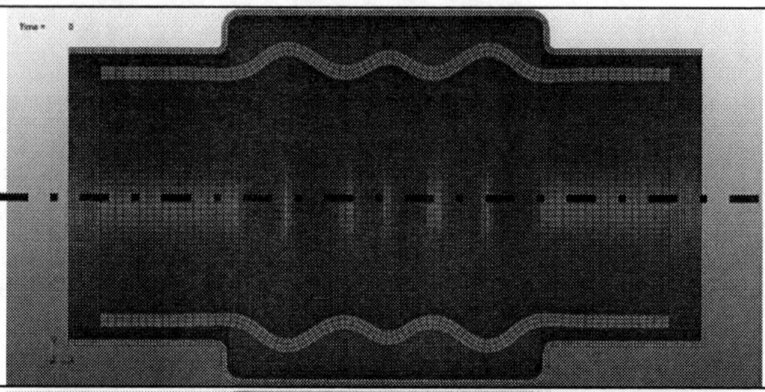

Result of optimized preform:

- uniform material thickness along part length,
- no significant thinning, even in conical transition areas
- full contact to cavity in cylindrical area, sufficient cooling rate is achieved

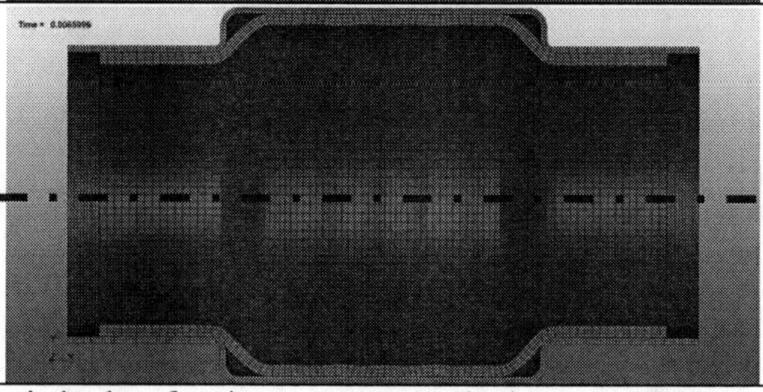

Figure 7. Simulation result of optimized preforming parameters and subsequent HMGF.

6. HMGF of preformed specimen

The existing tool was initially designed for a pipe outer diameter of 44 mm resulting in a circumferential elongation of 27%. This can be achieved without a preforming operation. By reduction of the pipe diameter to $d_0 = 2r_0 = 40\ mm$, the circumferential elongation develops towards 40%, which is no longer feasible without preforming and repeatedly results in a provoked rupture of the pipes. In a further experimental setup, a feeding motion was tested but did result in either a crumpling of the pipe ends or collapse of the pipe in the middle section. Accordingly, a preforming of the pipe is mandatorily required. The preformed specimen are heated up in a furnace at $\vartheta_{furn} = 950°C$ for a period of $t_{furn} = 300\ s$. The transfer operation was performed manually, but still in a repeatable manner. Thus a transfer time of $t_{transf} = 5.0 \pm 0.5\ s$ was ensured. Coming from the furnace, the forced convection due to the swift handling leads to a cooling rate of about 12 K/s, which further decreases with the temperature level of the part. Due to the wall thickness of $s_0 = 2.0\ mm$ and the predominant radiation merely on the outside of the specimen, the cooling rate is comparatively low. From the insertion temperature of about $\vartheta_{insert} = 890°C$ the temperature further drops due to radiation and convection on the upper side and the loose initial tooling contact on the lower side of the part. Regarding a time period of $t_{closing} = 2.5 \pm 0.5\ s$ for activation of the security curtain of the press and closing of the cavity, a starting temperature of $\vartheta_{start} = 840 \pm 25°C$ is achieved on the part. The utilized forming tool is shown in Figure 8, where the expansion in the middle area to $d_1 = 2r_1 = 56\ mm$ is obvious.

Figure 8. Successful HMGF process with expansion from $d_0 = 40\ mm$ to $d_1 = 56\ mm$. Preform shown on the left, finally achieved component on the right.

7. Characterization of the manufactured components

During the experimental trials, the optimized preforming proved robust in terms of feasibility. An interesting side effect in this case is the middle area of the preformed part which still is about 2 mm less in diameter and therefore does not yet get in direct contact to the die during the inserting of the part. Within CMM measurements, the shape of the cylindrical form feature was within a tolerance of 0/-0.05 from reference. Section cuts did show a complete martensitic transformation in the cylindrical section. Due to the lack of contact in the conical transition zone, a slight drop in micro hardness was observed in that area. If required, this can be achieved by an adaption in tool design.

8. Summary and outlook

Hot metal gas forming (HMGF) is a process well suitable for a variety of high-strength and lightweight pipe-typed or even profile components. Previously, the limitation in component design was severely restricted by the feasible circumferential elongation in the various sections along the components axis. Due to the challenging tribological conditions within the HMGF process, a notable axial feeding is not possible in particular within an industrial process environment. By an innovative preforming step based on a tool-supported bulging operation, significant material resources can be moved into the subsequent forming zone. By this precedent axial material feed and compression, the axial elongation portion is lowered. As a benefit, extraordinary circumferential strain values can be achieved. So far, the methodology has been applied on cylindrical form features. A transfer of the approach towards non-rotational, but arbitrary regular shaped cross sections is ongoing.

Acknowledgments

The content presented within this paper was funded within IGF 21359N.

References

[1] Müller R and Mosel A 2014 Characterisation of Tool Coatings for Press Hardening *Adv. Mater. Res.* **966–967** 259–69

[2] Alimov A and Haase R 2022 Upset bulging as a preforming operation for hot metal gas forming of 22MnB5 tubes *IOP Conf. Series: Mat. Sci. and Engineering*

Less damage accumulation of aluminum alloy sheet during electromagnetic forming based on Gurson-Tvergaard-Needleman model

W Liu[1,2]*, H Zhou[1], Z Meng[3], J Li[1] and S Huang[1]

[1] School of Materials Science and Engineering, Wuhan University of Technology, Wuhan, China
[2] State Key Laboratory of High Performance Complex Manufacturing, Central South University, Changsha, China
[3] School of Automotive Engineering, Wuhan University of Technology, Wuhan, China
* E-mail: weiliu@whut.edu.cn

Abstract. The formability of aluminum alloy sheet is usually improved during electromagnetic forming (EMF). Here, the ductility enhancement mechanism was investigated on the basis of Gurson-Tvergaard-Needleman (GTN) model. The parameters of GTN models under quasi-static tension and EMF conditions were respectively determined by inverse identification. For AA2024-O aluminum alloy sheet under quasi-static tension, the initial void volume fraction is identified as 0.006, the critical void volume fraction where voids begin to aggregate is 0.075, the void volume fraction of nucleating particles is 0.131 and the mean nucleation strain is 0.44. With the similar initial void volume fraction, the other parameters under EMF condition are determined as 0.045, 0.055 and 0.58. The evolution curve of void volume fraction with strain were compared under EMF condition and quasi-static tension. The results show that the total growth of void volume fraction is less under EMF condition than that under quasi-static tension. A box part with inclined flange was preliminary achieved by electromagnetic forming process. The GTN model was finally verified by comparing the numerical and experimental results.

1. Introduction

Electromagnetic forming (EMF) is a high speed forming process by using Lorentz force to make workpiece deform. When an impulse current runs along the coil, an eddy current is induced in the adjacent conductive workpiece, and the Lorentz force is consequently generated between the coil and workpiece. The forming speed of material can exceed 10^2 m/s and the strain rate can reach up to several 10^3 s^{-1}. Compared with that during the quasi-static forming processes at room temperature, the formability of aluminum alloy sheet can be effectively improved during EMF [1-3]. However, it is still needed to investigate the ductility enhancement mechanism of aluminum alloy sheet during EMF.

The ductile damage behavior of metals is summarized as nucleation, growth and coalescence of micro voids [4]. The ductile damage accumulation in sheet metal forming process will not only relate to the formability, but also affect the mechanical properties in service [5]. Gurson-Tvergaard-Needleman (GTN) model is one of the widely-used damage model [6-9] to predict the ductile fracture of sheet metal in numerical simulation. It uses void volume fraction to represent the damage degree of sheet metals. Imbert et al [10] compared the electromagnetic bulging processes of AA5754 aluminum alloy sheet without and with concave die by experiment and numerical simulation. The GTN model with the same parameters under quasi-static condition was used to numerically investigate the damage behavior of aluminum alloy sheet. The higher formability of EMF was attributed to the increased hydrostatic pressure and the decreased void volume fraction by the interaction between aluminum alloy sheet and concave die. Feng et al [11] used the GTN model to numerically predict the fracture behavior of 5052-O aluminum alloy sheet under high speed electromagnetic impaction. The parameters of GTN model were determined via scanning electron microscopy after EMF experiment. Zeng et al [12] investigated the influences of discharge voltage, impact distance and pre-strain level on the damage evolution of aluminum alloy sheet during EMF with uniform pressure actuator. The numerical simulation with GTN model showed that the damage accumulation was reduced by the high impact pressure between the die and sheet.

In order to inspect the ductile damage behavior of aluminum alloy sheet during EMF by numerical simulation with GTN model, the parameters should be determined at high strain rate. At present, there are three methods to determine the parameters of GTN model: metallographic analysis [13], representative volume element (RVE) [14] and inverse identification [15, 16]. Metallographic analysis is used to determine the damage parameters by observing the metallographic phase of tensile samples at different stages of tensile test. RVE is a method to obtain nucleation-related parameters through finite element analysis of a single cell. The inverse identification is performed by updating the material parameter in numerical simulation to minimize the discrepancy between the numerical and experimental results. Due to the advantages of global accuracy and good convenience, the inverse identification method is commonly adopted to determine the parameters of GTN model.

The GTN model parameters of AA2024-O aluminum alloy sheet were respectively identified with the quasi-static tensile test and electromagnetic hole flanging (EMHF) test. The damage evolution curves were compared under the quasi-static and EMF conditions to inspect the ductile damage behavior of aluminum alloy sheet during EMF. At last, an EMF experiment of a box part with inclined flange was carried out. The GTN model was used for numerical simulation, and it was validated by comparing with the experimental results.

2. Parameter identification of GTN model

2.1. GTN model

The GTN model is expressed as following:

$$\varphi=\left(\frac{\sigma_{\mathrm{M}}}{\sigma_{\mathrm{Y}}}\right)^2+2q_1f^*\cosh\left(\frac{3q_2\sigma_{\mathrm{H}}}{2\sigma_{\mathrm{Y}}}\right)-1-\left(q_1f^*\right)^2=0 \tag{1}$$

where σ_M is the von Mises equivalent stress, σ_Y is the yield stress, σ_H is the hydrostatic stress, q_1, q_2 are the correction coefficients (the values are usually defined as $q_1 = 1.5$ and $q_2 = 1$), and f^* is the equivalent void volume fraction. The overall rate of the void volume fraction in the GTN model is derived by the rate of the void volume fraction caused by the growth of existing voids and caused by void nucleation.

The parameters to be determined in GTN model include: the initial void volume fraction f_0, the void volume fraction f_N of nucleating particles, the critical void volume fraction f_c where voids begin to aggregate, the failure void volume fraction f_F, the mean nucleation strain ε_N, and the standard deviation s_N of the normal distribution of ε_N (the value is usually defined as $s_N = 0.1$). Here, GTN model is used to characterize the ductile damage evolution of the fully annealed AA2024-O aluminum alloy sheet with the initial thickness of 1mm. The failure of material is not considered. The parameters f_0, f_N, f_c and ε_N of GTN model are determined by inverse identification under quasi-static tension and EMF conditions.

2.2. Inverse identification of GTN parameters under quasi-static tension

The quasi-static tensile specimen was shown in Figure 1(a). The longitudinal direction of the specimen is consistent with the rolling direction of the sheet. Unlike the parallel section of standardized uniaxial tensile specimen, a large circular arc with a radius of 500 mm was designed to ensure that the localization occurred in the central region of the specimen, as shown in Figure 1(b). The quasi-static tensile test was carried out at room temperature and nominal strain rate of 0.001 s^{-1} to obtain the experimental force and elongation curve of 2024-O aluminum alloy sheet. The tests were repeated with three times to show a good agreement.

The 3D finite element model with the same geometry and constrains as the quasi-static tensile test was presented in Figure 2. The part was meshed with shell elements in software LS-Dyna. The element size was refined to be 0.5mm to guarantee the numerical accuracy of deformation in the central area. The fully fixed constraint was adopted at the left end and the displacement was applied at the right end. The simulated force and elongation curve was derived by the reaction force at the left end and the distance increment ΔL between the point A and B.

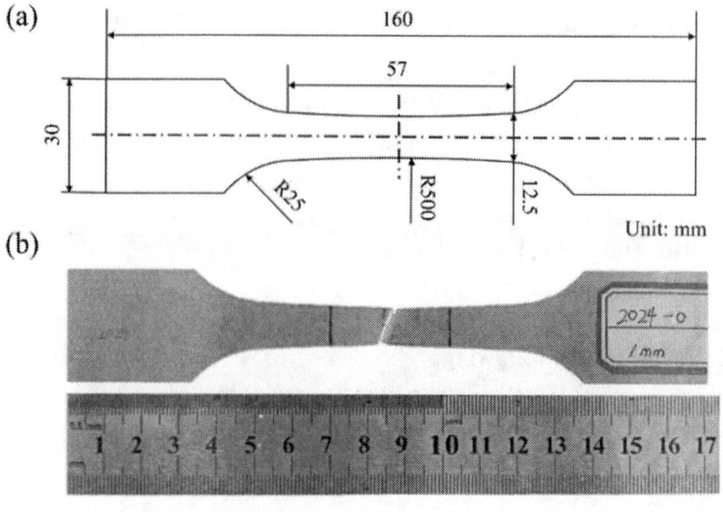

Figure 1. Quasi-static tensile specimen: (a) Geometry, (b) Fracture.

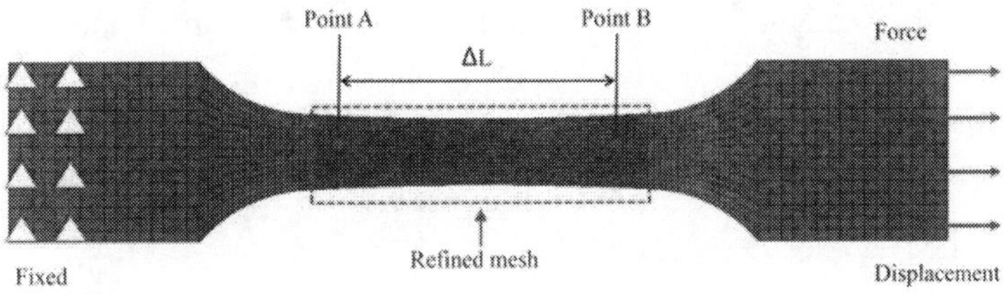

Figure 2. Finite element model of quasi-static tensile test.

The inverse identification can be summarized as numerical optimization problem of minimizing the discrepancy between the simulated and experimental results by updating the parameters of GTN model. Here, the successive response surface method (SRSM) was used to match the experimental and simulated curves of force and elongation for the parameter optimization. The identified parameters of GTN model were shown in Table 1. Finally, the simulated curve showed a good agreement with the experimental one, as shown in Figure 3.

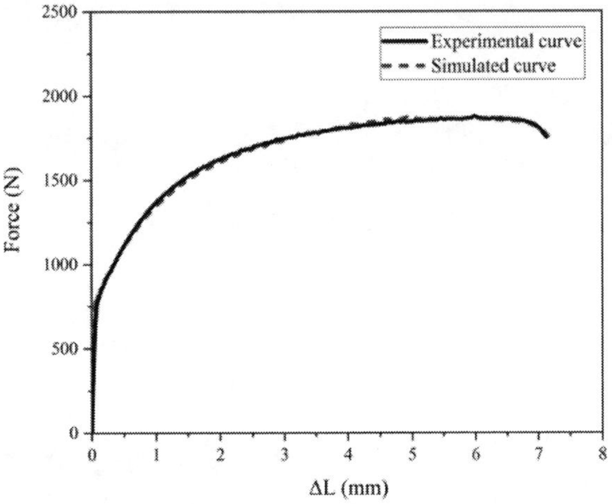

Figure 3. Experimental and simulated curves of quasi-static tensile test.

2.3. Inverse identification of GTN parameters under EMF condition

The principle of EMHF experiment was shown in Figure 4(a). It consists of charging circuit and discharging circuit. In this experiment, the capacitance and voltage were chosen to be 550μF and 3.25kV respectively. The capacitor tank C was previously charged by a power supply of high voltage U_0. When the discharging circuit was turn on, a transient electrical current flowed along the coil. Then, a transient magnetic field was generated around the coil and an eddy current was induced in the conductive sheet. Consequently, the Lorentz forces occurred between the coil and sheet. The sheet undergone elastoplastic deformation at high speed. A circular drawbead was used to prevent the outer margin from flowing into die. The experimental setup was shown in Figure 4(b). The die and base were made of steel. A flat spiral coil with winding number of 12 was used in the experiment, as shown in Figure 4(c). The interval between adjacent turns was 1.6 mm. The cross-section of each turn was rectangle of 2.5mm×10mm. The coil was made of copper by electrical discharge machining. It was fixed in polytetrafluoroethylene (PTFE) with epoxy to guarantee the strength and insulation. The distance between the coil and sheet was 2 mm. The outer and inner diameter of specimen were 180mm and 60mm, as shown in Figure 5(a). After EMHF, the displacement along Z direction and the thickness at the marked points of F, G, H, M and N along a generatrix were measured with accuracy of 0.01mm and 0.001mm, as shown in Figure 5(b). The height of workpiece reached 24.77mm, and the minimum thickness at the top was 0.840mm.

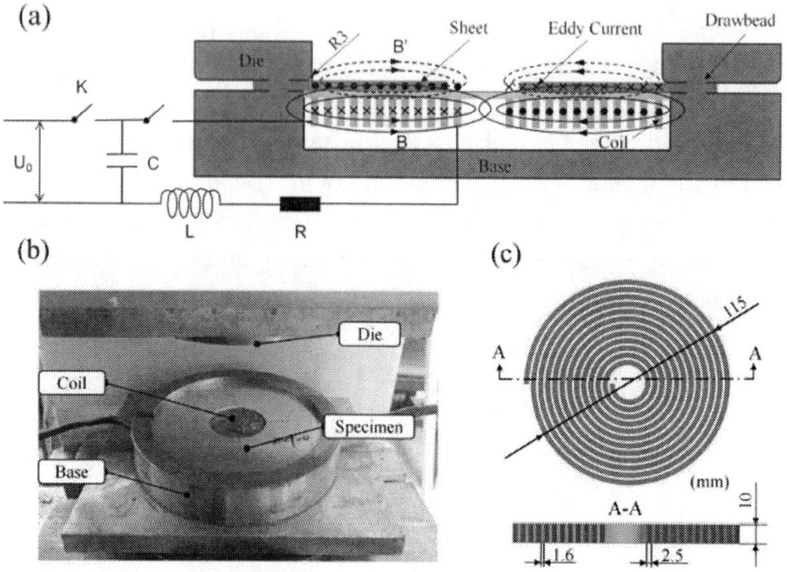

Figure 4. EMHF experiment: (a) Principle, (b) Setup, (c) Coil.

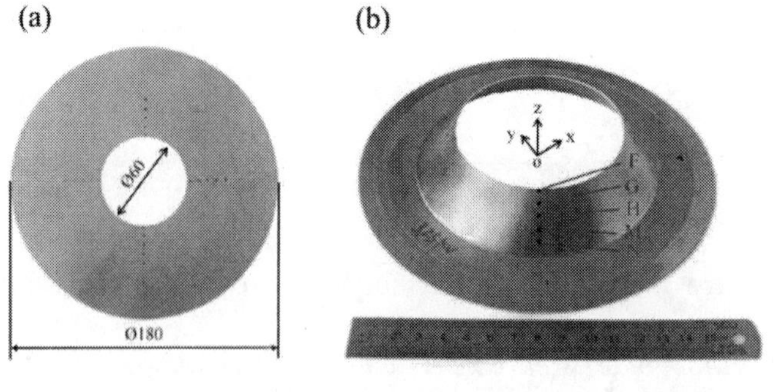

Figure 5. EMHF specimen: (a) Before deformation, (b) After deformation.

The numerical simulation of EMHF experiment was performed by sequential coupling of electromagnetic and mechanical fields. The electromagnetic field in air was calculated by boundary element method. It avoided the meshes of air and reduced the computation cost. The die, coil and base were defined to be rigid and fixed. The sheet and coil were discretized with hexahedral elements to accomplish the calculation of electrical currents, as shown in Figure 6. The aluminum alloy sheet was characterized by isotropic elasticity, Johnson-Cook (J-C) hardening model and GTN model. The elastic modulus and Poisson ratio were 68GPa and 0.33, respectively. The J-C model was defined as $\sigma = \left(74\text{MPa} + 362\text{MPa} \times \varepsilon_p^{0.51}\right)\left(1 + 0.0073 \times \ln(\dot{\varepsilon}/\dot{\varepsilon}_0)\right)$ with $\dot{\varepsilon}_0 = 10^{-3}s^{-1}$. A surface-to-surface contact with the friction coefficient of 0.2 was applied between the sheet and other components.

Figure 6. Numerical model of EMHF.

The parameters of GTN model under EMF condition were inversely identified by minimizing the discrepancy between experimental and simulated displacements along Z direction of workpiece. The simulated displacements showed a good agreement with experimental ones, as shown in Figure 7(a). The simulated and experimental thicknesses were compared to validate the accuracy of parameters. The numerical thicknesses almost matched with the experimental ones, as shown in Figure 7(b).

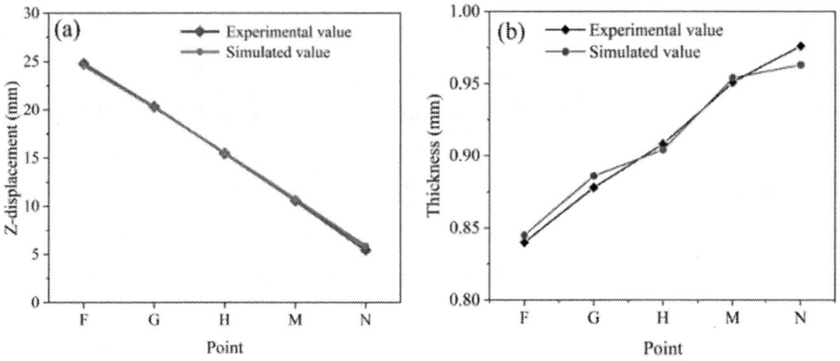

Figure 7. Experimental and simulated results of EMHF: (a) Displacement, (b) Thickness.

The identified parameters of GTN model were compared with the similar initial void volume fraction in Table 1. Considering that the deformation path at the edge of EMHF specimen is nearly the same with that at the central zone of quasi-static tensile specimen, the curves of void volume fraction (VVF) and strain at the two zones were compared to inspect the ductile damage evolution, as shown in Figure 8. The VVF increased with the deformation under the both conditions. However, the VVF increased significantly after the effective plastic strain of about 20% under quasi-static tension, while it increased slowly under EMF condition. With the deformation becoming larger, the induced current and Lorentz forces decreased quickly. The EMF deformation continued to increase by inertia effect, so the inertia effect was assumed to result in slow growth of VVF.

Table 1. Identified parameters of GTN model.

Parameters	f_0	f_c	f_N	ε_N
Quasi-static tension	0.006	0.075	0.131	0.44
EMHF	0.006	0.045	0.055	0.58

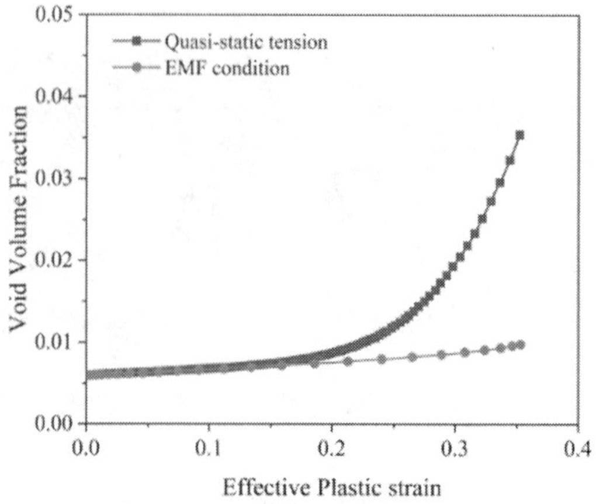

Figure 8. Comparisons of VVF and strain curves under EMF condition and quasi-static tension.

3. Application for EMF of a box part with inclined flange

A box part with inclined flange was shown in Figure 9(a). The maximum and minimum heights were 32mm and 17mm respectively. Due to the complex geometry, it is difficult to be deformed by the traditional forming methods such as deep drawing. Because of the high formability of aluminum alloy sheet during EMF, an EMF process with flat rectangular coil was applied for the box part, as shown in Figure 9(b). The coil was positioned at horizontal plane, and the sheet was placed at inclined plane. Where the sheet was closer to the coil, the Lorentz forces were larger and the deformed depth was larger. When the sheet contacted to the bottom of die, the rubber at the bottom absorbed the residual energy of sheet to form the flat bottom. The experimental setup was shown in Figure 10(a). The blank holder was installed on the hydraulic machine to press the flange of specimen with pressure of about 20MPa in order to avoid wrinkles. The flat rectangular coil was shown in Figure 10(b). The overall size of the coil is 110mm×96mm×10mm and the interval between adjacent turns was 3 mm. The capacitance and voltage were chosen to be 550μF and 6.5kV here. The specimens were shown in Figure 11. The principal strains were measured by a grid analysis software after deformation.

(a)

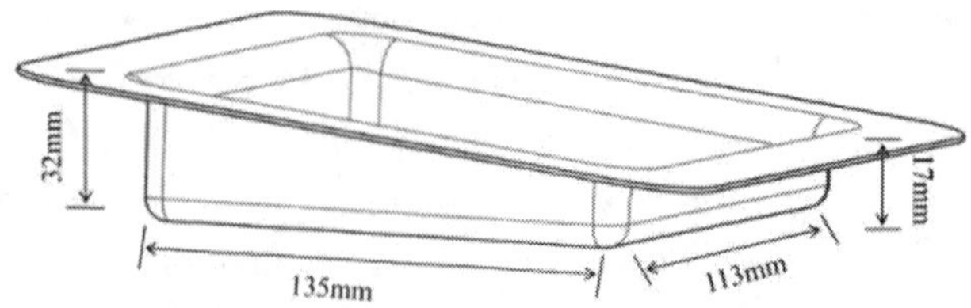

(b)

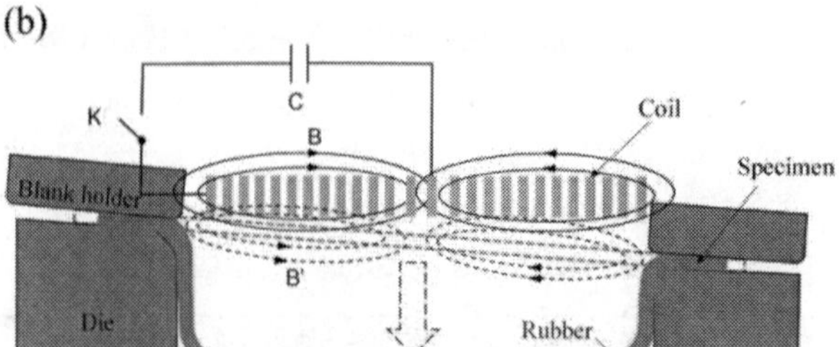

Figure 9. Box part with inclined flange: (a) Geometry (b)EMF principle.

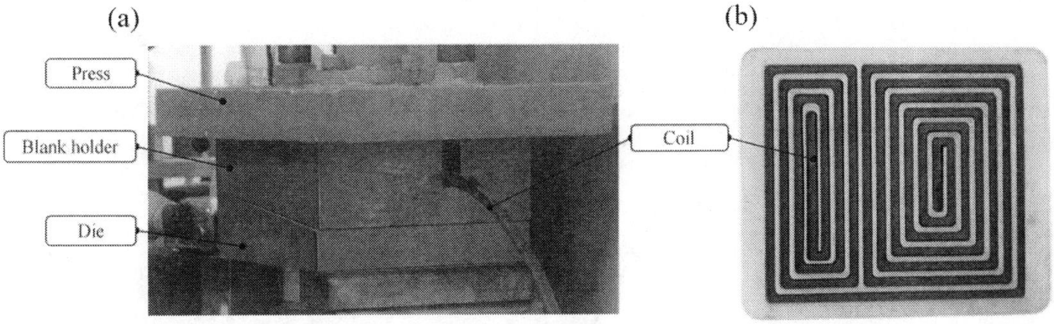

Figure 10. EMF experimental setup: (a) Die and tools, (b) Coil.

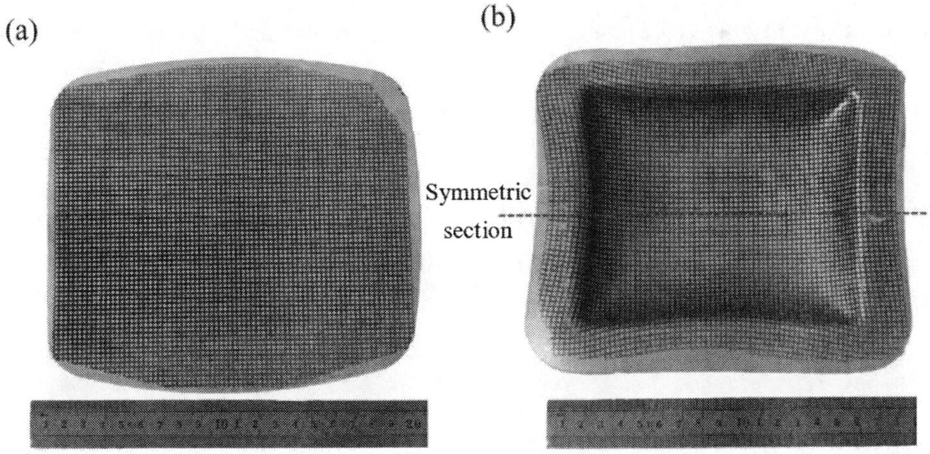

Figure 11. Specimens with inclined flange: (a) Before deformation, (b) After deformation.

The numerical model of EMF for the box part with inclined flange was shown in Figure 12. The GTN model was used for the AA2024 aluminum alloy sheet. The simulation results were shown in Figure 13. The thickness reduction of the box part was show in Figure 13(a). The maximum and minimum thickness reductions were 7.79% and -10.02%. The distribution of VVF was shown in Figure 13(b). The maximum value of 0.0072 appeared at the maximum depth.

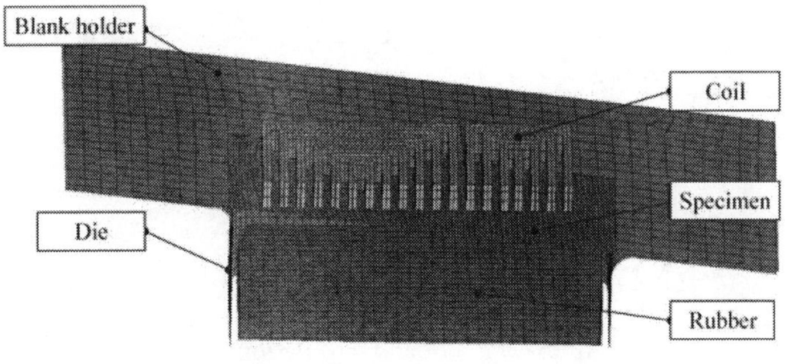

Figure 12. Numerical model of box part with inclined flange.

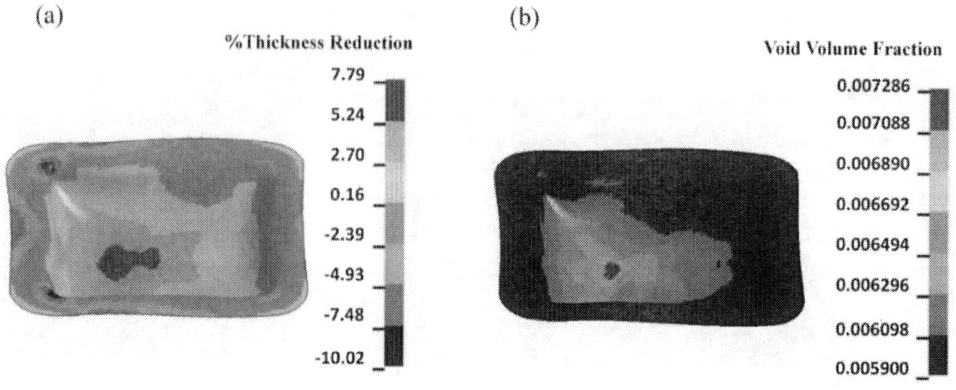

Figure 13. Numerical results: (a) Thickness reduction, (b) Void volume fraction.

In order to verify the GTN model, the results of experiment and simulation were compared along the symmetric section in Figure 11(b). The height, thickness, principal strains of specimen were compared with the numerical results respectively, as shown in Figure 14. The profile of the part was close to the simulated one, although the experimental height was slightly larger than the simulated value at the smaller depth, as shown in Figure 14(a). The maximum and minimum thicknesses were 1.090mm and 0.926mm, as shown in Figure 14(b). The first and second principal strains of numerical simulation were basically consistent with those of experimental part, as shown in Figure 14(c) and 14(d). The results showed that the parameters of GTN model is accurate for numerical simulation.

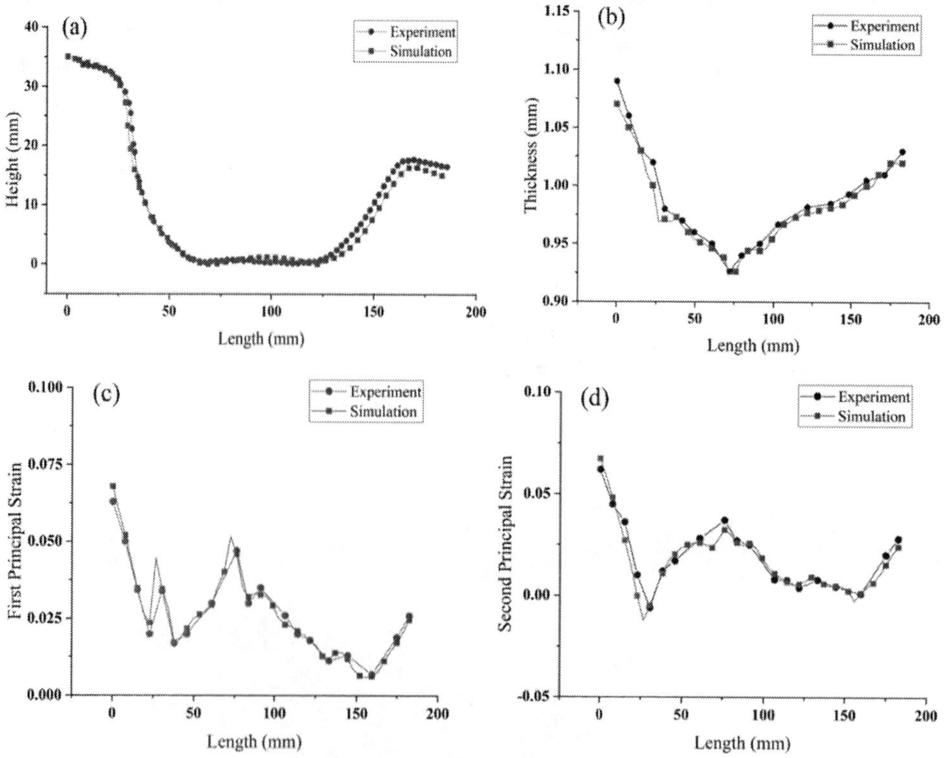

Figure 14. Comparisons of EMF experiment and simulation: (a) Height, (b) Thickness, (c) First principal strain, (d) Second principal strain.

4. Conclusion

The inverse identification of GTN model were performed under the quasi-static tension and EMF conditions for AA2024-O aluminum alloy sheet. With the similar initial void volume fraction, the other parameters are different under both conditions. The curve of void volume fraction and strain increased more significantly under quasi-static tension when compared with that under EMF condition. Therefore, the less damage accumulation was revealed for AA2024-O aluminum alloy sheet during EMF. It can be attributed to an important aspect of the ductile enhancement mechanism for EMF. The numerical simulation with GTN model was performed for EMF of a box part with inclined flange. The accuracy of GTN model was verified by comparing the simulation and experimental results.

5. References

[1] Psyk V, Risch D, Kinsey B L, Tekkaya A E and Kleiner M 2010 *J. Mater. Process. Technol* **211** 787–829
[2] Imbert J A, Worswick M A, Winkler S A, Golovashchenko S B and Dmitriev V B 2005 *SAE* **114** 6–18
[3] Su H, Huang L, Li J, Xiao W, Zhu H, Feng F, Li H and Yan S 2020 *J. Mater. Sci. Technol* **70** 125–35
[4] Tekkaya A E, Bouchard P O, Bruschi S and Tasand C C 2020 *CIRP Ann-Manuf Techn* **69** 600–23
[5] Tekkaya A *et al.* 2015 *CIRP Ann-Manuf Techn* **64** 629–53
[6] Gurson and A L 1977 *J. Eng. Mater. Technol* **99** 297–300
[7] Tvergaard V 1981 *Int. J. Fract* **17** 389–407
[8] Tvergaard V and Needleman A 1984 *Acta Metall* **32** 157–69
[9] Needleman A and Tvergaard V 1984 *J. Mech. Phys. Solids* **32** 461–90
[10] Imbert J M, Winkle S L, Worswick M J, Oliveira D A and Golovashchenko S 2005 *J. Eng. Mater. Technol* **127** 145–53
[11] Feng F, Li J, Yuan P, Zhang Q, Huang P, Su H and Chen R 2018 *Metals* **8** 761
[12] Zeng X, Meng Z, Liu W, Huang S and Lin Y 2020 *Int J Adv Manuf Tech* **109** 745–54
[13] Rakin M, Cvijovic Z, Grabulov V, Putic S and Sedmak A 2004 *Eng. Fract. Mech* **71** 813–27
[14] Steglich D, Siegmund T and Brocks W 1999 *Comput. Mater. Sci* **16** 404–13
[15] Ying L, Wang D, Liu W, Wu Y and Hu P 2018 *Int. J. Mater. Form* **11** 527–39
[16] Wang M and Li X 2012 *Rare Met. Mater. Eng* **41** 795–9

Acknowledgments

The work is financially supported by National Natural Science Foundation of China (grant number 52005374) and by open research fund of State Key Laboratory of High Performance Complex Manufacturing, Central South University (grant number Kfkt2021-04).

Numerical analysis of a backward flow forming operation of AA6061-T6 and comparison with experiments

K Mocellin[1], M A Vidal[1,2], F Frascati[2] and P O Bouchard[1]

[1] MINES Paris-Tech, PSL- Research University, CEMEF – Center for Material Forming, CNRS UMR 7635, BP 207, 1 rue Claude Daunesse, 06904 Sophia Antipolis Cedex, France

[2] MBDA France, Rond-Point Marcel Hanriot, 18020 Bourges Cedex, France

katia.mocellin@minesparis.psl.eu

Abstract. Incremental forming processes can be used to produce thin products (tubes or sheets)[1]. Very high deformation of the material can be reached taking advantage of the local and cyclic loading of the material. In this study we will focus on backward flow forming of aluminum tubes. In this process, tube thickness is reduced by the combined action of a rotating mandrel that imposes the inner radius and 3 rollers that decrease progressively the tube thickness. An experimental campaign is conducted on a laboratory device to study the influence reduction rate. The corresponding configurations are simulated to understand the mechanical loading path. The material characterization is presented to focus on the influence of the chosen behavior law on flow forming simulation results. Different damage criteria coming from the literature are studied to evaluate their capability to predict fracture and to compare the amount of damage reached for each process configurations. Even if none of them is able to predict accurately damaging configurations, the classical Cockroft and Latham seems to be the only capable to reflect the hierarchisation of configurations.

Introduction

Tube flow forming is an incremental forming process used to produce thin resistant tubes. The incremental loading condition allows the material to reach high deformation without exhibiting fracture. Inner diameter is controlled with a mandrel while thickness is reduced with rotating rollers. Several authors have studied the process from both experimental and numerical point of views [2, 3] comparing mainly global results as geometry or loads but not linking local mechanical variables with fracture prediction.

The objective, in this work, is to develop a numerical model able to understand material loading path, to predict fracture occurrence and to compare it with experimental results.

To save time and material, a lab scale device is used to study the process rather than an industrial device adapted to large tube dimensions. Smaller tubes are machined in the inner material waste coming from industrial preforms preparation as represented in **Figure 1**. Geometries of the rollers are defined homothetically from the industrial configuration. Dealing with smaller specimens allows also to reduce the associated CPU time.

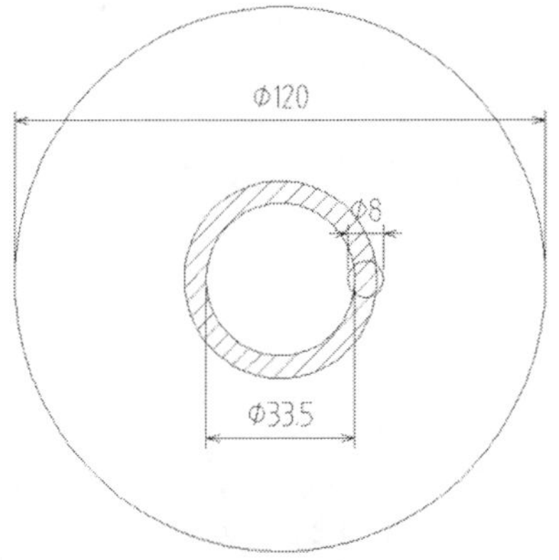

Figure 1. Dimensions of the initial extruded bar from which the industrial flow forming preforms, lab scale device preforms (hachured) and compression samples are machined.

Numerical model has to be fed with material behavior laws and damage criteria. The computational model is then used to reproduce an experimental campaign in which increasing reduction ratio is applied to exhibit the limitation of the process. Three damage models are then evaluated for the prediction of fracture.

1. Material characterization and experimental flow forming campaign

1.1. Material characterization

The behavior of the material was identified using compression tests. Compression samples were extracted from the initial bar along the extrusion direction, in the same zone were flow forming tubular preforms were also machined. Localization of both tubes and samples are represented in **Figure 1**. Samples are 12 mm height and 8 mm in diameter.

The samples were tested at different strain rates of 0.001, 0.1 and 10 s^{-1} and temperatures, 25, 70 and 100 °C to cover the large range of thermo-mechanical conditions encountered in the flow forming process. Load-displacement curves are shown in

Figure 2. The identification of the parameters of the behavior law was done in three steps using inverse analysis: strain hardening part first, dependence to the temperature and then dependence to strain rate using an in-house software of inverse analysis [4]. Because of heat generated by plastic deformation at high strain rate, the dependence to temperature must be identified prior to the dependence to strain rate.

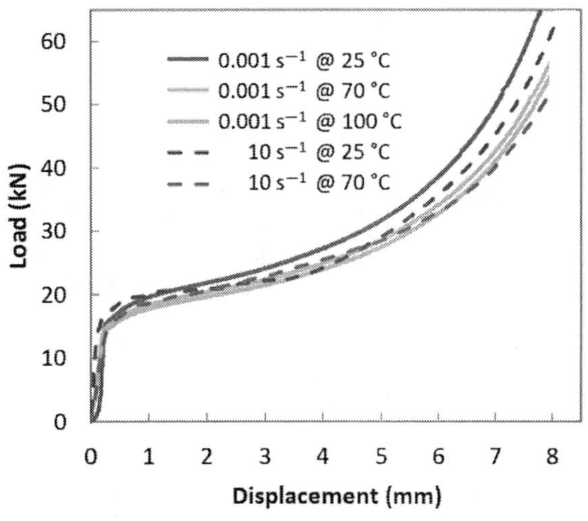

Figure 2. Experimental load-displacement curves of compression tests at 0.001 s−1 and 25, 70 and 100 °C (solid lines) and at 10 s−1 at 25 and 70 °C (dotted lines).

A material behavior model depending on both temperature and strain rate should be used in the flow forming simulation to account for heat generated both by plastic deformation and friction with tools. The chosen behavior law follows the Johnson-Cook model but with a Voce hardening part as described in equation (1):

$$\sigma = [\sigma_S - (\sigma_S - \sigma_0)\exp(-K\bar{\varepsilon})]\left[1 + C\ln\frac{\dot{\bar{\varepsilon}}}{\dot{\bar{\varepsilon}}_0}\right]\left[1 - \left(\frac{T - T_R}{T_M - T_R}\right)^m\right] \quad (1)$$

σ_0 is the yield stress (MPa), σ_S is the saturated stress (MPa), K is a dimensionless parameter describing the hardening effect, $\bar{\varepsilon}$ is the equivalent plastic strain, C is a dimensionless parameter describing the material sensitivity to strain rate, $\dot{\bar{\varepsilon}}$ is the equivalent strain rate (s^{-1}) of the test, $\dot{\bar{\varepsilon}}_0$ is the strain rate of the reference test (s^{-1}), T is the temperature of the test (°C), T_R is the reference temperature (°C), T_M is the melting point of the material and m is a dimensionless parameter describing the material sensitivity to temperature. The values of the identified parameters for AA6061-T6 alloy using compression tests are summed up in Table 1. This law will be denoted as VJC (Voce and Johnson Cook) law in the following.

Table 1. Identified values using inverse analysis of the parameters of the Voce-Johnson-Cook (VJC) behaviour law for AA6061-T6 tested in compression.

Voce hardening part	Temperature Johnson-Cook part	Strain rate Johnson-Cook part
$\sigma_0 = 301.16\ MPa$	$m = 0.81$	$C = 0.005$
$\sigma_S = 362\ MPa$	$T_R = 25\ °C$	$\dot{\bar{\varepsilon}}_0 = 0.001\ s^{-1}$
$K = 35{,}57$	$T_M = 632\ °C$	

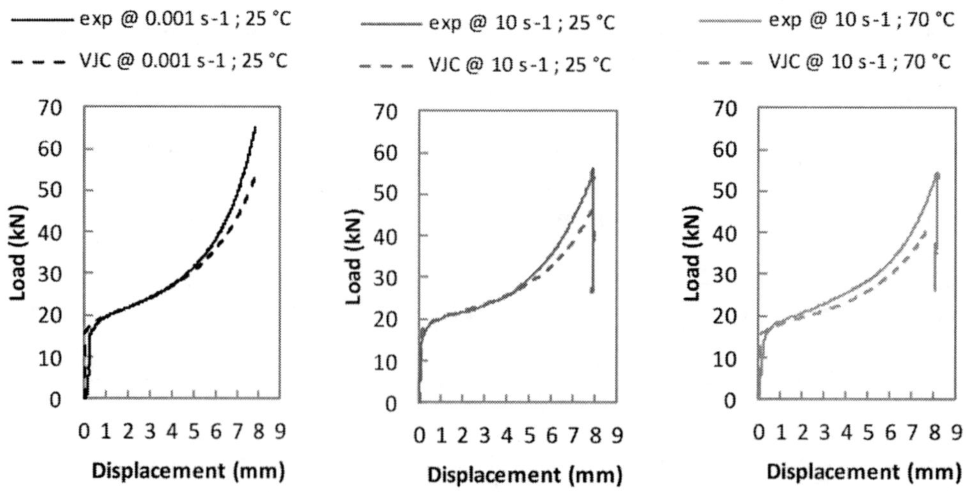

Figure 3 : Comparison of experimental and computed curves of force vs displacement

The comparison between the VJC law and the experimental results are compared on **Figure 3** for 3 different testing configurations. Correlation is better at the beginning and worse ate the end due to excessive large friction after 50% of reduction rate of the specimens.

1.2. Experimental flow forming campaign

Flow forming experiments are conducted in CEMEF on a lab sized flow forming machine [5]. It consists in a backward flow forming device with a rotating mandrel. The tubular preform is clamped on at its end and formed with 3 conical rollers. These rollers are arranged at 120° to each other, they are staggered along the Z axis -the revolution axis of the mandrel- and along the thickness of the preform. Each roller takes one third of the total depth (**Figure 4**). The loads can be recorded only on one roller per operation but on all the three directions (radial, tangential and axial).

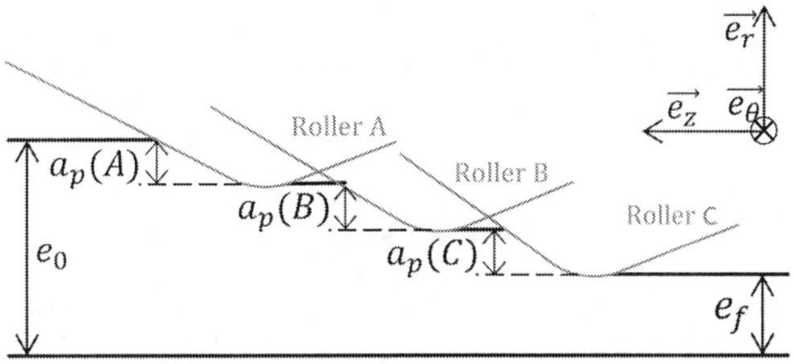

Figure 4.: Illustration of the three rollers staggered along Z if they are represented in the same plane $(\vec{e_r};\vec{e_\theta})$

An important parameter of the flow forming operation is the reduction ratio or thickness reduction ratio *TR*(%) which is defined by:

$$TR(\%) = 100 \times \frac{e_0 - e_f}{e_o} \tag{2}$$

where the reduction ratio is expressed in %, e_0 is the initial thickness of the preform (mm) and e_f is the final thickness of the tube (mm).

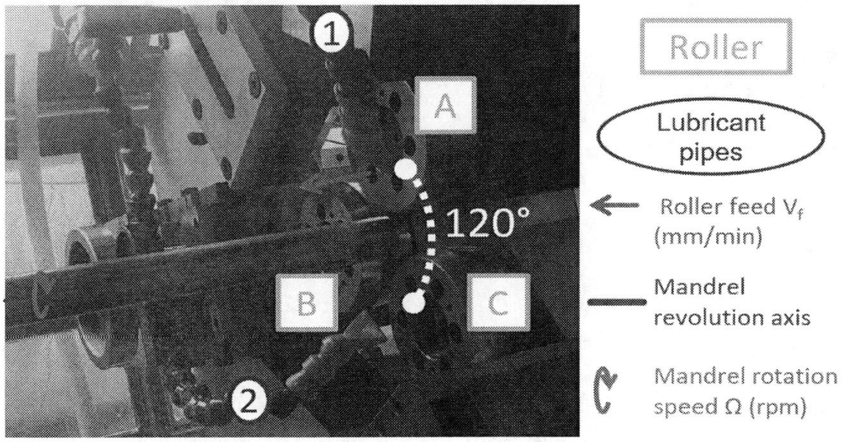

Figure 5. Experimental flow forming device in CEMEF.

For the study of the influence of the behaviour law on tool forces in simulation, we have chosen the following flow forming parameters: a mandrel speed of 50 rpm and an axial feed of 0.2 mm/rotation. Lubrication is performed with intensive flow of water-soluble cutting oil.

For the local analysis of the stress state, four flow forming experiments were conducted on 5 mm thick tubes at different reduction ratios using the process parameters listed below:

Tube 1: (TR60 P1): One pass of 60 % of reduction (expected final thickness is 2 mm)

Tube 2: (TR80 1P1): One pass of 80 % of reduction (expected final thickness is 1 mm)

Tube 3: 2 passes. The first one (TR80 2P1) of 40 % ratio (expected final thickness is 3 mm), then a second one on the resulting tube (TR80 2P2) with 66.67% reduction (expected final thickness is 1 mm)

In this study, the software Forge® was used. It is based on an implicit formulation with a mixed finite element in velocity and pressure [6]. Forge® is dedicated to the numerical simulation of forming processes implying large deformation. Automatic remeshing and parallel computation can be used to handled large plastic strain in the material and high number of nodes necessary to discretise accurately both the part and the local contact areas [7].

To control the volume variation of the tube during the simulation and to reduce the computation time [8], the description of kinematic interaction is modified. The {mandrel+tube} system is fixed whereas the rollers are rotating around the system at the actual mandrel rotation speed, Ω. The rotation of rollers around themselves due to friction with the tube is fixed during the simulation. The rotation speed is estimated with:

$$\omega_i R_i = \Omega_{mandrel}(R_{mandrel} + e_i) \quad (3)$$

where $\boldsymbol{i}$ is related either to the roller A, B or C, ω_i is the roller rotation speed (rpm), R_i is the roller radius (mm), $\Omega_{mandrel}$ is the mandrel rotation speed (rpm), $R_{mandrel}$ is the mandrel radius (mm) and e_i is the tube thickness after the roller i (mm). A global view of the 3D flow forming simulation explaining the specific kinematics is available in **Figure 6** (a).

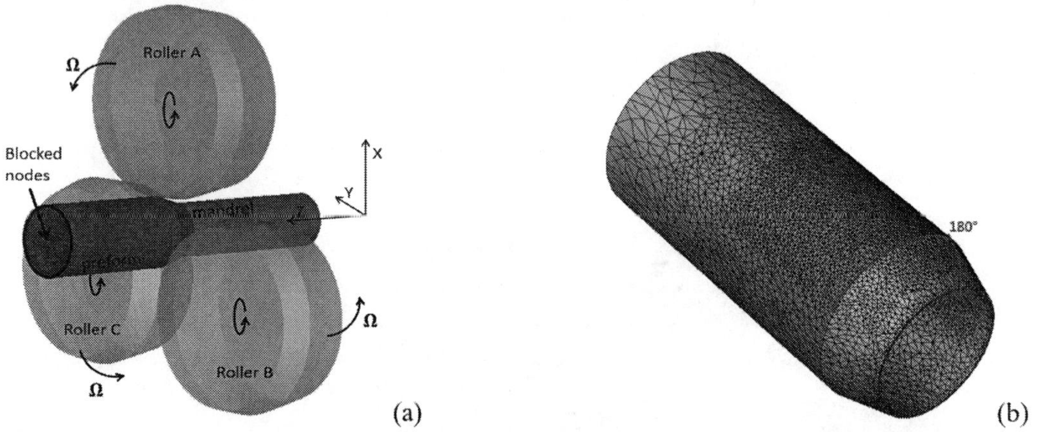

(a) (b)

Figure 6: (a) Global view of a 3D flow forming simulation using Forge®, (b) Meshed tube used in flow forming simulations in Forge®.

In the flow forming process, the contact zone between the rollers and the tube is very small, thus a small mesh size is required to properly describe the contact events. Moreover, the final thickness of the tube, depending on the reduction ratio, varies from 1 to 2 mm in this study. To capture the difference of stress state along the thickness of the tube, the mesh size is chosen to be lower than the half of the theoretical final thickness. However, having a fine mesh size on the whole tube would lead to very long computation time. To prevent that, only half of the tube is finely meshed as shown in **Figure 6** (b).

Each material point in the flow formed zone of the tube is deformed several times by each roller, at high strain rates and during short contact time, attesting the incremental deformation of the process. Small computation time steps are then required in the simulation, increasing the computation duration of the simulation. The computation time step is arbitrary chosen as half of the time that a roller needs to cover 1° of the tube.

As the tube is lubricated during the experimental flow forming operation, the heat exchange coefficient between the tube and its environment in simulation is set to 2100 $W.m^{-2}.K^{-1}$. The heat exchange coefficient between the tube and the tools is set to 2000 $W.m^{-2}.K^{-1}$. These values are classically used for cold forming lubricated processes.

As the rollers are rotating around themselves due to friction with the tube, a friction coefficient between the tube and the rollers is put in the simulation. A Coulomb-Tresca friction model is used with the respective following parameters: $\mu = 0.2$ and $\overline{m} = 0.4$.

To evaluate the capability to predict fracture of the tubes, damage criteria are computed. Several criteria can be found in the literature for analysis of flow forming [8, 9]. In the following we will focus on the Latham & Cockroft damage depending on the first principal stress, the Oyane based on triaxiality evolution and a variation of the Lou & Huh criterion [10] expressed with respect to local shearing conditions as described in equation (4).

Latham & Cockroft $$D_{CLN} = \int_0^{\bar{\varepsilon}} \langle \sigma_1 \rangle d\bar{\varepsilon}$$

Oyane $$D_{Oy} = \int_0^{\bar{\varepsilon}} \langle 1 + \frac{\eta}{A} \rangle d\bar{\varepsilon}$$

Shearing based on Lou & Huh $$D_{Shearing} = \int_0^{\bar{\varepsilon}} \langle \frac{\sigma_1 - \sigma_3}{\bar{\sigma}} \rangle d\bar{\varepsilon} \quad (4)$$

where σ_1 and σ_3 with first and last principal stress components and $\bar{\sigma}$ the equivalent von Mises stress. η the stress triaxiality A a parameter depending on the material here taken to 3. $\langle . \rangle$ a function defined as $\begin{cases} \langle x \rangle = x \text{ if } x > 0 \\ \langle x \rangle = 0 \text{ if } x \leq 0 \end{cases}$ and $\bar{\varepsilon}$ the plastic deformation

The local mechanical path undergone by the material in flow forming is complex, multiaxial and cyclic. It is then difficult to point out what are the most damaging stress states.

2. Experimental campaign results vs simulation modeling

2.1. Forming forces

Flow forming simulations is compared to experimental results for the 60% of programmed thickness reduction, in the same processing conditions. The forces on the first roller are measured and compared to the simulated ones on **Figure 7**.

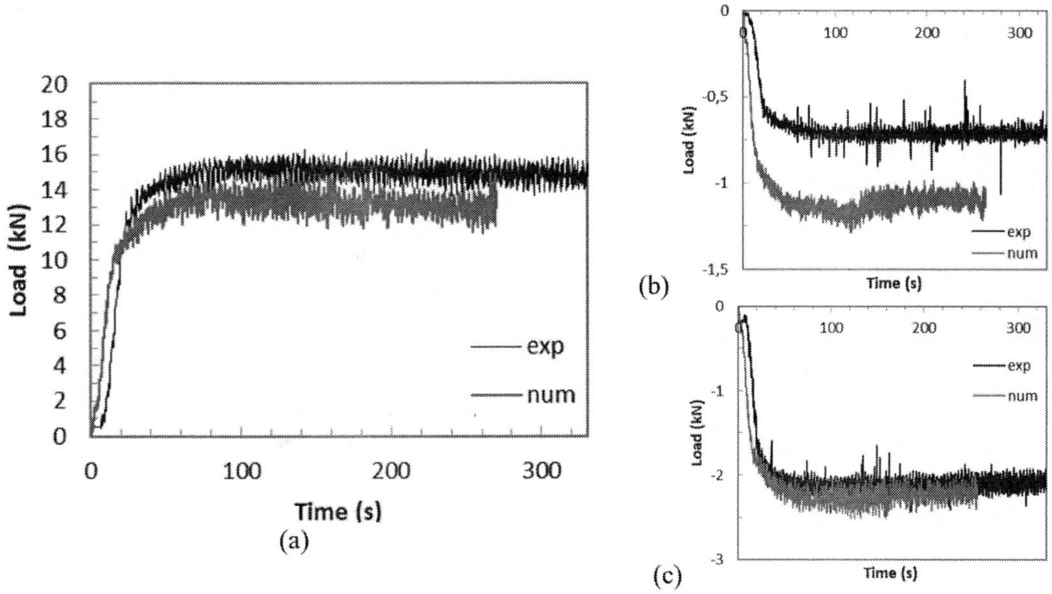

Figure 7. Radial(a), tangential (b) and axial (c) loads on roller A (kN) as a function of process time (s) experimentally and simulated by Forge®.

The force is first increasing when flow forming the conical part and reaching a plateau both in simulation and experimentally when the tubular part is flow formed at a constant reduction ratio.
Radial force is the highest and is well predicted by the numerical model. Axial force is also well predicted while radial component is underestimated by the simulation. It can be noticed that the value of this component is quite low and is experimental accuracy poor depending on the force sensor capacity.

2.2. Effect of the reduction ratio
We will focus on the radial component of the force that is the highest value and the most representative and limiting of the process (on industrial plant it is the only one recorded). The plateau values of radial forces are reported for different reduction ratio in **Figure 8**. The force is increasing with the reduction ratio.

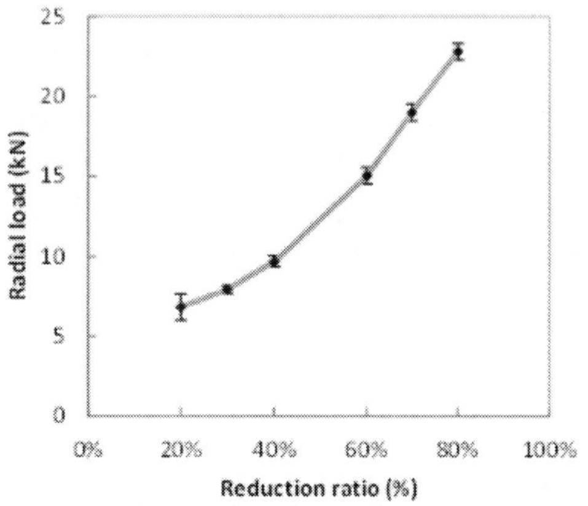

(a)

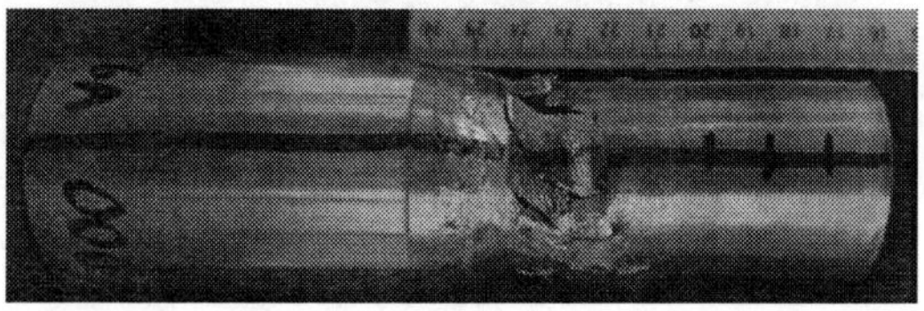

(b)

Figure 8. Evolution of forming load with the reduction ratio(a) and fracture observed for 80% TR

The increase of the reduction ratio leads to fracture for 80 % reduction in one step. However, for the third tube the same global reduction ratio is programmed in 2 flow forming steps as above. This tube did not show any fracture. Different computed damage criteria are then analyzed on 4 configurations to see which ones can predict the increase of the reduction ratio should lead to an increase of damage and a reduction ratio of 80% leads to more damage when achieved in one pass rather than in two.

2.3. Analysis of local mechanical variables
The plastic strain evolution is analysed with computational sensors on **Figure 9.**. Those sensors are recording of mechanical and thermal variables associated to given material points of the tube all along the simulation. One is located close to the outer surface, solicited by the roller, the second, close to the inner surface, nearby the mandrel.

The plastic strain is globally higher on the outer surface, values for 80 % (TR801P1) are higher than 60% (TR601P1) and 40% (TR80 2P1) but values for 80 % in 2 passes (TR802P2) is higher than the one for one pass (TR801P1). For 40%, the outer surface is highly deformed, this reduction ratio is not representative of "flow forming" conditions and then difficult to compare with values obtained with higher TR. However, the expected order is respected on the inner face of the tube.

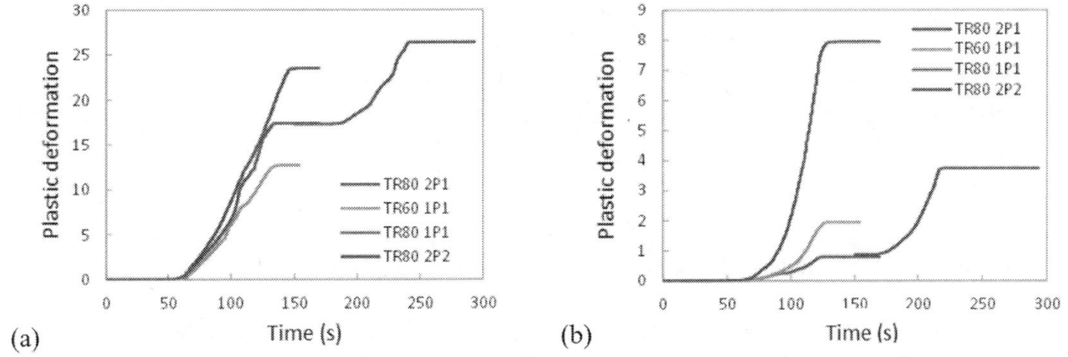

Figure 9. Evolution of plastic strain on sensors placed close to the outer surface (a) or the inner one (b) during flow forming in different configurations

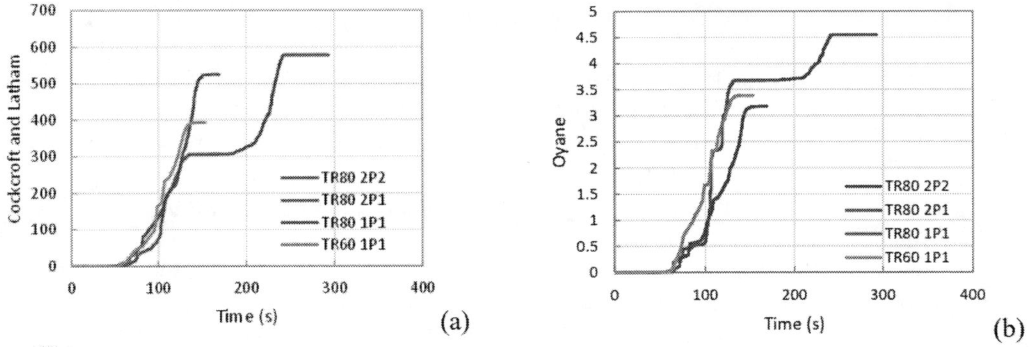

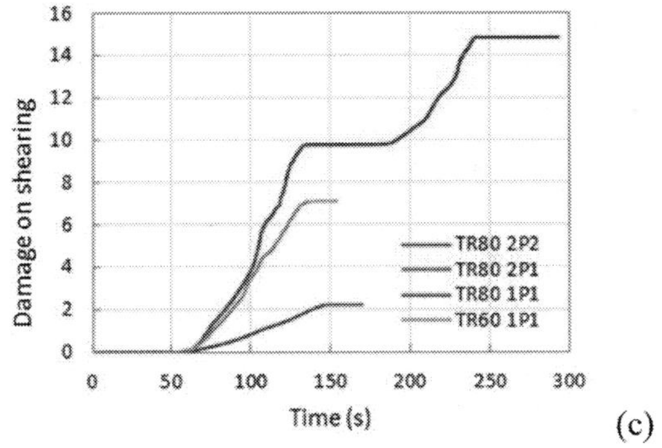

Figure 10. Evolution of damage criteria on sensors close to the outer surface in different flow forming conditions. (a) Latham & Cockroft criterion, (b), Oyane criterion, (c) criterion based on shear component of the stress tensor.

3. Results analysis

Evolution of plastic strain is used as a fracture criterion for monotonic strain path undergone during standard identification test. In flow forming the level of cumulated plastic strain withstood by the material is very high and coming from complex, cyclic and non monotonic loading path, it cannot then be used directly as a fracture criterion. It was thus interesting to notice that plastic strain is increasing with reduction ratio but for an equivalent final reduction rate of 80 % higher plastic strain evolution is different if one or two passes are used. The distribution of plastic strain within the tube thickness is then an interesting observation to study further in the future.

Damage criteria are giving different prediction. Oyane or the shear-based criterion are predicting decreasing damage with increasing reduction ratio. Only the Latham and Cockcroft based on the first principal stress and thus the tensile configuration allows to reach the expected hierarchy of the configurations. The difference between the 2 routes of 80 % is however small.

As a conclusion tube flow forming is complex to analyze because the material is incrementally and cyclically deformed reaching values of plastic strain almost impossible to reproduce with standard characterization testing. Classical damage criteria are not giving a clear answer, enhanced formulation must be considered. Mechanical analysis can provide information for the prediction of flow formability, but some metallurgical aspects are neglected in the approach and could help to understand more accurately the capability of the material to be deformed without fracture.

References

[1] Music O, Allwood JM, Kawai K-I. A review of the mechanics of metal spinning. *J Mater Process Technol* 2010; **210**: 3–23.
[2] Günay E, Fenercioğlu TO, Yalçinkaya T. Numerical analysis of thermo-mechanical behavior in flow forming. *Procedia Struct Integr* 2022; **35**: 42–50.
[3] Bylya OI, Khismatullin T, Blackwell P, et al. The effect of elasto-plastic properties of materials on their formability by flow forming. *J Mater Process Technol* 2018; **252**: 34–44.
[4] Roux É, Tillier Y, Kraria S, et al. An efficient parallel global optimization strategy based on Kriging properties suitable for material parameters identification. *Archive of Mechanical Engineering,* 2020,**67**-2,p 169
[5] Depriester D, Massoni E. On the Damage Criteria and their Critical Values for Flowforming of ELI Grade Ti64. *Key Eng Mater* 2014; **622–623**: 1221–1227.
[6] Chenot J-L, Bouchard P-O, Fourment L, et al. Numerical Simulation and Optimization of the Forging Process. *12th ICFC 2011*, May 2011, Stuttgart, Germany.
[7] Coupez T, Digonnet H, Ducloux R. Parallel meshing and remeshing. *Appl Math Model* 2000; **25** (2), p.153-175
[8] Ma H, Xu W, Jin BC, et al. Damage evaluation in tube spinnability test with ductile fracture criteria. *Int J Mech Sci* 2015; Complete: 99–111.
[9] Xu W, Wu H, Ma H, et al. Damage evolution and ductile fracture prediction during tube spinning of titanium alloy. *Int J Mech Sci* 2018; **135**: 226–239.
[10] Lou Y, Huh H. Prediction of ductile fracture for advanced high strength steel with a new criterion: Experiments and simulation. *J Mater Process Technol* 2013; **213**: 1284–1302.

Comparison of three different ductile damage models for deep drawing simulation of high-strength steels

B-A Behrens[1], D Rosenbusch[1], H Wester[1] and P Althaus[1]*

[1] Institute of Forming Technology and Machines (IFUM), Leibniz Universität Hannover, An der Universität 2, 30823 Garbsen, Germany

* althaus@ifum.uni-hannover.de

Abstract. High-strength steels are increasingly used in deep drawing for automotive parts due to their improved strength properties. To increase the formability and thus extend the process limits, a deep drawing process with additional force transmission has been developed. For a numerical optimisation of the considered process, an exact modelling of the failure behaviour is essential. The forming limit curve (FLC) is widely used to predict the onset of necking in sheet metal forming. However, the validity of the FLC is limited to the case of linear strain paths. Therefore, the scope of past investigations has been on failure modelling depending on the stress state. This article presents the experimental-numerical characterisation of the failure behaviour of the high-strength steels HCT600X and HX340LAD. Tensile tests with butterfly specimens were carried out under varying stress states and simulated with ABAQUS to parametrise the stress-based models Johnson-Cook (JC), Modified Mohr-Coulomb (MMC) and DF2016. An additional experiment was carried out to evaluate the models in terms of predicted outcome accuracies by comparing the onset of fracture in the simulation with the experimental findings. In future investigations, the improved damage modelling will be applied in a deep drawing simulation with additional force transmission to optimise the process design.

1. Introduction

Due to the ongoing demand for reduction of CO_2 emissions, lightweight construction is still one of the main research topics in the automotive sector [1]. A reduced overall vehicle mass can be achieved by the application of high-strength steels. The improved strength properties qualifies them for structural components with low weights [2]. To exploit the full potential of these steels, existing production processes must be continuously improved to increase productivity, material utilisation, energy balance and dimensional accuracy of the final parts.

Deep drawing of sheet metals is mainly used for the forming of car chassis parts. To expand the forming limits, a deep drawing process with an additional force transmission has been developed [3]. This technology is based on the local introduction of an additional force in the bottom of the part during forming, which initiates a pressure superposition in critical areas for fracture. By delaying the initiation of necking, the process window is extended, which was proven in experimental tests [4]. A useful tool to determine the extended process window for different materials is the finite elements (FE) method, which enables a process investigation without cost- and time-consuming trial and error test. However, for a numerical investigation, a precise modelling of the failure behaviour is important to detect fractures during forming. The FLC, which describes the forming limit as a function of major and minor strains, is a standard tool for the prediction of necking in sheet metals. Nevertheless, the range of validity is limited to linear strain paths. Necking that occurs during deep drawing in the bottom radius of the part could not be reliably predicted with the FLC [5]. Therefore, a better approach is to describe the failure behaviour by means of a stress-based damage model to improve the prediction accuracy.

In this work, three ductile damage criteria Johnson-Cook (JC), Modified Mohr-Coulomb (MMC) and DF2016 are applied to predict fracture for two high-strengths steels HCT600X and HX340LAD. Tensile test with butterfly specimen are carried out and simulated in ABAQUS to parametrise the coefficients of the criteria. The aim is to improve the damage modelling for a later application in the simulation of deep drawing with additional force transmission. The overall goal is to improve the process design and achieve a reduction of costs by avoiding time- and cost-consuming trail-and-error tests.

2. Stress based damage modelling

Modelling of the damage behaviour of ductile metals has been the focus of a wide range of researches in the last decades. A common approach is the use of an uncoupled fracture criterion, which is usually defined by an empirical formula. Since the multi-axial stress state has a great influence on the damage behaviour, current models take the stress triaxiality η into account, which is defined as the ratio of mean normal stress σ_m to the von Mises equivalent stress $\bar{\sigma}$ and the three principal stresses σ_1, σ_2 and σ_3:

$$\eta = \frac{\sigma_m}{\bar{\sigma}} = \frac{1/3\,(\sigma_1 + \sigma_2 + \sigma_3)}{\sqrt{1/2\,[(\sigma_1 - \sigma_2)^2 + (\sigma_2 - \sigma_3)^2 + (\sigma_1 - \sigma_3)^2]}} \tag{1}$$

Johnson and Cook published a criterion that calculates the equivalent plastic strain at fracture ε_{pl}^{f} as a function of stress triaxiality η, dimensionless plastic strain rate $\dot{\varepsilon}^*$ and homologous temperature T^* [6]:

$$\varepsilon_{pl}^{f} = [D_1 + D_2 \exp(D_3\eta)\,][1 + D_4 \ln \dot{\varepsilon}^*][1 + D_5 T^*] \tag{2}$$

T^* and $\dot{\varepsilon}^*$are defined by equation 3, where T_m is the melting temperature and $\dot{\varepsilon}_{ref}$ and T_{ref} are the reference strain rate and the reference deformation temperature, respectively.

$$\dot{\varepsilon}^* = \frac{\dot{\varepsilon}}{\dot{\varepsilon}_{ref}} \qquad T^* = \frac{T - T_{ref}}{T_m - T_{ref}} \tag{3}$$

Recent studies show that the JC criterion is still widely used today, e. g. for the simulation of blanking [7] or friction drilling [8]. However, it was found that the third stress invariant ξ, defined by equation (4), also has a great influence of the materials fracture characteristic.

$$\xi = \frac{27(\sigma_1 - \sigma_m)(\sigma_2 - \sigma_m)(\sigma_3 - \sigma_m)}{2(1/2\,[(\sigma_1 - \sigma_2)^2 + (\sigma_2 - \sigma_3)^2 + (\sigma_1 - \sigma_3)^2])^{3/2}} \tag{4}$$

Therefore, current models consider the third stress invariant by including the Lode angle θ or the normalised Lode angle $\bar{\theta}$ defined by equation (5) in the calculation [9].

$$\theta = 1/3 \arccos(\xi) \qquad \bar{\theta} = 1 - \frac{6\theta}{\pi} \tag{5}$$

Bai and Wierzbicki extended the Mohr-Coulomb criterion for ductile metals, which is widely known as MMC criterion [10]:

$$\varepsilon_{pl}^{f} = \left\{\frac{A}{c_2}\left[c_\theta^s + \frac{\sqrt{3}}{2 - \sqrt{3}}(c_\theta^{ax} - c_\theta^s)\left(\sec\left(\frac{\bar{\theta}\pi}{6}\right) - 1\right)\right]\left[\sqrt{\frac{1 + c_1^2}{3}}\cos\left(\frac{\bar{\theta}\pi}{6}\right) + c_1\left(\eta + \frac{1}{3}\sin\left(\frac{\bar{\theta}\pi}{6}\right)\right)\right]\right\}^{-1/n} \tag{6}$$

The MMC criterion has been successfully applied for various materials, e. g. aluminium alloys [11], advanced high strengths steels [12] and magnesium alloys [13]. Lou et al. published another ductile damage criterion, called DF2016, which is defined by the following equation [14]:

$$\left(\frac{2}{\sqrt{L^2+3}}\right)^{C_1}\left(\langle\frac{f(\eta,L,C)}{f(\frac{1}{3},-1,C)}\rangle\right)^{C_2} \varepsilon_{pl}^{f} = C_3 \qquad \text{with} \qquad f(\eta, L, C) = \eta + C_4\frac{(3-L)}{3\sqrt{L^2+3}} + C \tag{7}$$

L is the Lode parameter, which is defined as follows:

$$L = \frac{2\sigma_2 - \sigma_1 - \sigma_3}{\sigma_1 - \sigma_3} \quad (8)$$

The DF2016 criterion can be transformed into the space of $(\varepsilon, \eta, \overline{\theta})$ by the following relation:

$$\tan(\theta) = \frac{\sqrt{3}(L+1)}{3-L} \quad (9)$$

Due to their simple formulation, uncoupled damage criteria offer a simple implementation in existing FE codes. The parametrisation of the criteria usually requires an extensive number of experiments to ensure the consideration of sufficient stress states. Therefore, a special tool system will be used, which enables the induction of different stress states in a butterfly specimen.

3. Test procedure

For the tensile tests, a butterfly specimen was used following the geometry of Dunand and Mohr [15], which was geometrically modified so that the fracture appears more clearly in the centre of the specimen due to a stronger strain localisation [16]. Figure 1 (a) shows the geometry of the butterfly specimen.

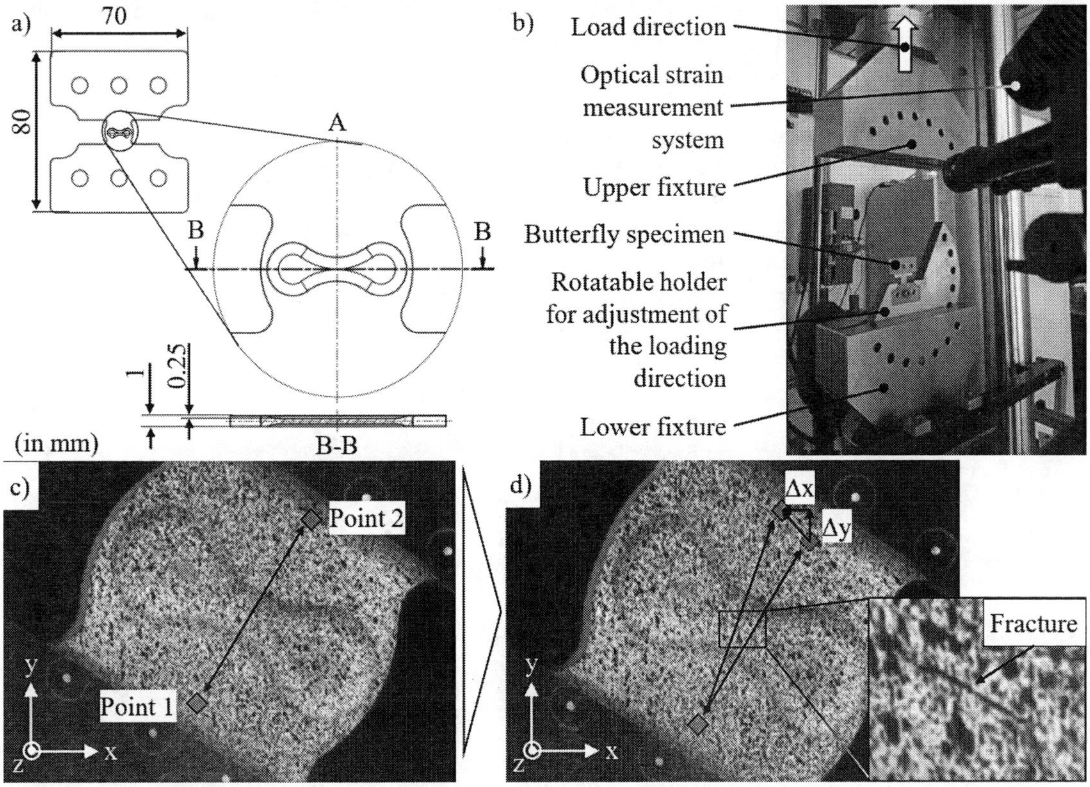

Figure 1. Tool setup for the tensile tests (a), geometry of the butterfly specimen (b), centre of the specimen before (c) and after the tensile test (43.5°) with visible fracture (d)

The specimens were cut by water jet cutting from sheet metals of the high-strength steels HCT600X and HX340LAD with a sheet thickness of 1.0 mm. On both sides the surfaces in the centre of the specimen were machined away with 25 % of the initial sheet thickness. It was ensured that the surface roughness remained under the tolerances of Ra < 0.8 and Rz < 6.3 to avoid local effects on the fracture initiation due to roughness peaks. For the tensile tests, a special tool system in a single-axis tensile testing machine from Zwick Roell was used, which is shown in figure 1 (b). The tool consists of two rotatable specimen holders as well as an upper and a lower fixture to fasten the holders. Due to the circular arrangement of the mounting holes, the specimen holders can be rotated in steps of 15.5°. This allows the testing under varying angles α of force application, which results in different stress states in the centre of the specimen ranging from a shear compression state to uniaxial tension. During the tests, the holders are fixed against rotation. The test were carried out quasi-statically at a constant tool speed of 0.02 mm/s until the onset of fracture appeared in the centre of the specimen. The optical measuring system Aramis from GOM was used to directly measure the displacements in x- and y-direction of the specimen by Digital Image Correlation (DIC). For this purpose, a stochastic pattern was applied on the specimen surface before testing. Furthermore, the optical measuring system was used to manually determine the onset of fracture, which was assumed when the first visible crack appeared on the surface of the specimen. The tests were performed at room temperature under the angles α of -3°, 12.5°, 28°, 43.5°, 59°, 74.5° and 90°. Each experiment was repeated five times. Subsequently, the experiments were evaluated in Aramis 2021 to obtain the displacements of the specimen in x- and y-direction. Therefore, two points were defined and the change of distance in x and y was measured, which are exemplary shown in figure 1 (c) and (d). Afterwards the average value of the five repetitions were calculated. These values were used as boundary conditions in the following numerical simulations.

4. FE modelling

Numerical simulations of the tensile test were carried out in ABAQUS to determine the equivalent plastic strain, stress triaxiality and normalised Lode angle Therefore, elastic-plastic material models with the yield criteria according to Hill were defined [17]. The characterisation and material model validation of the considered materials were performed in previous study [3]. For this reason, tensile tests were carried out at room temperature according to DIN EN ISO 10002 with the specimens aligned at 0°, 45° and 90° to the rolling direction. Additionally, hydraulic bulge tests were carried out to extend the flow curves and thus enhance the extrapolation accuracy. A combined approach of Swift and Hocket-Sherby was used for HCT600X and the approach according to Gosh was chosen for HX340LAD, because they provided the best approximation to the experimental data. The resulting flow curves and the yield locus with the corresponding parameters applied in the simulation are shown in Figure 2.

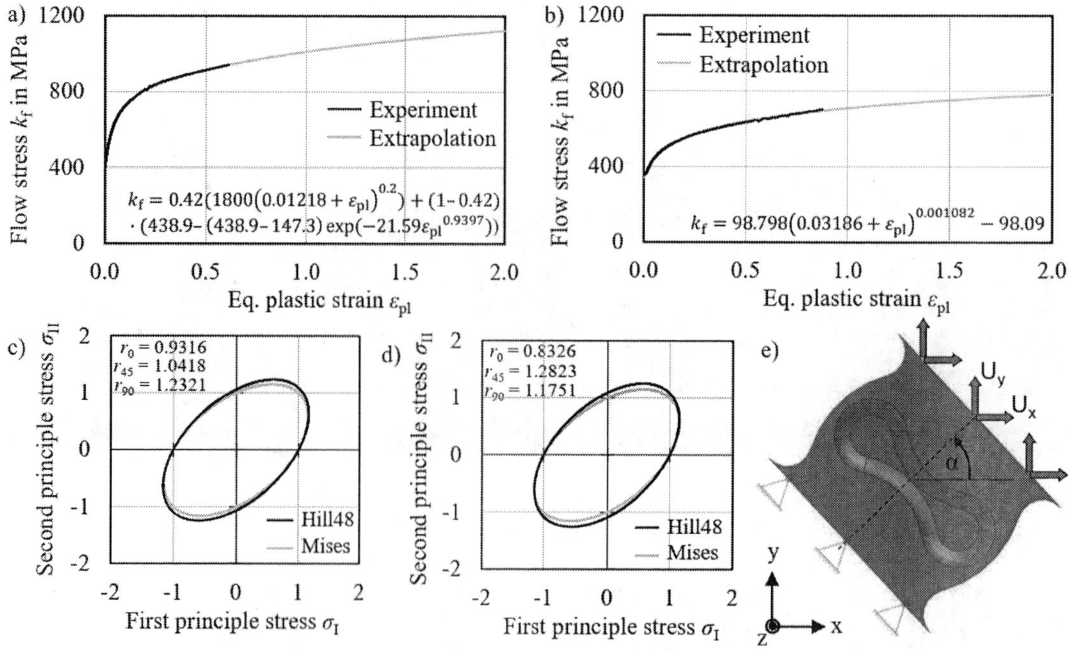

Figure 2. Flow curves and yield surfaces for HCT600X (a, c) [3] and HX340LAD (b, d) and the simulation model with the applied boundary conditions for α = 43.5° (e)

Since only the displacements of the specimen centre were measured during the experiments, only this section was modelled for the simulation. The centre was arranged in the x-y-plane analogous to the position in figure 1 (c). One-half of the geometry was modelled and a symmetry plane in z-direction was defined. A mesh size of 0.1 mm was applied. The measured displacements from the tensile tests were specified as the boundary conditions in x- and y-direction. The simulation model as well as the applied boundary conditions are shown in figure 2 (e). After simulation, the equivalent plastic strain and the three principal stresses in the centre of the specimen were evaluated to calculate the stress triaxiality and the normalised Lode angle. Therefore, the element with the maximum equivalent plastic strain was selected for evaluation. This element was chosen because it was always located in the centre of the investigation area of the specimen, which represented a good agreement to the position of the fracture in the experiments. Since the stress triaxiality and the Lode angle vary during the deformation of the specimen, they are normalized by the equivalent plastic strain up to the onset of fracture based on the following equation.

$$\eta_{norm.} = \frac{1}{\varepsilon_{\mathrm{pl}}^{f}} \int_0^{\varepsilon_{\mathrm{pl}}^{f}} \eta \, \mathrm{d}\varepsilon_{\mathrm{pl}} \quad \text{and} \quad \bar{\theta}_{norm.} = \frac{1}{\varepsilon_{\mathrm{pl}}^{f}} \int_0^{\varepsilon_{\mathrm{pl}}^{f}} \bar{\theta} \, \mathrm{d}\varepsilon_{\mathrm{pl}} \tag{10}$$

After determination of the parameters, the model coefficients of the damage criteria JC, MMC and DF2016 were calibrated. A and n of the MMC criterion were determined by fitting the flow curve to the Swift hardening model. For simplicity, the parameters c_{θ}^{ax} and c_{θ}^{s} were left at their default value of one analogous to [10]. The remaining parameters were fitted to the experimental data by reducing the sum of the least squares between experimental-numerical plastic strains and predicted plastic strains by the criteria. The results are given in table 1.

Table 1. Coefficients of the ductile fracture criteria and the average error of the fit

	JC			MMC				DF2016				
	D_1	D_2	D_3	A	n	c_1	c_2	C	C_1	C_2	C_3	C_4
HCT600	0.370	0.264	8.992	1018.6	0.143	0.166	560.033	1.949	5.521	5.482	0.616	1.734
Sum of least squares	0.03065			0.02452				0.01937				
HX340	0.256	0.576	3.486	713.0	0.165	0.208	401.934	3.115	1.821	6.918	0.522	0.479
Sum of least squares	0.00609			0.06465				0.02580				

The force application angle of 43.5° was not used for the parametrisation as this experiment will be used subsequently to validate the fracture criteria. While the criterion DF2016 achieved the best fit for HCT600X, the JC criterion showed the least sum of least squares for HX340LAD. In figure 3, the eq. plastic fracture strain calculated by the fracture criteria are shown as a function of stress triaxiality in the plane stress state.

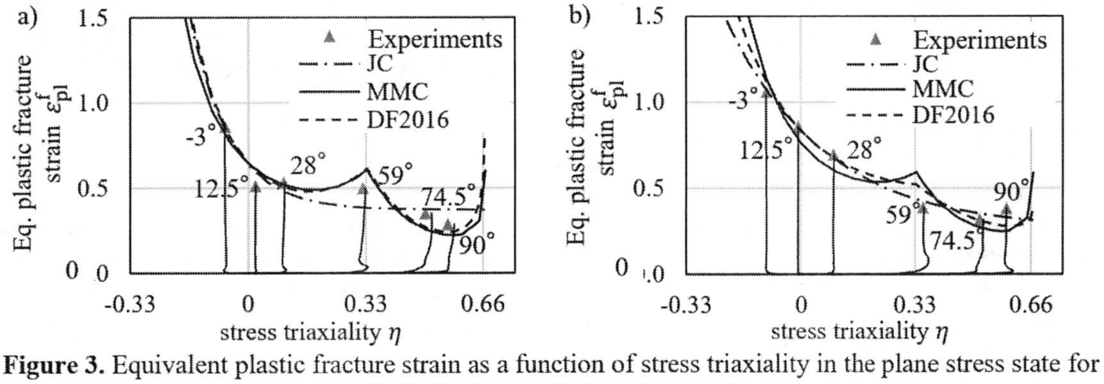

Figure 3. Equivalent plastic fracture strain as a function of stress triaxiality in the plane stress state for HCT600X (a) and HX340LAD (b)

While MMC and DF2016 show the characteristic peaks of the plastic fracture strain at a stress triaxiality of 0.33 and 0.66, the JC criterion assumes a falling slope of the fracture strain with an increasing stress triaxiality. The highest plastic fracture strains were determined at the angle of -3°, where a pressure superimposed state is present. The criteria assume a strongly increasing plastic strain when the stress triaxiality approaches the section of uniaxial pressure, which is in accordance to the existence of a cut off value at η = -0.33, which has been revealed by Bao et al. [18].

5. Verification of the ductile damage criteria

As mentioned previously, the test with the force application angle of 43.5° was not used to parametrise the fracture criteria. This experiment was used to validate and compare the fracture criteria in terms of prediction accuracy. For this purpose, the damage models were implemented in ABAQUS in the form of $\varepsilon(\eta, \bar{\Theta})$. The specimen was arranged at the angle of 43.5° and a multiple of the x- as well as y-displacement from the experiments were applied as boundary conditions. For the validation, the resulting displacement of the specimen at the onset of fracture was compared to the experimental findings. In the simulation, the fracture occurred when the output variable *DUCTCRT* reached the value of one. Figure 4 shows the displacements in x- and y-direction of the experiments as well as the simulations. The error bars of the experiments show the 95 % confidence interval of the repetitions.

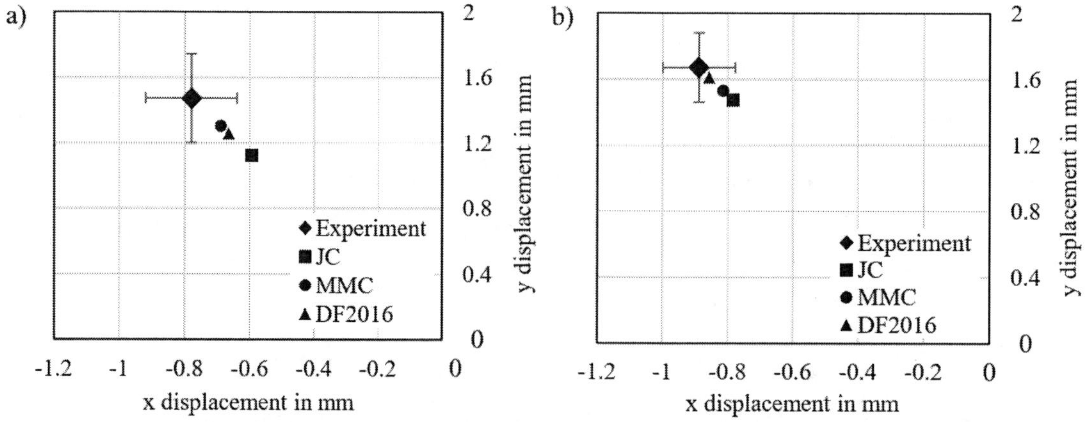

Figure 4. Displacement of the butterfly specimen in x- and y-direction at the onset of fracture from experiments and predicted by the damage criteria for HCT600X (a) and HX340LAD (b)

In all simulations, the onset of fracture occurred earlier than in the experiments. The best approximation for HCT600X was obtained with the MMC criterion with a deviation of 16.18 % from the mean value of the five experimental repetitions. A similar accuracy was achieved by the DF2016 criterion with a deviation of 19.23 %. Both models are within the 95 % confidence interval of the experiments. The largest deviation was reached by the JC criterion with 33.49 %. For HX340LAD the DF2016 criterion achieved the best approximation with a deviation of 4.9 %. MMC and JC showed deviations of 11.76 % and 16.52 %, respectively. It must be noted that the precise determination of the onset of fracture was difficult in the experiments, which can be one reason for the deviations.

In figure 5 the equivalent plastic strain at the onset of fracture measured in the experiments is compared to the simulation results. Since the best agreements are achieved by MMC for HCT600X and DF2016 for HX340LAD, only the plastic strains at fracture according to these criteria are shown. Since the onset of fracture occurs earlier in the simulation, the eq. plastic strain in the centre of the specimen is slightly lower than in the experiment, especially for HCT600X. However, the simulations show a good agreement regarding the localization of the maximal plastic strain as well as the overall deformation of the specimen geometry.

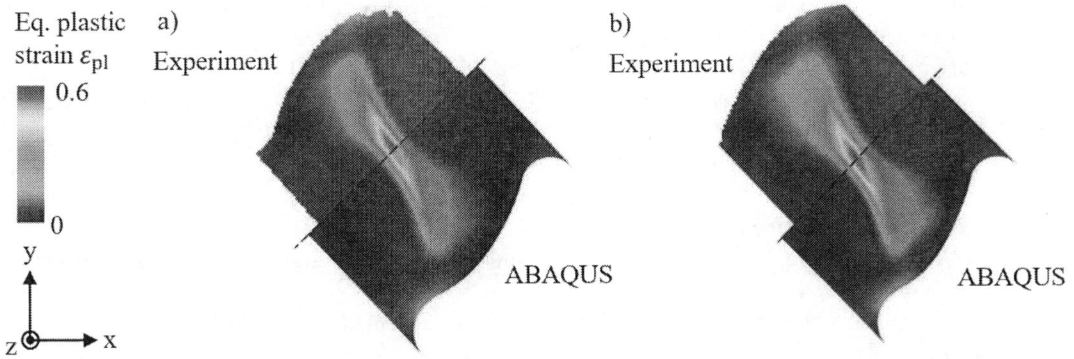

Figure 5. Comparison of the equivalent plastic strain at the onset of fracture from the experiment with the simulation predicted by MMC for HCT600X (a) and DF2016 for HX340LAD (b) for α = 43.5°

6. Conclusion and outlook

This research presents the application of three ductile fracture criteria for the prediction of failure in the two high-strength steels HCT600X and HX340LAD. Quasi-static tensile tests with butterfly specimen were carried out at room temperature under different angles of force application to induce varying stress states in the specimen centres. The tests were simulated with ABAQUS to identify the equivalent plastic strain as well as the normalised stress triaxiality and Lode angle. Subsequently, the parameters were used to calibrate the coefficients of the damage criteria JC, MMC and DF2016. Afterwards the criteria were implemented in ABAQUS and validated by simulating an additional test. In all simulations, the onset of fracture occurred earlier than in the experiment. The best agreements were achieved by the criteria MMC for HCT600X and DF2016 for HX340LAD, which shows that the models perform differently depending on the material considered. The JC criterion showed the greatest deviations to the experimental findings. This is attributed to the fact that the influence of the Lode angle is neglected and a falling course of the fracture strain is assumed with an increasing stress triaxiality.

In future investigations, the criterion MMC for HCT600X and DF2016 for HX340LAD will be applied in the simulation of a deep drawing process with additional force transmission. Due to the improved fracture characterisation, the process limits can be determined more precisely without cost- and time-consuming trial and error test. This will improve the process design and promotes the forming of high quality components.

7. Acknowledgement

The results presented were obtained in the project “Extension of the forming limits during deep drawing by additional force transmission” – 212270168. The authors thank the German Research Foundation (Deutsche Forschungsgemeinschaft, DFG) for their financial support.

References

[1] Tisza M and Czinege I 2018 Comparative study of the application of steels and aluminium in lightweight production of automotive parts *Int. J. Lightweight Mater. Manuf.* **1** (4) pp 229-238

[2] Schmitt J-H and Iung T 2018 New developments of advanced high-strength steels for automotive applications *C. R. Acad. Sci.* **19** (8) pp 641-656

[3] Behrens B-A, Bonk C, Grbic N and Vucetic M 2017 Numerical analysis of a deep drawing process with additional force transmission for an extension of the process limits *IOP Conf. Ser.: Mater. Sci. Eng.* **179 (1)** 012006

[4] Behrens B-A, Bouguecha A, Bonk C, Grbic N and Vucetic M 2017 Validation of the FEA of a deep drawing process with additional force transmission *AIP Conf Proc* **1896** 080024

[5] Behrens B-A, Bouguecha A, Bonk C, Rosenbusch D, Grbic N and Vucetic M 2017 Influence of the determination of FLC's and FLSC's and their application for deep drawing process with additional force transmission *Proceedings of 5th International Conference on Advanced*

Manufacturing Engineering and Technologies pp 405-417
[6] Johnson G R and Cook W H 1985 Fracture characteristics of three metals subjected to various strains, strain rates, temperatures and pressures *Eng. Fract. Mech.* **21** (1) pp 31-48
[7] Behrens B-A, Bouguecha A, Vucetic M, Krimm R, Hasselbusch T and Bonk C 2014 Numerical and Experimental Determination of Cut-edge after Blanking of Thin Steel Sheet of DP1000 within Use of Stress based Damage Model *Procedia Eng.* **81** pp 1096-1101
[8] Behrens B-A, Dröder K, Hürkamp A, Droß M, Wester H and Stockburger E 2021 Finite Element and Finite Volume Modelling of Friction Drilling HSLA Steel under Experimental Comparison *Materials* **14** p 5997
[9] Cao T S 2017 Models for ductile damage and fracture prediction in cold bulk metal forming processes: a review. *Int J Mater Form* **10** pp 139–171
[10] Bai Y and Wierzbicki T 2010 Application of the extended Coulomb-Mohr model to ductile fracture *Int. J. Fract* **161** pp 1-20
[11] Beese A M, Luo M, Li Y, Bai Y and Wierzbicki T 2010 Partially coupled anisotropic fracture model for aluminum sheets *Eng. Fract. Mech.* **77** (7) pp 1128-1152
[12] Li S, He J, Gu B, Zeng D, Xia Z C, Zhao Y, Lin Z 2018 Anisotropic fracture of advanced high strength steel sheets: Experiment and theory *Int. J. Plast.* **103** pp 95-118
[13] Lee J-Y, Steglich D, Lee M-G 2018 Fracture prediction based on a two-surface plasticity law for the anisotropic magnesium alloys AZ31 and ZE10 *Int. J. Plast* **105** pp 1-23
[14] Lou Y, Chen L, Clausmeyer T, Tekkaya A E and Yoon J W 2017 Modeling of ductile fracture from shear to balanced biaxial tension for sheet metals *Int J Solids Struct* **112** pp 169-184
[15] Dunand M and Mohr D 2011 Optimized butterfly specimen for the fracture testing of sheet materials under combined normal and shear loading *Eng Fracture Mech* **78** pp 2919-2934
[16] Peshekhodov I, Jiang S, Vucetic M, Bouguecha A and Behrens B-A 2016 Experimental numerical evaluation of a new butterfly specimen for fracture characterisation of AHSS in a wide range of stress states *IOP Conf Ser: Mat Sci and Eng* **159** p 12015
[17] Hill R 1948 A theory of the yielding and plastic flow of anisotropic metals *Proc. R. Soc. Lond.* **193** pp 281-297.
[18] Bao Y B and Wierzbicki T 2005 On the cut-off value of negative triaxiality for fracture *Eng. Fract. Mech.* **72** (7) pp 1049-1069

IOP Conf. Series: Materials Science and Engineering 1238 (2022) 012022 doi:10.1088/1757-899X/1238/1/012022

3D fracture modeling based on the coupling between damage criteria, phase field and crack propagation

P-O Bouchard[1], H Eldahshan[1,2], D Pino Munoz[1], J Alves[2] and E Perchat[2]

[1] MINES Paris-Tech, PSL- Research University, CEMEF – Center for Material Forming, CNRS UMR 7635, BP 207, 1 rue Claude Daunesse, 06904 Sophia Antipolis Cedex, France

[2] Transvalor SA, E-Golf Park, 950 Avenue Roumanille, 06410 Biot, France

pierre-olivier.bouchard@minesparis-psl.eu

Abstract. In addition to predicting the onset of fracture, some specific material forming processes require the modeling of crack propagation and the accurate prediction of the fracture surface. In such configurations, the use of the deletion element technique is not adapted since it suffers from major mesh dependency and spurious artificial volume loss. This work introduces a new alternative in which the phase field method is coupled to a 3D automatic crack initiation and propagation technique within the Forge® finite element software.

1. Introduction

In some given metal forming processes the modeling of fracture is mandatory to get the final shape of the manufactured component. This is the case for processes such as blanking, machining or cutting. In such configurations, ductile damage can be modeled either by uncoupled or coupled damage models (the readers can refer to [1] for a review on ductile damage in metal forming). Once a given threshold is reached, fracture must be modeled. Fracture creates a discontinuity which is often modeled using the deletion element technique. This is a robust technique that consists in deleting the elements from the mesh once a damage threshold has been reached. However, this technique suffers from several drawbacks such as dependency to the mesh size and spurious volume loss. In addition, the stress field at the crack tip cannot be captured accurately since it is based on the deletion of elements.
We propose here an alternative approach that combines the phase field for ductile fracture and a 3D crack insertion and propagation technique independent from the mesh.

Initially defined for brittle fracture problems [2], the phase field method was then extended to ductile fracture [3]. This work introduces new contributions to the field so as to be applied to metal forming processes:

- a new formulation in which the phase field-damage formulation can be coupled to ductile damage models;
- adaptive remeshing followed by the identification of the crack surface intersection with arbitrary mesh topologies based on the phase field evolution [4];
- fitting the crack surface within the mesh using local mesh partitioning operations;

- a nodal duplication strategy based on the local coloring algorithm in order to open the crack faces followed by a volume remeshing step [5].

The proposed framework offers a robust numerical tool for the modeling of damage to fracture transition in complex industrial processes. Metal sheet blanking and piercing processes will be used to show the accuracy and efficiency of this new approach compared to the conventional element deletion method.

2. The phase field method for ductile fracture

Initially based on a generalized formulation of the Griffith theory of brittle fracture [2], the phase field method was later on adapted to numerical computations through a regularized approximation presented in [6]. This new formulation consists in the minimization of the regularized total energy functional:

$$\mathcal{E}_{reg}(u,d) = \int_{\Omega_h} g_e(d)\psi_e(\mathcal{E}^e)d\Omega_0 + \int_{\Omega_h} G_c\gamma(d,\nabla d)d\Omega_0 \tag{1}$$

Where u and d are respectively the displacement field and the phase field (d=0 for an intact material and d=1 for a totally damaged state). Ψ_e is the elastic strain energy, G_c is the fracture toughness, Ω_h is the volume of the domain and $g_e(d)$ is a degradation function. The regularized fracture energy (crack surface density) is defined by:

$$\gamma(d,\nabla d) = \frac{1}{2l_c}\left(d^2 + {l_c}^2|\nabla d|^2\right) \tag{2}$$

Where l_c is a length scale parameter.

The elastic degradation function $g_e(d)$ is used to couple the effect of crack propagation on the bulk material. The following quadratic form is chosen $g_e = (1-d)^2$ in order to satisfy the following conditions: $g_e(0) = 1, g_e(1) = 0$ and $g_e'(1) = 0$.

The minimization of the regularized energy functional with respect to the phase field parameter conducts to the following evolution of the phase field equation:

$$d - l_c^2\nabla^2 d = -g_e'(d)H_e \tag{3}$$

Where H_e is the local history function that is defined by:

$$H_e = \frac{l_c}{G_c}\max_n \psi_e\left(\mathcal{E}^e(x,d_n)\right) \tag{4}$$

Where n is a unit vector normal to the surface. As defined in [3] a plastic degradation function can be added to local history function in order to handle ductile fracture.

In the present work, it is decided to define the local history function H (replaces H_e) so as to introduce a damage variable D and its threshold D_{th}:

$$H = \eta_c\langle D - D_{th}\rangle \tag{5}$$

Where η_c is an additional material parameter used to control the post-peak stress response of the material once D has reached its threshold D_{th}. And $\langle x\rangle = x \ if \ x \geq 0 \ and \ \langle x\rangle = 0$ otherwise. With such a formulation, it is possible to use different kinds of ductile fracture criteria such as the one presented in [1].

3. 3D modeling of fracture driven by the phase field variable

Once the damage variable has reached its threshold, the phase field variable increases leading to progressive degradation of the material's mechanical behavior. Finally, the phase field threshold is reached leading to fracture. The idea consists in using the phase-field variable (D) in order to identify regions where the transition from damage to fracture should occur. It is worth mentioning here that the non-local nature of the phase-field formulation leads to a mesh independent solution, which is one of the main drawbacks of the element deletion technique.

In [5], a new methodology was implemented in Forge® in order to propagate 3D cracks. This methodology, called CIPFAR (Crack Initiation and Propagation algorithm using the phase Field model and Adaptive Remeshing), and detailed in [5] can be summarized as follows:

- The phase field is computed as described in the previous section. At this stage, an adaptive remeshing strategy is used to ensure extreme mesh refinement in the area in which fracture is expected to initiate. This strategy is detailed and validated in [4].
- The phase field solution provides a highly concentrated damage field in which the crack surface can be identified as the points that have a phase field value of 1. In order to provide a unique and accurate fracture surface localization, the phase field solution is combined with its gradient. This approach enables to deal with multiple cracks.
- The intersection of the crack surface with the mesh edges is identified and new nodes are inserted. Local remeshing operations are carried out so as to fit the crack topology with the mesh.
- The nodes lying on the crack surface are duplicated in order to open the new crack. Finally a subsequent remeshing operation is carried out to insure mesh quality and mechanical fields are transported from the old mesh to the new mesh.

The whole methodology, illustrated in Figure 1 and validated in [4, 5], provides an efficient, robust and accurate 3D initiation and propagation technique driven by the phase field computation. The CIPFAR methodology is also conservative in terms of volume, solving in this way the other drawback of the aforementioned element deletion technique.

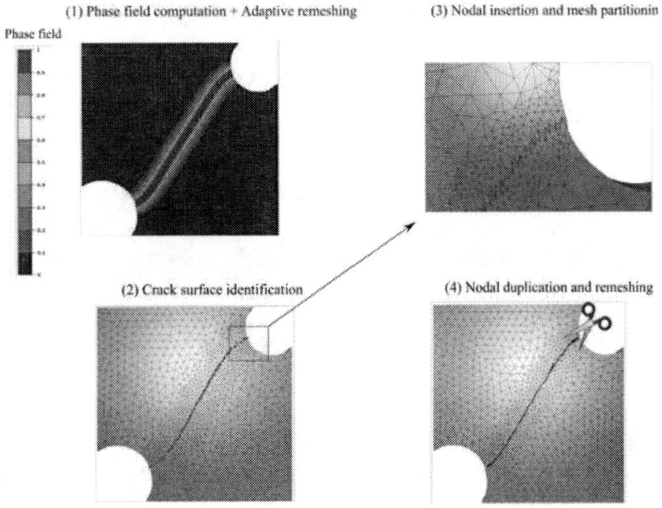

Figure 1: Summary of the main steps of 3D crack initiation and propagation based on the phase field computation and automatic remeshing

4. Application to a multi-stage forming process

The methodology was applied to several configurations of metal forming processes such as extrusion chevrons cracking or shear cutting (see [7] for more details). The accuracy was compared to the conventional kill element technique. The CIPFAR algorithm showed very good ability to model damage to fracture transition and the failure surface quality was very convincing with particular accuracy on predicting fracture zone and burr height in shear cutting processes. In the following, the methodology is applied to a 3 stages forming process that includes blanking, piercing and finally bending.

4.1 Multi-stage process definition

The three stages of the process together with the final shape of the component are detailed in Figure 2.

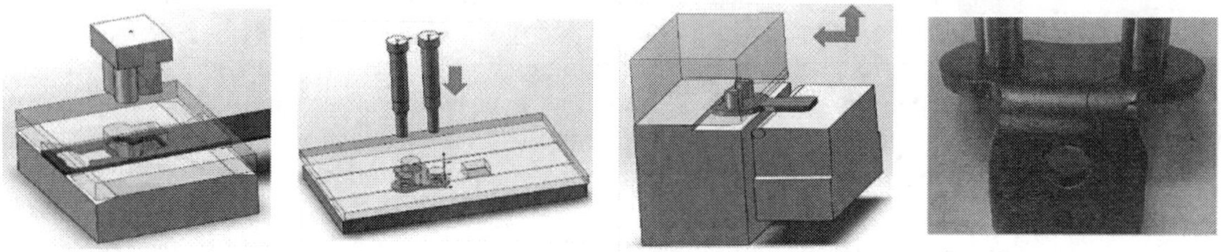

a) Stage 1: blanking b) Stage 2: piercing c) Stage 3: bending (90°) d) Final shape

Figure 2: Forming process split into a) blanking, b) piercing and c) bending. The shape of the final part is given in d) (scale not mentioned for confidentiality reasons)

4.2 Material and input process parameters

The material is a 36Mn5 steel grade with an elastic-plastic behavior with strain rate dependency. The Young modulus E and Poisson ratio ν are respectively 210 000 MPa and 0.3. The yield stress σ_y [MPa] is given by the following hardening law:

$$\sigma_y = 1011\bar{\varepsilon}^{0.316}\dot{\bar{\varepsilon}}^{0.012} e^{\frac{0.018}{\bar{\varepsilon}}} \tag{6}$$

Where $\bar{\varepsilon}$ and $\dot{\bar{\varepsilon}}$ are respectively the equivalent strain and equivalent strain rate.

The Coulomb limited by Tresca friction law is used for the tangential contact between the different dies and the billet. The Coulomb and Tresca friction parameters are respectively equal to 0.2 and 0.4

The normalized Latham-Cockcroft damage model is chosen to drive the phase field evolution all along the three stages (See Equation (5)):

$$D = \int_0^{\bar{\varepsilon}_f} \frac{\langle\sigma_1\rangle}{\bar{\sigma}} d\bar{\varepsilon} \tag{7}$$

Where σ_1 is the maximum principal stress and $\bar{\sigma}$ is the equivalent von Mises stress. Parameter η_c is fixed and equal to 200 whereas D_{th} will be varied. The characteristic length scale l_c is fixed to 1.2 mm.

An initial mesh size of 1 mm is used in the whole domain. The adaptive remeshing strategy is used to refine the mesh in the regions where the equivalent plastic strain exceeds a threshold of 0.05. Two mesh sizes are used in the numerical simulations: 0.6 mm and 0.3 mm respectively for the coarse mesh and the fine mesh. For the crack propagation simulations, the effective crack area is chosen as two times the minimum mesh size.

The accumulated plastic strain field is transferred from each step to the following one in order to reflect the effect of strain hardening on the material behavior. However, the damage field is not transferred from each stage to the next one. This assumption is necessary here in order to avoid immediate crack initiation at the very beginning of stages 2 and 3 in the immediate vicinity of the previous crack surface. This problem can be viewed as a consequence of the use of a coupled model to the material behavior at the macroscale that does not describe accurately the growth of voids near the cracked zone at the micro-scale. The size of this damaged zone should be much smaller than the characteristic length scale (and hence the smallest element size) in order to accurately capture the size of the damaged zone. However, the problem would be computationally very expensive especially in 3D calculations. It should be noted that neglecting the effect of damage might lead to inaccurate results in the cases where the damaged surface is subjected to another shearing process in the following stages. Indeed, a future investigation of this assumption is needed to have a better understanding of its limitations.

In the following, a comparison will be carried out between the conventional kill element technique and the new CIPFAR algorithm. The influences of the mesh size and of the damage threshold D_{th} are detailed in [8].

4.3 Results

Figure 3 shows the final shape of the part obtained after stages 1 (shear cutting) and 2 (piercing) with the new CIPFAR algorithm and with the conventional kill element technique.

At the end of stage 1, a better smoothness of the fracture surface can be observed with the CIPFAR algorithm. In addition, the formation of burrs can be accurately captured with the CIPFAR algorithm since the surface details are not eliminated during the simulation, contrary to the element deletion algorithm for which all burrs are removed from the mesh.

Regarding stage 2, two main observations can be made: (i). The discrete crack predicted by the CIPFAR algorithm leads to a better quality of the sheared surfaces as compared to the element deletion algorithm; (ii). the CIPFAR algorithm leads to an accurate prediction of the crack path which can be clearly seen in the exact contact between the punch and sheared edge. On the other hand, the deletion of elements results in a gap without any contact between the punch and the sheared edge in which the size of this gap is dependent on the mesh size.

It is interesting to notice as well that the CPU time for both methods are similar. Despite the coupling introduced by the CIPFAR algorithm, a reduced number of Newton-Raphson iterations compared to the deletion element technique conducts to a similar CPU time (see [8] for more details). In addition, the element deletion technique conducts to very rough fracture surface that can lead to degraded convergence of the mechanical resolution given the complex contact path at such rough surfaces.

Figure 4 shows the end of stage 4. In Figure 4.a two values of D_{th} are used in order to illustrate its influence on the final failure pattern. Figure 4.b shows the comparison with the experimental observation and a very good agreement both in terms of location of failure initiation and crack length.

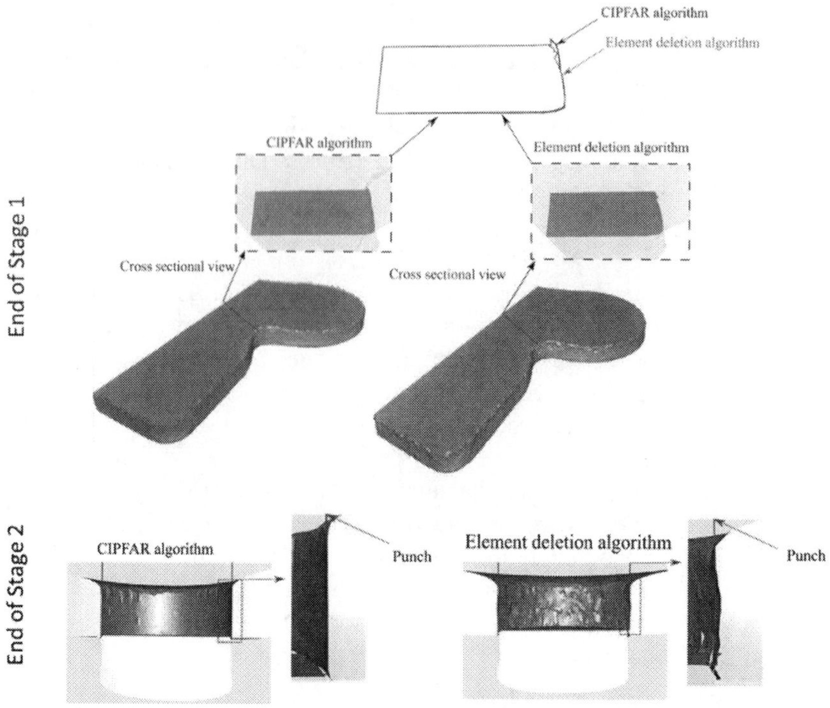

Figure 3: Comparison between the CIPFAR algorithm (left) and the conventional element deletion technique (right) at the end of stages 1 and 2.

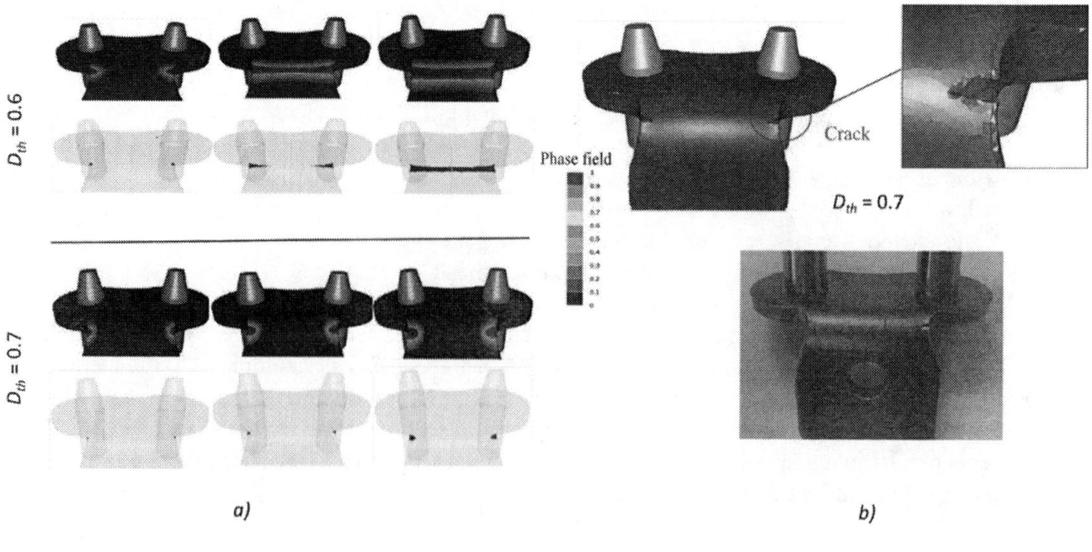

Figure 4: a) Influence of D_{th} on the final fracture pattern at the end of the 3rd forming stage (bending) and b) comparison with experimental observation.

5. Conclusion

In this work, ductile fracture is addressed by enhancing the phase field approach for ductile fracture through the coupling with damage criteria. The new developed approach also deals with the transition from continuum damage to discrete fracture. A specific methodology where the crack surface is localized within the mesh based on the phase field variable and its gradient is developed. Mesh adaptation techniques are then carried out in order to incorporate 3D crack surfaces in a very robust and efficient manner to model ductile fracture propagation. The method has been successfully applied to conventional ductile fracture problems as well as to ductile fracture occurring during material forming processes.

The multi-stages process shown in this paper illustrates the accuracy of the new approach compared to the classical element deletion technique. Fracture surfaces are really smooth which makes it easier to compute further forming process stages on the result of the previous forming stage. The new method is also very efficient since the CPU time is equivalent to the one of an uncoupled failure criteria using the deletion element algorithm.

The transfer of the phase field variable from one stage to the following remains to be discussed. This phase field variable is already very close to 1 in the failed area and in a thickness layer which depends on the characteristic length scale. The transfer of this variable from one stage to the following may conduct to premature cracking in this area whereas experimentally only micro-voids would be present in this area. The same issue would arise with non-local damage approaches.

The damage criteria used here is one of the simplest dedicated to ductile fracture. However, the way damage is introduced within the phase field formulation (See equation (5)) makes it appropriate for any other kind of damage criteria as shown in [8].

References

[1] Tekkaya A E, Bouchard P O, Bruschi S and Tasan C, Damage in metal forming 2020 *CIRP Annals* **69** 600-23.

[2] Francfort G A and Marigo J J, Revisiting brittle fracture as an energy minimization problem 1998 *J Mech Phys Solids* 46.

[3] Borden M J, Hughes T J, Landis C M, Anvari A and Lee I J, A phase-field formulation for fracture in ductile materials: Finite deformation balance law derivation, plastic degradation, and stress triaxiality effects 2016 *Comput Methods Appl Mech Eng* **312** 130–66.

[4] Eldahshan H, Bouchard P O, Alves J, Perchat E and Pino Munoz D, Phase field modeling of ductile fracture at large plastic strains using adaptive isotropic remeshing 2021 *Comput Mech* **67** 763–83

[5] Eldahshan H, Alves J, Bouchard P O, Perchat E and Pino Munoz D, CIPFAR: A 3D unified numerical framework for the modeling of ductile fracture based on the phase field model and adaptive remeshing 2021 *Comput Methods Appl Mech Eng* **387** 114171

[6] Ambrosio L, and Tortorelli V M, Approximation of functional depending on jumps by elliptic functional via t-convergence 1990, *Commun. Pure Appl. Math.* **43** 999–1036

[7] Eldahshan H, Pino Munoz D, Alves J, Perchat E and Bouchard P O, 3D crack initiation and propagation applied to metal forming processes 2022 (minor revisions) *J Mater. Proc. Tech.*

[8] Eldahshan H, 3D ductile damage to fracture transition based on phase field and mesh adaptation: application to metal forming processes 2022, PhD Mines ParisTech.

IOP Conf. Series: Materials Science and Engineering 1238 (2022) 012023 doi:10.1088/1757-899X/1238/1/012023

Experiments on forming behaviour of the aluminium alloy AA6016

R. Norz[1]*, F. R. Valencia[2], S. Gerke[2], M. Brünig[2] and W. Volk[1]

[1] Technical University of Munich, Chair of Metal Forming and Casting, Garching, Germany

[2] Universität der Bundeswehr München, Institut für Mechanik und Statik, Neubiberg, Germany

*roman.norz@utg.de

Abstract. Investigation of the sheet metal formability is an important issue in the design of components. During metal forming different deformation histories occur which may cause various damage and failure mechanisms on the micro-level affecting the quality of products. Therefore, description of the material characteristics by means of a linear limit change curve is not sufficient because the remaining formability and the fracture processes also depend on deformation history. For this purpose the aluminium alloy AA6016 is systematically investigated in order to obtain their forming capability as well as the damage and fracture behaviour after complex deformation paths. Different uniaxial tests with various initially unloaded and preloaded specimens have been performed to identify the effect of the forming history, the loading direction and the stress state on the formability and the mechanical parameters of the investigated material. The results can be used to propose a critical ratio between the pre-strain and the change in loading direction from which on shear failure occurs instead of necking behaviour leading to the loss of formability.

1. Introduction

The forming history significantly influences the remaining formability of sheet metals. For instance, Graf and Hoshford [1] found, that significant pre-forming affects the remaining formability of the aluminium alloy EN AW 2004-T4. Bergström and Ölund [2] received similar results for an aluminium-killed steel (AK-steel) and the aluminium alloy EN AW 2036-T4. Werber et al. [3,4] investigated three different pre-forming states (uniaxial, plane-strain and biaxial) as well as multiple pre-forming heights up to 50 % of the linear Forming Limit Curve (FLC) of an EN AW 6014 and found a significant effect of the pre-forming on the FLC. All mentioned publications coincide, that the pre-forming height significantly influence the remaining formability as well as the pre-forming state (e.g. uniaxial or biaxial). The pre-forming and the strain history also influences the fracture mode of materials. Gerke et al. [5] applied the X0-specimen to determine the dependence of non-proportional loading paths on damage evolution and fracture behaviour of an aluminium alloy EN AW 6082 and indicated the significant influence of the strain history on the fracture behaviour. In continuation Brünig et al. [6] confirmed these results with the H-specimen. Besides its influence on damage evolution and fracture behaviour the pre-loading history affects remarkably the load-displacement curves, while the final stresses are almost unaffected. Consequently, the different pre-forming states can change the failure as well as the force-displacement behaviour significantly. Uniaxial pre-forming can lead to more brittle behaviour in subsequent shear tests based on a change from shear dominated damage mechanisms to a more void-dominated failure modes. When conducting the reverse load change, namely pre-shear followed by an uniaxial tensile test, the damage behaviour is micro-shear-crack dominated. Next to the pre-forming state and level, also the loading direction of anisotropic sheet metals has a significant effect on the remaining formability and the failure mode. Volk et al. [7] analysed the influence of a change in loading direction for a micro-alloyed steel HC340LA in Nakajima experiments. A distinct interaction between pre-forming level and change in loading direction on the formability and failure mode is stated. Specimens with no change in loading direction show ductile necking behaviour, while changes in loading direction by 45° or 90° leads to a shear failure of the material with significant loss of formability under uniaxial stress states. Change in loading direction lead to a significant increase in void growth, resulting in shear fracture.

In this paper, the aluminium alloy EN AW 6016-T4 with an initial thickness of 1 mm is investigated. The material has been initially characterized by tensile, plane-strain and biaxial tests under different angles with respect to the initial rolling direction. To investigate the influence of the pre-forming height and changes in loading direction, Nakajima experiments with pre-formed specimens are conducted. After the experiments SEM pictures of the fracture surfaces are taken to visualise the dominant failure modes.

2. Experimental setup

The experiments for the initial characterization are carried out on a Zwick Z150 uniaxial testing machine. To ensure comparable results of all experiments the machine velocity has been adapted to ensure a strain rate of 0.004 1/s. This was done by using a Laserextensometer to measure the strain between the strain gauge length. The strain gauge length was set to 50 mm for the tensile tests and 30 mm for both plane-strain specimen geometries. The strain distribution at the other specimen surface have been extracted by digital image correlation (DIC) with the optical measurement system Aramis SRX, allowing its comparison with numerically obtained results. All experiments are conducted under 0°, 22.5°, 45°, 67.5° and 90° with respect to the rolling direction. The used geometries can be found in Figure 1.

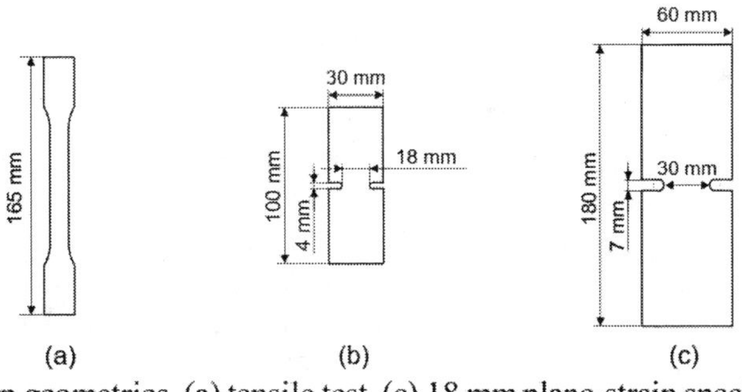

Figure 1. Used specimen geometries, (a) tensile test, (c) 18 mm plane-strain specimen, (c) 30 mm plane-strain specimen

The initial Nakajima and bulge experiments are carried out on a Zwick BUP1000. The Nakajima experiments are conducted according to ISO12002-2 with a constant punch velocity of 1 mm/s. The frame rate for the optical measurement system Aramis 4M is set to 10 Hz and the onset of necking is determined by the Time Dependent Evaluation Method (TDEM) by Volk and Hora [8]. A total of nine different specimen widths under 0°, 45° and 90° to the initial rolling direction are investigated, leading to a total of 25 data points corresponding to each other in all investigated directions. To reduce the friction between the punch and the specimen a PVC-pad with a thickness of 3 mm and lubrication paste are applied. The consecutive forming of the pre-strained specimens is conducted in the same way as the initial Nakajima experiments using four different specimen widths. The chosen specimen widths are 30 mm, 100 mm, 130 mm and 235 mm for the biaxial specimen. For all experiments at least three valid experiments are performed.

The pre-forming of the specimens has been realized on a hydraulic press using a modified Marciniak-tool, proposed by Weinschenk and Volk [9], a more detailed description of the tool can be found there. The punch speed is set to a constant value of 15 mm/s and the blank holder force is set to 1200 kN. In combination with the drawbead, the material flow is impeded. To allow high uniaxial pre-forming heights, the specimen geometry, as shown in Figure 2 (b), is used. The reduced width in the center of the specimen, where the punch is in contact with the specimen, ensures that the main deformation occurs there. To minimize a possible deviation in the pre-forming levels and pre-forming states an incremental pre-forming process is chosen. After a certain displacement the specimens will be re-lubricated with deep drawing oil to minimize the influence of friction on the results. The used tool, the specimens geometry and the process to change the loading direction are shown in Figure 2. The sample designation IF stands for the initial pre-forming direction, while PF stands for the post-forming direction. The introduced strains of the pre-forming process are measured by GOM Argus. A total of three different pre-forming levels, approximately 25 %, 35 % and 50 % of the linear FLC, are taken into account.

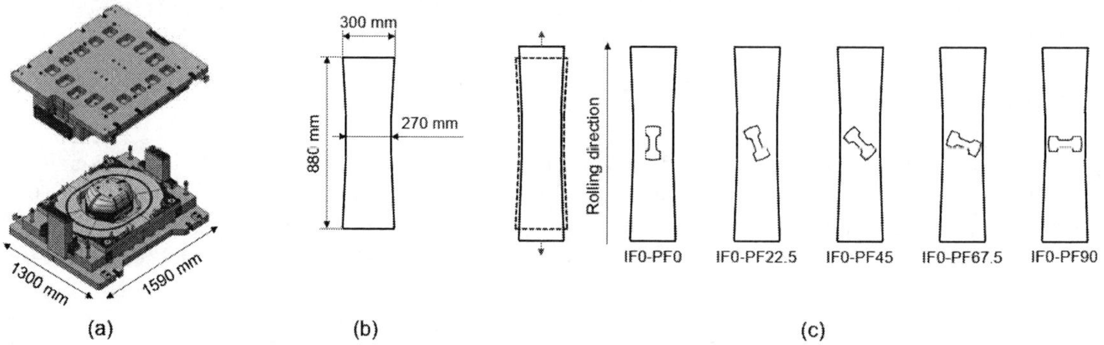

Figure 2. (a) Marciniak tool used to pre-form the specimens, (b) specimen geometry and (c) specimen extraction and nomenclature.

3. Results

3.1. Initial experiments

The tensile and plane strain tests indicated only a small material influence of the loading direction on the force-displacement curves, see Figure 3 (a) and (b). The force-displacement curve of the plane-strain specimen with a width of 18 mm the lower ones in Figure 3 (b) due to the smaller cross-section. It is also noted, that the deviation between the specimens is very small. The influence of the angle with respect to the rolling direction on the formability i.e. the Forming Limit Curves (FLC), is indicated in Figure 3 (c). The FLC under 0° with respect to the rolling direction displays the lowest formability under all investigated strain states. For the uniaxial Nakajima specimens with 30 mm width, brittle shear fracture was observed, the major strain distribution can be found in Figure 5 (a).

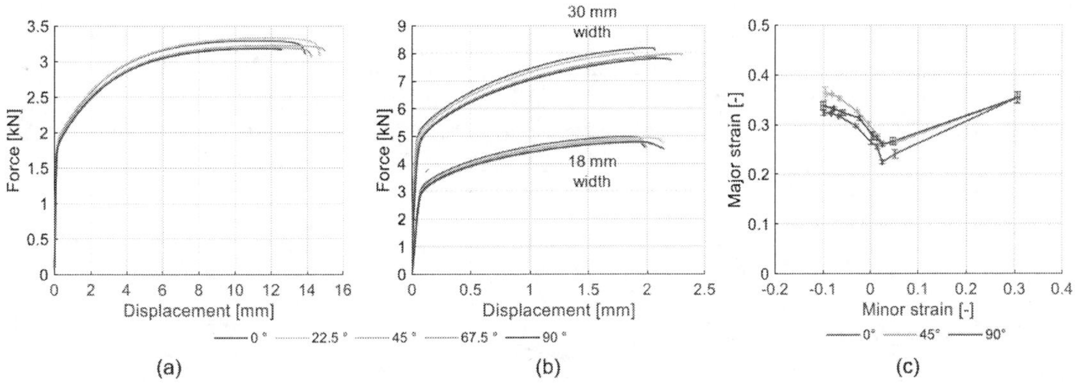

Figure 3. (a) Force-Displacement curves from tensile test, (b) force-displacement curves for the 18 mm plane-strain and the 30 mm plane-strain specimens and the (c) forming limit curve for different angles with regard to the rolling direction.

The Lankford-coefficients and their variation under different angles with regard to the initial rolling direction can be found in Table 1.

Table 1. Lankford-coefficients under different angles with respect to the initial rolling direction.

$r_{0°}$	$r_{22.5°}$	$r_{45°}$	$r_{67.5°}$	$r_{90°}$
0.672	0.593	0.464	0.529	0.611

All those experiments are used to determine and to optimize the yield criterion for this material. Best results could be obtained for the YLD2000-2d criterion by Barlat et al. [10]. The final determination of parameters has been realized by an inverse optimization procedure using the experiments used to characterize the material and was conducted with LS-Dyna using the R11.2.1 solver with double precision. As for example the plane-strain specimens or the tensile test under 22.5° and 67.5° are not used in the calibration process of the YLD2000-2d model and can therefore be used for the optimization. The optimization was done in LS-Dyna. The optimized material parameters, given in Table 2, are used in future numerical investigations of the experiments.

Table 2. Values for the YLD2000-2d model.

α_1	α_2	α_3	α_4	α_5	α_6	α_7	α_8	m
0.9764	0.9705	0.8999	1.0234	1.0249	1.0348	0.9452	0.1279	7

3.2. Pre-formed Nakajima experiments

The pre-formed Nakajima experiments reveal an influence of the loading direction on the formability of the material. The higher the pre-forming, the more a change in loading direction influences the remaining formability. Where all investigated post-forming angles show similar remaining formability for the 25 % pre-forming, the IF0°-PF45° and IF0°-PF90° are below the IF0°-PF0° curve for the 50 % pre-forming under uniaxial stress state, using the specimen with a specimen width of 30 mm. Therefore, a change in loading direction does influence the remaining formability in the uniaxial stress state. For all other specimen geometries, 100 mm, 130 mm and 235 mm, only a small effect can be observed, see Figure 4. Compared to the results of Volk et al. [11] for the micro-alloyed steel HC340LA the influence is less pronounced. This is possibly caused by the fact that the fracture curve, where the first crack in the specimen occur, and the forming limit curve, where necking begins, are close to each other and therefore only limited sliding is possible before the specimen shows the first cracks. This is not the case for the HC340LA steel, where the material can be further deformed after the onset of necking until the first cracks occur. A closer look on the experimental FLC indicates, that the uniaxial specimens stagnate in the major strains. This is caused by the non-linearity of the strain path in the experiment. When the pre-formed Nakajima experiments under a uniaxial stress state are conducted, the onset of necking is far beyond the linear FLC. The strain path has been linearized by the methodology of Volk and Gaber [12]. After linearization, the initial experimental FLC is in good agreement with the FLCs obtained from pre-formed experiments.

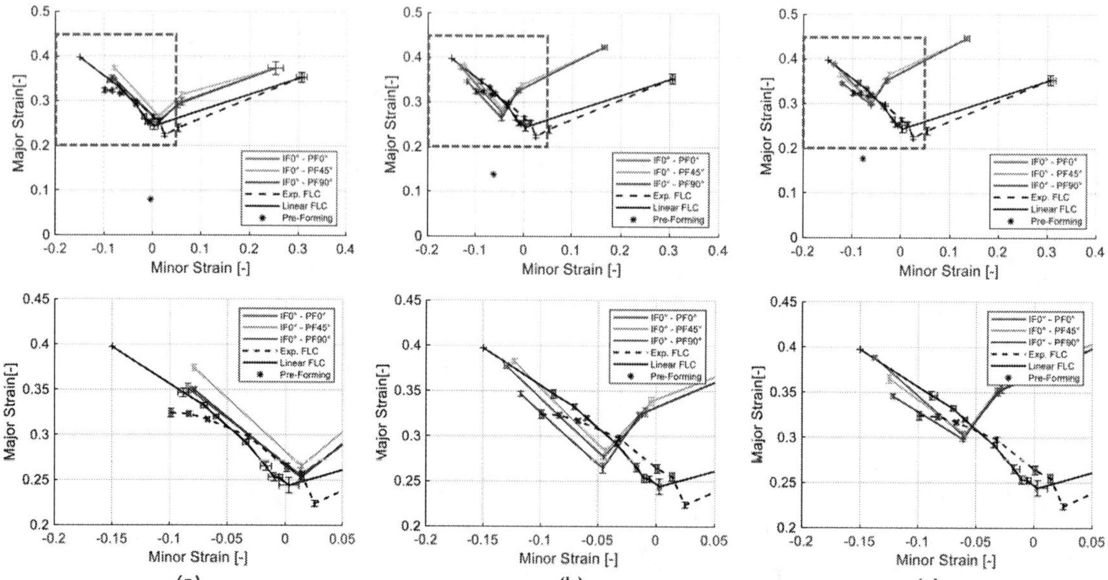

Figure 4. Forming Limit Curves for the three pre-forming levels, (a) 25 % of the linear FLC, (b) 35 % of the linear FLC and (c) 50 % of the linear FLC. The areas in the red-square are plotted in the lower figures with a higher resolution.

It is also found, that the failure mode has altered after the change in loading direction of 90° (IF0-PF90) and the pre-forming of 50 % of the linear FLC. While the initial material and the material with no change in loading direction (IF0-PF0) indicate a shear failure behaviour, the IF0-PF90 specimens show ductile failure behaviour with a necking zone in the middle of the specimens. This can be seen in contrast to the material behaviour of the HC340LA, where the initial material and the IF0-PF0 specimens revealed ductile failure behaviour in the centre of the specimens and the IF0-PF90 specimens showed brittle shear failure behaviour. DIC pictures of the major strain distribution for the initial specimen under 0° with respect to the rolling direction, a IF0-PF0 specimen and a IF0-PF90 specimen before fracture can be seen in Figure 5.

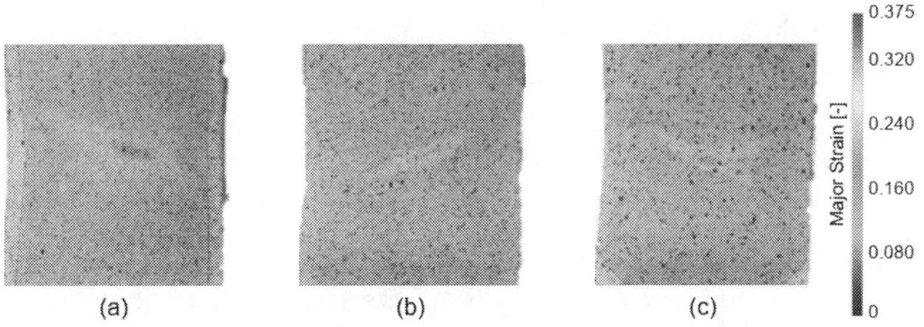

Figure 5. Major strain distribution at the stage before crack of the (a) initial 0° uniaxial Nakajima specimen, (b) of the IF0-PF0 uniaxial Nakajima specimen with a pre-forming of 50 % and (c) of the IF0-PF90 uniaxial Nakajima specimen with a pre-forming of 50 %.

Using scanning electron microscopy (SEM), fracture surfaces from Nakajima experiments with pre-formed specimens were analysed. The SEM images reveal that the IFO-PF90 fracture surface experiences ductile extrusion failure modes manifested in the greater detachment of material in the matrix, as observed in Figure 6 (c) (red circle). In addition, the generation of higher number of voids should have led to earlier failure as observed in the tension and plane strain tests, observed in Figure *3* (a) and (b). On the other hand, micrographs of the initial 0° uniaxial Nakajima experiments (Figure 6 (a)) and the pre-formed Nakajima experiments IF0-PF0, in Figure 6 (b) show similar behaviour. The fracture surfaces show stretched and oval voids, which means that the failure mode of the material was due to a sudden shear failure without any prior necking. The reason for these behaviours will be addressed in future work during this project.

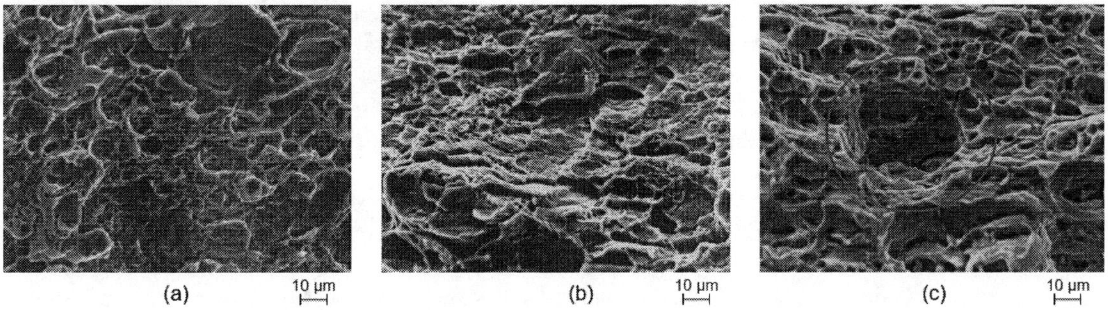

Figure 6. (a) Fracture surface of an initial 0° uniaxial Nakajima specimen, (b) fracture surface of a IF0-PF0 uniaxial Nakajima specimen and (c) fracture surface of a IF0-PF90 uniaxial Nakajima specimen.

4. Conclusion and Outlook

The material AA6016-T4 has been investigated in this paper, regarding the influence of uniaxial pre-forming on the remaining formability. It is found, that changes in loading direction influences the failure mode of the material after a pre-forming of 50 % of the linear FLC. The failure mode changes from brittle shear fracture to ductile necking fracture. Nevertheless, reduced formability can be observed, but is not as significant as for the HC340LA investigated in [11]. To determine the influence of different pre-forming states, also plane-strain pre-formed Nakajima experiments are planned. Next to the experimental investigations, a numerical damage model will be developed which should be able to capture the anisotropic damage behaviour of the investigated material. This model will give more information on the occurring stress states and damage mechanisms leading to the different fracture modes.

Acknowledgement

The authors would like to thank the German Research Foundation (DFG) for the financial support under the grant number 455960756. Furthermore, the support of Wolfgang Saur (Institut für Werkstoffe des Bauwesens, Universität der Bundeswehr München) in performing the scanning electron micrographs is gratefully acknowledged.

References

[1] Graf A. and Hosford W. 1993 Effect of Changing Strain Paths on Forming Limit Diagrams of Al 2008-T4. *Metall Mater Trans A* **24:**671. DOI: 10.1007/BF02646529
[2] Bergström Y. and Ölund S. 1982 The forming limit diagram of sheet metals and effects of strain path changes on formability: a dislocation treatment. *Materials Science and Engineering* **56:**47–61. DOI: 10.1016/0025-5416(82)90181-1
[3] Werber A., Liewald M., Nester W., Grünbaum M., Wiegand K., Simon J., Timm J., Bassi C. and Hotz W. 2012 Influence of Different Pre-Stretching Modes on the Forming Limit Diagram of AA6014. *KEM* **504-506:**71–76. DOI: 10.4028/www.scientific.net/KEM.504-506.71
[4] Werber A., Liewald M., Nester W., Grünbaum M., Wiegand K., Simon J., Timm J. and Hotz W. 2013 Assessment of forming limit stress curves as failure criterion for non-proportional forming processes. *Prod. Eng. Res. Devel.* **7:**213–221. DOI: 10.1007/s11740-013-0446-6
[5] Gerke S., Zistl M. and Brünig M. 2020 Experiments and numerical simulation of damage and fracture of the X0-specimen under non-proportional loading paths. *Engineering Fracture Mechanics* **224:**106795. DOI: 10.1016/j.engfracmech.2019.106795
[6] Brünig M., Zistl M. and Gerke S. 2021 Numerical Analysis of Experiments on Damage and Fracture Behavior of Differently Preloaded Aluminum Alloy Specimens. *Metals* **11:**381. DOI: 10.3390/met11030381
[7] Volk W., Norz R., Eder M. and Hoffmann H. 2020 Influence of non-proportional load paths and change in loading direction on the failure mode of sheet metals. *CIRP Annals* **69:**273–276. DOI: 10.1016/j.cirp.2020.03.009
[8] Volk W. and Hora P. 2011 New algorithm for a robust user-independent evaluation of beginning instability for the experimental FLC determination. *Int J Mater Form* **4:**339–346. DOI: 10.1007/s12289-010-1012-9
[9] Weinschenk A. and Volk W. 2017 FEA-based development of a new tool for systematic experimental validation of nonlinear strain paths and design of test specimens. *AIP Conference Proceedings* **1896**. DOI: 10.1063/1.5007966
[10] Barlat F., Brem J.C., Yoon J.W., Chung K., Dick R.E., Lege D.J., Pourboghrat F., Choi S.-H. and Chu E. 2003 Plane stress yield function for aluminum alloy sheets—part 1: theory. *International Journal of Plasticity* **19:**1297–1319. DOI: 10.1016/S0749-6419(02)00019-0
[11] Volk W., Gruber M. and Norz R. 2020 Prediction of limit strains during non-proportional load paths with a change in loading direction. *IOP Conf. Ser.: Mater. Sci. Eng.* **967:**12069. DOI: 10.1088/1757-899X/967/1/012069
[12] Volk W. and Gaber C. 2017 Investigation and Compensation of Biaxial Pre-strain During the Standard Nakajima- and Marciniak-test Using Generalized Forming Limit Concept. *Procedia Engineering* **207:**568–573. DOI: 10.1016/j.proeng.2017.10.1022

IOP Conf. Series: Materials Science and Engineering 1238 (2022) 012024 doi:10.1088/1757-899X/1238/1/012024

Analytical approach to damage prediction in incremental sheet metal forming

Sahil Bharti, Deepu Raneesh, Hariharan Krishnaswamy and Sushanta Kumar Panigrahi

Department of Mechanical Engineering, Indian Institute of Technology Madras, Chennai 600036, India

E-mail: me17d017@smail.iitm.ac.in

Abstract. Incremental sheet metal forming (ISF) is known to exhibit higher formability compared to conventional stamping. It is established that the mechanism of failure during ISF is by fracture occurring at higher effective strain than the local necking observed in traditional forming processes. The deformation limit in ISF is therefore estimated numerically using a suitable continuum damage model. However, simulation of incremental forming process including the damage model is often computationally expensive. In the present work, a simpler alternative combining finite element simulation and numerical solution outside finite element software is utilized. The finite element simulation is performed ignoring the damage model. The stress and strain history thus obtained in critical locations is utilized to estimate the damage evolution. The proposed method is useful when utilizing uncoupled continuum damage models. The proposed method is validated for typical cases by comparing it against the predictions from finite element method. Excellent correlation is observed between the proposed method and finite element simulation results.

Keywords: Incremental Sheet forming (ISF); Damage modelling ; Fracture Forming limit (FFL)

1. Introduction

Incremental sheet metal forming (ISF) is a highly versatile technique aimed at dieless sheet metal forming. It finds relevance in many industries, viz. automotive, aerospace, biomedical sectors for manufacturing complex sheet metal components [1, 2]. In ISF, the maximum strain region is dynamic due to the moving tool that postpone the localization of strains allowing larger overall deformation before complete fracture.

Failure during ISF is a combined effect of material behaviour and the stress states to which they are subjected to [3]. In ISF the deforming specimen endures a multiaxial state of stress. Depending upon the geometry of the component being formed the loading conditions and strain path (mode) vary at different locations within the material. Based on experiments and finite element (FE) simulations, it has been observed that the plastic deformation in ISF is limited to a narrow zone in the vicinity of tool-sheet contact area [4]. This can be understood by tracing the strain evolution in the contact region. Since the failure mechanism during ISF is by fracture rather than local necking, fracture forming limit curve (FFL) are used to represent deformation limits as opposed to the conventional forming limit curve (FLC) [5, 6]. Although the overall formability of ISF is higher than the conventional forming, the process is limited by gradual

IOP Conf. Series: Materials Science and Engineering 1238 (2022) 012024 doi:10.1088/1757-899X/1238/1/012024

thinning and subsequent fracture [7, 8]. An investigation of the failed parts confirms that the cracks in the part manufactured by ISF were ductile in nature [9]. It can be shown that typical FFL can be fit as a straight line (in the range of uniaxial to equibiaxial tension deformation mode as applicable to FLC) with a constant slope in the traditional two dimensional principal strain space [7].

An accurate prediction of fracture during ISF is essential for successful process planning. From the mechanics of materials perspective (based on stress strain distribution), continuum damage models can be effectively employed to obtain a theoretical estimate for the critical fracture limits [10]. The constitutive description of damage modelling is broadly classified as coupled and uncoupled models [11]. The damage variable evolves and mutually influences the plasticity relations in the former approach. In the case of uncoupled models such as modified Mohr-Coulomb [12] or Hosford Coulomb [13], the damage evolution is independent of the stress-strain relation and the onset of failure is estimated when the accumulated damage variable reaches a critical value. The uncoupled damage models are relatively simpler to implement numerically and are increasingly gaining interest in successfully modelling fracture prediction in ISF [14]. Huang et al. [10] provided an estimation of approximate stress distribution in the ISF deformation zone using force equilibrium. Martins et al. [15] estimated the damage in the material using a fracture criterion based on hydrostatic stress. Nguyen et al. [16] utilized Oyane fracture criterion to estimate damage in ISF for magnesium. Malhotra et al. [17] used damage approach in FE analysis of a simple cone. It was observed that the damage accumulates the most in the regions of local bending in the sheet during deformation. Similar results were shown by the study conducted by Shamsari et al. [18] using Modified Mohr-Coulomb fracture criterion. The numerical simulations indicated that significant damage accumulates in the transition region between the deformed and contact zone.

Determining the location of the damage initiation is critical and may help to analyze key factors leading upto material fracture. Although a great amount of research is reported on the deformation mechanism of ISF, initiation and evolution of the material fracture is scarcely discussed. Implementation of damage in FE simulations involves meticulously written user subroutines, which utilizes heavy computational resources. As mentioned earlier, damage accumulation in uncoupled approach is computed independent of the deformation behaviour of the material. This presents an opportunity to reduce the overall computational load by separating the FE analysis from the damage calculations. In this context, the current work presents a preliminary framework where the FE simulations are used to determine the deformation history, thereby damage progression is estimated via a standalone analytical code using the FE output. This procedure is validated using selected ISF formed components.

2. Methodology

Commercially pure aluminium alloy (AA1050) sheet of 1.2 mm thickness is considered in this study. A series of mechanical tests were conducted to record the mechanical behaviour of the material till fracture. Material response under different loading conditions was achieved by varying the specimen geometry (details are discussed in section 3). All the mechanical tests were performed using a 100 kN capacity (Zwick-Roell 100) universal testing machine (UTM) at an initial strain rate of $0.005s^{-1}$ and the displacements were recorded by video extensometer.

2.1. Experiments on single point incremental forming (SPIF)

Components with three geometrical variations namely, (i) variable wall angle conical frustum (VWACF), (ii) truncated pyramid and (iii) a hybrid shape with five lobes were designed and developed by incremental forming in the present work (Fig.1). SPIF experiments were conducted on CNC machining center ('Jyoti Huron KMill'). The sheet metal blank of dimensions (150 mm × 150 mm × 1.2 mm) was rigidly held by a customized fixture, such that the material flow is constrained along the periphery to avoid any draw-in during forming. Tool of 8 mm diameter was moved across the sheet with a constant feedrate of 750 mm/min. 3D spiral toolpath conformal to the designed component geometry were developed using commercially available CAM software, 'Mastercam' (ver.19.0.7874.0). Hydraulic oil was applied to the forming surface to reduce the contact friction during forming.

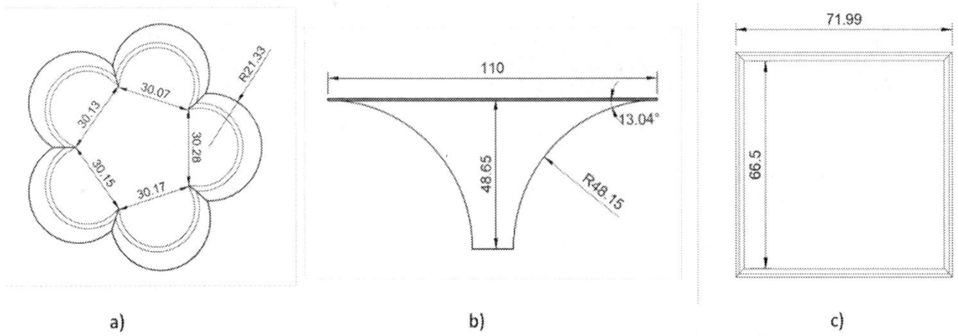

Figure 1. ISF formed geometries a) 5 lobe b) Variable wall angle conical frustum (VWACF) c) Truncated Pyramid

In order to characterize the forming limits by fracture all the component(ϵ_{2f}) at the vicinity of the crack were estimated using circle grid analysis. Thickness strains (ϵ_{3f}) are estimated by measuring final thickness at the crack location. Details of the procedure can be found in work by Cristino et al. [19]. Laser etched samples with grid of 1.5mm (d) diameter circles were used, and the shape change of circle to elliptical indicated the nature of strain induced.

$$\epsilon_{2f} = \ln\left(\frac{d_2}{d}\right) \tag{1}$$

where d_2 is the length of minor axis of the ellipses after deformation. The strain measurements are obtained using 3Com USB camera. The major inplane strains ϵ_{1f} at fracture were obtained by imposing volume constancy of plastic deformation.

$$\epsilon_{1f} = -(\epsilon_{2f} + \epsilon_{3f}) \tag{2}$$

2.2. Finite Element Simulations of tension tests and ISF
Commercially available explicit finite element code ABAQUS® was used to simulate the mechanical tests and SPIF. 3D solid elements C3D8R (8 node linear Brick with hourglass control) were chosen with minimum mesh size of 0.2 mm and 0.5 mm for mechanical tests and SPIF respectively. Mesh is refined in critical regions with a minimum of 3 elements along the thickness direction. The boundary conditions for UT, shear and notch tests and SPIF are depicted in Fig.2.

A user-defined VUMAT subroutine was implemented to simulate the mechanical behaviour. The plastic deformation was modelled assuming vonMises yield criterion and isotropic hardening. An uncoupled phenomenological damage criterion (Hosford-Coulomb) is implemented to simulate the damage evolution.

2.3. Proposed methodology for computation efficiency
It will be subsequently shown that the use of VUMAT increases the computation time significantly even for relatively simpler shapes. Unlike coupled damage models, the uncoupled

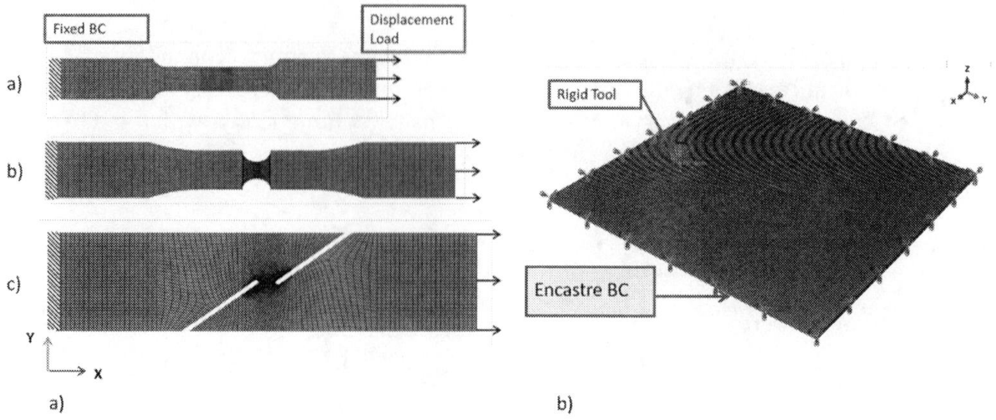

Figure 2. FE modelling , Boundary conditions for (a) Mechanical tests (b) SPIF

damage models such as that being investigated in the present work does not interfere with the plastic constitutive behaviour. Instead, the damage evolution is independently evaluated for each element based on the deformation history. Taking advantage of the uncoupled formulation, a combined approach is proposed in the present work to reduce the computation time significantly. The proposed methodology involves finite element simulation without the use of any damage model. The stress and strain history of the elements are exported and the damage evolution is estimated outside the framework of finite element code. In the present work, we have used commercial MATLAB code for estimating damage. The damage model of interest is independently coded in MATLAB and using the strain and stress history, the damage variable is computed. The deformation limit when the damage variable reaches a critical value is considered as failure. The procedure is shown in the flow chart below (Fig.3).

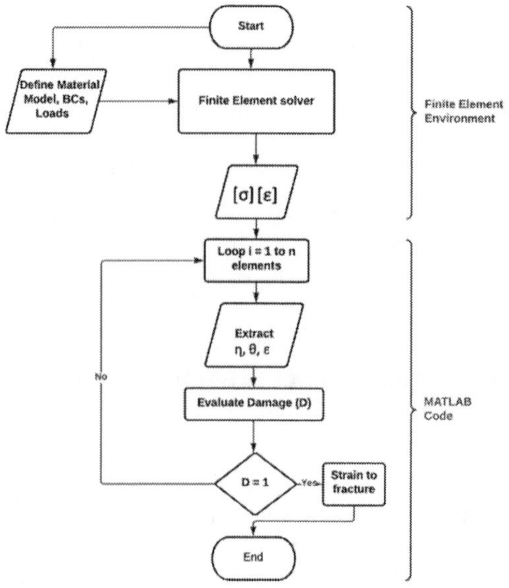

Figure 3. Flowchart depicting the integration of FE analysis and MATLAB output

Since damage is evaluated independently, the in-built material models in commercial software that are numerically efficient can be used to simulate ISF. In addition to the above, different combinations of yield criterion, hardening and damage models can be evaluated easily. In the previous approach, a separate VUMAT has to be written for each combination and is time consuming. One of the factors that contribute to increased computation time in VUMAT is the damage evaluation of all the elements in the model. In practice, only select few regions are subjected to large strain. Therefore, it is unnecessary to evaluate the entire model. In the proposed method, a minimum number of elements from the highly strained region is chosen and only these selected elements are tracked for damage accumulation.

3. Results and Discussion

The uniaxial stress- strain flow behaviour is described using a combined hardening law (Eq.3), as depicted in the Fig.(4). A linear combination of Swift and Voce hardening rules are used to describe the pre and post necking deformation of the material till the onset of fracture [13, 20].

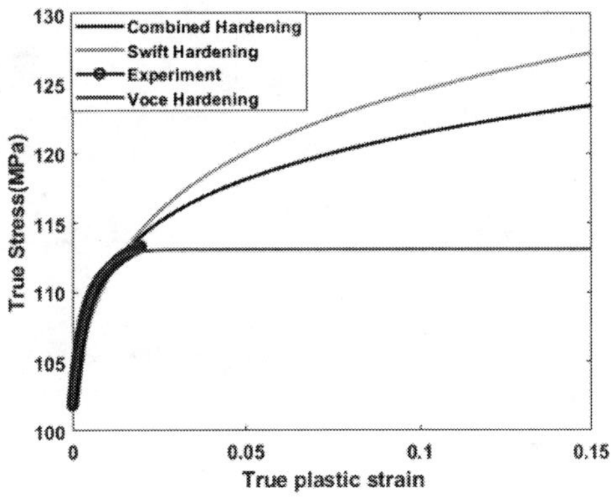

Figure 4. Experimental true stress - true strain response along with calibrated combined hardening curves based on Swift hardening, Voce hardening.

$$\sigma = QK(\epsilon_o + \epsilon_p)^n + (1-Q)[\sigma_y + A(1 - e^{-B\epsilon_p})] \quad (3)$$

Here K, ϵ_o,n,ϵ_p and σ_y denote strength coefficient, strain offset, strain hardening coefficient, equivalent plastic strain and yield stress respectively. A and B are fitting constants and Q is a post necking hardening parameter. Table 2 lists the mechanical properties of the material.

Table 1. Calibrated coefficients of the hardening law Eq.(3)

K(MPa)	ϵ_o	n	Q	A	B	σ_y
141.0736	0.002707	0.055	0.73	11.15	220.7	101.9138

3.1. Calibration of damage model

The stress state at a material point can be described in terms of dimensionless ratios of stress invariant, namely stress triaxility (η) and Lode angle parameter ($\bar{\theta}$). In terms of principal stress components(σ_1,σ_2,σ_3), vonMises effective stress ($\bar{\sigma}$), η and $\bar{\theta}$ can be defined as follows :

$$\bar{\sigma} = (\frac{1}{2}[(\sigma_1 - \sigma_2)^2 + (\sigma_2 - \sigma_3)^2 + (\sigma_3 - \sigma_1)^2])^{1/2} \quad (4)$$

Table 2. Mechanical properties of aluminum AA1050

E (GPa)	YS (MPa)	UTS (MPa)	Poisson's ratio	Elongation(%)	
				uniform	fracture
71.659	101.914	113.2987	0.33	2.31	9.96

$$\eta = \frac{\sigma_1 + \sigma_2 + \sigma_3}{3} \tag{5}$$

$$\bar{\theta} = 1 - \frac{2}{\pi} \arccos\left[\frac{-27}{\pi}\eta\left(\eta^2 - \frac{1}{3}\right)\right] \tag{6}$$

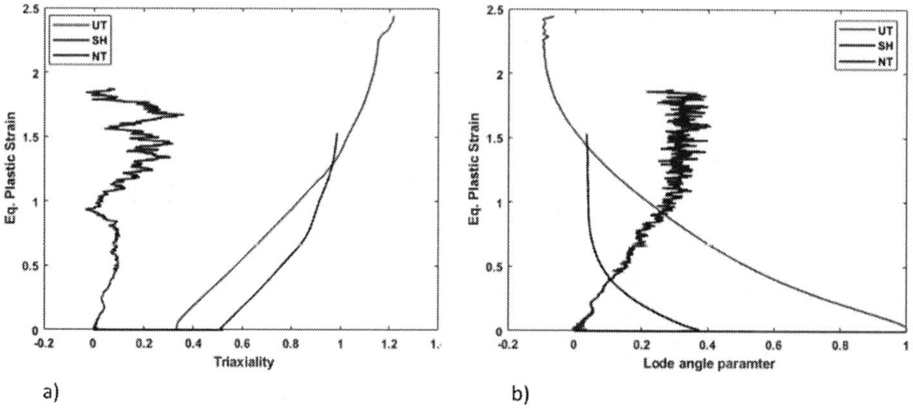

Figure 5. Evolution of eq.plastic strains with respect to a) triaxility and b) Lode angle parameter for mechanical tests

Fig.(5) depicts the evolution of (η) and ($\bar{\theta}$) with equivalent plastic strain(ϵ_p), obtained from the FE simulations of mechanical tests under uniaxial tension, notched tension and shear loading. For each loading condition, the stress state changes with the deformation of the material. It can be seen that the values of η and $\bar{\theta}$ are constant upto the onset of necking, and changes thereafter with further deformation.

In the present work, an uncoupled phenomenological fracture initiation model, Hosford Coulomb fracture criterion (HC) is used to predict the onset of ductile fracture in ISF. It is based on a localization criteria defined in terms of Hosford equivalent stress and normal stress on the plane of maximum shear, Eq(7) ([13]).

$$\left\{\frac{1}{2}((\sigma_1 - \sigma_2)^a + (\sigma_2 - \sigma_3)^a + (\sigma_3 - \sigma_1)^a)\right\}^{1/a} + c(\sigma_1 + \sigma_3) = b \tag{7}$$

HC model is a stress based criterion which is transformed from principal stress space, using Lode angle parameter dependent functions (f_1,f_2,f_3) as given below,

$$f_1[\bar{\theta}] = \frac{2}{3} cos[\frac{\pi}{6}(1 - \bar{\theta})] \tag{8}$$

$$f_2[\bar{\theta}] = \frac{2}{3}cos[\frac{\pi}{6}(3+\bar{\theta})] \tag{9}$$

$$f_3[\bar{\theta}] = -\frac{2}{3}cos[\frac{\pi}{6}(1+\bar{\theta})] \tag{10}$$

The principal stress may be expressed in (η,$\bar{\theta}$,$\bar{\epsilon}$) space, as given :

$$\sigma_1 = \bar{\sigma}(\eta + f_1) \tag{11}$$

$$\sigma_2 = \bar{\sigma}(\eta + f_2) \tag{12}$$

$$\sigma_3 = \bar{\sigma}(\eta + f_3) \tag{13}$$

Using Eqs.(11-13), in the Eq.(7) the stress at the onset of fracture (σ_f) can be formulated as:

$$\bar{\sigma} = \sigma_f(\eta, \bar{\theta}) = \frac{b}{[\frac{1}{2}((f_1 - f_2)^a + (f_2 - f_3)^a + (f_1 - f_3)^a)]^{\frac{1}{2}} + c(2\eta + f_1 + f_3)} \tag{14}$$

Parameter a controls the Lode angle sensitivity of the fracture strain, c is called the friction parameter and b is the strain to onset of fracture under uniaxial tension. Furthermore, strain at onset of fracture ($\bar{\epsilon}_f$) is obtained by taking inverse of isotropic hardening law (Eq.3), which can be expressed as, $\bar{\epsilon}_f = f^{-1}(\bar{\sigma}_f)$. Here, f signifies hardening function rewritten in terms of $\bar{\sigma}_f$."

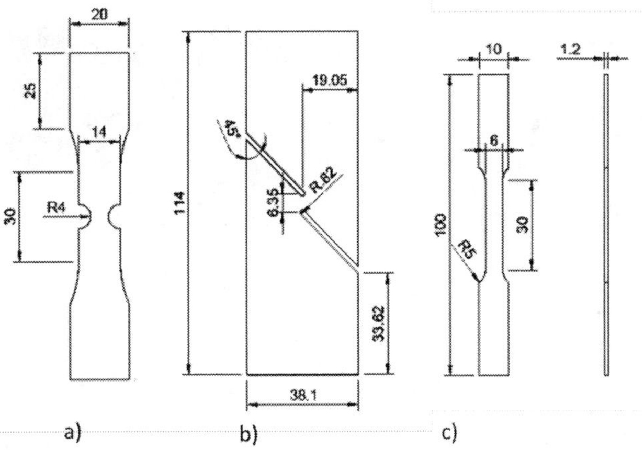

Figure 6. Specimen geometries a) Notch Test (NT) b) Shear Test (SH) c) Uniaxial Tension (UT)

Strain data at the onset of fracture for three monotonic tests namely uniaxial tension (UT), shear test (ST) and notch test (NT) are used to calibrate the damage model. The loading history denoted by (η_i,$\bar{\theta}_i$) of each calibration test is evaluated with help of FEA simulation. It is assumed that the displacement at the onset of fracture (δ_f) coincides with the location of highest equivalent plastic strain within the specimen near fracture Fig(7). The damage progression (D) for each loading case is calculated according to the Eq.(15):

$$D = \int_0^{\bar{\epsilon}_f} \frac{d\bar{\epsilon}}{\bar{\epsilon}_f(\eta, \bar{\theta})} = 1. \tag{15}$$

D is damage indicator, which attains unity at the onset of fracture. The unknown parameters of HC damage model obtained by fitting fracture surface to the experimental fracture strains for UT,SH NT tests is given in Table 3. The above formulation is coded in MATLAB environment for calibration and offline prediction of fracture strains for a given forming condition.

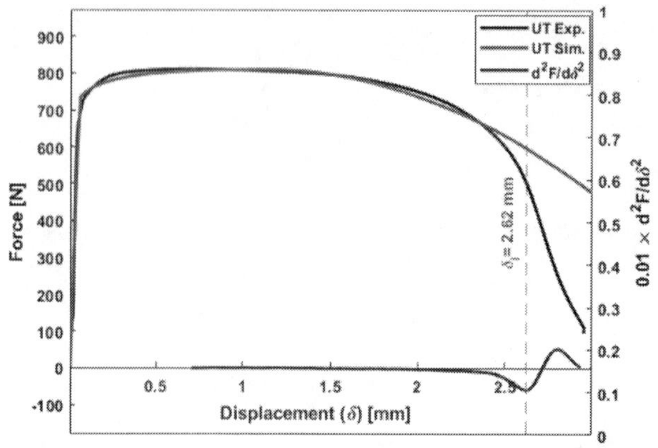

Figure 7. Experimental and numerical force -displacement response along with dashed line marking the onset of fracture in UT test.

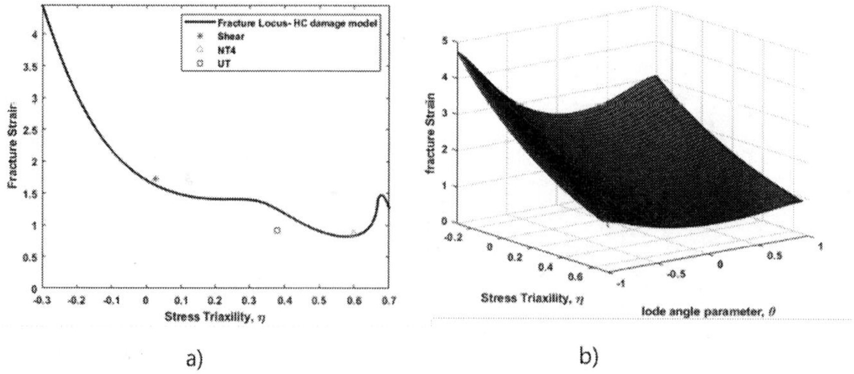

Figure 8. Fracture locus for HC damage criterion (a) 2D fracture locus and (b) 3D fracture surface in $\eta,\bar{,}\ \epsilon_f$ space

Table 3. Damage Parameters

a	b	c
1.7068	139.082	0.02765

3.2. Validation of numerical code

In order to test the validity of the code, the fracture strains obtained by FE simulations are compared with analytically estimated strain values for the mechanical tests. Fig.(9) shows the fracture strain distribution obtained by FE simulations and the analytical results for each mechanical test.

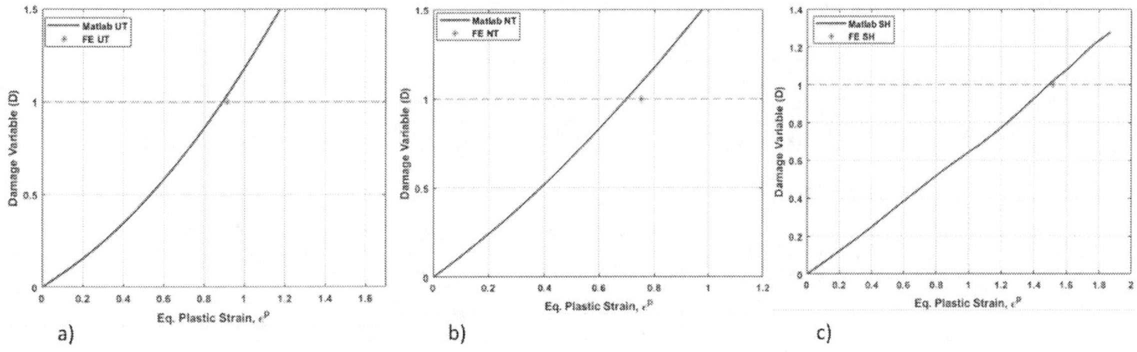

Figure 9. Fracture strains and evolution for damage estimated for a) Uniaxial tension test (UT) b) Notch test (NT) and c) Shear test (SH) using MATLAB code

Furthermore, to test the applicability of the analytical method to predict the onset of failure strains in ISF, strain path data (η,$\bar{\theta}$ and ϵ^p) were extracted for the critical elements from FE simulations and used as the inputs for the MATLAB code to estimate the fracture stains. It can be seen from Table 4 that the fracture strains under various loading conditions are well predicted by the coded damage theory in MATLAB.

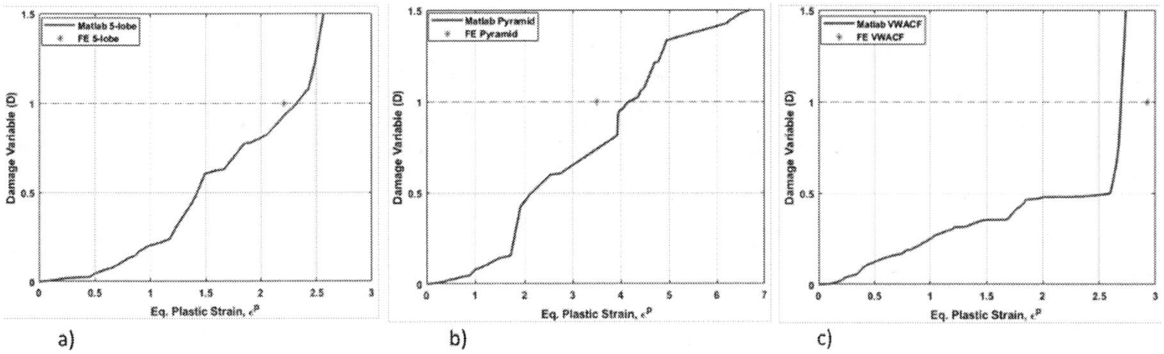

Figure 10. Fracture strains and evolution for damage estimated for (a) 5 - lobe (b) Pyramid and (c)VWACF using MATLAB code

Table 5, lists the CPU time recorded for FE analysis of ISF, when damage calculations are implemented with material subroutine (VUMAT), compared to analyses when damage is computed outside of FE environment. A workstation running on a windows platform, Intel(R) Core(TM) i7-7700 CPU @ 3.60GHz 32 GB RAM is used to run all the FE simulations. A significant difference in computational time can be noticed in both cases.

Fig.(11) depicts the inplane major and minor strain distribution, and the comparative placement of FFLs along with the evolution of strain path for the critical elements at fracture obtained by experiments, numerical simulations and analytical code. The experimental fracture forming limit is marked by interpolating a straight line from the values of inplane strains obtained by grid circle analysis. Similarly, fracture limit lines (FFLs) developed based on numerical prediction and MATLAB code represent the onset of fracture stains. The analytical and numerically estimated fracture strains are depicted by violet and yellow colored diamond marks respectively. Dashed lines marks the average strain path (strain ratio, β), recorded while

Table 4. FE and analytically computed fracture strains in mechanical and ISF tests for calibrated Damage models

ISF	Pyramid	VWACF	5 Lobe
Code (ϵ_f)	4.14	2.65	2.39
FEM (ϵ_f)	3.90	2.90	2.21
Abs. Error(%)	6.1	8.6	8.14
Mechanical Tests	UT	NT	SH
Code (ϵ_f)	0.916	0.702	1.49
FEM (ϵ_f)	0.913	0.75	1.51
Abs. Error(%)	0.03	6.4	1.1

Table 5. Computational time ISF tests

ISF	Pyramid	VWACF	5 Lobe
FE with Damage (hrs)	75	115	46
FE without Damage (hrs)	36	54	22
Difference (hrs)	39	61	24

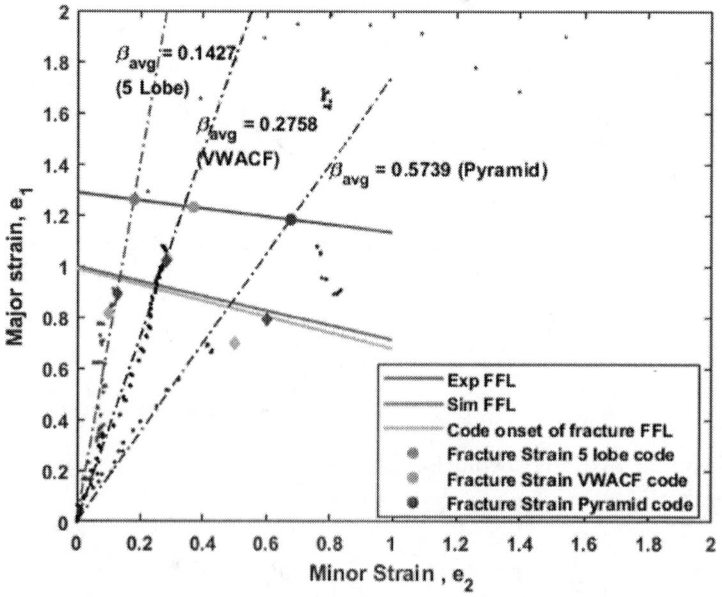

Figure 11. Fracture forming limits

forming three ISF parts i,e. 5 lobe, VWACF and pyramid. Furthermore, the damage evolution after onset is implemented by extrapolating the strain paths to predict the inplane strains, as function of average strain ratio corresponding to the loading, given as :

$$e_1 = \frac{1}{\beta} e_2 \tag{16}$$

here e_1 and e_2 are the major and minor inplane strains.

4. Conclusions

The present work demonstrated the analysis of incremental forming using uncoupled continuum damage model. The trend of fracture initiation predicted and experiments match well, though the simulation estimate were conservative. In order to reduce the computation time, we have proposed a hybrid strategy to separate the FE simulation without damage and later use the stress-strain history to analytically estimate the failure. It is demonstrated that the computation time reduced approximately by 50% when FE simulation is carried without damage model. The results were consistent with three different geometries analyzed for incremental forming. The deformation history of critical elements is analyzed using a MATLAB code after FE simulation. The predictions in the proposed approach is compared with that predicted using a user defined material model. It was shown that the results correlate very well. Therefore, the presented methodology can be utilized to significantly reduce the computation time. The added advantage is the flexibility of analyzing different combination of constitutive relations without detailed development of individual subroutines.

5. Acknowledgments

Authors would like to acknowledge Department of Heavy Industries, Government of India and Mahindra & Mahindra for funding the project (sanction number: 7(8)/2019-AEI (19310)).

References

[1] J Jeswiet, F Micari, G Hirt, A Bramley, Joost Duflou, and J Allwood. Asymmetric single point incremental forming of sheet metal. *CIRP annals*, 54(2):88–114, 2005.
[2] Tegan McAnulty, Jack Jeswiet, and Matthew Doolan. Formability in single point incremental forming: A comparative analysis of the state of the art. *CIRP Journal of Manufacturing Science and Technology*, 16:43–54, 2017.
[3] L. Filice, L. Fratini, and F. Micari. Fracture mechanics fundamentals and applications. *Boca Raton CRC Press*, 2017.
[4] Y. Fang, B. Lu, J. Chen, D.K. Xu, and H. Ou. Analytical and experimental investigations on deformation mechanism and fracture behavior in single point incremental forming. *Journal of Materials Processing Technology*, 214(8):1503–1515, 2014.
[5] J.M. Allwood, D.R. Shouler, and A. Erman Tekkaya. The increased forming limits of incremental sheet forming processes. In *Sheet Metal 2007*, volume 344 of *Key Engineering Materials*, pages 621–628. Trans Tech Publications Ltd, 8 2007.
[6] Kali Prasad, Hariharan Krisnaswamy, Dilip Banerjee, and Uday Chakkingal. An investigation into the influence of interrupted loading in improving the stretch-flangeability of dual phase steel. In *Defect and Diffusion Forum*, volume 414, pages 81–87. Trans Tech Publ, 2022.
[7] M B Silva, M Skjoedt, A G Atkins, N Bay, and P A F Martins. Single-point incremental forming and formability—failure diagrams. *The Journal of Strain Analysis for Engineering Design*, 43(1):15–35, 2008.
[8] Kali Prasad, B. Venkatesh, Hariharan Krishnaswamy, Dilip K. Banerjee, and Uday Chakkingal. On the interplay of friction and stress relaxation to improve stretch-flangeability of dual phase (dp600) steel. *CIRP Journal of Manufacturing Science and Technology*, 32:154–169, 2021.
[9] P. Gupta and J. Jeswiet. Observations on heat generated in single point incremental forming. *Procedia Engineering*, 183:161–167, 2017. 17th International Conference on Sheet Metal, SHEMET17.
[10] G. Hirt, J. Ames, M. Bambach, R. Kopp, and R. Kopp. Forming strategies and process modelling for cnc incremental sheet forming. *CIRP Annals*, 53(1):203–206, 2004.
[11] Vivek K Barnwal, Shin Yeong Lee, Jisik Choi, Jin Hwan Kim, and Frédéric Barlat. Performance review of various uncoupled fracture criteria for TRIP steel sheet. *International Journal of Mechanical Sciences*, 195(January), 2021.
[12] Keunhwan Pack and Dirk Mohr. Combined necking & fracture model to predict ductile failure with shell finite elements. *Engineering Fracture Mechanics*, 182:32–51, 2017.
[13] Dirk Mohr and Stephane J. Marcadet. Micromechanically-motivated phenomenological hosford–coulomb model for predicting ductile fracture initiation at low stress triaxialities. *International Journal of Solids and Structures*, 67-68:40–55, 2015.
[14] Sahil Bharti, Aishwary Gupta, Hariharan Krishnaswamy, S.K. Panigrahi, and Myoung-Gyu Lee. Evaluation

of uncoupled ductile damage models for fracture prediction in incremental sheet metal forming. *CIRP Journal of Manufacturing Science and Technology*, 37:499–517, 2022.

[15] P.A.F. Martins, Niels Bay, Martin Skjødt, and M.B. Silva. Theory of single point incremental forming. *C I R P Annals*, 57(1):247–252, 2008.

[16] Moriya Oyane, Teisuke Sato, Kunio Okimoto, and Susumu Shima. Criteria for ductile fracture and their applications. *Journal of Mechanical Working Technology*, 4(1):65–81, 1980.

[17] Rajiv Malhotra, Liang Xue, Ted Belytschko, and Jian Cao. Mechanics of fracture in single point incremental forming. *Journal of Materials Processing Technology*, 212(7):1573–1590, July 2012. Funding Information: The authors of this work gratefully acknowledge the support provided by the National Science Foundation (CMMI-0758607 and CMMI-0727843) for this work.

[18] Mohammad Javad Mirnia and Mohsen Shamsari. Numerical prediction of failure in single point incremental forming using a phenomenological ductile fracture criterion. *Journal of Materials Processing Technology*, 244:17–43, 2017.

[19] Valentino AM Cristino, Maria Beatriz Silva, Pak Kin Wong, and Paulo AF Martins. Determining the fracture forming limits in sheet metal forming: A technical note. *The Journal of Strain Analysis for Engineering Design*, 52(8):467–471, 2017.

[20] Ji Hyun Sung, Ji Hoon Kim, and R.H. Wagoner. A plastic constitutive equation incorporating strain, strain-rate, and temperature. *International Journal of Plasticity*, 26(12):1746–1771, 2010.

IOP Conf. Series: Materials Science and Engineering 1238 (2022) 012025 doi:10.1088/1757-899X/1238/1/012025

A case study of mechanical and thermal fatigue of press hardening dies

K Chantziara[1], F Javadzadeh Kalahroudi[1], J Bergström[1], M Grehk[1] and P Ulfberg[2]

[1] Karlstad University, Karlstad, Sweden
[2] Gestamp HardTech AB, Luleå, Sweden

katerina.chantziara@kau.se

Abstract. Press hardening provides ultra-high strength steel components, typically boron steels, of complex geometries. In the process, the steel sheet is heated in a furnace to the austenitization temperature, transferred to the press, then simultaneously formed at high temperature and cooled in the die. Life limiting factors for the press hardening dies are mechanical fatigue, thermal fatigue, and wear. In the present case study two die segments were selected where critical damages were mechanical and thermal fatigue, respectively. The dies were made of a H13 type premium hot-work tool steel with complex heated die technology, die design integrating an advanced cooling system, for pressing automotive frame parts.
The first die failed due to mechanical loading with a crack initiated from the ejector pin area. The die design, the mechanical loads, the elevated temperature, and the tool steel crack resistance are main factors to consider. In the second die cracks initiated from an ejector pin hole, as well, due to thermal cycles causing alternating compressive and tensile stresses at the surface, which led to crack nucleation because of the accumulation of local plastic strain in the surface.

1. Introduction

During the last decades press hardening (also called hot stamping) has become increasingly popular in automotive industry. This is due to the advantageous features involved as optimized processing, allowing for tailored component properties with high and optimized mechanical performance with improved high strength and high toughness. In the press hardening process blanks are both hot formed and quenched in one step in a water-cooled tool to achieve high strength [1]. The final steel component will have a martensitic microstructure with improved mechanical properties, and it allows using lower thickness components and in the end lighter final products. Evidently, it reduces both the amount of steel required per vehicle body shell and the fuel consumption when the vehicle is in use. In addition, the higher strength of components significantly improves the energy absorption capacity and therefore the crash performance of structural components [2]. This point plays an important role in the production of car body components in the automotive industry.

However, the high temperature forming process entails demanding working conditions to the forming steel dies. The main damage mechanisms leading to die failure in hot forging processes are abrasive and adhesive wear, mechanical and thermomechanical fatigue, plastic deformation and oxidation [3]. According to the statistics provided by many authors, 70 % of forging dies are taken out of service due to wear, 25% due to mechanical fatigue, and only 3-5% of the forging dies failed due to thermal cracking and plastic deformation [4]. All of the mechanisms occur simultaneously from the very beginning of the press hardening process, and they can be more or less intensive depending on factors as the design of the die, lack of hardness at elevated temperature, contact time, sliding distance, inadequate die material, work piece temperature and lubrication [3]. Moreover, high temperature processes involve changes in mechanical properties due to microstructural changes and thermal softening; surface chemical and morphological changes due to oxidation and diffusion; and deterioration of the surface and bulk material by adhesive/abrasive wear and thermal fatigue. Therefore, investigations on these processes are complex. The surface and near surface regions are points of attention since many of the initial damages occur there [5].

One of the most common materials used in press hardening is manganese–boron steels, as the 22MnB5 grade. The presence of boron in this steel helps to stabilize the austenite phase at elevated temperatures and limits the formation of ferrite and pearlite by promoting the formation of martensite during rapid quenching [6]. As a result, the martensitic microstructure provides the hardened final product with a high tensile strength up to 1500 MPa.

In the present case study the focus is on segments of damaged press hardening dies used for production of automotive cross member and side plates. The present effort is split into two sections for each die. First, macro-observations were performed to identify the different damages, then hardness measurements and micro-investigation was carried out by SEM for analyzing the occurred mechanisms. Finally, after detailed examination, mechanical and thermal fatigue were identified as the two main mechanisms of matrix damage.

2. Materials and methods

The press hardening dies were made of QRO 90 Supreme, a high performance chromium hot-work tool steel and specified to hold a hardness of 52 ± 1 HRC (~550 HV) for this application. The blank material was made of an uncoated boron steel, 22MnB5, with sheet thickness 5.9 mm. Subsequent to the press hardening process, it transformed to a martensitic microstructure with a hardness of about 500 HV from initial ferritic–perlitic structure of the blank material with hardness of 170 HV.

The nominal chemical composition of both the dies and the sheet metal are displayed in **Table 1**.

Table 1. Chemical composition, (%wt.), of dies and sheet metal, Fe-bal.

	C	Cr	Mo	Si	Mn	V	Al	Ti	B
QRO 90 Supreme	0.38	2.6	2.25	0.30	0.75	0.9			
22MnB5	0.22	0.25		0.20	1.25		0.04	0.035	0.003

The press hardening process involves several steps. First, the boron steel sheet is heated up in a furnace to its austenitization temperature (about 930 °C). Then, it is transferred by means of a robot arm into the press. During transferring the temperature drops and the sheets are about 850 °C when they are put in the press. In the next step, the sheets are simultaneously formed and quenched in a closed water-cooled die, and the contact time between the die and sheet metal is about 26 s. Finally, at about 220 °C the pressed sheets are ejected from the dies. In order to maintain a controlled temperature history throughout the die an intricate cooling and heating, channel system is perforating the die.

The overview images of the two case study dies are presented in **Figure 1** and **Figure 2**. Both of the dies considered as damaged after approximately 20000 number of press cycles. Cross member die with a mechanical fatigue fracture is displayed in the CAD-model in **Figure 1a**. The ejector-pin hole and the cooling channel are the main die features that influenced the fracture and both were designed with a 15mm margin from the die boundaries. The main crack of interest suffered from mechanical fatigue, **Figure 1b**, it originates from the ejector pin hole and grew to a length of 7 cm long on the die´s upper surface. Beneath the die surface cooling channels are running along the die.

The side-plate die suffered from thermal fatigue cracks around the ejector pin hole after 20000 press cycles. The ejector pin hole opens on the flat surface and there is a row of running cooling channels about 10 mm below and parallel to the surface as displayed in the CAD model in **Figure 2a**. **Figure 2b** shows the top view of the die surface. On the left part and around the ejector-pin hole the surface have

IOP Conf. Series: Materials Science and Engineering 1238 (2022) 012025 doi:10.1088/1757-899X/1238/1/012025

a roughened appearance. Various damages were found on the die, however the die failed due to the thermal fatigue cracks originating from the ejector pin hole (**Figure 2c**).

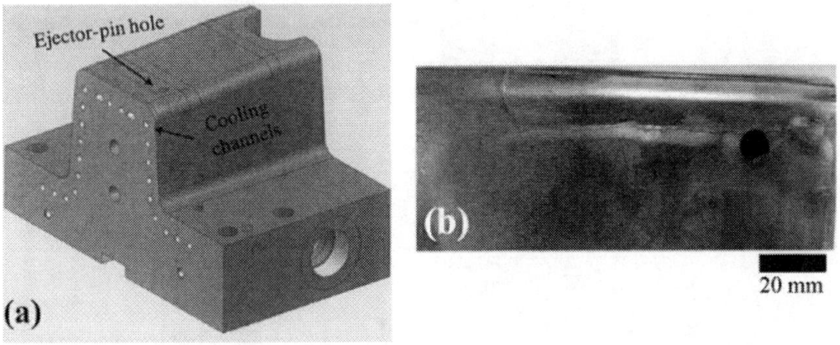

Figure 1. a) CAD model showing the design of the cross member die, and (b) overview of the main crack.

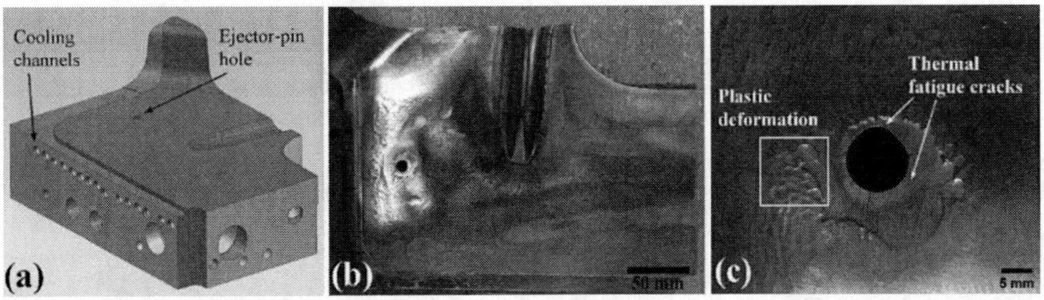

Figure 2. (a) CAD model showing the ejector pin hole and cooling channels, (b) overview of the side-plate die, and (c) thermal fatigue around the pin hole and plastic deformation damage on the die.

To identify failure mechanisms, detailed studies of the tool surface and polished cross-sections were performed by scanning electron microscopy (SEM) and energy dispersive X-ray spectroscopy (EDS) (JSM-7900F with X-MaxN EDX detector). Hardness measurements were carried out by macro- and microhardness Vickers indentations (Future Tech FM) using a load of 300 kg, 0.50 kg and 0.25 kg.

3. Results

3.1. Failure mechanisms in the press hardening die for automotive cross member

The main mechanical fatigue crack originates from the ejector pin hole growing in two opposite directions along the die, **Figure 3a**, initiated by the stress concentration of the pin hole and controlled by principal stresses along the blank sliding direction and perpendicular to the die length. The crack grows along the die and perpendicular to the die surface down to a depth of 1.5 mm. Here, it meets up with another crack following the die radius seemingly affected by the thermal stresses between the cooling channel and the curved die surface. The examination of the fractured surfaces, the die surface, cross-sections along and transverse to the die, and ejector pin hole and cooling channel, revealed different types of damage mechanisms that took place and lead to the degradation of the die and eventually its fracture. A grading of the main different damages is summarized, seen in **Figure 3b**, as; i) mechanical fatigue, ii) thermal fatigue, iii) blank - die surface press and oxidation damage, iv) internal creep damage, and v) cooling channel damage. Nevertheless, types i) and ii) are the two main damage mechanisms that led to the fatal failure of the die, and the mechanical and thermal fatigue evidence can be seen already by the naked eye on the fractured surface (**Figure 3a**).

SEM photos of the fractured surface imply that two different events took place, the mechanical and the thermal fatigue, even though coordinated by the press cycle. The fracture lines revealed that the damage of the events meeting points was approximately 1.5 mm beneath the die´s blank surface, **Figure 4a**.

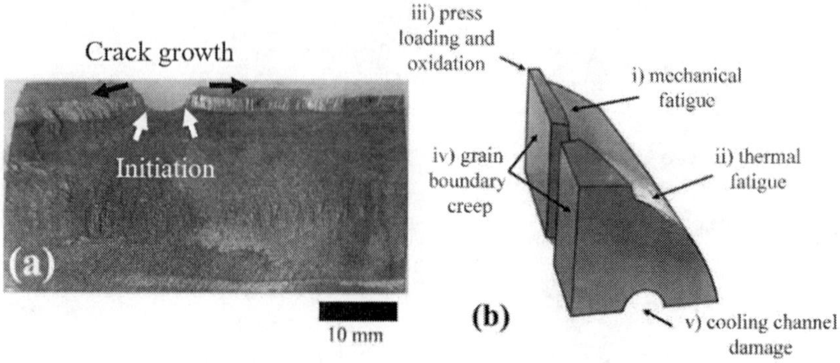

Figure 3. (a) Overview of the fractured surface (b) Grading of the different types of damages.

The assumption of the two mechanisms is being confirmed by the evidences of **Figure 4b and 4c**. **Figure 4b**, shows a cross-section transverse to the die length direction and perpendicular to the blank surface of the die. It seems that microcracks start from the upper surface and grow in a direction in accordance to a transverse stress responsible to the mechanical stress fracture. Concerning thermal stresses, thermal cracks were detected in a die curvature cross-section as viewed in **Figure 4c**.

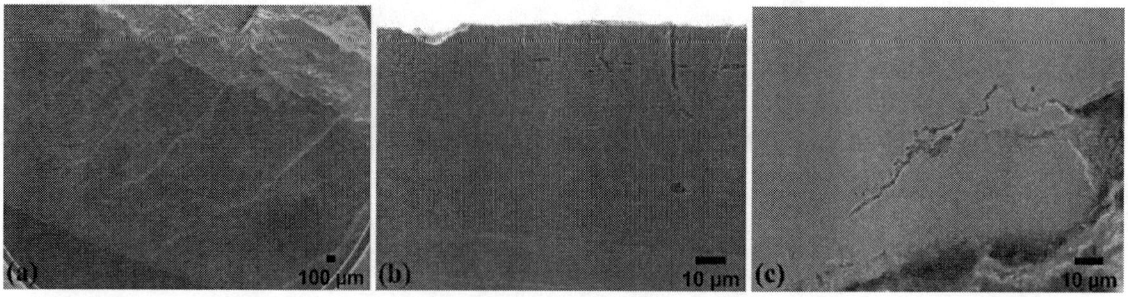

Figure 4. (a) Overview of the mechanical and thermal stress fractured surfaces meeting, (b) cracking due to mechanical loading in the die-blank surface contacts, (c) thermal crack in the die curvature cross-section.

Other secondary damage mechanisms than mechanical and thermal fatigue took place as well. SEM photos revealed phenomena of oxidation, especially beneath the blank-die surfaces **(Figure 5a)**. Oxidation cavities were detected not only beneath the blank surface of the die, but in most cross-sections beneath examined surfaces; ejector-pin hole surfaces, the fractured surface and in some cases different from the cracked area.

Internal cavitation with microvoid formation related to creep-fatigue mechanisms was observed at grain boundaries as is displayed in **Figure 5b**. The microcavities were found not only at samples directly associated to the main crack, but also at samples cut from the bulk of the die. Noteworthy was not the size of the cavities, but the extent of the phenomenon. Although, these types of creep induced cavities are small in size, they are high in their number and the number getting higher closer to the hot surfaces.

Last but not least, corrosion pits were found beneath the surfaces along the cooling channels, combined with cracks at their end (**Figure 5c**). The small cracks propagated in the channel perpendicular direction, a feature that may be attributed to preload stresses due to clamping of the die segments.

Macro- and microhardness values of the die along the depth of the ejector-pin hole, as well as close to the main crack path change to the depth of 1mm, in a cross-sectioned fracture sample are shown in **Figure 6**. In general, it seems that the hardness is being maintained during the press hardening and there is no significant softening effect taking place.

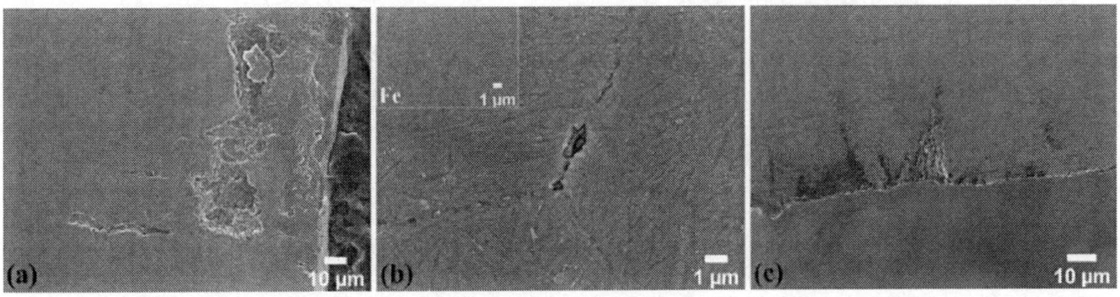

Figure 5. (a) Oxidation cavities beneath the surfaces, (b) grain boundary cavitation, and (c) pit at the cooling channel surface.

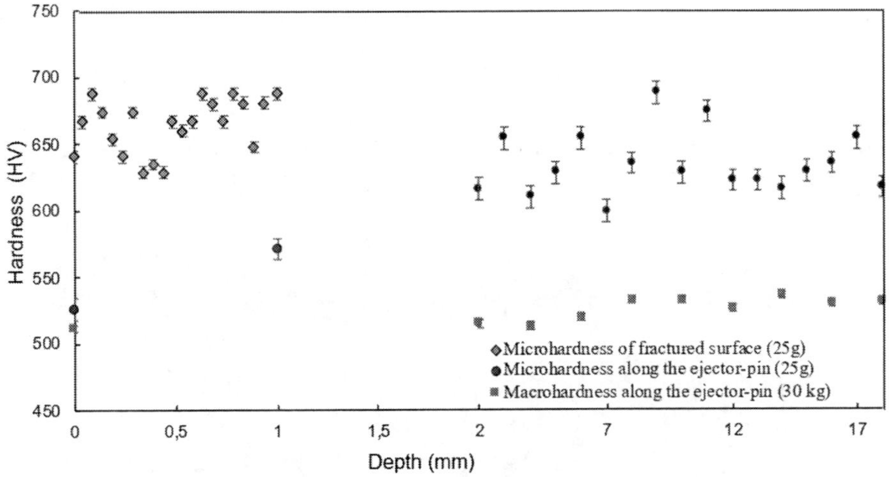

Figure 6. Microhardness measurements along the ejector-pin hole, 1 and 15 mm along the height of fractured surfaces´ cross section.

3.2. Failure mechanisms in the press hardening die for automotive side-plates

Observations revealed that the tool had several radial cracks initiating from the ejector pin hole, **Figure 2c**. Some of the cracks were relatively straight beneath the surface, and some were branched, **Figure 7a** and **Figure 7b**. Crack sizes along the die surface and ejector pin hole surface were measured displaying a maximum length of 8.8 mm and maximum depth of 4.5 mm, **Figure 7c**. There is an almost linear relation between crack depth (a) and length (c), with a/c= 0.41 indicating a semi-elliptical surface crack in the die as is visualized in **Figure 7d**.

In higher magnifications, **Figure 8**, SEM observations revealed networks of cracks in the surface oxide typical of thermal fatigue cracking. These networks were observed along the cracks, **Figure 8a**, on the tool surface away from the cracks, **Figure 8b**, and close to the die surface on the polished cross-section of the die. EDS results confirmed the presence of oxides rich in Cr in the crack networks.

Furthermore, oxide layers were observed inside the thermal fatigue cracks on the polished cross-section of the die, **Figure 9**. The EDS results confirmed the presence of oxide layers enriched in Ca in the cracks, and water leakage from the cooling channels may be a probable reason for the presence of Ca.

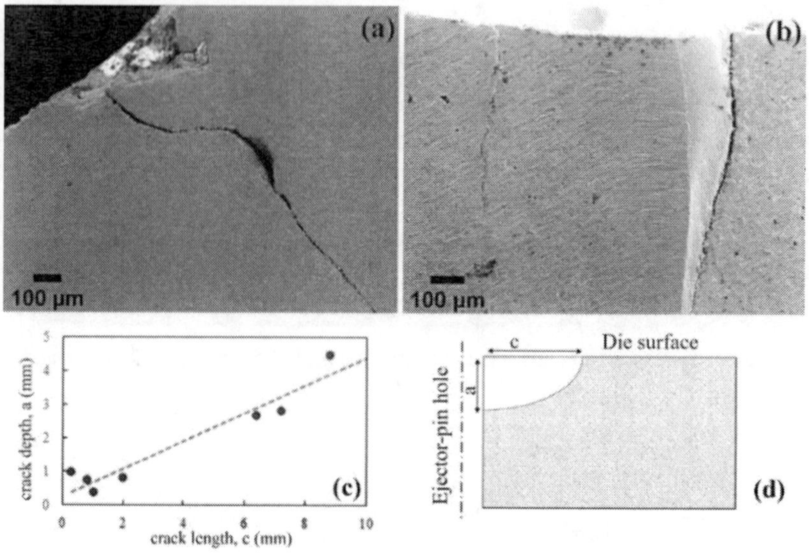

Figure 7. (a, b) Measured thermal fatigue cracks originated from the ejector pin hole, (c) crack depth, a, and length, c, of cracks initiated from the ejector pin hole.

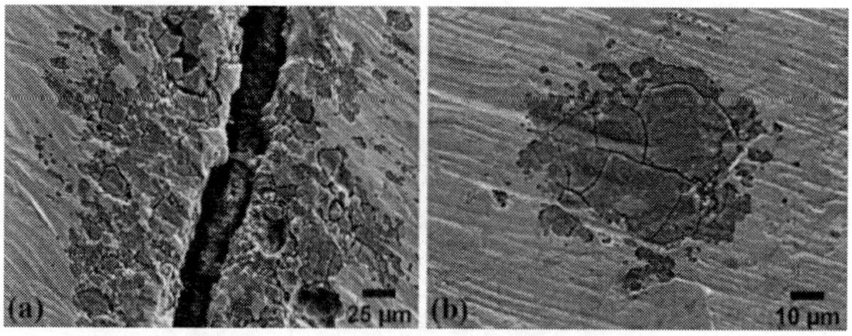

Figure 8. Oxide layer (a) along the cracks on the surface of the die (b) on the tool surface (c) near the die surface on a polished cross-section of the die.

Microhardness measurements were performed along the thickness of the die perpendicular to the surface in different locations including along the ejector pin hole and along two different cracks (1 and 2) up to the depth of 12 mm, **Figure 10**. For the cracks, the indents where recorded at polished cross sections (**Figure 10**) and at a distance corresponding to about 3 times the intent width. Measurements along the ejector pin hole were performed at a distance of 0.15 mm from hole.

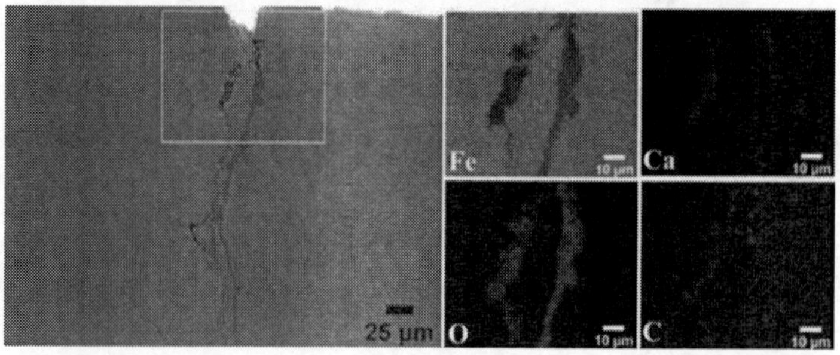

Figure 9. Oxide layers inside the cracks on a polished cross-section of the die. EDS result confirms the presence of Ca.

The microhardness values increases along the ejector pin hole when reaching the depth of about 10 mm from the surface. Moreover, considering the measured hardness values in the first 2 mm depth from the surface, a small reduction can be observed below the surface. The maximum difference in the measured hardness is about 90 HV in crack 1.

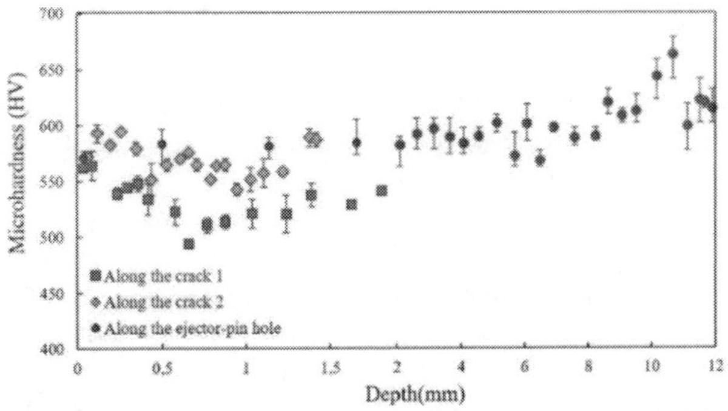

Figure 10. Microhardness measurements along the ejector pin hole and two different cracks using a load of 50 g. The diagram is magnified in the range of 0-2 mm.

Some evidence of plastic deformation was also observed on the surface of the die, **Figure 2b and c** and **Figure 11**. SEM images of the chemically etched part revealed three different zones near the surface comprising 1) a deformation zone, 2) a transition zone, and 3) the tempered martensitic base material zone, **Figure 11a**. The hardness values of these regions were measured as 634.8±12.3 HV, 584.4±8 HV, and 574.2±14.2 HV, respectively. Viewed in higher magnification the top layer displays a deformed microstructure, **Figure 11b**. Additionally, EDS results showed the same chemical composition in all the three regions; therefore, the possibility of adhesive wear occurrence from the sheet metal was rejected.

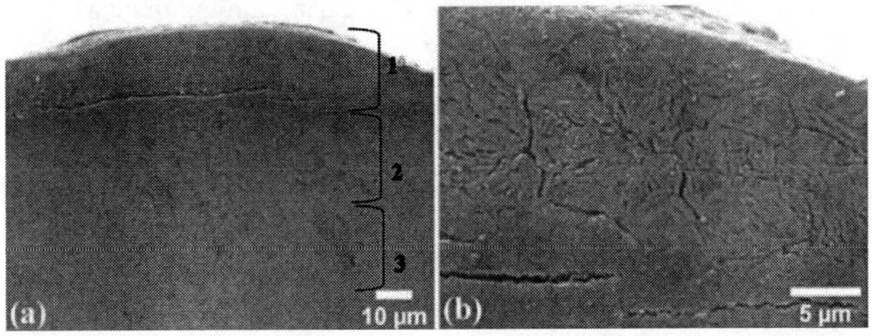

Figure 11. SEM images of the chemically etched part: (a) plastic deformation in surface layer of die comprising three different zones of 1) deformation zone, 2) transition zone, and 3) tempered martensitic zone, and b) the deformed microstructure of region 1 in higher magnification.

4. Discussion

The examination of the cross member die of this case study, revealed different kind of damages due to the press hardening process. Except the mechanical and thermal fatigue, which were the two main failure mechanisms, oxidation, corrosion and creep mechanisms took place due to the high stresses, the temperature and the environment.

Figure 3a and **Figure 4a, b**, revealed that the existence of high stresses led to the formation of long and straight to the depth cracks in this complicated die design. Furthermore, the characteristic fatigue feature of radial lines in the fractured surface starting from the upper blank surface to the depth confirmed the fact that mechanical fatigue was one of the damages that contributed to the final fracture of the die. The pattern and the direction of the fracture surface markings imply that the initiation points are at the ejector-pin hole, even though it was difficult to detect the exact initiation points due to the oxidized surface. From the ejector-pin hole the fatigue crack propagated at either side of the hole along the die segment. The main reason for the mechanical fatigue damage to happen in such an extent and type is believed to be the combination of the die´s design and the high stresses. In general, the design of the press hardening dies is always a case of discussion and constitutes one of the crucial factors that should be taken into account for a long die life [2]. In first case, the design of the die is challenging with the ejector-pin hole acting as a stress concentrator located close to the die segment curvature and side surface, raising the stress in this area.

Regarding the thermal fatigue damage, it seems that not only the temperature but the duration of the process and the heat transport from the blank through the die are critical factors. The design of the die seems to be critical in this case also, since the surfaces from where the heat can be transferred (cooling channel and outer surface of the die) are close to each other and accordingly a high thermal gradient is being created and repeated after each cycle. The curved crack around the cooling channel following the die segment curvature is a reflection of the thermal stress formation. Thus, the above phenomena led to the subsurface thermal fatigue and formation of thermal cracks in the die **Figure 4c**.

In addition, oxidation cavities located below the blank surfaces is one more damage mechanism that may become critical. The oxide layer that forms at elevated temperatures experiences a mechanical strain, which can result from one or a combination of the following mechanisms: 1) strain from the applied mechanical loading in the material, 2) mismatch in thermal expansion coefficients among the different stoichiometries of the material and the oxide or 3) viscous sliding of the oxide segments on the metal substrate [7]. On the die surface oxide pits and abrasive wear scratches were observed, and as viewed on the cross-sections internal oxidation was present, **Figure 5a**.

Furthermore, the combination of load with high temperature leads to the formation of creep microcavities. Even though thermal cycles are present, it can be considered that the temperature of the die remains high enough during the whole process since the production line works quick and the contact time is approximately 26 s. The die, except of the repeated mechanical load, experiences a constant loading due to die segment clamping as it is bolted to two other identical dies. Thus, the creep criteria temperature and stress may be fulfilled for microvoid creep damage to occur, **Figure 5b**. The microvoid and microcrack formation takes place through grain boundary sliding and carbides can concentrate the tensile stress at the interface at which the carbide and the matrix meet. Finally, the voids are getting bigger in size and as a result they meet each other around grains, forming microcracks. Although, these types of fatigue-induced cavities are small in size, they are also high in their number of density [8], which leads to degradation of the die.

Last but not least, pitting corrosion occurring in the cooling channels with cracks extending at the end of these pits was observed, **Figure 5c**. The existence of corrosion pits can be attributed to either a water leak or to the die cleaning process that is applied from time to time. The presence of Ca to the fractured surface implies that water is a part in the damage scenario.

The fact that the hardness was maintained, despite the extended damage that led to the cracking of the die, can be attributed both to the temper resistance of the material, but also acts like a validation that the combination of the design with stress and high temperature can result in low life; and also, demonstrates the importance and the challenge of a good and proper design for press hardening applications.

The side-plate die main damage comprised cracks initiated from around the ejector pin hole, **Figure 2c**. A limited number of cracks were present with a high total length, as local stress relief along each thermal crack may postpone the crack growth of adjacent cracks as larger ones propagate [9]. In addition, the appearance of the cracks may be strongly determined by the maximum temperature during each thermal cycle. In a study on QRO 90 steel samples subjected to thermal fatigue results revealed that the crack path was relatively straight in the thermal cycling up to 700 °C, whereas branched cracks in addition to the single cracks were present after cycling to 850 °C [10]. These results are in agreement with our observation in which the die was subjected to the blank sheet at temperature of 850 °C, and the produced cracks were both single and branched cracks.

The observed cracks initiating from the ejector-pin hole (**Figure 7**) with even distribution of radial cracks around the hole and a network crack pattern in the microscale are typical of thermal fatigue cracks. Thermal fatigue is one of the first damage mechanisms in die-casting, forging and press hardening dies. In press hardening, generally, there are two steps. In the first step, which is heating stage (closing and pressing), the thermal expansion of the surface layer is constrained by the cooler sublayers, and a compressive stress field develops at the surface. Next, during die opening and cooling the contraction of the surface is hindered by the heated the sublayers, and a tensile stress state appears at the surface. Under this condition the die material is subjected to a significant damage through a process of non-isothermal low cycle fatigue (LCF) [11]. Therefore, alternating compressive and tensile stresses at the surface of the tools that arises from differential thermal expansion/contraction during sudden transient temperature changes may lead to thermal fatigue damage in hot forming operations [12]. Additionally, even higher temperature gradients are produced at slim and sharp corners like the ejector-pin hole, which results in higher stresses and strains. Hence, low-cycle fatigue gradually accelerates thermal fatigue cracking of these regions [9], [10].

Some oxidized crack networks appeared along the main cracks on and right below the die surface, and oxide layers were, as observed on the polished cross-section, inside the deep cracks of the die. Although the oxidized surface crack networks, **Figure 8**, were rich in Cr and observed in many different regions, they seem to be not as harmful as oxide layers rich in Ca inside the cracks, **Figure 9**. The presence of oxides in the cracks may lead to an increase in compressive stresses and plastic yield during the heating phase of the thermal cycle, and during the cooling phase the tensile stresses are further increased and thus amplifies the continued crack growth [12]. The negative impacts of the oxide layers are due to their low thermal expansion, bigger volume and brittleness [13]. Moreover, the grain boundaries along the crack are oxidized easily and form brittle oxidized grain boundary zones subsequently adding paths of thermal fatigue crack propagation [11].

Microhardness measurement along the ejector-pin hole and the two cracks showed some variation in hardness, as shown in **Figure 10**. A lower hardness at the die surface of the ejector-pin hole compared to 10 mm below the surface, and a slight drop in hardness closer than 2 mm to the die surface along the two cracks indicates a slight softening close to the die surface. Generally, tool steel temper resistance to softening is one of the important factors in thermal fatigue crack resistance [14], however as the hardness reduction was slight in this die, its effect on the thermal fatigue damage may be little.

Figure 2c and **Figure 11** displays plastic deformation the die surface and parallel subsurface cracks, confirmed by the observed deformed microstructure and the higher hardness values in the deformed layer, **Figure 11**. Calculations based on the heat transfer from the blank to the die interior revealed that the temperature on the surface of the die may reach about 500°C. At this temperature the yield strength of the material is reduced to about 880 MPa, (**Table 2**), and mechanical and thermal stresses generated may exceed the yield point and plastic deformation is likely to occur. Moreover, shrinkage of the sheet metal in contact with the die during the 26 s long cooling step may cause high shear stresses on the surface of the die, and could be an important driving force for the occurrence of plastic deformation.

Table 2. Yield point values at different temperatures for QRO 90 Supreme tool steel [15].

Temperature (°C)	200	300	400	500	600	700
Yield strength (MPa)	1200	1100	1000	880	640	340

5. Conclusions

The failure case study on two press hardening dies for production of cross member and side-plate components analysed several die damage mechanisms. However, the most severe and life limiting failure mechanism in the study were mechanical fatigue and thermal fatigue. The mechanical fatigue is a results from loads during pressing and thermal fatigue depends on the thermal gradients occurring during pressing, thus both are related to the press hardening cycle. The press hardening components are made of thick boron steel sheets, thus with a large heat transfer into the die during press hardening. Therefore other elevated temperature dependant damages were found, e.g. oxidation, creep and corrosion, but mainly related to the time at high temperature. The case study is summarized by the following conclusions.

- Both the mechanical fatigue and thermal fatigue damages in both dies are related to ejector pin holes, and cooling channels, indicating a potential for design improvements.
- Other secondary phenomena, such as oxidation, creep and corrosion, degrades the quality of the die and at the end it may contribute with critical life limiting damages.
- Abrasive wear and oxidation pits may be observed on the die surface, but also severe plastic deformation on the side-plate die occurred during press hardening as a result of shear between the plate and die in combination with reduced die surface yield strength at the high temperatures.

References

[1] H. Hoffmann, H. So, and H. Steinbeiss, "Design of hot stamping tools with cooling system," *CIRP annals,* vol. 56, no. 1, pp. 269-272, 2007.

[2] R. Neugebauer *et al.*, "Press hardening—An innovative and challenging technology," *Archives of civil and mechanical engineering,* vol. 12, no. 2, pp. 113-118, 2012.

[3] Z. Gronostajski, M. Kaszuba, M. Hawryluk, and M. Zwierzchowski, "A review of the degradation mechanisms of the hot forging tools," *Archives of Civil and Mechanical Engineering,* vol. 14, no. 4, pp. 528-539, 2014.

[4] A. A. Emamverdian, Y. Sun, C. Cao, C. Pruncu, and Y. Wang, "Current failure mechanisms and treatment methods of hot forging tools (dies)-a review," *Engineering Failure Analysis,* vol. 129, p. 105678, 2021.

[5] J. Hardell, S. Hernandez, S. Mozgovoy, L. Pelcastre, C. Courbon, and B. Prakash, "Effect of oxide layers and near surface transformations on friction and wear during tool steel and boron steel interaction at high temperatures," *Wear,* vol. 330, pp. 223-229, 2015.

[6] S. Mozgovoy, "High temperature friction and wear in press hardening," Luleå tekniska universitet, 2014.

[7] R. Neu and H. Sehitoglu, "Thermomechanical fatigue, oxidation, and creep: Part I. Damage mechanisms," *Metallurgical transactions A,* vol. 20, no. 9, pp. 1755-1767, 1989.

[8] P. Rodriguez and K. B. S. Rao, "Nucleation and growth of cracks and cavities under creep-fatigue interaction," *Progress in materials science,* vol. 37, no. 5, pp. 403-480, 1993.

[9] D. Klobčar, J. Tušek, and B. Taljat, "Thermal fatigue of materials for die-casting tooling," *Materials Science and Engineering: A,* vol. 472, no. 1-2, pp. 198-207, 2008.

[10] V. Velay, A. Persson, G. Bernhart, J. Bergström, and L. Penazzi, "Thermal fatigue of a tool steel: experiment and numerical simulation," in *6th International Tooling Conference. Karlstad University, Sweden*, 2002, pp. 793-814.

[11] J. QC, S. HL, and G. QF, "Thermal fatigue behavior of new type high-Cr cast hot work die steel," *ISIJ international,* vol. 44, no. 6, pp. 1103-1107, 2004.

[12] D. Mellouli, N. Haddar, A. Köster, and H. F. Ayedi, "Hardness effect on thermal fatigue damage of hot-working tool steel," *Engineering Failure Analysis,* vol. 45, pp. 85-95, 2014.

[13] D. Klobčar, L. Kosec, B. Kosec, and J. Tušek, "Thermo fatigue cracking of die casting dies," *Engineering failure analysis,* vol. 20, pp. 43-53, 2012.

[14] H. A. Abdulhadi, S. Aqida, M. Ishak, and G. Mohammed, "Thermal fatigue of die-casting dies: An overview," in *MATEC Web of Conferences*, 2016, vol. 74: EDP Sciences, p. 00032.

[15] *Uddeholm QRO® 90 Supreme*, Uddeholms AB, 2021.

Formability of roll-formed carbon fibre reinforced metal hybrid components and its experimental validation

X Hu[1]*, C Creighton[1], P Zhang[1], N Müller[2], T Reincke[2], R Taube[3] and M Weiss[1]

[1]Institute for Frontier Materials, Deakin University, Waurn Ponds, Pigdons Rd. VIC. 3216, Australia

[2]Institute of Joining and Welding, Technische Universität Braunschweig, Langer Kamp 8, 38106 Braunschweig, Germany

[3]Ford Motor Company of Australia Ltd, 1735 Sydney Rd, Campbellfield VIC. 3061, Australia

* corresponding author's e-mail: huxin@deakin.edu.au

Keywords: CFRP, Hybrid, Lightweight, Roll Forming, Formability

Abstract: Carbon Fibre Reinforced Metal Hybrid (CFRMH) materials that combine a sheet metal substrate with a reinforcing carbon fibre patch represent a promising solution to reduce weight while increasing the structural and crash performance of future automotive vehicles. CFRMHs cannot be formed with conventional stamping processes and at high volumes. This currently reduces their widespread application. Roll forming is increasingly used in the automobile industry for the forming of lightweight and high-strength metal structural components. The major deformation mode in roll forming is simple bending and this reduces interlaminar shear and compressive stresses that lead to fibre failure and delamination issues when stamping CFRMH sheet materials. This work analyses the potential of applying the conventional roll forming process for the manufacture of a simple top hat section shape from CFRMH sheet. For this, first, the material is produced in a hot press and then formed to shape in a laboratory roll forming facility at room temperature. Different layup sequences are tested, and component quality is analysed after roll forming by visual inspection. The results suggest that roll forming presents a promising manufacturing method for the production of automotive components from CFRMH sheets. The roll formed open shells are joined to produce crash box structure and tested in 3-point bending. The results show that depending on the fibre orientation a significant increase in weight specific strength is achieved.

1. Introduction

Carbon Fibre Reinforced Metal Hybrids (CFRMH) contains a metal sheet substrate that is laminated to a carbon fibre reinforcement patch. CFRMH was first developed and applied in the aerospace industry to provide high specific strength and stiffness, good fatigue resistance and wear capacity. Recently, with the increasing focus on lightweight design lightweight CFRMHs have been introduced to the automobile industry[1].

Early studies on the manufacture of fibre metal hybrid laminate mainly focused on common autoclave curing processes and Vacuum-Assisted Resin Transfer Moulding (VARTM) and were aimed at the aerospace industry[2,3]. These methods require a long cycle time and high cost in the moulding process [4], which has become the main obstacle to promoting the application of CFRMH in the transportation industry. The existing direct forming technologies for fibre/metal laminates include incremental sheet forming, laser forming and die stamping. However, the formability of hybrid sheet materials is lower than that of conventional sheet metals. This is mainly due to the delamination at the metal/carbon fibre interface and the tearing or buckling of fibres when formed[5].

Roll forming is increasingly used for the manufacture of automotive structural and crash components from high-strength steel such as rocker panels and bumper beams. The major deformation in roll forming is pure bending and this allows the forming of sheet materials with high material strength and low ductility. The tool contact is reduced to point contact between the top and the bottom forming rolls which can reduce interlaminar shear stress when forming a hybrid composite structure [6]. Therefore, roll forming may present a promising alternative for the rapid forming of CFRMH components [7].

In this study, flat CFRMH plates containing a flat mild steel sheet substrate laminated to various carbon fibre lay-up directions are produced in a hot press and then roll-formed to a hat-shaped open-shell component. Part quality after roll forming is visually inspected, after that crash box type hollow structural components are produced by rivet joining two roll-formed open-shell components; this is then followed by three-point bending tests to investigate the structural performance and compare load-bearing ability between the metal and the hybrid crash box. Finally, the damage on the tested parts was visually examined.

2. Materials and method

2.1. Materials and CFRMH sheet preparation

A mild steel (MS) with a thickness of 0.8 mm was used as the substrate and standard tensile tests were conducted according to ASTM E8/E8M [8]. The engineering stress-strain curve of three samples oriented in the rolling direction is shown in Figure 1. A commercial thermoplastic PA6 based CFRP supplied by Celanese (Celstran® CFR-TP PA6 CF60-03) with a continuous unidirectional (UD) fibre tow and a thickness of 0.13mm[9], was used as the reinforcement patch. This was bonded to the steel substrate with a thermoplastic adhesive multilayer-film Nolax Cox 490-1 (Nolax) from the Nolax AG [10]. The material properties provided by the supplier are given in Table 1.

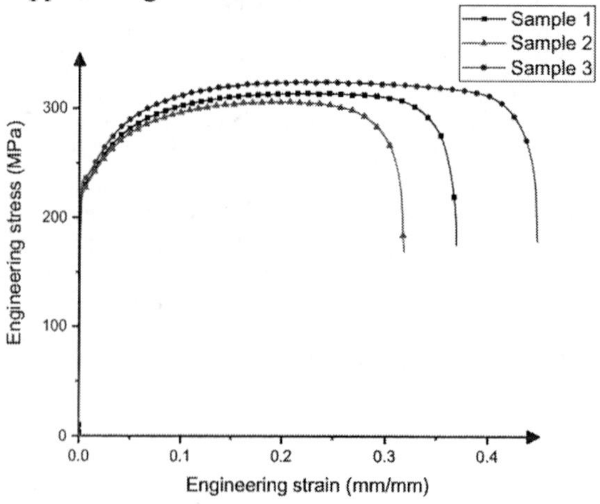

Figure 1 Engineering stress-strain curves for three samples of mild steel

Table 1: Material properties for CFR-PA6 and Nolax adhesive [9, 10]

Physical properties	CFR-PA6	Nolax Cox 490-1
Melting temperature (°C)	220	152-188
Glass transition temperature (°C)	57	-
Forming temperature (°C)	271	-
Tensile strength (MPa)	1909	-
Tensile Modulus (GPa)	100	-

The hot press process can be split into two steps. First, the Nolax film was trimmed to size and then fixed in the centre of the steel substrate. After this, the sheet was hot pressed at 200 °C and 1 bar for 2 mins. The 0.9 mm shims were used to prevent the foil from squeezing out and then the compound was taken out to cool down under a weight of 5kg. Subsequently, six layers of CFR-PA6 composite were stacked up the top surface of metal sheet facing the fixed Nolax film. The sheet was then hot pressed for a second time at 250°C, 1 bar and 50 seconds. This was followed by cooling to 50 °C with a weight of 5kg. In Figure 2a, the dimensions of the steel substrate, the adhesive and the of CFR-PA6 composite are shown.

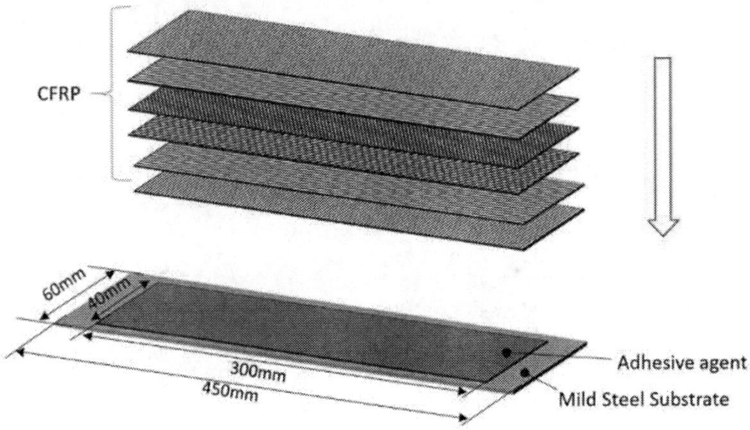

Figure 2: Schematic of CFRMH blank

CFRMH blanks with three different lay-up conditions were produced and those are shown in Table 2.

Table 2: Test setup

Description	Fibre orientation
$[0°]_6$ oriented in the rolling direction and the subscript means six layers of CFR-PA6	$[0°]_6$ $[90°]_6$ $[45°/-45°]_3$

2.2. Roll forming of the U-channels

The trapezoidal cross-section shape shown in Figure 3 was manufactured with the roll forming mill shown in Figure 4 and a strip feeding speed of 30mm/s was used. The roll former was limited to three roll stations and one feed-in guide. The roll gap was set as the total thickness of the hybrid blank (1.58 mm) and the station distance was 300 mm. The bend angle increment was 10°. To form the final bend angle of 50°, therefore, the material had to be first roll-formed to 30° followed by a retooling of stations and the re-feed of the preformed sheet into the roll former to manufacture the final 50° angle. The roll forming flower pattern is given in Figure 3. Forming defects such as fibre breakage and delamination were analysed by visual surface inspection.

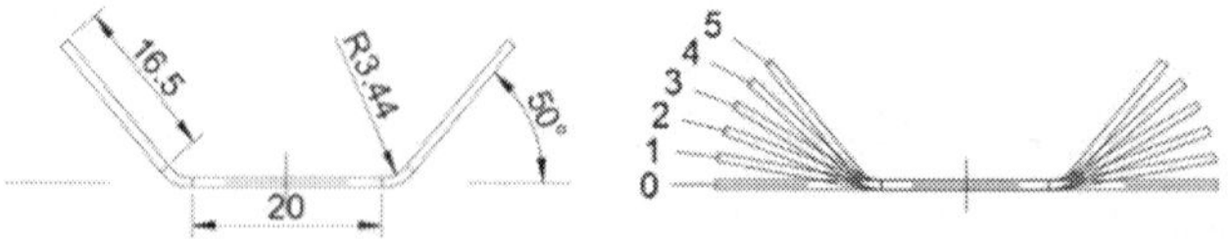

Figure 3 Profile geometry and the roll forming flower pattern

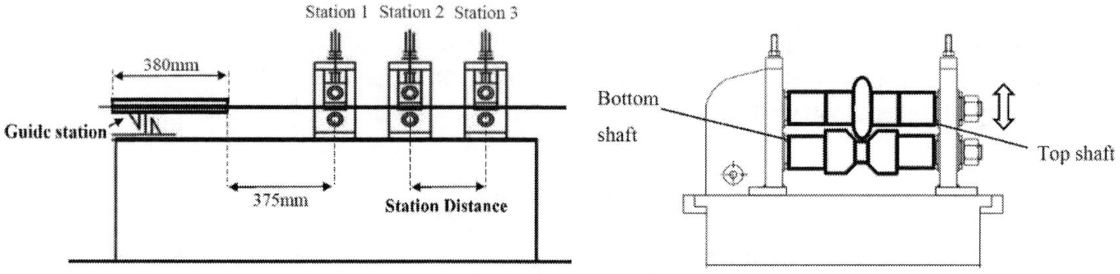

Figure 4 Schematic of the conventional roll forming mill

2.3. Crashbox assembly and flexural test

To test the structural performance, a crash box was created with two roll-formed trapezoidal parts. For this, the flange area was bent to form a top hat structure using a folding die. Then, two open shells were assembled using aluminium blind rivets in the top hat flange region (Figure 5a). In the centre section where the major deformation occurred the rivets were positioned close to each other with a distance of 12.5mm to ensure sufficient joint strength. The cross-section profile of the profile is shown in Figure 5a. After joining, three-point bending tests were performed based on DIN-EN-ISO 14125 [11] with the test setup shown in Figure 5b. The span length between two supporters is set as 105 mm. The tests were conducted at a rate of 5 mm/min and terminated at the maximum punch stroke of 19 mm.

After the test, the deformed crash box was opened up by removing the rivets and material damage analysed by visual surface inspection and compared to the as-roll formed condition.

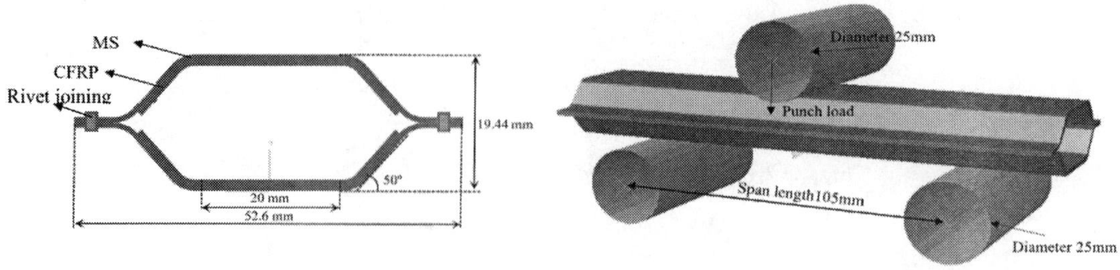

(a) (b)

Figure 5: (a) Cross-section profile of riveted double-hat crash structure and (b) flexural test setup

3. Results

3.1. Product quality of the roll formed parts

Images of the patch surface after roll forming are shown in Figure 6 for the 3 different fibre lay-up conditions. When the fibres are aligned with the profile radius direction, MS/CFRP $[0°]_6$, a successful hybrid patch is roll formed with no visible defects. For the MS/CFRP $[90°]_6$ specimens, fibre buckling perpendicular to the bend line was observed. The fibre patch is positioned on the inner side of the bend and this leads to compressive stresses perpendicular to the bend line and parallel to the fibre direction of the CFRP $[90°]_6$. The results suggest that the fibres do not compress sufficiently when formed at room temperature and instead buckle to accommodate the compressive deformation. There is also some fibre cracking 45° to the bend line direction in the CFRP $[45°/-45°]_3$ specimen. No obvious delamination between the CFRP and the metal substrate was observed for all fibre directions.

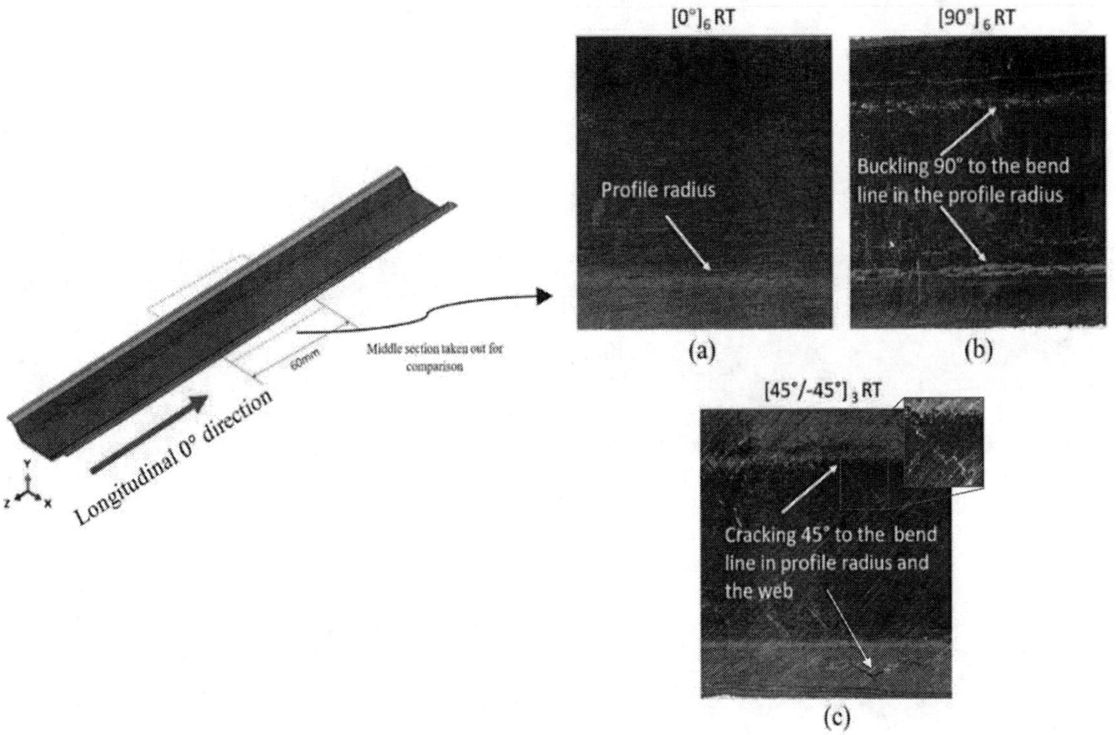

Figure 6 Top view: Surface quality of cold roll formed CFRMH parts with various layup orientations

3.2. Bending load and specific strength

The weight of all samples was estimated using the individual material densities and volumes. For the Mild Steel (MS), the CFRP and the Nolax adhesive densities of 7.86 g/cm3 and 1.447 g/and 1.10 g/cm3 were assumed respectively. This in combination with the individual material volumes gives an overall weight of 0.453 kg for the single mild steel crash box and of 0.501kg for the CFRMH crash box counterpart. Figure 7a presents the recorded load-displacement responses of the MS base condition and the three different CFRMH conditions. The peak values were taken as the maximum bending load and the specific strength was determined by dividing the maximum bending load by the sample weight. Figure 7a illustrates the highest maximum bending load for the MS/CFRP $[90°]_6$condition. This represents an increase of 44% compared to the sole MS base condition. Even though the MS/CFRP$[0°]_6$ showed the best roll forming result it only led to a load improvement of 24% compared to the sole metal base condition. When the weight factor is considered, specific strength comparison was given in (b) Figure 7b. All hybrid samples have a higher specific strength in comparison to the sole metal sample. The MS/CFRP $[90°]_6$ still shows the highest specific strength improvement which is 35.6% compared to the sole metal condition.

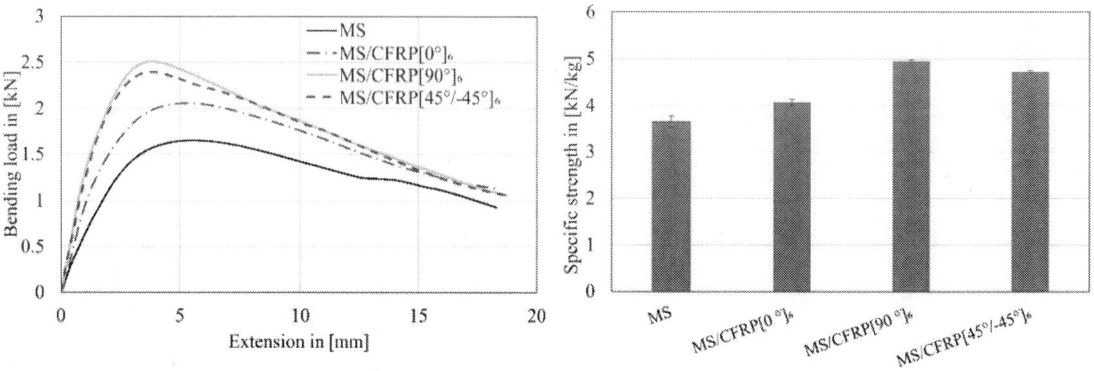

(a) (b)

Figure 7 Mechanical performace (a) Bending load-displacement response (b) Specific strength comparison

After the 3-point bending test, the crashboxes were removed from the tester and then separated into the top and bottom channels followed by visual inspection. Figure 8 shows that during three-point bending deformation a centre bend is formed into the web. This leads to stretch deformation perpendicular to the part length direction, i.e., parallel to the fibre direction of the MS/CFRP $[0°]_6$ and results in fibre breakage as can be seen in Figure 8a. In the MS/CFRP $[90°]_6$, the fibres are oriented perpendicular to the direction of stretch. This leads to matrix cracking and the spacing out of fibres (Figure 8b). Inter-ply delamination near the bending line is obvious for the MS/CFRP $[45°/-45°]_3$ sample in Figure 8c. Overall, the occurrence of fibre cracking suggests that interface lamination after roll forming is sufficient to enable the loading of fibres during the structural 3-point bend test loading.

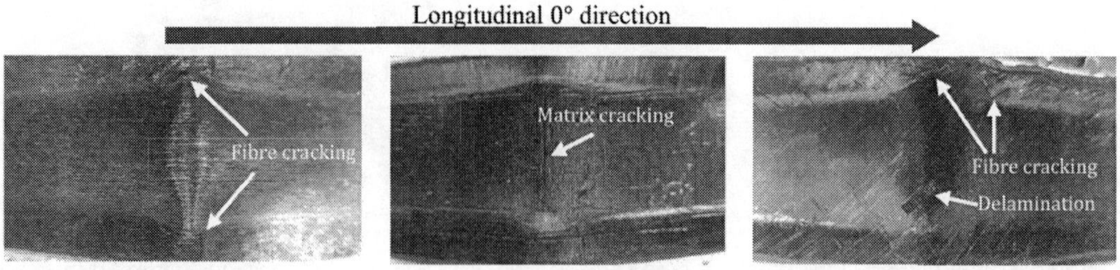

(a) Steel/CFRP $[0°]_6$ (b) Steel/CFRP $[90°]_6$ (c) Steel/CFRP $[45°/-45°]_3$

Figure 8 Failure status after 3-point bending

4. Conclusion

A Carbon Fibre Reinforced Metal Hybrid (CFRMH) component consisting of a sheet metal substrate laminated to a carbon fibre patch was successfully produced by hot pressing and conventional roll forming at room temperature. Subsequent 3-point bending trials performed on crash boxes produced from the roll-formed hybrid shells indicate an increase in specific strength of up to 35% compared to the sole metal sheet solution depending on the fibre orientation in the carbon fibre patch.

In the future, the hot-pressing step may be replaced by a continuous in-line hybrid sheet production process. If this can be achieved the roll forming process may represent a cost-effective solution to produce lightweight CFRMH components for the transport industry. The following conclusion can be made:

- The roll forming process allows the forming of metal/carbon fibre hybrid at room temperature.
- The roll formed component quality depends on the fibre orientation in the carbon fibre re-enforcement patch. If the fibres are oriented parallel to the roll formed bend line (MS/CFRP $[90°]_6$) and then the best part quality is achieved.
- The roll formed CFRMH components showed a higher specific strength compared to the sole steel counterpart. The level of the increase in specific strength depends on the fibre orientation.

5. Acknowledgements

The authors would like to thank ARC Training Centre in Lightweight Automotive Structure (project number IC160100032) and the Ford Motor Company for their financial support. The authors further acknowledge the support provided by Nikolai Müller from the Technische Universität Braunschweig in performing the experimental work of this study.

References

[1] Su X and Wagner D 2019 Integrated computational materials engineering development of carbon fiber composites for lightweight vehicles

[2] Bienias J, Majerski K, Jakubczak P and Mania R 2018 Forming of thin-walled profiles made of FML in autoclave process *J. Aircr. Engi. Aeros. Tech* **90** 506-14

[3] Botelho E, Silva R, Pardini L and Rezende M 2006 A review on the development and properties of continuous fiber/epoxy/aluminum hybrid composites for aircraft structures *Materials Research* **9** 247-56

[4] Park S, Choi W and Choi H. S 2010 A comparative study on the properties of GLARE laminates cured by autoclave and autoclave consolidation followed by oven postcuring *Int. J. Advan. Manu. Tech* **49** 605-13

[5] Jong T. W 2004 *Forming of Laminates* (Delft: Delft University of Technology)

[6] Weiss M, Abeyrathna B and Pereira M 2018 Roll formability of aluminium foam sandwich panels *Int. J. Adv. Manu. Tech* **97** 1-13

[7] Bhattacharyya D and Martin T. A 1999 Roll forming of composite sheets ed Karger-Kocsis (Dordrecht: Springer Netherlands) pp 721-7

[8] *Standard Test Methods for Tension Testing of Metallic Materials* 2016

[9] C. Corporation. CELSTRAN® CFR-TP PA6 CF60-03 [Online]. Available: https://tools.celanese.com/products/datasheet/SI/CELSTRAN%C2%AE%20CFR-TP%20PA6%20CF60-03

[10] Reincke T, Hartwig S and Dilger K 2019 Improvement of the adhesion of continuously manufactured multi-material joints by application of thermoplastic adhesive film *Int J. Adhe. A* **93** 102321

[11] *Fibre-reinforced plastic composites - Determination of flexural properties* 1998

Improvement of shear cutting process by stress superposition via cross-elastic partholder

D Briesenick[1]*, M Liewald[1] and S Senn[1]

[1] Institute for Metal Forming Technology, University of Stuttgart, Germany

*david.briesenick@ifu.uni-stuttgart.de

Abstract. High quality requirements for shear-cut workpieces made of advanced sheet metal materials nowadays pose highest demands on cutting technologies. In this context, stress superposition in shear zones represents a common approach to adjust cutting surface quality or reduce trim load, but usually requires sophisticated process technologies. For example, fine blanking technology allows to increase clean-cut portion, but requires advanced tool and press technology. Given this background, the present paper deals with a novel, low cost and straightforward approach of stress superposition in conventional shear cutting processes and tools. For this purpose, the partholder was geometrically modified to achieve vertical and horizontal force transmission into the blank. Presented investigations with implemented Z-shaped partholder and its cross-elasticity show a significant impact on cutting surface quality. The results were obtained by a numerical study using the FEA-Software DEFORM 2D, wherein geometry and process parameters were varied. By superposing compressive stresses, clean-cut portion could be increased up to 43 % compared to conventional shear cutting of DP600 having a sheet thickness of 1 mm. Additionally, rollover height was significantly reduced by up to 50 % due to the transversal forces of the cross-elastic partholder, while burr height could be kept constant.

1. Introduction

Material separation by shear cutting technologies is commonly used in production processes. Many technical products made of metal are separated or pierced by shear cutting at least once in its production chain [1]. Here, the most frequently used method in metal forming industry is the conventional shear cutting, which can be performed on single-acting presses. Those separation processes with stroke rates up to 3,000 strokes per minute offer considerable cost advantages, particularly in large-scale production of sheet metal components [2]. Nowadays, the quality requirements of manufactured parts with regard to the cutting surface are constantly increasing and process results of conventional shear cutting are often no longer sufficient [1]. High-quality cutting surfaces, which can be used without additional rework, are characterized by a small rollover, high clean-cut portion and marginal burr heights [3]. With ordinary shear cutting tools and processes, however, only a clean-cut portion of up to 50 % is achievable [4], [5]. Therefore, for producing parts with higher quality requirements, subsequent reworking or finishing steps are necessary, which reduce the overall economic efficiency of the production chain.

For this reason, various special cutting tools and processes have evolved over the last decades. Two-stage shear cutting, for example, combines conventional shear cutting with a subsequent trimming operation to produce precise and functional part contours [6], [7]. Counter cutting avoids burr formation and improves cutting surface quality by a two- or three-stage, counteracting embossing and shearing process [8]. These multi-stage shear cutting processes may increase part contour quality, but are accompanied by high demands on multiple axes tool design, part positioning issues, tooling cost or multiple acting presses. Therefore, single stage shear cutting technologies are more likely implemented to reduce complexity and cost of the manufacturing tools and processes. For example, conventional cutting of high-strength steels shows a tremendous reduction of clean-cut portion where a compressive stress superposition in the shear zone can counteract and again extend the formability of the material [9]. In this context, Senn et al. [10] investigated the impact of a new sheet metal cutting tool design using a concave nose punch geometry to superpose compressive stresses and thus increase the clean-cut portion up to 73 % for various steel grades. However, filigree punch edges reduce tool life in series production. In general, fine blanking technology is commonly applied to achieve highest part contour quality and precision [11]. This cutting process is characterized by a minimized cutting clearance, a special partholder geometry with a vee-ring indented along the cutting contour of the blank and an additional counter-acting pad. Due to this, the tool and process design of fine blanking generates a beneficial stress state in the shear zone and prevents an early fracture of material, which leads to clean-cut portions up to 100 % [12]. However, the complex and multi-acting tool and press design, lower stroke rates and a reduced tool life make fine blanking a cost-intensive process compared to conventional shear cutting.

Against this background, the present paper deals with a new approach for stress superposition in shear zones via a cross-elastic partholder geometry. Supported by numerical results, the proposed and new partholder design transforms the vertical load of the press into a horizontal and vertical load towards the blank, which superposes stresses in the cutting area and improves shear cutting processes.

2. Concept of cross-elastic partholder

In general, this paper pursues the research hypothesis of using a geometrically controlled blankholder deformation in combination with tool material elasticity for a simultaneous reversible horizontal and vertical partholder force transmission on blanks. Herein, the lateral contraction of elastic-plastic material, described by Poisson's ratio [13], is used to generate a force transmission in the sheet metal plane with the help of static friction. Simultaneously, the vertical force transmission still provides the general functionality of a partholder with positioning and lift-off prevention of the blank. This tool setup is shown in figure 1 a) as a cross section for piercing rectangular or circular cut-outs.

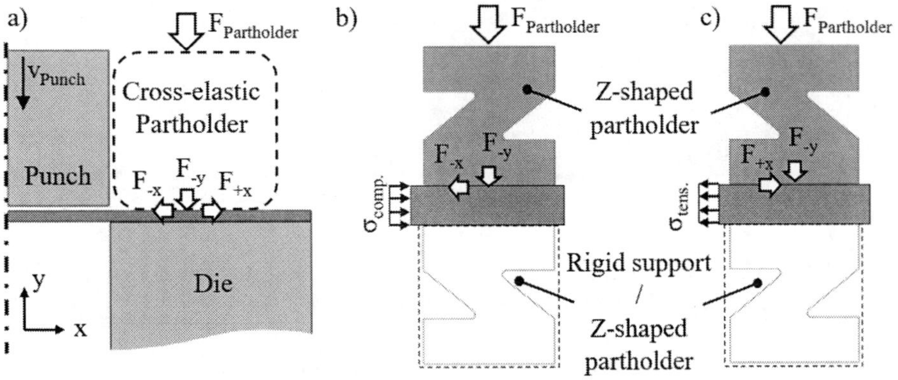

Figure 1: a) Concept of a cross-elastic partholder in shear cutting tools and derived design variants with one- or double-sided Z-shaped partholders for b) compressive and c) tensile stress superposition.

However, a symmetrical vertical force transmission on the blank by a lateral material contraction ($F_{-x} = F_{+x}$) would not sufficiently impact the stress state in the shear zone. One solution for this is the concept of a Z-shaped partholder, derived from a single spring winding and shown in figure 1 b) and c). It enables a geometrically controlled and targeted cross-elasticity in a desired lateral working direction. Thereby, the diagonal alignment of the Z-shaped element defines the deformation and working direction of the partholder. If the diagonal points top down towards the punch, a horizontal compression force on the blank is transferred and leads to a compressive stress superposition in the shear zone. As explained in the introduction, such a compressive stress superposition can increase the cutting surface quality. Here, it is conceivable to use one Z-element, only acting on one blank side, or a double-sided solution with two horizontally symmetrically located Z-shaped partholders (see figure 1 b)). If the diagonal points top down away from the punch, a horizontal tensile force on the blank is transferred and leads to a tensile stress superposition. In particularly, this is important for mechanical separation processes of high-strength steels to decrease cutting forces [14]. Likewise, this can be realized by one Z-shaped partholder or a double sided-solution (see figure 1 c)). The numerical results presented in this paper deal with the approach of compressive stress superposition to increase cutting surface quality, measured by a high clean-cut portion and a minimum of rollover height.

3. Numerical study on a partholder design for compressive stress superposition in shear zones

The virtual tool design presented in this paper was carried out using the FEA-Software DEFORM 2D. All numerical investigations and simulations refer to the dual-phase steel DP600 with a thickness of t = 1 mm. For the blank, an elastic-plastic material model was chosen, whose material properties were determined by means of tensile tests in previous work of authors (see figure 2) [10]. Damage modelling used, refers to the normalized Cockroft & Latham criterion with a cumulative real-time value $C_{normC\&L}$. This criterion is commonly applied in shear cutting simulation and integrates the ratio of maximum principal stress and the equivalent stress up to the limit fracture strain. Thereby, critical limit of plastic work (C_{crit}) is locally considered by deleting adjacent mesh elements. The determination of the critical limit of plastic work was done by conventional shear cutting experiments and an afterwards iterative calibration of C_{crit} = 2.8 of simulated cutting surface quality. Here, conventional shear cutting simulation with a cutting clearance of 10 %, a standard partholder geometry and rigid tools with sharp punch and die radii (0.05 mm) were used and served additionally as a valid reference for the new partholder design. All other settings like mesh sizes and boundary conditions are described in the following and were kept constant to ensure comparable results. The implemented elastic-plastic and damage modelling led to a simulated clean-cut portion of 27 % and a rollover height of 0.08 mm, which is in a good agreement with the experimental data (see figure 2 a)).

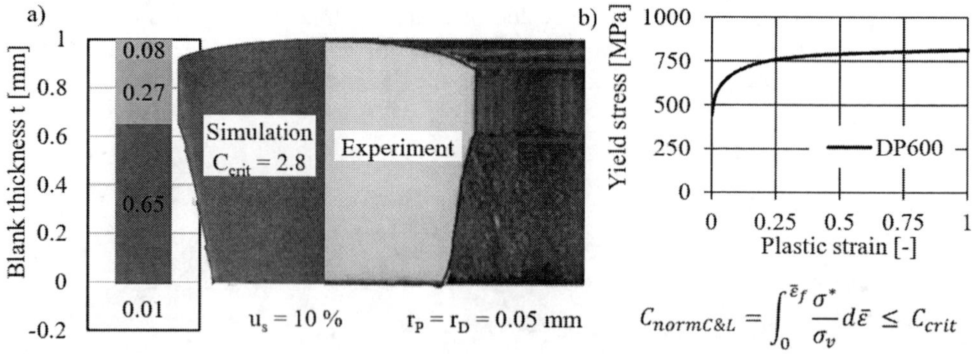

Figure 2: a) Validation of simulated surface quality for conventional shear cutting of DP600 and b) corresponding stress-strain curve and damage modelling of selected material.

3.1. Modelling approach of shear cutting with cross-elastic partholder

A parametrized 2D FE-model of shear cutting with cross-elastic partholder was built with DEFORM 2D Software in an integrated pre-processor. To reduce computational time, plane strain symmetric boundary conditions were applied to model shear cutting of rectangular cut-outs, see figure 3 a). The 2D finite elements were extruded in Z-direction with a thickness of 1 mm. The elastic-plastic modeled blank was discretized by 15,000 elements with an adaptive meshing and refinement towards the shear zone, limited by a minimum edge length of 0.006 mm. The nominal Z-shaped partholder geometry and its dimensions were conceptually designed and also discretized by 2D finite elements and an elastic tool steel material model (AISI-D2) was assigned. For calculation, punch and die were defined as rigid bodies and friction coefficients were set locally for blank and partholder at μ = 0.2 and globally at μ = 0.1.

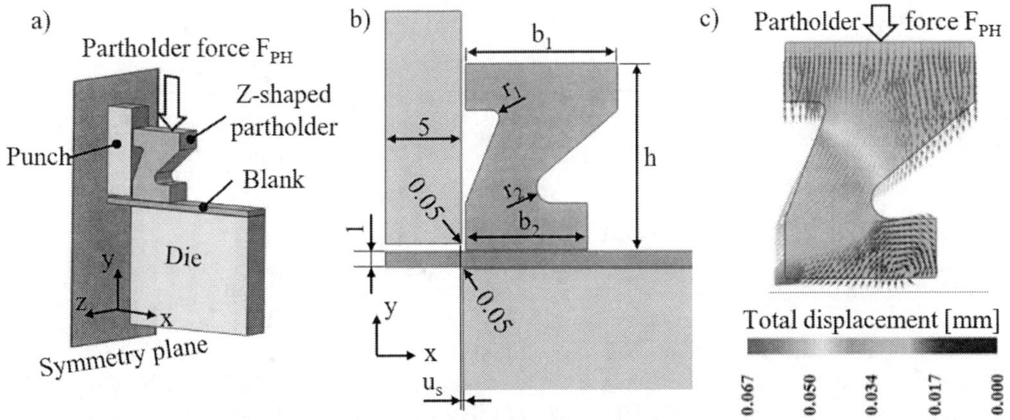

Figure 3: a) 2D FE-model with plane strain symmetry and b) dimensions of Z-shaped partholder with c) the cross-elastic deformation for nominal simulation setup.

For the initial assumed design, the principal mechanism of the targeted cross-elasticity by Z-shaped partholders and two-dimensional force transmission on blanks can be illustrated by the total nodal displacements of the respective elements, see figure 3 c). The vertical applied partholder force F_{PH} of 5,000 N compresses the Z-shaped component and results in a lateral deformation of the partholder towards the shear zone at its contact area with the blank. These minor elastic deformations of the Z-shaped partholder towards the downward moving punch do result into an improved cutting surface quality compared to conventional shear cutting, see figure 4. Even for the initial process design with a reduced cutting clearance and comparable partholder forces, the rollover height was reduced by 50 % due to the lateral stress superposition. Furthermore, the clean-cut portion increases from 27 % to 47 %. Regarding the fracture surface of conventional and shear cutting with Z-shaped partholder, the fracture height and fracture angle was reduced because of the stress superposition together with a smaller cutting clearance (6 %) of initial simulated design. In contrast, a slight increase of the burr height from 0.01 mm to 0.03 mm was numerically predicted when cutting with Z-shaped partholder.

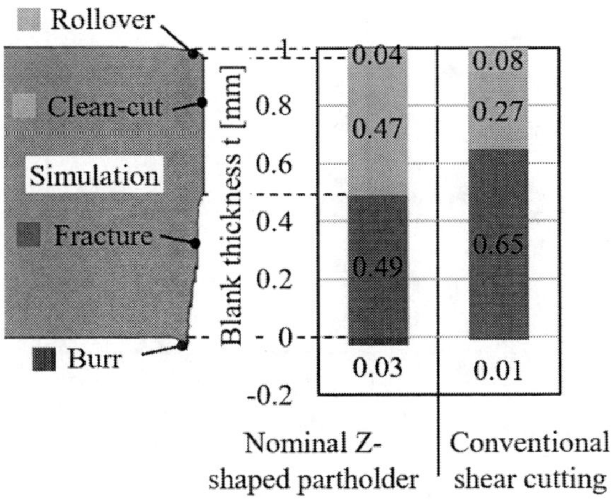

Figure 4: Comparison of simulated surface quality for cutting with nominal Z-shaped partholder and conventional shear cutting.

3.2. Sensitivity analysis

With this 2D FE-model, a numerical sensitivity analysis was carried out regarding the influence of the Z-shaped partholder geometry, the cutting clearance u_S and the partholder force F_{PH} on the cutting surface quality. The initial designed geometries and tool setup is shown in figure 3 b) together with the varied parameters. In addition, the nominal settings as well as the considered parameter space are given in table 1. The sensitivity analysis, presented in the following, based on a Latin-Hypercube (LHC) parameter sampling with more than 50 runs and different parameter combinations.

Table 1: Summary of investigated process and geometry parameters.

Parameter	b_1 [mm]	b_2 [mm]	r_1 [mm]	r_2 [mm]	h [mm]	u_s [%]	F_{PH} [kN]
Initial	10	8	0.5	1	12	6	5
LHC-interval	8 - 12	6 - 10	0.3 – 0.7	0.8 – 1.2	10 - 14	1 - 11	3 -7

In the course of the sensitivity analysis, the Z-shaped partholder geometry, the cutting clearance u_s and the partholder force F_{PH} were varied using a LHC parameter sampling. This enabled the determination of the impact of these parameters on the cutting process and the cutting surface quality, represented by estimated linear correlation coefficients ρ. Due to an internal and dependent scaling of the Z-shaped partholder geometry parameters by DEFORM 2D software, all varied dimensions could be summarized as one geometry parameter. Subsequently, correlations between this geometry parameter as well as the other input parameters with the cutting surface quality (burr, clean-cut, rollover and blank imprint) and process conditions (cutting surface and partholder stress) were identified. These correlations are shown in figure 5 a). Here, a positive value ρ of up to 1 indicates a rise of result variable when the input parameter is increased, and vice versa ($-1 \leq \rho \leq 0$).

Due to the dependent scaling of the Z-shaped partholder geometry size within narrow boundaries, only minor correlations can be directly observed in cutting quality. It shows a certain tendency, that increasing the size of the partholder reduces the burr formation, increases the clean-cut portion, requires more cutting force but reduces the stress of the partholder itself. The elaborated correlation between a reduced cutting clearance, an enhanced surface quality and increased process loads are in a good agreement with the state of the art for conventional shear cutting [1]. It is thereby conformed that a minimized cutting clearance increases the clean-cut portion, reduces burr and rollover height but requires significantly higher cutting forces.

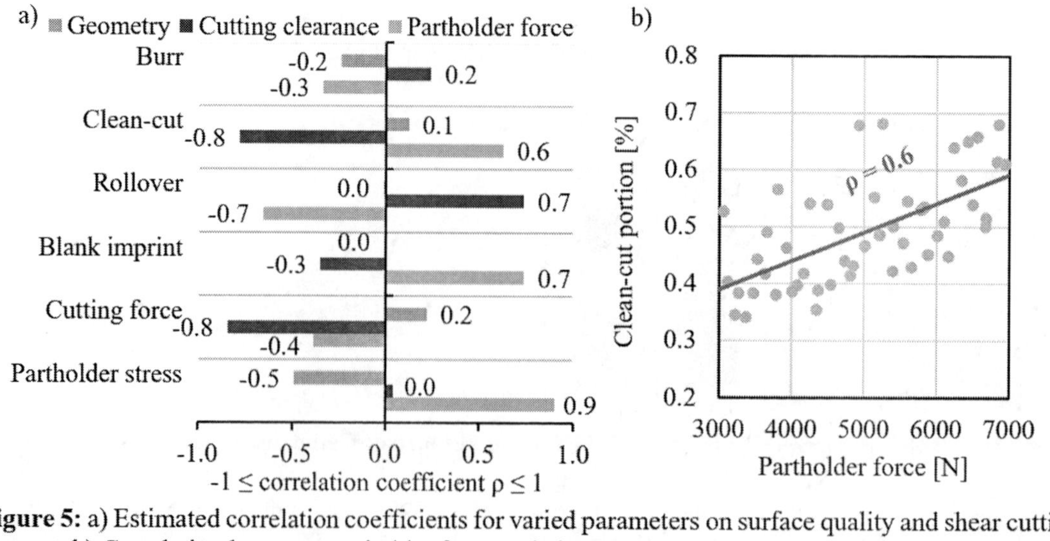

Figure 5: a) Estimated correlation coefficients for varied parameters on surface quality and shear cutting process. b) Correlation between partholder force and clean-cut portion due to cross-elastic partholder.

In general, it can be stated that the Z-shaped partholder geometry enables the partholder force as an influencing factor on cutting surface quality and process. Compared to a partholder or stripper of conventional shear cutting tool design, which only serves as a rigid blank fixture or lift-off prevention, the new partholder design impacts the cutting surface quality and the process parameters. Due to the cross-elastic behaviour of Z-shaped partholder, an increase of partholder force tremendously enhances the cutting surface quality. It directly impacts the clean-cut portion and reduces burr formation and rollover height. It is limited by the formation of blank imprints (< 0.001 mm) at the contact zone between partholder and blank surface to ensure an overall sufficient part quality for cut-outs. It also affects the load on partholder, represented by an evaluation of maximum partholder stress, which is discussed in the following section.

3.3. *Best-case design of DoE*

Within the investigated parameter space (see table 1) and the total of 50 simulation runs, a best-case design of the Z-shaped partholder was selected. This selection was made by taking into account the maximum achievable clean-cut portion and a marginal imprint of the partholder on the blank surface. The resulting partholder dimensions are shown in figure 6. In addition, the partholder stress distribution occurring and the achievable cutting surface quality are depicted in this figure. Thus, with the presented geometry, a cutting clearance u_s of 1 % and a partholder force F_{PH} of 5.1 kN, a clean-cut portion of 67 % and a neglectable blank imprint is obtained. However, the improvement of cutting surface quality is accompanied by a critical effective stress for tool steel (σ_{max} = 2,600MPa) at the lower radius (r_2 = 1 mm). On contrary, selected high carbon and chromium steel AISI-D2 exhibits a compressive strength up to 2,500 MPa and a bending strength up to 4,000 MPa [15]. Thus, even for static failure evaluation, critical loads on Z-shaped partholder geometry can be stated. In terms of an application with cyclic loading by thousands of strokes and a considerably reduced fatigue limit (~ 1,000 MPa) [16], current design shows an extensive risk of premature fracture. This needs to be tackled in a future re-design work to reduce the notch effect in respective area and increased tool strength with improved material properties by surface or heat treatment [16], [17] and later on verified in experimental endurance tests.

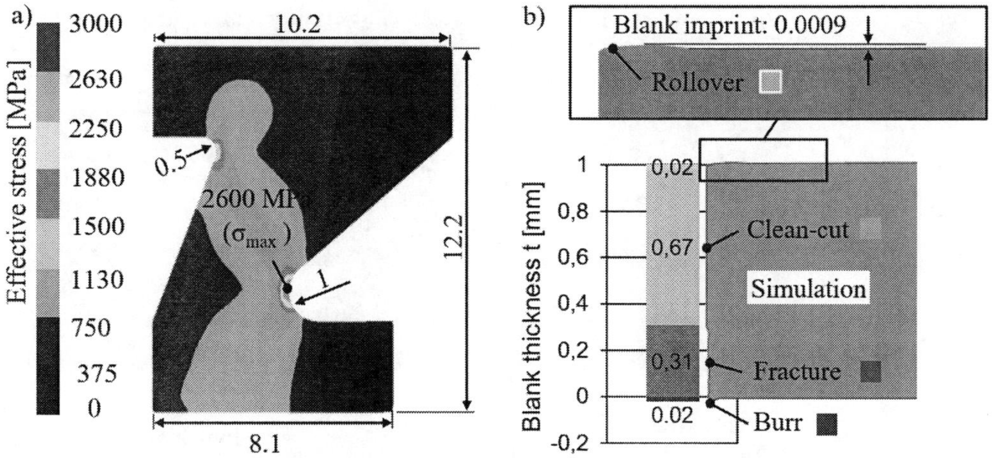

Figure 6: Numerical results for a) effective stress on selected best-case Z-shaped partholder geometry with corresponding dimensions and b) simulated cutting surface quality.

4. Classification of shear cutting results

In the following chapter, the cutting results obtained with the devised best-case design of the Z-shaped partholder is numerically compared to those of a conventional shear cutting tool design as well as a state of the art fine blanking setup, but without utilizing a counter-acting pad. The comparability is ensured by defining the same cutting clearance, material, simulation settings as well as punch and die radii. In this context, figure 7 shows the mean stress distributions obtained with the three different shear cutting processes, each at a punch intrusion depth of 0.3 mm. For conventional shear cutting (figure 7 a)), the formation of tensile stress bands can be observed, which provoke an early failure of material. On contrary, the other two process variants (figure 7 b) and c)) show a fully compressive stress coverage at the shear zone, which supports an extended plastic material flow and thus high clean-cut portions.

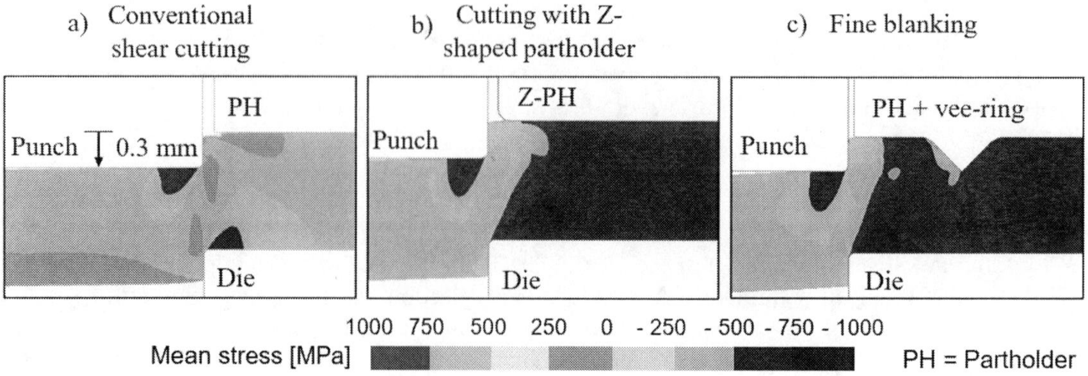

Figure 7: Mean stress distribution at a punch intrusion of 0.3 mm for a) conventional shear cutting, b) cutting with Z-shaped partholder and c) fine blanking.

Figure 8 shows a comparison of the cutting surfaces obtained with the three investigated shear cutting methods and illustrates the resulting formations of rollover, clean-cut, fracture and burr. Additionally, these numerical results show the emerging damage in the cutting area, which represents the edge hardening and provides information of the formability in subsequent forming operations. Here, conventional shear cutting reveals the lowest cutting surface quality with a clean-cut portion of 47 % and severe edge hardening along more than 50 % of blank thickness. Simulation results for the new approach using the Z-shaped partholder offers an increased clean-cut portion of 67 %, reduced rollover and burr height as well as minor edge hardening phenomena. At current state of work, fine blanking still provides highest cutting surface quality (73 % clean-cut portion) with lowest material damage in shear zones. Nevertheless, rollover and burr formation are found at the same level as cutting with Z-shaped partholder.

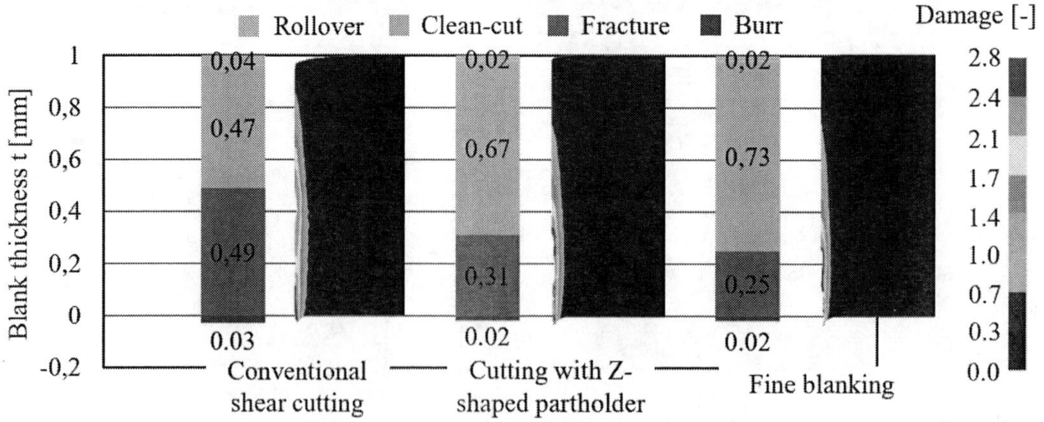

Figure 8: Comparison of simulated cutting surface quality and cutting edge hardening (damage) with constant cutting clearance ($u_s = 1$ %), punch radii ($r_P = 0.05$ mm) and die radii ($r_D = 0.05$ mm).

Figure 9 presents numerically determined force-stroke curves for conventional cutting (black), fine blanking (blue) and cutting with Z-shaped partholder (grey). The calculated forces relate to the 2D-FE-model at a section depth of 1 mm. For all processes, maximum cutting force occurs at a punch intrusion depth of 0.1 mm at the end of the elastic deformation phase of blank material. Here, maximum cutting force of conventional and fine blanking amounts to 442 N and a slightly higher force (453 N) is required for cutting with Z-shaped partholder. The subsequent plastic deformation phase leads to a decrease of cutting forces for all three investigated tool designs. It can be stated, that fine blanking and cutting with Z-shaped partholder do exhibit same flow behavior whereas conventional shear cutting shows a slightly higher decrease of forces. The abrupt force drop marks the seperation of material when the shearing stress finally exceeds its shear fracture limit. This correlates with the calculated clean-cut portions of all three tool designs (figure 8). After material separation, the push-phase proceeds at different constant force levels. Here, the compressive stress superposed by fine blanking and cutting with Z-shaped partholder leads to an increased contact pressure between cutting and lateral punch surface and thus to marginally increased push forces.

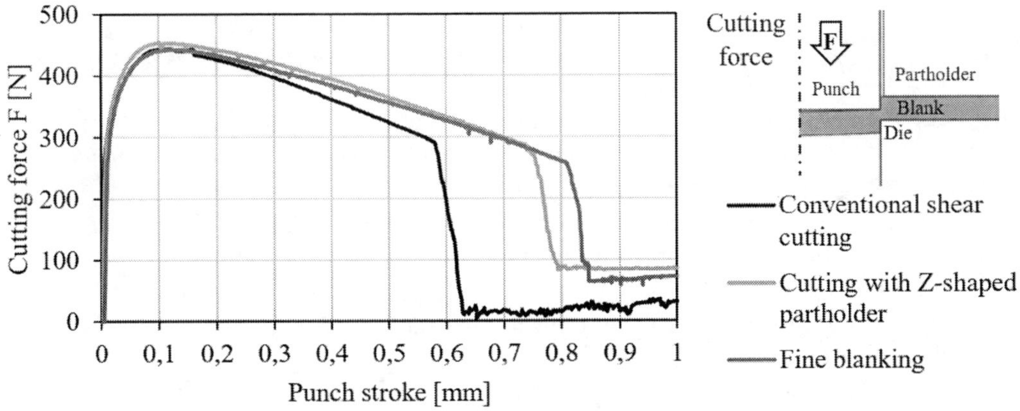

Figure 9: Numerically calculated cutting force curves for conventional shear cutting, cutting with Z-shaped partholder and fine blanking.

5. Conclusion and outlook

In the numerical study presented in this paper, a new design of a cross-elastic partholder was analysed to superpose stresses in shear cutting zones. First, suggested Z-shaped partholder geometry for compressive stress superposition was modelled in DEFORM 2D FE-software and a parametric shear cutting simulation was set up. The numerical sensitivity analysis revealed correlations of the partholder geometry, cutting force and clearance on the process and the cutting surface quality. Due to the targeted cross-elasticity of Z-shaped partholder, clean cut-portion could be controlled by normally applied partholder force. Second, the best-case design of the Z-partholder, leading to the highest achievable clean-cut portion (67 %), was compared to the numerical results of conventional shear cutting (47 %) and fine blanking (73 %).

These first prospectful results for stress superposition via cross-elastic partholder geometry to enhance cutting surface quality by minor tool and process modifications of a conventional shear cutting tool motivates for future research work. Currently, a topology optimization of partholder geometry with wider boundaries and different target options such as cutting surface quality or tool load is set up to achieve multiple geometries for both tensile and compressive stress superposition. Furthermore, physical implementation of new partholder design into a prototype tool and an experimental verification of the numerical results is necessary. Here, partholder geometry and cutting process have to be analysed in terms of static and fatigue strength as well as achievable cutting surface quality.

References

[1] Klocke, F 2017 *Fertigungsverfahren 4 Umformen* Springer Berlin Heidelberg

[2] Hörmann F 2008 *Einfluss der Prozessparameter auf einstufige Scherschneidverfahren zum Ausschneiden mit endkonturnaher Form* Dissertation Technical University of Munich

[3] Sachnik P 2017 *Methodik für gratfreie Schnittflächen beim Scherschneiden* Dissertation Technical University of Munich

[4] Lee T C, Chan L C, Zheng P F 1997 Application of the finite-element deformation method in the fine blanking process *Journal of Materials Processing Technology* **63** pp 744–749

[5] Kräusel V 2013 *Gestaltung und Bewertung einhubiger Scherschneidverfahren mit starren Werkzeugen unter besonderer Berücksichtigung der Schnittflächenqualität an Blechbauteilen* Habilitation Technical University of Chemnitz

[6] Kühlewein R 2003 *Einfluss der Prozessparameter auf das Nachschneiden schergeschnittener Konturen* Dissertation Technical University of Munich

[7] Paetzold I, Feistle M, Golle R et al. 2018 Determination of the minimum possible damage due to shear cutting using a multi-stage shear cutting process IOP Conf. Ser.: *Materials Science and Engineering* **418**

[8] Liewald M, Kappes J, Hank R 2010 Schnittgratfreies Scherschneiden mittels Konterschneiden Verlag Meisenbach *utfscience* pp 1–8

[9] Klocke F, Sweeney K, Raedt H W 2001 Improved tool design for fine blanking through the application of numerical modeling techniques *Journal of Materials Processing Technology* **115** Nr. 1 pp 70–75

[10] Senn S and Liewald M 2019 Investigation of a new sheet metal shear cutting tool design to increase the part quality by superposed compression stress. IOP Conf. Ser.: *Materials Science and Engineering* **651**

[11] Birzer F 2014 Feinschneiden und Umformen Wirtschaftliche Fertigung von Präzisionsteilen aus Blech Landsberg Verlag *Moderne Industrie*

[12] Feuerhack A, Trauth D, Mattfeld P et al. 2015 Fine Blanking of Helical Gears *60 excellent inventions in metal forming* pp 187–192

[13] Greaves G N, Greer A L, Lakes R S et al. 2011 Poisson's ratio and modern materials *Nature Materials* **10** Nr 11 pp 823–837

[14] Panico T 2012 *Beurteilung der Kantenbelastung beim Scherschneiden höchstfester Stahlbleche mit Hilfe der Finiten-Element-Methode* Dissertation Technical University of Munich

[15] FIBRO GmbH 2022 Ground Precision Components - Comparative Graphs Standard Parts pp. E16-E17

[16] Betzwar Kotas A, Danninger H, Weiss B et al. 2017 Fatigue testing and properties of hardmetals in the gigacycle range *International Journal of Refractory Metals and Hard Materials* **62** pp 183–191

[17] GuuY H and Hocheng H 2001 Improvement of fatigue life of electrical discharge machined AISI D2 tool steel by TiN coating *Materials Science and Engineering* **318** Nr. 1 pp 155–162

IOP Conf. Series: Materials Science and Engineering 1238 (2022) 012028 doi:10.1088/1757-899X/1238/1/012028

Determination of temperature dependence in Modified-Mohr-Coulomb failure model for process simulation of shear cutting

B-A Behrens[1], K Brunotte[1], H Wester[1], C Kock[1] and D Kildonaviciute[1]

[1] Institute of Forming Technology and Machines, Leibniz Universität Hannover, An der Universität 2, 30823 Garbsen, Germany

E-mail: kildonaviciute@ifum.uni-hannover.de

Abstract. Shear cutting is a proven process for chip-less separation of metals and used as a cost-effective production method. The process design of shear cutting is usually based on time and cost intensive experimental tests. Therefore, numerical depiction of the process offers great potential to reduce these practical tests. An important aspect hereby is the representation of temperature dependency of the material. However, the classical Modified-Mohr-Coulomb (MMC) failure model does not depict this effect directly. Therefore, the objective of this paper is to investigate the temperature dependency of MMC failure model for simulating shear cutting. The required data is obtained from tensile tests on miniaturised specimens under variation of temperature on dual-phase steel DP1000, using an additional optical measurement system. To determine stress triaxiality, Lode angle parameter and plastic strain at failure, the experimental tests are simulated using the Finite-Element simulation program ABAQUS. The MMC failure surfaces are fitted using the least squares method in Matlab. With this approach, temperature dependent MMC failure model was built to be used for shear cutting simulation of DP1000 steel.

1. State of the Art

Increasing quality demand on sheet metal components using shear cutting requires high precision, which is achieved with a suitable process design [1]. Standard simulations based on Finite-Elemente (FE) method use simplified calculation algorithms to describe stress dependent material failure [2]. Due to the strong temperature increase in shear cutting [3], it is necessary to develop and integrate a temperature dependent failure model, so that the numerical representation of these processes can accurately predict the temperature influence on material failure [4]. The classical Mohr-Coulomb (MC) model predicts material failure at a certain combination of normal stresses σ_1, σ_2 and σ_3 as shown in [5]:

$$c_2 = max\left\{\frac{1}{2}|\sigma_1 - \sigma_2| + c_1(\sigma_1 + \sigma_2), \frac{1}{2}|\sigma_2 - \sigma_3| + c_1(\sigma_2 + \sigma_3), \frac{1}{2}|\sigma_3 - \sigma_1| + c_1(\sigma_3 + \sigma_1)\right\} \quad (1)$$

The constant c_1 is referred to as coefficient of friction and c_2 as shear resistance in MPa. Through geometric relations in the deviator plane, σ_1, σ_2 and σ_3 can be given as functions of mean stress σ_m, stress triaxiality η and Lode angle parameter θ [5]:

$$\sigma_1 = \left(1 + \frac{2\cos(\theta)}{3\eta}\right)\sigma_m, \sigma_2 = \left(1 + \frac{2\cos\left(\frac{2}{3}\pi - \theta\right)}{3\eta}\right)\sigma_m, \sigma_3 = \left(1 + \frac{2\cos\left(\frac{4}{3}\pi - \theta\right)}{3\eta}\right)\sigma_m \quad (2)$$

In [5] the MC criterion was represented as a function of the equivalent stress $\bar{\sigma}$:

$$\bar{\sigma} = c_2 \left[\sqrt{\frac{1+c_1^2}{3}} \cos\left(\frac{\pi}{6} - \theta\right) + c_1 \left(\eta + \frac{1}{3} \sin\left(\frac{\pi}{6} - \theta\right)\right) \right]^{-1} \quad (3)$$

Furthermore, Lode angle parameter can be reduced to a range of $-1 \leq \bar{\theta} \leq 1$ by normalizing it [5]:

$$\bar{\theta} = 1 - \frac{6\theta}{\pi} \quad (4)$$

However, in [5] it was also shown that ductile fracture detection improves by taking the strain into account. Thus, by transforming the MC fracture criterion from stress based into stress and strain based space under condition of monotonic loading, the Modified-Mohr-Coulomb (MMC) failure model was built in terms of equivalent plastic strain at fracture $\bar{\varepsilon}_p^f$:

$$\bar{\varepsilon}_p^f = \left\{ \frac{A}{c_2} \left[c_3 + \frac{\sqrt{3}}{2-\sqrt{3}} (1-c_3) \left(\sec\left(\frac{\bar{\theta}\pi}{6}\right) - 1 \right) \right] \left[\sqrt{\frac{1+c_1^2}{3}} \cos\left(\frac{\bar{\theta}\pi}{6}\right) + c_1 \left(\eta + \frac{1}{3} \sin\left(\frac{\bar{\theta}\pi}{6}\right) \right) \right] \right\}^{-\frac{1}{n}} \quad (5)$$

with A and n as strength and work hardening exponent in Swift hardening law, respectively, as well as c_3 as material constant [6]. Various authors have successfully implemented the MMC failure model for aluminum [6] as well as for steel [7]. Xiao et al. in [8] extended the MMC failure model to integrate the strain rate and temperature dependency for aluminum according to the Johnson-Cook failure model. However, the data shown only portrayed results made under constant temperature.

Based on the literature it is clear that the failure models are affected by material temperature dependency. Therefore, this paper conducts a solution approach using MMC failure model on DP1000 to improve the calculation of ductile material failure by taking the temperature into account.

2. Experimental tests

2.1. *Tensile tests*

The tensile tests were carried out on a quenching and forming dilatometer DIL 805A/D+T from TA Instruments, shown in figure 1 (a), at a quasi-static strain rate of 0.001 s^{-1}. The length change of the specimen was recorded by two quartz push rods. A more detailed description of dilatometer can be found in [9] and [10]. Uniaxial specimens were used to evaluate flow behaviour in uniaxial tension conditions.

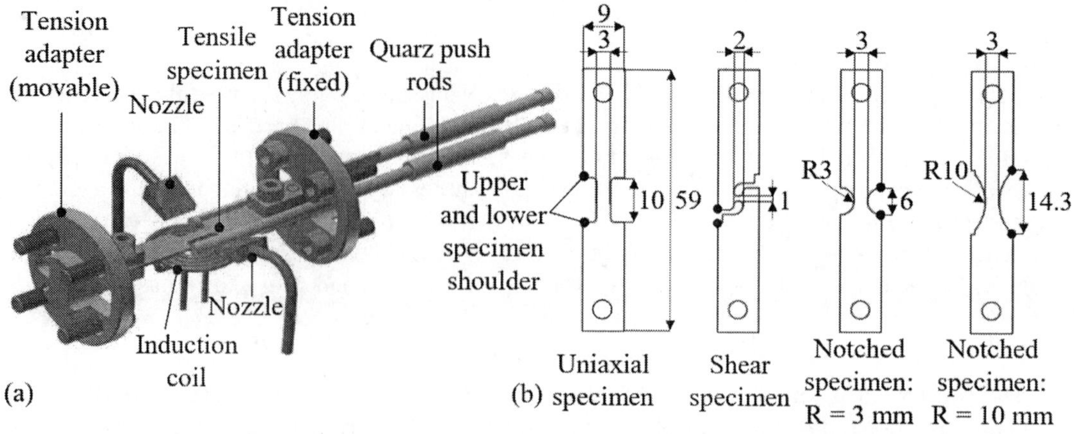

Figure 1. Schematic setup of dilatometer (a), sample geometries and dimensions (b).

In addition to room temperature of 20 °C, tests were performed at elevated temperature of 100 °C, 200 °C, 300 °C and 500 °C. Since the failure model represents varying stress states and its accuracy depends on the number of support points for the calibration of the failure surfaces, tensile tests on shear and notched specimens were performed to generate different stress states, shown in figure 1 (b). In shear tensile tests, triaxialities of $\eta = 0$ were to be induced. Notched tensile specimens with radius R = 3 mm and R = 10 mm were to produce higher triaxiality values than uniaxial tensile test. Since in [2] and [4] the determined stress triaxiality during shear cutting approximated the range of $-0.2 \geq \eta \geq 0.5$, it was decided to use such specimen geometries to ideally cover the region of stress triaxialities in shear cutting. The specimens were cut along the rolling direction by means of a water jet cutter of a 1 mm thick DP1000 sheet. The tests were repeated three times for each temperature and each specimen geometry.

A calculation of the flow curves determined from dilatomter tests was required for the simulation, in which a maximal forming degree of $\varepsilon = 0.06$ was reached. Since this was not sufficient for the numerical simulation, an extrapolation was done using a combined approach of the Swift [11] and Voce [12] (eq. 6) hardening law in equation 7. These methods contained the lowest values in the least squares method to the flow curves obtained from dilatometer tests compared to Ludwig and Hockett-Sherby approaches.

$$k_{Swift}(\varepsilon) = A(\varepsilon_0 + \varepsilon)^n,\ k_{Voce}(\varepsilon) = \sigma_S + (\sigma_0 - \sigma_S)e^{\left(-\frac{\varepsilon}{E}\right)} \quad (6)$$

$$k_{\mathrm{Comb}}(\varepsilon) = \alpha \cdot A(\varepsilon_0 + \varepsilon)^n + (1 - \alpha)(\sigma_S + (\sigma_0 - \sigma_S))e^{\left(-\frac{\varepsilon}{E}\right)} \quad (7)$$

k stands for flow stress, ε_0 for starting plastic strain, σ_0 for yield stress, A, n, σ_S, E for material parameters and α for a weighting factor. For each temperature considered, material parameters were obtained from dilatometer tensile tests. To validate the accuracy of the extrapolation methods, a tensile testing machine Dynamess S100/ZD with a mechanical extensometer and GOM optical measurement system ARAMIS was used. Hereby the change in length and width of tensile specimens in 0 °, 45 ° and 90 ° rolling direction was recorded. The digital extensometer of the optical measurement system allowed to obtain the flow curve beyond the uniform strain. Therefore, flow curve of a uniaxial specimen at room temperature was built up to a plastic strain of $\varepsilon = 0.2$ at a quasi-static strain rate. Moreover, using tensile specimens in different rolling directions, the anisotropic parameters r_0, r_{45}, r_{90} were determined.

2.2. *Design of the simulation model*

Specimen models were created in ABAQUS 6.14 from dilatometer samples presented in chapter 2.1. To numerically investigate material anisotropy, ABAQUS parameters R_{11}, R_{22}, R_{33}, R_{12}, R_{13}, R_{23} were determined from anisotropy parameters r_0, r_{45}, r_{90}, together with Hill'48 coefficients [13]. The Hill'48 model was prior used to describe the forming and failure behavior of DP980 sheet in literature [14, 15].

Solid eight-node 3D elements C3D8R were used for the implicit static simulation. Direct method was used as a solver with a full Newton solution technique. The specimens were meshed with an element edge length of 0.1 mm. In the forming area Arbitrary Lagrangian Eulerian (ALE) adaptive mesh domain was activated for the steps of displacement. No mass nor time scaling was implemented. Symmetries were used for the uniaxial and both notched specimens, therefore only quarter models were simulated. Due to the geometry of the shear specimen only the symmetry relative to the central plane of the sheet was exploit. A time-length change curve was applied as a boundary condition for axial displacement to the upper end of each specimen model, while the lower end was fixed. To identify the length change in ABAQUS, the distance between the upper and the lower shoulder of each specimen was measured, shown in figure 1 (b). When this area reached the length change at fracture from dilatometer experiments, the simulation was declared finished, thus indicating fracture. Of the three curves of time-length change at fracture, which resulted from the test repetition, the intermediate one was taken as the specimen fracture condition. Such technique was used simulating each specimen. With this approach, $\varepsilon_{\mathrm{p}}^{\mathrm{f}}$, η and $\bar{\theta}$ were determined. While $\varepsilon_{\mathrm{p}}^{\mathrm{f}}$ was the equivalent of the deformation degree at maximal length change, the stress state varied during the deformation, thus impacting η and $\bar{\theta}$ [16]. For this reason, [16] suggested to use average values of η_{av} and $\bar{\theta}_{\mathrm{av}}$, which are defined in equation 8.

$$\eta_{\mathrm{av}} = \frac{1}{\varepsilon_{\mathrm{p}}^{\mathrm{f}}} \int_0^{\varepsilon_{\mathrm{p}}^{\mathrm{f}}} \eta(\varepsilon) d\varepsilon_{\mathrm{p}},\ \bar{\theta}_{\mathrm{av}} = \frac{1}{\varepsilon_{\mathrm{p}}^{\mathrm{f}}} \int_0^{\varepsilon_{\mathrm{p}}^{\mathrm{f}}} \bar{\theta}(\varepsilon_{\mathrm{p}}) d\varepsilon_{\mathrm{p}} \quad (8)$$

3. Results and Discussion

3.1. Results of the tensile tests

Results depicted from dilatometer tests on the uniaxial specimen are shown as a force-length change curve in figure 2 (a) for various temperatures and as a length change at fracture-temperature diagram for various specimen geometries in figure 2 (b). In literature, steels are assumed to have decreasing strength with increasing temperature. The so-called blue brittleness is considered to be an exception. If the steel material contains a certain mass of dissolved foreign atoms such as carbon and nitrogen, they diffuse to the dislocations at particular temperatures and block them. This means the tensile strength increases [17]. In figure 2 (a) a decrease of force was noticed from 25 °C to 100 °C, caused by a thermal softening effect. When considering the length change versus temperature of the uniaxial specimen in figure 2 (b), a blue brittleness range was identified, in which the length change decreases at 100 °C and 200 °C compared to 20 °C, whilst the force stays nearly constant in figure 2 (a). Moreover, a blue brittleness range was identified between 200 °C and 300 °C at which the force remains almost constant as well.

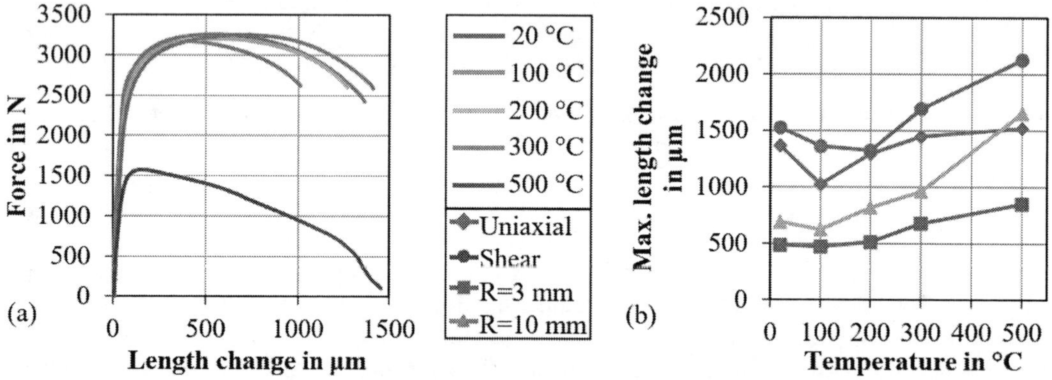

Figure 2. Force-length change of the uniaxial specimen at various temperatures (a) and length change at fracture-temperature diagram for each specimen geometry (b).

Figure 3 (a) shows flow curves of uniaxial specimen at various temperatures from dilatometer testing. Figure 3 (b) compares the flow curves of dilatometer and Dynamess tests as well as the extrapolation approaches of Swift, Voce and the combined method at 20 °C, shown in equations 6 and 7.

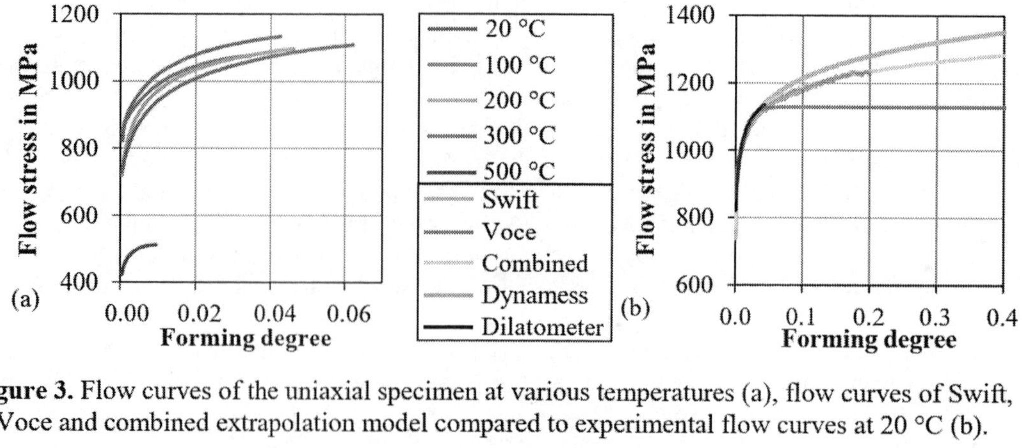

Figure 3. Flow curves of the uniaxial specimen at various temperatures (a), flow curves of Swift, Voce and combined extrapolation model compared to experimental flow curves at 20 °C (b).

The value of $\alpha = 0.7$ was fitted as the weighting factor. It can be depicted, that the combined extrapolation approach was the most suitable. The same weighting factor α was used for extrapolation of the flow curves for all further temperatures in figure 3 (a), since the parameters of Swift and Voce hardening law were determined for each temperature separately. The parameters are shown in table 1.

Table 1. Swift and Voce hardening law parameters for each temperature.

	Swift			Voce		
T in °C	A in MPa	ε_0	n	σ_S in MPa	σ_0 in MPa	E
20	1450	0.0002	0.0773	1128	850	93.8
100	1410	0.0008	0.0772	1083	850	92.3
200	1480	0	0.0933	1083	748	111.8
300	1460	0	0.0951	1098	760	70.9
500	640	0	0.0473	511	407	471.1

Table 2 shows the anisotropy parameters r_0, r_{45}, r_{90}, determined from Dynamess tensile tests on uniaxial specimens, together with ABAQUS parameters R_{11}, R_{22}, R_{33}, R_{12}, R_{13}, R_{23} and the Hill'48 coefficients, which were used to portray material anisotropy in the simulations.

Table 2. Anisotropy parameters and Hill'48 coefficients for input in ABAQUS.

r_0	r_{45}	r_{90}	F	G	H	N	R_{11}	R_{22}	R_{33}	R_{12}	R_{13}	R_{23}
0.86	1.27	1.13	0.41	0.54	0.46	1.68	1.00	1.07	1.03	0.95	1.00	1.00

3.2. Results of the simulations

The combined extrapolation approach, described in section 2.1, was evaluated by comparison of the force-length change diagrams from simulations with the experimentally determined results. Figure 4 (a-d) shows the findings for uniaxial, shear and both notched specimens, correspondingly.

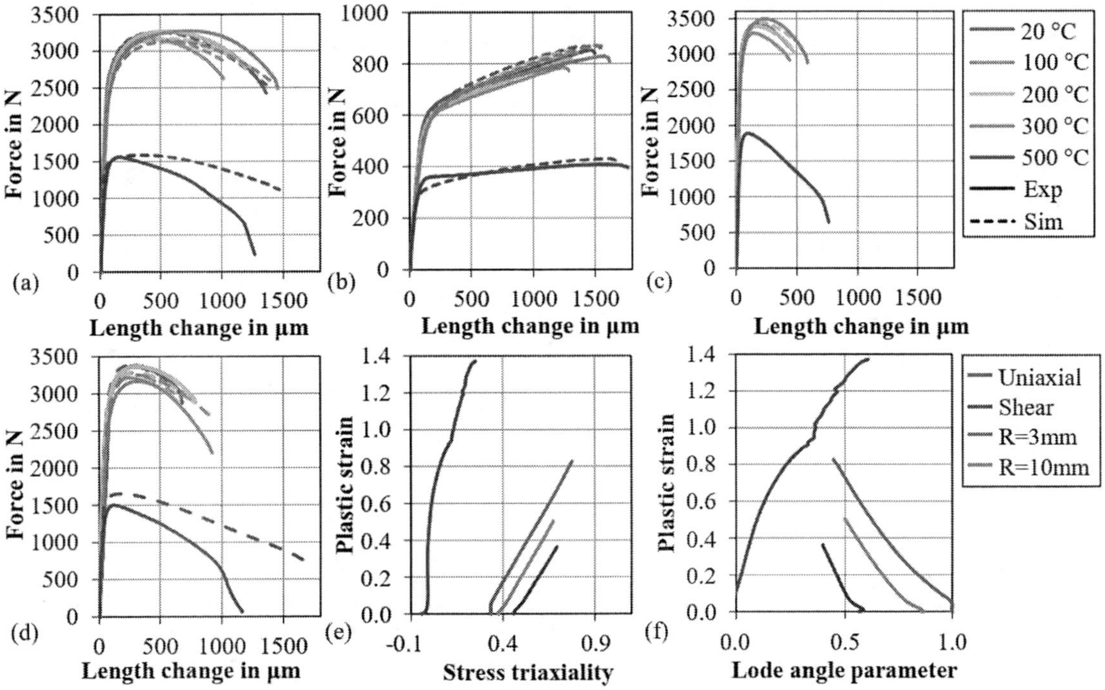

Figure 4. Force-length change curve of uniaxial (a), shear (b), notched R = 3 mm (c) and R = 10 mm (d) specimen from experiment (solid line) and simulation (dashed line) at temperatures. Plastic strain-stress triaxiality (e) and plastic strain-normalized Lode angle parameter (f) at 20 °C for all specimens.

Overall, good agreement was observed between experiment and simulation results, in particular at temperatures up to 300 °C. At 500 °C the force-length change curves held the largest deviation between numerical and experimental results for all specimens in that the force was overestimated in the simulation. Since at 500 °C a strong thermal softening effect occurred in the experimental testing for all specimens, which lead to a partial halving of tensile strength compared to lower temperatures, the temperature-dependent weighting factor α should be considered for subsequent simulations to depict this effect correctly. Figures 4 (e) and (f) show exemplary curves of plastic strain-triaxiality and plastic strain-normalized Lode angle parameter for all specimens at 20 °C, respectively. The values of the stress triaxiality and normalized Lode angle change over the plastic strain. While the stress triaxialities show values close to each other, the normalized Lode angles cover a wider range of stress states.

3.3. Generation of failure surfaces

As described above, only the average values over the degree of deformation $\varepsilon_{\mathrm{p}}^{\mathrm{f}}$ were calculated for the stress values η and $\bar{\theta}$ to build the MMC failure model using equation 8, which are summarized in table 3 for all specimen geometries at various temperatures. As expected, the average triaxiality of the shear specimen was at about $\eta_{\mathrm{av}} = 0$. The notched specimens showed slightly higher values regarding the average stress triaxiality, but lower plastic strain at fracture compared to uniaxial and shear specimens, with an exception at 500 °C of the notched specimen with R = 10 mm.

Table 3. Characteristic values η_{av}, $\bar{\theta}_{\mathrm{av}}$ and $\varepsilon_{\mathrm{p}}^{\mathrm{f}}$ at various temperatures of each specimen.

	Uniaxial			Shear			R = 3 mm			R = 10 mm		
T in °C	η_{av}	$\bar{\theta}_{\mathrm{av}}$	$\varepsilon_{\mathrm{p}}^{\mathrm{f}}$	η_{av}	$\bar{\theta}_{\mathrm{av}}$	$\varepsilon_{\mathrm{p}}^{\mathrm{f}}$	η_{av}	$\bar{\theta}_{\mathrm{av}}$	$\varepsilon_{\mathrm{p}}^{\mathrm{f}}$	η_{av}	$\bar{\theta}_{\mathrm{av}}$	$\varepsilon_{\mathrm{p}}^{\mathrm{f}}$
20	0.53	0.72	0.78	0.01	0.12	1.08	0.57	0.48	0.33	0.54	0.63	0.56
100	0.43	0.84	0.43	0.01	0.06	0.97	0.59	0.47	0.32	0.50	0.68	0.41
200	0.46	0.80	0.59	0.01	0.05	0.97	0.57	0.49	0.36	0.54	0.64	0.59
300	0.50	0.76	0.72	0.02	0.09	1.16	0.63	0.45	0.57	0.57	0.63	0.68
500	0.64	0.63	1.07	0.07	0.25	1.68	0.77	0.35	0.96	1.01	0.36	1.97

The determination of the material parameters c_1, c_2 and c_3 in the MMC model was carried out with the curve fitting tool in Matlab using the least squares method. For each temperature the determined parameters are shown in table 4.

Table 4. MMC failure model parameters c_1, c_2 and c_3 for each temperature.

T in °C	c_1	c_2 in MPa	c_3
20	0.08	775	0.92
100	0.12	735	0.90
200	0.11	800	0.95
300	0.09	795	0.92
500	0.05	340	0.89

Figure 5 (a-e) shows the failure surfaces for temperatures 20 °C, 100 °C, 200 °C, 300 °C and 500 °C, correspondingly. All failure surfaces held the lowest plastic strain at fracture for stress triaxiality $\eta = 0.8$. A minimal influence of stress triaxiality can be depicted with respect to normalized Lode angle parameter at values of $\eta \approx 0.8$, but an increasing influence at $\eta \approx 0$. For the stress state parameters $\eta = 0$ and $\bar{\theta} = -1$, the plastic strain at fracture was at its highest for all temperatures. At 100 °C the stress state seemed to have less influence on the plastic strain at fracture, which made the MMC surface to appear flatter. When compared to 100 °C, the curvature of the surface at 200 °C increased and thus the influence

IOP Conf. Series: Materials Science and Engineering 1238 (2022) 012028 doi:10.1088/1757-899X/1238/1/012028

of the stress state to the plastic strain at fracture. The failure surfaces at 200 °C showed a higher level than at 100 °C due to the increasing forming capacity. Same behaviour was observed for failure surfaces at 300 °C and 500 °C. Since the plastic strain at fracture was well above the maximal axis value of $\varepsilon_p^f = 3.0$ at $\eta \approx 0$ and $\bar{\theta} = -1$ at 500 °C, the failure surface was cut off from the axis in figure 5 (e).

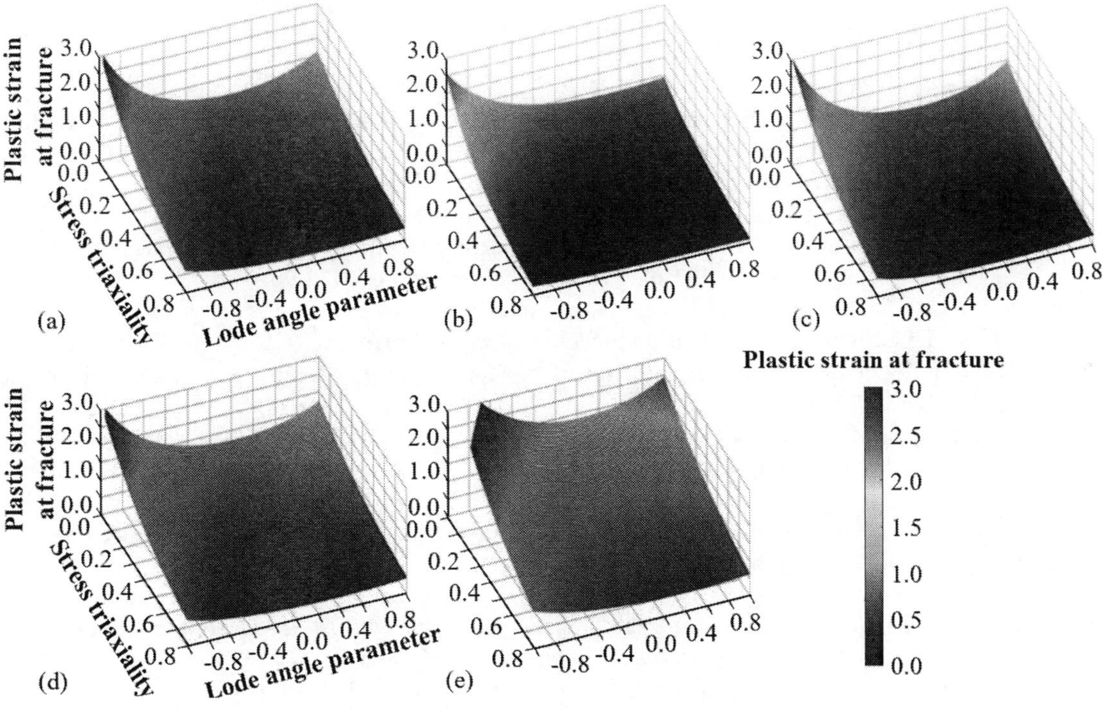

Figure 5. MMC failure surfaces at (a) 20 °C, (b) 100 °C, (c) 200 °C, (d) 300 °C and (e) 500 °C.

The failure surfaces exhibited different behaviour at varying temperatures, thus implying that a temperature dependent adjustment of the curvature is necessary. It can be therefore concluded that a simple multiplicative scaling of the failure surface represented at room temperature to obtain the MMC damage model at elevated temperatures is not useful for the correct representation of the plastic strain at fracture.

4. Conclusion

Due to the temperature increase in shear cutting, the aim of this paper was to investigate the temperature effect on MMC failure model, which is to be implemented into a numerical simulation model of shear cutting. The first step hereby was to characterize the temperature dependency of a DP1000 sheet material by the use of miniaturized tensile specimens on a dilatometer. An abnormal behavior caused by blue brittleness was depicted. The data from tensile tests was successfully imported into ABAQUS simulation using a combined extrapolation method of Swift and Voce to obtain the required values for the determination of MMC failure model surfaces. It was successfully shown, that the temperature dependent material behavior highly influences the MMC damage model. Therefore, future investigations should be carried out considering and extension and parametrization of the damage model to incorporate the temperature dependency directly into the equation, thus increasing the accuracy when simulating the shear cutting process. In further publications, the failure model will be applied for simulating shear cutting process.

Acknowledgement
The authors are much obliged to the DFG (German Research Foundation) for the financial support of the project "Improved FE simulation of the shear cutting process using a temperature and strain rate-dependent extension of the MMC model" (project number: 199808648).

References

[1] Schneider M, Peshekhodov I A, Bouguecha A and Behrens B-A 2016 A new approach for user-independent determination of formability of a steel sheet sheared edge *J. Prod. Eng. Res. Devel.* **10** 241-252

[2] Gutknecht F, Steinbach F, Hammer T, Clausmeyer T, Volk W and Tekkaya A E 2016 Analysis of shear cutting of dual phase steel by application of an advanced damage model *Procedia Struct. Integr.* **2** 1700–1707

[3] Behrens B-A, Bouguecha A, Vucetic M, Krimm R, Hasselbusch T and Bonk C 2014 Numerical and Experimental Determination of Cut-edge after Blanking of Thin Steel Sheet of DP1000 within Use of Stress based Damage Model *Procedia Eng.* **81** 1096 – 1101

[4] Behrens B-A, Bouguecha A, Peshekhodov I and Bonk I 2013 Comparison between the Lemaitre and a modified Lemaitre damage model in sheet steel blanking *IDDRG 2013 Zurich* 201–206

[5] Bai Y and Wierzbicki T 2009 Application of extended Mohr–Coulomb criterion to ductile fracture *Int. J. Fract.* **161** 1-20

[6] Mirnia M J and Shamsari M 2017 Numerical prediction of failure in single point incremental forming using a phenomenological ductile fracture criterion *J. Mater. Process. Technol.* **244** 17–43

[7] Cheloee D A, Kadkhodapour J, Pourkamali A A, Khoshbin M, Alaie A and Schmauder S 2021 Micromechanical modeling of damage mechanisms in dual-phase steel under different stress states *Eng. Fract. Mech.* **243** 107520

[8] Xiao X, Pan H, Bai Y, Lou Y and Chen L 2019 Application of the Modified Mohr-Coulomb fracture criterion in predicting the ballistic resistance of 2024-T351 aluminum alloy plates impacted by blunt projectiles *Int. J. Impact Eng.* **123** 26–37

[9] Behrens B-A, Dröder K, Hürkamp A, Droß M, Wester H and Stockburger E 2021 Finite Element and Finite Volume Modelling of Friction Drilling HSLA Steel under Experimental Comparison *J. Mater.* **14** *5997*

[10] Behrens B-A, Rosenbusch D, Wester H and Stockburger E 2022 Material Characterization and Modeling for Finite Element Simulation of Press Hardening with AISI 420C *J. Mater. Eng. Perform.* **31** 825–832

[11] Swift H W 1952 Plastic instability under plane stress *J. Mech. Phys. Solids* **1** 1–18

[12] Voce E The relationship between stress and strain for homogeneous deformations *J. Inst. Met.* **74** 537–62

[13] Dassault Systèmes Simulia Corporation 2017 *Abaqus Analysis User's Manual: Anisotropic yield/creep* https://abaqus-docs.mit.edu/2017/English/SIMACAEMATRefMap/simamat-c-anisoyield.htm

[14] Li S, He J, Gu B, Zeng D, Xia Z C, Zhao Y and Lin Z 2018 Anisotropic fracture of advanced high strength steel sheets: Experiment and theory *Int. J. of Plast.* **103** 95–118

[15] Gu B, He J, Li S and Lin Z 2020 Anisotropic fracture modeling of sheet metals: From in-plane to out-of-plane *Int. J. Solids Struct.* **182-183** 112–140

[16] Bai Y and Wierzbicki T 2008 A new model of metal plasticity and fracture with pressure and Lode dependence *Int. J. Plast.* **24** 1071–1096

[17] Liu W and Lian J 2021 Stress-state dependence of dynamic strain aging: Thermal hardening and blue brittleness *Int. J. Miner. Metall.* **28** 854

IOP Conf. Series: Materials Science and Engineering 1238 (2022) 012029 doi:10.1088/1757-899X/1238/1/012029

Sheared edge formability characterization of cold-rolled advanced high strength steels for automotive applications

A Lara[1], D Frómeta[1], S Parareda[1], D Casellas[1,2], P Larour[3], J Hinterdorfer[3], E Atzema[4] and M Heuse[5]

[1] Eurecat, Centre Tecnològic de Catalunya, Plaça de la Ciència, 2, Manresa 08243, Spain
[2] Luleå University of Technology, 971 87 Luleå, Sweden
[3] voestalpine Stahl GmbH, voestalpine-Straße 3, 4020 Linz, Austria
[4] Tata Steel, P.O. Box 10.000, 1970 CA IJmuiden, The Netherlands
[5] Faurecia Autositze GmbH, Garbsener Landstraße 7, 30419 Hannover, Germany

E-mail: toni.lara@eurecat.org

Abstract. Edge cracking has become a limiting factor in the use of some advanced high strength steels (AHSS) for high-performance automotive applications. This fact has motivated the development of a multitude of experimental tests for edge formability prediction over the last years. In this sense, the Hole Expansion Test (HET) according to ISO16630 has been established in the automotive industry as a standard procedure for edge cracking sensitivity ranking. However, whereas it may be useful for rapid material screening, the results are often not accurate and reliable enough. Consequently, alternative methods based on Digital Image Correlation (DIC) have been proposed aimed at improving the prediction of edge cracking occurrence during forming and obtaining useful strain data that can be implemented in forming simulations. This paper explores the applicability of different DIC-based methods, such as Half-Specimen Dome Tests, Sheared Edge Tensile Tests, and KWI hole expansion tests with a flat nosed punch, for characterizing the edge formability of three cold-rolled AHSS sheets. The results obtained from the different testing methods are compared and validated with a laboratory-scale demonstrator. Finally, the limitations and advantages of the different methods are discussed.

1. Introduction

Advanced High Strength Steels (AHSS) have become the dominant material choice for lightweight construction in automotive. Steel is affordable, has good in-service performance, manufacturability, and recyclability. AHSS allow down gauging and, therefore, weight saving, making them the preferred choice over conventional steels. Today, a modern vehicle body contains about 73% of these steels and this percentage will continue to grow in the coming years. In the body structure weight is saved by employing cold rolled galvanised AHSS such as dual phase (DP) and complex phase (CP) steels.

The main drawback is that the high strength achieved in these steel grades is usually associated with limitations in cracking resistance and formability. As a result, during their processing, unexpected defects (cracks, stretching, thinning, etc.), not predicted by traditional experimental or computational approaches at the product design stage, can occur. Because of this, forming parameters and raw material quality must be carefully controlled to assure defect-free part production and guarantee part quality.

Typical metal sheet forming processes for manufacturing AHSS parts includes cutting operations like shearing, blanking or punching. It is well known that these operations produce a shear-affected zone in the cut edge, with material deformation and damage (work hardening, residual stresses, voids and cracks) [1].

Such damage degrades the edge mechanical properties which locally reduces formability, needed for post-cutting forming operations. In materials with limited ductility such as AHSS, edge damage may trigger edge cracks in sheared areas that expand during forming operations involving stretch flanging or hole expansion. This cracking phenomenon is known in the automotive industry as edge-cracking. Edge-cracking compromises part quality because it cannot be predicted from the part design stage and cracks could appear in part production when cutting parameters or coil properties change.

A significant amount of research work has been done on sheet metal mechanical shearing for conventional steel grades. However, the observed behaviour of AHSS is not always following the same trend as conventional steels where higher ductility means higher edge-crack resistance. It is accepted that edge-cracking resistance in AHSS is related to both the microstructure and the cut edge integrity (presence of cracks) [2, 3]. AHSS with high strength (≥800 MPa) are especially sensitive to the presence of cracks or defects at the sheared edge.

Traditional approaches to characterize formability and mechanical properties in sheet materials do not give sufficient accuracy and reproducibility to properly describe edge fracture and laboratory results cannot be transferred to industrial parts. FLC curves and tensile tests provide only predictions on global formability for restricted conditions like dome stretching but not for local formability modes such as stretch flanging, hole extrusion and tight-radius bending [4]. Local formability failure modes are an entirely different failure condition, where fractures occur from applied concentrated (local) deformation. AHSS grades showing high global formability (DP, TRIP) do not always show high local formability as well.

From the multiple tests available in the literature, the most common and the only standardized one to assess the formability of cut edges on punched sheets is the Hole Expansion Test, ISO16630 [5]. The value obtained in this test is the Hole Expansion Ratio (HER). Large variations in the HER of the same AHSS grades are well known from practical testing experience [6] and have become an obstacle in objectively evaluating the edge-cracking resistance of AHSS. Thus, ISO HET is not suitable for quality control method in a production line and other alternative edge-crack tests as KWI test, hole tension test, or others with the aid of Digital Image Correlation (DIC) systems have been developed. An extensive overview of edge-crack tests found in the literature is given in reference [7]. Nevertheless, and despite all the uncertainties about the applicability of ISO HET to characterize the edge-cracking resistance, the HET closely resembles industrial cases and the method is very easy and fast to perform.

In the present work, the applicability of different DIC-based methods, such as Half-Specimen Dome Tests, Sheared Edge Tensile Tests, and KWI hole expansion tests with flat nosed punch is explored and compared with classical Hole Expansion Tests. All these tests are also compared with a laboratory-scale demonstrator. Finally, the limitations and advantages of these different methods are discussed for the investigated steels.

2. Materials and methods

2.1. Materials

Three cold-rolled AHSS grades with thicknesses between 0.8 and 1.5 mm are analysed in the study: a dual-phase steel (DP1000) and two complex phase steels, CP1000HD and CP1200. Table 1 shows their mechanical properties extracted following ISO6892-1 standard [8] and using specimens type 2 with a reference gage length of 80 mm.

2.2. Experimental methods

2.2.1. Hole Expansion test (HET). These tests were performed according to ISO16630. Squared samples with an initial punched hole of 10 mm diameter in the centre were used (Figure 1a). According to standard recommendations, the hole was punched using a punch-to-die clearance of 12%. The hole expansion was performed using a conical expansion tool with a top angle of 60° (Figure 1b).

Table 1. Mechanical properties of the investigated steels. t, thickness; YS, yield stress; UTS, ultimate tensile strength; UE, uniform elongation; TE, total elongation; n, strain hardening coefficient at uniform elongation (UE); r, anisotropy at 4% elongation.

Steel	t [mm]	Direction	YS [MPa]	UTS [MPa]	UE [%]	TE [%]	n	r_4
DP1000	0.8	Rolling	762	1066	7.7	11.7	0.074	0.84
		45°	736	1048	7.9	11.8	0.076	1.10
		90°	735	1074	7.5	10.5	0.073	0.94
CP1000HD	1.5	Rolling	893	1052	7.3	11.1	0.071	0.91
		45°	905	1052	7.0	10.6	0.068	1.02
		90°	909	1062	7.1	10.7	0.068	0.96
CP1200	1.5	Rolling	1075	1208	5.2	8.3	0.051	0.88
		45°	1076	1206	5.2	7.8	0.051	0.96
		90°	1079	1215	5.1	8.3	0.049	0.92

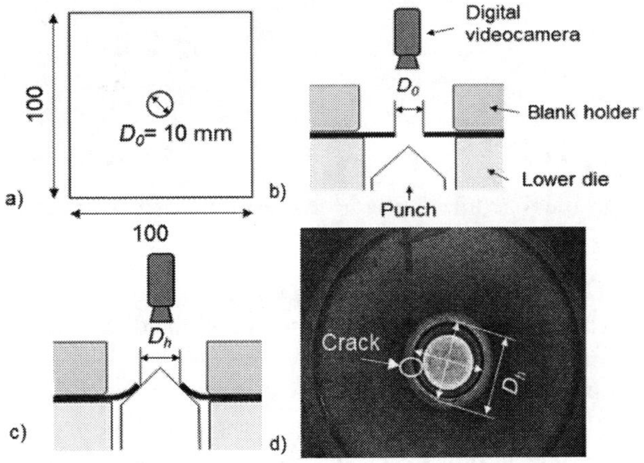

Figure 1. a) Typical specimen geometry used for hole expansion tests. Experimental setup for the determination of the HER before (b) and after (c) the test. d) Image for evaluation of D_h.

The tests were performed simultaneously in three laboratories to show the variability of the test due to coaxiality of the punching tool and the method for crack detection. In Lab1 the tests were conducted in a drawing press at a displacement rate of 0.3-0.5 mm/s without video instrumentation. The tests were stopped manually by naked eye and the measure was done after springback. Tests at Lab2 were performed in a universal testing machine at a displacement rate of 1 mm/s and were stopped after the first trough-thickness crack was observed. Crack occurrence was detected by using a high-resolution video camera (Figure 1b). Finally, a sheet metal testing machine was used by Lab3 at a displacement rate of 0.25 mm/s using a USB camera to judge the appearance of the first crack by the operator. A minimum of 3 specimens per steel were tested. The hole expansion ratio was obtained as follows:

$$\lambda = \frac{D_h - D_0}{D_h} \times 100$$

where λ is the HER, D_h is the hole diameter at crack detection and D_0 is the initial hole diameter. D_h was measured using a calliper by Lab1 and Lab3 in unloaded condition, and from the images of the video camera using a digital image analysis software by Lab 2 in loaded condition (Figure 1d).

2.2.2. KWI Hole Expansion test with flat nosed punch. The ISO HET, though well established and widely used, does by design exhibit a contact force on the edge, which is expected to influence the results, at least for some materials [9]. Sometimes by sheer geometrical influences there is a stretched flange which has no contact force on the flat surface of the sheet, like in the corner of a bodyside of a car. For that reason, the three materials were also subjected to flat nosed punch test.

The same squared samples of standard HET were used (Figure 2a). The hole expansion was performed using a flat nosed punch (Figure 2b). These tests were performed simultaneously in Lab1 and Lab3. In Lab1, same punch and press equipment as ISO HET were used and tests were also stopped manually by naked eye. The same speed as for conical punch was used (0.5 mm/s). Lab 3 performed all the tests in a universal sheet metal testing machine at a displacement rate of 0.1 mm/s. The tests were stopped after the first trough-thickness crack was observed. In this case, crack initiation was detected by using a high-resolution video camera and strain with DIC.

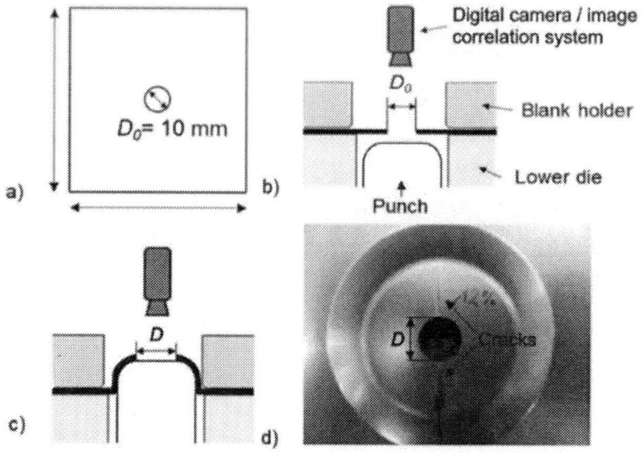

Figure 2. a) Typical specimen geometry used for KWI hole expansion tests. Experimental setup for the determination of the HER before (b) and after (c) the test. d) Image for evaluation of D.

It should be noted that the failure in a flat nosed punch test may originate in the bulk of the material rather than on the edge. This depends on the size of the hole, the clearance, and material properties.

2.2.3. SETi test. Sheared edge tensile improved, where improved is referring to the DIC measurement, was introduced in 2015 [10]. In these uniaxial tensile tests, the sample was a straight strip with both sides sheared, enabling the preparation easier and allowing much more freedom in setting the cutting clearance. The specimen width was 20 mm and the free length was 50 mm. The crosshead displacement rate was 0.5 mm/s and the DIC frequency was 15 Hz.

The tooling was laid out to function on a universal testing machine with good control over force and displacement as well as having good data acquisition. Furthermore, the results could be also compared to ISO and flat nosed punch hole expansion tests. As in previous cases, the cutting clearance used in the test was 12% to compare all the tests carried out. The shearing operation has been carried out at two speeds: 1 mm/s (low); and 100 mm/s (high).

2.2.4. Half Specimen Dome test (HSDT). This test was developed by Shih et al. [11] as an additional approximation to determine the edge cracking behaviour during flanging operations. Shih used it to analyze the edge quality of sheared, waterjet and laser cutting specimens. Compared to hole flanging tests, the half dome test is a more time and cost-saving test for evaluating edge cracking behaviour. HSDT is also unaffected by aspects relating to the manufacture of the tools for punching the holes, as the coaxiality between punch and die. Shih et al. also concluded that shearing angle, tool clearance and anisotropy will also affect edge cracking behaviour.

Figure 3a illustrates the HSDT tools, the same used to determine the FLC. To perform the tests, a DIC system has been used to detect the onset of the first through-thickness crack. The main outputs of HSDT are the dome height (or drawing depth/punch stroke), Figure 3b, and fracture strain of the sheared edge in the previous stage to first crack appears, Figure 3d. To perform HSDT tests, a double-acting hydraulic press was used. The press-holder was equipped with a circular rib to avoid material sliding. All tests were performed at 1 mm/s punch speed.

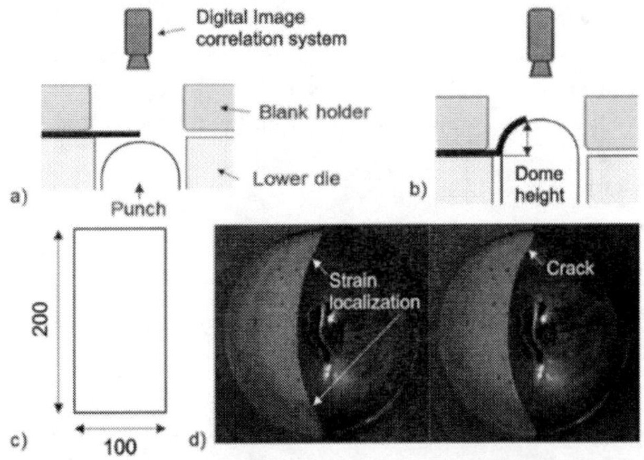

Figure 3. Nakajima tool before (a) and after (b) the test. Specimen geometry used for HSDT (c). DIC image used for evaluation of dome height and strain at previous stage to the appearance of crack (d).

During the tests, it was detected that, although the rib to prevent the specimen from slipping, in some cases this was not effective. In these cases, the dome height obtained is considered to be above the real one, so these values are not always reliable. Thus, in order to have results less dependent on the possible slipping of the specimen, in this article only the values of strain at fracture have been considered.

Regarding specimen identification, the code used for specimen identification is in accordance with the defined in ASTM E399 [12]. Three orientations were performed: L-T (0°), D-D (45°) and T-L (90°). The first letter indicates the orientation of the specimen geometry with respect to the rolling direction, and the second letter indicates the crack propagation direction.

2.2.5. Laboratory-scale demonstrator. The semi-industrial tests have been performed using a S-shape contour tool to test the crack sensitivity of the steels after an open cut. This test was performed in two steps. First, a rectangular blank is cut and after, a curved flange is formed, Figure 4. Depending on the position of the blank, different flange heights and elongations could be achieved. Bending and cutting were performed in the same tool changing the punch and die. Due to the concave contour of the central part of the tool the crack appeared in this area.

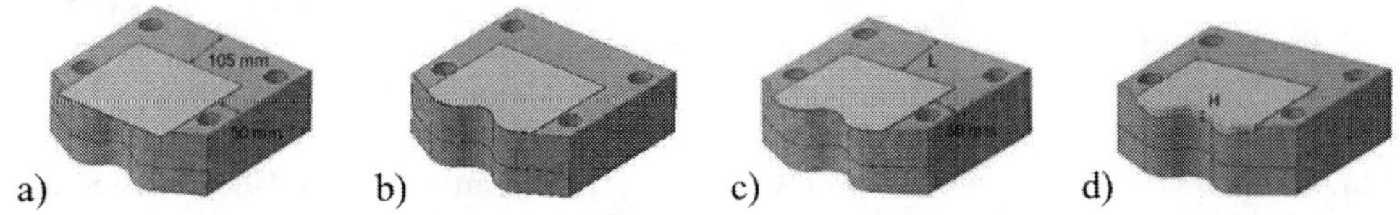

Figure 4. (a) Position of the rectangular blank on the cutting tool, (b) after cutting, (c) positioning the cutting specimen on the bending tool, (d) finished specimen (after flanging).

During cutting the blank is always positioned in the same way on the tool to have the same shaped blank for the further flanging operation. To achieve different flanging heights H, the distance L will be varied in steps of 0.5-1 mm until a crack occurred. To measure the elongation on the edge after forming, the blanks are marked with parallel lines (distance 5 mm) by etching, Figure 5a.

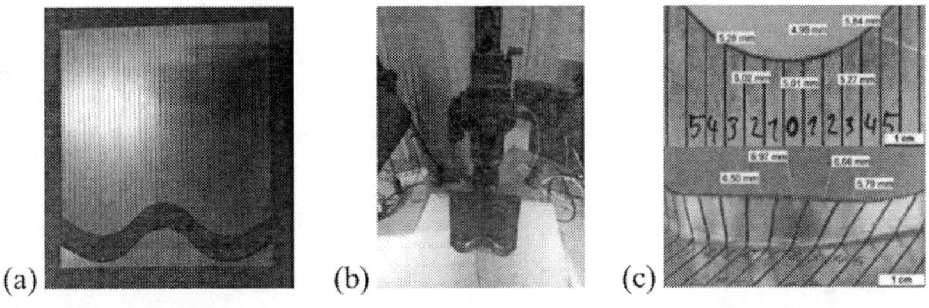

Figure 5. Marked blank with parallel lines (a). Camera setup (b) and optical measurement (c).

After cutting the edge, the length was measured on the blank between these lines. After the test, the deformed edge length was measured again. With these values, the elongation of the edge was calculated. The accuracy of these measurements is limited. The deformed specimen has a 3D contour which will affect the optical measurement. To minimize this failure the specimens were measured in different angles. Figure 5b shows the camera setup and Figure 5c the measurement before and after flanging.

3. Results and discussion

3.1. Hole Expansion tests (HET)

Table 2 shows the hole expansion ratio of the investigated materials in punched 12% clearance for the three laboratories and drilled conditions for Lab1. As clearly seen in this table, drilled HER values are significantly higher than punched ones.

Table 2. Hole Expansion Ratio for the investigated steels.

Steel	Driller conical HER, λ [%]		Punched Conical HER, λ [%]					
			Lab1		Lab2		Lab3	
	Mean	Std. dev.	Mean	Std. dev.	Mean	Std. dev.	Mean	Std. dev.
DP1000	48	7	37	4	27	2	34	3
CP1000HD	126	2	86	5	70	3	-	-
CP1200	137	8	67	5	65	5	59	9

Regarding the values of the punching HER, there are important differences between laboratories. The measurement in Lab2 was carried out with the sample loaded and probably overestimated the elastic contribution in comparison to the fully plastic unloaded condition (Lab1 and Lab3). On the other hand, by measuring the diameter directly on the tool with the video camera, the tests were stopped before in comparison to bare eye operator manual stopping. Thus, both effects should be compensated, and the results can be considered comparable, whereby this is not the main reason for deviations, and scattering probably comes mostly from punching operation.

3.2. KWI Hole Expansion test with flat nosed punch

These tests show that the three steels are not edge crack sensitive at analysed clearance, following in all cases the fractures plane strain paths. In all the tests, the crack starts after some pronounced necking away from cut hole edge, then it propagates to the hole edge. This is however not a classical edge crack. Additionally, the hole diameters have been measured using a calliper, following the ISO way, resulting in mild overshoot in case of large fractures, table 3.

As just said, table 3 illustrates the considerable difference between HER values obtained with ISO and flat nosed punch tests, and that the values are much more similar between the two laboratories than the standard HER tests. In this case, the values obtained with the flat nosed punch test are much lower than those of the standardized HET.

Table 3. Results of flat nosed punch tests with 12% clearance for the investigated steels.

Steel	HER flat nosed punch, λ [%]			
	Lab1		Lab3	
	Mean	Std. dev.	Mean	Std. dev.
DP1000	21	2	22	3
CP1000HD	31	1	-	-
CP1200	21	1	25	1

3.3. SETi test

The SETi tests are evaluated by finding the strain on the edge in the stage just before failure at the position of failure. The failures classified as abrupt are those where failure occurred in one stage of DIC system measurement, and it cannot be firmly considered as clear edge failure, table 4. Thus, if the strain on the edges for the abruptly failing specimens is trusted, then it seems fast cutting reduces edge ductility. The fuzziness of the results was judged to be mainly due to the uncertainty of where fracture has originated. This is inherent in the simple tensile loading of the test combined with material with relatively good edge ductility, i.e. high HET. The SETi test only worked well for DP1000.

Table 4. Edge failure (major) strain (including abrupt failures).

Steel	Slow cutting				Fast cutting			
	Logarithmic		Technical [%]		Logarithmic		Technical [%]	
	Mean	Std. dev.	Mean	Std. dev.	Mean	Std. dev.	Mean	Std. dev.
DP1000	0.336	0.029	40	3	0.327	0.009	39	1
CP1200	0.567	0.031	76	3	0.428	0.062	53	6

3.4. Half Specimen Dome tests (HSDT)

All HSDT values were extracted from the analysis of the results of the DIC measurements. To obtain the strain coordinates in the sheared edge a curve was made following them to determine the strain as a function of position. The maximum strain values were extracted from this curve, thus locating the two necking zones on the edge of the specimen during the HSDT test, just before the fracture, Figure 3d.

To analyse the strain path followed by the necking zone of HSDT specimens, also a point was situated in the cutting edge. By this point, it was observed that the strain path of the necking zone was uniaxial. Major strain extracted from HSDT is shown in table 5. The burr position was in die side (UP) in the same way as in the HER tests.

Table 5. Edge failure (major) strain in HSDT for the investigated steels.

Steel	Edge direction		Logarithmic		Technical [%]	
	Angle	E399	Mean	Std. dev.	Mean	Std. dev.
DP1000	0°	L-T	0.186	0.038	20	4
	45°	D-D	0.222	0.013	25	1
	90°	T-L	0.151	0.038	16	4
CP1000HD	0°	L-T	0.370	0.016	45	2
	45°	D-D	0.457	0.057	58	6
	90°	T-L	0.417	0.050	52	5
CP1200	0°	L-T	0.400	0.012	49	1
	45°	D-D	0.354	0.037	42	4
	90°	T-L	0.393	0.021	48	2

Table 5 shows the effect of anisotropy in major strain values. In all steels, there is a clear anisotropy effect. Thus, the higher values for DP1000 and CP1000HD are obtained with trimmed edge oriented at 45° from RD. For CP1200 steel, the results of 0° and 90° are quite similar and higher to those of 45°.

3.5. Laboratory-scale demonstrator

The average height and elongation over the rolling direction were calculated to compare the steels. The results of the flange height show a clear ranking. The DP1000 achieve the lowest and the CP1000HD the highest flange height without crack. The same trend is visible in the elongation, Table 6. This table does not show deviation in the results since only one specimen was tested per material and orientation, due to the arduous manual work involved in these measurements.

DP1000 shows the clearest crack occurrence of all materials. The achievable flange height is in diagonal the highest, as already stated by Unruh et al. for this type of steels [13]. Parallel and perpendicular have the same height. Overall, the achievable flange height is the lowest for DP1000.

CP1000HD achieves parallel to the rolling direction the biggest height. Elongation without crack is in diagonal direction slightly higher than parallel. CP1200 shows in the flange height for all three directions comparable values.

Table 6. Flange height and elongation for the investigated steels.

Steel	Edge direction		Results without crack		Results with crack	
	Angle	E399	Height [mm]	Elongation	Height [mm]	Elongation
DP1000	0°	L-T	6.67	0.166	7.53	0.195
	45°	D-D	7.35	0.169	7.70	0.174
	90°	T-L	6.70	0.146	7.40	0.168
CP1000HD	0°	L-T	17.70	0.502	18.80	0.760
	45°	D-D	16.80	0.521	17.95	0.588
	90°	T-L	15.61	0.391	15.95	0.600
CP1200	0°	L-T	13.40	0.391	14.30	0.512
	45°	D-D	12.94	0.321	13.30	0.468
	90°	T-L	13.25	0.320	13.35	0.330

3.6. Test comparison

To compare the different tests aimed at characterizing edge formability, the only one that is standardized, the ISO HET, was initially taken as a reference. To avoid introducing differences in tool design and measurements between the different laboratories, the results of the different tests carried out by the same laboratory are compared, except in the case of semi-industrial tests.

If the ISO HET test is compared with the HET flat nosed punch, the relationship is far from being linear, Figure 6a. This seems logical if one considers that in KWI test the fracture never appeared by the edge. When the comparison is made between the ISO HET and the SETi test, since data is only available for two steels, it is not possible to clearly analyse whether the results show similar trends, or not, Figure 6b. However, it is clear that the cutting speed affects to the maximum edge strain obtained for CP steel. For DP steel this effect is not observed.

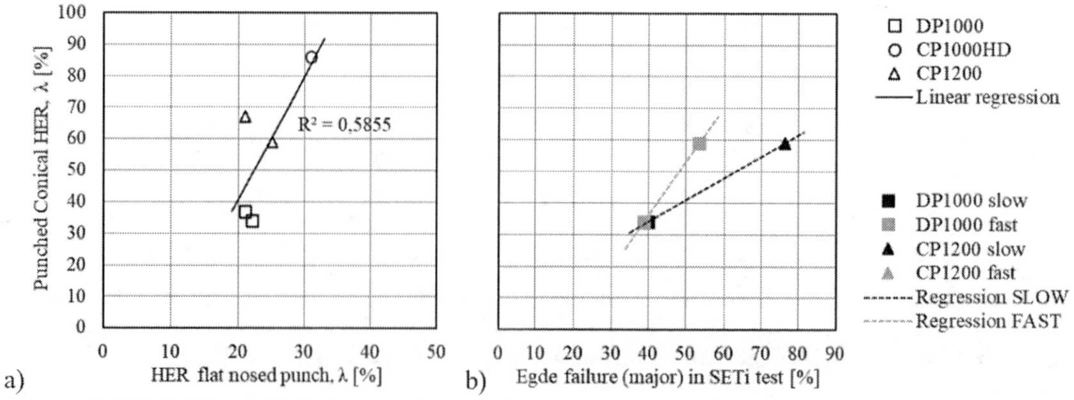

Figure 6. ISO HER vs. HER flat nosed punch (a). ISO HER vs. edge failure strain at SETi test (b).

In the case of the relationship between the ISO HET and HSDT tests (Figure 7), the linear relationship is clear, and it is even more evident when only the minimum strain obtained in the HSDT test for each material is represented, Figure 7b. The latter would better represent what actually occurs in the ISO HET, in which the crack appears in the orientation that presents the lowest resistance to edge-cracking.

Since the results between the ISO HET and the HSDT can be considered equivalent, the laboratory-scale demonstrator results are compared only with the HSDT values, Figures 8 and 9. Thus, same edge orientations relative to the rolling direction can be directly compared.

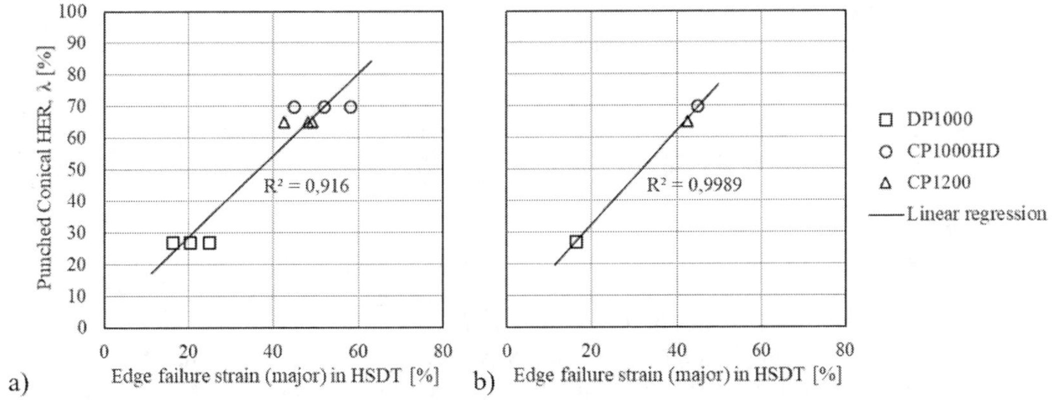

Figure 7. ISO HER vs. edge failure major strain at HSDT test with all data points (all orientations) (a) and only with the lowest value (one orientation) (b).

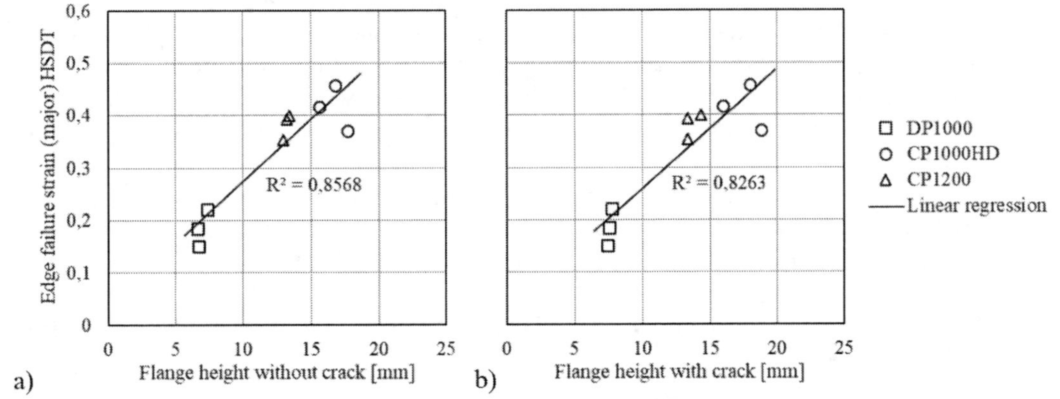

Figure 8. Edge failure major strain at HSDT vs. flange height without crack (a); and with crack (b).

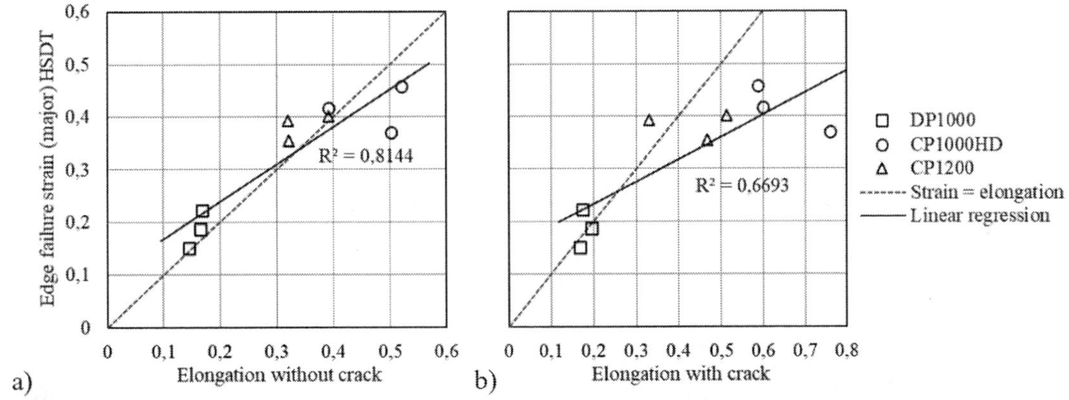

Figure 9. Edge failure major strain at HSDT vs. elongation without crack (a); and with crack (b).

This last comparison, made between the demonstrator specimens with and without cracks, highlights two facts. On the one hand, the relationship between the results of these two tests is clearer when comparing the values of the demonstrator without crack, such as the HSDT results, taken in the previous stage to fracture, Figures 8a and 9a. And on the other, that in this case, the greatest strain obtained in the HSDT is similar to the elongation obtained in the laboratory-scale demonstrator, Figure 9a. It should be considered, however, that the strain measurement and the detection of the previous stage to fracture are sufficiently different between both tests, which would justify the greater dispersion of results in the more industrial test, more evident as the strain increases.

4. Conclusions

After analysing the sheared edge formability of three cold-rolled advanced high strength steels through different tests and a demonstrator, the main conclusions that can be drawn are the following.

- The ISO HET presents notable differences in the results reported by different laboratories for the same steel. These differences, despite the different methods to detect the crack and measure the diameter, can be attributed to punching tool design and wear, nominal tool clearance, cutting punch speed and differences in punch and die coaxiality.
- The flat nosed punch test may be more applicable to certain processes than ISO HET, but failure occurs in the bulk material for the steels and clearance analysed, and then consequently edge condition cannot be assessed. If they compare to each other, seems to be the HER results in flat nosed punch tests are much lower than in ISO HET.
- HET flat test requires therefore more careful sample analysis in comparison to ISO conical test where crack almost always start at hole edge. By post-mortem sample inspection with or without necking it can be roughly estimated if the crack starts from the edge or from bulk material.
- SETi test is unsuitable to obtain clear local fracture strains in case the materials exhibit relative good local ductility. Due to this it could not be established if the trend in local fracture strains coincides with that in global fracture strains such as provided by ISO HET and flat nosed punch tests. To determine the influence of cutting speed further testing is required.
- HSDT test allows detecting the effect of material anisotropy in edge-cracking, which cannot be measured by HET. The ISO HET and HSDT show a linear relationship. Linearity that is more evident when only the minimum strain obtained in the HSDT for each material is represented.
- The results of the HSDT and laboratory-scale demonstrator tests show a clear relationship, especially when comparing of the demonstrator samples before the appearance of cracks. The lower-than-expected linearity may be caused by the difficulties with the measurement of the elongation on the edge due to the 3D contour of the flanged specimen. In this case, in which the strain measurement is manual and not automatic as in the HSDT, the accuracy depends on the experience of the operator.

It is necessary to comment, however, that these results are obtained from only 3 cold-rolled steels, for which it is necessary to extend them to more steels to confirm the observed trends.

Acknowledgments

The authors gratefully acknowledge the funding received by the Research Fund for Coal and Steel programme (847213-CuttingEdge4.0-RFCS-2018).

References

[1] Levy B S and Van Tyne C J 2012 *J. Mater. Eng. Perform.* **21** pp 1205-13
[2] Gutiérrez D, Escaler J, Lara A, Casellas D and Prado J M 2011 *Proc. IDDRG2011 Conf.* (Bilbao)
[3] Lara A, Frómeta D, Molas S, Rehrl J, Suppan C and Casellas D 2016 *Proc. IDDRG2016 Conf.* (Linz) pp 499-509
[4] World Auto Steel 2021 *AHSS Application Guidelines* version 7.0
[5] ISO16630 2017 Metallic materials – Method of hole expanding test
[6] Larour P, Pauli H, Freudenthaler J, Lackner J, Leomann F and Schestak G 2016 *Proc. IDDRG2016 Conf.* (Linz) pp 480-98
[7] Schneider M, Geffert A, Peshekhodov I, Bouguecha A and Behrens B A 2015 *Mater. Sci. Technol.* **46** pp 1196-217
[8] ISO6892-1 2019 Metallic materials - Tensile testing - Part 1: Method of test at room temperature
[9] Ooms B R J 2017 *MSc thesis, University of Twente*
[10] Atzema E and Seda P 2015 *Forming Technology Forum 2015* (Zurich)
[11] Shih H C, Hsiung C K and Wendt B 2014 *SAE Technical Paper* **2014- 01-0994**
[12] ASTM E399 Standard Test Method for Plane-Strain Fracture Toughness of Metallic Materials
[13] Unruh K and Heuse M 2017 *Steels in Cars and Trucks Conf.* (Amsterdam)

IOP Conf. Series: Materials Science and Engineering 1238 (2022) 012030 doi:10.1088/1757-899X/1238/1/012030

Blanking of Stainless Steel

I Pätzold, P Tröber, M Welm* and W Volk

Technical University of Munich, TUM School of Engineering and Design, Chair of Metal Forming and Casting, Walther-Meissner-Strasse 4, 85748 Garching, Germany

E-mail: markus.welm@tum.de

Abstract. Slug pulling, adhesion formation and edge fracture are major challenges in the sheet metal processing industry. They lead to a strong reduction in part quality, process stability as well as profitability. In this study, investigations are carried out on stainless steel X5CrNi18-10 to address the previous mentioned challenges. While slug pulling and edge fracture strongly depend on the geometric characteristics of active elements as well as the selection of process parameters, adhesion formation is mainly determined by temperature and thermoelectric currents. In this publication, the influence of the die channel geometry on the slug pulling effect and the part quality is investigated. Furthermore, the temperature profile over the shear cutting process as well as the resulting thermoelectric currents are determined for the test material. The relationship between edge crack sensitivity, shear cutting parameters and strategies is examined. These investigations thus form the basis for an improved understanding of the shear cutting of stainless steel.

1. Introduction

Blanking is the most widely used method for cutting sheet materials in the serial process due to its profitability. However, depending on the materials to be processed, special aspects must be taken into account. The blanking and forming of stainless steel for example poses great challenges to the sheet metal processing industry, for example slug pulling, adhesive wear and edge-fracture sensitivity.

Starting with the punching of small holes, slugs can be pulled out of the die channel after the separation process, especially when using high cutting rates [1]. In series production, this leads to marks on components [2] and the break-out of active element edges due to the blanking of double sheets [3]. The forces influencing slug pulling can be divided in two categories: Causing and hampering slug pulling. The former include burr clamping, vacuum and adhesive effects [4], the latter gravitation and friction. [5] was able to show, that the adjustment of the geometric properties of the active elements as well as the choice of process parameters can reduce slug pulling effectively.

Since stainless steel strongly tends to form adhesions, galling and adhesive wear are additional major challenges in its processing. A reduction of part quality, process stability as well as an increased risk of tool damage are possible consequences [6]. Therefore, adhesion formation has to be minimized, which is only possible by a profound knowledge of wear causing interactions as well as the associated influencing factors. In addition to the surface roughness of the punch, temperature plays an important role. Both should be as low as possible in order to reduce the amount of adhesions [7]. However, the dissipation of at least 60 % of the plastic work [8] leads to temperatures of up to 230 °C during blanking of stainless steel with a thickness of 2.5 mm [9].

Due to the Seebeck effect, the resulting temperature gradient in the tool causes thermoelectric currents, which also show a significant influence on adhesion formation [10]. Both the current strength and its direction, which are determined by the Seebeck coefficients of tool and sheet material as well as the temperature, show a significant influence on the amount of adhesion on the lateral surface of the punch [11].

The cut surface qualities resulting from the selected process parameters contributes to a large extent to the further processability of stainless steel. Component edges that are severely damaged by shear cutting often fail in further forming process steps due to edge cracks. Especially in blanking tools, which have low stiffness and manufacturing accuracy, displacements of the active elements occur under load, resulting in large deviations of the nominal die clearances and thus of the circumferential cutting surface qualities [12]. An adaptation of the shear cutting process by means of suitable process parameters or the application of two-stage shear cutting enables process-reliable further processing of shear cut edges [13].

The present investigations comprise experimental studies which, on the basis of adapted shear cutting parameters, offer easy-to-apply solutions for the processing of stainless steel under the aspects of slug pulling, adhesive wear and edge fracture.

2. Material

For the investigations, a stainless steel X5CrNi18-10 (1.4301) in different sheet thicknesses was used which, due to its resistance to acids and corrosion, is often used for the processing of components on white goods, but also in the pharmaceutical, food, construction or automotive sectors [14]. The metastable austenitic stainless steel stands out for its good formability, however high plastic deformations lead to a significant increase in strength [15]. The microstructure consists of a homogeneously distributed phase of austenite with linearly embedded delta ferrite and small amounts of deformation martensite. It has a yield strength $R_{p0,2}$ of a minimum of 230 MPa, a tensile strength R_m in a range of 540 to 570 MPa, and a minimum total elongation A_g of 45 % [16].

3. Tools and Methods

3.1. Shear Cutting Tools and Presses

For the experimental investigations several modular shear cutting tools and presses were used. The modular design of the tools allows the variation of a variety of process parameters. [4] gives a detailed description of the tool design with its special features for measuring the vacuum, burr clamping and adhesive force. The investigated die channels used in this experimental study are illustrated in Figure 1.

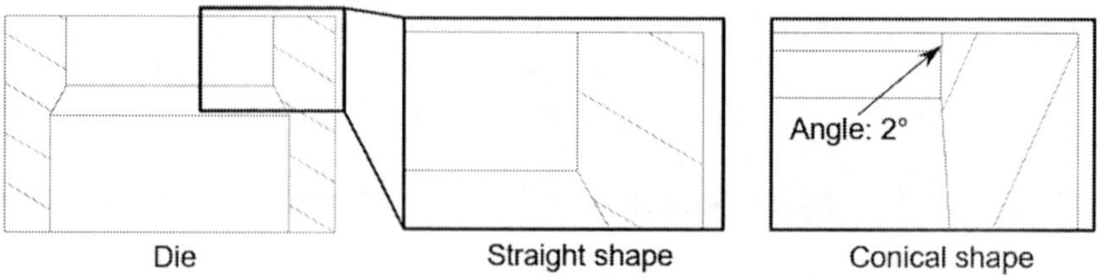

Figure 1. Investigated die channel geometries.

All experiments were conducted on servo-controlled modules of the stamping and forming machine GRM-NC of the company Otto Bihler Maschinenfabrik GmbH & Co. KG, Germany, which features high flexibility in adjustable stroke height and cutting speed.

The experiments on thermoelectric currents were performed on a special blanking tool, where the active elements and the sheet metal are electrically insulated by ceramic applications to prevent interfering signals. Its schematic structure is illustrated in Figure 2. In all experiments, the active elements consist of the cold working steel X155CrVMo12-1 (1.2379). The circular punch as well as the counter punch exhibited a diameter of 70 mm. The cutting edge radius was 50 μm. The HFA 3200 plus from Feintool AG, Switzerland, a hydraulic fine blanking press with a maximum press force of 3200 kN, was used for these experiments.

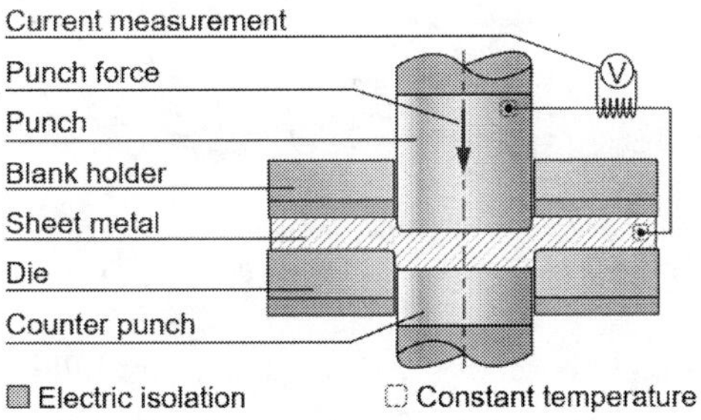

Figure 2. Schematic diagram of the blanking tool for investigating thermoelectric currents.

The edge crack investigations were performed on two different shear cutting tools, which differ mainly in the cutting line geometry. The one for producing the *Edge-Fracture-Tensile-Test* (*EFTT*) specimen according to [17] is shown in Figure 3 a) and described in detail in [19]. Based on the specimen geometry, both an open and closed cutting line can be realized in the tool. The initial geometry and the process steps for manufacturing the specimens are shown in Figure 3 b).

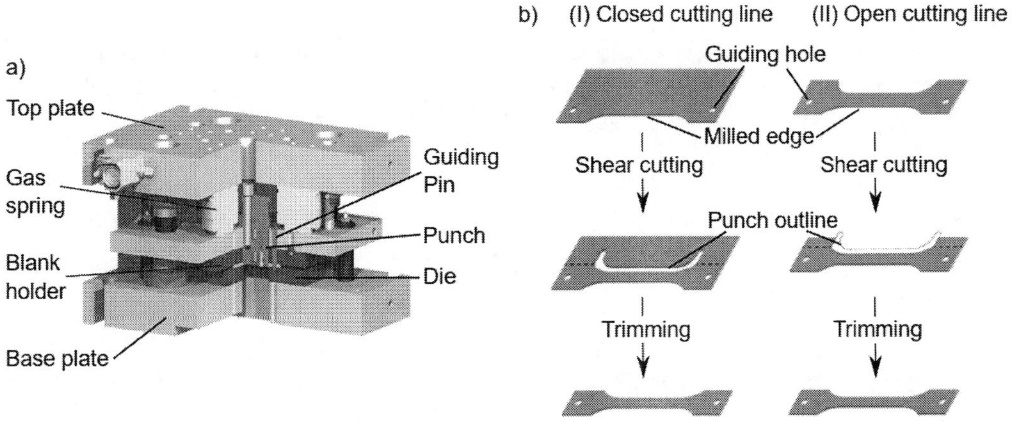

Figure 3. a) *EFTT* Tool design according to [17]. b) Manufacturing of *EFTT* specimen with an open (I) and closed (II) cutting line according to [18].

The second tool, shown in Figure 4 allows the centric placement of a hole in the very middle of the specimen for the hole tensile test as specified in [20]. Both tools were operated on a high-performance automatic stamping press BSTA 1600 from Bruderer AG, Switzerland, with a maximum press force of 1600 kN.

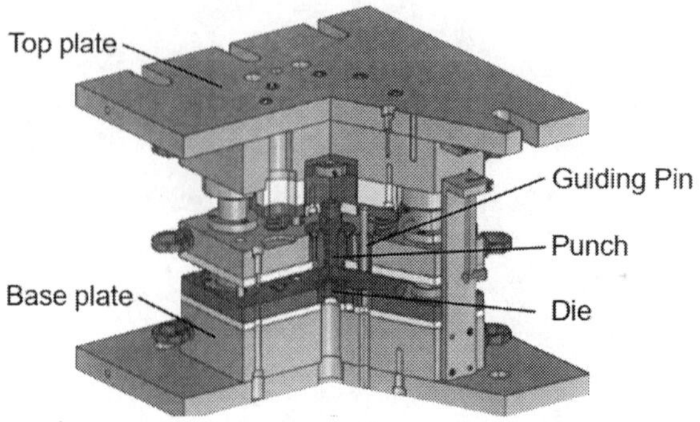

Figure 4. Shear cutting tool for manufacturing Open Hole Tensile Test specimen.

3.2. Measurement Equipment

An assessment of the die channel geometry with regard to its influence on slug pulling requires several data from the tool and the produced parts. The punch force as the most important value for determining frictional forces between slug and die is measured with a piezoelectric sensor in the cutting unit. A parallel tactile travel sensor gauges the punch travel. Both signals are recorded with a sample rate of 100 kHz. For classifying the part quality, the geometric characteristics of the shear cut surfaces are recorded by the digital microscope VKX-100, from Keyence, Japan. This microscope was also used for the initial inspection of the active elements.

Both the direction and the strength of the thermoelectric currents were measured using the K2 current clamp from Chauvin Arnoux, France, in a closed electrical circuit, consisting of punch, sheet metal and a connecting wire. In order to relate this data to the entire blanking process, occurring forces and the punch travel were also recorded.

To evaluate the edge crack sensitivity, the Aramis SRX contactless optical deformation analysis system from GOM mbh, Germany was used to capture the strain distribution which appears on the specimen surface during testing. This enables to differentiate between the failure types deformation fracture and edge crack [21]. Furthermore, it was used to determine the maximum major strain before failure of the specimen. The evaluation of the geometric characteristics of the shear cut surfaces was performed using the digital microscope VHX 2000, from Keyence, Japan.

3.3. Experimental Design

Figure 5 illustrates the experimental procedure, including all blanking parameters for determining the guidelines for processing stainless steel with regard to blanking. During the slug pulling experiments, the influence of two different die channel geometries on the cutting forces and the friction between slug and die, as well as the part quality, was investigated on 0.7 mm thick sheet material. While slug pulling is a common problem when blanking small geometries, the shear cut surface characteristics of the stamping grid serves as quality criterion. The two examined die geometries are a common one with a straight channel and a conical one with a channel angle of 2 °, corresponding to a compression of the slug of 0.14 mm at the maximal immersion depth of 1.85 mm. Every experiment is divided in three strokes: The first is the typical stroke up to an immersion depth of 1.4 mm, in which the material is separated. For the second stroke the stamping grid is removed and inserted distance sheets push the slug to a depth of 2 mm. For determining the frictional force between punch and guide bush, the third stroke is performed without stamping grid and slug, but with distance sheets. The force of the third stroke is then subtracted from the one of the second stroke. This enables an investigation of the frictional forces between slug and die without the influence of other friction during the blanking process.

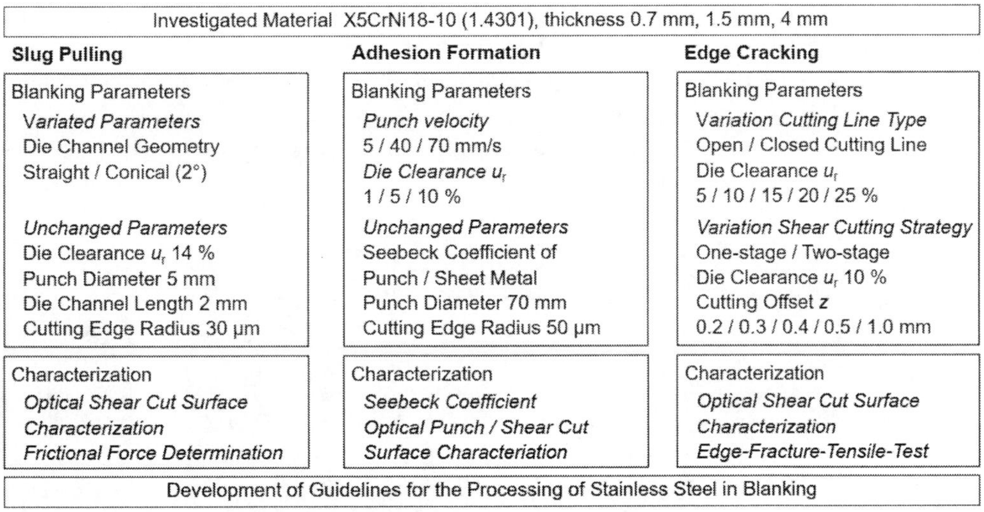

Figure 5. Experimental procedure.

Since the amount of adhesion formation increases with the naturally arising thermocurrent in case of the same current direction [11], this investigation focuses mainly on the influence of process parameters on these currents. Therefore, both die clearance and punch velocity was varied and the occurring thermoelectric current was measured. In order to relate the curve of the latter to the entire blanking process, the punch-, counterpunch- and blank holder forces as well as the punch travel were measured simultaneously. After nine consecutive strokes, three each per punch velocity, the amount of adhesions on the lateral surface of the punch was documented optically. All tests were repeated several times so that all values given were averaged over at least five strokes. In all the presented experiments, no lubricants were applied.

For the edge crack investigations the influence of cutting line type in combination with different die clearances as well as the shear cutting strategy, examining the influence of the width of cutting offset was varied on 1.5 mm thick stainless steel. To determine the influence of the cutting line type the *EFTT* was used. In order to investigate the influence of two-stage shear cutting on the residual forming capacity, hole tensile tests were performed with a hole diameter of 10 mm and a specimen width of 30 mm. All experiments were carried out with sharp-edged ($r < 10\,\mu$m) active elements. The large range of die clearances was chosen due to possible displacements in the series process. In a worst-case scenario, when the nominal die clearance is 10 %, this means that 0 % die clearance can occur on one side of the contour and 20 % on the other side. For the two-stage shear cut specimens, the die clearance for the first and second step was selected to be $u_{1st} = u_{2nd} = 10\,\%$. All specimens were tested transverse to the rolling direction. For both methods the tests were performed using optical measurement with a measuring frequency of 10 Hz. The frequency was doubled shortly before fracture (20 Hz), which enables to determine the crack initiation even more precisely.

4. Results and Discussion

4.1. Slug Pulling

4.1.1. Frictional forces

Varying the die channel geometry has a significant influence on the frictional forces between slug and die, shown in Figure 6 a). The straight die geometry triggers a maximum force of 495 N, right after the punch gets in contact with the slug. Subsequently, from 1.47 mm the force slightly declines due to leveling of roughness asperities and the slug starting to leave the die channel, which causes the area of contact between clean shear and die getting smaller. From 1.75 mm to 1.85 mm the slug is completely pushed out of the die channel and thus the force decreases rapidly.

With a conical channel geometry the force reaches 1127 N when the punch starts pushing the slug. In contrast to the straight geometry the force raises up to 1252 N at 1.69 mm. After that, the decreasing area of contact exceeds the effect of slug compression and thus the force drops. The overall higher friction forces with the conical die are caused by higher radial forces due to two mechanisms: First, a compressive stress is induced in the slug during the cutting process which causes a higher elastic spring back after material separation. Second, the compression of the slug increases with higher immersion depth, which enhances the slug oversize. With higher frictional forces between slug and die, the conical die provides a higher probability of preventing slug pulling.

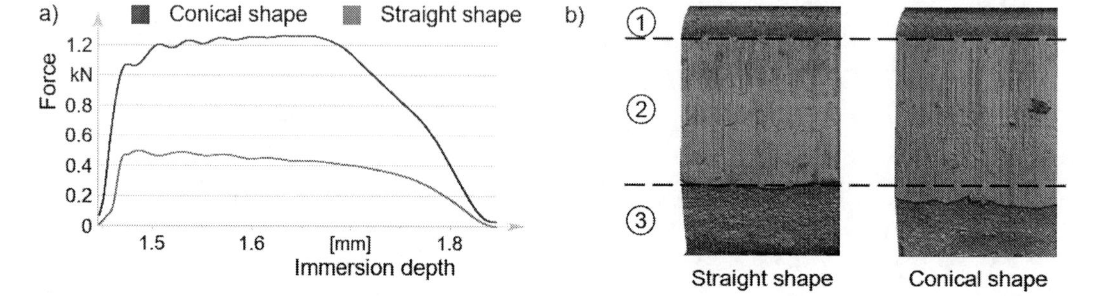

Figure 6. a) Frictional forces between slug and die from an immersion depth of 1.45 mm to 1.85 mm where the clean shear leaves the die channel. b) Shear cut surface characteristics for both die variants with roll-over (1), clean shear (2) and fractured surface (3).

4.1.2. Part Quality

A higher compression stress for the conical shape of the die channel also affects the shear cut surface characteristics, as illustrated in Figure 6 b). While the rollover is nearly constant with 14.7 % for the straight shape and 14.9 % for the conical shape, respectively, the clean shear proportion differs. It is 54.0 % for the straight channel compared to 60.8 %. Also the percentage of the fractured surface changes from 31.3 % to 24.4 %. This can be explained by the higher compression stress retarding the crack initiation. Furthermore, this stress condition changes the fracture angle from 23.5 ° for the straight geometry to 20.2 °. This shows that changing the geometry of the die to avoid slug pulling can significantly influence the component properties.

4.2. Adhesion Formation

Figure 7 illustrates the maximum thermoelectric currents, arising during blanking 4 mm stainless steel with the above-mentioned setup. Furthermore, it shows the lateral surface of the punch next to the cutting edge for the respective die clearance.

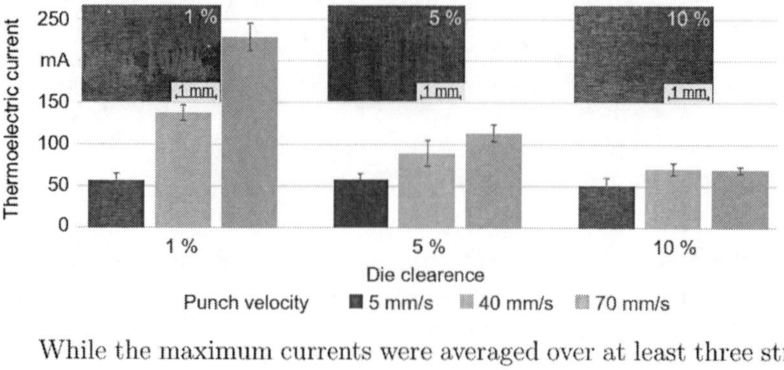

Figure 7. Thermoelectric current depending on the die clearance and punch velocity as well as the lateral surface of the punch

While the maximum currents were averaged over at least three strokes, the punch surface was photographically documented after nine consecutive strokes. The positive current values indicate a technical current direction from punch to sheet metal. This can be traced back to the Seebeck coefficients of punch and sheet metal since these alone determine the thermoelectric current direction. With regard to adhesion formation, this direction has a reinforcing effect [11]. The maximum thermoelectric current rises significantly with increasing punch velocity and decreasing die clearance. For a die clearance of 1 %, the maximum thermocurrent in case of the lowest punch velocity amounts to 56.8 mA. The maximum current rises to 137.5 mA for 40 mm/s and to 228.6 mA for 70 mm/s, which can be attributed to the increasing temperature due to a reduced time for heat equalization [10]. In sum, the lowest velocity reduces the thermoelectric current by 75.2 % compared to the highest one. However, the influence of punch velocity decreases with an increasing die clearance. While the reduction amounts to 49.5 % for a die clearance of 5 %, it is only 26.4 % for 10 %. Furthermore, almost no difference in the thermocurrent occurs between a velocity of 40 mm/s and 70 mm/s in the latter case. The decreasing impact of velocity can be attributed to the lower clean cut portion, which results in a decreasing amount of conducted forming work and thus a lower process temperature. This effect is enhanced by the growing shear zone in which the dissipated energy is distributed. When comparing the lateral surfaces of the punch between all die clearances, it can be seen that a die clearance of 1 % leads to the highest amount of galling. Almost the entire area at the cutting edge is covered with adhesion. In case of 5 % die clearance, a considerably less amount of adhesion occurs. However, adhesions, which are formed primarily on the grinding marks, can still be recognized. A die clearance of 10 % exhibits the lowest amount of galling, with only a few local adhesions. The results show that the thermoelectric current strength qualitatively correlates with the amount of adhesion, which increases significantly with smaller die clearances due to higher surface pressures and temperatures.

4.3. Edge Fracture

4.3.1. Variation of Cutting Line Type

Figure 8 shows the influence of an open and closed cutting line as well as different die clearances on the major strain distribution of the specimen surface before failure and the shape of the cut surfaces by means of single-stage *EFTT*. It can be seen that the way of strain distribution strongly depends on the shear cutting parameters used. All specimens produced with a closed cutting line fail from the edge in the range of die clearances from 5 to 25 %. Specimens with die clearances $u_r < 15\,\%$ are particularly sensitive and show typical edge cracks. Here, a strain hot spot forms at the shear cut edge and the specimen fails at very low strains without global necking. For the open cutting line, only specimens produced with die clearances $u_r > 15\,\%$ fail from the edge, specimen with $u_r = 25\,\%$ due to edge crack.

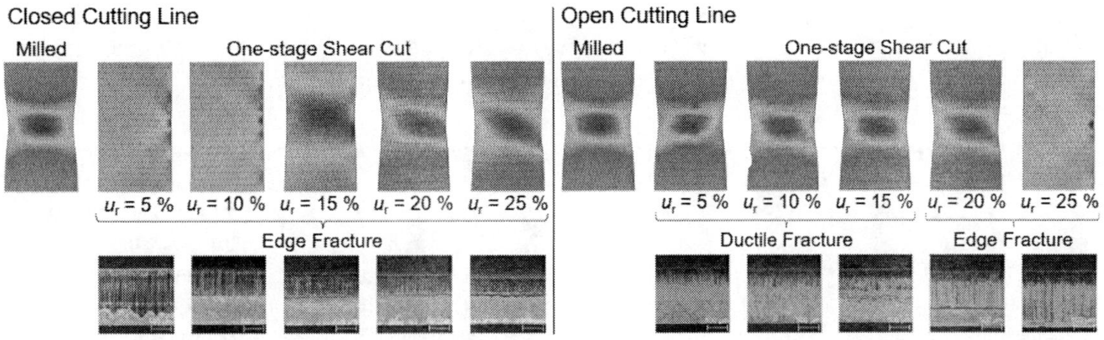

Figure 8. Strain distribution before fracture and photographic documentation on *EFTT* specimens for open and closed cutting line with variable die clearance.

The reason for this is the high proportion of clean shear as well as the strong burr formation, which causes high notch stresses. The advantage of the open cut over a closed cutting line is in particular the lower stiffness of the slug. The slug can bend downwards the punch face and thus provokes a tensile stress at the punch edge. This results in an earlier fracture, a reduced roll-over and clean shear zone and, consequently, a significant reduction in strain hardening and damage within the shear affected zone. As Figure 9 shows, selecting an open cutting line with adapted shear cutting parameters ($u_r = 5\,\text{to}\,15\,\%$) leads to almost no reduction in the forming capacity of the component edge.

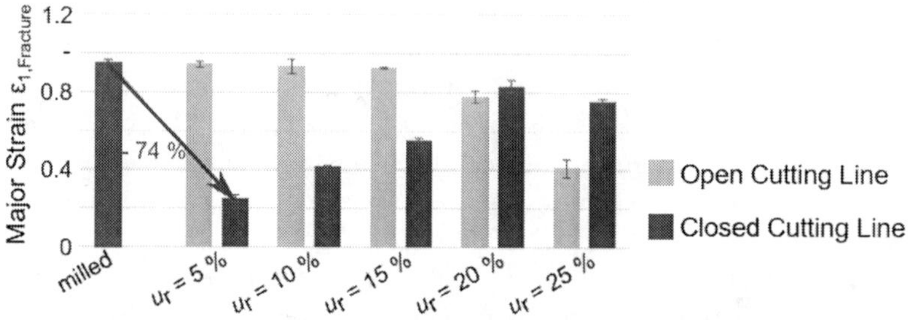

Figure 9. Major strain before failure determined in *EFTT* for different cutting line types and die clearances.

In the case where closed cutting lines are required for production purposes, further processing of the shear-cut specimen edges is considered critical in the case of one-stage shear cutting. Deviations from the die clearance in the tool under load can reduce the edge forming potential by up to 74 %.

4.3.2. Variation of Shear Cutting Strategy

To avoid the risk of edge cracks, especially in the closed cut, two-stage shear cutting has already been successfully applied for high strength steels in [13]. When using two-stage shear cutting, in the first shear cutting process, a hole smaller by the width of the cutting offset z is precut at critical points of a punching process. The scrap in the first step is fixed circumferentially and is therefore displaced mainly in vertical direction. This leads to a strong work hardening in the area of the later shear cut edge. In the second shear cutting process, only a ring with the width of the cutting offset z is removed. The scrap is free to move in vertical and radial direction. The hardening resulting from the first step is thus removed, and the deformation occurring in the shear affected zone of the second step is almost completely transferred to the scrap if the process parameters are selected appropriately.

For the stainless steel X5CrNi18-10, the residual forming capacity could also be increased in the investigations by applying a two-stage shear cutting process. For single-stage punching, the residual forming capacity for the investigated range increases with larger die clearances. Figure 10 shows that with sharp active element edges, die clearances in the first and second step of $u_{1st} = u_{2nd} = 10\,\%$ and cutting offsets in the range between 0.3 mm and 1.0 mm, the forming capacity of two-stage shear cut edges almost reaches that of the milled specimens. The photographic documentation shows that for single-stage shear-cut specimens with hole diameters of $D = 10\,\text{mm}$, the clean shear portion decreases with increasing die clearance.

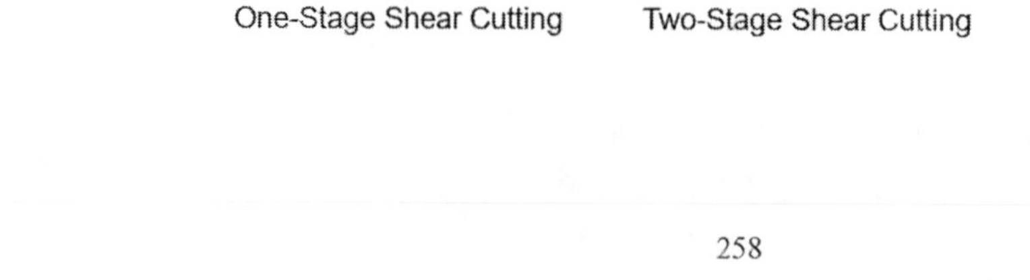

Figure 10. Major strain before failure determined in hole tensile tests for different shear cutting strategies and cutting offsets.

IOP Conf. Series: Materials Science and Engineering 1238 (2022) 012030 doi:10.1088/1757-899X/1238/1/012030

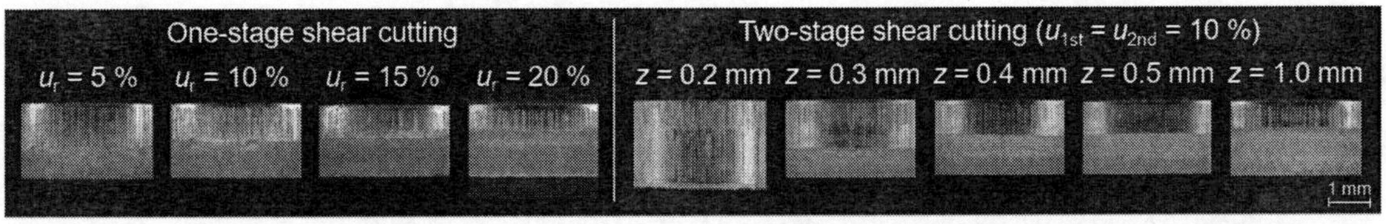

Figure 11. Photographic Documentation of punched specimen with a diameter of $D = 10$ mm under one-stage and two-stage shear cutting.

This leads to a lower work hardening in the shear affected zone and thus to a higher residual forming capacity. The shear cut surface of the two-stage shear cut specimen with a cutting offset of $z = 0.2$ mm has a very high proportion of clean shear as well as a large burr, which leads to a strong reduction in the achievable forming capacity. Although the proportion of clean shear is higher than in the single-stage punching, a cutting offset of $z = 0.3$ mm allows an increase in the residual forming capacity, which is due to the fact that most of the deformation from the first step of punching ends up in the slug of the second process step, as described in [13]. For increasing widths of the cutting offsets, a lower clean shear percentage is achieved than in the single-stage trim. This also correlates with the achievable major strains in the hole tensile test.

5. Conclusions

The following conclusions can be drawn for the investigations carried out on slug pulling, adhesion formation and edge fracture using the stainless steel X5CrNi18-10 (1.4301):

The die channel geometry has significant influence on the frictional forces between slug and die. Consequently, the right choice of the geometry helps to reliably prevent slug pulling. A conical shaped die channel with an angle of 2° enhances the frictional force by a factor of 2.5. Changing the stress conditions during the cutting process due to other die channel shapes affects the shear cut surface characteristics and thus part quality. Especially the crack initiation is affected and thus the amounts of clean shear and fractured surface.

With regard to the thermoelectric current, a strong influence of the punch velocity could be revealed in case of a small die clearance. For a relative die clearance of 1 %, the current strength is about four times higher in case of the highest punch velocity, compared to the lowest one. With an increasing die clearance, this influence decreases and becomes negligible in case of 10 %. With regard to adhesion formation, the current strength qualitatively correlates with the amount of adhesion on the lateral surface of the punch. Therefore, the process parameters should be chosen appropriately in order to keep thermoelectric currents as low as possible. In the presented setup, the results suggest a die clearance of at least 5 %. Compared to 1 %, similar shear cut surface qualities could be reached, while adhesion formation and thus galling on the lateral surface of the punch could be reduced significantly.

Single-stage shear cutting of stainless steel with open cutting line should be performed with die clearances of 5 to 15 % to reduce the risk of edge cracks occurring. The reason for this is a bending of the waste and thus tensile stresses at the punch edge which cause an earlier crack initiation and consequently lower work hardening in the shear affected zone. In single-stage shear cutting, the lowest possible smooth cut percentage should be achieved. In the event that a closed cut is required for production reasons, it is advisable to use a two-stage shear cutting process to increase the edge formability. The two-stage shear cutting allows the material to flow radially and vertically, removing the hardening from the first step and shifting nearly all of the work hardening from the second shear cutting process to the scrap. For a hole diameter of $D = 10$ mm cutting offsets between 0.3 and 1.0 mm allow a significant increase in residual formability.

References

[1] Tittel V and Bernadic L (2012) *A review of methods and precautions against slug pulling*; International Journal of Pure and Applied Sciences and Technology **12(1)**

[2] Dannemann E, Sugundo S, Lange K (1981) *On the Adhesion of Blanks to the Punch in Blanking and Piercing* Annals of the CIRP **30/1/1981** doi:10.1016/S0007-8506(07)60917-2
[3] Strasser F (1963) *Wie vermeidet man das Hoch-kommen von Butzen und Ausschnitten?* Werkstatt und Betrieb **96** pp. 319-320
[4] Kindsmüller A, Schrepfer A, Stahl J, Pätzold I, Nürnberger A, Golle R, Volk W (2021) *Influence of cutting parameters on mechanisms causing slug pulling* Production Engineering **15** doi:10.1007/s11740-021-01061-3
[5] Stahl J,Pätzold I, van den Bosch L, Kindsmüller A, Golle R, Volk W (2021) *The Frictional Force between Slug and Die in Shear Cutting after Material Separation* Key Engineering Materials **833** doi:10.4028/www.scientific.net/KEM.883.285
[6] Golovashchenko S (2006) *A study on trimming of aluminum autobody sheet and development of a new robust process eliminating burrs and slivers* International Journal of Mechanical Sciences **12** pp. 1384–1400 doi:10.1016/j.ijmecsci.2006.07.007
[7] Tröber P, Welm M, Weiss H A, Demmel P, Golle R, Volk W (2021) Temperature, thermoelectric current and adhesion formation during deep drawing WEAR **477** doi: 10.1016/j.wear.2021.203839
[8] Talonen J, Hanninen H (2007) *Energetics of plastic deformation of metastable Austenitic stainless steel* Steel Res. Int. **78** pp. 260–265 doi:10.1002/srin.20070588
[9] Tröber P, Welm M, Weiss H A, Demmel P, Golle R, Volk W (2019) The influence of process parameters and sheet material on the temperature development in the forming zone Manufacturing Review **6** 9 doi:10.1051/mfreview/2019005
[10] Tröber P, Demmel P, Golle R, Volk W (2017) On the influence of Seebeck coefficients on adhesive tool wear during sheet metal processing CIRP Ann. -Manuf. Technol. **66** pp. 293–296 doi:10.1016/j.cirp.2017.04.116
[11] Welm M, Tröber P, Weiss H A, Demmel, P, Golle R, Volk W (2020) Thermoelectrically Based Approaches to Reduce Adhesive Wear During Blanking JOM - The Journal of The Minerals, Metals Materials Society (TMS) **72**
[12] Larour P, Freundenthaler J, Essbichl R, Kerschbaum M (2017) *Influence of shear cut edge condition on conical hole expansion ratio* SCT2017 5th International Conference on Steels in Cars and Trucks 2017, Amsterdam-Schiphol, Netherlands
[13] Gläsner T, Sunderkötter C, Hoffmann H, Volk W, and Golle R (2017) *Development of a 2-stage shear cutting-process to reduce cut-edge-sensitivity of steels* Journal of Physics Conference Series **896** doi:10.1088/1742-6596/896/1/012104
[14] Rana R (2018) *High Performance Ferrous Alloys* Springer doi:10.1007/978-3-030-53825-5
[15] N N (2013) *Handbook of Stainless Steel* Outokumpu Oyj, Finland
[16] DIN EN 10088-2 (2022) *Stainless steels - Part 2: Technical delivery conditions for sheet/plate and strip of corrosion resisting steels for general purposes*; Beuth Verlag GmbH: Berlin, Germany
[17] Feistle M, Krinninger M, Pätzold I, Volk W (2015) *Edge-Fracture-Tensile-Test, 60 Excellent Inventions in Metal Forming* Springer Chapter 30, pp. 193-198 doi:10.1007/978-3-662-46312-330
[18] Feistle M, Golle R, Volk W (2015) *Determining the influence of shear cutting parameters on the edge cracking susceptibility of high-strength-steels using the edge-fracture-tensile-test* 48th CIRP Conference on Manufacturing Systems - CIRP CMS 2015
[19] Feistle M, Krinninger M, Pätzold I, Stahl J, Golle R, Volk W (2017) *Design and conceptualization of a cutting tool to investigate the influence of the shear cutting process on edge crack sensitivity* Journal of Physics **896** doi:10.1088/1742-6596/896/1/012106
[20] Feistle M, Pätzold I, Golle R, Volk W (2019) *Open hole tensile tests for the determination of the edge-crack sensitivity of sheared holes dependent on specimen geometry, cutting parameters, and the notch factor* 18th International Conference on Sheet Metal **29** pp. 412-419 doi:10.1016/j.promfg.2019.02.156
[21] Pätzold I, Feistle M, Golle R, Hoffmann, Volk W, Frehn A, Ilskens R (2017) *Identification of Edge Crack Sensitivity on Shear Cut Component Edges made of High-Strength Materials using suitable Testing Methods* 5th International Conference on Steel in Cars and Trucks

A Study of the Boundary Conditions in the ISO-16630 Hole Expansion Test

A Barlo[1], M Sigvant[1,2], L Pérez[3], M S Islam[1], and J Pilthammar[1,2]

[1]Department of Mechanical Engineering, Blekinge Institute of Technology, Karlskrona, Sweden
[2]Volvo Cars, Dept. 81110 Strategy & Concept, Olofström, Sweden
[3]RISE IVF AB, Component Manufacturing Unit, Olofström, Sweden

E-mail: `Alexander.Barlo@bth.se`

Abstract. As new and more advanced sheet metal materials are introduced to the market, more accurate techniques for determination of failure limits are needed. One area that needs attention is edge formability, where the ISO-16630 standardized Hole Expansion Test currently is used to express this through the Hole Expansion Ratio. Over the years, this standard has been criticized for producing a large scatter in repeated tests. This paper investigates a new setup for the Hole Expansion Test which introduces draw beads into the setup to ensure sufficient restraining of the specimen during the test in an effort to reduced the scatter. In total 62 tests of a DP800 steel alloy were executed, but a large scatter in the results were still seen. It was therefore concluded that a lack of restraining force in the Hole Expansion Test was not the primary cause of the reported scatter seen in other tests.

1. Introduction

With the increased focus on combating the rising environmental issues in the world today, a key goal for most automotive manufacturers is to reduce the carbon-footprint of a vehicle throughout its lifetime. One way to do so is by introducing lightweight and eco-friendly materials into their products. These new materials are however much more advanced than their predecessors and therefore much more difficult to validate in the design and manufacturing feasibility analysis stages. One of the large challenges these new materials bring, is how to accurately predict edge fractures caused by straining of punched holes during the forming process.

In the past decade and more, the Hole Expansion Ratio (HER) has been used as a measure for the edge formability of materials. The HER value is determined through the ISO standardized Hole Expansion Test [1] where a conical punch is driven through a hole in the test specimen, with a diameter of 10 mm, until a through-thickness crack appears. The HER value is then determined as presented in Eq. 1, where D_h is the average hole diameter after rupture, and D_o is the original hole diameter.

$$\lambda = \frac{D_h - D_o}{D_o} \times 100 \tag{1}$$

Over the years, the Hole Expansion Test has however been criticized by several authors, including Schneider et al. [2] and Larour et al. [3], for producing a high scatter when repeating the test on identical material from the same batch. Some authors, including Schneider et al. [2] and Chiriac & Chen [4], have attributed this to the operator reliant post-processing of the test, however, this paper work with a different hypothesis: That the scatter is caused by insufficient restraining of the material during the test.

2. Experimental Setup and Specimen Preparation

In order to test if the restraining of the material during the Hole Expansion Test has an impact on the reported scatter, a new experimental setup was designed. In order to restrain the material, draw beads were introduced to the setup, where a die set from the Nakajima test setup, normally used to determine Forming Limit Curves, was used. With the application of an already existing die set for the Nakajima test setup, another benefit is also that the 3D Digital Image Correlation (DIC) system ARAMIS from GOM GmbH is embedded in the setup. According to Larour et al. [3] it should however be noted, that 3D DIC cannot be applied to materials where $\lambda > 70\%$ since the cameras in the ARAMIS 3D DIC setup fails to focus once a punch displacement of 50 mm has been passed.

The main criteria to the punch according to the ISO standard is that the cone of the punch should have a tip angle of $60.00 \pm 1°$. This criteria was taken into the design of the new punch, where the punch was scaled up to have a diameter of Ø100 mm to fit in the Nakajima die. Another change that was implemented in the new punch is that the tip of the punch is cut away, so that the top of the punch is circular with a diameter of Ø8 mm. This was done in order to provide the cameras for the 3D DIC with the most optimal conditions, and since the initial hole diameter of the test specimen is Ø10 mm, this should have no influence on the test results. The new experimental design can be seen in Figure 1.

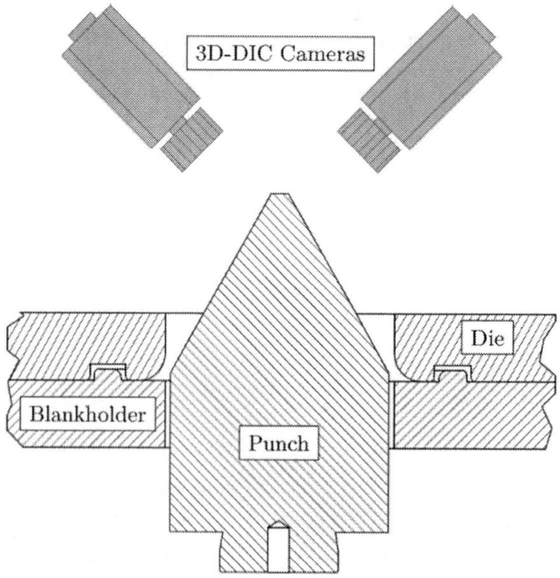

Figure 1. Cross-section view of the new experimental setup introducing draw beads and DIC system to the Hole Expansion Test.

Another change that was made from the ISO standard Hole Expansion Test is the shape of the blank. In the standard, the specimen is exemplified with a square blank, but in order to fit into the new setup with the die and blankholder from the Nakajima setup, the shape of the blank is changed to be circular. A requirement for the blank geometry is that "*The test piece shall be flat and of such dimensions that the centre of any hole is not less than 45 mm from any edge of the test piece...*" [1]. For this reason, the full-blank Nakajima geometry (Ø200 mm) was chosen, which ensures that the centre of the hole is 100 mm from the edge of the specimen. The geometry of the modified blank geometry can be seen in Figure 2.

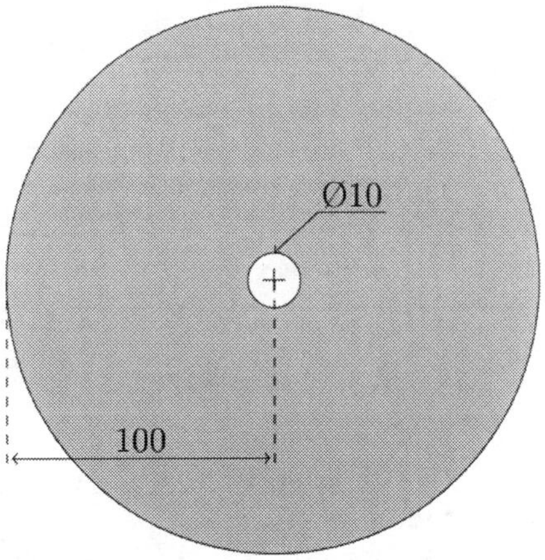

Figure 2. Modified blank geometry. All measurements in millimetres.

The cutting of the centre hole was performed using the punching operation, as specified in the ISO standard. This study focuses on a DP800 steel alloy with a sheet thickness of 1.2 mm, why a cutting clearance of 12 ± 1 % was used as in agreement with the standard. This resulted in an inside diameter of the die of $d_d = 10.288$ mm. Once the hole has been punched, a stochastic pattern is applied to the specimen to prepare it for DIC measurements. The pattern is applied so that the exit surface of the punched hole faces the die, thereby ensuring that the burrs created by the punching process faces away from the punch.

In order to ensure that the centre of the hole was aligned with the centre of the punch, a guide pin, mounted in a drilled hole on top of the punch, was used. The guide pin system can be seen in Figure 3 and a blank with formed draw bead can be seen in Figure 4.

Figure 3. Guide pin used for ensuring the centering of the hole.

Figure 4. Pre-formed blank after removal of guide pin and with draw beads formed.

In order to form the draw bead, the rule of volume consistency dictates that the material used for this must be taken from somewhere else. Therefore, in order to ensure that the dimension of the centre hole was not changed radically during this "pre-forming" a test was conducted without the punch, so that only the draw bead was formed. A manual measurement of the blank before the forming showed that the initial hole diameter was $D_o = 10.013$ mm and the hole diameter after forming of the draw bead was $D_h = 10.047$ mm. This results in an increase of the diameter of $\lambda_{pre} = 0.34$ %, which is deemed negligible compared to the spread of the HER-values and to not influence the results of the Hole Expansion Test.

3. Experimental and Post-Processing Procedure

By moving the setup into a single-action Wemhöner hydraulic press with a press capacity of 1300 tons, another deviation from the ISO standard needs to be done. In the standard, a testing speed of 1 mm/s is specified, however, the minimum speed of the utilized press is 5 mm/s. Another challenge when moving to a hydraulic press is how to determine when a through-thickness crack has appeared. In the Nakajima test setup, the operator does not visually observe the crack, but detects it by listening, and once the crack is heard, the test is terminated. The edge cracks in the Hole Expansion Test are not possible to detect by listening for the particular material tested, why a certain point in time where the test should be terminated is difficult to determine. This causes a risk that too deep of a draw could damage the tooling. Therefore, a mechanical stop was implemented in the press ensuring a fixed depth of the drawing. Based on initial experiments, a maximum drawing depth of 28 mm was chosen.

Another factor, and an important one in this study, is the restraining force applied to the specimen. In the standard, the following is specified about the clamping force: "*Apply sufficiently high clamping force to the test piece to prevent any material draw-in from the clamping area during the test*"[1]. The term 'a sufficiently high clamping force' is open to interpretation, but an example is given in the standard stating that a clamping force of 50 kN or higher should be sufficient for a test piece with the dimension 150×150 mm. As this investigation focuses on a DP800 steel grade, the blank holder force used in the press was increased significantly to a starting force of 330 kN, and the force increases as the tests proceeds to more than 380 kN. This scenario is exemplified in Figure 5. With the combination of the draw bead and this level of blank holder force, it is believed that no draw-in will occur during the testing.

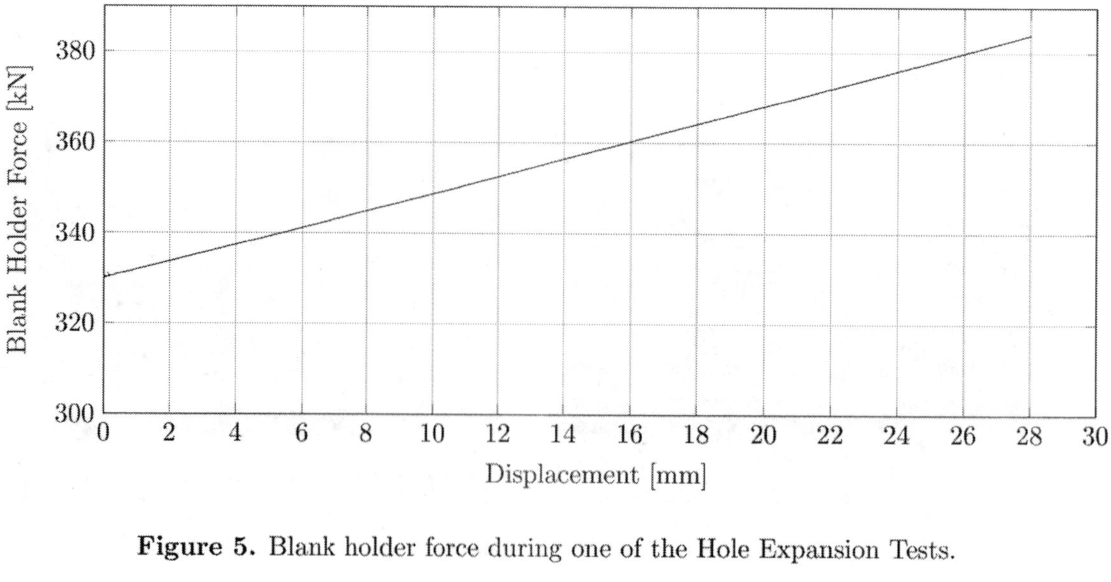

Figure 5. Blank holder force during one of the Hole Expansion Tests.

Fixing the draw depth is not without consequences. The obvious issue is that the final specimen will in the vast majority of the cases have passed a point where a through-thickness crack first appears. In the same way, cases where a through-thickness crack has yet not presented itself could occur. In the case of the latter, not much can be done in the setup, except changing the mechanical stop, and try again. In the first case, the general methodology presented by Chiriac & Chen [4] can be utilized, where the Hole Expansion Test was recorded with a digital camera, and a digital measurement of the hole diameter occurs at the frame that shows the first through-thickness crack.

Therefore, by utilizing the images captured by the 3D DIC system (with a sampling frequency of 60 Hz), a measurement of the hole diameter of any given stage in the test is possible, and should result in a reduction of the uncertainties introduced by the operators reaction time. Figures 6 and 7 presents a close up of the collar on a Hole Expansion Test at two different stages with a draw depth difference of approximately 0.17 mm. In Figure 6 it can be seen that the crack has been initiated in the outer diameter, but has not yet propagated all the way through the thickness. Figure 7 presents the state of the collar two stages later, and here a full through thickness crack is present.

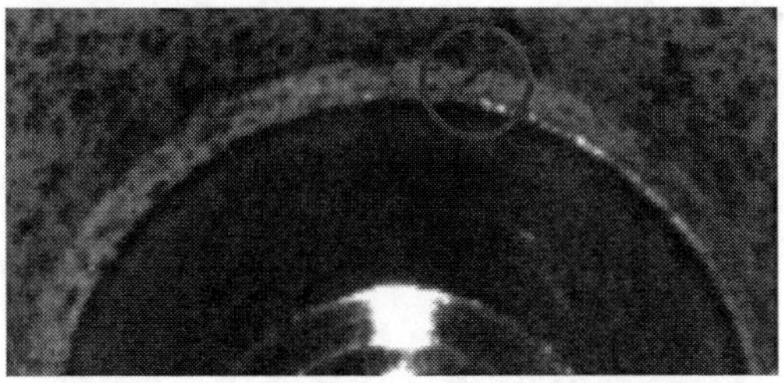

Figure 6. Close up of collar on test exhibiting crack initiation.

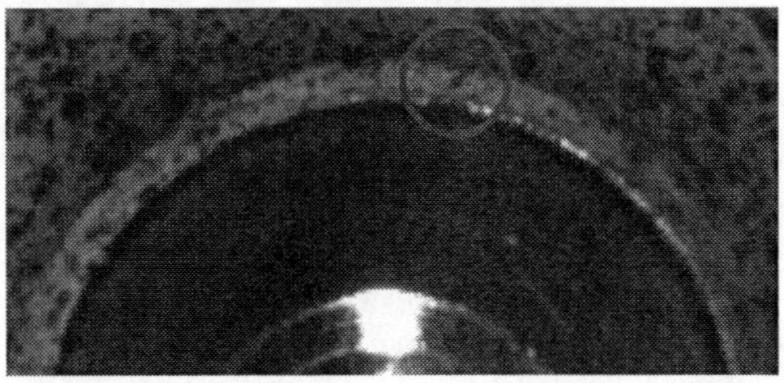

Figure 7. Close up of collar on test exhibiting through thickness crack.

Once the stage where the initial through thickness crack appears had been identified, the diameter of the hole is measured based on a method mapping a certain number of pixels in the image to a reference length. In total, four measurements of the diameter were taken, two from the image of the left camera (D_{h1} and D_{h2}) and two from the right camera (D_{h3} and D_{h4}). The known reference length in the image is the hole made for mounting the guide pin presented in Figure 3, where the length is known to be 6.004 mm. Figure 8 presents where the reference geometry was taken, and how the two measurements are placed in relation to each other. Once the four measurements were obtained, the average Hole Expansion Ratio $\overline{\lambda}$ was calculated using the average hole diameter $\overline{D}_h$ as presented in Eq. 2.

$$\overline{\lambda} = \left(\frac{\left(\frac{D_{h1}+D_{h2}+D_{h3}+D_{h4}}{4} \right) - D_o}{D_o} \right) \times 100 \tag{2}$$

4. Results

In total 62 test specimens were tested in the new experimental setup, where 59 of the executed tests were successful, and a through thickness crack was detected before the mechanical stop was reached. Three of the executed tests showed no evidence of fracture or only partial fracture (not through the thickness) at the point where the mechanical stop was reached. Figure 9 presents the $\overline{\lambda}$-values of the 62 tests, and the three unsuccessful tests have been marked separately. A mean value of the 59 successful tests was found to be 31.184 %, with a standard deviation of $\sigma = 4.2910$.

As it can be seen from the results presented in Figure 9, the test still produces a large scatter, indicating that the lack of restraining force in the test is not the primary cause of the reported scatter. An interesting trend that can be observed from the results, is the decrease of the $\overline{\lambda}$-values beginning with sample number 48. No apparent reason stands out when looking at the

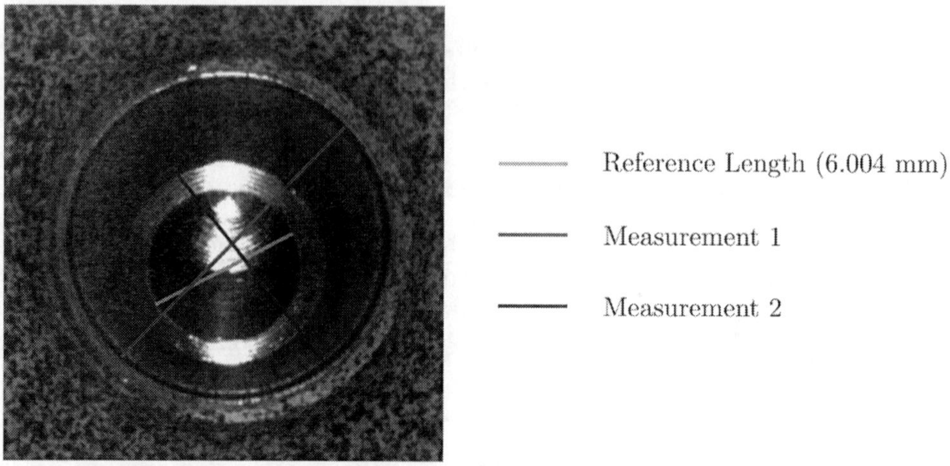

Figure 8. Illustration of the measurement of the hole diameter. Two measurements are taken on each of the images recorded by the DIC system.

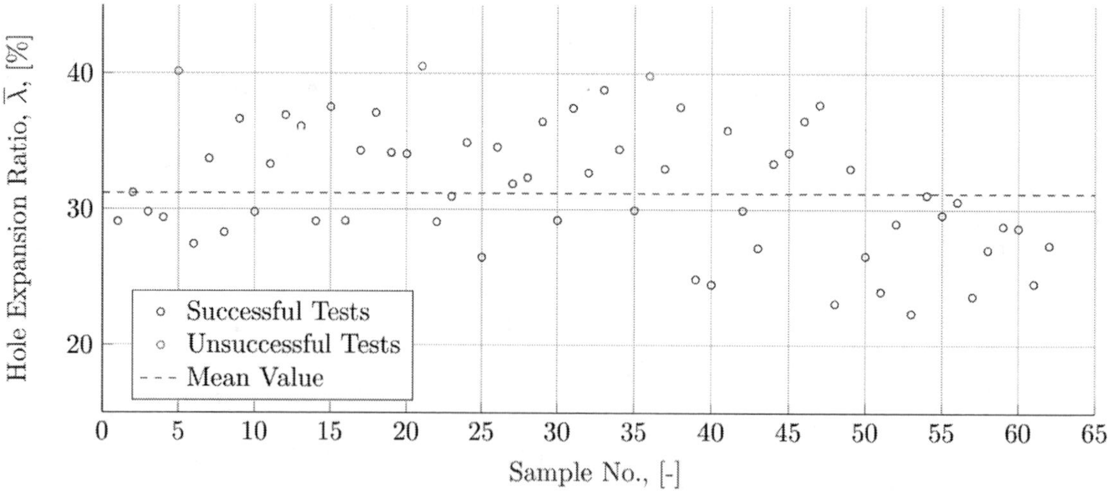

Figure 9. Hole Expansion Ratios of the 62 tests conducted. In total three tests were unsuccessful due to no presence of through-thickness cracks.

DIC stages at which the specimens failed, why a more in-depth discussion of the results must take place.

5. Discussion

As presented, the results of the 62 Hole Expansion Tests still produce a significant scatter, which indicates that the boundary conditions of the ISO-16630 test is not the primary cause for the scatter. Another factor that could highly influence the results of the test is the stochastic behaviour of fracture. Historically, the Hole Expansion Test has been evaluated once a through-thickness crack has appeared in the sample. With the previous setup this was the only way to proceed, but with the new setup presented in this paper, it is possible to evaluate the test in an arbitrary point in time due to the images recorded by the DIC system. Therefore, it could be argued that the test should be evaluated at a stage before the first surface crack appears. The problem with edge cracks is that they are not induced by necking, but are subject to the direct fracture phenomena as described by Manopulo & Carleer [5]. Therefore, the current methods of determining the onset of localized necking as presented by e.g. Sigvant et al. [6] cannot be applied, and a different approach must be found.

Another interesting trend that was observed in the results is the downwards trend of the $\overline{\lambda}$-values on the last specimens tested (sample number 48-62). Several authors, including Yoon et al. [7] and Karelova et al. [8], have stressed the importance of the edge quality of the punched hole in the Hole Expansion Test. One possible explanation for the downwards trend in the results could be due to wear of the tool used for punching the holes in the test specimens. The tool used for punching in this series of experiments was new and never before used, why it is not unlikely what is seen is the tool experiencing some initial wear before settling into a stable condition. Another possible explanation for the trend could be an unintended wear of the conical punch. The scaled up punch used for the new setup was manufactured from a tool steel and subsequently vacuum hardened to have a hardness of 61±1 HRC. In order to get the punch to fit with the specifications, a grinding operation was performed after the hardening. If this grinding operation has removed most of the hardened material, it is possible that the conical punch has experienced some wear, which has resulted in a change in the friction conditions between the punch and the test specimen. Figure 10 presents a comparison of the Force-Displacement curves for test 15 and 49. This comparison shows that there are differences between the two curves, especially in the low displacement area, indicating that friction conditions might have changed between the punch and the sheet, as well as a difference in the total displacement accounting for the difference in the $\overline{\lambda}$-values of the two tests.

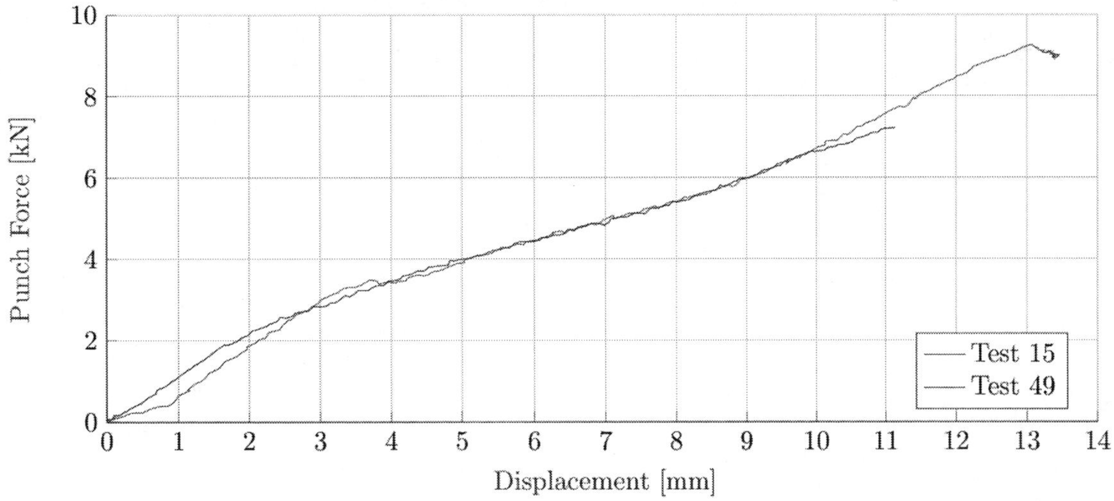

Figure 10. Comparison of the Force-Displacement response of two tests from each end of the test series.

6. Conclusion

A new experimental setup aiming to reduce scatter in the results of the ISO-16630 Hole Expansion Test was tested. The new setup was based on a hypothesis that the scatter reported by several authors was due to an insufficient restraining of the material during the test. Therefore, a conical punch was developed according to the specifications of the ISO-16630 standard and a die set from a Nakajima test was used due to the presence of a draw bead. For the evaluation of the test, images from a 3D DIC system was analysed in a third party software where a mapping of pixels between a known distance and the hole diameter was used to determine the final hole diameter.

In total 62 tests of a DP800 steel alloy were performed, where 59 of these successfully presented a through-thickness crack, and the remaining three did not due to the test begin terminated by a mechanical stop in the hydraulic press used. The results of these 62 tests still presented a large scatter with a standard deviation of 4.291 % why it can be concluded that restraining force of the sample during the test is not the primary cause of the scatter in the Hole Expansion Test.

Acknowledgements
Open Access funding was provided by Blekinge Institute of Technology. This study was also funded by VINNOVA in the Sustainable Production sub-program within Vehicle Strategic Research and Innovation (FFI) program (grant number 2020-02986).

References
[1] International Standard Organization 2018 Metallic materials – sheet and strip - hole expanding test (ISO 16630)
[2] Schneider M *et al* 2015 Overview and comparison of various test methods to determine formability of a sheet metal cut-edge and approaches to the test results application in forming analysis *Mat.-wiss u. Werkstofftech* **46**(12) pp 1196-1217
[3] Larour P *et al* 2014 Evaluation of alternative stretch flangeability testing methdos to iso 16630 standard *IDDRG 2014 Conference Proceedings*
[4] Chiriac C and Chen G 2008 Local formability characterization of ahss - digital camera based hole expansion test development *IDDRG 2008 Conference Proceedings*
[5] Manopulo N and Carleer B 2020 A new workflow for the effective distinction between necking induced splits and direct fracture phenomena *IOP Conf. Series: Mater. Sci. Eng.* **967** 012066
[6] Sigvant M *et al* 2008 The definition of incipient necking and its impact on experimentally or theoretically determined forming limit curves *IDDRG 2008 Conference Proceedings*
[7] Yoon J I *et al* 2016 Factors governing hole expansion ratio of steel sheets with smooth sheared edge *Met. Mater. Int.* **22**(6) pp 1009-1014
[8] Karelova A *et al* 2009 Hole expansion of dual-phase and complex-phase ahs steels - effect of edge conditions *Steel Research International* **80**(1) pp 71-77

IOP Conf. Series: Materials Science and Engineering 1238 (2022) 012032 doi:10.1088/1757-899X/1238/1/012032

Modelling stamped edges in FEM breakage analyses of high-strength steel safety components

R A Lingbeek[1]*, D Schalk-Kitting[2], P Gruschka[3], M Uurike[4], P Peetsalu[5], P Blome[1]

[1]Autoliv B.V. & Co. KG, Otto-Hahn-Str. 4, 25337 Elmshorn, Germany
[2]VIRTUAL VEHICLE Research GmbH, Graz, Austria
[3]AUDI AG, Ingolstadt, Germany
[4]AS Norma, Tallinn, Estonia
[5]Tallinn University of Technology, Tallinn, Estonia
* e-mail: roald.lingbeek@autoliv.com

Abstract. The development process for new safety components includes break load tests, where the component undergoes very large strains. Besides the mechanical properties of the steel, which are changed by the forming process, the break load depends on the quality of the stamped edges. Target of this paper is to investigate whether a hole-tensile test with a stamped hole can demonstrate the effects of stamped-edge quality on the load bearing capacity – and secondary, how such a test procedure can be reproduced with a multi-scale FEM procedure. A series of Hole Tensile Tests (HTT) has been performed with a HSLA steel grade. Here, only minor differences in break load and total elongation were found. The local strain just prior to breakage at the edge of the stamped hole shows remarkably high values, but little difference is found when comparing a machined sample to stamped samples. The multi-scale FEM approach was demonstrated using literature data for dual-phase steel, which is known to be more sensitive to edge cracks. Indeed, a stamped edge, which includes significant hardening and pre-damage, shows earlier fracture in this FEM calculation – however, the crack propagation, which is needed to capture the full breakage of the HTT sample, is not modelled correctly – this limits the application of the method on safety components.

1. Introduction

The influence of the blanking or shear-cutting process on the failure behaviour of sheet metal material edges has been widely investigated for deep drawing operations and is also partially simulated with sufficient accuracy [1-5]. However, the influence of the stamping process on the breaking load behaviour of stamped safety-related components in belt systems is less frequently explored in literature. In a dynamic loading scenario, safety components may undergo very large deformations: Clearly, a premature initiation of failure at the stamped edge is unacceptable.

Figure 1 shows high tensile strain in the bent area (point A) of a simplified example bracket. Two distinct failure modi, also known from physical break load tests [6], have been reproduced in FEM by setting the limit strain A_{local} at a lower and higher value in the material model – failure at the bending (Figure 1 – centre) reduces the break load significantly [6]. The maximum bending strain at point A, is on a stamped edge of the bracket. Hence, it is natural to assume an influence of the stamped edge stretchability on the break mode and break load.

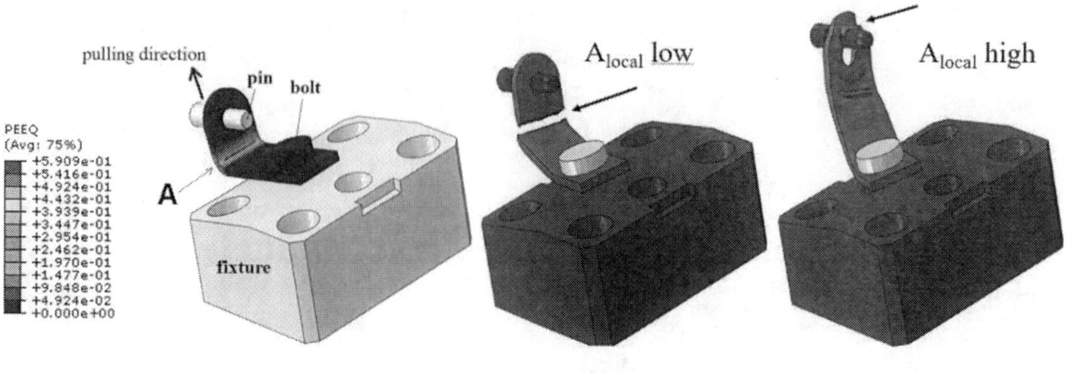

Figure 1. Pre-bent example bracket in FEM break load simulation, with two failure modes

To achieve no-compromise quality for safety related components, both the material selection and the design of the shear-cutting process must be carefully assessed in the component design phase already. This requires that the edge properties resulting from material selection and process design are known/understood and added as an extension to the above presented FEM break load model.

1.1. Literature on stamped edge stretchability

A shear-cutting process typically uses a punch, die and blank holder. The cutting clearance, which is the gap between die and punch (in this work represented by the character Q and expressed as % of the steel thickness), as well as the sharpness of the tool (e.g., the radius of the punch edge) define the characteristics of the shear face and the characteristics of the shear-affected zone (SAZ) for the sheared material [4]. To investigate the influence of the shear-cutting parameters on edge characteristics and consequently on the failure behaviour of the stamped edges in forming related operations, various experimental tests have been developed, such as Hole-Expansion Test (HET) [ISO 16630], Diabolo-test [19] or Half-dome-test [1]. The tensile test with one-sided stamped edge [7] and Hole-Tensile-Test [8-11] allow free loading of the stamped edge (e.g. not involving tool contact and friction) and thus better represent the loading situation in a safety component break load test.

The common finding from experimental tests is that the deformation in the SAZ is the dominant factor in controlling failure during stamped edge stretching [4, 11]. The presence of voids and microcracks plays a secondary role according [4]. As a side remark: [11] points at the differences in residual stresses after drilling or stamping as a future research topic.

While shear face characteristics and characteristics of the shear-affected zone correlate to the stretchability of the edge, one cannot determine the stretchability of the edge directly from the morphology of the shear face. Furthermore, the different types of steels, especially of different AHSS families (Complex-phase, Ferritic-Bainitic, Dual-phase) have different sensitivities to parameters in the stamping process [2, 11]. [17] compares hole expansion ratios for DP780 steel from 6 different suppliers: The authors found some supplier's steel to show significant differences in Hole-Expansion Ratio (HER) when shear-cut with various clearances, the samples of other suppliers proved insensitive. In contrast to DP steels, only few experimental studies can be found in literature for HSLA steel, which is often used for safety components in belt systems (e.g., see example in [6] for a seatbelt bracket: S550MC).

1.2. Edge stretchability in FEM simulations

FEM simulations may be the key technique to fundamentally understand the stamped edge stretchability of various steel families (and differences between different suppliers) and to assess component performance in industrial context. [8-10] combine the shear-cutting step and Hole-Expansion Test in a single FEM model. This requires enormous computational effort since the stamping process requires high-resolution modelling, which limits industrial application. Two more efficient modelling approaches to model stretchability of stamped edges are often found in literature: The first approach is used in the context of sheet metal forming. It applies an empirical edge strain limit with fixed values to edge elements, such as the "edge forming limit" [5] or the commercially available option in AutoForm, demonstrated by [12]. In industrial practice, this requires a large library of material and process combinations. It is not clear how to generalize such an edge crack criterion to be included in a break-load simulation chain as presented by [6], which uses solid elements. How to set an edge forming limit strain for point A in Figure 1, or the limit strain on the surface of a hole-tensile test?

The second (less empirical) approach maps accumulated damage and hardening from a high-resolution shear-cutting simulation (in plane-strain) to the coarse mesh in the macro scale load scenario. This approach is called "multi-scale approach" throughout this paper and can be included into the sequential forming and break load simulation chain such as presented in [6].

FE simulations of the shear-cutting process can be found in various literature (for example [8-10, 12, 14, 15, 18]). The simulations are generally in good agreement with physical tests, specifically in terms of shear face characteristics and hardening distribution in the SAZ, even for simplified material models. The challenge is, to capture the stretchability of this edge in the secondary forming or break load (or dynamic loading) scenario, using the same material model. [8-10] and [18] report that the multiscale approach in HET give good results for HER, where edge failure (initiation of a crack) is defined by the deletion of first edge elements. This paper focuses on the multiscale approach in the context of safety component breakage tests, summarized in two research questions:

1. *Can a hole tensile test, using HSLA steel, be used to assess a potential drop in break load on a stamped component with decreasing edge quality? How can the maximum strain at the stamped edge be measured in-situ?*
2. *Can the influence of shear-cutting be accurately covered by a sequential (multiscale) shear-cutting and break load simulation, using the simplified material model from [6]?*

2. Experimental testing

This paper uses the HTT to show differences in break load and elongation from macro-viewpoint (over the sample as a whole), but also to measure the maximum local strain at the hole just prior to breakage, using digital image correlation (DIC). In this section, results from HTT experiments are reported, using a high quality HSLA steel and various cutting clearances on the stamped hole. Unlike the HET, the HTT shows friction-free loading of the stamped edges which matches the loading situation of safety components in break load tests well.

2.1. Material and sample geometry

The used material is an EN 10149 S700MC hot-rolled HSLA steel with sheet thickness 3 mm and following mechanical properties: Rm = 789 MPa, Rp0.2 = 735 MPa and A5 = 21.7 %. There is no generally accepted standard for the HTT: Various specimen geometries can be found in the literature (for example, [8] and [11]). The sample shape used in this work is shown in Figure 2. A hole is stamped in the centre of the specimen with a diameter of d0 = 10 mm, using a HET stamping toolset (i.e. the die diameter is varied to achieve different cutting clearances - this has been found to have just minor effect on the load-bearing cross-section and hence on the maximum load of the HTT and is therefore neglected in this work.). A cutting clearance. Q = 5.8% is representative for stamping of HSLA in serial production. It produces a large burnish zone but induces considerable cold work hardening in the SAZ. Q = 19 % leads to a smaller burnish zone and more microcracks but reduces the cold work hardening in the SAZ. Reference specimens (hole edge in initial material conditions) have been obtained from stamped hole HTT by milling the shear-affected zone (i.e. 500 µm on the circumference) away. The amount of 500 µm for the SAZ was determined from hardness measurements in the cross-section perpendicular to the punched edges. The hole for the milled HTT has consequently a larger diameter of 11 mm. The resulting reduced load bearing cross-section in the HTT due to the milling has been taken into account in the given results. The test program includes three samples each for Q = 5.8 %, Q = 19 % and for milled hole.

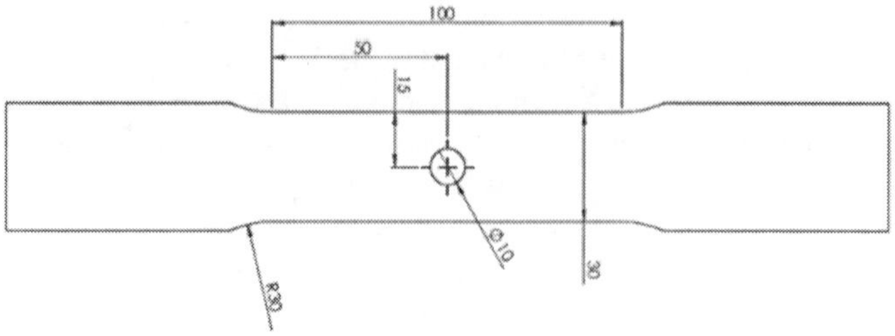

Figure 2. Specimen geometry for the Hole Tensile Test.

2.2. Break load testing and DIC measurement

During the HTT, strains are measured locally with a GOM ARAMIS® optical system. The resulting data files contain information about the deformation on the surface as well as the measured pulling force from the universal testing machine (UTM). The time of crack initiation is determined visually for all tested specimens (i.e., last picture frame before a visible crack occurs). Figure 3 shows exemplarily contour plots (technical strains) for an HTT sample with Q = 19 % in three different deformation stages (i.e., early deformation stage in a); last frame before fracture is detected in b); visible fracture in c)).

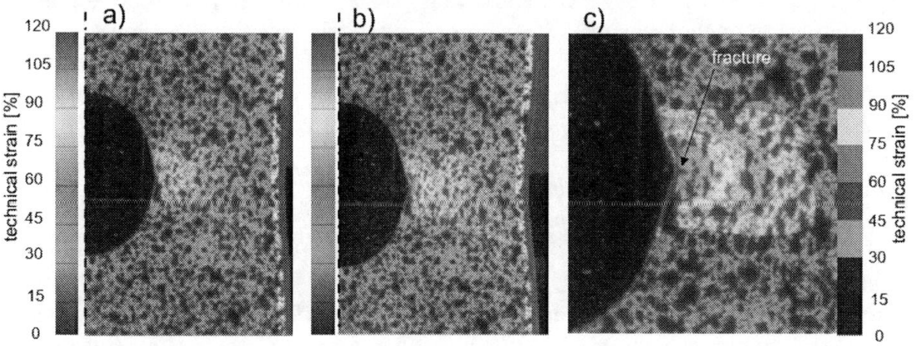

Figure 3. Exemplary Aramis results for HTT with Q = 19 %: Contour plots of technical strains at three different deformation stages.

2.3. Applied methodology for limit strain analysis

In the initial (undeformed) stage, an ellipse was fitted along the points of the inner boundary (i.e., the hole – see Figure 4 a) of the ARAMIS® generated mesh. Multiple parallel sections at a distance of 0.125 mm were generated, with the mid-section passing through the centre of the ellipse (see also Figure 4 a). On the 5 inner cross sections a total of 10 inspection points were set symmetrically on both sides of the hole at a distance of approximately 0.4 mm from the edge (see Figure 4 b). The technical strains are exported for all available stages. The maximum strains from the 10 inspection points of the left and right side are taken at the last picture frame before a visible crack occurs to represent failure of the specimen (see Figure 4 c).

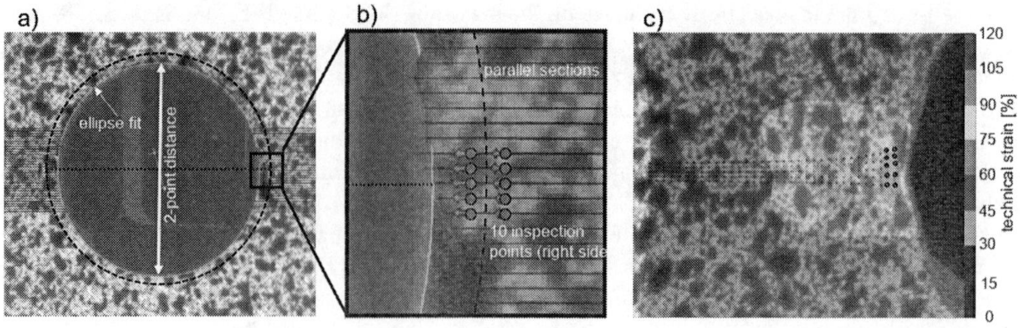

Figure 4. Schematic representation of the applied methodology for strain analysis procedure for HTT.

2.4. Experimental Results

The results of the strain analysis described above for all HTT tested in this work are given in Figure 5 in terms of technical strains in loading direction of the specimen. A box plot format is used to capture the statistic characteristic of the determined fracture strains (note: 20 strain data points are evaluated for each single specimen). The results in Figure 5 shows that there is no significant difference on the average technical strains at fracture for the two tested cutting clearances (i.e., Q = 5.8 % and Q = 19 %). Furthermore, it can be seen from Figure 5 that the strain values for HTTs with Q = 19 % show higher variances. For the milled specimens, a trend towards higher average technical strains in the critical regions seems to be present.

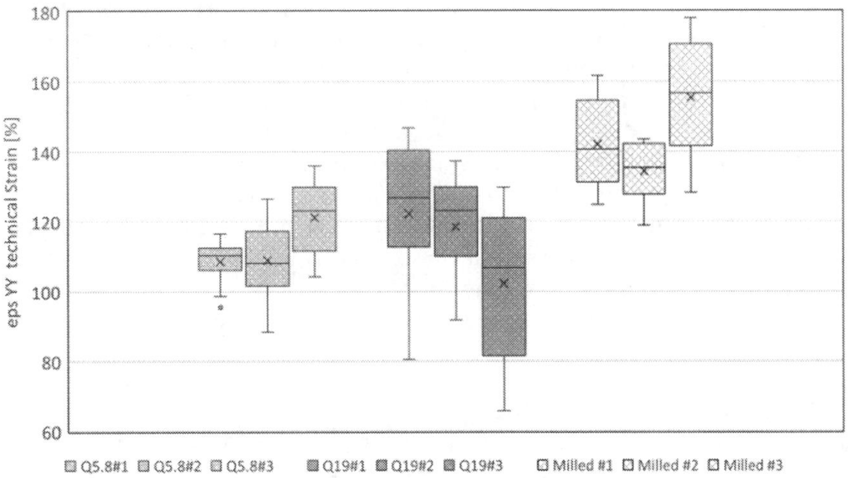

Figure 5. Results of fracture strain evaluation for the HTT specimen with cutting clearance of Q = 5.8 % and Q = 19 %. 20 strain data points are evaluated for each single specimen and are given in a box plot format for each specimen to capture the statistic characteristic of the determined fracture strains. The results for HTT with milled hole are given as a reference.

The synchronized force signal from the ARAMIS® result file was exported for all available stages and related to initial critical cross-section to account for the different hole diameters for stamped and milled hole. For the displacements, two-point distance at the stamped hole in loading direction (see Figure 4 a) – marked with white arrow) was used. Figure 6 shows results in terms of engineering stress over a relative measure for displacement (i.e. relative Y distance).

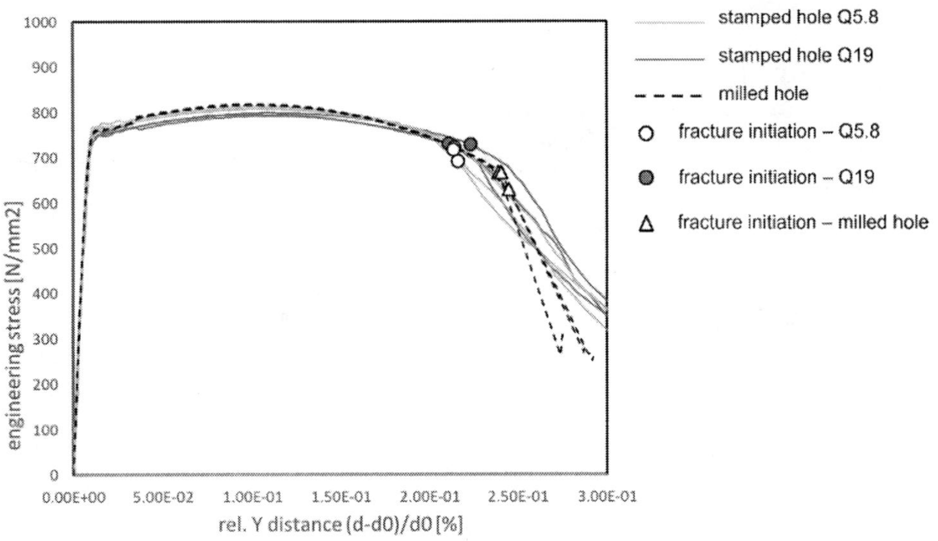

Figure 6. Engineering stress over corrected relative distance Y for the tested HTT specimen. The point of fracture initiation is marked for all specimens. (Note: The corrected relative distance Y on the abscissa is based on the elongation of the punched hole in loading direction and was chosen to account for different initial diameters d0.)

From Figure 6 can be seen that stamped as well as milled HTT specimens show very good reproducibility. There are no significant differences between the achieved maximum engineering stresses for stamped and milled holes in HTT. Between Q = 5.8 % and Q = 19 % specimens a slight difference in maximum stress can be observed. This is probably due to a differently pronounced shear-affected zone as well as the load-bearing cross sections due to the change in cutting clearance. Regarding the maximum displacements before visible cracks occur (marked with circles for stamped holes in Figure 5) there is no significant difference between the specimens with cutting clearances 5.8 % and 19 % that could be attributed to differences in the shear-affected zones. The maximum displacements for the milled specimens (see triangles in Figure 6) exceed those seen in the specimens with stamped holes.

3. Multiscale FEM modelling of HTT with stamped edge

This section describes an investigation into a FEM multiscale approach for the hole tensile test, with stamped edges. The study uses HTT data for DP780 steel with 1.6 mm sheet thickness, taken from literature [8]. This basis for the data has been chosen, and not the produced results from the presented HSLA test campaign, because neither the force-displacement (Figure 6), nor the DIC strain measurement (Figure 5) tests showed a clear dependency on the edge properties (milled, stamped, clearance).

3.1. Model description and material modelling

Figure 7 a and 7 b show the dimensions and meshing details of the LS-DYNA FEM simulation model (half-model) for the HTT according to [8]. In the HTT FEM model, solid elements with element size 0.1 mm are used. The meshing directly at the edge of the hole in the HTT model is obtained with 5 element ring layers which allow to represent the so-called shear-affected zone of a stamped edge (see Figure 7 b).

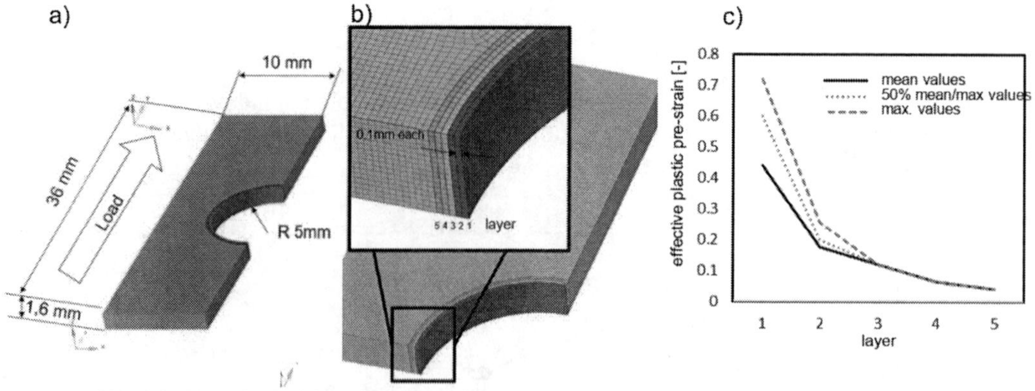

Figure 7. Dimensions and meshing details of the LS-DYNA FEM simulation model for the HTT (half-model).

A simplified material model for plasticity and fracture has been derived from the DP780 mechanical properties in [8] and [10]. The methodology used generates an elastic-plastic FEM material model including breakage from tensile test values (e.g. a material certificate) alone. In this method, there is no accumulation of damage in the form of stiffness degradation – when the limit strain (in dependency of the triaxiality) is reached, failure is instantaneous, and the element is deleted. The details on the methodology are given in [6]. The validation of the simplified plasticity and fracture model for the DP780 shows good correlation with the reported experimental values in [8]: Slightly higher force values

IOP Conf. Series: Materials Science and Engineering 1238 (2022) 012032 doi:10.1088/1757-899X/1238/1/012032

are obtained in the force/displacement results which may be attributed to the isotropic plasticity in the material model. The maximum displacement at failure from the FEM HTT simulation with the simplified material model falls well into the range of the experimental results given in Figure 14b in [8].

3.2. Multiscale approach – mapping 2D cutting simulation results onto 3D model

In [10] detailed axis-symmetric 2D cutting simulations have been conducted for the investigated DP780 grade for various cutting clearances. The pre-damage and plastic strain values obtained from 2D cutting simulations for different cutting clearances show relatively small variations compared with the overall pre-damage. Also, pre-damage and effective plastic strain have a nearly identical distribution and can be scaled by a factor into each other [10]. Fundamentally, it may make sense to analyse pre-damage and cold work hardening individually (and pre-damage as a tensorial parameter [20]), however, this does not match with the simplistic material model [6] and should be focus of future work.

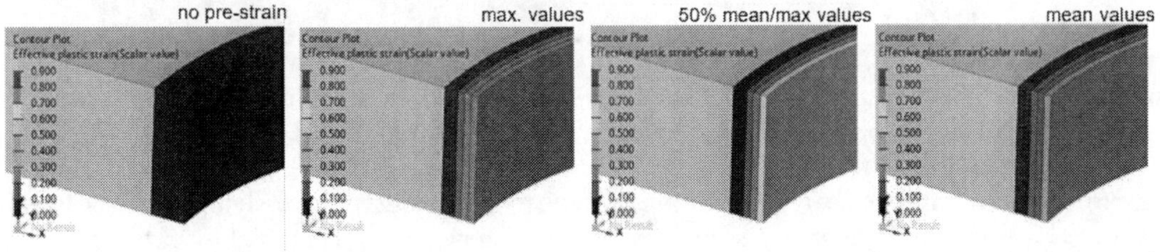

Figure 8. Initial conditions of the hole tensile test FEM model – mapping of pre-strain values onto the 3D model (no pre-strain for "milled" hole; max. value; 50 % mean/max. values; mean value).

For demonstration reasons, an average pre-damage continuous distribution is obtained for all die clearances in [10] (the reader is referenced to Figure 19 in [10] – this distribution serves as the master curve representing a shear-affected zone. Superimposing this curve on the element discretization at the edge in the 3D model (the 5 rings), the question arises as to which discrete curve values should be transferred to the integration points of the individual elements of each ring as the initial condition. In this work, the mean values, the maximum values, and values in between the aforementioned values (50 % mean/max. values) from the master curve have been explored in the multiscale FEM concept (see Figure 7 c for the three different effective plastic strain distribution that serve as initial condition in the FEM model to represent the SAZ). Figure 8 shows the HTT FEM model with the different pre-strain distribution as initial condition in the SAZ.

3.3. Simulation results

Figure 9 shows the numerical force/displacement results for the HTT were pre-damage and pre-strain values have been considered at the holes edge as described above. The force-displacement results from the HTT simulation model representing edge conditions for milled edges (i.e. no shear-affected zone) is given as a reference (i.e. grey solid curve in Figure 9). The results show that earlier fracture occur in the HTT FEM models where stamped edges are considered. The numerically predicted total elongation of the sample lie clearly above the experimental range for stamped edges (i.e. light-grey marked area in Figure 9 - the reader is referred to Figure 14b in [8] for the detailed experimental results). Furthermore, it can be seen from Figure 9 that there is no significant difference in the force/displacement result for the three different degrees of the pre-damage/pre-strain at the holes edge (i.e. the black solid curve, grey dotted curve and grey dashed curve in Figure 9 to represent different degrees of pre-damage).

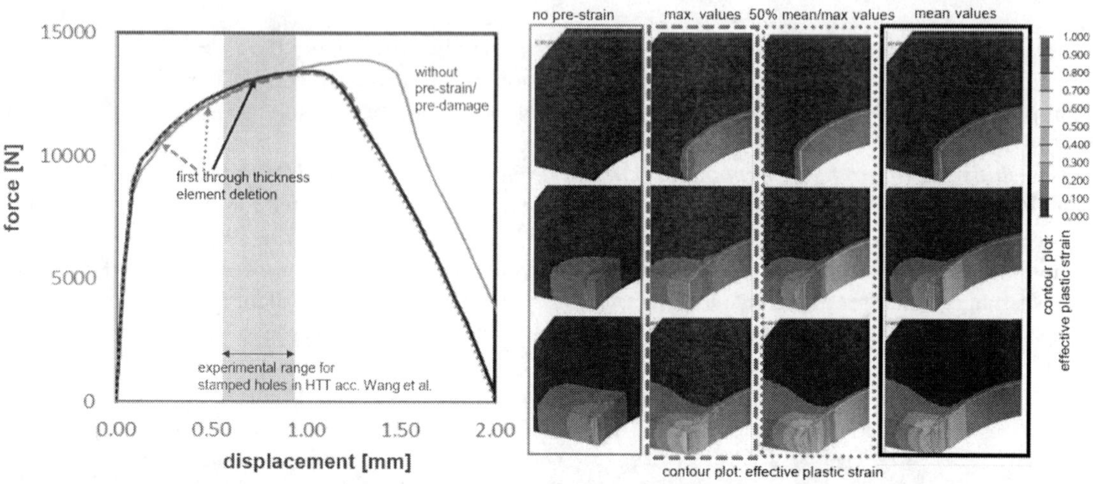

Figure 9: Numerical force/displacement result for HTT with different pre-damage and pre-strain values in the SAZ (left). Evolution of element deletion in the HTT models (right).

A more detailed look on the failure evolution in the FEM model where stamped edges have been considered via initial pre-damage and pre-strain values shows that for the highest pre-damage degree (grey dashed curve in Figure 9 (left), and the grey dashed column (right)) a first through thickness element deletion happens already in the very beginning of the component stretching. In the force/displacement curve this event is visible through very small kink in the curve (indicated in Figure 9 with the grey dashed arrow). The further evolution of the force/displacement curve after first through thickness element deletion follows qualitatively the force/displacement behaviour of the "pre-damage mean value" model (i.e. black solid curve) but a slightly lower force level. The slightly lower force level can be attributed to the reduced cross-section at the critical area through element deletion. For the grey dotted and black solid curve similar behaviour is observed in terms of element deletion but at later simulation stages. However, it becomes clear from this results that a first through thickness element deletion with element size of 0.1 mm does not trigger macro scale fracture in the HTT model making the overall force/displacement answer independent of the edges pre-damage values for the investigated pre-damage degrees.

4. Discussion

Can a hole tensile test, using HSLA steel, be used to assess a potential drop in break load on a stamped component with decreasing edge quality?

Firstly, in the maximum values of engineering stress in the engineering stress/displacement curves (Figure 6) of the hole-tensile-tests for the S700MC (HSLA) samples, no clear distinction can be made between milled and stamped specimens, and obviously, no difference can be observed between the two cutting clearances either. More interestingly, the overall elongation-before-breakage of the samples has only a minor advantage for the milled sample. This insensitivity of the HSLA samples is in sharp contrast to the published curves for DP780 by [8]. Secondly, the mean value of the measured (eps yy technical) strains are, for all samples, well above 100%. Whereas HSLA steel is known for superior local ductility, the authors interpret the measurement values as proof of an exceptionally "damage tolerant" steel. The strain at breakage is higher for the milled samples, hinting at an influence of the stamped edge, but not significantly so.

Hence, for the measured batch of steel, the HTT is unable to link stamped edge quality to macro-level breakage. The authors recommend analysis for a series of HSLA steels from various suppliers, similar to [17] to learn whether the insensitivity of HSLA in an HTT is fundamental.

Looking back at the example bracket in Figure 1, the HTT procedure lacks the typical strain reversal of the safety part application [6], where a flat, stamped blank is bent into shape as a bracket which is then pulled flat again in a subsequent load scenario: The test procedure should include shear-cutting, forming and then the break load situation as a third step. Future analyses could extend the stamp+bend test in [16] with a subsequent tensile stage.

Can the influence of shear-cutting be accurately covered by a sequential (multiscale) shear-cutting and break load simulation, using the simplified material model from [6]?

Because of the insignificant differences in the experimental HTT results for HSLA, the originally intended combination of physical tests and FEM analysis was not followed up in this paper. When published [8] DP780 plastic strain and pre-damage in the SAZ are mapped onto the mesh of an HTT sample, the early onset of a crack is indeed reproduced in FEM. However, the crack initiates at the wrong time, and the propagation is captured incorrectly. This is easily understood as the used [6] material model does not include crack propagation physics: element deletion is instantaneous once the strain limit is reached. Also, the element-wise deletion is likely to be much too coarse to capture the crack propagation.

To assess the break load of safety components with stamped edges, crack propagation is likely the key aspect. Basically, any stamped edges will have some level of microcracks: The main question is whether these propagate into a macro breakage or not.

5. Acknowledgements

The publication was prepared with the cooperation of Virtual Vehicle Research GmbH in Graz, Austria. The authors would like to acknowledge the financial support within the COMET K2 Competence Centers for Excellent Technologies from the Austrian Federal Ministry for Climate Action (BMK), the Austrian Federal Ministry for Digital and Economic Affairs (BMDW), the Province of Styria (Dept. 12) and the Styrian Business Promotion Agency (SFG). The Austrian Research Promotion Agency (FFG) has been authorized for the program management. They would furthermore like to express a thank-you to Thyssenkrupp Hohenlimburg for supporting the experimental work.

6. References

[1] Gu J, Kim H, Shih H-C and Dykeman J 2018 Effects of Blanking Conditions to Edge Cracking in Stamping of Advanced-High Strength Steels (AHSS), *SAE Technical Paper* 2018-01-0626

[2] Konieczny A 2008 On the Formability Behavior of Advanced High Strength Steels, *Steel Research Int., Special Edition Metal Forming Conference* 2008, **1**

[3] Konieczny A and Henderson T 2007 On Formability Limitations in Stamping Involving Sheared Edge Stretching, *SAE Technical Paper* 2007-01-0340

[4] Levy B S and Van Tyne C J 2012 Review of the Shearing Process for Sheet Steels and Its Effect on Sheared-Edge Stretching, *J. Materials Engineering and Performance*, **21(7)**, pp 1205-1213

[5] Gläser T, Sunderkötter C, Hoffmann H, Volk W and Golle R 2017 Development of a 2-Stage Shear-Cutting-Process to Reduce Cut-Edge-Sensitivity of Steels, *IDDRG Proceedings*

[6] Lingbeek R, Aldemir B, Soni A, Jagalur S and Yao L 2020 Material Spread and Local Failure in Breakage Modeling for Steel Safety Components, *23rd Int. Conf. on Material Forming, ESAFORM 2020, Procedia Manufacturing*

[7] Feistle M, Pätzold I, Golle R and Volk W 2019 Open Hole Tensile Tests for the Determination of the Edge-Crack Sensitivity of Sheared Holes Dependent on Specimen Geometry, Cutting Parameters, and the Notch Factor, *18th Int. Conf. on Sheet Metal, SHEMET 2019, Procedia Manufacturing* 29

[8] Wang K, Luo M and Wierzbicki T 2014 Experiments and Modeling of Edge Fracture for an AHSS Sheet, *Int. J. Fract.* **187**, pp 245-268

[9] Wang K and Wierzbicki T 2015, Experimental and Numerical Study on the Plane-Strain Blanking Process on an AHSS Sheet, *Int. J. Fract.* **194**, pp 19-36

[10] Wang K, Greve L and Wierzbicki T 2015 FE Simulation of Edge Fracture Considering Pre-Damage from Blanking Process, *Int. J. Solids Structures* **71**, pp 206-218

[11] Pathak N, Butcher C, Worswick M J, Bellhouse E and Gao J 2017 Damage Evolution in Complex-Phase and Dual-Phase Steels during Edge Stretching, *Materials* **10**, 346

[12] Falk J 2017 Fracture Prediction of Stretched Shear Cut Edges in Sheets Made of Dual-Phase Steel, *Master thesis, Blekinge Institute of Technology, Karlskrona, Sweden*

[13] Gutknecht F, Steinbach F, Clausmeyer T and Tekkaya A E 2015 Advanced Material Model for Shear Cutting of Metal Sheets, *XIII Int. Conf. on Computational Plasticity: Fundamentals and Applications, COMPLAS*

[14] Krinninger M, Opritescu D, Golle R and Volk W 2017 On the Opportunities of Problem- and Process-Adapted Shear Cutting Simulations for Effective Process Design, *Int. Conf. on the Technology of Plasticity, ICTP 2017, Procedia Eng. 207*

[15] Krinninger M, Steinlehner F, Opritescu D, Golle R and Volk W 2017 On the Influence of Different Parameters on the Characteristic Cutting Surface when Shear Cutting Aluminum, *50th CIRP Conf on Manufacturing Systems, Procedia CIRP 63*

[16] Tsoupis I, Hildering S and Merklein M 2014 Bending of High-Strength Low-Alloyed Steel with Respect to Edge Crack Sensitivity Caused by Shearing Operations, *11th Int. Conf. on Technology of Plasticity, ICTP 2014, Procedia Eng. 81*

[17] Kim H 2021 A New Testing Method for Evaluating Edge Cracking of AHSS, *Great Design in Steel (GDIS) Symposium*

[18] Bhattarai Y 2019 Numerical Edge Failure Prediction of Stamped Parts: Analysis of Influence of Mapping Approach on the Prediction Quality, *Master thesis, Technical University of Dortmund, Germany*

[19] Held C, Liewald M and Sindel M 2010 Erweiterte Werkstoffprüfverfahren zur Charakterisierung von Leichtbaublechwerkstoffen im Hinblick auf die Kantenrisssensitivität, *Materials Testing* **52**, pp 596-602

[20] Heath A, Gese H, Oberhofer G and Dell H 2013 , Modeling failure for nonlinear strain paths with CrachFEM, *AIP Conference Proceedings* **1567**, 587-590

Reduction of adhesive wear with use of tool coating reducing thermoelectric currents

A Schrepfer[1], A Schott[2], P Tröber[1], M Keunecke[2], M Welm[1], F Steinlehner[1], R Golle[1] and W Volk[1]

[1] Chair of Metal Forming and Casting, Technical University of Munich, 85748 Garching, Germany
[2] Fraunhofer Institute for Surface Engineering and Thin Films IST, 38108 Braunschweig, Germany

Agnes.Schrepfer@tum.de

Abstract. Blanking is one of the main processes in producing sheet metal components. The impact of adhesive wear of the blanking tools on the cost efficiency is high, especially when processing aluminum. As an influencing factor on adhesive wear in blanking, thermoelectric currents have been considered to a very small extent. Recent investigations have shown that thermoelectric currents occur in every blanking tool due to the rise of temperature in the shear zone in interaction with a difference of the material-specific Seebeck coefficients of tool and sheet metal. A reduction in the amount of wear was observed in case of a smaller difference of Seebeck coefficients and therefore lower levels of flowing thermoelectric currents. This paper follows the approach of adjusting the Seebeck coefficient of the tool material CF-H40S+ to the one of the sheet material by means of coating with CrAlN, in order to reduce the flowing thermoelectric currents and consequently adhesive wear. For this purpose, the effect of coating on the Seebeck coefficients, the thermoelectric currents during blanking as well as resulting wear quantities were investigated. Basic correlations could be revealed.

1. Introduction

The chemical and physical processes that take place in the contact zone of two components in interaction, leading to a progressive loss of material from the surface of a solid body are referred to as wear mechanisms [1]. Adhesive wear, as one of the most common wear mechanisms, predominates in the processing of aluminum alloys, stainless steel and titanium [2]. According to Czichos, adhesions are caused by contact between two asperities which initially deform plastically due to high surface pressure. Therefore, fragments of the protective oxide layer are pulled off and bare metallic surfaces come into contact. As the resulting chemical bonds between the contact partners are stronger than the cohesive bonds within the materials, a relative movement between the two partners causes a separation within the cohesively weaker bonded partner and a material transfer to the stronger bonded one [3]. These adhesions formed lead to a reduction in component quality. During cyclic tool engagement, material transfer builds up and consequently comes off again as sliver particles [4]. These free particles lead to an increased risk of damages of the tool, the peripheral electronics or the press [5]. Currently lubricants and coatings are used in the industrial environment to reduce adhesive wear, as they are minimizing the friction between the contact partners and therefore also the resulting frictional heat [6].

However thermoelectric currents were considered to a minor extent. While blanking, a temperature rise can be observed in the shear zone, which occurs due to three reasons: Besides macroscopic friction between tool and sheet metal and microscopic friction on an atomic level, the conversion of plastic work into heat contributes most to the temperature rise. For the aluminum alloy BS 2011, about 90 % of the work dissipate into heat [7] and only a small portion is stored in the structure of the material [8]. Therefore, the occuring temperature is mainly influenced by the thickness and the mechanical properties of the sheet metal. It is also basic reason for the emergence of thermoelectricity during metal forming. This phenomena, called Seebeck effect, occurs when two different conductors come into contact and are subject to a temperature gradient. It leads to electric voltages that are measurable in an open circuit. In addition to the temperature gradient, the difference between the material-specific Seebeck coefficients (SCs) influences the thermoelectric voltage. The SC describes the thermoelectric behavior of materials as a measure of the increase in thermoelectric voltage depending on the temperature and is mainly influenced by the chemical composition. However, different crystal structures and grain sizes of the materials also change the SC [9]. Therefore, it must be determined experimentally for each material. In a closed conductor circuit, thermoelectric circular currents are generated due to the existing thermoelectric voltage. The strength of the current also depends on the resistance R of the entire conductor circuit [10]. Considering the temperature increase in the blanking tool, the conditions for the Seebeck effect to occur are also fulfilled during cold forming and cutting as shown in figure 1. Tool and sheet material are representing two conductors which are subject to a temperature gradient.

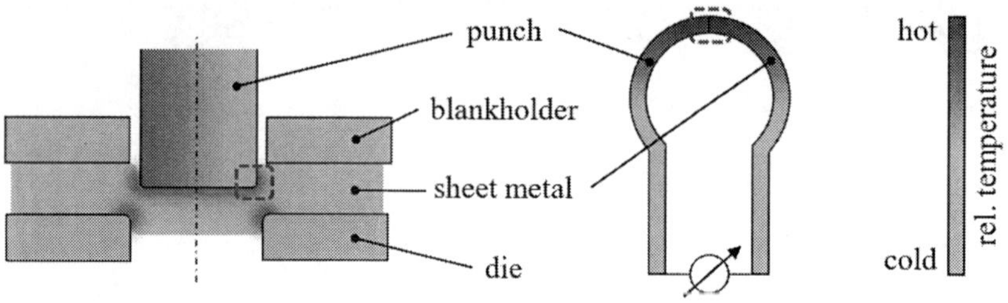

Figure 1. Schematic representation of the Seebeck effect during shear cutting [15].

At first the influence of thermoelectric currents on adhesive wear has mainly been considered in the field of machining. Several studies have shown that introducing a current that counteracts the natural thermoelectric current reduces wear and thus increases the service life of tools. In addition an insulation of the active elements in contact can also have the same effect [11]. Uehara et al. showed that the influence of tool insulation depends on the materials used [12]. Then Tröber et al. examined the influence of thermoelectric currents on the wear of tools during blanking. For this purpose, shear cutting tests were carried out with three different punch materials (1.2379, 1.4301, CF-H40S) and the sheet material EN AW 5083. A determination of the SCs showed that the stainless steel alloy 1.4301 and the sheet material have very similar coefficients values [13]. With this material combination, almost no thermoelectric current flowed between tool and sheet [14]. In contrast to the sheet material, the carbide CF-H40S has a negative SC causing the electrons to flow from punch to sheet metal. Due to the larger difference of the SCs of punch and sheet metal when using CF-H40S a larger amount of wear was formed on the punch due to the stronger thermoelectric currents in comparison to the punch out of 1.4301. However, since the SC of 1.2379 is significantly higher than that of the aluminum alloy, the direction of the current is reversed. A determination of the amount of wear showed that despite a lower current, the amount of adhesion was greater compared to the carbide punch, which showed that not only the amount but also the flow direction of the current has an influence on adhesive wear. The height and direction of the thermoelectric current and thus also the amount of wear can therefore be influenced by a suitable choice of material combinations [13]. In addition, the effects of the natural thermoelectric currents can be influenced by means of an externally controlled current, which is applied into the contact zone between the punch and the sheet [15].

In this work to reduce adhesive wear, the Seebeck coefficient of the tool material is to be adapted to that of the sheet material by means of specially developed coating in order to reduce the flowing thermoelectric current. The influence of the coating on the resulting thermoelectric current and the resulting adhesions at the punch will be investigated in experimental shear cutting tests.

IOP Conf. Series: Materials Science and Engineering 1238 (2022) 012033 doi:10.1088/1757-899X/1238/1/012033

2. Experimental Setup

2.1. Determination of the thermoelectric properties

For the experimental determination of SCs of conductive and solid materials a special test rig was developed at the Chair of Metal Forming and Casting [9]. A schematic sketch of the setup is shown in figure 2. Based on an integral method, the material sample is subjected to a defined temperature difference. While the one end of the sample is kept at a constant temperature of 0 °C with the aid of an ice bath, the other end is sequentially heated to 500 °C and then cooled back down to 0 °C. Thermocouples measure the temperature of the sample at each end. The warming and cooling process is carried out by contacting the sample with a copper block, which is heated by a heating cartridge and cooled again by applying compressed air and dry ice. The measurement of the arising thermoelectric voltage is recorded via two wires attached to the ends. Platinum wires with a purity of 99.99+ % are used as a reference material for the thermoelectric characterization because of its high inertness and temperature resistance. In order to minimize environmental influences on the measurement signal, the thermoelectric voltage, which is usually only in the range of µV, is amplified by a factor of 1000 inside the test chamber. To counteract corrosion of the materials due to the high temperatures, the entire test chamber is flooded with the inert gas argon. The SC is then calculated from the recorded thermoelectric voltage signal.

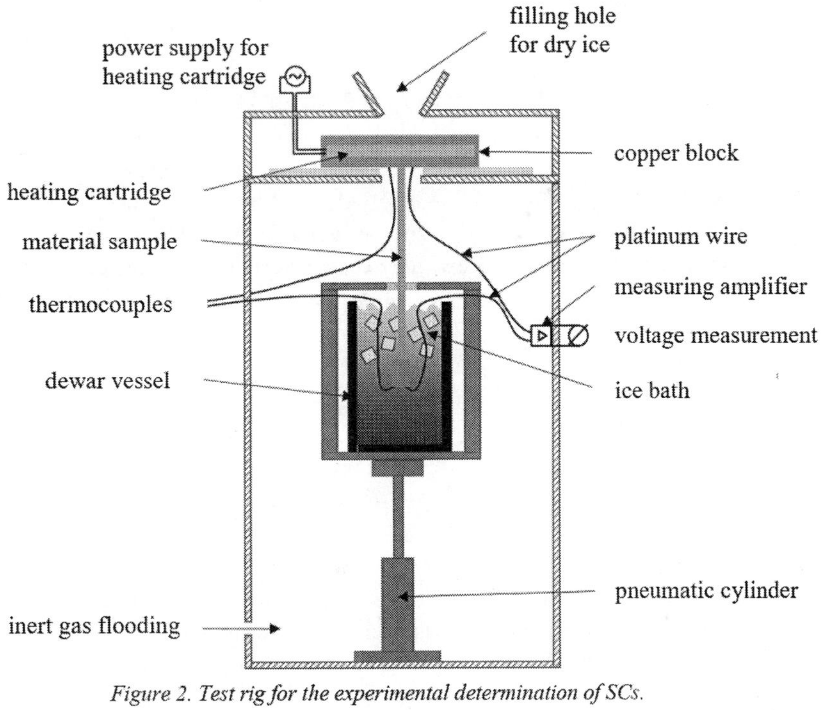

Figure 2. Test rig for the experimental determination of SCs.

2.2. Blanking tool and process parameters

The blanking tool for carrying out the experimental investigations has been developed specifically for measuring thermoelectricity during forming and blanking. Due to the four-track design, a simultaneous investigation of different test configurations is possible. In order to obtain accurate measurement results, both the sheet metal strip and the active elements are electrically insulated. This prevents interfering signals. To measure thermoelectric currents, a cable is attached on each punch and the sheet material, building a short circuit between them. The current flows through this connection, where a Hall probe determines both the strength and the direction of the measurable component of the thermoelectric current from the change in the electric field around the conductor. The single stroke tests were carried out on the high performance stamping press BSTA 1600-181, Bruderer, Frasnacht (Switzerland). In order to be able to make statements about the wear status after only a few strokes an absolute cutting gap of 125 μm was selected. With a sheet thickness of 2.5 mm, this corresponds to a relative cutting gap of 5 %. All tests were carried out with 150 strokes per minute, which corresponds to a punch impact speed of 120 mm/s, and without applying lubricants.

2.3. Investigated materials

The Seebeck coefficients of the tool material CF-H40S+ as well as the stainless steel material 1.4301 (X5CrNi18-10) and the aluminum alloy EN AW 5754 (AlMg3) were measured. For the investigation of the wear reduction potential of the coating, single stroke tests were carried out with the aluminum alloy. The chemical composition of these materials can be found in Table 1, 2 and 3.

Table 1. Chemical composition of 1.4301 in wt%.

	C	Si	Mn	P	S	Cr	Ni	N	Fe
1.4301	-	0.4	1.7	-	-	18.5	8.0	-	Balance

Table 2. Chemical composition of EN AW 5754 in wt%.

	Si	Fe	Cu	Mn	Mg	Cr	Zn	Ti	Al
EN AW 5754	0.1	0.3	-	0.1	3.1	0.1	-	-	Balance

Table 3. Chemical composition of CF-H40S+ in wt%.

	WC	Co	Other
CF-H40S+	86.6	11.8	-

2.4. Measurement of electrical resistivity

To characterize the thermoelectrical behavior of the layers, their electrical resistivity is determined by four-point measurement [16]. For this purpose, precisely defined geometries are inscribed by means of laser structuring in the layers coated on insulating substrate material. The used design can be described with width (w = 0.2 mm) and length (l = 3.88 mm) of the structure. Using equation 1 and 2 with thickness t of the thin film the surface resistivity $R_{\square}$ can be determined. The electrical resistivity for thin films can be calculated with equation 3.

$$R_{\square} = \frac{\rho_{Thin\,film}}{t} \tag{1}$$

$$R_{\square} = R_{4Point} \times \frac{w}{l} \tag{2}$$

$$\rho_{Thin\,film} = R_{4Point} \times \frac{w}{l} \times t \tag{3}$$

2.5. Coating deposition procedure by physical vapor deposition (PVD)

Hard coatings and their modifications were deposited using a PVD magnetron sputtering system CemeCon 800/9. The system is equipped with 4 sputtering cathodes with a size of 500 mm x 88 mm and was operated in d.c. and pulsed d.c. mode. The chamber dimensions are 800 mm x 800 mm x 900 mm. The residual pressure before starting the process was $<10^{-3}$ Pa. Prior to coating depositions substrates were cleaned with a water-based cleaning procedure. Instantly after the cleaning procedure samples were mounted into the vacuum chamber. Prior to the coating process an ion etching step with argon ions for additional surface cleaning was applied. Different coatings could be deposited by combination of different target materials on the sputtering cathodes and reactive gases like nitrogen during coating process. By varying the applied power to the sputter cathodes different coating compositions and layer sequence could be obtained. Coatings were deposited on substrates as well as ceramic plates for electrical characterization. The substrates and tools were fixed at sample holders, which are charged with a negative voltage for increased ion bombardment, and coated while rotating through the vacuum chamber.

3. Coating development

3.1. Strategy for modification of the Seebeck coefficient by adjusting the electrical resistivity

In order to minimize the wear occurring during blanking, the pairing of the SCs of the workpiece and the tool has to be adapted to each other. For this purpose, the SC of the tool has to be modified by means of coating. The SC is a material constant and is influenced by free electrons. Consequently, the electrical conductivity respectively electrical resistivity has a significant influence on the thermoelectric properties of materials. Depending on the charge carrier density in a material, the electrical conductivity shows an opposite course to the SC. As the charge carrier concentration decreases, the SC increases. Thus, electrically insulating materials have a very high SC. [17] For the development and study of thermoelectric properties of thin films, the method of 4-point measurement for the determination of electrical resistivity was used. Due to the existing correlation of electrical conductivity and the SC, the approach is to adjust the thermoelectric properties of the thin films to the desired target range based on measured electrical resistivity by modifying the coatings structure by different phases. [18] Similar approaches are also described by [19], where the thermoelectric properties of two individual materials can be modified in combination as a nanocomposite. Most of the standard hard coating used for blanking tools, like TiN, CrN etc. have low electrical resistivities. To attain electrical resistivity and thermal conductivity in an appropriate range, such coatings has to be modified with other elements or phases. The used technique of PVD magnetron sputtering allows to deposit coatings in a wide composition range. Due to the possibility to use different target materials like Cr and Al in one coating process, it is easy to affect the coating structure and phase composition too.

3.2. Development of modified CrN coatings

Very good effects for the adaptation of the SC to the sheet material were achieved by a modification of CrN coatings with aluminum respectively AlN. The specific electrical resistivity can be influenced by orders of magnitude (see table 4) by addition of AlN to CrN.

Table 4. Specific electric resistivity of applied coatings.

Coating	Specific electric resistivity [Ωcm]
CrN	~ 4.62×10^{-4}
CrAlN_1	~ 6.08×10^{-2}
CrAlN_2	~ 4.56×10^{-1}

Different CrAlN coatings were prepared by power variation of the cathodes. Due to significant different sputter rates of Cr and Al in reactive nitrogen gas mixtures three Al targets and one Cr target were used. Typical target power was between 0.5 – 2 kW per target. The negative bias voltage was in range of 30-60 V typically. The process pressure was in the range of 0.4 Pa. The analysis of the chemical composition (figure 3 a) was investigated by electron probe micro analysis (EPMA). It shows that for CrAlN_1 an Al content of 30±1 at%, a Cr content of 18±1 at% and 50±1 at% nitrogen was obtained. For CrAlN_2 an Al content of 35±1 at%, Cr content of 11±1 at% and 50±1 at% nitrogen was measured. By reducing the Cr content and simultaneously increasing the Al content the electrical resistivity was enhanced by an order of magnitude from appr. 10^{-2} Ωcm to the range of 10^{-1} Ωcm (figure 3 b). Because of pure Al and Cr target materials with nitrogen as reactive gas were used for coating deposition, it is reasonable that the coating structure is affected by the rotational movement of the tools through the vacuum chamber and along different coating material and resulting in a layered coating structure of alternating AlN rich and CrN rich layers. These coating structure of multilayers or nanolayers for modified nitride coatings is comparable with the developments of modified nitride hard coatings for nanostructured high-performance tool coatings [20]. Therefore, this approach seems to be promising for blanking tools coatings as well. Due to the fact that relative high Al contents are needed to attain the desired electrical resistivity range and thus the SC, it is recommended to use combinations of AlCr-mixture and Al-targets for coating deposition. For TiAlN coatings the use of mixture or alloyed targets provide good and well adherent hard coatings for tools [21] but lower specific electrical resistivity values compared to CrAlN (figure 3 b).

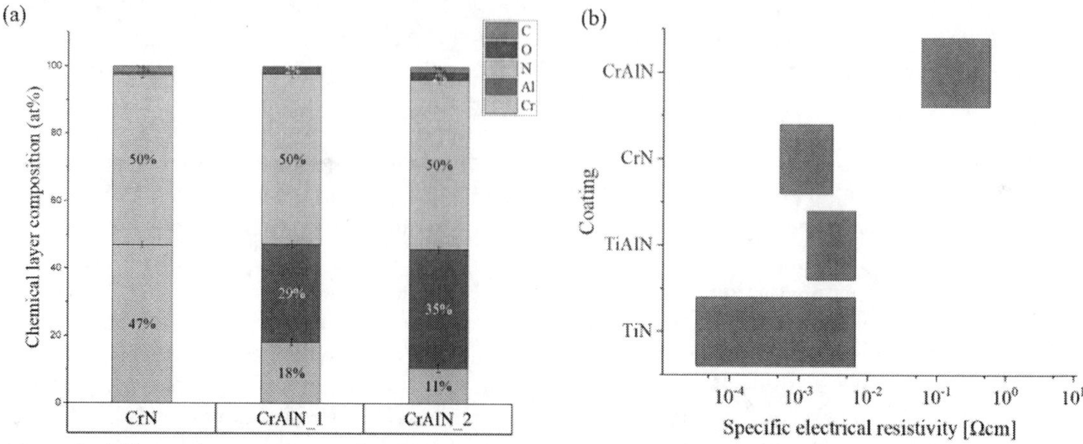

Figure 3. a) Chemical composition results of EPMA analysis for CrN and modified CrAlN layers b) Specific electrical resistivity of different hard coatings.

In combination with an adapted pre-treatment and surface finish of the blanking tools it should be possible to increase the coating performance and maintain the desired electrical resistivity and SC range. Furthermore it could be possible to adjust the coating resistivity and SC to the tool and workpiece material. For this purpose further development work is needed.

4. Results and Discussion

4.1. *Seebeck coefficients of carbide before and after coating and of selected sheet materials*

The SC of CF-H40S+ is negative in the temperature range from 0 °C to 300 °C due to the high cobalt content. In comparison, the SCs of the sheet materials EN AW 5754 and 1.4301 are positive, resulting in a high difference in the SCs of the sheet and the uncoated tool material as shown in figure 4 a. By means of coating with CrAlN, the SC difference between the carbide and the sheet material can be significantly reduced. Best results were obtained using the chemical composition of the CrAlN_1-layer (see 3.2), which is also shown in figure 4 a as a layer on CF-H40S+. A concomitant reduction in the

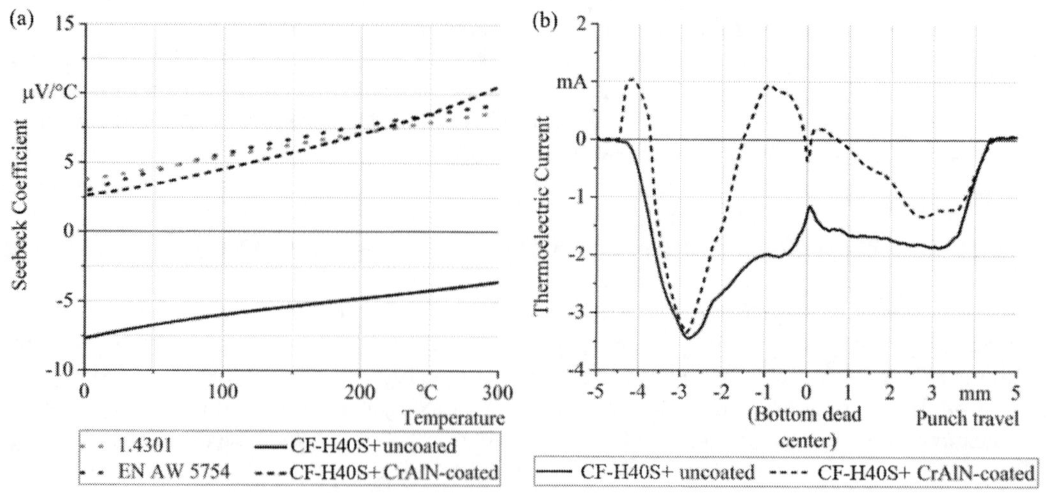

Figure 4. a) Seebeck coefficients of CF-H40S+ uncoated and CrAlN-coated, EN AW 5754 and 1.4301 b) Thermoelectric currents over punch travel for CF-H40S+ uncoated and CrAlN-coated while cutting EN AW 5754.

amount of adhesions on the punch should therefore be possible with the use of this coating with regard to literature.

4.2. Impact of Seebeck coefficients on thermoelectric currents

Figure 4 b shows the course of the measured thermoelectric currents averaged over 9 strokes when blanking the aluminum alloy EN AW 5754. It can be seen that the maximum thermoelectric current, flowing between sheet and punch, has a negative sign, which by definition corresponds to an electron flow from punch to sheet metal. At the beginning of the contact between the punch and the sheet metal, the current increases as the punch travel progresses until the moment of crack formation in the sheet metal. Afterwards, it decreases until the bottom dead centre is reached. When the punch movement is reversed, the current rises again slightly until it drops to zero when the punch leaves the punching grid. The maximum absolute value of the averaged thermoelectric current in case of an uncoated punch is -3.49 mA. Due to the change in SC (figure 4 a), a reduction in the flowing current is possible when a CrAlN layer is used. It can be seen, that as well as at the beginning and near the bottom dead center the thermoelectric current showed even positive values, at the exact bottom dead center it drops back into negative values. This can be attributed to the fact that when measuring the SC, a slow heating of the entire sample takes place over a larger volume, resulting in a greater influence of the base material. During blanking, the temperature in the shear zone increases rapidly, which mainly heats the layer and the layer material thus dominates in terms of the SC. Therefore, the SC of the CrAlN-coated punch is higher than that of the sheet material at higher temperatures, resulting in reversed current directions. A shift of the current curves in a positive direction over a great part of the punch travel can be observed. However at a punch position of -3 mm, the absolute maximum values of both punches are in a similar range. This can be attributed to the fact that the coating started to peel off from the stamp surface. Since the adhesion of the coating fails primarily at the cutting edge of the punch, as it is case with most coatings used in industry, similarly high thermoelectric currents can be measured during the cutting process. When the cutting process is finished and the cutting edge loses contact with the sheet metal, there is only contact between the coated lateral surface of the punch and the sheet metal. During this phase and the return stroke, the effects of the changed Seebeck coefficient are clearly visible in the thermoelectric current curve.

4.3. Thermoelectric currents and adhesive wear

Conclusions about the amount of adhesion that has been formed on the punch over the first 9 strokes can be drawn by looking at the recorded punch force. Figure 5 a shows the measured punch force during stroke 1, stroke 5 and stroke 9 over the punch travel while cutting EN AW 5754. All curves show a rapid increase when the punch hits the sheet metal at -4.1 mm, a sharp drop in the force when cracks occur and a force required to push the slug through the die channel at -1.5 mm to 0 mm. During the return stroke tensile forces act on the punch in opposite direction, which indicate friction between punch and punching grid. When the punch leaves the punching grid, the force reaches a value of zero. It can be seen, that maximum force, push through force and pull back force are higher when using the CrAlN-coated punch which can be attributed to the higher surface roughness of the coated punch. A comparison of the maximum push-through force and the maximum retraction force after stroke 1 and stroke 9 shows that both forces increase with the number of strokes performed. This indicates that the amount of adhesion on the punch increases noticeably, causing high friction. A thermal expansion of the punch can be excluded, as the individual strokes were performed with a time interval of several minutes. Comparing the punch force curves from 1 to 9 strokes in figure 5 b gives indication of the wear-reducing effect of the CrAlN coating. An increase in the push-through force and retraction force is significantly less pronounced here than with the previously considered uncoated punch.

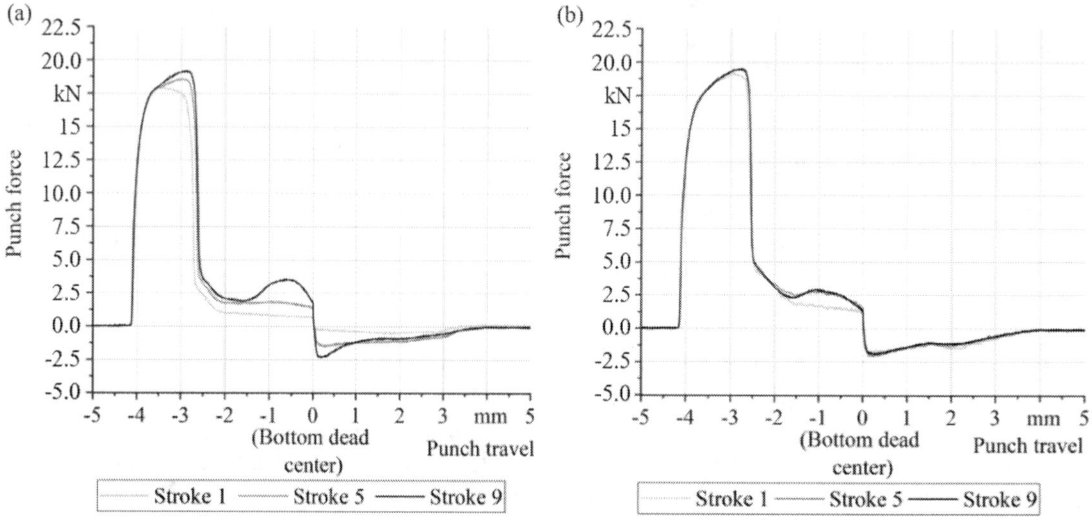

Figure 5. Punch forces over punch travel for CF-H40S+ (a) uncoated and (b) CrAlN-coated when cutting EN AW 5754.

The assumption of a reduced amount of adhesive material transfer with CrAlN-coated punch is confirmed when comparing the punch surfaces before the first stroke and after the ninth stroke. Using a laser microscope, the punches worn lateral surface was recorded in at least four locations evenly distributed around the circumference, where an area with a width of 1 mm was recorded. For analysing the wear quantity the curvature of the punch was compensated and the adhesive wear volume was measured. In order to exclude the influence of manufacturing disparities, the material volume resulting from the surface roughness was subtracted. Finally, the wear coefficient is calculated by dividing the remaining volume by the surface area under consideration. Therefore, the wear coefficient indicates the height of transferred material assuming an even distribution. Already after nine strokes, a formation of adhesions on the uncoated punch surface is clearly visible. An evaluation of the height of the adhesions on the punch surface revealed a wear coefficient (VQ) of 0.35 µm. In contrast, the laser microscope measurements of the punch surface of the CrAlN-coated punch after 9 strokes in figure 6 show only slight wear. An evaluation of the wear coefficient resulted in a value of 0.21 µm, which represents 60 % of the amount of adhesions of the uncoated punch. A close look at the surface of the CrAlN-coated punch shows that the coating started to detach as the recorded thermoelectric currents suggested, therefore, the adhesion of the coating needs to be further improved for use in an industrial environment.

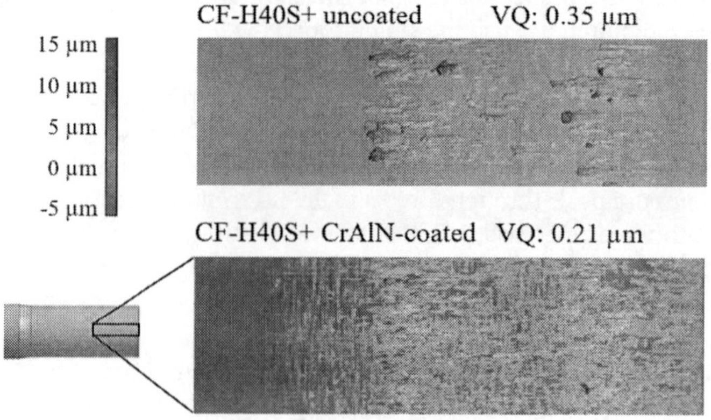

Figure 6. Laser microscopic pictures of the adhesive wear distribution on the punches lateral surface after cutting EN AW 5754.

In order to be able to exclude an influence of the temperature development in the shear zone as well as different friction conditions between punch and sheet metal as a possible reason for the reduced thermoelectric currents, the cutting surfaces of the punching grids were also considered. Figure 7 shows the individual zones of a characteristic shear cut surface. While cutting at first the roll over and clean-cut form. With advancing punch position material failure begins and the sheet metal tores. The torn zone and the burr arise. Figure 8 shows that the CrAlN-coated punch leads to a higher proportion of clean-cut with 57 % in comparison to 45 % of clean-cut using the uncoated punch. This shows that cracking occurs earlier with progressive punch stroke in the uncoated punch, suggesting that the temperatures in the shear zone are lower here. Thus, although the temperatures in the shear zone are higher with the CrAlN-coated punch, the thermoelectric current was lower and so were the resulting amounts of adhesions to the punch. A lower measured thermoelectric current when using the CrAlN-coated punch is therefore due to the adjusted Seebeck coefficient. Likewise, the surface of the clean-cut after only one stroke is significantly rougher when using the CrAlN-coated punch than when using an uncoated punch, because of the higher roughness of the coated punch. Although the higher roughness normally leads to higher adhesions on the punch due to the surface pressure, lower adhesions were measured here. This improvement can therefore also be attributed to the Seebeck coefficient.

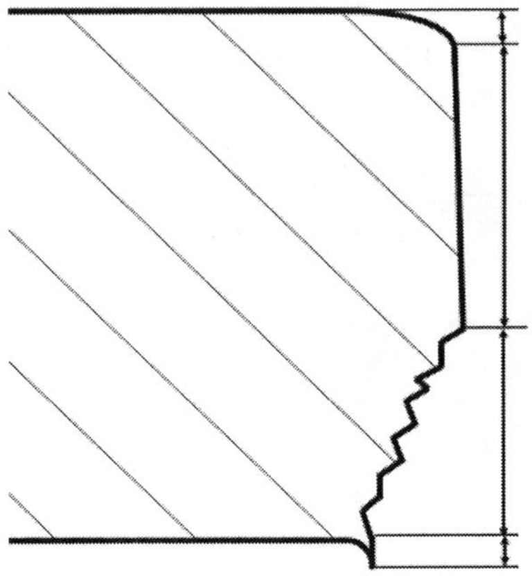

Figure 7. Cutting surface characteristic values.

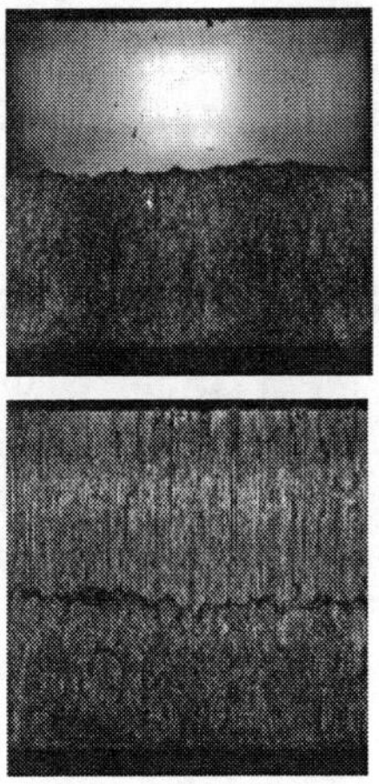

Figure 8. Cutting surface of EN AW 5754 after cutting with (a) uncoated and (b) CrAlN-coated punch.

5. Conclusion

In this paper, it could be shown that it is basically possible to influence and adjust the Seebeck coefficient by a coating and its modification respectively. By adjusting the Seebeck coefficient of the tool material to the one of the sheet metal, the flowing thermoelectric current and consequently the amount of adhesive wear could be reduced. Furthermore, the findings will help to improve the understanding of the wear causing interactions. The improvement of the mechanical coating properties as well as its adhesion will be part of future projects.

Acknowledgement

The authors would like to thank the German Federation of Industrial Research Associations (AiF) for the financial support under the grant number 20693N.

References

[1] Czichos H and Habig K-H 2010 *Tribologie Handbuch – Tribometrie, Tribomaterialien, Tribotechnik* (Wiesbaden: Vieweg + Teubner-Verlag) pp 117-30

[2] Groche P, Nitzsche G and Elsen A 2008 Adhesive wear in deep drawing of aluminum sheets

Annals of the CIRP **57** pp 295-8.
[3] Buckley D H 1981 Surface effects in adhesion, friction, wear and lubrication *Elsevier Tribology Series 5*
[4] Krinninger M, Steinlehner F, Opritescu D, Golle R and Volk W 2017 On the influence of different parameters on the characteristic cutting surface when shear cutting aluminum *Procedia CIRP* **63** pp 230-5
[5] Schilp H, Hoffmann H, Golle R and Hoogen M 2004 Flitterreduzierung beim Schneiden von Aluminiumblech *Blech Rohre Profile* **51**, pp 78-9
[6] Schulz J, Brinksmeier E, Meyer D and Schultalbers M 2015 Vertiefende Untersuchungen zu Verschleißschutzadditiven und deren Wechselwirkung mit Metalloberflächen *56. Tribologie-Fachtagung GfT-Tagungsband*
[7] Macdougall D 2000 Determination of the plastic work converted to heat using radio-metry *Experimental Mechanics* **40/3** pp 298-306.
[8] Demmel P, Hoffmann H, Golle R, Intra C, Volk W 2015 Interaction of heat generation and material behaviour in sheet metal blanking *Annals of the CIRP* **64**, pp 249-52
[9] Demmel P, Tröber P, Kopp T, Golle R, Volk W and Hoffmann H 2013 Characterization of the thermoelectric behavior of plastically deformed steels by means of relative Seebeck coefficient *Materials Science Forum* **755** pp 1-7
[10] Barnard R D 1972 *Thermoelectricity in metals and alloys*
[11] Shan H and Pandey P 1975 Wear of Cutting Tools: Thermoelectric effects *Wear* **32** pp 167–79
[12] Uehara K, Sakurai M and Ikeda T 1992 On the problem of thermoelectric current in metal cutting *Annals of the CIRP* **41** pp 75–8
[13] Tröber P, Demmel P, Hoffmann H, Golle R and Volk W 2017 On the influence of Seebeck coefficients on adhesive tool wear during sheet metal processing *CIRP Annals* **66/1** pp 293-6
[14] Welm M, Tröber P, Weiss H, Demmel P, Golle R and Volk W 2020 Thermoelectrically based approaches to reduce adhesive wear during blanking *JOM* **72/7** pp 2025-35
[15] Tröber P, Weiss H A, Kopp T, Golle R and Volk W 2017 On the correlation between thermoelectricity and adhesive tool wear during blanking of aluminum sheets *International Journal of Machine Tools and Manufacture* **118-119** pp 91-7
[16] Chen X, Dai W, Wu T, Luo W, Yang J, Jiang W and Wang L 2018 Thin film thermoelectric materials: Classification, characterization, and potential for wearable applications *Coatings* **8** *p* 244
[17] Rowe D M 2006 *Thermoelectrics handbook: Macro to nano* (CRC Press Taylor & Francis Group)
[18] Gharavi M A, Kerdsongpanya S, Schmidt S, Eriksson F, Nong N V, Lu J, Balke B, Fournier D, Belliard L and le Febvrier A 2018 Microstructure and thermoelectric properties of CrN and CrN/Cr2N thin films *Journal of Physics D: Applied Physics* **51/53**
[19] Dehkordi A M , Zebarjadi M, He J and Tritt T M 2015 Thermoelectric power factor: Enhancement mechanisms and strategies for higher performance thermoelectric materials *Materials Science & Engineering R* **97** pp 1-22
[20] Stein C 2016 Entwicklung von nanostrukturierten Nitriden und c-BN als Verschleißschutzschichten für Zerspanwerkzeuge (Stuttgart: Fraunhofer Verlag) **160**
[21] M. Keunecke et al 2010 Modified TiAlN coatings prepared by d.c. pulsed magnetron sputtering *Surface & Coatings Technology* **205** pp 1273–8

IOP Conf. Series: Materials Science and Engineering 1238 (2022) 012034 doi:10.1088/1757-899X/1238/1/012034

Prediction of sheared edge characteristics of advanced high strength steel.

O Sandin[1], S Hammarberg[1], S Parareda[2], D Frómeta[2], D Casellas[1,2], P Jonsén[1]

[1] Division of Solid Mechanics, Luleå University of Technology, 971 87 Luleå, Sweden
[2] Eurecat, Centre Tecnològic de Catalunya, Unit of Metallic and Ceramic Materials, Plaça de la Ciència, 2, Manresa 08243, Spain

E-mail: olle.sandin@ltu.se

Abstract. In the present work, numerical models are developed for the shearing and cutting process of advanced high strength steel-blanks which can predict the edge morphology in the shear effected zone. A damage model, based on the modified Mohr-Coulomb fracture surface, is calibrated. To increase the predictability of the numerical models, the fracture surface is fine-tuned in areas corresponding to the stress-state of cutting, a methodology called Local calibration of Fracture Surface (LCFS). Four cutting cases with varying clearance are simulated and verified with experimental tests, showing good agreement. It is thus found that the suggested methodology can simulate cutting with adequate accuracy. Furthermore, it is found that solely using plane-stress tensile specimens for calibrating the fracture surface is not enough to obtain numerical models with adequate accuracy.

1. Introduction

Advanced High Strength Steels (AHSS) have become the dominant material choice for lightweight construction in the automotive and transport sector for its good in-service performance, manufacturability and recyclability. However, AHSS have some limitations regarding crack resistance and formability, which may trigger edge-cracking problems. The cause of the edge-cracking phenomena, that arises during forming or crash, have in several publications been traced to the damage induced by shear cutting processes used in early stages of the part manufacturing [1–6]. Numerical modelling methods serve as important tools during the product development, but the damage from shear cutting is seldom accounted for in traditional forming simulations. The commonly used forming limit diagram (FLD) defines the failure through thinning and localised necking and is extensively used for mild steels, but for damaged edges of AHSS materials failure tends to occur well before FLD criteria are fulfilled. To numerically model the shear cutting of AHSS can therefore aid in understanding and evaluating the edge-cracking in forming processes. This work presents an inverse modelling method to define material hardening and calibration of a Generalised Incremental Stress State Dependent Damage (GISSMO) failure- and damage model for an AHSS grade, based on plane stress tensile testing. Initial results of punching simulations showed deviating agreement in terms of cut edge shapes compared to experimental results. The authors of this article then developed an iterative approach of locally scaling the fracture locus in stress states corresponding to punching failure in order to obtain closer agreement to experimental edge morphology. The material investigated in this article is a complex phase AHSS grade called CP1000HD of 1.5mm thickness, commonly used in automotive crash applications.

2. Constitutive modelling

Quasi-static tensile testing of four different specimen geometries enabled evaluation of the material properties under varying stress states. The results from the tensile testing was used for calibrating a plastic hardening model and to define a fracture locus, both to be used for the shear cutting FE-model presented in Section 3. The tensile specimen geometries were presented by Sjöberg et al. [7] and are schematically shown in Figure 1.

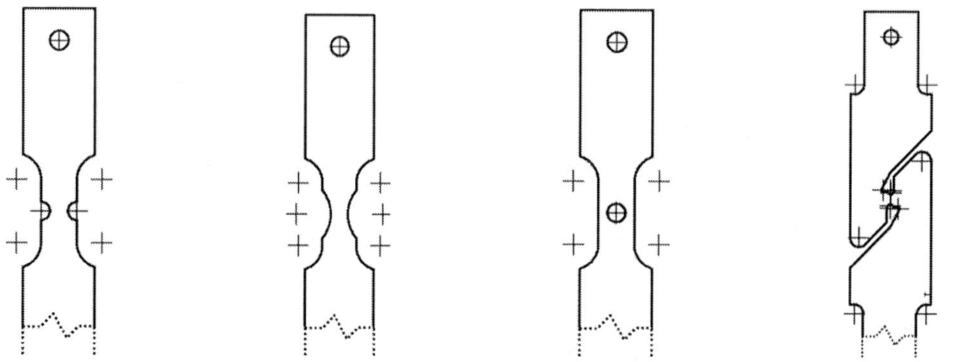

(a) R3.75. (b) R15. (c) Hole. (d) Shear.

Figure 1: Tensile specimen geometries.

2.1. Plasticity

An inverse modelling scheme was used to calibrate the plastic hardening model of the CP1000HD material. The inverse modelling approach consisted of calibrating the parameters of a Stiebler hardnening model such that the force - displacement results from a FE-model of the R3.75 specimen would match the corresponding experimental results. The Stiebler hardening model was described by Stiebler et al. [8] and shown in Equation (1), where parameters c_1 to c_4 were adjusted through the inverse modelling scheme shown in Figure 2. Along with the Stiebler model parameters, the Young's modulus E was calibrated. With the hardening models calibrated, a plastic hardening curve was written to an isotropic piecewise plasticity model (called MAT_24) for the FE-software LS-Dyna.

$$\sigma(\varepsilon_p) = c_1 + c_2\varepsilon_p + c_3(1 - \exp(-c_4\varepsilon_p)). \tag{1}$$

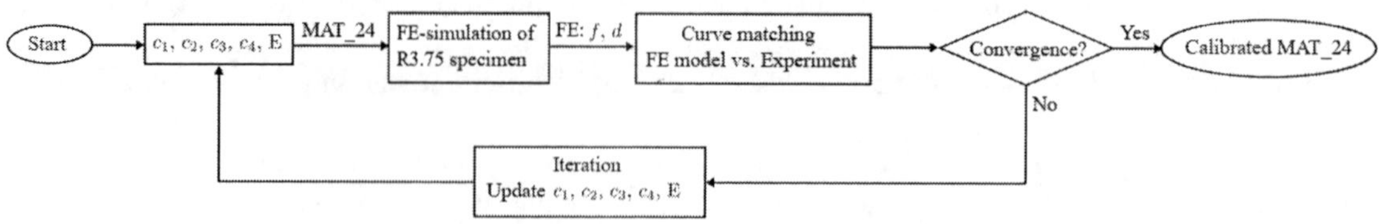

Figure 2: Calibration flowchart for the Stiebler hardening model.

2.2. Damage- and failure

In this work, modeling of damage and failure was carried out using the Generalized Incremental Stress State Dependent Damage Model (GISSMO), initially developed by Neukamm et al. [9]

and further improved by Basaran et al. [10]. GISSMO, implemented in LS-DYNA [11], is a phenomenological damage and failure model, using effective plastic strain, ϵ^{p}_{eff}, as a failure criteria. To account for stress-state dependent failure, the failure criterion may be expressed as a function of stress-state triaxiality, η, and lode angle parameter, $\bar{\theta}$, i.e., $\epsilon^{p}_{eff} = \epsilon^{p}_{eff}(\eta, \bar{\theta})$. Using tensor notation, the stress triaxiality and lode angle parameter may be expressed in accordance with Equations (2) and (3), where s_{ij} refers to the deviatoric part of the Cauchy stress tensor.

$$\eta = \frac{\sigma_{kk}}{3\sqrt{\frac{3}{2}s_{ij}s_{ij}}} \tag{2}$$

$$\bar{\theta} = \frac{1}{3}\arccos\left(\frac{3\sqrt{3}}{2}\frac{\frac{1}{3}s_{ij}s_{jk}s_{ki}}{(\frac{1}{2}s_{mn}s_{mn})^{3/2}}\right). \tag{3}$$

In Equation (4), the accumulative damage rule, implemented in GISSMO, is defined. $\dot{D}$ is the incremental damage value, n is a damage exponent, $\epsilon_f\left(\eta, \bar{\theta}\right)$ is the stress-state dependent failure strain, and $\dot{\epsilon}^{p}_{eff}$ is the increment of effective plastic strain.

$$\dot{D} = \frac{n}{\epsilon_f\left(\eta, \bar{\theta}\right)} D^{(1-\frac{1}{n})}\dot{\epsilon}^{p}_{eff} \tag{4}$$

By coupling the damage parameter, D, to the flow stress, σ, material degradation may be modeled, see Equation (5).

$$\sigma^* = \sigma\left(1 - \left(\frac{D - D_{crit}}{1 - D_{crit}}\right)^m\right) \tag{5}$$

Initiation of the coupling between damage and stress is governed by the material instability parameter, F, accumulated according to Equation (6). At $F = 1$, the coupling to the stress is initiated. ϵ_i refers to the instability strain, which may be prescribed as a function of both stress triaxiality, η, and the lode parameter, $\bar{\theta}$.

$$\dot{F} = \frac{n}{\epsilon_i(\eta, \bar{\theta})} F^{(1-\frac{1}{n})}\dot{\epsilon}^{p}_{eff} \tag{6}$$

2.3. Modified Mohr-Coloumb fracture locus

From the four tensile test geometries described in Section 1, fracture strains for four unique stress-states were obtained and used for fitting a modified Mohr-Coulomb (MMC) fracture locus, first presented by Bai and Wierzbicki [12], see Equation (7). The determined values of constants $C1 - C5$ are presented in Table 1. Assuming plane stress, the lode angle parameter, $\bar{\theta}$, may be expressed as in terms of stress triaxiality, according to Equation (2.3).

$$\varepsilon_f(\eta, \bar{\theta}) = \left\{C_2\left[C_3 + \frac{\sqrt{3}}{2-\sqrt{3}}(C^{ax} - C_3)\left(sec\left(\frac{\bar{\theta}\pi}{6}\right) - 1\right)\right]\cdot \right.$$
$$\left. \cdot\left[\sqrt{\frac{1+C_1^2}{3}}cos\left(\frac{\bar{\theta}\pi}{6}\right) + C_1\left(\eta + \frac{1}{3}sin\left(\frac{\bar{\theta}\pi}{6}\right)\right)\right]\right\}^{-\frac{1}{C_5}}, \tag{7}$$

where

$$C^{ax} = \begin{cases} 1 & \text{for } \bar{\theta} \geq 0 \\ C_4 & \text{for } \bar{\theta} < 0. \end{cases} \tag{8}$$

$$\bar{\theta} = 1 - \frac{2}{\pi}\cos^{-1}\left(-\frac{27}{2}\eta\left(\eta^2 - \frac{1}{3}\right)\right) \tag{9}$$

Table 1: MMC parameters.

Parameter	C_1	C_2	C_3	C_4	C_5
Value	0.0992	7.7814	1.0247	1.0479	0.2127

The MMC fracture locus was used as initial estimate for an inverse modelling scheme that scaled the fracture surface and calibrated the GISSMO parameters n and m. The aim of the inverse modelling was to obtain correlation between simulation and experiments in terms of material failure of the R3.75 tensile specimen. A more detailed description of the inverse modelling procedure for the hardening curve in Section 2.1 and the GISSMO damage- and failure model was given by Sandin et al. [13].

3. Numerical shear cutting models

To evaluate the effect of different tool clearances for shear cutting, two-dimensional axisymmetric FE-models of the shear cutting processes were developed in LS-Dyna. The models consisted of rigid tools (punch, die and blank holder) and the blank to which the isotropic piecewise plastic model and the GISSMO model was assigned. Spring elements were defined to model the blank holder spring and to account for the punch tool stiffness. Figure 3 shows the mesh of one of the FE-models simulated.

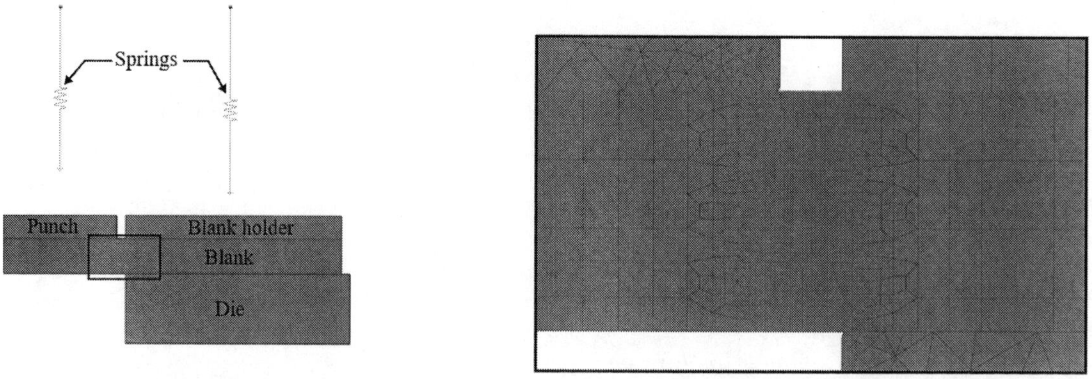

(a) Overview. (b) Close-up on the shear affected zone.

Figure 3: 2D axisymmetric FE-model setup.

Different tool clearances were obtained through horizontal offsetting of the axisymmetrixal die mesh, which corresponds to changing the die diameter. Table 2 shows the tool clearance and the corresponding die radius r_{Die} for the different load cases. The variation of clearances where chosen to include both optimal industrial cutting clearances (12-20%) and the extreme cases of 5.3% and 27%. Table 3 states the geometric properties for the punching models that remained constant at all load cases, i.e. the punch diameter r_{Punch}, the blank thickness t_{Blank}, the punch edge radius $r_{\text{Punch edge}}$ and the die edge radius $r_{\text{Die edge}}$. The geometric properties of the experimental shear cutting setup were measured at the time of conducting the experiments. The tool clearance was calculated according to Equation (10).

Table 2: Clearance and corresponding die radius of the shear cutting processes.

Clearance [%]	5.267	12.133	20.533	27.000
r_{Die} [mm]	5.077	5.180	5.306	5.403

Table 3: Constant geometric properties of the shear cutting FE-models.

t_{Blank} [mm]	r_{Punch} [mm]	$r_{\text{Punch edge}}$ [μm]	$r_{\text{Die edge}}$ [μm]
1.50	4.998	30	30

$$\text{Clearance [\%]} = \frac{r_{die} - r_{punch}}{t_{Blank}} * 100 \tag{10}$$

4. Local calibration of fracture surface (LCFS)

As stated in Section 1 were initial FE-results of the shear cutting processes inaccurate in terms of cut edge morphology. The FE-results were not able to capture the crack initiation following the shearing process, why the burnish surface length was over-predicted. The inaccurate modelling results were considered to originate from the damage- and failure model described in Section 2.3, where the fracture locus consequently lacked failure data that corresponds to the stress state of interest. As the initial FE-results over-predicted the length of the burnish surface, the authors of this article could conclude that the failure strain values of the MMC surface was too large at the point of crack initiation following the burnish. In order to increase the predictability of the damage- and failure model, an iterative approach was applied of fine tuning the fracture strain in the $(\eta,\bar{\theta})$-space corresponding to the crack initiation stress state.

Through numerical modelling of the shear cutting processes were the values of η and $\bar{\theta}$ at crack initiation identified for each cutting clearance investigated. In a Matlab script, a square box of $\pm 0.1\ \epsilon^p_{eff}(\eta,\bar{\theta})$ was removed from the original MMC fracture locus around the stress state point of crack initiation and replaced with the scaled failure strain value. Continuity of the fracture locus was preserved through spline interpolation with 100 point interpolation density between the original MMC surface and the scaled failure strain value. Figure 4 shows the CP1000HD fracture locus after applying the LCFS methodology on the MMC surface presented in Section 2.3. In Figure 4 the $(\eta,\bar{\theta})$-points corresponding to the stress state at crack initiation for different punching clearances are shown with colored dots. As the calibration points for 5.3% and 12.1% lay close to each other, the 12.1% calibration point was omitted. Figure 4 also shows the MMC plane stress failure curve defined by Equation 7, which was calibrated by the tensile tests in Figure 1. Consequently, the calibration technique means local adjustments to the original MMC failure surface in regions corresponding to crack initiation for a wide range of cutting clearances, while retaining the areas of the original MMC surface calibrated by tensile testing.

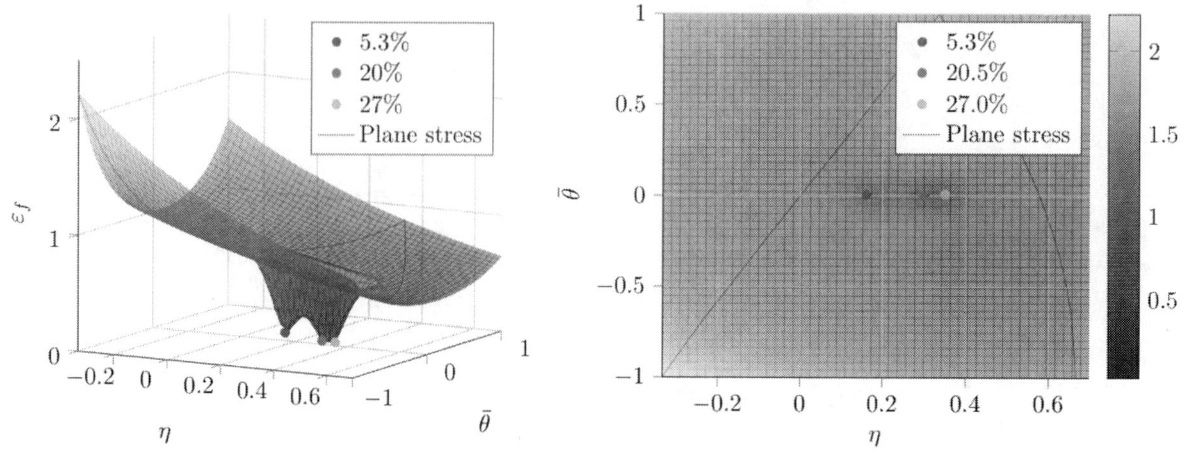

Figure 4: The fracture locus of CP1000HD after LCFS.

5. Results

In Figure 5, the cut edges from experiments and simulations are presented. Figure 5a contains the cut edge from the experiment, using a clearance of 5.3 %, and Figure 5b the corresponding simulation, using the MMC fracture locus. The simulation is able to predict the primary and secondary burnish but not the second zone of fracture, present in the experiments. Furthermore, the location of the secondary burnish deviate from experimental results. In Figure 5c, the cut edge of the simulation, using the LCFS, is presented. A closer agreement is now found between the experimental- and numerical results, as the model now predicts two burnish zones and two fracture zones, approximately at the correct locations.

In Figure 5d, the experimental cut edge, using a clearance of 12.1 % is presented, whereas Figures 5e and 5f contain the corresponding simulations, using the MMC surface and the LCFS, respectively. The effect of the LCFS is evident, as it predicts burnish and fracture zones agreeing with the experiments, whereas the model using the regular MMC does not.

Further increase of clearance, to 20.5 and 27 %, reduces predictability of the simulations using the regular MMC, seen in Figure 5g and 5h, as well as Figures 5j and 5k. For both clearances, the numerical models based on the LCFS produce cut edges in closer agreement with experimental data, see Figure 5g and 5i, as well as Figures 5j and 5l. Table 4 shows the percentual distribution of the cut edge zones over the blank thickness from experiments, as well as comparison with numerical results from cutting simulations with both original MMC failure model and calibrated failure model using LCFS.

Table 4: Cut edge zones from experiments and numerical modelling, expressed in terms of percentual distribution of the cut edge zones over the blank thickness. Numerical results are presented as results from original MMC failure model and from calibration using LCFS, separated by a forward slash symbol.

Clearance	Roll over [%]		Burnish [%]		Fracture [%]		Burr [%]		2nd burnish [%]	
	Exp.	MMC/LCFS	Exp.	MMC/LCFS	Exp.	MMC/LCFS	Exp.	MMC/LCFS	Exp.	MMC/LCFS
5.3%	3.6	2.9/2.9	22.4	47.4/48.6	22.9+20.9	16.1/16.0	0.9	0.7/0.7	30.2	33.6/32.5
12.1%	6.3	5/4.8	17.0	42.9/19.3	76.7	52.1/75.9	-	0.3/0.7	-	-
20.5%	9.4	8.2/8.1	19.7	91.8/21.0	70.9	-/70.9	-	18.7/4.9	-	-
27.0%	10.9	11.9/8.5	22.0	88.1/20.6	67.1	-/70.9	-	22.7/6.5	-	-

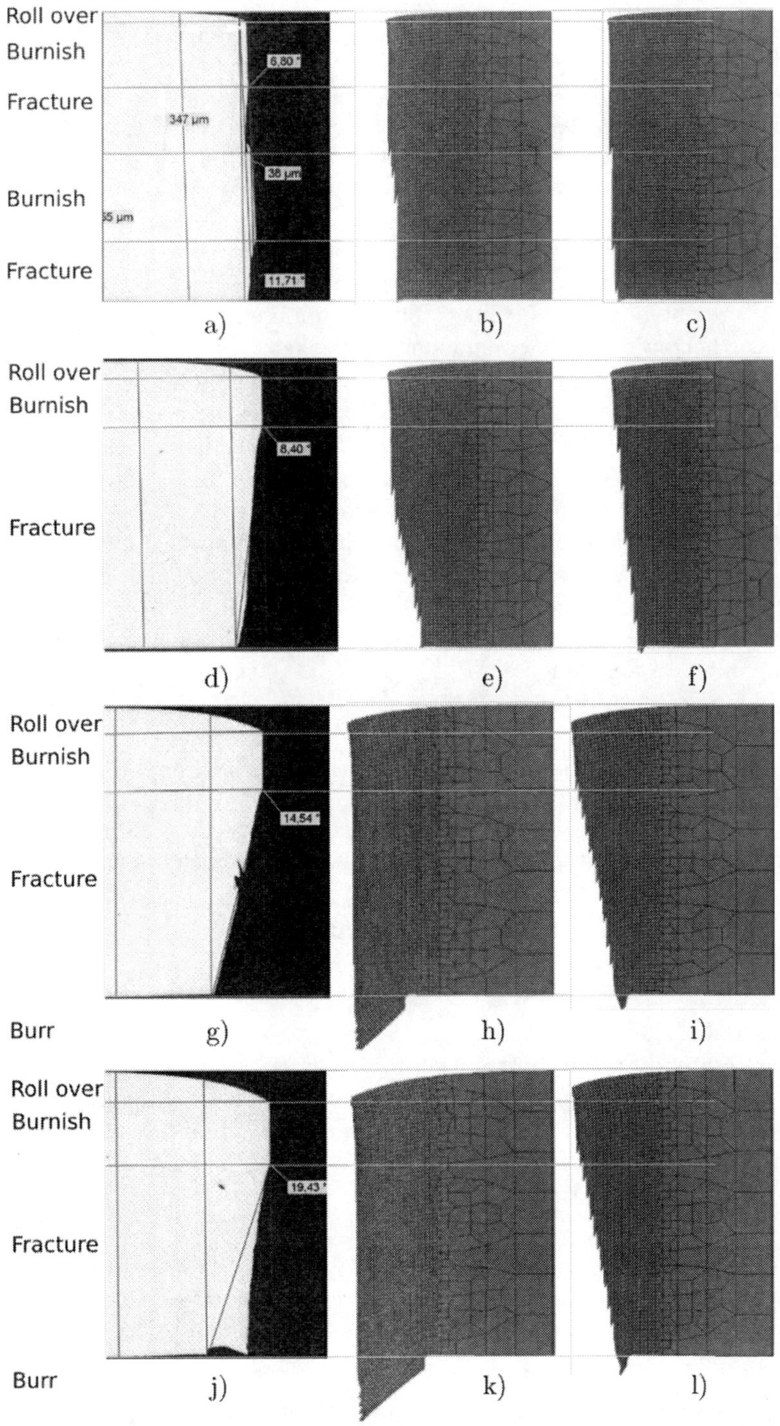

Figure 5: Cut edges from experiments and simulations are compared. Rows correspond to clearances of 5.3, 12.1, 20.5 and 27 %, respectively. In each row, the experiments are compared with simulated cut edges for the original and scaled failure surface, respectively.

6. Discussion and conclusions

The results presented in Figure 5 shows that utilisation of LCFS on the MMC fracture locus yields improved modelling results in terms of cut edge morphology. The enhanced modelling accuracy of LCFS is emphasised as it improves the modelling results for a large variation of punching clearances, ranging from 5.3% to 27.0%. As was done in this article, comparing experimental and numerical cut edge results by a single cross section, shows the possibility to improve the modelling results using LCFS. However, further work should the comparison between experimental and numerical cut edge results include effects from the circumferential variation often present in hole punching. The circumferential variation is caused by multiple factors such as orthotropic material behaviour, non-coaxial tool alignment and uneven tool ware.

However, the LCFS methodology can also be considered as a consequence of deficient calibration of the fracture locus. The inverse calibration procedure of the CP1000HD GISSMO model presented in Section 2.3 utilised a MMC fracture locus, which in turn was calibrated based on plane stress DIC measurements from tensile testing. Considering the stress state identification mentioned in Section 4 and shown in Figure 4 it can be concluded that the stress state corresponding to crack initiation during punching is not on the plane stress curve and rather found in the plane strain valley under mixed shear/uniaxial tension. The inverse modelling of the GISSMO model regarding correlation to the R3.75 tensile testing might furthermore impair the punching modelling accuracy, as the failure model risks to be sub-optimized towards a stress state not comparable to punching.

Ideally would the calibration of the MMC fracture locus include failure data from stress states corresponding to the different stages of the shear cutting process. However, such calibration experiments are particularly challenging for thin sheets as few experimental procedures exists that can control both trixiality and lode angle at material failure and that represents the plain strain condition that prevails during hole punching. Traditionally have this kind of trixiality/lode angle testing used both tensile testing of thin sheets and solid cylindrical specimens under compression, as presented by Bai and Wierzbicki [12, 14]. The results from the LCFS procedure can therefore aid in calibration of failure models where punching-like stress states are unable to be generated experimentally.

Acknowledgments

The authors are grateful for the contribution by Dr. Patrick Larour and Ing. Josef Hinterdorfer at voestalpine Stahl GmbH for providing experimental results of cut edges and for providing the sheet steel on which the numerical model was based. This research was partially funded by the European Commission, Research Fund for Coal and Steel, under grant agreement 847213 and the results are part of the CuttingEdge4.0 project.

References

[1] Konieczny A and Henderson T 2007 *SAE Tech. Pap. 2007-01-0340* pp 1–12

[2] Shih H C, Chiriac C and Shi M F 2010 *ASME Int. Manuf. Sci. Eng. Conf. MSEC 2010* **1** p 599–608

[3] Dykeman J, Malcolm S, Yan B, Chintamani J, Huang G, Ramisetti N and Zhu H 2011 *SAE Tech. Pap. 2011-01-1058* pp 1–10

[4] Thomas D J 2013 *J. Fail. Anal. Prev.* **13** pp 451–462

[5] Sigvant M, Falk J and Pilthammar J 2017 *J. Phys. Conf. Ser.* **896**

[6] Frómeta D, Tedesco M, Calvo J, Lara A, Molas S and Casellas D 2017 *J. Phys. Conf. Ser.* **896**

[7] Sjöberg T, Marth S, Kajberg J and Häggblad H-Å 2017 *Eur. J. Mech. A/Solids* **66** pp 279–286

[8] Stiebler K, Kunze H-D and El-Magd E 1991 *Nucl. Eng. Des.* **127** pp 85–93

[9] Neukamm F, Feucht M, Haufe A and Roll K 2008 *10th International LS-DYNA Users Conference* (Dearborn, MI)

[10] Basaran M, Wölkerling S D, Feucht M, Neukamm F and Weichert D 2010 *LS-DYNA Anwenderforum* (Bamberg)

[11] Livermore software technology 2020 Tech. rep.

[12] Bai Y and Wierzbicki T 2008 *Int. J. Plast.* **24** pp 1071–1096

[13] Sandin O, Jonsén P, Frómeta D and Casellas D 2021 *Materials* **14** p 7821

[14] Bai Y and Wierzbicki T 2010 *Int. J. Fract.* **161** pp 1–20

IOP Conf. Series: Materials Science and Engineering 1238 (2022) 012035 doi:10.1088/1757-899X/1238/1/012035

An approach to describe edge ductility

V Kesti[1*], M Folmerz[2], R Viereliä[3], P Rautio[3], R Ruoppa[3], P Plosila[4] and A Kaijalainen[4]

[1] SSAB Europe, Raahe, Finland
[2] SSAB Europe, Borlänge, Sweden
[3] Lapland University of Applied Sciences, Kemi, Finland
[4] Materials and Mechanical Engineering, CASR, University of Oulu, Oulu, Finland
*vili.kesti@ssab.com

Abstract. One of the challenges of utilizing advanced high-strength steels is their limited ability to withstand the forming of cut edges. Large production quantities of parts often lead to mechanical punching/shearing processes in blank preparation, providing a challenging starting point for forming processes. The most commonly used edge ductility test is the ISO 16630 hole-expansion test to describe stretch-flangeability properties. However, this method has been widely criticized for its large-scatter, unreliable results and the fact that it covers only a certain stress-strain state of cut edge forming. In addition, it does not provide enough data to be reliably used in forming simulations to predict edge failures. This paper presents an approach to create a more comprehensive way of describing overall edge ductility. Multiple edge forming test methods coupled with digital image correlation (DIC) were selected in order to investigate different edge loading scenarios in both open and closed trim line forming situations. Data regarding limiting local strain before cracking was collected for several steel grades, and results are gathered in 2D and 3D "Edge FLC" -figures. Attempts to utilize these results for simulation purposes are also presented. Results indicate that this approach can be useful to evaluate overall edge forming limits.

1. Introduction

The usage of advanced high-strength steels (AHSS) in car and truck parts has significantly increased in the last 15 years. Simultaneously, the complexity and overall demand for parts, their production and final performance has been increasing. For example, hot-rolled steel grades are often used in complex chassis components with mechanically cut blanks, followed by challenging forming. Cutting creates a work-hardened zone, that is, a shear-affected-zone (SAZ) [1], with high local strains, initial voids and even microcracks due to the intense mechanical shearing process. Therefore, the sheared edge often shows much lower formability than the base material. One of the major challenges to fully utilize advanced high-strength steels in these applications has been their limited ability to withstand the forming of mechanically cut edges. Another challenge is the difficulty of comprehensive edge formability characterization and lack of reliable tools to predict failures in simulation. For example, the traditional forming limit curve (FLC) can overestimate material properties in edge formability [2,3], leading to crack formation in production. Therefore, it is important to have a proper method of characterizing edge formability in a reliable and comprehensive way.

Edge ductility is most often described as a hole-expansion ratio according to the ISO 16630 hole-expansion test [4]. During the past 15 years, this method has been comprehensively investigated but also majorly criticized due to its large scatter, unreliable results and the difficulty of obtaining the same results between different laboratories [5-9]. One major limitation of the ISO 16630 test is that it covers only one stress-strain state of cut edge forming [9,10]. Depending on part design, sheared edges can be formed in a variety of different loading scenarios during actual part production and thus multiple different test methods have been created over the years. Schneider et al. [11] have gathered an overview of different methods, including an approach to describe different loading scenarios in a variety of tests by strain gradient. One approach has been presented by Frometa et al. [12], who showed that during edge forming, steels with greater fracture toughness create higher strain at fracture, together with a lower cutting clearance sensitivity.

In this paper, an approach is presented to describe overall edge ductility, both from the perspective of the steel development phase in order to compare different steel concepts as well as the part design phase. Multiple steel grades are studied in various loading situations, together with the possibility of utilizing the results for simulation purposes. One major point of interest is to study the possible differences between materials at the same strength level in different loading scenarios: is the ranking of the materials the same in all tests or will it vary depending on the loading case? Five different edge forming test methods coupled with digital image correlation (DIC) were selected in order to investigate multiple edge loading scenarios in both open and closed trim line forming situations. The main idea is to cover a wide scope of loading situations, including forming in- and out-of-plane. In addition to different loading situations and the possibility of DIC, the tests were selected for easy accessibility (no special tools for cutting) and flexibility of sample preparation (the possibility to use different cutting clearances, sheet thicknesses and steel grades). Data concerning the limiting local strain before cracking was collected for several steel grades, and the results are shown in a maximum major strain comparison chart together with 2D and 3D "Edge FLC" -figures. These figures can also be used to evaluate the formability against simulation results. The strain path was also detected for each test setup to describe the whole loading scenario.

2. Experimental

In total six hot-rolled 3 mm steels were selected for the testing, including three grades for 355 MPa and three for 700 MPa yield strength level. For the minimum 355 MPa level, a standard S355J2 was selected as the basic standard material together with two different concepts of thermomechanically (TM) hot-rolled S355MC steels, named S355MC(1) and S355MC(2). The two latter materials exhibit an approx. 50 MPa higher yield strength than the standard material, while elongation levels are very similar for all three steels. For the 700 MPa materials, three different TM hot-rolled steels were investigated: standard S700MC(1), trial S700MC(2) and the recently developed Docol 800HHE advanced high-strength steel. The mechanical properties of the investigated materials are shown in Table 1.

Table 1. Mechanical properties of test materials.

Material	$R_{p0.2}$ (MPa)	R_m (MPa)	A_5 (%)	A_g (%)
S355J2	370	540	29	14
S355MC(1)	430	500	31	13
S355MC(2)	420	495	31	14
S700MC(1)	750	850	19	9
S700MC(2)	720	840	15	6
Docol 800HHE	705	800	21	9

Five different test methods were selected for the actual test matrix: hole-expansion with spherical Nakajima punch (NHE) and flat punch (KWI), tensile test with punched hole (HTT), a double-bending test (DBT) [13] and a Diabolo-test [14] (Figure 1), with the three first having a closed trim line and the latter two an open trim line for the cut edge. The NHE, KWI and Diabolo tests were conducted in an Erichsen sheet metal testing machine with 600 kN blank holder force and 0.5 mm/s punch speed. NHE tests were performed with a hemispherical punch (diameter 100 mm) with multipurpose grease lubrication on the punch side and a blank diameter of 215 mm. KWI tests were conducted with a flat punch (diameter 50 mm) and sample size of 150 x 500 mm (three tests per sheet with 150 mm separation). The KWI test had 0.1 mm Teflon foil with multipurpose grease on both sides of the foil as lubricant. The Diabolo test had a special diabolo-shaped punch with a 20 mm radius, a blank size of 30 mm x 215 mm and no lubrication. DBT was performed with an Amada HFP 130-3 press brake with a punch speed of 1.0 mm/s, lower tool width of 40 mm and punch radius of 10–20 mm depending on material and flange height (10–18 mm). Different flange height/radius setups were used in order to obtain a wider scope of strain concentration situations, aiming for cracked specimens. The blank size in DBT was 145 mm x 145 mm. HTT was conducted with a Zwick Z250kN tensile testing machine, and the loading rate on the yielding area was 30 MPa/s and the speed at the end of the test was 0.005/s. Sample dimensions were 40 mm x 290 mm. For the sake of comparison, an ISO 16630 standard HER test without DIC measurement was conducted for all materials.

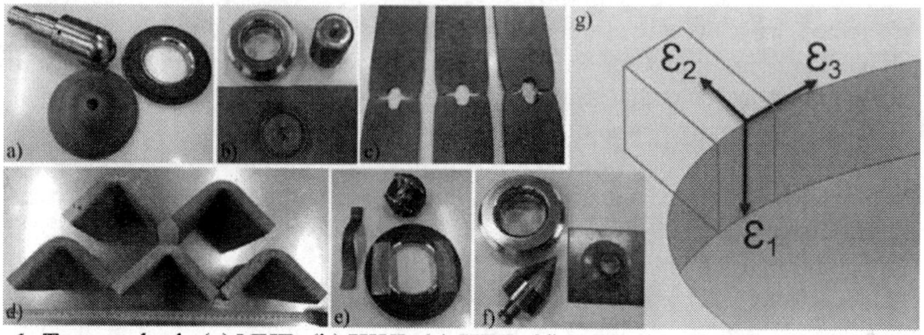

Figure 1. Test methods (a) NHE, (b) KWI, (c) HTT, (d) DBT, (e) Diabolo, (f) ISO 16630 and (g) edge strain components.

The holes for NHE, KWI and HTT were punched with the same tools as the ISO 16630 standard requires, with a 10 mm hole diameter and 12 % cutting clearance in an Erichsen forming machine. The Diabolo and DBT samples were cut in a CNC power shearing machine with 20 % cutting clearance. For each test, a minimum of five repetitions were performed, and for some tests, even more data is included into the analysis. All test samples were painted with a stochastic paint pattern and recorded with the GOM Aramis optical strain measurement system (frame rate of 10 Hz). Local point strains just before cracking were measured for three different edge strain components according to Schneider et al. [11]: ε_1 (thickness strain), ε_2 (minor strain) and ε_3 (major strain), shown in Figure 1(g). Strain analysis was conducted as close to the edge as possible with a facet size of 2.1 mm x 2.1 mm.

Different loading cases of these five tests were introduced and computed by Schneider et al. [11] and are schematically presented in Figure 2. Strain components ε_1, ε_2 and ε_3 are presented as simplified strain gradients starting from the edge, with the Y-axis showing the major strain component and X-axis the distance from the edge. The ISO 16330 test is also shown (Figure 2(f)) for comparison. The strain gradients are simplified, meaning that the absolute value of the major strain together with the depth of the gradient can vary from case to case, but currently only shows the basic trend. For example, the NHE and ISO tests show similar strain gradient profiles, but it is evident that, for example, the ε_2 component will have a distinct depth of strain gradient due to different test punch geometry. Tests were selected to provide different loading scenarios for the edge, including both in- and out-of-plane situations. In addition to different strain gradient profiles, different strain concentration profiles for ε_3 components along the cut edge were also selected by the test method. DBT, Diabolo and HTT tests force the strain to localize geometrically due to test geometry, while NHE and KWI create an initially uniform strain concentration around the hole, similar to the ISO16630 test. The strain will localize in these tests as well, but it will happen due to the strain localization tendency of each material, not by forcing the strain to localize by tool geometry as in the three tests mentioned earlier. One aim of the tests was also to include both open and closed trim lines into the test matrix: DBT and Diabolo representing the former and NHE, KWI and HTT the latter.

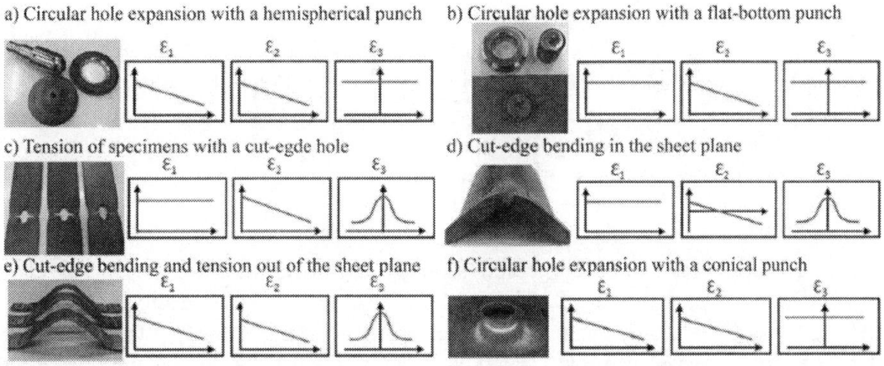

Figure 2. Theoretical strain gradients due to test geometry in three different directions on the edge area for (a) NHE, (b) KWI, (c) HTT, (d) DBT, (e) Diabolo and (f) ISO 16630. [11]

3. Results and discussion

3.1. Max local major strain

First, only data regarding major strain component ε_3 was collected from the tests to describe the absolute maximum local strain before crack formation. One goal of this approach was to simplify the results and to find out if the materials would end up being ranked in the same order in different tests when considering just maximum local strains. Figure 3 presents the results for maximum local major strain, except for ISO 16630, for which results are presented as basic hole-expansion ratios (with log scale) due to lack of DIC measurements. For materials in the 355 MPa level, it is evident that standard S355J2 creates much lower ISO HER and maximum local strain values compared to TM hot-rolled materials. When comparing two TM hot-rolled 355 MPa materials, it can be concluded that maximum local strains in NHE and KWI are ranked in a similar order than in the ISO HER test, with version 1 being better than version 2. However, in the rest of the tests, S355MC(2) shows higher local strains than S355MC(1). This indicates that a certain material could be optimal in one loading scenario, but for another scenario, another material would be a better choice. To conclude: if relying only on one test method, materials could be ranked in the "wrong" order. When comparing 700 MPa strength level materials (Figure 3(b)), the same ranking of the materials stays through all test methods: Docol 800HHE showing superior maximum strains compared to S700MC(1) and S700MC(2). It is still important to note that HTT provides a statistically similar level of results for all 700 MPa grades. This leads to the fact that if these materials are only tested by HTT, the excellent edge ductility behavior of Docol 800HHE in all other situations could be completely missed. According to the results, it is very difficult to create a comprehensive description of the edge ductility property by performing only one test.

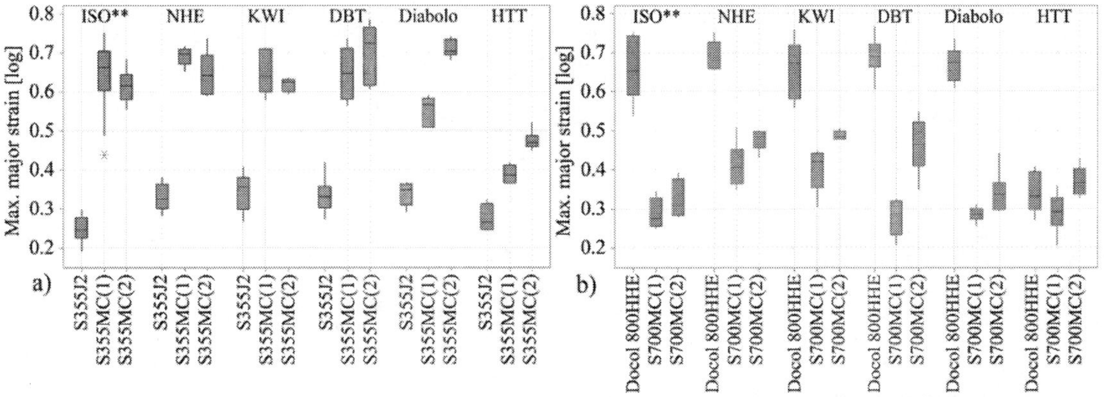

Figure 3. Max local major strain [ε_3] values for different tests: (a) 355 MPa and (b) 700 MPa materials. **ISO HE test values are converted into log scale from [%].

3.2. 2D Edge FLD

In order to use the minor strain component in evaluation as well, "2D Edge FLD" was created to visualize the data in a similar manner to traditional forming limit diagrams. The thickness direction strain component ε_1 is still neglected and only the ε_3 and ε_2 components were used as major and minor strains, respectively. Results for both material groups are presented in Figure 4 (a-f). This approach provides a larger amount of information about material behavior in the different tests. As can be seen from the results, all selected tests provide compressive strains for minor strain. It is also evident that relative scatter in minor strain values is much higher than for the major strain within the same test method. For the very moderate edge ductility material S355J2, the major strain for all tests seems to achieve a similar level, even though the test methods create different minor strain values, varying from -0.01 to -0.21. This indicates that the S355J2 material tolerates only a certain level of strain along the edge in different loading scenarios. On the other hand, S355MC(2) shows a trend of increasing major strain with higher compressive minor strains. It can be noted that HTT creates the lowest level of major-minor strain combination for S355MC(1), S355MC(2) and Docol 800HHE. For the rest of the materials, HTT results are not so distinctly low when compared to other test results. DBT and Diabolo tests create the highest combination of major (tensile) and minor (compressive) strain in general, with S700MC(1) being the only material that creates a rather uniform scatter and results for all tests. The result pattern in 2D Edge FLD shows the level where the combination of major and minor strain become limiting. It can be presumed that forming beneath this level should create crack-free components on the edge areas. This approach creates a good visualization of the qualities of the different materials when comparing overall edge ductility.

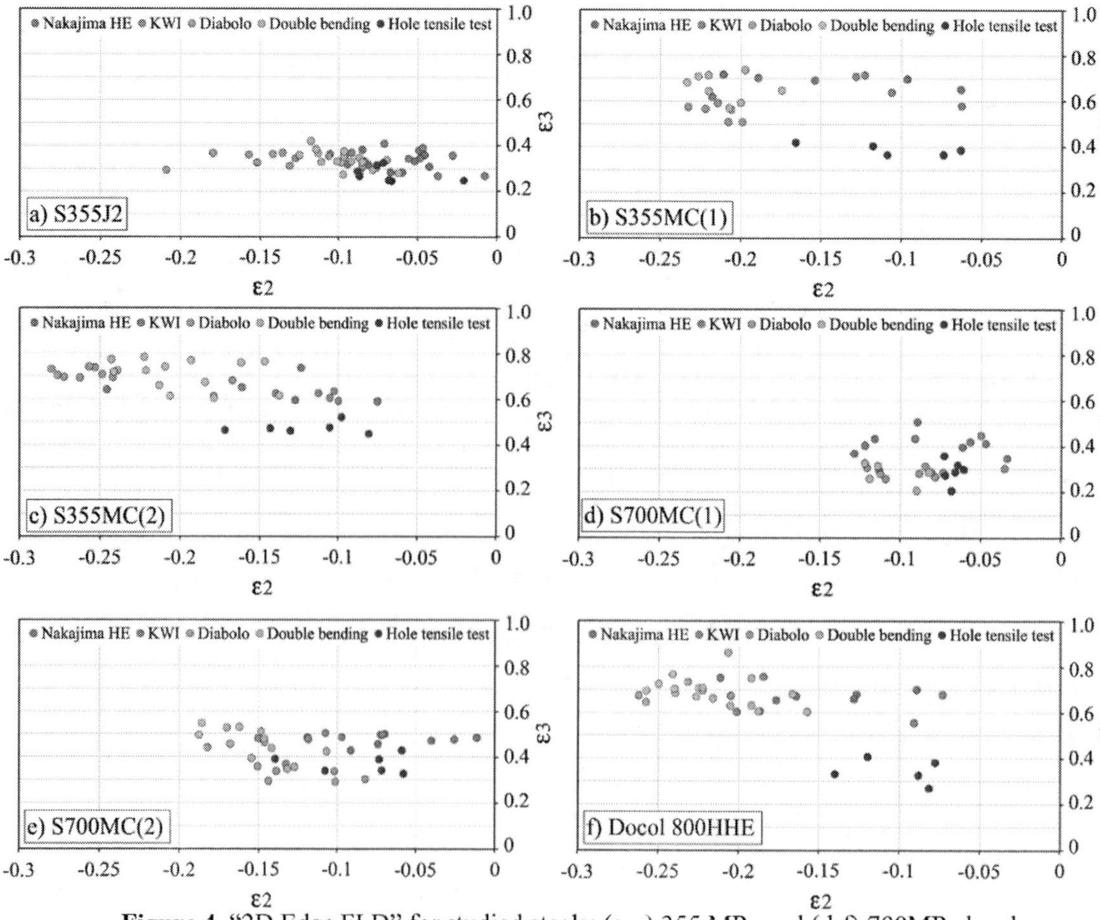

Figure 4. "2D Edge FLD" for studied steels: (a-c) 355 MPa and (d-f) 700MPa level.

2D Edge FLD approach could be too conservative in some cases. For example, in a situation that S355MC(1) or S700MC(1) materials are formed with specific NHE test loading situation. In this case, the limiting curve would be considerably lower than the results of NHE test indicate, as seen in Figure 4(b) and (d). If part design includes only one specific loading case, materials could be optimally tested only by one test to describe edge ductility. However, industrial parts often include multiple different loading cases, creating the need of simultaneous judgement of different stress-strain situations. This will make it very difficult or even impossible to choose only one test to describe edge ductility in a proper way. In order to complement the approach of major and minor strain, it is important to take the complete strain state situation into consideration, including thinning tendency and strain path.

3.3. *3D Edge FLD*

Going even further with analysis, an attempt was made to describe the final situation just before crack formation, including thinning as thickness reduction strain component ε_1. Results are presented as "3D Edge FLD" surface plots for 355 MPa and 700 MPa materials in Figure 5 and Figure 6, respectively. The 2D Edge FLD has now been placed into the Y- and X-axes, including the maximum local strain before cracking for the ε_3 and ε_2 components. The thickness reduction strain, ε_1, is in the Z-axis, creating the "height" of the surface. The idea is to visualize the limiting strains, not only for the major and minor strains but also for the maximum thinning before crack occurs. As can be seen in Figure 5, steel S355J2 creates a much lower and smaller limiting surface than the two other 355 MPa steels, indicating that the material does not exhibit much necking of the edge before cracking. Therefore, it can be said that the material is very prone to edge cracking but without the propensity to necking. With the "3D Edge FLD" approach, it is also possible to evaluate the thinning tendency in more detail. With similar levels of major and minor strain, a very different level of thinning of the edge area is seen for different materials. For example, with rather high minor and major strain levels, the Docol 800HHE material creates more thinning than the other two 700 MPa grades. It is also evident that when comparing the two 700 MPa grades, standard S700MC(1) and Docol 800HHE, the overall "edge ductility surface" is much wider and higher for the latter one. It can be concluded that the amount of necking on the edge area can be very different from test to test and from material to material.

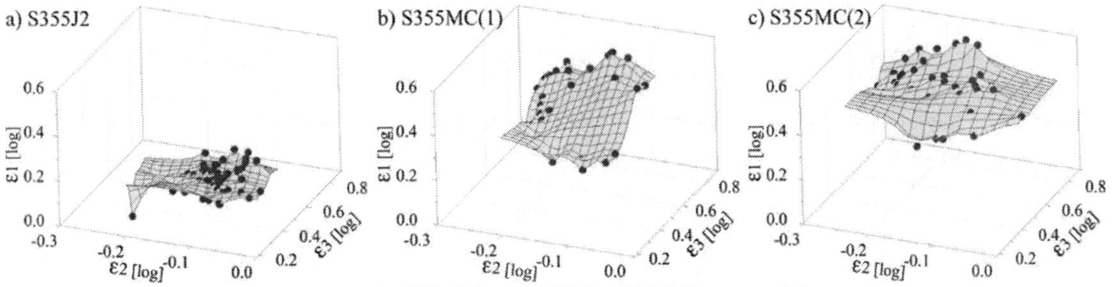

Figure 5. "3D Edge FLC" for 355 MPa materials.

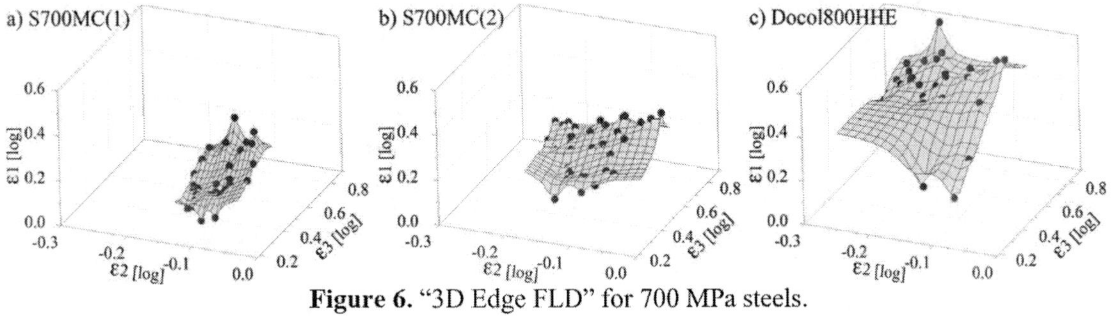

Figure 6. "3D Edge FLD" for 700 MPa steels.

There are often limits for the maximum thinning of a material when it comes to part production and simulations. According to these results, it could be stated that the thinning limit, when edge cracking is used as a failure criterion, could be very different in different materials. For example, a 0.2 [log] thickness strain could be too much for some materials, while for others, it would be underestimating the material ability. This could lead to a situation where full material performance is not utilized if the thinning of the edge does not reach critical limits. When comparing different materials, 3D Edge FLD visualizes and describes the overall edge ductility in a comprehensive way and makes the comparison between materials easy. This would also allow for more information during the steel development phase when comparing different steel concepts.

3.4. Simulation case

The simulation case was run with the Autoform R10 forming simulation software for the S700MC(1) steel. Two hypothetical parts were designed to create good and bad situation in the edge forming area. The aim is to present simplified example on how Edge FLD could be used for cracking risk evaluation. The schematic presentation of the parts production process for both good and intentionally bad design is shown in Figure 7. Blanks (500mm x 450mm) are initially cut into similar shape followed by a drawing operation (depth of 100mm), trim cutting and flanging operation. It can be noted that there is a risk of increased cut edge strain in the concave area of the part. Hence, it is important to be careful with the trim line in this area. For the good design, the trim line in the cutting operation is smooth and has no clear localization points. Whereas for the bad design, a cut-out is intentionally added right into the concave area (arrow) to act as a localization point for the strains in the following flanging operation. The more intensive localization behavior of the bad design can be clearly seen in Figure 7(c), showing major strain concentration along the edge in the critical area.

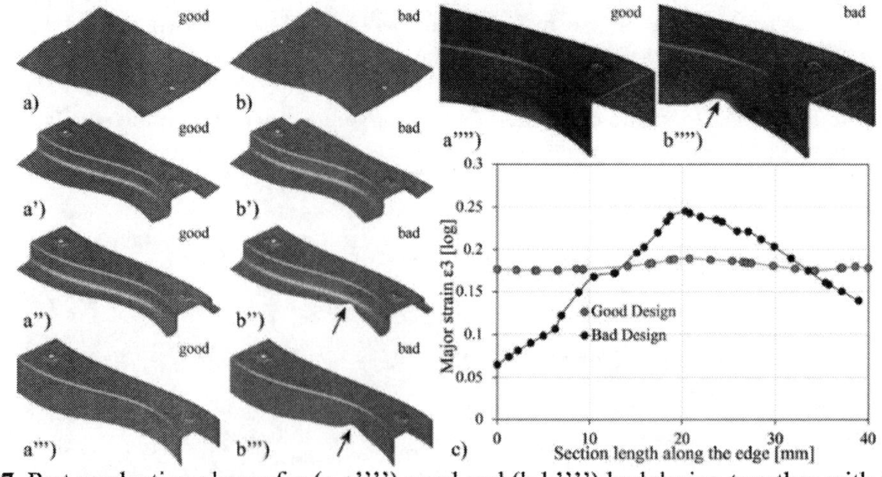

Figure 7. Part production phases for (a-a'''') good and (b-b'''') bad design together with (c) strain concentration along the critical area.

Strains are simulated on the critical area for all three strain components and presented in 2D Edge FLD in Figure 8a. It can be concluded that the part with good design could be possible to produce with S700MC(1) as it is just on the forming limit curve of the 2D Edge FLD. On the other hand, the part with bad design creates a major-minor strain combination which is in the crack risk area above the forming limit curve in the 2D Edge FLD, leading to a high risk of crack formation in the forming process. Thickness reduction was also analyzed in the simulation process. The results for both the good and bad design are put into 3D Edge FLD in Figure 8b. It can be seen that the good design (green dot) is located below the limiting strain surface while the bad design (black dot) creates strain combination above the strain surface. This indicates that the thickness reduction might also be too much for the bad design, leading to problems in part production.

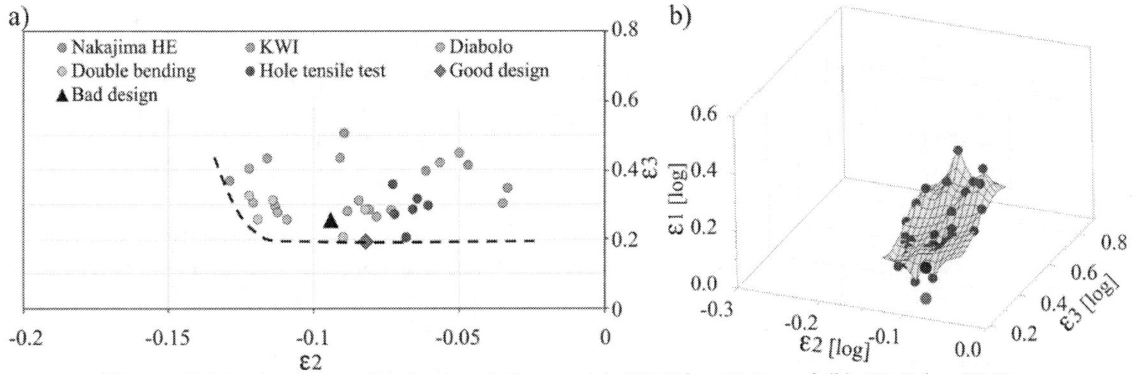

Figure 8. Maximum strains in simulation vs (a) 2D Edge FLD and (b) 3D Edge FLD.

In this study, the limiting strain was determined as maximum local strain just before cracking from the very edge of the samples. Atzema et al. [10] indicated that this approach can be feasible to describe limiting failure criteria for edges and could be utilized for edge strain criteria for simulation purposes. However, there is a possibility that this approach could create too optimistic an expression of failure criteria in real industrial applications. Therefore, fine-tuning of failure criteria should be performed through validation tests in the future. One possibility could be to use a similar best-fit parabola process that is used for FLD determination [15], which is an approach based on necking behavior. This approach could lead to better scatter behavior control, as the cracking phenomenon itself tends to be a more or less random event, especially in sheared edges. However, as seen from the results, the necking phenomenon could be suppressed in certain situations and may be difficult to use for edge cracking. In future work, proper limiting strain criteria needs to be validated through modeling and actual verification tests for edge areas.

3.5. Strain path

In order to identify the characteristic loading of each case, strain paths for different test setups are shown in Figure 9 for S355J2 and Docol 800HHE. These materials were selected to describe the low and high end of studied materials, from the perspective of both the strength as well as edge ductility. One representative test was selected for each case, and the strain history was plotted for each frame for the major ε_3 and minor ε_2 strains. It can be seen that the strain path is rather linear in all cases, taking into account that in most cases there seems to be some scatter in the final, unstable region before fracture. For S355J2, the path seems to slightly curve toward the plane strain region at the end stages of the DBT and Nakajima tests, indicating that the compressive minor strain slightly decreases. The same behavior was not seen for the Docol 800HHE material, and strain paths follow a linear path. This indicates that the strain paths can also be affected by the material properties, not only by test geometry. The effect of linear vs. non-linear strain path during edge forming could be an interesting research field in the future, together with different pre-strain conditions.

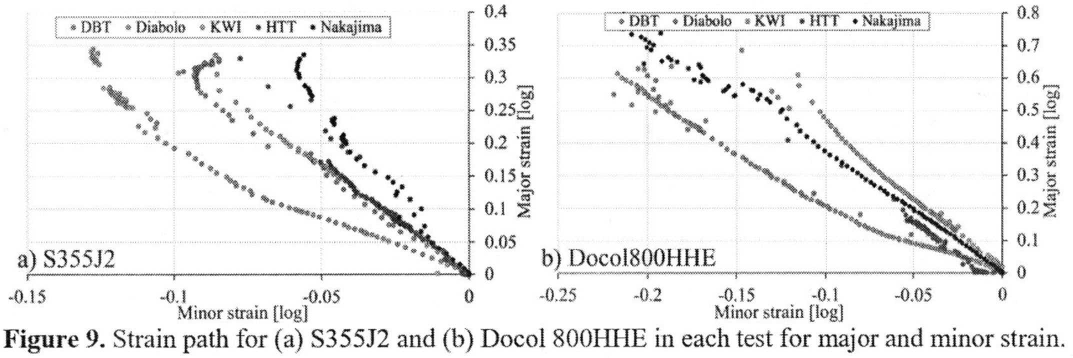

Figure 9. Strain path for (a) S355J2 and (b) Docol 800HHE in each test for major and minor strain.

4. Summary

In this study, several known edge ductility tests were selected to study the varying loading cases of sheared edges. Results are presented in multiple different ways: maximum local major strains, 2D Edge FLD and 3D Edge FLD, including presentation of the strain path in different test setups. According to the results, relying only on one test method can lead to erroneous choices during steel development or when selecting material for specific applications. Therefore, multiple different test methods are needed in order to describe material edge ductility in a comprehensive way. Approaches with 2D and 3D Edge FLDs were presented, and they created a good visualization of overall edge ductility, providing more information about material behavior in different tests. It is shown that Edge FLDs could be used to evaluate limiting strains also for simulation purposes. More work needs to be done to set and validate the limiting failure criteria for sheared edges in different loading situations.

Acknowledgments

Financial assistance of the Business Finland, projects ISA – Intelligent Steel Applications and FOSSA–Fossil-Free Steel Applications, are acknowledged.

References

[1] Levy B S and Van Tyne C J 2008 Failure during sheared edge stretching *J. Mater. Eng. Perform.* **17** 842–8
[2] Shi M and Chen X 2007 Prediction of stretch flangeability limits of advanced high strength steels using the hole expansion test *SAE Technical Paper* 1693
[3] Nitta J, Yoshida T, Hashimoto K and Kuriyama Y 2008 Development of the practical evaluation test and a study of numerical evaluations of edge fracture for stretch flangeability of sheet metal forming *Proc. Int. Conf. IDDRG (Olofström)* pp 93–101
[4] ISO 16630 2017 Metallic Materials - Sheet and Strip - Hole Expanding Test
[5] Atzema E, Borsutzki M, Braun M, Brockmann S, Bülter M, Carlsson B, Larour P and Richter A 2012 European round robin European round robin test for the hole expansion test according to ISO 16630 *Int. Conf. Proc. Neuere Entwicklungen in der Blechumformung (Fellbach)* pp 171-84
[6] Hance B 2017 Practical application of the hole expansion test *SAE Int. J. Engines* **10** 247-57
[7] Schneider M and Eggers U 2011 Investigation on punched edge formability *Int. Conf. Proc. IDDRG (Bilbao)*
[8] Larour P, Pauli H, Freudenthaler J, Lackner J, Leomann F and Schestak G 2016 Experimental artefacts on ISO 16630 hole expansion ratio *Int. Conf. Proc. IDDRG (Linz)* pp 480-98
[9] Larour P, Schauer H, Lackner J and Till E T 2016 Edge crack simulation with the modular "smiley" forming tool *Int. Conf. Proc. IDDRG (Linz)* pp 457–79
[10] Atzema E and Seda P 2015 SETi, sheared edge tensile test improved *Int. Conf. Proc. IDDRG (Shanghai) pp* 984–91
[11] Schneider M, Geffert A, Peshekhodov I, Bouguecha A and Behrens B A 2015 Overview and comparison of various test methods to determine formability of a sheet metal cut-edge and approaches to the test results application in forming analysis *Materialwissenschaft und Werkstofftechnik* **46** 1196-217
[12] Frómeta D, Lara A, Parareda S and Casellas D 2019 Evaluation of edge formability in high strength sheets through a fracture mechanics approach *AIP Conf. Proc.* **2113**, 160007
[13] Bouaziz O, Douchamps S, Durrenberger L and Bui-Van A 2010 The double bending test: a promising new way for an optimal characterization of cut-edges ductility *Int. Conf. Proc. IDDRG (Graz)* pp 575–82
[14] Held C, Schleich R, Sindel M and Liewald M 2009 Investigation on advanced forming limit prediction techniques for high strength steels *Int. Conf. Proc. IDDRG (Golden)* pp 461–70
[15] ISO 12004-2 2021 Metallic materials - Sheet and strip - Determination of forming-limit curves Part 2: Determination of forming-limit curves in the laboratory

IOP Conf. Series: Materials Science and Engineering 1238 (2022) 012036 doi:10.1088/1757-899X/1238/1/012036

Design Guidelines for interlocked stator cores made of CoFe sheets

D M Martin[1], F Bäcker[2], B Deusinger[3], C Kubik[1], P Gehringer[1], J Schröder[2] and P Groche[1]

[1] Technical University of Darmstadt, Institute for Production Engineering and Forming Machines Darmstadt, Germany
[2] VACUUMSCHMELZE GmbH & Co. KG, Hanau, Germany
[3] Technical University of Darmstadt, Institute of Electrical Energy Conversion Darmstadt, Germany

E-mail: daniel.martin@ptu.tu-darmstadt.de

Abstract. CoFe lamination stacks used for high-performance electric motors can be manufactured economically in high volumes by interlocking. In order to ensure sufficient joint strength with minimized sheet thickness, a comprehensive knowledge of the influences of various process parameters, such as embossing depth, clearance and counter punch force, is essential. To analyze these parameters, which also influence the magnetic properties, experiments are carried out and resulting joint strengths are determined in top tensile tests. The negative influences of the cutting process on magnetic conductivity and thus hysteresis losses due to residual stresses and plastic deformation are well known. In the subsequent stacking step, an influence of embossing and pre-stresses on the material properties is expected. In addition, local electrical contacts between the sheets may occur due to the interlocking process, causing additional eddy currents. Loss measurements are conducted to investigate the effect of the joining process on the magnetic properties of the stack. In doing so, the influence of process parameters such as the embossing depth and clearance on eddy current power losses is analyzed.

1. Introduction

As a result of progressive electrification, particularly in the mobility sector, the importance of economical manufacturing processes for components of electric drives is increasing while product quality requirements remain high. In vehicles that have to meet high performance requirements, lamination stacks made from CoFe alloys are used in the manufacturing of stator cores for electric motors. Compared with conventional FeSi electrical steel, these are characterized by a higher saturation flux density, allowing higher torques to be achieved within the same assembly space and higher power density at high rotational speeds.

In addition to the hysteresis losses, which depend on the material properties and are dominant at low frequencies, power losses in the stators of electrical machines consist mainly of eddy current losses, which represent a significant share of the power losses at medium and especially at high speeds [1]. These arise from eddy currents, caused by an induced voltage perpendicular to a time-varying magnetic field in the soft-magnetic core. The thickness t of the individual sheets enters quadratically into the specific mass-related eddy current power loss p_{W}. Therefore, iron cores do not consist of a solid body, but are made of individual laminations electrically insulated from each other. For the same height of the stator or rotor, the added losses of the laminated core are thus lower. For this reason, efforts are being made to further reduce the thickness of the laminations in order to reduce iron losses. In the present work, CoFe laminations with a thickness of 0.2 mm are investigated.

One manufacturing process for stacking sheets, which represents an economical alternative for the production of larger quantities, is interlocking. After the cutting step (usually a shear cutting process), the process is subdivided into the process steps embossing (Figure 1(b)) and stacking (c), in which the laminations are pressed together with the stacking force F_s. The mechanism of action in interlocking is based on an interference fit between the outer and inner diameter of the punching nubs [2], see Figure 1(a). The resulting joint strength F_j must be sufficiently high to ensure safe handling during transport and further processing as well as the necessary strength during operation. In addition, high geometric accuracy is required to avoid stacking errors like curved or skewed stacks.

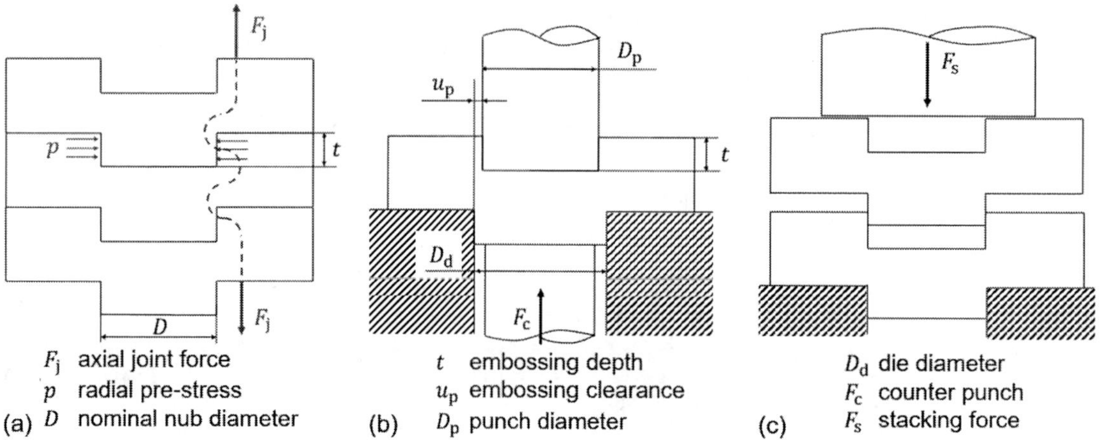

Figure 1. Schematic representation of the process steps in.

To ensure reliable joining of the individual sheets into a stack and at the same time minimize the sheet thickness to reduce eddy current losses, comprehensive knowledge of the process variables in the design process is essential. To enable prediction of joint strength, the interrelationships of numerous process parameters must be analyzed.

Thereby, the aim is to quantify the influence of the parameters embossing depth t, clearance u_p, counter punch force F_c and stacking force F_s on the resulting joint strength by means of experiments. However, different effects act along the process chain, which influence not only the mechanical strength but also the magnetic properties of the lamination stacks leading to increased iron losses [3]. Depending on the cutting process, thermal and/or mechanical residual stresses are introduced into the material during the cutting step [4, 5], and plastic deformations can cause changes in the microstructure [6].

A negative influence of these mechanisms on the electro-magnetic conductivity and thus the iron losses has been known for FeSi electrical sheet for a long time [7, 8, 9] and can be described with various modeling approaches [10], but are still the subject of current research activities [11]. During the stacking step, local electrical contacting between the layers may occur, which causes additional eddy currents. Furthermore, the deformations caused by the embossing of the nubs and the pre-stressing generated during stacking influence the material properties analogously to the cutting process [3, 12].

In previous experimental investigations, the significance of eddy current losses in the area of the punching nubs has been demonstrated [13]. This is due to the electrical contacting (damage to the insulation induces additional eddy current paths) as well as the change in the magnetic flux density in the area of the formed nub [1]. As a result, field displacement effects lead to a local reduction of the magnetic flux density and thus to a reduction of the available mechanical power. Furthermore, the influence of the nub shape on the joint strength was considered [13]. An investigation of the influence of the embossing clearance of cylindrical nubs on the deformation of the contact zone and the joint strength was carried out by Lin et al. [14] and Park et al. [15].

The presented work investigates which parameters are relevant for the design of interlocked joints for use in electric motors. Experimental investigations are used to determine the influence of these parameters on the target variables joint strength and iron losses in a stator core.

2. Methods and Materials

The subject of the investigations are test stacks made of 10 layers of CeFe test rings of sheet thickness 0.2 mm with the geometry shown in Figure 2(a) which are stacked using interlocking. Each layer, including the bottom one, is provided with one cylindrical punching nubs with a diameter of 1 mm. The interlocking process takes place in the unannealed state. Top tensile tests are used to identify which influencing variables have the greatest impact on mechanical joint strength. To determine the influence of the most significant design parameters on the component properties with regard to the subsequent application, an annealing process is carried out in the further test series after the stacking process before the mechanical and electromagnetic target values of the stacks are determined experimentally.

In the annealed condition, the test material has the properties listed in Table 1. The annealing process after stacking is performed for optimum magnetic properties at 880 °C for 6 h. The sheets are coated with a magnesium ethylate coating which is transformed to MgO during annealing. The influence of annealing on the joint strength of interlocked components has not yet been investigated so far. In the case of CoFe alloys, however, it is known that the mechanical and magnetic material properties can be influenced by an annealing process, with lower coercivity being associated with lower yield strength and vice versa. Also for FeSi alloys the influence of the annealing process on the magnetic properties due to the strain distribution has been proved [3]. Thus, the mechanical and magnetic properties of the material behave in opposite ways. The investigations presented are based on material that has not yet been heat treated prior to processing and is annealed for optimum magnetic properties following the interlocking process.

Table 1. Material properties of the CoFe alloy (*measured at a sheet thickness of 0.35 mm; [1]annealed for optimum magnetic properties) [16].

composition by weight	saturation magnetization B_s	mass density ρ	coer-civity $H_c^{*,1}$	yield strength $R_{p0.2}^{*,1}$	tensile strength $R_m^{*,1}$	modulus of elasticity $E^{*,1}$
49 % Fe, 49 % Co, 2% V and Nb	2.35 T	8.12 g/cm^3	50 A/m	210 MPa	400 MPa	200 GPa

The test tools shown in Figure 2 are used to produce the interlocked lamination stacks. In order to reduce complexity and avoid unwanted interdependencies between the process steps, embossing and stacking are implemented in separate tools. Figure 2(b) shows the tool for the embossing process. This has a three-part design and consists mainly of an upper tool carrying the punches, a lower tool containing the dies and the counter punch mechanism, and a guidance plane inbetween that carries the blank holder. The embossing tests are carried out on an eccentric press in single-stroke operation, whereby the embossing depth t is specified by setting the bottom dead center. Figure 2(c) shows the stacking tool, in which the lamellae are pressed into a stack by applying a defined force in the axial direction. The individual test rings are inserted manually and centered radially via a guidance component in the middle. The tangential centering is achieved via the nubs themselves.

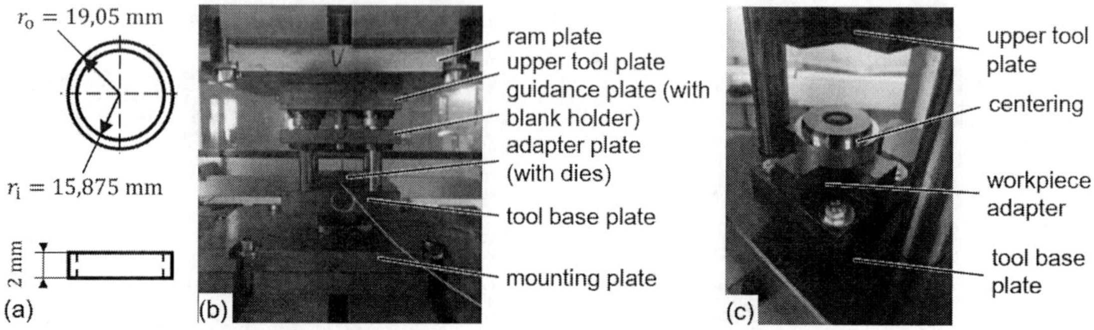

Figure 2. Test stack geometrie (a), embossing tool (b) and stacking tool (c).

The joint strength F_j of the interlocked specimens is evaluated by means of top tensile tests on a tensile testing machine, whereby a force in the axial direction is applied and slowly increased continuously until failure of the joint by separation of the stack occurs. As shown in Figure 3, the force is applied via disposable adapters made of plastic (PLA), which are bonded both to the test rings and to receptacles made of aluminum, which are clamped in the testing machine.

To investigate the magnetic properties, iron loss measurements are performed according to [17] in the frequency range 50 Hz to 1000 Hz at field strengths of 0.5 T to 2 T. For this purpose, the test stacks are placed in a trough and mechanically wrapped with a copper coil, as shown in Figure 4. To evaluate the influence of the joining process on the iron losses in the core, reference measurements are made for all frequencies and field strengths on a loosely stacked package of sheets without punching nubs, which serve as a reference for the additional losses due to interlocking.

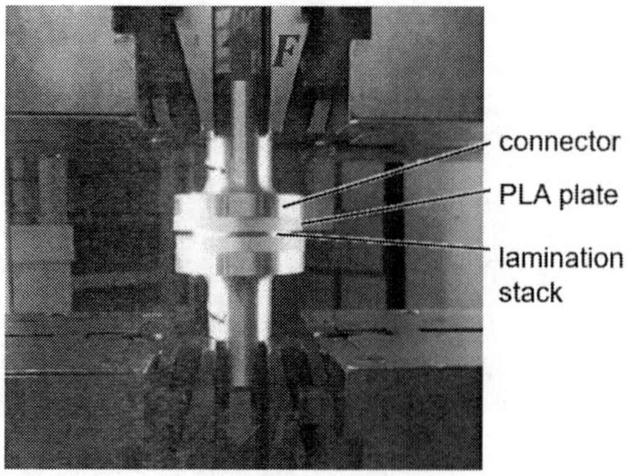

Figure 3. Test setup for determination of the axial joint force.

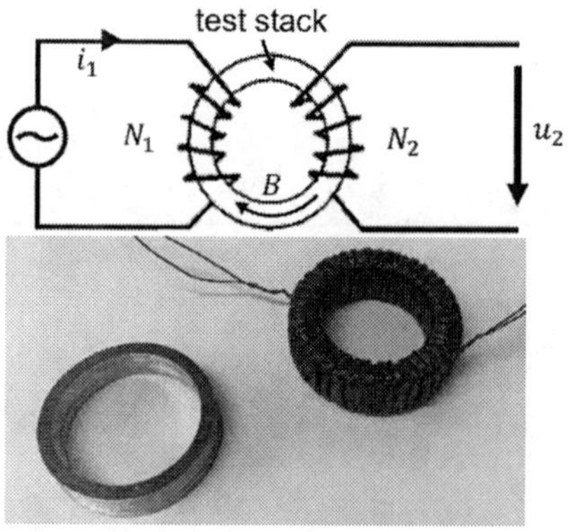

Figure 4. Test setup for measuring iron losses of test rings according to [17].

In a preliminary test, it is first checked whether the number of interlocking spots has a significant influence on the joint strength of the stacks. A comparison of the joint strength of a stack provided with 5 nubs distributed over the circumference with a stack with only one nub per layer shows that the achieved strength scales in good approximation with the number of nubs. The result of this comparison normalized to the joint strength of the one-nub stack is shown in Figure 5. For this reason, the following investigations can be carried out representatively using rings with only one nub. In this way, the influence of the varied process variables on the individual nubs can be considered.

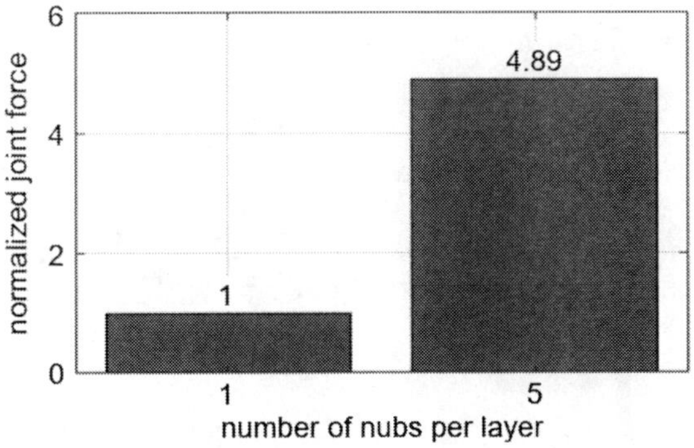

Figure 5. Influence of the punching nub number on the joint force of the lamination stack normalized to the joint strength of a stack with one interlocking spot.

3. Results

3.1. Joint Strength

To identify the main influencing variables on the target quantity joint strength, a full factorial 2^3 DOE is carried out varying the embossing depth, the embossing gap and the blank holder force. Previously, a variation of the stacking force was done to evaluate its influence on the joint strength. The results of this is shown in Figure 6(a). It should be mentioned that all results are normalized to the maximum achieved joint force in this test series. Accordingly, the joint strength increases slightly with increasing stacking force. For the conducted experiments, a stacking force of 1500 N was considered to be sufficient to investigate the influence of the relevant process parameters of the embossing step. The blank holder force was set to 30% of the maximum punch force during emobssing, according to preliminary experiments. The embossing depth is set to 50% or 90% of the sheet thickness s by adjusting the lower dead center. The embossing clearance is varied between 0.5% and 1.5%s by changing the punching dies. The counter punch force is set to 15% or 25% of the maximum punch force by varying the pre-stress of a spring in the counter punch mechanism located in the tool base plate. As can be seen in Figure 6(b), an increase in the embossing depth leads to a significant increase in the joint strength achieved. An increase in the embossing clearance leads to a moderate decrease (Fig. 6(c)) while increasing the counter punch force only leads to a slight drop in the joint strength (Fig. 6(d)).

Using the normalized step values listed in Table 2, the effect of the varied parameters and their combinations on the resulting joint strength is calculated. These are shown in Figure 7. It can be seen that embossing depth and clearance have the greatest influence on the joint strength, while the counter punch force has no direct influence. However, it can be observed that a higher counter punch force helps to increase the maximum achievable embossing depth, without failure due to nub break-off or local fracture.

Based on these findings, the influence of stamping depth and embossing clearance is further investigated in another fully factorial 3^2 series of experiments. Figure 8(a) shows the joint strength as a function of embossing depth at an embossing clearance of 1%, and Figure 8(b) shows the joint strength as a function of embossing clearance at an embossing depth of 70%. Analogously to the results in Fig. 7, all results in Fig. 8 are normalized to the maximum joint strength achieved in this test series. From these results, a regression model was used to derive the diagram shown in Figure 8(c) to predict the achievable joint strength when the embossing

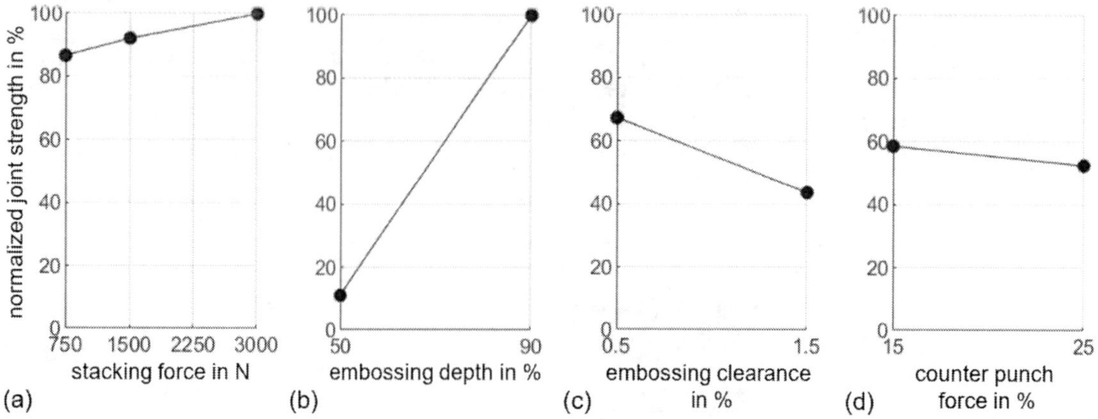

Figure 6. Influence of the embossing depth and clearance on the joint strength of a single interlocking nub.

Table 2. Standardized step values of variation parameters for calculating the effects on the joint force.

standardized step values	-1	1
embossing depth t in $\%s$	50	90
embossing clearance u_p in $\%s$	0.5	1.5
counter punch force F_c in $\%F_\mathrm{s}$	15	25

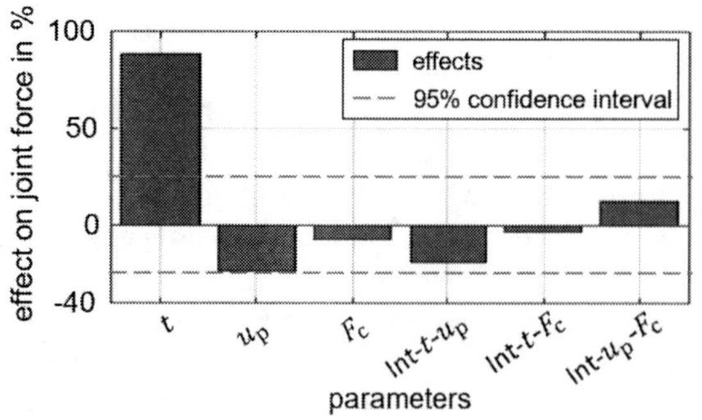

Figure 7. Effects of process parameters on joint force; entries labelled with 'Int-...' denote the effects of the interaction of two parameters.

depth and clearance are varied. Increasing the embossing depth from 50% to 70% results in a significant increase in joint strength. A further increase in the embossing depth only causes a smaller increase. Up to an embossing depth of 80%, reliable stacking was possible in all configurations. With further increase, a critical range was identified. Above an embossing depth of 80%, there is an increased risk for failure of the nubs during the embossing process and thus a safe joining process is not guaranteed.

Counter punch force and embossing clearance have been shown to influence the maximum achievable embossing depth. Both factors make it possible to further increase the embossing depth before the residual cross section of the nub fails. The embossing clearance also has a direct influence on the joint strength. As the embossing clearance decreases, an increase in joint strength can be observed. Furthermore, the stacks were subjected to a final annealing to achieve the optimum magnetic properties. This also has an influence on the joint strength

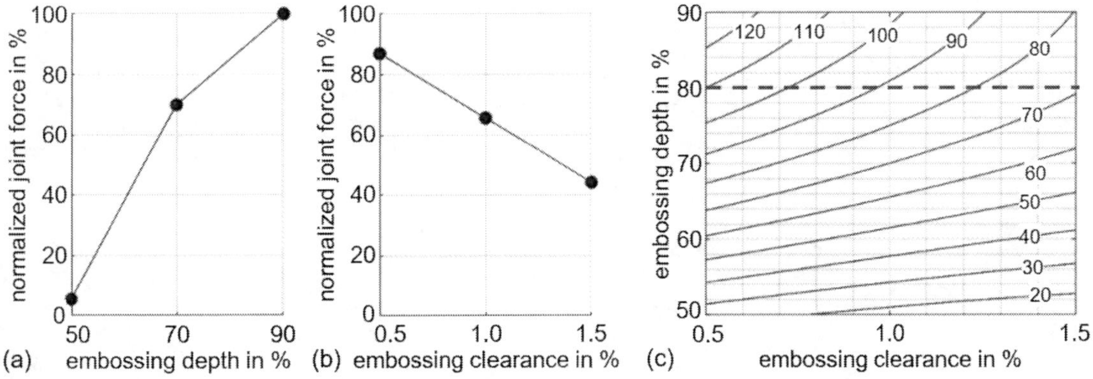

Figure 8. Influence of the embossing depth and clearance on the joint strength of a single interlocking nub: (a) variation of the embossing depth at an embossing clearance of 1.0%; (b) variation of the embossing clearance at an embossing depth of 70%; (c) regression map for variation of embossing depth and clearance

as in all parameter configurations the joint force of the annealed stacks was at least two times higher than the joint force of the unannealed stacks. The explanation for this can be assumed to be locally occurring thermal welding at the contact surfaces of the nubs during the annealing process.

3.2. Iron Losses

Figure 9 shows the results of the iron loss measurements for different embossing depths (a-c) under variation of the embossing clearance. The graphs show the relative change of the total iron losses in the stack compared to the reference measurement on loosely stacked samples. No significant influence of the stamping depth on the core losses can be detected.

In Figure 9, the 95% confidence intervals of the plotted curves are shown in dashed lines. The large measurement uncertainty is due on the one hand to the small number of samples (3 repetitions) and on the other hand to the fact that test rings with only one interlocking spot were used, which cause only small amounts of additional iron losses, so that the percentage deviations are highly afflicted with uncertainty.

However, it can be observed that the losses of the interlocked samples are always higher than those of the loose lamination stacks for all configurations in the entire frequency range. This leads to the conclusion that interlocking causes an increase in hystereses as well as eddy current losses. At high frequencies, there is a tendency for the iron losses to be higher the smaller the embossing clearance becomes. This indicates that the eddy current losses in particular are influenced by the embossing clearance. One explanation for this is the contact area at the nub flank, which is shown in Figure 9(d-f) as an example for an embossing depth of 80%. In the case shown, the contact area is significantly larger for an embossing clearance of 0.5 % than for a clearance of 1.0 % or 1.5 %.

4. Discussion and Conclusion

The presented investigations show that a reliable interlocking process is possible for 0.2 mm thin sheets of cobalt-iron, where the embossing depth and the embossing clearance can be identified as significant influencing factors on joint strength. Entscheidend für die Verbindungsfestigkeit sind die radialen Spannungen zwischen den Fügepartnern und die Höhe der Kontaktfläche. A larger embossing clearance causes a higher deformation of the joining zone during the stacking step due to the interference fit and a higher draft angle. This leads to a reduction in the

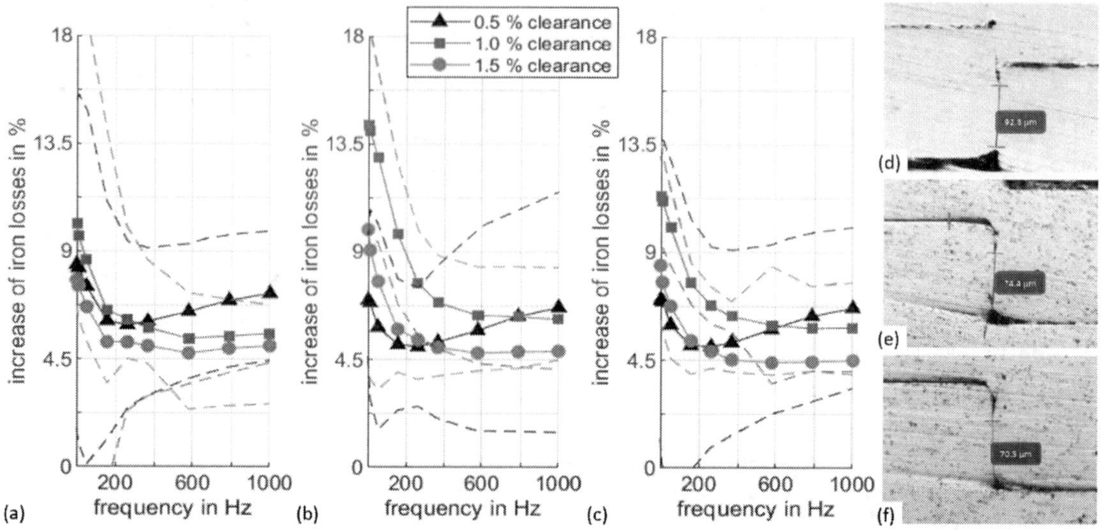

Figure 9. Experimentally determined interlocking-induced iron losses under variation of the embossing clearance at embossing depths of 70% (a), 80% (b), and 90% (c) and contact area height at the nub flank for a embossing depth of 80% at a clearance of 0.5% (d), 1.0% (e), and 1.5% (f).

effective contact area heigth and a lower joint force after removal of the punch. This observation is consistent with the investigations of Lin et al. [14]. A clear influence of embossing depth and clearance on the core losses cannot be proved.

The counter punch force shows no direct influence on the joint strength but is necessary to increase the maximum achievable embossing depth before failure occurs. An increase in the pressing force in stacking leads to a slight increase in the joint strength.

The annealed stacks show at least a doubling of the joint strength compared to the unannealed stacks. The comparison of a stack that has five interlocking spots with the stacks that have only one seems to show good scalability in terms of joint strength.

The interlocking of test stacks consisting of 10 layers of 0.2 mm thick individual sheets of the selected annular geometry made of a CoFe alloy with 49% cobalt using single cylindrical nubs of 1 mm diameter leads to a measurable increase in the core losses in all combinations of embossing depth and clearance considered under variation of the excitation frequency in the range from 0 to 1000 Hz. This increase was demonstrated experimentally. The measured increase in power loss is due to an increase in both hysteresis (offset between the curves) and eddy current losses (higher slope of the curve). However, the difference between the iron losses in the entire stack of sheets with only one interlocking spot compared with the reference measurement on a loose stack of sheets without interlocking spots is small. Taking into account the present measurement uncertainty in the investigated range of the process variables embossing depth and embossing clearance, no clear relationship to the iron loss performance can be described.

This suggests that when designing an interlocked joint in which the area of the nubs is small compared with the total stack, as in the case studied, the embossing depth and embossing clearance can be selected for the maximum achievable joint force of the individual nubs. Accordingly, the number of nubs required to achieve the necessary joint strength of the stack can be minimized. However, this initially applies only to a constant nub diameter. A statement on the influence of the nub diameter and thus an evaluation of whether many smaller nubs should be preferred over a few large ones with the same overall joint strength is not yet possible.

Therefore, additional stack properties such as vibrational behaviour will have to be taken into account in further investigations. In addition, the influence of different nub shapes on the iron losses needs to be further investigated.

Acknowledgments
The research presented here was carried out at the Institute of Production Engineering and Forming Machines (PtU) at Technical University of Darmstadt in close cooperation with Vacuumschmelze GmbH, Hanau. We would like to thank Vacuumschmelze GmbH for financing the tools, providing the experimental equipment for the magnetic measurements, and providing industrial expert advice. Values of experimental results have been normalized for this paper as the exact values are considered intellectual property by Vacuumschmelze GmbH. The authors would like to thank the German Federation of Industrial Research Associations (AiF) and the German Federal Ministry for Economic Affairs and Climate Action (BMWK) for funding for personnel resources within the framework of project no. IGF 22036 N. Additionally, thanks go to Mr. Robin Löw, who provided a valuable contribution to the execution and evaluation of the experiments. Furthermore, the authors would like to thank Bruderer AG for providing the high-speed press BSTA 810-145 on which the tests were carried out. 15:30

References

[1] Lamprecht E 2014 *Der Einfluss der Fertigungsverfahren auf die Wirbelstromverluste von Stator-Einzelzahnblechpaketen für den Einsatz in Hybrid- und Elektrofahrzeugen: Zugl.: Erlangen, Nürnberg, PhD thesis* (*Fertigungstechnik - Erlangen* vol 247) (Bamberg: Meisenbach) ISBN 9783875253627
[2] Groche P, Wohletz S, Brenneis M, Pabst C and Resch F 2014 *Journal of Materials Processing Technology* **214** 1972–1994 ISSN 09240136
[3] Senda K, Toda H and Kawano M 2015 *IEEJ Journal of Industry Applications* **4** 496–502 ISSN 2187-1094
[4] Leuning N, Steentjes S, Stöcker A, Kawalla R, Wei X, Dierdorf J, Hirt G, Roggenbuck S, Korte-Kerzel S, Weiss H A, Volk W and Hameyer K 2017 *AIP Advances* **8** 047601 URL `https://aip.scitation.org/doi/10.1063/1.4994143`
[5] Schauerte B, Leuning N, Vogt S, Moll I, Weiss H, Neuwirth T, Schulz M, Volk W and Hameyer K 2020 *Journal of Magnetism and Magnetic Materials* **504** 166659 ISSN 03048853
[6] Senda K, Ishida M, Nakasu Y and Yagi M 2006 *Journal of Magnetism and Magnetic Materials* **304** e513–e515 ISSN 03048853
[7] Nakata T, Nakano M and Kawahara K 1992 *IEEE Translation Journal on Magnetics in Japan* **7** 453–457 ISSN 0882-4959
[8] Hubert O and Hug E 1995 *Materials Science and Technology* **11** 482–487 ISSN 0267-0836
[9] Emura M, Landgraf F, Ross W and Barreta J 2003 *Journal of Magnetism and Magnetic Materials* **254-255** 358–360 ISSN 03048853
[10] Bali M and Muetze A 2017 *IEEE Transactions on Industrial Electronics* **64** 2547–2556 ISSN 0278-0046
[11] Leuning N, Elfgen S, Weiss H A, Volk W and Hameyer K 2019 *e & i Elektrotechnik und Informationstechnik* **136** 184–194 ISSN 0932-383X
[12] Vogt S, Neuwirth T, Schauerte B, Weiss H A, Falger P M, Gustschin A, Schulz M, Hameyer K and Volk W 2019 *Production Engineering* **13** 211–217 ISSN 0944-6524
[13] Nakayama T and Kojima H 2007 *Journal of Materials Engineering and Performance* **16** 7–11 ISSN 1059-9495
[14] Lin H S, Fu H C, Liu L H, Huang Y K and Fang D H 2017 *Procedia Engineering* **207** 992–997 ISSN 18777058
[15] Park K and Choi S R 2002 *Journal of Materials Processing Technology* **130-131** 477–481 ISSN 09240136
[16] VACUUMSCHMELZE GmbH & Co KG Soft magnetic cobalt-iron alloys: Vacoflux and vacodur URL `https://vacuumschmelze.com/products/soft-magnetic-materials-and-stamped-parts/49-Cobalt-Iron---VACOFLUX-and-VACODUR`
[17] DIN Deutsches institut für Normung eV 2020-01 DIN EN IEC 60404-6:2020-01; vde 0354-6:2020-01: Magnetic materials - part 6: Methods of measurement of the magnetic properties of magnetically soft metallic and powder materials at frequencies in the range 20 hz to 100 khz by the use of ring specimens (iec 68/595/fdis:2018)

Optimum blanking clearance choice method by an approach coupling experimental trials and simulations

P Schreiber[1], J Rizk[2], M Rachik[2], and A Maillard[1]

[1]CETIM – Technical Centre for Mechanical Industries – 52 avenue Felix Louat – 60300 Senlis – France
[2]Laboratoire Roberval, Sorbonne universités, Université de Technologie de Compiègne, Centre de recherche Royallieu, CS60319, 60203 Compiègne, France

pierrick.schreiber@cetim.fr

Abstract. Blanking process of sheet metal is accomplished by a shearing action on a thin plate placed between two sharp cutting edges, known as "punch" and "die". The main problem of this sheet metal cutting technology, that causes major shutdowns of the press machine, is the punch wear. Among all blanking parameters, the clearance set between the punch and the die has the major influence on tool wear. In this context, our study focuses on identifying the optimum blanking clearance that guarantees a maximum abrasion and chipping wear reduction for different sheet metal materials and different blanking operations. The developed approach combines experimental blanking tests and numerical simulation of blanking processes. Based on assumptions made in this paper, links are established between the stripping force and the abrasive wear behaviour of punching tools and also between the maximum pressure on the face of the punch and the risk of chipping. Using these links, criteria have been developed allowing to propose an optimum blanking clearance for all the configurations tested in this paper. A next step of this study consists on applying the optimum punch-die clearances in a variable way along a complex blanking shape and realizing experimental tests to compare the results (tool life and edge quality) with a constant blanking clearance.

1. Introduction

The mechanical blanking process on a press is commonly used by manufacturers in sheet metal cutting operations. The high productivity rate of this process allows it to intervene at different stages of the sheet metalworking process: at the beginning of the process to blank the sheet or pierce the pilot holes, during the process to prepare for bending or drawing operations and at the end of the process to trim or separate the part from the strip. Tool life is a primary concern for manufacturers in order to guarantee the quality of the manufactured parts while maintaining the productivity, and among all the sheet metal cutting operations, the blanking operation induces the most significant tool wear which is the primary cause of press shutdowns.

The main consequences of wear formation on blanking tools are listed below:

- Dimensional variations or distortions of parts,
- Increase in the size of the burrs on the part produced,

- Increased blanking forces and more particularly the so-called "stripping" force (Figure 1) which strongly contributes to the wear of the punch [1].

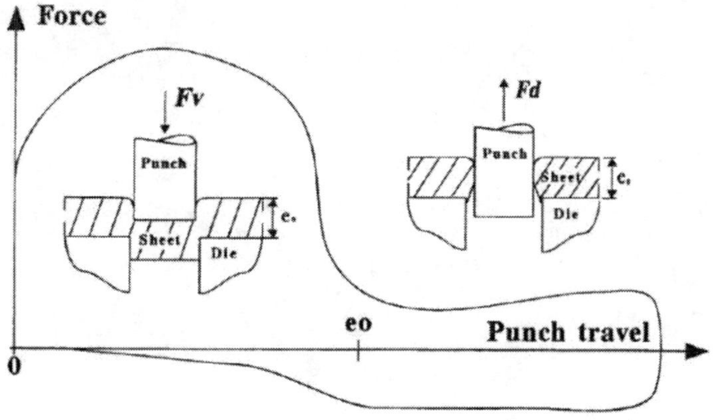

Figure 1. Blanking force during punch travel [1]

Among all blanking parameters, the clearance set between the punch and the die has the major influence on tool wear formation. Several studies had been carried out on the subject. Bell [2] and Högman [3] worked on the influence of the blanking clearance on the punch wear for the blanking of a 1 mm Dual Phase high strength steel. The results of their experiments showed that small clearances cause galling while high clearances cause chipping risks due to high bending stresses in the blanked edge. Thus, they found that there was an optimum clearance that minimises the punch wear formation. Another study implementing variable blanking clearance was reported by Högman for the blanking of a 1 mm thick DP800 sheet metal [4]. This study has shown that variable blanking clearances can be efficient in reducing tool wear formation. An increase of 10% in the clearance on the corner of the square punch results in an increase of 45 000 strokes to 200 000 strokes in the tool life. To improve the choice of an optimum clearance on the local punch shape, Subramonian et al. used a numerical method that consists of analysing the stresses developed at the punch corner during blanking operation and thus applying a more uniform stress distribution on the punch [5]. This method has been validated by experiments and the results show that punch life has increased from 126000 strokes to 350 000 by using a variable clearance. More recently Bohdal *et al.* [6] and Yousefi *et al.* [7] reported extensive studies on tool wear processes in nibbling and punching. The authors combined numerical simulation and experimental investigations to analyse the wear formation mechanisms.

Our study focuses on identifying the optimum blanking clearance that guarantees a maximum abrasion and chipping reduction for different sheet metal materials. We will analyse more particularly the abrasion wear that takes place on the sides of the punch. Different simple shapes of blanking were used to evaluate their influences. The method used is based on experimental trials combined with a simulation campaign carried out in order to choose a blanking clearance providing resistance to chipping wear. An analysis is then carried out with the prospects for further actions.

2. Experimental aspects

The experiments were carried out on a 200T crankshaft press equipped with a blanking tool. This blanking tool, presented on Figure 2, is provided with guiding columns on which the blank holder is also connected to ensure proper centering of the whole set (the punch, the guiding bush and the die). We also observe on Figure 2, the pretensioned load washer on the top of the punch.

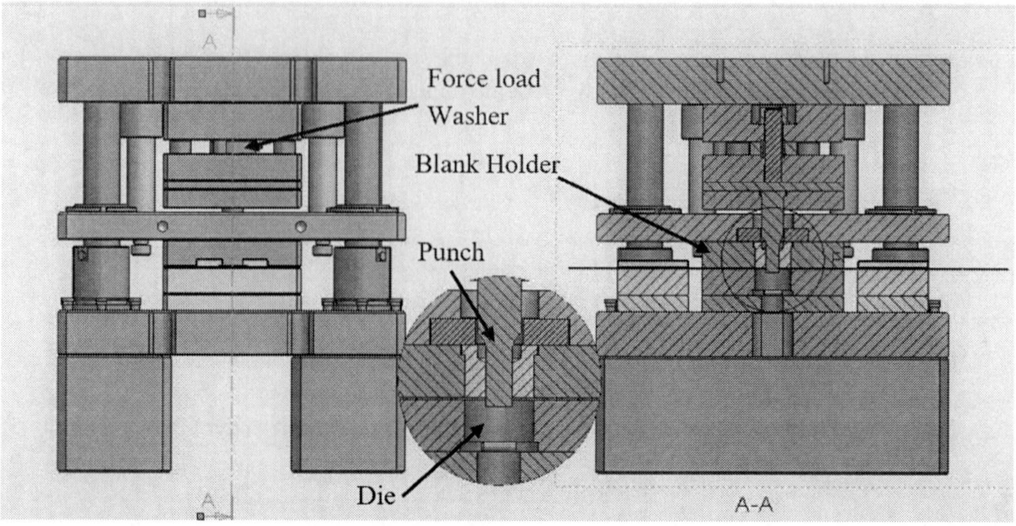

Figure 2. Blanking tool

The blanking process has been characterised by the following measurements :

- Displacement – Load – Time curve of the blanking process. A load washer on the top of the punch allows measuring the blanking force and a displacement wire sensor between the bed and the ram of the press allows measuring the displacement of the ram. Figure 3 illustrates a displacement – load – time curve measured with this set up. With this method, it is possible to link the load with the displacement of the ram through the synchronisation made.
- Variation of the blanked edge dimension after extraction of the punch. This variation is due to the springback of the hole blanked edge [1] and is measured with calibrated rods whose diameter varies with a step of 0,01 mm. The precision of the calibrated rods diameter is equal to ± 3µm.
- Height of the burnished zone of holes. The observation of the blanked edge is made using an optical microscope. The analyse and the measurement of the burnished zone is made through image post processing with a precision of ± 0,01 mm.

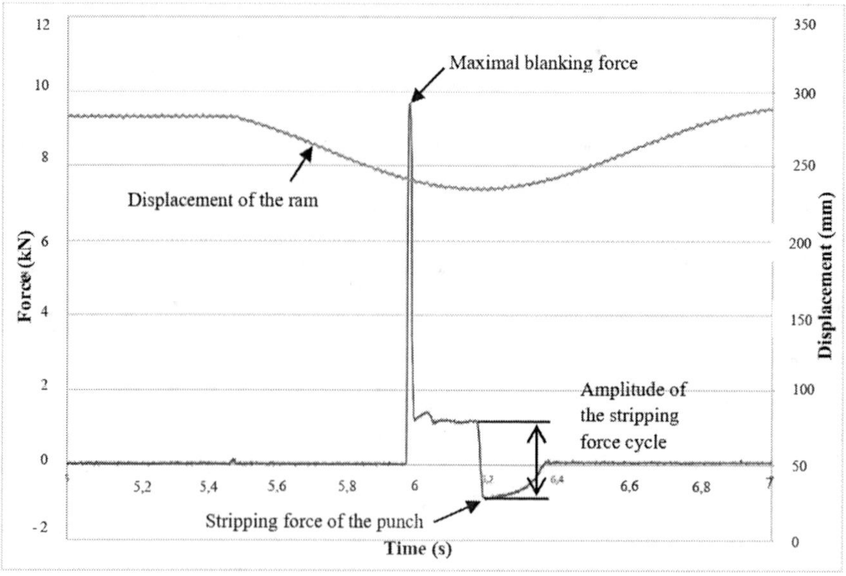

Figure 3. Displacement – Load – Time curve

The blanking experiment was realized in dry conditions. Five samples have been blanked for each configuration.

A map experiment was carried out, varying the following parameters:

- Two tested materials: high strength steel HC380LA and stainless steel X2CrNi18-09, with 1 mm thickness.
- Six radial blanking clearances: 2,5; 5; 7; 10; 15 and 20 % of the initial thickness. The clearance has been transferred on the die.
- Three simple shapes (Figure 4), to represent the following localized geometry: straight edges and different values of fillet radius that we can encounter on complex shapes. A rectangular (18 x 15 mm) punch witch a 1 mm radius and two circular (Ø8 and 16 mm) punches.

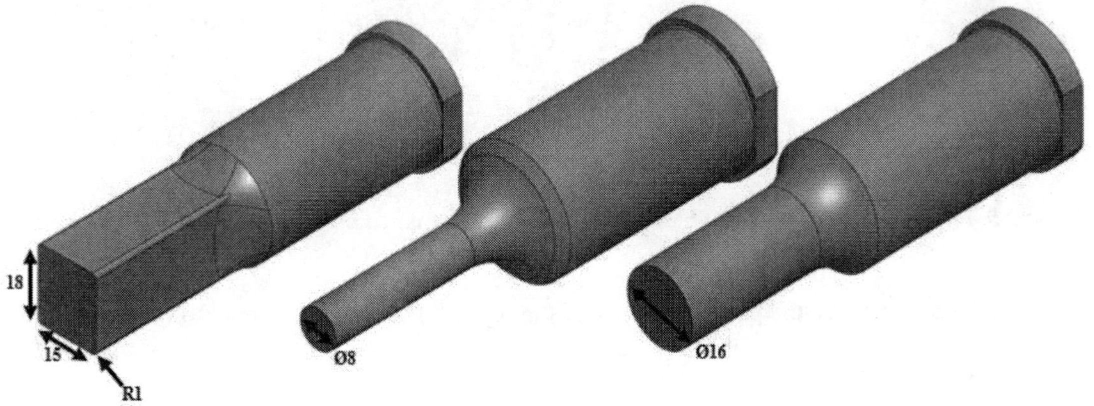

Figure 4. Punches shape

The measured outputs through these experiments are the blanking force, the dimensional variation of the blanked edges, and the distribution of the burnish zone of the hole. All values expressed in this article are averages of five measurements.

3. Numerical aspects

The various blanking tests carried out under this research program are simulated with the help of Abaqus software using a solid axisymmetric finite element model. The sheet metal is discretized using 3875 nodes and 3674 axisymmetric solid elements with reduced integration (CAX4R). The calculations involve three steps. In the first step, the blanking process is simulated using Abaqus/Explicit. The results are then transferred from Abaqus/Explicit to Abaqus/Standard for the subsequent calculations. In the second step, the calculations are achieved using Abaqus/Standard. The fracture line is then created, the slug elements are deactivated as well as the contact between the sheet metal, the die and the blank holder. This way, the blanked edge springback and the resulting pressure on the punch side are computed. In the third and final step, the contact between the sheet metal and the punch is deactivated and the blank edge geometric changes are computed. The calculations steps are illustrated in Figure 5.

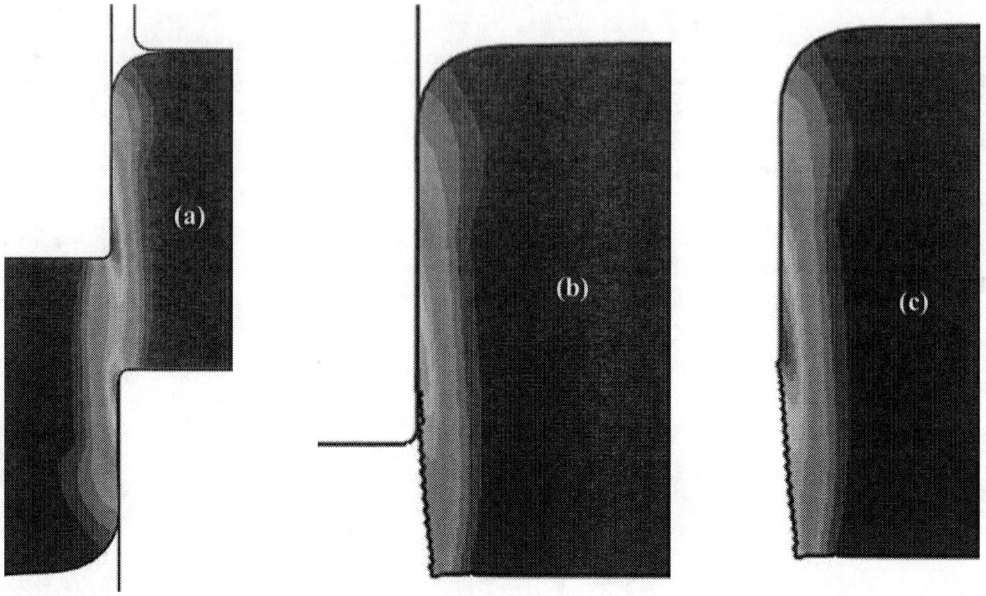

Figure 5. Finite element model, (a) blanking simulation, (b) punch pressure due to spring back, (c) cut edge changes due to spring back

In all the calculations, the punch, the die and the blank holder are considered to be rigid while the sheet metal behaviour is described using a *J2* plasticity model. The material flow stress curve is described using Krupkowski strain-hardening law:

$$\sigma = K(\bar{\varepsilon} + \varepsilon_0)^n \tag{1}$$

σ is the yielding stress, $\bar{\varepsilon}$ is the equivalent plastic strain, K, ε_0 and n are material parameters. The material input data for the two investigated steels are reported in table 1. To complete these data, the Young modulus E is equal to 210 000 MPa and the Poisson ratio ν is equal to 0,3.

Table 1. Mechanical characteristics and data for hardening laws for the materials used

Material	$Rp_{0,2}$*(MPa)*	*Rm (MPa)*	*A (%)*	r	K (MPa)	ε_0	n
HC380LA	383.3	510.2	22.8	0.5797	800.09	0.008	0.167
X2CrNi18-09	292.9	678.9	52.6	0.773	1442	0.0252	0.404

The parameters of the hardening law are obtained from trials on a tensile test machine equipped with cameras and by performing a DIC (Digital Image Correlation) method.

In this article, the fracture is simulated when the penetration of the punch reaches the value of the penetration measured experimentally corresponding to the effective rupture. It should be noted that the fracture line is created based on experimental observation because the damage modelling in blanking process and in particular the damage model calibration requires significant effort [8]. But in the future, we plan to take damage into account with the help of simple models like other researchers [9] who successfully used a simple Cockroft-Latham model to investigate damage induced by shearing process. They showed that the model can be calibrated using blanking tests and finite element modelling.

4. Experimental results

An assumption has been carried out in this paper by considering that the most important abrasion wear occurs during the stripping phase of the blanking operation. The contact between the burnished zone of the blanked edge and the side of the punch (Figure 6) is considered to be the main mechanism of abrasion wear formation. That assumption supposes that this contact is established on the entire burnish zone [1].

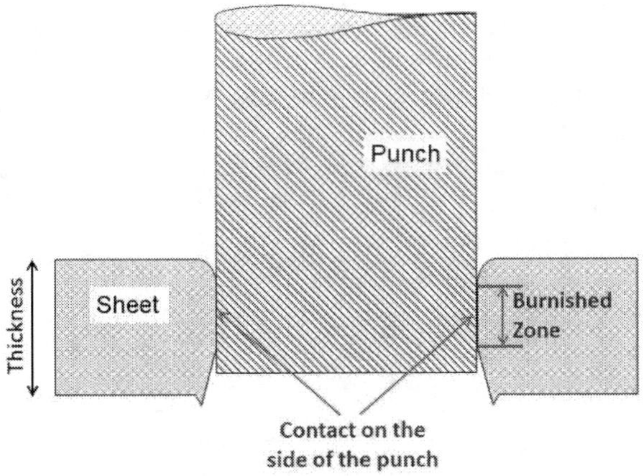

Figure 6. Presentation of the contact between the burnished zone and the side of the punch during punching

As stated before, the characterised outputs of interest using this experiment are the stripping force, the dimensional variations of the blanked edge and the height of the burnished zone of this edge.

Figure 7 shows the normalised stripping forces as a function of the blanking clearance for both tested materials. The stripping force has been normalised by the blanked perimeter so comparisons between the samples can be held independently of the dimension of the blanked perimeter (relation (2)).

$$f_{ns} = \frac{f_s}{P} \quad (2)$$

f_{ns} is the normalised stripping force, f_s is the stripping force and P is the blanked perimeter.
General trends are observed for both tested materials:

- The normalised stripping force decreases when the blanking clearance increases,
- The normalised stripping force increases when the size of the punch decreases,
- The normalised stripping force is relatively small for a punch shape mostly made of straight edges (rectangular punch).

Figure 8 shows the dimensional variations expressed as a function of the blanking clearance. The dimensional variation of the hole has been expressed using the diameter of the punch and the obtained value has been divided by two in order to obtain springback value of the blanked edge of the hole. On Figure 8, D is the measured diameter and $D_{nominal}$ is the diameter of the punch. For comparison purposes, the straight edges could be considered as circles with an infinite radius. The results in Figure 8 show that an increase in the blanking diameter led also to an increase in the springback of the sheet. Indeed, there is more springback on straight edges because the deformation is not hindered, unlike the curved edges where the deformation closes the hole and is therefore limited by shrinkage.

On the contrary, the results in Figure 8 for the two circular punches show that the springback is higher for a smaller diameter. This result can be explained by the fact that the springback depends also from the distribution of the elastic stresses generated during the blanking of the edges which is higher for smaller punching diameter.

In Figure 9, the burnish zone of the blanked edges is expressed as function of the blanking clearance. It is observed that the burnish zone decreases with an increasing blanking clearance (this observation is in agreement with the rules of the blanking profession). And finaly, by characterising the different shapes by their radius of curvature, the burnish zone is found to have an increasing trend when the radius of curvature decreases.

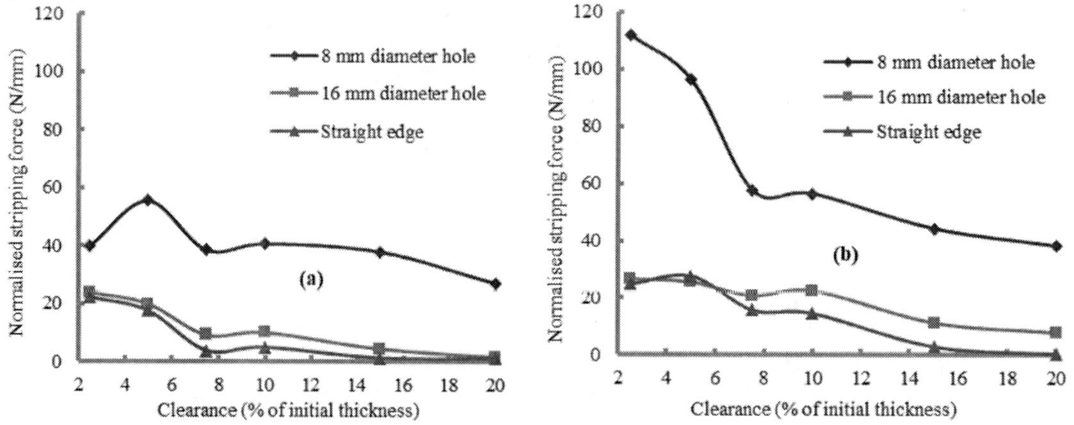

Figure 7. Normalised stripping forces as a function of blanking clearance, (a) HC380LA steel, (b) X2CrNi18-9 steel

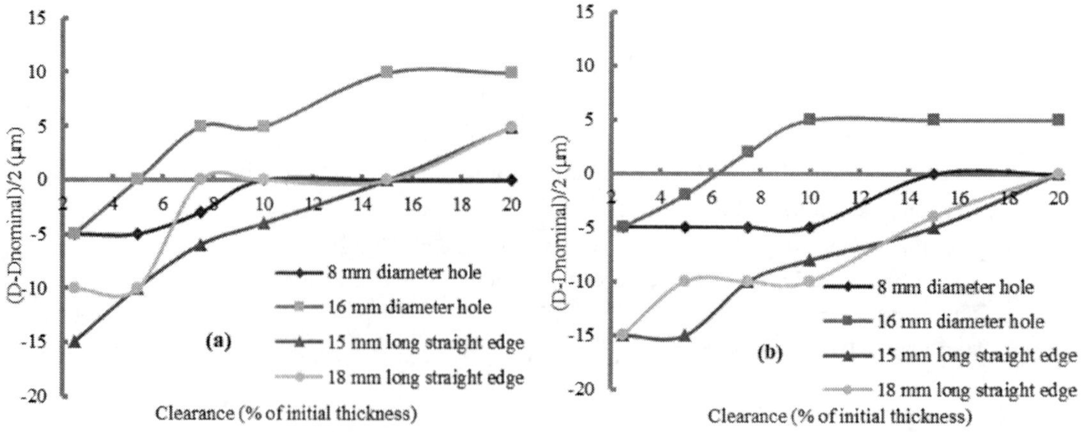

Figure 8. Dimensional variation as a function of blanking clearance, (a) HC380LA steel, (b) X2CrNi18-9 steel

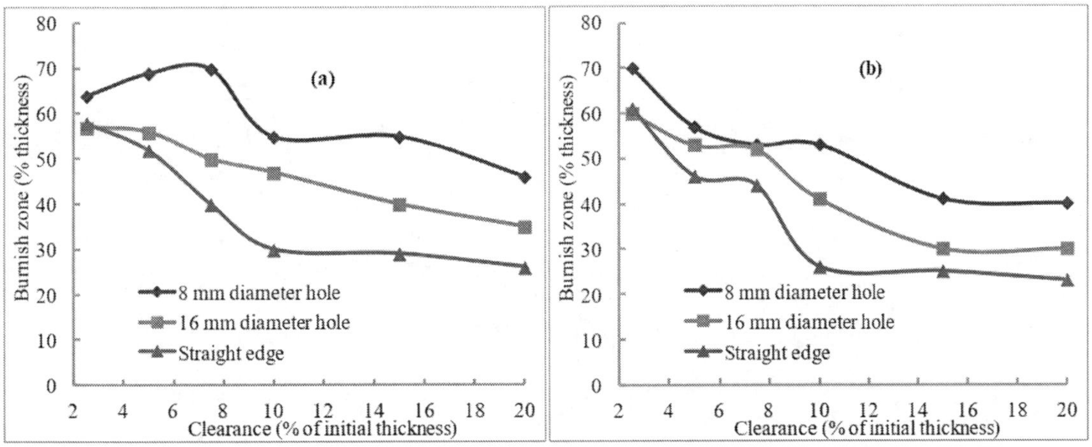

Figure 9. Burnish zone evolution with clearance, (a) HC380LA steel, (b) X2CrNi18-9 steel

5. Numerical results

In addition to the experimental tests, numerical simulations of the blanking operation were carried out in order to deduce the following measures:

- Maximal pressure on the face of the punch (Pmax),
- Contact surface and the side pressure of the blanked edge on the side of the punch.

Each simulation has been evaluated through a comparison with the experimental maximum blanking force. The calculation parameters have been ajusted so as to minimise the differences between the maximum blanking force obtained experimentally and numerically. Figure 10 shows an example of a comparison made to evaluate the simulation of one tested configuration. In that evaluation, it can be estimated that the difference between the two forces is acceptable and the analysis of the pressure is therefore allowed.

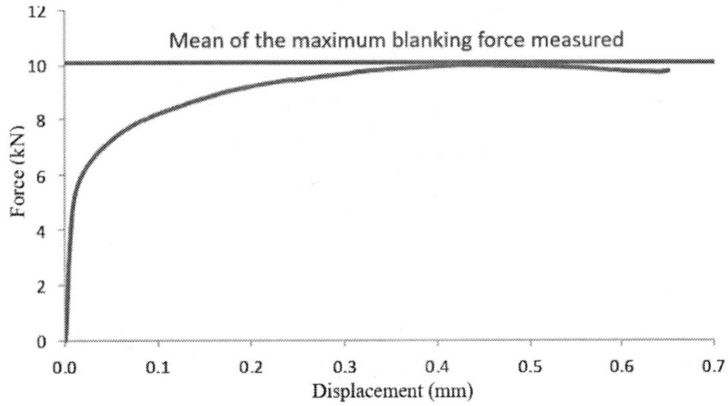

Figure 10. Validation of the simulation through the maximum blanking force (HC380LA, thickness 1mm, Ø8 mm, clearance=20 %).

Figure 11 shows a typical example of the distribution of the pressure on the face of the punch expressed as a function of the radius of the punch. This value is extracted just before rupture.

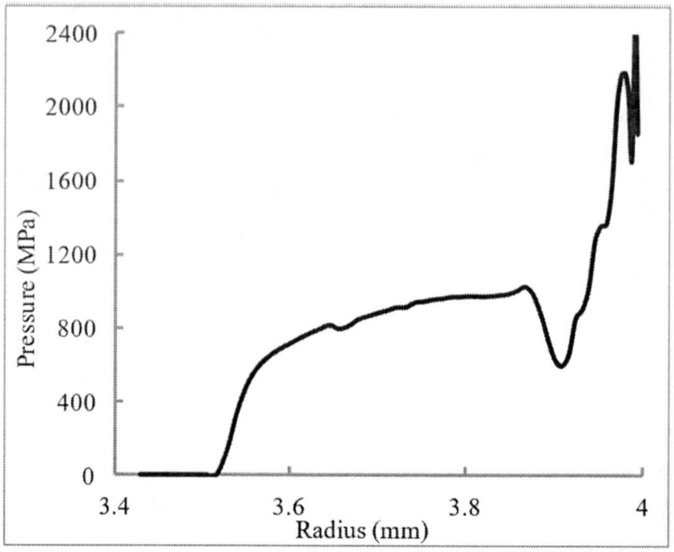

Figure 11. Pressure distribution on the face of the punch as afunction of the radius of the punch (HC380LA, thickness 1mm, Ø8 mm, clearance=2,5%).

Figure 11 shows that the maximal pressure is located close to the blanking edge. Its position coincides with the initiation of the chipping which is shown as an example on Figure 12. Within the scope of this article, none chipping case has been met during the experiments.

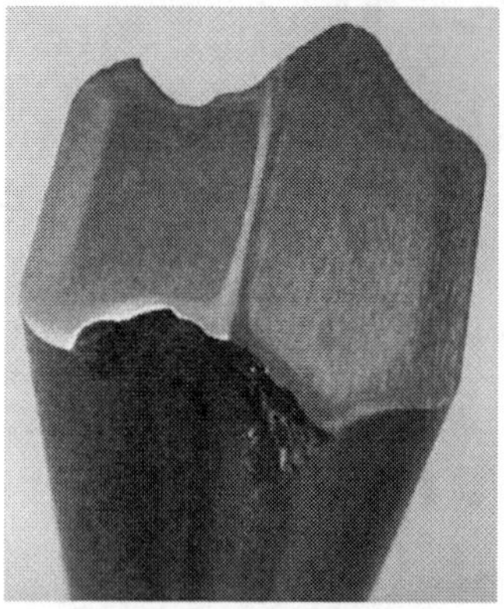

Figure 12. Example of chipping on the blanking edge of the punch (CETIM results met outside the presented work)

Based on the results obtained from Figure 11 and 12, one can assume that the risk of chipping is related to the maximum pressure on the face of the punch.

Figure 13 shows the numerical maximum pressure on the punch as a function of the blanking clearance.
The following trends are observed:

- The maximum pressure decreases when the blanking clearance increase regardless of the shape of the punch except for the case of stainless steel with clearance of 20% (Figure 13b).
- The maximum pressure decreases when the diameter of a circular punch increases with the same exception said in the line above.
- The maximum pressure increases with the maximum tensile strength of the material.

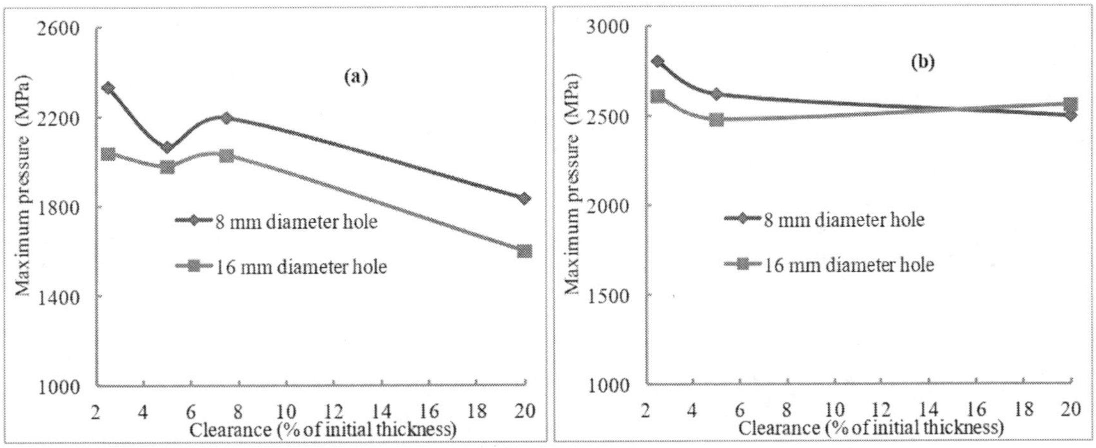

Figure 13. Predicted maximum pressure, (a) HC380LA steel, (b) X2CrNi18-9 steel

Figure 14 shows an example of a contact between the side of the punch and the simulated blanked edges, in which the contact doesn't seem to apply on a surface but on two lignes instead. This result contradicts the rubbed length assumption made in the forth paragraph which considered that the entire burnish zone was in contact with the side of the blanking tool. Nevertheless, this contradiction could be explained by the fact that the simulation doesn't take into account neither the evacuation of the slug, and therefore the springback of the blanked edge, nor the stripping of the punch.

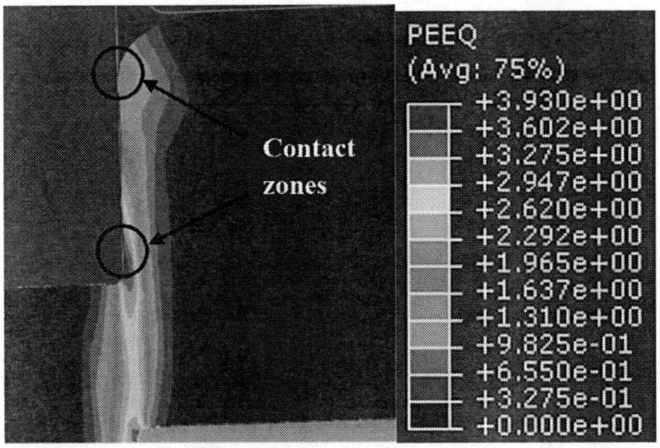

Figure 14. Contact example between the punch and the blanked edge during a punching operation (HC380LA, thickness 1mm, Ø8 mm, clearance=2,5%).

The blanked edge springback and the pressure on the punch side obtained after the second step of the calculations (Figure 5b) haven't been analysed because of aberrant results which need an improvement of the numerical model.

6. Analysis and discussions

All results presented in the paragraphs 4 and 5 will be used, in the following, to choose an optimum blanking clearance that reduces abrasion and chipping wear. The assessment of abrasion wear is made by studying the variation of the stripping forces as a function of the blanking clearance. Indeed, an analysis of the parameters of the Archard Law [10] leads to conclude that the abrasion wear is linked to the normal stress, the friction coefficient, the hardness of the materials facing each other's and the rubbed length. The study of the blanking clearance influence fixes the following parameters: the coefficient of friction (μ) and the hardness of the materials facing each other. In that case, we can compare the abrasion wear corresponding to several blanking clearances by establishing the following relation:

$$\frac{V_{c1}}{V_{c2}} = \frac{\sigma n_{c1}}{\sigma n_{c2}} \cdot \frac{rl_{c1}}{rl_{c2}} \tag{3}$$

V represents the wear volume, c represents the blanking clearance, σn represents the normal stress and rl represents the rubbed length.
rl is equivalent to the height of the burnish zone (hb). The normal stress σn can be also expresses by the following relation:

$$\sigma n = \frac{fns}{hb} \cdot \frac{1}{\mu} \tag{4}$$

Combining relations (3) and (4) leads to:

$$\frac{V_{c1}}{V_{c2}} = \frac{fns_{c1}}{fns_{c2}} \tag{5}$$

Knowing that this abrasion wear criterion doesn't always allow to decide between several clearances, a criterion based on the quality of the blanked edge is chosen. This criterion is based upon the height of the burnish zone (hb). Indeed, in industry, a compromise between the tool life and the part quality is aimed.

The chipping risk of the punch blanking edge can be evaluated by comparing the yield strength of the tool steel to the maximum pressure which occurs on the punch. Thus, the criterion can be written as:

$$C.P_{max} \leq Re\ (Tool\ steel) \tag{6}$$

C is a dimensionless security constant that includes the chipping risk by fatigue ($C<1$), P_{max} is the maximum pressure in the blanking direction and Re is the yield limit of the tool steel.
While the abrasion wear criterion concerns the side of the punch, the chosen chipping risk criterion concerns the face of the punch pressed against the blank. This criterion indicates a minimal resistance of the tool to exceed and can integrate a fatigue limit.

From these different approaches, it is possible to establish recommendations on optimum blanking clearances to use on different simple shapes regarding the risk of wear and to temper these choices in the aim of respecting the quality of the edge. These optimum clearances are determined by an analysis of the abrasion wear risk and then controlled regarding the chipping risk. A table of recommended variable clearances is therefore proposed for both materials tested (Table 2).

Table 2. Recommended variable blanking clearances

	Ø16	Ø8	Straight Edge
HC380LA	5 %	10 %	5 %
X2CrNi18-09	2,5 %	7,5 %	2,5 %

7. Conclusion

In the present paper, an optimum blanking clearance choice method by an approach coupling experimental trials and simulations has been investigated. This method has been applied to two materials and three blanking shapes to define an optimum blanking clearance that is aiming to reduce the tool wear. It was highlighted that the stripping force would be the major parameter used to identify an optimum blanking clearance for reducing abrasion wear. It was also found that this parameter has a direct link with the Archard Law to estimate the wear volume. Also, the risk of chipping has been defined by a criterion involving the maximum pressure on the punch and the yield limit of the tool steel. Finally, the maximum pressure used to establish the previous criterion was obtained by a numerical simulation of the blanking operation. The next steps of this study are to apply the optimum punch-die clearances in a variable way along a complex blanking shape and realizing experimental tests to compare the results (tool life and edge quality) with a constant blanking clearance and to improve the numerical model by including damage modelling.

References

[1] Maillard A 1995 *Proc. Int. Deep Drawing Research Group Conf. (Colmar)*
[2] Bell T and Krause B 2005 *Proc. Great Designs in Steel Sym. (Livonia)*
[3] Högman B 2002 *Proc. 6th Int. tooling conf. (Karlstad)*
[4] Högman B 2004 *Proc. Int. Conf. Recent Advances in Manuf. & Use of Tools & Dies (Olofström)*
[5] Subramonian S and Ciocirlan B 2016 *Proc. Int. Deep Drawing Research Group Conf. (Linz)*
[6] Bohdal L, Kukielka L, Patyk R, Loska K, Chodor J and Czyrewski K 2022 *Materials* **15** 107
[7] Yousefi M and Pervaiz S 2022 *Simulation Modelling Parctice and Theory* **114** 102415
[8] Basak S, Kim C, Jeong W, Jung Y I and Lee M G 2022 *Int. J. Mech. Sci.* **219** 107109
[9] Han S, Chang Y, Wang C Y and Dong H 2022 *J. Mat. Proc. Technol.* **299** 117377
[10] Archard J F 1953 *J. Appl. Phys.* **24** 981

Evaluation of hole expansion formability of high strength AA7075 alloy under varying temper conditions

Kali Prasad[a*], Abrar Salam Ebrahim[b], Hariharan Krishnaswamy[a], Uday Chakkingal[b] and Dilip K. Banerjee[c]

[a]Manufacturing Engineering Section, Department of Mechanical Engineering, IIT Madras, Chennai-600036, India
[b]Metal Forming Laboratory, Department of Metallurgical and Materials Engineering, IIT Madras, Chennai-600036, India
[c]Material Measurement Laboratory, National Institute of Standards and Technology (NIST), Gaithersburg, MD 20899, USA

E-mail: Corresponding author(*): kali.iitm@gmail.com

Abstract. There has been a widespread increase in the use of aluminum alloys in automotive industries for meeting ever-growing light-weighting requirements. However, edge formability is a critical manufacturing challenge that restricts their widespread use. Edge formability of sheet metal is determined using a hole expansion test (HET) and is evaluated by the hole expansion ratio (HER). The present study investigates the effect of temper conditions on the edge formability of AA7075 alloy sheets. Hole expansion tests were conducted in different temper such as W-temper (super saturated solid solution followed by water quenching), under aged (UA), and peak aged (PA) conditions. Two different hole preparation techniques, a punching and a drilling process, were used to prepare samples with varying edge conditions. The results demonstrate that the W-temper has the highest edge formability irrespective of hole edge conditions. Researchers have reported that uniaxial stress state prevails at the hole edge during the HET. Consequently, uniaxial tensile tests were conducted on for each temper condition and various tensile properties such as yield stress (YS), ultimate tensile strength (UTS), ratio of yield stress to ultimate tensile strength (YS/UTS) were determined to evaluate edge formability. Furthermore, microstructural and failure analysis of the failed specimens were performed to explain the deformation behavior during the HET.

keywords: Stretch-flangeability; Hole expansion; Aluminium AA7075 alloy; Formability; Heat treatment

1. Introduction

In recent years, aluminum alloys have gained considerable attention in the automotive and aerospace industries for light weighting applications [1, 2, 3, 4]. Among the various aluminum alloys, precipitate hardened aluminum alloys are widely used due to their high strength to weight ratio, good wear and corrosion resistance [1, 5]. However, one of the major technological problem that restricts their widespread usage is their inferior room temperature formability, especially in precipitate hardened conditions [3, 6]. To address these manufacturing challenges, the alloys are often formed at elevated temperatures [7]. However, elevated forming temperatures pose significant manufacturing and tooling difficulties, such as reduced die life and poor surface quality of formed components [8]. To circumvent these technological challenges, cold forming technique is currently employed. In this method, the part is subjected to solid solution treatment and then quenched to attain super saturated solid solution (S.S.S) phase and thereafter cold formed [9]. The strength of the formed part is increased by subsequent artificial aging post forming. This process is beneficial as lower forming forces are required, and it improves the die service life and formability[9].

In automotive industries, the formability measurement of sheet metal is a routine task. An accurate formability assessment of sheet metal is very important before the component is subjected to forming operation. Edge formability of sheet material is one of the important formability parameters which is evaluated using the hole expansion test. Typically, the traditional forming limit diagram (FLD) is used to evaluate formability. However, FLD is not suitable for evaluating the edge formability of sheet metal through hole expansion test (HET) data, since HET results are sensitive to hole edge conditions [10, 11, 12]. The test data are used to evaluate a dimensionless parameter called hole expansion ratio (HER), which is defined as the ratio of the change in the hole diameter of the specimen at failure to the initial hole diameter. A higher HER value indicates that the sheet has superior edge formability. A high value of edge formability is desired for sheet materials, especially when forming flanges in complicated shapes. Hole expansion deformation behaviour has been widely studied by various groups [10, 11] and it is well known that the edge formability during HET depends on the punch profile. In the case of conical punch, the hole edge deforms nearly in uniaxial condition [10, 11, 12]. Following this observation, various researchers have correlated the tensile properties like yield strength (YS), uniform elongation (UE) and total elongation (TE) with HER [10]. Majority of these studies were performed on ferrous alloys [10, 11, 12, 13]. Aluminum alloys received less attention in this regard. Despite the importance and potential use of AA7075, only limited studies have been carried out to understand the edge formability, particularly with varying temper conditions. Therefore, the objective of the present study is to investigate the edge formability of AA7075 alloy with varying temper and edge conditions. A systematic approach is used to investigate the influence of temper condition on edge formability of the AA7075 alloy sheet, and results are correlated with uniaxial tensile properties.

2. Materials and methods

2.1. Experimental method

In the present work AA7075 with sheet thickness 2 mm, obtained from Bharat Aerospace, India [1] was used. The chemical composition of the as-received AA7075 alloy sheets was measured using optical emission spectroscopy (OES) and is listed in Table:1. Square sheets of dimensions 90×90 mm^2 with a center hole of diameter 10 mm were made using drilling and punching process from the as received AA7075 sheets. Two stage drilling of ϕ 5 mm followed by ϕ 10 mm at 630 rpm (lip angle of drill bit 59°) was utilized to prepare drilled edge sample. The punching operation was performed at punch velocity of 10 mm/s. All the heat treatments and aging were performed after specimen preparation. In addition, uniaxial tensile specimens as per the ASTM E8 standard [14] were fabricated using wire cut electric discharge machining (WEDM) process. These sheet specimens (hole expansion and tensile specimens) were solution treated at 480 °C for 2 hours followed by water quenching to obtain S.S.S condition. This treatment is referred

[1] Certain commercial equipment, instruments, software or materials are identified to describe a procedure or concept adequately. Such identification is not intended to imply recommendation, endorsement or implication by NIST that the equipment, instruments, software or materials identified are necessarily the best available for the purpose.

to as W Temper condition. Thereafter, these sheets were artificially aged at 120 °C for 1 hour and 24 hours to obtain under aged (UA) and peak aged (PA) conditions, respectively. Fig. 1 shows the heat treatment cycle followed for obtaining various temper conditions. Tensile tests were performed using a Zwick/Roell 100 kN machine. Specimens were tested with a nominal strain rate of 0.001 s^{-1}. All the experiments were repeated three times to ensure repeatability of test data.

Table 1: Chemical composition of as-received AA7075 alloy sheet.

Elements	Zn	Mg	Cu	Cr	Ni	Si	Ti	Fe	Al
Contents (%)	6.02	1.3	1.43	1.4	1.3	1.43	1.43	1.43	balance

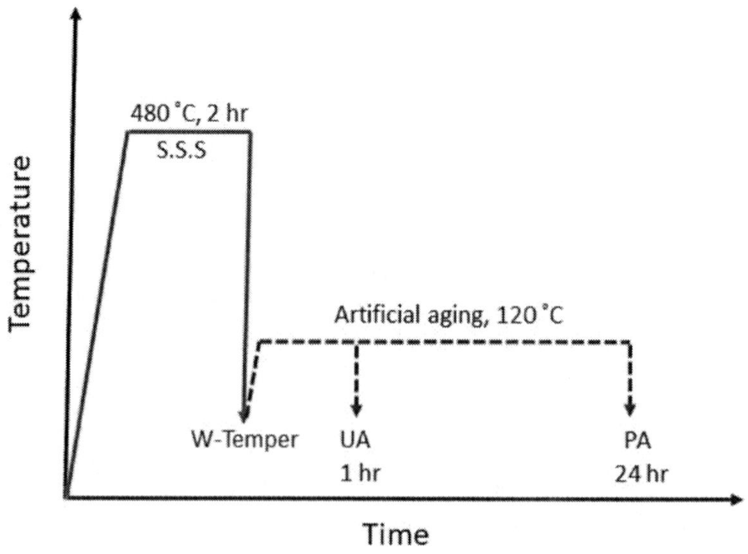

Figure 1: Heat treatment cycle followed for obtaining various temper conditions used for both sheet and uniaxial tensile specimens.

2.2. Hole expansion test

Standard hole expansion tests (HET) were performed on the heat treated specimens using a conical punch with a cone angle of 60°. HET was conducted as per the ISO 16630 2009 standard [15] (Fig.2). A blank holding force of 65 kN was applied in order to prevent drawing in of the sheet specimens. The test was continuously monitored with the help of a camera with an external light source. The test was stopped once through-thickness cracks appeared. The hole expansion ratio (HER) value was calculated using equation (1).

$$HER(\%) = \frac{D_f - D_i}{D_i} \times 100 \tag{1}$$

where, D_f and D_i refers to the final and initial diameter of the central hole. D_i is calculated using the average value of four diameters measured at angles of 45°.

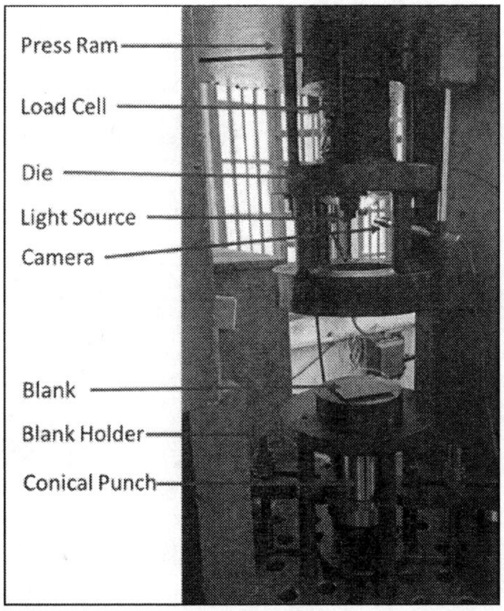

Figure 2: Experimental setup used for hole expansion deformation process.

3. Results and discussion

3.1. Correlating: HER with tensile properties

The shape of the HET specimen before and after the hole expansion deformation process is shown in Fig.3(a,b) respectively. The values of HER in different temper conditions are tabulated in Table 2. It is observed that the HER values significantly varied with hole edge and temper condition. Maximum HER was obtained in W-temper, followed by UA and PA for both the edge conditions [16].However, the drilled edge was found to exhibit higher HER compared to the punched edge. The dependence of HER with the hole edge can be attributed to difference in the surface roughness for the two edges, which is discussed in Section 3.2. It is known that, the hole edge in HET deforms in uniaxial tensile deformation conditions [10, 11]. Therefore, to better understand the edge formability characteristics of this alloy, HER values are correlated with the uniaxial tensile properties of the sheet with varying temper conditions.

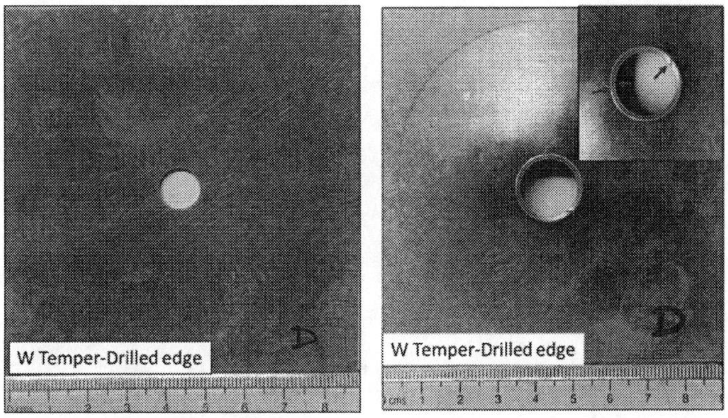

(a) Before HET (b) After HET

Figure 3: A typical HET specimen (a) before deformation, (b) after deformation (arrows in the insert figure showing the through thickness crack).

Table 2: Evaluated HER values for AA7075 sheet in different temper and hole edge conditions.

Condition	HER values (%)	
	Punched edge	Drilled edge
W-Temper	36.08 ± 1.35	43.52 ± 2.26
UA	20.74 ± 1.37	24.25 ± 1.86
PA	14.20 ± 2.69	18.62 ± 1.42

Fig.4 shows the stress strain response of AA7075 alloy in different temper conditions. As the aging proceeds from solutionized (W-Temper) to PA condition, significant changes in the deformation behaviour are observed. Serrated deformation behavior is observed in the W-temper condition. This is characteristic of occurrence of Portevin–Le Chatelier effect (PLC) [3, 17]. In the W-temper condition, solute elements present in the aluminum matrix, interact with the dislocations during plastic deformation resulting in the serrated flow behavior. However, when the material is subjected to artificial aging, the solute elements precipitate out of the aluminum matrix. The depletion of solute elements in the aluminum matrix, is manifested by the suppression of serrated flow behavior in UA and PA condition. During deformation, dislocations interact with the precipitates, and this interaction is responsible for the increase in strength levels in the aged condition. The tensile properties obtained in different temper conditions are tabulated in Table 3. The yield strength (YS) and ultimate strength (UTS) significantly increased upon aging. It is also important to note that the strain hardening capacity of the material significantly decreased as the alloy is subjected to increased aging treatment. In case of W-temper condition, the ratio of yield strength to ultimate tensile strength (YS/UTS) is 0.42 whereas, as the material is aged the ratio of yield strength to ultimate tensile strength (YS/UTS) increased from 0.66 to 0.85 for UA and PA respectively, indicating significant loss in the ductility. This is evident from considerable decrease in uniform elongation (UE) and total elongation (TE) values obtained in UA and PA compared to W-temper condition.

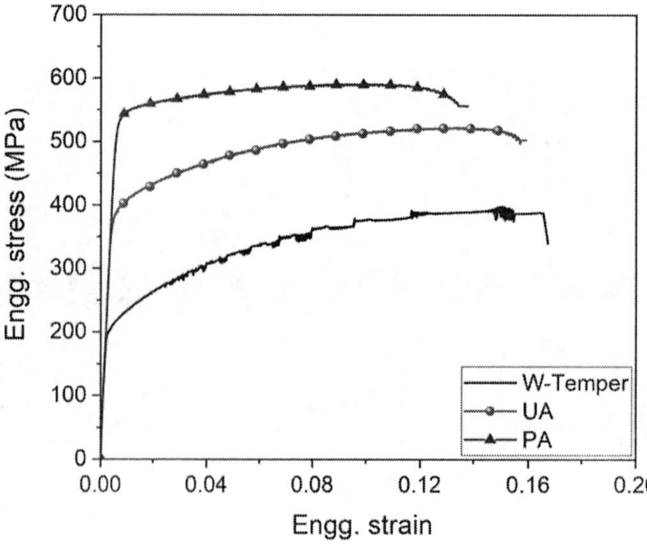

Figure 4: Stress-strain response of AA7075 alloy in different temper conditions.

IOP Conf. Series: Materials Science and Engineering 1238 (2022) 012038 doi:10.1088/1757-899X/1238/1/012038

Table 3: Uniaxial tensile properties of AA7075 alloy in various temper conditions.

Temper condition	YS (MPa)	UTS (MPa)	YS/UTS	UE	TE
W-Temper	182.91 ± 2.07	492.08 ± 3.27	0.37 ± 0.001	0.13 ± 0.003	0.17 ± 0.002
UA	394.65 ± 3.44	591.10 ± 3.43	0.66 ± 0.002	0.11 ± 0.003	0.15 ± 0.001
PA	549.29 ± 2.57	641.27 ± 2.58	0.85 ± 0.0005	0.07 ± 0.004	0.13 ± 0.002

Fig.5(a-e) presents plots of variation of HER with tensile properties (YS, UTS, YS/UTS, UE, TE) (Table. 3). It is observed that irrespective of tensile properties, the HER values for drilled edge are higher than those obtained from the punched edge tests. HER values are largest for the W-temper followed by UA and PA temper condition as evident from Fig.5(a-e). These figures also show that the HER values continued to decrease with aging. HER values ultimately decreased by 60.6% and 57.2% from corresponding values for the W-temper condition for drilled edge and punched edge hole tests respectively for the PA condition. As observed from Fig.5a and Fig.5b, there exists a negative correlation between HER and YS and UTS. In the W-Temper condition, the material exhibits the least values of both YS and UTS. However, as the aging continues from UA to PA condition, both the YS and UTS values considerably increase, thereby decreasing the HER values. To further understand, we plotted the HER variation with the YS/UTS ratio. In the W-Temper condition, the ratio of YS/UTS is 0.37, which increases to 0.66 and 0.85 in UA and PA temper conditions respectively. The increase in the ratio of YS/UTS indicates the decrease in material formability, which is also manifested by lower HER values as shown in Fig.5c. Similar negative correlation between HER and YS and UTS values were reported by Paul et al.[18] for wide grades of automotive steels. HER values were found to be positively correlated with UE and TE for both the hole edge conditions, as shown in Fig.5d and Fig.5e. The correlation is limited to the given edge condition. Varying the edge condition during sample preparation could lead to a different correlation, which is a limitation when attempting to correlate the HER with the uniaxial behavior.

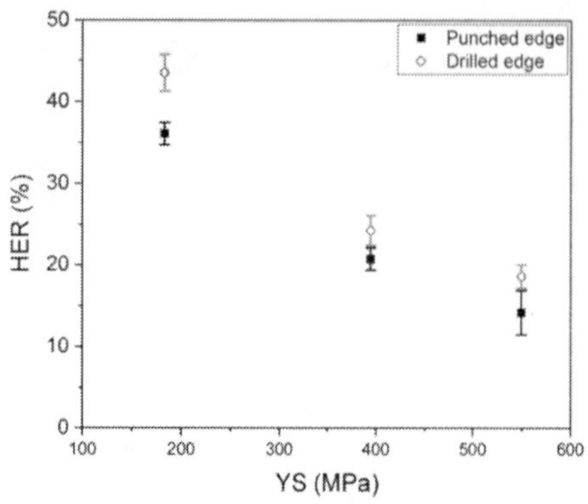

(a) Yield strength (YS)

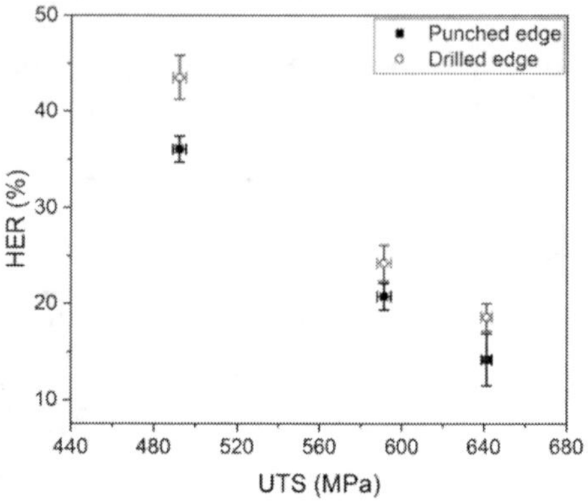

(b) Ultimate tensile strength (UTS)

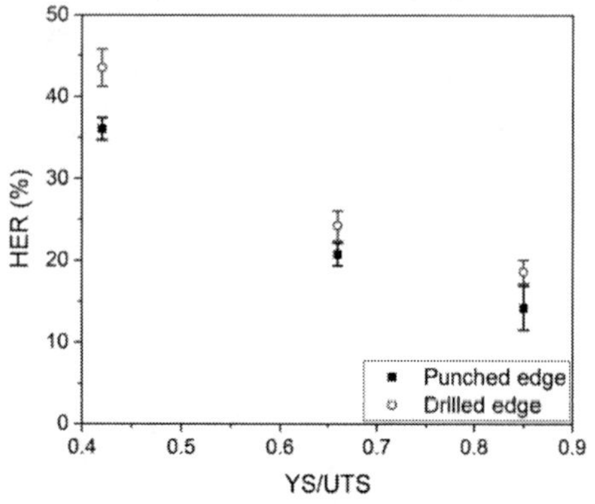

(c) YS/UTS ratio

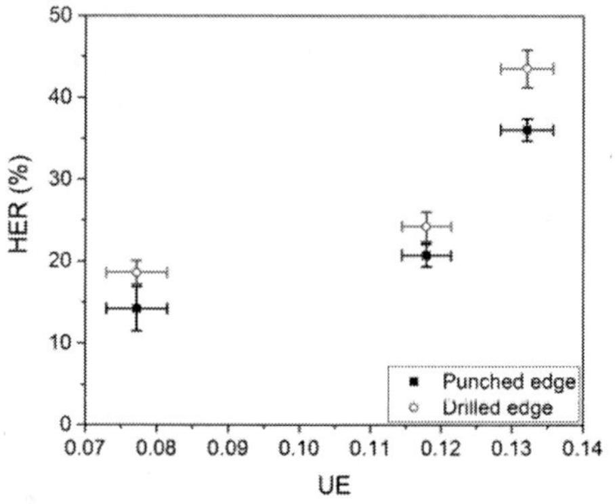

(d) Uniform elongation (UE)

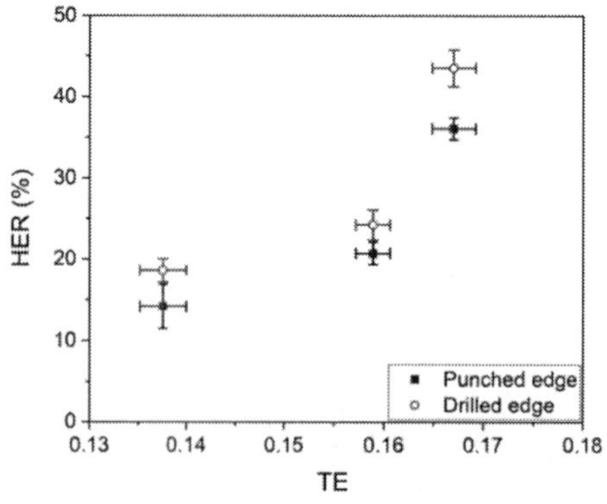

(e) Total elongation (TE)

Figure 5: Variation of HER with (a) Yield strength (YS), (b) Ultimate tensile strength (UTS), (c) Yield stress to Ultimate tensile strength ratio (YS/UTS), (d) Uniform elongation (UE) (e) Total elongation (TE) in different temper and edge conditions.

3.2. Effect of edge condition

Fig.6a shows a schematic of HET specimen position during hole edge preparation using punching and drilling process. The measured average surface roughness (Sa) for drilled and punched hole is 1858 ± 35 nm and 2981 ± 27 nm respectively. It is well established that edge preparation technique plays a critical role in HER values obtained in HET [10, 11]. During the hole edge preparation method, significant plastic deformation takes place near the hole edge. As a result, initial micro-cracks are generated due to punching and drilling process which is shown in Fig.6b and Fig.6c. These initial micro-cracks serve as crack initiation sites during HET. It is important to note that during punching, the initial micro-cracks are aligned along the through thickness direction whereas, in the drilling process, the initial micro-cracks are mostly in transverse direction. During HET the through thickness micro-cracks tend to open up easily, resulting in lower values of HER in punched specimens. On the other hand, the transverse direction micro-cracks for the drilled edge conditions provide some resistance to the crack progression. Consequently, higher HER values are obtained for the drilled edge conditions.

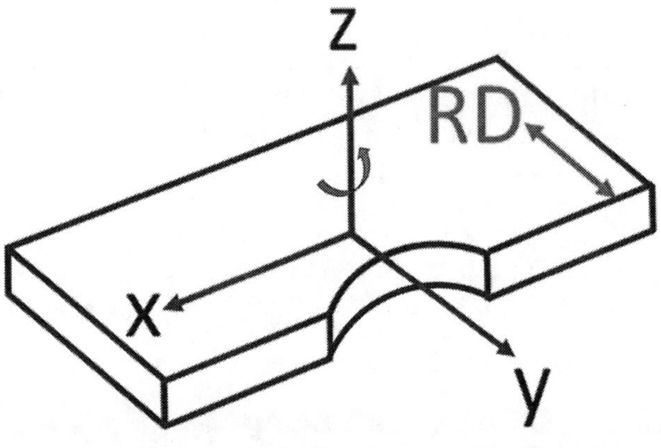

(a) Schematic-Specimen preparation and orientation

(b) Punched edge

(c) Drilled edge

Figure 6: (a) Schematic of specimen preparation and orientation (punching and drilling is done along the "z" axis. Arrow represent drill axis). SEM micrographs showing the presence of micro-cracks in the HET specimens (b) Punched edge (c) Drilled edge

4. Conclusions

The objective of the present study was to investigate the effect of temper condition on the edge formability of aluminum alloy AA7075 sheets using hole expansion tests. From this study, the following are major findings.

1. HER was found to be sensitive to the hole preparation method and temper conditions.

2. W-temper condition was found to be a promising method for improving the edge formability of AA7075 alloy.
3. HER correlate well with the tensile properties. HER is inversely proportional to YS and UTS and directly proportional to UE and TE.
4. Through-thickness micro-cracks formed during punching tend to open up easily during HET resulting in lower HER values. On the other hand, transverse micro-cracks are seen to form along the plane of the sheets during drilling. These cracks appear to delay the eventual failure during HET resulting in higher HER values for drilled hole specimens.

5. Acknowledgment

Authors would like to thank Prof. V. Subramanya Sarma, Department of Metallurgical and Materials Engineering, IIT Madras, for giving access to lab facilities for carrying out heat treatment work. One of the authors, Kali Prasad, acknowledges Mr. Amit Singh (M.S research scholar) and Mr. Achintya Patra (Ph.D research scholar), Department of Metallurgical and Materials Engineering, IIT Madras for helping in conducting heat treatment.

References

[1] Dursun T and Soutis C 2014 *Materials Design (1980-2015)* **56** 862–871 ISSN 0261-3069
[2] Tiwari J, Pratheesh P, Bembalge O, Krishnaswamy H, Amirthalingam M and Panigrahi S 2021 *Journal of Materials Research and Technology* **12** 2185–2204 ISSN 2238-7854
[3] Prasad K, Balaji V, Krishnaswamy H, Phani P S and Carlone P 2022 *Mechanics of Materials* 104279 ISSN 0167-6636
[4] Prasad K, Krishnaswamy H and Jain J 2018 *Journal of Materials Processing Technology* **255** 1–7 ISSN 0924-0136
[5] Lee H, Chae H, Kim Y S, Song M J, Lim S, Prasad K, Krishnaswamy H, Jain J, An K and Lee S Y 2021 *Mechanics of Materials* **158** 103899 ISSN 0167-6636
[6] Zheng K, Politis D J, Wang L and Lin J 2018 *International Journal of Lightweight Materials and Manufacture* **1** 55–80 ISSN 2588-8404
[7] Huo W, Hou L, Zhang Y and Zhang J 2016 *Materials Science and Engineering: A* **675** 44–54 ISSN 0921-5093
[8] Ghiotti A, Bruschi S and Medea F 2017 *Wear* **376-377** 484–495 ISSN 0043-1648 21st International Conference on Wear of Materials
[9] Choi Y, Ha J, Lee M G and Korkolis Y P 2021 *Scripta Materialia* **205** 114178 ISSN 1359-6462
[10] Paul S K 2020 *Materialia* **9** 100566 ISSN 2589-1529
[11] Prasad K, Venkatesh B, Krishnaswamy H, Banerjee D K and Chakkingal U 2021 *CIRP Journal of Manufacturing Science and Technology* **32** 154–169 ISSN 1755-5817
[12] Prasad K, Krishnaswamy H, Banerjee D and Chakkingal U 2022 *Tribology in Manufacturing Processes* (*Defect and Diffusion Forum* vol 414) (Trans Tech Publications Ltd) pp 81–87
[13] LChen, Kim J K, Kim S K, Kim G S, Chin K G and BC De Cooman 2010 *steel research international* **81** 552–568
[14] ASTM E8 / E8M-16a, standard test methods for tension testing of metallic materials, 2016 Tech. rep. ASTM International West Conshohocken, PA
[15] 2009-07 ISO 16630:2009-metallic materials-sheet and strip-hole expanding test Tech. rep.
[16] Choi Y, Ha J, Lee M G and Korkolis Y P 2021 *Journal of Materials Processing Technology* **296** 117211 ISSN 0924-0136
[17] Leacock A G, Howe C, Brown D, Lademo O G and Deering A 2013 *Materials Design* **49** 160–167 ISSN 0261-3069
[18] Paul S 2014 *Journal of Materials Engineering and Performance* **23** 3610–3619

IOP Conf. Series: Materials Science and Engineering 1238 (2022) 012039 doi:10.1088/1757-899X/1238/1/012039

Measurement of edge fracture strain of dual-phase steels by in-plane bending test

M Masoumi Khalilabad[1], ES Perdahcıoğlu[1], EH Atzema[2] and AH van den Boogaard[1]

[1] Chair of Nonlinear Solid Mechanics, Faculty of Engineering, University of Twente, Enschede, The Netherlands

[2] Tata Steel,Research & Development, PO Box 10000, NL-1970 CA IJmuiden, The Netherlands

E-mail: m.masoumikhalilabad@utwente.nl

Abstract. Edge ductility of dual phase steels is highly affected by the preparation of the edge and especially shear cutting has a detrimental effect on the formability. In this study, a novel in-plane bending test is used to characterize the edge ductility of dual-phase steels. Digital image correlation technique is employed to calculate the fracture strain along the edge in the desired material orientation. The proposed test setup isolates the material-dependent aspects, namely damage and hardening, from those resulting from testing procedure factors such as contact stresses and friction compared to conventional hole expansion capacity (HEC) test. The results show that the edge crack phenomenon depends on the material orientation.

1. Introduction

Dual-phase (DP) steels provide excellent formability and high strength to weight ratio [1]. They have attracted particular attention due to new environmental initiatives adopted by car industry to reduce the weight and to increase the safety of the car [2]. However, DP steels suffer from the low edge ductility or premature fracture at shear cut edges below the forming limit curves (FLC) [3].

The shear cutting process (such as blanking or punching) severely deforms the edge of material and forms a shear-affected zone (SAZ) at the edge of the material. The fracture surface of the shear cut edge consists of the rollover, burnish, fracture, and burr regions, schematically shown in Figure 1a.

Several tests are available in the literature to assess edge ductility. The hole expansion capacity (HEC) test based on ISO 16630 is the most common test to investigate whether a sheet metal is suitable for forming process or not concerning edge ductility. A schematic of the HEC test can be seen in Figure 1b. It was widely shown that HEC results are sensitive to test parameters such as punch geometry [4], initial hole diameter [5], hole curvature variation [6], edge cutting method [7], direction of strain gradient *i.e.* perpendicular [8] or parallel [9] to edge direction. One reason for this observation can be the complexity of stress state at the edge of the test specimen that is impacted by friction, contact forces, and out of plane deformation. Thus the HEC value could not solely be considered a material characteristic. In other words, the variation in the measured fracture strain [10] significantly impacts engineering interpretation of the results [11]. Furthermore, anisotropy in edge crack sensitivity cannot be captured due to inherent rotational symmetry in the test. Also, performing digital image correlation (DIC) for local measurement of strain along the edge in an HEC test is not straightforward due to out-of-plane deformation [12]. Alternatively, to remove the contact stress and friction at the edge caused by the conical punch of the HEC test, a hollow cylindrical punch with a flat bottom can be used [13]. This punch geometry produces in-plane deformation without contact and friction between the edge of the hole and the punch [14]. While it seems that a cylindrical punch may provide edge material properties independent of test parameters [15], it also has been reported that fracture can occur far from the edge due to different stress state compared to the the material edge [16]. The material can endure less deformation in the plane-strain condition (lowest point on FLC) than uniaxial stress state. Whereas the hole edge is in a uniaxial stress state, further from the edge it shifts towards a plane strain state, where failure may occur.

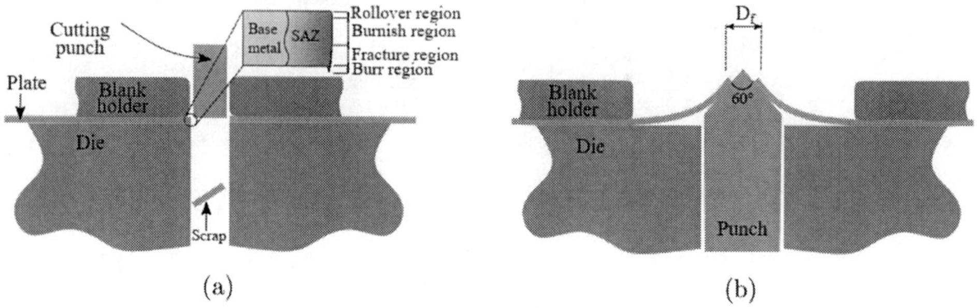

Figure 1: Schematic of a) the punching process and shear affected zone (SAZ) and b) the hole expansion capacity (HEC) test.

Wang *et al.* [17] and Atzema *et al.* [18] introduced the sheared edge tensile test improved (SETi) technique that utilizes a conventional tensile test with sheared edges equipped with DIC for assessing edge ductility. SETi test can be performed in the desired material orientation; However, a tensile test to evaluate edge ductility works only for materials sensitive to edge cracks. Otherwise, failure will occur after necking, which in return makes it challenging to isolate edge ductility. Kobe hole tensile test [19] and side bending test [20] are other alternative tests; independent of the burr orientation, friction, and contact stress. However, the sample geometry used in this study localizes the strain on a tiny portion of the edge, which may not represent the whole sample due to local property variations.

A reliable characterization method for assessing edge ductility of high-strength steel is required to isolate the material-dependent factors such as damage and hardening from structural elements such as contact stresses and friction. It is a prerequisite for understanding the physics of the problem and the root cause of edge cracks. As described earlier, this task is unachievable by current approaches such as the HEC test. In the present research, this challenge is addressed by introducing a novel in-plane bending test to investigate edge ductility. The following sections begin with a comprehensive explanation of the proposed in-plane bending setup, including using digital image correlation in the measurements. Subsequently, two common grades of advanced high-strength steels used in the car industry were selected for the edge ductility investigation by the in-plane bending test.

2. In-plane bending setup

2.1. Fixture design and sample geometry

Recently, an in-plane bending test for sheet metals (Figure 2a) was developed to offer more stability in finding hardening curves at strains above the uniform strain [21]. The strain gradient that occurs in bending delays the necking, making it ideal for investigating edge ductility. The sample geometry and the location of applied forces for deforming the sample are shown in Figure 2b. The sample design is such that the maximum tension occurs at the top edge of the sample. This provides easy edge preparation by any cutting method for edge ductility study, considering that the upper edge is straight. As it can be seen, there is no friction and no contact stress on the edge at the beam area as opposed to the HEC test.

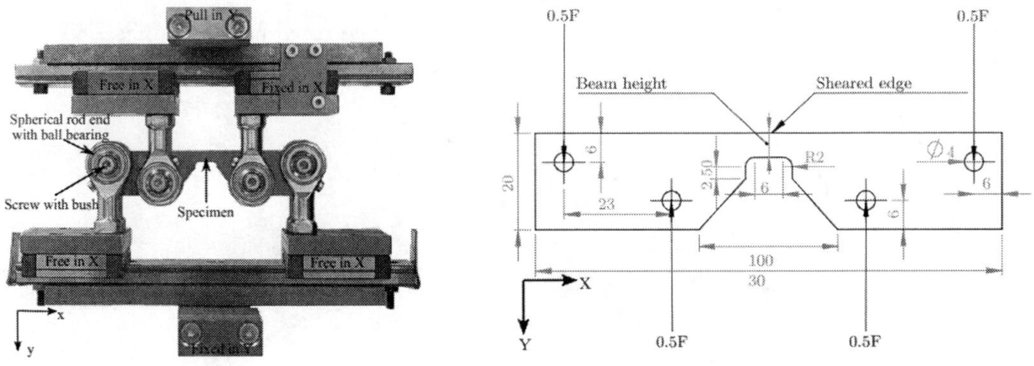

(a) (b)

Figure 2: In-plane bending test; a) The test fixture, b) Sample geometry. The dimensions are in millimeter.

2.2. Fracture strain measurement

The fracture strain is required for edge ductility studies, which can be measured by the DIC technique. This technique works based on processing images taken from the beam section of the in-plane bending test during deformation. The method can provide a relatively accurate strain distribution over the edge compared to the HEC value measured by the caliper. DIC commonly uses a white painted background with black speckles on the surface of the sample for creating the best contrast. However, the layer of paint makes crack detection difficult. Thus, a pattern was applied directly on the material surface with the help of spraying the black color without white background. An IDS camera model UI-35910CP-C-HQ revision two equipped with RICOH lens model FL-BC7528-9M was employed for recording a sequence of images. A frame rate of 10 fps with an image resolution of 4912×3684 pixel was chosen. Subsequently, image analysis was done by the open-source DIC code, Ncorr [22]. The subset with a radius of 20 pixels was selected as the DIC parameter. The numerical mean of the strain distribution along the edge is called 'edge strain at fracture'. This value could be used for comparison purposes with the HEC value, considering that the HEC value is also an average quantity over the whole circumference. All reported strains from different tests in the current investigation refer to the logarithmic strain measure.

2.3. Moment-curvature measurement

In order to calculate the moment *vs.* curvature diagram, the applied force (F) was measured by a Zwick tensile machine of 5.0 kN capacity. Then, the moment was calculated by Equation (1).

$$M = \frac{F \cdot d \cdot \cos(\alpha - \beta)}{2} \tag{1}$$

Where d and α, shown in Figure 3, are constant values during rotation, related to the position of the holes on the bending sample, and β is the angle of rotation which was measured by image processing in GOM correlate software. In this method, a local coordinate system was defined with the help of three points located in the rigid part of the sample far from the deformation zone. Subsequently, the rotation (β) during bending was tracked. To calculate the curvature by Equation (2) [23], the angle (θ) between two imaginary lines in the beam area, shown in Figure 3, was followed during bending.

$$\kappa_{\mathrm{beam}} = \frac{\theta}{l_0} \tag{2}$$

where, l_0 is the initial distance between the two lines. The maximum distance between the two lines is 4.5 mm.

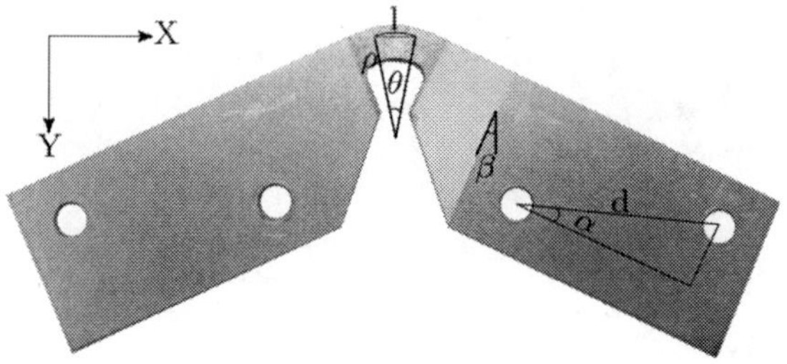

Figure 3: The deformed sample after in-plane bending test.

2.4. Hole expansion capacity test as a reference of the edge fracture strain

Hole expansion capacity tests were done based on ISO 16630. In this test, a hole with a diameter of 10 mm was punched in a rectangular steel plate (90 mm × 90 mm); subsequently, the hole was expanded by a conical punch until a through-thickness crack was detected. Then, the engineering fracture strain (e_{fracture}) or hole expansion capacity (HEC value) was calculated based on the original and final hole diameter (D_0 and D_{f}) by using Equation (3). The test was repeated three times.

$$e_{\mathrm{fracture}} = \frac{D_{\mathrm{f}} - D_{\mathrm{o}}}{D_{\mathrm{o}}} \tag{3}$$

3. Design of experiment

Dual-phase steels were used as the test material in this research. A commercial-grade DP800 hot-dip galvanized (EN 10346:2015: HCT780X+Z) and a trial grade aimed at UTS 1000 MPa, taken for early evaluation of edge ductility, referred as DP1000. Both were produced by Tata Steel with a similar thickness of 1.3 mm, as two common grades of advanced high-strength steels in the car industry.

In order to prepare the top edge of samples for edge ductility investigation in the rolling direction (RD) or transverse direction (TD), a shearing tool with adjustable cutting clearance of 2 µm precision was used. A cutting clearance of 12% was chosen, similar to the punching clearance of the hole expansion capacity test. Then, laser cutting was employed for cutting the rest of the edges and holes of the geometry. Also, it was demonstrated by Ref.[24] that the edge fracture strain is influenced by the time between cutting and testing, and it stabilizes after two weeks. Therefore, all tests in this research were performed more than two weeks after cutting the samples to mitigate the time dependency effect. Each experiment was repeated at least five times and at most ten times to ensure the test repeatability.

IOP Conf. Series: Materials Science and Engineering 1238 (2022) 012039 doi:10.1088/1757-899X/1238/1/012039

4. Results

The following sections elaborate on the effect of material orientation on edge ductility by the in-plane bending test. Also, it shows the limitation of the HEC test for measuring the fracture strain in different material orientations.

4.1. The effect of material orientation on fracture strain

The stress-strain curves measured by a tensile test (ISO 6892) in rolling (RD) and transverse (TD) directions of the as-received DP800 material are reported in Figure 4a. The achieved moment-curvature diagrams of the in-plane bending tests based on the methods explained in Section 2.3, are shown in Figure 4b. The overall material anisotropy of DP800 is already observed in the tensile test results, where the uniform elongation in RD is around 2.4 % higher than in TD (Figure 4a). The same trend is observed for edge ductility (Figures 4b), which confirms the dependency of edge ductility to material orientation.

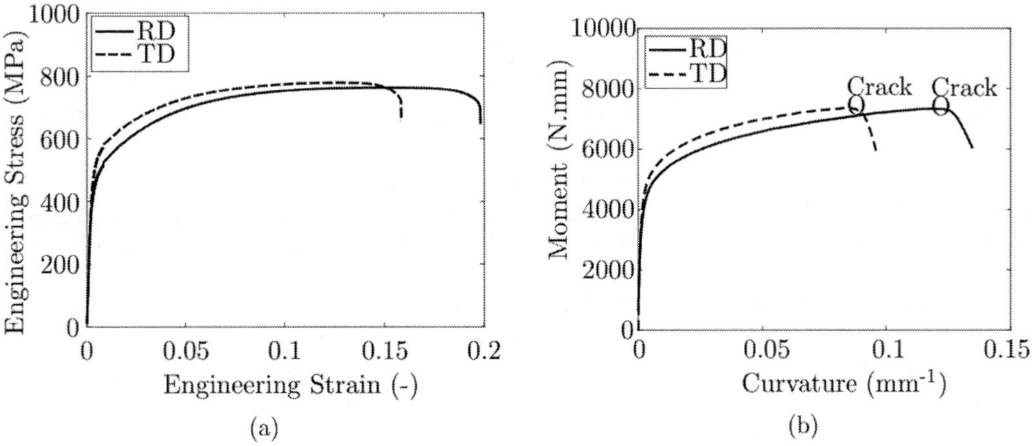

Figure 4: The characterization of DP800 in rolling and transverse directions. a) Stress-strain curve measured by tensile test and b) Moment-curvature diagram measured by bending test.

The strain distribution of the in-plane bending test at at the presence of the edge crack is shown in Figure 5a. It can be seen that the maximum stretch is on the sheared edge of the material. The average strain distribution on the edge is extracted from DIC measurement at the moment of fracture and depicted in Figure 5b, in RD and TD material orientations for DP800 and DP1000. The result shows that the rolling direction performs better with respect to the edge ductility issue for both materials.

The HEC values of DP800 and DP1000 are 0.21 ± 0.005 and 0.22 ± 0.01 respectively; which are close to each other. The direct comparison of the HEC test with the in-plane bending test is not straightforward, considering the complex stress state at edge material during the HEC test due to friction, out-of-plane deformation, and contact stress. The edge crack in HEC test occurred in the rolling direction as indicated in Figure 5c, where the major principal stress direction is parallel to TD. This implies that the transverse direction is the weakest material orientation for edge cracks. This shows that it is not possible to measure edge ductility of the material in the rolling direction with the HEC test due to inherent rotational symmetry in the test, which means that the weakest direction determines the result.

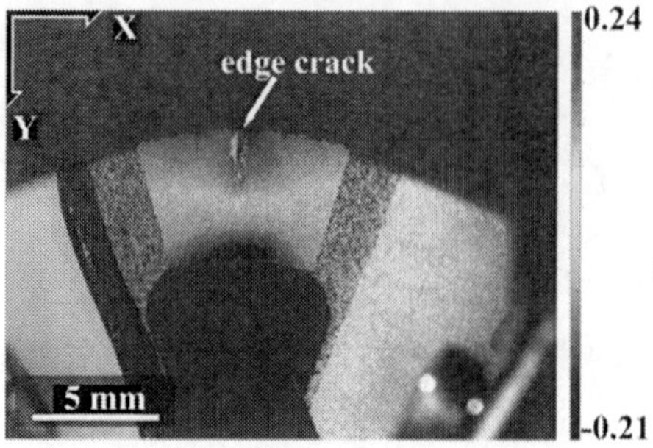

(a)

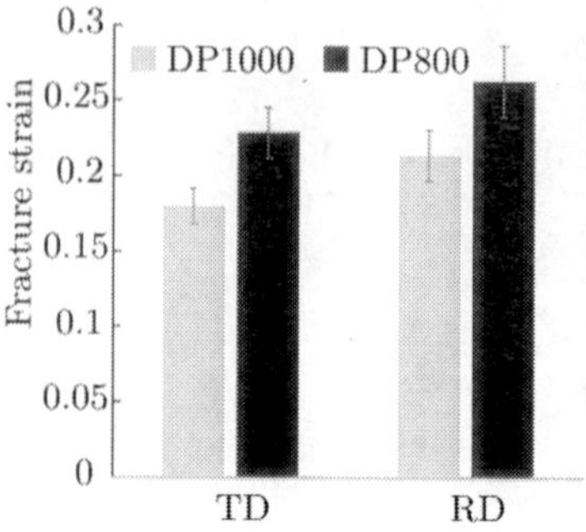

(b)

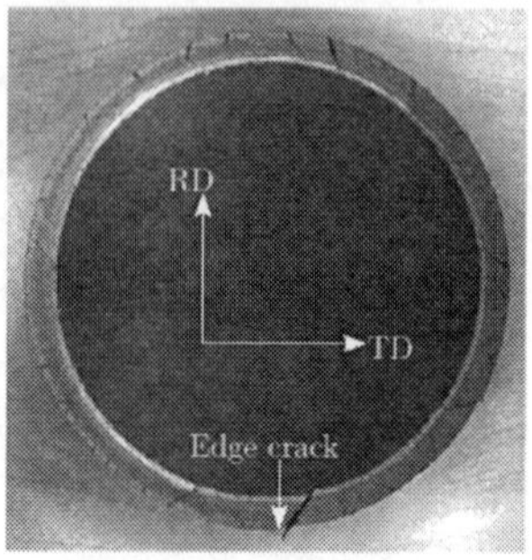

(c)

Figure 5: a) The edge strain distribution after fracture of in-plane bending test of DP800 material. b) The fracture strain in transverse and rolling directions of DP800 and DP1000. C) Hole expansion capacity test of DP800. The major vertical cracks confirm that the weakest direction is TD.

4.2. The edge crack initiation and growth

The edge of the material after the bending test with vertical and horizontal cracks is shown in Figure 6a. The horizontal crack in region 'A' was created due to shear cutting and not the bending test. Interestingly, it does not play a role in the initiation of the edge crack during bending. Figure 6b and 6c show that multiple vertical cracks initiated in the burr region, and there are a few other cracks that encompass the whole thickness. These micro-cracks are not present in the rollover region. It is in agreement with the higher HEC value reported by [16] for cases with the burr in direct contact with the punch of the HEC test than the specimen with the burr away from the punch. Actually, the punch contact in the HEC test delays the crack initiation at the side of contact by applying compressive stresses. Also, the out-of-plane deformation imposes more deformation at the side of the plate away from the punch. In other words, the material does not necessarily fail at the weakest location and the result of HEC test can be influenced by contact stress, punch type and friction [4, 14]. This is not the case for the in-plane bending test, and hence, the in-plane bending test gives a better representation of pure material behavior.

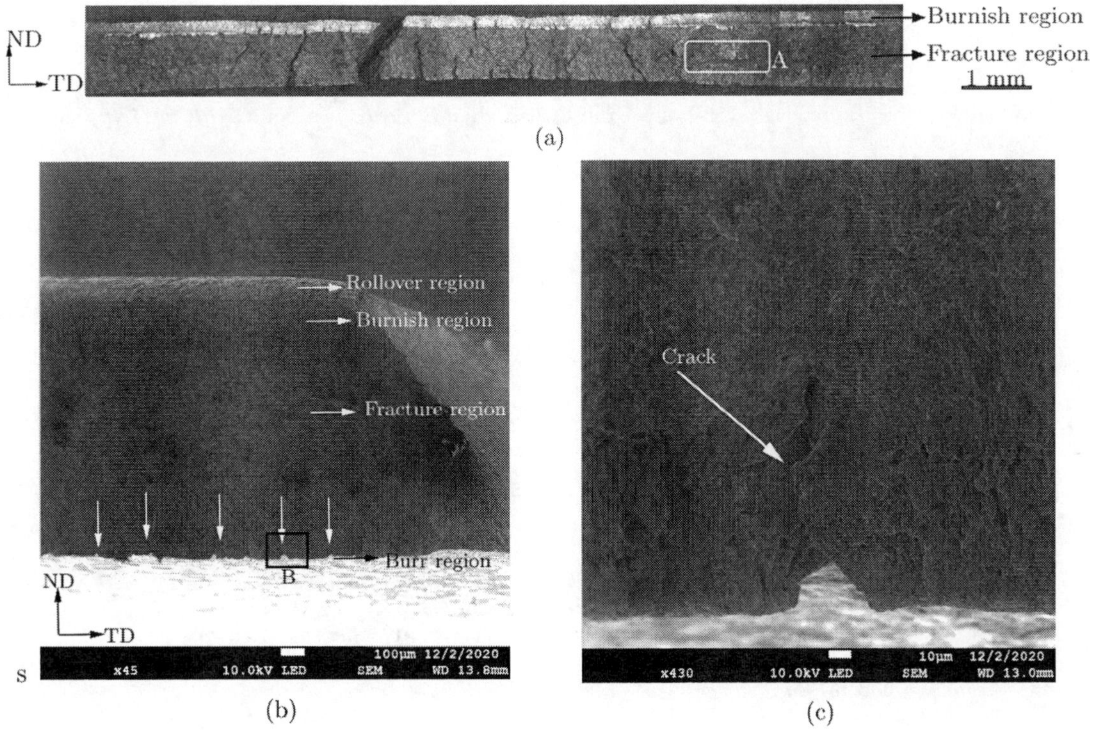

Figure 6: The edge appearance after bending test. a) Multiple vertical cracks are visible. Region A shows a horizontal crack created by shear cutting process. b) The microcracks initiated at burr region, while there is no visible micro crack at the roll over region. c) higher magnification of rectangle B.

5. Conclusions

In this study, an in-plane bending test was employed as a novel method for assessing edge ductility of two common steel grades in the automobile industry. The investigation proved that the edge crack phenomenon depends on material orientation. A sheared edge parallel to the transverse direction has a lower fracture strain. Therefore, the in-plane bending test is reliable for measuring edge ductility, considering that it provides more details of fracture strain than the HEC test in the desired material orientation. Also, the edge properties along the thickness direction are inhomogeneous; many micro-cracks have been observed on the burr side while only a few spread the entire thickness.

6. Acknowledgment

This research was carried out under project number T17019c in the framework of the Research Program of the Materials innovation institute (M2i) (www.m2i.nl) supported by the Dutch government.

References

[1] Ghoncheh M, Sengupta J, Wu N, Gao J and Phillion A 2021 *Journal of Materials Processing Technology* **289** 116936
[2] Han Y, Chu X, Kuang S, Li T, Xie C and Teng H 2019 *Journal of Materials Engineering and Performance* **28** 372–381
[3] Frómeta D, Tedesco M, Calvo J, Lara A, Molas S and Casellas D 2017 *Journal of Physics: Conference Series* vol 896

[4] Thesing L, Boff U, of Materials L S I J and 2016 U 2016 *International Journal of Materials Engineering and Technology* **15** 159–170
[5] Kim H, Shang J, Beam K, Samant A, Hoschouer C and Dykeman J 2016 *Journal of Physics: Conference Series* vol 734
[6] Yu X, Chen J and Chen J 2016 *International Journal of Advanced Manufacturing Technology* **86** 1083–1094
[7] Larour P, Pauli H, Freudenthaler J and Grünsteidl A 2011 *Proceedings of the IDDRG2011 (Bilbao, Spain, June 05-08)*
[8] Iizuka E, Urabe M, Yamasaki Y and Hiramoto J 2017 *Journal of Physics: Conference Series* vol 896 (IOP Publishing) p 012008
[9] Yoshida H, Yoshida T, Sato K, Takahashi Y, Matsuno T and Nitta J 2013 *Nippon Steel Technical Report* **1035** 18–24
[10] Hance B M 2017 *SAE International Journal of Engines* **10** 2017–01–0306
[11] Atzema E, Borsutzki M, Braun M, Brockmann S, Buelter M, Carlsson B, Larour P and Richter A 2012 *Proceedings of the Neue Entwicklungen in der Blechumformung, Fellbach, Germany* 23–25
[12] Larour P, Freudenthaler J, Grünsteidl A and Wang K 2014 188–193
[13] Konieczny A and Henderson T 2007 *SAE Transactions* 20–29
[14] Stachowicz F 2008 *Archives of Civil and Mechanical Engineering* **8** 167–172
[15] Paul S K 2020 *Proceedings of the Institution of Mechanical Engineers, Part B: Journal of Engineering Manufacture* **234** 671–676
[16] Pathak N, Butcher C and Worswick M 2016 *Journal of Materials Engineering and Performance* **25** 4919–4932
[17] Wang J, Link T M and Merwin M J 2011 *Proceedings of the International Conference on New Developments in Advanced High-Strength Sheet Steels* International Conference on New Developments in Advanced High-Strength Sheet Steels, 2008 (Warrendale: AIST) pp 361–366
[18] Atzema E and Seda P 2015 *Forming Tech. Forum, Zurich, Switzerland* pp 984–991
[19] Pathak N, Butcher C, Worswick M J, Bellhouse E and Gao J 2017 *Materials* **10** 1–29
[20] Matsuno T, Sato K, Okamoto R, Mizumura M and Suehiro M 2016 *Journal of Materials Processing Technology* **230** 167–176
[21] Naseem S, Perdahcıoğlu E S, Geijselaers H J M and van den Boogaard A H 2020 *Experimental Mechanics*
[22] Blaber J, Adair B and Antoniou A 2015 *Experimental Mechanics* **55** 1105–1122
[23] Hu J, Marciniak Z and Duncan J 1979 *Mechanics of Sheet Metal Forming* (Elsevier)
[24] Atzema E and Seda P 2017 *5th SCT2017 - Steels in Cars and Trucks* (Netherlands)

IOP Conf. Series: Materials Science and Engineering 1238 (2022) 012040 doi:10.1088/1757-899X/1238/1/012040

Sheared Edge Formability Characterization of High Strength Aluminum Alloys at Room and Elevated Temperatures using Hole Expansion Tests

A Narayanan[1*], R George[1], K Cheong[1], A Mohamadizadeh[1], C Shi[2], R Long[3], E Boettcher[4], A Weinschenk[5], S Huhn[5], M J Worswick[1] and C Butcher[1]

[1]University of Waterloo, 200 University Avenue West, Waterloo, ON N2L 3G1, Canada

[2]Magna International, 750 Tower Drive, Troy, MI 48098, United States

[3]Arconic Technology Center, 100 Technical Drive, New Kensington, PA 15069, United States

[4]Honda Development & Manufacturing of America, 21001 OH-739, Raymond, OH 43067, United States

[5]Hexagon Canada, 555 Michigan Drive, Suite 200, Oakville, ON L6L 0G4, Canada

anadvait@uwaterloo.ca

Abstract. High strength 6000- and 7000-series aluminum alloys have significant potential for automotive light weighting due to their high strength-to-weight ratios. Alternate thermal processing routes and warm forming operations with rapid heating to 150 - 300°C are being explored to expand their forming process windows. While the constitutive and formability behavior of these alloys at elevated temperatures is an active area of research, limited data is available on their sheared edge stretchability, particularly under elevated temperature forming conditions. To this end, the influence of cutting clearance on the edge fracture limits of two precipitation hardening alloys, AA6013-T4 and AA7075-T6, and a non-heat treatable AA5182-O were first investigated at room temperature using the conical hole expansion test. A 12% clearance was found to be suitable for the three materials and used to process the hole expansion samples prior to warm forming. Microhardness tests were used to characterize the depth and severity of the shear affected zone (SAZ) for the AA7075-T6 before and after the warm forming cycle. The forming temperature was found to increase the hole expansion ratio by approximately 400%, 150%, and 520% for the AA5182-O, AA6013-T4, and AA7075-T6 respectively relative to the room temperature tests. The edge stretchability during W-temper forming of the AA6013 and AA7075 at room temperature was also assessed. It was observed that W-temper forming negatively influenced the hole expansion of AA6013 by approximately 30% relative to the T4 condition. Conversely, the hole expansion ratio for AA7075 when formed in W-temper increased by over 35% with respect to the T6 temper.

1. Introduction

The need for lighter and fuel efficient vehicles has driven the increased adoption of high strength 6000- and 7000-series alloys in the automotive stamping industry. However, additional heat treatment and elevated temperature forming processes are often necessary to enable formation of complex parts since these age hardenable alloys have limited ductility at room temperature [1]. One approach is to solutionize and quench the material followed by immediate cold forming in the ductile W-temper [2]. Although promising, the W-temper is unstable at room temperature due to natural aging such that the time between quenching and stamping becomes critical. Alternatively, heating the material in the T4 or T6 temper to temperatures above 150°C and warm forming has been found to enhance the material ductility and part depths considerably for 5000- and 7000-series alloys [3]. On the contrary, limited improvements in forming depth and limit strains have been observed in general during warm forming of 6000-series alloys [3, 4]. The available literature has mainly addressed the drawability and stretchability of high strength aluminum alloys. Studies focused on sheared edge formability of these materials at room or elevated temperature have been scarce despite hole punching followed by extrusion being commonplace in automotive panels. Several studies have been conducted with regards to dependence on cutting clearance for 5000- and 6000-series aluminum at room temperature wherein a gap equal to 20% of the sheet thickness or lower was identified to be appropriate [5, 6]. The effect of heat treatment and warm forming on sheared edge stretchability remains unexplored.

The objective of the present study is to evaluate the influence of cutting clearance, hole expansion temperature, and W-temper forming on sheared edge fracture limits of aluminum alloys using the conical hole expansion (HX) test. Three commercial aluminum alloys are studied: a non-precipitation hardenable AA5182-O, and two heat treatable alloys, AA6013-T4 and AA7075-T6. The room temperature sheared edge formability was first characterized to select the best cutting clearance for each material. Blanks with a 10 mm diameter central hole were then prepared by either drilling and reaming or punching, after which the hole expansion test was conducted at elevated temperatures of 250°C, 300°C and 175°C for the AA5182-O, AA6013-T4 and AA7075-T6 samples, respectively. In addition, the sheared edge stretchability of AA6013 and AA7075 at room temperature were also characterized in the W-temper condition. Finally, the severity of work hardening near the edge after punching and before the forming cycle were characterized for AA7075-T6 using microhardness tests to correlate with the hole expansion capabilities at room and elevated temperature.

2. Materials and Experiments

2.1. Mechanical Properties

The room temperature tensile mechanical properties of the three materials at a strain rate of 0.001 s^{-1} are listed in Table 1. The AA5182-O sheet was supplied in the annealed condition while the AA6013 and AA7075 were provided in the T4 and peak-aged T6 tempers respectively. The plasticity response characterization of the same lot of AA5182-O and AA7075-T6 was performed in prior work by the authors [7, 8].

Table 1. Mechanical properties for the studied materials along the rolling direction. The virtual extensometer lengths used to evaluate the uniform and total elongation values of AA5182-O, AA6013-T4, and AA7075-T6 were 50 mm, 25 mm, and 12.5 mm respectively due to different tensile geometries.

Material	Thickness (mm)	Yield Strength (MPa)	Ultimate Tensile Strength (MPa)	Uniform Elongation (%)	Total Elongation (%)
5182-O	1.5	146	302	20.6	25.7
6013-T4	2.0	156	314	20.8	30.6
7075-T6	2.0	483	566	11	16.0

2.2. Hole Processing Techniques

Circular holes of 10 mm diameter were punched with three different cutting clearances (c) of 5%, 12%, and 19% with respect to the sheet thickness. The 10 mm hole size was chosen in accord with the ISO16630 standard [9]. A 203.2 mm diameter blank was positioned on lifter springs with its edges in contact with guide stops on the tooling to ensure accurate positioning of the hole during cutting. The

IOP Conf. Series: Materials Science and Engineering 1238 (2022) 012040 doi:10.1088/1757-899X/1238/1/012040

holding force and strip removal was achieved via a punch stripping system with steel springs designed for compact space requirements as shown in Figure 1 (b). Monarch Klir 2 cutting oil was applied on the specimen surface before punching to minimize burr formation. The punching speed was set to 125 mm/s and conducted at room temperature for all test conditions considered. The dimensions and the punching tool setup are depicted in Figure 1. Additional samples were also fabricated by drilling a hole of 9.5 mm followed by reaming to the final 10 mm diameter to consider the baseline material condition with minimal strain hardening at the edge.

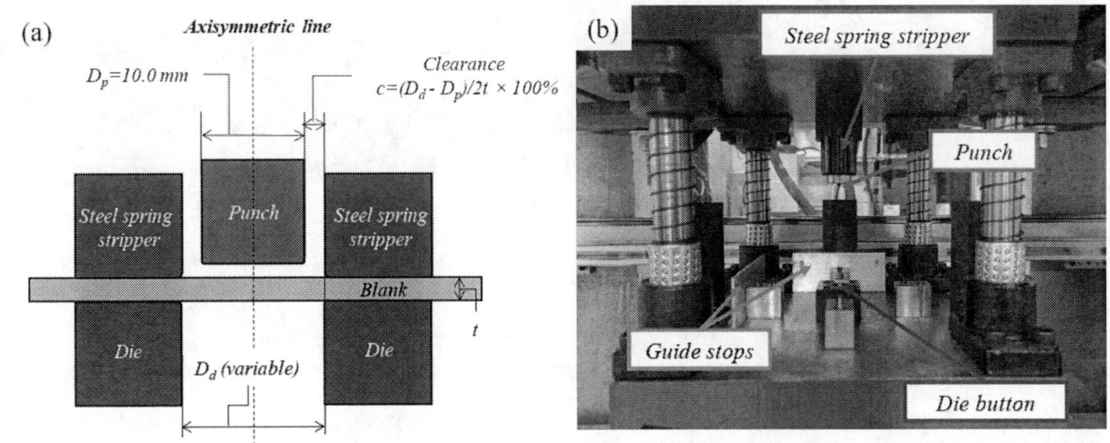

Figure 1. (a) Schematic of the punching process, and (b) punching tool at the University of Waterloo.

2.3. Conical Hole Expansion at Room and Elevated Temperatures

The Fast Forming System (FFS) developed at the University of Waterloo and shown in Figure 2 was employed to perform the warm hole expansion tests [11]. The FFS system utilizes an ABB® robotic arm to maneuver the specimen from the loading station to the contact furnace and the hole expansion tooling. The specimens were heated to the forming temperature rapidly before removal and transfer to press by the robot within approximately 12 seconds. The forming temperature chosen for the materials and corresponding heating times in the furnace are tabulated in Table 2. The hole expansion temperature for each alloy was selected based upon the literature and/or prior Nakazima dome test results that resulted in good sheet forming limit strains. Thus, the temperatures chosen for each material may not necessarily be the most favorable in terms of maximizing the sheared edge fracture limits.

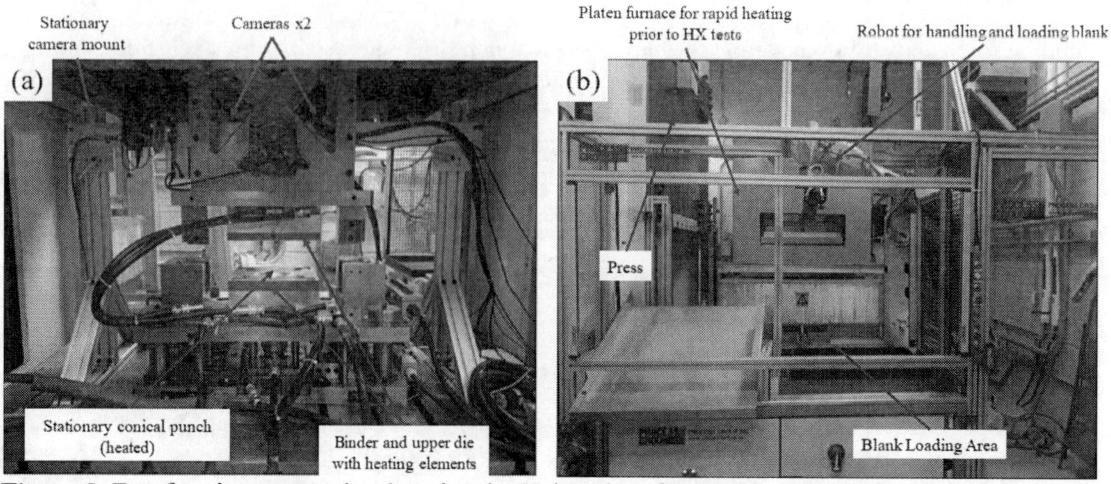

Figure 2. Fast forming system developed at the University of Waterloo [11]. (a) Tooling setup and (b) blank heating and handling system for warm hole expansion tests

Table 2. Blank forming temperature and heating time in contact furnace prior to hole expansion. The tools are heated to the same temperature as the specimens to facilitate a near isothermal test condition.

Material	HX temperature (°C)	Contact furnace temperature (°C)	Hold time in contact furnace (s)
AA5182-O	250	300	7
AA6013-T4	300	350	10
AA7075-T6	175	220	8

The conical punch had a minimum and maximum diameter of 4 mm and 40 mm, respectively, with a taper angle of 60° in accordance with the ISO16630 standard. The hydraulic binder imposed a clamping force of approximately 400 kN on the specimen. No lubricant was applied to the punch or the specimen before the tests due to its potential breakdown at elevated temperatures. After sample placement, the binder and die clamped the specimen and moved downward in concert at a speed of 1.0 mm/s to extrude the hole while the punch was stationary. Hole deformation was continuously captured using two high speed stereo cameras having a frame rate of 750 frames per second and equipped with Sigma® macro lenses with 105 mm focal length. The test was terminated as soon as a through-thickness crack was detected in the monitoring system. Subsequently, the first image that showed the crack was identified and the corresponding hole geometry was evaluated using the ImageJ software. A schematic of the conical hole expansion tooling and a recorded image of an AA5182-O sample at the first appearance of a through-thickness crack are shown in Figure 3. The room temperature hole expansion tests were performed on an MTS formability press with a clamping force of 640 kN in dome test tooling without lockbeads. The tests were conducted without any lubricant to be in line with the elevated temperature cases. Details regarding the tooling, test setup and data acquisition are discussed in Noder *et al.* [10]. A minimum of five repeats were conducted for each test condition.

IOP Conf. Series: Materials Science and Engineering 1238 (2022) 012040 doi:10.1088/1757-899X/1238/1/012040

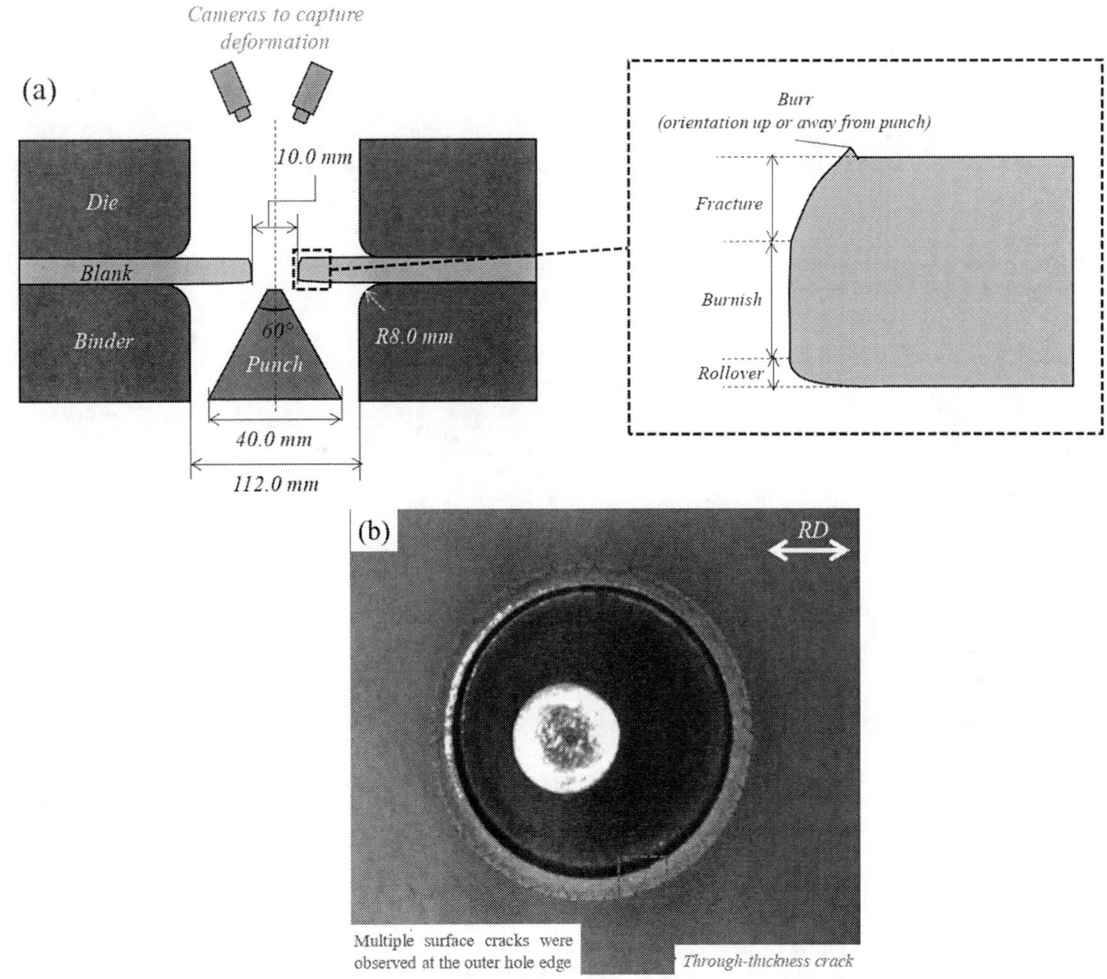

Figure 3. (a) Schematic and dimensions of the tooling employed in the FFS for warm hole expansion tests, and (b) a representative image of an AA5182-O sample at instant of through-thickness crack after punching with 12% clearance and hole expansion at room temperature.

The average inner hole diameter at fracture, D^f_{inner}, was determined from four measurements of the image at fracture and used to evaluate the hole expansion ratio (HER) defined as

$$HER(\%) = \frac{D^f_{inner} - D_0}{D_0} \times 100 \tag{1}$$

where D_0 is the initial hole diameter equal to 10 mm.

The HER represents deformation under a non-linear strain path due to the initial compression of the inner hole edge when in contact with the conical punch before gradually transitioning to uniaxial tension [12]. Conversely, the outer hole edge remains in a uniaxial tensile stress state throughout the test and is utilized to determine the fracture strain from the initial and final diameters [10, 12]. The average major and equivalent strain in terms of the average outer diameter at fracture (D^f_{outer}) are

$$\varepsilon^f_{eq} = \varepsilon^f_1 = \ln\left(\frac{D^f_{outer}}{D_0}\right) \tag{2}$$

3. Results and Discussion

3.1. Edge Sensitivity to the Cutting Process at Room Temperature (RT)

The effect of cutting process and punching clearance on the room temperature HER and fracture strains for the studied materials are depicted in Figure 4. The AA5182-O and AA7075-T6 samples were found to be relatively insensitive to the cutting clearance. Interestingly, the measured HER for AA6013-T4 samples after shearing with 12% clearance was significantly higher at 46% compared to the other two punching conditions. A cutting clearance of 12% was then selected for the W-temper and warm forming conditions. All three alloys were moderately sensitive to the cutting process since the HER of the reamed specimens was at least 50% higher than that of the corresponding edges sheared with 12% punch-die clearance.

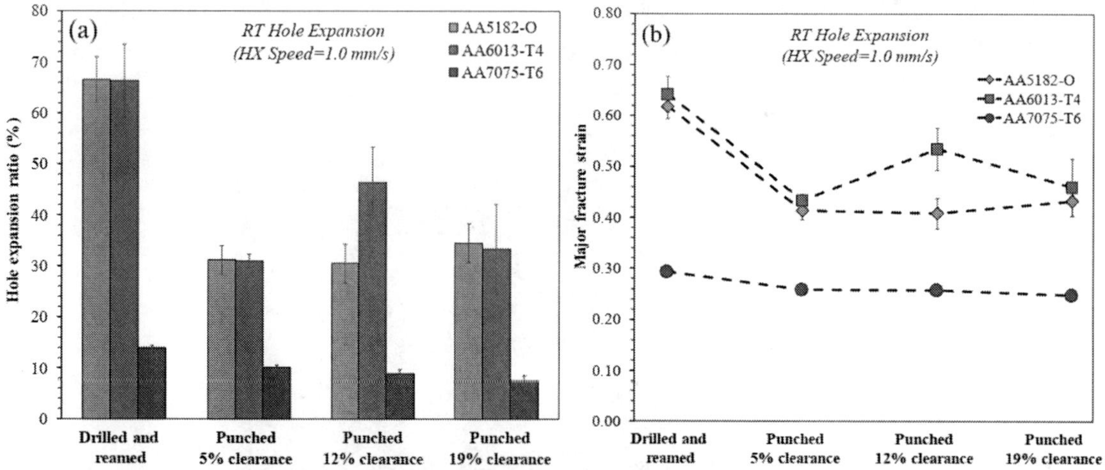

Figure 4. (a) Measured hole expansion ratios and (b) average major fracture strains from RT hole expansion tests of reamed and sheared edges. The error bars denote the standard deviation in the measured data.

3.2. Effect of W-Temper Forming on Sheared Edge Fracture Limits

The AA6013 and AA7075 specimens were solutionized for 30 minutes at 570°C and 470°C, respectively, followed by rapid quenching between flat dies, RT hole punching and extrusion within 6 minutes for the W-temper tests. The comparison of the W-temper hole expansion with the as-received material is shown in Figure 5. The sheared edge formability of AA6013 sheet was found to decrease in W-temper forming by approximately 30% relative to the T4 condition. On the contrary, the HER increased in W-temper forming of AA7075 and was comparable to that of the drilled and reamed test conducted in the T6 temper.

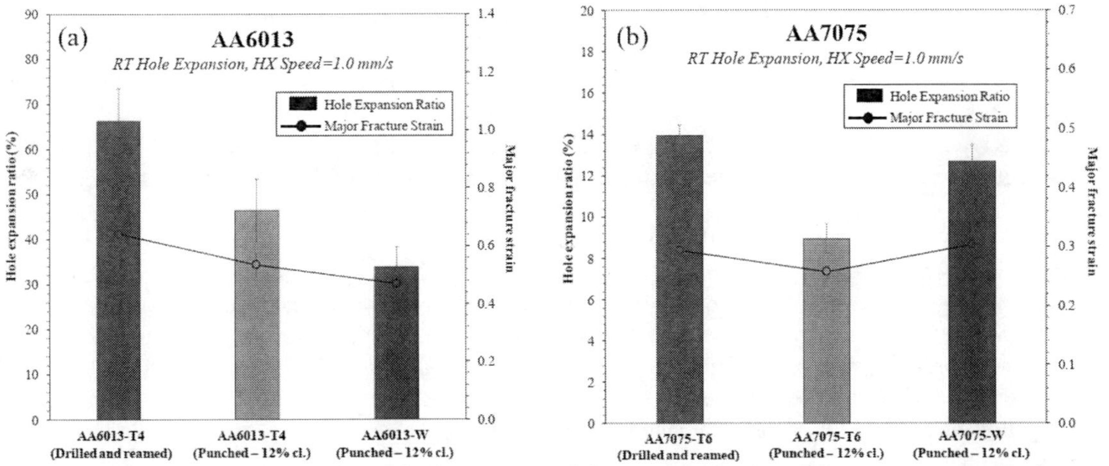

Figure 5. Comparison of the sheared edge HER and fracture strain from W-temper tests with the as-received condition for a) AA6013 and (b) AA7075. A cutting clearance of 12% was used for the punched conditions.

3.3. *Influence of Forming Temperature on Sheared Edge Fracture Limits*

A significant influence of hole expansion temperature on the edge fracture limits was observed for the three materials, as can be seen in Figure 6 for both reamed and punched holes. The observed HER at 250°C for AA5182-O was at least five times higher than that of the room temperature tests. Fracture was not achieved in the AA5182-O tests at 250°C so the HER of 150% is a lower bound estimate. The sheared edge HER at elevated temperatures was approximately 2.5 and 6.2 times higher than the RT test for the AA6013-T4 and AA7075-T6 samples, respectively. A similar trend is also noted in the fracture strains.

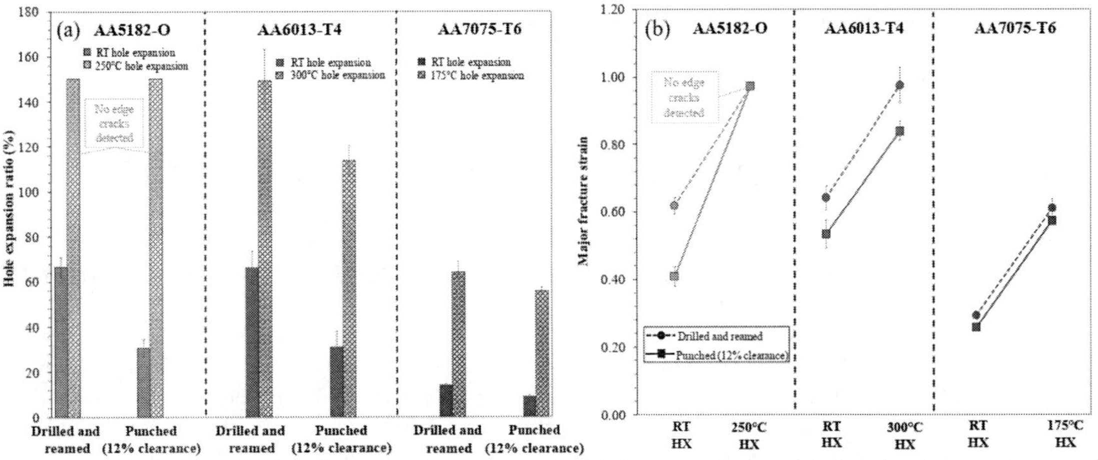

Figure 6. Variation of (a) hole expansion ratio and (b) fracture strain with forming temperature and cutting process for the three aluminum alloys. No fracture was observed in the AA5182-O tests conducted with both the reamed and punched specimens at 250°C. The HX tests were conducted at a speed of 1.0 mm/s.

3.4. *Hardness Distribution After Hole Punching for AA7075-T6*

The reamed and sheared AA7075-T6 specimens were sectioned along the rolling direction (RD) for hardness measurements since through-thickness cracks predominantly occurred in this direction during hole expansion. A few samples were also subjected to a rapid heating cycle at 175°C for 8 seconds after hole processing but prior to sectioning and mounting to gauge the hardness change due to temperature. The specimens were then cold mounted in the puck using a mixture of epoxy resin and hardener (25:3

ratio by weight) due to heat sensitive nature of the AA7075-T6. The microhardness tests were conducted after a resin curing time of approximately 12 hours. One representative sample was considered for each condition. The microhardness measurements were conducted across the SAZ along three lines located at approximately 25%, 50%, and 75% through the sheet thickness, as shown in Figure 7(a). The first indent was placed at approximately 0.15 mm away from the edge while the distance between two consecutive indents was roughly 2.5 times the indenter size or 0.20 mm. A notable SAZ with a hardness gradient near the edge was not observed for the punched specimen at RT with the values comparable to a reamed specimen as shown in Figure 7 (b). This lack of SAZ hardness change is also reflected in the small absolute differences in the HER and fracture strain between the reamed and punched specimens for the AA7075-T6 shown in Figure 4. Treating the specimen to a heating cycle at 175°C after hole punching slightly reduced the hardness but does not explain the significant difference in the sheared edge fracture limits. It is hypothesized that the reduction in the hardness is due to potential over ageing that is accelerated by the deformation induced near the edge despite only 8 seconds of hold time in the furnace at 175°C. The increased sheared edge formability at 175°C is presently attributed to a reduction in the flow stress [11] that delays damage nucleation near the edge.

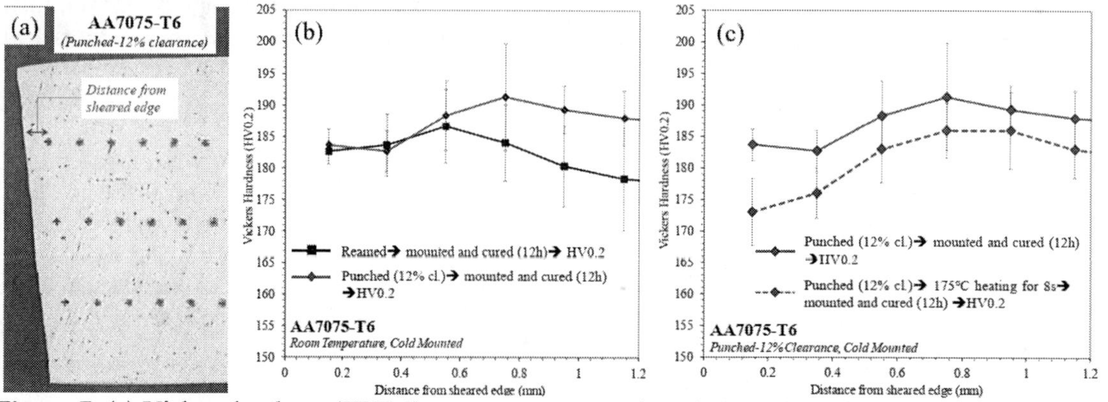

Figure 7. (a) Vickers hardness (HV0.2) measurements conducted along three lines for an AA7075-T6 sample, (b) average hardness *versus* the distance from the hole edge for a reamed and punched specimen at RT, and (c) hardness distribution for a punched specimen at RT and held at 175°C for 8 seconds.

4. Conclusions

The sheared edge formability of three different aluminum alloys was experimentally characterized at room and elevated temperature. At RT, the AA5182-O and AA7075-T6 did not exhibit a strong sensitivity to the cutting clearance whereas 12% clearance performed the best for the AA6013-T4. Cold punching and HX of the W-temper at room temperature negatively affected the fracture strains for the AA6013, whereas a fracture strain increase of over 10% was seen for AA7075-W with respect to the peak-aged condition. With respect to the RT sheared edge HER values of 31%, 46%, and 9% for a 12% cutting clearance, the hole expansion capability at elevated temperature were higher by approximately 5.0, 2.5, and 6.2 times for the AA5182-O, AA6013-T4, and AA7075-T6 specimens, respectively. Finally, microhardness tests conducted for the punched AA7075-T6 specimens did not indicate the presence of any hardness gradient within the SAZ and likely accounts for its low edge sensitivity to the cutting process. However, observed hardness distribution could not explain the increase in HER and fracture strain with temperature. Future work will address the influence of other conditions such as pre-aging, hot forming, and stroke rate on the sheared edge stretchability of these alloys.

Acknowledgements

Financial support from the Natural Sciences and Engineering Research Council of Canada, the Canada Research Chairs Secretariat, the Ontario Advanced Manufacturing Consortium, the Ontario Research Fund and Ontario Centres of Excellence is gratefully acknowledged. The studies on the AA6013 and

AA7075 aluminum alloys were also supported by Honda R&D Americas, Promatek Research Centre (Cosma International) and the Arconic Technical Center. The AA5182 studies received additional support from Hexagon Canada.

References

[1] Liu Y, Zhu Z, Wang Z, Zhu B, Wang Y and Zhang Y 2017 *Procedia Eng.* **207** p 723-728.
[2] Sáenz de Argandoña E, Galdos L, Ortubay R, Mendiguren J and Agirretxe X 2015 *Key Engineering Materials* **651** p 199-204.
[3] Kim H, Hahnlen R, Feister T and Tunga V 2018 *IOP Conf. Ser.: Mater. Sci. Eng.* **418** p 012029.
[4] Li D and Ghosh A K 2004 *J. Mater. Process. Technol.* **145**(3) p 281-293.
[5] Tsoupis I and Merklein M 2016 *IOP Conf. Ser.: Mater. Sci. Eng.* **159**(1) p 012017.
[6] Hu X H, Sun X and Golovashchenko S F 2016 *Finite Elem Anal Des.* **109** p 1-13.
[7] Noder J and Butcher C 2019 *Int. J. Mech. Sci.* **163** p 105138.
[8] Rahmaan T, Noder J, Abedini A, Zhou P, Butcher C and Worswick M J 2020 *Int. J. Impact Eng.* **135** p 103390.
[9] International Organization for Standardization 2017 'Metallic Materials—Sheet and Strip—Hole Expanding Test' 16630 p 1-7.
[10] Noder J, Gutierrez J E, Zhumagulov A, Dykeman J, Ezzat H and Butcher C 2021 *Materials* **14**(17) p 4970.
[11] Pishyar H 2021 Master's thesis, *University of Waterloo.*
[12] Narayanan N, Abedini A, Khameneh F and Butcher C 2022 An Experimental Methodology to Characterize the Uniaxial Fracture Strain of Sheet Metals using the Conical Hole Expansion Tests *J. Mater. Eng. Perform.* Accepted.

Stretch flangeability of AHSS automotive grades versus cutting tool clearance, wear, angle and radial strain gradients

P Larour[1], J Hinterdorfer[1], L Wagner[1], J Freudenthaler[1], A Grünsteidl[1], M Kerschbaum[1]

[1]voestalpine Stahl GmbH, voestalpine-Straße 3, A-4020 Linz, Austria.

E-mail: patrick.larour@voestalpine.com

Abstract. Stretch flangeability tests are performed on cold rolled AHSS automotive steel grades for crack sensitivity evaluation. The ISO16630 tests procedure is used with different cutting edge conditions including cutting clearance, cutting tool wear, cutting tool angle and hole expansion punch tool geometry. Among all influence parameters, the cutting clearance is considered to be the most critical to control. Low (<10%) and high (>20%) clearances are significantly detrimental to hole expansion ratio (HER). An optimum in HER is reached around 15% clearance. Both cutting punch wear and cutting die wear each affect negatively the HER values in comparison to new sharp cutting tools. Varying the cutting angle with concave and convex hole punching tool geometries instead of orthogonal cutting has a rather negative effect on stretch flangeability due to irregular cut edge quality along hole perimeter and excessive tool wear in the concave configuration. The influence of hole expansion punch geometry (flat R25, biaxial Nakajima R25 vs. ISO16630 60° conical) on the HER values is also investigated. The (true) HER values are proportional to the FE simulated logarithmic radial strain gradient at fracture.

1. Introduction

Advanced High Strength Steels (AHSS) are widely used in the automotive industry, driven by vehicle weight reduction and crash safety requirements. Their reduced shear edge formability however is a growing concern [1]. Shear edge failure may occur for strain levels much lower than the forming limit curve (FLC) depending on the steel grade considered [2]. The conical hole expansion test, as described in the ISO 16630 procedure [3], is suitable to characterize the stretch flangeability of steel grades.

Among all influence parameters, the cutting clearance is considered to be the most critical for stretch flangeability control. High cutting clearances are advised to avoid crack formation after hole expansion test in AHSS [4]. Nominal clearances ≥ 12% can be considered as long as burr formation is not occurring [4]. Own investigations on AHSS confirm an optimal clearance between 12% and 20% for optimum hole expansion ratio (HER) [5]-[6]. Current AHSS processing guidelines suggest adapted clearances according to sheet strength and thickness level [7]. Optimal HER values for higher strength and thickness sheet steels require an increase of cutting clearance from typically ≈8% for mild steels up to ≈15% for AHSS [7].

Close monitoring of cutting tool wear is also necessary for edge crack minimizing [7]. A significant decrease of HER values from 25% with a sharp punch down to 16% for a worn out cutting punch has been observed in [8] for a DP980 steel grade. Significant negative tool wear effects on stretch flangeability of AHSS are also found in [9] for hot rolled HDT780C steel. An increase of punch radius as well as die radius up to 250μm both decrease the HER values. A combination of both punch and die tool wear together has an even worse influence on HER values degrading from 80% down to 30% [9].

HET investigations have been performed in [10] for DP980 and DP1180 with flat, conical and roof top cutting punches. Flat expansion HER values are similar for flat and conical cutting punch geometries and worse for roof top cutting punch configuration. Investigations in [11] for DP600 and DP980 show the positive influence of angled straight edge cutting on edge crack strain values with an optimum for 6° lower/upper blade angle as well as 3° rake shear angle at 15% clearance. Similarly a 6° bevel or conical cutting punch yield the best conical HER values for DP600, DP800 and DP1000 steels in [12].

Some investigations also suggest that the strain gradient (radial curvature of hole edge during expansion) influences significantly positively the HER values [13]-[17]. The radial strain gradient can be influenced by varying the conical expansion tool angle and initial hole diameter [13]. The use of hemispherical Nakajima punches influences also the radial strain gradient amount at cut edge [14]. According to [15] this strain gradient might effectively suppress local neck formation and have a positive influence on the limiting edge crack strain. This radial strain gradient is stronger for 60° conical and hemispherical punches compared to a flat punch [15]. A linear positive dependency of edge crack strain versus radial strain gradient is clearly shown in [16]-[17]. Different hole diameters from 10 to 50mm combined with 50-60° conical punches as well as Nakajima and Marciniak punches are varied to achieve a large range of strain gradient at hole edge in [16]-[17]. Even elliptical holes have been used to clarify the (small) influence of tangential major strain gradients along hole edge on HER values [17].

The stretch flangeability of AHSS cold rolled steels cut edges is determined in the present work based on the 60° conical ISO16630 hole expansion test set up (HET) with different cutting clearance (5-40% range), cutting tool wear conditions (sharp/worn shear punch/die) and punch cutting angle (concave, convex, orthogonal). Additionally flat and hemispherical hole expansion punch geometries have been compared with the ISO standard conical punch geometry for strain gradient analysis.

2. Experimental procedures

2.1. Investigated materials

Following cold rolled AHSS grades (CR) in the thickness range between 0.8mm and 1.5 mm have been investigated: a dual-phase steel (DP1000) and two complex phase steels, CP1000HD and CP1200. Table 1 shows their transverse mechanical properties according to ISO6892-1 standard [18] using specimens type 2 with a reference gage length of 80 mm.

Table 1. Transverse mechanical properties of the investigated steels (ISO 6892-1 type 2).

Steel grade designation	t	R_{p02}	R_m	A_g	A_{80}	n_{Ag}	r_4
	mm	MPa	MPa	%	%		-
DP1000 (CR700Y980T-DP-UC-U)	0.8	735	1074	7.5	10.5	0.073	0.94
CP1000HD (CR780Y980T-CH-UC-U)	1.5	909	1062	7.1	10.7	0.068	0.96
CP1200 (CR900Y1180T-CP-GI50/50-U)	1.5	1079	1215	5.1	8.3	0.049	0.92

2.2. ISO 16630 standard conical hole expansion test

Figure 1 shows schematically the 60° conical ISO 16630 hole expansion test (HET) setup. The limiting hole expansion ratio at first through-thickness crack (HER) is expressed in % as the ratio of hole diameter expansion to the original hole diameter as follows [3]:

$$HER = \frac{D_h - D_0}{D_0} x100 \tag{1}$$

with D_h: average hole diameter at first through-thickness crack; D_0: initial (10mm) hole diameter.

All tests are performed in the burr-up configuration (burr on die side) at 12% cutting clearance with a 60° conical punch and 10mm initial hole diameter according to ISO 16630 standard. The tests are stopped manually without optical magnifying device at first through-thickness crack (Figure 1). The final hole diameter is measured with a caliper in unloaded condition at two locations away from edge crack according to ISO 16630 procedure.

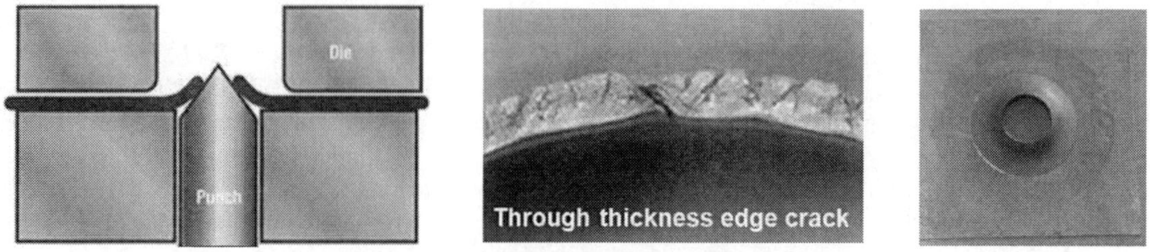

Figure 1. ISO 16630 Hole expansion HET test setup.

Table 2 shows the hole expansion ratio in punched 12% clearance and drilled condition of the investigated materials (mean and standard deviation). HER values are higher in the drilled than punched condition, since this process minimizes material damage around hole perimeter in comparison to shear cut edges.

Table 2. Hole expansion ratio for the investigated steels (punched vs. drilled conditions).

Steel type	Steel grade	Punched Conical HER, λ [%]		Drilled conical HER, λ [%]	
		Mean	Standard Deviation	Mean	Standard Deviation
Cold rolled	DP1000	37	4	48	7
	CP1000HD	86	5	126	2
	CP1200	67	5	137	8

2.3. Alternative hole expansion test punch geometries

Additional hole expansion tool geometries have also been investigated (flat R25, hemispherical Nakajima R25 vs. 60° conical ISO16630 punch) with a 10mm hole diameter in punched vs. drilled condition (Figure 2). This way some more in plane hole expansion load configuration can be simulated.

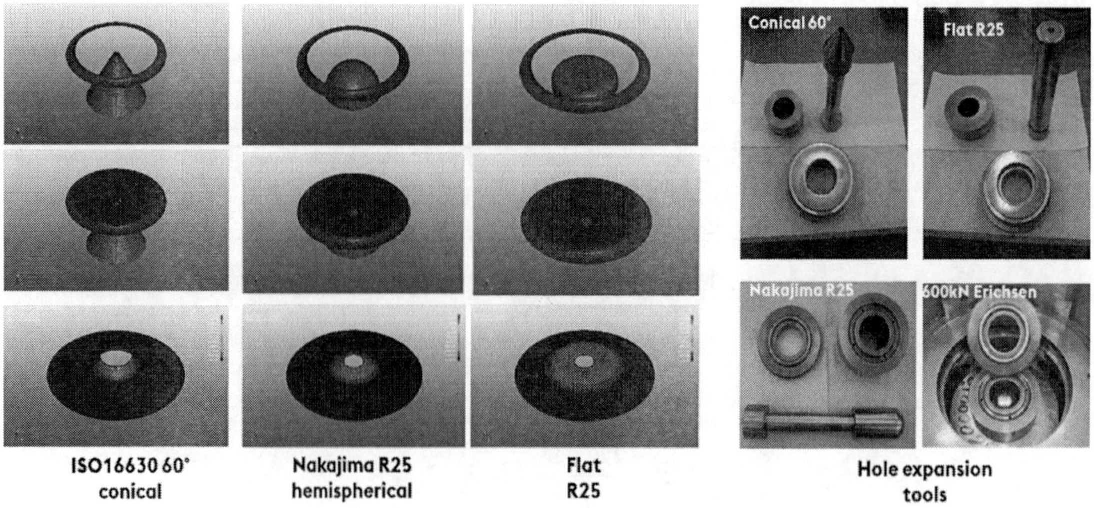

Figure 2. Hole expansion punch geometry (conical, hemispherical, flat).

2.4. Test program overview

In the following chapter different parameters on conical hole expansion ratio have been investigated such as cutting clearance (chapter 3.1), cutting tool wear (chapter 3.2) and cutting tool angle (chapter 3.3). The influence of strain gradient during subsequent hole expansion has also been investigated with different punch geometries (chapter 3.4). Wherever suitable an investigation on cut edge parameters with a 360° panoramic optical hole edge inspection system is also given. All chapters should be considered as independent investigations and are not necessarily interconnected.

3. Results and discussion

3.1. ISO16630 conical hole expansion ratio vs. cutting clearance

The influence of punching clearance on hole expansion ratio (HER) based on ISO16630 test procedure is shown in Figure 3 in the 5-40% cutting clearance range. The drilled HER values are given as reference. Low (<10%) and high (>20%) clearances are significantly detrimental to hole expansion ratio. An optimum in HER is reached around 15% clearance. Those results are consistent with previous HER vs. clearance findings in [4]-[6] and AHSS processing guidelines for stretch flangeability optimization [7].

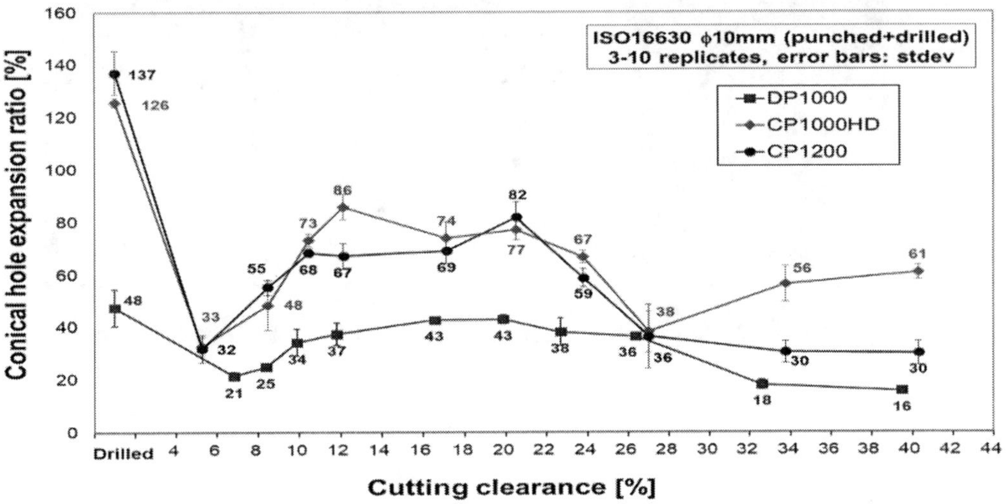

Figure 3. ISO 16630 hole expansion ratio vs. cutting clearance and for drilled condition.

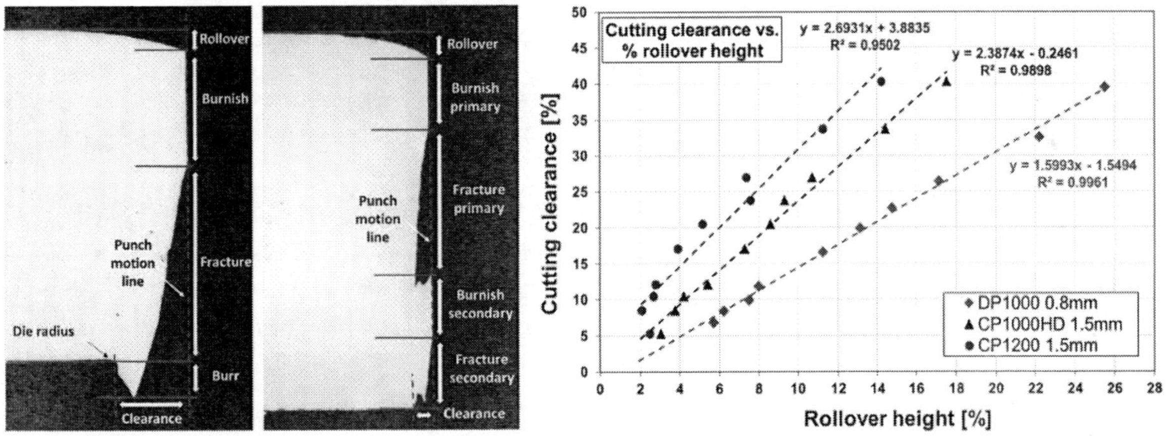

(a) (b)

Figure 4. (a): Cut edge parameters definition; (b): Cutting clearance vs. rollover height.

Figure 4a shows the definition of cutting parameters such as rollover, burnish, fracture and burr expressed in % fraction of material thickness. At low clearance a distinction is made between primary and secondary burnish (as well as primary and secondary fracture).

The cutting clearance is directly linear proportional to rollover height clearance (and vice versa) for a given material, thickness and cutting tool (Figure 4b). The higher the material stiffness (strength and thickness), the lower the rollover bending amount. The proportionality of rollover and clearance maybe helpful in industrial condition. The effective cutting clearance in components could then be calibrated easier directly from rollover height without the need for metallographic sections. Since cutting clearance is a key parameter in stretch flangeability control, such rollover-clearance correlation should be helpful.

Cut edge investigations vs. cutting clearance show the emergence of local geometrical notches within punched edge such as secondary burnish at low clearance (Figure 5a) or excessive burr combined with a rough stair step like fracture shape at high clearance (Figure 5b,c).

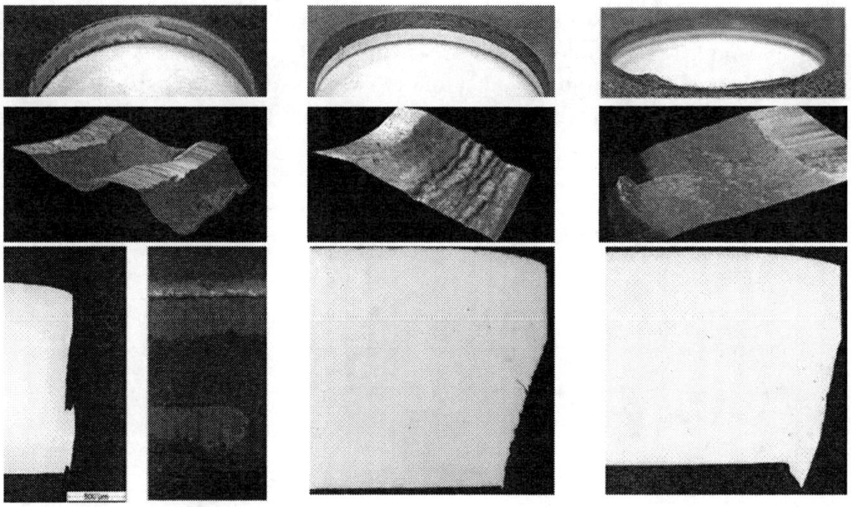

(a) (b) (c)

Figure 5. Cut edge defects illustration. (a): secondary burnish, (b): rough fracture, (c): burr.

Secondary burnish, as defined in Figure 4a, occurs for the investigated AHSS steels at clearance below ≈8% and is correlated with a decrease in HER values at low clearances (Figure 6a). Burr is found above ≈25% clearance and is also linked to a reduction in HER values at high clearances (Figure 6b).

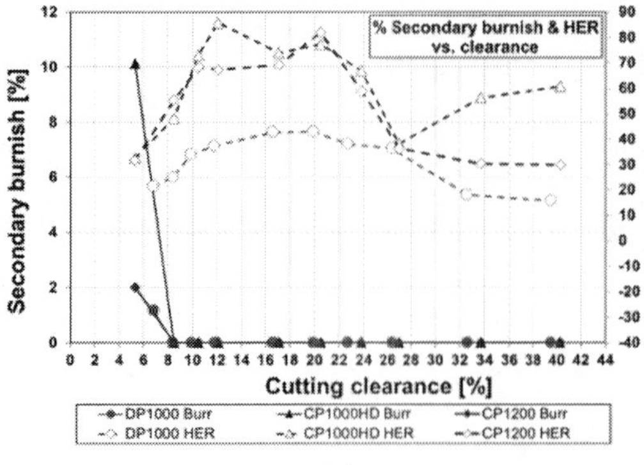

(a)

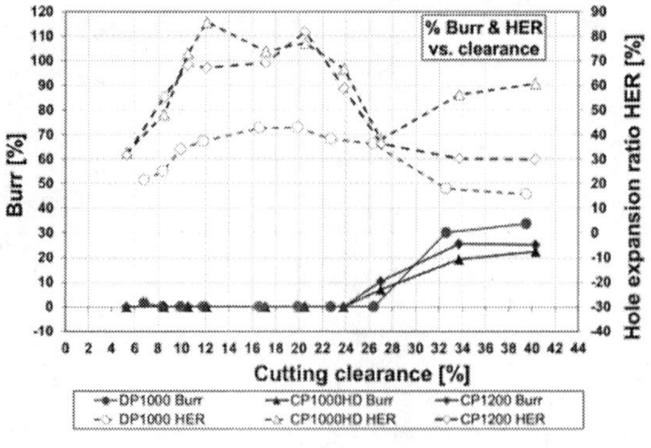

(b)

Figure 6. (a): Secondary burnish & HER vs. clearance; (b): Burr & HER vs. clearance.

HER values are maximum at 15 ± 5% clearance when burnish values are minimum (Figure 7a), fracture height values are maximum (Figure 8a) and burnish to fracture ratio are minimum (Figure 9a).

HER values generally increase with decreasing burnish height (Figure 7b), increasing fracture height (Figure 8b) as well as decreasing burnish to fracture ratio (Figure 9b).

HER values are maximum for burnish values around 15% for DP1000, 20% for CP1000HD and 25% for CP1200 (Figure 7). HER values are optimum for fracture height values around 75% for all investigated steels (Figure 8). HER values are also maximum for burnish to fracture ratio values around 20% for DP1000, 30% for CP1000HD and 35% for CP1200 (Figure 9).

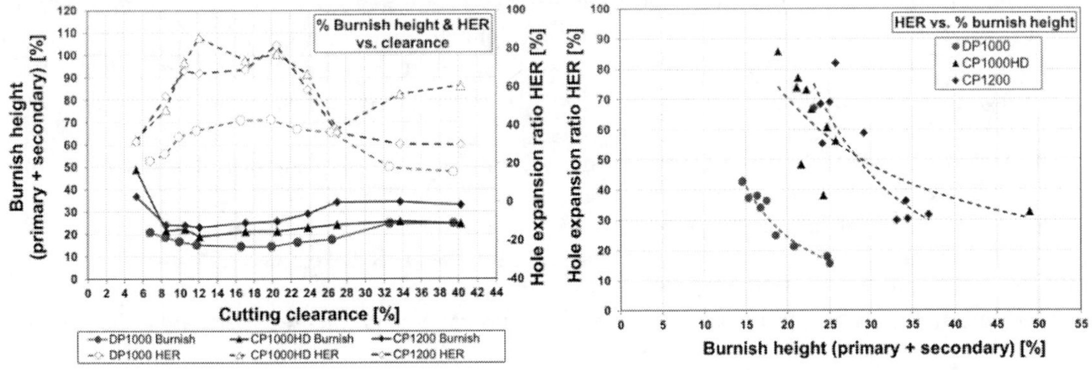

(a) (b)

Figure 7. (a): Burnish height & HER vs. clearance; (b): HER vs. burnish height.

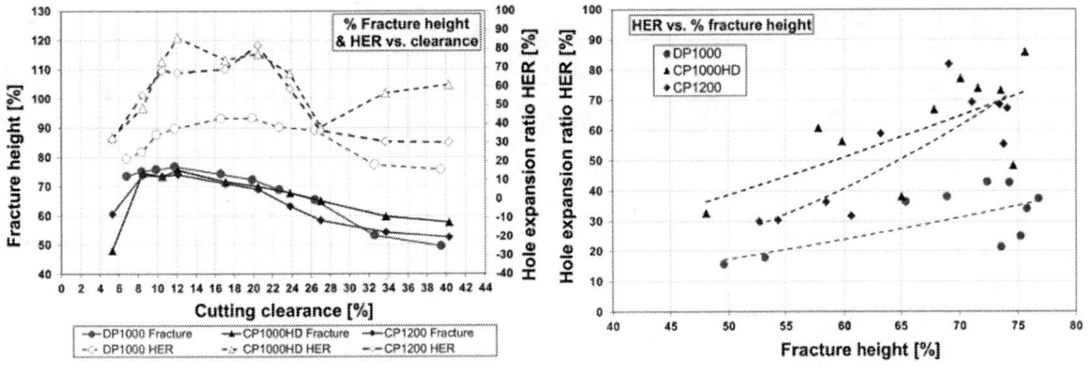

(a) (b)

Figure 8. (a): Fracture height & HER vs. clearance; (b): HER vs. fracture height.

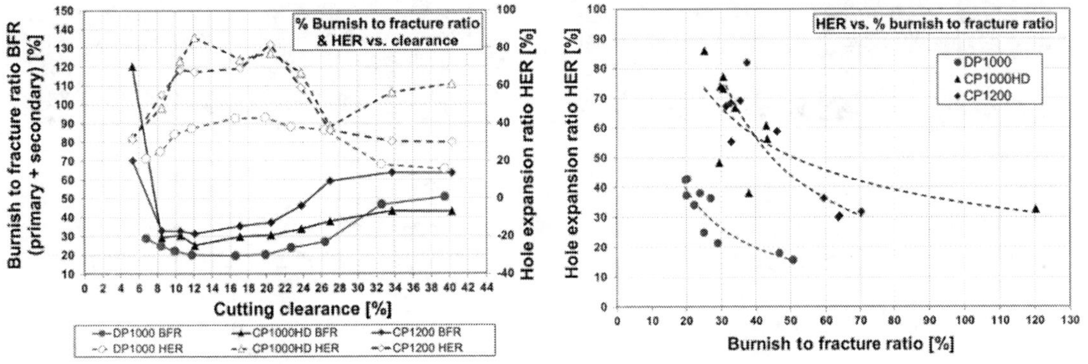

(a) (b)

Figure 9. (a): Burnish to fracture ratio & HER vs. clearance; (b): HER vs. burnish to fracture ratio.

3.2. ISO16630 conical hole expansion ratio vs. cutting tool wear

The cutting tool wear has been experimentally reproduced for CP1000HD and CP1200 at 9.6% and 12.2% clearance with various tool wear conditions. Two punching tool configurations have been chosen: a new punch with 30±10µm radius and a worn out punch with 50±10µm punch radius naturally abrasively rounded off after ≈ 3000 strokes. Two cutting die configurations have also been investigated: a new cutting die with 20±10µm radius and a manually grinded cutting die with 200±30µm radius. According to industrial feedback a large 200µm die radius is a realistic worst case. The punch tool wear with a 50µm edge radius seems rather representative of natural abrasive wear after a few 1000 strokes.

Figure 10 shows the corresponding hole expansion ratio (HER) values results. Both cutting punch wear and die wear affect negatively the HER values. This effect seems to increase with increasing cutting clearance (9 to 12%). A combination of both punch and die wear is additive and can lead to a significant cumulative worst case HER decrease in comparison to new sharp cutting tools configuration (≈90% to 55% for CP1000HD, ≈70% to 50% for CP1200 for 12% clearance). Similar effects have been reported in [7]-[9] for AHSS grades.

Cutting tool wear with increased punch and/or die radius leads to a minimal increase in rollover, a significant increase in burnish height and a decrease in fracture height. Tool wear is also inducing burr generation at large cutting die radii (200µm).

As shown in Figure 11 for CP1000HD and CP1200, the HER values correlate with the amount of burnish or fracture height. Burnish increases and fracture height decreases with increasing tool wear for a given clearance. Burr appears only for large die radii, which additionally lowers the HER values.

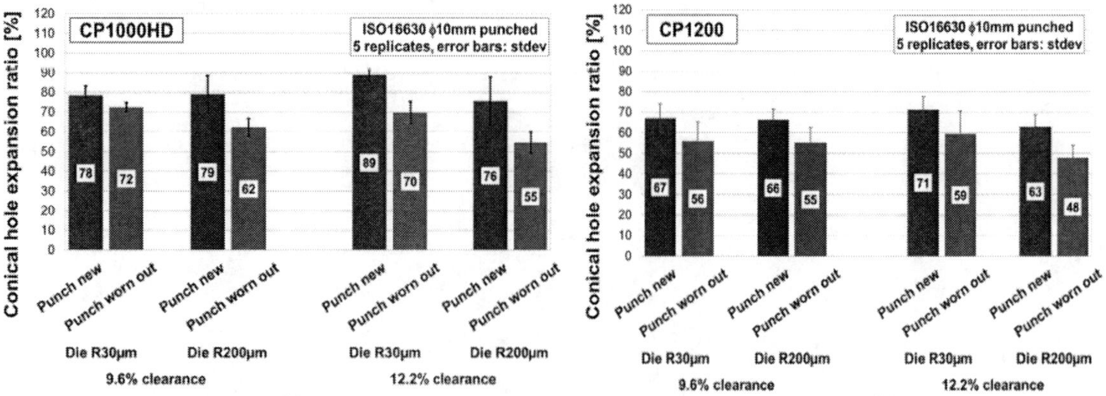

(a) (b)

Figure 10. HER vs. cutting die & punch radius and clearance. (a): CP1000HD; (b): CP1200.

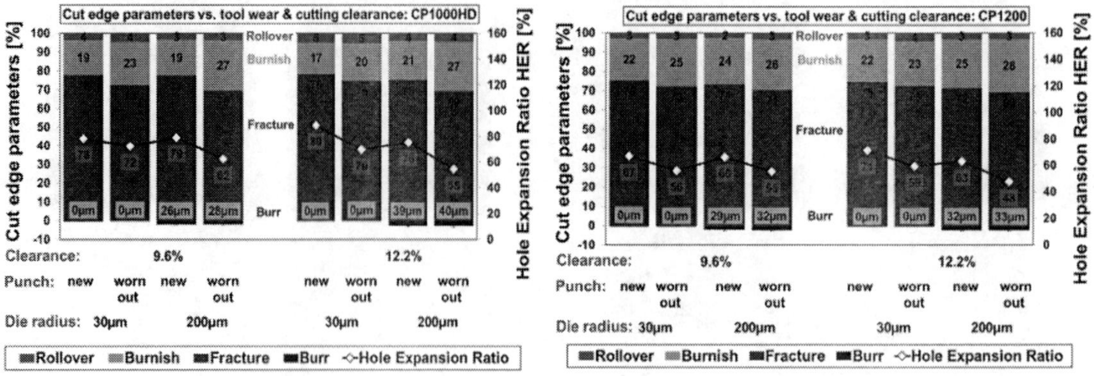

(a) (b)

Figure 11. Cut edge parameters vs. tool wear & clearance. (a): CP1000HD, (b): CP1200.

Figure 12 shows a clear linear dependency of HER values vs. burnish or fracture height (Figure 12a) and burnish to fracture ratio (Figure 12b) when all tool wear data are plotted together (CP1000HD vs. CP1200, 9.6% vs. 12.2% clearance, for all investigated tool wear conditions). Such cut edge parameters could be therefore suitable indicators for tool wear monitoring in production lines.

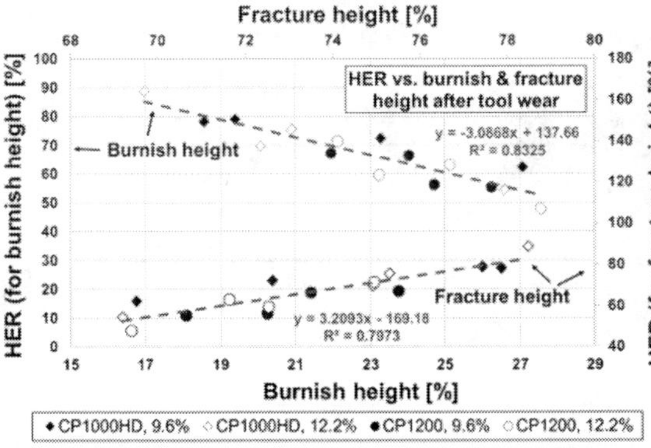

(a)

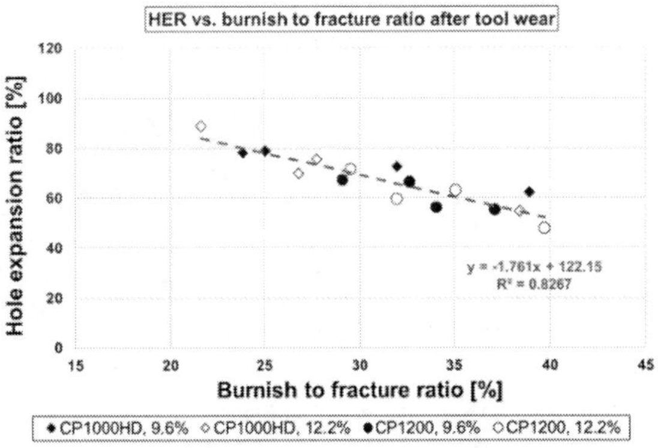

(b)

Figure 12. HER for all tool wear conditions vs. (a) burnish or fracture height (b): burnish to fracture ratio.

3.3. ISO16630 conical hole expansion ratio vs. cutting angle

Literature work is increasingly found to improve HER values by punching tool geometry optimization [10]-[12]. The ISO16630 classical punching configuration prescribes a 10mm hole diameter punched with an orthogonal 90° punching angle [3]. Additionally, concave and convex punch geometries have been investigated as shown in Figure 13. The convex roof top punch shape is flat in the bottom middle with a 17° angle on both sides. The concave bottom punch shape is flat on the outer regions with a 5mm radius hollow shape in the center. Figure 14 shows the influence of punching tool geometry on conical HER values. Figure 15 shows exemplarily for CP1000HD the effect of orthogonal, concave and convex punch on cut edge (360° panoramic pictures with an optical cutting edge inspection tool).

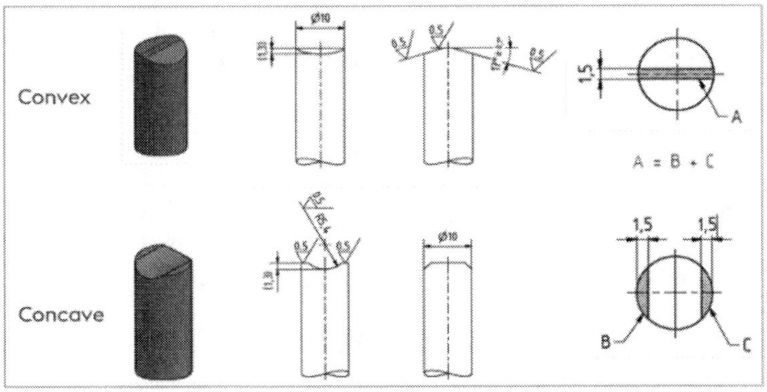

(a)

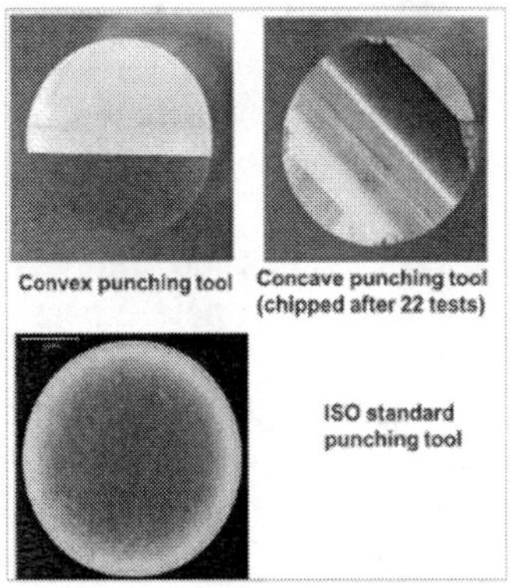

(b)

Figure 13. (a): Cutting tool convex/concave geometry; (b): Cutting tools bottom face view.

A positive effect on HER values could not be found using such alternative shear punch geometries (Figure 14). Contrary to orthogonal cutting, the cutting clearance is inhomogeneous along hole perimeter for concave and convex punch geometries, the amount of rollover and burnish varying sharply along hole perimeter (Figure 15). At some symmetric cut edge locations burnish vanishes with an unusual direct transition between rollover and fracture zone. The local cutting clearance is undefined either too low or too high, which is in both cases detrimental to HER ability of the punched material in comparison to an homogeneous orthogonal cutting tool.

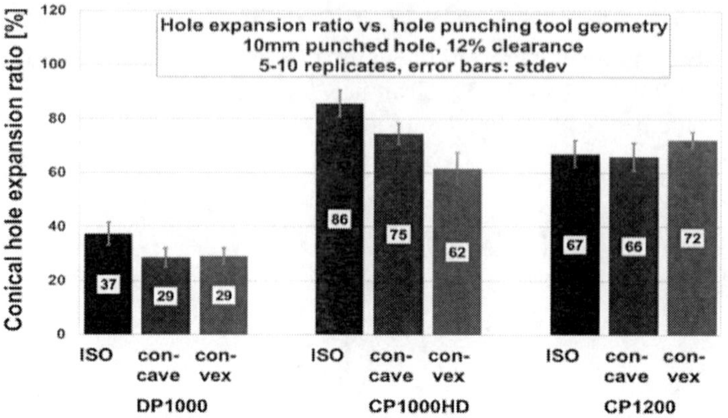

Figure 14. ISO 16630 HER ratio vs. cutting tool geometry (orthogonal, concave, convex).

The punch tool wear is also massively increased through early chipping after only 22 strokes for the concave hollow punch shape as shown in Figure 13b. This is due to excessive notch stress concentration level in this hollow concave configuration at the transition between flat and curved punch regions. Higher tool material grades and heat treatment are needed for such elaborated punch geometries [12].

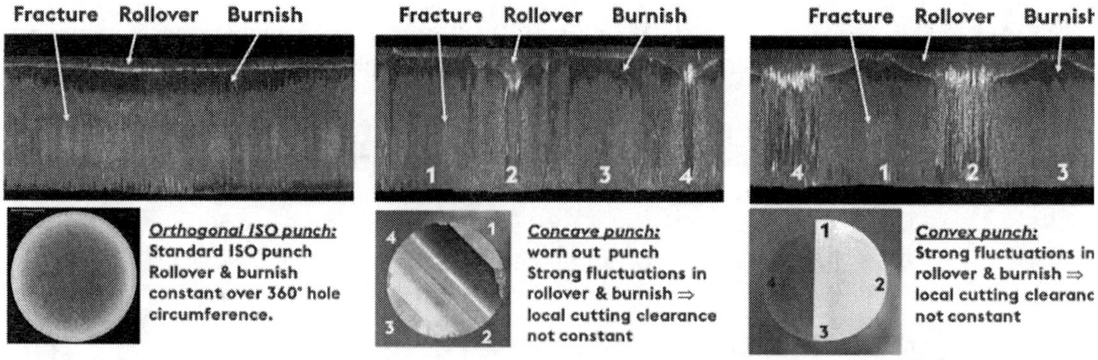

Figure 15. Cut edge 360° morphology vs. cutting tool geometry (CP1000HD).

3.4. Hole expansion ratio vs. expansion punch geometry and radial strain gradient

The influence of hole expansion punch geometry (flat R25, Nakajima R25 vs. ISO16630 60° conical) on the HER values in punched (12% clearance, 10mm hole diameter) vs. drilled condition has been investigated (Figure 2). The tests are simulated with LS DYNA (shell elements, friction coefficient 0.1, Barlat anisotropy model, flow curves based on tensile and bulge tests, no damage model). The experimental HER values are chosen as fracture criteria for the FE simulation and the linear radial logarithmic major strain gradient up to 2-2.5mm transverse from hole edge are determined.

The HER results are given in Figure 16. Figure 17a shows exemplarily the method for determining the logarithmic radial strain gradient at HER level. Figure 17b shows an overview of testing results including all materials, hole conditions and punch geometries. The logarithmic HER values are linear proportional to the logarithmic radial strain gradient, in agreement with previous literature investigations [13]-[17], the strain gradient sensitivity being defined as the slope of log. HET vs. log. radial strain gradient is highest for CP1200 and CP1000HD with values around 4 double as high as for DP1000 (≈ 2). CP grades are known to be more sensitive to sheet curvature bending effects [5], [13]-[14]. Such HER-strain gradient information in Figure 17b may be implemented in FE postprocessing for safe assessment of HER values depending on the local flange geometry of components.

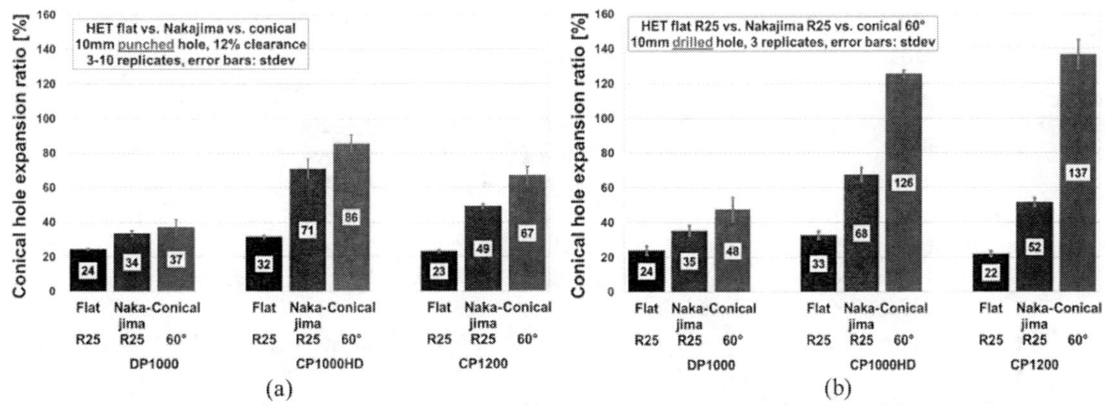

Figure 16. Conical HER vs. expansion punch geometry. (a): punched hole; (b) drilled hole.

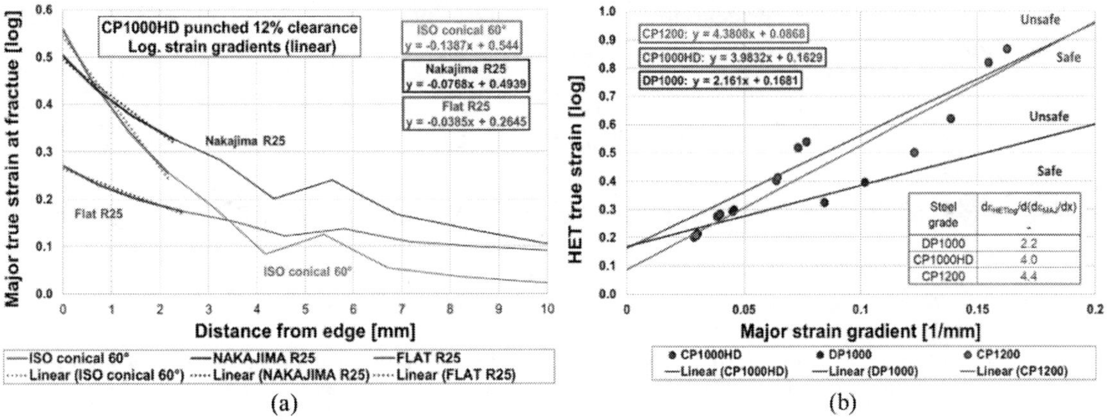

Figure 17. FEM Log. major strain gradient vs. HER level. (a): strain gradient method from last 2-2.5mm $d\varepsilon_{MAJ}/dx$ from edge; (b): results overview in punched and drilled conditions.

4. Conclusions

Following conclusions can be drawn from this investigation:

- **Conical ISO16630 HET:** The superiority of complex phase with regard to conical hole expansion ratio is clearly shown both in punched (12% cutting clearance) and drilled conditions.
- **HER vs. cutting clearance:** Among all influence parameters, the cutting clearance is considered to be the most critical factor to consider and control. Low (<10%) and high (>20%) clearance are significantly detrimental to hole expansion ratio. An optimum in HER is reached around 15% clearance for AHSS. Sharp local geometrical notches within cut edge such as secondary burnish at low clearance or excessive burr, combined with a rough fracture shape at high cutting clearance should be avoided for stretch flangeability improvement.
- **HER vs. tool wear:** Both cutting punch wear and cutting die wear affect negatively the HER values. A combination of both punch and die cutting tool wear can lead to an even more significant HER decrease in comparison to a new sharp cutting tool configuration (HER ≈90% to 55% for CP1000HD, HER ≈70% to 50% for CP1200 at12% clearance). Cutting tool wear with increased punch and/or die radius leads to a minimal increase in rollover, a significant increase in burnish and a decrease in fracture height. Some burr appears at high cutting die radii. The influence of cutting tool wear (punch and die) together with the knowledge of exact cutting clearance should therefore be considered in

the early component design phase. A close online monitoring of burnish to fracture ratio may deliver some information about cutting tool wear evolution in production lines.

- **HER vs. cutting angle:** A positive effect on HER values could not be found using alternative concave and convex hole punching tool geometries. Contrary to orthogonal cutting, the cutting clearance is irregular along hole perimeter for concave and convex punch geometries, the amount of rollover and burnish varying sharply along hole perimeter, which is detrimental to HER ability. The cutting punch tool wear is also massively increased for the concave hollow punch shape.
- **HER vs. radial strain gradient:** The influence of hole expansion punch geometry (flat R25, Nakajima R25 vs. ISO16630 60° conical) on the HER values in punched (12% clearance, 10mm hole diameter) vs. drilled condition has been investigated. The varying radial strain gradient at fracture has been determined with FE Simulation. The (true) hole expansion ratio values are proportional to the logarithmic radial strain gradient. The strain gradient sensitivity is the highest for CP1200 and CP1000HD twice as high as for DP1000. This non negligible radial strain gradient effect should be therefore implemented in stretch flanging FE simulations for early edge crack detection warning.

Acknowledgments
The authors gratefully acknowledge the funding received by the Research Fund for Coal and Steel programme (847213-CuttingEdge4.0-RFCS-2018).

References
[1] Konieczny A and Henderson T (2007) *SAE Tech. Paper 2007-01-03*. DOI: 10.4271/2007-01-0340.
[2] McEwan et al. (2009) A new approach to predicting edge splits-the combined FLC/HEC diagram. IDDRG 2009 Conference 1-3 June 2009, Golden, Co, USA.
[3] ISO 16630:2017(E) Metallic Materials - Sheet and Strip - Hole Expanding Test.
[4] Gutierrez D et al. (2011) Strategies to improve stretch flangeability in AHSS. IDDRG2011 conference, Bilbao, Spain.
[5] Larour P et al. (2011) Alternative stretch flangeability characterization methods for AHSS steel grades. IDDRG2011 conference, Bilbao, Spain.
[6] Larour P et al. (2021) *Mater. Sci. Eng.* **1157** 012054. DOI: 10.1088/1757-899X/1157/1/012054.
[7] AHSS Application Guidelines. Cutting, Blanking, Shearing & Trimming. World Steel Association (2021). https://ahssinsights.org/forming/cutting
[8] Chiriac C and Shi M F (2013) *SAE Tech. Paper* 2013-01-1774. DOI: 10.4271/2013-01-1774.
[9] Gläsner T et al. (2015) *Key Eng. Materials* **651–653** 1261-70. DOI:10.4028/www.scientific.net/kem.651-653.1261.
[10] Pu C et al. (2019) *SAE Technical Paper* 2019-01-1086. DOI: 10.4271/2019-01-1086.
[11] Shih H et al. (2014) *SAE Technical Paper* 2014-01-0994. DOI: 10.4271/2014-01-0994.
[12] Shi H et al. (2017) *SAE Technical Paper* 2017-01-1705. DOI: 10.4271/2017-01-1705.
[13] Krempaszky C et al. (2014) Towards more efficient hole expansion testing. IDDRG 2014 Conference, Paris, France.
[14] Larour P et al. (2014) Evaluation of alternative stretch flangeability testing methods to ISO16630 standard. IDDRG 2014 Conference, Paris, France.
[15] Ilinich A et al. (2011) *SAE Technical Paper* 2011-01-1062. DOI:10.4271/2011-01-1062.
[16] Beier T and Woestmann S (2016) Berücksichtigung von schergeschnittenen Blechkanten zur Auslegung von Formgebungsprozessen höherfester Stahlwerkstoffe in der FEM Umformsimulation. 14. LS-DYNA Forum 2016, Bamberg, Germany.
[17] Iizuka E et al. (2017) *J. Phys. Conf. Ser.* **896** 012008. DOI: 10.1088/1742-6596/896/1/012008.
[18] ISO 6892-1:2016(E) Metallic materials-Tensile testing - Part 1: Method of test at room temperature.

IOP Conf. Series: Materials Science and Engineering 1238 (2022) 012042 doi:10.1088/1757-899X/1238/1/012042

Adjustment of fracture locus to improve edge crack resistance

Niloufar Habibi[1*], Sebastian Münstermann[1]

[1]Steel Institute, RWTH Aachen University, Aachen, Germany

niloufar.habibi@iehk.rwth-aachen.de

Abstract. Edge crack sensitivity of advanced high strength steels is posing challenges in forming the sheets with shear-cut edges. In fact, cutting process reduces formability of material by leaving residual damage at the cut edges and makes it prone to edge cracking. The difficulty of the problem mainly lies in the complex loading path that sheets undergo through shear-cutting and the subsequent forming processes. In other words, the stress state evolves from pure-shear to plane-strain during cutting process and changes to mostly uniaxial tension during the following hole-expansion test. The present study aimed to investigate the physical mechanism underlying the formation of edge cracks. The overall aim of the study is to tailor an improved microstructural configuration leading to a significantly reduced edge crack sensitivity. As a first step in this process, an artificial fracture locus is iteratively identified which promises to improve edge crack resistance. In this regard, a dual phase steel was considered as reference material and its fracture locus was calibrated based on uncoupled Bai-Wierzbicki fracture model. The artificial fracture locus representing a virtual material with improved edge crack resistance changed in a way that lower residual damage was inherited from the cutting process, which consequently improved the hole-expansion ratio.

1. Introduction

Dual phase (DP) Steel sheets are widely used in automotive industries. Their complex microstructure with soft ferrite matrix and hard martensite islands, leads them to represent good combination of strength and formability and have high potential for reducing weight components [1]. However, some challenges still remain in their forming processes. They mostly undergo several complicated forming processes to obtain their final shape. Shear-cutting is a common forming process which is used for the initial cutting, such as door bumpers or cut-out the corner of the window. This process imposes some adverse effects on the shear cut edge, which become profound during the subsequent forming processes and cause edge cracking. This type of failure, which happens in DP steels, cannot be predicted by using conventional methods like forming limit diagrams or material properties included yield strength, tensile strength and elongation. For investigating into edge crack sensitivity, the forming capacity at the manufactured edge should be considered, i.e. the residual damage at the edge and edge quality [2]. Although some methods have been proposed to prediction of edge cracking, the physical insight of this phenomenon have remained unknown yet.

The defects introduced by the pre-damage from the initial punching process and the damage accumulation during the hole expansion test (HET) determine the formation of edge crack for DP steel sheets. Wang et al. [3,4] reported that the punched edges were experienced the highest level of pre-damage which considerably influenced on the subsequent hole-expansion forming test. Habibi et al. [5] numerically investigated the effects of pre-damage and roughness of an edge from the punching process on prediction of hole expansion ratio. The results showed that the pre-damage play the most significant role in the hole expansion ratio and its prediction. Furthermore, Park et al. [6] established a dual-scale finite element simulation to study the effect of microstructure on the punching and subsequent HET. They revealed that more homogenous microstructure led to the lower damage at the cut edge and higher hole expansion ratios.

The present work aims to a thorough study on stress state evolution during a punching and the subsequent hole expansion processes. For this purpose, the fracture strain differed at specific stress states and new fracture loci were created. This study can propose the best fracture locus configuration to design a material with higher edge crack resistance.

2. Materials and methods

Dual-phase CR590Y980T-DP steel sheets, DP1000 according to DIN EN 10346, were received with the thickness of 1.5 mm. Bai-Wierzbicki uncouple fracture model [7], was applied to describe the fracture behavior in different stress states (equation 1). It is worth mentioning that the model is defined in terms of stress-triaxiality, η, and normalized Lode angle parameter, $\bar{\theta}$, to describe stress-state of the material during the deformation.

$$\int_0^{\varepsilon^f} \frac{d\varepsilon^p}{(C_1 e^{-C_2\eta} - C_3 e^{-C_4\eta})\bar{\theta}^2 + C_3 e^{-C_4\eta}} = 1 \qquad (1)$$

Herein, ε^P is equivalent plastic strain, ε^f is equivalent plastic strain at fracture, and C_i is material constant. The material constants were determined by a combined experimental and numerical approach. The mechanical properties were evaluated through shear, plane-strain tension, uniaxial tension, and biaxial tension loading conditions along with ARAMIS digital image correlation (DIC) analysis. The parallel numerical simulations were performed by using finite-element software Abaqus/Explicit 2017. In all the simulations, eight-node reduced-integration solid elements of type C3D8R were employed. In order to minimize the computational time, a mesh size of 0.1 mm was applied only on those regions of samples which were exposed higher plastic straining. The material parameters are summarized in table 1.

Table 1. Mechanical properties of the studied material [8]

C_1	C_2	C_3	C_4
0.15	1.5	0.08	1.5

Moreover, the numerical simulations of a two-stage shear-cutting_ subsequent hole-expansion test were performed for a set-up with sharp cutting edges tool (edge radii of 0.05 mm) and a die clearance of 10 %. In order to reduce the computation time, only a quarter of the symmetric model was considered, figure 1. In both steps the tools were considered as discrete rigid parts, and the sheet was modeled as a deformable homogenous solid. Friction between the tools and the sheet was considered as the Coulomb model with a coefficient of 0.1 [3]. The contact pairs were defined as a surface-to-surface algorithm. However, the contacts between new surfaces which are created during cutting process and the tool were defined as a node-to-surface algorithm. Since the applied fracture model is a local model and influenced by the mesh size, mesh size of 0.1 mm was used in the critical regions of the sheet like the models which were used for the parameter calibration as well. In order to apply homogenous force distribution on the sheet, finer elements were defined in the contact regions of the tool. An optimal clamping force was chosen such that the blank was neither drew in nor torn improperly during the processes. The edge crack sensitivity was assessed based on hole expansion ratio. The conventional HER is measured when a crack passed throughout the thickness of the sheet. But here HER was measured as a crack penetrated through the plane of the sheet. The crack initiation and propagation were defined by element deletion method.

3. Stress states throughout the processes

In order to design new fracture loci, the elaborate understanding of shear-cutting and hole expansion processes is required. In this regard, the stress-state during these processes were studied at different stages by tracking stress-triaxialty and Lode-angle parameter. For better illustration each steps is presented in a separate part.

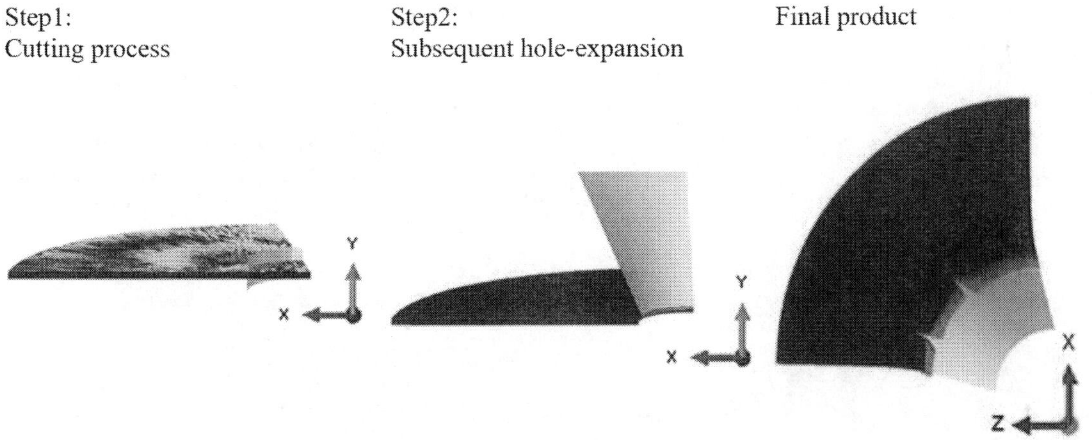

Figure 1. The two-step shear-cutting_hole-expansion test.

3.1. Shear-cutting process

Figure 2 illustrates the cutting process and changes in stress-state variables, i.e. stress-triaxiality and normalized Lode angle parameter. According to the simulation results, throughout the cutting process $\bar{\theta}$ in the critical elements remained about zero, while η increased from zero to 0.8, i.e. the stress-state changed from pure shear (η=0 and $\bar{\theta}$=0) to plane-strain tension (η=0.8 and $\bar{\theta}$=0). Therefore, the stress-state varied only on the line $\bar{\theta}$=0, the valley of the fracture locus.Note that ultimate separation was caused by meeting of two cracks in the fracture part of the edge. In fact, first a crack initiated near the punch and propagated through the thickness. Meanwhile another crack initiated from the other side of the thickness and propagated towards the first crack.

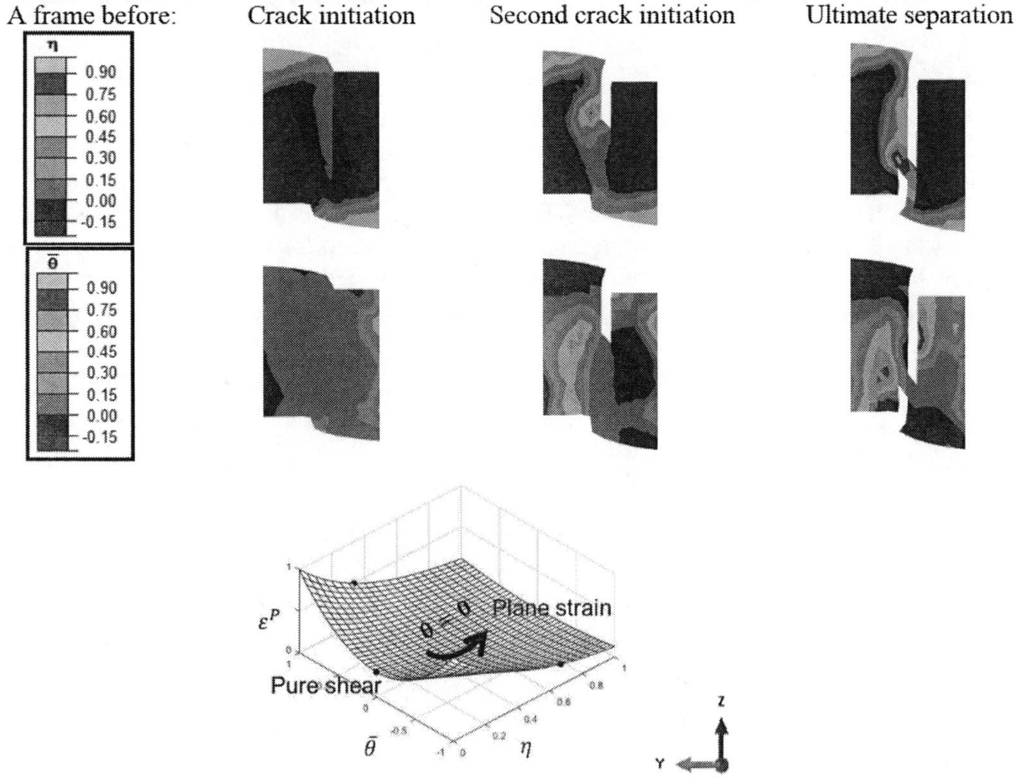

Figure 2. The evolution of stress-state at critical elements through the shear-cutting process.

IOP Conf. Series: Materials Science and Engineering 1238 (2022) 012042 doi:10.1088/1757-899X/1238/1/012042

3.2. Subsequent hole-expansion test

The evolution of stress-state at the hole edge during the hole-expansion test is depicted in figure 3. The results reveal that the elements at the edge experienced initially uniaxial tension state. As any crack appeared at the edge and propagated through the thickness, the ligaments between the cracks underwent pure shear deformation. Furthermore, the adjacent elements in the sheet (far from the edge) experienced relatively higher stress-triaxiality and plane-strain to biaxial tension stress-states. Therefore, when a crack is propagating through the thickness, the stress states of critical elements differs from uniaxial tension to plane strain tension, and when it is propagating in to the plane, the stress states changes from plane strain tension to biaxial tension.

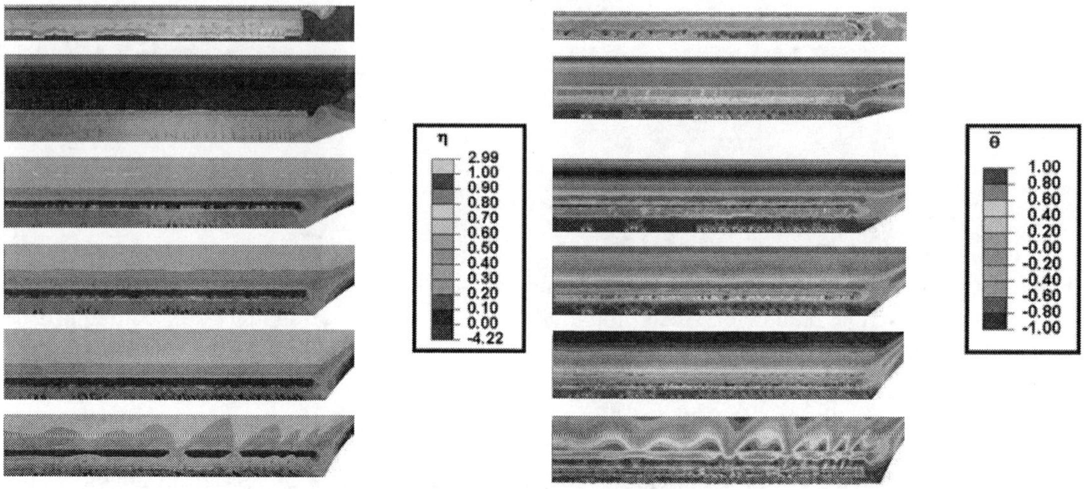

Figure 3. The evolution of stress-state at edge elements through hole expansion test.

4. Fracture locus design

The effects of level and shape of fracture locus on edge crack sensitivity were studied by designing different loci. For this aim, the values of equivalent plastic strain in different stress-states were individually multiplied once by ½ and once by 2. Afterwards, the fracture locus was plotted for each condition by fitting the best surface on the fracture points of pure shear (Sh), plane-strain tension (PS), uniaxial tension (UT), and biaxial tension (BT) stress-states. Note that all new fracture loci were different from the original one. The edge crack sensitivity of each fracture locus was investigated by simulating the shear-cutting process and subsequent hole-expansion test and the results are illustrated in the following parts.

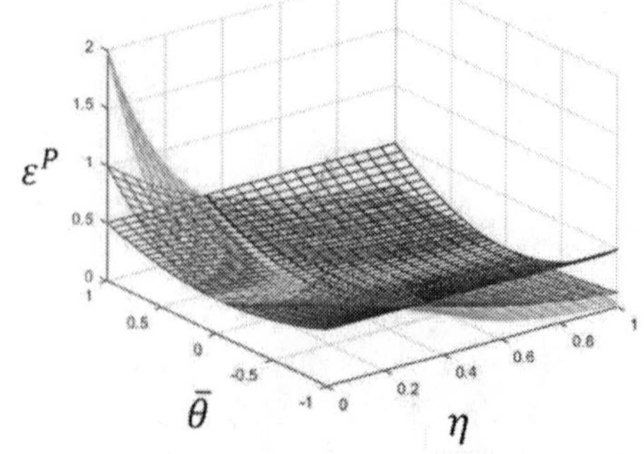

Figure 4. Example of fracture loci design, here by multiplying the equivalent plastic strain at the fracture of biaxial tension states by 1/2 (green locus) or 2 (blue locus).

IOP Conf. Series: Materials Science and Engineering 1238 (2022) 012042 doi:10.1088/1757-899X/1238/1/012042

4.1. Effect of fracture locus on shear-cutting process

Figures 5 and respectively compare the residual damage at the cut edge and the stroke of the cutting punch until the total separation for the original material with the imaginary fracture loci. Note that the residual damage is shown by failure indicator which varies from 0 (element with no damage) to 1 (failed element). As mentioned, the stress-state through shear-cutting process changed in the valley of the fracture locus ($\bar{\theta}$=0). In addition, the shape and level of the valley were calibrated by shear and plane-strain tension modes. In other words, by changing the value of fracture strain in uniaxial or biaxial tension, the valley still remains the same. Thus, the material showed similar behavior during the cutting process, while it depended on fracture strain of pure shear and plane-strain tension. By multiplying ε_{Sh}^{f} to half, the crack initiated earlier and propagated rapidly before any second crack can initiate. In contrast to multiplying by 2, where the first crack initiated later, which made the roll-over part bigger. Also, a second crack created, propagated further and met the first crack in upper part of the edge. This behavior induced more damage than the former condition. The influence of changing the fracture strain of plane-strain condition, damage at the edge became smaller by making ε_{PS}^{f} double and higher by making it half. However, the changes are not significant regarding induced damage by the cutting process.

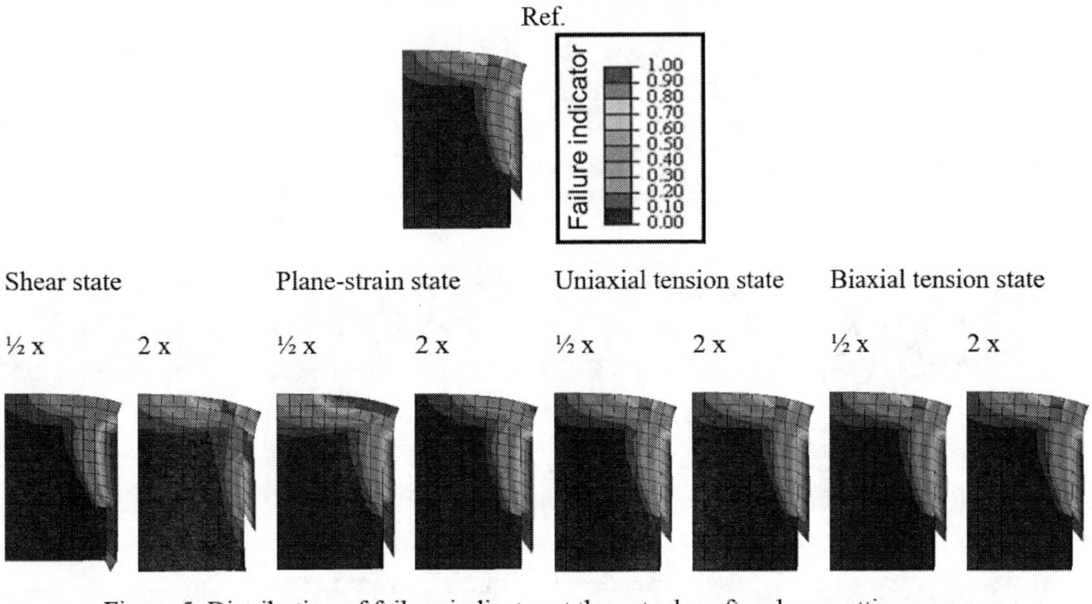

Figure 5. Distribution of failure indicator at the cut edge after shear-cutting process

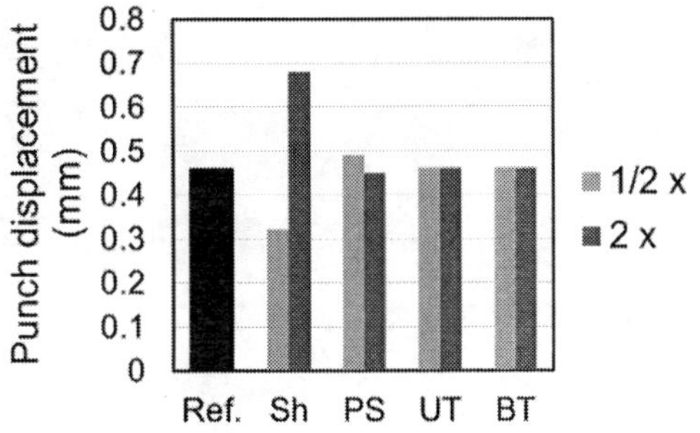

Figure 6. Comparison of punch displacement of ultimate separation in shear-cutting process.

4.2. Effect of fracture locus on hole-expansion ratio

By differing the fracture locus, the edge behaved differently through the hole-expansion test. In some conditions, the edge was more sensitive to created multiple cracks through the thickness before a crack

passed through the specimen, such as ½ ε^f_{Sh}, 2 ε^f_{PS}, ½ ε^f_{UT}, 2 ε^f_{BT}. Whereas in the other conditions, cracks were prone to propagates towards the specimen much earlier and led to smaller hole-expansion ratio, figure 7 and 8. It is worth mentioning, in case the fracture strain decreases at the edge (with uniaxial tensile mode), the formation of cracks at the edge becomes easier and increases. While increasing of fracture strain for elements far from the edge which were underwent plane strain and biaxial tension modes, suppresses propagation of the crack through the plane, gives the chance and encourage more cracks forming at the edge.
This behavior depended more on the stress-state through the process than the pre-damage at the edge. However, for this material and the defined materials with this small die clearance, the applied damage from the cutting process varied negligibly for all the cut edges. Also, the effect of fracture strains in high stress-triaxiality conditions was more profound and the HER became much bigger by increasing them. Moreover, by plotting the volume under the fracture locus for high stress-triaxiality versus HER, a linear relation was detected, figure 9. As mentioned, high stress-triaxialities are the reason of crack propagation into the materials.

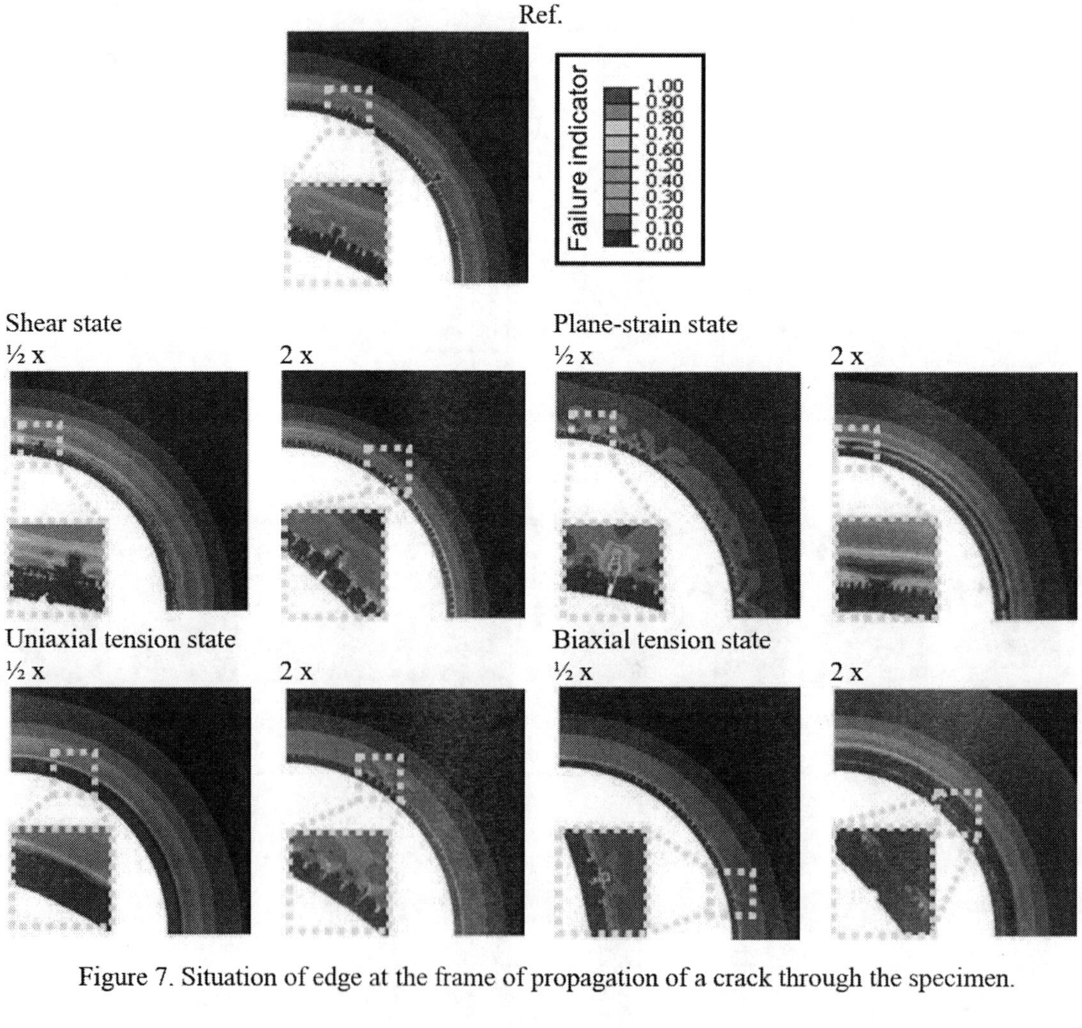

Figure 7. Situation of edge at the frame of propagation of a crack through the specimen.

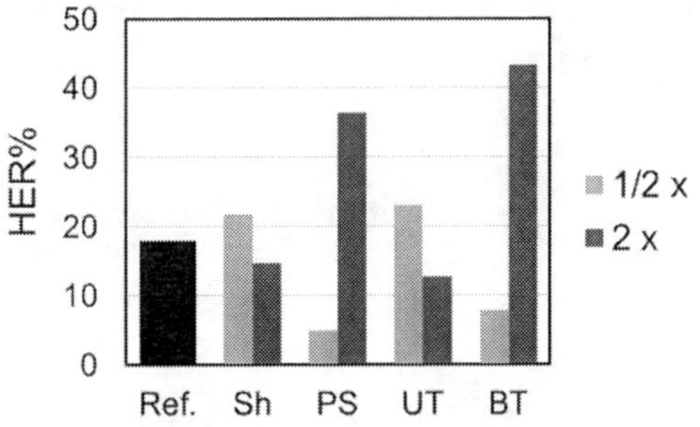

Figure 8. Comparison of hole expansion ratio between different conditions

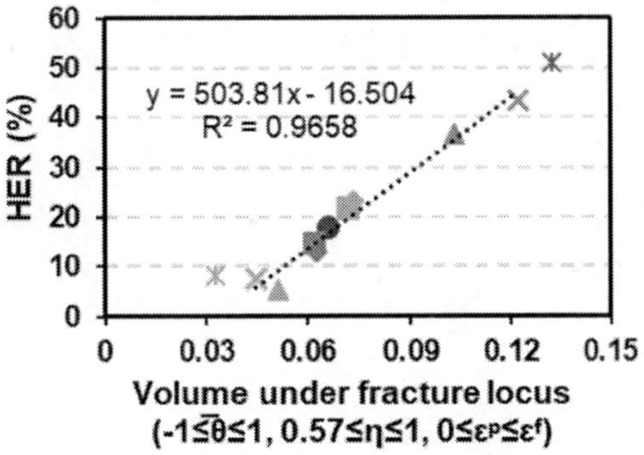

Figure 9. Relation between hole expansion ratio and volume under fracture surface for high stress-triaxiality. Red, green, and blue marks represent original, ½ x, and 2 x conditions, respectively (● original, ■ shear, ▲ plane-strain, ◆ uniaxial tension, ^{x} biaxial tension, and * all-four equivalent plastic strains).

5. CONCLUSIONS

The level and shape of fracture surface influences on the behavior of the material through shear-cutting and the subsequent hole-expansion processes. The results show that increasing the level of fracture surface in high stress-triaxiality conditions significantly increases the resistance to edge cracking by measuring of HER as a crack penetrate through the sheet plane. Whereas, increasing the level of fracture surface in relatively low stress-triaxiality slightly degrades this resistance. As the material experiences plane strain and biaxial tension at the crack propagation into the material plane, the changes in these two stress states caused huge difference in the cracking.

6. ACKNOWLEDGEMENT

Simulations were performed with computing resources granted by RWTH Aachen University under project <rwth0459>.

REFERENCES

[1] Nasheralahkami S, Golovashchenko S, Pan K, Brown L and Gugnani B 2016 *SAE Int.* 2016-01-0358
[2] Casellas D, Lara A, Frómeta D, Gutiérrez D, Molas S, Pérez L, Rehrl J, Suppan C 2017 *Metall. Mater. Trans. A* **48** 86
[3] Wang K, Luo M and Wierzbicki T 2014 *Int. J. Fracture* **187** 245
[4] Wang K, Greve L, Wierzbicki T, 2015 *Int. J. Solids Struct.* **71** 206
[5] Habibi N, Beier T, Richter H, Könemann M, Münstermann S 2019 *IOP Conf. Ser.: Mater. Sci. Eng.* **651 (1)** 012073
[6] Park S, Cho W, Jeong B-S, Jung J, Sung S, Na H, Kim S-I, Lee M-G, Han HN 2021 *Int. J. Solids Struct.* **236-237** 111312

[7] Bai Y and Wierzbicki T 2008. *Int. J. Plasticity* **24** 1071
[8] Habibi N, Pütz F, Könemann M, Brinnel V, Münstermann S, Feistle M, Volk W 2018 *IOP Conf. Ser.: Mater. Sci. Eng.* **418 (1)** 012058

IOP Conf. Series: Materials Science and Engineering 1238 (2022) 012043 doi:10.1088/1757-899X/1238/1/012043

Modeling tension-compression asymmetry and failure anisotropy in bending operations of a magnesium alloy

D Steglich[1] **and J Besson**[2]

[1] Helmholtz-Zentrum Hereon, Institute of Material Systems Modeling, D-21502 Geesthacht, Germany, dirk.steglich@hereon.de
[2] Centre des Matériaux, PSL Research University, Mines ParisTech, UMR CNRS 7633, 91003 Evry Cedex, France, jacques.besson@mines-paristech.fr

Abstract. The deformation behavior of the magnesium sheet alloy AZ31 at room temperature is investigated using different standard and non-standard mechanical tests as well as respective finite-element simulations. In order to address the bendability, three-point bending tests were performed for different orientations of the sample with respect to the sheets axes of orthotropy. The elastic-plastic deformation was modeled using a two-yield-surface model, which is able to describe the evolving strength differential effect as well as the anisotropy of the material. For the prediction of material failure, a numerically efficient damage criterion based on transformation of strain rates is calibrated and applied. It is shown that this model is able to predict the observed edge failure of specimens with good accuracy.

1. Introduction

Despite the high strength-to-weight ratio of magnesium alloys in general, the application of rolled sheets and extruded profiles is still restricted. Challenges to face for applications are its planar anisotropy, the tension-compression asymmetry (the so-called strength differential effect, SD effect) [1] as well as its limited formability at room temperature. Strong and efficient prediction tools for forming operations, for structural responses to loads and for lifetime analyses are required on the macro scale to extend the use of magnesium alloys.

The tension-compression asymmetry requires special (not von Mises) yield functions to describe plastic deformation. Formulation of the yield function including the third stress invariant naturally account for this asymmetry, but are not necessarily convex. Formulations using linear transformations of the stress deviator lead to convex yield functions [2, 3]. For describing the evolution of anisotropy/SD effect, either the coefficients of anisotropy can be defined as functions of an equivalent strain [4, 5], or a two-surface model [6, 7] can be used.

In most industrial applications, the use of shell or membrane elements is preferred due to their computational efficiency and robustness. In these cases plane stress conditions are considered. Therefore, stress triaxiality and Lode angle depend on each other. This prevents the use of two-parameter approaches triggering the evolution of a damage variable to address material failure. Hence, common models applied in sheet metal forming to predict the forming limits are based on through-thickness necking of the materials, e.g. [8]. However, for some e.g. advanced high-strength steel (AHSS) and magnesium alloys, fracture occurs without preceding necking. In these cases, failure models waiving an evolution equation for damage appear suitable. These uncoupled damage models can be used to signpost a mechanical stage, for which either the overall forming is limited, or structural instability sets in. These models are computationally more efficient compared to coupled damage models. The required damage parameters indicating critical stages are derived from the elasto-plastic behavior of the material by a hybrid approach involving mechanical testing and numerical analyses. Focusing on magnesium sheets in the following, successful predictions are reported by several authors, e.g. [9-11].

The method described in the following reproduces the direction-dependent irreversible deformation as well as the failure behavior. Alongside (smooth) tensile and notched samples, Arcan-type samples with emerging shear stress states are incorporated in the parameter optimisation strategy. Failure points are analysed as global quantities. Fracture is predicted in terms of crack initiation location and local mechanical state. An uncoupled failure model is proposed and exploited for this purpose. It is shown that regardless of its phenomenological character, the proposed model predicts the deformation and failure at monotonous loading of the magnesium sheets investigated.

2. Material and calibration tests

The material under investigation is the magnesium alloy AZ31 (Mg + 3%Al + 1%Zn) in its as-received condition. It was provided by POSCO in sheets of 1.17 mm thickness. The microstructure is fully recrystallized, evidencing an average grain size of 11 μm. The matrix of mechanical tests conducted comprises tensile test following the DIN standard along the rolling direction (RD), along the transverse direction (TD), and along 45° to the RD (DD), compression tests along the RD and TD as well as through-thickness (TT), Arcan-type samples cut along the RD and TD, with loading angles of 0°, 15°, 30°, and 45°, as well as notched bars with a notch radius of 2 mm (NB2) and 6 mm (NB6), respectively. Details of the tests are given in previous publications [12, 13]. Table 1 summarises the tests conducted.

Table 1. Test matrix for model calibration.

	tension			compression			Ar 0°		Ar 15°		Ar 30°		Ar 45°		NB2		NB6	
orientation	RD	TD	DD	RD	TD	TT	RD	TD	RD	TD	RD	TD	RD	TD	RD	TD	RD	TD
# tests	4	4	4	4	4	4	3	3	3	3	3	3	3	3	4	4	4	4

3. Models

3.1. Plasticity model

A phenomenological two-yield surface model [7] is adopted here to numerically describe the deformation behaviour. This model principally captures the tension-compression asymmetry and describes the evolution of the strength differential effect. A symmetric (for glide mechanisms) and an asymmetric yield surface (for twinning mechanisms) are combined by two interacting plastic potentials. Assuming the additive decomposition of the strain increment, the total strain increment consists of the elastic and parts related to glide and twinning, respectively:

$$d\boldsymbol{\varepsilon} = d\boldsymbol{\varepsilon}^e + d\boldsymbol{\varepsilon}^p = d\boldsymbol{\varepsilon}^e + d\boldsymbol{\varepsilon}_g^p + d\boldsymbol{\varepsilon}_t^p. \quad (1)$$

The plastic strain increments are calculated from the two yield functions, namely Bron & Besson [14] and CPB [2] using normality rule. Two coupled isotropic hardening functions are used:

$$R_g(p_g, p_t) = R_{0g} + H_g p_g + Q_{1g}(1 - \exp[-b_{1g} p_g]) + Q_{2g}(1 - \exp[-b_{2g} p_g]) + H_{gt} p_t, \quad (2)$$

$$R_t(p_g, p_t) = R_{0t} + Q_{1t}(\exp[-b_{1t} p_t] - 1) + H_{tg} p_g, \quad (3)$$

with p_g and p_t being the plastic multipliers of the glide and twinning mechanism, respectively. The determination of the numerous parameters was realized by running FE-simulations of the mechanical tests, constructing a target function from the differences between simulation and experimental results and minimizing the target function.

3.2. Failure model

An approach based on linear transformations of the plastic strain rate tensor referred in the orthotropic material axes is chosen for modelling of directionality of failure. A local fracture criterion is formulated as an integral over the loading history:

$$D = \int_0^t f(\dot{\boldsymbol{\varepsilon}}_{pl})dt = D_c \tag{4}$$

In order to address the anisotropy in failure strains, the strain rate components in the material frame need to be mapped to a "corresponding" strain rate tensor. This is done by a linear transformation rendering a deviatoric deformation measure $\dot{\boldsymbol{\varepsilon}}^d_{pl}$,

$$\dot{\boldsymbol{\varepsilon}}^d_{pl} = \boldsymbol{L} \cdot\cdot\ \dot{\boldsymbol{\varepsilon}}_{pl} \tag{5}$$

with

$$\boldsymbol{L} = \begin{pmatrix} \frac{1}{3}(L_{TT}+L_{SS}) & -\frac{1}{3}L_{SS} & -\frac{1}{3}L_{TT} & 0 & 0 & 0 \\ -\frac{1}{3}L_{SS} & \frac{1}{3}(L_{LL}+L_{SS}) & -\frac{1}{3}L_{LL} & 0 & 0 & 0 \\ -\frac{1}{3}L_{TT} & -\frac{1}{3}L_{LL} & \frac{1}{3}(L_{LL}+L_{TT}) & 0 & 0 & 0 \\ 0 & 0 & 0 & L_{LT} & 0 & 0 \\ 0 & 0 & 0 & 0 & L_{TS} & 0 \\ 0 & 0 & 0 & 0 & 0 & L_{SL} \end{pmatrix}, \tag{6}$$

which introduces six adjustable parameters L_{ij}. A material point is assumed to be broken if

$$D = \int_0^t \left(\langle \dot{\varepsilon}^d_I \rangle + \langle \dot{\varepsilon}^d_{II} \rangle\right)dt = 1, \tag{7}$$

with $\langle . \rangle$ being the positive part function. The damage indicator D can be evaluated as a local quantity in a post processing step of the FE-simulation for any forming process. By the relation equation (5) a tensor transformation is applied, similar to what is common in stress tensor transformations used in yield functions for anisotropic media.

3.3. Model simplifications and parameter calibration

Previous investigations [15] revealed that for the present material, the mechanical anisotropy is primarily manifested by its basal texture, which renders a plastic deformation behaviour similar to planar anisotropy. The failure strain quantified by tensile tests, however, differs significantly between RD, TD and DD orientation. In order to reduce the models complexity, the following supposition is made: Coefficients in the linear transforms of the plasticity model affecting the RD and TD orientation were set to unity – RD and TD deformation behaviour become hence identical. Parameters affecting the out-of-plane direction (plasticity and failure model) were set to unity, since testing in this orientation is not possible for sheets. Isotropic hardening functions for glide- and twinning mechanisms are taken from [13]. All other plasticity parameters were re-calibrated using the data provided in the test matrix shown in table 1. The fits are visualised here using the force-elongation record of tensile and compression tests as an example, see figure 1. The resulting yield functions are shown for the space of plane principal stresses in figure 2. Note that due to the simplification, the simulated curves compromise RD and TD behavior with respect to their plastic response.

The parameters of the failure model are determined using the global failure strains measured during the tensile tests, see figure 3. Local strains recorded by digital image correlation are generally higher due to strain localization, but show more scatter. The model is apparently able to capture the direction-dependent fracture of tensile samples with model parameters L_{LL}=10.0, L_{TT}=1.3, L_{SS}=3.0, and L_{LT}=1.4.

Note that in this approach the anisotropy in failure is captured, although the anisotropy with respect to deformation is not.

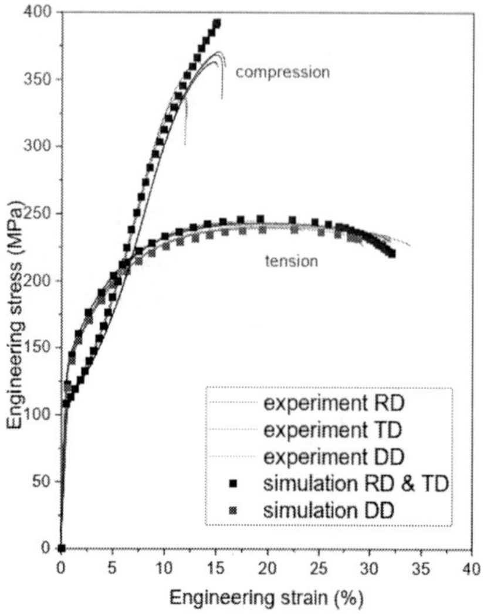

Figure 1. Stress-strain data for uniaxial tension and compression.

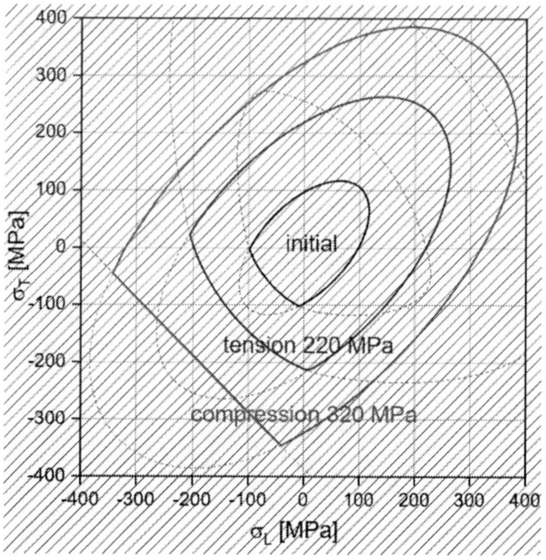

Figure 2. Plane stress representation of the 2-yield-surface model.

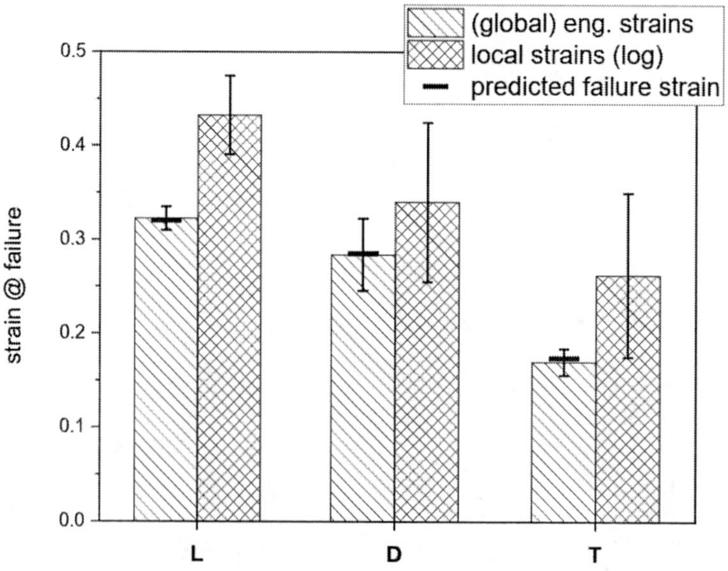

Figure 3. Verification of the failure model: predicted global failure strains compared to experimental counterparts.

4. Three-point bending

4.1. Experiments

For the three-point bending tests, rectangular strips with the as-received thickness of 1.17 mm, 60 mm in length and 10 mm in width are used. The tests are conducted with an in-house tool with a punch radius of 2 mm and circular supports with a radius of 5 mm, see figure 4. 15 samples each from orientations along the RD, along the TD, and along the DD are fabricated and tested. The bending tests are performed with spans of 18 mm, 20 mm, and 25 mm between the supports. These dimensions are set with the help of an in-built scale on the support platform. Furthermore, the samples are coated with the use of a thin Teflon sheet between the supports and the beam. Figure 5 displays two bent strips of the 18 mm and 25 mm width configuration. Details are reported in [16].

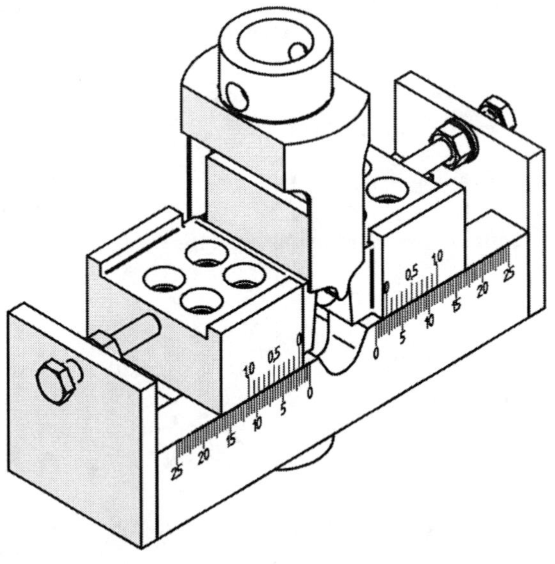

Figure 4. Adjustable bending apparatus.

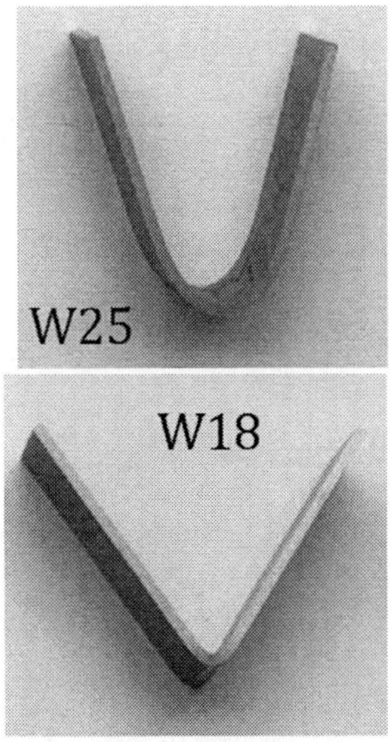

Figure 5. Bending samples after testing.

4.2. Modeling results

The simplified two-yield surface plasticity model is used to simulate the three-point bending tests by conducting 3D FE-simulations using twofold symmetry conditions. Frictionless contact between sheet, punch and support is assumed. Figure 6 shows the results in terms of global punch force as a function of displacement for all tested spans for the three orientations. The fit is excellent over the whole range of deformation and demonstrates the high predictive power of this model.

Failure during bending was evidenced in the experiments for the configurations W18 (TD and DD) and for W20 DD. Samples in RD orientation never failed. The same holds true for the wide configuration W25, where all samples remained intact. This result is captured by the failure model. Failure points are indicated by symbols in figure 6. They meet with the first occurrence of a crack in the sheets, which is associated with a load drop of the force-displacement record.

5. Discussion and Conclusion

The formability here is explored here by means of bending tests, in which commonly the deformation state is between uniaxial stress and plane strain deformation. While the deformation model is calibrated using test experiencing a variety of stress triaxialities ranging from -1/3 (uniaxial compression) to close to 2/3 (round notched bar tension), the failure model is tuned on the basis of tensile test only. This appears inconsistent. However, trials for finding elements of the failure transformation matrix, equation 6, meeting the failure point of all tests given in table 1 gave less satisfying results than those depicted in figure 6. A possible reason for this is the high ductility of tensile test along the RD, which is detailed in [13].

It can be concluded, that the simplified two yield surface model together with the phenomenological failure model provide a powerful tool for the prediction of formability of hcp metals, for which a strength differential effect is reported. For the prediction of bendability, uniaxial tests for calibration of the failure model appears sufficient as it provides accurate results.

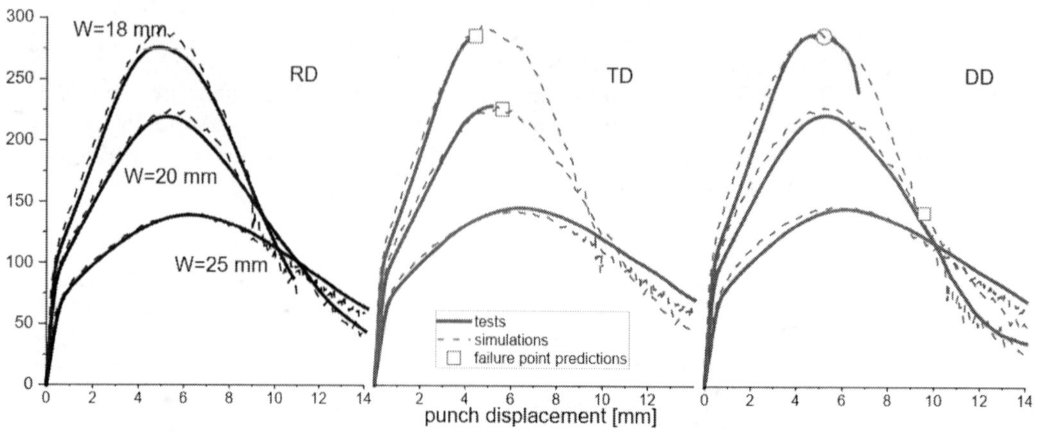

Figure 6. Prediction of the deformation and failure behaviour during bending.

Acknowledgments
The authors thank POSCO for providing the material. They furthermore appreciate the assistance of Abhishek Nagendra during mechanical testing.

[1] Hosford W F 1993 *The mechanics of crystals and textured polycrystals* (New York: Oxford University Press)
[2] Cazacu O, Plunkett B and Barlat F 2006 Orthotropic yield criterion for hexagonal closed packed metals *Int. J. Plast.* **22** 1171-94
[3] Plunkett B, Cazacu O and Barlat F 2008 Orthotropic yield criteria for description of the anisotropy in tension and compression of sheet metals *International Journal of Plasticity* **24** 847-66
[4] Yoon J, Cazacu O and Mishra R K 2013 Constitutive modeling of AZ31 sheet alloy with application to axial crushing *Materials Science and Engineering: A* **565** 203-12
[5] Mekonen M N, Steglich D, Bohlen J, Letzig D and Mosler J 2012 Mechanical characterization and constitutive modeling of Mg alloy sheets *Materials Science and Engineering: A* **540** 174-86
[6] Kim J H, Kim D, Lee Y-S, Lee M-G, Chung K, Kim H-Y and Wagoner R H 2013 A temperature-dependent elasto-plastic constitutive model for magnesium alloy AZ31 sheets *International Journal of Plasticity* **50** 66-93
[7] Steglich D, Tian X and Besson J 2016 Mechanism-based modelling of plastic deformation in magnesium alloys *European Journal of Mechanics - A/Solids* **55** 289-303
[8] Marciniak Z and Kuczynski K 1967 Limit strains in the processes of stretch-forming sheet metal *International Journal of Mechanical Sciences* **9** 609-20
[9] Jia Y and Bai Y 2016 Ductile fracture prediction for metal sheets using all-strain-based anisotropic eMMC model *International Journal of Mechanical Sciences* **115-116** 516-31
[10] Lee J-Y, Steglich D and Lee M-G 2018 Fracture prediction based on a two-surface plasticity law for the anisotropic magnesium alloys AZ31 and ZE10 *International Journal of Plasticity* **105** 1-23
[11] Li F-F, Fang G and Qian L-Y 2019 Forming limit analysis of Mg-2Zn-1.2Al-0.2Ca-0.2RE alloy sheet using ductile fracture models *International Journal of Damage Mechanics* **29** 1181-98
[12] Steglich D 2020 Mixed-Mode Deformation and Failure of a Magnesium Sheet Quantified using a Modified Arcan Fixture *Experimental Mechanics* **60** 109-18
[13] Steglich D and Besson J 2021 Prediction of deformation and failure anisotropy for thin magnesium sheets under mixed-mode loading *Mechanics of Materials* **163** 104064
[14] Bron F and Besson J 2004 A yield function for anisotropic materials: Application to aluminium alloys *Int. J. Plast.* **20** 937-63

[15] Jeong Y and Steglich D 2020 Modelling-assisted description of anisotropic edge failure in magnesium sheet alloy under mixed-mode loading *International Journal of Mechanical Sciences* **181** 105680
[16] Nagendra A and Steglich D 2020 Experimental and Numerical Bendability Analysis of a 3rd Generation Magnesium Alloy *Procedia Manufacturing* **47** 1274-80

IOP Conf. Series: Materials Science and Engineering 1238 (2022) 012044 doi:10.1088/1757-899X/1238/1/012044

Prediction of forming limit curve for AA6061-T6 at room and elevated temperatures

Ahmed Kacem*, Hervé Laurent, Sandrine Thuillier

Univ. Bretagne Sud, UMR CNRS 6027, IRDL, F-56100 Lorient, France

E-mail: *ahmed.kacem@univ-ubs.fr

Abstract. Forming Limit Curve (FLC) has been widely adopted as a practical criterion for evaluating the formability of sheet metals. Predicting a reliable FLC by a virtual methodology could lead to robust process optimization before expensive tool manufacturing. In order to increase the predictive capabilities of the virtual forming tools, an accurate modeling of the forming limit curve should be considered at room and elevated temperatures. In this work, the isothermal forming limit curves of 6000 series aluminum alloy sheet metal are predicted by performing numerical simulations of Nakajima test. A stress triaxiality and Lode angle based ductile fracture criterion is used to determine the forming limit curve. Also, the ductile fracture criterion is extended to add the impact of temperature on ductile fracture prediction. The hybrid experimental-numerical approach is used to calibrate the ductile fracture criterion. The forming limit curve of AA6061-T6 sheet metal, with a thickness of 1 mm, is predicted using the calibrated ductile fracture criterion at room and elevated temperatures. Numerical simulations are performed in 3D with the finite element code Abaqus. The limit strains are determined for specimens undergoing deformation under different strain paths. Influence of temperature on the predicted forming limit curve is discussed.

1. Introduction

Forming Limit Curves (FLC) are widely used to evaluate the formability of sheet metals. FLC displays in principal strain space in the sheet plane the combinations of strains either at the onset of local necking (FLC at necking) or at the onset of fracture (FLC at fracture). FLC at fracture are very useful to evaluate the forming limits of metals when ductile fracture is induced without onset of localized necking (e.g. material with minor necking, bending [1], incremental sheet metal forming processes [2]). To predict numerically FLC at fracture, accurately calibrated ductile fracture (DF) criterion is strongly recommended.

Jain et al. [3] have predicted FLC at fracture of AA6111-T4 aluminum alloy by means of various ductile fracture criteria. It is found that the maximum shear stress criterion by Tresca predicts reasonably well the fracture limits over a large range of strain ratios. Takuda et al. [4] have successfully predicted the FLC at fracture of aluminum alloy sheets with Oyane's ductile fracture criterion [5]. Han and Kim [6] proposed a new ductile fracture criterion to better predict the experimental FLC at fracture. Lou et al. [7] have applied a new stress triaxiality and Lode angle based ductile fracture criterion to construct the FLC at fracture. Comparison between the experimental and predicted FLCs demonstrates the accuracy of this new ductile fracture criterion. Yang et al. [8] have effectively obtained FLC at fracture of AA7075-T6 aluminum alloy with Lou's ductile fracture criterion. However, the above-mentioned studies refer to FLC at fracture obtained only at room temperature (RT) whereas FLC at fracture for elevated temperature is scarcely reported. Recently, Heidari et al. [9] have obtained experimentally and numerically FLCs at fracture for the AA6063 alloy at RT and 200°C, by using several ductile fracture criteria without considering the dependency of the ductile fracture criterion to the temperature. It is found that increasing the temperature moved the FLC upward, about 15%. In this study, a stress triaxiality and Lode angle based ductile fracture criterion is used to investigate the effect of temperature on the prediction of FLC at fracture for aluminum alloy. The ductile fracture criterion is extended to include the effect of temperature on ductile fracture prediction. Numerical simulations of Nakajima test with various specimen geometries are conducted until fracture to obtain the FLC at fracture of 1 mm thickness AA6061-T6 alloy at RT and 200°C.

2. Material and methods

2.1. Material

AA6061-T6 sheet aluminum alloy with 1 mm thickness is used in this work. The material parameters which have been obtained from uniaxial tensile test are given in Tab.1.

Table 1. Mechanical properties in the rolling direction and plastic anisotropy coefficients of AA6061-T6 at RT and 200 °C.

	YS (MPa)	UTS (MPa)	r_0	r_{45}	r_{90}	$\bar{r}$	Δr
RT	270	318	0.59	0.78	0.81	0.74	-0.08
200 °C	205	211	0.64	0.71	0.72	0.7	-0.03

$\bar{r} = (r_0 + r_{90} + 2r_{45})/4$ $\Delta r = (r_0 + r_{90} - 2r_{45})/2$

For the finite element (FE) simulation, isotropic hardening with Hill48 yield criterion is chosen to model the mechanical behavior of this material. The yield function is given by:

$$f = \bar{\sigma} - \sigma_y\left(\bar{\varepsilon}_p\right) \tag{1}$$

where $\bar{\sigma}$ is the Hill48 equivalent stress that is defined as follow, from the components of the Cauchy stress tensor:

$$\bar{\sigma} = \sqrt{F\left(\sigma_{yy} - \sigma_{zz}\right)^2 + G\left(\sigma_{zz} - \sigma_{xx}\right)^2 + H\left(\sigma_{xx} - \sigma_{yy}\right)^2 + 2L{\sigma_{yz}}^2 + 2M{\sigma_{xz}}^2 + 2N{\sigma_{xy}}^2} \tag{2}$$

F, G, H, L, M and N are the Hill's coefficients. Due to the low value of the planar anisotropy coefficient Δr, transverse isotropy is assumed. Therefore, the values of Hill's coefficients are calculated considering only $\bar{r}$ (i.e. assuming $r_0 = r_{45} = r_{90} = \bar{r}$) according to these equations:

$$F = G = \frac{1}{1+\bar{r}} \qquad H = 1 - G \qquad N = \frac{1+2\bar{r}}{1+\bar{r}} \tag{3}$$

Due to the lack of available data regarding the mechanical behavior in the thickness of the sheet, L and M are kept equal to their isotropic value (i.e $L = M = 1.5$).

$\sigma_y\left(\bar{\varepsilon}_p\right)$ is the hardening function that is chosen with a saturation form of Hockett-Sherby modified to include the strain rate sensitivity at elevated temperature [10].

$$\sigma_y(\bar{\varepsilon}_p, \dot{\varepsilon}) = \left[\sigma_0 + Q\left(1 - \exp\left(-b\left(\bar{\varepsilon}_p\right)^n\right)\right)\right]\left[\frac{\dot{\varepsilon}}{\dot{\varepsilon}_0}\right]^m \tag{4}$$

where $\bar{\varepsilon}_p$ is the equivalent plastic strain, σ_0 is the initial yield stress, Q represents the maximum change in the size of the yield surface, b defines the growth rate of the yield surface, n is the strain hardening coefficient, $\dot{\varepsilon}$ is the strain rate, $\dot{\varepsilon}_0$ is a constant strain rate normalization factor and m is the strain rate sensitivity coefficient.

Lou's ductile fracture criterion [7, 11] is used to predict the onset of fracture. This original criterion is expressed as follows:

$$D = \frac{1}{C_3} \int_0^{\bar{\varepsilon}_f} \left(\frac{2}{\sqrt{L^2+3}} \right)^{C_1} \left(\frac{\langle 1+3\eta \rangle}{2} \right)^{C_2} d\bar{\varepsilon}_p \tag{5}$$

where D is a failure indicator that gives the onset of fracture when it reaches unity. $\bar{\varepsilon}_f$ is the equivalent plastic strain at fracture. C_1, C_2 and C_3 are material parameters that need to be calibrated. η and L are the stress triaxiality and the Lode parameter, respectively.

An extended form of Lou's ductile fracture criterion which includes the impact of temperature on ductile fracture criterion is also used. The extended form recently proposed by Kacem et al.[12] is expressed as:

$$D(T) = \frac{1}{C_3(1+C_4\, T^*)} \int_0^{\bar{\varepsilon}_f} \left(\frac{2}{\sqrt{L^2+3}} \right)^{C_1} \left(\frac{\langle 1+3\eta \rangle}{2} \right)^{C_2} d\bar{\varepsilon}_p \tag{6}$$

where C_4 is an additional material parameter to include the influence of temperature T. T^* is the homologous temperature, defined by :

$$T^* = \frac{T - T_{ref}}{T_m - T_{ref}} \tag{7}$$

T_{ref} and T_m are respectively the reference and melting temperature ($T_{ref} = 293\,\text{K}$ and $T_m = 873\,\text{K}$).

2.2. Nakajima test

Nakajima forming method is used to obtain the FLC. Fig.1 shows the forming tools and the specimen geometries used to obtain three different strain paths. A hemispherical punch of $\phi\, 33$ mm diameter is used to deform the specimen with an outer diameter of $\phi\, 60$ mm into the die cavity with $\phi\, 35.25$ mm diameter corresponding to a ratio of 1/3 to the standard dimensions [13]. A blank-holding force of 100 kN is set to clamp firmly the sheet between the die and blank-holder. The punch speed is set at $1.5\,\text{m.s}^{-1}$ according to the standard ISO 12004-2 [13].

2.3. Numerical model

Isothermal numerical simulations of Nakajima test are conducted in 3D using Abaqus/Standard. The tools are modeled as rigid surfaces whereas the workpiece is modeled as a deformable body as shown in Fig.2. One quarter of the workpiece is meshed with 3D solid hexahedral elements with reduced integration (C3D8R). A fine mesh size of approximately 0.2 mm is used in the local zone where fracture is likely to occur and the size is increased to approximately 0.6 mm at sections away from this zone. Moreover, 4 elements are used through the thickness. The workpiece is assumed to have elastic-plastic behavior with the material properties specified in Section 2.1 and given below. Since, experimentally, an optimal lubricant system is strongly recommended to meet the requirement of a fracture on the top of the dome [13], the friction between the punch and the workpiece is neglected for the sake of simplicity. For the other tool-workpiece interfaces, Coulomb's friction law is used with a friction coefficient of 0.2. In a first step, the force is applied to the blank-holder. Then, the punch is constrained to move in the axial direction. The ductile fracture criterion is implemented in the FE code through UVARM subroutine which calculates the damage indicator D (Eq.6) at each increment. The major and minor values in the sheet plane of the first element reaching the critical value of $D = 1$ (critical element) are recorded to plot the FLC at fracture. For a first approach in this study, isothermal conditions are considered and the blank temperature is supposed uniform. FLC at fracture are predicted at RT and 200°C.

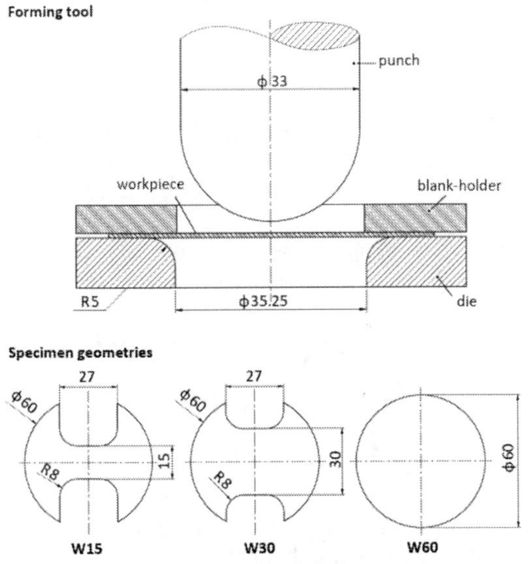

Figure 1. Geometry and dimensions (in mm) of forming tool and specimens for Nakajima test.

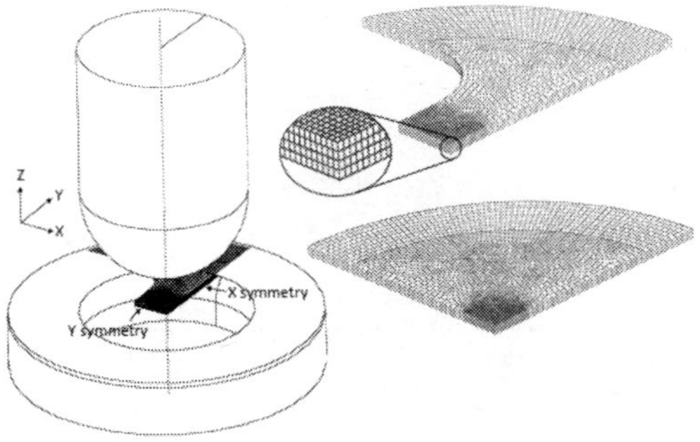

Figure 2. FE model of Nakajima test and meshing for specimens W15 and W60.

3. Results

3.1. Calibration of material parameters

In a recent study conducted by the same authors [12], experiments and simulations of tensile tests up to fracture on dog-bone specimens, notched specimens with different radius (NR5 and NR15), specimens with a central hole (HR4) and shear specimens (SH) are used to calibrate the hardening and fracture parameters of AA6061-T6 at RT and 200°C. Hardening law parameters have been identified manually through several trial and error iterations until the numerical and experimental load-displacement curves obtained with all fracture specimens show good agreement. Hybrid experimental-numerical approach have been used to determine the fracture strain and the corresponding stress state parameters for each specimen geometry as described by Kacem et al. [14]. The fracture parameters are then optimized by minimizing a cost function. Though the process is considered isothermal, the two DF criteria are used out of comparison's sake. Tabs.2 and 3 show the identified hardening law parameters and calibrated fracture parameters, respectively. Figs.3 and 4 show the hardening curves predicted by the modified Hockett-Sherby model and the fracture locus for AA6061-T6, respectively.

Table 2. Hardening law parameters identified at RT and 200 °C.

	σ_0 (MPa)	Q (MPa)	b	n	$\dot{\varepsilon}_0$ (s^{-1})	m
RT	244	135	7	0.68	–	–
200 °C	130	76	7	0.36	10^{-4}	0.04

Table 3. DF criterion parameters calibrated at RT and 200 °C.

	Lou's original DF criterion			Lou's extended DF criterion			
	C_1	C_2	C_3	C_1	C_2	C_3	C_4
RT	0.640	0.250	0.612	0	0.201	0.573	1.321
200 °C	0	0.166	0.801				

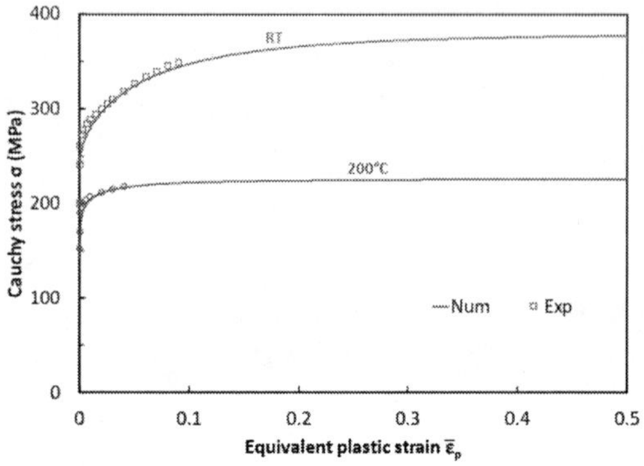

Figure 3. Hardening curves of AA6061-T6 predicted at RT and 200°C. Symbols represent experimental points obtained from uniaxial tension of dog-bone specimen at $\dot{\varepsilon} = 10^{-3}\,\mathrm{s}^{-1}$.

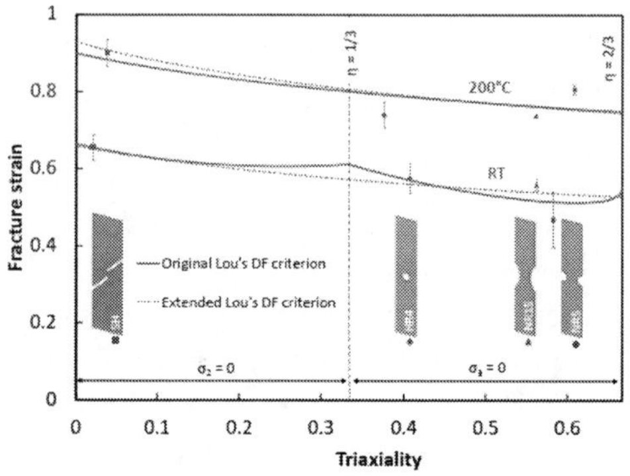

Figure 4. Fracture locus of AA6061-T6 constructed in the space of stress triaxiality and equivalent plastic strain under plan stress condition at RT and 200°C.

3.2. Ductile fracture prediction

The parameter D is used to indicate the onset of ductile fracture when it reaches $D = 1$ and as an indicator to assess the damage state in the specimen. Fig.5 shows the distribution of the parameter D in the deformed specimen at fracture initiation (i.e. when a first element reaches the critical value $D = 1$) for both ductile fracture criteria at RT and 200°C.

As expected, the critical zone where fracture initiates is found on the top of the dome for all specimens. The critical zone is more localized for W15 and W30 specimens. However, W60 specimen shows no signs of localized necking prior to fracture making the determination of necking point required to establish the FLC at necking very difficult for this kind of specimen. The original and extended fracture models show a similar distribution of the parameter D. To compare more quantitatively between both fracture criteria, Tab.4 shows the equivalent plastic deformation at fracture $\bar{\varepsilon}_f$ predicted by both models at RT and 200°C.

It can be seen that the values of $\bar{\varepsilon}_f$ predicted by the original DF criterion are very close to the ones predicted by the extended form, especially at elevated temperature where the maximum difference is found about 0.5 %. At room temperature, the maximum difference between both models is about 3.7 %. Therefore, it is better to use the extended form instead of the original form to predict the forming limits since the extended form is more useful to predict ductile fracture initiation under isothermal and non-isothermal conditions whereas the original can be only used under isothermal conditions.

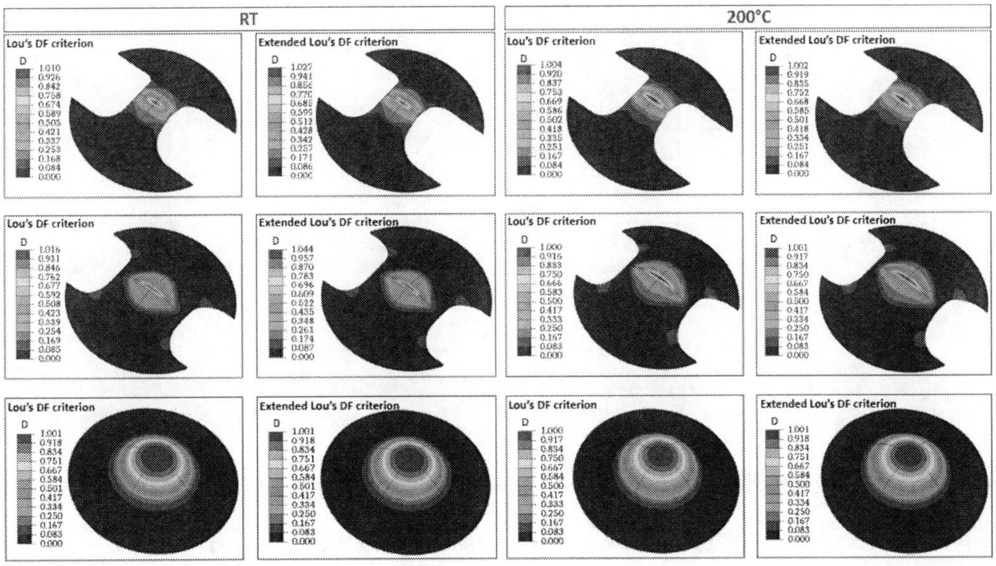

Figure 5. Damage distribution at the onset of fracture at RT (left hand side) and 200°C (right hand side)

Table 4. Equivalent plastic strain at fracture $\bar{\varepsilon}_f$ predicted at the critical element when $D = 1$ by the original and extended DF criteria at RT and 200°C.

		Lou's original DF criterion	Lou's extended DF criterion
W15	RT	0.542	0.550
	200 °C	0.775	0.777
W30	RT	0.521	0.539
	200 °C	0.765	0.764
W60	RT	0.550	0.530
	200 °C	0.750	0.746

3.3. Forming limit curves at fracture

The histories of strains (i.e. major and minor strains) are extracted in the critical element to obtain the strain path and the fracture point as shown in Fig.6. Only results obtained from the extended DF criterion are considered since both criteria give similar results.

It can be seen that the FLC at fracture tends to have an approximately linear shape as was also found experimentally at RT in the work of Takuda et al. [4] for AA1100-O and AA5182-O aluminum alloy sheets with a thickness of 1 mm. As can be observed, the temperature has a high effect on the major strain limit with an increase from RT to 200°C of about 40 % that is higher than that found by Heidari et al. [9] for AA6063.

The same linear strain path is obtained for W60 specimen for RT and 200°C corresponding to equi-biaxial tensile state. However, for W15 and W30 specimens due to the punch geometry, the strain path are non-linear as was widely reported in previous works, e.g.[15, 16]. The deformation starts with bending, leading to an initial balanced biaxial strain path, which is the major cause of the non-linear strain path in Nakajima test. It can be seen that the non-linear strain path effect is more pronounced at 200°C.

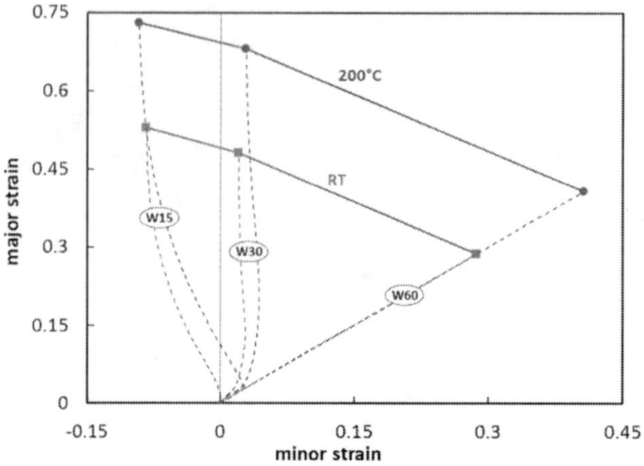

Figure 6. FLC at fracture at RT and 200°C. The dashed lines represent the strain path of each specimen.

4. Conclusion

To investigate the effect of temperature on the forming limits of aluminum alloy, FLC at fracture of AA6061-T6 are determined numerically at RT and 200°C using Nakajima test. To predict the onset of fracture, Lou's DF criterion is used in numerical simulation. An extended form of the DF criterion which include the influence of temperature on ductile fracture prediction is also used. Comparison between the original and extended forms shows almost similar results making the use of the extended form more recommended especially in non-isothermal conditions. The same approximately linear shape of FLC at fracture is obtained at RT and 200°C. It is found that the temperature has a strong influence on the forming limits. The limit strain may increase by 40 % with the raise of temperature from RT to 200°C. It is also found that the non-linear strain path effect inherent to Nakajima test increases with the raise of temperature.

References

[1] Thuillier S, Le Maoût N and Manach P Y 2011 *Materials & Design* **32** 2049–2057
[2] Silva M B, Skjoedt M, Atkins A G, Bay N and Martins P A F 2008 *The Journal of Strain Analysis for Engineering Design* **43** 15–35
[3] Jain M, Allin J and Lloyd D J 1999 *International Journal of Mechanical Sciences* **41** 1273–1288
[4] Takuda H, Mori K, Takakura N and Yamaguchi K 2000 *International Journal of Mechanical Sciences* **42** 785–798
[5] Oyane M, Sato T, Okimoto K and Shima S 1980 *Journal of Mechanical Working Technology* **4** 65–81
[6] Han H N and Kim K H 2003 *Journal of Materials Processing Technology* **142** 231–238
[7] Lou Y, Huh H, Lim S and Pack K 2012 *International Journal of Solids and Structures* **49** 3605–3615
[8] Yang Z, Zhao C, Dong G and Chen Z 2021 *Journal of Materials Processing Technology* **291** 117044
[9] Heidari A, Ghassemi A and Atrian A 2019 *The International Journal of Advanced Manufacturing Technology* **106** 2595–2611
[10] Laurent H, Coër J, Manach P Y, Oliveira M C and Menezes L F 2015 *International Journal of Mechanical Sciences* **93** 59–72
[11] Lou Y and Huh H 2013 *International Journal of Solids and Structures* **50** 447–455
[12] Kacem A, Laurent H and Thuillier S 2022 Experimental and numerical investigation of ductile fracture for AA6061-T6 sheets at room and elevated temperatures (Submitted for publication)
[13] ISO 12004-2 2008 *Metallic materials — Determination of forming-limit curves for sheet and strip — Part 2: Determination of forming-limit curves in the laboratory*
[14] Kacem A, Laurent H and Thuillier S 2021 *Engineering Fracture Mechanics* **248** 107686

[15] Leppin C, Li J and Daniel D 2008 Application of a method to correct the effect of non-proportional strain paths on nakajima test based forming limit curves *Proc. of the NUMISHEET 2008 Conf., Interlaken, Switzerland* pp 217–221
[16] Zhang L, Min J, Carsley J E, Stoughton T B and Lin J 2017 *International Journal of Mechanical Sciences* **133** 217–226

IOP Conf. Series: Materials Science and Engineering 1238 (2022) 012045 doi:10.1088/1757-899X/1238/1/012045

An alternative time-based strategy for the evaluation of forming limits using optical experimental measurements

Celalettin Karadogan[1]*, Marcel Görz[1], Mathias Liewald[1] and Marco Bühler[1]
[1] Institute for Metal Forming Technology, University of Stuttgart, Holzgartenstr. 17, 70174, Germany

*celalettin.karadogan@ifu.uni-stuttgart.de

Abstract. Optical measuring systems allow to observe the sheet metal instability more precisely. The section method was first standardised and made available in commercial measurement systems in order to evaluate the forming limit curves almost automatically. The insignificant contribution of the necking area to the evaluated critical strains compared to the neighbouring volume of material led to alternative approaches such as the time-based evaluation criteria. The majority of these criteria take into account the strain evolution at the material necking location and are also suitable for evaluating other types of limits such as edge-crack sensitivity. Some advanced high-strength steels typically exhibit very short deformation after the onset of instability, which complicates the application of time-based approaches because they require sufficient amount of deformation after necking. The proposed approach is an attempt to reduce the dependence of the time-based approaches on the amount of deformation after necking by taking into account the local characteristics of the measurement curves. The approach is applied to some advanced high-strength steels to evaluate the forming instability and edge-crack sensitivity. The superiority of this identification approach is demonstrated in this paper by comparison with classical time-based approaches.

1. Introduction

Simplicity of the strain-based forming limit diagrams (FLD) makes them the most used tool in the assessment of the sheet metal formability. The probability of defects in terms of strain localization or rupture can be determined by the locations of principal strains relative to the forming limit curve (FLC). The implementation of digital measurement system took a long time in the online measurement of strains during a sheet formability test. With the use of digital online measurement systems, the forming limit curve is classically determined according to DIN EN ISO 12004-2 in a Nakajima or Marciniak test. The experimental determination of the critical deformation points is carried out according to DIN 12004-2 using the so-called line section method. In this method strain distributions along a section perpendicular to the rupture orientation at a measurement state just before the material separates is used to calculate the limit strains. As alternative evaluation scheme time-dependent methods have been extensively investigated to remedy the adverse effect of problematic strain distributions along sections [1], [2]. These however have their own difficulties especially in determining a clear and unique inflection point in time as the initiation of the instability. The majority of these criteria take into account the strain evolution at the material necking location and are also suitable for evaluating other types of limits such as edge-crack sensitivity.

The processing of high-strength and ultra-high-strength sheet metal steel grades poses new challenges for the automotive industry. In particular, high-strength dual-phase steels increasingly exhibit cracks at the component edges due to collar forming and body drawing processes [3]. The tendency of a material to develop early edge cracks due to extreme hardening effects while transversal shearing is also referred to as *edge crack sensitivity (ECS)*. However, ECS is not solely dependent on the sheet materials used. The manufacturing process by which the edge of the semi-finished product was produced, as well as the associated cutting process parameters, also have a major influence on ECS [4]. If a component edge is produced by a shear cutting process, the formability of the sheet material at this edge degrades, hence the sheet metal edge exhibits increased ECS [3]. Those cracks are not directly predictable by common failure criteria e.g., the forming limit curve (FLC). Therefore, several approaches to test and predict edge cracks in simulation were developed in recent years [4], [5].

To critically analyze the most known testing methods for instability and ECS the German Association of the Automotive Industry (VDA) formed a group on a project called "Local formability and edge crack sensitivity". Among many research institutes, sheet metal producers and sheet metal users, authors took part in performing the edge fracture tensile tests and the hole tensile tests. As decided by the project group, the maximum principal deformation in these tests was determined at three different points in time: At the time of the 1% force drop relative to the maximum force $\varphi_{1,1\%Force}$, at the time of the observed crack initiation $\varphi_{1,Riss}$, and at the time of evaluated critical thinning of the material $\varphi_{1,Inst}$ (localized instability).

In the tests performed to determine ECS, the maximum principal deformation change is preferably located directly at the specimen edge. Since the optical measurement data at the specimen edges are usually error-prone, a safety distance to the component edges is used to create the virtual mesh of the optical measurement to avoid possible errors. To generate an experimental dataset, demonstrating repeatability (intra-institution) and reproducibility (inter-institutions), various institutions took part in performing the same experiments. The special type of specimen for the edge-fracture-tensile-tests was proposed by the Chair of Metal Forming and Casting (UTG) of the Technical University of Munich [6], and these specimens are cut also by UTG for the mentioned VDA-Project using the introduced punching tool setup. In addition, the time-force-strain measurements recorded by all project partners are evaluated for the localized instability ($\varphi_{1,Inst}$) by UTG using a script based on the method developed collaboratively by Volk and Hora [1] and probably further developed at UTG (see Figure 1). The most classical time-based approach proposed in this study uses the thinning rate slopes before and after the initiation of the instability to find the transition point.

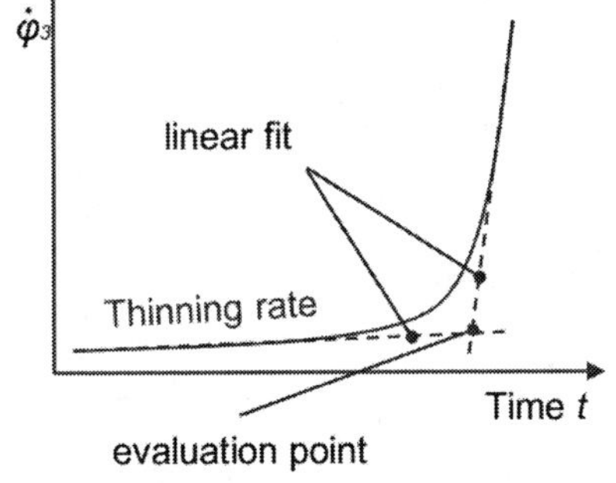

Figure 1: A time-based evaluation method to identify the localized instability [6].

To justify the use of localized instability ($\varphi_{1,Inst}$) in this study as a reliable approach, it is necessary to first discuss the alternative evaluation strategies mentioned above. Use of the maximum principal strain at a force drop of 1% beyond the maximum force ($\varphi_{1,1\%Force}$) has the advantage that it is clearly defined by the force curve. This improves the repeatability and reproducibility of the tests and also allows the use of a fully automated evaluation strategy. The disadvantage of this strategy is that a drawing force reduction is not necessarily coupled with the localized instability or edge fracture. The determination of the maximum principal deformation when the specimen is cracked $\varphi_{1,Riss}$ has the advantage that there is a direct correlation of the measured value with the localized instability or edge failure. The disadvantage of this approach is that the point of time of the crack is specified visually by the operator and therefore is not automated and objective. This causes a degradation in repeatability and reproducibility.

As an interesting example on the use of the time-based localized instability ($\varphi_{1,Inst}$), Figure 2 shows the thinning rate and the principal major strain as a function of time recorded at the most-critical point (for this case the most critical interior point and not a point on the edge of specimen) on a classical tensile test of the material HR660Y760T-CP. The shown thinning rate has several straight regions, resulting in two possible evaluation times (Point 1 and Point 3). The evaluation performed by UTG determined for this case the Point 1 as the point of time of critical thinning and beginning of instability. This results in a relatively low prediction for the principal deformation of approx. $\varphi_1 = 0.12$. Since these reference specimens, having both edges milled, are cracked in the center of the specimen, this value should be approximately located on the FLC of the material in the forming limit diagram (see Figure 3). The values for major and minor strain at the evaluation times are plotted in the forming limit diagram as shown in Figure 3. The FLC data was taken from the material libraries provided by AutoForm and Dynamore simulation programs and provided by the manufacturer of this material. It can be seen in this figure that the Point 1 is located significantly below the FLC. With manual intervention the same algorithm can produce Point 3 or even Point 2 (as the closest point to the intersection of the linear interpolations on first and third segments of the curve) which are almost located on the FLC and thus better represents the state of the critical thinning. This example shows the need to reconsider the assessment strategies for the instabilities within or at the edge of the specimen. To obtain a conservative, reliable estimate of the failure time, the determined time of the critical thinning should be defined at the beginning of the significant increase of the thinning rate, which is the Point 2, for this specific case.

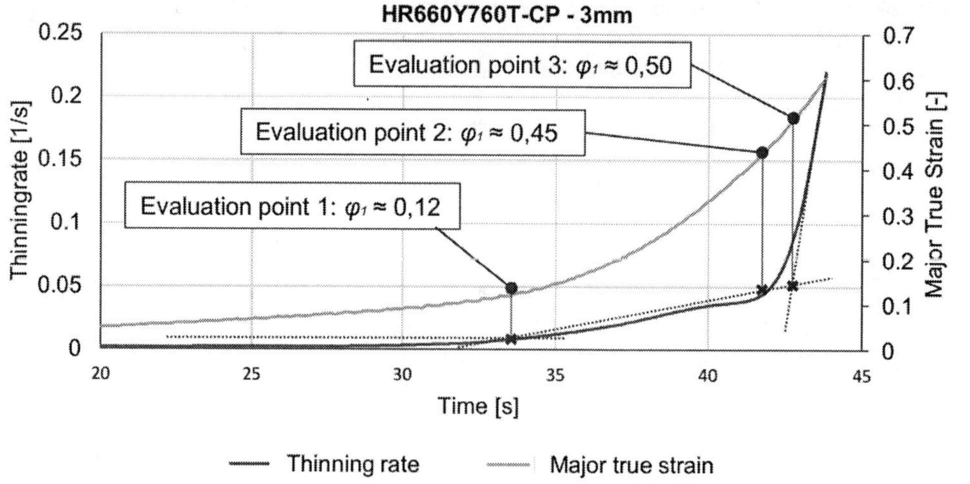

Figure 2: Evaluation points for the time-based method using a reference tensile test (HR660Y760T-CP).

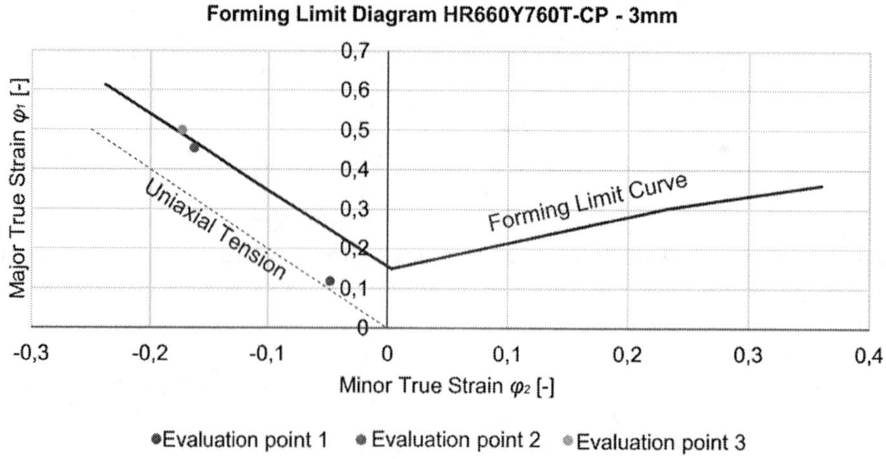

Figure 3: Evaluated points in the forming limit diagram of the material HR660Y760T-CP.

2. New approach for the automated evaluation of the point of critical thinning

This section develops an alternative strategy for the automated evaluation of the point of critical thinning based on the time-strain method with data obtained during stretching of conventional tensile test specimen. The strain values are to be measured at the most critical point of the specimen, i.e. at the failure point or at the nearest measurement point if the failure is at the edge of the specimen. If this point lies on the edge of the optical measurement mesh, the criterion provides the strain values for the ECS of the tested material. This strategy is tested for the hole tensile test and the edge-fracture-tensile-test but also generally applicable to other procedures. The algorithm for determining the critical thinning strain value is given in Figure 4 and the details are explained afterwards on the basis of a punched hole tensile test for the material CR780Y980T-CP and the and edge fracture tensile test for the material HR660Y760T-CP.

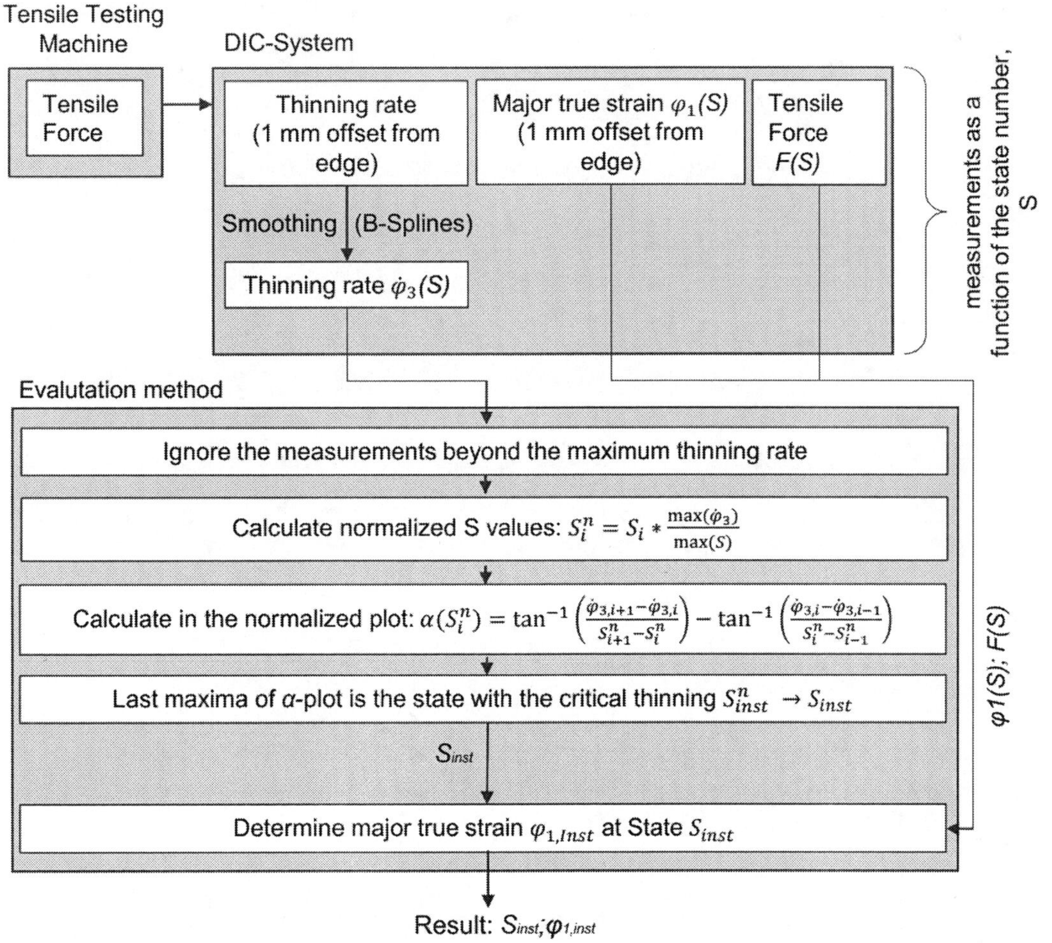

Figure 4: Flow chart for determining the time of critical thinning.

The thinning rate $\dot{\varphi}_3$ is used to determine the correct point of time of critical thinning. The states of the optical measurement system are used as the quasi-time in the approach and all measurements are made as a function of this integer number. In a first step, the thinning rate is subjected to smoothing using B-Splines. The maximum achieved thinning rate refers to the state number, up to which the calculations explained below are based on. In other words, data measured beyond this state is not considered. In the next step, the state (i.e., time) of the critical thinning or the instability point is found. Looking at the thinning rate curve in Figure 2, a human observer searches for the point of instability in the region of maximum curvature. For this purpose, the scales of the axes are usually normalized optically. For the same reason, the calculated values of curvature in such a plot do in fact make more sense after numerically normalizing the scale of the one axis with respect to the one of the other. Therefore, we re-scaled the x-axis, the integer numbers corresponding to the state number. Numerous test calculations using the curvature κ of the modified graph of the thinning rate as a function of re-scaled state numbers revealed the fact that not only the curvature of the imagined continuous curve but also the distances between the points of the underlying discrete representation are influential in the curvature calculation. This phenomenon is explained in Figure 5, in which the use of curvature is shown on the left side and the alternative use of angular divergence is shown on the right. The greater distance (usually observed with increased thinning rate) between the second and third points of interest inversely affects the calculated curvature value. This conclusion demonstrates that human perception can be better represented by angular divergence then curvature on the discrete representation. As shown in the Figure 5, the angular divergence can be calculated in the simplest way by calculating the angle between the two line-segments connecting the first and second points and the second and third points. An extended version for the calculation of the angular divergence can be achieved by considering many before and after points and fitting lines to these before and after points. It has been shown that the approach based on angular divergence remains stable and always indicates the same critical point, up to 5 before and after points.

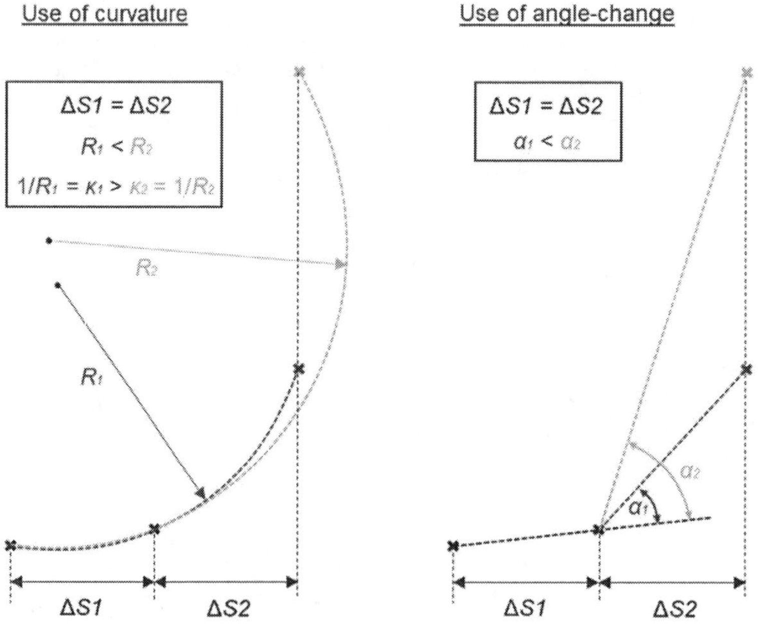

Figure 5: Use of curvature vs. angle change in finding the instability point based on three measurement points.

The calculated α values are then used to evaluate the instability point, being the state corresponding to the last peak of the α-plot. This state number is associated with the measured major and minor strains, which can then be used as the critical strain point.

3. Experimental Setup

The edge fracture tensile tests and hole tensile tests were carried out based on the tensile test according to DIN EN ISO 6892-1. The specimen geometry corresponds to that being used for the tensile test according to ASTM standard. The hole tensile test specimen ere punched with a hole diameter of 10 mm in a specimen of 20 mm width. In case of an edge-fracture-tensile-test, the samples were milled on one edge and shear cut on the other edge by UTG for all project partners, so that a study of the repeatability and reproducibility of the evaluation of the edge cracking behavior on a shear cut edge became feasible. The relative cutting gap was reported as 12.0% of the sheet thickness. Reference tests are also carried out with edges milled on both sides. The drawing direction during the tests was perpendicular to the rolling direction (90°), since this is where the smallest changes in shape occur when the specimen cracks and a conservative assumption is therefore made [7]. The drawing speed during the tests was constant and is 12.0 mm/min as specified by the VDA project team. For statistical validation, seven specimens per test material and edge condition were tested. The 5M GOM-ARAMIS was used as a DIC system for capturing strain values from test specimens. The reference length for strain calculation was recorded to be 0.6 mm. The shape changes are recorded on the burred face of the cutting edge in case of the shear-cut specimens. The materials used in this study are CR780Y980T-CP and HR660Y760T-CP being only a subset of the whole materials considered in the scope of the VDA project. The material parameters of the aforementioned materials are presented in Table 1.

Table 1: Material properties of CR780Y980T-CP s and HR660Y760T-CP

Material	t [mm]	Yield Strength [MPa²]	UTS [MPa]	Frature Strain min. [%]
CR780Y980T-CP	1,5	889.8	987.9	8.2
HR660Y760T-CP	3,0	759.8	889.6	11.6

4. Results and Discussion

The use of the proposed approach is shown in Figures 6 and 7 for the materials CR780Y980T-CP and HR660Y760T-CP with the classical ISO tensile specimen geometries used also for the edge-fracture-tensile-test and the hole tensile specimen. The plot of the angle α calculated as explained above is given in Figure 6 for a classical tensile test. The last highest point of the plot of angle α then corresponds to the point in time of the critical thinning of the specimen. This plot also demonstrates the estimation by the method proposed by [1] using exactly the same measurement data performed by the owner of the script. The major strains corresponding to the evaluated states are comparatively shown below in Figure 6. A further comparison is performed using a hole tensile test with a punched hole as shown in **Figure 7**, demonstrating also the calculated critical major strain values near the punched edge. In Figures 6 and 7, the divergence angle α is also calculated using 5 pre- and post-points. As explained in section 2, the specified critical point remains unchanged regardless of the number of pre- and post-points used for the angle calculation. Such critical major strain is important in evaluating an extended FLC curve valid for the punched edges of a blank.

Figure 8 shows the simplified forming limit diagram for the material HR660Y760T-CP as a comparison for the evaluated critical major strains φ1,Inst,IFU and φ1,Inst,UTG. The results of the proposed criteria are labeled with "IFU" and the results of the method introduced by [1] are labeled with "UTG". Using the provided experimental data, UTG evaluated the critical values for all project partners using the script explained in [1]. Reference tests referred to the specimens with both edges milled. Shear-cut specimens are labeled as "Punch". The principal strain values evaluated by the proposed approach φi,Inst,IFU for the reference specimens lie close to the FLC of the material. The shear-cut specimens, however, are found significantly below the FLC curve, showing a remarkably reduced edge crack sensitivity of this material. In comparison, the principal strains φi,Inst,UTG determined by UTG were calculated significantly below the forming limit curve.

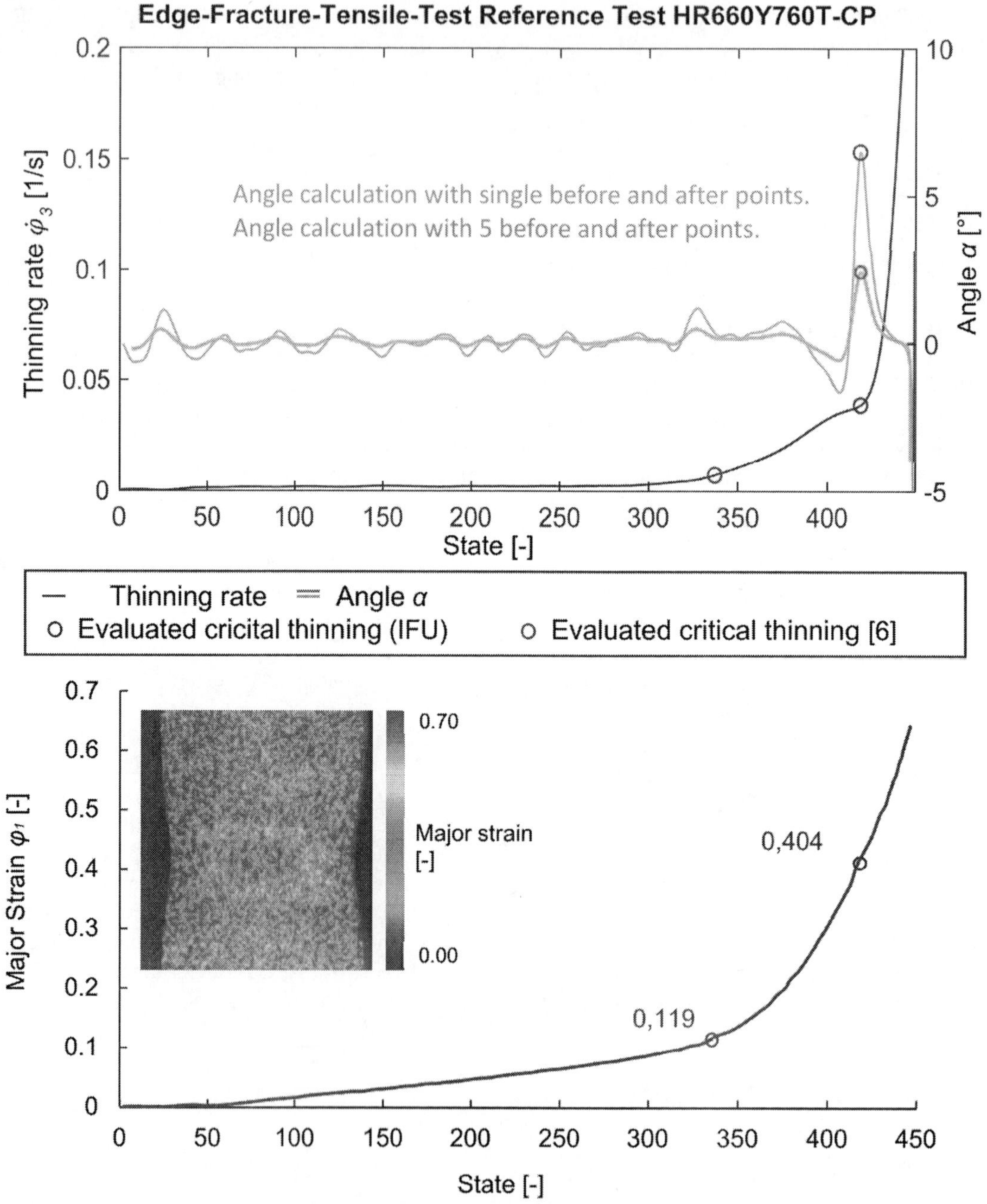

Figure 6: Calculated values for critical thinning and the major strain in case of a classical tensile specimen with even edges. The used material is HR660Y760T-CP.

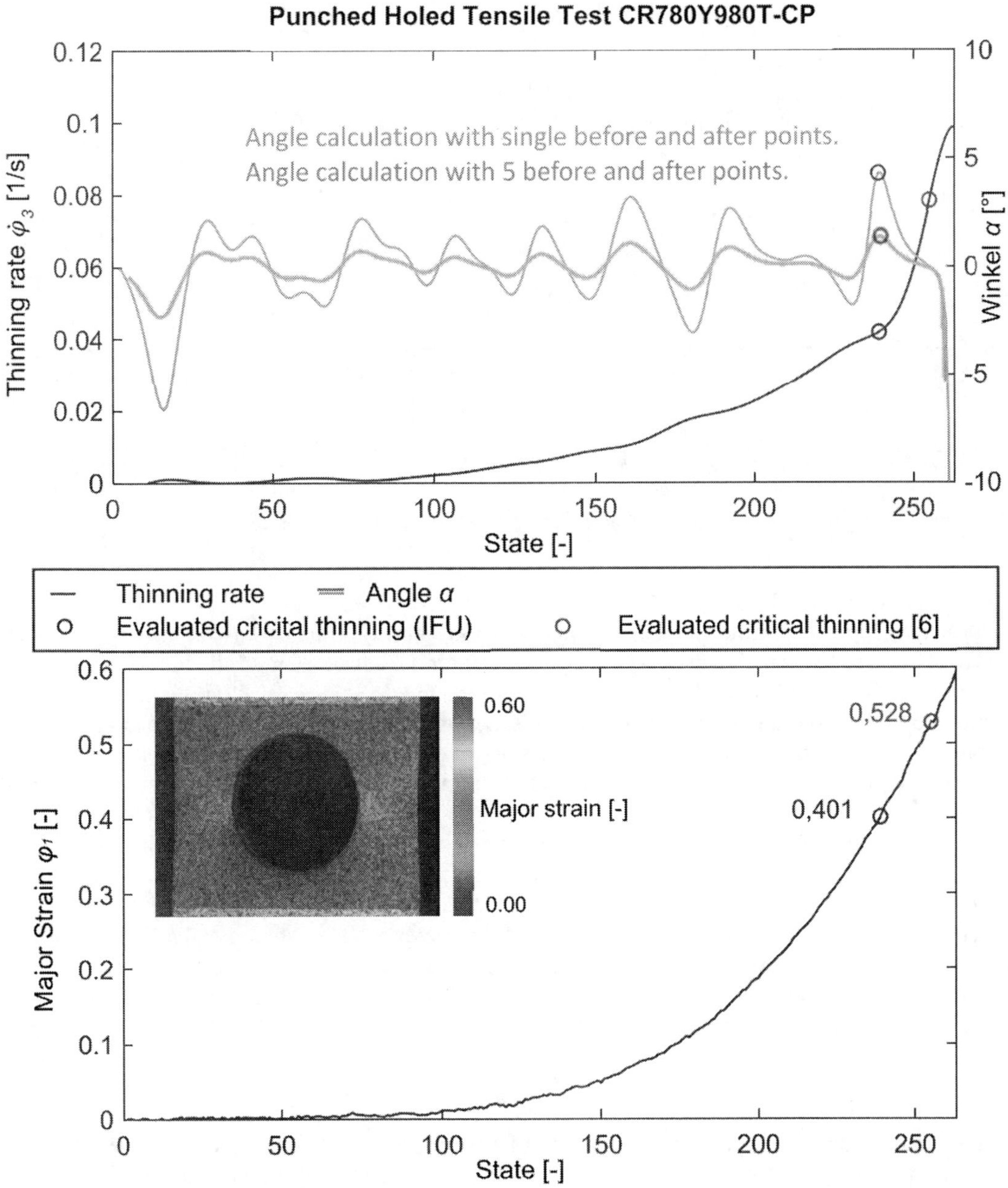

Figure 7: Calculated values for critical thinning and the major strain for punched hole tensile test specimen. The used material is CR780Y980T-CP.

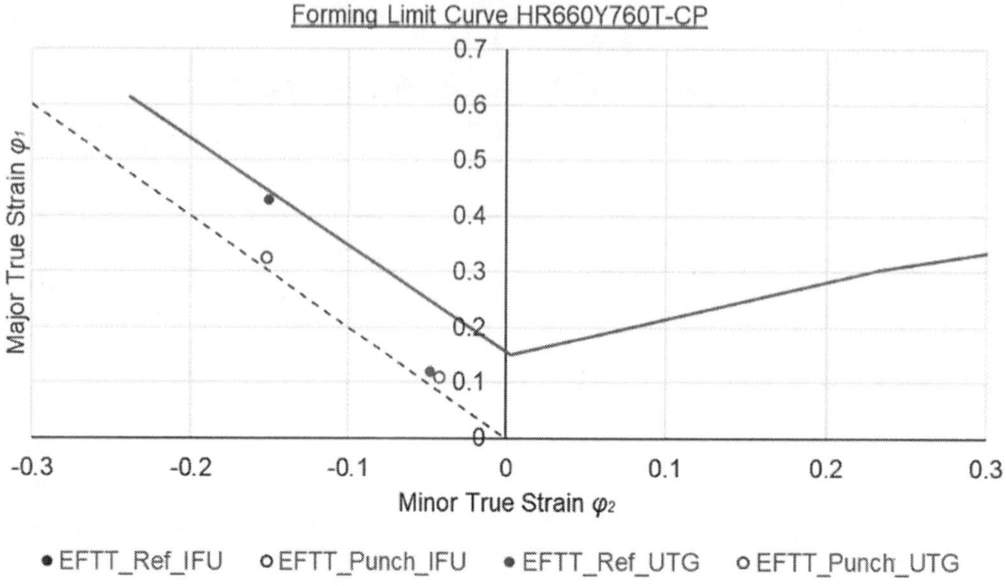

Figure 8: Forming limit diagram for the material HR660Y760T-CP with determined forming limit strains

5. Conclusion and Outlook

The proposed approach for the time-based evaluation of instability strains in case of far-edge localisation or edge crack sensitivity is shown to yield proper results for the advanced high strength steels considered in the scope of the VDA Project on "Local formability and edge crack sensitivity". The shortcomings arising due to the three stable regions for the evolution of the thinning rate, compared to the two stable regions assumed by the most classical time-based strategies, are eliminated. The plausibility of the results is demonstrated by comparing the evaluated strain values with the forming limit curves of these materials. The introduced evaluation procedure until now is fully automated, it does not require manual interventions. The required strain values are measured at the most critical point, and the time history of the entire strain distribution on the specimen is not required for the calculations. As it relies on a discrete calculation strategy, the proposed approach also requires a sufficient amount of measurement points properly representing the continuous behavior of the material. This strategy, developed within the framework and time span of mentioned VDA project, has not yet been tested for the classical evaluation of FLC curves with Nakajima test setups. A comparison with the classical approaches in this context also is planned as a follow-up study by the authors.

Acknowledgements

The material provided in the scope of the VDA Project "Local formability and edge crack sensitivity" is highly appreciated. The preparation of the EFTT specimens and the evaluation of the critical strains using a proprietary time-based calculation strategy by UTG is gratefully acknowledged.

References

[1] Volk W and Hora P 2011 New algorithm for a robust user-independent evaluation of beginning instability for the experimental FLC determination International *Int. J. Mater. Form.* **4** pp.339–46

[2] Merklein M, Kuppert A and Geiger M 2010 Time dependent determination of forming limit diagrams *CIRP Annals* **59** pp. 295–98

[3] Beier T and Woestmann S 2016 Berücksichtigung von schergeschnittenen Blechkanten zur Auslegung von Formgebungsprozessen höherfester Stahlwerkstoffe in der FEM-Umformsimulation *14. LS-DYNA Forum Proceedings*

[4] Feistle M, Golle R, and Volk W, 2022, Edge crack test methods for AHSS steel grades: A review and comparisons *J. Mater. Process. Technol.* **302** Elsevier B.V.

[5] Feistle M, Pätzold I, Golle R and Volk W 2016 Predicting Edge Cracks on Shear-Cut High-Strength Steels by Modified Uniaxial Tensile Tests *Key Eng. Mater* **703** pp. 49–55
[6] Feistle M, Golle R, and Volk W 2016 Determining the Influence of Shear Cutting Parameters on the Edge Cracking Susceptibility of High-strength-steels Using the Edge-fracture-tensile-test Procedia CIRP **41** pp. 1078–83
[7] Gläsner T, Imat V, Schneider M, Sunderkötter C, Theiss M and Westhäuser S 2015 Berücksichtigung der Kantenrissempfindlichkeit in der Umformsimulation Bamberg: Verlag Meisenbach

Investigation on the formability of friction stir welded Al-TWB through incremental forming

Shubham[1], Kiran K Nayak[1], Pilli J Teja[1], Rahul Jain[1] and Kaushik Bandyopadhyay[1]

[1] Indian Institute of Technology Bhilai, Raipur-492015, Chhattisgarh, India

Email: kaushik@iitbhilai.ac.in

Abstract. In the quest for weight reduction, automobile industries introduced the technology of tailor welded blanks (TWB) which offered utilization of material as per functional requirement. In general, friction stir welding (FSW) is mostly used for fabricating aluminum based TWBs as fusion welding of Al—alloys are problematic. Recently, single point incremental forming (SPIF) gained popularity among researchers as well as in industry as a dieless flexible forming technique. Hence, in this study two aluminum sheets namely AA6061 and AA7075 of similar thickness were welded together using FSW with square pin. The formability of the TWBs were evaluated in terms of cup height, wall angle and fracture location through SPIF incorporating different tool geometries. To assess the effect of weld, similar welded Al blanks were also deformed in the similar condition. The formability of the similar welded blanks and corresponding parent metals did not differ much and fracture did not initiate at the weld. However, cup depth for the TWB was in between the cup heights of the parent metals and failure was not observed at the weld region. TWB cups did not show any weldline shift.

1. Introduction

In the case of incremental forming, complete deformation of sheets occurred through gradual localised deformation of the same, instead of simultaneous deformation of the complete sheet by the application of punch, die and binder. In general, small hemispherical tools are used to cause localised plastic deformation in the sheets, which is then extended all over the clamped sheet along the best suitable path to achieve the desired deformed shape [1,2]. On one hand, being a dieless forming process, incremental forming offers the flexibility to fabricate any shape by just changing CNC/robot programming to design the suitable tool path without the necessity of designing new die-punch sets. On the other hand, localised nature of deformation enables the sheet to sustain more amount of deformation compared to the conventional sheet metal forming. However, the major drawback of the process involves longer processing time of the components. Therefore, the process is being adapted in the industries for customised product development, batch production or rapid prototyping purposes [3]. Among the incremental forming processes, single point incremental forming is more flexible and popular due to involvement of less amount of tooling [4]. A detailed discussion on deformation mechanisms, springback and surface roughness in SPIF process was reported by Gatea et al. [2]. In general, it is observed that formability improves with increase in tool rotation speed and decreases with feed rate. The fracture is mostly observed at the corner region of deformed part due to equi-biaxial stretching condition. In a recent review article, Duflou et al. [4] discussed several other advancements in the SPIF process which enabled the SPIF to overcome its limitation and is being considered for applications like cranial implants, self-bearing roof structure, die making etc. Application of multistep tool path [5], heat assisted SPIF [6], electrically assisted SPIF [7] are the main advancements which further enhances the performance of SPIF process.

In the past decades, one of the major technology enhancements in the field of automobile industry to address the ever-restricting fuel efficiency norms was the development of tailor welded blanks (TWB). The TWBs are single blanks fabricated by joining two or more sheet metals with different grades, gauges or surface properties (coated/uncoated) in butt welding configuration mostly by laser/friction stir welding [8]. TWBs offers the advantage of placing right materials at the selected position as per the functional/design requirements. Thus weight reduction, optimised material utilisation, cost effectiveness etc. can be achieved. However, presence of different materials in terms of thickness or properties on either side of the weld makes forming of TWBs a challenging job. Several studies have already been reported on formability evolution of TWBs in various mode of deformation. Panda et al. [9] concluded that the formability of DP980-HSLA TWB reduced during stretch forming operation due to non-uniform deformation which led to failure in the weaker HSLA side. Reduction in formability of DP980-IFHS TWBs were also reported in case of deep drawing operation by Bandyopadhyay et al. [10]. The failure was always observed at the cup bottom, in the weaker material and parallel to the weld in case of deep drawing of TWBs. However, not only the reduction in formability, but also weld line movement due to non-uniform deformation is another typical problem associated with forming of TWBs. Bandyopadhyay et al. [10] also reported the severity of weld line movement in case of DP980-IFHS and DP980-DP600 laser welded blanks for deep drawing operations. It is observed that during stretching weldline shifts towards the stronger material as can be seen at the cup bottom for deep drawing [10], also at the pole for LDH testing [11] and weldline shifts towards the weaker material where compressive stress is predominant as can be seen at the flange or cup wall portion in case of deep drawing. Therefore, several techniques were proposed to restrict weldline movement and as a consequence to improve formability like split binder [12], clamping pin [13], proper placement of initial weldline [14]. However, most of the research on formability study of TWBs involves conventional sheet metal forming, and SPIF of TWBs are relatively rare in open literature. One of the preliminary study on SPIF of TWBs was successfully conducted by Silva et al. [15]. In the study, AA1050-H111sheets with different thicknesses were combined to fabricate the TWBs which were then deformed under SPIF process using experimental benchmark tests. Tayebi et al. [16] also successfully performed SPIF on TWBs fabricated by joining AA6061 to AA5083 using FSW process. It was concluded that reduction in formability of TWBs in SPIF process can be avoided by selecting proper welding and SPIF parameters.

Therefore, in the present study, formability behaviour of a particular Al-TWB with dissimilar material combination were studied under SPIF process. Unlike the previous studies, the sheet thicknesses of the materials are selected to be 1 mm. Further, in the present study a complex shape is chosen as the deformed geometry to understand the extent of formability under complex strain state. Finally, different shape and size of tools are employed to understand the effect of tool geometry on formability and corresponding force requirement.

2. Material and methodology

In the present study, two aluminium alloys AA6061 and AA7075 were selected for fabrication of the TWB and further formability study. Both the Al alloys were of 1 mm thickness. The mechanical properties of both the aluminium alloys are compared in the table below (Table 1) in terms of yield strength, and ultimate tensile strength.

Table 1: Comparison of the tensile test properties

Material	Yield Strength (MPa)	Ultimate Tensile Strength (MPa)
AA6061	290	327
AA7075	280	304

2.1. Friction stir welding process

Three sets of welded blanks were created using FSW process for which a linear CNC machine (make: ETA) was employed. Among them two were similar welded blanks fabricated by combining AA6061 to AA6061 and AA7075 to AA7075, respectively. The third one is the TWB fabricated by combining AA6061 with AA7075. The fixture and the tool geometry for the FSW process are shown in figure 1. The major challenge during welding of 1 mm blank was buckling of the sheet which was overcame by providing proper clamping. The shoulder diameter of the FSW tool was 15 mm and the cross-section of the tool pin was square with the side of 3.5 mm. The height of the tool pin was 0.9 mm. For all the three types of welding, the FSW parameters were kept same; the tool rotation was 500 rpm, the welding speed was kept at 100 mm/min and plunge depth was 0.05 mm with a tilt angle of 1°.

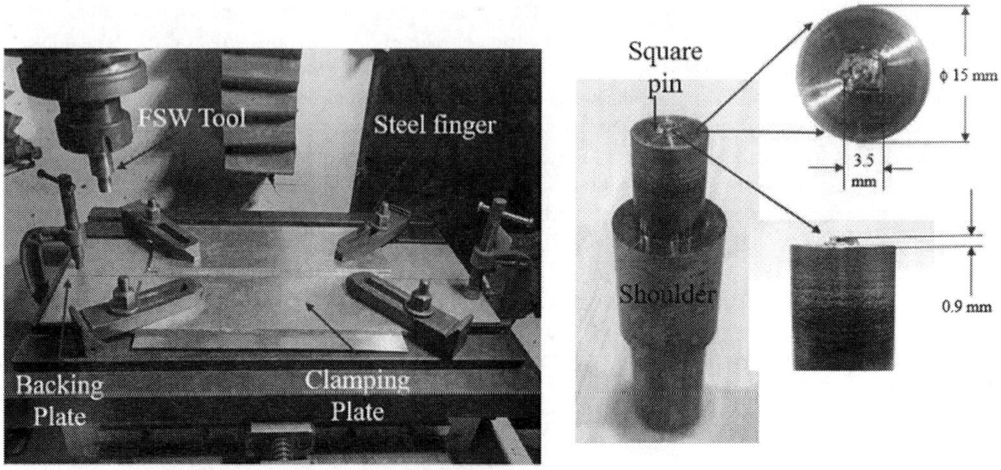

Figure 1. The fixture and tool for FSW process

2.2. Weld quality evaluation

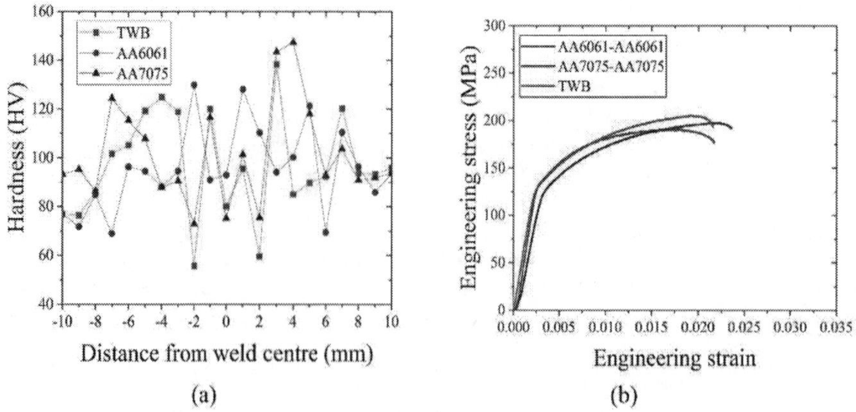

Figure 2. Assessment of weld quality in terms of (a) cross weld hardness (b) tensile testing

The weld quality was assessed in terms of cross weld hardness test (figure 2a) and tensile test was performed keeping the weld perpendicular to the loading axis (figure 2b). Hardness in the weld zones were measured with a load of 100 g and dual time of 10s. The hardness at the weld was observed to be increased compared to the corresponding parent metals as can be seen from figure 2a. It is because finer grain size is obtained in FSW process and as per Hall-Petch equation hardness is inversely proportional to the grain size. Tensile test shows the UTS of all the three-weld samples is around 200 MPa. The failed sample revealed that failure occurred from the heat affected region. In all the three welds, the weld efficiency was more than 60%. FSW process has successfully welded one mm thin aluminium alloy (similar and dissimilar) and with optimization of process parameters, weld efficiency can be further improved.

2.3. *Incremental forming*

The SPIF process was performed using a laboratory scale fixture with a forming window of 110×110 mm. The blank size required for the fixture was 150mm×150mm which had to be clamped between the blank holder and the backing plate using 12 numbers of screws. During experiment the complete SPIF set up was placed on a dynamometer (make: Kistler) for measuring the deformation force. In this study, two different shape of tools were used; one with hemispherical tip another with flat bottom.

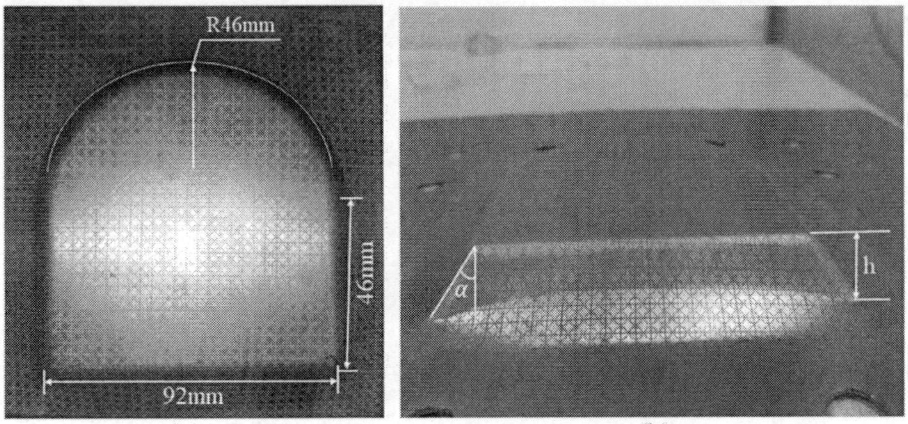

(a) (b)

Figure 3. True shape of the deformed geometry

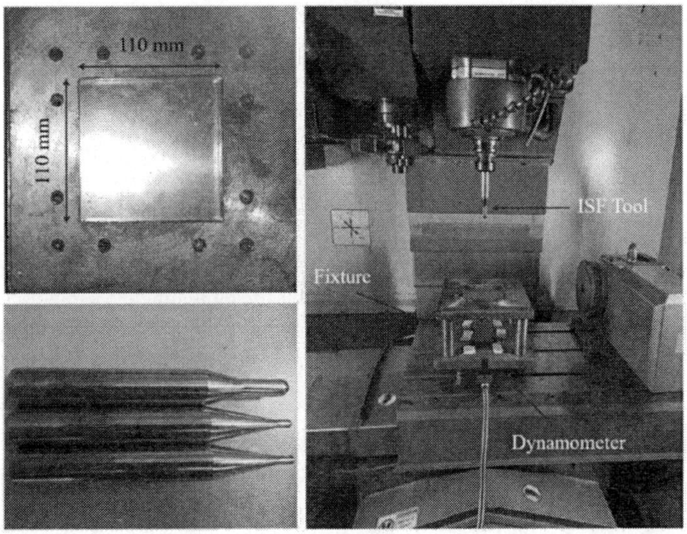

Figure 4. The fixture, tools and complete set up for the SPIF process

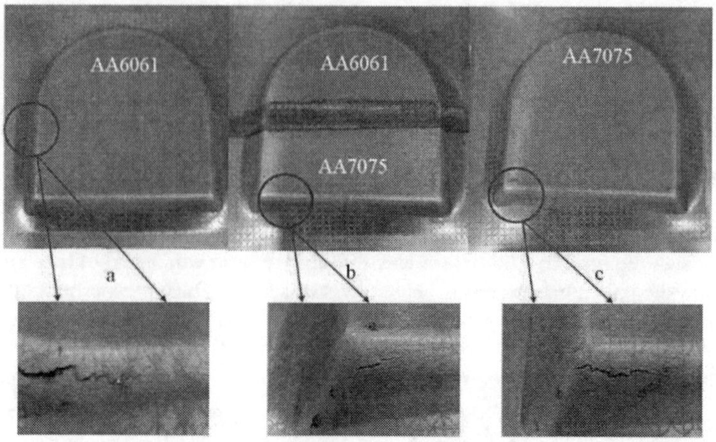

Figure 5. The deformed cups obtained after SPIF process using a hemispherical tool with 10 mm diameter along with the failure location

For hemispherical tool two different tool diameter were considered; φ5 mm and φ10 mm. Flat bottom tool was of φ5 mm with corner radius of 1 mm. The tool path was designed and codified using Mastercam software which were later incorporated in the CNC system through memory card. In this study, a complex geometry (figure 3) developed by combining pyramid and conical shape was selected to be the deformed shape to understand the formability of welded blanks under more complex strain state with flat and curvilinear surfaces. The complete SPIF process was performed using a 4-axis CNC system (make: AMS V540). For the SPIF process, 4500 rpm was used as the tool rotation speed, feed rate was selected as 1000 mm/min with a step depth of 0.1 mm. The fixture, tools, and the complete experimental set up can be seen in figure 4. The SPIF process was continued till any appearance of crack in the deforming sheet. During the process, a mirror was placed under the deforming sheet to check appearance of crack. All the blanks were etched with circular grid marks and the grid marked surface was placed opposite to the tool facing surface. Circular grid marks were used to measure the strain of the deformed surfaces. During experimentation with parent metal, it was observed that formability drastically reduced at forming angle of 60°. Therefore, all the experiments were conducted at this angle. Some of the deformed cups after the SPIF process are shown in figure 5. Deformed cups from TWBs did not show any weldline movement.

3. Results and discussions

3.1. Forming height

The forming heights of the deformed cups were measured using a 3D scanner (Make: Artec Eva) after dismounting the deformed samples from the fixture. All the cup heights are reported in Table 2. It can be observed that tool with hemispherical tool tip delivered higher cup height compared to that obtained by the flat bottom tool. In case of AA6061 cups, almost 72% cup height reduction was observed for flat bottom tool compared to the highest cup height that was achieved for the material. Similarly, 71.5% reduction was observed for AA7075 material. The highest cup heights were observed in case of hemispherical punch with 10 mm diameter for both the parent materials. This results corroborated well with previous studies [17]. Application of hemispherical tool with 5 mm diameter resulted in reduction of around 20% cup height for AA661 and 10% cup height for AA7075 compared to the corresponding maximum cup height. In case of the welded blanks, the cup heights were almost similar to the corresponding parent metals, without any significant reduction in cup height due to presence of the weld. In this regard it was also observed that none of the failure in the deformed cups were in the weld or in the vicinity of the weld though failure was observed at the weld zone during tensile testing. Similar to the observation made for the parent metals, the maximum cup heights for the welded blanks were obtained with hemispherical punch with 10 mm diameter and the reduction in cup height due to application of hemispherical tool with 5 mm diameter followed the similar trend as reported for the parent metals. For the case of TWB, the cup height was observed to be in between the cup heights reported for the combining parent metals whenever hemispherical tools were used.

Table 2: Comparison of experimental cup heights obtained after SPIF process

Material	Tool Geometry	Tool Diameter (mm)	Forming Angle	Fracture Height (mm)
AA6061	Hemispherical	10	60°	11.075
	Hemispherical	5	60°	8.875
	Flat Bottom	5	60°	3.075
AA7075	Hemispherical	10	60°	12.575
	Hemispherical	5	60°	11.275
	Flat Bottom	5	60°	3.575
AA6061+AA6061 (Welded blanks)	Hemispherical	10	60°	11.375
	Hemispherical	5	60°	8.575
	Flat Bottom	5	60°	2.875
AA7075+AA7075 (Welded blanks)	Hemispherical	10	60°	11.875
	Hemispherical	5	60°	10.575
	Flat Bottom	5	60°	3.475
AA6061+AA7075 (TWB)	Hemispherical	10	60°	11.575
	Hemispherical	5	60°	9.575
	Flat Bottom	5	60°	3.075

3.2. Force analysis

During the SPIF experimentation, the force measurements were performed employing a dynamometer for comparing the force requirement with different tool shapes and sizes. The obtained Z-direction force data with all the three types of tools are plotted for both the parent metals in figure 6a, b. For the TWB, force data are plotted for the case of 5 mm hemispherical tool (figure 6c). It can be delineated from the plots that force required for the deformation increased when hemispherical tool diameter increased from 5 mm to 10 mm. This can be related to the localized area under the tool which is being deformed. In case of 10 mm tool diameter, the localized deformation zone was higher compared to that of the 5 mm tool diameter. Therefore, higher force was recorded with the 10 mm diameter tool. The lowest force was recorded for the flat bottom tool. The deformation region was quite small in this case as the corner diameter of the flat bottom tool was 1 mm. It can also be observed that the force requirement for AA6061 was relatively higher compared to that of AA7075. This can be correlated to the corresponding material strength reported in section 2. In case of the TWB, the force data reflected the load requirement for either side of the parent metal for which two distinctive force data is visible. This indicates the force requirement corresponds to that of the AA6061 while the tool travel in the AA6061 side, and vice-versa. Similar observation was also made for hemispherical tool with 10 mm diameter and the flat bottom tool.

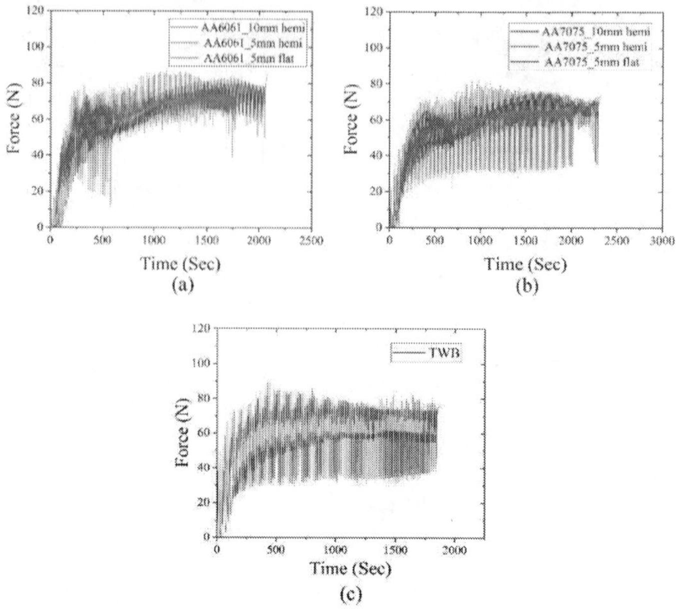

Figure 6. Comparison of forming forces for (a, b) parent metals and (c) TWB

3.3. *Strain analysis*

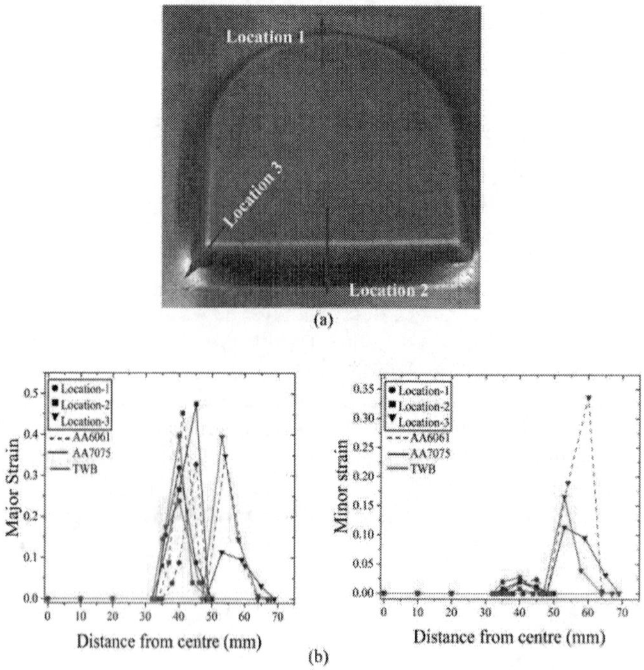

Figure 7. (a) The location and measurement direction for strain, (b) the measured major and minor strain

The complex geometry selected for the deformed shape, in this study, has three distinct regions; three flat surface (part of the pyramid shape), one conical face, and the corner regions. Therefore, the strain measurement was performed at three different regions as can be seen in the figure 7a. Strain values were measured from the cup center to the flange region along three different lines to cover all the distinct region. The measured major and minor logarithmic strain data for the samples deformed with 10 mm hemispherical tool are plotted in figure 7b. It can be inferred from the figure that regions with flat surfaces (location 2) follows a plane strain condition during deformation as minor strain was negligible. However, the strain measured across the conical face (location 1) showed major strain to fall between 0.4-0.5 and the minor strain to be around 0.02, indicating strain state to be in the biaxial stretching zone. Significant rise in both major and minor strain was observed in the corner region (location 3) which indicates a biaxial mode of stretching. The failure location of the various deformed samples as was shown in figure 5 can well be corroborated with the measured strain state. The failure was always observed close to the cup corner which experienced biaxial mode of stretching. The results can be well correlated to the results reported by Basak at al. [18] where it was clearly shown that fracture limiting strain are lower in the biaxial region. For other deformed samples, which were deformed using the 5 mm diameter hemispherical tool and the flat bottom tool, fracture was observed to initiate at the corner and then propagated to the flat surfaces. In case of the TWB, no strain differences across the weld were observed in the flat un-deformed portion and the weld remained at initial position. From this it can be inferred that localized mode of deformation did not affect the complete blank in SPIF process and the weldline shift was not observed. However, in the case of conventional forming, complete TWB deforms simultaneously in eccentric manner due to strength difference and weldline shifts.

4. Summary

In this study, TWB fabricated with 1 mm thick AA6061 and AA7075 were successfully deformed using SPIF process. From this study following conclusions can be drawn

1. With selection of proper parameters, and proper fixture design, sound friction stir welded blanks can be obtained even with sheet thickness of 1 mm. A weld efficiency of more than 60% was achieved in the study which can be further enhanced with parametric optimisation.

2. In case of SPIF process the tool geometry, and tool size have significant influence on force requirement during deformation. It was observed force requirement was more for hemispherical tool tip with 10 mm diameter compared to 5 mm diameter. Force requirement for deformation was lowest with flat end tools. For TWB two distinct force curve were observed which corresponds to the force related to the component blanks.

3. Cup heights for the similar welded blanks were similar to the corresponding parent metals and no failure was observed at the weld zone. Hence, presence of friction stir weld did not affect the cup height. However, cup height for the TWB was in between that of the component blanks.

4. In case of SPIF process of TWB, no weld line movement was observed due to localised nature of deformation. Hence, no complex set up development is necessary for weld movement restriction like conventional sheet metal forming processes.

References

[1] Silva MB, Skjoedt M, Martins PAF, Bay N. *Int J Mach Tools Manuf.* 2008;**48**(1):73–83.

[2] Gatea S, Ou H, McCartney G. *Int J Adv Manuf Technol.* 2016;**87**(1–4):479–99.

[3] Emmens WC, Sebastiani G, van den Boogaard AH. *J Mater Process Technol.* 2010;**210**(8):981–97.

[4] Duflou JR, Habraken A-M, Cao J, Malhotra R, Bambach M, Adams D, et al. *Int J Mater Form.* 2018;**11**(6):743–73.

[5] Verbert J, Belkassem B, Henrad C, Habraken AM, Gu J, Sol H, et al. *Int J Mater Form.* 2008;**1**(1):1203–6.

[6] Ambrogio G, Filice L, Manco GL. *CIRP Annals* 2008;**57**:257–60.

[7] Adams D, Jeswiet J. *Proc Inst Mech Eng Part B J Eng Manuf.* 2014;**228**(7):757–64.

[8] Merklein M, Johannes M, Lechner M, Kuppert A. *J Mater Process Technol.* 2014;**214**(2):151–64.

[9] Panda SK, Hernandez VHB, Kuntz ML, Zhou Y. *Metallurgical and Materials Transactions A.* 2009;**40**(8):1955-67.

[10] Bandyopadhyay K, Panda SK, Saha P, Padmanabham G. *J Mater Process Technol.* 2015;**217**.

[11] Kumar A, Gautam V, Kumar DR. *J Manuf Sci Eng.* 2018;**140**(3)

[12] Ahmetoglu MA, Brouwers D, Shulkin L, Taupin L, Kinzel GL, Altan T. *J Mater Process Technol.* 1995;**53**(3–4):684–94.

[13] Kinsey B, Liu Z, Cao J. *J Mater Process Technol.* 2000;**99**(1–3):145–53.

[14] Mennecart T, Güner A, Khalifa N Ben, Tekkaya AE. *Key engineering materials* 2014; **611**;955-962.

[15] Silva MB, Skjoedt M, Vilaça P, Bay N, Martins PAF. *J Mater Process Technol.* 2009;**209**(2):811–20.

[16] Tayebi P, Fazli A, Asadi P, Soltanpour M. *CIRP J Manuf Sci Technol.* 2019;**25**:50–68.

[17] Kim YH, Park JJ. *J Mater Process Technol.* 2002;**131**:42–6.

[18] Basak S, Prasad KS, Sidpara AM, Panda SK. *Int J Mater Form.*. 2019; **12**(4):623-42

IOP Conf. Series: Materials Science and Engineering 1238 (2022) 012047 doi:10.1088/1757-899X/1238/1/012047

Deep drawing of DC06 at high strain rates

M Tulke[1*], E Galiev[2], V Psyk[2], V Kräusel[2] and A Brosius[1]

[1] Technische Universität Dresden, Chair of Forming and Machining Processes, 01062 Dresden, Germany

[2] Fraunhofer Institute of Machine Tools and Forming Technology, 09126 Chemnitz, Germany

*marc.tulke@tu-dresden.de

Abstract. Velocity effects in production technology provide technological and economic advantages, but they require consideration of strain rate dependent material behaviour in numerical process modelling. This paper proposes a deep drawing process with DC06 at high strain rates of up to 100 s^{-1}. Basis of the investigation are the results of the inverse parameter investigation at high strain rates in [1]. In addition to the velocity dependent flow curve in numerical modelling, the damage parameters are also taken into account. The drawing ratio and the punch velocity that affects the strain rate and the stress state in the deep drawing process are varied while the blank holder force is constant. Due to the high punch speed and the pronounced inhomogeneity of the strain rate over the work piece volume, the punch speed and punch force is very well suited to verify the numerical results. Additionally, the mesh dependency of the temperature evolution in the sheet metal is examined for further investigations of the adiabatic heating of the work piece during deformation in experiment and simulation. In this context, the deep drawing tools as well as the results of the numerical and experimental investigations are presented and discussed.

1. Introduction

The use of speed effects in production engineering enables the exploitation of technological and economic advantages [2], which also applies to high speed forming and high-speed cutting processes [3]. High-speed effects enable manufacturing with higher quality and without additional finishing in contrast to conventional processes [4]. In order to describe and model high-speed processes in detail, suitable data on the flow curve and failure behaviour for different strain rates that occur are of central importance, since materials subjected to high-speed forming often exhibit extended forming limits [5]. Furthermore, it was shown in [6] that the coefficient of friction decreases with increasing velocity in strip tensile tests with minimum quantity lubrication. The same result was obtained in an analysis of the influence of punch speed on the process window in micro deep drawing [7] by demonstrating the reduction of the friction coefficient with increasing punch speed.

The challenge of material characterisation at high velocities is the accurate measurement of the material parameters due to the high testing speeds, small specimen size and limited accessibility of the sample. An alternative approach to direct measurement of stress and strain for determining material properties at high strain rates is an inverse parameter identification. In [1] Psyk et al. successfully developed an approach using an accelerated hammer in order to implement high speed tensile and shear tests. They used a pneumatic acceleration to achieve hammer speeds of up to 8 m/s and corresponding strain rates in the test specimen of up to 10^3 s^{-1} and an electromagnetic acceleration for realizing even higher hammer speeds (up to 30 m/s) and corresponding strain rates of up to 10^5 s^{-1} in the specimen. The method is based on measuring auxiliary test parameters and using them as input data for the parameter identification in an inverse simulation. More recently, the identified material data was refined and supplemented with values considering more moderate strain rates (10^{-1} s^{-1} to 10^2 s^{-1}) [8].

In this study, the pneumatic test setup is modified in such way, that it can be used for investigating deep drawing at high strain rates. In this context, the punch acceleration and the punch force are measured in order to characterize the process. The research results presented in [1] and [8] will be used to model the process numerically. This will provide deeper knowledge of high-speed deep drawing and enable a more comprehensive verification of the determined material parameters for strain rates in the range of up to $\dot{\varepsilon} = 100$ s^{-1}.

2. Experimental setup

The pneumatically driven accelerator based on the principle idea suggested in [9] enables high-speed deep drawing on a conventional press. It is presented in Figure 1 together with the deep drawing tools and the measuring sensors for the punch force and displacement. The main structure of the pneumatic device is taken from [10] and has been supplemented by the deep drawing tools. As detailed in [10] the piston of the accelerator is first filled with compressed air and when the press pushes the piston downwards, the air in the piston and the hammer chamber is compressed. While the compressed air in the hammer chamber is acting on the hammer flange, the hammer moves upwards into its starting position. In the further course of the stroke the highly compressed air in the piston streams into the upper part of the hammer chamber through the space between hammer cover and lifting rod. Consequently, the hammer moves downwards and when it passes the holes in the hammer cover, highly compressed air rushes into the hammer chamber and accelerates the hammer up to 6 m/s. In contrast to the testing device for the material characterisation, the hammer in the accelerator is equipped with a punch and the piezo load cell (9041A, KISTLER) is mounted between the punch and hammer for measuring the punch force during the deep drawing process. The punch acceleration is measured by an accelerometer sensor (8044, KISTLER) with a measurable acceleration up to 30,000 m/s². Subsequently, the punch displacement is calculated by integrating the acceleration signal of the punch twice. The investigated material is a deep drawing steel DC06 (1.0873) with a sheet thickness of $t = 1$ mm that has a significant strain rate dependency and is widely used in industry. The punch diameter is $d_{punch} = 40$ mm, the punch profile radius amounts $r_{profil} = 5$ mm, the inner die radius is $d_i = 42$ mm and the die edge radius is $r_{die} = 5$ mm. Depending on the initial blank diameter of $D = 52$ mm up to $D = 68$ mm, different drawing ratios from $\beta = 1.3$ up to $\beta = 1.7$ are investigated. The distance between blank holder and die is 1.15 mm for realising a distanced deep drawing.

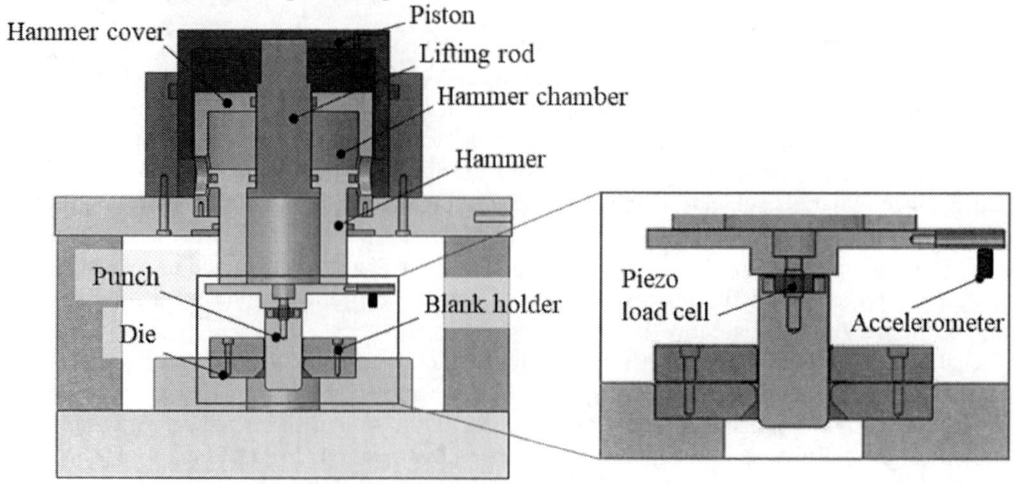

Figure 1. Test setup for deep drawing with high strain rates according to [9]

3. Numerical investigation

3.1. High-speed tensile test

As already mentioned, the material characterisation from [1] and [8] provides the basis for the numerical investigations. Here, the determined material properties are initially verified by tensile tests at a strain rate $\dot{\varepsilon} = 50\ s^{-1}$. For this purpose, the simulation with a quasi-static flow curve for a speed of $\dot{\varepsilon} = 0.001\ s^{-1}$ and the simulation with strain rate dependent flow curves are each compared to the results of experimental tensile tests are carried out on the high-velocity testing machine ZWICK HTM 16020. The specimen geometry and the strain rate-dependent flow curves for strain rates $\dot{\varepsilon} = 0.001\ s^{-1}$ up to $\dot{\varepsilon} = 180\ s^{-1}$ are summarised in Figure 2. Flow curves for higher strain rates are neglected due to the maximum considered strain rate of $\dot{\varepsilon} = 100\ s^{-1}$ in the presented investigations.

The finite element model of the tensile test is built in LS-Dyna R12.0.0. The blank is meshed with fully integrated solid elements (ELFORM 2) with an initial element size of 0.5 mm x 0.5 mm and 0.2 mm in the sheet thickness direction without remeshing. The material behaviour is defined by material type MAT24 (piecewise linear plasticity) according to the flow curves in Figure 2. The damage description is based on the GISSMO model using MAT_ADD_EROSION and is defined by the instability strain ε_{crit} and failure strain ε_{fail} versus triaxiality, shown in Figure 2 and the damage exponent DMGEXP = 2.0 and the fading exponent FADEXP = 2.0.

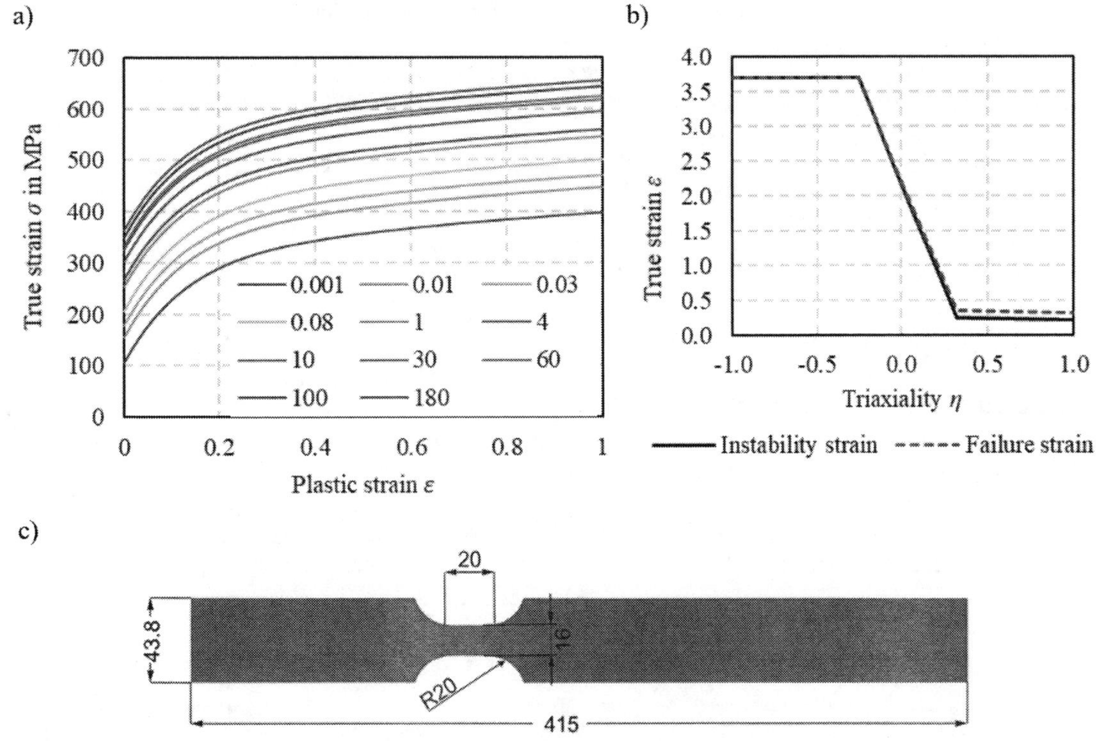

Figure 2. Input parameters for numerical tensile test: Strain rate dependent flow curves (a), instability and failure strain for GISSMO-model (b) and geometry of the specimen (c) according to [8]

3.2. High-speed deep drawing

The FE-model of the deep drawing process is also built in LS Dyna R12.0.0 and the blank is meshed with fully integrated 3D shell elements (ELFORM -16) with nine integration points in sheet thickness, because in the case of deep drawing, through-thickness stresses are negligible. The initial element size amounts 0.2 mm with adaptive refinement. The minimal allowed element size is 0.05 mm. The calculation was performed using an explicit model with a mass scaled solution. The FE-model of the rotational symmetrical cup is reduced to a quarter for saving resources and time in calculation. The deep drawing tools punch, blank holder and die are defined as rigid bodies and the friction is modelled by a Coulomb friction model with a friction coefficient of $\mu = 0.1$. In addition to the velocity dependent material parameters, adiabatic heating during the forming process is also taken into account. Similar to the simulation of the tensile test, the material behaviour in the deep drawing simulation is also defined by the flow curves presented in Figure 2 combined with the GISSMO model using MAT_ADD_EROSION.

3.3. Mesh sensitivity of temperature

For future consideration of the temperature development during forming of the cup, the influence of the element size on the temperature development is investigated. For this purpose, a square specimen with element sizes of 0.05 mm, 0.01 mm and 0.005 mm is modelled. The specimen is generally assigned room temperature $T_{RT} = 295$ K, with a small defined region being assigned an elevated starting temperature $T_i = 495$ K. The volume that is assigned the elevated temperature T_i is the same for all three simulations. Here, a thermal-only calculation is performed to determine the mesh sensitivity dependency of the temperature independent of the forming process. The temperature distribution is calculated via material model T03 (THERMAL ISOTROPIC TD) based on the temperature depending parameters listed in Table 1.

Table 1. Parameters for thermal modelling

Temperature T in K	293	373	473	573
Heat capacity c in J/(kg · K)	460	480	510	550
Thermal conductivity λ in W/(m·K)	57.5	55.7	43.4	39.6

4. Results and discussion

4.1. High-speed tensile test

In order to validate the material properties, the engineering stress-strain curve determined via numerical simulation of the tensile test was compared to corresponding measurements. This verification clearly shows the necessity of considering the strain rate dependency of the flow curve in numerical investigations, see Figure 3. While the simulation with the strain rate dependent material model shows good agreement with the experimental result, the engineering stress in the simulation with the quasi-static flow curve is clearly underestimated and does not correspond to experimental results. It is noticeable that the oscillations at the beginning of the stress curve are more pronounced in the numerical results than in the experiment, although here, the simulation with strain rate dependency reproduces the experimental results better. The oscillation is due to the impact as a result of the high velocity

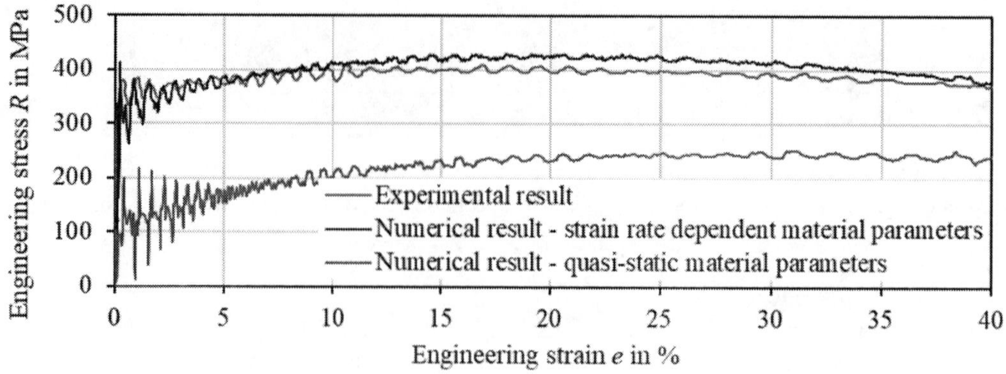

Figure 3. Results of experimental and numerical tensile test with a nominal strain rate of 50 s^{-1}

The strain in the experimental setup is optically recorded by an extensometer Rudolf 200XR. Therefore, the front side of the specimen is primed white and two sharply drawn black lines define the measuring area. In numerical simulation, the strain is measured by a virtual extensometer that is modelled with two lines similar to the measuring principle of the experimental extensometer.

4.2. High-speed deep drawing

Based on the knowledge of the verification the strain rate dependent material model is used for numerical simulation of the deep drawing tests with punch velocities up to $v_p = 9$ m/s, but for the sake of completeness, also in this case additional comparative simulations with quasi-static material parameters were carried out. Here the punch force and punch displacement are evaluated as relevant process variables, since these can be measured in the experiments. Similar to the tensile test, the numerical results differ depending on punch velocity and resultant strain rate. Figure 4 shows exemplarily the required experimentally and numerically determined punch forces for different drawing ratios $\beta = 1.3$ and $\beta = 1.5$ at a fixed punch velocity of $v_p = 4$ m/s. Due to the larger blank diameter $D_0 = 60$ mm for the drawing ratio $\beta = 1.5$, there is a larger punch displacement, and due to the larger forming, the punch force is also significantly larger than for the drawing ratio beta=1.3. The resultant cup height is 13.8 mm for a drawing $\beta = 1.5$ and in case of $\beta = 1.3$ the cup height amounts 7.7 mm.

There is a good agreement between the experimentally determined force curve and the numerically determined force curve with strain rate-dependent material modelling. In contrast, the required punch force is significantly underestimated by up to 50 % in the simulation with quasi-static material modelling due to the lower flow stress at low strain rates. The cell data was smoothed by a moving average with a smoothing width of seven data points.

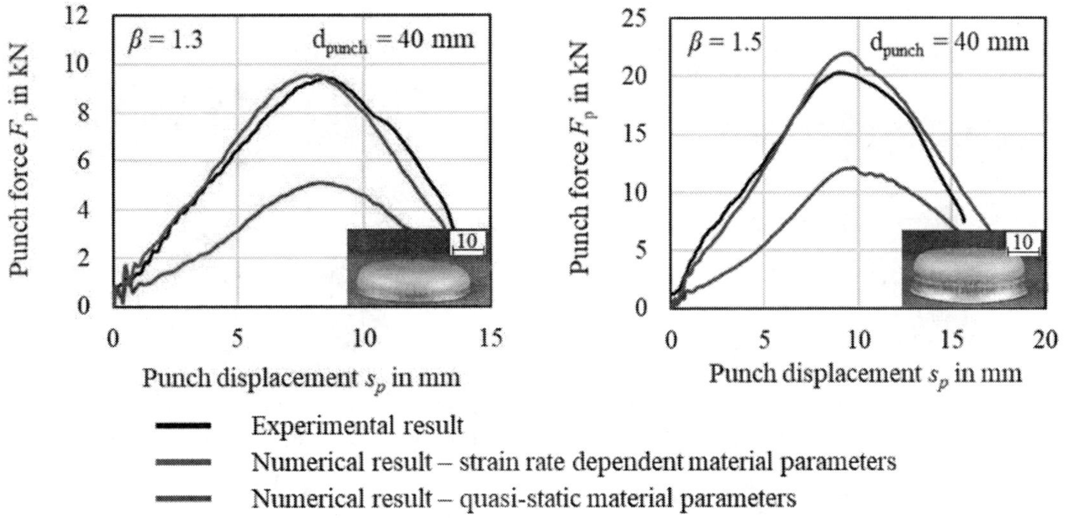

Figure 4. Experimentally and numerically determined punch force for drawing ratios of $\beta = 1.3$ (left) and $\beta = 1.5$ (right)

Numerical results show that an average strain rate $\dot{\varepsilon}_{average} = 37$ s^{-1} is obtained, with extreme strain rate values between $\dot{\varepsilon}_{min} = 20$ s^{-1} and $\dot{\varepsilon}_{max} = 70$ s^{-1} for a punch velocity of $v_p = 4$ m/s. The distribution and development of the strain rate for three different states during the deep drawing process for $\beta = =1.5$ is shown in Figure 5. Based on the different process stages, it can be seen that the areas with the highest strain rate shift according to the forming zone in the sheet and that the strain rate varies locally within a process stage.

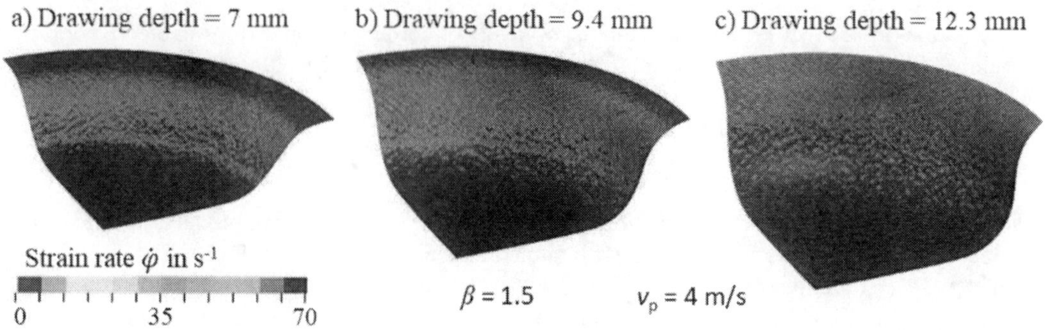

Figure 5. Strain rate distributions at different states of the deep drawing process for a punch velocity of $v_p = 4$ m/s and a drawing ration $\beta = =1.5$.

While at punch velocities up to $v_p = 7$ m/s only the punch force varies as a function of the strain rate, the simulation with a punch velocity of $v_p = 9$ m/s that results in strain rates up to 100 s^{-1}, also shows a change in the damage of the material. If the strain rate dependence of the flow curve is not taken into account at these high velocities, the simulation shows a material failure, depicted in Figure 6, which is caused by the impact of the punch at very high velocity. In contrast, a cup is completely drawn, if the material is modelled in a strain rate dependent mode. This shows very clearly that for technological forming processes with high process speeds, it is very crucial to determine the material properties for high strain rates so that material behaviour and failure are modelled realistically and a process simulation enables realistic predictions.

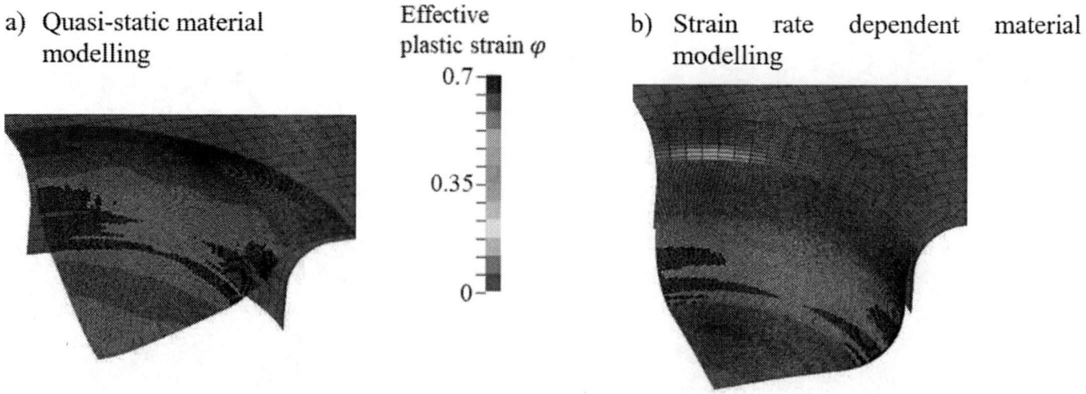

Figure 6. Numerically determined material behaviour at high strain rates due to different material modelling: a) Quasi-static material modelling and b) strain rate dependent material modelling

4.3. Mesh sensitivity of temperature

The results of the sensitivity study of the temperature regarding the mesh size shows a significant difference in the temperature development. The temperature evolution for the investigated element sizes is shown in Figure 7. The final temperature of coarse meshing differs from the final temperature of fine meshing. Already after the short time of 0.6 ms the temperature difference amounts 36.1 K between the finest mesh and the coarse mesh with an element size of 0.05 mm. As a large part of the mechanical work is converted into heat during forming with high strain rates and thus high temperatures are reached in the work piece, a significant influence of the mesh size on the temperature calculation is to be expected.

If the temperature is predicted incorrectly, this can have a negative effect on the temperature depending flow stress and thus on the calculated process prediction. Furthermore, it has been shown that the mesh dependency is more pronounced for irregular meshes, which usually cannot be avoided along curved edges, as in the investigated specimen in Figure 7. Comparison tests with regular rectangular elements with element lengths of 0.05 mm and 0.005 mm show only a temperature difference $\Delta T = 24$ K between both cases compared to irregular meshes.

Accordingly, fine meshing in the forming area is necessary to realistically represent the development and distribution of the forming heat during the process. Consequently, fine elements are necessary in numerical investigations with high strain rates and temperature dependency. In the case of deep drawing with high strain rates, a sub-modelling with fine meshing of selected areas is conceivable in order to save calculation time and resources.

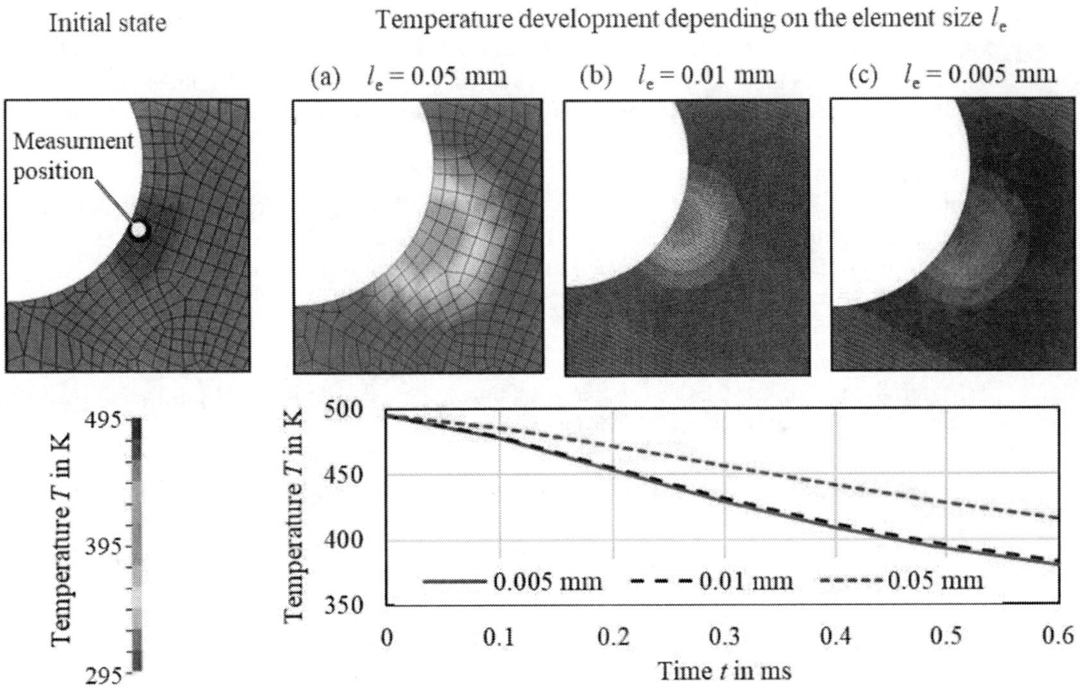

Figure 7. Influence of the element size on the temperature development for an element length $l_e = 0.05$ mm (a), $l_e = 0.01$ mm (b) and $l_e = 0.005$ mm (c).

For the presented numerical investigations on deep drawing, a maximum temperature $T = 395$ K within the cup is obtained at an average strain rate of 100 $s^{-1}$and a drawing ratio $\beta = 2.0$. The course of the temperature in the cup is shown in Figure 8. However, the element size of 0.5 mm in the modeled cup is comparatively large with respect to the results on the mesh sensitivity of the temperature. In future investigations, suitable modeling needs to be developed in order to correctly represent the temperature in the cup and, at the same time, to keep the computational time low.

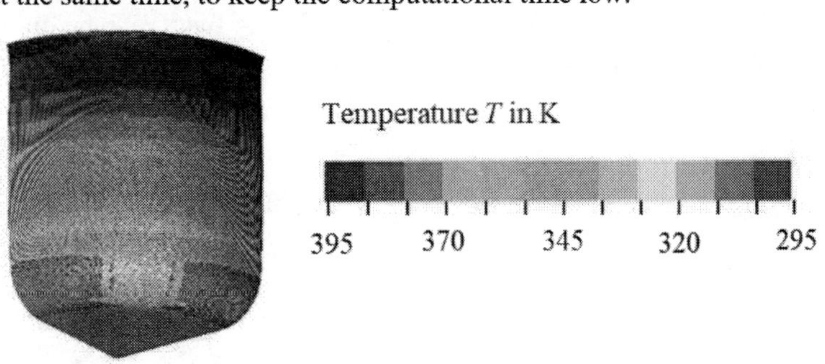

Figure 8. Numerical determined temperature development in deep drawing

5. Conclusion and outlook

The investigations show an approach for deep drawing of DC06 with maximum strain rates up to 100 s^{-1} in numerical studies. The material parameters determined in earlier investigations [1] were verified in experimental tests with a local maximum strain rate of $\dot{\varepsilon}_{max}$ = 70 s^{-1}. Due to the time and position dependant distribution of the strain rate during the deep drawing process, the average strain rate in the forming areas of the deep drawn cups amounts to $\dot{\varepsilon}_{average}$ = 37 s^{-1}. It was shown that the determination of the strain rate dependent flow curve is essential for realistic numerical results. In comparative calculations with quasi-static material parameters, the punch force was underestimated by 50 % compared to the experimental result.

Further investigations concern the integration of temperature measurement and increasing the punch velocity up to v_p = 10 m/s in the experimental setup to verify the numerically determined material damage at strain rates higher above 100 s^{-1} and to realize a drawing ratio up to β = 2.0. In addition, the modeling of the cup needs to be adapted in such a way that the temperature development can be correctly predicted taking into account the mesh sensitivity of the temperature with a suitable computation time.

6. Acknowledgements

The presented work was performed within the project "Material- and failure characteristics for high-speed forming and cutting", which is funded by the German Research Foundation (Deutsche Forschungsgemeinschaft, DFG) under reference number PS 68/2-3 and BR 3500/19-3.

References

[1] Psyk V, Scheffler C, Tulke M, Winter S, Guilleaume C and Brosius A. 2020, Determination of Material and Failure Characteristics for High-Speed Forming via High-Speed Testing and Inverse Numerical Simulation, JMMP 2020, **4(2)**. 31.

[2] Neugebauer R, Bouzakis K-D, Denkena B, Klocke F, Sterzing A, Tekkaya A E and Wertheim, R. 2011, Velocity effects in metal forming and machining processes. CIRP Annals – Manufacturing Technology, **60**, S. 627-650

[3] Barthel T and Drossel W-G. 2014, Energieeffizienz beim Scherschneiden. Internationale Konferenz "Neuere Entwicklungen in der Blechumformung", 283-292

[4] Davies R., Austin ER. 1970, Development in high speed metal forming, New York: Industrial Press Inc

[5] Psyk V, Kurka P, Kimme S, Werner M., Landgrebe D, Ebert A and Schwarzendahl M. 2015, Structuring by electromagnetic forming and by forming with an elastomer punch as a tool for component optimisation regarding mechanical stiffness and acoustic performance. Manufacturing Rev. **2.23**

[6] Mehic B, Engler S and Huskic A. 2015, Entwicklung einer Streifenziehversuchsanlage zur Untersuchung des Reib- und Verschleißverhaltens von tribologisch beanspruchten Oberflächen, Forschungsforum Österreichischer Fachhochschulen **9**

[7] Vollertsen F and Hu Z, 2010 Analysis of punch velocity dependent process window in mirco deep drawing, Production Engineering **4**, 553-559

[8] Galiev E, Winter S, Reuther F, Psyk V, Tulke M, Brosius A and Kräusel V. 2022, Applied Science. MDPI, Local temperature development in the fracture zone during uniaxial tensile testing at high strain rate experimental and numerical investigations, (submitted)

[9] Yanagihara N, Saito H, Nakagawa T. Proc. 1980 Conf. 21st International Machine Tool Design and Research **21,** 123-128

[10] Tulke M, Scheffler C, Psyk V, Landgrebe D, Brosius A. 2017, Procedia Engineering, Conf. ICTP **207**, 2000-2005

IOP Conf. Series: Materials Science and Engineering 1238 (2022) 012048 doi:10.1088/1757-899X/1238/1/012048

Post necking evaluation of the tensile test using artificial neural networks

F Hakenbeck, R Lafarge and A Brosius

felix.hakenbeck@tu-dresden.de

remi.lafarge@tu-dresden.de, alexander.brosius@tu-dresden.de, ff-if@mailbox.tu-dresden.de

Abstract. This paper introduces a new method for evaluating the tensile test using finite element simulations and data-driven artificial neural networks. For this purpose, a synthetic data set was generated by finite element simulations using LS-DYNA. Artificial neural networks of two different topologies were trained and tested on parts of this synthetic data set. The networks use geometry information of the necking area as input data to predict a correction factor to convert the stress obtained from the tensile test to the equivalent flow stress. The best models are evaluated on the test set and good results are achieved.

1. Strategies for evaluating the tensile test

The flow properties of a given material can be represented by means of a flow curve. A flow curve shows the relationship of the stress at which the material begins or continues to flow versus the degree of deformation. One method to determine the flow curve is the uniaxial tensile test, an inexpensive and relatively simple method. Up to the beginning of specimen necking, the stress is uniaxial. Afterwards the stress state becomes multiaxial. If the stress state is uniaxial the flow curve can be determined using the stress-strain curve. After the initiation of necking, the evaluation becomes much more complex and inaccurate. For many materials, the beginning of the necking begins at relatively small degrees of deformation. However, degrees of deformation for industrial forming processes are significantly greater. For this reason, the stress-strain curve is extrapolated with different methods which results in inaccurate stress-strain curves.

Tu et al. divide strategies for evaluating the uniaxial tensile test into three different classes: analytical methods, iterative experimental-numerical methods and inverse methods [1] which will only be discussed briefly here. For further information [1] can be reviewed.

Analytical methods determine a correction factor to convert the true stress into a uniaxial equivalent stress. These methods are easy to use and quite simple, however compared to other methods not as accurate [1] when the existing stress state diverges from the uniaxial assumption. One analytical method is the formula introduced by Bridgman [2]. A more recent method is introduced by Paul et al. which aims to use strain data measured with digital image correlation (DIC) to correct the local stress and strain states in the necking area [3].

Finite element method simulations are conducted for iterative experimental-numerical methods. Furthermore, real tensile test experiments are carried out. Afterwards the parameters of the simulation are iteratively adjusted to minimize the deviation between the results of the simulation and the experimental test. Iterative experimental-numerical methods can produce very accurate results. However, the better accuracy comes with greater complexity as well as a higher demand on resources and expertise [1].

Dunand et al. use an iterative procedure to adjust local deformations [4]. Marth et al. use a full-field method to characterize strain hardening parameters even after the localized constriction has developed [5].

The third category encompasses the inverse methods. For these methods a tensile specimen with a predefined equivalent stress-strain curve is modelled. Afterwards the equivalent strain after diffuse necking is characterized. The relationship between the true stress from numerical modelling and the equivalent stress at the equivalent strain is investigated, which is afterwards put into a formula used to correct the stress [1].

Schneider et al. study the stress distribution in the necking using FE simulation and an alternative correction formula for rectangular cross-section specimens is developed [6]. Coppieters et al. use the balance between internal and external work and an optimization based on the Levenberg-Marquardt method to estimate the parameter values of an associated hardening law [7].

The existing methods for evaluating the tensile test after necking are either easy to apply but less accurate or complex to apply with higher possible accuracy. Currently, there seems to be no solution for a method that is as simple to use and fast as well as accurate at the same time.

The method presented in this paper is intended to fill the gap by using a data driven approach. Such approaches, including artificial intelligence, enable finding patterns in large amounts of data. As a result, less expertise regarding the evaluation of tensile tests is required. However, the successful application of data-driven methods essentially depends on the quality and quantity of a representative database. Furthermore, because simulations are used to generate the given database for the case at hand, it is necessary to validate methods developed solely on synthetic data on data from real experiments such as the tensile test or bulge test [1].

2. Artificial neural networks

For this paper two of the commonly used topologies for artificial neural networks, multilayer perceptrons (MLP) and convolutional neural networks (CNN), were used to train models for the prediction of a correction factor to accurately determine the flow stress after necking.

MLPs are the default type of artificial neural network consisting of multiple fully connected layers which results in perceptrons in each layer being fully connected to all perceptrons in the previous and following layer. Therefore, the amount of parameters in a MLP model grows exponentially with the amount of neurons in a model. A high amount of neurons in a model is necessary if the patterns hidden in the data are complex [8].

The other investigated topology are convolutional neural networks. Neurons of resulting models can consider data in a specified window [9]. For this reason, neurons are only sparsely connected which results in less complex models with similar performance compared to MLP models.

Jeong et al. train an artificial neural network on data from finite element simulations which is then used to convert indentation load-depth curves into hardening parameters required to describe uniaxial tensile flow [10]. Jeník et al. determine hardening curves using an artificial neural network. However, this solution does not take the influence of geometry on the tensile behavior of the tested material into account. Therefore the method is limited to materials with no strongly pronounced strain hardening and anisotropy [11].

3. New method for stress analysis

The aim of this paper is to use Artificial Intelligence and geometry information of a specimen during the tensile test to predict the quotient α_{corr} of flow stress σ_y and the true stress in tensile direction σ_{true}.

Flow stress and true stress are frequently used expressions. For every timestep of every simulation this factor is calculated independently.

$$\alpha_{corr} = \frac{\sigma_y}{\sigma_{true}} \qquad (1)$$

The multiplication of this correction factor with the stress obtained from a tensile test or simulation the flow stress can be calculated. The task at hand is a regression for which the dataset must contain the input as well as the output data. The factor to be predicted contains information that cannot be extracted from experimental tensile tests, but from finite element simulations. For this reason, simulations of the uniaxial tensile test are carried out and afterwards the geometry information used as input and the correction factor used as output data are extracted. The principal progression of the correction factor over a simulation is shown in figure 1.

The input data for the artificial neural networks are geometrical information of the specimen for each timestep of each simulation. The used geometrical data are the width differences between the current timestep and the first timestep in relation to the distance to the narrowest point of the specimen. The narrowest cross-section of the specimen is identified and using interpolation, 32 points in a given distance from the narrowest cross-section are extracted from the simulations. From each of the points the specimen width at this point is extracted and compared to the width of the same point at the initial state of the simulation.

Within one given simulation, the input data was extracted for every 20th timestep and visualized as shown in figure 2a. Markers of the same color are points extracted from the same timestep. The earlier the timestep the smaller is the width difference compared to the initial state of the simulation. With an increasing time step the necking area forms. Figure 2b shows a timestep of a specimen when the necking already started. The extracted points at that timestep are comparable to the 280th timestep shown in figure 2a.

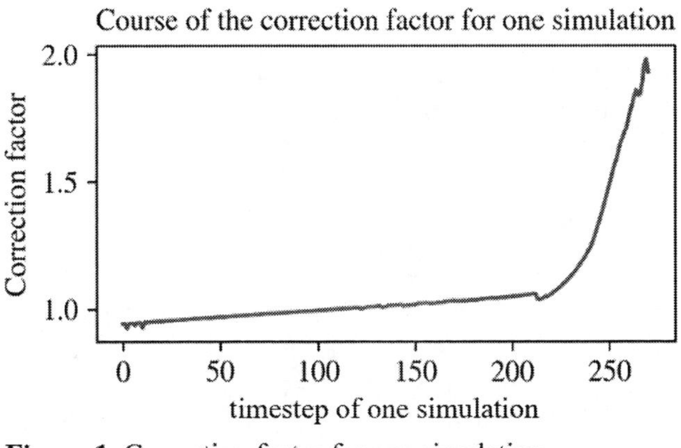

Figure 1. Correction factor for one simulation

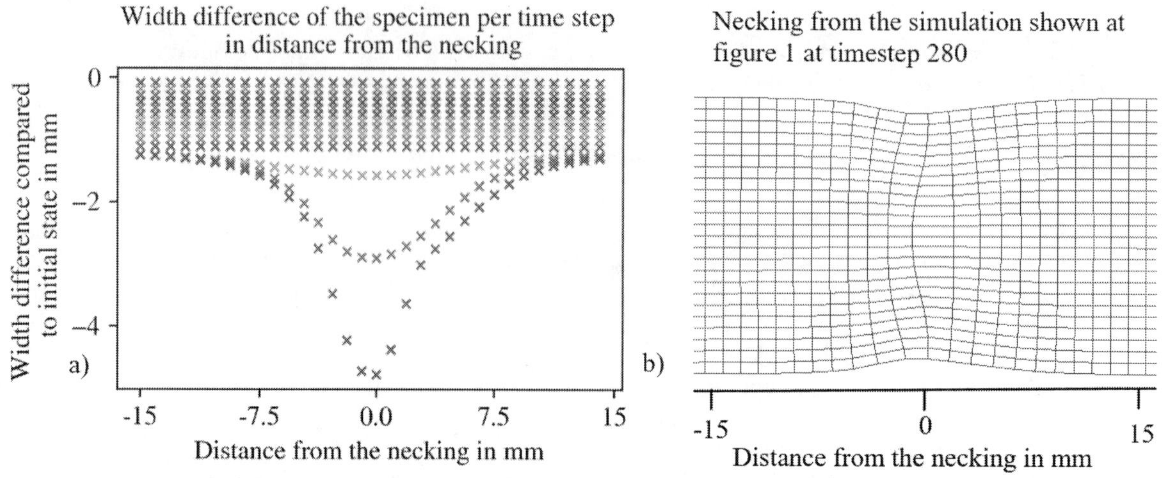

Figure 2. Input data generated from one simulation for every 20th timestep (a) and example of simulated specimen at one timestep (b)

4. Virtual Experiments

4.1. Database

The database used for the training and testing of the artificial neural network models are generated using finite element method simulation. The simulations are carried out using only the Swift hardening law to sufficiently represent the simulated materials. The flow stress can be calculated with the Swift hardening law as shown in formula (2) [12].

$$\sigma_y(\varepsilon_p) = k(\varepsilon_0 + \varepsilon_p)^n \quad (2)$$

σ_y and ε_p are the equivalent stress and strain and k, ε_0 and n are parameters that assume material-dependent values normally determined by uniaxial tensile tests [13]. Furthermore, only isotropic hardening is simulated. A two-dimensional finite element method of a tensile test specimen is created. Afterwards a pretest is conducted to reduce the necessary computing power for resources necessary to create the database is minimized while its accuracy is as high as necessary. The parameters under study for this pretest are the amount of time between two states of the simulation and the amplitude height of the perturbation shell thickness.

The limits of the parameters of the selected hardening laws for steel as shown in formula 2 are derived from literature and adapted to the given task [14][15][16]. The upper and lower parameter limits are shown in table 1. From the resulting parameter space, 500 samples are taken using Latin Hypercube Sampling, a sampling method optimal for numerical experiments. Compared to the random Monte Carlo method, the necessary effort can be reduced by up to 50 % without having to forego the advantages of the method [17].

These samples are used to create the necessary input files for the simulations. The simulations are then carried out and used as a database for training and testing the artificial intelligence models. For this purpose, for every timestep in every simulation the previously mentioned input and output data is extracted automatically using a Python script.

For reproducibility reasons, a random seed was determined to generate the same outputs for functions that use random elements with a pseudo number generator. The data set is divided into a training and a testing set in the ratio of 4:1. Subsequently, the training set is divided into five subsets for cross validation. Four of these subsets are used to form a slightly smaller training set. The last subset is used

IOP Conf. Series: Materials Science and Engineering 1238 (2022) 012048 doi:10.1088/1757-899X/1238/1/012048

as validation set resulting in each model being trained five times on different data. The average performance of each model is calculated and the model with the best average performance of each topology is then trained on the entire training set and afterwards evaluated on the testing set. The metric to evaluate the performance of each model is the mean squared error often used for regression.

Table 1. Swift hardening law parameters

Parameter	Minimum	Maximum
k in MPa	600	1200
ε_0	0.02	0.05
n	0.05	0.3

4.2. Artificial Intelligence

Before the hyperparameter optimization and cross validation are carried out, pretests were conducted to estimate the range of complexity in which the best model of each topology might be situated. Instead of training many models with a wide range of complexity on the entire dataset, few manually chosen models with different complexities were tested to estimate the range of complexity in which the best models might be situated. Once the range was determined many models with their complexity inside the given range were trained and tested. Without these pretests more models need to be trained and tested to get an accurate model. By manually reducing the range of model complexity beforehand, the necessary computing power for solving the task could be reduced.

In addition, various pretests were carried out for the CNN topology to determine a range in which the optimal structure of the identification part is located. The optimal structure for the identification part for the task at hand consist of five one dimensional convolutional layers. In each layer the product of the number of filters and kernel size is the same. From one layer to the next the number of filters is halved and the kernel size is doubled. Afterwards the influence of this identification part on the complexity of the detection part as well as the model's performance is investigated. The identification part of the CNN are the convolutional layers and the detection part is the MLP part of the CNN.

The models are trained on a system with 128 GB RAM, an Intel® e5-2667 v4 and a NVIDIA® Quadro M5000. The training process takes several days. The studied hyperparameters for the MLP and the CNN topology with their researched limits are shown in table 2 and table 3.

Table 2. Studied hyperparameters of the MLP topology

Hyperparameter	Minimum	Maximum
Learning rate	0.00001	0.1
Amount of trainable parameters	351	15,145

Table 3. Studied hyperparameters of the CNN topology

Hyperparameter	Minimum	Maximum
Learning rate	0.0001	0.1
Amount of trainable parameters	2,143	13,444

5. Evaluation

In this section the results of the conducted experiments are presented. Figure 3 shows the deviations between the predictions of the best models for each topology and the true values for the correction factor.

The performance of the best model of each topology is nearly identical. The predictions of the best MLP model have a 5 % relative deviation with a confidence interval of 95 % and the predictions of the best CNN model have a 6 % relative deviation with a confidence interval of 95 %. However, the best MLP model has roughly eight times the number of parameters of the best CNN model. Also the best MLP model is the most complex model possible within the examined hyperparameter space. In contrast the best CNN model is in terms of model complexity in the bottom 20 %.

If the expected yield stress is to be 350 MPa, then there is a 95 % probability that the final corrected stress predicted by the best CNN model will be between 329 MPa and 371 MPa.

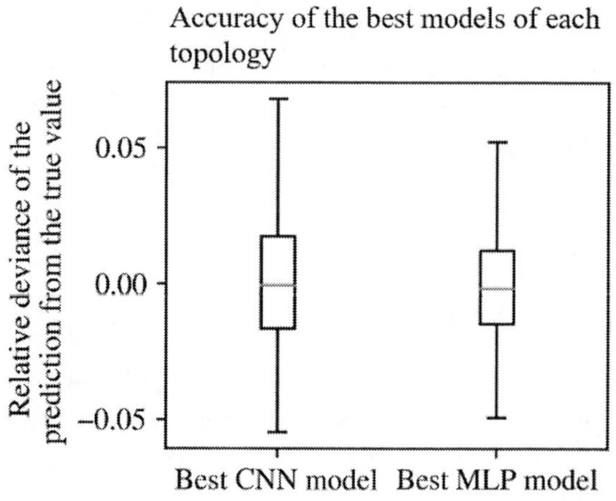

Figure 3. Boxplots of the relative deviance of the prediction of the best models for each topology from the true values

The fact that the best MLP models is the most complex suggests that an even more complex model could achieve better performance. So it stands to reason that since this topology has achieved better results, it should be investigated further.
But even though the best CNN model has a slightly worse performance than the best MLP model, further research should focus on this topology, as similar performance compared to the MLP topology could be achieved with 1/8 of the model complexity.

6. Discussion and further research

The proposed method shows accurate results on the synthetic data generated from finite element simulations. However, the resulting models need to be validated on real experiments like the tensile test or bulge test. Furthermore, the database used to train the models can be made more diverse by using different hardening laws, different approaches to determine the flow curve, different flow curve models and wider ranges of material parameters. This can result in more robust models for stress correction and further validate the accuracy of the method presented in this paper. However, the more diverse database could mean that models have to become more complex to still achieve the same performance.

For this publication only one-dimensional data (of the width of the specimen) were used as input data. Instead, the depth of the specimen could be used in combination with the width, or the cross-sectional area could be used. Using these kinds of input data could result in more robust but more complex models as well.

The deviations in the predictions can be reduced in multiple ways. Each prediction of a model is for one time step in a simulation or a real tensile test. One possibility to reduce the impact of outliers on one prediction is to use a Gaussian filter or another topology that regards the input and output data as time series.

The prototype presented in this paper shows promising accuracies for a possible application on data from real experimental tensile tests and bulge tests.

7. Summary

A new method for stress correction based on artificial intelligence and finite element simulations is proposed. An artificial intelligence model predicts a stress correction factor on the basis of geometric data of the necking area of a tensile specimen. 500 finite element simulations using the Swift hardening law were carried out and used as database for the artificial neural network models. For the network topologies multilayer perceptrons and convolutional neural networks a hyperparameter optimization was carried out. The performance of the best model of each topology was evaluated and compared. The proposed method shows promising results on the synthetic data and is also usable on data from real tensile tests.

References

[1] Tu S, Ren X, He J and Zhang Z 2020, *Stress–strain curves of metallic materials and post-necking strain hardening characterization: A review*, *Fatigue Fract Eng Mater Struct*, **43**, no. 1, pp. 3–19, doi: 10.1111/ffe.13134.

[2] Bridgman P W 1964, *Studies in Large Plastic Flow and Fracture: With Special Emphasis on the Effects of Hydrostatic Pressure*. Accessed: Jan. 20, 2022. [Online]. Available: https://doi.org/10.4159/harvard.9780674731349

[3] Paul S K, Roy S, Sivaprasad S, Bar H N and Tarafder S 2018, *Identification of Post-necking Tensile Stress–Strain Behavior of Steel Sheet: An Experimental Investigation Using Digital Image Correlation Technique*, *J. of Materi Eng and Perform*, **27**, no. 11, pp. 5736–5743, doi: 10.1007/s11665-018-3701-3.

[4] Dunand M and Mohr D 2010, *Hybrid experimental–numerical analysis of basic ductile fracture experiments for sheet metals*, *International Journal of Solids and Structures*, **47**, no. 9, pp. 1130–1143, doi: 10.1016/j.ijsolstr.2009.12.011.

[5] Marth S, Häggblad H-Å, Oldenburg M and Östlund R 2016, *Post necking characterisation for sheet metal materials using full field measurement*, *Journal of Materials Processing Technology*, **238**, pp. 315–324, doi: 10.1016/j.jmatprotec.2016.07.036.

[6] Scheider I, Brocks W and Cornec A 2004, *Procedure for the Determination of True Stress-Strain Curves From Tensile Tests With Rectangular Cross-Section Specimens*, *Journal of Engineering Materials and Technology*, **126**, no. 1, pp. 70–76, doi: 10.1115/1.1633573.

[7] Coppieters S, Cooreman S, Sol H, Van Houtte P and Debruyne D 2011, *Identification of the post-necking hardening behaviour of sheet metal by comparison of the internal and external work in the necking zone*, *Journal of Materials Processing Technology*, **211**, no. 3, pp. 545–552, doi: 10.1016/j.jmatprotec.2010.11.015.

[8] Géron A 2019, *Hands-on machine learning with Scikit-Learn, Keras, and TensorFlow: concepts, tools, and techniques to build intelligent systems*, Second edition. Beijing [China] ; Sebastopol, CA: O'Reilly Media, Inc.

[9] Lecun Y, Haffner P and Bengio Y 2000, *Object Recognition with Gradient-Based Learning*.

[10] Jeong K, Lee H., Kwon O M, J. Jung, Kwon D and H. N. Han 2020, *Prediction of uniaxial tensile flow using finite element-based indentation and optimized artificial neural networks*, *Materials & Design*, **196**, p. 109104, doi: 10.1016/j.matdes.2020.109104.

[11] Jeník I, Kubík P, Šebek F, Hůlka J and J. Petruška 2017, *Sequential simulation and neural network in the stress–strain curve identification over the large strains using tensile test*, *Arch Appl Mech*, **87**, no. 6, pp. 1077–1093, doi: 10.1007/s00419-017-1234-0.

[12] Swift H W 1952, *Plastic instability under plane stress*, *Journal of the Mechanics and Physics of Solids*, **1**, no. 1, pp. 1–18, doi: 10.1016/0022-5096(52)90002-1.

[13] Fernandes J 1998, *A modified swift law for prestrained materials*, *International Journal of Plasticity*, **14**, no. 6, pp. 537–550, doi: 10.1016/S0749-6419(98)00027-8.

[14] Panich S, Barlat F, Uthaisangsuk V, Suranuntchai S and Jirathearanat S 2013, *Experimental and theoretical formability analysis using strain and stress based forming limit diagram for advanced high strength steels*, *Materials & Design*, **51**, pp. 756–766, Oct. 2013, doi: 10.1016/j.matdes.2013.04.080.

[15] Pham Q-T, Lee B-H, Park K-C, and Kim Y-S 2018, *Influence of the post-necking prediction of hardening law on the theoretical forming limit curve of aluminium sheets*, *International Journal of Mechanical Sciences*, **140**, pp. 521–536, doi: 10.1016/j.ijmecsci.2018.02.040.

[16] Reis L C, Prates P A, Oliveira M C, Santos A D and Fernandes J V 2017, *Inverse identification of the Swift law parameters using the bulge test*, *Int J Mater Form*, **10**, no. 4, pp. 493–513, doi: 10.1007/s12289-016-1296-5.

[17] Siebertz K, van Bebber D and Hochkirchen T 2017, *Statistische Versuchsplanung: Design of Experiments (DoE)*, 2. Auflage. Berlin [Heidelberg]: Springer Vieweg, doi: 10.1007/978-3-662-55743-3.

Development of a high-temperature double-layer bulge test for failure prediction in gas-based hot forming of a high-strength aluminium alloy

T Teeuwen[1*], N K Baru[2], K J Tilly[3], D Bailly[4] and G Hirt[5]

[1-5]Institute of Metal Forming (IBF), RWTH Aachen University, Intzestraße 10, 52072 Aachen, Germany

*tobias.teeuwen@ibf.rwth-aachen.de

Abstract. In order to meet the continuously increasing environmental concerns, automotive lightweight concepts of replacing steels with high strength aluminium alloys are one promising solution. Therefore, complex automotive structural components manufactured with new processes like rapid gas-based hot sheet metal forming can become a key factor from a forming technology point of view. A product and process development phase, which is mainly assisted by numerical simulations, mandates knowledge of the material forming limits under the process conditions. The aim of this work is to determine the FLD of a 6xxx precipitation-hardening aluminium alloy within the solution heat treatment temperature range under gas-based hot forming process conditions. For this purpose, a hydraulic double-layered bulge test, as introduced by Banabic, is used as the basis and a high-temperature test setup as well as a characterization methodology are established. FE simulations are utilized to optimize the specimen geometries aiming at specific strain paths. Experiments are conducted with the optimized specimen geometries and the high-temperature FLD is extracted. The FLD is further used in simulations for process parameter identification for failure-free gas-based hot forming of a laboratory scale benchmark component. Experimental validations showed the reliability of the methodology and the determined FLD for failure prediction in the novel forming process.

1. Introduction

Automobile and aerospace industries are experiencing enormous pressure to design the mobility of the future in a resource-efficient and eco-friendly manner. To enhance the automotive lightweight design, hot forming processes like hot stamping have been invented and developed for high strength manganese-boron steels. Nowadays, the high strength lightweight alloys such as aluminium alloys of 6xxx and 7xxx series and titanium alloys that can potentially replace steel from a mechanical point of view are being considered. However, due to their limited formability at room temperature, several new forming processes at elevated temperatures are being invented. Due to their improved ductility at elevated temperatures, relatively complex geometries can be realized, especially by using active media for forming operations instead of/in combination with traditional deep drawing processes. The current study is related to the gas-based hot forming of EN AW-6010-S precipitation hardening aluminium alloy.

The novel forming processes are nowadays studied with the help of Finite Element Analysis (FEA) to enable an effective and efficient process development. In order to efficiently model the limits of material formability and predict the failure within FE simulations, forming limits of the material are input to the simulation. Commonly, Forming Limit Diagram (FLD), which quantitatively describes the sheet metal formability under different strain states is used. For this purpose, the FLD needs to correspond to the process conditions and it should cover the strain states that occur within the component. The gas-based hot forming process currently under investigation is characterized by high temperature and strain rate and the strain paths are unlike punch-based deep drawing processes. Therefore, a novel high temperature testing concept and a setup that represents the strain states and paths as in gas-based forming processes are necessary.

2. State of the art

2.1. Failure prediction during hot forming

Several conventional punch-based and nonconventional active-media-based hot sheet metal forming processes are being developed since the last few decades [1]. For failure prediction during the process layout, a knowledge of the material forming limits is necessary. This can be done with help of formability parameters obtained from qualitative process-specific tests such as limiting drawing ratio from drawing tests [2] or limiting bulge height from a bulge test [3]. Some studies like [4] involved development and calibration of complex damage models based on these tests and their implementation in process simulations to model the failure. Unlike the process-specific tests, the material formability can be quantitatively described with help of a FLD comprising the forming limits under different strain states. FLDs are therefore widely and most commonly used for failure prediction in sheet metal forming. Apart from traditional strain-based FLDs, stress-based FLDs also exist and these are better at prediction of failure especially in active media based forming processes due to their strain path independency [5]. However, mostly strain-based FLDs are used because the stresses cannot be experimentally measured and require additional plasticity model based correlation procedures from strain measurements [6].

2.2. High temperature FLDs and characterization methods

For FLD determination, punch-based Nakajima and Marciniak tests are widely used. However, the high temperature setup with hot tools gets more complicated compared to traditional setups. In some studies like [7] and [8], an elevated temperature Nakajima test setup was realised to determine the warm and hot formability of aluminium. The FLDs determined by these tests are generally influenced by friction between the specimen and the punch. Moreover, they usually show strain paths similar to those of punch-based deep drawing processes.

When it comes to active media based forming processes, these frictional effects are absent and the strain paths are different. Therefore several active media-based tests for determination of FLDs are being developed. Hydraulic bulge test [9] is a classical example, but it gives the formability of the material at lower temperatures under biaxial scenario only. Hot gas bulge test [10] is a modified version of hydraulic bulge test which is capable of determining the biaxial forming limits at high temperatures. Some works, for instance [11], enhanced hot gas bulge test by modifying the ellipticity of the die as in [9] to obtain different strain paths and succeeded in determining the right side of the FLD of Magnesium alloy AZ31B sheets at 400°C. Apart from punch-based tests and bulge tests, another recent study [12] investigated in-plane test concepts for FLD determination of AA7075 alloy in between 200 and 400 °C.

Banabic et al. developed a novel hydraulic double-layer bulge test [13] to determine the entire FLD of AA6016-T4 at room temperature. Others also determined FLDs at high temperatures by using double layered gas based test methods. For example in [14] the specimen has two slits instead of circular holes to reduce the stress concentration. Another example is [15] where FLDs of a titanium alloy were determined at 970°C under constant strain rate by controlling the gas pressure rate. In this study, bi-layered gas bulge tests were performed with circular and elliptical dies and specimen geometries with elliptical holes. Moreover, the study [16] involved tests with similar specimen geometries with holes for FLD determination for electrohydraulic forming.

Several studies investigated the hot sheet formability through advanced test setups. But, applying these techniques to aluminium alloys at their solution heat treatment temperature complicates the DIC analysis due to the ductility and extreme surface enlargement of the material. This is a major point that makes the extraction of FLD quite challenging that is less emphasised on. Moreover, currently no studies known to the authors exist, that employ the extracted FLD to predict failure during a gas-based forming process. The current study aims to fill these gaps by realising a test setup and a systematic pattern for DIC analysis for hot FLD determination and its validation by laboratory scale benchmark tests.

3. Test setup and materials

3.1. Material

This study aims to determine the FLD of the 1.5 mm thick EN AW-6010-S aluminium alloy sheet at 565°C. During the experiments, the material is heat-treated according to the gas forming process route which involves solutionising at nearly 565°C for 20 minutes and subsequent specimen deformation until failure. Since the material is annealed at solution heat treatment temperature, complete recrystallization took place. Therefore, no significant anisotropic effects were observed and hence neglected in this study.

3.2. Test concept and setup

3.2.1 Test concept. The current test concept is based on the hydraulic double-layer bulge test [13] in which the test specimen with two holes is deformed with help of a carrier sheet and fluid pressure. This technique is adapted to high temperature conditions with hot tools and gaseous active medium. As shown in Figure 1a, the setup involves four major parts: heated die and blank holder, carrier blank and the specimen. The carrier blank gradually bulges and acts as a punch where it transmits the deformation load to the specimen. By varying the hole geometry and distance between the holes of the specimen, different strain paths can be obtained. The specimen geometries suggested by Banabic et al. [13] are shown in the Figure 1b. This comprises of five different specimen geometries resulting in two strain paths on the left side (uniaxial-like, intermediate (-)), one on the vertical axis (plane strain) and two strain paths on the right side (intermediate (+), biaxial).

3.2.2 Test setup. The test setup designed for realising the above mentioned test concept shown in Figure 1a is realised by modifying the existing biaxial hot gas bulge test [10], which was earlier developed for high temperature material characterization at the Institute. The heated die and blank holder have an outer diameter of 206 mm and the die has an inner opening hole diameter of 100 mm. The outer diameter of the specimen and the carrier blank correspond to the tools, i.e., 206 mm. The geometry and size of the specimens' holes vary according to the aimed strain path. In order to simplify the experiments, carrier sheets are also made of same alloy as the specimen to be consistent with the solution heat-treatment temperature and time. The carrier sheet thickness is taken to be 4 mm, which is significantly higher than the specimen thickness of 1.5 mm. By this, the carrier sheet should remain intact until the thinner specimen fails.

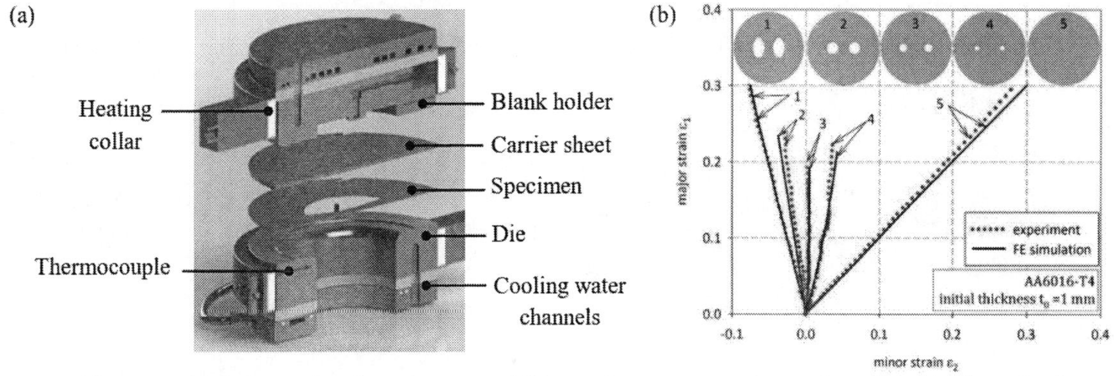

Figure 1. (a) Schematic of test setup and (b) specimen geometries and strain paths from [13]

The die and blank holder are heated via circumferential heating collars and appropriately insulated to avoid heat loss to the environment. The heat transfer to the press is compensated with help of integrated cooling water channels at the interfaces between the press and tools. Therefore, the tools can be maintained at a nearly constant test temperature of 565°C throughout the experiments with help of two separate PID temperature controllers and type K thermocouples. During the experiment, the specimen and the carrier blank are placed between the hot tools, where they get conductively heated up to the test temperature and undergo solution heat treatment for 20 minutes before the deformation. The forming medium is high-pressure nitrogen gas that passes through the gas distribution hole at the centre of the blank holder and acts on the carrier sheet, which bulges and simultaneously deforms the specimen. During the experiment, the carrier sheet and the specimen might undergo slight cooling (max. 10 K) because of convection due to pressurized gas. But, due to conduction between the hot dies and specimen sheets the temperature loss is compensated and therefore near isothermal conditions can be assumed. The rate of gas pressure can be controlled linearly with help of an integrated valve-setup and is set to 2 bar/s. The blank holder force is set to be constant at 250 kN, which ensures a complete sealing throughout the experiment.

3.3. DIC and pattern design with preliminary biaxial hot gas bulge tests

The DIC based GOM Aramis 4M setup is used to track the specimen deformation and the local major and minor strains. As a part of this, two cameras continuously record the specimen in the deformation zone through the die opening. Since the 6xxx alloys exhibit high ductility and high surface enlargement under the solution heat treatment temperature conditions, a temperature resistant and stable pattern that adheres well to the surface of the specimen is necessary for DIC analysis. Preliminary studies showed that the typical manually created stochastic pattern where a secondary contrasting colour is finely sprayed on an initial primary black or white coated specimen surface disintegrates during the deformation. The patches of newly generated aluminium surface at high specimen surface enlargement are moreover highly reflective and inhibit the GOM Aramis system from pattern deformation detection until specimen failure. Therefore, the existing methods of stochastic pattern creation are not very useful under current conditions and a stable pattern needs to be developed through further preliminary studies before proceeding with the actual experiments for FLD determination.

A significant part of the current study deals with development of a new multi-layer pattern concept in a step-by-step manner to improve the deformation analysis via DIC technique. As a first step, the initial specimen surface roughness is increased by sand blasting which further improved the adhesion of the paint to the specimen. After this, different stochastic pattern combinations are tested in order to check the stochastic pattern quality and integrity. It was observed that the conventional method of spraying stochastic pattern onto a primary coating did not give satisfactory results. Therefore, alternative methods of stochastic pattern generation were evaluated. Finally, a random Matlab-generated stochastic pattern was chosen and applied via laser marking. Marking this on an initially roughened specimen coated with high temperature resistant white furnace paint enhanced the pattern detectability as shown in Fig. 2c.

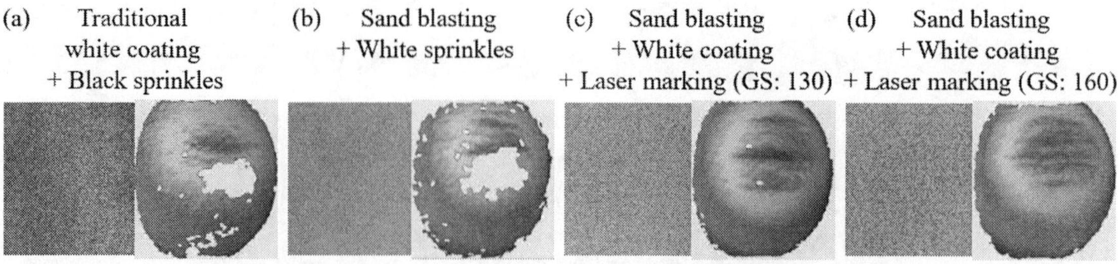

Figure 2. (a – d) Selected instances of tested stochastic pattern combinations for the DIC analysis showing the pattern quality and the amount of detectable region of the biaxial specimen at 565 °C

The laser marking results in local surface modification as the initial white coating is removed and the underlying aluminium surface is selectively oxidised. In contrast to manual spraying method, this is advantageous because a stable stochastic pattern can be applied via a reproducible and robust method by modifying the aluminium surface layers. Finally, increasing the grey-scaling value (GS) during laser-marking to 160 recorded a major strain of up to 83 % at the dome region as shown in Fig. 2d and this turned out to be the best pattern among all. Adjusting the grey-scaling changes the sensitivity of the laser at which it considers a grey value as to be marked or not.

4. Simulation-aided verification and optimization of the specimen geometries

Following the test conception and setup, it should first be investigated if the specimen geometries proposed in [13] result in the desired strain paths with current material under the high temperature conditions. Therefore, the FE models are setup, the simulations are performed starting with the existing specimen geometries ranging from uniaxial to biaxial strain states and the resulting strain paths are analysed.

4.1. Model setup and definition of boundary conditions

With help of symmetry boundary conditions a quarter FE model according to the experimental concept is setup as shown in the Fig. 3a. The geometries of the tools, carrier sheet and the specimen with notches are designed in Solidworks 2018, meshed in Hypermesh and the FE model is setup in LS-PrePost V4.3 and simulated in LS-Dyna V9.0. The simulations are carried out using an explicit FE solver with an automatic time step increment.

4.1.1 Modelling and discretization of parts. The die and blank holder are modelled as rigid bodies with Type-2 Belytschko-Tsay shell elements of 1 mm element size. The elastic deformation of tools is neglected and only the surfaces of the tools are modelled with shell elements in order to minimise the computational effort. The die is fixed and the blank holder is free to only translate in the vertical direction under action of blank holder force. The carrier sheet and specimen are modelled as elastoplastic deformable bodies with Type-2 shell elements of section thicknesses 4 mm and 1.5 mm respectively. The average mesh size for discretization of these deformable bodies is finalised to be 1 mm based on a mesh convergence analysis. For thickness discretization, nine integration points are taken in shell normal direction. The material properties for deformable bodies are taken from author's previous work [17] and the plasticity modelled using the piecewise_linear_plasticity card in LS-Dyna. The simulations are isothermal at a temperature of 565°C.

4.1.2 Interactions and loads. The three different interactions between the tools and deformable bodies are modelled by forming_surface_to_surface contact algorithm. For the time being, friction is modelled according to the Coulomb model with the friction coefficients of 0.17 and 0.1 corresponding to the Boron Nitride and Graphite sprays taken from literature [18]. Two different loads corresponding to the experiment should be represented in the model, the gas pressure and the blank holder force. The pressure load is modelled as a surface load acting on the elements of the carrier sheet with help of the card load_segment_set. The linearly increasing pressure load is taken as in the experiments. As the carrier sheet starts bulging, the pressure load is transmitted to the specimen that also deforms along with the carrier sheet. A constant blank holder force of 250 kN is applied on the blank holder in the vertical direction as a rigid body force in order to hold the carrier sheet from drawing into the die during process.

4.2. Simulated strain paths and modification of geometries

The simulated strain paths are shown in the following Fig. 3b and it can be seen that all the geometries exhibit the strain paths as intended. The plane strain geometry is further investigated in terms of the size of the notches and the distance between the notches. Based on this simulation study, the specimen dimensions are finalised.

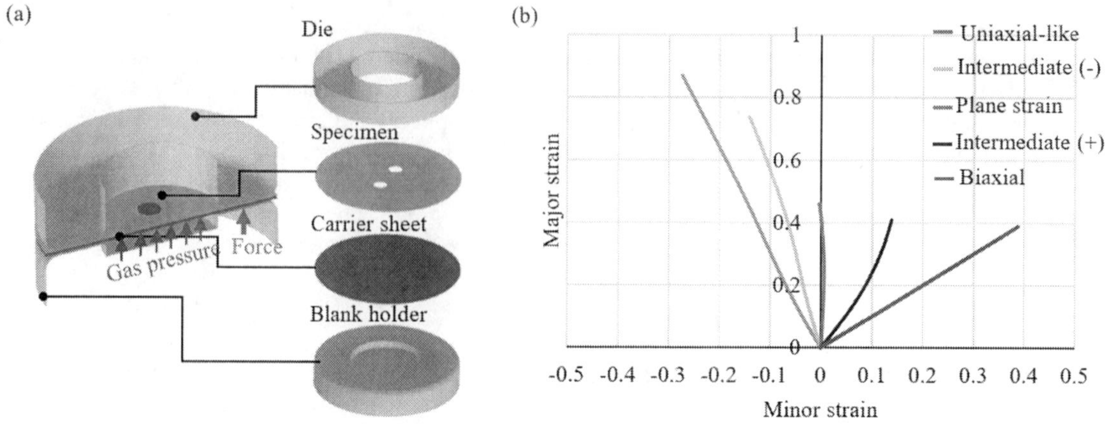

Figure 3. (a) FE-Model setup and (b) simulated strain paths

5. Experimentation and determination of FLD

5.1. Specimen preparation

The experiments are performed with the finalized specimen geometries. To track the evolution of the strain paths and to identify the instance of failure, the previously finalised stochastic pattern combination is applied on the specimens as shown in Fig. 4.

5.2. Experimentation and strain measurement via DIC technique

During the experimentation, as the gas is gradually released the carrier blank gradually bulges simultaneously deforming the specimen along with itself. As the specimen deforms, the strain increases at the polar region. The evolution of the strains is continuously monitored by the GOM-Aramis. The major strain distributions before the specimen failure are also shown in the following Fig. 4.

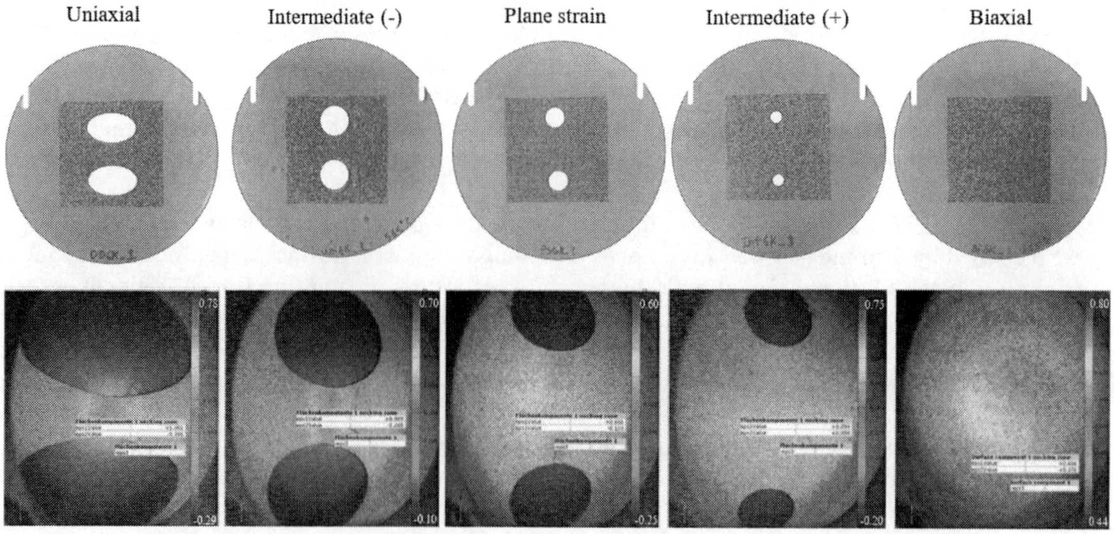

Figure 4. Initial specimens with laser-marked stochastic pattern and the corresponding DIC maps during the experimentation at 565 °C showing the major strain distribution

5.3. *Analysis of strain paths and extraction of FLD*

The standards (ISO 12004) of Nakajima tests are used to analyse the formability and to analyse the strain paths and extract the limiting major and minor strains at the instance of failure. Comparison of progression of the simulated and experimental strain paths in Fig. 5 shows that they are almost similar with very slight deviations. These deviations can be reasoned in differences between experiment and simulation, for example, temperature drop because of cooling due to gas, evolving friction coefficient, variation in chamber pressure evolution, etc. The limiting strains at the dome at the instance of crack propagation are extracted and these are connected to construct the FLC. It can be seen that the plane strain forming limit is below all the remaining specimens and as the plane strain is the critical point and breaks at minimum. This FLC is further evaluated with help of a cross-die benchmark test as described in the following chapter.

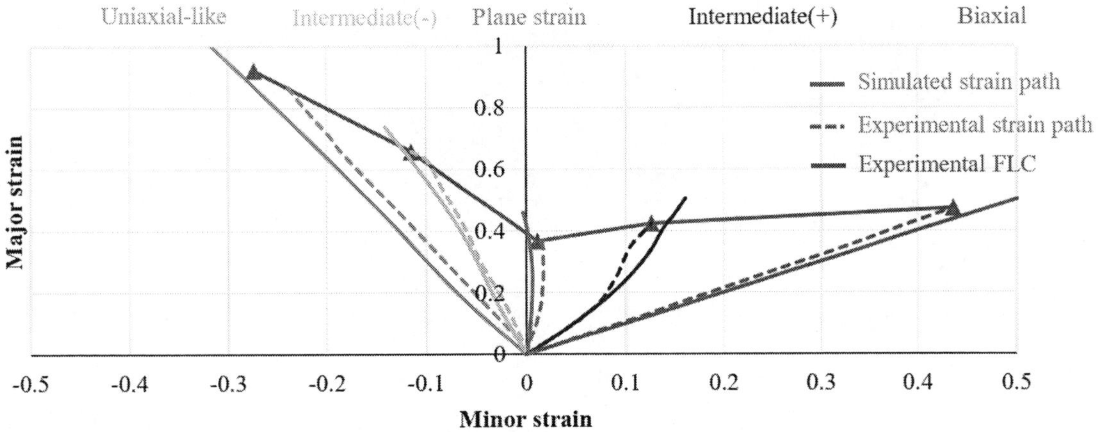

Figure 5. Simulated and experimental strain paths and the FLC of EN AW 6010-S 1.5 mm at 565°C

6. Validation with a cross-die benchmark test

6.1. *Benchmark test setup and experimentation*

An existing cross-die benchmark specimen shown in Fig. 6a, formed via pure gas-based forming process is used to validate the determined FLC. Similar to the setup in [17], this benchmark test consists of hot die and blank holder at 565°C as shown in the Fig. 6b. The 1.5 mm thick EN AW-6010-S initial blank whose shape is already optimized for the current gas-based hot forming process is placed in between the die and blank holder, solutionised and formed with help of continuously increasing gas pressure by simultaneous regulation of the blank holder force.

For validation of the current FLC, two different scenarios with different values of blank holder force are chosen so that in the first scenario a successful component is manufactured and in the second scenario a failure in the component is recorded. The blank holder force in the first scenario is varying from 10 kN to 70 kN during the forming process whereas in the second scenario it is initially 14 kN which increases to 70 kN during forming. The rate of increase of gas pressure is 5 MPa/s in both of these experiments.

Figure 6. (a) Cross-die specimen, (b) experimental setup for forming the cross-die benchmark specimen and (c) corresponding simulation model

From the resulted cross-die components in Figure 7 it can be observed that a failure-free component with complete form-filling resulted in first scenario when the initial blank holder force is lower, whereas in the second scenario the specimen cracked and could not deform further due to gas leakage. The second scenario also shows an insufficient material draw-in in the flange region.

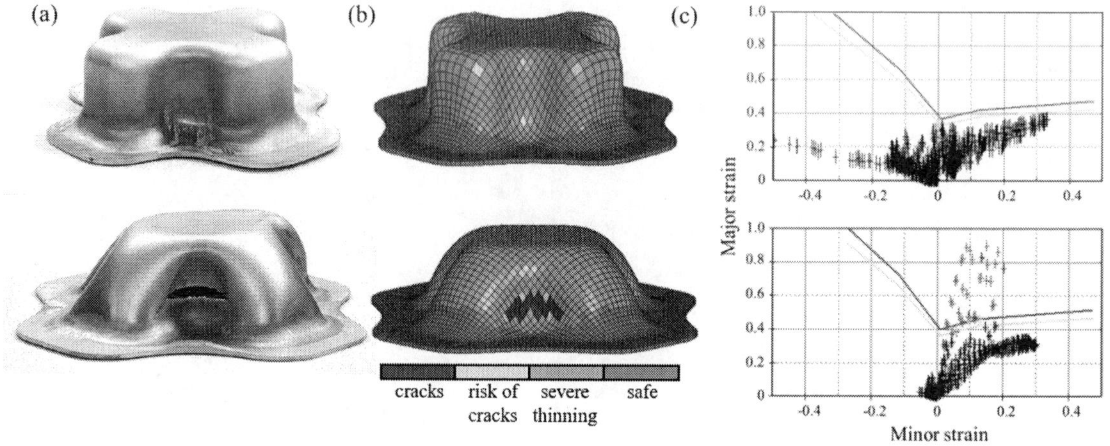

Figure 7. (a) Experimental, (b) simulation results and (c) corresponding strain states of first scenario (upper) with 10 kN and second scenario (lower) with 14 kN initial blank holder force

6.2. FE modelling, simulation and validation

The FE model of cross-die benchmark forming process consisting of rigid tools and deformable blank is setup according to the simulation methods in [17] with Type-2 shell elements. All the boundary conditions in the simulation are modelled as close to the experiment as possible. Simulations with aforementioned values of blank holder force are carried out. By taking the experimental FLC previously shown in Fig. 5, formability investigations are carried out using the 'Formability-Key' option in LS-Dyna during post processing and the resulting strain-states (Fig. 7c) are analysed. The line indicating 'risk of cracks' region is taken to be 10% below the FLC.

It can be observed that the simulation corresponding to first scenario shows only thinning, but no cracks or no risk of cracks. In simulation of the second scenario, cracks are observed in the concave wall region of the cross-die as highlighted in Fig. 7b. This is because of high initial blank holder force, which restricts sufficient material flow from flange region that ultimately resulted in risk of cracks and subsequent cracks. The first instance of occurrence of crack is depicted here. Once the cracks are observed, the further deformation of the specimen is not realistic as the gas escapes through the cracks. Even though a slight temperature drop is observed in the benchmark test due to convection and the FLC is determined at nearly isothermal conditions, the results of crack prediction are quite reliable. So, it can be said that the FLC is validated with the cross-die simulation model successfully.

7. Summary, conclusions and future work

A test concept for extracting the high temperature FLD for failure prediction during gas-based hot forming processes is devised, developed and taken into operation. Experiments are carried out and the high temperature FLD of ductile aluminium alloy is determined. The FLC is further used in simulations of a laboratory scale cross-die benchmark component for failure prediction and the simulation results are experimentally validated. Based on the work done, the following key findings can be reported:

- The existing hydraulic double layer bulge test [13] has been successfully adapted for testing ductile aluminium alloys under high temperature conditions by substituting the liquid medium with gaseous medium and integrating appropriate heating and cooling elements.

- A promising method for generation of heat resistant and mechanically stable DIC patterns for highly ductile materials has been developed. This method involves surface roughening by sand blasting, primary coating with temperature resistant white paint and application of stochastic pattern via optimised laser marking.
- The specimen geometries from [13] resulted in nearly linear strain paths for the investigated EN AW 6010-S alloy even under high temperature conditions. Accordingly, a high temperature FLD at 565°C is successfully determined.
- The FLC applicability is partially validated for rapid gas-based hot forming process with help of a laboratory-scale cross-die component. Failure occurrence during the forming process is reliably predicted within corresponding FE simulations.

In future, the simulation models could be further improved so that the experimental and simulated strain paths correspond even more to each other by further investigations on the themo-mechanical and frictional boundary conditions. Following this, the specimen geometries could be further optimized to obtain the ideal strain paths and additionally the strain path corresponding to deep drawing. Moreover, FLD determination under process relevant non-linear strain paths could be investigated. Also, strain rate control could be implemented within the experiments to determine FLDs at different and constant strain rates.

Acknowledgements
The authors gratefully acknowledge the financial support by ZIM (Zentrales Innovationsprogramm Mittelstand) within the Project ZF4791302LL9 "Hot-Alu-Testing". Moreover, the authors would like to thank the project partner Dirkra Sondermaschinenbau GmbH & Co. KG for their contribution.

References
[1] Neugebauer R, Altan T, Geiger M, Kleiner M and Sterzing A 2006 *CIRP Annals* **55** (2), 793-816
[2] Wang H, Luo Y, Friedman P, Chen M and Gao L 2012 *Trans. Nonferr. Met. Soc. China* **22** 1-7
[3] Banabic D, Vulcan M and Siegert K 2005 *CIRP Annals* **54** (1), 205-208
[4] Rong H, Hu P, Ying L, Hou W and Zhang J 2019 I. *J. Mech. Sci.* **156** 59-73
[5] Gajjar N, Modi B and Digavalli RK 2019 *J Fail. Anal. Preven.* **19** 1792–1800
[6] Hu Q, Zhang F, Li X and Chen J 2018 *Acta Mechanica Solida Sinica* **31** No. 3 259-289
[7] Wang N, Ilinich A, Chen M, Luckey G and D'Amours G 2019 *I. J. Mech. Sci.* **151** 444-460
[8] Zhu M, Lim YC, Liu X, Cai Z, Dhawan S, Gao H and Politis DJ 2021 *I. J. Lightweight Mat. & Man.* **4** 269-280
[9] Lazarescu L, Comsa DS, Nicodim I, Ciobanu I and Banabic D 2012 *Trans. Nonferrous Met. Soc. China* **22** 275-279
[10] Braun A, Storz J and Hirt G 2015 *METEC & 2nd ESTAD*
[11] Abu-Farha F, Verma R and Hector Jr. LG 2012 *J. Mat. Pro. Tech.* **212** 1414– 1429
[12] Reuther F, Lieber T, Heidrich J and Kräusel V 2021 *Materials* **14** 1882
[13] Banabic D, Lazarescu L, Paraianu L, Ciobanu I, Nicodim I and Comsa DS 2013 *CIRP Annals – Man. Tech.* **62** 255-258
[14] Mitukiewicz G, Anantheshwara K, Zhou G, Mishra RK and Jain MK 2014 *J. Mat. Pro. Tech.* **214** 2960-2970
[15] Wu Y, Liu G, Liu Z and Wang B 2016 *Materials & Design* **108** 298–307
[16] Maris C, Hassannejadasl A, Green DE, Cheng J, Golovashchenko SF, Gillard AJ and Liang Y 2016 *J. Mat. Pro. Tech.* **235** 206-219
[17] Baru NK, Teeuwen T, Teller M, Hojda S, Braun A and Hirt G 2021 *IOP Conf. Ser.: Mater. Sci. Eng.* **1157**
[18] Liu Y, Zhu B, Wang K, Li S, Zhang Y, 2020 *Tribology International* **151** 106504

Accurate plane strain compression test validation

Adriana Neag[1,2], Qifan Hu[1] and Tudor Balan[1]

[1] Arts et Metiers Institute of Technology, Université de Lorraine, LCFC, HESAM Université, F-57070 Metz, France

[2] Dept of Materials Science and Engineering, Technical University, Cluj-Napoca 400641, Romania

E-mail: tudor.balan@ensam.eu

Abstract. Large strain characterization of sheet metals has become increasingly important with the generalization of advanced high strength steels, for which the tensile test provides data over a very reduced strain range. Among the numerous alternative characterization tests, the plane strain compression test (PSCT) requires a small amount of material and classical testing machine and acquisition. PSCT was mainly used for hot forming characterization, but recently it has been proved sufficiently accurate for application in cold metal forming. This work provides an in-depth validation of the PSCT by means of the finite element method. When converting the PSCT force-displacement curve into a stress-strain curve (flow curve), several analytical corrections are applied. Several sets of such corrections terms were proposed in the literature, some of which are consistently used by all authors, while others being only used in some papers. The FE simulation of the test was used in order to validate these correction terms and their hypotheses. The originality of the approach is the design of a sequence of test configurations which allow for the individual validation of each and every one of the correction terms concerning the effect of several test parameters. The FE simulations showed that the analytical exploitation of the PSCT provides a very good accuracy. They helped identifying the most suitable correction terms to consider.

1. Introduction

Accurate stress-strain curves of sheet metal are required for realistic simulations, for example in the automotive industry. According to the ISO 6892-1 standard, stress-strain curves are traditionally built from the uniaxial tensile test. However, this test cannot provide data at strains beyond uniform elongation within the ISO norm specifications. Therefore, it is necessary to find other alternative methods to obtain an accurate flow curve. Over the past decades, many alternative methods have been developed. Some of the more commonly used in the literature are: bulge test, simple shear test, stack compression test, in-plane torsion test, and plane strain compression test. Only the hydraulic bulge test gave rise to an ISO standard. Most of these experiments require dedicated testing devices, sometimes dedicated testing machines, and image correlation-based acquisition. The plane strain compression test is an exception to this rule, since it makes use of a simple test device and a universal tension-compression machine and standard force-displacement acquisition. From this point of view, it is thus very appealing for industrial application, especially as it requires small amounts of sample material. While it was traditionally applied in the context hot rolling applications, several authors seem to indicate that the test is sufficiently accurate for the flow curve determination in cold forming conditions [1][2].

In the plane strain compression test (PSCT), a rectangular sample is compressed between two rectangular dies, wider than the part, and extend the part in the direction of its length, as illustrated in Figure 1. The displacement of one of the tools is imposed and the force is measured from which a stress-strain curve under the plane strain stress state is derived. The friction between the tool and the workpiece is reduced using lubricants made of different materials such as graphite, glass, polymer (Teflon), oil, which are chosen mainly according to the temperature of the test. Several drawbacks were summarized in the literature [3]: lateral misalignment of the tool, stress concentration at tool edge, heterogeneous distribution of deformations and friction. Lateral tool misalignment reduces the area of the workpiece in contact with the tool and affects the measured force. The concentration of stresses on the edges of the tool can cause crack initiation at relatively low tool penetration. And the distribution of stresses in the part is heterogeneous, which poses problems of interpretation of the results of the plane strain compression test. To improve the reliability of the plane compression test, Becker and Pöhlandt [1] used a tool with a radius and proposed an analytical method for the determination of the effective tool width. However, friction cannot be entirely eliminated in the test. Thus, the correction of friction has been studied by several researchers, who applied Tresca's law and Coulomb's law or a combination of the two [4] to the friction between the tool and the test piece. These studies provide friction correction equations for each case. Finally, optimization of the specimen geometry was also proposed [5].

The purpose of this paper is to validate the various correction terms proposed in the literature for the PSCT flow curve calculation, by means of finite element simulations. The research methodology adopted in this work is described in Section 2. Section 3 summarizes the main results of the study, which are further discussed in Section 4.

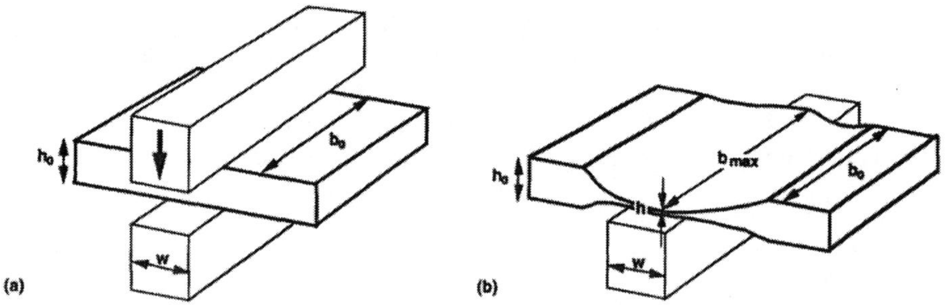

Figure 1. Schematic description of the PSCT test configuration a) before and b) after test and main notations [6].

2. Research methodology

An analytical analysis of the results of the plane strain compression test is required to extract the material's flow curve from the PSCT. Several correction terms are involved in these calculations, whose validation is insufficient in the literature. Therefore, an efficient and reliable verification method is necessary. To validate these corrections, FE simulation with the Forge® software were used. Such simulations allow for individual and combined analyzes of the impact of these corrections.

The method proposed is summarized in Figure 2:

- First, a material law is chosen, corresponding to a cold formed steel sheet. Here, a C35 steel material was considered; its properties were selected from the Forge® software database.
- Then PSCT simulations are performed under different conditions. The applied force as a function of tool movement is recorded. The geometry of the test is defined by the sheet thickness (2 mm), tool width (6 mm), sample width (40 mm).

- Finally, stress-strain curves are calculated by applying the PSCT analytical formulas to the FE-based force-displacement recordings. These curves can be used to compare with the curve generated by the constitutive equation, which serves as reference. The analytical analysis equations used in this work were taken from [2].

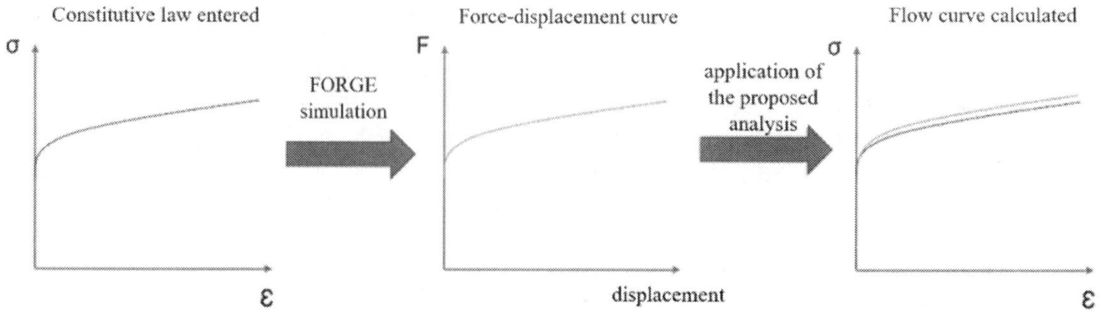

Figure 2. Research methodology used for the numerical validation of the PSCT.

This approach allows one to rigorously compare the result of the PSCT procedure to a rigorously known reference. Moreover, simulation further allows one to decompose the problem in sub-problems in order to decouple the different correction terms. Consequently, several 2D and 3D test configurations were simulated, as schematically shown in Figure 3. Two-dimensional simulations did not need to be compensated for the sample width variation while frictionless simulations avoid the influence of friction, and the small length configuration is not affected by shear side effect. Simulation configurations were designed to emphasize only one single effect (width variation; friction; shear…) and thus validate individually each correction term, which would be impossible experimentally. The tools were described as rigid, analytical surfaces, while the material was considered elasto-plastic and meshed with triangular / tetrahedral elements. According to a convergence study, a fine element size of 25µm was used for all simulations. This was important in particular for a correct estimation of the contact area. Given the large strains, remeshing was used in order to maintain a good mesh quality. It was verified that strain rate and heat transfer during the process do not have any influence on the obtained force-displacement curves.

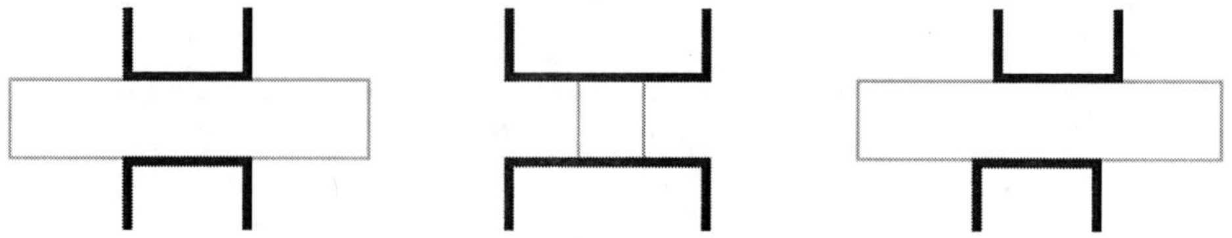

Figure 3. Simulated PSCT configurations within the validation procedure: reference configuration (l>w); small length configuration (l<w); and misalignment configuration.

3. Simulation results

The small length configuration was used in order to explore the corrections related to friction correction and to the yield surface model employed in the analytical calculation of the flow stress. Indeed, in this configuration the shape of the sample geometry remains perfectly rectangular thus involving less approximations.

Figure 4 shows predicted flow curves using 2D PSCT simulations with the small length configuration. Various friction models were used for the correction. The figure clearly shows that the friction correction improves the accuracy of the flow stress calculation, in particular at large strains. In turn, these simulations showed an equivalent accuracy for all the friction models tested: Coulomb friction, Tresca friction and combined friction. This may explain why recent publications [2] still use a friction correction based on the Coulomb friction model, although this model is not acceptable for this test on physical grounds.

The same analysis using the reference configuration, with a sample longer than the tool width, led to the same conclusions. With the reference configuration, the accuracy of the free boundary condition at the die exit surfaces was questions by several authors. Becker and Pöhland [1] proposed a so-called shear correction to address this problem. Figure 5 shows that this shear correction does not improve flow curve calculation, provided that a von Mises yield surface is used in the analytical calculation of the flow stress. In contrast, the usage of Tresca's yield surface as was done in [1] significantly overestimates the flow stress; this may explain why these authors found the shear correction useful, as its effect is a decrease in the flow stress.

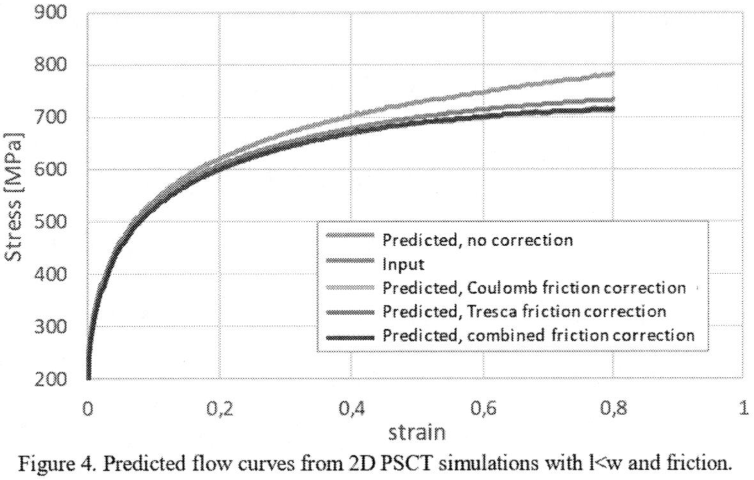

Figure 4. Predicted flow curves from 2D PSCT simulations with l<w and friction.

A classical correction term [2-5] aims to take into consideration the increase of the sample width b, which induces an increase in the contact surface between dies and sample, thus affecting the calculation of the average normal stress. The simulation of the width increase required a full 3D simulation. The correction term makes use of the maximum sample width measured at the end of the experiment. As it is made clear in Figure 6, this correction term compensates with very good accuracy the influence of the width increase on the flow curve calculation, although the corresponding error is not very large. In the last two simulations the friction was set to zero in order to avoid superposition of several errors. Nevertheless, the same conclusions were also obtained with similar simulations where small to moderate friction was considered. Also, the three different friction correction terms were again confronted to each other using 3D simulations of the PSCT. The same conclusions are drawn as in two dimensions: the friction correction is necessary, the three proposed correction terms are accurate and there is almost no difference between their predictions.

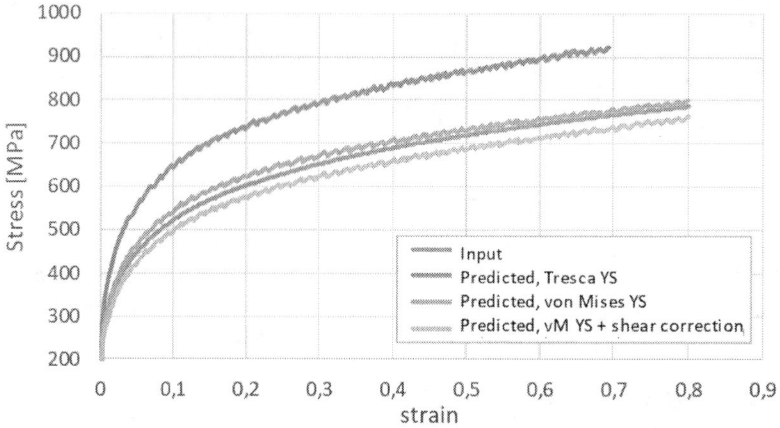

Figure 5. Predicted flow curves from the frictionless 2D PSCT simulations with l>w.

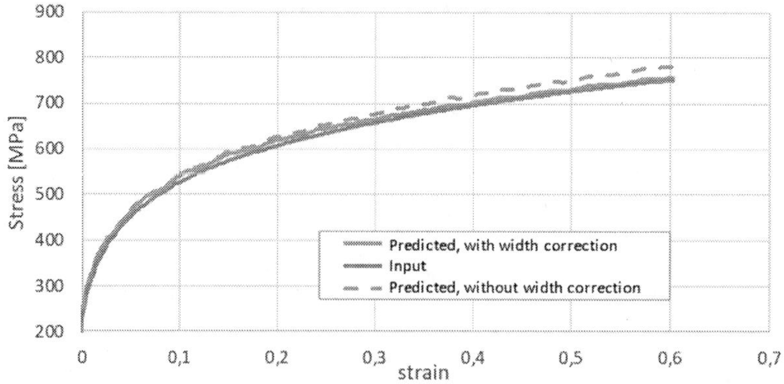

Figure 6. Predicted flow curves from the frictionless 3D PSCT simulations; influence of width correction.

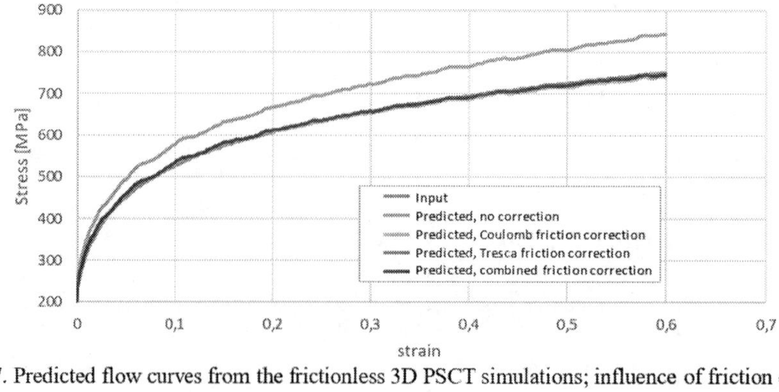

Figure 7. Predicted flow curves from the frictionless 3D PSCT simulations; influence of friction correction.

Finally, the influence of an accidental tool misalignment was investigated with 2D simulations. No correction term was proposed, but the influence of a misalignment imposed in the simulations on the resulting flow curve was determined. Figure 8 plots the errors between flow curve with misalignment and the input flow curve. These results show that the error is negligible for tool misalignments inferior to 0.2 mm. However, for a tool misalignment of 0.5 mm the error becomes very significant, approaching 10%. It is thus recommended to carefully consider the tool alignment when building PSCT devices.

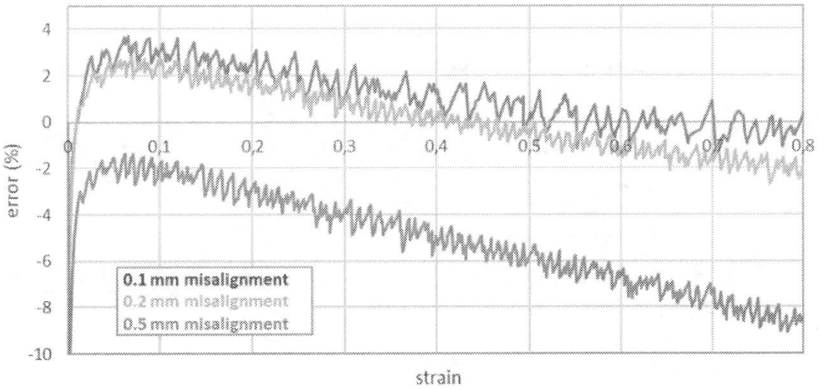

Figure 8. Predicted flow curves from the 2D PSCT simulations; influence of tool misalignment.

4. Discussion and conclusions

The plane strain compression test provides an economical means for flow curve determination of sheet metals at large strains. The analytical calculation of the flow curve from force-displacement measurements involves a number of corrections to compensate for edge effects, friction effects etc. While it is very difficult or even impossible to validate these correction terms independently with respect to experimental data, the numerical simulation of the test provides a rigorous means for validation, in particular when several test configurations are used. In this work, several analytical correction factors were analysed. These are the effect of the plasticity surface, shear-type edge effect, effect of friction and effect of the specimen width correction. The investigation conforms that for materials without plastic anisotropy, the von Mises yield surface should be preferred to Tresca's. In this case, the so-called shear correction of the edge effects does not improve the results and should be avoided. In turn, the classical compensation terms for the specimen width increase and the friction influence proved both necessary and accurate. In particular, the performed simulations could not show any difference between the correction terms using Coulomb and Tresca friction laws. Some of these conclusions could be further explored with more detailed simulations.

References

[1] Becker N and Pöhlandt K 1989 *CIRP Annals* **38** 227
[2] Chermette C, Unruh K, Peshekhodov I, Chottin J and Balan T 2020 *Int. J. Material Forming* **13** 269
[3] Aksenov S, Kliber J, Puzino Y and Bober S 2015 *J. Chem. Techn. Metallurgy* **50** 644
[4] Lacey A J, Loveday M S and Mahon G J 2002 *Materials at High Temperatures* **23** 85
[5] Semiatin S L and Altan T 2010 Measurement and interpretation of flow stress data for the simulation of metal-forming processes *Air Force Research Laboratory Report*, USA
[6] Gelin J C, Ghouati O and Shahani R 1994 *Int. J. Mech. Sci.* **36** 773

IOP Conf. Series: Materials Science and Engineering 1238 (2022) 012051 doi:10.1088/1757-899X/1238/1/012051

Influence of criteria on the determination of forming limits in thickness reduced cruciform specimens

Z Wang [1], D Guines [1] and L Leotoing [1]*

[1] Univ Rennes, INSA Rennes, LGCGM (Laboratoire de Génie Civil et Génie Mécanique) - EA 3913, F-35000 Rennes, France

* lionel.leotoing@insa-rennes.fr

Abstract. Biaxial tensile test methods using cruciform specimens have been drawing attention as an effective approach to characterize sheet metal formability. The cruciform specimens usually adopt a thickness reduction at the central area to achieve a relatively large strain gradient for localized necking and fracture investigation. For conventional tests, many criteria have been proposed to determine the forming limits at necking, however, none has been widely accepted and applied to the cruciform specimens with non-uniformed thickness reduction. Therefore, in this work, different existing necking criteria are applied and discussed for numerical procedure in order to define an appropriate necking criterion for the thickness reduced cruciform specimens. AA6061-T4 sheet with a thickness of 2mm is used as a target material. The predicted forming limit curves are compared for different criteria, and the feasibility of each criterion is discussed in detail. Thanks to two dedicated specimens proposed by the authors, forming limits are determined for a wide range of strain states.

1. Introduction

The forming limit curve (FLC) is the common tool for assessing the formability of sheet metals. It's represented in the space of principal major and minor strains and illustrates the limit strains at the onset of localized necking under various linear strain paths [1]. The two primary experimental methods recommended by the international standard ISO 12004-2 [2], the Nakazima and Marciniak tests, have been standardized to construct the FLC of sheet metals. Both tests use punches to form a set of specimens with varying widths to obtain limit strains under different strain paths. It's necessary to have a nearly frictionless state in the zone of evaluation, since friction can influence the position of necking and the linearity of the strain path [3]. Thus, the lubrication measures are commonly employed between the punch and specimen. However, expending the applications of the tests to determine the sheet metal formability at high temperatures remains a challenge, since the limited thermal stability of lubricants [4]. In recent years, biaxial tensile test with a dedicated cruciform specimen has become an interesting alternative to determine the FLC for sheet metals, especially at high temperatures and large strain rates. The test is frictionless and the strain path can be directly controlled by adjusting the biaxial tensile ratios [5].

Significant efforts have been made to design cruciform specimens for the sheet metal formability evaluation. A common feature of these cruciform specimens is the use of a progressive thickness reduction in the specimen central area, and the aim is to achieve a relatively large strain gradient for localized necking and fracture. For example, Zidane et al. [6] proposed a cruciform specimen with a two-step thickness reduction at the central area to determine the FLC of the AA5086 sheet. Song et al.

[7] proposed a cruciform specimen with slits in each arm and progressive thickness reduction at central area to determine the FLC of the DP600 sheet. However, this kind of non-uniformed thickness reduction introduces uncertainty in determining the forming limits at necking, since almost all of the existing necking criteria are developed for flat specimens. Therefore, it's necessary to investigate the feasibility of the existing criteria in determining the forming limits based on the thickness reduced cruciform specimen.

The necking criteria can be classified into three categories: position-dependent method, time-dependent method and position-time-dependent method. As defined in ISO 12004-2, the position dependent method evaluates the distribution of the principal main strains along the cross section perpendicular to the crack, and the necking strain corresponds to the highest point of an inverse parabolic curve which is fitted to the main strains [8]. The time-dependent method is based on the evaluation of strain history. The forming limits at necking are determined at the moment of the abrupt change in the first or the second time derivative of strains [9]. For example, Iquilio et al. [10] proposed a necking detection method by means of thickness variation and applied it to determine the forming limits for AISI 430 stainless steel. In this method, the center and edge of the necking area were identified, and the onset of localized necking is considered as the moment when the thickness variation rate curve at the edge of necking area changes its slope drastically. For the position-time-dependent method, the basic concept is the analysis of the difference in the strain evolution between necking and the adjacent areas. The strain increment ratio method is a typical position-time-dependent method and is frequently adopted in the M-K model [11]. In this method, the localized necking is assumed to occur when strain increment ratio between a point located in the necking area and a point in the adjacent area reaches the critical value [12]. In addition, Zhang et al. [13] proposed a necking detection method by analyzing the evolutions of the average thickness strain for two rectangular zones, and applied it to determine the forming limits of AA5754 in the biaxial tensile test.

In this paper, four different necking criteria were applied to detect the forming limits in the numerical simulation of the in plane biaxial test. The predictions of all the criteria are discussed in order to recommend an appropriate necking criterion for the thickness reduced cruciform specimens. Two dedicated cruciform specimens were employed to cover the strain paths from equi-biaxial to uniaxial tension. Finite element (FE) simulations coupled with the Gurson–Tvergaard–Needleman (GTN) model were conducted to predict failure behaviours of AA6061 sheets.

2. FE model construction and FLC prediction

2.1. Numerical simulation of the biaxial tensile test

AA6061-T4 sheets with 2mm thickness were selected for the study. The FE models based on two dedicated cruciform specimens (shown in Figure 1) were built in ABAQUS/Explicit. By adjusting the biaxial tensile ratios between the L1 and L2 directions, Specimen I [7] was used to obtain the strain paths from equi-biaxial to plane strain, and specimen II [14] was used to obtain the strain paths from plane strain to uniaxial tension. The FE models were modeled by eight-node brick solid elements with reduced integration (C3D8R), and the element size was refined in the thickness reduced zone to ensure the accuracy of the predictions, with a minimum mesh size of 0.2mm. In addition, 3 elements were adopted in the thickness direction.

The basic mechanical properties of the AA6061-T4 sheet used in simulations were obtained from uniaxial tensile tests at room temperature and quasi-static strain rate ($0.002s^{-1}$) along rolling direction. The strain hardening of the material was described by the Voce hardening function, as shown in equation (1).

$$\sigma_{Hard} = \left[166.6 + 191.3 \times \left(1 - e^{-13.3\bar{\varepsilon}^{p}}\right)\right]\text{MPa} \tag{1}$$

In order to consider the damage effects and to predict the failure behavior of the material, the GTN model [15] was employed and implemented in ABAQUS by the user subroutine VUMAT. The yield function of the GTN model takes the following form:

$$\Phi = \left(\frac{\sigma_e}{\sigma_{\text{Hard}}}\right)^2 + 2q_1 f^* \cosh\left(\frac{3q_2\sigma_m}{2\sigma_y}\right) - 1 - \left(q_1 f^*\right)^2 = 0 \tag{2}$$

$$f^* = \begin{cases} f & f \le f_c \\ f_c + \dfrac{f_u - f_c}{f_f - f_c}\left(f - f_c\right) & f > f_c \\ f_u & f = f_f \end{cases} \tag{3}$$

$$\dot{f} = \dot{f}_{grow} + \dot{f}_{nucl}; \quad \dot{f}_{grow} = (1-f)\dot{\varepsilon}^p : \boldsymbol{I}; \quad \dot{f}_{nucl} = A_N \cdot \dot{\bar{\varepsilon}}^P \tag{4}$$

$$A_N = \frac{f_n}{S_n\sqrt{2\pi}} \exp\left(-\frac{1}{2}\left(\frac{\bar{\varepsilon}^P - \varepsilon_n}{S_n}\right)\right) \tag{5}$$

Where σ_{Hard} is the yield stress evolving with the material hardening, σ_e is the von Mises equivalent stress, $\sigma_m = \sigma_{ii}/3$ is the mean stress. q_1 and q_2 are material coefficients. f is the void volume fraction of the material and f^* is the effective void volume fraction. f_c is the void volume fraction to trigger coalescence, $f_u = 1/q_1$ is the ultimate volume fraction and f_f represents the void volume fraction at fracture. f_n represents the volume fraction of the nucleated void, ε_n is the mean value of the normal distribution of the nucleation strain, and S_n is the standard deviation. The main parameters in the GTN model are listed in Table 1 [16].

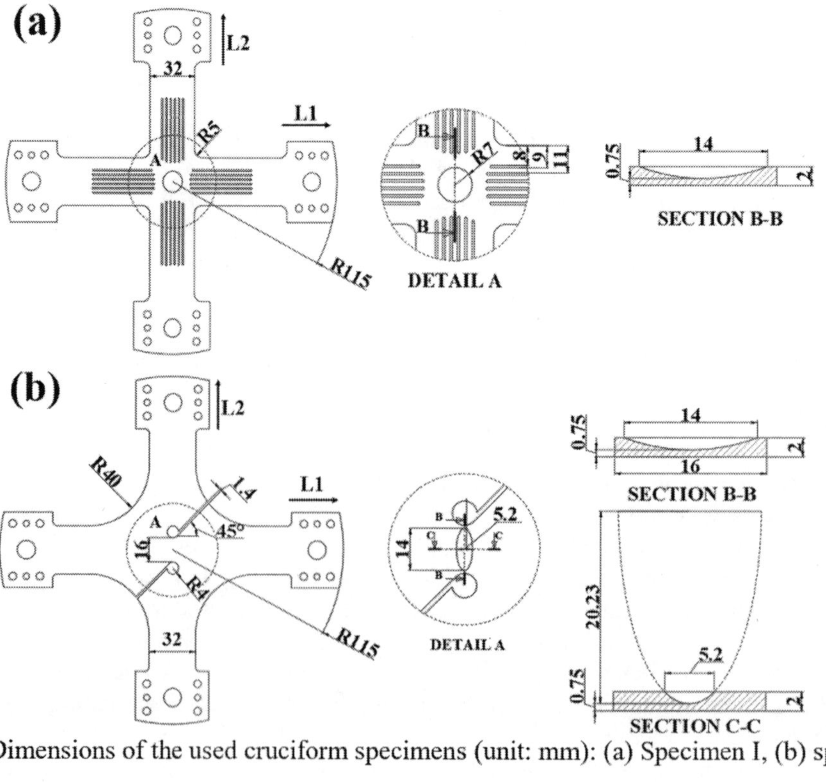

Figure 1. Dimensions of the used cruciform specimens (unit: mm): (a) Specimen I, (b) specimen II

Table 1. Main parameters in the GTN model

Parameters	q_1	q_2	f_n	S_n	ε_n	f_0	f_c	f_f
Value	1.5	1	0.01285	0.07	0.168	0.00025	0.01755	0.02887

2.2. FLC prediction with different necking criteria

2.2.1. Position dependent method (CRIT1). As the method is explicitly explained in ISO 12004-2 [2], a brief introduction is given here and the simulation results of specimen I under equi-biaxial loading condition are used as an example. As shown in Figure 2(a), the major strain field just before the onset of initial fracture was employed for the forming limits determination. Since a significant strain gradient is generated at the specimen thickness reduced area and the maximum strain was located at the specimen center, only one cross section passing through the specimen center and perpendicular to the crack was considered. As shown in Figure 2(b), the major strains along the cross section were fitted by an inverse parabolic curve ($f(x)=1/(ax^2+bx+c)$). The inner boundary of the fit window was determined by the maximum value of the second derivative of the major strains, while the width of the fit window was calculated by the function $W=10\times(1+\bar{\varepsilon}_2/\bar{\varepsilon}_1)$. The major limit strain was represented by the highest point of the inverse parabolic curve, and the minor limit strain was identified from the strain path at the corresponding major limit strain [17]. By adjusting the biaxial tensile ratios on the two cruciform specimens, the predicted FLC on the basis of CRIT1 is presented in Figure 6.

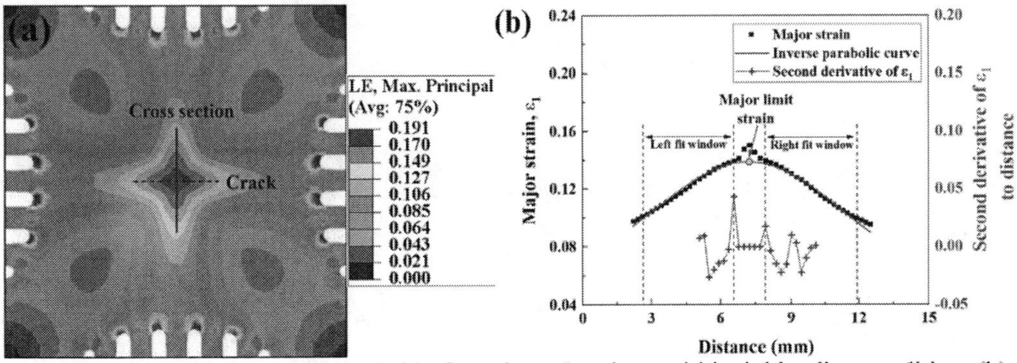

Figure 2. (a) The major strain field of specimen I under equi-biaxial loading condition, (b) determination of the major limit strain based on the CRIT1.

2.2.2. Thickness variation method (CRIT2). Once the localized necking occurs, material points within the necking area will show a significant difference in thickness variation rate compared to the points at the edge and outside of the necking area. Thickness variation method is based on this strain localisation, and the key is the identification of the center and edge of the necking area [10]. As shown in Figure 3, five points were created distant by 0.2mm along the direction perpendicular to the crack, starting from the center of the necking area. The thickness variation rates of the five points were analysed to identify the edge of the necking area and then determine the forming limits at necking. It can be seen, in Figure 3(c), that the curves of thickness variation rate for points P1, P2 and P3 decreased monotonically, while the slopes of the curves for P4 and P5 changed from negative to positive after the localized necking occurred. Thus, the edge of necking area was identified at P4, and the moment of localized necking was determined at the time when the slope of the thickness variation rate curve for P4 equal to zero. Subsequently, the forming limits at necking (point P1) can be identified according to the determined necking time. With the thickness variation method, CRIT2, the predicted FLC is presented in Figure 6.

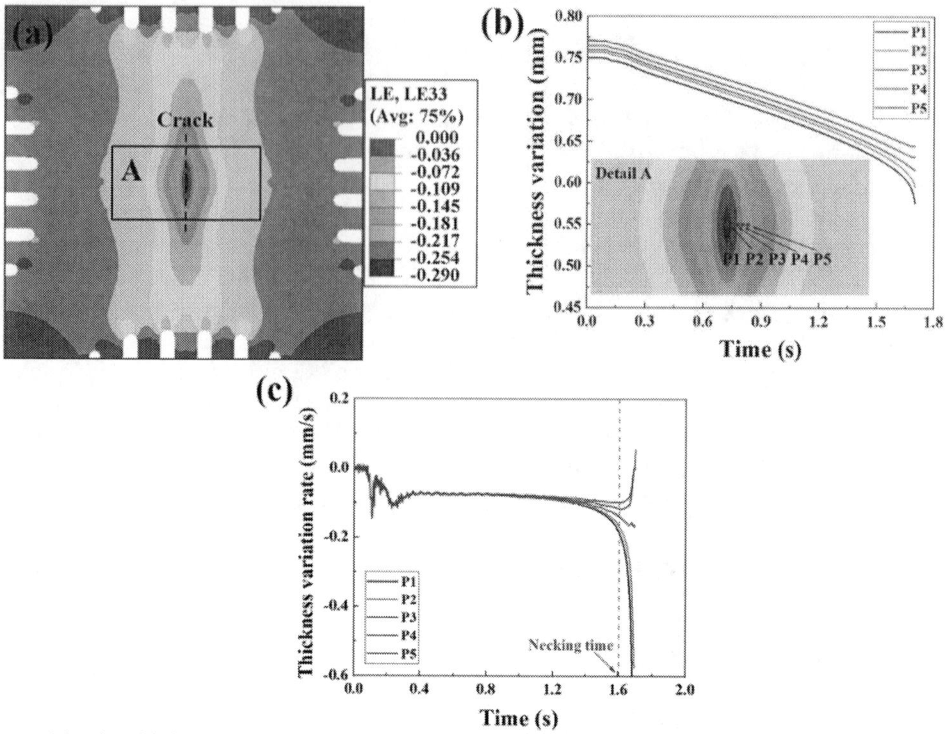

Figure 3. (a) The thickness strain field of specimen I under the biaxial tensile ratio of V_{L1}/V_{L2} = 1/0.4 (mm/s), (b) thickness variations of the five points, (c) determination of necking time based on the CRIT2.

2.2.3. Equivalent plastic strain increment ratio method (CRIT3). Figure 4 presents the application of CRIT3 to determine the forming limits at necking with specimen II under the plane strain condition. Two elements were selected on the surface of specimen II, where element A was at the center of the necking area and element B was outside the necking area at a distance of 1mm from element A. It can be seen in Figure 4(b), as entering the plastic domain, the equivalent plastic strain increase rate of element A was significantly higher than the one of element B. The ratio of the equivalent plastic strain increment between the two elements was calculated by the function ($r=\Delta\bar{\varepsilon}_A^p/\Delta\bar{\varepsilon}_B^p$), with a time interval of 0.025s. As presented in Figure 4(c), when the ratio reaches the critical value (in this work, the common value of 7 was selected), localized necking appears. Then, according to the found localized necking time, the forming limits at necking are given by the corresponding strain pairs. With the CRIT3, the predicted FLC is shown in Figure 6.

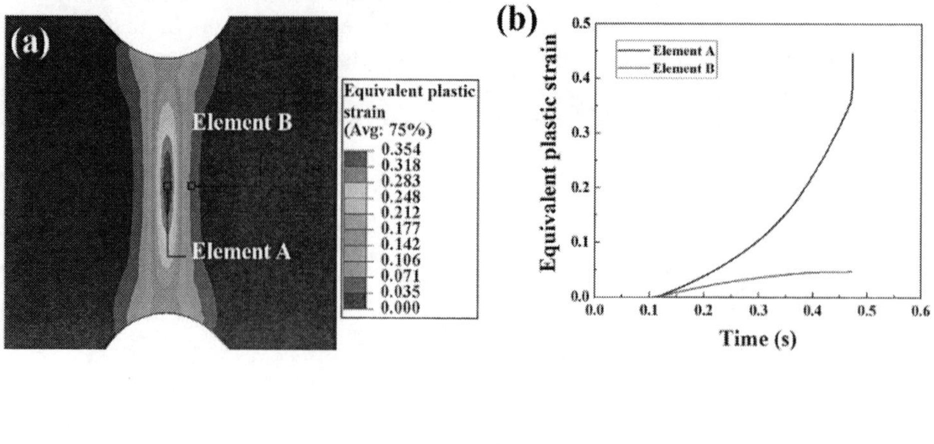

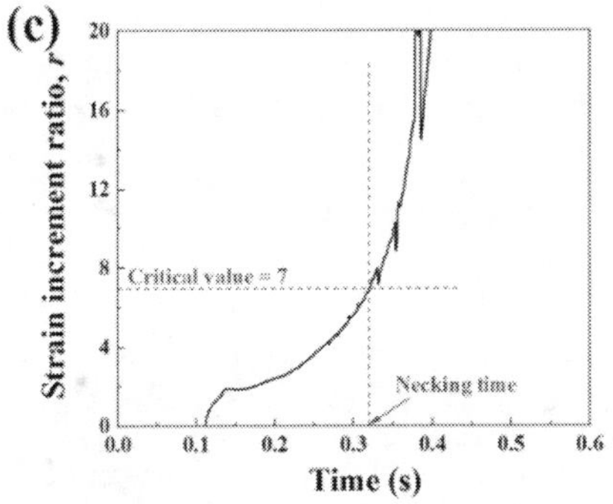

Figure 4. (a) Equivalent plastic strain field of specimen II under plane strain condition, (b) evolutions of the equivalent plastic strain in Elements A and B, (c) evolution of the equivalent strain increment ratio.

2.2.4. Spatio-temporal method (CRIT4). As shown in Figure 5(a), two rectangular zones (BZ and RZ) were defined on the surface of specimen II with the same symmetric line located at the center of necking area. The size of the BZ zone was 3.3×1.2 (mm) and the size of the RZ zone was 3.3×3 (mm). Since the localized necking induces an increase in the non-uniformity of the thickness strain distribution, more thickness strain is accumulated in the BZ zone, so that the ratio of the average thickness strain between the two zones develops nonlinearly. In this method, the localized necking is assumed to appear at the beginning of an increasing difference between the average thickness strain within the two zones. In figure 5(b), two straight lines were used to fit the uniform deformation stage and the final stage of the average thickness strain curve, respectively. The intersection point of the two fitted lines was recognized as the onset of localised necking. The FLC determined by CRIT4 is shown in Figure 6.

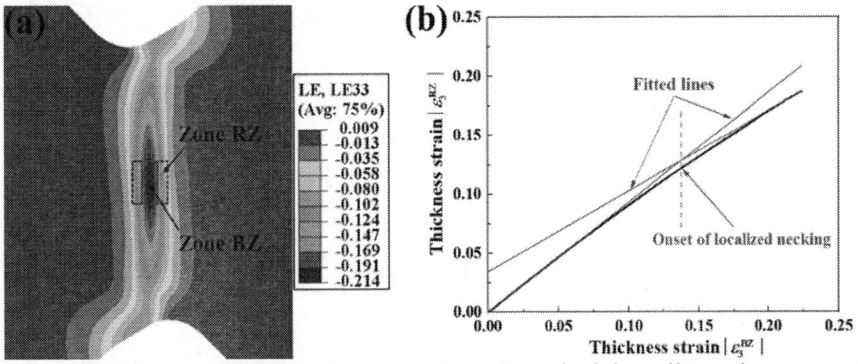

Figure 5. (a) Thickness strain field of specimen II under uniaxial tensile strain state, and selection of zones BZ and RZ. (b) Determination of onset of localised necking based on the CRIT4.

3. Comparison of the predicted FLCs and experimental validation

Figure 6(a) presents the FLCs determined by the four necking criteria. The right branch of the FLC was obtained from specimen I and the left branch from specimen II. The two specimens provided an overlapping area near the plane strain state, and the forming limits of the two specimens under the plane strain state were in good agreement. For the right branch, the forming limits determined by CRIT2 were slightly higher than the results of the other criteria. As CRIT2, CRIT3 and CRIT4 did not detect the occurrence of necking under the equi-biaxial loading condition, the fracture strain was used as the forming limits at necking. For the left branch, significant differences can be observed in the predicted FLCs, especially close to the uniaxial tension state. The reason for the difference might result from the large strain gradient due to the non-uniformed thickness reduction of specimen II.

Therefore, the numerical simulation of a conventional uniaxial tensile test was carried out. Then, the forming limits were determined by the four necking criteria respectively, and the results were utilized as reference values for the comparison. As shown in Figure 6(b), the major forming limit strains determined by CRIT3 and CRIT4 were close to the reference values, while the CRIT1 and CRIT2 provided significantly higher results than the reference values. It can be concluded that CRIT1 and CRIT2 are greatly influenced by the non-uniformed thickness reduction, and both criteria overestimate the forming limits in specimen II. In contrast, CRIT3 and CRIT4 are more stable and therefore more suitable for application to the specimen with non-uniformed thickness reduction.

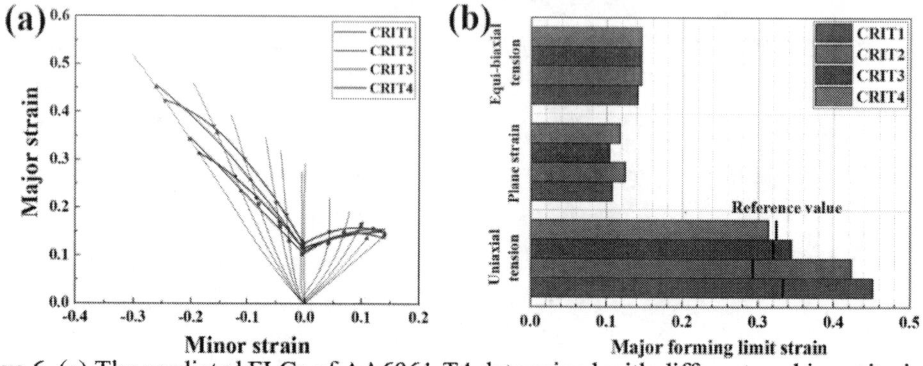

Figure 6. (a) The predicted FLCs of AA6061-T4 determined with different necking criteria, (b) the major forming limit strains for three representative strain states.

4. Conclusion

In this work, four existing necking criteria are applied and discussed for numerical procedure in order to evaluate their implementation for the specimens with non-uniformed thickness reduction. The results show that the non-uniformed thickness reduction causes CRIT1 and CRIT2 to overestimate the forming limits of the material. In contrast, CRIT3 and CRIT4 are more stable and therefore more suitable for the thickness reduced cruciform specimen. In addition, from an experimental point of view, the forming limit determination by CRIT1 relies on the best fit of an inverse parabolic curve; however the large strain gradient in the specimen with non-uniformed thickness reduction brings the challenge to the best fit. CRIT2 is greatly influenced by the noise in the strain measurement and therefore requires smoothing of the strain-time data. CRIT3 and CRIT4 are easy to use for experimental determinations of forming limits at necking. However, the critical value used in CRIT3 is usually empirical and lacks physical meaning. The width of the fitting window used to fit the straight lines in CRIT4 has a large influence on the results of the forming limits determination. A complete comparison between predictive and experimental results for all the strain paths will confirm the choice of the appropriate criterion.

References

[1] Keeler S P and Backhofen W A 1963 Plastic instability and fracture in sheet stretched over rigid punches. *Trans. Am. Soc. Met.* **56** 25–48

[2] BS EN ISO 12004, Metallic materials - Sheet and strip - Determination of forming limit curves Part 2: Determination of forming limit curves in the laboratory, 2008

[3] Min J, Stoughton T B, Carsley J E and Lin J 2016 Compensation for process-dependent effects in the determination of localized necking limits. *Int. J. Mech. Sci.* **117** 115-134

[4] Zhang R, Shi Z, Shao Z, Yardley V A and Lin J 2021 Biaxial test method for determination of FLCs and FFLCs for sheet metals: validation against standard Nakajima method. *Int. J. Mech. Sci.* **209** 106694

[5] Leotoing L, Guines D, Zidane I and Ragneau E 2013 Cruciform shape benefits for experimental and numerical evaluation of sheet metal formability. *J. Mater. Process. Tech.* **213** 856-863

[6] Zidane I, Guines D, Leotoing L and Ragneau E 2010 Development of an in-plane biaxial test

for forming limit curve (FLC) characterization of metallic sheets. *Meas. Sci. Technol.* **21**(5) 055701

[7] Song X, Leotoing L, Guines D and Ragneau E 2017 Characterization of forming limits at fracture with an optimized cruciform specimen: Application to DP600 steel sheets. *Int. J. Mech. Sci.* **126** 35-43

[8] Chu X, Wang Z, Chen C, Lin S and Leotoing L 2020 Experimental investigation of punch curvature influence in Nakazima test and numerical FLD prediction of AA5086. Ferroelectrics **565** 12-25

[9] Chu X, Leotoing L, Guines D and Ragneau E 2014 Temperature and strain rate influence on AA5086 Forming Limit Curves: experimental results and discussion on the validity of the MK model. *Int. J. Mech. Sci.* **78** 27-34

[10] Iquilio R A, Castro Cerda F M, Monsalve A, Guzman C F, Yanez S J, Pina J C, Vercruysse F, Petrov R H and Saavedra E I 2019 Novel experimental method to determine the limit strain by means of thickness variation. *Int. J. Mech. Sci.* **153-154** 208-218

[11] Marciniak Z, Kuczynski K 1967 Limit strains in the processes of stretch-forming sheet metal. *Int. J. Mech. Sci.* **9** 609-620

[12] Zhang C, Leotoing L, Zhao G, Guines D and Ragneau E 2011 A Comparative Study of Different Necking Criteria for Numerical and Experimental Prediction of FLCs. *J. Mater. Eng. Perform.* **20** 1036-1042

[13] Zhang R, Shi Z, Shao Z, Dean T A and Lin J 2021 A novel spatio-temporal method for determining necking and fracture strains of sheet metals. *Int. J. Mech. Sci.* **189** 105977

[14] Wang Z, Guines D, Chu X and Leotoing L 2022 Characterization of forming limits at fracture from shear to plane strain with a dedicated cruciform specimen. *Int. J. Mater. Form.* **15** 7

[15] Gurson A L 1977 Continuum theory of ductile rupture by void nucleation and growth: Part I - Yield criteria and flow rules for porous ductile media. *J. Eng. Mater. Technol.* **99** 2

[16] Yildiz R A and Yilmaz S 2020 Experimental Investigation of GTN model parameters of 6061 Al alloy. *Eur. J. Mech. A-Solid.* **83** 104040

[17] Noder J and Butcher C 2019 A comparative investigation into the influence of the constitutive model on the prediction of in-plane formability for Nakazima and Marciniak tests. *Int. J. Mech. Sci.* **163** 105138

IOP Conf. Series: Materials Science and Engineering 1238 (2022) 012052 doi:10.1088/1757-899X/1238/1/012052

Axisymmetric V-Bending of Sheet Metal: Determining the Fracture Strain and the Weakest Material Direction for Plane Strain Tension in One Test

T Beerli[1*], V Grolleau[1,2], D Mohr[1], C C Roth[1]

[1] Department of Mechanical and Process Engineering, Swiss Federal Institute of Technology (ETH), Zurich, Switzerland
[2]Univ. Bretagne Sud, Laboratoire IRDL UMR CNRS 6027, F-56100 Lorient, France
*beerlit@ethz.ch

Abstract. Plane strain tension is one of the most critical loading conditions leading to ductile failure in forming and crash applications. Hence knowing the fracture strain and weakest orientation for this stress state is crucial for safe design. A novel axisymmetric V-bending experiment is proposed to identify the strain to fracture and the weakest sheet material orientation for plane strain tension under proportional loading. After clamping its inner and outer boundaries, a disc-shaped specimen with an annular gage section is bent over a tubular knife of diameter 54mm. A single camera takes images of the top surface of the specimen throughout the experiment. It allows for timely crack detection and digital image correlation-based strain measurements. Experiments are performed on a range of modern engineering materials comprising two aluminium alloys (AA7075 and AA6014) as well as two steels (DP1180 and CR4). The experiments are complemented by Finite Element simulations to assess the robustness of the novel experimental approach.

1. Introduction

Plane strain tension is one of the most critical loading conditions for ductile failure in forming and crash applications. Most fracture models, whether they are phenomenological or micromechanics-based, assume a minimum in failure strain under plane strain tension. For a Levy von Mises solid, a triaxiality of $\eta = 1/\sqrt{3} \approx 0.58$ and a Lode angle parameter of zero characterizes this stress state.

Existing experimental testing techniques for this stress state can be divided into in-plane and out of plane experiments. One of the most common in-plane experiment uses symmetrical notched tensile specimens [1-3]. These specimens expose severe limitations in the strain field and thus stress state homogeneity, which is partially due to through thickness necking [4, 5]. In addition, early onset of fracture might occur from the free edges in a tensile stress state, thereby leading to an underestimation of the fracture strain for plane strain tension.

Out of plane experiments on the other hand are usually performed on flat specimens, for example in Keeler or Nakazima experiments [6]. The most common out of plane testing technique is V-bending of a flat sheet strip coupon over a knife [7, 8], which has also been included in several norms, e.g. VDA 238-100, DIN 50111, ASTM E290, ISO 7438, and JIS Z2248. The major limitation of the V-bending experiment is the maximum strain that can be obtained, once full fold over is reached. Dihedral Mini-Nakazima experiments [9] can be used to overcome this shortcoming by taking advantage of a stretch-bending type of mechanism.

The present paper is an extension of [10], in which a novel experimental technique is presented which allows to determine the weakest material orientation from a single test. A wide range of material thicknesses between 0.5 to 2.5mm can be assessed, allowing for fracture strains of at least 0.8. The main idea revolves around a disk shaped sheet specimen clamped on the inside and outside being bent over an axisymmetric knife. To validate the novel experimental setup, four new sheet metals are tested with up to five repeats: 1.2mm thick aluminium alloy AA6014, 2mm thick aluminium alloy AA7075, 1mm thick dual phase steel DP1180 and a 0.6mm thick deep drawing steel CR4. The results are compared to Dihedral Mini-Nakazima experiments as well as to numerical simulations.

2. Design

In most plane strain tension fracture experiments the material is tested in one orientation at a time, hence several experiments with different orientations have to be performed per material. The idea of the novel axisymmetric V-bending technique is to test all material orientations at the same time and determine the orientation with the lowest fracture strain under plane strain tension. Figure 1a shows the setup used for the test. It comprises a lower part (dark) which is fixed to the machine and does not move during the experiment. It consists of an inner shaft ② which clamps the disk shaped sample on the inside (Figure 1d) with a single fine pitch M16-12.9 screw. The outside of the specimen is clamped by a top and bottom ring ③ connected with eight M10 screws. The outer clamping of the specimen is connected to the lower fixture with the help of four rods ⑤. The upper part (light gray) is connected to the load cell and the moving crosshead of the testing machine as well as to the tubular knife ④ (Figure 1c). The camera with attached ring light ⑥ is connected to the upper part to maintain a constant focal distance from the specimen surface. The clamping solution chosen for this setup allows an unobstructed view on the whole specimen surface (Figure 3c) enabling planar Digital Image Correlation.

Figure 2 exemplarily shows the mechanism of the experiment with a 1.5mm thick generic material (DP 780 hardening from [11] with von Mises constitutive framework). The ring shaped knife moves upwards and thereby deforms the disk specimen. From the beginning of the experiment, the strain concentrates in a toroidal prism above the knife with the highest strains on the top surface. To assess the quality of the stress state, the evolution of the stress triaxiality in the element with maximum equivalent plastic is evaluated (Figure 2b). For the whole experiment, the stress triaxiality of the element stays between 0.568 and 0.575, which is well within ±1.5% of the ideal value of $\eta = 1/\sqrt{3}$ for plane strain tension for a Levy-von Mises material. To quantify the amount of deformation we introduce the effective strain based on the principal strains as

$$\bar{\varepsilon} = \frac{2}{\sqrt{3}}\sqrt{\varepsilon_I^2 + \varepsilon_I \varepsilon_{II} + \varepsilon_{II}^2} \ . \qquad (1)$$

IOP Conf. Series: Materials Science and Engineering 1238 (2022) 012052 doi:10.1088/1757-899X/1238/1/012052

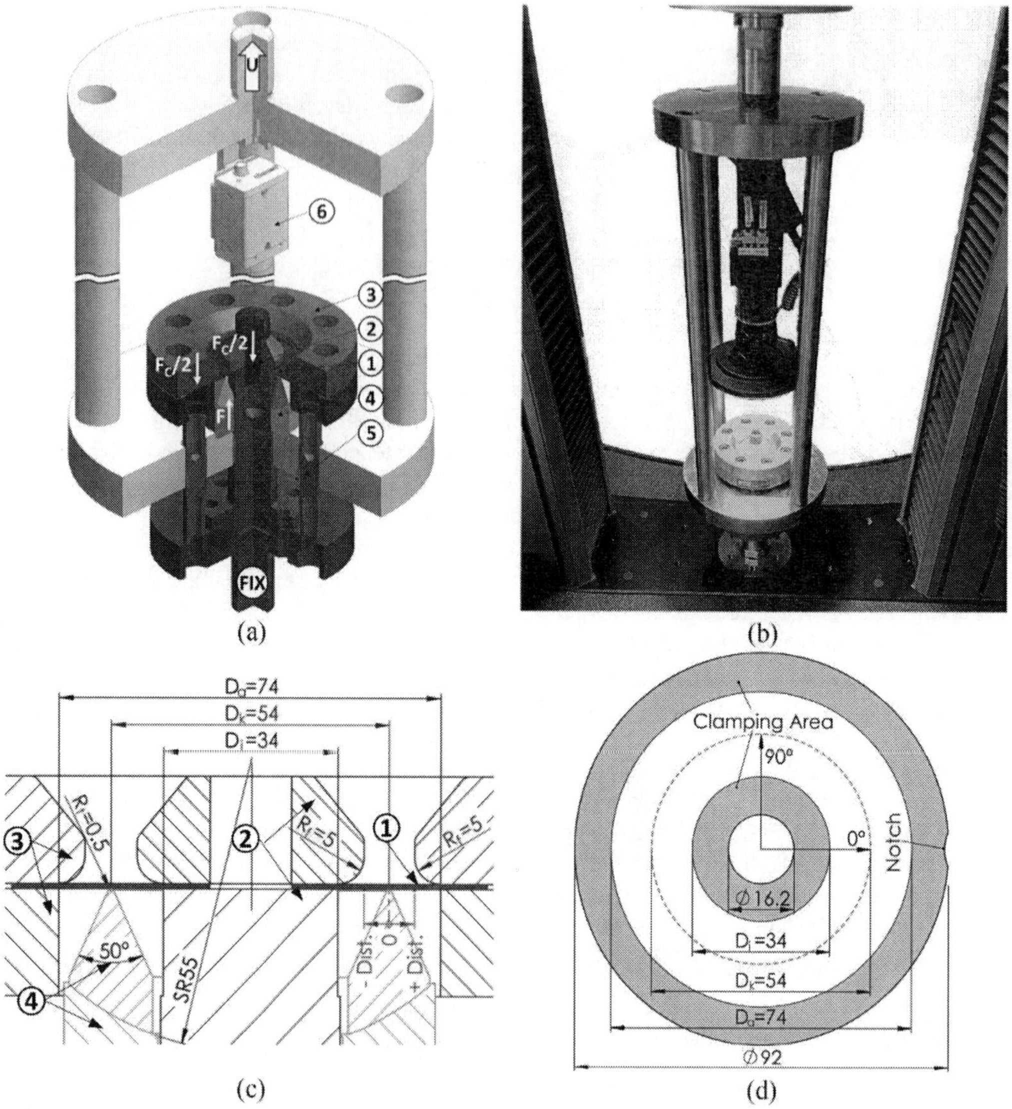

Figure 1. Experimental Setup: (a) axisymmetric V-bending device and (b) picture of setup with ① specimen (red), ② inner clamping with center screw, ③ outer clamping, ④ knife-like circular punch (green), ⑤ lower fixture (dark grey), ⑥ camera with ring light and upper fixture (light grey). (c) Technical drawing of the central part and (d) of the specimen. A notch is introduced to mark the rolling direction.

To determine the optimal specimen dimension a parametric study is performed. The diameter of the disk is limited by the maximum force of the machine. To be able to test sheets up to 2.5mm on a 250kN electro-mechanical testing machine (Instron 5985), the outer diameter is fixed to 92mm (Figure 1d) and the hole diameter to 16.2mm allowing a M16 screw to fit. The single screw allows for a maximum clamping force of 145kN in the centre, which is matched by eight M10 screws on the outside. With the constraint of keeping the clamping width and the distance from the knife on the inside and outside equal, the remaining dimensions of the specimen are selected. Numerical studies show that a sharper knife leads to lower forces at fracture, thus a tip radius of 0.5mm is chosen (Figure 1c). To compensate any misalignment in the setup, the circular knife is set on a spherical joint.

IOP Conf. Series: Materials Science and Engineering 1238 (2022) 012052 doi:10.1088/1757-899X/1238/1/012052

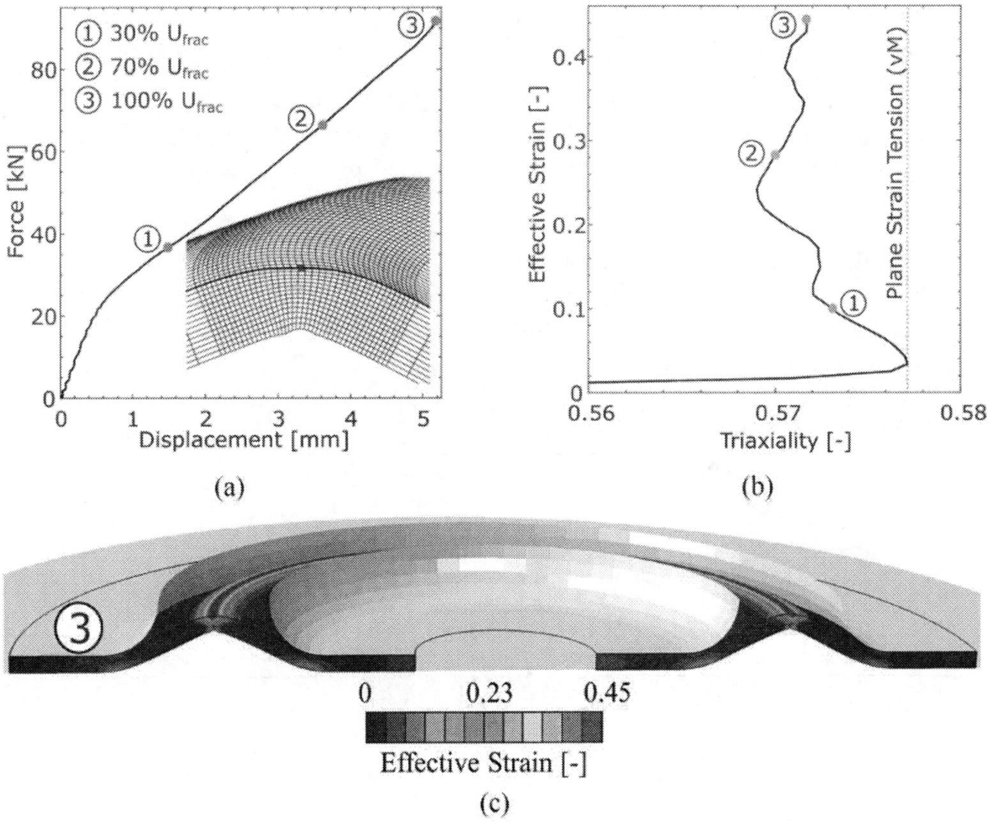

Figure 2. Numerical results for the generic material. (a) Force-displacement curve, (b) loading path for surface element (denoted by red dot in (a)) and (c) contour plot of a section cut of the specimen at time point 3.

3. Computational models

For the design and validation of the testing technique, finite element simulations are performed. For the steels, a rate independent non-associated quadratic plasticity model as proposed in [12] is used, whereas for the aluminium alloy a 3D extension of the YLD2000-2D proposed in [13, 14] is implemented. A self-similar hardening law is used to define the relationship between the equivalent plastic strain and the deformation resistance [13]. All simulations are carried out with the commercial FE solver Abaqus/Explicit (6.13-2) and the respective material models are implemented in VUMAT user subroutines. A model of the full disk is created, with 14 elements over the thickness and an element size in the critical area of 0.1mm (Figure 2a), resulting in more than 600'000 reduced integration solid brick elements (C3D8R). The clamping area as well as the knife are modelled as analytical rigid bodies and a penalty contact friction coefficient of 0.1 is used. Uniform mass scaling is applied to the model so that a simulation is completed within 500'000 explicit time steps, ensuring that inertia effects can be neglected.

4. Experiments

The performance of the new experimental setup is assessed on four engineering materials each with at least 3 repeats:

- 1.2mm thick aluminium alloy AA6014
- 2mm thick aluminium alloy AA7075
- 1mm thick dual phase steel DP1180
- 0.6mm thick deep drawing steel CR4

The samples for the experiments are waterjet cut as shown in Figure 1d. To allow for Digital Image Correlation (DIC), a random speckle pattern with a white background and black speckles of a size of around 50µm is applied. The axisymmetric V-bending setup is mounted in the testing machine and all tests are carried out with a crosshead speed of 2mm/min. A SVS-Vistek hr25CCX (5120 x 5120 pixels) camera with a 55mm f2.0 lens is used to take images of the test at 2fps and with spatial resolution of 14 µm/pixel. A LED ring light is mounted on the lens, evenly lighting the whole specimen surface during the experiment. Images from the experiments are post processed with the commercial DIC software Vic 2D (Correlated Solutions). A subset size of 25 and a step size of 6 are selected and the strain fields are computed using a Gaussian filter of size 5.

Table 1. Yld2000-3D model parameters for aluminium AA6014 and AA7075

	α_1 [-]	α_2 [-]	α_3 [-]	α_4 [-]	α_5 [-]	α_6 [-]	α_7 [-]	α_8 [-]	ρ [kg/m^3]
AA6014	0.915	1.041	0.768	1.003	0.9626	0.7501	0.9305	1.276	2780
AA7075	0.8991	1.039	0.9980	1.008	1.017	0.9519	1.0298	1.052	2780
	A [MPa]	ε_0 [-]	η [-]	k_0 [MPa]	Q [MPa]	β [-]	α [-]	E [GPa]	ν [-]
AA6014	469.3	0.00295	0.259	135.5	179.2	12.169	0.008	73	0.33
AA7075	855.4	0.00262	0.151	502.4	194.1	10.08	0.5	69	0.33

Table 2. Non-associated Hill'48 model parameters for MP980 and CR4 steels

	P_{12} [-]	P_{22} [-]	P_{44} [-]	G_{12} [-]	G_{22} [-]	G_{44} [-]	A [MPa]	ε_0 [-]
CR4	-0.5923	1.021	3.139	-0.6558	0.9081	3.033	553	0.00417
DP1180	-0.4948	0.9896	3.095	-0.4573	0.9023	2.946	1917	0
	η [-]	k_0 [Mpa]	Q [MPa]	β [-]	α [-]	E [GPa]	ν [-]	ρ [kg/m^3]
CR4	0.2704	158.3	236.3	9.427	0.99	200	0.3	7850
DP1180	0.1168	717.1	484.9	169.3	0.4485	202	0.3	7850

4.1. Results for the 2mm thick aluminum alloy AA7075

The results from the three experiments on the 2mm thick aluminum AA7075 are shown in figure 3. Excellent repeatability is obtained on the force displacement curves (black solid lines). All tests expose an initial elastic response (up to 0.7mm and 37kN), followed by a plastic regime that is indicated by a change in slope from 52kN/mm to 14kN/mm before a force maximum is reached (Figure 3a). As the slope of the force-displacement response flattens after transition to plastic deformation, the slope of the average effective strain around the sample (blue solid line) increases. A first crack occurs on the specimen surface at an effective strain between 0.18 and 0.2.

Figure 3b shows the strain evolution along the specimen circumference. A uniform evolution is observed exposing only minor deviations which are attributed to clamping errors. At 90% of the total displacement the first crack appears, which subsequently grows until the sample fails. It is not possible to detect the crack from a change in the evolution of the effective strain or the force displacement curve for this material. Only with optical methods the crack can be detected. To evaluate the quality of the plain strain condition during the experiment, we also introduce the absolute strain ratio

$$\lambda = \frac{\varepsilon_I}{|\varepsilon_{II}|}. \qquad (2)$$

In theory, for plane strain tension this ratio would increase to infinity as the limit of the second principal strain approaches zero. For the selected experiments (Figure 3d) the absolute strain ratio increases with the effective strain. Throughout the test it stays well above 20, reaching more than 60 at the location of fracture.

4.2. Results for the 1mm thick dual phase steel DP1180

The force-displacement and average effective strain curves of the four repeats of the DP1180 experiments are shown in figure 4a. Very good repeatability is observed for both measurements. An initially elastic response (28kN/mm) is followed by a plastic regime (22kN/mm). After reaching a force maximum at about 80kN, a slight decrease in the force level is observed for all specimen before complete loss of load carrying capacity occurs.

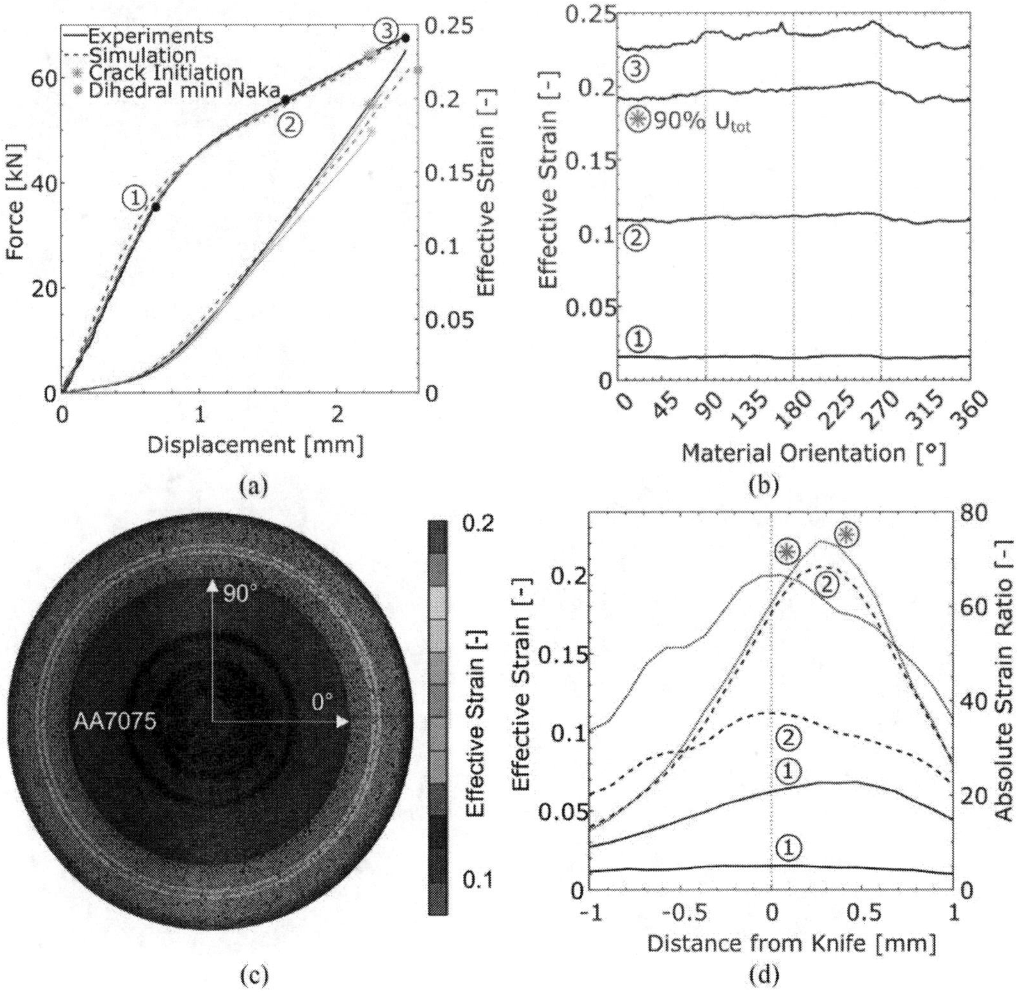

Figure 3. Results for Aluminium AA7075 experiments. (a) Force-displacement curves (solid grey lines) and average effective strain curves (solid blue lines) of the experiments, the corresponding simulation (dashed red line) and the fracture strain for the weakest material orientation of the dihedral mini Nakazima experiment (green dot). The chosen reference experiment is plotted darker and used in (b-d) (b) Evolution of the effective strain around the specimen, (c) Plot of the effective strain at the last instant before failure. (d) Circumferential average effective strain (black) and strain ratio (blue) along a radial path shown in green in figure 1c at the time points denoted in (a).

However, the initial onset of fracture (determined from visual inspection) occurs much earlier at a force level of about 72 kN and an effective strain of 0.26±0.01. The evolution of the crack is shown in figure 4b-d on a two quarter painted sample to better observe the crack. At step I (figure 4b) with U= 0.86Ufmax and only 75% of the strain at force maximum, the first two cracks develop in the negative 90°-orientation. These initial cracks continue to grow and at step II additional cracks start to appear in the 90°-orientation (Figure 4c). At step III, the instant before complete fracture, i.e. full specimen separation, the cracks in 90° and -90°-orientation have grown to ±22.5° each. This underlines the importance of evaluating crack initiation with optical methods – force maximum criteria or slope change criteria might lead to significantly overestimating the fracture strain; in this case by 36%.

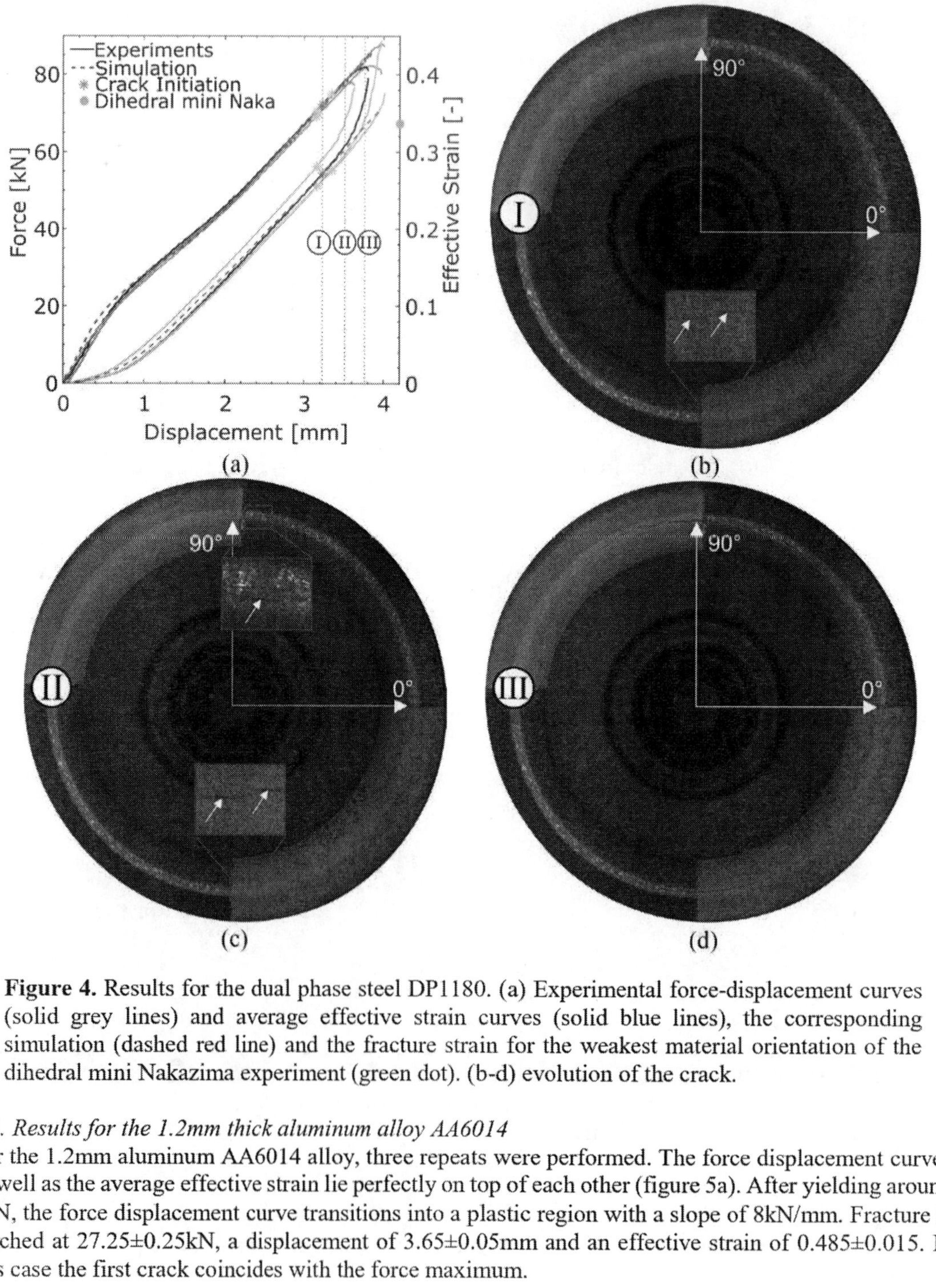

Figure 4. Results for the dual phase steel DP1180. (a) Experimental force-displacement curves (solid grey lines) and average effective strain curves (solid blue lines), the corresponding simulation (dashed red line) and the fracture strain for the weakest material orientation of the dihedral mini Nakazima experiment (green dot). (b-d) evolution of the crack.

4.3. Results for the 1.2mm thick aluminum alloy AA6014

For the 1.2mm aluminum AA6014 alloy, three repeats were performed. The force displacement curves as well as the average effective strain lie perfectly on top of each other (figure 5a). After yielding around 4kN, the force displacement curve transitions into a plastic region with a slope of 8kN/mm. Fracture is reached at 27.25±0.25kN, a displacement of 3.65±0.05mm and an effective strain of 0.485±0.015. In this case the first crack coincides with the force maximum.

4.4. Results for the 0.6mm thick deep drawing steel CR4

As a fourth material a 0.6mm thick deep drawing steel CR4 is tested. Excellent repeatability is obtained for the three repeats (figure 5b), all of which expose only a small elastic region before the transition to the plastic region occurs. The first crack appears in all experiments at the force maximum of 31.6±0.3kN, a displacement of 4.8±0.08mm and an average effective strain of 0.65±0.03. After force maximum the effective strain continues to rise up to 0.8 before the force abruptly drops upon final fracture.

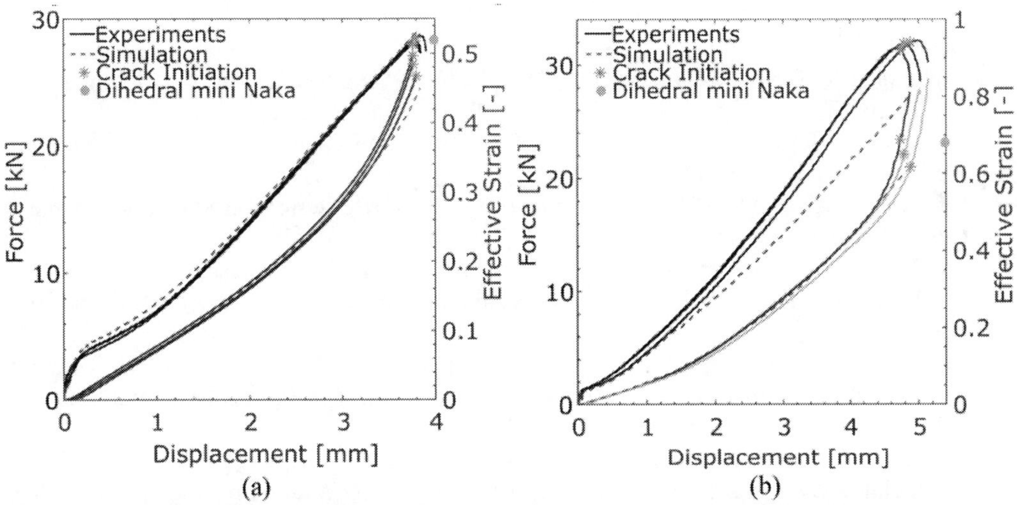

Figure 5. Results for the (a) aluminium alloy AA6014 and (b) the deep drawing steel CR4. Experimental force-displacement curves (solid grey line) and average effective strain curves (solid blue lines), the corresponding simulation (dashed red line) and the fracture strain for the weakest material orientation of the dihedral mini Nakazima experiment (green dot).

5. Discussion

5.1. Comparison with numerical results

For each of the four materials a simulation of the axisymmetric V-bending experiment is carried out using a full finite element model as described in section 2. For each material the calibration is done as described in [9], the model parameters are given in tables 1 and 2. For all but the CR4 the force displacement response from the simulations (red dashed curves in Figures 3a, 4a and 5) show excellent agreement with the experiment, assuming a machine stiffness of 100kN/mm. For the CR4 the simulation underestimates the force at fracture by 15%. This is tentatively attributed to an underestimation of the friction coefficient for the thin sheet. The estimation of the effective strain from the numerical simulations also works well for all materials, matching the fracture strain within less than 5% error.

5.2. Comparison with other plane strain tension experiment

All axisymmetric V-bending experiments are compared to the fracture strains obtained from Dihedral Mini-Nakazima (DMN) experiments in the weakest direction (green dots in Figures 3a, 4a and 5) performed as in [9]. For all experiments the fracture strain from the V-bending is slightly lower than in the DMN experiment with the biggest difference in the DP1180 with a difference of 0.08 and the aluminum AA7075 alloy with a difference of 0.05, while the other two materials have less than 5% difference. This is attributed to the challenging crack detection in DMN experiments which might lead to a late identification of a cracked specimen. Another possible reason might be the difference in measurement technique, as for the axisymmetric V-bending the average effective strain is taken, which might be slightly lower than the strain at the location where the first crack occurs.

6. Conclusion

A novel axisymmetric V-bending experiment is develop that allows characterizing the fracture strain in the weakest direction of sheet metal under proportional plane strain tension loading. The sample has a disk shape which is bent over a tubular knife of 54mm diameter. The design of the device allows to monitor the whole specimen surface with a single camera used to determine the surface strains with the help of planar DIC. The setup allows to probe all material directions in one single experiment and determine the least ductile material orientation for low strain as well as strains up to 0.8. Due to the axisymmetric shape of the specimen, early fracture from free (i.e. machined) edges is not possible.

References

[1] Wagoner R H 1980 Measurement and Analysis of Plane-Strain Work-Hardening *Metall Trans A* **11** 165-75
[2] Gruben G, Hopperstad O S and Børvik T 2013 Simulation of ductile crack propagation in dual-phase steel *Int J Fracture* **180** 1-22
[3] Lou Y and Yoon J W 2017 Anisotropic ductile fracture criterion based on linear transformation *Int J Plasticity* **93** 3-25
[4] Dournaux J, Bouvier S, Aouafi A and Vacher P 2009 Full-field measurement technique and its application to the analysis of materials behaviour under plane strain mode *Mat Sci Eng a-Struct* **500** 47-62
[5] Flores P, Tuninetti V, Gilles G, Gonry P, Duchêne L and Habraken A M 2010 Accurate stress computation in plane strain tensile tests for sheet metal using experimental data *Journal of Materials Processing Technology* **210** 1772-9
[6] Banabic D 2010 Sheet Metal Forming Processes: Constitutive Modelling and Numerical Simulation *Sheet Metal Forming Processes: Constitutive Modelling and Numerical Simulation* 1-301
[7] Roth C C and Mohr D 2016 Ductile fracture experiments with locally proportional loading histories *Int J Plasticity* **79** 328-54
[8] Noder J, Dykeman J and Butcher C 2021 New Methodologies for Fracture Detection of Automotive Steels in Tight Radius Bending: Application to the VDA 238–100 V-Bend Test *Exp Mech* **61** 367-94
[9] Grolleau V, Roth C C, Lafile V, Galpin B and Mohr D 2019 Loading of mini-Nakazima specimens with a dihedral punch: Determining the strain to fracture for plane strain tension through stretch-bending *Int J Mech Sci* **152** 329-45
[10] Beerli T, Grolleau V, Roth C C and Mohr D (submitted) Axisymmetric V-Bending: a Single Experiment to Determine the Fracture Strain and Weakest Sheet Material Direction for Plane Strain Tension.
[11] Li X, Roth C C, Bonatti C and Mohr D 2022 Counterexample-trained neural network model of rate and temperature dependent hardening with dynamic strain aging *Int J Plasticity* **151** 103218
[12] Mohr D, Dunand M and Kim K H 2010 Evaluation of associated and non-associated quadratic plasticity models for advanced high strength steel sheets under multi-axial loading *Int J Plasticity* **26** 939-56
[13] Roth C C, Fras T and Mohr D 2020 Dynamic perforation of lightweight armor: Temperature-dependent plasticity and fracture of aluminum 7020-T6 *Mech Mater* **149**
[14] Dunand M, Maertens A P, Luo M and Mohr D 2012 Experiments and modeling of anisotropic aluminum extrusions under multi-axial loading - Part I: Plasticity *Int J Plasticity* **36** 34-49

IOP Conf. Series: Materials Science and Engineering 1238 (2022) 012053 doi:10.1088/1757-899X/1238/1/012053

On the use of elliptical bulge tests in material characterization through inverse methodologies

M Rossi and A Lattanzi

Department of Mechanical Engineering and Mathematical Science, Università Politecnica delle Marche, Ancona, Italy

E-mail: m.rossi@staff.univpm.it

Abstract. The hydraulic bulge test is an experimental technique that can be used to identify the properties of sheet metals up to large strain. The hydraulic bulge test, usually, is a free expansion of a sheet blank through a circular hole, driven by the pressure of a fluid, where the ends of the blank are blocked using a suitable die. The state of stress during the bulge test is mainly equi-biaxial, however, using elliptical holes with different aspect ratios, it is possible to produce heterogeneous stress-strain fields that can be used to identify the properties of the material through an inverse method; in this case, the non-linear Virtual Fields Method (VFM) was adopted. With respect to the traditional tests employed with the VFM in plasticity, i.e. double notched specimens or specimens with holes subjected to tensile tension, elliptical bulge tests allow to investigate a different zone in the stress-strain space, with a higher triaxiality. The capability of this type of test to calibrate the parameters of advanced anisotropic models is here assessed using simulated experiments.

1. Introduction

The hydraulic bulge test is a well established experimental protocol generally used to investigate the mechanical response of thin bodies under multiaxial loading conditions. The test finds several examples of application for the characterization of the plastic behaviour of sheet metals thanks to the production of a biaxial stress state on the dome apex. This also allows to achieve larger plastic deformations compared to the standard uniaxial tensile test, providing important information on the material formability.

Circular bulge tests has been already used to characterize the plastic anisotropy of sheet metals [1, 2] and for the study of the fracture behaviour of advanced high strength steels [3]. The working principle of bulge test was also used to develop other multiaxial tests such as the multiaxial tube expansion in [4] or the use of elliptical dies for the bulge test [5], in order to investigate the mechanical response of the material under un-balanced biaxial conditions up to large plastic deformations.

The computation of the biaxial stress and strain curve is generally performed by using kinematic relationships between the dome height, the bulge curvature and the forming pressure: for instance, in the work of Rees [6] is presented an analytical method to predict the hoop and longitudinal stress at the ellipsoidal dome on the basis of the Hill48 anisotropy model (i.e. the method requires to know yield function coefficients), while a more general formulation based only on kinematic relations was recently proposed in [7].

IOP Conf. Series: Materials Science and Engineering 1238 (2022) 012053 doi:10.1088/1757-899X/1238/1/012053

2. Inverse methodology

Usually, the bulge test is used to identify the stress-strain curve in a single point of sheet metal, however, using inverse methods, it is possible to include in the identification algorithm the whole full-field measurement obtained with a suitable optical system, e.g. digital image correlation.

One of the most used inverse methods is the non-linear virtual fields method (VFM). the VFM can be used to calibrate a set of constitutive parameters $\boldsymbol{\xi}$ by a minimization of a cost function $\psi(\boldsymbol{\xi})$ that represents the balance between the internal and the external virtual work, according to the principle of virtual work (PVW). Using the finite strain theory the PVW can be written in the reference configuration using the 1st Piola-Kirchhoff tensor $\mathbf{T}^{1PK}$, so $\psi(\boldsymbol{\xi})$ becomes:

$$\psi(\boldsymbol{\xi}) = \left| \underbrace{\int_{\mathscr{B}_0} \mathbf{T}^{1PK} : \delta\mathbf{F}^{\bullet}\, dV_0}_{\text{IVW}} - \underbrace{\int_{\partial\mathscr{B}_0} \left(\mathbf{T}^{1PK} \cdot \mathbf{n}_0\right) \cdot \delta\mathbf{v}\, dA_0}_{\text{EVW}} \right| \tag{1}$$

where the first integral is the internal virtual work (IVW) and the second is the external virtual work (EVW); $\delta\mathbf{v}$ is a kinematically admissible virtual fields and $\delta\mathbf{F}^{\bullet}$ is the corresponding gradient. The parameters $\boldsymbol{\xi}$ are used to compute the stress from the measured strain field using a suitable stress integration algorithm (e.g. radial return, backward Euler, etc.). This approach can be applied to the bulge test using the plane stress assumption. In fact, Eq. 1 is valid for any virtual field and it is written in the undeformed configuration, that, in this case, is a planar sheet. Thus, selecting virtual fields of this form:

$$\delta\mathbf{v} = \begin{bmatrix} \delta v_x \\ \delta v_y \\ 0 \end{bmatrix} \quad \text{and} \quad \delta\mathbf{F}^{\bullet} = \begin{bmatrix} \delta F^{\bullet}_{xx} & F^{\bullet}_{xy} & 0 \\ \delta F^{\bullet}_{yx} & F^{\bullet}_{yy} & 0 \\ 0 & 0 & 0 \end{bmatrix} \tag{2}$$

all components in z are equal to zero and the problem reduces to two dimensions. A thorough description of the application of VFM in case of shell structures where the bending stress is negligible is given in [8].

3. Numerical model

Nowadays, simulated experiments based on numerical simulations are becoming the state-of-art for the study of identification strategies through heterogenous test, allowing to explore different fundamental aspects of the inverse procedure such as the specimen geometry [9] or the error associated to the experimental setup [10].

In this work, numerical simulations of elliptical bulge test by using different test configurations were performed to define the mechanical states produced on the specimens. All the analysis were performed by means of the commercial implicit FE solver Abaqus/Standard®. As illustrated in Figure 1, the numerical model is composed by two main parts: the elliptic die, which drives the deformed shape of the specimen, and the sheet-metal specimen. Due to the double symmetry of the problem, only one quarter of the specimen was modelled.

Here we considered two principal elliptical die openings of the bulge, regulated through the major radius r_x and the minor radius r_y and expressed by means of the aspect ratio $\beta = r_y/r_x$. The specimen geometry is also defined according to β, assuming a constant thickness of 0.8 mm. All the details about the die and specimen dimensions are listed in Table 1. Note that the die's fillet radius r_f was kept constant and equal to 8 mm in all the analyses, according to the study presented in [7].

The specimen was discretized by using solid 8-node linear elements with reduced integration and hourglass control (C3D8R), employing an average mesh size of 2 mm and setting-up 6 elements along the thickness. Besides, the die was modelled as discrete rigid shell, imposing an

IOP Conf. Series: Materials Science and Engineering 1238 (2022) 012053 doi:10.1088/1757-899X/1238/1/012053

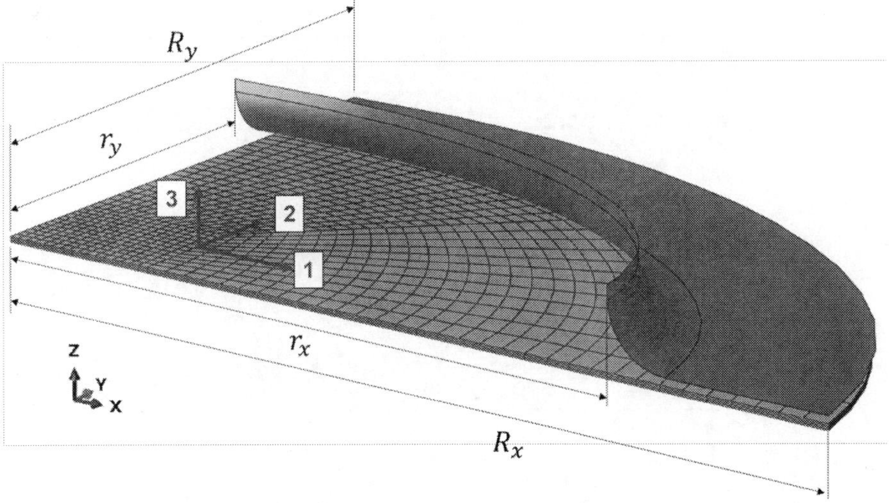

Figure 1. Schematic view of the numerical model adopted for the virtual experiment. The coordinate system in red indicates the material orientation of the specimen, where 1 indicate the rolling direction RD, 2 the transverse direction TD and 3 the normal direction ND.

Table 1. Elliptic bulge aspect ratios and specimen dimensions numerically simulated.

β	r_x	r_y	R_x	R_y
0.5	100	50	130	80
0.25	100	25	130	55

average mesh size of 2 mm. To avoid a disproportionate radial sliding far from the gauge zone and to simulate the effect of the blank-holder, all nodes on the outer side of the specimen were fixed, while the pressure load was imposed on all the bottom surface. The contact between the blank specimen and the die was modelled as frictionless hard contact with finite sliding, enforced by using the Augmented Lagrange Method. Such boundary condition is an approximation of what occurs in a real test, however, it can be considered acceptable in this case because the strain maps used for the VFM identification are far from the edge of the die.

The material behaviour introduced in the FE modelling was taken from previous investigations of the authors on a BH340 bake-hardenable steel [11], adopting the hardening curve obtained from circular bulge tests on the same material and here described through the Swift law:

$$\bar{\sigma}(\bar{\varepsilon}_p) = K(\varepsilon_0 + \bar{\varepsilon}_p)^N \tag{3}$$

where $\bar{\sigma}$ is the flow stress, $\bar{\varepsilon}_p$ is the accumulated equivalent plastic strain and K, ε_0 and N are material parameters. Plastic anisotropy was introduced through the Hill48 material model [12], whose coefficients are listed in Table 2.

Table 2. Material parameters adopted in the FE model.

Elastic Properties					
E/GPa			ν		
200			0.3		
Plastic Properties					
Swift law			Lankford coeff.		
K/MPa	ε_0	N	R_0	R_{45}	R_{90}
709.2	0.0021	0.259	1.52	0.89	1.67
Hill48 coefficients					
f	g	h	l	m	n
0.3612	0.3968	0.6032	1.5	1.5	1.0536

4. Results and discussions

A fundamental characteristic of inverse methods based on full-field measurements is the capability to extract more information from a single test compared to standard calibration methods. Therefore, one of the basic ideas is to generate heterogeneous mechanical states on the specimen to provide a comprehensive description of the material behaviour during the identification process. An effective strategy to increase the amount of material data – especially in the case of anisotropic materials – is also to feed the identification with test on specimen obtained at different material orientations.

Following this approach, we here considered two different tests, one with the RD and the other with the TD directions aligned with the specimen major axis (R_x), for each elliptical geometry. Figure 2 displays the displacement field along the Z direction and the corresponding equivalent plastic $\bar{\varepsilon}_p$ strain distribution at the last timestep during the bulge test with $\beta = 0.5$. For both specimen orientations, the maximum dome height exceeded 40 mm, while the maximum equivalent plastic strain was 0.57% and 0.67% for the RD-parallel and TD- parallel configurations, respectively. It is worth mentioning that we considered as last step the last converged increment before plastic instabilities and excessive elements distortions on the dome height occurred during the analysis. The effect of anisotropy is also showed by the different maximum value of forming pressure obtained for the two tests: 92 bar for the the RD-parallel and 82 bar for the TD-parallel specimens.

Reducing the aspect ratio of the ellipse to $\beta = 25\%$ had the primary effect to drastically increase the forming pressure to 200 bar for the RD-parallel configuration and 180 bar for the TD-parallel one. Nonetheless, the maximum dome height was 25 mm and 27 mm for the two specimens, achieving a maximum plastic strain of 0.46% per the RD-parallel specimen and 0.62% for the TD-parallel specimen (Figure 3).

To evaluate the heterogeneity of the mechanical states generated by the elliptical bulge tests, Figure 4 illustrates the stress distribution on the normalised π-plane combined with the contours of the reference yield surface. Note that all the data are reported with respect the pure equi-biaxial stress state, where the yield surface is equal to 1.

The elliptical bulge with aspect ratio $\beta = 0.5$ produced large amount of mechanical states capable of covering a wide portion of the yield surface that includes the equi-biaxial stess condition, and mostly all the generalised plane strain condition for material orientations from 0° to 90° with respect to the RD (Figure 4a). The two specimen orientations with respect the ellipse major axes does not show significant differences in stress heterogeneity, however, a

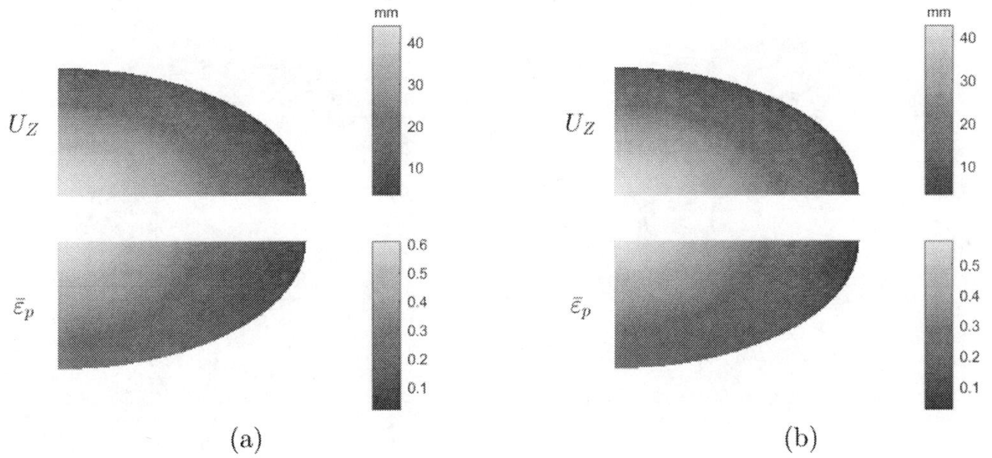

Figure 2. Full-field maps of the out-of-plane displacement and equivalent plastic strain from elliptic bulge tests with $\beta = 0.5$: (a) RD parallel to the ellipse major axis, (b) TD parallel to the ellipse major axis.

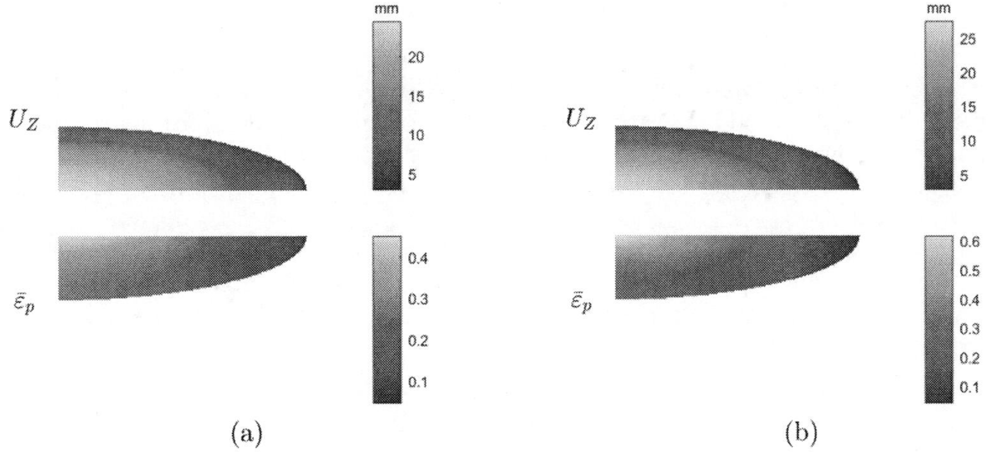

Figure 3. Full-field maps of the out-of-plane displacement and equivalent plastic strain from elliptic bulge tests with $\beta = 0.25$: (a) RD parallel to the ellipse major axis, (b) TD parallel to the ellipse major axis.

more dense number of points is observed on the yield surface hemisphere corresponding to the material orientation parallel to the ellipse minor axis.

On the other hand, reducing the bulge aspect ratio to $\beta = 0.25$ polarizes the stress distribution towards the uniaxial tensile condition (Figure 4b) of the material orientation parallel to the ellipse minor axis. Although there is also here large number of datapoints in the plane strain condition, there are no information about the pure equi-biaxial stress state, moreover, only few points get close to the plane strain along the 45° direction with respect to the RD.

From this analysis it turns out that the elliptical bulge test with $\beta = 0.5$ represents the best candidate for the identification of the anisotropic behaviour of materials.

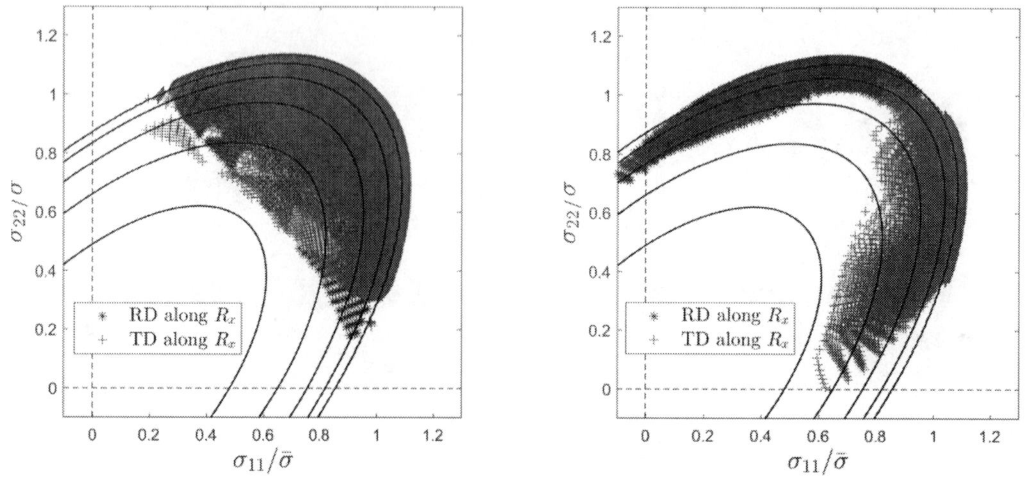

(a) (b)

Figure 4. Distribution of the stress states with yield surface for: (a) $\beta = 0.5$, (b) $\beta = 0.25$. The contour lines of the yield surface are plotted every $\sigma_{12}/\bar{\sigma} = 0.1$

4.1. Parameter identification

A detailed study of the VFM application to elliptical bulge test is beyond the scope of the paper. The intent here is instead to compare the potentiality of the elliptical bulge test with respect to the classical geometries used to characterize anisotropic plasticity model with VFM, i.e. double notched specimen.

To this purpose, the elliptical bulge test with RD along x and $\beta = 0.5$ is used to identify the coefficients of the Swift law and the Hill48 criterion using a single simple virtual field, which is based on the equilibrium of the forces along the z-direction. The internal virtual work (IVW) and external virtual work (EVW) are, respectively:

$$IVW = p \int_{\mathscr{A}_0} F_{33}^{-T} \, dA_0 \tag{4}$$

and

$$EVW = \int_{\partial \mathscr{A}_0} \left(T_{31}^{1PK} \cos\alpha + T_{32}^{1PK} \sin\alpha \right) \, s_0 \, dl_0 \tag{5}$$

where p is the fluid pressure, F_{ij}^{T} is the inverse of the deformation gradient and T_{ij}^{1PK} is the first Piola-Kirchhoff stress computed from the strain data and the constitutive parameters $\boldsymbol{\xi}$; the area A_0 is a circle placed at the top of the elliptical bulge, in the undeformed configuration, α is the angle of normal vector with respect to the circumference $\partial\mathscr{A}_0$, and s_0 is the metal sheet thickness. The different steps to obtain Eqq. 4 and 5 are given in [8].

The 6 constitutive parameters are extracted minimizing the difference between IVW and EVW using a standard Nelder-Mead simplex algorithm, the results are listed in Table 3. The algorithm is able to identify correctly the hardening law and R_{90}, however, large errors are obtained for R_0 and R_{45}. To compare the elliptical bulge test with other tests, Table 4 reports the identification error obtained in [13] for the same constitutive model using a double notched specimen cut along the rolling direction. Using a single VF the level of error is much larger with respect to the elliptical bulge test, showing that, potentially, the latter has a richer information.

Table 3. Results of identification.

	Constitutive parameters					
	Swift law			Lankford coeff.		
	K/MPa	ε_0	N	R_0	R_{45}	R_{90}
Reference	709.2	0.0021	0.259	1.67	0.89	1.52
Identified	717.3	0.0024	0.266	1.29	0.41	1.53
Error	1.2%	12.9 %	2.8 %	-22.5 %	-54.3%	0.90 %

Table 4. Error obtained with a double notched specimen in [13]

		Swift law			Lankford coeff.		
Used VF		K/MPa	ε_0	N	R_0	R_{45}	R_{90}
1	Error	-20.8%	14.5%	4.0%	67.5%	70.0%	81.3%
2	Error	-1.0%	15.5%	3.7%	-7.4%	4.7%	52.0%
3	Error	-1.7%	12.2%	2.6%	0.7%	-6.7%	41.2%

However, the experimental complexity of the bulge test is more critical than that of the double notched specimen, this aspect must be also taken into consideration when comparing the two methods.

Eventually, in order to reduce the error obtained with the elliptical bulge test it is necessary to implement more VFs and check the influence of the material orientation, this will be addressed in future works. Moreover, it is necessary to extend the analysis to other anisotropic yield criteria, for instance Yld2000-2D, where more constitutive parameters are involved and allow to better reproduce real materials.

5. Conclusion

This paper presents a preliminary study to assess the effectiveness of elliptical bulge tests in the inverse identification of the anisotropic behaviour of sheet metals. In particular, the study focuses on the stress distribution achievable using elliptical bulge tests with different aspect ratios and present a first identification example using a single user defined virtual field. The main outcomes are:

- the elliptical bulge test allows to cover a large portion of the stress space and the distribution is strongly affected by the aspect ratio β;
- the largest heterogeneity is obtained using an aspect ratio $\beta = 0.5$;
- the identification with the VFM, using a single user defined VF and a single test with $\beta = 0.5$, allows to retrieve correctly the hardening parameters and one of the three anisotropic coefficients;
- although the identification is not perfect, comparatively, the elliptical bulge test leads to better results with respect to the same VFM approach applied to a standard double notched specimen.

In the future it is necessary to implement for the elliptical bulge test case the routines already used to generate automatic virtual fields for the 2D case. Then the developed numerical models will be used to conduct and extensive validation and to properly quantify the accuracy of the VFM applied to elliptical bulge tests.

Acknowledgement

This project has received funding from the Research Fund for Coal and Steel under grant agreement No. 888153.

References

[1] Barlat, F., Aretz, H., Yoon, J. W., Karabin, M., Brem, J., Dick, R., 2005. Linear transfomation-based anisotropic yield functions. *Int. J. Plasticity*, 21(5):1009–1039.

[2] Chen, K., Scales, M., Kyriakides, S., Corona, E., 2016. Effects of anisotropy on material hardening and burst in the bulge test. *Int. J. Solids Struct.*, 82:70–84.

[3] Barnwal, V. K., Lee, S.-Y., Choi, J., Kim, J.-H., Barlat, F., 2021. On the fracture characteristics of advanced high strength steels during hydraulic bulge test. *Int. J. Mech. Sci.*, 190:106032.

[4] Kuwabara, T., Sugawara, F., 2013. Multiaxial tube expansion test method for measurement of sheet metal deformation behavior under biaxial tension for a large strain range. *Int. J. Plasticity*, 45:103–118.

[5] Williams, B. W., Boyle, K. P., 2016. Characterization of anisotropic yield surfaces for titanium sheet using hydrostatic bulging with elliptical dies. *Int. J. Mech. Sci.*, 114:315–329.

[6] Rees, D. W. A., 1995. Plastic flow in the elliptical bulge test. *International journal of mechanical sciences*, 37, 4:373-389.

[7] Chen, K., Scales, M., Kyriakides, S., 2020. A general and unified theory of the kinematic relationships in bulge tests. *J. Mech. Phys. Solids*, 143:104086.

[8] Rossi, M., Lattanzi, A., Barlat, F., J.-H. Kim, 2022. Inverse identification of large strain plasticity using the hydraulic bulge-test and full-field measurements. *Int. J. Solids Struct.*, 111532.

[9] Barroqueiro, B., Andrade-Campos, A., Dias-de-Oliveira, J., Valente, R. A. F., 2020. Design of mechanical heterogeneous specimens using topology optimization. International Journal of Mechanical Sciences, 181, 105764.

[10] Rossi, M., Lava, P., Pierron, F., Debruyne, D., Sasso, M., 2015. Effect of DIC spatial resolution, noise and interpolation error on identification results with the VFM. *Strain*, 51(3), 206-222.

[11] Lattanzi, A., Barlat, F., Pierron, F., Marek, A., Rossi, M. 2020. Inverse identification strategies for the characterization of transformation-based anisotropic plasticity models with the non-linear VFM. *International Journal of Mechanical Sciences*, 173, 105422.

[12] Hill R, 1948. A theory of the yielding and plastic flow of anisotropic metals. P Roy Soc A-Math Phy, 193:281–297

[13] Rossi, M., Pierron, F., Štamborská, M., 2016. Application of the virtual fields method to large strain anisotropic plasticity. *Int. J. Solids Struct.*, 97:322–335, 2016.

Hot tensile and expansion tests of Ductibor®1000 steel

N Demazel[1], A Boyer[1,2], H Laurent[1] and M C Oliveira[2]

[1]Univ. Bretagne Sud, UMR CNRS 6027, IRDL, F-56321 Lorient, France
[2]CEMMPRE, Department of Mechanical Engineering, University of Coimbra, Polo II, Rua Luís Reis Santos, Pinhal de Marrocos, 3030-788 Coimbra, Portugal

E-mail: nathan.demazel@univ-ubs.fr

Abstract. The boron steels present high mechanical properties which are of great interest for lightening the vehicle parts. But these steels, as the Ductibor®1000, need to be hot formed. The hot stamping process consists of a heating/austenization step leading to an increase of the formability, followed by simultaneous forming and quenching steps. In order to understand the thermomechanical and metallurgical cycles occurring during this process, this study focuses on two different test devices at high temperatures using direct resistance heating: a Gleeble 3500 machine for tensile tests and a new bulge test device. The two different tests were performed on the Ductibor®1000 steel for a temperature of 900°C (with temperature holding during deformation) after an austenization step, which consists of heating up to 900°C and a temperature holding stage. The stress-strain curves of the tensile tests obtained using Digital Image Correlation (DIC) are compared to the stress-strain curves of the bulge tests obtained using a laser profilometer and the membrane theory.

1. Introduction

In order to reduce greenhouse gas emissions, the car manufacturers look for lighter vehicles while maintaining high safety standards. The quenchable steels with their high mechanical properties can respond to these requests. The hot stamping process allows to produce parts with these quenchable steels, as described in detailed in [1,2,3]. It consists of an austenization step at high temperature follow by simultaneous forming and quenching steps in a press to obtain high tailored properties [4].

Some devices were developed to perform tests over the temperature range in the austenitic state, mainly to evaluate the forming limit curve (FLC). In their work, Shao et al. [5] proposed an adaptation to the Gleeble machine using cruciform specimens. Thanks to the resistance heating integrated in the Gleeble machine, the test can be performed up to a maximum temperature of 1000 °C. Li et al. [6] used Nakazima tests to characterize the 22MnB5 steel at 600 and 700 °C. Turetta et al. [7] and Bariani et al. [8] developed a device to simulate thermomechanical cycles occurring during hot stamping process thanks to an induction heating of the sheet, a punch with cartridges heaters and compressed air jets for cooling and quenching. This device enables evaluating the FLC of high-strength steels (HSS) at temperatures up to 900 °C under hot stamping conditions.

The major mode encountered in many sheet forming processes is the biaxial stress mode. Some works proposed to evaluate the stress-strain curves in warm and hot conditions in this mode using hydraulic bulge test devices. Li et al. [9] investigated the formability of the AZ31 magnesium alloy using a bulge test device and pulse current to heat a rectangular sheet at 400 °C. In their work, Braun et al. [10,11,12] developed a hot gas bulge test device with resistance heating of rectangular sheets to characterize quenchable steels up to 900 °C.

In this paper, a bulge test device using resistance heating to characterize metal sheets up to 900 °C is presented. This bulge test device differs from others reported in literature because it is possible to test specimens at very high temperature considering a constant temperature during the deformation, which is carried out by pressure of a gas and therefore without friction. The material studied is a Ductibor®1000 steel. Since it is still little studied, this material is succinctly described. Contrary to Usibor® steels, Ductibor® steels offer a high ductility which leads to a very good impact behaviour. Then, the hot bulge test device developed and its principle are presented. The method using a laser profilometer to evaluate the stress-strain curves is described. This method is validated with a comparison to a digital image correlation method (DIC method) on a EN AW6061-T6 aluminium alloy. Finally, the stress-strain curves of the Ductibor®1000 resulting from hot gas bulge tests and a tensile test are given for a forming temperature of 900 °C.

2. Method

2.1. Material
The material studied is the Ductibor®1000 sheet with a thickness of 0.8 mm and a protective layer composed of 90 wt. % aluminium and 10 wt. % silicon. Its chemical composition is given in Table 1.

Table 1: Chemical composition (wt. %) of Ductibor®1000 (ArcelorMittal)

Max. C	*Max. Si*	Max. Mn	Max. P	Max. S	*Al*	*Max. Ti*	*Max. Nb*	*Max. Cu*	*Max. B*	*Max. Cr*
0.10	0.6	1.8	0.03	0.01	0.01 - 0.1	0.05	0.10	0.20	0.005	0.20

2.2. Experimental device
The basic diagram of the hot bulge test principle is shown in Figure 1.a. First, the circular blank is clamped on its periphery between the die and the blank-holder. Then, the blank is heated using resistance heating. Finally, the gas pressure is applied to form the bulge while the temperature is maintained [13].

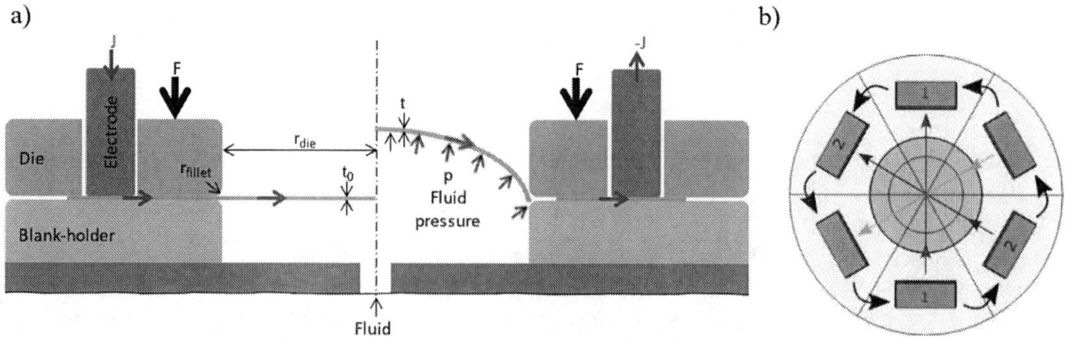

Figure 1: Basic diagram of a) the hot gas bulge test and b) the rotary permutation of active electrode pair.

The resistance heating system is composed of a multi-electrodes system to apply the electrical current to the blank. This system allows a good temperature homogeneity thanks to a rotary permutation of the electrical field, as shown in Figure 1.b. Pneumatic cylinders associated to each pair of electrodes ensure the permutation. The alternative current is provided by an electrical generator, which is supplied by two phases of the three-phase network at 400 V, 200 A, i.e. 80 kVA. The measured root mean square current (RMS current) during heating does not exceed 6000 A and the maximum measured voltage at the electrodes is about 5 V for a maximum power of 22 kW. The applied current is managed by a temperature control loop using a 250 µm diameter K-thermocouple welded at 10 mm from the blank centre. Others K-thermocouples can be welded at 20, 30 and 40 mm from the centre to monitor the temperature uniformity. The tools are placed in the Instron 8803 tensile machine, which applies a clamping force of 300 kN. These tools are electrically isolated from the blank and the electrodes with Miglasil® mica sheets and Kapton® adhesive tapes. The gas pressure is ensured by a 200 bar argon tank associated with a control system, composed of a TESCOM ER5000 controller and a 0-70 bar pressure sensor placed in the cavity under the blank. Concerning the evaluation of the stress-strain curve, the use of a DIC system is more difficult at high temperature, due to the pattern deterioration and contrast losses during bulge test. Thus, the KEYENCE LJ-V7200 laser profilometer is used to monitor the maximum dome height of a line passing through the blank centre and the stress is calculated using the membrane theory. More details about the stress-strain evaluation are given in the next section.

2.3. Stress-strain curves determination

To evaluate the stress-strain curve based on the results obtained with the laser profilometer, there are three key variables, the maximum height of the dome monitored by the laser profilometer, the measured pressure and the initial thickness. The membrane theory is used to determine the stress.

According to the ISO 16808:2014 standard [14], the bending can be neglected if the ratio of the thickness to the diameter of the tool is less than $1/33$ (r_{die} = 60 mm and t_0 = 0.81 mm), leading to the assumption that the stress component normal to the surface is equal to zero at the centre of the blank. Another assumption used in the ISO 16808:2014 standard, is that, at the dome apex, the stress state is assumed equibiaxial. Moreover, the curvature radii at mid-thickness are the same in all directions. Thus, the biaxial stress σ_b is determined using the following form of the membrane theory:

$$\sigma_b = \frac{p\rho}{2t} \tag{1}$$

where p is the applied pressure, t is the thickness of the sheet at the dome apex, which is calculated by the Hill relation [15]:

$$t = t_0\left(1 + \left(\frac{h}{r_{die}}\right)^2\right)^{-2} \tag{2}$$

and ρ is the curvature radius of the dome, defined by the Panknin formula [16] using the dome height h, the die radius r_{die} and the fillet radius of the die r_{fillet} = 5 mm (see these dimensions in Figure 1.a):

$$\rho = \frac{(r_{die} + r_{fillet})^2}{2h} + \frac{h}{2} - r_{fillet} \tag{3}$$

The biaxial strain is calculated from the thickness strain:

$$\varepsilon_b = -\,ln\left(\frac{t}{t_0}\right) \tag{4}$$

In the case of the complete DIC method, the curvature radius is determined via a sphere fitted to the surface and the thickness is calculated using the volume constancy condition during plastic deformation. More details are given in [17].

3. Results and discussion

3.1. Validation of the method using laser profilometer

As written in the first section, the DIC system is hard to use at high temperature, because of the pattern deterioration and the loss in the contrast quality. Thus, to avoid these issues, another system using a laser profilometer is used to determine the stress-strain curve. The method proposed to be used with the laser profilometer was tested for the EN AW6061-T6 at 150 °C. The results are compared to those calculated using two other methods. One is referred as the complete DIC method, since both the radius of curvature and the thickness are evaluated based on data from DIC. The other, based on the method using laser profilometer, uses the evolution of the dome height extracted from DIC data and equations (2-4). Finally, the results obtained from a tensile test performed on a Gleeble machine using a DIC system are also discussed. All these results are shown in Figure 2. Globally, there is a good agreement between the stress-strain curves resulting from the bulge tests, i.e. using the three methods. Concerning the tensile test results, the stress level is lower than the one resulting from the bulge tests. Nevertheless, this can be due to the difference on the strain rate levels attained for each type of test, given that the EN AW6061-T6 presents a positive strain rate sensitivity at high temperature [18]. The tensile test is performed with a strain rate of 2.10^{-3} s^{-1} while the bulge tests start at a strain rate of 1.10^{-3} s^{-1} that increases up to 8.10^{-3} s^{-1}.

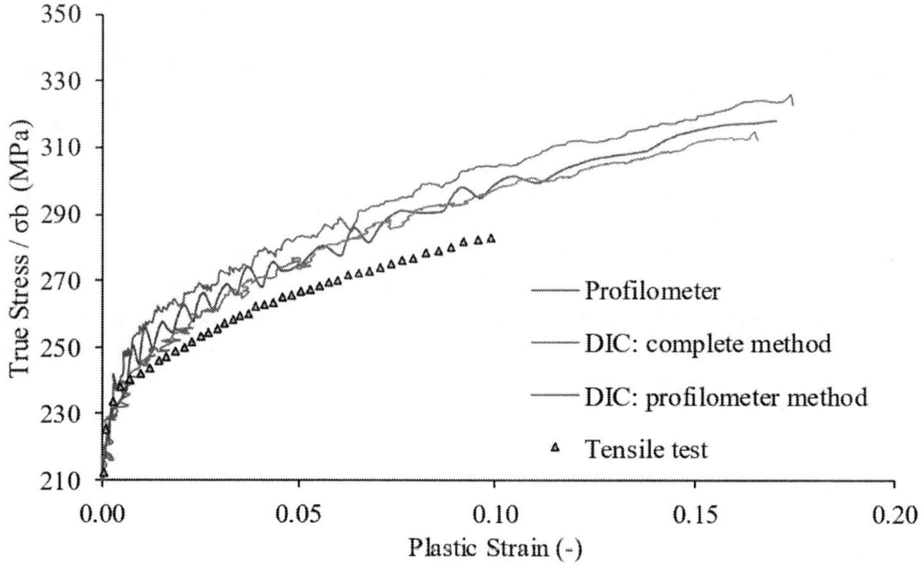

Figure 2: Stress-strain curves of the EN AW6061-T6 at 150°C, obtained from a bulge test and the proposed method using a laser profilometer, complete DIC method, the evolution of the dome height from the DIC system and equations (2-4), and from a tensile test using DIC system.

3.2. Temperature uniformity during the bulge test

Figure 3 shows the temporal evolution of the temperature at 10, 20, 30 and 40 mm from the blank centre, the set temperature and the pressure applied to form a Ductibor®1000 blank during bulge test at 900°C. The blank centre reached 900 °C in 60 s and the bulge test occurred between 80 and 107 s with a good temperature uniformity. The maximum temperature difference did not exceed 30 °C up to 40 mm radius. The temperature drop at 100 s for thermocouple located at 40 mm (TC40mm) was due to its dissociation from the blank.

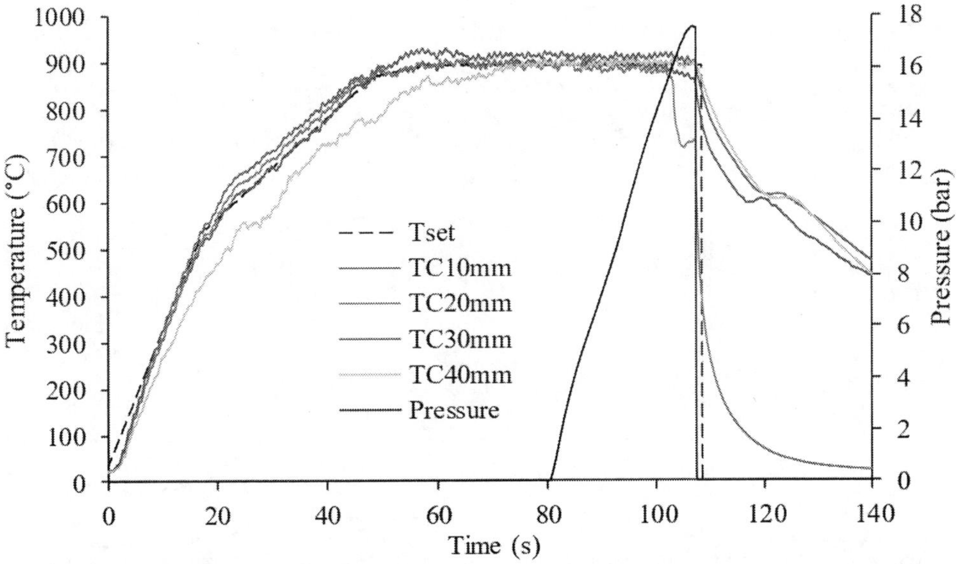

Figure 3: Temporal evolutions of the temperature at 10, 20, 30 and 40 mm from the blank centre and temporal evolution of the pressure for a forming test at 900 °C.

3.3. Stress-strain curves at 900 °C

The stress-strain curves of three bulge tests and a tensile test performed at 900 °C are presented in Figure 4. The bulge tests 2 and 3 were aborted contrary to the bulge test 1, for which the ramp pressure (shown in Figure 3) led to a burst. The uniaxial test was performed until the rupture. At the austenitic state, the quenchable steel presents an isotropic behaviour [19], thus the uniaxial tensile test and the bulge tests results are compared with the same assumption. The biaxial stress level is lower than the uniaxial stress level and the strain level before rupture is higher for the bulge test 1. However, the comparison of the stress-strain curves between the bulge and the uniaxial tests is biased by the strain rate difference. A positive strain rate sensitivity was observed for the Ductibor®1000 at high temperature, using tensile tests performed on a Gleebe machine and as for the Usibor®1500P [19]. In the case of the tensile test, the strain rate is around 2.0 s^{-1} while for the bulge tests the strain rate begins at around 1.10^{-3} s^{-1} and increases up to 8.10^{-2} s^{-1}, as shown in Figure 5 for the bulge test 1.

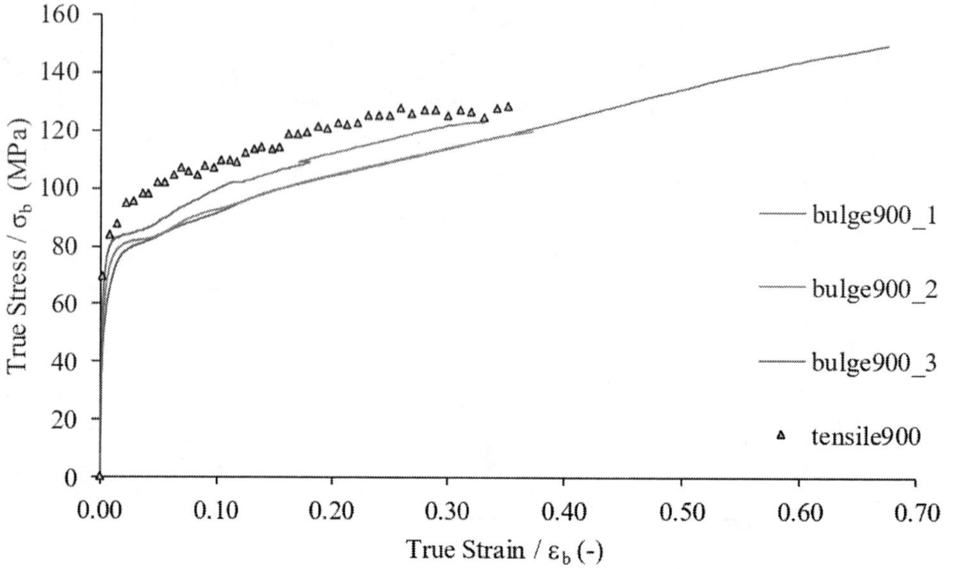

Figure 4: Stress-strain curves of Ductibor®1000 for three bulge tests and a tensile test at 900°C.

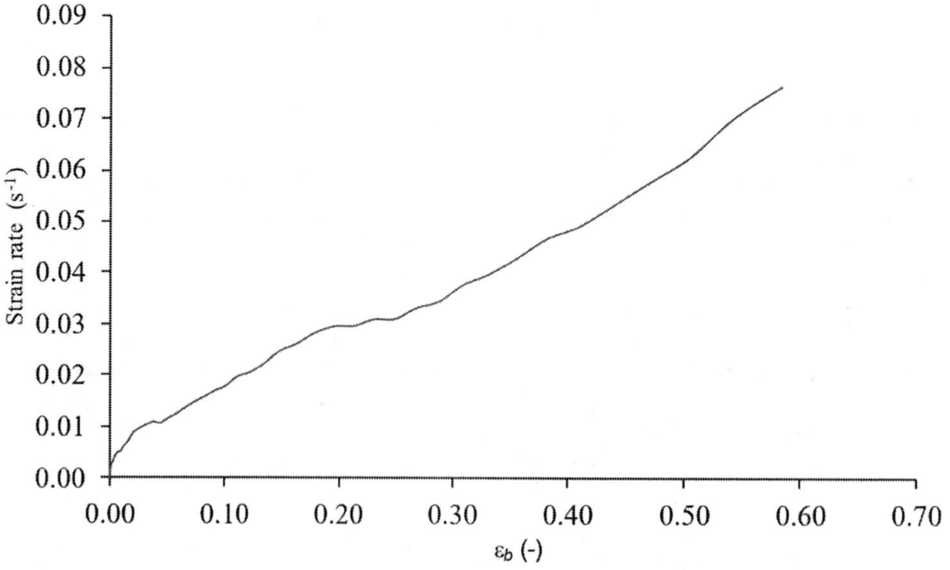

Figure 5: Strain rate as a function of the biaxial strain during the bulge test 1.

4. Conclusion

This paper presents an innovative hot gas bulge test, which allows to perform tests at high temperature up to 900 °C, with a good temperature uniformity at the blank centre thanks to a multi-electrodes system for resistance heating. Due to difficulties to obtain the deformations at high temperatures with a DIC system, another method to determine stress-strain curves using the membrane theory and a laser profilometer was validated with bulge tests performed with EN AW6061-T6 blanks at 150 °C. The proposed method is used to determine stress-strain curve at high temperatures. The results presented show a good agreement between stress-strain curves of Ductibor®1000 blanks resulting from bulge tests and a tensile test performed at 900 °C.

Acknowledgments
The authors would like to thank SREM Technologies-France (http://www.srem.fr/fr/), for the de-sign and the conception of the device and especially for their skills in the electrical domain. They also thank ArcelorMittal-France for the providing of Ductibor®1000 steel. The authors would also like to acknowledge the funding that sponsored this research work: the national funds from the French Ministry of Higher Education and the financial support of the Portuguese Foundation for Science and Technology (FCT) under project with reference PTDC/EME-EME/30592/2017 and by European Regional Development Fund (ERDF) through the Portugal 2020 program and the Centro 2020 Regional Operational Programme (CENTRO-01-0145-FEDER-031657) under the project UIDB/00285/2020.

References
[1] Karbasian H, & Tekkaya A E 2010. A review on hot stamping. *Journal of Materials processing Technology*, **210**(15), 2103-2118.
[2] Merklein M, Wieland M, Lechner M, Bruschi S, & Ghiotti A 2016. Hot stamping of boron steel sheets with tailored properties : A review. *Journal of Materials Processing Technology*, **228**, 11-24.
[3] Mori K, Bariani P F, Behrens B-A, Brosius A, Bruschi S, Maeno T, Merklein M, & Yanagimoto J 2017. Hot stamping of ultra-high strength steel parts. *CIRP Annals*, **66**(2), 755-777.
[4] ChiuHuang C-K, Wang S-W, & Lee P-K 2019. Investigation of the influence of forming parameters on the springback of hot-stamped hat-shaped parts. *IOP Conference Series: Materials Science and Engineering*, **651**(1), 012029.
[5] Shao Z, Li N, Lin J, & Dean T A 2016. Development of a New Biaxial Testing System for Generat-ing Forming Limit Diagrams for Sheet Metals Under Hot Stamping Conditions. *Experimental Mechanics*, **56**(9), 1489-1500.
[6] Li J, Min J, Qin K, Lin J, Liu F, & Liu L 2011. Investigation on the Effects of Sheet Thickness and Deformation Temperature on the Forming Limits of Boron Steel 22MnB5. *Key Engineering Materials*, **474–476**, 993–997.
[7] Turetta A, Bruschi S, & Ghiotti A 2006. Investigation of 22MnB5 formability in hot stamping opera-tions. *Journal of Materials Processing Technology*, **177**(1-3), 396-400.
[8] Bariani P F, Bruschi S, Ghiotti A, & Turetta A 2008. Testing formability in the hot stamping of HSS. *CIRP Annals*, **57**(1), 265-268.
[9] Li C, Zhang K F, Jiang S S, & Zhao Z P 2012. Pulse current auxiliary bulging and deformation mechanism of AZ31 magnesium alloy. *Materials & Design*, **34**, 170-178.
[10] Braun A, Bambach M, & Hirt G 2014. Investigation of a bulge test at high temperatures and high strain rates using a finite-element simulation study. *Metal Forming*, **9**.
[11] Braun A, Storz J, Bambach M, & Hirt G 2014. Development of a Pneumatic Bulge Test for High Temperatures and Controlled Strain Rates. *Advanced Materials Research*, **1018**, 245-252.
[12] Braun A, Waerder M, Hirt G 2016. Forming limit and Flow curve determination of hot stamping steels using a hot-gas-bulge-test, Forming Technology Forum 2016, Munich, 2016.
[13] Demazel N 2018. Développement de solutions de chauffage par conduction électrique pour l'emboutissage des aciers trempants : Approche numérique et expérimentale. PhD Thesis; Université Bretagne Sud
[14] ISO 16808:2014 - Metallic materials - Sheet and strip - Determination of biaxial stress-strain curve by means of bulge test with optical measuring systems 2014.
[15] R Hill M A 1950 C A theory of the plastic bulging of a metal diaphragm by lateral pressure, *The London, Edinburgh, and Dublin Philosophical Magazine and Journal of Science,* **41**(322) 1133-1142.
[16] Panknin W 1959. Der hydraulische Tiefungsversuch und die Ermittlung von Flisskurven (The hydraulic bulge test and the determination of the flow stress curves), Ph.D. thesis, University of Stuttgart.
[17] Boyer A, Laurent H, & Oliveira M C 2019. Evaluation of the stress vs strain curve using a high

temperature bulge test device. *IOP Conference Series: Materials Science and Engineering*, **651**, 012048.

[18] Simões V M, Laurent H, Oliveira M C and Menezes L F 2019. The influence of warm forming in natural aging and springback of Al-Mg-Si alloys. *International Journal of Material Forming* **12**(1), 57–68.

[19] Merklein M & Lechler J 2006. Investigation of the thermo-mechanical properties of hot stamping steels. *Journal of Materials Processing Technology*, **177**(1-3), 452-455.

IOP Conf. Series: Materials Science and Engineering 1238 (2022) 012055 doi:10.1088/1757-899X/1238/1/012055

On the topology design of a mechanical heterogeneous specimen using geometric and material nonlinearities

M Gonçalves[1], A Andrade-Campos[1] and S Thuillier[2]

[1] Centre for Mechanical Technology and Automation (TEMA), Department of Mechanical Engineering, University of Aveiro, Portugal

[2] Univ. Bretagne Sud, UMR CNRS 6027, IRDL, F-56100, Lorient, France

E-mail: mafalda.goncalves@ua.pt, gilac@ua.pt, sandrine.thuillier@univ-ubs.fr

Abstract. Material characterization and the calibration of constitutive models play an important role in the majority of the forming processes nowadays. For an accurate virtualization of the processes, the material mechanical behavior needs to be known a priori. The classical characterization procedure involves carrying out several standard mechanical tests to identify the required information about the material. However, this procedure turns out to be expensive and time-consuming. Heterogeneous mechanical tests have been used to overcome this issue. By providing richer information with a reduced number of tests, their use can enhance the actual material characterization process. This work aims at making a significant contribution towards this goal by designing a heterogeneous mechanical test using topology optimization. A specimen topology is obtained with a heterogeneous displacement field by applying the theory of compliant mechanisms. Due to the large displacements considered, a geometrically nonlinear finite element analysis and a topology optimization procedure are proposed. Material nonlinearity is taken into account as well, to design solutions closer to reality. An optimal mechanical test with a highly heterogeneous strain field is obtained and evaluated, considering its diversity, using mechanism theory and a mechanical indicator.

1. Introduction

The virtualization of sheet metal forming processes is proceeding at an increasing pace. Material characterization and model calibration procedures play a key role in the accuracy of that process. A whole range of classical mechanical tests is usually carried out to identify the required parameters to characterize a specific material behavior. In order to avoid the costs and time associated with this procedure, heterogeneous mechanical tests have been used. Several authors have already addressed this topic by proposing nonstandard configurations for mechanical tests towards the improvement of the material characterization process. The design of a mechanical test based on the empirical knowledge of the authors was addressed in [1–3], for example. The strain field heterogeneity was used as a criterion to lead the design process using optimization methodologies [4–7]. A step further relies on evaluating the uncertainty of the identified parameters [8–10]. Adding the errors from full-field measurements techniques, the full measurement chain can be considered to evaluate the identification quality [11, 12].

Although several methodologies have already been proposed, to the best of the authors' knowledge, there is still a need for a systematic design methodology. An optimization approach is required to achieve an optimal solution for the test design. Also, the obtained designs need to be independent on the initial guess of the design. Therefore, this work aims at filling this gap by proposing an nonlinear topology-based optimization methodology for designing specimen geometries with heterogeneous displacement and strain fields. The goal is to provide a relevant quantity and quality of information about the material behavior with a single test. The design methodology is based on a nonlinear topology optimization problem. Geometric and material nonlinearities are considered in order to account for the large displacements and the nonlinear material behavior. Several works have already addressed the nonlinear topology design of structures and compliant mechanisms. Topology optimization of structures with both geometric and material nonlinearities was addressed by Jung *et al.* [13]. Huang *et al.* [14] focused also on structural design under displacement loading. Capasso *et al.* [15] used nonlinear mechanics to analyze the stress-based topology optimization of compliant mechanisms. A similar work was proposed by Leon *et al.* [16], who also addressed the topology design of compliant mechanisms with material and geometric nonlinearities using stress constraints.

This work proposes a methodology to design specimen geometries with strain heterogeneities by coupling the design by topology optimization and an extended version of the theory of compliant mechanisms [17]. On this basis, non standard specimen designs can be obtained with highly heterogeneous displacement fields. Heterogeneity is introduced through the displacement field by applying two displacements in specific locations of the specimen. Therefore, several mechanical states can be induced depending on the applied displacements. A previous work already accounted for the large deformations that the specimen is submitted to [18]. In this work, material nonlinearity is introduced, being the nonlinear material behavior described by von Mises yield criterion and the Swift law. An optimal specimen geometry is obtained with a nonconventional geometry. Its performance in inducing a diversity of stress states is evaluated using a mechanical indicator.

2. Heterogeneous test design

Heterogeneous mechanical tests are a step ahead in material characterization due to the heterogeneous displacement and strain fields that are induced in them. These provide relevant information about the material behavior that is crucial for the calibration of the constitutive models. To induce such heterogeneity in a specimen is the goal of some of the test design methodologies that have been developed. In this work, it is proposed to introduce heterogeneity in the displacement field by coupling the design by topology optimization and an extended version of the compliant mechanisms. The former is used to obtain nonstandard specimen geometries, which on its own already can generate nonhomogeneous displacement fields. The latter is added to the approach to create more heterogeneity by applying displacements in specific locations of the specimen. Depending on the chosen locations, specific strain/stress states can be induced in the specimen and, consequently, lead to the enrichment of the strain field.

The design by topology optimization starts from a predefined design domain to find the optimal material distribution inside the boundaries of that domain. The starting point chosen in this work is represented in Figure 1. A schematic representation of the initial design domain is shown, which geometry is chosen in order not to restrict the material distribution update. A uniaxial tensile loading test is reproduced. Only one quarter of the design domain is represented. The load applied by the grips of the testing machine is represented by $\mathbf{F}_{\text{in}}$. The height of the design domain is more than twice than the height of the grips since it is assumed that this height is enough to provide the optimal solutions. Based on the theory of compliant mechanisms, two displacements are applied in specific locations of the specimen. Two locations, input and output, are chosen empirically by the authors. The first one corresponds to the displacement of the grips and the second is applied preferentially far from the specimen boundaries. It can be applied in different directions, being responsible for the way the specimen deforms.

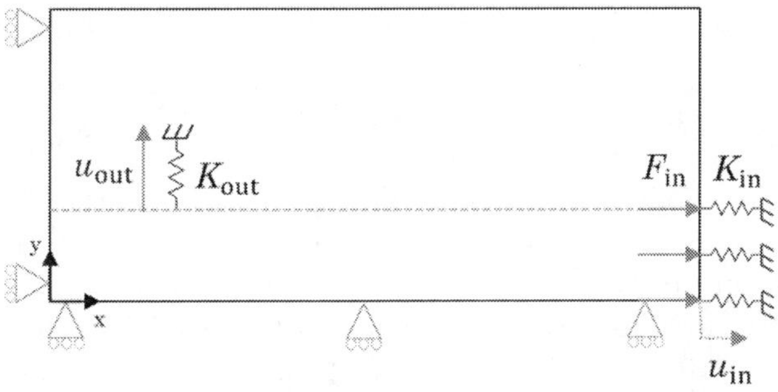

Figure 1: Schematic representation of the topology design domain subjected to a uniaxial tensile loading.

2.1. Problem formulation

The design optimization problem aims at finding the optimal material distribution for the specimen. The material layout, $\mathbf{X}$, is defined by the design variables, X_e, that represent the relative density of each element. These variables characterize the quantity of material in each element, taking the value 0 when the element is empty or the value 1 when the element is full. The density of each element is updated at each iteration of the optimization process until the optimum of the objective-function is found. In this work, it is proposed to maximize the ratio between the displacements in the output, u_{out}, and input, u_{in}, locations in order to enhance the heterogeneity of the displacement field of the specimen. Therefore, the optimization problem is defined as follows:

$$\begin{aligned} \underset{\mathbf{X}}{\text{maximize}} \quad & T(\mathbf{X}) = \frac{u_{\text{out}}(\mathbf{X})}{u_{\text{in}}(\mathbf{X})}, \\ \text{subject to} \quad & \mathbf{R} = \mathbf{0}, \\ & \frac{\sum_{e=1}^{M} X_e V_e}{\sum_{e=1}^{M} V_e} - V^* \leq 0, \\ & 0 \leq \rho_{\min} \leq X_e \leq 1, \;\; e = 1, 2, .., M. \end{aligned} \tag{1}$$

where M is the number of elements and V_e the volume fraction of each one. V^* corresponds to the volume fraction required for the final specimen design. $\rho_{\min}$ represents a minimum value for the element density to avoid numerical issues. A nonlinear finite element analysis is proposed for the design procedure. When the equilibrium is achieved, the balance between the external and internal loads of the specimen, $\mathbf{R} = 0$, is required to be zero.

2.2. Nonlinear Finite Element Analysis

To reproduce accurately the specimen behavior when submitted to a tensile loading test, it is proposed to carry out a nonlinear finite element analysis. Both material and geometric nonlinearities are considered and the latter aims at accounting for the large displacements that the specimen is submitted to. For that purpose, the Green-Lagrange strain tensor is used, being composed of linear and nonlinear parts that can be represented as follows:

$$\mathbf{E} = \mathbf{E}_{\text{L}} + \mathbf{E}_{\text{N}} = \mathbf{B}_{\text{L}}\mathbf{U} + \mathbf{B}_{\text{N}}\mathbf{U}, \tag{2}$$

where $\mathbf{B}_{\text{L}}$ and $\mathbf{B}_{\text{N}}$ correspond to the linear and nonlinear transformation matrices between nodal displacements, $\mathbf{U}$, and element strains, $\mathbf{E}$, respectively. Along with the strain measure, the second Piola-Kirchoff stress tensor, $\mathbf{S}$, is also employed in the system analysis.

The equilibrium of the system is one of the constraints of the optimization procedure and can be defined as

$$\mathbf{R} = \int \mathbf{B}^{\mathrm{T}}\mathbf{S}\, dV - \mathbf{F}_{\mathrm{ext}} = \mathbf{F}_{\mathrm{int}} - \mathbf{F}_{\mathrm{ext}}, \tag{3}$$

where $\mathbf{B}$ is composed of both the linear and nonlinear matrices, $\mathbf{B}_{\mathrm{L}}$ and $\mathbf{B}_{\mathrm{N}}$, respectively. An incremental-iterative approach is used to solve the nonlinear finite element analysis. The external load is applied in increments $\Delta\mathbf{F}_{\mathrm{ext}}$ during which the displacements $\Delta\mathbf{U}$ are found iteratively to achieve the equilibrium for the current configuration. At each load increment, the equilibrium is achieved when the residual $\mathbf{R}$ is equal to zero. Its computation considers the current state of the topology, being $\mathbf{F}_{\mathrm{int}}$ the internal load of the system considering its history and $\mathbf{F}_{\mathrm{ext}}$ the load that has already been applied. The iterative procedure is solved using the Newton-Raphson algorithm and requires the determination of the tangent stiffness matrix defined as

$$\mathbf{K}_{\mathrm{T}} = -\frac{\partial \mathbf{R}}{\partial \mathbf{U}}, \tag{4}$$

that can be written as follows

$$\begin{aligned} \mathbf{K}_{\mathrm{T}} &= \mathbf{K}_{\mathrm{L}} + \mathbf{K}_{\mathrm{N}} + \mathbf{K}_{\mathrm{S}} \qquad (5) \\ &= \int \mathbf{B}_{\mathrm{L}}^{\mathrm{T}}\mathbf{D}_{\mathrm{ep}}\mathbf{B}_{\mathrm{L}} dV + \int \mathbf{B}_{\mathrm{L}}^{\mathrm{T}}\mathbf{D}_{\mathrm{ep}}\mathbf{B}_{\mathrm{N}} dV + \int \mathbf{B}_{\mathrm{N}}^{\mathrm{T}}\mathbf{D}\mathbf{B}_{\mathrm{L}} dV + \int \mathbf{B}_{\mathrm{N}}^{\mathrm{T}}\mathbf{D}_{\mathrm{ep}}\mathbf{B}_{\mathrm{N}} dV + \int \mathbf{G}^{\mathrm{T}}\mathbf{M}\mathbf{G} dV. \end{aligned}$$

The first two terms, $\mathbf{K}_{\mathrm{L}}$ and $\mathbf{K}_{\mathrm{N}}$, are the stiffness matrices due to the small and large displacements, respectively. The latter term, $\mathbf{K}_{\mathrm{S}}$, is related to the initial stress state, in which $\mathbf{M}$ is composed of elements of the stress tensor $\mathbf{S}$ and $\mathbf{G}$ stands for a derivative matrix of shape functions with respect to coordinates. The consistent tangent matrix, represented as $\mathbf{D}_{\mathrm{ep}}$, improves the overall convergence of the equilibrium equations when the Newton-Raphson method is used for the latter [19]. It is computed during the material behavior analysis that will be described in the following section, considering plane stress conditions.

2.3. Material behavior

The elastoplastic material behavior is taken into account in this work. Both the elastic and the plastic behaviors are considered isotropic, being the former described by Hooke's law. The yield function can be represented as follows:

$$f(\boldsymbol{\sigma}, \bar{\varepsilon}^{\mathrm{p}}) = \sigma_{\mathrm{vm}}(\boldsymbol{\sigma}) - \sigma_y(\bar{\varepsilon}^{\mathrm{p}}) \tag{6}$$

where $\sigma_{\mathrm{vm}}(\boldsymbol{\sigma})$ stands for the von Mises equivalent stress defined based on the stress tensor $\boldsymbol{\sigma}$. The yield stress, represented as σ_y, is defined by the equivalent plastic strain, $\bar{\varepsilon}^{\mathrm{p}}$. Thus, the isotropic hardening behavior of the material is characterized by the Swift Law, which can be computed as

$$\sigma_y(\bar{\varepsilon}^{\mathrm{p}}) = K(\varepsilon_0 + \bar{\varepsilon}^{\mathrm{p}})^n \tag{7}$$

where K, ε_0 and n are the constitutive model parameters. The yield surface is described by the von Mises yield criterion [20] considering plane stress conditions as

$$\sigma_{\mathrm{vm}}(\boldsymbol{\sigma}) = \sqrt{\sigma_{xx}^2 + \sigma_{yy}^2 - \sigma_{xx}\sigma_{yy} + 3\tau_{xy}^2} \tag{8}$$

where σ_{xx}, σ_{yy} and τ_{xy} correspond to the components from Cauchy stress tensor $\boldsymbol{\sigma}$.

In the finite element analysis, the second Piola-Kirchhoff stress tensor, $\mathbf{S}$, is used (work-conjugate to the Green's strain tensor) in contrast to the Cauchy stresses, $\boldsymbol{\sigma}$, that are applied in the analysis of the material behavior. While the first one is related to the original configuration, the latter relates to the current configuration [19]. Therefore, it is needed to establish a relationship between them that is given by

$$\mathbf{S} = J\mathbf{F}^{-1}\boldsymbol{\sigma}\mathbf{F}^{-\mathrm{T}}, \tag{9}$$

where $\mathbf{F}$ is the deformation gradient between the original and the new configurations and J its determinant.

2.4. Sensitivity analysis

The objective-function of the topology optimization problem evaluates the ratio between the displacements in the output and input locations. Therefore, the sensitivity of the objective-function can be defined as

$$\frac{dT}{d\mathbf{X}} = \frac{\frac{du_{\text{out}}}{d\mathbf{X}} u_{\text{in}} - \frac{du_{\text{in}}}{d\mathbf{X}} u_{\text{out}}}{u_{\text{in}}^2}, \tag{10}$$

where the derivative of the output and input displacements can be derived using the adjoint variable method in the following manner,

$$u_{\text{in}} = \mathbf{L}_{\text{in}}^{\mathrm{T}}\mathbf{U} + \boldsymbol{\gamma}^{\mathrm{T}}\mathbf{R}, \tag{11}$$

$$u_{\text{out}} = \mathbf{L}_{\text{out}}^{\mathrm{T}}\mathbf{U} + \boldsymbol{\lambda}^{\mathrm{T}}\mathbf{R}, \tag{12}$$

where $\mathbf{L}_{\text{in}}$ and $\mathbf{L}_{\text{out}}$ stand for two vectors with the value of one at the input and output locations and zeros in the remaining ones, respectively. Assuming that the equilibrium has been found, the terms $\boldsymbol{\gamma}^{\mathrm{T}}\mathbf{R}$ and $\boldsymbol{\lambda}^{\mathrm{T}}\mathbf{R}$ are equal to zero and, therefore can be added to the displacements. The derivative of the input and output displacements can be written as

$$\frac{\partial u_{\text{in}}}{\partial X_e} = -p\left[(1 - X_e)\,\rho_{\min}^{p-1} + X_e\right]\boldsymbol{\gamma}_e^{\mathrm{T}}\mathbf{F}_{\text{int}}^e, \tag{13}$$

and

$$\frac{\partial u_{\text{out}}}{\partial X_e} = -p\left[(1 - X_e)\,\rho_{\min}^{p-1} + X_e\right]\boldsymbol{\lambda}_e^{\mathrm{T}}\mathbf{F}_{\text{int}}^e, \tag{14}$$

respectively. The adjoint vectors $\boldsymbol{\lambda}$ and $\boldsymbol{\gamma}$ can be easily obtained as the solutions to the linear adjoint equations $\mathbf{K}_{\mathrm{T}}\boldsymbol{\lambda} = \mathbf{L}_{\text{out}}$ and $\mathbf{K}_{\mathrm{T}}\boldsymbol{\gamma} = \mathbf{L}_{\text{in}}$. The tangent stiffness matrix has already been found during the equilibrium equations. The material properties are determined using the Solid Isotropic Material with Penalization (SIMP) method [21], in which the element densities are penalized using a penalization factor, p. The adopted methodology [22] establishes a non-homogenized combination of solid and void in intermediate elements, being required a linear interpolation between the two phases, solid and void.

2.5. Solution evaluation

From the design optimization procedure, an optimal specimen geometry is obtained. Based on the stress/strain fields that are induced after a uniaxial tensile loading test, the specimen is evaluated using a mechanical indicator [7]. It analyzes the performance of the specimen in inducing a high diversity of stress states and can be computed as

$$id = \prod_{s=1}^{3}\left[\frac{3}{\sum_{e=1}^{n} X_e}\sum_{e=1}^{n}\left({}^{s}\delta_e\ Z_e\ X_e\right)\right]. \tag{15}$$

Solutions with stress concentrations or unstressed material are penalized based on the von Mises stress using the parameter Z_e. The index s indicates the stress state (tension, compression, and shear) that the element e is subjected to. According to the principal strains relationships, the operator ${}^s\delta_e$ assumes the value of one or zero if the element is in the stress state s or not, respectively. Based on a generalization of the smooth Heaviside function, the parameter ${}^s\delta_e$ can defined as

$$ {}^s\delta_e = \begin{cases} \frac{1}{2}\Big(1 - \tanh\big(\beta\left(\varepsilon_e^1 + 0.75\varepsilon_e^2\right)\big)\Big), & s = 1 \\ \frac{1}{4}\Big(1 + \tanh\big(\beta\left(\varepsilon_e^1 + 0.75\varepsilon_e^2\right)\big)\Big)\Big(1 - \tanh\big(\beta\left(\varepsilon_e^1 + 1.5\varepsilon_e^2\right)\big)\Big), & s = 2 \\ \frac{1}{2}\Big(1 + \tanh\big(\beta\left(\varepsilon_e^1 + 1.5\varepsilon_e^2\right)\big)\Big), & s = 3 \end{cases} . \tag{16} $$

where the parameter β controls the "agressiveness" of the Heaviside function. ε^1 and ε^2 stand for the major and minor principal strains, respectively. The ideal solution would present the same amount of material in the three stress states (tension, compression, and shear) without stress concentrations or unstressed material.

3. Implementation

The design procedure is based on a nonlinear topology-based optimization methodology from which an optimal specimen is obtained. Figure 2 represents the structure of the developed methodology.

At the beginning of the design optimization, the design domain is established along with the volume fraction and the filtering parameters [22]. The material distribution that is being evaluated is submitted to a nonlinear finite element analysis, from which the displacements and internal forces are obtained. The sensitivity analysis is performed as well as the objective-function computation. A filtering technique is applied to the sensitivities and the design variables are updated by the Method of Moving Asymptotes (MMA) [23]. A new iteration starts with the updated topology in case the convergence criterion is not reached. Otherwise, an optimal solution is obtained. At the end of this process, the solution is submitted to a performance evaluation using the mechanical indicator.

A nonlinear finite element analysis is proposed to account for the nonlinear behavior of the specimen. Since material and geometric nonlinearities are considered, an incremental analysis is proposed to evaluate the structure behavior as the load level evolves. The equilibrium $\mathbf{R} = \mathbf{0}$ is found, at each increment, iteratively using the Newton-Raphson method [19]. The tangent stiffness matrix describes the structure current configuration, being updated at each iteration. The equilibrium at each load increment is achieved when a balance between the external and internal loads is obtained. For determining the last ones, the computation of the stresses is required. The elastoplastic behavior of the material makes it necessary to use an yield criterion to predict when yielding occurs and, consequently, to compute correctly the stresses. If the stresses are computed by an elastic relationship and they are found to lie within the yield surface, the material is found to remain elastic. Otherwise, these are already in the plastic regime, being no longer determined by a linear assumption. The Backward-Euler return method [19] needs to be applied to determine the plastic strain increment and, consequently, the updated stresses and the consistent tangent matrix.

4. Results

The developed methodology is applied to the design of an optimal heterogeneous mechanical test. From the initial design domain, represented in Figure 1, three specimen geometries are obtained considering material and geometric nonlinearities. These solutions are shown in Figure 3. A

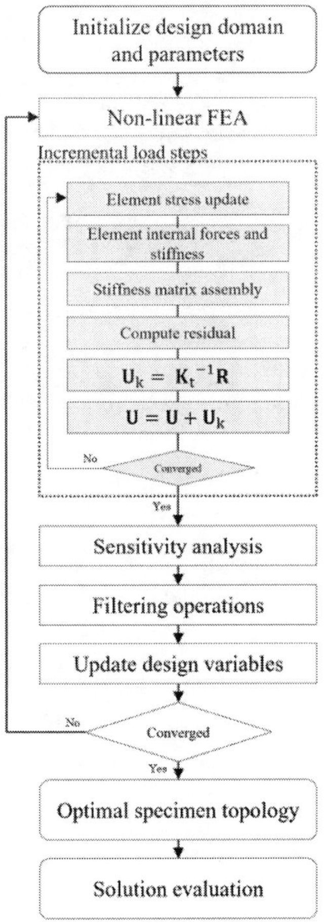

Figure 2: Flow diagram of the nonlinear topology optimization methodology.

volume fraction of 35% of the total initial volume is required for the specimen designs. The location of the output displacement is chosen to be established in the left symmetry boundary of the specimen and it is represented by the red arrow. A mesh of 50 × 50 elements is used. These parameters are the ones that led to the best results [24]. The elastic material properties are represented in Table 1 as well as the constitutive model parameters related to the Swift Law.

Table 1: Elastic properties and constitutive model parameters for Swift's isotropic hardening law of DP600 [25].

	Elastic		Swift		
	E [GPa]	ν	K [MPa]	ε_0	n
DP600	210	0.3	979.46	0.00535	0.194

Three different assumptions have led to the topologies presented in Figures 3(a), (b) and (c), that are entitled as solutions A, B and C, respectively. The first one considered the material behavior linear elastic. With the same material behavior, solution B was obtained assuming

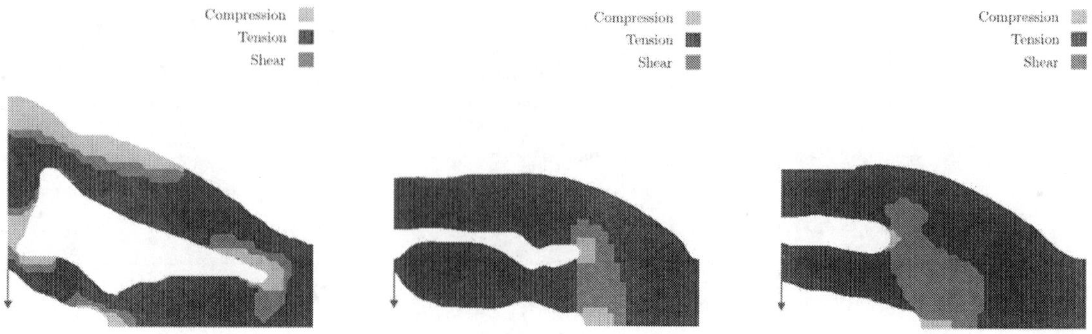

(a) Solution A (b) Solution B (c) Solution C

Figure 3: Obtained specimen designs with a topology-based optimization methodology considering: (a) a linear material and geometric behavior, (b) geometric nonlinearities with a linear elastic material and (c) geometric and material nonlinearities. The stress states induced in the specimen are represented.

geometric nonlinearities in order to account for the large displacements that the specimen is submitted to. Material and geometric nonlinearities were considered to obtain the last topology. The elastoplastic material behavior influences the design optimization process and, consequently, the obtained specimen designs.

Regarding the obtained specimen geometries, although the material is similarly placed in the locations where the displacements and loads are being applied, the overall material distribution is rather different. Their performance in inducing a high diversity of stress states is evaluated using the performance indicator. For the solutions A, B and C, the values of the performance indicator are 0.0359, 0.0190 and 0.0215, respectively, being the solution A the one with the highest heterogeneity of stress states. However, this scalar indicator does not take into consideration the equivalent plastic strain through the specimen or the duration of the test, since an early rupture of the specimen limits the quantity and quality of the information that can be extracted from the specimen. Therefore, a Finite Element Analysis (FEA) was conducted in ABAQUS/Standard using a forming limit diagram to predict the moment which the specimen rupture occurs. In Figure 4, the obtained results are represented. The ratio of the major and minor principal strains ($\varepsilon^1/\varepsilon^2$), the equivalent plastic strain ($\bar{\varepsilon}^{\mathrm{P}}$) and the von Mises stress (σ_{VM}) are depicted for each solution at the moment just before rupture.

Concerning the ratio between the major and minor principal strains, it can be noticed that due to the applied tensile load, most of the specimen is subjected to tension. However, a significant area of the specimens under other mechanical states, such as compression and shear (the mechanical states considered in this work) can be observed, particularly, in solution C. This validates the purpose of this methodology that aims at creating heterogeneity of mechanical states. The equivalent plastic strain is represented along the specimen at the moment just before rupture. In solution A, the rupture occurs too early in the testing procedure due to its complex geometry. Therefore, plastic strains only exists in small areas of the specimen. A similar issue arises in solution B. The solution C consists in the specimen geometry with the largest area in the plastic regime. Regarding the von Mises stress, the maximum values of each solution are similar, being around 700 MPa. Solution A, due to its early rupture in specific locations of the geometry, presents the maximum value in those locations, being most of the specimen area under low values of the von Mises stress. In contrast, the specimen C presents higher values of the von Mises stress over its surface, leading to a higher duration of the test and, consequently, providing more information about the material behavior.

IOP Conf. Series: Materials Science and Engineering 1238 (2022) 012055 doi:10.1088/1757-899X/1238/1/012055

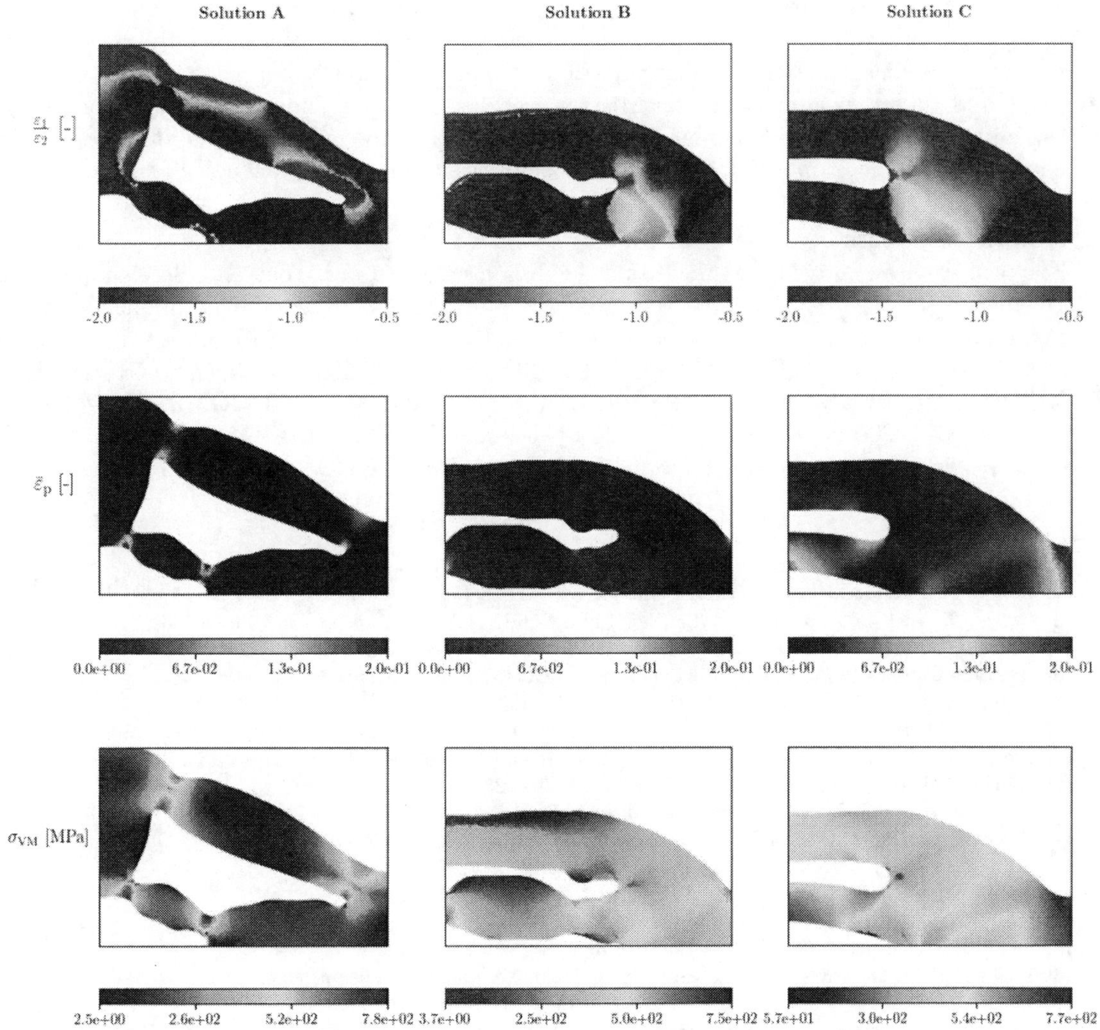

Figure 4: Obtained results from FEA: the ratio between major and minor principal strains ($\varepsilon^1/\varepsilon^2$), the equivalent plastic strain ($\bar{\varepsilon}^{\mathrm{p}}$) and the von Mises stress (σ_{VM}) for each solution (A, B and C).

Based on these results, it is possible to conclude that the solution C, obtained with the methodology proposed in this work, consists in the specimen design that better suits for the improvement of the material characterization process. Therefore, this methodology should be used in order to obtain specimen geometries that take into account the real nonlinear material and geometric behavior.

5. Conclusions

In this work, a nonlinear topology-based methodology is proposed for the design of a heterogeneous mechanical test. Geometric and material nonlinearities are considered for the design of test in order to lead to more realistic results. Large deformations are taken into account in the finite element analysis as well as the elastoplastic material behavior, characterized by the von Mises yield criterion and the Swift law. An optimal specimen geometry is obtained.

The influence of the material and geometric nonlinearities in the test design is analyzed by comparing to other solutions. Several differences were noticed between the topologies related to the material layout and, consequently, to their performance. Solution C, the one obtained assuming a nonlinear material and geometric behavior, presented the largest area in the plastic regime and the more interesting von Mises stress distribution. The obtained specimen design has the potential to provide a higher quality and quantity of information about the material behavior, being the proposed methodology of major relevance for an accurate test design.

Acknowledgments

This project has received funding from the Research Fund for Coal and Steel under grant agreement No 888153. The authors also gratefully acknowledge the financial support of the Portuguese Foundation for Science and Technology (FCT) under the projects CENTRO-01-0145-FEDER-029713, POCI-01-0145-FEDER-031243 and POCI-01-0145-FEDER-030592 by UE/FEDER through the programs CENTRO 2020 and COMPETE 2020, and UIDB/00481/2020 and UIDP/00481/2020-FCT under CENTRO-01-0145-FEDER-022083. M. Gonçalves is grateful to the FCT for the Ph.D. grant Ref. UI/BD/151257/2021.

References

[1] Haddadi H and Belhabib S 2012 *Int. J. Mech. Sci.* **62** 47–56
[2] Belhabib S, Haddadi H, Gaspérini M and Vacher P 2008 *Int. J. Mech. Sci.* **50** 14–21
[3] Pottier T, Vacher P, Toussaint F, Louche H and Coudert T 2012 *Exp. Mech.* **52** 951–963
[4] Souto N, Andrade-Campos A and Thuillier S 2016 *Int. J. Mech. Sci.* **107** 264–276
[5] Andrade-Campos A, Aquino J, Martins J M and Coelho B 2019 *Metals* **9**
[6] Aquino J, Andrade-Campos A, Martins J M and Thuillier S 2019 *Strain* **55** 1–18
[7] Barroqueiro B, Andrade-Campos A, Dias-de Oliveira J and Valente R A 2020 *Int. J. Mech. Sci.* **181** 105764
[8] Pierron F, Vert G, Burguete R, Avril S, Rotinat R and Wisnom M R 2007 *Strain* **43** 250–259
[9] Bertin M B, Hild F and Roux S 2016 *Strain* **52** 307–323
[10] Chamoin L, Jailin C, Diaz M and Quesada L 2020 *Int. J. Solids Struct.* **193-194** 270–286
[11] Rossi M, Badaloni M, Lava P, Debruyne D and Pierron F 2016 *AIP Conf. Proc.* vol 1769
[12] Wang P, Pierron F, Rossi M, Lava P and Thomsen O T 2016 *Strain* **52** 59–79
[13] Jung D and Gea H C 2004 *Finite Elem. Anal. Des.* **40** 1417–1427
[14] Huang X and Xie Y M 2008 *Eng. Struct.* **30** 2057–2068
[15] Capasso G, Morlier J, Charlotte M and Coniglio S 2020 *Mech. Ind.* **21**
[16] De Leon D M, Gonçalves J F and de Souza C E 2020 *Struct. Multidiscip. Optim.* **62**
[17] Zhu B, Zhang X, Zhang H, Liang J, Zang H, Li H and Wang R 2020 *Mech. Mach. Theory* **143** 103622
[18] Gonçalves M, Andrade-Campos A and Thuillier S 2022 *Proceedings of the 25th ESAFORM conference*
[19] Crisfield M A 1991 *Non-Linear Finite Element Analysis of Solids and Structures, Essentials* (Wiley)
[20] Mises R v 1913 *Nachr. Ges. Wiss. Göttingen, Math.-Phys. Kl.* **1913** 582–592
[21] Rozvany G, Zhou M and Birker T 1992 *Struct. Multidiscip. Optim.* **4** 250–252
[22] Fu Y F, Rolfe B, Chiu L N N, Wang Y, Huang X and Ghabraie K 2020 *Adv. Eng. Softw.* **150** 102921
[23] Svanberg K 2009 *Encyclopedia of Optimization* ed Floudas C A and Pardalos P M (Boston, Massachusetts: Springer) pp 3832–3834
[24] Gonçalves M, Andrade-Campos A and Dias-de Oliveira J A 2021 *XVI International Conference on Computational Plasticity. Fundamentals and Applications* (Barcelona, Spain)
[25] Ozturk F, Toros S and Kilic S 2014 *Procedia Manuf.* **81** 760–765

IOP Conf. Series: Materials Science and Engineering 1238 (2022) 012056 doi:10.1088/1757-899X/1238/1/012056

A relative approach for uncertainty quantification of inversely identified material behavior using a heterogeneous test for sheet metal

M Conde[1], S Coppieters[2] and A Andrade-Campos[1]

[1] Centre for Mechanical Technology and Automation (TEMA), Department of Mechanical Engineering, University of Aveiro, Campus Universitário de Santiago, 3810-193 Aveiro, Portugal

[2] Department of Materials Engineering, KU Leuven, Ghent Technology Campus, Gebroeders De Smetstraat 1, Ghent, 9000, Belgium

E-mail: marianaconde@ua.pt

Abstract. The process of calibrating the material parameters of a constitutive model has inherent uncertainty sources. These can lead to inadequate material behavior characterization and, consequently, errors in the numerical simulations. A robust mechanical test is necessary for accurate material mechanical characterization. In this work, it is relatively quantified the uncertainties of the parameter identification of the Swift hardening law and the Yld2000-2d yield function using Digital Image Correlation (DIC) and a DIC-levelled approach. The levels of uncertainty are calculated with a derivative-based local method, having into consideration the Karush-Kuhn-Tucker (KKT) conditions for optimization problems. The mechanical test used for the parameter calibration was an optimized interior notched specimen that presents several strain and stress states simultaneously. The synthetic images were generated based on the numerical data, and the strain fields were accessed using DIC techniques. It was concluded that the step size is the DIC parameter that most influences the material parameter identification.

1. Introduction, framework and literature review

Nowadays, non-homogeneous mechanical tests came to replace the conventional tests in material mechanical characterization and constitutive parameters identification. These tests, called heterogeneous mechanical tests in this paper, present several strain states simultaneously that can be measured using optical techniques, such as DIC. This large data acquisition can then feed an inverse analysis for parameter identification, such as the Finite Element Model Updating (FEMU) method. Yet, it is of utmost importance to estimate the errors and uncertainties of these identifications. The quantification of the associated errors and the influence of chosen experimental or numerical parameters are essential for parameter identification and, consequently, the accurate simulation and engineering decision making. The DIC procedure alone presents a vast number of sources of uncertainty, namely, the setup misalignment, the sample and camera motion, the air turbulence, the image quality, the setup calibration, the speckle size and shape, the illumination, the image resolution and noise, the aperture, the exposure, the lens distortions, the coordinate system, the correlation for the displacement measurements and the strain and virtual strain gauge (VSG) analysis. Some of these have been evaluated in [1, 2, 3]. Moreover, the material parameter identification itself produces error propagation since it usually compares experimental data measured with optical techniques and numerical data [4]. So, when using optical cameras to measure displacement fields, it is possible to link image noise to displacement uncertainties, and the latter ones with material parameter fluctuations [5]. Several studies have identified material parameters using FEMU [6, 7, 8, 9, 10]. Also, for model calibration, the Virtual Fields Methods (VFM) is often used, as in [11, 12, 13, 14]. Nevertheless, the vast majority of the mentioned studies did not account for uncertainty measurements in the parameter identification.

The present work aims at quantifying the uncertainties in material parameter identification. Uncertainties in a system can be quantified using probabilistic approaches, such as Monte Carlo simulations, importance sampling, perturbation methods, polynomial chaos expansion (PCE), or non-probabilistic approaches, such as interval analysis and Fuzzy theory [15]. However, the theories and methodologies for uncertainty propagation and inverse uncertainty quantification are computationally expensive and increase dramatically with the dimensionality of the problem (i.e. the number of input variables and/or the number of unknown parameters). Moreover, these suffer from identifiability issues, i.e. multiple combinations of unknown parameters can yield the same prediction, being different parameters indistinguishable. Therefore, knowing that an identification procedure in plasticity can take up to hours, an expedited method is required. Here, considering that the identification problem is solved using optimization techniques and when using virtual experiments the global minimum is known, the KKT conditions [16] can be used for uncertainty quantification and clearness of solution. This approach substantially reduces the computational time.

The identification regarded is a FEMU DIC-levelled technique as in [17], which was already employed with a heterogeneous test in [18]. This FEMU technique compares both the experimental and numerical strain fields using the DIC subset-based method for the strain measurements, minimizing the error between the two. The mechanical test that feeds the FEMU is a heterogeneous test proposed in [19], which was originated by a shape optimization design approach based on a heterogeneity indicator. The constitutive model under consideration is the Swift hardening law and, for the anisotropy behavior, the Yld2000-2d yield function. The sources of uncertainty here analyzed are the subset size, step size and strain window size. Furthermore, the uncertainty technique used is based on the perturbation of the sources of uncertainty and its derivative quantification.

The paper is divided into four main sections. In Section 2, the methodology and implementation, used to quantify the uncertainties of the material parameter identification, are explained in detail. It starts by a simplification of the material calibration problem having in consideration theoretical concepts of optimization problems. The uncertainty technique is described along with (i) the identification problem itself, (ii) the mechanical test in consideration, (iii) the material under analysis, (iv) the simulation and strain calculation and (v) the analyzed sources of uncertainty. In Section 3, the calculated levels of errors in each material parameter and due to each source of uncertainty are presented and discussed. Finally, Section 4 presents the main conclusions.

2. Methodology and implementation

2.1. Uncertainty quantification and problem approach

Model calibration with a large number of material parameters is an optimization problem with a large number of optimization variables that requires severe computational efforts. Furthermore, uncertainty quantification of this type of problem is even more time-consuming. Thus, a simplification of the problem is necessary.

Having into consideration that the identification of material parameters is a non-linear unconstrained optimization problem, the problem's solution satisfies the KKT necessary conditions [16]. These are first-order necessary conditions for a solution to be optimal. This means that for a cost function f and the optimum solution x^* of the problem, the stationary KKT condition has to be satisfied and can be written as follows:

$$\nabla f(x^*) = 0. \tag{1}$$

In theory, for a continuous cost function f without the presence of noise, these conditions are

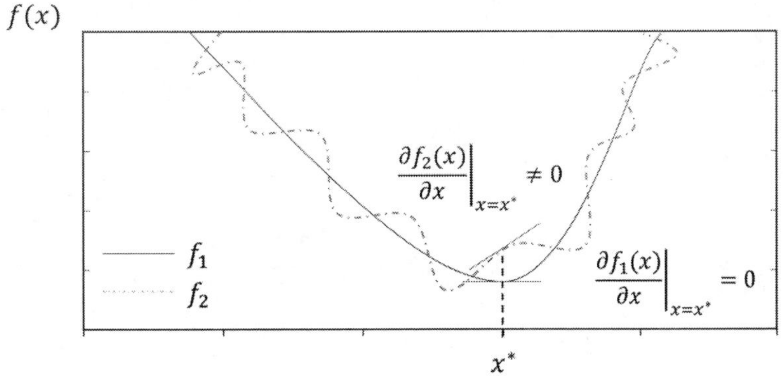

Figure 1. Illustration of the KKT conditions in theory and with a noisy function.

perfectly verified. The partial derivative of the cost function in relation to the optimization variable in the optimal solution is equal to zero (see f_1 in Figure 1). Yet, when f_1 suffers some kind of transformation $f_2 = g(f_1)$ and shows a noisy behavior, the gradient can become different from zero (see f_2 in Figure 1). This transformation is a possible source of errors that must be taken into consideration. Therefore, the level of noise and uncertainties can be evaluated by means of the KKT conditions and it is assumed that the levels of uncertainties increase with the deviation to the KKT conditions.

Assuming that the equality in Equation 1 is verified, it is possible to analyse specific sources of uncertainty. These levels of uncertainty can be estimated with the derivative-based local method that can be written as:

$$\frac{\partial}{\partial \boldsymbol{\theta}}\left(\frac{\partial f}{\partial x}\right), \tag{2}$$

for a $\boldsymbol{\theta}$ representing a vector of analysed sources of uncertainty. Although not yet demonstrated as an accurate method for uncertainty analysis, this approach can be used as a relative method and as a support tool for engineering decision making.

2.2. Identification problem under analysis

The identification approach adopted was the FEMU that is driven by the minimization of the optimization function written as:

$$\begin{aligned}\varphi(\boldsymbol{\chi}) =& \frac{1}{n_\mathrm{t}}\sum_{i=1}^{n_\mathrm{t}}\left\{\frac{1}{3n_\mathrm{p}}\sum_{j=1}^{n_\mathrm{p}}\left[\left(\frac{\varepsilon_{xx}^{\mathrm{num}}(\boldsymbol{\chi})-\varepsilon_{xx}^{\mathrm{exp}}}{\varepsilon_{\mathrm{max}}^{\mathrm{exp}}}\right)^2+\left(\frac{\varepsilon_{yy}^{\mathrm{num}}(\boldsymbol{\chi})-\varepsilon_{yy}^{\mathrm{exp}}}{\varepsilon_{\mathrm{max}}^{\mathrm{exp}}}\right)^2+\left(\frac{\varepsilon_{xy}^{\mathrm{num}}(\boldsymbol{\chi})-\varepsilon_{xy}^{\mathrm{exp}}}{\varepsilon_{\mathrm{max}}^{\mathrm{exp}}}\right)^2\right]_j\right.\\ &\left.+\left(\frac{F^{\mathrm{num}}(\boldsymbol{\chi})-F^{\mathrm{exp}}}{F_{\mathrm{max}}^{\mathrm{exp}}}\right)^2\right\}_i,\end{aligned} \tag{3}$$

where χ is the vector of the unknown material parameters, F is the load value and ε are strain components of the mechanical test. The superscripts "num" and "exp" refer to the numerical data obtained iteratively during the optimization process and the experimental data, respectively. The number of time instances is n_t and the number of in-plane measurement points is n_p. $F_{\max}^{\exp}$ is the maximum load value and $\varepsilon_{\max}^{\exp}$ is the maximum strain value of all in-plane components of the test.

Applying the methodology mentioned in Section 2.1, the partial derivatives of the cost function φ with respect to the material parameters χ that should satisfy the KKT conditions and Equation 1 can be calculated as:

$$\begin{aligned}\frac{\partial \varphi}{\partial \chi} = & \frac{1}{3 n_t n_p \cdot (\varepsilon_{\max}^{\exp})^2} \sum_{i=1}^{n_t} \left\{ \sum_{j=1}^{n_p} \left[\left(2 \cdot (\varepsilon_{xx}^{\text{num}}(\chi) - \varepsilon_{xx}^{\exp}) \cdot \frac{\partial \varepsilon_{xx}^{\text{num}}(\chi)}{\partial \chi} \right) + \right.\right. \\ & \left.\left. \left(2 \cdot (\varepsilon_{yy}^{\text{num}}(\chi) - \varepsilon_{yy}^{\exp}) \cdot \frac{\partial \varepsilon_{yy}^{\text{num}}(\chi)}{\partial \chi} \right) + \left(2 \cdot (\varepsilon_{xy}^{\text{num}}(\chi) - \varepsilon_{xy}^{\exp}) \cdot \frac{\partial \varepsilon_{xy}^{\text{num}}(\chi)}{\partial \chi} \right) \right]_j \right\}_i + \\ & \frac{1}{n_t \cdot (F_{\max}^{\exp})^2} \sum_{i=1}^{n_t} \left(2 \cdot (F^{\text{num}}(\chi) - F^{\exp}) \cdot \frac{\partial F^{\text{num}}(\chi)}{\partial \chi} \right)_i ,\end{aligned} \tag{4}$$

where the partial derivatives of each numerical strain component $\left(\frac{\partial \varepsilon^{\text{num}}(\chi)}{\partial \chi}\right)$ and of the load value $\left(\frac{\partial F^{\text{num}}(\chi)}{\partial \chi}\right)$ with respect to the material parameters can be quantified using the finite difference method (FDM).

2.3. Heterogeneous mechanical test in consideration

An innovative heterogeneous mechanical test proposed in [20, 19] was the base to generate synthetically deformed images that were used as reference virtual material data in the approach, as well as used for the iterative analysis. The geometry of the heterogeneous specimen is depicted in Figure 2. This interior notched specimen was designed by shape optimization, based on the maximization of the heterogeneity. The heterogeneity of the test was evaluated using several heterogeneity indicators that favor solutions that exhibit several strain states simultaneously with large magnitudes. The shape of the interior notch follows a third-degree spline defined by a total of 20 points. This specimen has 2 symmetries, ensuring the balance during the experimental testing. The specimen is tested under uniaxial loading conditions up to rupture.

2.4. Constitutive model and material definition

Hooke's law was used to describe the linear elastic material behavior of DP600 steel, with a modulus of elasticity of 210 GPa and Poisson's ratio of 0.3. The Swift law [21] was defined for the hardening behavior, and the Yld2000-2d yield criterion [22] for the plastic anisotropy. The reference material parameters used in the virtual experiment of the DP600 steel with 0.8 mm thickness are presented in Table 1. The rupture criterion implemented was based on the FLD (Forming Limit Diagram) [23]. The exponential of the Yld2000-2d function nominated as a is equal to 6 since the material has a BCC type of crystal structure.

2.5. FEA and DIC test simulation: DIC-levelled approach

To model and analyze the heterogeneous mechanical test described in Section 2.3, both the Abaqus [24] and MatchID 2D [25] software were used. A DIC-levelling approach was used to

IOP Conf. Series: Materials Science and Engineering 1238 (2022) 012056 doi:10.1088/1757-899X/1238/1/012056

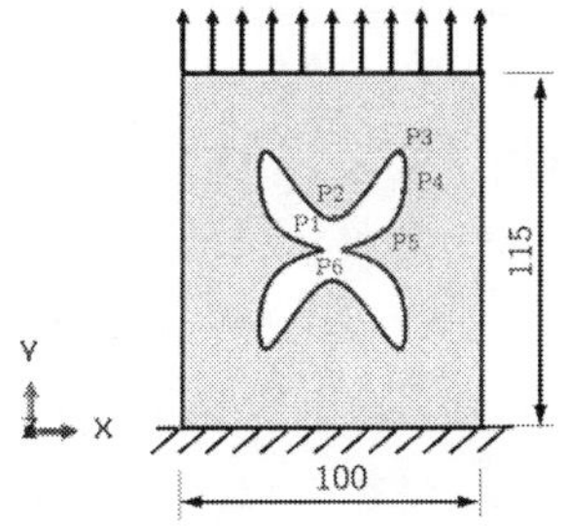

	P1	P2	P3	P4	P5	P6
xx [mm]	0.000	3.708	23.092	23.505	15.889	3.086
yy [mm]	10.000	11.412	31.784	17.078	5.163	0.000

Figure 2. Geometry, dimensions (in mm) and loading conditions of the heterogeneous mechanical test used in the parameter identification approach. Coordinates of the spline considering the origin located in the center of the specimen, in the first quadrant.

Table 1. Reference material parameters of the Swift hardening law and Yld2000-2d yield criterion used for the virtual experiment [23].

Swift law	K [MPa]			ε_0			n		
	979.460			0.00535			0.194		
Yld2000-2d	α_1	α_2	α_3	α_4	α_5	α_6	α_7	α_8	a
	1.011	0.964	1.191	0.995	1.010	1.018	0.977	0.935	6

determine the strain gradients. This was already tested in [4] and implemented in a FEMU method in [17, 18]. First, for the FE analysis, the 2D test was conducted in Abaqus under plane stress state conditions, using 2975 CPS4R elements. The material behavior was implemented using the UMMDp developed by JANCAE (Japan Association of Nonlinear CAE) [26]. A displacement was imposed in the top edge of the specimen, in the vertical direction, up to rupture. The bottom edge of the specimen was kept fixed. A total of 19 time increments presenting elastic and plastic regimes of deformation were used to extract the nodal coordinates and displacements of the mesh. Secondly, these results were used to synthetically deform a speckle pattern to be analyzed using the DIC method. The pattern was numerically generated with 3 px dot size, to avoid aliasing [27] and printed in true dimensions. The hardware and DIC analysis parameters are shown in Table 2. Moreover, a performance analysis was conducted to access the adequate DIC parameters, which are presented in the same table.

2.6. Analyzed sources of uncertainty

Several sources of uncertainty could be evaluated since both the FEA and DIC approaches can influence the outcome of the problem. In the present paper, only the subset size, step size and strain window size were analyzed and discussed. These are the elements of the vector $\boldsymbol{\theta}$ in Equation 2. These are parameters from the DIC analysis that directly influence the strain gradients observed in the specimen and, thus, are very important in parameter identification problems. In order to determine the result of Equation 2, a perturbation of approximately 20 % was applied for each of the parameters that were considered the most suitable for the DIC analysis, after the DIC performance analysis. Although it seems an excessively large value for perturbation to determine the derivatives through the FDM, this is because these DIC parameters are integers that can only vary within specific values and intervals. Therefore, the subset sizes of 21 px and 25 px, the step size of 5 px and 6 px and the strain window size of 11

Table 2. Hardware and DIC analysis parameters using MatchID 2D software [25].

Parameters	
Camera	Flir Blackfly BFS-U3-51S5M-C
Image resolution [px^2]	2448×2048
Focal length [mm]	12.5
Average speckle size [px]	3
Image filtering	Gaussian, 5 px kernel
Image conversion factor [mm/px]	0.07241
Subset size [px]	21
Step size [px]	5
Subset shape function	Quadratic
Matching criterion	ZNSSD
Interpolation	Bi-cubic spline
Strain window [data points]	11
Strain formulation	Green-Lagrange
Post-filtering of strains	Spatial
Displacement noise-floor [px]	0.009
Strain noise-floor [mm/mm]	1.209×10^{-4}

and 13 data points were evaluated.

3. Obtained results and analysis

First, the size of the material parameter perturbation was analyzed in order to better determine the partial derivatives of the FEMU cost function in relation to the material parameters calculated with Equation 4. Perturbation sizes of 10%, 5%, 1% and 0.1% were studied. The results are presented in Figure 3 and they revealed a similar tendency for each parameter. Ideally, a smaller perturbation would be more adequate for the FDM. Though, the 0.1% perturbation shows discrepancies in some anisotropy parameters compared to the other perturbation sizes. Thus, 1% material parameter perturbation was chosen for the following analysis since it is the smaller perturbation that is consistent in relation to the others.

Completing the calculus according to the methodology present in Section 2, with a perturbation of 1 % in the material parameters, it was possible to investigate the influence of each material parameter in the FEMU DIC-levelled problem and the verification of the KKT conditions.

Figure 4 shows the absolute value of the partial derivatives of the cost function with respect to the material parameters (as in Equation 4) in two sets of material parameters. The first is the optimum set of material parameters. Theoretically, its derivatives should be zero, satisfying the KKT conditions. Yet, practically, some of the derivatives are larger than zero. This quantifies the inherent errors caused by the interpolations, transformations, filters and noise and should be accounted for in the interpretation of the following results. The second set of material parameters is obtained randomly. It is presented to understand if the parameter influences the identification process. This is because when starting a FEMU identification, the initial set of material parameters usually is different from the optimum set and each parameter should show influence in the cost function for the optimisation process to flow. If a material parameter is not activated in the mechanical experiment, the value of $\left|\frac{\partial \varphi}{\partial \chi}\right|$ would be zero and the methodology presented could not lead to any conclusions. It can be observed that all the parameters are

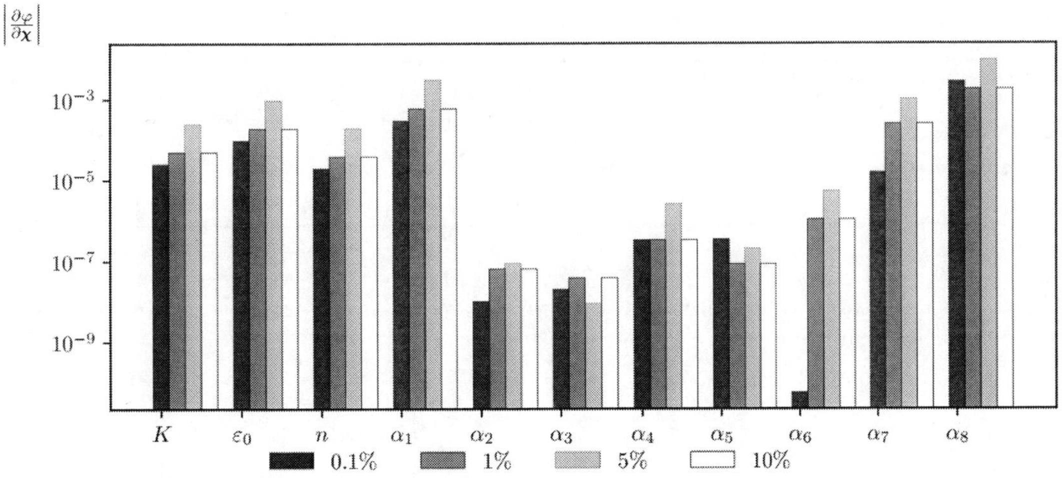

Figure 3. Absolute value of the partial derivatives of the cost function (φ) with respect to the material parameters ($\boldsymbol{\chi}$) with different perturbation sizes.

activated and, thus, influence the identification process. Yet, some present a smaller influence in the FEMU cost function, and, consequently, their derivatives are closer to zero with the optimum set of material parameters. These can present a smaller sensitivity in the methodology. The material parameters of the random set can be observed in Table 3.

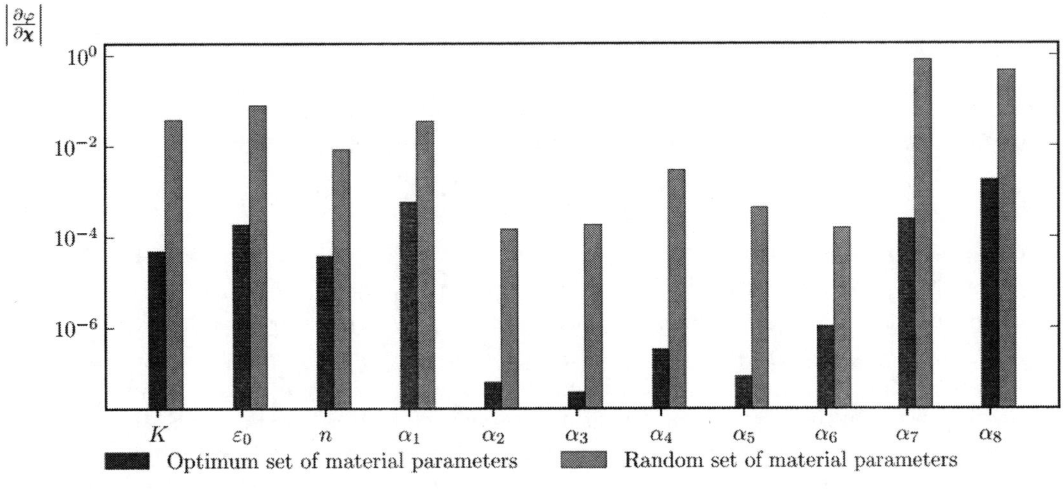

Figure 4. Absolute value of the partial derivatives of the cost function (φ) with respect to the material parameters ($\boldsymbol{\chi}$) in the optimum and random sets of parameters.

The influence of each source of uncertainty in the problem can be seen in Figure 5. It can be stated that α_7 and α_8 are more sensitive to uncertainties. This can be both from the propagation of errors that are observed in Figure 4 and from the influence of each studied source of uncertainty (subset, step and strain window sizes). Furthermore, the step size is the analyzed source of uncertainty that largest influences the FEMU DIC-levelled parameter identification.

Table 3. Random set of material parameters of the Swift hardening law and Yld200-2d yield criterion used to determine $\left|\frac{\partial \varphi}{\partial \boldsymbol{\chi}}\right|$ in Figure 4.

Swift law	K [MPa]			ε_0			n		
	960.511			0.00559			0.189		
Yld2000-2d	α_1	α_2	α_3	α_4	α_5	α_6	α_7	α_8	a
	1.123	1.310	0.899	0.905	0.974	1.200	1.076	0.869	6

The subset and strain window sizes have a similar impact on the problem and in the search for the identification solution.

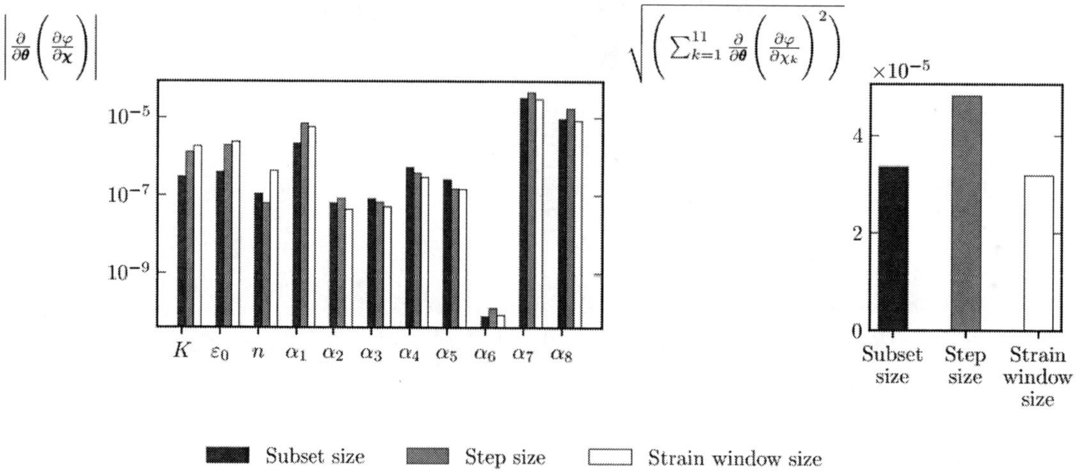

Figure 5. Influence of the subset size, step size and strain window size in each material parameter and in the overall parameter identification approach.

4. Main conclusions

In the present paper, a methodology was developed to quantify the uncertainties in material parameter identification, using the assumptions of the KKT necessary conditions of non-linear unconstrained optimization problems. This way it was possible to simplify the identification problem and estimate the uncertainties with less computational effort, leading to faster engineering decision making. The calibration problem considered was the FEMU DIC-levelled approach and the uncertainty technique was the derivative-based local method by means of the finite difference method. The mechanical test used in the inverse identification process was an interior notched specimen for uniaxial loading conditions that presents several strain and stress states simultaneously. The calibration of the Swift hardening law and Yld2000-2d yield function was evaluated. The sources of uncertainty analyzed were the subset, step and strain window sizes.

It was concluded that the obtained errors in the material parameter identification using a FEMU DIC-levelled technique were originated by the propagation of errors of the measurements, interpolation and discretization of the problem itself and from the influence of the sources of uncertainty. Moreover, the step size has a larger impact on the identified material parameters, compared to the subset size and strain window size for the studied speckle pattern and strain gradients. Special attention must be paid when choosing the DIC settings for an inverse identification of material parameters since they have a direct effect on the strain gradients and measurements and also a different impact in the inverse identification of the material parameters.

Acknowledgments

This project has received funding from the Research Fund for Coal and Steel under grant agreement No 888153. The authors also acknowledge the financial support of the Portuguese Foundation for Science and Technology (FCT) under the project PTDC/EME-APL/29713/2017 by UE/FEDER through the programs CENTRO 2020 and COMPETE 2020, and UID/EMS/00481/2013-FCT under CENTRO-01-0145-FEDER-022083. Mariana Conde is grateful to the Portuguese Foundation for Science and Technology (FCT) for the PhD grant 2021.06115.BD.

References

[1] Lava P, Coppieters S, Wang Y and Houtte P V 2011 *Applied Mechanics and Materials* **70** 165–170
[2] Lava P, Coppieters S, Wang Y, Houtte P V and Debruyne D 2011 *Optics and Lasers in Engineering* **49** 57–65 ISSN 0143-8166
[3] Rossi M, Lava P, Pierron F, Debruyne D and Sasso M 2015 *Strain* **51** 206–222 ISSN 14751305
[4] Lava P, Jones E M C, Wittevrongel L and Pierron F 2020 *Strain* **56**
[5] Roux S and Hild F 2020 *International Journal of Solids and Structures* **184** 14–23
[6] Haddadi H and Belhabib S 2012 *International Journal of Mechanical Sciences* **62** 47–56 ISSN 00207403
[7] Peixoto J, Andrade-campos A and Thuillier S 2020 Calibration of Johnson-Cook Model Using Heterogeneous Thermo- Mechanical Tests *Procedia Manufacturing* vol 47 (Elsevier B.V.) pp 881–888 ISBN 3512343708 ISSN 2351-9789
[8] Pereira A F G, Prates P A, Sakharova N A, Oliveira M C and Fernandes J V 2014 *Engineering with Computers*
[9] Pottier T, Toussaint F and Vacher P 2011 *European Journal of Mechanics, A/Solids* **30** 373–382 ISSN 09977538
[10] Souto N, Andrade-campos A and Thuillier S 2015 *Journal of Materials Processing Tech.* **220** 157–172 ISSN 0924-0136
[11] Martins J M P, Andrade-campos A and Thuillier S 2019 *International Journal of Solids and Structures* **172-173** 21–37 ISSN 0020-7683
[12] Jones E M, Carroll J D, Karlson K N, Kramer S L, Lehoucq R B, Reu P L and Turner D Z 2018 *Computational Materials Science* **152** 268–290 ISSN 09270256
[13] Wang P, Pierron F, Rossi M, Lava P and Thomsen O 2015 *Strain* 1–44
[14] Zhang S, Xing T, Zhu H and Chen X 2020 *Materials* **13** 1–12 ISSN 19961944
[15] Christian S 2017 *Uncertainty Quantification* 1st ed (Springer, Cham) ISBN 978-3-319-54338-3
[16] Kuhn H and Tucker A 1956 *Princeton University Press, Princeton, NJ*
[17] Henriques J, Conde M, Andrade-Campos A and Xavier J 2022 Identification of Swift law parameters using FEMU by a synthetic image approach based on digital image correlation *Esaform 2022 - 25th International Conference on Material Forming* (Braga, Portugal)
[18] Conde M, Henriques J, Coppieters S and Andrade-Campos A 2022 Parameter identification of Swift law using a FEMU-based approach and an innovative heterogeneous mechanical test *Esaform 2022 - 25th International Conference on Material Forming* (Braga, Portugal)
[19] Conde M, Andrade-Campos A, Oliveira M G and Martins J M P 2021 Design of heterogeneous interior notched specimens for material mechanical characterization *Esaform 2021 - 24th International Conference on Material Forming* (Liège, Belgique)
[20] Conde M 2020 *Design of a heterogeneous interior notched specimen using shape optimisation approach* Masters thesis University of Aveiro, Portugal
[21] Swift H W 1952 *Journal of the Mechanics and Physics of Solids* **1** 1–18
[22] Barlat F, Brem J C, Yoon J W, Chung K and Dick R E 2003 *International Journal of Plasticity* **19** 1297–1319
[23] Ozturk F, Toros S and Kilic S 2014 Effects of anisotropic yield functions on prediction of forming limit diagrams of DP600 advanced high strength steel *Procedia Engineering* vol 81 (Elsevier B.V.) pp 760–765 ISSN 18777058
[24] Dassault Systèmes 2014 Abaqus 6.14 Online Documentation

[25] MatchID 2021 New MatchID release 2021.1
[26] Takizawa H, Kuwabara T, Oide K and Yoshida J 2016 Development of the subroutine library 'UMMDp' for anisotropic yield functions commonly applicable to commercial FEM codes *Journal of Physics: Conference Series* vol 734 ISSN 17426596
[27] International Digital Image Correlation Society, Jones E and Iadicola M 2018 *A Good Practices Guide for Digital Image Correlation*

Characterization of elasto-plastic transition of sheet metal by using large-scale four-point bending test

W Liegard[1,2], **L Tabourot**[1] **and P Balland**[1]

[1] Univ. Savoie Mont Blanc, SYMME, FR-74 000 ANNECY, France
[2] Bystronic Group, Niederönz Switzerland

Abstract. The elastic properties of a material do play an important role in the spring back behaviour of sheet metal and have a dominant influence on the depth of where the bending tool has to go. In industry, the elastic behaviour is often approximated to a pure linear behaviour defined by two coefficients: Young's modulus (E) and yield stress (Re). This modelling hypotheses result in geometrical errors in the angle and flange lengths of bended part. In order to better determine material behaviour at very small strains, a large-scale four-point bending test machine is presented in this study. Generating pure bending and measuring the residual radius allow observation of the transition between elastic and plastic deformation behaviour. By deforming the test beam in the supposed pure elastic area, it appears that the pure elasticity behaviour of the material is indeed non-existent. Plastic strains can be detected as soon as the very first bending of the sample, at very low equivalent stress well under usual accepted elastic yield stresses. Perspectives are addressed in order to solve this difficulty, which hinders the possibility of correctly simulating the bending process, particularly with regard to tight geometry requirements.

1. Introduction

In sheet metal forming industry, it is essential to reduce non-productive time but also to reduce the quantity of wasted material per produced part. The former objective is of paramount importance in forming processes like bending that allows the production of small series of parts. In such cases, the price itself and the time cost of trial and error become non-negligible compared respectively to the market cost of the part and its production time. These difficulties also go hand in hand with the generation of a lot of wasted metal and result in an over consumption of raw material.

Three-point bending is now seen as an easy and reliable process that can soundly be integrated in general production. Indeed, in a growing number of places, it is becoming more and more common to see machined parts being replaced by bended parts. In these cases, bended parts have to support extra engineering functionalities which can be fulfilled with the control of supplementary geometrical specifications. Main target is not anymore to aim for an angle with a given tolerance but also for defined sides' dimensions with strict tolerance regarding process' usual capabilities.

Almost all bending machine vendors are investing into systems that allow directly measuring the spring back on the fly in order to correct it in real-time. However, this solution does not provide the means to respect the supplementary constraints imposed on the formed part, namely the length of its sides. Therefore, one way to control both the angle and sides' lengths is to conduct an optimisation that aims to define the initial width of the flank and the machine control parameters, using some design of experiments. This optimisation can be carried out using finite element tool to simulate the behaviour of the flank under the effect of boundary conditions modelling machine thrust. To generate accurate

IOP Conf. Series: Materials Science and Engineering 1238 (2022) 012057 doi:10.1088/1757-899X/1238/1/012057

predictions, it is essential to have an accurate modelling of the material behaviour. Nevertheless, until now and as far as finite element simulations are concerned, the constitutive behaviour models do not allow for a prediction with sufficient accuracy. Air bending induces important spring back that can be affected by many parameters of the material [1]. Among all of them, so-said elastic parameters, Young's modulus and yield stress are first order determinants of the elasticity behaviour of the blank.

2. Influence of Young's modulus and yield stress on bending simulation

To study the bending process sensibility upon the elastic parameters, FEM Abaqus software was used to simulate a bend obtained with a 12 mm opening die and a punch with a radius of 1.5 mm on a 2 mm steel sheet metal. As three-point bending is used on parts with an important length and width relative to the thickness dimensions, it allows to consider plane strain in the bended section. Therefore, 2D simulation and symmetrical conditions are acceptable simplifications to optimize time simulation vs. accuracy insofar as the aim was not to simulate any 3D effect such as anticlastic curvature [2] but to evaluate at least the influence of each elastic material parameters taken separately. Different courses of the punch were imposed to reproduce real cases where the corresponding angle was measured. The modelling of the material was characterized by a Young's modulus of 210 GPa and a yield stress Re set at 130 MPa. The elastic behaviour is therefore assumed to be linear isotropic. The Von Mises criterion is used to determine the yield strength and isotropic hardening based on the equivalent stress-strain curve described by the true stress-true strain curve obtained in tension.

Mean corresponding material parameters were identified on a mild steel DC03 sheet metal used to produce the real part. A first set of simulations (Figure 1) was provided to study the dependence of the final angle with Young's modulus and yield stress parameters. The variations on these parameters are introduced to reflect the experimental measurement uncertainties that prevail when they are identified.

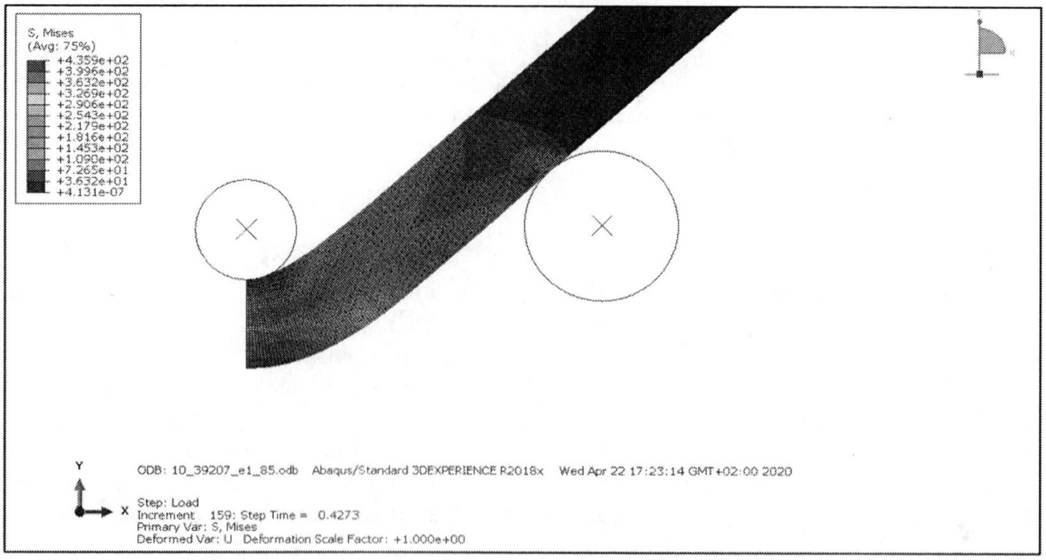

Figure 1. Stress mapping obtained during the 2D simulation of a bend on DC03 metal sheet of 2 mm with a 12 mm opening, a 3 mm diameter punch and a 12 mm stroke after tooling removal.

Plotting of the variation between final angles obtained with two numerical simulations with a different E value (Figure 2Error! Reference source not found.) shows that variation on E modulus alters the prediction enough to limit its predictive reliability. Rather small, this variation is already too big to be consistent with imposed tolerances on the industrial process. In fact, the targeted tolerance on the angle is ±0.2°. As the two parameters are independent, the error can be cumulated, and it becomes clear that the prediction cannot be seen as reliable.

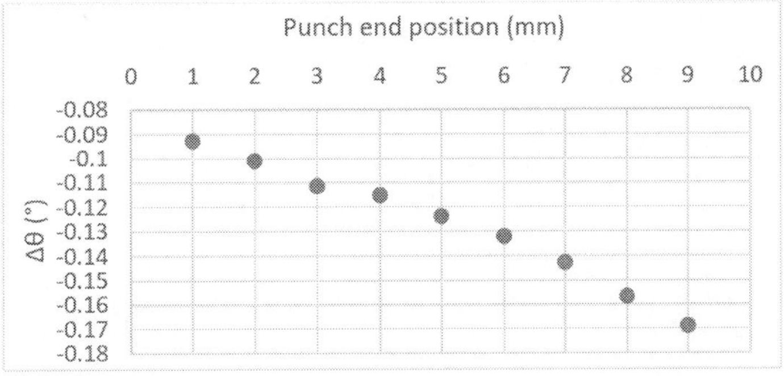

Figure 2. Plot of the angle variation after a variation of E modulus from 200 GPa to 220 GPa

Simulations have also been performed to determine the influence of yield stress Re (Figure 3). The yield stress uncertainties also have similar effects on the angle determination.

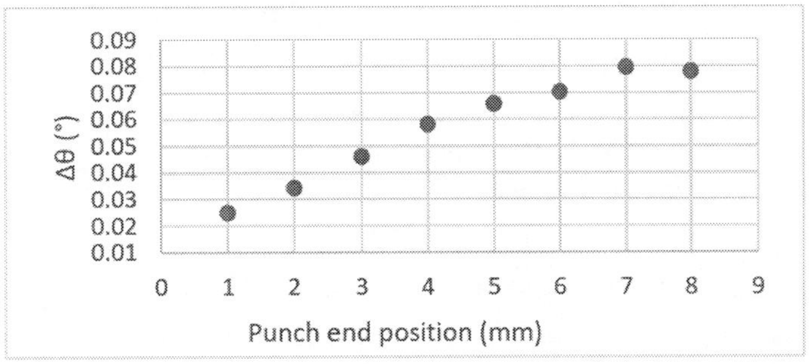

Figure 3. Variation of the angle after a variation of yield stress from 130 MPa to170 MPa

The effects of uncertainties on the elastic parameters can be cumulatively added resulting in hazardous estimation of bending angle on which it will be impossible to base the industrial bending operation optimisation.

3. Definition of parameters based on tensile test

The above results show the necessity to provide unquestionable identification of elastic parameters to provide, with the help of the simulation, an accurate prediction of bending operations. For this purpose, two materials were selected - DC04 and DC03, steels defined for cold forming processes. The objective is to focus the identification process of elastic parameter and to determine a possible deterministic way of obtaining univocally elastic parameters that would increase reliabilities of finite element simulation.

The stress-strain curve, obtained by tensile test (Figure 4), for the two selected materials is the usual support of elastic parameters identification. These tensile tests have been run a 5569 Instron machine and a 12.5 mm width extensometer to measure the strain, the strain rate was set at 7 mm/min. For this we use sample from the sheet metals cut in different directions, such as, L for longitudinal (in the laminated direction), D for diagonal (45° to the laminated direction) and T for transversal (90° to the laminated direction). Each time, three samples were tested. In the figures under, only one of the samples is represented as the results were consistent.

The identification process of the elastic parameters is rather trivial. After identifying a linear stress-strain dependency, two yield stress values would respectively be set – 190 MPa for DC03 and 135 MPa for DC04.

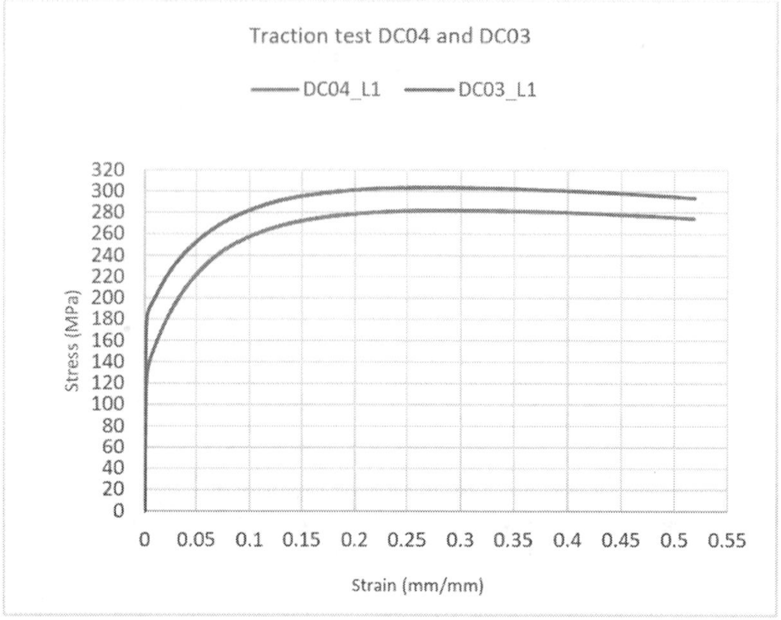

Figure 4. Comparison between DC03 and DC04 tensile curves. Trivial determination of yield stress provides Re=140 MPa for DC04 and 190 MPa for DC03.

A closer examination of the strain region between 0 and 0.004 (Figure 5), gives another possibility to define elastic limit Re0.2 of a material. This value is provided first by sheet metal suppliers but could also be determined directly by the experimenter. The method for determining the elastic limit Re0.2 [3] is to determine the portion of the curve on which elastic behaviour is considered. In one way or another, a linear regression is performed to obtain the slope of the line that approximates this portion of the curve. This slope is therefore the considered Young's modulus. To obtain the elastic limit, a straight line is drawn through (0.002) with the same slope and its intersection with the stress-strain curve is determined. Corresponding stress is the Re0.2 value that can be used as yield stress. This is nevertheless a little fuzzy insofar as plasticity events have already occurred in material. Note also that the use of such value in Abaqus will immediately introduce errors due to the fact that as long as the elastic limit is not reached, the behaviour is assumed to be perfectly elastic.

This induces that the determination of yield stress for DC03 is very dependent on the identification process. For DC03, the determined yield stress is 190 MPa and for DC04, yield stress is 140 MPa. Thus, for DC03, if we take the value from Figure 5, it appears that Re0.2 would be 182 MPa, that is to say under the value of Re that has been defined in the Figure 4 (190 MPa). Furthermore, the linearisation approximation induces Young's modulus approximation that also depends on the experimenter and has a direct effect on the yield stress determination.

Furthermore, even approximating these curves as linear seems inappropriate with respect to the accuracy to conduct a simulation of bending. Usual method to define Young's modulus in the industry for those materials is to take the so-called part of the curve, here between 0 and 100 MPa for DC04 and between 0 and 120 MPa for DC03, and to use the directing coefficient of the approximate line as the E modulus of the material. This has been represented in Figure 6a and Figure 6b.

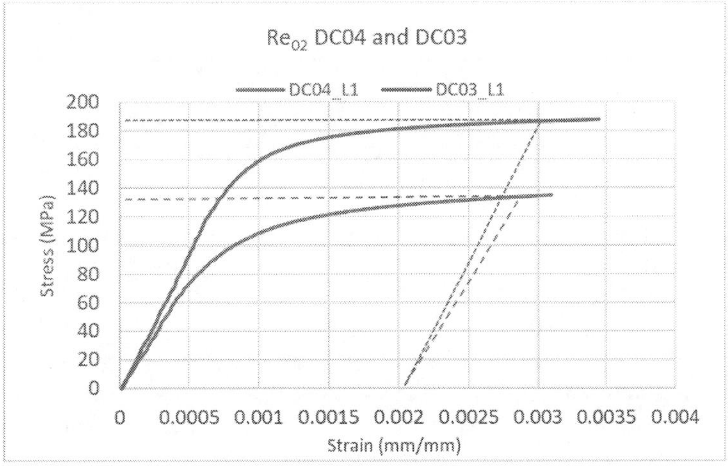

Figure 5. Magnification of the curve at the transition from purely elastic to elasto-plastic behaviour to determine Young's modulus and yield stress at 0.002 strain for DC03 and DC04.

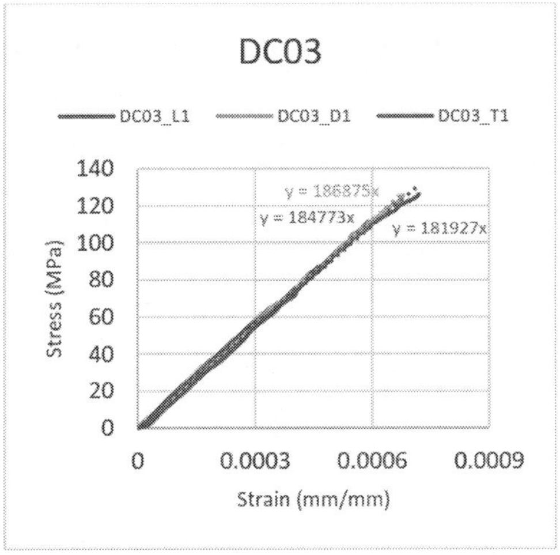

Figure 6a. Determination of Young's modulus based on the measures between 0 MPa and 130 MPa.

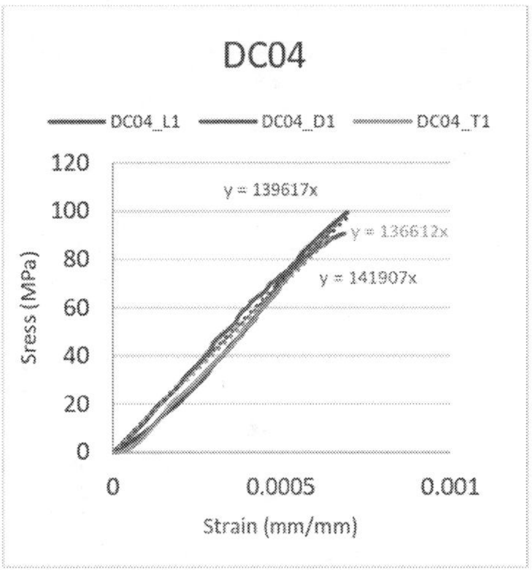

Figure 6b. Determination of Young's modulus based on the measures between 0 MPa and 100 MPa.

Young's modulus is usually not provided by sheet metal suppliers, but as mentioned previously is a necessary parameter to be defined in the simulation. If we compare the estimated Young's modulus with what is often given as a standard for steel, 210 GPa, measured in i.e. 184.5 GPa ±2.5 GPa for DC03 and 139.4 GPa ±2.6 GPa for DC04, it can be concluded that a simulation ran with standard values adjusted to the generic parameters provided by suppliers' certificates will not lead to an exploitable prediction of the results for optimising industrial process.

Table 1: Comparison of the different values of Young's modulus and yield stress obtained with the different methods.

Method	DC03 Young's modulus GPa	DC04 Young's modulus GPa	DC03 Yield Stress MPa	DC04 Yield Stress MPa
Tensile curve determination	Not evaluated	Not evaluated	190	130
Using strain of 0.2 %	182	168	182	135
Linear approximation between 0 MPa and 100 or 120MPa	184.5	139.4	Not evaluated	Not evaluated

Table 1 shows how the values defining the elastic behaviour of a sheet metal is questionable and can be considered as random. The values summarised in Table 1 are varying sometimes by almost 17% for the Young's modulus and by around 5% for the yield stress. The tendency of approximations is also not all the time an over neither an under estimation. This may have a real impact on the quality of the predictions and may lead to wrong estimations of the bend results.

4. Observation of the elasto-plasticity with a large-scale bending bench

Showing all the possible approximations of the different elasticity parameters with a tensile test leads to finding an unequivocal way to determine them. An in-depth understanding of physical processes occurring at very small stresses and strains states are therefore necessary to achieve high accuracy in the predictions.

In that purpose, a large-scale bending bench has been developed. With this bench, it is possible to generate small stress into the sample while generating a large global displacement which is easier to observe and measure.

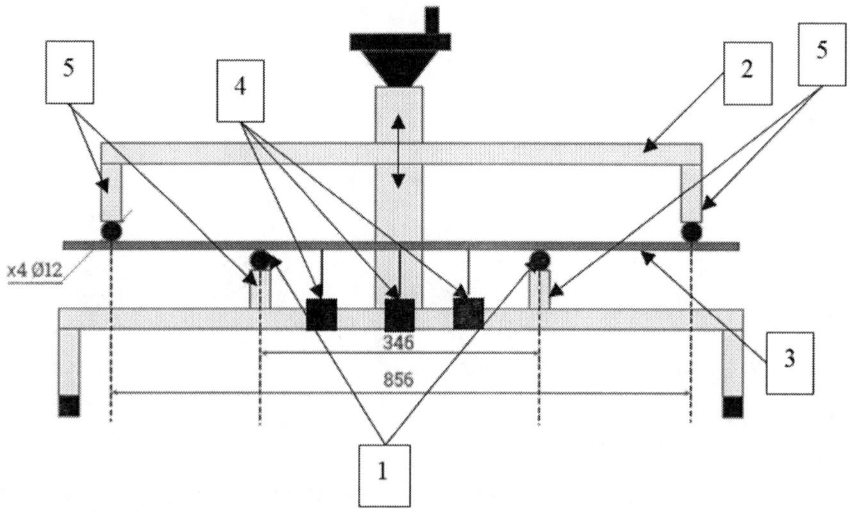

Figure 7. Concept of the first version of the large-scale bending bench. 1 Punches, 2 V-bock, 3 Sheet metal sample, 4 Laser sensors, 5 force sensors. Dimensions in millimetres.

The bench is composed of two fixed punches, denoted (1) in Figure 7. The upper beam playing the role of a large V-block and denoted (2) will be moved up and down. The displacements will be made in steps. In the first step, the two points are moving to a given position and in the subsequent step, the points are moved back to the starting position. The punches strokes are incrementally increased after each step to increase the deformation. After each step, the residual displacement of the sample part (3) is finely measured with three laser sensors (4). During the tests force sensors (5) will measure the force applied on each point (fix and moving). This will allow to evaluate the symmetry of the system and to calculate the applied moment on the sample.

The primary reason for choosing this type of test is because it provides the advantage of generating a constant load over a large portion of the test sample. The motion of the two points from the upper beam generates two forces that are equal and symmetrical. The resulting moment exerted on the sample is then constant over the distance between the two fixed points, as shown in Figure 8.

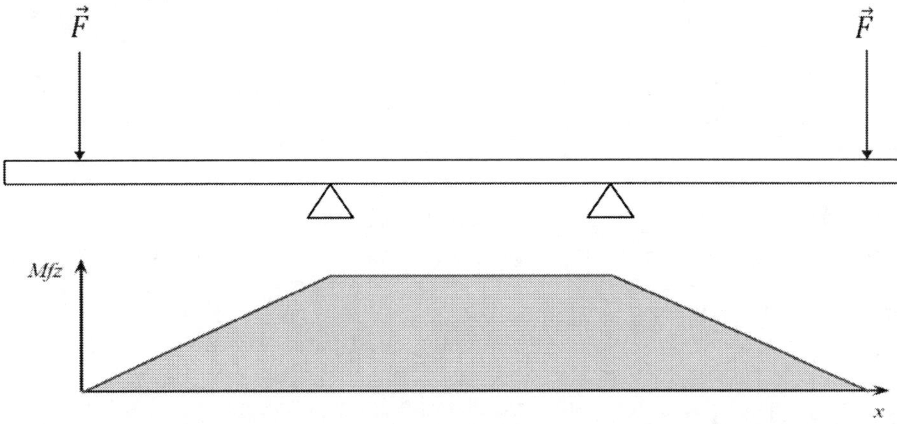

Figure 8. Evolution of the moment along the sample. The constant moment is between the two fixed punches (corresponding the dimension 346 mm) and the moment is null at the contact of the two moving points (at 0 mm and 856 mm).

Preliminary tests were performed on different samples of a DC04 sheet metal (Figure 9), for which residual displacements measured by the central sensor were plotted versus the calculated applied stress for a given stroke of the punches.

The "applied" stress, used as the horizontal axis of the graph (Figure 9) was calculated based on the beam theory [4] and the geometry of the bending bench.

$$\sigma = -\frac{Mfz}{I_{gz}}.y \tag{1}$$

Mfz is the moment calculated based on the force measured by the sensors and the distance taken to calculate the moment is the one between a moving a fix and a moving point, the y distance corresponds to a point in the height of the sample that needs to be evaluated. The zero value corresponds to the middle layer of the sample. I_{gz} is calculated according to equation (2) where b is the width of the sample (35 mm) and the h is the thickness of the sample (2 mm).

$$I_{gz} = \frac{bh^3}{12} \tag{2}$$

The value of the stress plotted on Figure 9 corresponds to the value for $y = h / 2$. This corresponds to the higher stress applied on the sample.

It is then possible to observe residual displacement that is the consequence of plastic strain obtained even if the average stress in the sample is far below yield stress whatever the method to obtain it. Here Re = 125 MPa is considered for DC04. These results confirm that considering the material as purely elastic before reaching the yield stress has to be questioned in the objective to obtain accurate simulation of bending starting with the prediction of spring back.

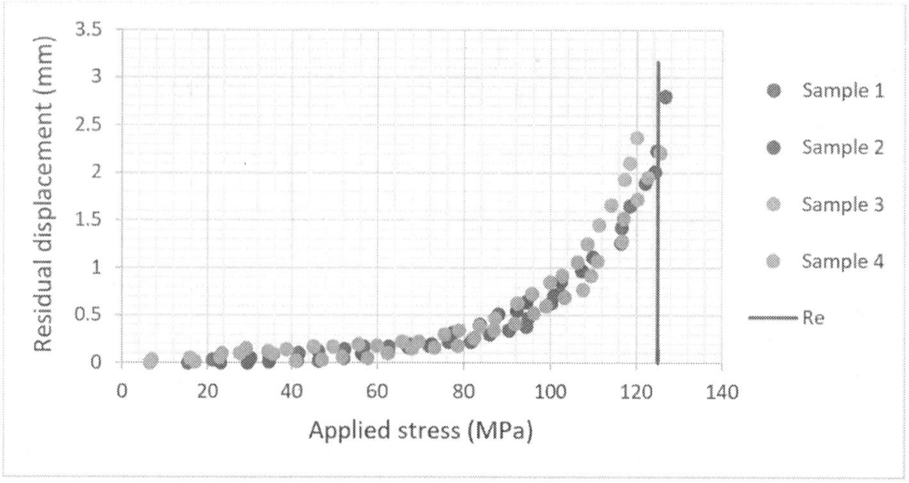

Figure 9. Residual displacement of the central point measured by the central laser after each descent of the crossbar versus tensile stress at the sample surface for DC04.

5. Conclusion and perspectives

This study aims to answer the problem of obtaining good parts immediately in 3-point air bending industrial process.

The answer to this problem lies in the use of finite element simulations. It is then shown that the necessary accuracy on the geometry of the simulated part cannot be obtained if one uses the parameters of the elastic behaviour obtained by the usual processes of identification from tensile tests or by being satisfied with the values provided on the material files by the raw material suppliers. In fact, considering the material as purely elastic in this region of deformation, suggests that all parts of the sheet metal are deforming in the same way.

In order to closely study the behaviour of the so-called and therefore hypothetic purely elastic stage, a bending bench has been developed. Several tests have been conducted below the 'apparent' elastic yield stress with results showing already permanent changes of shapes.

Therefore, another way to describe the material behaviour is to consider very low strains since the beginning of the deformation. Considering the heterogeneity of the sheet metal is certainly the prerequisites for defining models that can reproduce the plasticity that occurs at the smallest deformations.

Sheet metal is composed of grains with different orientations [5], it is then highly improbable that all the grains are behaving uniformly under an applied stress. This heterogeneity should be considered in the modelling in order to achieve the targeted accuracy.

As a follow-up study, a hybrid compartmented model [6] which takes into account this heterogeneity will be integrated in the bending simulation. The model mitigates the uncertainties of characterising the elastic part of the material by approaching the material as full elasto-plastic. It is believed that it will allow us to reach a satisfactory description of the first part of the curve to be able to correctly simulate residual deformation phenomena even at very small deformations.

References

[1] Garcia-Romeu M L, Ciurana J and Ferrer I 2007 Springback determination of sheet metals in an air bending process based on an experimental work *J. Mater. Process. Technol.* **191** 1–3 174–177

[2] Shi J, Pan S, Markum R E and Good J K 2021 Anticlastic bending impact on webs transiting rollers *Mech. Res. Commun.* **111** 103661

[3] Lemaître J et Chaboche J L 1985 *Mécanique des matériaux solides, DUNOD*

[4] Bauchau O A and Craig J I 2009 *Structural Analsyis* Springer
[5] Benson S 2019 Grain size, Part I: Material grain size matters in sheet metal bending *The Fabricator*
[6] Charleux L, Tabourot L, Roux E, Issack Farah M and Bizet L 2020 Dependency of the Young's modulus to plastic strain in DP steels: A consequence of heterogeneity? *Mater. Today Commun.* **24** 100972

On the use of the Gleeble® test as a heterogeneous test: sensitivity analysis on temperature, strain and strain rate

B. Coelho[1,*] **and S. Thuillier**[1]

[1]Univ. Bretagne Sud, UMR CNRS 6027, IRDL, F-56100 Lorient, France

E-mail: *bernardete.coelho@univ-ubs.fr

Abstract. The development of hot and warm forming processes for the manufacture of complex-shaped alloy components has been increasing. At the same time, the knowledge of the effect of certain variables involved in such processes such as temperature and strain rate is essential for an accurate modelling of the materials and processes. This work provides a sensitivity study on the effect of different testing conditions variables on beta titanium-molybdenum (Ti-Mo) alloys. Data from quasi-static uniaxial tensile tests performed on a dog bone shaped specimen using a Gleeble® machine is post-processed with Aramis digital image correlation (DIC) software and numerical models of the tests are developed using Abaqus® finite element analysis (FEA) software. Results considering different temperature conditions, different applied strain rates and different gauge lengths for the determination of the mechanical properties of the material are discussed.

1. Introduction

Titanium alloys have been of high interest for industries such as aeronautical, biomedical, luxury and marine due to their lightweight, high structural properties, excellent corrosion resistance, biocompatibility and also their mechanical properties that allow flexible forming processes for the manufacturing of thin and light parts. However, under certain conditions, such materials may present plastic instabilities (serrated yielding and macroscopic localised bands of plastic strain) associated to the Portevin-Le Chatelier (PLC) effect. Luo *et al.* [1] have performed an experimental study based on digital image correlation (DIC) to evaluate the effect of temperature and applied strain rate on the PLC bands that appear in Ti-xMo alloys. They performed uniaxial tensile tests using a Gleeble® 3500 system coupled with Aramis (GOM™'s DIC software) [2]. This work intends to provide a first step in the numerical modelling of such experiments initially investigated by Luo *et al.*. More precisely, an analysis on the use of the Gleeble® test to extract the material properties to be used as inputs in numerical models is here presented. It should be noticed that, at this stage, the focus is not on the PLC effect but the mechanical tests themselves and their interpretation for an accurate numerical modelling.

Gleeble® thermal-mechanical testing systems [3] are commonly used to extract constitutive data of sheet metals that are used in hot and warm forming processes and, therefore, experience complex temperature profiles. Such systems are known for enabling accurate control of temperature and testing speed. However, the heating by Joule effect and the heat loss that occurs due to the water cooling of the grips holding the specimen during the tensile tests may become a limitation due to the generation of temperature gradients and non-homogeneous deformation in the specimen [4, 10, 5, 6, 7]. At the same time, even performing such tests at constant quasi-static applied strain rates, spatial variations on the strain rate along the specimen are also experimentally observed [8, 9]. This means that the material properties characterisation may be affected as each part of the material deforms differently due to the variation of both temperature and strain rate dependent flow properties. Thus, for an accurate estimation of the material properties, such heterogeneities may need to be taken into consideration. This is particularly important for the case of titanium alloys given their low thermal conductivity and high sensitivity of deformation resistance on temperature when compared with other alloys [7].

A number of works in the literature have reported the experimental flow stress sensitivity of metal alloys to both temperature and strain rate. Typically, for metallic materials such as aluminum alloys [10], magnesium alloys [11], boron steels [12, 13, 6] and dual-phase steels [9] a negative sensitivity to temperature and positive sensitivity to strain rate has been reported. However, the reported flow stress sensitivity of titanium alloys to temperature and strain rate does not follow the "traditional" behaviour due to the effect of PLC bands [1, 14, 7]. Although in compression tests (not tension) some titanium alloys have presented as well a negative sensitivity to the temperature and positive to the strain rate [14, 7]. Others, in tension, have shown a negative sensitivity to strain rate and a variable sensitivity to the temperature, depending on the chemical composition of the alloy [1].

Considering the flow stress sensitivity to such variables may, however, not be enough to validate the numerical prediction of Gleeble® tests. The heterogeneous spatial distribution of both temperature and strain rate may need to be taken into account [8, 9]. Nonetheless, investigations measuring and quantifying the influence induced by both temperature and strain rate heterogeneities in the testing procedures and in their numerical modelling are still lacking.

To replicate Gleeble® tests through FE models, DIC information as well as the test conditions are used to define the model geometry, boundary conditions and material properties [15, 8, 7]. However, the most adequate procedure on how to extract and use such information is not clearly defined. That is the case, for instance, of (1) the choice of the adequate gauge length to calculate the average strains, (2) the decision of modelling the full specimen or just a part of the gauge area or even (3) the type of numerical model, *e.g.*, as thicker specimens may present temperature gradients in thickness some authors have implemented a coupled electro-thermo-mechanical model considering the full specimen and the grips to capture such gradients [7].

This work suggests a procedure to extract the material properties from full field data considering the smallest and the largest gauge lengths. Simulations of the largest section possible of the Ti-Mo specimen at different conditions of temperature and strain rate are analysed.

2. Experimental procedures and calculation methods

The nominal composition of the tested material is Ti-18Mo and its chemical composition (wt.%) is: Mo 17.87 % and Ti 82.13 %. The material density ($\rho_{Ti-18Mo} = 11.91$ g/cm^3) was calculated as the sum of the weighted densities of each component. The tensile specimens are cut from a rolled sheet with 1 mm thickness according to the dimensions provided in figure 1. The longitudinal direction (x direction in Fig. 1) is aligned with the rolling direction of the sheet.

For the uniaxial tensile tests at elevated temperatures, the specimens were heated to 250 and 350 °C (temperature imposed in the centre) using a constant heating rate of 10 °C/s, followed by a holding time of 10 s prior to the test [1]. The temperature during each tensile process is considered constant as the fluctuations measured by three thermocouples, TC1, in the centre of the specimen, TC2 and TC3, both ±5 mm distant from the centre, as shown in Fig. 1, are maintained within 2 °C (see examples of temperature evolution in Fig. 2). As shown in Fig. 2, the first step consists in the heating of the specimen. When the imposed temperature is reached in the centre of the specimen, the mechanical load is applied.

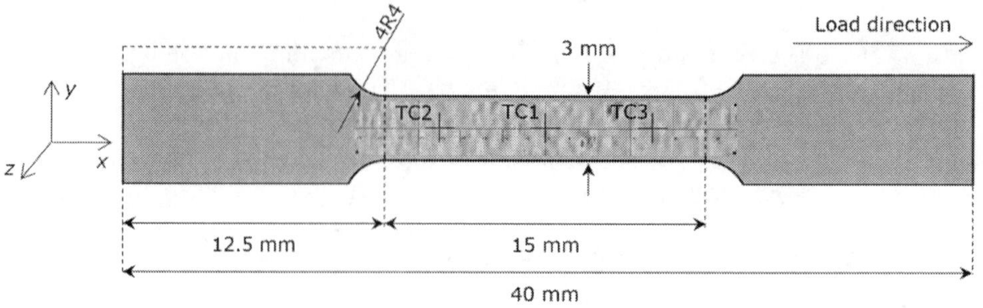

Figure 1: Dimensions of the tested specimens and DIC region of interest (ROI).

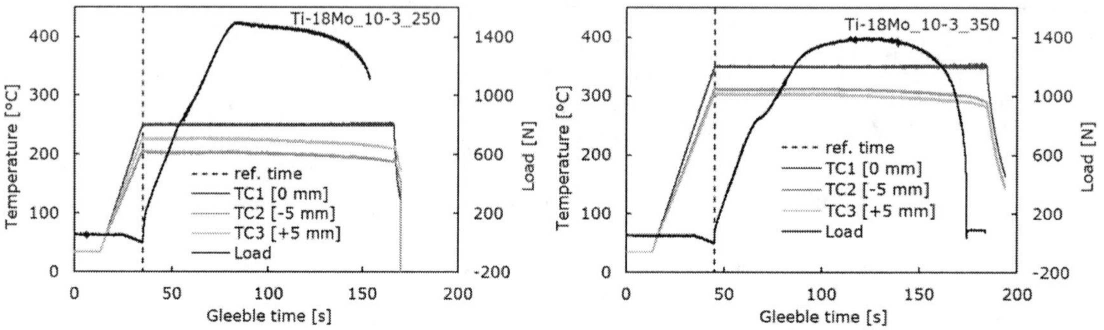

Figure 2: Temperature measured by each thermocouple, TC1, TC2 and TC3, and applied load for the tests at a strain rate of 10^{-3} s^{-1} and temperatures of 250 °C [left] and 350 °C [right].

Temperature and load data are recorded by the Gleeble® machine while the DIC full-field displacement and strain data are obtained by Aramis (using two digital cameras with a rate of 10 frames per second). However, Aramis and Gleeble® data are not synchronised and are recorded at different time intervals. Thus, for the synchronisation, the first instant of positive load, ref. time in Fig. 2, is used as reference. Additionally, the Gleeble load-time curves are used to plot the Aramis load-, displacement- and strain-time curves at the same time intervals as the Gleeble® data. This is performed using linear interpolations. The initial time $t_{num} = 0$ s shown in Fig. 3 is defined at the beginning of the load increase observed in Fig. 2.

The variation of temperature along the specimen length (longitudinal direction) is determined by fitting a curve (2nd order polynomial function) to the three points of average temperature measured by the thermocouples during the test, as shown in Fig. 3. In the region where the sectional area of the specimen is constant,*i.e.* within the 15 mm gauge length, the temperature from the centre to the extremities decreases from 350 to around 250 °C, which corresponds to a temperature gradient of 13 °C/mm. Variations of temperature in the transversal direction (y direction) are not measured. However, according to the literature, for specimens of such small dimensions, those variations are usually not significant in quasi-static processes [4, 9].

For the DIC analysis on the performed tests, the selected region of interest (ROI) is indicated in Fig. 1 by the coloured region, which has an approximate area of 18×2.5 mm^2. Since this area is manually selected, it slightly varies between tests. For the strain measurement, a subset window or facet size of 9 px and a step size of 4 px is used. A sensitivity study was performed in advance to determine such parameters.

The average displacements in x, y and z were extracted from the boundaries of the ROI, *i.e.* at $x \approx -9$ mm, here called fixed side, and $x \approx 9$ mm, here called loaded side. The obtained displacement-time curves are provided in Fig. 3. In order to consider zero displacements at the

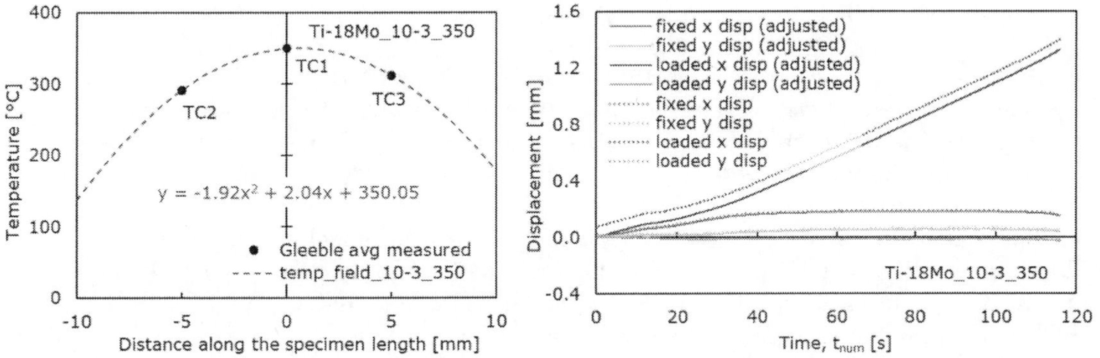

Figure 3: Average temperature along the specimen [left] and average displacements in x, y and z directions obtained by DIC in the left/fixed and right/loaded sides of the ROI [right].

beginning of the test, the observed initial displacements associated with dilatation effects are removed (see adjusted displacement curves in Fig. 3).

To obtain the true stress-strain curves, the Cauchy stress tensor component σ_{xx}^{C} is calculated by the equation:

$$\sigma_{xx}^{\mathrm{C}} = (F/S_0)\exp(\epsilon_{xx}), \tag{1}$$

where F represents the applied load, S_0 the initial cross-sectional area and ϵ_{xx} the average logarithmic strain. In this work, two different stress-strain curves are obtained by considering two different gauge lengths to calculate the average strain: (1) a larger gauge length of approximately 15 mm, which corresponds to the maximum length with constant cross-sectional area and (2) a smaller gauge length of less than 2 mm, selected where the strain localisation occurs (different for each test) with the aim of capturing larger strain values. Here, the different analysed datasets are mentioned as GLOBAL and LOCAL, for the larger and the smaller localised gauge areas, respectively. Those two different stress-strain curves are used to obtain the material hardening curves by visually fitting a curve between them using the Swift-Voce hardening law that describes the isotropic yield stress as:

$$\sigma_Y = \delta\left[K(\epsilon_0 + \epsilon_{\mathrm{p}})^n\right] + (1-\delta)\left[\sigma_{Y_0} + R_{\mathrm{sat}}\left(1 - exp(-\beta\epsilon_{\mathrm{p}})\right)\right], \tag{2}$$

where the first and second terms of the equation correspond, respectively, to the Swift and to the Voce law and δ is the weighting factor. ϵ_{p} is the equivalent plastic strain and the parameters K, ϵ_0, n, σ_{Y_0}, R_{sat} and β are material parameters to identify. The plastic strain is derived from the Hook's law and the total strain decomposition assumption:

$$\epsilon^{\mathrm{p}} = \epsilon - \epsilon^{\mathrm{e}} = \epsilon - \sigma_{xx}^{\mathrm{C}}/E, \tag{3}$$

where ϵ, ϵ^{p} and ϵ^{e} are, respectively, the total, plastic and elastic strains and E is the Young's modulus, which is obtained from the slope of the stress-strain curves.

For the calculation of the strain rate, $\dot{\epsilon}_{xx} = \partial\epsilon_{xx}/\partial t$, the forward finite differences method is used considering time intervals of 10 s. Before such calculations, in order to minimise errors from noisy data, the strain-time curves where smoothed using the exponential smoothing method:

$$\epsilon_t^{\mathrm{s}} = \alpha\epsilon_{t-1} + (1-\alpha)\epsilon_{t-1}^{\mathrm{s}}, \tag{4}$$

where ϵ_t^{s} and $\epsilon_{t-1}^{\mathrm{s}}$ represent the forecasted/smoothed strain for the current and previous time period, respectively; α is a smoothing constant between 0 and 1 (here, $\alpha = 0.2$ is used) and ϵ_{t-1} is the actual value of strain for the previous time period.

3. Numerical modelling: geometry, boundary conditions and material properties

FE simulations using explicit dynamic analysis are performed to study the effect of temperature and strain rate in the replication of the quasi-static tensile tests on Ti-18Mo alloys. For that, a 3D model of the ROI shown in Fig. 1 is developed using Abaqus®/Explicit 2018 [16]. The model geometry is defined by the experimental coordinates (x, y) of the black points located in the vertices of the ROI represented in the DIC mesh of Fig. 4, which corresponds to the coloured region of Fig. 1. For the model thickness, only half is considered (5 mm) and a z-symmetry boundary condition is imposed. The x and y displacement curves, adjusted curves of Fig. 3, are imposed to all nodes of the corresponding side of the model as shown in Fig. 4. As a displacement $\Delta u_x = 1.18$ mm is applied in 115 seconds (0.01 mm/s), a mass scaling factor of 1×10^6 is used to speed up the analysis. It should be noticed that the influence of simulating a ROI longer than the constant width area has not been investigated yet. However, it is used only to apply displacement boundary conditions. The average strains, as explained in the previous section, are calculated only over gauge areas with a constant width.

For the simulations requiring to account with the longitudinal temperature profiles, a temperature analytical field using the equation provided in Fig. 3 is also imposed as an initial boundary condition to all nodes of the model. It should be noticed that Abaqus® only takes into account this boundary condition when temperature-dependent material properties are provided. Given the reduced thickness of the specimen, the variation of the temperature along the y axis is assumed constant. The generated mesh is composed of eight-node linear brick reduced integration elements (C3D8R) with enhanced hourglass control (default) and an average size of 0.14 mm, which is approximate to the dimensions of the mesh from the DIC analysis, as can be seen in Fig. 4, in order to obtain a similar number of measurement points as in the experiments.

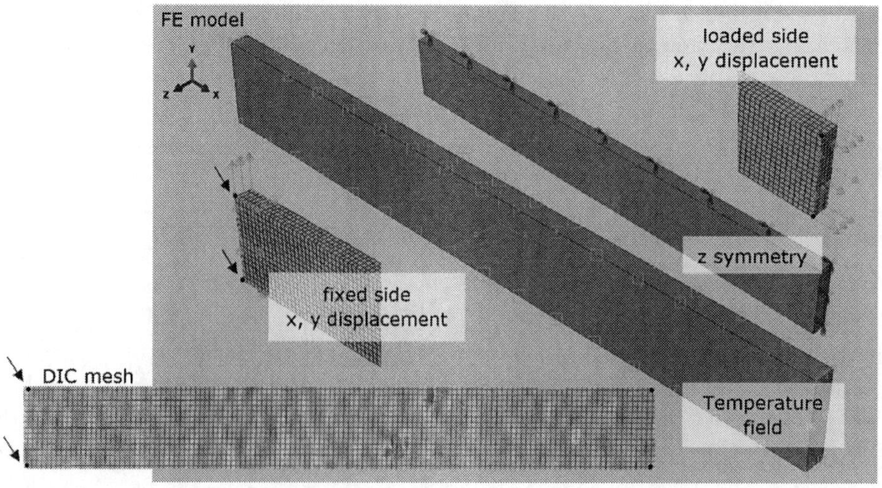

Figure 4: DIC mesh and 3D FE model showing the imposed boundary conditions.

Isotropic elasto-plasticity is assumed for the material mechanical behaviour. The material properties to be used as input in the numerical model include the material density, the Young's modulus, the Poisson's ratio and the hardening curves. Both the density and the Poisson's ratio ($\nu = 0.33$) are considered to be the same for different temperatures and strain rates. The Young's modulus at a certain temperature and strain rate is determined by the slope of the Cauchy stress - logarithmic strain curve of the test performed under such conditions. The same for the hardening curve at each specific condition of temperature and strain rate. Considering such properties, four different scenarios are simulated in this work:

num: a material with no temperature nor strain rate dependency, so a single value for the Young's modulus and a single hardening curve are considered;

num_SR: a strain rate dependent material, where an additional hardening curve obtained from a test at a lower strain rate is considered;

num_T: a temperature dependent material, where an additional value for the Young's modulus and an additional hardening curve obtained from a test performed at a lower temperature are considered; and

num_T_SR: a temperature and strain rate dependent material, which includes an additional Young's modulus at a lower temperature and 3 additional hardening curves, 1 at a lower temperature, same strain rate, and 2 at a lower strain rate and two different temperatures.

To define the isotropic hardening model in Abaqus®, the yield stress σ_Y is given as a tabular function of the plastic strain and, depending on the scenario, as a function of temperature and/or strain rate. The yield stress at a given state is simply interpolated from this table of data, and it remains constant for plastic strains exceeding the last value given as tabular data.

For strain rate dependent materials, the hardening curve at static strain rate is also a required input. Since such data is not available for the studied material, the same hardening curve obtained at the lowest strain rate is used, meaning the material is considered to not change its plastic mechanical behaviour for lower applied strain rates.

4. Results and discussion

In this section, the material properties extracted from tests performed at two different temperatures (250 and 350 °C) and two different strain rates (10^{-4} and 10^{-3} s^{-1}) are provided. After that, the results of the numerical simulation of the tensile test performed at 350 °C and 10^{-3} s^{-1} considering the four different scenarios described in section 3 are provided and compared with the experimental results. It should be noticed that the presented results correspond to selected tests at each different condition of temperature and strain rate, however, the analysis of two tests at each different condition was performed to verify the repeatability of the results.

Fig. 5 provides the Cauchy stress - logarithmic strain curves obtained from different average strains according to the selected gauge length (GLOBAL/LOCAL). It can be observed that smaller gauge lengths allow to significantly obtain larger values of strain. However, smaller gauge lengths, when centred in the localised strain, also increase the noise in data. For this reason, the Young's modulus is calculated from the slope of the GLOBAL stress-strain curves. The flow stress curves also demonstrate that the studied Ti-18Mo alloy has a negative sensitivity to both temperature and strain rate, *i.e.*, the stress decreases when the temperature increases from 250 to 350 °C and/or when the strain rate increases from 10^{-4} to 10^{-3} s^{-1}. As initially introduced in this paper, this behaviour is not a classical behaviour of metallic materials and is justified by the PLC effect. The temperature sensitivity is more accentuated at a lower strain rate, with larger stress differences at 10^{-4} s^{-1} than 10^{-3} s^{-1}, and the strain rate sensitivity is more accentuated at a lower temperature, with larger stress difference at 250 °C than 350 °C.

The hardening curves obtained from both the GLOBAL and LOCAL average strains are shown in Fig. 6. As can be observed, the Swift-Voce law, adjusted by a *trial-and-error* visual fitting approach, provides a satisfactory fit with the experimental data. At a first stage of this work, the use of just the LOCAL data to get the hardening curve was tested. However, it demonstrated not to be the best approach as the numerical flow stress curves would present lower values of stress compared with the real/experimental curves. The same was observed when using just the GLOBAL data, with resulting stress values higher than expected and lower strain ranges. Adjusting a curve between the GLOBAL and LOCAL experimental curves allow to obtain the closest approximation to the real material behaviour as it is demonstrated ahead. The Swift-Voce parameters of each curve, as described in equation 2, are provided in Table 1, where

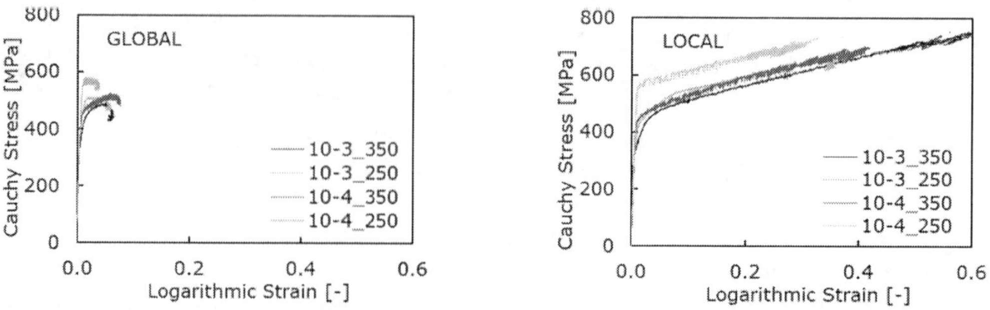

Figure 5: Cauchy stress - logarithmic strain curves for the selected tests using the average strain from a large (GLOBAL) and localised (LOCAL) gauge lengths.

10-3_350, 10-3_250, 10-4_350 and 10-4_250 correspond to the tests at the imposed conditions of 10^{-3} s^{-1}/350 °C, 10^{-3} s^{-1}/250 °C, 10^{-4} s^{-1}/350 °C and 10^{-4} s^{-1}/250 °C, respectively.

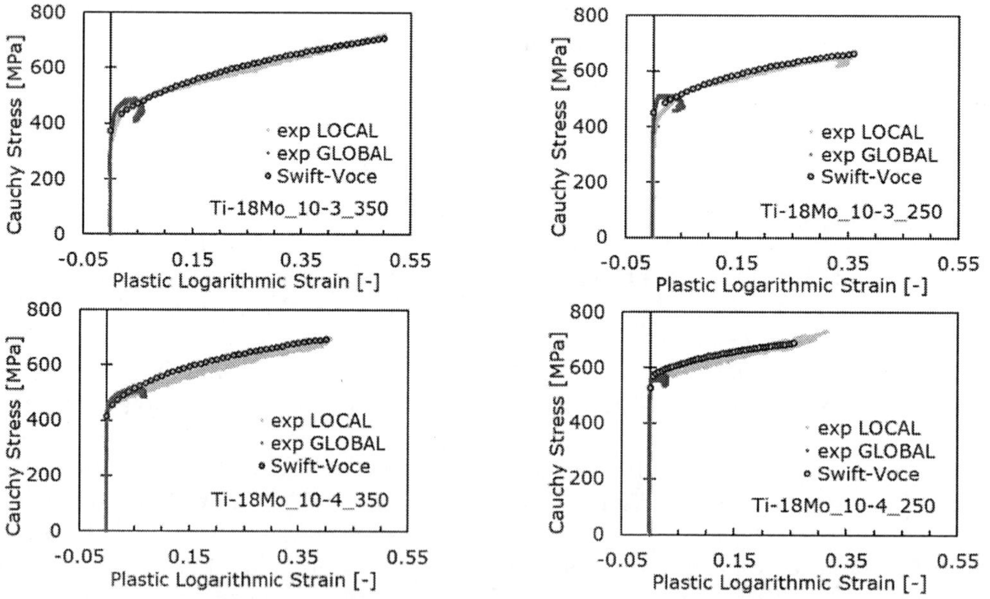

Figure 6: Hardening curves obtained experimentally and fitted Swift-Voce hardening for the selected tests performed at 4 different conditions of temperature and strain rate.

Table 2 presents the material properties extracted from the analysed tests. The values of the Young's modulus and the initial yield stress correspond to the average of two tests performed under the same conditions of temperature and strain rate. The maximum logarithmic strains before rupture correspond to the values obtained for the selected tests.

To analyse the influence of temperature and strain rate in the FE replication of the tensile tests, 4 different scenarios are simulated. Fig. 7 shows the material inputs (hardening curves and values of Young's modulus) considered in each different scenario. For the scenarios considering material strain rate dependency (SR), the 10^{-4} s^{-1} curves are used for the static strain rate.

The validation of the numerical simulations with the experimental results obtained by DIC is presented in Fig. 8 through the representation of the nominal stress - relative elongation curves. Fig. 9 shows, for each case (the experimental and the four numerical scenarios), the strain maps in the longitudinal direction (ϵ_{xx}) at two different stages of the tensile tests: (1) at

Table 1: Swift-Voce hardening parameters for the tests performed at 4 different conditions.

Param. \ Test	10-3_350	10-3_250	10-4_350	10-4_250
δ [-]	0.88	0.40	0.96	0.50
K [MPa]	840	700	800	990
ϵ_0 [-]	0.050	0.015	0.030	0.040
n [-]	0.28	0.20	0.19	0.18
σ_{Y_0} [MPa]	450	550	500	500
R_{sat} [MPa]	250	230	500	80
β [-]	200	4	1000	400

Table 2: Young's modulus E, initial yield stress σ_{Y_0}, and maximum logarithmic strain before rupture ϵ_{max} obtained for the 4 different testing conditions. E and σ_{Y_0} are average values from two tests.

Test	E [GPa]	σ_{Y_0} [MPa]	ϵ_{max} [-]
10-3_350	74	409	0.54
10-3_250	64	450	0.59
10-4_350	65	413	0.62
10-4_250	63	514	0.46

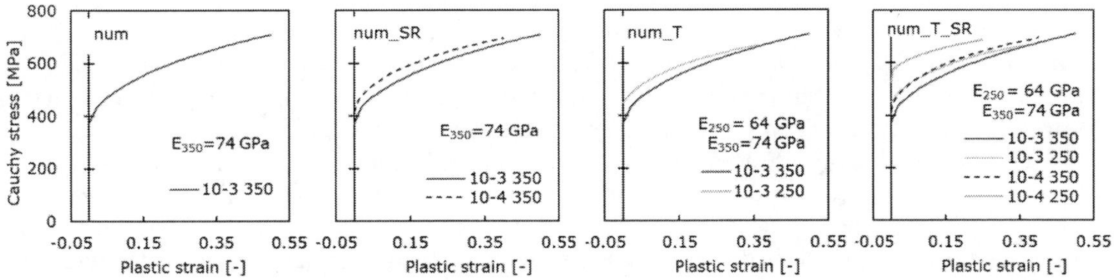

Figure 7: Difference in plastic and elastic properties considered in each simulated scenario.

the final stage before rupture and (2) at the stage of observed maximum load, which occurs at approximately $t_{num} = 70$ s.

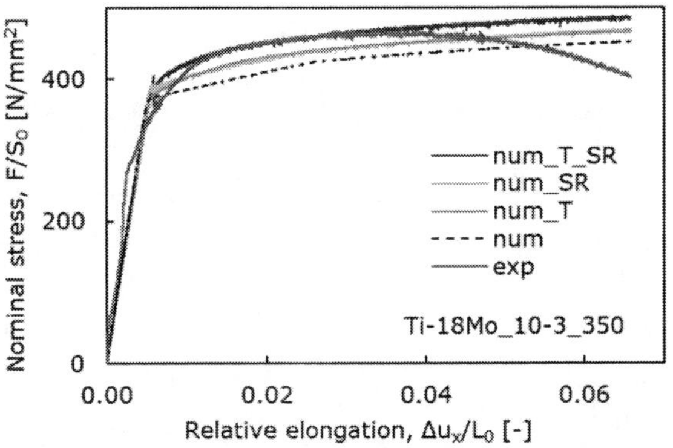

Figure 8: Experimental and numerical results for the nominal stress - relative elongation curves.

Overall, the FE models approximate well the DIC results up to the maximum load. After that, the numerical model is not capable of capturing the material softening.

Comparing the most simple model (num) with the models assuming a temperature dependent material (num_T and num_T_SR), it is clear that adapting the boundary conditions (specifically, in this case, considering the temperature spatial variation), allows to optimise the simulation of the mechanical test. In fact, only the temperature dependent models (T) achieve the expected strain localisation, as can be observed in Fig. 9. At the same time, only these two models result in a wider range of strain. However, the maximum strain value is only achieved by the model with both temperature and strain rate dependency (num_T_SR) and uniquely at the instant of

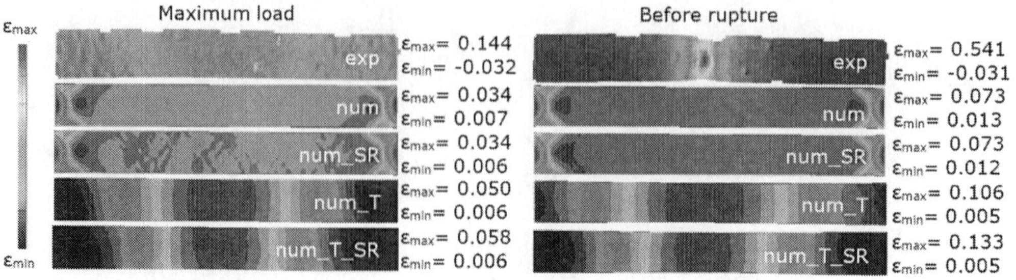

Figure 9: Experimental [Aramis] and numerical [Abaqus®] strain fields (ϵ_{xx}) obtained at an instant near the maximum load ($t_{\text{num}} = 70$ s) and before rupture.

maximum load ($\epsilon_{\text{max}} \approx 0.06$). The experimental value of ϵ_{max} indicated in Fig. 9 is 0.144 but this corresponds to a measurement in two specific points in the specimen containing a defect in the speckle pattern (red-yellow points observable in the figure). Indeed, the localised region is represented by the green colour which corresponds to $\epsilon_{\text{max}} \approx 0.06$. At the same time, for the experimental scenario before rupture, the indicated value of maximum strain, in the red region, is 0.541 but the green region presents an approximate value of 0.26.

A more detailed description of the DIC and numerical (num_T_SR) strain profiles along the specimen during the tensile test is shown in Fig. 10. The presented data is extracted from a line parallel to the tensile direction x in the middle of the specimen ($y = 0$). The superimposed experimental and numerical data up to the maximum load demonstrates that the numerical model is capable of predicting not only the maximum value of strain, as previously observed, but also the shape of spatial variation of strain along the specimen length.

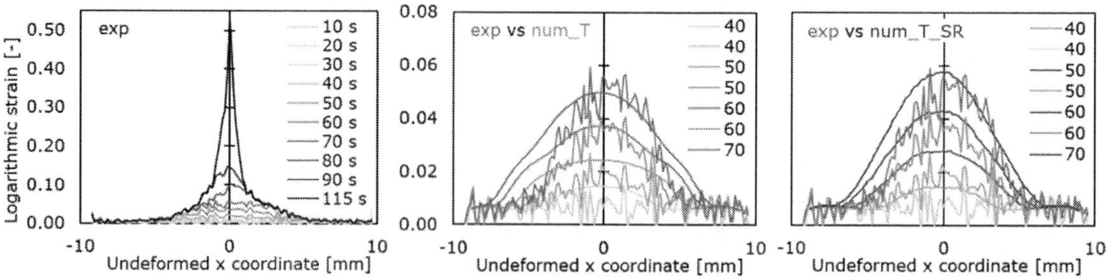

Figure 10: Strain spatial distribution obtained by DIC (exp) and superimposed numerical results from the two temperature dependent models up to the instant of maximum load.

Since the strain rate gives a more accurate picture of the instantaneous evolution of the structure, its spatial distribution at different stages of the tensile test up to the maximum load obtained experimentally (exp) and by the two numerical models assuming a temperature dependent material (num_T and num_T_SR) is presented in Fig. 11. The num_T_SR is capable of predicting well the maximum strain rate that occurs in the middle of the specimen ($x = 0$ mm) when reaching the maximum load. However, the strain rate decreases faster than desired when going far away from the centre. Surprisingly, the num_T model is capable of predicting, at the instants near the maximum load, the (almost) constant strain rate from the centre up to around ± 3 mm but at a lower value of strain rate as expected. On the other hand, the results for this particular model present not only a single strain rate "peak" in the centre of the specimen but also two additional smaller "peaks" that start near the centre of the ROI (even before the main "peak" starts appearing at 40-50 s) and move in the direction of the extremities of the ROI as the deformation proceeds (similar with the numerical observations reported in [8]).

This also demonstrates the importance of modelling a larger region of the specimen than the commonly used small lengths where the temperature remains approximately constant. The latter may omit important information concerning the accuracy in predicting the evolution of the structure. Such additional peaks are not observed by the num_T_SR model, which makes it a better approximation of the DIC results.

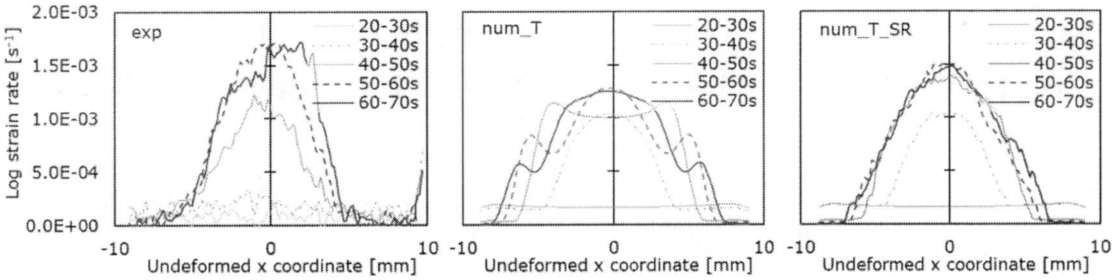

Figure 11: Evolution of the strain rate profiles along the specimen length (at y=0) up to the instant of maximum load obtained by DIC (exp) and by the two temperature dependent models.

5. Conclusions

The present work demonstrates that the temperature gradients and the strain rate spatial distribution along the specimens in Gleeble® tensile tests that are observed experimentally have a significant impact in the deformation behaviour of the material and should be taken into consideration in the numerical simulations to ensure an adequate replication of the tests. Moreover, the presented results show that the use of the Gleeble® test, if performed at two different conditions of both temperature and strain rate, can provide rich information for a satisfactory numerical prediction of the material behaviour over a large gauge length. However, specimens of different materials, size and shapes (as suggested by [4] and [5]), different gauge lengths (also suggested by [4] and [6]), as well as different testing conditions (such as higher temperatures and different strain rates) should be analysed in order to validate any suggestion for a standardisation of the Gleeble tensile testing for mechanical properties measurement.

Acknowledgments

The authors would like to acknowledge Région Bretagne's support with the post-doctoral research grant ref. 20003859 under the ViFoMeTi project.

References

[1] Luo S, Castany P and Thuillier S 2019 *Mater Sci Eng A* **756** 61–70
[2] GOM, Aramis www.gom.com/nl/3d-software/gom-system-software/aramis-professional.html Accessed: 2022
[3] Dynamic Systems Inc, Gleeble www.gleeble.com Accessed: 2022
[4] Norris S and Wilson I 1999 *Model Simul Mat Sci Eng* **7** 297
[5] Haddadi H and Belhabib S 2012 *Int J Mech Sci* **62** 47–56
[6] Zhang R, Shao Z, Lin J and Dean T A 2020 *Exp Mech* **60** 1289–1300
[7] Xiao H, Fan X, Zhan M, Liu B and Zhang Z 2021 *J Mater Process Technol* **288** 116868
[8] Labergere C, Guelorget B and Francois M 2014 *Int J Solids Struct* **51** 3944–3961
[9] Martins J, Thuillier S and Andrade-Campos A 2021 *Int J Mech Sci* **202** 106511
[10] Coër J, Bernard C, Laurent H, Andrade-Campos A and Thuillier S 2011 *Exp Mech* **51** 1185–1195
[11] Ulacia I, Salisbury C, Hurtado I and Worswick M 2011 *J Mater Process Technol* **211** 830–839
[12] Li H, He L, Zhao G and Zhang L 2013 *Mater Sci Eng A* **580** 330–348
[13] Li N, Sun C, Guo N, Mohamed M, Lin J, Matsumoto T and Liu C 2016 *J Mater Process Technol* **228** 2–10
[14] Wan P, Zou H, Wang K, Zhao Z, Lu S and Nie H 2020 *Metals* **10** 956
[15] Zhang C, Bellet M, Bobadilla M, Shen H and Liu B 2011 *Inverse Probl Sci Eng* **19** 485–508
[16] Dassault Systèmes, Abaqus FEA www.3ds.com/products-services/simulia/products/abaqus/ Accessed: 2022

On the inverse identification methods for forming plasticity models using full-field measurements

A Andrade-Campos[1], N Bastos[1], M Conde[1], M Gonçalves[1], J Henriques[1], R Lourenço[1], J M P Martins[1,2], M G Oliveira[1,2], P Prates[1] and L Rumor[1]

[1] Centre for Mechanical Technology and Automation (TEMA), GRIDS Research Unit, Mechanical Engineering Department, University of Aveiro, 3810-193 Aveiro, Portugal
[2] Université de Bretagne Sud, UMR CNRS 6027, IRDL, F-56100 Lorient, France

E-mail: gilac@ua.pt

Abstract. The simulation of deep drawing processes and its quality is intrinsically dependent on the accuracy of the constitutive model in reproducing the mechanical behaviour of the sheet metal material. Today, the calibration of elastoplastic models – correspondent to the inverse identification of the material parameters – often uses full-field measurements, through Digital Image Correlation (DIC) techniques, to capture non-homogeneous strain fields and states, coupled with non-straightforward numerical inverse methodologies. In the last decade, new parameter identification methodologies, such as the Finite Element Model Updating (FEMU), the Constitutive Equation Gap (CEG) method, the Equilibrium Gap Method (EGM) and the Virtual Fields Method (VFM) have been developed and have proven to be effective for non-linear plasticity models. Nonetheless, the FEMU and the VFM have distinguished themselves from the others. More recently, supervised Machine Learning (ML) techniques have been also used as an inverse identification method. These artificial intelligence-based methods use large datasets of numerical tests to train an inverse model in which the input is the history of the strain field and loads during the test, and the output are directly the material parameters.

The goal of this paper is to analyse, compare and discuss these inverse identification methods, with particular focus on the FEMU, VFM, and ML methodologies. A heterogeneous tensile-load test is considered to compare in detail the FEMU, VFM, and ML strategies.

1. Introduction

Nowadays, modelling and simulation have become a critical stage in product design. Even in the development of new materials, engineers resort to simulation for fast progress and comprehension of the intrinsic phenomena. However, in Solid Mechanics, the simulation is fully dependent on the numerical methods and the material models. Although the numerical methods and their approximations can be evaluated using benchmarks and convergence analysis (e.g. numerical mesh analysis), the material models can only be evaluated when confronted with experimental tests. Yet, each classical experimental test represents a simple phenomenon and a single stress/strain state, making a rigorous experimental campaign very laborious. Moreover, complex states as the ones experienced in deep drawing are not represented and taken into account in these experimental campaigns, therefore a correct material behaviour characterisation is not guaranteed.

IOP Conf. Series: Materials Science and Engineering 1238 (2022) 012059 doi:10.1088/1757-899X/1238/1/012059

Recently, experimental full-field measurements, particularly using Digital Image Correlation (DIC) techniques, opened large potentialities in material characterization in the sense that experimentally full-field displacements and strain maps can be measured. The advantage of local strain analysis of the Finite Element Analysis (FEA) was therefore reduced because these can now be experimentally measured. However, local stress analysis is still is not measurable experimentally, and is only possible to obtain with FEA (called numerical stress reconstruction).

A phenomenological macroscopic material constitutive model, in particular in elastoplasticity, relates the strain with stress. Therefore, using the strain knowledge coming from full-field measurements, the constitutive model computes the stress field, which could be easily calibrated and compared with the experimental stresses, again, if these could be measured. However, during mechanical tests, independently of homogeneous or heterogeneous, only the load (which corresponds to the stress surface integration) can be measured. Therefore, the calibration of a material constitutive model must be done indirectly because the model's output is not locally and (yet) experimentally known. Nevertheless, knowing the model mathematical structure is already a great help[1] since the model calibration reduces to a parameter identification problem.

In the last decades, several strategies have emerged to solve the inverse parameter identification problem in solid mechanics, inclusively, using full-field measurements [23, 11]. Clearly the first and the most known is the Finite Element Model Updating (FEMU) [14] followed by the Virtual Field Method (VFM) [10, 12, 28, 26]. Other methods must be also highlighted, such as the Constitutive Equation Gap Method (CEGM) [19], the Equilibrium Gap Method (EGM) [8, 9], the Constitutive Compatibility Method (CCM) [24], the Dissipation Gap Method [4], the Self-Optimizing Method (Self-OPTIM) [32] and the Integrated Digital Image Correlation Method (Integrated-DIC) [29]. Other methods can be retrieved from a literature review, however, these seem to be variations of the previous methods. From this list, only the FEMU, the VFM, the CEGM and the EGM proved to be effective in the parameter identification of non-linear models [23]. Very recently, with the explosion of data-driven methods, Machine Learning (ML) and Artificial Intelligence (AI) techniques, supervised learning was used to identify the parameters of constitutive models [5, 6, 17].

The distinction between the parameter identification methods is in the use of the strain and load as an inverse method. While the FEMU method uses the numerical strain field and the load, and directly and iteratively compares them to the correspondent experimental values, other methods such as the VFM use equilibrium equations to guide the identification process. In these equilibrium equations, the experimental strain and load are used in both sides of the equilibrium equation, the stress field is reconstructed using the experimental strain field and a residual is minimised. The difference is then:

- For the CEGM, the equilibrium equation is energetic (energy norm) and the objective function to minimise is based on the difference between an updated statically admissible stress field and the stress field reconstructed using the experimental strain field and the constitutive model [19]. The particularity of this method is the statically admissible stress field, which must verify the boundary conditions of the experimental test, as well as the equilibrium equation [23]. An FEA model can be used to calculate this latter stress field.
- For the EGM, a local equilibrium is assessed using the continuity of the stress vector at the interfaces [8, 9]. This local equilibrium can be calculated using finite differences between the elements interfaces, and their gap must be minimised. Here again, the stress update algorithm uses the (measured) experimental strain. For the boundary interfaces, the

[1] Although here the previous knowledge of the mathematical structure of the model is considered an advantage for the material characterisation because it simplifies the identification problem, for a general discussion it can be seen as a disadvantage because this formulation can work as a constraint for the reproduction of the material behaviour.

equilibrium is prescribed using the external load. For the case of a free boundary interface, the load is zero. Therefore, this method minimises the local stress equilibrium into the domain and the boundary to find the parameters.

- The equilibrium in the VFM method is taken from the principle of virtual work, which balances the internal virtual work with the external virtual work in the absence of body-forces [27], i.e. verifies if the stress fields are statically admissible. The external work uses the experimentally measured load and the internal work is found using the stress reconstructed from the experimental strain field and the constitutive model. The VFM has the advantage of producing both local and global equilibrium, depending on the virtual field used. These virtual fields act as filters to obtain different work equilibrium equations and can be obtained manually or automatically. There are no limitations on the number of virtual fields and there is no need for a FEA analysis.

All the previous inverse techniques require an optimization algorithm for an efficient search of the parameters while minimizing its objective functions (equilibrium conditions or external "numerical *vs* experimental" differences).

Machine learning methods, particularly the well-known Artificial Neural Networks (ANN), use data to build a model. These are (data-driven) regression models in which the regressor (or its mathematical formulation) is not initially known and can be found together with its parameters during a fitting operation, which is called supervised training. For the parameter identification problem, the goal is to develop a model that, with the knowledge of the strain field and load[2], can give suitable numerical parameters. This approach makes use of the inverse problem formulation to directly build an inverse model. No stress update algorithm is required nor any equilibrium conditions. However, its black-box characteristic frightens the engineers because there is no guarantee of equilibrium or any other physics. Only data is used and all Key Performance Indicators (KPI) metrics are statistical, as in regressions.

The goal of this paper is to promote a discussion on the parameter identification methods, with particular emphasis on the FEMU, VFM and ML methods, representing the external, equilibrium and data-driven approaches, respectively. For that purpose, the paper is divided into two parts. The first part, which includes sections 2-4, presents the methods, their particular formulations and their implementation. The second part (section 5) presents an example to discuss the performance of each approach.

2. Parameter identification as an inverse problem

In Solid Mechanics, FEA simulation requires the geometry of the solid, boundary conditions including forces, and the material behaviour, which is reproduced by the material constitutive model and its parameters and properties. The solution of an FEA simulation includes the displacements, strains, and stresses. This problem is considered the *direct problem* and its solution is obtained by FEA simulations using the Finite Element Method (FEM). The parameter identification problem is one of the *inverse problems* of the previous direct problem, where the displacement and strain fields become known and the material parameters are unknown. This problematic is illustrated in Figure 1.

In a general formulation, the inverse parameter identification problem can be written as the

[2] It is important to recall that the stress field is not directly available from the mechanical tests. Only the global force can be measured.

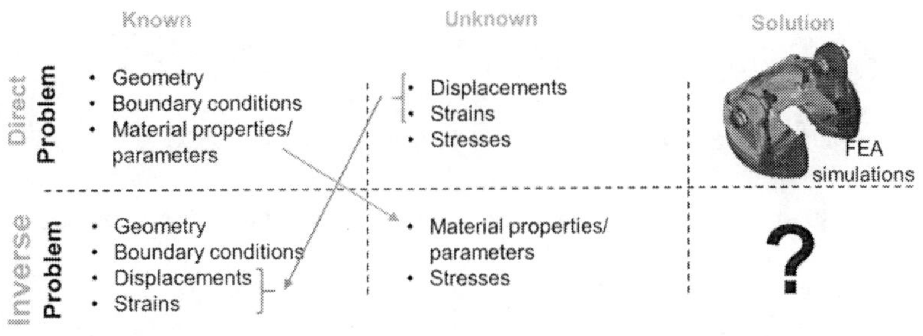

Figure 1. Direct problem versus inverse problem in the material characterisation and parameter identification.

search for the set of the r parameters $\mathbf{A}$ that

$$\begin{aligned} \underset{\mathbf{A}}{\text{minimise}} \quad & \mathcal{L}(\boldsymbol{\sigma}(\mathbf{A}, \boldsymbol{\varepsilon}^{\text{exp}}, \boldsymbol{\varepsilon}^{\text{num}}), \boldsymbol{\varepsilon}^{\text{exp}}, \mathbf{u}^{\text{exp}}, \mathbf{F}^{\text{exp}}, \ldots) & \\ \text{subject to} \quad & g_m(\mathbf{A}) \leq 0, & m = 1, \ldots, M \\ & \mathbf{K}(\mathbf{A})\mathbf{u} = \mathbf{F}, & \\ & A_i^{\min} < A_i < A_i^{\max}, & i = 1, ..., r \ , \end{aligned} \tag{1}$$

where $\mathcal{L}$ is a function that relates the measured experimental displacements $\mathbf{u}^{\text{exp}}$, strains $\boldsymbol{\varepsilon}^{\text{exp}}$ and loads $\mathbf{F}^{\text{exp}}$ with their numerical correspondent. This function must use the constitutive model to reconstruct the stress, however, the input of the stress update algorithm changes from one method to other. The inverse problem can be subjected to some constraints. Inequality constraints are represented by M functions g and the side constraints depend on the maximum and minimum of each variable, represented by $A^{\max}$ and $A^{\min}$, respectively. An equality constraint $\mathbf{Ku} = \mathbf{F}$ is also defined, however, it is generally satisfied if an FEA is used or other static equilibrium condition. For non-linear constitutive models, which depend on the deformation history, $\mathcal{L}$ must be integrated in (virtual) time, meaning that it must be evaluated for n_s time instants:

$$\mathcal{L}(\mathbf{A}) = \sum_{i=1}^{n_s} \mathcal{L}_i(\mathbf{A}). \tag{2}$$

3. Inverse methodologies for full-field measurements

The definition and solving procedure for $\mathcal{L}(\mathbf{A})$ is the main difference of the methods. For external and equilibrium methods, the parameters are found iteratively by $\mathbf{A} = \min \mathcal{L}(\mathbf{A})$. The Machine Learning method build the inverse model $\mathcal{L}^{\text{ML}}$ and obtain directly the parameters solving $\mathbf{A} = \mathcal{L}^{\text{ML}}(\boldsymbol{\varepsilon}^{\text{exp}}, \mathbf{F}^{\text{exp}})$.

3.1. External methods and the Finite Element Model Updating (FEMU)

The external methods are the ones that the experimental values are external to the constitutive model and the stress reconstruction. The numerical calculations are completely separated from the experimental measurements and their comparison is made with a kind of numerical-experimental gap metric. The FEMU is the most widely used method for parameter identification and reportedly first employed by Kavanag and Clough in 1971 [14].

The FEMU method generally uses a FEA model[3] of the mechanical test to calculate the displacements, strain and stress fields with the assumption that the experimental boundary

[3] Although FEA is clearly the most used simulation method in solid mechanics, other numerical methods for space and time integration can be used, such as IGA, finite differences, etc.

conditions are fully reproduced numerically. This assumption is one of the major disadvantages of the FEMU method because some boundary conditions and phenomena are very difficult to reproduce. Contact and friction (resulting in sliding) in the grips is an example of this difficulty. Today, these boundary conditions can be retrieved from the DIC displacement measurements and input in the numerical model, however, the measurement noise can produce stress concentrations at the boundaries.

The stress integration of the constitutive model is made inside the FEA. However, only strains and global loads are compared. For the case of full-field measurement[4], the objective function $\mathcal{L}$ can be written as

$$\mathcal{L}^{\text{FEMU}}(\mathbf{A}) = \sum_{i=1}^{n_s}\left[\sum_{j=1}^{n_p}\left(\varepsilon_{xx}^{\text{exp}} - \varepsilon_{xx}^{\text{num}}(\mathbf{A})\right)_j^2 + \left(\varepsilon_{yy}^{\text{exp}} - \varepsilon_{yy}^{\text{num}}(\mathbf{A})\right)_j^2 + \left(\varepsilon_{xy}^{\text{exp}} - \varepsilon_{xy}^{\text{num}}(\mathbf{A})\right)_j^2\right]_i + \sum_{i=1}^{n_s}\left(\mathbf{F}^{\text{exp}} - \mathbf{F}^{\text{num}}(\mathbf{A})\right)_i^2 . \quad (3)$$

The xx, yy and xy subscripts denote the strain components at each of the n_p space integration points in a 2D plane stress formulation. The minimization of this function leads to a perfect match on the strain field and on the global load. A scheme of this approach is depicted in Figure 2.

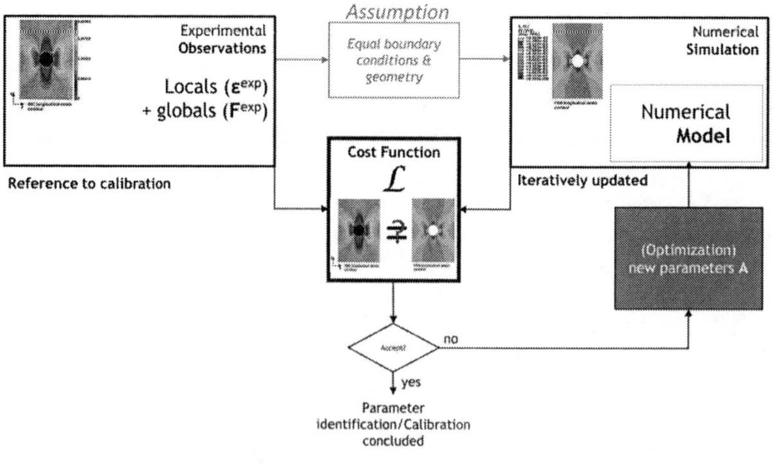

Figure 2. Schematic flowchart for the FEMU method.

One recent discussion concerns strain interpolation. Although there is a consensus for time interpolation, because it is rather easy to use the same time instants, the space interpolation can lead to smoothing or discontinuity errors if the numerical mesh is quite different from the experimental DIC subset mesh. A DIC-leveled approach [18, 13] can be the solution, however, accompanied by a large increase in computational effort.

It is also good practice to normalize the objective function 3 and give equal weight to both the strain and forces. However, generally, the choice of the weights is made intuitively and can lead to unsuccessful results.

[4] One advantage of the FEMU method is that it is easily adapted to the available experimental data and can be used with partial measurements or even with a single average value.

3.2. Equilibrium methods and the virtual field method (VFM)

Equilibrium methods solve the parameter identification problem with the verification of balance/equilibrium conditions. For the case of the VFM, the equilibrium condition comes from the Principle of Virtual Works, which states that, in the absence of acceleration and body forces, the internal virtual work must be equal to the external virtual work, i.e.

$$W^{\star}_{\text{internal}}(\mathbf{A}) = W^{\star}_{\text{external}} \Leftrightarrow \underbrace{\int_{\Omega} \boldsymbol{\sigma}(\mathbf{A}, \boldsymbol{\varepsilon}^{\text{exp}}) : \boldsymbol{\varepsilon}^{\star} dV}_{\text{Internal virtual work}} = \underbrace{\int_{\Gamma_f} \bar{\mathbf{f}}^{\text{exp}} \cdot \mathbf{u}^{\star} dS}_{\text{External virtual work}} \ , \tag{4}$$

for a body Ω and a surface density of the force $\bar{\mathbf{f}} = \mathbf{F}/S$ applied over the boundary Γ_f. It must be noted that the principle of virtual work is independent of the constitutive model, therefore, it can be widely used. The static equilibrium between the external and internal work is a global equilibrium and requires the full knowledge of all forces, including the reactions of the boundary conditions, which is very difficult to measure. However, the major advantage of this method is the use of a kind of filter functions called virtual fields, defined by the virtual strain $\boldsymbol{\varepsilon}^{\star}$ and displacement $\mathbf{u}^{\star}$, which allow to obtain multiple local and global equilibrium equations and, therefore, filter unknown boundary forces[5]. For more details concerning the demonstration of the VFM, please refer to [28].

Although the major advantage of this method comes from the virtual fields (VFs), these are also its main drawback. The choice of suitable VFs is not straightforward and is a critical stage in the identification process. These VFs must be chosen satisfying (i) zero virtual displacement on the constrained boundary condition, due to the unknown forces on these boundaries, (ii) be constant on the load boundary condition, due to the knowledge of the resultant of the applied force, instead of its distribution, and (iii) C^0 continuity. Today, the drawback of VFs selection was clearly minimized with its calculation by stiffness-based [26, 1] and sensitivity-based [20] procedures. For non-linear elastoplasticity, the objective function is then,

$$\mathcal{L}^{\text{VFM}}(\mathbf{A}) = \sum_{i=1}^{n_s} \left(\int_{\Omega} \boldsymbol{\sigma}(\mathbf{A}, \boldsymbol{\varepsilon}^{\text{exp}}) : \boldsymbol{\varepsilon}^{\star} dV - \int_{\Gamma_f} \bar{\mathbf{f}}^{\text{exp}} \cdot \mathbf{u}^{\star} dS, \right)_i^2 \tag{5}$$

which should be minimised. Figure 3 illustrates this approach.

3.3. Data-driven and Machine Learning (ML) methods

When solving inverse parameter identification problems, data-driven methods are similar to response surface methods in the sense that these use the data to make the relation between the input and output. This relation is built as a regression. Therefore, data-driven methods do not solve an inverse problem but previously build directly the inverse model as a regression. Machine learning methods, where ANNs, support vector machines and random trees are included, are the new competitors for solving the parameter identification problem. This ML approach uses direct problem solutions to fit (here called supervised learning) the regression model. Although this approach does not inversely and iteratively identify the parameters of the constitutive model, it still solves an identification problem but for the hyper-parameters of the ML model. However, this problem is much simpler because the regressor structure (the topology or architecture of the ML model) is well known and can be easily differentiated. After proper learning, the solution for the material model identification is promptly obtained. The disadvantage of this method is the

[5] Although these virtual displacement and strain seem to have the meaning and units of length, in fact, these functions cannot be seen as real displacement and strain. Nevertheless, these remain with the units of displacement and strain. It is the use of virtual quantities that makes the principle of virtual work dissimilar to the energy balance. The VFM grounds on the local equilibrium equations and force boundary equations.

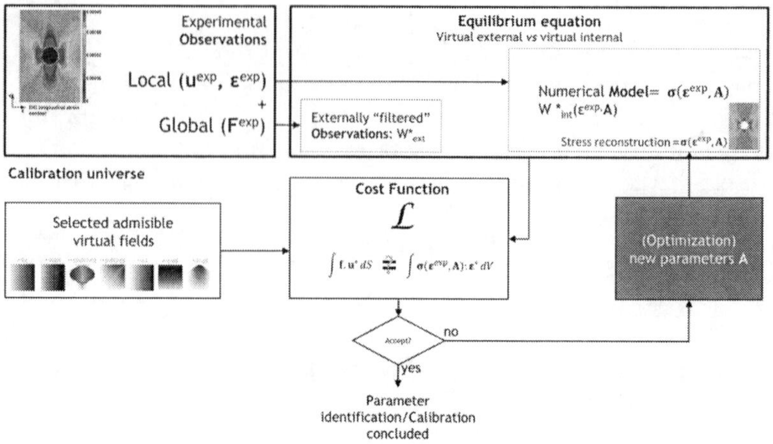

Figure 3. Schematic flowchart for the VFM method.

computational effort to build the direct problem database. For numerical material constitutive models, which are going to be used in FEA, the database can be built using the same FEA simulator considering the strain field and the load as input feature and the parameters as output features. This method is illustrated in Figure 4.

The formulation for a ML approach can be written as

$$\mathbf{A} = \mathcal{L}^{\mathrm{ML}}(\boldsymbol{\varepsilon}^{\mathrm{exp}}, \mathbf{F}^{\mathrm{exp}}, \mathbf{W}, \mathbf{b}), \tag{6}$$

where

$$(\mathbf{W}, \mathbf{b}) = \min \mathcal{F}\left(\mathbf{A}_i, \boldsymbol{\varepsilon}_i, \mathbf{F}_i\right)_{\mathrm{FEA}, i=1,\ldots,ndata}. \tag{7}$$

The hyper-parameters $\mathbf{W}$ and $\mathbf{b}$ of the activation functions[6] are found by minimising a given loss function $\mathcal{F}$, which is generally the mean square error (MSE) of all *ndata* residuals with a regularization term. The training database $(\mathbf{A}_i, \boldsymbol{\varepsilon}_i, \mathbf{F}_i)_{\mathrm{FEA}, i=1,\ldots,ndata}$ is used once for the training, then the same ML model architecture with the found hyper-parameters can be used for different experimental sets $(\boldsymbol{\varepsilon}^{\mathrm{exp}}, \mathbf{F}^{\mathrm{exp}})$. However, it is essential that the experimental input set has the same number and, for the strain field, location of the set $(\boldsymbol{\varepsilon}_i, \mathbf{F}_i)_{\mathrm{FEA}, i=1,\ldots,ndata}$ used for training. A DIC-levelled approach can be also used to create the training database. Here, similarly to the FEMU approach, the experimental geometry, loading, and boundary conditions must be very well reproduced by the FEA simulation used for training.

4. The role and dependency on the optimization technique

All procedures for inverse parameter identification require optimization techniques and all formulations lead to least-square optimization problems (see equations 3,5 and 7). Therefore, optimization algorithms such as Levenberg-Marquardt [21] or Trust Region Reflective algorithms are expected to be used. However, these are gradient-based methods and, consequently, are sensitive to the objective function behaviour and noise, and the initial set of parameters. Alternatives to gradient-based least square algorithms are derivative-free least-square algorithms. Nonetheless, unexpectedly, this type of algorithms is not used by the scientific community. Instead, the Nelder-Mead is commonly used in the field. An analysis concerning some optimization algorithms can be seen in [25], however, this is still an open question.

[6] These activation functions depend on the ML model. In the case of ANN, these functions are related to neurons. However, for random trees or support vector machines, these have different meanings.

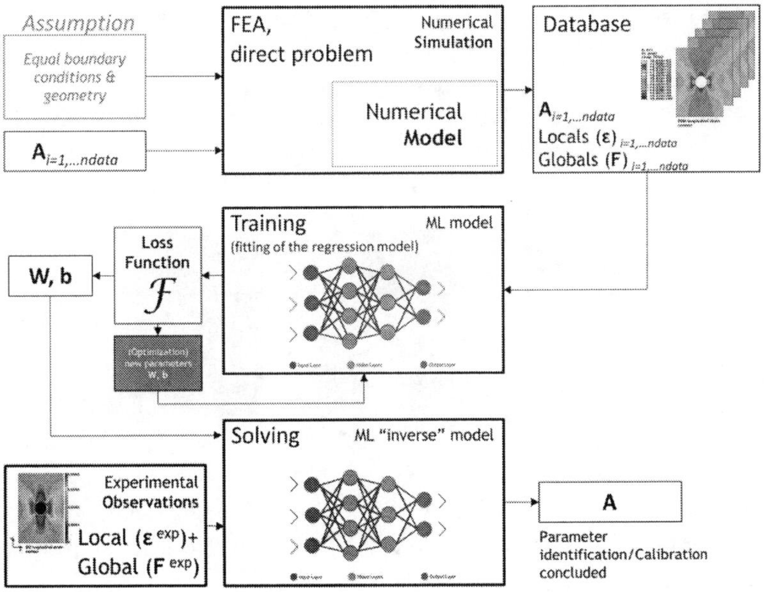

Figure 4. Schematic flowchart for the ML method.

ML models are fitted (the training phase) with gradient-based algorithms because their topology is well known and their derivatives can be easily calculated by Automatic Differentiation (AD). ADAM, a stochastic gradient descent algorithm, is one of the most used optimization techniques [16]. Nevertheless, least-square algorithms are still being used.

5. Comparison of the parameter identification in non-linear elastoplasticity

In order to compare the different presented approaches, a simple mechanical heterogeneous specimen is used for the identification of the hardening of a isotropic constitutive model. It is assumed that the elastoplastic material model (i) has the elastic behaviour described by the linear Hook law, where Young's modulus and Poisson's ratio are defined by $E = 210$ GPa and $\nu = 0.3$, and (ii) the plastic yield is governed by the von Mises yield criterion and hardening is described by the Swift's law, given by

$$\sigma_{\rm Y} = k\,(\varepsilon_0 + \bar{\varepsilon}^{\rm p})^n\,, \quad \text{with} \quad \varepsilon_0 = \left(\frac{\sigma_0}{k}\right)^{1/n}, \tag{8}$$

where $\sigma_{\rm Y}$ is the flow stress, σ_0 is the initial yield stress, k and n are the hardening coefficient and exponent, respectively, and $\bar{\varepsilon}^{\rm p}$ is the equivalent plastic strain. The three material parameters to be identified are $\mathbf{A} = \{k, n, \sigma_0\}$.

5.1. *Heterogeneous dog-bone test*

For this study, synthetic data from a virtual experiment is used for all three approaches. This procedure has the advantage of knowing *a priori* the solution, as ground truth, although some would suggest not to use this methodology [31].

A heterogeneous specimen with smooth cross-section change is selected. This test, here called heterogeneous dog bone, was retrieved from [22, 15] and can be seen in Figure 5. For the problem, only one quarter is used due to symmetries conditions in $x = 0$ and $y = 0$. The model, numerically created using Abaqus standard, assumes the plane stress formulation and uses a regular mesh of 112 CPS4 (bilinear and full integration) elements. A vertical displacement of

3 mm is applied on the top of the specimen and 30 time increments were used. The virtual material is characterised by the parameter set $\mathbf{A} = \{565, 0.26, 160\}$. An equivalent plastic strain of 0.25 can be observed in the specimen's centre and the deformation gradient is observed mainly in the vertical direction. Although this test is heterogeneous, only hardening heterogeneity is observed. No anisotropy is accounted and all specimen yields in the same $\sigma_1 - \sigma_2$ yield locus direction.

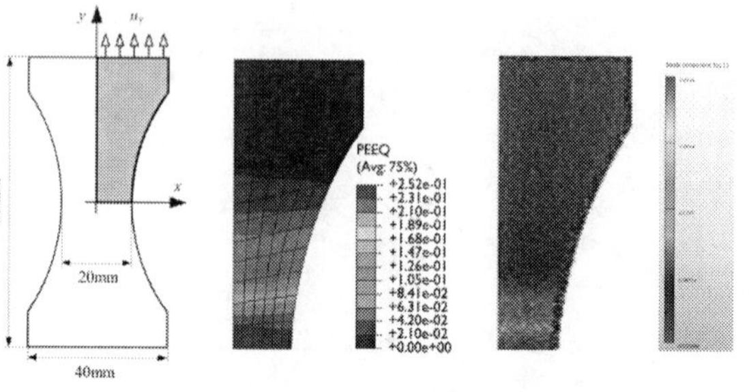

Figure 5. Specimen geometry and corresponding finite element model and DIC measurement. Equivalent Plastic strain distribution after a displacement of 3 mm. The test was retrieved from [22].

5.2. Results and discussion
Implementation details for the three inverse problem approaches can be seen in [22, 23] and [2]. For all approaches, the logarithmic strain was retrieved from the centroid of the element, and the identical coordinates for the strain points are considered for the experimental and the numerical values, avoiding interpolation errors. The hardening model is implemented in both FEA (for the FEMU and the training in ML) and the VFM by a user subroutine. For the VFM, the simplest virtual fields were chosen: $u*_x = 0$ and $u^*_y = 0$. These only two VFs were sufficient for the correct determination of the parameter. Nevertheless, multiple VFs or special determined VFs (e.g. using the sensitive-based VFs) would result in even better results, with the cost of larger CPU time.

For the ML approach, a gradient boosted trees algorithm, called XGBoost and proposed by Chen and Guestrin [7], is used. It can be shown that this model is more efficient than the ANN, which is the most used in ML [30]. However, it still needs a training data set for the three material parameters. Here, the database was limited to 5000 combinations of the parameters, which were retrieved using the Latin Hypercube Sampling (LHS) [3], and using the following universe: $\mathbf{A} \in \{[280, 700], [0.1, 0.3], [80, 300]\}$.

Considering that the Levenberg-Marquardt optimization algorithm is starting set dependent, two initial parameters sets are used for both the FEMU and the VFM approaches. These can be seen in Table 1 and are denoted as superior and inferior due to their values. However, from the obtained results, it can be seen that the VFM method is not as sensitive to the initial parameters set as the FEMU approach. Both starting sets have led the VFM to the same final parameters, whose errors are very low. These are not zero because the virtual experiment was retrieved using FEA and not using the VFM itself. But the fact that both starting sets lead to the exact same results shows that it is the global minimum for the VFM.

The FEMU has also reached good results for both initial sets, but their difference shows the method's sensitivity to the initial dependence and more prone to local minima. The magnitude of the parameter difference to the virtual material parameters of the FEMU is similar to the VFM, however, it took more than ten times to reach the final results, confirming the large CPU efficiency of the VFM method.

Table 1. Initial sets and identified parameters for the three approaches (FEMU, VFM and ML) presented.

$\mathbf{A} =$	k [MPa]	n	σ_0 [MPa]	Obj. function $\mathcal{L}(\mathbf{A})$
$\mathbf{A}^{\text{sup}}$	965	0.35	234	—
$\mathbf{A}^{\text{inf}}$	165	0.08	100	—
$\mathbf{A}^{\text{VFM}}$	564.24	0.265	160.72	2.29×10^{-5}
Error [%]	0.134	1.923	0.450	—
$\mathbf{A}^{\text{FEMU, sup}}$	568.74	0.2633	160.36	5.67×10^{-6}
Error [%]	0.662	1.269	0.225	—
$\mathbf{A}^{\text{FEMU, inf}}$	552.77	0.250	159.7	7.95×10^{-5}
Error [%]	2.164	3.729	0.187	—
$\mathbf{A}^{\text{ML}}$	564.67	0.259	160.12	1.46×10^{-5}
Error [%]	0.058	0.385	0.075	—

The ML approach seems to be effective, and the errors were not larger than the ones from the other competitor approaches. It has the advantage of total non-dependence of the starting set because it does not require an optimization procedure for the identification. If the time for training is not considered, the ML approach is, without doubt, the most efficient because the solution is obtained instantaneously. However, training requires a quite large database, here with 5000 direct problem solutions, which took around 7 days to be fulfilled. The training itself took only 50 minutes for 5000 samples and 6020 features.

Nevertheless, it should be noticed that the influence of the size of the database was analysed, using database dimensions of 750, 1000, 1500, 2500, and 4500, and it was verified that a database of only 1500 samples could deliver MSE smaller than 2% in training. This result is even more enhanced if PCA analysis is introduced and only 450 samples could be used without loss of quality. For more details, see [2].

Figure 6 shows that the hardening with the parameters obtained by the FEMU, VFM and ML approaches is similar, however, some small differences can be seen for the hardening identified by the ML approach.

6. Concluding remarks

In this paper, three approaches for parameter identification in non-linear elastoplasticity were compared and discussed. Although the FEMU and the VFM are regularly used in the scientific communities, ML approaches are not so usual. The results here presented are from the identification of a 3-parameter non-linear hardening model. Even though the case study is not very demanding, the advantages and disadvantages of each approach could be disclosed. The advantage of the ML is not being starting-set and optimization dependent but requires large computational effort for the database building and training. The VFM is clearly very computational effective and robust, yet it requires a wiser implementation. The FEMU is, without doubt, the simplest and wide-range method to work with. But its optimization dependence and effort are major drawbacks. Therefore, the three methods present disadvantages that maybe could be eliminated if a hybrid method or redundant strategy would be used.

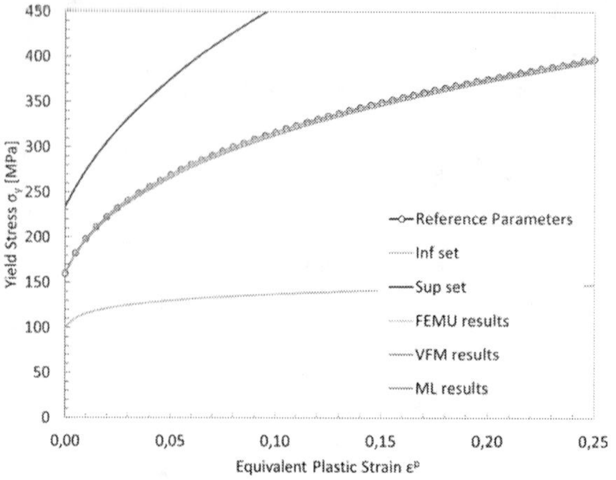

Figure 6. Hardening curve for the parameters obtained with the FEMU, VFM and ML approaches. Comparison with the virtual material Swift hardening law.

Acknowledgements
This project has received funding from the Research Fund for Coal and Steel under grant agreement No 888153. The authors also gratefully acknowledge the financial support of the Portuguese Foundation for Science and Technology (FCT) under the projects PTDC/EME-APL/29713/2017 (CENTRO-01-0145-FEDER-029713), PTDC/EME-EME/31243/2017 (POCI-01-0145-FEDER-031243), PTDC/EME-EME/31216/2017 (POCI-01-0145-FEDER-031216) and PTDC/EME-EME/30592/2017 (POCI-01-0145-FEDER-030592) by UE/FEDER through the programs CENTRO 2020 and COMPETE 2020, and UIDB/00481/2020 and UIDP/00481/2020-FCT under CENTRO-01-0145-FEDER-022083.

References
[1] S. Avril, M. Grédiac, and F. Pierron. Sensitivity of the virtual fields method to noisy data. *Computational Mechanics*, 34(6):439–452, 2004.
[2] N. Bastos, P. Prates, and A. Andrade-Campos. Material parameter identification of elastoplastic constitutive models using machine learning approach. *Proceedings of the 25th ESAFORM conference, Key Engineering Material*, pages 1–8, 2022.
[3] M. Baudin. pydoe - design of experiments for python, 2015.
[4] B. Blaysat, E. Florentin, G. Lubineau, and A. Moussawi. A dissipation gap method for full-field measurement-based identification of elasto-plastic material parameters. *International journal for numerical methods in engineering*, 91(7):685–704, 2012.
[5] A. Chamekh, H. Bel Hadj Salah, and R. Hambli. Inverse technique identification of material parameters using finite element and neural network computation. *The International Journal of Advanced Manufacturing Technology*, 44(1):173, Nov 2008.
[6] A. Chamekh, H. BelHadjSalah, R. Hambli, and A. Gahbiche. Inverse identification using the bulge test and artificial neural networks. *Journal of Materials Processing Technology*, 177(1):307–310, 2006. Proceedings of the 11th International Conference on Metal Forming 2006.
[7] T. Chen and C. Guestrin. Xgboost: A scalable tree boosting system. In *Proceedings of the 22nd ACM SIGKDD International Conference on Knowledge Discovery and Data Mining*, KDD '16, page 785–794, New York, NY, USA, 2016. Association for Computing Machinery.
[8] D. Claire, F. Hild, and S. Roux. Identification of damage fields using kinematic measurements. *Comptes Rendus Mécanique*, 330(11):729 – 734, 2002.
[9] D. Claire, F. Hild, and S. Roux. A finite element formulation to identify damage fields: the equilibrium gap method. *International Journal for Numerical Methods in Engineering*, 61(2):189–208, 2004.
[10] M. Grédiac. Principe des travaux virtuels et identification. *Comptes rendus de l'Académie des sciences. Série 2, Mécanique, Physique, Chimie, Sciences de l'univers, Sciences de la Terre*, 309(1):1–5, 1989.

[11] M. Grédiac and F. Hild. *Full-field measurements and identification in solid mechanics*. John Wiley & Sons, 2012.
[12] M. Grediac, F. Pierron, S. Avril, and E. Toussaint. The virtual fields method for extracting constitutive parameters from full-field measurements: a review. *Strain*, 42(4):233–253, 2006.
[13] J. Henriques, M. Conde, A. Andrade-Campos, and José Xavier. Identification of swift law parameters using femu by a synthetic image dic-based approach. *Proceedings of the 25th ESAFORM conference, Key Engineering Material*, pages 1–8, 2022.
[14] K.T. Kavanagh and R.W. Clough. Finite element applications in the characterization of elastic solids. *International Journal of Solids and Structures*, 7(1):11 – 23, 1971.
[15] J.-H. Kim, J.-S. Park, F. Barlat, and F. Pierron. Characterization of dynamic hardening behavior using acceleration information. *Procedia Engineering*, 207:245–250, 2017. International Conference on the Technology of Plasticity, ICTP 2017, 17-22 September 2017, Cambridge, United Kingdom.
[16] D. P. Kingma and J. Ba. Adam: A method for stochastic optimization, 2017.
[17] Z. Ktari, C. Leitão, P.A. Prates, and A. Khalfallah. Mechanical design of ring tensile specimen via surrogate modelling for inverse material parameter identification. *Mechanics of Materials*, 153, 2021. cited By 4.
[18] Pascal L., E. M. C. Jones, L. Wittevrongel, and F. Pierron. Validation of finite-element models using full-field experimental data: Levelling finite-element analysis data through a digital image correlation engine. *Strain*, 56(4):e12350, 2020. e12350 10.1111/str.12350.
[19] P. Ladeveze and D. Leguillon. Error estimate procedure in the finite element method and applications. *SIAM Journal on Numerical Analysis*, 20(3):485–509, 1983.
[20] A. Marek, F. M. Davis, and F. Pierron. Sensitivity-based virtual fields for the non-linear virtual fields method. *Computational Mechanics*, Apr 2017.
[21] D.W. Marquardt. An algorithm for least-squares estimation of nonlinear parameters. *Journal of the society for Industrial and Applied Mathematics*, 11(2):431–441, 1963.
[22] J. M. P. Martins, S. Thuillier, and A. Andrade-Campos. Identification of material parameters for plasticity models: A comparative study on the finite element model updating and the virtual fields method. *AIP Conference Proceedings*, 1960(1):110007, 2018.
[23] J.M.P. Martins, A. Andrade-Campos, and S. Thuillier. Comparison of inverse identification strategies for constitutive mechanical models using full-field measurements. *International Journal of Mechanical Sciences*, 145:330–345, 2018.
[24] A. Moussawi, G. Lubineau, E. Florentin, and B. Blaysat. The constitutive compatibility method for identification of material parameters based on full-field measurements. *Computer methods in applied mechanics and engineering*, 265:1 – 14, 2013.
[25] M.G. Oliveira, J.M.P Martins, B. Coelho, S. Thuillier, and A. Andrade-Campos. On the optimisation efficiency for the inverse identification of constitutive model parameters. *ESAFORM 2021 [Online]*, 1(1486):1–13, 2021.
[26] F. Pierron, S. Avril, and V. Tran. Extension of the virtual fields method to elasto-plastic material identification with cyclic loads and kinematic hardening. *International Journal of Solids and Structures*, 47(22):2993–3010, 2010.
[27] F. Pierron and M. Grédiac. *The virtual fields method: extracting constitutive mechanical parameters from full-field deformation measurements*. Springer Science & Business Media, 2012.
[28] F. Pierron and M. Grédiac. *The virtual fields method: extracting constitutive mechanical parameters from full-field deformation measurements*. Springer Science & Business Media, 2012.
[29] J. Réthoré. A fully integrated noise robust strategy for the identification of constitutive laws from digital images. *International Journal for Numerical Methods in Engineering*, 84(6):631–660, 2010.
[30] R. Shwartz-Ziv and A. Armon. Tabular data: Deep learning is not all you need, 2021.
[31] A. Wirgin. The inverse crime, 2004.
[32] G.J. Yun and S. Shang. A self-optimizing inverse analysis method for estimation of cyclic elasto-plasticity model parameters. *International Journal of Plasticity*, 27(4):576 – 595, 2011.

Investigation on tool properties for the production of components with micro textured surfaces

M Reck and M Merklein

Institute of Manufacturing Technology, Friedrich-Alexander-Universität Erlangen-Nürnberg, Egerlandstr. 13, 91058 Erlangen, Germany

manuel.reck@fau.de

Abstract. Defined surface micrcotextures offer potential for the friction reduction in elastohydrodynamic contacts and can thus increase the energy efficiency of technical systems. The production of prostheses, where microtextured surfaces can provide anti-bacterial effects is another application area. To manufacture these components economically in large quantities, forming processes are to be aimed at. Since lightweight parts are crucial in most application areas, the forming of sheet metal is the subject of current research. A major aspect of the presented investigation is the behavior of tool materials for the production of such microtextured sheet metal parts. Applying microtextures of varying geometries to the tools is a key challenge. Limitations in the accuracy of electrical discharge machining processes for the fabrication of micro cylinders, prisms and cuboids are therefore analyzed on the high-speed steels 1.3343 and 1.3244. In addition, the tool surfaces are characterized regarding their friction properties with case hardening steel as a contact partner. In this paper, the tool friction is analyzed for the steel 1.7131 (16MnCr5) with an initial sheet thickness of $t_0 = 2.4$ mm, which is often utilized for high-wear applications due to its high strength and toughness. In this context, the friction behavior of the wax-containing lubricant Beruforge 150 DL made for bulk forming processes is compared with zinc-free high-performance lubricants Raziol CLF 65-400 of varying viscosity.

1. Introduction

Applying defined microscopic structures on metal surfaces have the potential to reduce friction between highly loaded lubricated components under relative motion [1]. In bioengineering, micro textured surfaces can be implemented to provide antibacterial effects [2]. Recent investigation approaches for the single stage fabrication of micro textured metal components are based on a cup backward extrusion process in combination with micro coining [3]. In this process, the high lateral material flow of the specimen impedes a precise imprint of the micro textured coining punch onto the semi-finished material [3]. A promising method for overcoming this process limitation is the combination of sheet-bulk metal forming processes with micro coining. In sheet-bulk metal forming, which is characterized by the application of bulk forming operations on sheet metal semi-finished parts [4], the global adaption of the tribological conditions enables a targeted material flow control. Besides the use of tailored tool and workpiece surfaces [5], the lubricant influences the tribological system and therefore the friction. The lubricants are mostly designed for their specific application, either in sheet metal forming or bulk metal forming. Their properties are thus adapted to long sliding path for sheet metal forming or high contact normal stresses for bulk metal forming [5]. Due to the appearance of both described load cases in sheet-bulk metal forming, the suitability of lubricants for sheet-bulk metal forming is to be investigated.

2. Methodology and experimental setup

Different zinc-free sheet metal forming lubricants are investigated and compared to a bulk metal forming lubricant. In drawing processes, the lubricant viscosity is a major factor to adapt the friction system. This can be explained by the velocity dependant friction conditions between the forming tool and the sheet blank [6]. The examined lubricants are therefore chosen according to its varying viscosity. In case of the Raziol CLF lubricants, the corresponding viscosities at 40°C are 65 mm^2/s, 100 mm^2/s and 400 mm^2/s, whereby an increasing viscosity positively influences the load capacity of the lubricant [7]. Common application fields are drawing, forming and coining processes of construction steel, stainless steel and aluminum [7]. Due to the addition of extreme pressure additives, a high pressure absorption capacity and a good tool wear protection is reached [7]. The behaviour is furthermore compared to the bulk forming lubricant Beruforge 150 DL, which is a wax-containing and water-miscible oil [8]. Since it was identified in [9] as suitable for the use in sheet-bulk metal forming, it is used as reference lubricant. In accordance with [9], the pin-extrusion and the ring-compression test are applied to characterize the friction conditions for sheet-bulk metal forming processes. Figure 1 shows the tool setup of both processes.

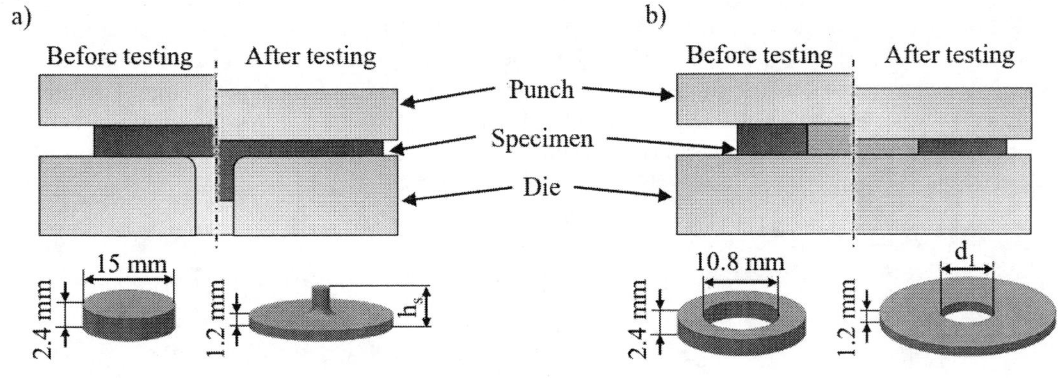

Figure 1. Tool setup for the friction characterization using the pin-extrusion test (a) and the ring-compression test (b)

Due to its frequent application in high-wear applications, the steel 1.7131 (16MnCr5) is used as blank material. The tests are conducted with tools from the high-speed steel 1.3343. In order to ensure universally applicable results, the tests are conducted on polished tools without microtextures. Cutting of the specimens in the desired geometry is realized using a CO^2 laser, which are subsequently processed in a hydraulic press Lasco 100SO. During the pin-extrusion test, the initial sheet thickness of 2.4 mm is reduced by 50 %, causing a radial material flow as well as a flow into the die cavity. The forming can be described as a combination of forward extrusion and upsetting and is therefore especially suitable for the friction characterization for loads comparable to those in bulk forming processes. With higher friction between the tools and the specimen, more material flows into the cavity and induces a larger pin height h_s. For processes with lower contact normal stresses, the ring-compression test can be applied [9], where a sheet metal ring is upset between two parallel tools. Under higher friction, the outward oriented radial material flow is hindered which results in inward oriented material flow of the material. As proposed in [10], the inner diameter d_1 is selected as the decisive factor for the characterization of the tribological conditions. Numerical models for both processes are furthermore developed to quantify the corresponding friction factor m. Since the finite element method is a common approach to numerically investigate forming processes [11], the forming software Simufact.forming 16.0 is chosen. Recommendations of [12] are moreover used to prove the validity of the numerical models.

Since the final aim of this investigation is to evaluate the suitability of the two high-speed steels 1.3343 and 1.3244 for micro texturization, they are analysed regarding their behaviour in the fabrication of microtextures, using the electronic discharge machining (EDM). For the production of high precision

IOP Conf. Series: Materials Science and Engineering 1238 (2022) 012060 doi:10.1088/1757-899X/1238/1/012060

products of elevated hardness, the production via EDM is a usual strategy [13]. The suitability is classified by the analysis of the form accuracy of the texture elements. Cuboids, prisms and cylinders with a height of 15µm and varying lateral dimensions are therefore eroded on the tool surface. Table 1 gives an overview of the analysed microstructure elements in the scope of this research.

Table 1. Dimensions of cuboid, prism and cylinder microstructures

Cuboid / Prism		Cylinder	
[h] µm	[l] µm	[h] µm	[Ø] µm
15	15	15	15
	20		20
	25		25
	30		30

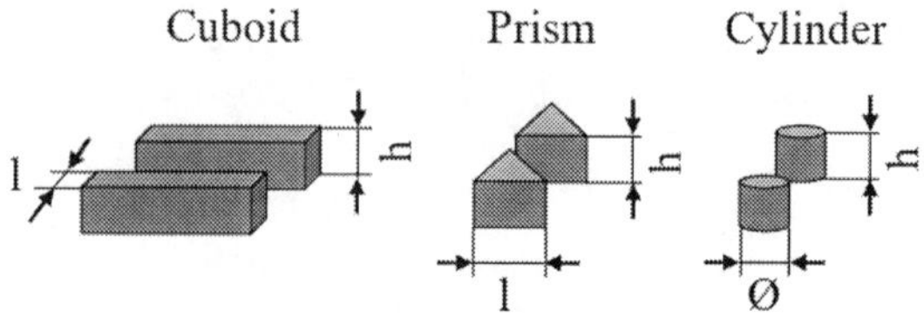

3. Results and discussion

The friction factors are determined by the combination of experimental results from the pin-extrusion and the ring-compression test with numerical results from the simulation models. On the one hand, the process force from the simulation and the experiment are compared to provide a process-related validation. The tested sheet metal of 16MnCr5 is modeled using an experimentally determined initial yield stress of 508 MPa. On the other hand, the experimentally obtained microhardness is compared to the calculated true strain distribution. Using the Fischerscope HM2000 from the Helmut Fischer GmbH, the hardness distribution is obtained in accordance with ISO 14577-1 at a testing load of 500 mN. The comparison of numerical and experimental results is given in Figure 2.

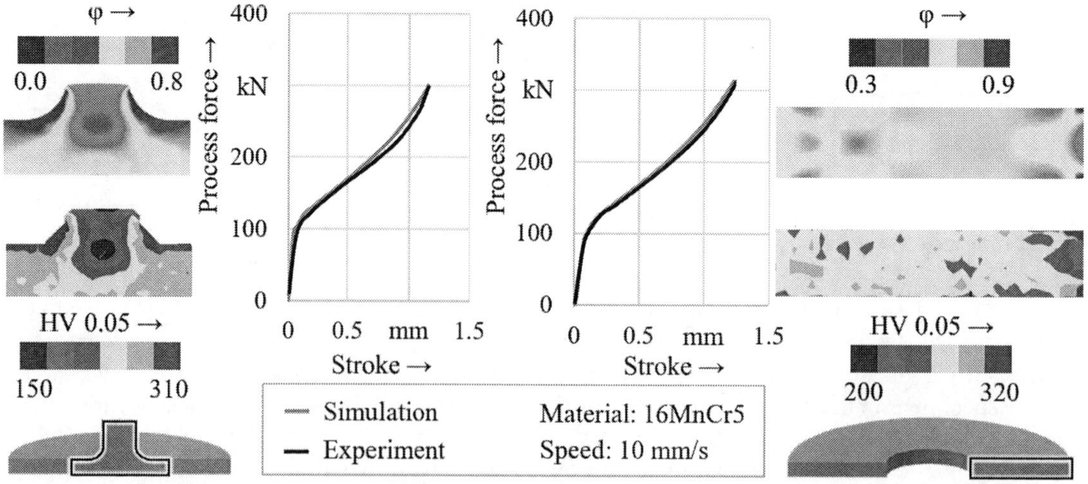

Figure 2. Validation of numerical models for the pin-extrusion and the ring-compression test

For the process forces, only small deviations are visible. The force curves show good compliance for the linear elastic forming at initiation of the process as well as in the phase of plastic material deformation. The ability of the simulation model to realistically predict the material's mechanical properties is proven by correlating the degree of deformation φ with the hardness values. These two parameters are connected due to an increased hardness induced by a high degree of deformation, which causes strain hardening [12]. Both simulation models predict the strain hardening behaviour of the material in a precise way. In conclusion, the good concordance of numerically calculated values with experimentally measured parameters enables the use of both simulations models for the determination of the friction factors.

In the experimental procedure for pin-extrusion and ring-compression, the four lubricants are applied onto the specimen surface in a quantity of 10 g/m^2. Figure 3a shows the height of the specimen h_s after the pin-extrusion test with n = 4 experiment replications, distinguishing between four applied lubricants.

A lubrication system using Beruforge 150DL results in the lowest height with 2.13 ± 0.002 mm. For all Raziol lubricants, the specimen and therefore the extruded pin is higher than for Beruforge 150DL. With rising viscosity, the height rises from 2.26 ± 0.003 mm to 2.48 ± 0.004 mm.

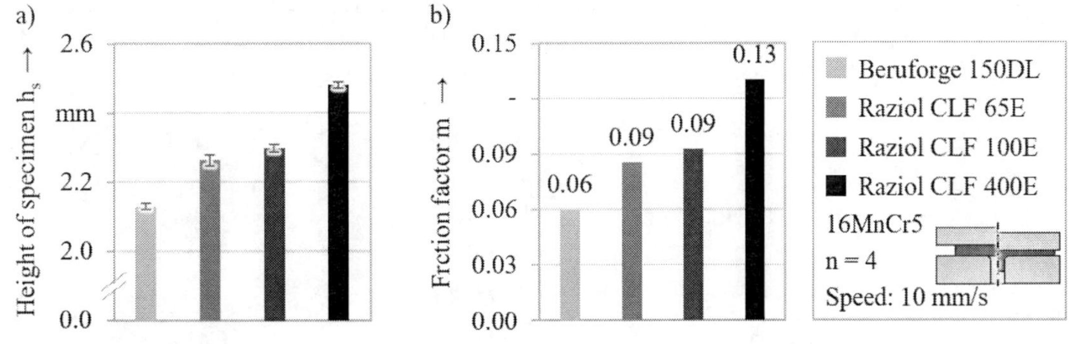

Figure 3: Inner specimen diameters (a) and corresponding friction factors (b) in the pin-extrusion test

The larger specimen heights and therefore higher friction factors in Figure 3b of all Raziol lubricants in comparison to Beruforge 150DL is justified by the additive substances. At elevated surface pressure, the wax particles, included in Beruforge 150CL, lead to a friction reduction in the pin-extrusion test. Since all three Raziol lubricants are designed for sheet metal forming at low surface pressure, the oil is pressed out of the forming zone during testing. The increased friction is therefore caused by a direct contact between the tools and the specimen, resulting in friction factors of up to 0.13.

For the characterization of the friction system of 16MnCr5 specimens under lower contact pressure, the ring-compression test is used. The boundary conditions regarding lubricant types, amount of oil applied and testing speed are chosen in accordance with the pin-extrusion test to ensure the comparability of the results. Figure 4a shows the measured inner diameters of the ring-compression specimens after the upsetting process whereas Figure 4b presents the corresponding friction factors, which are identified using the numerical model.

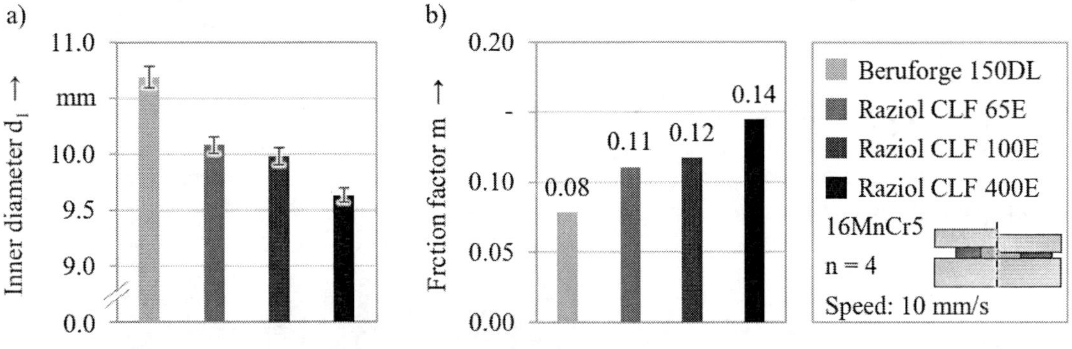

Figure 4: Specimen diameters (a) and corresponding friction factors (b) in the ring-compression test

For the ring-compression test with the Beruforge 150DL lubricant, an inner diameter of 10.69 ± 0.09 mm can be observed. The use of Raziol lubricants leads to smaller inner diameters with its minimum at 9.63 ± 0.06 mm for Raziol CLF 400E. A friction reduction with the use of wax containing lubricants is detected. This is similar to the ring-compression test, however the resulting friction factors are higher in the ring-compression test. The ring-compression test enables a free material flow in two directions, whereas in the pin-extrusion test, only a defined material flow into the die cavity is induced [9]. At an equal thickness reduction of 50 % in both test variants, the contact normal stresses are consequently lower in the ring-compression test [9], leading to higher friction factors.

Besides its friction system, the suitability of the tool material 1.3343 for the fabrication of micro texturized punches is analysed. In the design of micro textured surfaces for the use in friction reduction applications of highly loaded lubricated body contacts, the height is a decisive parameter to influence the hydrodynamic pressure and therefore the lubricant's film thickness [1]. The heights of cuboid textures, fabricated by EDM, are accordingly analysed for the high speed tool materials 1.3323 and 1.3244. As can be seen in Figure 5, the aimed height of each texture element is 15 μm while the target edge length varies between 15 μm and 30 μm.

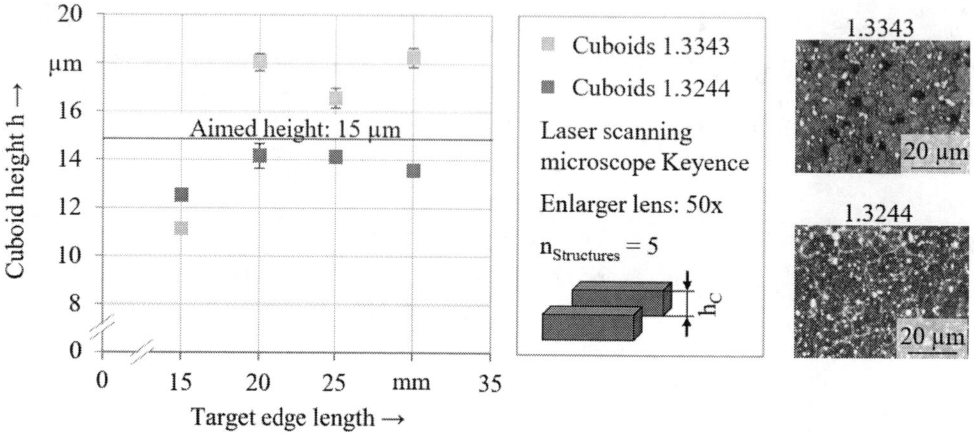

Figure 5: Heights of cuboid textures of the tool materials 1.3343 and 1.3244 fabricated by EDM

It can be observed for both materials, that a higher edge length has a beneficial influence on the height accuracy. This is due to a larger top surface, when fabricating textures with a larger edge length, which is not affected by unintentional detaching of particles in direct contact to the carbide electrode. For textures with a smaller edge length, the electrode is closer to the top surface centre and therefore negatively affects the overall height accuracy more than for textures with a higher edge length. Nevertheless, the height accuracy reaches saturation at an edge length of 20 μm. Furthermore, a material-dependent difference in the height accuracy is visible. The absolute height deviation for 1.3343 results in range between 1.7 μm and 3.8 μm whereas the range for 1.3244 is found between 0.8 μm and 2.3 μm. The use of powder metallurgical steels like 1.3244 can thus be a potential approach to enhance the height accuracy in the fabrication of micro textured forming tools. A comprehensive analysis of the manufacturing limits demands the investigation of additional structural variants. For this purpose, the heights of three texture variants are compared in Figure 6. Besides the cuboid structures, cylinders and prisms with concurring target edge length and diameter are selected.

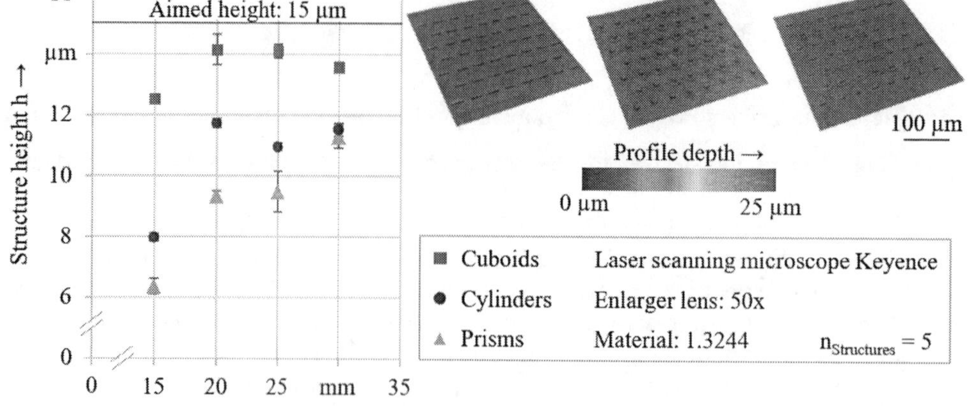

Figure 6: Heights of cuboid, cylinder and prism textures of 1.3244 fabricated by EDM

All four size variants show a form-dependent height accuracy behaviour. Cuboid structures fit the target height of 15 μm best, followed by cylinder elements. In all size variants, the lowest height accuracy is detected for prism texture elements. This is due to the larger separation of the carbid electrode from the element's top surface when performing the layered contour erosion. Structures with a target edge length respectively diameter of 15 μm are furthermore found to be influenced most by production inaccuracies, regardless of the geometrical form. This is in line with the findings for Figure 5. Regarding the element height, the selection of elements with a minimum edge length or diameter of 20 μm is therefore beneficial to narrow the manufacturing deviations.

4. Summary and outlook

The investigation has shown a methodology for the qualification of tool materials for the production of components with micro structured surfaces. To characterize the friction system of the high-speed steel 1.3343, the pin-extrusion test and the ring-compression test were used. In the combined experimental and numerical approach, the friction factors with four lubricant variants were identified. In both tests, the application of the bulk forming lubricant Beruforge 150DL leads the lowest friction factors, in comparison to the tested sheet metal forming lubricants from Raziol. Furthermore, the tool materials 1.3342 and 1.3343 were examined, regarding the suitability for the fabrication of microscopic texture elements by electrical discharge machining. In general, the powder metallurgical steel 1.3244 meets the target dimension better than 1.3343. A minimum edge length for cuboids and prisms and a diameter of 20 μm has a beneficial influence on the structure accuracy. In further investigations, measures to enhance the overall accuracy of the process are to be developed. In this context, an overcompensation of the target value as well as an adjustment of the electrode diameter is conceivable.

5. References

[1] Zahner M, Lentz L, Steinlein F and Merklein M: Investigation of production limits in manufacturing microstructured surfaces using micro coining. *Micromachines*, **11**(2017). 8: pp. 82-85
[2] Vadakkumpurath S, Venugopal A and Ullattil S: Influence of micro-textures on antibacterial behaviour of titanium-based implant surfaces: In vitro studies. *Biosurface and Biotribology*, **1**(2019). 5: pp 20-23
[3] Zahner M and Merklein M.: Analysis of combined extrusion micro coining process to manufacture microstructured tappets.17th International Conference on Metal Forming, 15, Toyohashi, Japan: 2018
[4] Merklein M, Allwood J, Behrens BA, Brosius A, Hagenah H, Kuzman K, Mori K, Tekkaya AE and Weckenmann A: Bulk forming of sheet metal. *CIRP Annals of Manufacturing Technolgoy*, (2012). 61: pp 725-745
[5] Löffler M, Gröbel D, Engel U, Andreas K and Merklein M: Analysis of effectiveness of lcally adapted tribological conditions for improving product quality in sheet-bulk metal forming. *Applied Mechanics and Materials*, (2015). 794: pp. 81-88
[6] Jiang Z, Zhao J and Xie H: Practice of micro deep drawing, in *Microforming Technology*. (2017). S. pp. 369-390
[7] Technisches Datenblatt "Raziol CLF E", Raziol Zibulla & Sohn GmbH.
[8] Datenblatt "Technisches Datenblatt Beruforge 150 DL", Carl Bechem GmbH.
[9] Vierzigmann U: Beitrag zur Untersuchung der tribologischen Bedingungen in der Blechmassivumformung - Bereitstellung von tribologischen Modellversuchen und Realisierung von Tailored Surfaces. Dissertation, (2019)
[10] Male A and Cockcroft M: A method for the determination of the coefficient of friction of metals under conditions of bulk plastic deformation. *Journal of the Institute of Metals*, **65**(1964). 93: pp. 38-46
[11] Oh S: Simulation of metal flow and tool deflection during cold extrusion process. in ASME Annual Winter Meeting. 1990.
[12] Tekkaya AE: A guide for validation of FE-simulations in bulk metal forming. *Arabian Journal for Science and Engineering* (2005). 30: pp. 113-136
[13] Ali M and Banu A: Electric discharge machining (EMD): A review. *International Journal of Engineering Materials and Manufacture*, **1**(2016). 1: pp. 3 -10
[14] Löffler M and Kolja A: Einfluss des Schmierstoffs auf die Werkstückqualität in der Blechmassivumformung. in Tagungsband XXXV. Verformungskundliches Kolloquium. 2016.

Acknowledgement
This work was supported by the German Research Foundation (DFG) within the scope of project 426217784.

An industrial application case to predict galling in hot stamping processes

S Berahmani[1], L Bruinekreeft[1], A Güner[2], J Venema[3] and M Sigvant[4]

[1] TriboForm Engineering B.V., Enschede, Netherlands
[2] AutoForm Engineering Deutschland, Dortmund, Germany
[3] Tata Steel, Research & Development, IJmuiden, Netherlands
[4] Volvo Cars, Olofstrom, Sweden

Abstract. Severe tool wear is an unwanted phenomenon which occurs widely during hot stamping processes due to extreme process conditions like high temperatures and the absence of lubricant. Galling is a wear mechanism in the form of adhesive wear in which some material from the sheet transfers to the tooling. In a longer term, the build-up of material on the tool can damage the sheet in the form of scratches and can negatively affect the heat transfer between sheet and tool. Therefore, it is important to develop advanced models to predict and control tool wear and galling during hot stamping processes. More recently an advanced friction model for hot stamping processes has been introduced to accurately describe frictional behavior of 22MnB5-AlSi. This study aims to further extend the advanced friction model of 22MnB5-AlSi into a galling prediction tool by first evaluating possible galling initiation models and next assessing growth models. These models are calibrated using experimental data from hot strip draw tests performed at Tata Steel. This results in a multi-dimensional galling model as a function of temperature, pressure, strain and also on the relative sliding distance in contact between the tool and the sheet. Finally, the predicted galling distribution on the tooling surfaces for two industrial parts from Volvo Cars are verified. The galling locations are accurately found on the parts by the applied galling model.

1. Introduction

Hot stamped parts in body-in-white applications are of a great interest due to their superior mechanical properties [1]. A commonly used material in direct hot stamping process is the ultra-high boron strength steel 22MnB5 with an aluminum silicon (+AlSi) coating to improve its corrosion resistance. For the tooling, the tool steel material with a coating layer to improve wear and frictional behavior is often utilized, which together with the 22MnB5 (+AlSi) form the most commonly used tribology system for hot stamping processes in the automotive industry. However, still one of the known issues during hot stamping processes is galling, which is a type of adhesive wear [2,3].

Galling occurs mainly due to the extreme process conditions during hot stamping with very high temperatures and the absence of lubricant, which causes particles from the softer sheet surface to transfers to the relatively hard tool surface which can affect the tribological behavior [2,3]. In a longer term, over multiple production cycles, the particles accumulated at the tool surface can eventually damage the sheet surface, observed in the form of scratches, and can negatively influence the heat transfer between sheet and tool during e.g. the quenching process. These unwanted phenomena cause a disturbance in the production batch, because the tooling needs to be cleaned or replaced [4]. Hence, developing advanced models to predict galling in forming simulations can significantly benefit achieving a better understanding of the process and tribological conditions which result in initiating galling in the forming process.

More recently, to describe the frictional behavior in forming simulations of hot stamping processes, an advanced friction model has been introduced by TriboForm® [4]. This friction model can predict friction under different process conditions such as a range of pressure and temperature and basically can substitute the conventional constant Coulomb friction modelling in hot stamping simulations [4,5].
For this study, the contact model implemented in the TriboForm advanced friction model is further adopted to predict galling. Also, a galling initiation model and a lump growth model are implemented, which are collectively calibrated based on the results from a series of hot strip-draw experimental testing by Tata Steel. Finally, the galling model is validated by comparing the severity of galling behavior observed for two industrial hot stamping parts, an A-pillar reinforcement and a side member from Volvo Cars, with the predicted galling distribution on the tooling surfaces.

2. Modelling approach

The 3D surfaces topographies from 22MnB5 (+AlSi) blank and tooling surfaces are the basis for both friction and galling models. The 3D surface of the sheet is captured after initial heating of the blank to a temperature of 930°C and then cooling it down, which resulted in a surface roughness of around 2 μm (Figure 1). The tooling surface belongs to a standard tool steel tooling from production with a typical roughness value in the range of 0.6 μm [6]. Furthermore, the material model according to hardening law proposed by Abspoel and van Liempt for hot forming is considered [7].

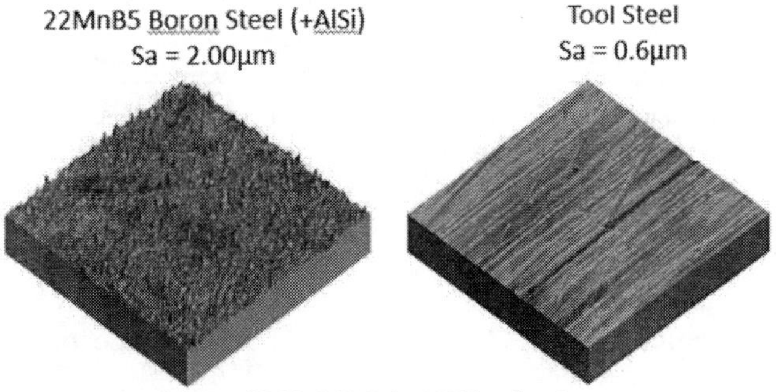

Figure 1. Surface topographies of 22MnB5 (+AlSi) after heat treatment and tooling surface.

2.1. *Galling models*

The contact area, coefficient of friction and asperity shape from the contact model of TriboForm Solver® [5] are used as inputs to the implemented galling models, namely the galling initiation model and a lump growth model. The galling initiation model shows the criteria for which an asperity will be a contributing factor to galling. This is followed by the lump growth model, which shows how adhered material will be deposited on the galling initiated asperities.

Before an asperity grows in size according to the lump growth model, it needs to be determined if the asperity is initiated for galling. A criterion needs to be set to check whether the composition of the asperity in combination with the input parameters of the process is suited to be initiated. In this study, the coating fracture initiation (CFI) criteria is used which is based on the plane strain fracture toughness of the coating as a threshold for galling initiation of an asperity.

In the next step, the lump growth model is implemented by following a couple of steps, first the separation height is determined to know which tool asperities are in contact and what their penetration is into the softer counter surface. Then, the material transfer from the sheet to the tool is determined in two steps: 1) determine transferred volume and 2) determine deposit layer. Then the stability is evaluated, because the material that is able to transfer only adheres when the asperity can hold the forces during sliding: 1) determine stresses within the asperity and 2) evaluate stability cases.
Finally, the adhered layer, according to one of the stability cases in the step before, is included in the new dimensions of the asperity. This cycle is repeated, which leads to multiple extra layers on the initial asperity, called growth of the asperity. Full details of the lump growth models and procedure can be found in the previous studies by van der Linde et al. [8].

2.2. Calibration

To simulate the conditions of a hot stamping process a series of hot strip draw tests are performed by Tata Steel. Using a roller heath furnace the blank is heated to a temperature of 930°C for 6 minutes to ensure full austenitization of the material [8]. The specification of strip and tool are given in the Table 1. The strips are then transferred to the stripdraw setup and tested with an interface temperature range from 450-750 °C under the nominal pressure of 2.5 and 5 MPa and a velocity of 100 mm/sec. Furthermore, a series of line measurements and 3D optical confocal measurements are performed using a Mahr PKG120 (cut-off 0.8mm) to determine the amount of area and volume of adhered wear at the locations with the greatest amount of wear (Figure 2).

The line measurements contains worn and unworn sections of the tooling surface and the confocal images provide a 3D height distribution of the tooling surface with the highest amount of observed wear. The line and 3D measurement are used in combination to determine the adhered volume, by first defining the reference height from the line measurement and then overlapping the 3D surface data on the reference line to quantify the adhered volume. Finally, the experimentally measured data is used for calibrating the lump growth model by using the correlation between worn volume and transferred volume.

Table 1: Specification of tested strip and tooling.

Strip		Tool	
Substrate	22MnB5	Material	DIN 1.2344
Coating	Al-Si (7-11% Si)	Finish	Milled
R_a (Rolling Direction)	0.15 μm	Hardness	48±2 HRC
R_a (Transverse Direction)	0.04 μm	S_a of testing tool	0.20± 0.05 μm
Dimensions	800 x 50 x 1.5 mm	Dimensions	90 x 11 mm
S_a (after heat treatment)	1.9 μm	S_a of production tool	0.60 μm

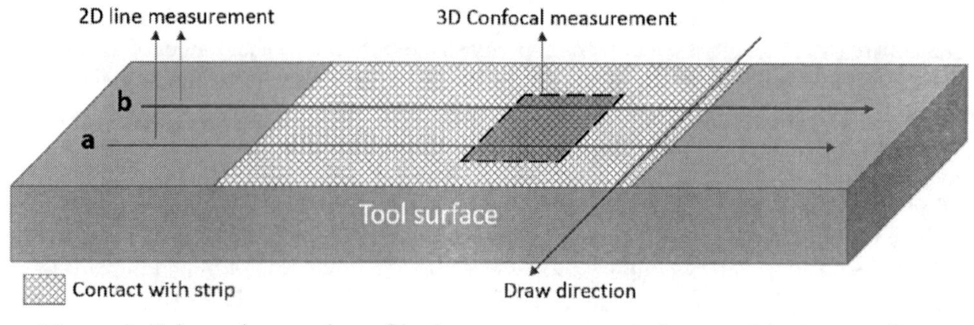

Figure 2. Schematic overview of both measurement techniques and its intersection.

IOP Conf. Series: Materials Science and Engineering 1238 (2022) 012061 doi:10.1088/1757-899X/1238/1/012061

3. Stamping simulations

Two industrial parts from Volvo cars, an A-pillar lower reinforcement and a side member, were studied (Figure 3 and 4). The simulation settings were following the standard settings at Volvo Cars to model hot stamping process using AutoForm® R8.0, including cyclic simulations considering the cooling channels to calculate the 3D heat conduction in the tool volumes. This leads to a more accurate surface temperature calculation.

The multi-dimensional friction model for the studied tribology system of 22MnB5 (+AlSi) and tool steel, is incorporated into the AutoForm simulation by using the TriboForm Plug-In®. The friction model also includes the galling model as a function of pressure, temperature, and strain over the sliding length. After running the simulations, the galling amount is projected back on the tooling surfaces using an in-house software.

The simulation results concerning the A-pillar reinforcement has been validated in the previous study by comparing the experimental and predicted thickness results [4]. For the current study, mainly the observed locations on the tooling surfaces with severe galling are compared with the predicted galling on the tooling surfaces.

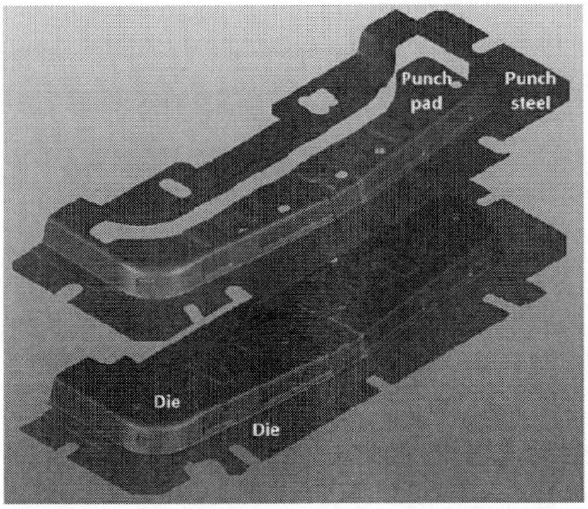

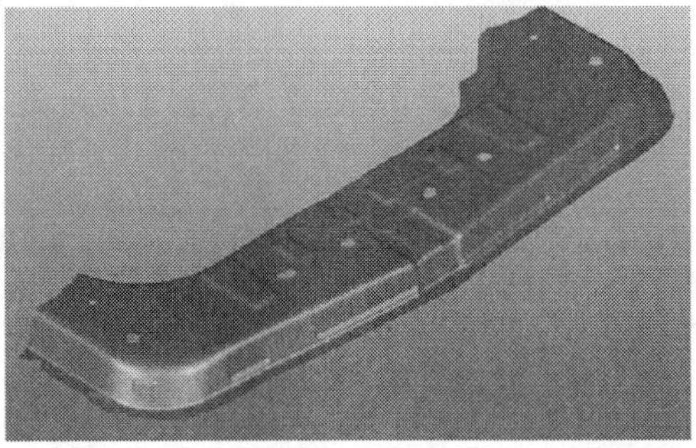

Figure 3. (Left) The tooling surfaces for the A-pillar lower reinforcement and (right) the final product.

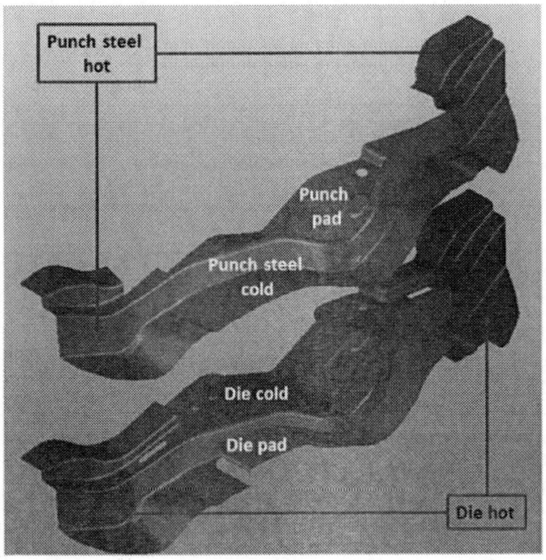

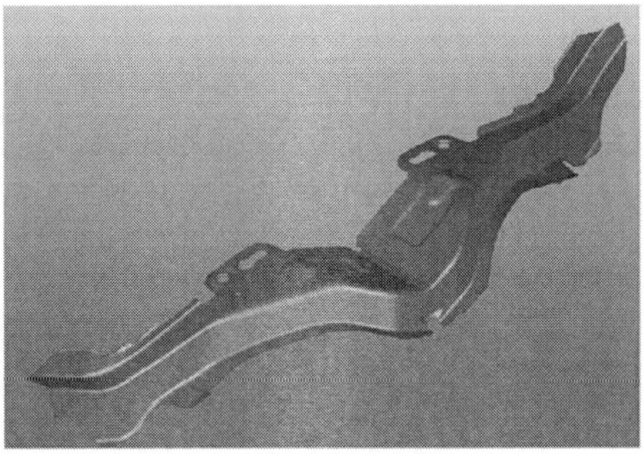

Figure 4. (Left) The tooling surfaces for the side member and (right) the final product.

4. Results and discussion

4.1. Calibration the galling models and calculating galling

The measured transferred volume from hot strip draw tests for a range of temperatures versus predicted worn volume by the galling model are shown in the Figure 5. Only experimental data points at the pressure of 5 MPa and the temperature from 500° C to 700° C are included as at 450° C no clear correlation between line and 3D measurements can be found. Also, it was observed that at the highest temperature of 750° C probably large lumps were removed that are also excluded from further analysis. The contact fracture galling initiation model (red line in Figure 5) shows an increase of adhered volume till a temperature of 625° C and then a decrease is predicted, which corresponds to the experimental findings.

The obtained calibration parameters were then used to extend the galling model for a wider range of pressure, temperature and strains as shown in the Figure 5 (Right). To predict the galling amount for these ranges, the galling initiation model was applied within the TriboForm Solver®. To check whether a single asperity is able to initiate lump growth for galling the equation below is used. The initiation according to the CFI model uses the nominal pressure to calculate the stress intensity factor for its fracture criterion. The equation states that if the stress intensity factor (K_I) of the single asperity in question exceeds the plane strain fracture toughness of the coating layer that the asperity will initiate fracture. Subsequently, the adhered volume is calculated over sliding length for the range of pressure, temperature and strain, which results in the galling model as a function of these parameters.

$$K_I \geq K_{IC}$$
$$K_I = Y\tau\sqrt{\pi a}$$

One important limitation is that during sliding contact both adhesive and abrasive wear are present, which are temperature dependent [9,10]. Therefore, the experimental data for the transfer volume shows adhered volume affected by abraded volume to an unknown extend. Also, the growth model is based on the adhesive galling, while compaction galling might also have a significant influence. However, the experimental adhered volume cannot distinguish between these two. Also, the procedure (position, measurement size) to determine the wear volume and/or wear height has a large effect on the results due to the inhomogeneity of the wear over the tool surface. Fracture of particles plays an important effect in compaction galling, another tests set up including deformation (for example strip drawing over a radius) could be interesting for calibration of the models.

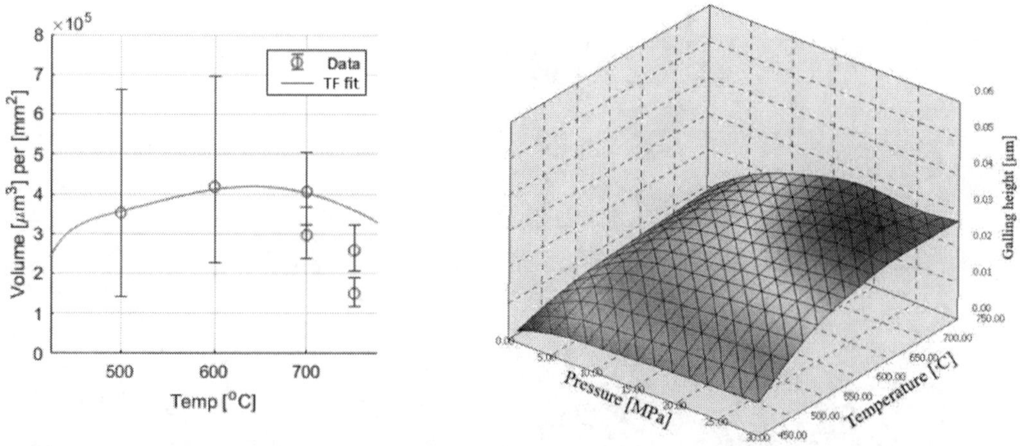

Figure 5. (Left) The transferred volume measured from strip draw tests versus predicted worn volume by the galling models. (Right) The predicted galling height as a function of pressure and temperature.

4.2. Galling prediction for the industrial parts

As the galling amount or galling height is not exactly known, it is mainly possible to compare the location with severe galling on the real tooling with the predicted locations on the tooling surfaces. The A-pillar part has shown a high amount of galling at the outer radii of punch which is also predicted correctly using the galling model (Figure 6). Less amount of galling was predicted for the punch pad and die which corresponds with the observation on the real tooling by Volvo cars.

However, there were two over-predicted points at the inner radii which was not observed in practice. A further analysis showed no significant difference in sliding length compared to the neighbouring edges, hence, these locations are attributed to a numerical error during the projection of galling on the tooling and not to real galling predictions.

It is important to be noted that the prediction of galling amount projected on the tool depends on the temperature, pressure and strain of the sheet during forming and on the relative sliding distance in contact between the tool and the sheet. This explains then the difference between different tool parts due to the difference in cumulative sliding distance, as the sliding length for the die and punch pad is significantly less than the sliding length on the punch.

Regarding the side member, the punch has shown a severe galling in reality (Figure 7). The simulation results show that the galling model can predict the location of galling for the side member to a very good extent (Figure 7). The severe galling location occurs when the tooling is heated up to 525° C, though, this high tool temperature is not directly applied within the galling model, still the tooling temperature does influence the sheet temperature, which is used to predict the galling amount per time increment.

Further evaluation demonstrates that the combination of pressure, temperature and strain does not indicate severe galling, hence the long sliding distance on these spots causes the severe galling. The punch pad and the die do not suffer from galling, which is in line with the experience at Volvo Cars. More interestingly, the simulation results of the side member show a higher prediction of galling amount compared to the A-pillar, which is also in accordance with the experience from Volvo Cars.

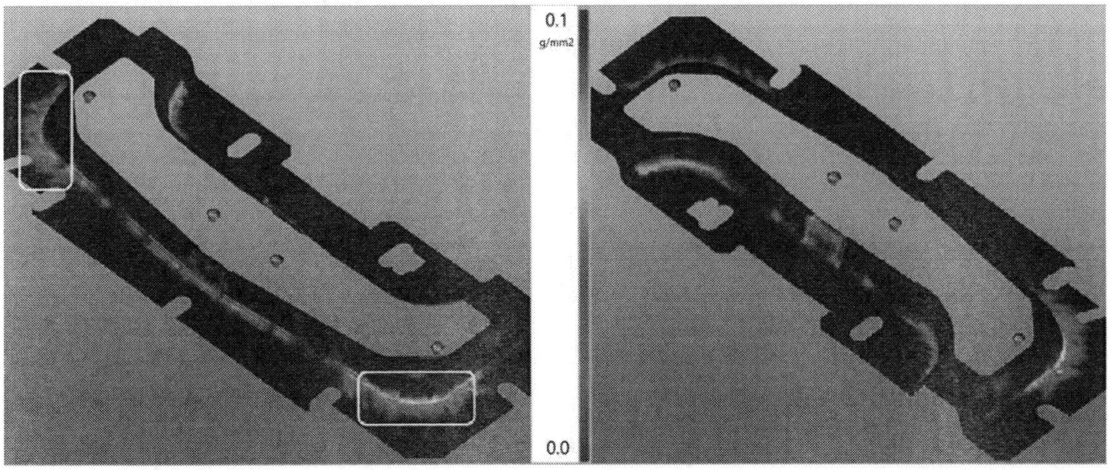

Figure 6. The distribution of galling (g/mm^2) on the outer radii of punch (right) and (left) inner radii views.

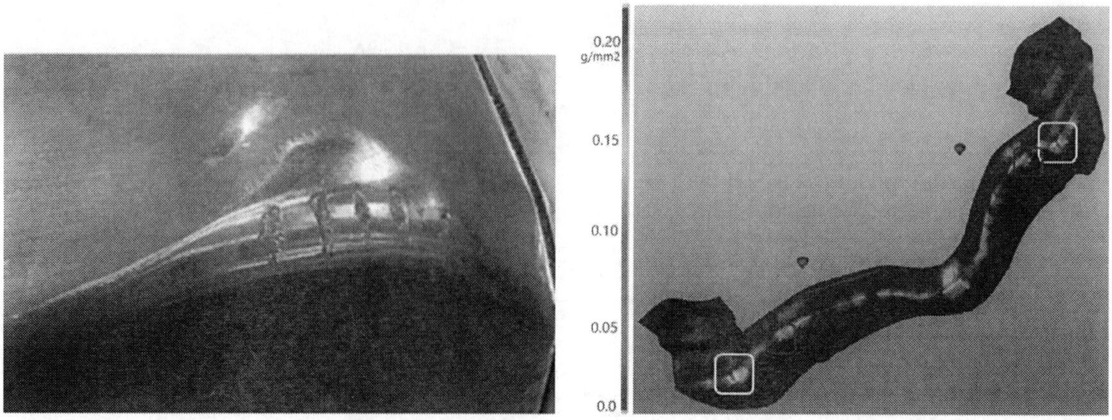

Figure 7. (Left) The observed galling at the tooling surface of side member which corresponds to (right) the distribution of predicted galling (g/mm^2) in the white boxes. Red boxes show the second location with high galling.

5. Conclusion and future work

In this study, the prediction capability of a galling model during hot stamping process has been studied for two industrial parts. To achieve this goal, first the original contact model of the TriboForm friction model is further developed and expanded with a galling initiation model and a lump growth model. The galling model is then calibrated by using the experimental data from hot strip draw tests and finally validated by comparing predicted galling locations with real parts. The calibration process was performed under a series of limitations as the exact amount of adhesive wear cannot be isolated from other forms of wear, namely compaction wear and abrasive wear.

Despite these limitations, a good match was found in terms of predicting the galling locations in both industrial parts which shows the high potential of the current model for predicting galling. However, as the severity of galling has not been measured in practice, it was not possible to compare the predicted galling amount with reality. The next step should focus on determining a quantitative approach to give an indication of galling risk based on the predicted galling amount.

References

[1] Karbasian H and Tekkaya A E 2010 A review on hot stamping *J. Mater. Process. Technol.* **210** 2103–18

[2] Schedin E and Lehtinen B 1993 Galling mechanisms in lubricated systems: A study of sheet metal forming *Wear* **170** 119–30

[3] Pelcastre L, Hardell J and Prakash B 2013 Galling mechanisms during interaction of tool steel and Al–Si coated ultra-high strength steel at elevated temperature *Tribol. Int.* **67** 263–71

[4] Güner A, Hol J, Venema J, Sigvant M, Dobrowolski F, Komodromos A and Tekkaya A E 2021 Application of an Advanced Friction Model in Hot Stamping Simulations: A Numerical and Experimental Investigation of an A-Pillar Reinforcement Panel from Volvo Cars *IOP Conf. Ser. Mater. Sci. Eng.* **1157** 012020

[5] Hol J, Meinders V T, Geijselaers H J M and Boogaard A H van den 2015 Multi-scale friction modeling for sheet metal forming: The mixed lubrication regime *Tribol. Int.* **85** 10–25

[6] Sigvant M, Pilthammar J, Hol J, Wiebenga J H, Chezan T, Carleer B and Boogaard T van den 2019 Friction in sheet metal forming: influence of surface roughness and strain rate on sheet metal forming simulation results *Procedia Manuf.* **29** 512–9

[7] Abspoel M, Neelis B M and Liempt P van 2016 Constitutive behaviour under hot stamping conditions *J. Mater. Process. Technol.* **228** 34–42

[8] van der Linde G 2011 *Predicting galling behaviour in deep drawing processes* PhD Thesis (Netherlands: University of Twente)

[9] Venema J, Hazrati J, Matthews D T A, Stegeman R A and Boogaard A H van den 2018 The effects of temperature on friction and wear mechanisms during direct press hardening of Al-Si coated ultra-high strength steel *Wear* **406–407** 149–55

[10] Venema J 2019 *Tribological interactions and modelling of friction in hot stamping* PhD Thesis (Netherlands: University of Twente)

TriboForm® software evaluation for ArcelorMittal Steel Products

A BOUZID[1, a*], H BOUR[1]

[1]ArcelorMittal Maizières, voie Romaine, BP30320, F-57283 Maizières-lès-Metz, France
a* aymen.bouzid@arcelormittal.com,

Abstract. In numerical simulations, input data should be correctly defined to enable reliable numerical predictions, close to experimental observations. The purpose of this study is to better take into account the tribological behaviour of ArcelorMittal steel products during the stamping process. TriboForm® is an important additional software solution for the simulation of tribology, friction, and lubrication in sheet metal forming process and enables the generation of multiple friction models. A comparison between experimental observations and numerical predictions was investigated for three ArcelorMittal steel grades (CR4-GI, CR180BH-GI+NIT and CR440Y780T-DH-GI) on several stamping parts. This study illustrates the interest of TriboForm® friction law compared to a standard and constant predefined friction coefficient.

1. Introduction

In stamping numerical simulations, the blank rheological behavior and the sliding properties of its surface should be described. These properties are respectively defined by a constitutive hardening law of the steel, a yield surface and a friction coefficient generally considered as a constant. The ArcelorMittal identification for hardening models is only based on stress-strain curves determined in uniaxial tension. The "combined Swift and Voce (S-V)" law is recommended and was used for this study. It offers good prediction beyond uniform elongation (UE) for different experimental tests [1].

It is already well known that the friction coefficient is highly dependent on several parameters like the contact pressure, material in contact, oil type, quantity, tool roughness, ... [2]. However, the friction properties are generally still not considered in detail in stamping numerical simulations. A constant Coulomb friction coefficient is often used and may limit the overall simulation accuracy. For example, a friction coefficient value equal to μ=0.15 is recommended by AutoForm® software.

TriboForm® software allows to simulate the effects of tool coatings, lubricants, material surface characteristics and new sheet materials. The objective of this study is to evaluate the effectiveness of TriboForm® software on ArcelorMittal steel grades during stamping process by comparing experimental observations to numerical predictions with AutoForm® using the standard and constant friction coefficient (μ=0.15) recommended by AutoForm® and the friction model proposed by TriboForm®.

2. TriboForm® friction model

The TriboForm® friction law is obtained from physical measurements performed on the surfaces in contact (Fig 1). These measured parameters are the following:

- The sheet and tool roughness,
- The lubrication type and amount,
- Stamping process such as pressure, velocity and temperature.

By generating these measured parameters, the TriboForm® friction law can be illustrated as a function of the contact pressure, strain and other parameters (Fig 2).

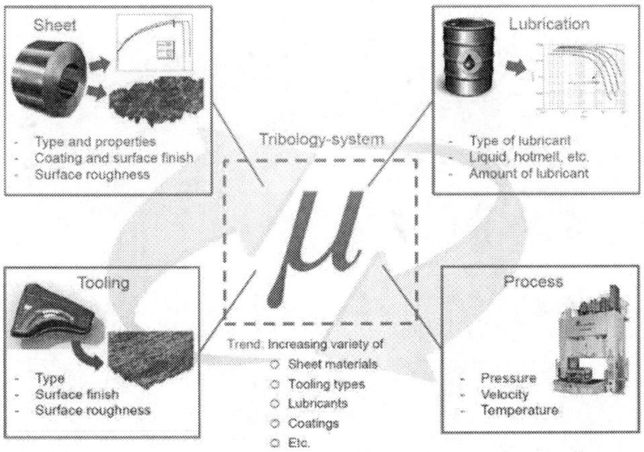

Figure 1: Influencing parameters on the tribological behaviour during stamping process [3]

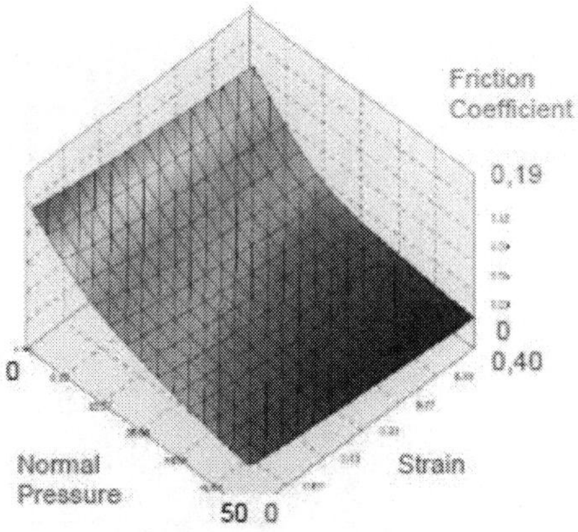

Figure 2: TriboForm® friction law [3]

The friction model proposed by TriboForm® can be imported in several stamping software such as AutoForm®, PamStamp® and LS-Dyna® using a TriboForm® FEM plug-in. In this study, all numerical simulations were conducted with AutoForm®, where an effective friction coefficient is calculated at each position of the blank and tools (upper and lower sheet sides, die, punch) for each calculation step (Fig 3).

The evolution of the friction coefficient depends on the current values of plastic strain, pressure, velocity and temperature [4] during the stamping process simulation.

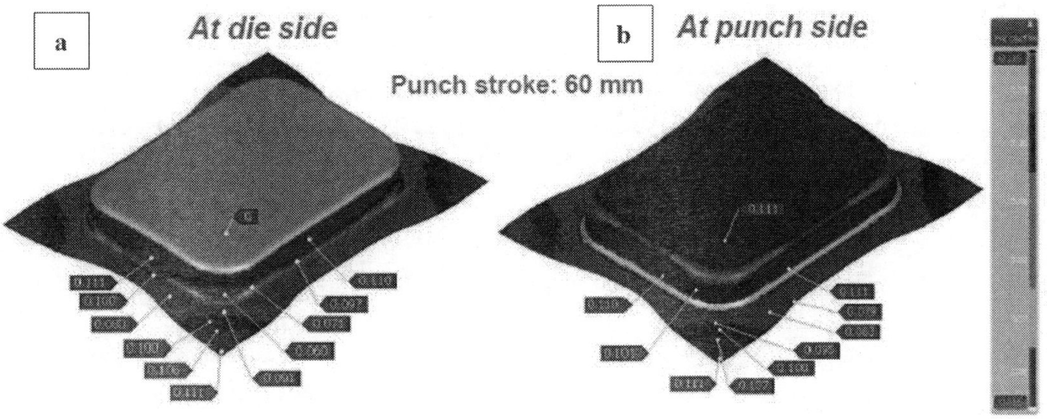

Figure 3: Display of friction coefficient cartography for a rectangular part simulated by AutoForm® (a) at the die side (b) at punch side

3. Experimental and numerical approaches

3.1. Selected Materials

This study was conducted with three ArcelorMittal steel grades: CR4, CR180BH and CR440Y780T-DH, two surfaces types: Galvanized and Galvanized + **N**ew **I**norganic **T**reatment (NIT) and three different tool shapes (Rectangular, Cross, and L-shape), Fig 4.

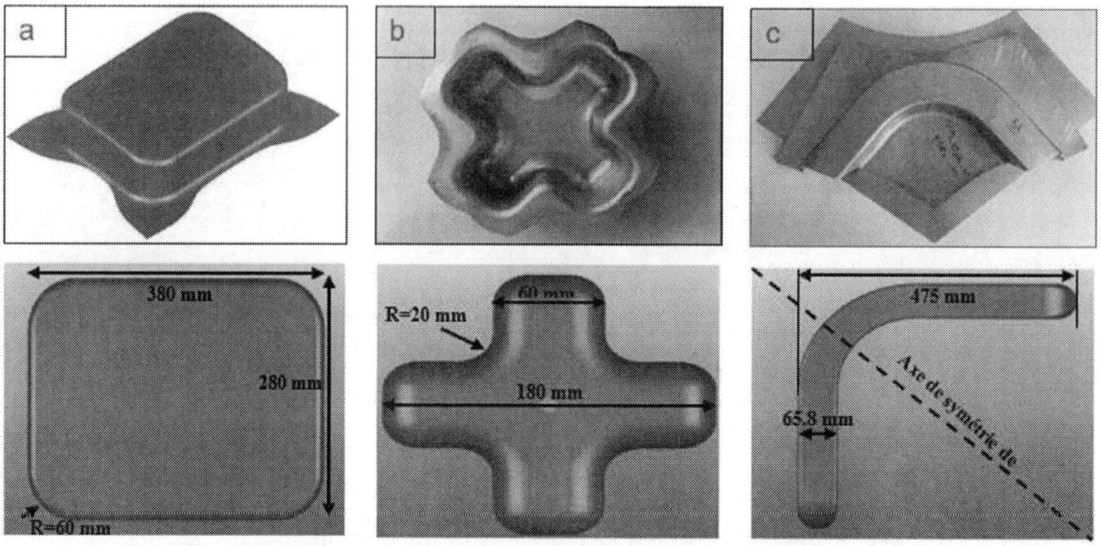

Figure 4: Shapes of the different parts (a) rectangular, (b) cross and (c) L shape

The metal sheet used for the rectangular part is a CR4-GI and a Quaker 6130 lubricant is applied to the sheet material with a constant amount of 1g/m². The metal sheet used for the cross-part made of CR180BH-GI treated with NIT and a Fuchs 3802-39S lubricant with 0.95g/m² is applied to the sheet material. And finally, for the L-part, a CR440Y780T-DH -GI is applied oiled with Quaker 6130 lubricant with a constant amount of 1.24g/m².

An overview of the applied tribology system of the different stamping parts is summarized in Table 1.

Table 1: Tribology system for rectangular, cross and L-parts

Sheet material	Sheet roughness	Sheet thickness	Lubricant	Tool material	Tools roughness
CR4-GI	1,25 µm	0.7 mm	Quaker 6130 (1 g/m²)	DIN X160 CrMoV12 (rectangular tool)	0.31 µm
CR180BH-GI+NIT	1.03 µm	0.65 mm	Fuchs 3802-39S (0.95 g/m²)	DIN X160 CrMoV12 (cross tool)	0.34 µm
CR440Y780T-DH-GI	0.87 µm	1.22 mm	Quaker 6130 (1.24 g/m²)	DIN 45 CrMnMoS8 (L-tool)	0.22 µm

3.2. Comparison between experimental observations and numerical predictions (AutoForm®)

In order to evaluate the interest of TriboForm®, numerical simulations were also conducted with a constant friction coefficient (μ=0.15). All numerical predictions were confronted to experimental observations in terms of stamping force, final sheet thickness after forming and plastic strain on localized areas.

Thinning measurements were done using an ultrasonic device. Experimental thinning and numerical predictions are illustrated in Figure 5 for the rectangular box made of CR4-GI 0.7mm. The Blank holder force is equal to 494 kN. A constant friction coefficient of μ=0.15 induces much higher thinning in the wall than those measured experimentally. The use of the TriboForm® friction model allows to predict more accurately experimental observations even if thinning is slightly underestimated.

To confirm the previous results, experimental thinning and numerical prediction for a rectangular box made of CR4-GI were done also with blank holder forces of 659 kN and 786 kN. The results show the same prediction accuracy than those obtained with a blank holder force equal to 494 kN.

Experimental thinning and numerical prediction were confronted for L-part and cross-part made of respectively CR440Y780T-DH-GI 1.22mm and CR180BH-GI+NIT 0.65mm. The comparison confirms that the use of TriboForm® friction law allows to predict more accurately the experimental observations than the standard Coulomb friction coefficient (μ=0.15).

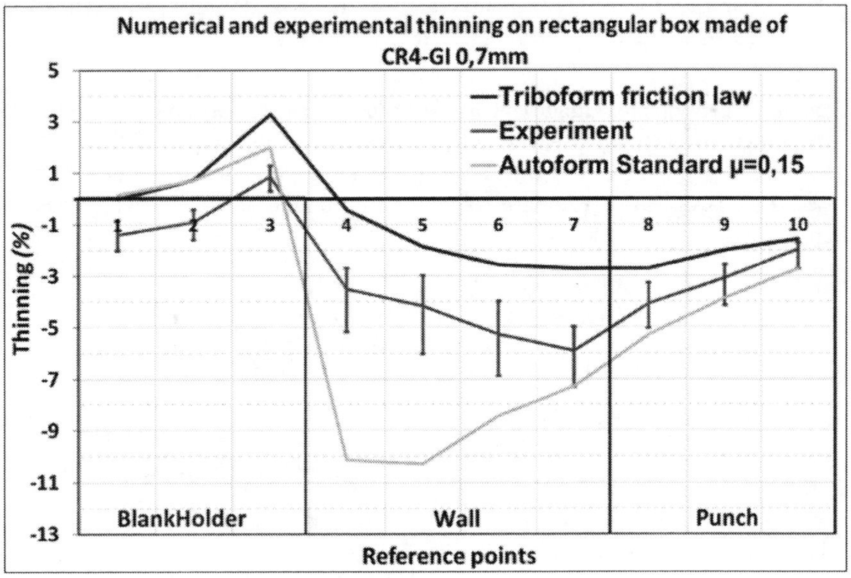

Figure 5: Experimentally measured and simulated thinning results for the Rectangular box made of CR4-GI 0.7mm

The stamping force evolution during the cross-shape forming of a CR180BH-GI and pretreated with NIT is illustrated in Figure 6. It can be clearly seen that the stamping force evolution including the TriboForm® friction model predict well the observations in try-out. However, the stamping force determined with a constant friction coefficient (μ=0.15) overestimates the experiments.

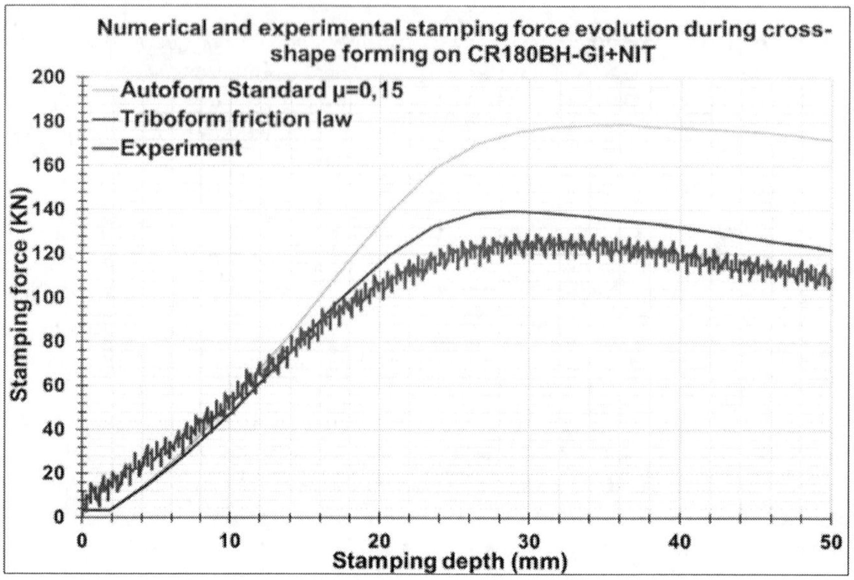

Figure 6: Experimentally measured and simulated stamping force results for the cross-shape made of CR180BH-GI+NIT 0.65mm

An element of the L-Shape part made of CR440Y780T-DH-GI 1.22mm has been selected to compare the plastic strain predictions to an experimental measurement using Digital Image Correlation (Fig 7). The predicted couple major-minor plastic strains are again better estimated when TriboForm® friction model is used compared to a standard friction coefficient (μ=0.15) where plastic strains are overestimated.

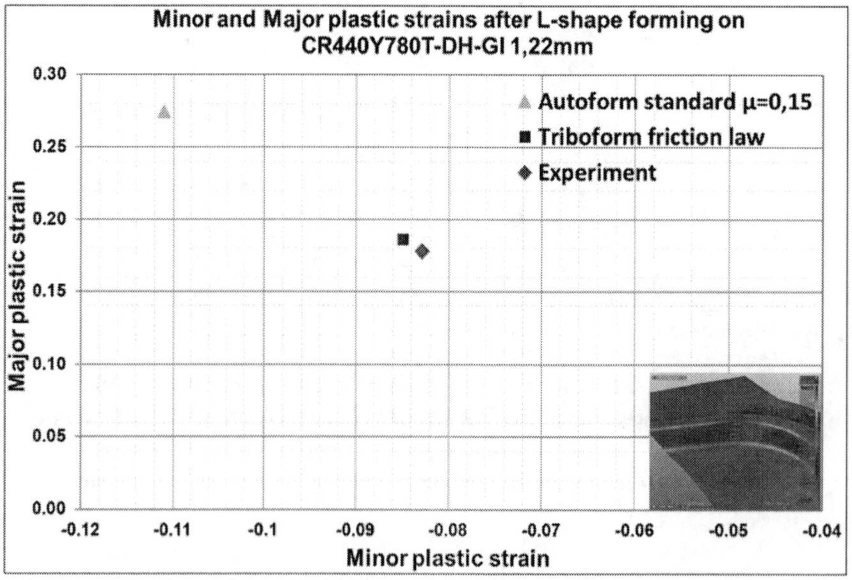

Figure 7: Experimentally measured and simulated plastic strain results for a localized are of L-shape part made of CR440Y780T-DH-GI 1.22mm

4. Conclusion

This study reveals the interest of the TriboForm® friction law compared to a standard and constant predefined friction coefficient (μ=0.15). A strong influence of the tribology, friction and lubricants conditions on the stamping process is observed.

Numerical predictions are improved for three stamping parts, in terms of sheet thickness reparation after forming, stamping force evolution and plastic strain field, when TriboForm® friction model is used.

The effects of type and amount of lubricant can also be taken into account in TriboFrom. This functionality allows to determine the optimal stamping process parameters (like drawing velocity and blank holder force) to enhance the part formability [5].

References

[1] X.Lemoine, S.Sriram and R.Kergen *Flow curve determination at large plastic strain levels to accurately constitutive equations of AHSS in forming simulation* Conference IDDRG 2011, 5-8 June, Bilbao, Spain

[2] M Sigvant, J Pilthammar, J Hol, J H Wiebenga, T Chezan, B Carleer and A H van den Boogaard *Friction in Sheet Metal Forming Simulations: Modelling of New Sheet Metal Coatings and Lubricants*

[3] www.triboform.com

[4] Grueebler R., Hora P. *Temperature dependent friction modeling for sheet metal forming. International Journal of Material Forming*, 2:251–254, 2009

[5] Sigvant M., Pilthammar J., Hol J., Wiebenga J.H., Chezan T., Carleer B., Boogaard T. van den *Friction and lubrication modeling in sheet metal forming simulations of a Volvo XC90 inner door*, IOP Conference Series: Material Science and Engineering 159 (1), 2016.

IOP Conf. Series: Materials Science and Engineering 1238 (2022) 012063 doi:10.1088/1757-899X/1238/1/012063

Investigation of frictional behavior of steel and aluminum at different temperatures

G Huang, F Fenton

ArcelorMittal Global R&D, East Chicago, IN 46321 USA

gang.huang@arcelormittal.com

Abstract. In order to evaluate the performance of the press line, it is essential to investigate the frictional behavior and properties of materials under production temperatures. In servo press line production conditions, the tooling temperature can reach 45-55 °C after around 150 stampings. In this investigation, a clamping fixture with drawbead and flat plate inserts was employed to conduct the friction tests at room temperature and an elevated temperature (54 °C). Three mild steels and two aluminum grades (5000 and 6000 series) were used in the tests. The results from drawbead tests showed that friction force on the two aluminum materials increased by 10% at 54 °C compared to that at room temperature. However, for the 3 steels the change of friction force was negligible from 22 °C to 54 °C. This implies that for exposed panel applications, the selected steel grades might have a more controlled sliding behavior potentially resulting in better control of formability than aluminum grades for the range of temperatures observed during production stamping.

1. Introduction

Cold stamping has been widely applied by a variety of manufacturing industries to form metal parts of different shapes. After workpieces are pushed into a die cavity for flowing through drawbead during a stamping process, they can experience relative movement at the metal/tooling interface that will be impacted by the tribological behavior, which together with the dissipated energy from the plastic deformation of the parts can lead to appreciable increase of temperature of the dies. Therefore, the cold stamping might not stay cold after many stamping strokes. Accordingly, metal flow control is an important aspect of stamping process optimization as it is essential to assure part quality.

Several tests are used in the stamping industry to determine the frictional behavior of the sheet metal, such as the drawbead simulator (DBS) test [1], bending under tension (BUT) test [2], and a simple flat plate sliding test to determine the frictional behavior typically characterized by a constant friction coefficient as in the Coulomb's model. These tests were designed to simulate the material experiencing friction at different locations of the stamping die. While frictional behavior of metals at temperature higher than 200 °C during warm forming has been extensively studied experimentally or numerically [3-5], there are very few investigations on the frictional behavior for steel and aluminum at a moderately elevated tooling temperature during cold stamping. Since temperature might have different effect on friction for steel and aluminum, the control of formability is important to manufacturer to ensure the outcome quality of cold stampings.

In this study, the frictional behavior using drawbead and flat binder (strip drawing) tests was characterized for different mild steel and aluminum grades at different temperatures. The friction performance was compared between steel and aluminum through studying the frictional behavior at room and elevated temperature.

2. Experimental method

2.1. *Materials*

Three mild steels with same thickness and two aluminum grades with different thicknesses were used in the friction tests. The two Extra Deep Draw Steel (EDDS) materials have very similar mechanical properties, except that EDDS-2 has a pre-phosphate layer on the Galvannealed (GA) coating. The mechanical properties of the four materials were listed in Table 1.

Table 1. Mechanical Properties of tested materials

Material	Coating	Thickness (mm)	YS (MPa)	TS (MPa)	TEL (%)
EDDS-1	GA	0.63	162	313	45
EDDS-2	GA (with pre-phosphate)	0.63	149	303	47
BH340	GA	0.63	240	365	28
Al-5xxx	None	0.76	94	203	26.7
Al-6xxx	None	0.90	111	211	22

2.2. *Experiment*

2.2.1. Friction test with drawbead. The clamping fixture housing the drawbead insert was used to conduct the friction tests, as shown in Figure 1. The drawbead set geometries are shown in Figure 2. To achieve a stable and accurate temperature control, an electrical heating system with a new die set was developed. Two independent resistance heating units were installed in the drawbead holders (for the punch and die of the drawbead set) and were controlled separately by the control unit. Thermal Insulators made of Mica were imbedded outside the drawbead holders to minimize the heat loss, so as to reach a stabilized temperature during the test. The heating system was able to render the temperature control between 22 °C and 200 °C. In this study, room temperature (22 °C) and the elevated temperature, 54 °C, were applied in the friction tests. The testing parameters are listed in Table 2.

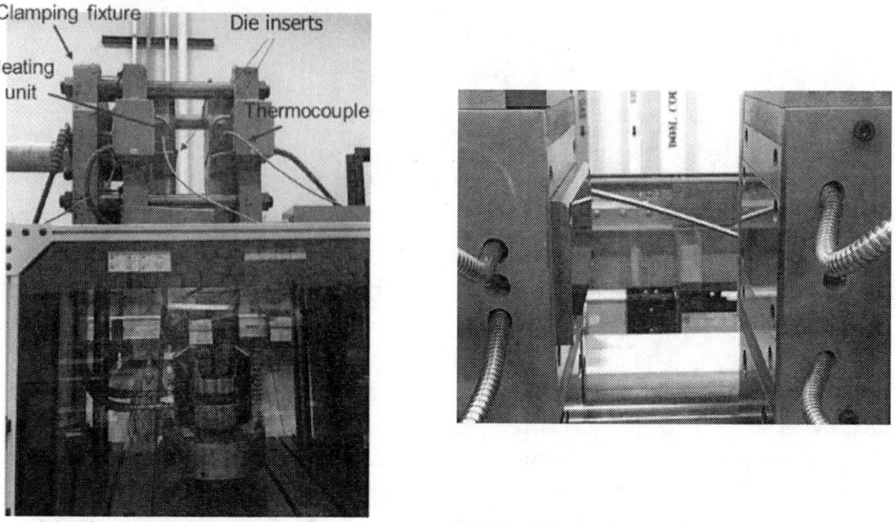

(a) Clamping fixture used to evaluate the frictional behavior

(b) Flat binder insert for strip drawing test

Figure 1. Setup of draw bead friction test

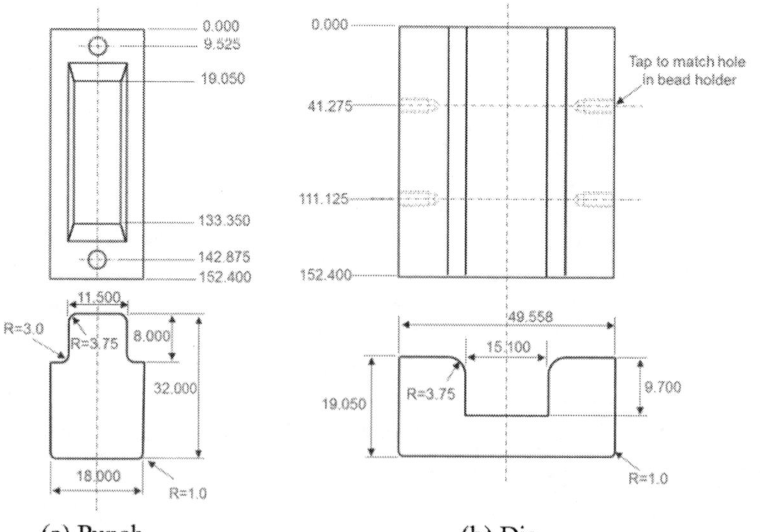

(a) Punch

(b) Die

Figure 2. Geometry of drawbead inserts

Specimens were cut with same dimensions of 889 mm × 101.6 mm, where the length of 889 mm is in the rolling direction. Thereafter, the edges were deburred thoroughly. No lubricant was applied on the specimen surface. Before testing, both steel and aluminum specimens were uniformly and thoroughly wiped with cloth towels to remove any surface debris, while no cleaning chemicals was applied. For tests at the elevated temperature, the punch and die of drawbead set were heated to reach the same temperature (54 °C) before starting the test. During the test, the specimen was first gripped below by the hydraulic gripping fixture which was connected to the actuator; then the top clamping fixture closed the drawbead set, followed by the pulling of the specimen through the drawbead right after the clamping load was reached. In this process, the data of pulling force, displacement and clamping load were recorded.

2.2.2. Friction test with binder (strip drawing test). The friction test with the flat binder is similar to the conditions using the drawbead, except that the die insert was replaced with flat plates, as shown in Figure 1b. As the specimen was in direct contact with the flat binders, no clearance was set. For the current tests, a pressure of 5 MPa was applied to the specimen. Table 2 also shows the testing parameters of the flat binder tests which were conducted on the 5 materials at room temperature and 54 °C.

Table 2. Friction testing parameters using two different inserts

Test parameters	Drawbead	Flat binder
Specimen size	101.6 mm x 889 mm	101.6 mm x 889 mm
Testing Temperatures	22 °C and 54 °C	22 °C and 54 °C
Pulling speed	200 mm/s	200 mm/s
Pulling distance	150 mm	150 mm
Clearance	10%	0%
Clamping pressure	5 MPa	5 MPa

3. Results and discussions

3.1. Results from drawbead friction tests

In the drawbead friction tests, the specimens were subjected to bending and unbending deformation mode when pulled through the drawbead, which results in springback. Figure 3 shows pictures of the tested specimens of EDDS1 and the 5xxx aluminum, where more severe springback was exhibited for the aluminum due to its lower modulus.

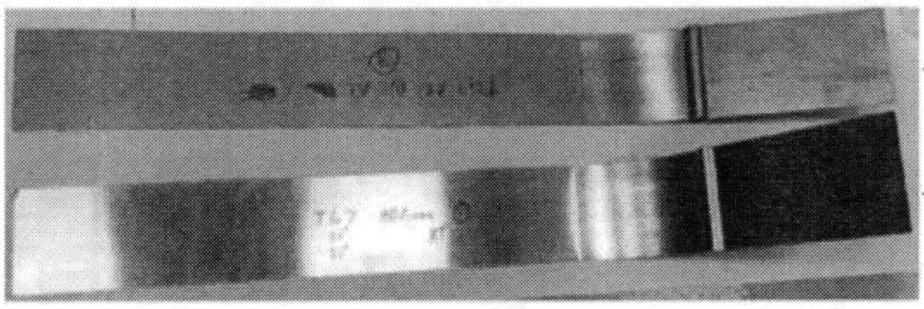

Figure 3. Example of specimens after tests

For tests at room temperature, 10% of material thickness was used for drawbead clearance by adding or removing shims. When the drawbead reached 54 °C for elevated temperature testing, it was found that the gap between the two flat blocks (at the location of drawbead and elsewhere) was reduced by about 0.08 mm, which is significant compared with the thickness of the mild steels. To study the effect of the change in clearance due to the thermal expansion from the tooling on the frictional behavior, two different types of tests were conducted: one with temperature compensation to maintain the same clearance, the other without temperature compensation or any clearance adjustment.

For the tests at 54 °C with temperature compensation, shims with thickness around 0.08 mm were used to retain the 10% of clearance. Figure 4 gives an example of comparison of pulling force on EDDS2 and Al6xxx as function of pulling distance. Other grades of steel and aluminum showed similar trend where with the maintained clearance, the friction force for 3 steels was reduced by around 10% at the elevated temperature, as compared with for room temperature, while for the two aluminum grades the friction force remained the same at the elevated temperature with the clearance adjustment.

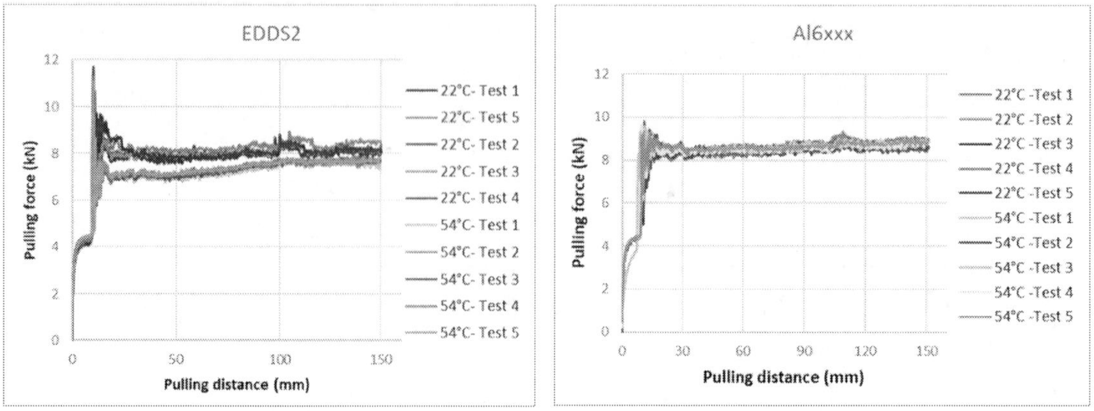

(a) EDDS2 (b) Al 6xxx

Figure 4. Example of comparison of frictional behavior between steel and aluminum with temperature compensation for clearance

For the tests at 54 °C without temperature compensation, no adjustment for clearance was applied, even though the clearance was significantly reduced by the thermal expansion of the tooling. Figure 5 summarizes the averaged pulling force along pulling distance for the 5 materials. The trend is clear that for three mild steels, the friction force was almost the same between 22 °C and 54 °C, even though the clearance at 54 °C was much lower than at 22 °C. However, for the two aluminum grades the friction force increased by 10% at 54 °C due to the clearance shrinkage.

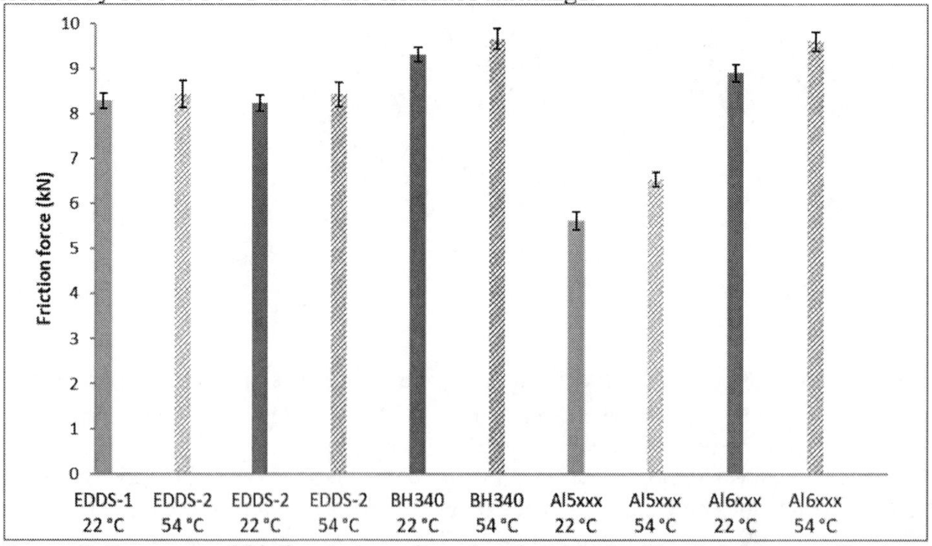

Figure 5. Summary of friction test results using drawbead without clearance adjustment

An example of the result of pulling force as a function of pulling distance is shown in Figure 6 for steel and aluminum. From Figures 5 and 6, it is obvious that the friction force exerted by the drawbead was not affected by the increase of the temperature when no extra adjustment was applied at the elevated temperature, which was somehow reflecting with clearance compensation for temperature the friction force of two aluminum grades increased by about 10%. This indicated that the tested steels had better control of formability in terms of friction performance than aluminum.

It should be noted that at pulling distance around 100 mm, there was a spike of friction force for the steels. This might be mostly caused by the roughening of the surface at the contact location away from the drawbead which was induced by the closing of the drawbead. Since the tooling gap for the steels are much lower than those of the aluminum grades due the thickness difference (while the clearance of 10% is the same), the surface roughening could be more significant than for the aluminum grades.

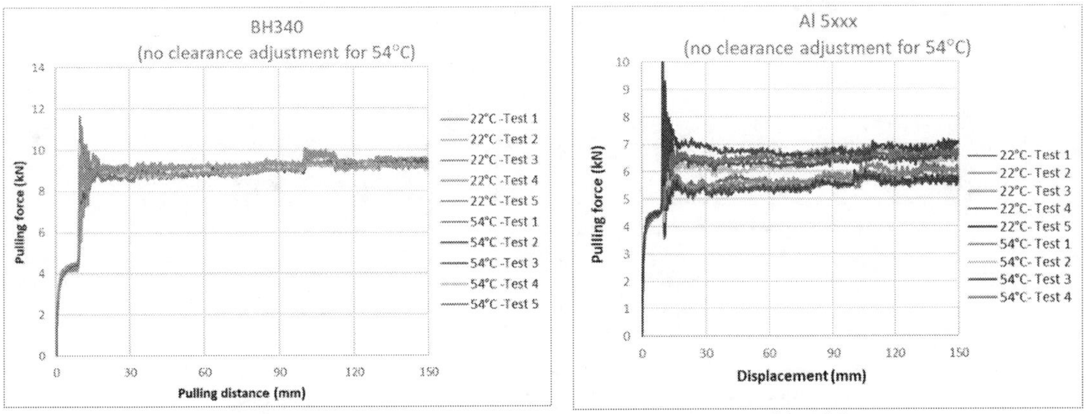

(a) BH340 (b) Al 5xxx

Figure 6. Example of comparison of frictional behavior between steel and aluminum with no temperature compensation for clearance

3.2. Results from strip drawing friction tests with flat binder

The drawbead friction test was able to mimic the material flow through the drawbead for the stamping shop, however it could not directly lead to the determination of correlation of coefficient (COF) as the friction force was dependent on the bending and unbending moment, which was greatly influenced by the thickness of the material. Therefore, an inverse method is required to determine the COF from the drawbead friction test [6]. In contrast, the strip drawing test, which forced the specimen directly contact with the flat binder, enables the direct determination of COF using Equation (1):

$$f = \frac{F}{2N} = \frac{F}{2pS} \qquad (1)$$

where f is the COF, F the pulling force, N the normal force applied on the specimen, p the pressure and S the contact area.

Figure 7 shows the pulling force as function of displacement for the two EDDS steels at room temperature and 54 °C. The COF calculated based on Equation (1) is summarized in Table 3. It can be seen that, in general the three mild steels have similar behavior of friction between room temperature and 54 °C, except that there was slight increase of COF for EDDS-1 (without pre-phosphate). Furthermore, the repeatability of tests was fairly high for all the tested steels. Table 3 indicated that the COF of EDDS with pre-phosphate was lower (around 10%) than that EDDS without pre-phosphate. This demonstrated that pre-phosphate has positive effect on the reduction of COF.

Figure 8 shows the pulling force as function of displacement for the two aluminum grades. Large variation was exhibited between test repeats. Meanwhile, a steady state of friction force was not reached for nearly all test repeats, as depicted in Figure 8. The highly scattered data prevented COF from being calculated for the two aluminum grades. The variability of the data was caused by the frequent breakage of aluminum specimens during the tests: splitting over 50% of the specimens was observed for both materials.

Even though the reduction of binding pressure might lead to testing data of better quality and determination of COF, the results from the current friction tests using flat binder (with 5 MPa binding pressure) again implied that the tested steels had better friction performance than the tested aluminum grades.

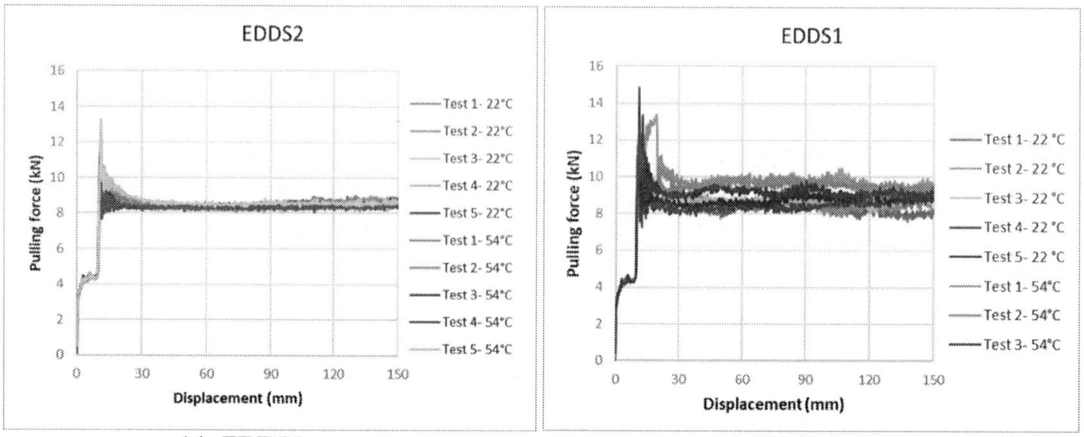

(a). EDDS2 (b). EDDS1

Figure 7. Result of pulling force versus displacement for two EDDS grades from strip drawing tests

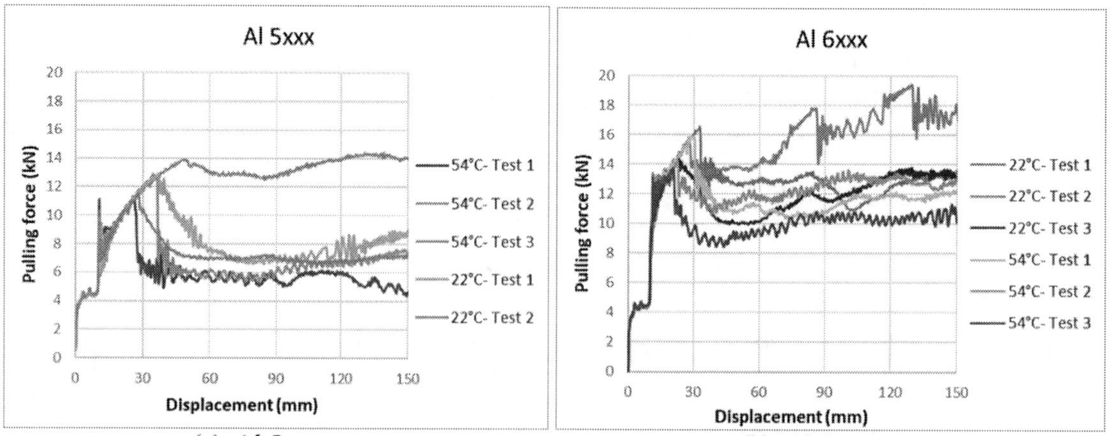

(a). Al 5xxx (b). Al 6xxx

Figure 8. Result of pulling force versus displacement for aluminum grades from strip drawing tests

Table 3. COF from strip drawing tests

Material		EDDS-2	EDDS-1	BH340	Aluminum 5000	Aluminum 6000
COF	22 °C	0.11	0.12	0.135	NA	NA
	54 °C	0.11	0.125	0.138	NA	NA

It is noted that all friction tests of both types were performed at the speed of 200 mm/s, while the machine actuator requires some time to accelerate to the relatively high speed. Therefore, it is essential to verify whether the stead speed of 200 mm/s was reached and maintained during the test. Figure 9 describes the history of displacement of specimens under testing. It shows that specimens reached the speed of 200 mm/s before 20 mm, and the speed was well maintained thereafter. Since the steady speed was aligned with the steady state of friction (Figures 4, 6 and 7) which started after 20 mm for both drawbead and strip drawing tests, the condition of valid speed for friction test was well satisfied.

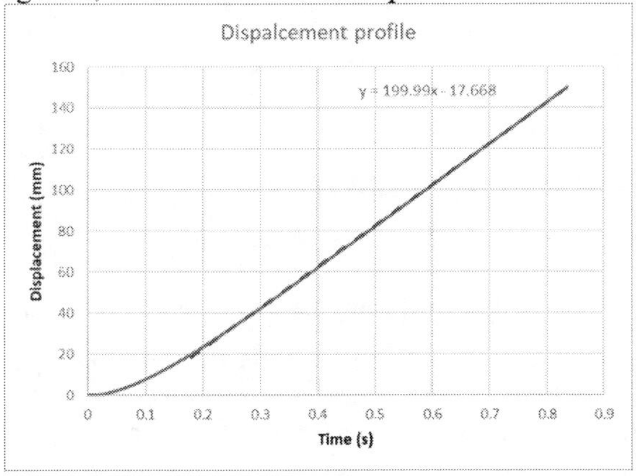

Figure 9. History of displacement (pulling distance)

4. Conclusions

Friction tests were conducted on three mild steels and two aluminum grades using drawbead and flat binder (strip drawing) conditions at room temperature and an elevated temperature. The results from drawbead friction tests showed that, when there was no temperature compensation for clearance, for steels the friction force did not change after the temperature was elevated from 22 °C to 54 °C, while for the aluminum grades the friction force increased by 10% at 54 °C compared to at 22 °C. The results from strip drawing friction tests showed that, the COF could be readily determined for three steels with the testing data, while for aluminum grades the variability was too high to determine the COF. This indicated that the tested steels might have better control of formability in terms of friction performance.

Acknowledgments

The authors would like to sincerely thank Mr. Robert Houser for conducting the friction tests. The constructive suggestions and advice from Dr. Sriram Sadagopan and Mr. Chris Brown are highly appreciated. The authors would also like to highly appreciate the encouragement from Dr. Jayanth Chintamani and Dr. Mark Atkinson.

References

[1] Nine H D 1978 *Mechanics of Sheet Metal Forming, Material Behavior and Analysis*, ed D P Koistinen and N-M Wang (New York: Plenum) pp 179-211

[2] Wilson W R D, Maikani H G and Saha P K 1991 Boundary Friction Measurements Using a New Sheet Metal Forming Simulator *Proc. NAMRC* **XIX** p 3742

[3] Hardell J and Prakash B 2008 High-temperature friction and wear behaviour of different tool steels during sliding against Al–Si-coated high-strength steel *Tribol. Int.* **41** pp 663-671

[4] Singh S K Mahesh K Kumar A Swathi M 2010 Understanding formability of extra-deep drawing steel at elevated temperature using finite element simulation *Materials & Design* **31** pp 4478-4484

[5] Liu Y Zhu Z Wang Z Zhu B Wang Y Zhang Y 2018 Flow and friction behaviors of 6061 aluminum alloy at elevated temperatures and hot stamping of a B-pillar *Int. J. Adv. Manufacturing Tech.* **96** pp 4063–4083

[6] Huang G, Yao H, Sriram S and Zhu H 2016 Use of an Inverse Method to Determine the Coefficient of Friction for Flow of Sheet Steel through a Drawbead IDDRG pp 697-711

IOP Conf. Series: Materials Science and Engineering 1238 (2022) 012064 doi:10.1088/1757-899X/1238/1/012064

Metamodeling of a deep drawing process using conditional Generative Adversarial Networks

Patrick Link*[1], Johannes Bodenstab[1], Lars Penter[1,2] and Steffen Ihlenfeldt[1,2]

[1] Fraunhofer Institute for Machine Tools and Forming Technology IWU, 09126 Chemnitz, Germany

[2] Institute of Mechatronic Engineering, Technische Universität Dresden, 01062 Dresden, Germany

*patrick.link@iwu.fraunhofer.de

Abstract. Optimization tasks as well as quality predictions for process control require fast responding process metamodels. A common strategy for sheet metal forming is building fast data driven metamodels based on results of Finite Element (FE) process simulations. However, FE simulations with complex material models and large parts with many elements consume extensive computational time. Hence, one major challenge in developing metamodels is to achieve a good prediction precision with limited data, while these predictions still need to be robust against varying input parameters. Therefore, the aim of this study was to evaluate if conditional Generative Adversarial Networks (cGAN) are applicable for predicting results of FE deep drawing simulations, since cGANs could achieve high performance in similar tasks in previous work. This involves investigations of the influence of data required to achieve a defined precision and to predict e.g. wrinkling phenomena. Results show that the cGAN used in this study was able to predict forming results with an averaged absolute deviation of sheet thickness of 0.025 mm, even when using a comparable small amount of data.

1. Introduction

In the past decade machine learning for optimization of processes became a widely used method in manufacturing [1]. Machine learning algorithms are often used to develop data-driven metamodels, which predict product quality as a function of relevant process parameters. In deep drawing these metamodels can be used for process optimization [2] or can be applied as fast prediction models in process control schemes [3]. The amount and quality of available data for training significantly affects the predictive power of the metamodel. The prediction of entire part geometry is of particular interest for optimizing and controlling deep drawing processes. For instance, the optimization of local blank properties, such as friction (based on the amount of oil for lubrication) and sheet thickness, requires the computation of the final geometry including blank and these local properties. Such investigations require metamodels that represent the underlying deep drawing process appropriately. Typically, enhanced metamodels comprise a large number of parameters, and consequently, need a large amount of training data. Data acquisition in manufacturing in general but specifically in sheet metal forming is time consuming and costly before start of production. Hence, a common approach to train metamodels is to generate data with FE simulation models. However, data generation by means of numerical simulations can as well be costly and time consuming when dealing with complex material and contact models or large parts with a great number of elements. It is desirable to keep the amount of FE simulations to a minimum. Therefore, it is necessary to develop metamodels for limited training data that still result in high predictive power.

Approaches reported in earlier literature can be distinguished into metamodels that predict scalars and metamodels that predict physical fields. For instance, Bonte et al. [2] described a methodology to optimize forming processes by using response surface methods and kriging models. Opposed to that, other authors described methods to predict physical fields. For example, Schwarz et al. used principle component analysis [4] and Fourier-based transformation methods [5] for dimension reduction with subsequent linear regression to predict physical fields of different sheet metal forming processes. Furthermore, Senn et al. [6] used a neuronal network to predict physical fields of a cup drawing process in dependency of the current time step. Moreover, Pfrommer et al. [7] proposed a deep multilayer perceptron as a metamodel for process optimization in manufacturing. Whereby, the network was applied to predict shear angle fields of a composite textile draping part as a function of 50 process parameters based on 584 training samples. However, beside these previously used dimension reduction techniques and deep multilayer perceptrons, recent publications suggest to apply CNNs as metamodels of physical processes. The capability to consider and recognize spatial information is especially promising in focusing on physical processes.

In order to predict results from FE simulations, Zimmerling et al. [8] proposed to represent all results as images in a Cartesian grid so CNNs with convolution and transposed convolution layers could be used for metamodeling. Thereby, the metamodel predicts the resulting shear angle fields of a composite textile draping process depending on different geometries to be formed using 10,000 training samples. Furthermore, Jiang et al. [9] suggested that a cGAN is able to predict a stress field of a two dimensional structure in dependence of loads and boundary conditions based on FE simulation data 38.400 FE simulation samples. Attar et al. [10] applied CNNs with U-Net architecture developed for image segmentation (see [11]) to separately predict displacement and thinning fields of a hot sheet metal forming process. Thereby, to develop fast metamodels during component design, critical radii and drawing depth of a square cup tool geometry were varied. Results show good predictive precision of the metamodels trained with 1.800 samples. In line with that, Zhou et al. [12] also applied CNNs with U-Net architecture to predict FE simulation results of a sheet metal forming process. Thereby, besides varying radii of the square cup and varying binder force additionally, sheet thickness has been varied to generate 1.080 samples.

Existing literature shows that applying CNN based approaches leads to good performance in both precision and generalization of the trained metamodels. Thereby, two metamodels were trained separately to predict displacement and thinning fields. Nevertheless, the amount of data used for training differs in literature. Previous work demonstrated that cGANs achieve high performance in computer vision applications [13], in modelling aerodynamic data [14] as well as in predicting stress fields of FE simulations with high accuracy [9]. Therefore, in this study one cGAN architecture is used to train a metamodel that predicts both displacement and thinning fields of FE results as a function of process parameters of a sheet metal forming process. Furthermore, the effect of amount of data used to train the metamodels on predictive power will be investigated.

2. Method

cGANs are commonly used for tasks in computer vision, for instance image generation or image to image translation [13]. In comparison to conventional GANs [15], a condition is added to the training process to specify the picture generated by the GAN. Using cGAN architectures to predict results of FE simulations in sheet metal forming requires to represent results as images (see [8–10, 12]). Hence, results have to be structured in a Cartesian grid with a same number of pixels for each result.

2.1. *Application example*

The methodology was explored on the example of a sheet metal forming process with a cross cup geometry. Since the cross cup is double symmetric, only one quarter of the geometry is modeled in FE simulation, see Figure 1.

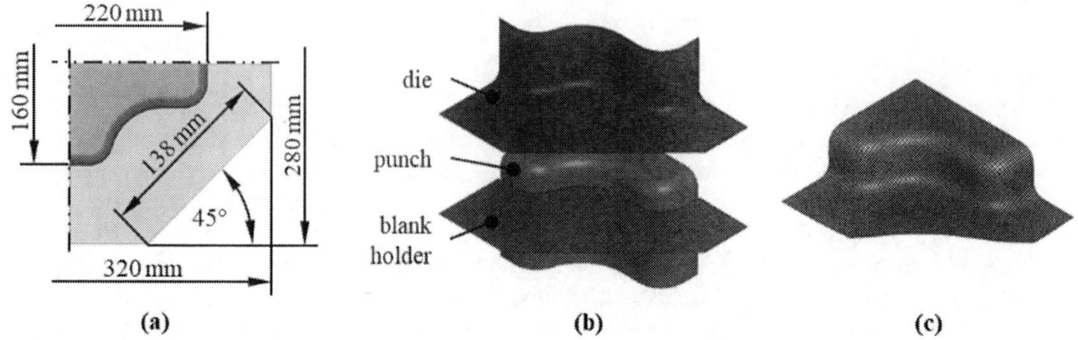

Figure 1. Form and dimensions of double symmetric (a) blank and punch of cross cup, (b) tool, and (c) exemplary deep drawn part.

2.2. *FE simulation model*

LS-DYNA explicit solver was used for numerical simulations of the deep drawing process. The blank was modeled with fully integrated Belytschko-Tsay shells. The tool was modeled with rigid shell elements. The blank mesh contains 4,844 elements, the die was meshed with 5,586 elements, the punch with 5,920 elements and the blank holder with 2,870 elements. The material used in this study was standard deep drawing steel DC01. The stress-strain curve $\sigma(\bar{\varepsilon}_p)$ in Figure 2 was used for modeling the material behavior during forming with material type 24 in LS-DYNA. Isotropic material behavior was assumed. Coloumb's law of friction with friction coefficient μ was applied for the contact between blank and tool.

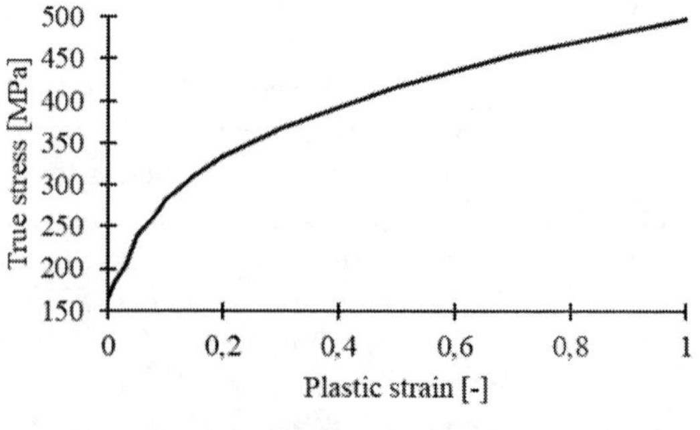

Figure 2. Stress-strain behaviour $\sigma(\bar{\varepsilon}_p)$ of DC01.

2.3. *Data acquisition and preparation*

The input space covers five parameters. Beside the process parameters drawing depth h and blankholder force F_{bh}, boundary conditions were taken into account by varying the friction coefficients μ and different material properties were considered by varying the initial sheet thickness s_0 and scaling the stress-strain-curve with a factor f_s:

$$\hat{\sigma}(\bar{\varepsilon}_p) = f_s\,\sigma(\bar{\varepsilon}_p) \tag{1}$$

To investigate the predictive power of the metamodel, large parameter ranges were used to ensure high variance of displacement and thickness fields, see Table 1.

Table 1. Parameters with ranges for data generation.

Parameter	Initial sheet thickness	Drawing depth	Blankholder force	Friction coefficient	Scale factor for $\sigma(\bar{\varepsilon}_p)$
Symbol	s_0	h	F_{bh}	μ	f_s
Unit	mm	mm	kN	-	-
Minimum	1.2	45	1	0.05	0.8
Maximum	1.8	80	360	0.20	1.2

To generate data for training the metamodels, a Latin hypercube sampling (LHS) was used. A total number of 10,000 samplings were created. As a first step of preprocessing data, 149 samples were excluded because of outliers and error terminations in FE simulation. 9,851 were stored in the database for training. In a second step, element results of the FE simulations were averaged to the corresponding nodes. Since FE simulations in solid mechanics mostly result in unstructured grids, an interpolation to a uniform Cartesian grid is required to use an image based cGAN. The results considered in this study contain node coordinates $\boldsymbol{x} = (x, y, z)$, node displacement $\boldsymbol{u} = (u_x, u_y, u_z)$ and averaged element thickness s at each node of the blank and the parts of the tool, respectively. This information is available for the configuration before and after forming. Linear interpolation in SciPy [16] was used to interpolate the results $\boldsymbol{u}$ and s into the Cartesian grid with 256×256 pixels. This results in a pixel size of 0.625 mm. Areas on the Cartesian grid without blank or tool geometry were set to -1 and channels were normalized between 0 and 1, see Figure 3.

2.4. Metamodel

In general, GANs consist of a generator network and a discriminator network [15]. During GAN training, these networks are trained based on a minimax game, in which the generator learns to generate images as similar as possible compared to the training data. The discriminator learns to distinguish between real and fake images. Thereby, both networks are trained alternately until Nash equilibrium is reached.

Isola et al. [17] proposed a general solution for image to image translation problems with cGANs. It is based on a U-Net (see [11]) for the generator and a patch wise discriminator. Thereby, only patches of a full generated image are evaluated to be real or fake. The architecture of the generator is extended by adding residual squeeze and excitation (RES-SE) blocks in the bottleneck of the U-Net (see [18, 19]), because this architecture for the generator achieved good results when working with FE simulation data [10, 12]. Furthermore, SE blocks [19] are used to recalibrate the channels of the skip connections and in the decoder of the generator to improve generalization capability [20], see Figure 3. The reduction ratio for all SE blocks was set to 8. In this study the discriminator architecture is based on [17]. The input pixel grid consists of the target values with four channels (256,256,4) and the conditional information with three channels (256,256,3). The scalar values are concatenated after layer 5, see Figure 3.

Each convolution operation uses a kernel with size of 4, and a stride of 2. Padding was set to 1. Convolution to layer 5 and 6 of the discriminator uses stride of 1 and padding of 0. Kernel size of convolution to layer 6 of the discriminator was set to 1. Additionally, after each convolution and transposed convolution operation ReLU was chosen as activation function for the generator. LeakyReLU with a negative slope of 0.2 for values < 0 was used for the discriminator, respectively. Except for the last layers of generator and discriminator, there was hyperbolic tangent (tanh) activation function set for generator and no activation function for discriminator, respectively. To provide noise into the training of the generator, dropout with a probability of 0.5 for an element to be zero was applied after transposed convolution to layer 7.

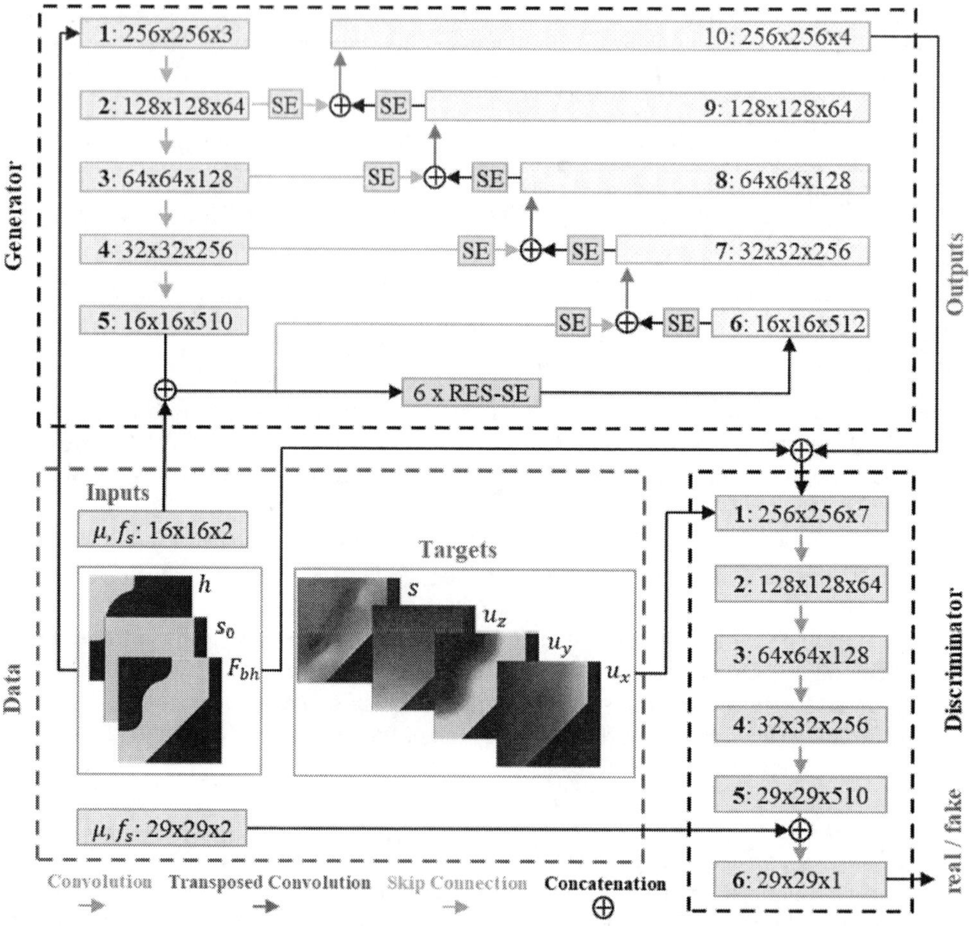

Figure 3. Inputs, outputs and architecture of the discriminator and generator of the cGAN.

The inputs for the metamodel are represented in both in an image and in a scalar manner. The initial sheet thickness s_0, drawing depth h and blankholder force F_{bh} are represented as images in the top view of the blank, punch and blankholder, respectively. This results in a pixel grid with three channels (256,256,3). Friction coefficient μ and the scale factor f_s of the stress-strain-curve were added to the feature map, see Figure 3. Thereby, each channel contains 16^2 equal values of μ and f_s for the generator and 29^2 equal values for the discriminator, respectively. Targets and outputs are represented as an image in form of a pixel grid (256,256,4) with four channels (u_x, u_y, u_z, s). The cGAN is implemented using PyTorch [21]. In order to train the cGAN Adam algorithm was used with a learning rate of 0.00025 and a batch size of 128. The loss function from [17] for cGAN was used to alternately train the generator and the discriminator.

3. Results and discussion

The cGAN was trained with a different number of training samples and corresponding epochs to achieve the same number of training steps according to Table 2. Calculation was executed on one GPU (NVIDIA Quadro RTX 5000) (approx. 4.5h for each metamodel).

Table 2. Number of samples used for training of cGAN.

Samples	**125**	**500**	**2,000**	**8,000**
Epochs	16,000	4,000	1,000	250
Best epoch	14,385	3,971	349	186

The best metamodel was chosen based on generator loss to generate the results shown in Figure 4. Figure 4 depicts randomly picked predictions for three examples: standard part without failure, part with high thinning, and part with wrinkles. Thereby, high thinning and wrinkling are represented in the dataset as a course of the large process parameters chosen in for the sampling. Predictions show that the networks could generally be able to predict deep drawing results for standard parts without any failure as well as for parts with high thinning or wrinkling.

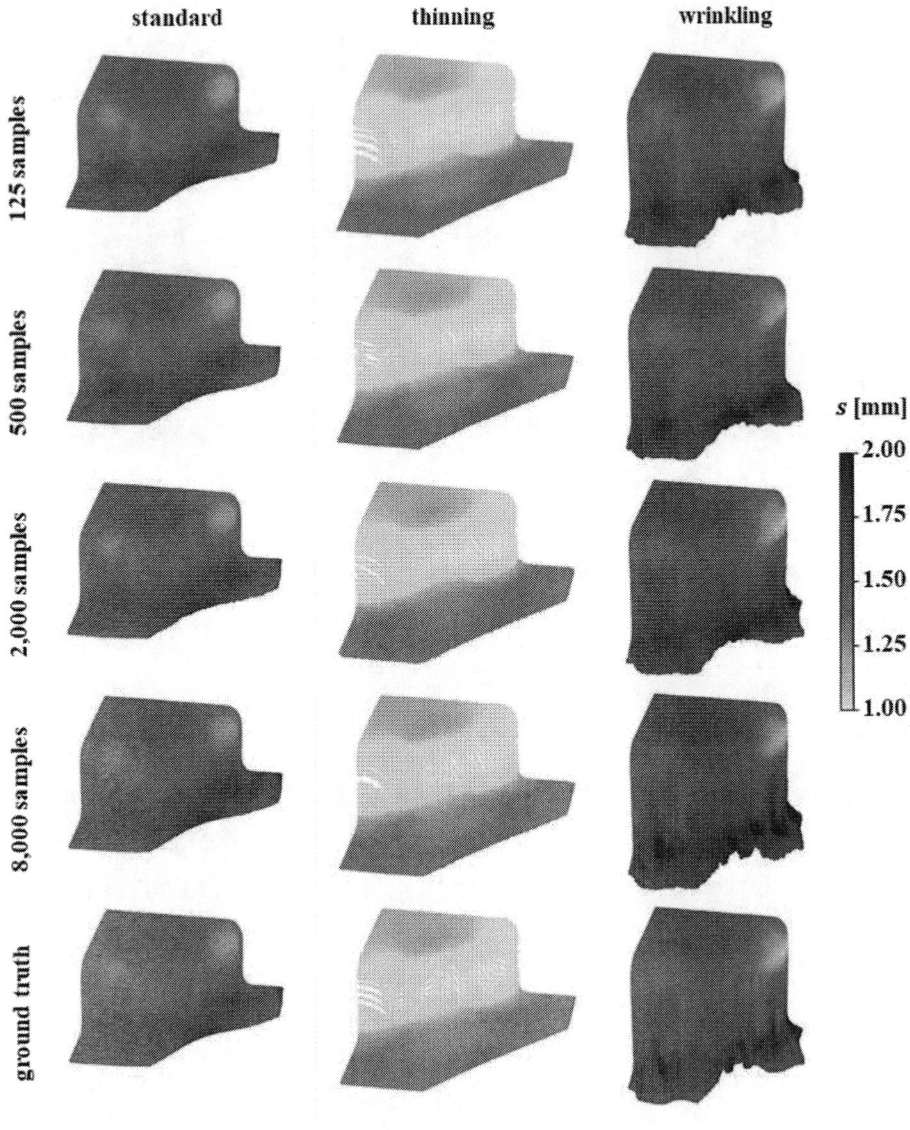

Figure 4. Three exemplary predictions from the test set (deformation and thickness) of the four trained metamodels and the corresponding ground truth.

Furthermore, the network achieved good predictions for standard parts and parts with high thinning, even with only 125 samples for training. Moreover, the draw-in is also mainly predicted correctly in examples shown in Figure 4 and Figure 5, even for small data sets. Opposed to that, the prediction of wrinkling only tended to be acceptable when using 8,000 samples for training, see Figure 4 and Figure 5 (a) and (c). Similar to that, 8,000 samples were needed for an acceptable prediction of the thickness field of the part with wrinkles.

In order to show the performance of the cGANs, Figure 5 depicts additional exemplary predictions with high and low predictive power of parts were high thinning and wrinkling occurred. The high predictive power results are mostly independent of the data used for training, except for predicting wrinkles. In contrast, the results with low predictive power show dependency on data used for training and blurriness.

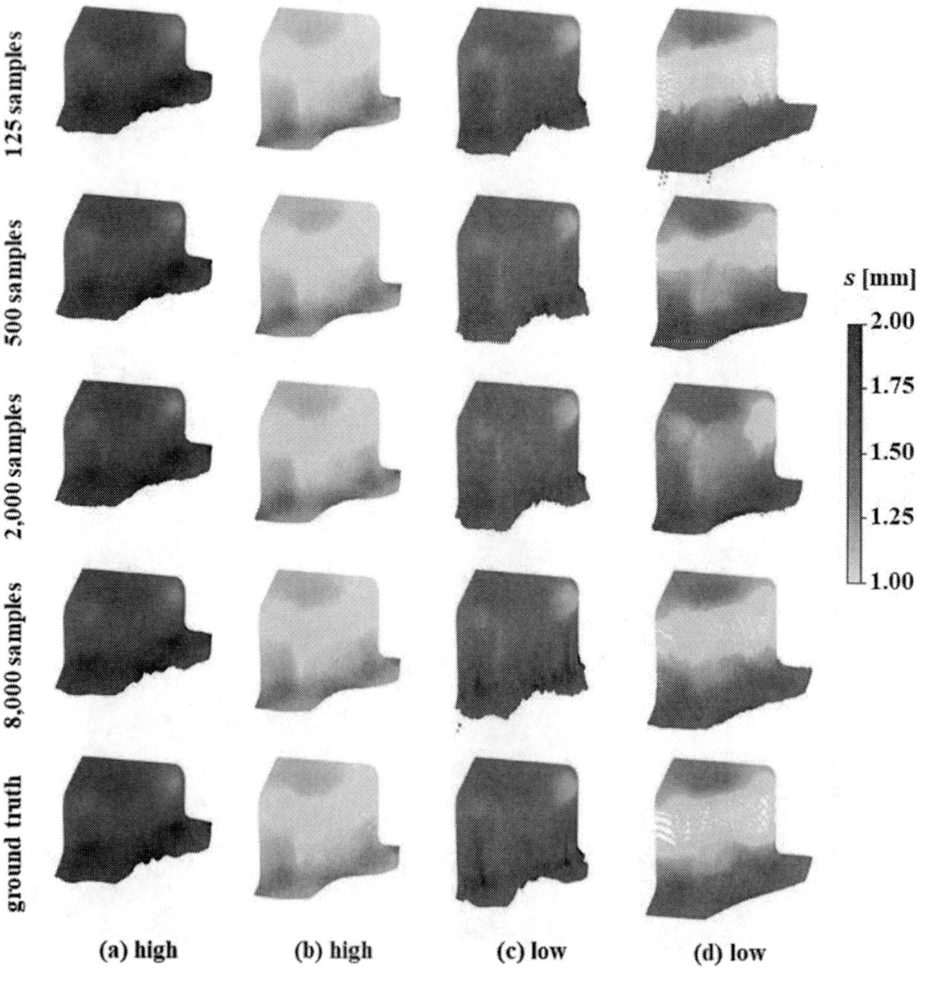

Figure 5. Exemplary predictions with high and low predictive power.

In general, most predictions were of high predictive power, as shown in Figure 5 (a) and (b). Nevertheless, some predictions were only of lower predictive power that can be observed in Figure 5 (c) and (d) showing deviations between predictions and ground truth. However, the overall high predictive power of the models can be seen in Table 3. Table 3 shows the average over all 1,951 test samples of the mean absolute error (MAE) between all pixels of the ground truth and the predictions from the metamodels. This shows, that with increasing number of training samples the deviation can be reduced. The largest effects are shown for u_y and u_z. Moreover, the values for z displacement are comparably high. In comparison to that, the average MAE of the sheet thickness is small and does not significantly decrease with increasing number of samples.

Table 3. Average MAE over all pixels for all four trained metamodels on test set with 1,951 samples.

Samples	**125**	**500**	**2,000**	**8,000**
Average MAE u_x [mm]	0.482	0.391	0.406	0.364
Average MAE u_y [mm]	0.727	0.584	0.578	0.504
Average MAE u_z [mm]	0.870	0.724	0.688	0.649
Average MAE s [mm]	0.027	0.024	0.024	0.022

The predictions as well as the performance indices of all four networks show overall good results. The comparably high average MAEs for all four channels may result from the large range of the process parameters and from the consequential samples with high thinning as seen in Figure 5 (d). Furthermore, it is also possible that there are some parameter combinations of the test dataset that caused large deviations between predictions and the ground truth, for instance the predictions from the cGAN trained with 500 and 2,000 samples in Figure 5 (d). Since the accuracy of predictions always depends on the application, the suggested approach is suitable to predict FE results of a sheet metal forming process with cGAN training and the suggested architecture in general. The predictions of the metamodels in dependency of the amount of training data show that even with very small datasets a good quality prediction can be made for parts without failures. In this context, it is questionable how robust the predictions of the metamodels are against a test dataset containing more than 1,951 samples, as neuronal networks can respond very sensitively. The metamodels based on small datasets are only able to predict tendencies of occurring wrinkling. To address these effects, further investigations on different architectures of both generator and discriminator have to be conducted. Especially, the complexity of the generator in comparison with the discriminator is of high importance to ensure maximum benefit of the adversarial training.

4. Conclusion

The investigations prove that cGANs are capable of predicting deep drawing results. The proposed architecture achieves good results for calculating sheet draw-in and thinning based on a small amount of training data. The authors show that the prediction quality of wrinkling depends much stronger on the amount of available data for training than does the forecasting of thinning and draw-in. If wrinkles are to be accurately mapped, more data is required than if only the tendency towards wrinkling is to be predicted. Future research will include adapting the architecture of the generator and discriminator. In addition, a more detailed investigation of the generated metamodels will take place.

Acknowledgements

We gratefully acknowledge the funding provided by the Fraunhofer Society as part of the lighthouse project *Machine Learning for Production (ML4P)*.

References

[1] Weichert D, Link P, Stoll A, Rüping S, Ihlenfeldt S and Wrobel S 2019 A review of machine learning for the optimization of production processes *Int J Adv Manuf Technol* **104** 1889–902
[2] Bonte M H A, van den Boogaard A H and Huétink J 2008 An optimisation strategy for industrial metal forming processes *Struct Multidisc Optim* **35** 571–86
[3] Allwood J M, Duncan S R, Cao J, Groche P, Hirt G, Kinsey B, Kuboki T, Liewald M, Sterzing A and Tekkaya A E 2016 Closed-loop control of product properties in metal forming *CIRP Annals* **65** 573–96

[4] Schwarz C, Ackert P and Mauermann R 2018 Principal component analysis and singular value decomposition used for a numerical sensitivity analysis of a complex drawn part *Int J Adv Manuf Technol* **94** 2255–65
[5] Schwarz C, Link P, Ihlenfeldt S and Drossel W-G 2021 Application of Fourier-related data reduction methods in sheet metal forming *Procedia CIRP* **99** 260–5
[6] Senn M, Jöchen K, Van T P, Böhlke T and Link N 2013 In-depth online monitoring of the sheet metal process state derived from multi-scale simulations *Int J Adv Manuf Technol* **68** 2625–36
[7] Pfrommer J, Zimmerling C, Liu J, Kärger L, Henning F and Beyerer J 2018 Optimisation of manufacturing process parameters using deep neural networks as surrogate models *Procedia CIRP* **72** 426–31
[8] Zimmerling C, Trippe D, Fengler B and Kärger L 2019 An approach for rapid prediction of textile draping results for variable composite component geometries using deep neural networks *AIP Conference Proceedings* **2113** 20007
[9] Jiang H, Nie Z, Yeo R, Farimani A B and Kara L B 2021 StressGAN: A Generative Deep Learning Model for Two-Dimensional Stress Distribution Prediction *Journal of Applied Mechanics* **88**
[10] Attar H R, Zhou H and Li N 2021 Deformation and thinning field prediction for HFQ® formed panel components using convolutional neural networks *IOP Conf. Ser.: Mater. Sci. Eng.* **1157** 12079
[11] Ronneberger O, Fischer P and Brox T 2015 U-Net: Convolutional Networks for Biomedical Image Segmentation *Medical Image Computing and Computer-Assisted Intervention – MICCAI 2015 (Lecture Notes in Computer Science)* ed N Navab *et al* (Cham: Springer International Publishing) pp 234–41
[12] Zhou H, Xu Q, Nie Z and Li N 2022 A Study on Using Image-Based Machine Learning Methods to Develop Surrogate Models of Stamp Forming Simulations *Journal of Manufacturing Science and Engineering* **144**
[13] Creswell A, White T, Dumoulin V, Arulkumaran K, Sengupta B and Bharath A A 2018 Generative Adversarial Networks: An Overview *IEEE Signal Process. Mag.* **35** 53–65
[14] Hu L, Zhang J, Xiang Y and Wang W 2020 Neural Networks-Based Aerodynamic Data Modeling: A Comprehensive Review *IEEE Access* **8** 90805–23
[15] Goodfellow I, Pouget-Abadie J, Mirza M, Xu B, Warde-Farley D, Ozair S, Courville A and Bengio Y 2014 Generative Adversarial Nets *Advances in Neural Information Processing Systems* ed Z Ghahramani *et al* (Curran Associates, Inc)
[16] Virtanen P, Gommers R, Haberland M and Hilboll A 2020 SciPy 1.0: Fundamental algorithms for scientific computing in Python *Nature methods*
[17] Isola P, Zhu J-Y, Zhou T and Efros A A 2017 Image-To-Image Translation With Conditional Adversarial Networks *Proceedings of the IEEE Conference on Computer Vision and Pattern Recognition (CVPR)*
[18] He K, Zhang X, Ren S and Sun J 2016 Deep Residual Learning for Image Recognition *Proceedings of the IEEE Conference on Computer Vision and Pattern Recognition (CVPR)*
[19] Hu J, Shen L and Sun G 2018 Squeeze-and-Excitation Networks *Proceedings of the IEEE Conference on Computer Vision and Pattern Recognition (CVPR)*
[20] Rundo L *et al* 2019 USE-Net: Incorporating Squeeze-and-Excitation blocks into U-Net for prostate zonal segmentation of multi-institutional MRI datasets *Neurocomputing* **365** 31–43
[21] Paszke A *et al* 2019 PyTorch: An Imperative Style, High-Performance Deep Learning Library *Neural Information Processing Systems* **32**

A data-driven methodology for separately quantifying the effects of tool wear of upper and lower tool on the quality of cut surfaces in shear cutting processes

S Niessner[1] and M Liewald[2]

[1] TRUMPF Werkzeugmaschinen GmbH + Co. KG, Johann-Maus-Straße 2, 71254 Ditzingen, Germany
[2] University of Stuttgart, Institute for Metal Forming Technology, Holzgartenstr. 17, 70174 Stuttgart, Germany

E-Mail: stephan.niessner1@trumpf.com

Abstract. Shear cutting processes are characterised by physical parameters that change in time and space because of influencing process effects. Those parameters partwise can be measured and recorded by sensors and assigned to causal process effects through detailed evaluation based on analytical models. If the analytical model can classify the causal process effects in detail, the quality of the processed sheet metal component can be monitored on the basis of knowledge about tool wear influences and so a permanent part quality only can be ensured by a preventive tool change. In detail, the quality of the cut surface of a shear cut component to be assessed in this process is influenced by both the tool wear effect of the upper and lower tool, whereby both in fact show a significant influence on the different dimensional and surface characteristics of the cut surface. Analytical models for such analysis often are not available, or, if available, it may deliver only proximate results that needs improvement for the respective process effects. Previous procedures for such prediction purpose are frequently based on large and complex neural networks that require as well as process a huge amount of data.

In this regard, a novel evaluation method has shown great potential for data-driven quantification of tool wear effects and looks capable of far outperforming previous process evaluation methods in its performance under different process conditions. In previous studies, specific upper and/or lower tool wear was not considered in a separate manner and a detailed inspection of the finished cut surface there for could not be carried out. This paper investigates how a data-driven approach can be used to evaluate specific wear effects of upper and lower tools in a differentiated manner, and in which degree of detail the main quality characteristics of the cutting surface in a shearing process thus can be assessed.

1. Introduction

Due to the expected increase in complexity in production and the customer requirement to deliver specific products in ever shorter production times, operators must be increasingly supported by adaptive assistance systems in the execution of their work. This calls the continuous enhancement of the degree of automation of corresponding manufacturing processes. One of the leading tasks in production technology is to continuously develop digital networking and make it even more economical, precise and future proof. By collecting and analysing process-related data in the immediate manufacturing process, it is possible to observe important information about the current quality of the manufactured parts at any time through data-driven monitoring systems, or even to use it for the implementation of adaptive control strategies. If the analytical model is able to classify the causal process effects in detail, the quality of the processed sheet metal component can be monitored on the basis of knowledge about tool wear influences and so a permanent part quality only can be ensured by a preventive tool change. Through the targeted use of assistance systems, both the flexibility of a system and its efficiency can be significantly increased, resulting in a decisive competitive advantage.

In the case of shear cutting processes, the product quality in the manufacturing process is particularly influenced by deviations in the cutting geometry of the tools caused by built up edges, rounding and edge breakouts [1]. The influences of the cutting geometry deviations, that negatively affected the area of the cut surfaces, can be detected with suitable measurement technology in temporal and spatial changes of the characterising physical process variables. The machine tool TRUMPF TruPunch 5000 used here sets new standards in terms of productivity and process reliability. In this respect, up to 1,600 cutting operations per minute are realised in the machine during machining, which poses special challenges for the evaluation speed.

In this context, time series that are composed of consecutive measurements over time are gaining popularity and are increasingly becoming the focus of data analysis. Time series are therefore the values of variables in a temporal sequence. Depending on the number of variables, time series can be divided into univariate time series and multivariate time series. The added value of the acquired process signals depends on the signal quality, the measurement concept and the localization [2]. Previous research has shown that sensor signals for the process force and the stroke movement in time series show strong deviations at constant process conditions [3]. The main reasons are time delays due to the drive concept, influences by machine dynamics and cycle times in the machine control. As cumulative disturbance influences, these lead to latency times that cause deviations of more than 30 per cent in the time series data of the sensor signals under constant conditions. The bandwidth of these disturbances prevents a classification use of models that are not very computationally intensive and can be trained fast or do not have to be pre-computed, such as the KNN algorithm. Previous methods, with which a classification for this prediction purpose was possible, are based on large and complex neural networks that require and process a huge amount of data. Current approaches therefore require enormous computing capacities and long computing times for the model training with slightly better performance [4]. Furthermore, they make an embedded approach in the machine control of the machine tool impossible and because of the extensive model complexity it is nearly impossible to explain or reconstruct misjudgement of the model. To solve these problems, a new method based on automatic feature extraction has recently been proposed to quantify the effects of tool wear [5]. By linking the recorded process signals, the inherent information content about the tool wear condition can be recovered, and signal interference is significantly reduced. The linking approach pursued describes a two-stage procedure that considers the physical origin of the individually recorded process signals from the multivariate time series data and combines them into a new physical quantity through smart pairing. The procedure not only reduces the necessary amount of data of the signals enormously, but also surpasses previous work in classification by far. However, the issue of how suitable this proposed method is for separately assessing the tool wear effect of the upper and lower tool from the process signals is still open. Die wear was not considered in this series of tests. For this purpose, each class of the cutting edge preparation on the punch was used in combination with an unused die. However, this differentiated evaluation is necessary if the resulting cut surface quality is to be assessed based on the tool wear effects. In detail, the quality of the cut surface of a shear cut component to be assessed in this process is influenced by both the tool wear effect of the upper and lower tool, whereby both in fact show a significant influence on the different dimensional and surface characteristics of the cut surface. This paper investigates how a data-driven approach can be used to evaluate specific wear effects of upper and lower tools in a differentiated manner, and in which degree of detail the main quality characteristics of the cutting surface in a shearing process thus can be assessed. Furthermore, a metamodel is presented from which quality maps can be derived that describe and evaluate the relationship between tool wear and cutting surface quality. According to the classification approach presented, these can be used as a link for a direct quality assessment based on process signals.

IOP Conf. Series: Materials Science and Engineering 1238 (2022) 012065 doi:10.1088/1757-899X/1238/1/012065

2. Overview of time series classification

Comprehensive data collection and analysis are basically no longer novel trends and can be considered as standard practice in many fields. A large part of this practice also dealt with time series data from sensor measurements or other types of monitoring, which are generally the most significant data sources for automation and the vision of "Industry 4.0". However, due to the nature of time series data; high dimensionality, large data size – especially for multivariant time series, this often remains in a difficult problem. Learning of representations and classification of time series continue to receive attention and can be divided into three categories according to the classification scheme: model based, distance based, and feature based [6].

2.1. Distance based classification

The first category is based on distance. After defining a distance function, the similarity between two time series is determined. Thus, the key point of these approaches is how to define the distance function. The most significant representatives of this category are the KNN and the support vector machine (SVM) classifier that have already been used successfully in a wide variety of applications [7] [8] [9] [10]. Both offer the possibility to use them directly for classification on the raw data. The basic idea in using SVM is to map a sequence in a feature space and find the hyperplane with the largest distance to separate two classes. The kernel function specifies the higher-dimensional feature space. Depending on the selected kernel function, SVM can be used for two categories, feature extraction and measure a distance [6]. The KNN classifier is a lazy learning method and does not pre-compute a classification model. For the KNN classifier, a variety of possibilities to define the distance function are already known [11]. However, this method has the serious disadvantage that the length of the two series must be equal, and it is sensitive to distortions in the time dimension [6]. To deal with this disadvantage, an alternative distance function is proposed, the DTW (Dynamic Time Warping) distance [12]. For this purpose, the sequences are non-linearly "warped" in the temporal dimension to determine a measure of their similarity independent of certain non-linear deviations in the temporal dimension. An overview of the different classifiers with the aim of classifying time series using a distance-based approach as well as a discussion of the strengths and weaknesses of each method can be found in [13]. Overall, it has been shown that compared to a variety of classifiers, such as neural networks, SVM and HMM, the KNN-classifier with dynamic time-warping distance generally has higher accuracy [7]. Today these approaches are commonly used until they fail with long or noisy time series.

2.2. Model based classification

The second category of time series classification approaches is based on models. The basis for this is the assumption that the time series in a class are generated by an underlying model, and then the model parameters are adjusted and determined by the training examples in that class. As a result, different models are generated for the respective classes. Subsequently, new time series can be compared with the models to determine to which class it belongs. The best-known representatives in this category are autoregressive models [14] [15] [16], the Markov models and hidden Markov models [17] [18].

2.3. Feature based classification

The selection of temporal features is a difficult task, as numerical data does not allow for an enumeration of features and is the key aspect of feature-based classification in time series data. Therefore, these methods are mainly divided into two categories: handcrafted feature methods and learned feature methods. An overview of feature based classification algorithms for multivariate time series represented in [19]. The principles and procedures of these methods are showed, and the advantages and disadvantages of each method are discussed. In the category of learned features, neural networks, especially those based on deep learning, have gained popularity in recent years [20] [21] [22]. One of the deep learning architectures successfully used in computer vision is convolutional neural networks (CNN). Besides their original domain of computer vision, CNNs are increasingly used in the classification of time series data with good performance [23]. Inspired by the CNN-framework for image recognition, these approaches are based on the transformation of a multivariate time series into a vector by singular value decomposition and other transformations [24]. CNNs can discover and extract the appropriate internal structure to automatically detect and generate deep features of inputs through convolution and pooling operations. In contrast, handcrafted methods require experience and domain knowledge, and are a complex and challenging task, which kindles the popularity of learned feature methods.

3. Experimental setup

3.1. Machine design and tooling concept

In the investigations in this paper, a hydraulically driven punching machine with a flexible tool system was used. The tool concept on upper (a) and lower (b) side are shown in figure 1. The force input into the punching process is via a hydraulic unit. For the vertical movement of the stroke, pressure is applied to the two differently sized surfaces on the upper and lower side of the ram in the punching head of the machine. Depending on the pressure ratio applied, the ram moves up or down.

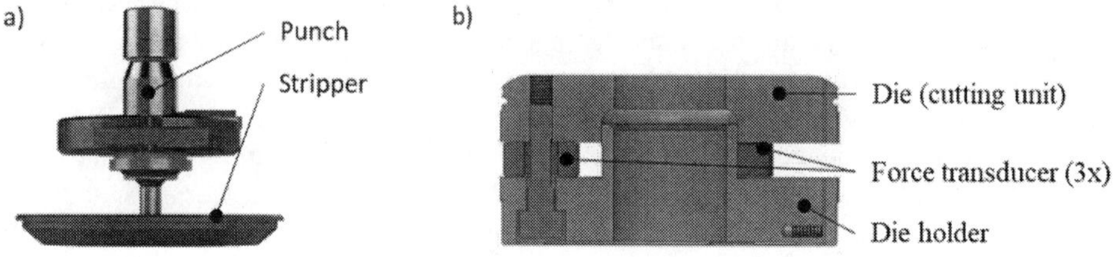

Figure 1. a) Tooling concept on upper side; b) Sectional view of the segmented die on the lower side

3.2. Applied Material

The test material investigated in this study was the deep drawn steel DC01 (EN-AW 1.0330) having nominal sheet thickness of 1.0 mm and 3 mm and a sheet format of 1000 mm * 1000 mm. To characterise the mechanical properties of the material, uniaxial tensile tests were carried out with various orientations to rolling direction. The resulting mechanical properties are shown in table 1. The lowest value of ultimate tensile strength occurred against rolling direction. Diagonally to the rolling direction, the value was significantly higher.

Table 1. Mechanical properties of the test material gained by tensile tests.

Material	Rolling direction [°]	E-Modul E [MPa]	Yield strength $R_{p0,2}$ [MPa]	UTS R_m [MPa]	Uniform elongation A_g [%]	Fraction elongation A_{100} [%]
	0	201027	184	308	23.0	40.8
EN-AW 1.0330	45	217253	202	325	20.5	36.5
	90	207657	193	304	22.4	39.2

3.3. Measurement methods and setup of the experiments

The data sets used in the experiments were a selection of three sensor signals captured at a fixed sampling rate between 2 and 20 kHz for each sensor in the manufacturing process. The recorded signals were collected by a measurement amplifier and were recorded time synchronously. The sensor signals used include the process force, the movement of the tool and the suspension behaviour of the machine frame. The tool movement, called stroke movement of the punch, was measured using a linear measuring system on the ram rod and were sampled at 2 kHz. The force measurements of the required force F_{Cut} for the cutting process were carried out via three rotationally symmetrically arranged load cells (KISTLER type 9021A), that were integrated in the die between die holder and cutting unit. The construction of the segmented die for force measurement is shown in figure 1 b). The piezoceramic sensors were installed with a preload force of 20 percent of the measuring range and were sampled at

20 kHz. For measuring the elastic components of the machine and the tools, strain gauges were applied to the C-Frame of the machine. The strain gauge sensors were calibrated in advance to reflect the resulting distance between punch and die, which occur due to the softness in the machine design and the tool components because of the applied process force, as a function of the measured strain in the C-frame.

For the entire investigation, round punches with a diameter of 8 mm were used, whereby the cutting clearance to the die was always 10 percent of the sheet thickness. For the wear classification, the individual classes of upper and lower tools are considered individually with a non-preparation counterpart, as well as continuous combined stages for the respective classes of the two tools. This results in a total of 13 different combinations that are considered within the scope of the examination. For each class of wear conditions investigated, 100 machining operations were performed on a separate sheet. The blank holder was positioned 0.6 mm above the surface of the sheet and acted only as a stripper. The investigated edge preparation of the tools for the experiments are summarised in table 2.

Table 2. Summary of the investigated cutting edge preparations.

Factors	Settings
Cutting edge punch (x_P mm * 60°)	Chamfer x_P: 0.000, 0.250, 0.500, 0.750, 1.000
Cutting edge die (x_D mm * 60°)	Chamfer x_D: 0.000, 0.025, 0.050, 0.075, 0.100

4. Data pre-processing and feature extraction

4.1. Data pre-processing

First, a pre-processing step is required in which the raw data is cropped to obtain a multitude of equal samples for each class of tool wear from the recorded data set. The time interval of the multivariate time series data to be cropped is triggered by the initial stroke movement of the tool. The end is defined by a fixed period, and the start value of all cropped intervals are reset to zero. If a predefined limit value is reached, all process signals within these common time stamps are cropped out. This converts a long time series with sequential cutting processes into many individual data sets. Each of these data sets contains exactly one cutting process that always starts at the same time.

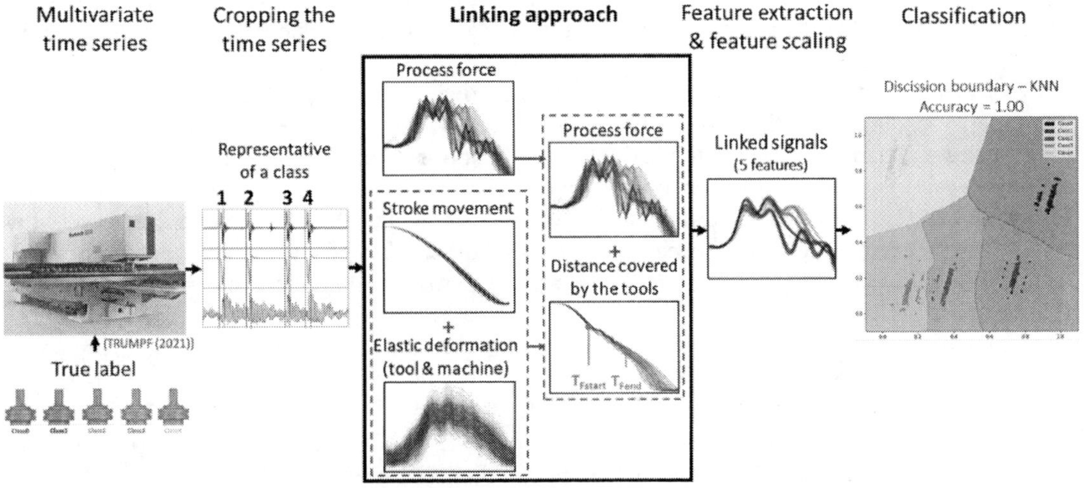

Figure 2. Flow chart of the used linking approach [5].

The linking approach proposed in [5] is used for the investigation, which starts from the generally known assumption that the mechanical work required for identical processes must also be identical. In the literature, for mechanical manufacturing processes, the representation of the energy transferred to or from the tool by the application of force along the stroke movement is recommended for this intention

[25]. For an actual implementation of this proposal, all tool and machine components in the force flow that are elastically deformed along the stroke movement must be considered. The implementation of the linking approach was carried out in a two-stage process and is shown in figure 2.

In the first step, the elastic components of the machine and the tool are extracted from the measured stroke movements on the ram rod. For the implementation of the first step, the considered time series must be available in the same sampling frequency, i.e. for each of the common time values in the multivariate time series, a measured value must be stored. For the execution of this requirement and for the consideration of non-linear effects in the signals, the series, which was recorded at a sampling rate of 2 kHz, is increased in their point density by a factor of ten using the scikit-learn class splinetransformer. The result is the actual distance covered by the upper and lower tools in relation to the workpiece to be processed. In the second step, the signals are converted into the mechanical work, i.e. the energy transferred to or from the tool via the application of force along the stroke movement.

4.2. Feature extraction

In the next step, the features are extracted from the cropped data sets. Good features for classification are characterised by the fact that the process is described completely and the changes in the signals that can be found within the classes to be recognised are emphasised as far as possible. In this context, the features of S_{Fstart}, S_{Fend}, S_{Fi50}, S_{Fd50}, S_{Fi80} and S_{Fd80} were found to be excellent representatives, where the corresponding measured values show the smallest variance within a class. The start S_{Fstart} and end value S_{Fend} are determined at 10% of F_{max} occurring in the increase and decrease. The extraction of the other features follows the same principle. The values in the increase and decrease were extracted at 50 and 80 percent of the maximum process force (i = increase, d = decrease), which was necessary for the respective cutting process. These values have proven to be a good trade-off to keep large differences as small as possible and still be able to describe the entire process to the extent possible. An additional significant feature is the work W_{total}, that must be expended for separating the sheet. It is equal to the area under the signals that lies between the values S_{Fstart} and S_{Fend}. A total of 7 features are thus extracted from the multivariate time series for each cutting process.

5. Experiments and results

5.1. Classification of wear effects

After defining the features, a normalisation is carried out in a final step. For uniformly distributed data, minmax scaling is generally recommended. First, the entire dataset is divided into training and test data, as the test data should only be scaled to the min-max values of the training data. The ratio of training to test data is set at 80 to 20. To visualise the decision boundaries of the model, the 7 features are reduced to two dimensions by a dimension reduction step with principal component analysis (PCA). The sum of the variance explained by the two remaining components is over 0.99. The minmax scaler, which is used, scales and translates each individual feature so that it is given in a range between zero and one in the training set. Therefore, the estimator is fitted to the training data and then transferred to the test data. Hence, the feature values of the test data can be larger or smaller than the specified interval between [0,1]. This procedure is preferred because it guarantees the independence of the models from the test data. The first investigation was carried out for a DC01 material with a sheet thickness of s = 1 mm. To find the best performing model for the classification, a grid search was implemented, in which the evaluation of the models was carried out using a cross-validation with k-folds = 5. The hyperparameters for the used grid search were weights ∈ [uniform, distance] and n neighbours ∈ [1, 10]. For the implementation of the dimension reduction, the division into training and test data, the normalisation and the classification, the standard methods from the scikit-learn library were used. For all hyperparameters of the applied classes that are not explicitly specified, the implemented default values were adopted. The best performance in the classification is achieved by the model with n neighbours = 3 and weights = "distance" with 95.4 percent and is shown in figure 3 with the decision boundaries as well as the data distribution (o = training, * = test). On the right side of figure 3, the heatmap for the respective tool wear classes has been added. It is easy to see that the misinterpretations of the model are more often found in neighbouring die wear (Class1 - Class4). Due to the ten times smaller chamfers on the cutting edge of the dies (in contrast to the punches), it is also much more challenging for the model to extract the necessary information content for different wear conditions from the process signals. However, the exception is the max. punch-side wear Class8 and the 4th combined wear Class11. By far the falsest imputations are recorded for these two classes. This suggests that a very similar feature image is generated for the 4th class of combined tool wear (Class11) and the max. wear allowance on the punch (Class8). A similar result was obtained for the sheet thickness of s = 3 mm, whereby the accuracy score of the best model with was slightly higher at 96.2 percent. However, the result of 95.4 – 96.2 percent accuracy for 13 wear classes to be classified shows the excellent capability of the presented procedure to evaluate specific wear effects of upper and lower tools in a differentiated manner.

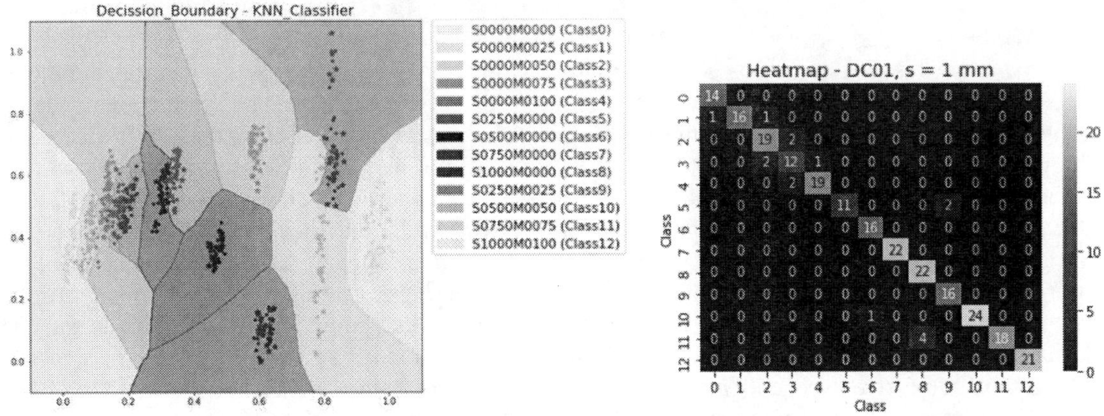

Figure 3. Decision boundary of the classification procedure and the resulting heatmap (s = 1 mm).

5.2. Influence of tool wear effect on the characteristics of the cut surface

Of the one hundred cutting operations processed in a wear class, every tenth perforation was cut out of the processed sheet for sample preparation and grinding for the cut surface assessment. To avoid irregularities and value distortions of the characteristics on the cut surfaces due to deviations in the concentric alignment of the upper and lower tools, it is expedient to characterise the samples taken from the perforations at two opposite points regarding their topology – 20 assessments for each class of tool wear. The VDI 2906 [26] was used to determine the cut surface characteristics, whereby only those parameters were considered that were relevant for the vertical topology structure – rollover height, clean cutting height, fracture height and burr height. Figure 4 summarises the results of the cutting surface measurement with the characteristic micrographs (s = 1 mm) for the respective wear classes.

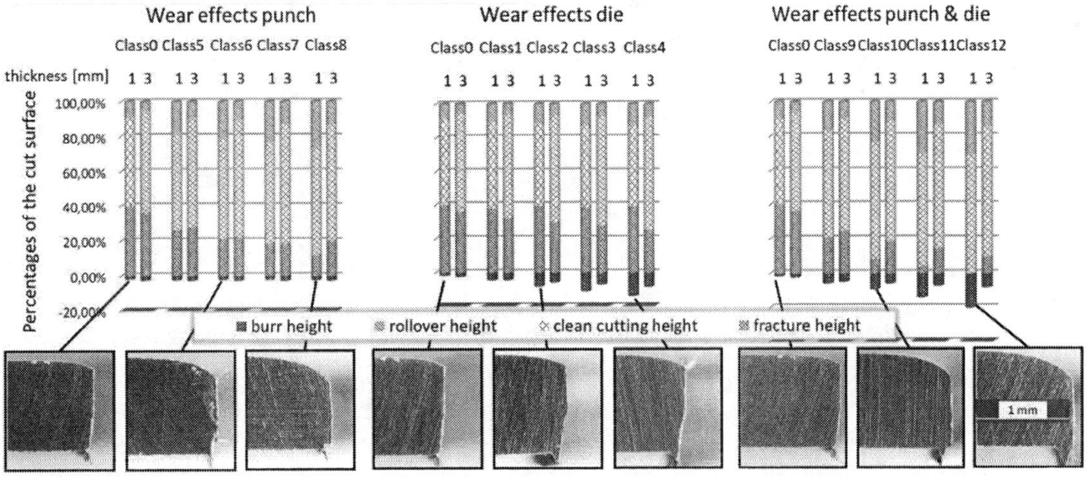

Figure 4. Characteristic parameter distribution of the shear cutting surface, given as a percentage of the sheet thickness. The micrographs shown were obtained for a sheet thickness of s = 1 mm.

For the punch-sided wear, it can be stated unambiguously that with increasing size, it is mainly the rollover height, that increases. In contrast, the resulting burr height remains almost identical across all wear levels. For the wear on the die side, it can first be stated that an increase in the wear effects leads to a strong increase in the burr height. In contrast to the wear on the punch side, where the rollover increased strongly with rising wear effects, the rollover remains almost constant with increasing wear effects on the die. When looking at the cutting surface characteristics for the combined wear, it is first noticeable that the behaviour of the rollover is almost identical to what could be determined for the punch-side wear. The same can be noted for the combined wear conditions in comparison to the wear conditions on the die side for the burr height. Overall, it can also be observed that the influences on the changes in the cut surface are more pronounced with thinner sheet thicknesses than with thicker ones.

5.3. Metamodel and quality plot for the surface quality parameter burr height

The metamodel is intended to provide the still missing bridge between the separate quantification of the effects of upper and lower tool wear from the process signals and the correlated formation of the burr height on shear cut components. For the resulting model, the total characteristic values of the determined burr heights, as determined in chapter 5.2, were entered in a diagram as a function of the two wear conditions on the punch and die. The x-axis shows the wear condition on the punch side and the y-axis the wear condition on the die side and the z-axis represents the burr height. Subsequently, the min. (green diamond), max. (red diamond) and average values (blue diamond) of the measured burr heights were determined. For the modelling of the burr topology, mesh grids are first generated for all three burr levels, which are then approximated with quadratic functions and multiplied in their point density. The result is the min., max., and average burr topology as a function of the wear effects on the upper and lower tools. To derive the quality maps for the ridge height, the isohypsis (height level lines) are entered in 0.02 mm steps in the respective surfaces of the model. From the burr height quality maps, the expected burr heights can be read quickly and easily depending on the different effects of upper and lower tool wear. Figure 5 shows the metamodel and the derived quality maps for a sheet thickness of 1 mm.

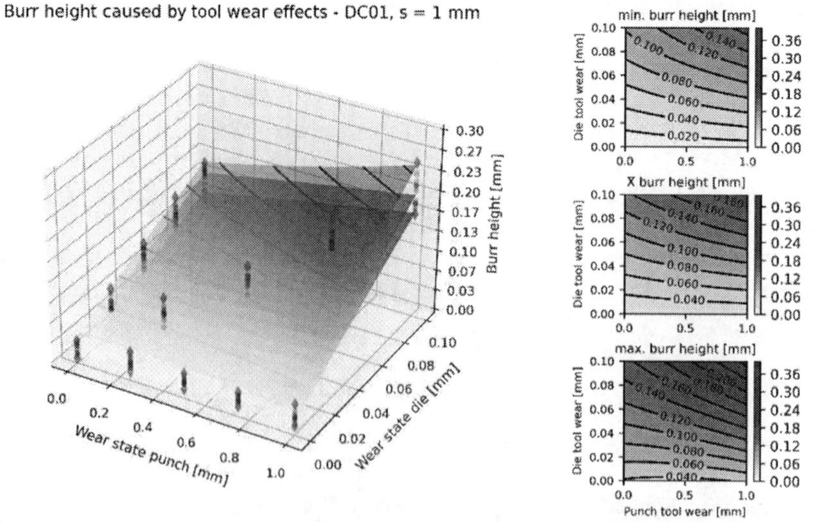

Figure 5. The metamodel for the burr height of the material DC01 and s = 1 mm as a function of the wear effects on the punch and die side is shown on the left and the derived quality maps for the assessment of the burr height on the right.

The metamodel is intended to provide the still missing bridge between the separate quantification of the effects of upper and lower tool wear from the process signals and the correlated formation of the burr height on shear cut components. In principle, this procedure can be applied to each of the characteristic quality parameters and the entire topology of the cutting surface can be described and evaluated after the shearing process. The metamodel and the three derived quality plots for the burr height assessment as a function of the wear effects on both tools for the sheet thickness of 3 mm is shown in figure 6.

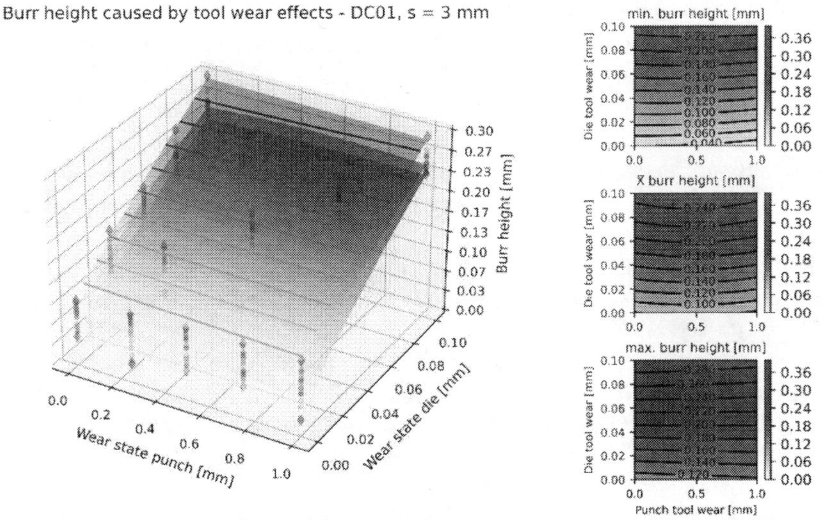

Figure 6. Metamodel for the material DC01 (s = 2 mm) and the corresponding quality maps.

Compared to figure 5, the isohypsis of all three quality maps in figure 6 increase almost in parallel with the wear effects on the die side. While the wear on the punch side still had a significant influence on the burr height with a sheet thickness of 1 mm, the resulting burr height is only influenced by the wear effects on the die side for a sheet thickness of 3 mm. In addition, it can be seen by the higher density of the burr level lines that the burr height increases much more in its absolute value with increasing wear effects on the die side. In reference to the sheet thickness, however, thinner sheets tend to cause higher burr levels.

6. Conclusion

In the experimental investigations carried out as part of this work, a data-driven methodology for separately quantifying the effects of upper and lower tool wear was presented for the first time, and how this can be used to evaluate the cut surfaces quality during the machining process. First, it could be shown that with the linking approach proposed in [5] for pre-processing the process signals, not only the wear condition on the punch side of the tools can be uniquely identified, but also the tool wear effects on the die side can be clearly assigned. Moreover, it could be shown that not only one-sided wear effects can be clearly assigned, but also combined wear effects on the upper and lower tool can be identified with the applied method. Afterwards, the tool wear effects used in the experimental investigations were analysed regarding their impact on the characteristics of the shear cut surfaces. For the distinctive quality characteristics, a procedure for the development of metamodels was presented, which can describe and evaluate the correlation behaviour of tool wear, both on punch and die, separately for the cutting surface quality. The proof for the procedure has been demonstrated exemplarily on the most significant characteristic, the burr height. By extending the study to a further sheet thickness, the recommended method could be validated, and the effectiveness of the approach was demonstrated as part of a proof of concept. This offers the unambiguous opportunity for the first data driven quantification of the quality of shear-cut components during the shear cutting process. It could also be established that the influence of the punch-sided wear effects on the burr height decreases with increasing plate thickness. Currently, the proof of concept has only been applied to one material and one tool contour, therefore it is recommended to falsify the transferability of the proposed method for other materials and tool contours in further investigations.

7. References

[1] Birzer F 1996 *Feinschneiden und Umformen: Wirtschaftliche Fertigung von Präzisionsteilen aus Blech* Verlag Moderne Industrie Band **134**

[2] Kolbe M and Hellwig W 2018 Überwachung von Stanzwerkzeugen. In Kolbe M and Hellwig W *Spanlose Fertigung Stanzen* pp 337-347 Springer Vieweg Wiesbaden

[3] Niessner S Liewald M 2020 Identification of methods for the in-situ measurement of cutting forces in a tool-bound punching machine *IOP Conf. Ser.: Mater. Sci. Eng.* **967** 012025
[4] Nießner S Khanguieldyan A and Liewald M 2020 Data-driven tool wear classification with a convolutional neural network in punching machines *International Conference on "New Developments in Sheet Metal Forming"* pp 101-108
[5] Niessner S Liewald M and Weyrich M 2022 A methodology to quantify tool wear effects in a shear cutting process based on an automatic feature extraction *14th IFAC Workshop on IMS*
[6] Xing Z Z Pei J and Keogh E J 2010 A brief survey on sequence classification *ACM SIGKDD Explorations Newsletter* **12(1)** pp 40-48
[7] Xi X Koegh E J Shelton C et al. 2006 Fast time series classification using numerosity reduction *ICML '06: Proceedings of the 23rd international conference on Machine learning* pp 1033-1040
[8] Hearst H Dumais S Osman E Platt J et al. 1998 Support Vector Machines *IEEE Intelligent Systems and Their Applications* **13(4)** pp 18-28
[9] Burian A Takala J 2003 On signal detection using support vector machines *Proceedings of the international symposium on signals, circuits and systems* **2** (July) pp 609-612
[10] Vapnik V N 1998 *Statistical learning theory* Wiley New York
[11] Wang X Mueen A Ding H et al. 2013 Experimental comparison of representation methods and distance measures for time series data *Data Mining and Knowledge Discovery* **26(2)** pp 275-309
[12] Koegh E J and Pazzani M J 2000 Scaling up dynamic time warping for datamining KDD *Proceedings of the sixth ACM SIGKDD international conference on Knowledge discovery and data mining* pp 285-289
[13] Wang X Mueen A Ding H et al. 2013 A review on distance based time series classification *Data Mining and Knowledge Discovery* **33 Issue 2** pp 378-412
[14] Kini, B V and Sekhar C C 2013 Large margin mixture of AR models for time series classification *Applied Soft Computing* **13(1)** pp 361-371
[15] Bagnall A Janacek G 2014 A run length transformation for discriminating between auto regressive time series *Journal of Classification* **31 (October)** pp 274-295
[16] Corduas M Piccolo D 2008 Time series clustering and classification by the autoregressive metric *Computational Statistics and Data Analysis* **52(4)** pp 1860-1872
[17] Antonucci A De Rosa R Giusti A et al. 2015 Robust classification of multivariate time series imprecise hidden markov models *International Journal of Approximate* Reasoning **56(B)** pp 249-263
[18] Smyth P 1997 Clustering sequences with hidden Markov models *Advances in Neural Information Processing Systems* **9** pp 648-654
[19] Wu J Li Y and Liu B 2018 An overview on feature-based classification algorithms for multivariate time series *IEEE 3rd International Conference on Cloud Computing and Big Data Analysis (ICCCBDA)*
[20] Maschler B Knodel T and Weyrich M 2021 Towards deep industrial transfer learning for anomaly detection on time series data *26th IEEE ETFA* Vienna
[21] Längkvist M Karlsson L Loutfi A 2014 A review of of unsupervised feature learning and deep learning for time-series modeling *Pattern Recognition Lett*ers 2014 **42(1)** pp 11-24
[22] Schmidhuber J 2015 Deep learning in neural networks: an overview *Neural Networks* 2015 **61(1)** pp 85-117
[23] Zhao B Lu H Chen S et al. 2017 Convolutional neural networks for time series classification *Journal of Systems Engineering and Electronics* **28 Issue 1** pp 162-169
[24] Li C Kahn L and Prabhakaran B 2005 Real-time classification of variable length multi-attribute motions *Knowledge and Information Systems* **10(2)** pp 163-183
[25] Liewald M and Wagner S 2018 S4.5 Scheren und Schneiden In Grote K-H Bender B Göhlich D *Dubbel* pp 70-79 Springer-Verlag Berlin Heidelberg
[26] The Association of German Engineers 1994 *VDI 2906-2: Quality of cut faces of (sheet) metal parts after cutting, blanking, trimming or piercing* Berlin: Beuth (Mai 1994)

Optimisation of deep drawn corners subject to hot stamping constraints using a novel deep-learning-based platform

H R Attar*, N Li

Dyson School of Design Engineering, Imperial College London, London SW7 2DB, UK

*corresponding author E-mail: h.attar19@imperial.ac.uk

Abstract. State-of-the-art hot stamping processes offer improved material formability and therefore have potential to successfully form challenging components. The feasibility of components to be formed through these processes is dependent on their geometric design and its complex interactions with the hot stamping environment. In industrial practice, trial-and-error approaches are currently used to update non-feasible designs where simulation runs are needed each time a design change is made. These approaches make the design process resource intensive and require considerable numerical and process expertise. To demonstrate a superior approach, this study presents a novel application of a deep-learning-based optimisation platform which adopts a non-parametric geometric modelling strategy. Here, deep drawn corner geometries from different geometry subclasses were optimised to minimise wasted volume due to radii while avoiding excessive post-stamping thinning. A neural network was trained to generate families of deep drawn corner geometries where each geometry was conditioned on an input latent vector. Another neural network was trained to predict the thinning distributions obtained from forming these geometries through a hot stamping process. Guided by these distributions, the latent vector, and therefore geometry, was iteratively updated by a new gradient-based optimisation technique. Overall, it is demonstrated that the platform is capable of optimising geometries, irrespective of complexity, subject to imposed post-stamped thinning constraints.

1. Introduction

Lightweighting is one of the most effective strategies to reduce vehicle use-phase emissions [1] and therefore address the ever-growing concerns surrounding global environmental impact [2]. The latest hot stamping processes (e.g., the HFQ® process for high strength aluminium alloys [3]) are key enablers for realising vehicle lightweighting. However, there are several inherent manufacturing complexities associated with these processes. These include uneven cooling during stamping [4] and viscoplastic effects on the material flow stress at elevated temperatures [5]. These complexities create unfamiliarity among industrial designers and constrain the space of feasible component designs.

Published literature focuses on the development of advanced material models [5–8] that predict the material constitutive behaviour under hot stamping conditions. These models are then used to simulate hot stamping processes with high accuracy using Finite Element (FE) simulations [4]. However, these simulations usually take place late in design processes when the component geometry is near completion and require considerable numerical and process expertise.

Recently, researchers have started exploring time-efficient alternatives to FE simulations for component feasibility assessment. For simple approximations of complex geometries, Attar *et al.* [9] developed guidelines for quickly identifying if a designed geometry is likely to fall within a forming process's window of capability. For more complicated geometries with many geometric dimensions, non-parametric modelling representations offer a promising route. From the domain of machine learning, researchers have explored using Convolutional Neural Networks (CNNs) as surrogate models of stamping processes. These models were able to rapidly predict post-stamping physical fields (e.g., thinning strain distribution) associated with a given die geometry as input [10–13]. Attar *et al.* [11] found that high accuracy in the predicted physical fields was achievable from CNNs when compared to FE simulations even for a complex hot stamping process.

However, these promising surrogate models are only capable of forward predictions without any backward feedback or guidance towards optimum designs. To address this drawback, Attar *et al.* [14] developed a novel deep-learning-based optimisation platform for optimising geometries subject to manufacturing constraints. The optimisation platform featured the interactions of two neural networks: 1) a geometry generator f_{θ_1} and 2) a manufacturing performance evaluator g_{θ_2}. The geometry generator took advantage of recently developed implicit neural representations [15] but adapted for representing stamping geometries with high accuracy [14]. These representations made use of Signed Distance Fields (SDFs) which implicitly represent arbitrary geometries by their zero-level-sets and these geometries were extracted using Marching Cubes [16]. The manufacturing performance evaluator was a CNN-based surrogate model which used a 2D projection of the extracted geometries to predict their post-stamping physical fields. Manufacturing constraints were imposed based on these fields and were used to formulate objective functions to be minimised. The minimisation was performed by using the CNN-predicted manufacturability information to iteratively update the geometries until convergence.

The purpose of this paper is to present a novel engineering application of the aforementioned optimisation platform. This application offers practical guidance for researchers and industrial engineers who are planning to use the platform to develop geometries that are optimised for stamping applications.

2. Development of the geometry generator model

The procedures developed in [14] were employed here to establish a geometry generator model f_{θ_1}. The definition of the geometries considered are first given followed by model details.

2.1. Geometry definition

The high level geometry class considered in this study was quarter deep drawn box corners. These corners are widely adopted on many rectangular or square shaped components, e.g., door inners in [17], but are challenging to successfully form [9]. Within this geometry class, three different geometry subclasses were defined and parameterised individually. These were named standard corners, chamfer corners and stepped sidewalls and can be seen in Figure 1. Different subclasses were used to showcase the non-parametric capability of the optimisation platform, as presented later in the paper.

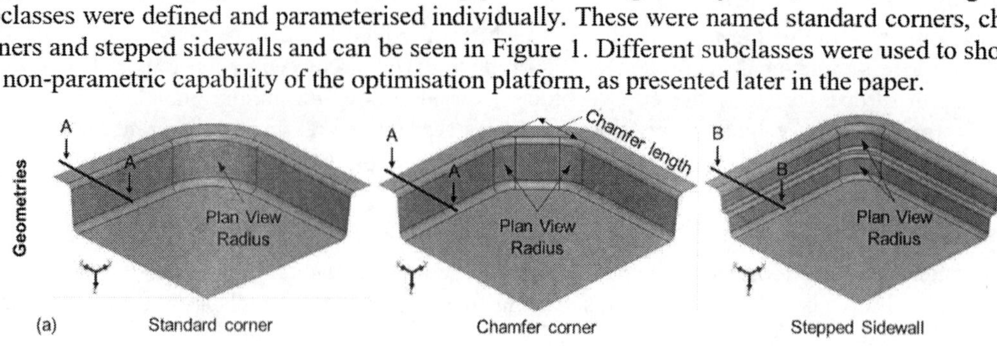

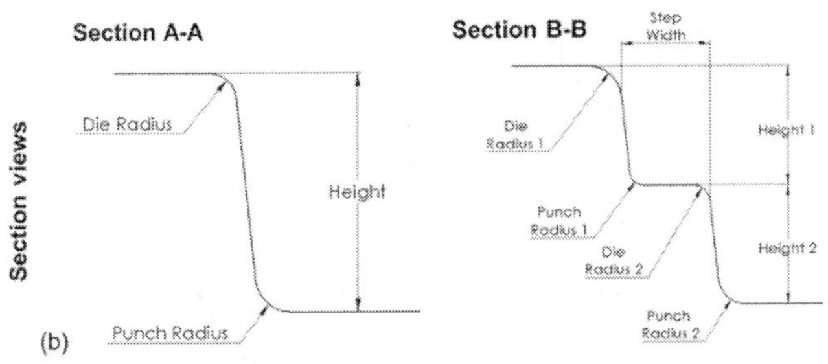

Figure 1. Definition of the three deep drawn box corner geometry subclasses considered in this study. Figure shows (a) 3D CAD views with corner parameters labelled and (b) section views with sidewall parameters labelled. Box half side length of 500 mm for all geometries.

2.2. *Data preparation*

The geometry generator model was trained to generate variants of die geometries from the geometry subclasses and this training process required example geometries. Using the CAD parameters shown in Figure 1, the Latin Hypercube design of experiments technique [18] and SolidWorks VBA was employed to generate these examples. For training data, a total of 1200 geometries were generated, 400 from each subclass. For test data, 300 geometries were generated, 100 from each subclass.

The generated geometries were then processed into a suitable form for neural network training. The top surface from each generated example die geometry was extracted in CAD software and exported as an STL surface mesh file, as shown in Figure 2(a). Each mesh was scaled such that the die edge was of unit length. The scaled meshes were then discretised into the point sets shown in Figure 2(b).

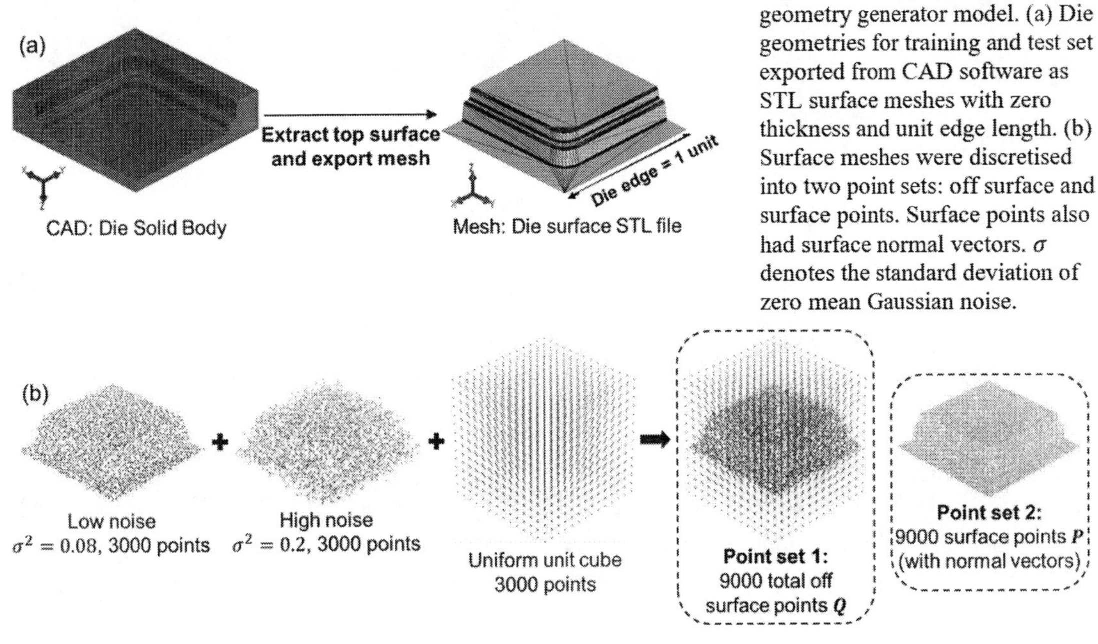

Figure 2. Data preparation for the geometry generator model. (a) Die geometries for training and test set exported from CAD software as STL surface meshes with zero thickness and unit edge length. (b) Surface meshes were discretised into two point sets: off surface and surface points. Surface points also had surface normal vectors. σ denotes the standard deviation of zero mean Gaussian noise.

2.3. *Model architecture*

The architecture of the neural network model f_{θ_1} used as the geometry generator is shown in Figure 3. This model was based on an Auto-Decoder multilayer perceptron, recently introduced by Park *et al.* [15]. The model took as input a latent vector $\boldsymbol{z} \in \mathbb{R}^{128}$ which was concatenated with the coordinates of a 3D query point in ambient space $\boldsymbol{x} \in \mathbb{R}^3$. The output was a SDF value at the query point. Querying f_{θ_1} on a 3D grid of points enabled a volumetric SDF to be generated over that grid and this SDF was conditioned on the latent vector $\boldsymbol{z}$. Changing $\boldsymbol{z}$ during optimisation would change the generated SDF and therefore change the geometry implicitly represented by the zero-level-set of the SDF [15].

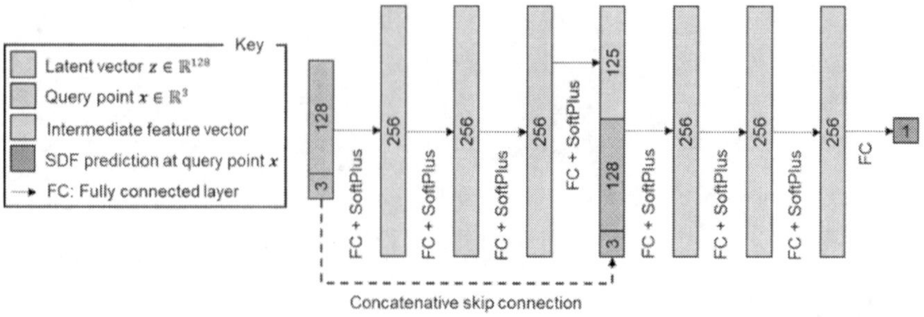

Figure 3. Architecture of the geometry generator neural network model f_{θ_1}. Numbers denote vector length.

2.4. Training and inference

The training process of the model shown in Figure 3 followed the developments presented in [14]. The model was trained to generate an SDF given only the 3D coordinates of input points. Ground truth SDF values were not needed, and the model was instead trained to obey the geometric properties of SDFs. Attar *et al.* [14] found that this approach resulted in networks that were capable of high accuracy reconstructions of small geometric features on stamping geometries (e.g., tight radii).

To train the model in this way, a loss function to be minimised was formulated in terms of the network parameters θ_1 and the latent vectors for geometries in each training batch $\{\boldsymbol{z}_i\}_{i=1}^{B}$. This function is shown in Equation (1)

$$\underset{\theta_1,\{\boldsymbol{z}_i\}_{i=1}^{B}}{\arg\min}\frac{1}{B}\sum_{i=1}^{B}\mathcal{L}_{\text{training}}(\theta_1,\boldsymbol{z}_i) = \underset{\theta_1,\{\boldsymbol{z}_i\}_{i=1}^{B}}{\arg\min}\frac{1}{B}\sum_{i=1}^{B}\left(\mathcal{L}_{\text{surface}}(\theta_1,\boldsymbol{z}_i)+\mathcal{L}_{\text{off surface}}(\theta_1,\boldsymbol{z}_i)+\mathcal{L}_{\text{reg}}(\boldsymbol{z}_i)\right) \quad (1)$$

where

$$\mathcal{L}_{\text{surface}}(\theta_1,\boldsymbol{z}_i)=\frac{1}{|\boldsymbol{P}|}\sum_{x_j\in\boldsymbol{P}}\left(\lambda_1\left|f_{\theta_1}(\boldsymbol{z}_i,\boldsymbol{x}_j)\right|+\lambda_2\alpha_j\left\|\nabla_{x_j}f_{\theta_1}(\boldsymbol{z}_i,\boldsymbol{x}_j)-\boldsymbol{n}_j\right\|_2\right) \quad (2)$$

$$\mathcal{L}_{\text{off surface}}(\theta_1,\boldsymbol{z}_i)=\frac{\lambda_3}{|\boldsymbol{Q}|}\sum_{x_k\in\boldsymbol{Q}}\left(\left\|\nabla_{x_k}f_{\theta_1}(\boldsymbol{z}_i,\boldsymbol{x}_k)\right\|_2-1\right)^2 \quad (3)$$

$$\mathcal{L}_{\text{reg}}(\boldsymbol{z}_i)=\lambda_4\|\boldsymbol{z}_i\|_2 \quad (4)$$

A brief breakdown of Equation (1) is now given, and further details can be found in the original paper by Attar *et al.* [14]. The $\mathcal{L}_{\text{surface}}(\theta_1,\boldsymbol{z}_i)$ term was formulated on the surface point set $\boldsymbol{P}$ and contained two terms weighted by the scalars λ_1 and λ_2 as shown in Equation (2). Minimising the first term encouraged the generated SDF value at the geometry surface $f_{\theta_1}(\boldsymbol{z}_i,\boldsymbol{x}_j)$ to be zero, since SDFs should represent the surface geometry by their zero-level-sets [15]. Minimising the second term encouraged the spatial gradients of the generated SDF at the surface $\nabla_{x_j}f_{\theta_1}(\boldsymbol{z}_i,\boldsymbol{x}_j)$ to be aligned with the surface unit normal vectors $\boldsymbol{n}_j$. This alignment of normals at the geometry surface is a geometric property of SDFs [14]. The α_j term gave a greater weighting to surface points belonging to tight geometry regions (e.g., tight radii) to encourage their accurate reconstruction.

The $\mathcal{L}_{\text{off surface}}(\theta_1,\boldsymbol{z}_i)$ term weighted by λ_3 was formulated on the off surface point set $\boldsymbol{Q}$. Minimising this term encouraged the spatial gradient of the generated off surface SDF $\nabla_{x_k}f_{\theta_1}(\boldsymbol{z}_i,\boldsymbol{x}_k)$ to have unit magnitude, which is another property of SDFs [14]. Minimising the $\mathcal{L}_{\text{reg}}(\boldsymbol{z}_i)$ term weighted by λ_4 encouraged latent vector magnitudes that were concentred near the origin. This concentration enabled similar latent vectors for similar geometries and helped training convergence [15]. After training, high quality geometric reconstructions were enabled from f_{θ_1} on unseen test set geometries.

At inference time, the determined network parameters θ_1 were fixed and an optimum latent vector for an unseen geometry was inferred through an iterative optimisation procedure. This inference process involved minimising the expression shown in Equation (5) to obtain an optimal latent vector which best describes the unseen geometry.

$$\begin{aligned}\arg\min_{z} \mathcal{L}_{\text{inference}}(\boldsymbol{z}) = \arg\min_{z} \frac{1}{|\boldsymbol{P}|} \sum_{\boldsymbol{x}_j \in \boldsymbol{P}} \left(\lambda_1 \left| f_{\theta_1}(\boldsymbol{z}, \boldsymbol{x}_j) \right| + \lambda_2 \alpha_j \left\| \nabla_{\boldsymbol{x}_j} f_{\theta_1}(\boldsymbol{z}, \boldsymbol{x}_j) - \boldsymbol{n}_j \right\|_2 \right) \\ + \frac{\lambda_3}{|\boldsymbol{Q}|} \sum_{\boldsymbol{x}_k \in \boldsymbol{Q}} \left(\left\| \nabla_{\boldsymbol{x}_k} f_{\theta_1}(\boldsymbol{z}, \boldsymbol{x}_k) \right\|_2 - 1 \right)^2 + \lambda_4 \|\boldsymbol{z}\|_2 \end{aligned} \quad (5)$$

3. Development of the manufacturing performance evaluator model

An overview of the manufacturing performance evaluator model g_{θ_2} is presented and further details can be found by Attar *et al.* [11,12]. This model was a CNN based surrogate of the HFQ process [3] and was able to predict the post-stamped thinning distribution associated with an input die geometry.

3.1. HFQ stamping simulations

Simulations in PAM-STAMP were conducted to obtain target thinning distributions for the training and testing datasets of g_{θ_2}. The CAD geometries described in Section 2.1. were converted into FE meshes for these simulations. The Python programming language was used to automate the loading of these meshes into PAM-STAMP and launch each simulation. The processing parameters were kept constant and set according to Attar *et al.* [14]. The temperature and strain rate dependent material model for AA6082 under HFQ conditions which was presented in [19] was used.

3.2. Model architecture and training

The Res-SE-U-Net architecture presented in Attar *et al.* [11,12] was used for g_{θ_2} and an overview of the manufacturability assessment using this model is presented in Figure 4. Specific details on the architecture of Res-SE-U-Nets can be found in the cited literature [11,12].

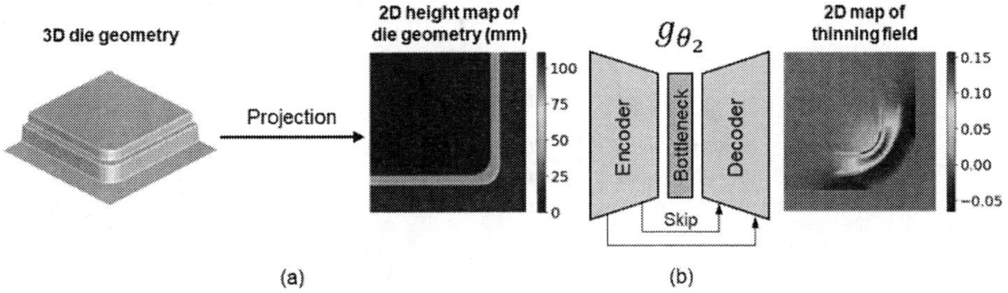

Figure 4. Overview of the manufacturing performance assessment approach. (a) 2D projection of 3D die geometry. (b) Thinning field prediction using CNN based surrogate model given the input height map.

The training data for g_{θ_2} consisted of input-target image examples like the one seen in Figure 4. For the input images, the 3D CAD geometries described in Section 2.1. were projected onto images to obtain 2D height maps. For the target images, the thinning distributions were first extracted from the HFQ stamping simulations and plotted onto the undeformed 2D blank nodes from the FE mesh. These 2D distributions were then linearly interpolated onto a uniform image grid.

g_{θ_2} was then trained in a supervised manner using the training data. Through an iterative optimisation process, the optimiser sought to find the combination of network parameters θ_2 such that the mean squared error (MSE) between target thinning distributions and network predictions was minimised. After training, high accuracy in thinning distribution predictions were seen from g_{θ_2} on unseen test set geometries.

4. Design optimisation

The optimisation platform was setup after f_{θ_1} and g_{θ_2} were established. An overview of the setup of the platform is presented in this section as well as details on two optimisation tasks.

4.1. Optimisation platform set up

Figure 5 shows the optimisation platform overview. The platform is based on the interaction of f_{θ_1} and g_{θ_2} and these networks are used to iteratively update the latent vector $\mathbf{z}$ to minimise $\mathcal{L}_{\text{task}}$.

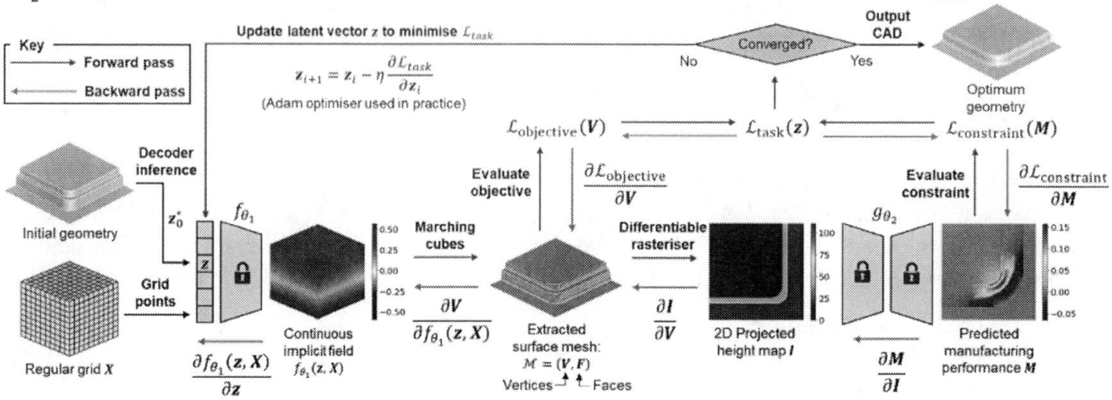

Figure 5. Optimisation platform overview. Padlocks on f_{θ_1} and g_{θ_2} denote fixed network parameters.

$\mathcal{L}_{\text{task}}$ was a function designed to be minimised and consisted of the summation of an objective function $\mathcal{L}_{\text{objective}}$ and a constraint function $\mathcal{L}_{\text{constraint}}$. These functions are task specific and were formulated based on the optimisation tasks considered in this study (see below). A forward pass through the platform occurred each optimisation iteration to compute $\mathcal{L}_{\text{task}}$ and update $\mathbf{z}$, as shown by the red arrows in Figure 5. $\mathbf{z}$ was iteratively updated by the Adam optimiser [20] by using a variation of the simplified gradient descent shown in Equation (6) where the subscript i denotes the ith optimisation iteration and η is a constant.

$$\mathbf{z}_{i+1} = \mathbf{z}_i - \eta \frac{\partial \mathcal{L}_{\text{task}}}{\partial \mathbf{z}_i} \tag{6}$$

A backward pass occurred each iteration to compute $\partial \mathcal{L}_{\text{task}}/\partial \mathbf{z}$ which was needed to perform the $\mathbf{z}$ updates as shown by the blue arrows in Figure 5. The chain rule was used to compute $\partial \mathcal{L}_{\text{task}}/\partial \mathbf{z}$ as shown in Equation (7). Details on the computation of these gradients are given by Attar *et al.* [14].

$$\frac{\partial \mathcal{L}_{task}}{\partial \mathbf{z}} = \left(\frac{\partial \mathcal{L}_{objective}}{\partial \boldsymbol{V}} + \frac{\partial \mathcal{L}_{constraint}}{\partial \boldsymbol{M}} \cdot \frac{\partial \boldsymbol{M}}{\partial \boldsymbol{I}} \cdot \frac{\partial \boldsymbol{I}}{\partial \boldsymbol{V}}\right) \cdot \frac{\partial \boldsymbol{V}}{\partial f_{\theta_1}(\mathbf{z}, \boldsymbol{X})} \cdot \frac{\partial f_{\theta_1}(\mathbf{z}, \boldsymbol{X})}{\partial \mathbf{z}} \tag{7}$$

4.2. Optimisation tasks

Two optimisation tasks were performed using the optimisation platform, one for a standard corner geometry and another for a stepped sidewall geometry. These tasks both involved optimising an initial geometry to minimise the wasted volume due to corner geometric features. Ideal geometries which have zero wasted volume would have infinitely sharp corners as shown in Figure 6. However, these geometries would impose large thinning defects and fractures during stamping [9]. To form these geometries successfully, their corner features must be carefully designed and optimised.

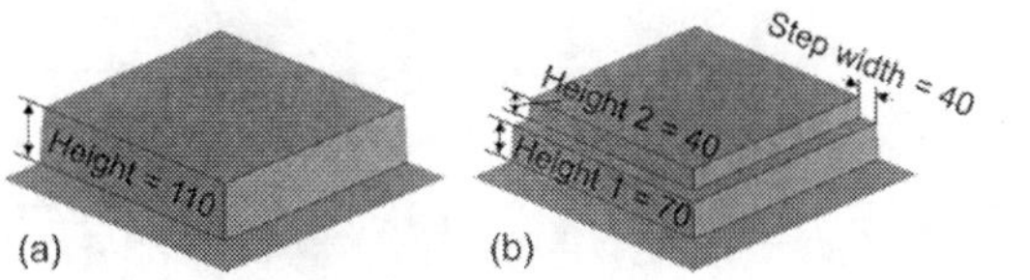

Figure 6. Ideal geometries for the optimisation tasks in this study. (a) Task 1 ideal geometry: Standard corner with no radii. (b) Task 2 ideal geometry: Stepped sidewall with no radii. Dimensions in mm.

The objective function $\mathcal{L}_{\text{objective}}$ and the constraint function $\mathcal{L}_{\text{constraint}}$ were first formulated for this optimisation problem. The formulation for $\mathcal{L}_{\text{objective}}$ is shown in Equation (8)

$$\mathcal{L}_{\text{objective}} = \lambda_1 \left(\sum_{x_i \in P_I} \min_{x_j \in P_R} \|x_i - x_j\|_2^2 + \sum_{x_j \in P_R} \min_{x_i \in P_I} \|x_i - x_j\|_2^2 \right) + \lambda_2 |H - H_{\text{ref}}| \tag{8}$$

where the bracketed term is the chamfer distance between point sets sampled from the ideal geometry $\boldsymbol{P}_I$ and the reconstructed geometry which is being optimised $\boldsymbol{P}_R$. These point sets are illustrated in Figure 7. A generous 30,000 points were sampled to approximate continuous surfaces. It is noteworthy that the generation of P_R occurred at each optimisation iteration and had to be done in a differentiable way to allow for $\partial\mathcal{L}_{objective}/\partial\boldsymbol{V}$ to be computed for Equation (7). In practice, the differentiable point sampler from the Pytorch3D framework was used for this purpose [21]. Weighting the chamfer distance by the scalar λ_1 formed the chamfer loss. Minimising the chamfer loss encouraged the geometry being optimised to be as close as possible to its idealised counterpart in Figure 6.

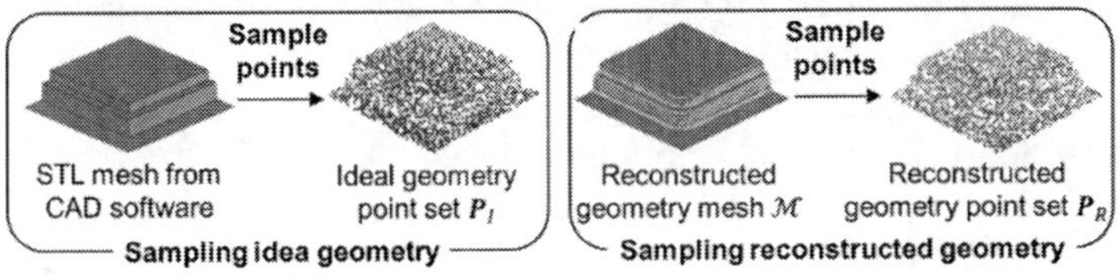

Figure 7. Illustration of point sets sampled from ideal geometry $\boldsymbol{P}_I$ and from the reconstructed geometry which is being optimised $\boldsymbol{P}_R$. These sets were used to compute the chamfer loss in Equation (8).

The second term in Equation (8) weighted by the scalar λ_2 was the height loss. Minimising this loss encouraged the height of the geometry being optimised H to be as close as possible to a desired reference height H_{ref}. Therefore, this loss ensured that where possible, optimisation of corner geometries were favoured over height. Here, $H_{\text{ref}} = 110$ mm as shown in Figure 6.

The formulation for $\mathcal{L}_{\text{constraint}}$ is shown in Equation (9)

$$\mathcal{L}_{\text{constraint}} = \lambda_3 \text{ReLU}\big(\max(\boldsymbol{M}(\mathbf{z})) - t_{\max}\big) \tag{9}$$

where λ_3 is a weighting scalar, $\boldsymbol{M}(\mathbf{z})$ is the generated post-stamping thinning field defined as

$$\boldsymbol{M}(\mathbf{z}) = g_{\theta_2}\left(\phi\left(f_{\theta_1}(\mathbf{z}, \boldsymbol{X})\right)\right) \tag{10}$$

where ϕ is the combined marching cubes [16] and differentiable rasteriser [21] steps in Figure 5. Here, ϕ can be thought of as a differentiable function that converts the volumetric SDF $f_{\theta_1}(\mathbf{z}, \boldsymbol{X})$ into a 2D height map of the geometry encoded in $\mathbf{z}$. The non-differentiability of marching cubes [22] has recently been addressed by Remelli *et al.* [23,24] and applied by Attar *et al.* [14]. The function $\text{ReLU}(x) := \max(0, x)$ was used in Equation (9) to make $\mathcal{L}_{\text{constraint}}$ inactive when the maximum thinning $\max(\boldsymbol{M}(\mathbf{z}))$ did not exceed a maximum allowable thinning constraint $t_{\max}$ and active otherwise.

4.2.1. Task 1 description and results. Task 1 was to optimise the initial chamfer corner geometry shown in Figure 8(b)(top). The objective of task 1 was to modify this geometry so that it is as close as possible to the ideal geometry shown in Figure 6(a) but subject to a 0.1 max thinning constraint. The results are shown in Figure 8.

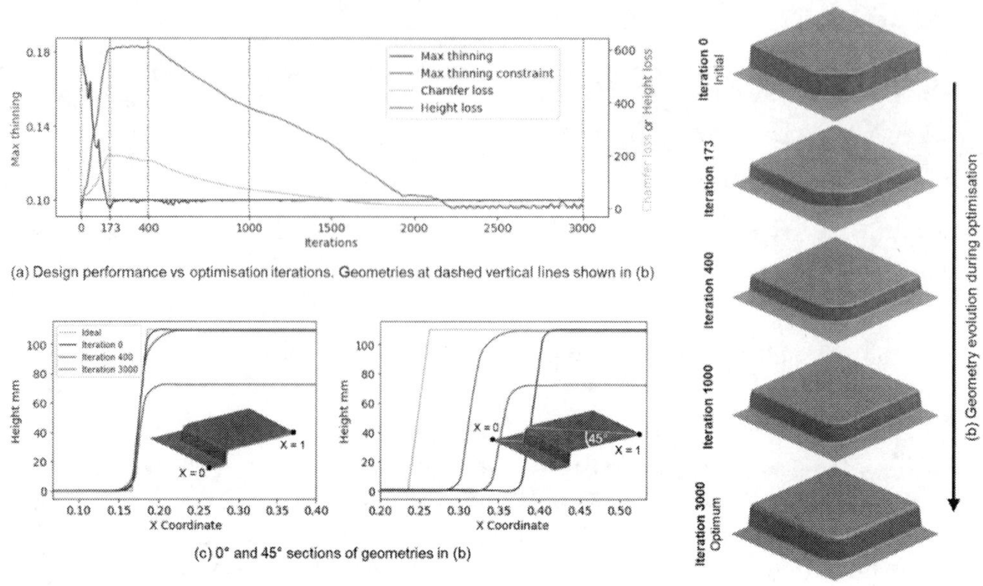

Figure 8. Optimisation task 1 results.

4.2.2. Task 2 description and results. Task 2 was to optimise the initial shallow standard corner geometry shown in Figure 9(b)(top). This shallow geometry could be, for example, one that might have been previously designed for a standard cold stamping process. Therefore, this task could be interpreted as optimising this geometry to take advantage of hot stamping capabilities that enable challenging features such as stepped sidewalls, tight radii and large draw depths [5,9]. The objective of task 2 was to modify this geometry so that it is as close as possible to the ideal geometry shown in Figure 6(b) but subject to a 0.2 max thinning constraint. The results are shown in Figure 9.

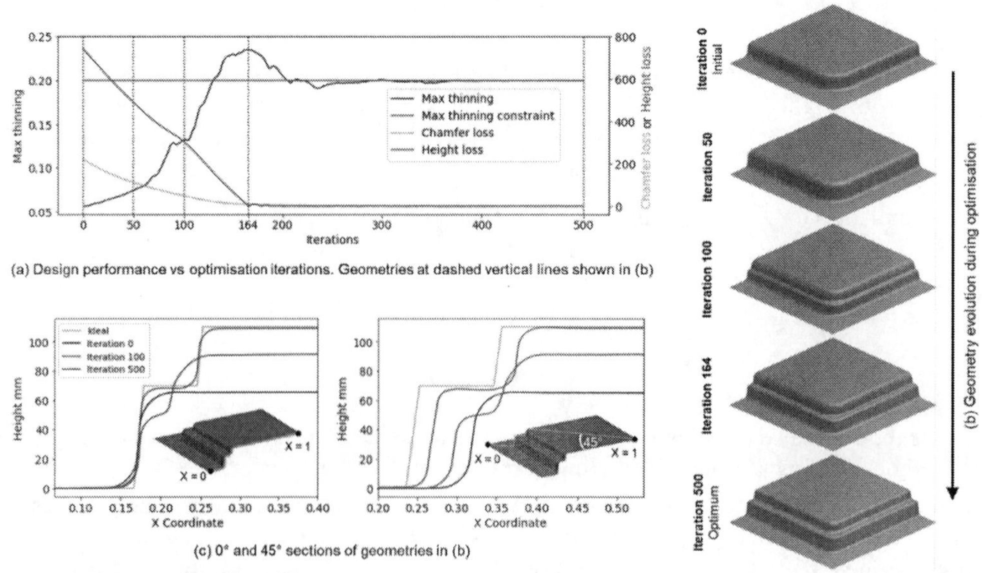

Figure 9. Optimisation task 2 results.

4.2.3. Discussion. The results for both tasks showed that expressive changes in geometries are possible. The optimisation history of task 1 is plotted on Figure 8(a). The maximum thinning was seen to decrease from 0.18 to 0.1 through the first 173 iterations and this decrease was mainly due to a reduction in geometry height as seen in Figure 8(b). At iteration 173, the constraint became inactive.

This formulation of Equation (9) prevented the constraint from becoming active again, as further explained by Attar *et al.* [14]. The remainder of the optimisation process was driven by minimising Equation (8). Any further geometric changes after iteration 173 occurred under constant maximum thinning at the constraint level. The final geometry at iteration 3000 was as close as possible to the ideal geometry in Figure 6(a) but with optimised corner features. The height loss in Equation (8) enabled the desired 110 mm height to be achieved but at the cost of larger radii.

An opposite scenario was seen for task 2. Figure 9(a) shows that the initial geometry had a maximum thinning of around 0.06 which was overly conservative since the allowable was set at 0.2. In this case, both the chamfer loss and the height loss proceeded to decrease with no manufacturing constraint since the constraint function in Equation (9) was inactive. Consequently, from iterations 0 to 164, height was increasing, the stepped feature was being introduced, and all radii were getting tighter. These changes resulted in increases in the maximum thinning which reached a maxima at iteration 164 but went back down to 0.2 shortly after. The final geometry at iteration 500 was as close as possible to the ideal geometry in Figure 6(b) but with optimised corner and stepped sidewall features.

The optimisation tasks were repeated with different constraint levels and the optimum geometries are shown in Figure 10. Noticeably tighter optimised corners can be seen when increasing maximum thinning constraint levels. At the strictest constraint level of 0.05, shallower geometries were obtained with larger radii and with no stepped feature for task 2. These results communicate understanding of feasible designs and design capabilities of a particular stamping process to industrial designers.

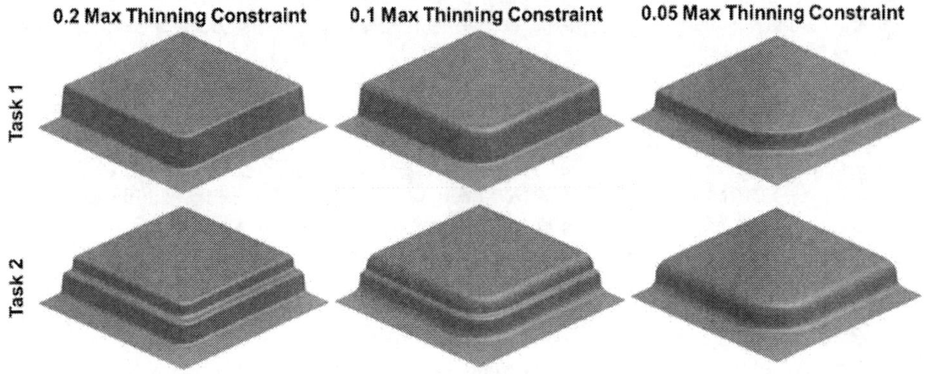

Figure 10. Optimum geometries from task 1 (top row) and task 2 (bottom row) for various maximum allowable thinning constraint values.

5. Conclusion

A novel application of a deep-learning-based platform for optimising sheet stamping geometries subject to manufacturing constraints was presented. Manufacturability-constrained optimisation of deep drawn corner geometries was performed to minimise wasted volume due to designed radii. Training details for two neural networks needed for the platform were given and these networks enabled non-parametric modelling and evaluation of sheet stamping geometries. The first network generated geometries and the second evaluated their manufacturing performance. New objective and constraint functions were formulated, and optimisation was performed by minimising these functions. It was demonstrated that expressive changes to geometries are possible using the optimisation platform. These changes were driven by achieving close to target designs while satisfying manufacturability and were irrespective of geometric complexity due to the deep-learning-based non-parametric approach. The presented application offers practical guidance for researchers and engineers who are planning to use the platform to develop optimum geometries for stamping processes.

Acknowledgments

The authors thank Impression Technologies Ltd and the UK EPSRC for funding support.

References

[1] Raugei M, El O, Wang L, Lin J and Morrey D 2014 Life cycle assessment of the potential environmental benefits of a novel hot forming process in automotive manufacturing *J. Clean. Prod.* **83** 80–6

[2] UK gov 2020 *2018 UK Greenhouse Gas Emissions, Final figures*

[3] Lin J, Dean, Trevor A, Garrett, Richard P and Foster, Alistair D 2008 Process for forming aluminium alloy sheet component

[4] El Fakir O, Wang L, Balint D, Dear J P, Lin J and Dean T A 2014 Numerical study of the solution heat treatment, forming, and in-die quenching (HFQ) process on AA5754 *Int. J. Mach. Tools Manuf.* **87** 39–48

[5] Mohamed M S, Foster A D, Lin J, Balint D S and Dean T A 2012 Investigation of deformation and failure features in hot stamping of AA6082: Experimentation and modelling *Int. J. Mach. Tools Manuf.* **53** 27–38

[6] Shao Z, Li N, Lin J and Dean T 2017 Formability evaluation for sheet metals under hot stamping conditions by a novel biaxial testing system and a new materials model *Int. J. Mech. Sci.* **120** 149–58

[7] Lin J, Mohamed M, Balint D and Dean T A 2014 The development of continuum damage mechanics-based theories for predicting forming limit diagrams for hot stamping applications *Int. J. Damage Mech.* **23** 684–701

[8] Wang A, Zheng Y, Liu J, El Fakir O, Masen M and Wang L 2016 Knowledge based cloud FE simulation a multi-objective FEA system for advanced FE simulation of hot stamping process *The 10th International Conference and Workshop on Numerical Simulation of 3D Sheet Metal Forming Processes*

[9] Attar H R, Li N and Foster A 2021 A new design guideline development strategy for aluminium alloy corners formed through cold and hot stamping processes *Mater. Des.* **207**

[10] Zimmerling C, Trippe D, Fengler B and Kärger L 2019 An approach for rapid prediction of textile draping results for variable composite component geometries using deep neural networks *AIP Conf. Proc.* **2113**

[11] Attar H R, Zhou H, Foster A and Li N 2021 Rapid feasibility assessment of components to be formed through hot stamping : A deep learning approach *J. Manuf. Process.* **68** 1650–71

[12] Attar H R, Zhou H and Li N 2021 Deformation and thinning field prediction for HFQ® formed panel components using convolutional neural networks *IOP Conf. Ser. Mater. Sci. Eng.* **1157** 012079

[13] Zhou H, Xu Q, Nie Z and Li N 2022 A Study on Using Image-Based Machine Learning Methods to Develop Surrogate Models of Stamp Forming Simulations *J. Manuf. Sci. Eng.* **144** 1–15

[14] Attar H R, Foster A and Li N 2022 Development of a deep learning platform for optimising sheet stamping geometries subject to manufacturing constraints *arXiv*

[15] Park J J, Florence P, Straub J, Newcombe R and Lovegrove S 2019 DeepSDF: Learning continuous signed distance functions for shape representation *arXiv*

[16] Lorensen W E and Cline H E 1987 Marching cubes: A high resolution 3D surface construction algorithm *Proc. 14th Annu. Conf. Comput. Graph. Interact. Tech. SIGGRAPH 1987* **21** 163–9

[17] Politis D J, Li N, Wang L, Lin J, Foster A D and Szegda D 2016 Prediction of Thinning Behavior for Complex-Shaped, Lightweight Alloy Panels Formed Through a Hot Stamping Process 395–401

[18] Bonte M H ., van den Boogaard A and Huétink J 2007 A Metamodel Based Optimisation Algorithm for Metal Forming Processes *Adv. Methods Mater. Form.*

[19] El Fakir O 2015 *Studies on the Solution Heat Treatment , Forming and in-Die Quenching Process in the Production of Lightweight Alloy Components* (Imperial College London)

[20] Kingma D P and Ba J L 2015 Adam: A method for stochastic optimization *3rd Int. Conf. Learn. Represent. ICLR 2015 - Conf. Track Proc.* 1–15

[21] Ravi N, Reizenstein J, Novotny D and Gordon T 2020 Accelerating 3D Deep Learning with PyTorch3D *arXiv*

[22] Liao Y, Donne S and Geiger A 2018 Deep Marching Cubes: Learning Explicit Surface Representations *Proc. IEEE Comput. Soc. Conf. Comput. Vis. Pattern Recognit.* 2916–25

[23] Remelli E, Lukoianov A, Richter S R, Guillard B, Bagautdinov T, Baque P and Fua P 2020 *MeshSDF: Differentiable Iso-Surface Extraction*

[24] Guillard B, Remelli E, Lukoianov A, Richter S, Bagautdinov T, Baque P and Fua P 2021 DeepMesh: Differentiable Iso-Surface Extraction *arXiv*

Towards a real-time tool state detection in sheet metal forming processes validated by wear classification during blanking

C Kubik[1,*], D A Molitor[1], M Rojahn[1] and P Groche[1]

[1] Institute for Production Engineering and Forming Machines, Darmstadt, Germany

[*] kubik@ptu.tu-darmstadt.de

Abstract. The potential of data for inline detection of changes in the physical state of sheet metal forming processes has been proven over the last decade. However, with production rates exceeding 300 parts per minute the time available for a workpiece-related processing of sensor data is reduced. Therefore, the analysis of large data sets is outsourced to the cloud taking advantage of the high computing power provided there. But within this cloud-based computing paradigm, the speed of data transmission hinders real-time analysis of data and causes latency between fault detection and its occurrence. To overcome this bottleneck, this study aims to evaluate a data-based monitoring (DBM) approach for estimating process states in high speed sheet metal forming in terms of their suitability for a decentralized analysis at the edge. Thereby, the DBM is evaluated according to the model accuracy and the absolute computing time. In order to quantify these key performance parameters and the applicability of the DBM on edge devices, a classification of 16 wear states during blanking is considered. Based on the key performance parameters, an optimal DBM approach for decentralized analysis is proposed and an empirical formulation is provided to estimate the absolute computing time depending on the computational resources used for data processing.

1. Introduction

The increasing availability of data enables companies to master challenges of modern manufacturing processes. Therefore, manufacturers aim to use knowledge gained from data to fulfill the increased demands on product quality by ensuring simultaneously an economical production [1]. In this context, process-related uncertainties, such as varying material properties, temperature conditions or wear, must be controlled to ensure this specified product quality. Especially in the manufacturing process of blanking, unexpected wear phenomena are a major reason for insufficient product quality, forcing an increase of scrap. Due to the strong tribological loads resulting from high contact pressure and sliding distances, blanking tools are exposed to abrasive and adhesive wear in the forming zone [2]. Even though wear has a significant impact on the blanking process and the quality of the processed workpieces, determining the state of wear requires additional effort by interrupting the process, removing and manually inspecting the tool [3]. This leads to a conflict between minimizing machine downtimes (manufacturer side) and maximizing product quality (customer side). For this reason, real-time estimation of wear states during blanking is mandatory for scheduling maintenance intervals and maximizing the machine utilization while ensuring a desired product quality.

In this context, DBM represents a successful technique for in-line estimation of process states and can reduce or prevent unplanned machine downtime. A study by McKinsey & Company shows that

DBM reduces machine downtime by 30 to 50 percent and extends the machine life by 20 to 40 percent [4]. Here, DBM aims to automatically extract hidden information from a heterogeneous database and quantify the current process state based on this knowledge. However, manually processing of the huge amount of available data in high speed blanking processes complicates this knowledge discovery procedure and requires advanced analysis techniques such as machine learning (ML) [5] or feature-based supervision (FS) [6]. In the context of in-line wear estimation during blanking, DBM approaches such as ML or FS are able to quantify complex and non-linear interdependencies between wear phenomena and a sensorial acquired variable [7].

Next to the discovery of knowledge from data, data management plays a crucial role, especially in blanking operations, where production rates of over 300 stroke per minute (spm) are achieved [8]. In this process, large amounts of measurement data are generated in a short time, which must be analyzed in relation to the workpiece. For a blanking process at 300 spm, and a material thickness of 0.3 mm, this results in a tool engagement time of less than 334 ms. In this case, the time frame of 334 ms for data acquisition, data processing and adjusting the process must be smaller than the time for one stroke due to the knowledge gained [9].

In practice, the data processing step is outsourced to external computational sources, due to the high computing power provided, hereinafter referred to as cloud. However, compared to the rapidly developing data processing speed and computing power in the cloud, bandwidth improvements are at a standstill. With the increasing amount of data, the data transmission speed becomes a bottleneck in the cloud-based computing paradigm [10]. To ensure a real-time DBM, data must be processed at the origin where it is gathered. A decentralized data processing leads to better bandwidth availability, a faster data transmission and reduced latency. Therefore, it is necessary to shift the procedure of DBM from the cloud to the edge [11,12]. In this study, edge is defined as a device that enables decentralized data processing directly at the source of data generation [13]. In contrast to the cloud, edge devices rely on their local computing capacities. Therefore, it is important to keep the required computing power for edge devices as low as possible. At the same time, the quality of the generated knowledge and thus the reliability of the inline estimated process state should not be limited. Figure 1 compares the paradigm of edge and cloud computing, adapted to a manufacturing process. Thereby, data is provided by the process via sensors and centrally (a) or locally (b) processed. As a result of the analysis, knowledge is provided, that allows the process to be adapted to varying boundary conditions. The quality of the discovered knowledge and, related to this, the identification of the current process state is decisive for an appropriate adaptation of the process.

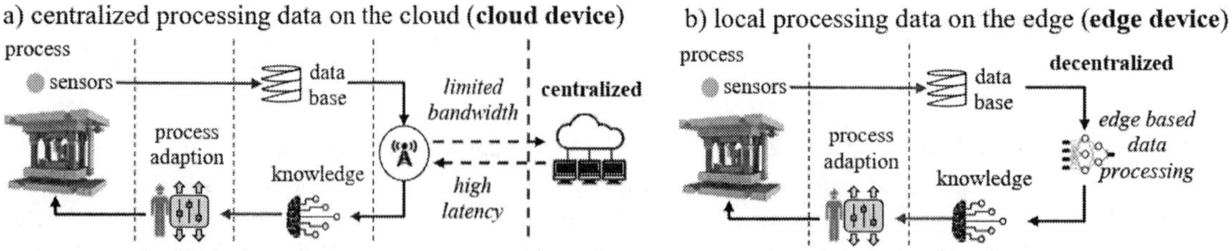

Figure 1: Comparison between cloud- and edge-based DBM related to manufacturing processes

Therefore, the objective of this work is to evaluate DBM for real-time estimation of process states in high speed sheet blanking with regard to their suitability for decentralized analysis at the edge. Thereby, the efficiency of DBM is quantified by the model accuracy (performance) and the absolute computing time calculated from the computational effort of the central processing unit (CPU). To quantify the efficiency of the DBM approaches, a classification of sixteen wear states is considered during a blanking operation at production rates up to 300 spm. Based on the two key parameters (model performance and absolute computing time) an optimal DBM for decentralized analysis at the edge is proposed. Therefore, sections 2 and 3 provide an introduction to blanking and the relevance of wear for this forming operation. Furthermore, an overview of existing DBM in sheet metal forming is given. Section 4 presents the experimental data sets with a total of about 4.000 strokes, the two key parameters and the training procedure for the model. Afterwards in section 5, the efficiency of the DBM approaches for a decentralized estimation of 16 abrasive wear state on the edge is quantified. In order to demonstrate the performance of the classification marginally differing wear conditions states are defined in 16 narrow intervals. Finally, section 6 summarizes the available results and provides a roadmap for future work.

2. Wear phenomena during blanking

As a result of the relative motion of the tool and the sheet metal at high stroke speed combined with high tribological loads, wear significantly affects the conditions of the blanking process. Thereby, the motion of the tool is characterized by three phases as shown in Figure 2 (a). During the punch-phase (I), the tool hits the sheet, causing an abrupt increase in the blanking force. The material changes from a linear elastic phase to a plastic deformation. In the process, the blanking force decreases until the material breaks completely. While in the push-phase (II) the blanked workpiece is ejected from the die, in the withdraw-phase (III) the tool is pulled out of the die. In both phases, the blanking force deviates from the zero level due to jamming between the blanked workpiece and the die and between the punch and the sheet metal strip. In addition to the relative motion, high contact stresses, reduced lubrication, large sliding distances and the processing of high-strength materials accelerate wear [14,15].

During blanking, wear is initiated by surface adhesions in the area of the forming zone. Due to the high contact stresses in the forming zone, the surfaces of the tool and sheet stick together. The relative motion and the jamming between the tool and the sheet cause frictional forces, that tear out small parts of the tool and cause the formation of wear particles. High abrasion is to be expected, especially near the cutting edge of the punch. According to Archard, the wear volume linearly depends on the contact normal stress and causes decreasing wear from the bottom of the punch upwards along the lateral surface [16]. In preliminary studies, a correlation was found between the contact normal stress and the distance from the bottom of the punch, as shown in Figure 2 (b). Thereby, the black line describes the outer contour of the cutting edge of the punch and the blue line the contact normal stresses along these outer contour.

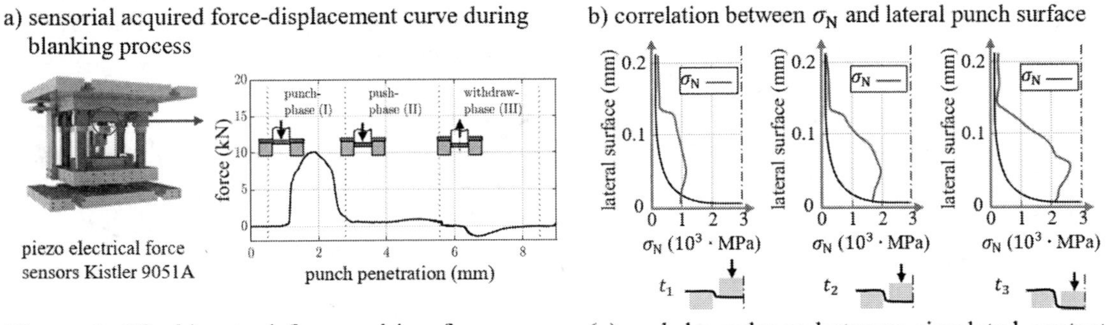

Figure 2: Blanking tool for acquiring force curves (a) and dependency between simulated contact normal stresses σ_N and the distance from the bottom of the punch (b)

As shown in the literature, the onset of wear as well as the type of wear are influenced by the material (sheet metal and tool), tool parameters (clearance and edge radii) [17,18] stroke speed, lubrication, and material properties (sheet thickness and alloy constituents) [19,20]. In particular, the material properties play a decisive role in the type of wear that occurs. According to Hohmann et al., soft graded steels (R_m < 350 MPa) tend to adhere to the harder surface of the tool. In contrast, high strength steel grades (R_m > 600 MPa) tend to cause abrasive wear on the edge of the blanking tool [21]. Despite this tendency, soft graded steels also show abrasive wear phenomena [22,23]. In this case, excessive wear occurs after a longer sliding distance (number of strokes) due to the lower contact normal pressure of the soft graded steels. Since the trend is towards processing increasingly high-strength materials, abrasive wear is occurring more frequently [24]. This leads to shorter maintenance intervals and a deterioration of the tool geometry in the form of a rounded cutting edge. In addition, the change in the tool geometry correlates with the quality of the blanked parts [3].

3. Data-based prediction of tool wear in sheet metal forming

Real-time DBM is mandatory to detect wear states inline during blanking or to quantify the actual tool geometry. Therefore, three monitoring approaches can be found in the literature: 1. binary fault detection (BFD), 2. feature-based supervision (FS) and 3. machine learning-based supervision (MLS).

BFD is mainly used in industrial practice and is limited to the identification of binary fault classes. By setting thresholds, envelopes and windows based on experiential knowledge, a distinction is made between process states that are acceptable or unacceptable (OK / NOK) [25,26].

In the case of FS, features are extracted from sensorial acquired data based on engineering knowledge. In their study, Hohmann et al. presented such a FS approach and showed that engineered features extracted from a force signal correlate with occurring abrasive and adhesive wear [21]. Klingenberg and de Boer extracted a feature describing the length of the punch penetration into the sheet until the maximum force is reached, as well as the work done in the punch phase, and correlated them with the rounding of the cutting edge [27]. A similar approach was chosen by Kubik et al, who extracted four features from the force signal and correlated them with the cutting edge radii of the punch [3,28].

In the MLS approach, features are extracted without the engineering or experiential knowledge. This so called transformation extracts features by an algorithm from the time domain, the frequency domain, the time-frequency domain or it is based on a model [29]. In the following step, these features serve as input for the training procedure of a model. Model types range from simple regression models to complex deep neuronal networks. One of the first applications of the MLS is described by Lee et al. [30]. They used an autoregressive time-series model to extract coefficients from blanking force. Based on these coefficients, the tool state (sharp vs. worn) was predicted using a linear discriminant function. Jin and Shi extracted features from the tonnage signal of a stamping process using a model-based transformation approach. These features served as an input for a hierarchical classifier to detect abnormal process states [31]. Hambli presented a backpropagation neural network for predicting burr height formation of blanked workpieces. Here, the inputs for the network are generated by a finite element analysis of the blanking process [32]. Ge et al. used a support vector machine (SVM) in their study to detect abnormal health conditions [33]. The entire signal of the stamping process serves as input for the SVM. In his work, Zhou et al. identified missing workpieces in one of the die stations of a progressive tool based on features extracted by principal component analysis (PCA) and SVM [34]. Asahi et al. present a Temporal Convolutional Network autoencoder for extracting features from raw time series and a SVN to classify 3 discrete wear states [35]. A similar framework is presented by Niemietz et al., leveraging the architecture of an autoencoder (AE) to track tool wear progression during fine blanking [36]. Another approach is shown by Molitor et al. who determine the wear state of a blanking tool inline by capturing images of the processed parts. The authors use Deep Convolutional Neural Networks to estimate 16 wear classes with classification accuracies of up to 99% [37].

A summary of the state of the art shows that wear significantly affects the efficiency of blanking processes. This applies in particular to abrasive wear, which occurs more and more frequently due to the processing of high strength materials. Though techniques such as FS and MLS are described in the literature for in-line wear state estimation, monitoring approaches in practice are limited to manual inspections and BFD. Although procedures for inline estimation of the actual wear condition based on sensor-acquired data can be found in the literature, a concrete specification (transformation and model) for the MLS approach is not given. In particular, the combination of extracted features and the chosen model and their influence on the estimation quality is missed. In addition, there is no information on the absolute computing time of the different FS and MLS approaches. However, this is important for high speed forming process such as blanking, where the time to process a single product is even less than 100 ms. In order to realize an inline and workpiece related tool state estimation on an edge device, the absolute computing time must to be minimized by maximizing the performance of the model.

4. Methodology

In order to estimate the actual abrasive wear state during blanking via DBM inline, three principal steps must be completed. In the first step, data is acquired to form the process as a basis for a further transformation and modelling procedure. Process data (time series: force) and quality data (label: wear state), which represent the label for each time series, are acquired. In the second step, the amount of data is reduced by a transformation (feature extraction) and serves as input for the modelling procedure. Since the BFD does not inherently allow a classification of multiple instances, this approach is not considered for further investigations in this study. For the FS approach, 10 engineering features are combined with 10 classification models. For the MLS approach, 23 transformation algorithms are combined with 10 classification models. As a result, 10 trained models for the FS approach and 230 trained models for the MLS approach are available, allowing classification of 16 wear classes based on the blanking force. In the third step, each model from the FS and MLS approaches is evaluated based on the key parameters showing its suitability for a decentralized monitoring of blanking process at the edge. Figure 3 summarizes the procedure.

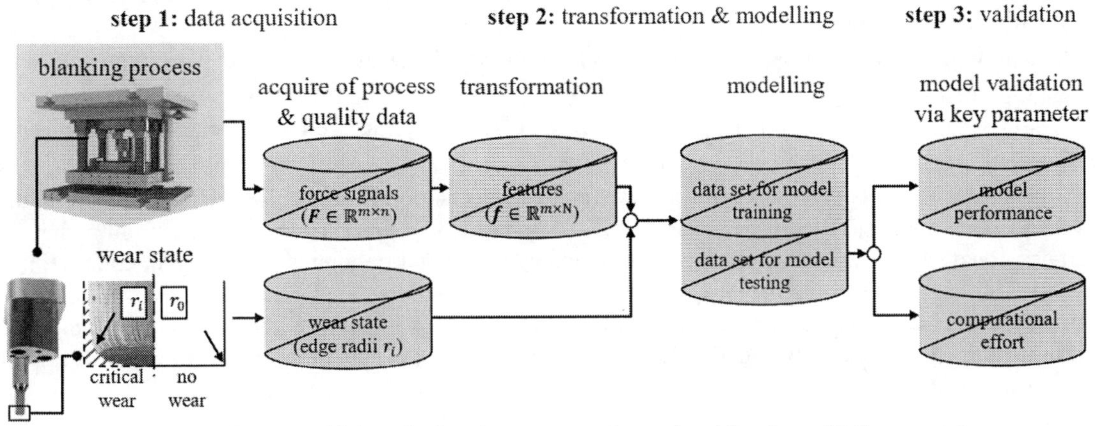

Figure 3: Procedure for quantifying the key parameters for a classification of 16 wear states

4.1. *Step 1: Experimental procedure and data acquisition*

All experiments were carried out on a Bruderer AG (BSTA 810) high-speed press at a stroke height of 35 mm and stroke speed of speeds of 100 spm and 300 spm. The blanking tool selected is a cylindrical punch with a diameter of 6 mm and a clearance of 0.15 mm. A cold rolled steel DC03 (1.0347) with a sheet thickness of 2 ± 0.02 mm and a tensile strength of 299.8 ± 2.9 MPa is selected as the test material. As shown in Figure 2 (a), the blanking tool consists of a lower and an upper part, which are connected by four guiding columns. Thereby, the blanking force is acquired via a piezoelectric force washer (Kistler 9051A) integrated into the direct force flux. The voltage signal of the sensor is digitized by a CompactRIO (NI cRIO 9047) with integrated measuring modules (NI 9215 – analog voltage input ± 10 V) at a sampling frequency of $f_s = 50$ kHz. The final measuring range during the blanking cycle is limited to an angle from 150° to 210° of the ram. This leads to approximately 10550 / 3520 data points per time series at a stroke speed of 100 / 300 spm. Since the data set for each wear state consists of approximately 123 observations, the force matrix is described by $\boldsymbol{F} \in \mathbb{R}^{m \times n}$ ($\boldsymbol{F}_{100} \in \mathbb{R}^{1968 \times 10550}$ and $\boldsymbol{F}_{300} \in \mathbb{R}^{1968 \times 3520}$), where m is the total number of observations for all wears states and n is the number of data points per time series. The labels for the modeling step are described by the cutting edge radii r_i of the punch which quantifies the actual abrasive wear state. They are varied in steps of 0,05 mm, resulting in 16 labeled classes $r_i \in \{0, \dots, 0.75\}$ mm. Since a critical wear state at the cutting edge has to be specified individually for each tool configuration and depends on the resulting geometry of the blanked workpiece (e.g. burr height [38]), a range up to $r_{16} = 0.75$ mm shall be estimated by the model. This maximum limit of excessive wear is slightly above the cutting edge radii of up to 0.4 mm shown in the literature [27]. However, it allows the classification model to be adapted to tool configuration with a higher tolerable cutting edge radii.

4.2. Step 2: Modelling and transformation

Based on the acquired data, features are extracted from the matrix **F**, leading to a feature vector $\boldsymbol{f} \in \mathbb{R}^{m \times k}$, where m is the total number of observations and k the number of extracted features with $k \leq n$. Afterwards, the feature vector serves as an input for the classification model, estimating the 16 abrasive wear states. However, to ensure a comparability between each transformation algorithm combined with a classification model, the number of features must be constant. Since for the FS approach ten engineering features are available, the total number of features for the MLS approach is fixed to 10, leading to a feature vector $\boldsymbol{f} \in \mathbb{R}^{1968 \times 10}$. While for the FS approach only engineering features are extracted for the MLS approach, 23 transformation algorithms are used. Figure 4 (a) summarizes the transformation algorithms for each approach and gives an overview of the applied classification models. In the MLS approach, features are extracted in the time domain, time-frequency domain or model-based. In the time domain, 10 engineering features according to Hohmann et al. are directly extracted from each time series [21]. Therefore, a force signal is divided into three phases, characteristic points are identified within each phase and features are derived describing the length (l_{punch}, l_{push}, l_{withdraw}), the maximum force ($F_{\text{max,punch}}$, $F_{\text{max,push}}$) and minimum force ($F_{\text{min,withdraw}}$), the work done (W_{punch}, W_{push} and W_{withdraw}) as well as the elastic gradient during the punch phase (δ_{punch}).

From the time-frequency domain, the wavelet transformation (WT), Hilbert Huang transformation (HHT) and Wigner Ville distribution (WVD) are used to extract features, transforming the one-dimensional signals into a two-dimensional function of time and frequency [39,40]. While the HHT and the WVD have a fixed mathematical algorithm, the wavelets in the WT are varied according to the coefficients Haar (WThaar), Daubechies with momentum 1 (WTdeb1), momentum 10 (WTdeb10) and momentum 20 (WTdeb20) and Coiflet with momentum 1 (WTcoi1), momentum 3 (WTcoi3) and momentum 5 (WTcoi5) [40].

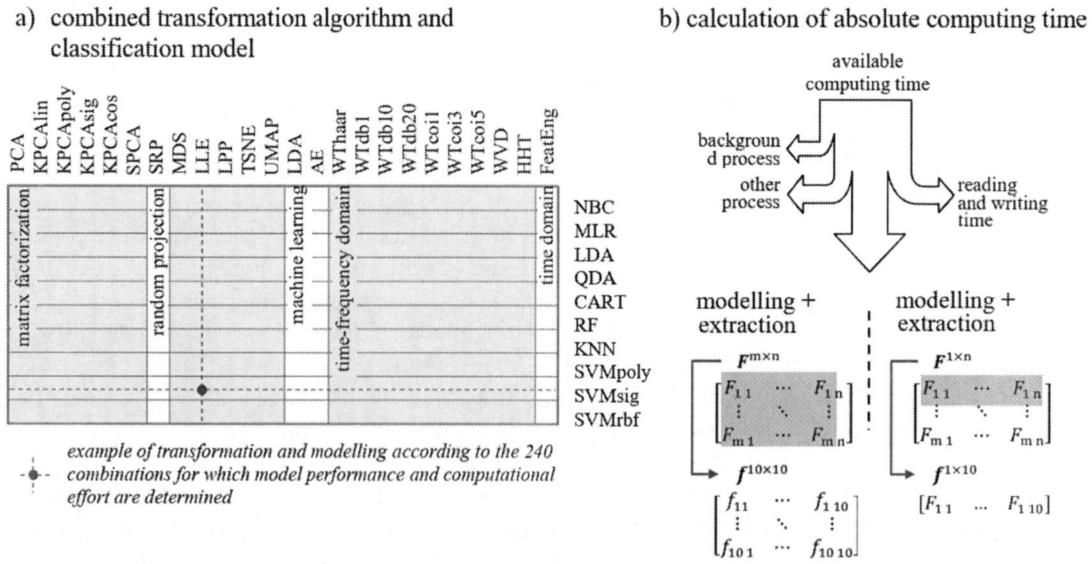

Figure 4: 240 combinations of transformation algorithms and classification models spread over FS and MLS approaches (a) and procedure for calculating the absolute computing time depending on the configuration T_{total} (transformation based on the total data set) and T_{time} (transformation based on a single time series) (b)

Matrix factorization, manifold learning, random projection, as well as ML based models discriminant analysis and AE are used for the model-based transformation. The matrix factorization finds latent structures in a data base and represents them in a lower dimensional space. Representatives of matrix factorization techniques are the principal component analysis (PCA) and its variants kernel PCA

(KPCA) and sparse PCA (SPCA) [41]. While ordinary PCA is a linear method describing its principal components by linear combinations of all input variables, SPCA overcomes this disadvantage by finding linear combinations that contain only a few input variables by adding a constraint on the number of non-zero elements in the matrix [42]. In KPCA, the data is projected by a kernel function into a higher dimension where it is linearly separable. In this study linear (KPCAlin), quadratic (KPCApoly), sigmoid (KPCAsig) and cosine (KPCAcos) kernels are used.

Manifold learning techniques transfer data from a multidimensional space into a two dimensional space via a non-linear mapping. The algorithms considered in this paper are multidimensional scaling (MDS), linear locally embedding (LLE), locality preserving projection (LPP), t-distributed stochastic neighbor embedding (TSNE) and the uniform manifold approximation and Projection (UMAP) [43].

In addition to the model-based approaches, a sparse random projection transformation (SRP) [44], a linear discriminant analysis (LDA) [45] and an AE are used for feature extraction in this study.

As there are many different types of ML methods used for various applications, the transformation techniques are combined with 10 classification models. Therefore, in this study a Naive Bayes classifier (NBC), classification and regression trees (CART), random forests (RF), linear (LDA) and quadratic discriminant analysis(QDA), multiple logistic regression (MLR), k-nearest neighbor (kNN) and support vector machines (SVM) are used to estimate the abrasive wear state with a minimal error. Since the SVM depends on a kernel function that transfers the input into a higher dimensional space, a polynomial (SVMpoly), a sigmoid (SVMsig) and a radial basis (SVMrbf) function are used in this study.

4.3. Step 3: Training procedure

For the training process of the 240 classification models, the sensorial acquired data is divided into 80 % training data and 20 % test data and statistically validated by 50 training cycles. It shows that the time for the training procedure of the classification models is independent of the transformation step which is demonstrated by a row dominance in Figure 5. In particular, models such as LDA and QDA as well as KNN and NBC require less training time. Moreover, they are easy to implement since no hyperparameter optimization has to be conducted. In contrast, SVMs show a training effort that is up to 10 times higher and requires a more or less complex hyperparameter optimization depending on the kernel function used. However, the training process for the investigated classification models is associated with low computational effort and training times of less than 1.5 seconds.

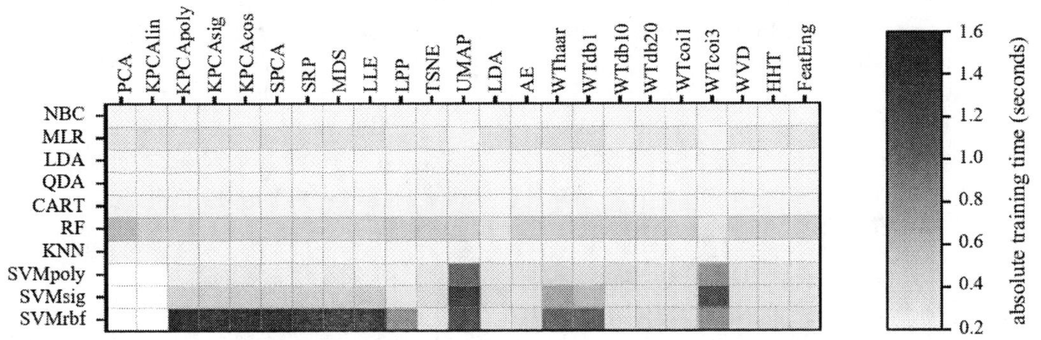

Figure 5: Summarizing the absolute time required to train the classifications model depending on the transformed data

4.4. Step 4: Evaluating efficiency of the DBM via key parameters

In order to compare the efficiency of the DBM, the key parameters model performance and absolute computing time are considered. Since various classification models determine a distribution of classes (e.g. NBC), the receiver operating characteristics (ROC) is used to calculate the model performance depending on its probability distribution over the classes [46]. To compare the ROCs for different models, the area under the curve (AUC) is calculated as a key parameter for the classification model performance. A higher value for the AUC close to 1 indicates that the classes are highly separable by the model. Since a multiple classification problem is considered in this study, N number of AUC are set up for N classes considering the one versus all methodology. The resulting performance parameter is the mean value of the AUC for each class.

Since the absolute computing time is affected by a number of parameters, the strategy shown in Figure 4 (b) is used to quantify this value. Thus only a part of the total computing time provided by the central processing unit (CPU) is available for the execution of the DBM procedure. Other running programs, such as the task manager, browser or background processes, are excluded from the computational time. This also includes the time for reading and writing to the random access memory (RAM), which must be taken into account due to the memory size of the input excess 16 MB of the CPU's internal memory. This makes 50 % of the CPU power available for the DBM process consisting of transformation and modelling. The CPU time is calculated by Equation (1), where n_{IC} is the number of clock cycles for executing the DBM and f_{CPU} is the cycle time of the CPU. Since the absolute CPU time $t_{\mathrm{CPU,abs}}$ depends on the number of cores used in parallel n_{C} and the CPU load A it can be assumed as follows:

$$t_{\mathrm{CPU}} = \frac{n_{\mathrm{IC}}}{f_{\mathrm{CPU}}} \qquad t_{\mathrm{abs}} = \frac{t_{\mathrm{CPU}}}{n_{\mathrm{C}} \cdot A} \tag{1}$$

In order to transfer the results obtained in this study to other computers, the following empirical approach (2) is chosen. It allows the estimation of the required resources, defined by the absolute computing time t_{abs} of the edge device for inline estimation of wear in high speed sheet metal forming processes:

$$t_{\mathrm{abs}} = t_{\mathrm{abs,study}} \cdot \frac{n_{\mathrm{C,study}}}{n_{\mathrm{C}}} \cdot \frac{A_{\mathrm{study}}}{A} \cdot \frac{f_{\mathrm{CPU,study}}}{f_{\mathrm{CPU}}} = t_{\mathrm{abs,study}} \cdot \frac{3.4}{n_{\mathrm{C}} \cdot A \cdot f_{\mathrm{CPU}}} \cdot [\mathrm{GHz}] \tag{2}$$

Table 1: Computer resources used for the examination in this study

type	cores $n_{\mathrm{C,study}}$	clock $f_{\mathrm{CPU,study}}$	physical RAM	average load A_{study}
Intel core i7 4770	2	3.4 GHz	16GB	50 %

Thus, the calculation of the absolute computing time in this study $t_{\mathrm{abs,study}}$ for the inline estimation of the wear state is composed of the transformation and insertion of these features into a trained classification model. The training process and the transformation matrix $\boldsymbol{R}$ generated from the model-based transformation is assumed to be completed. This results in two methods to determine the absolute computing time for the DBM procedure. If the feature extracting time is based on the whole matrix $\boldsymbol{F}$, this is called a transformation from the total data set (T_{total}). If extraction time is based on each time series, this is called a time series based transformation (T_{time}). In this case, $\boldsymbol{R}$ is provided within the training procedure, resulting in a feature vector for the n-the processed workpiece $\boldsymbol{f}^{(\mathrm{n})} = \boldsymbol{R} \cdot \boldsymbol{x}^{(\mathrm{n})}$ with $\boldsymbol{f} \in \mathbb{R}^{(\mathrm{n})\,10\times 1}$, $\boldsymbol{R} \in \mathbb{R}^{10\times 10}$ and $\boldsymbol{x} \in \mathbb{R}^{(\mathrm{n})\,10\times 1}$ (see Figure 4 (b)).

5. Results

The following results quantify the key parameters for real-time estimation of the wear state during blanking considering the absolute computing time, represented by the absolute computing time $t_{\mathrm{abs,study}}$, and the model performance, represented by the classification accuracy AUC. As explained in section 4.3, the absolute time is summed from the time to transform the entire data set (T_{total}) and the time to refeeding it into the model or from the time to transform a single time series (T_{time}) and the time to refeeding it into the model. The configuration T_{time} does not take into account the computing time for the modelling procedure as well as the generation of the model-based transformation matrix $\boldsymbol{R}$. Figure 6 (a) shows the absolute computing time plotted against model performance for a sampling frequency of 50 kHz for the configuration for T_{total} at 100 spm (10550 data points for each time series). Regardless of the combination of transformation algorithm and classification model, most MLS and FS approaches reach an AUC value above 97 %. In contrast, the absolute computing time depends on the combination of transformation algorithm and classification model, and Figure 6 (b) shows the transformation time is higher by 10^3 on average. The least extraction time is required for a random projection. Here, $\boldsymbol{F} \in \mathbb{R}^{m \times n}$ is multiplied by a randomly generated matrix $\boldsymbol{R} \in \mathbb{R}^{n \times k}$ and the dimension is reduced from n to k, so that the computing time is limited to a single matrix multiplication. The computing time for the matrix factorization approach based on an PCA rises since $\boldsymbol{F}$ needs to be decomposed by the sum of eigenvectors of the covariance matrix of $\boldsymbol{F}$. In this case, the eigenvalue problem must be solved, which leads to a higher absolute computing time. In the case of a KPCA, in a previous operation $\boldsymbol{F}$ is transposed into a higher dimensional space to generalize the approach to a non-linear method. This increases the computing time by solving the eigenvalue problem and transferring $\boldsymbol{F}$ to a higher dimensional space.

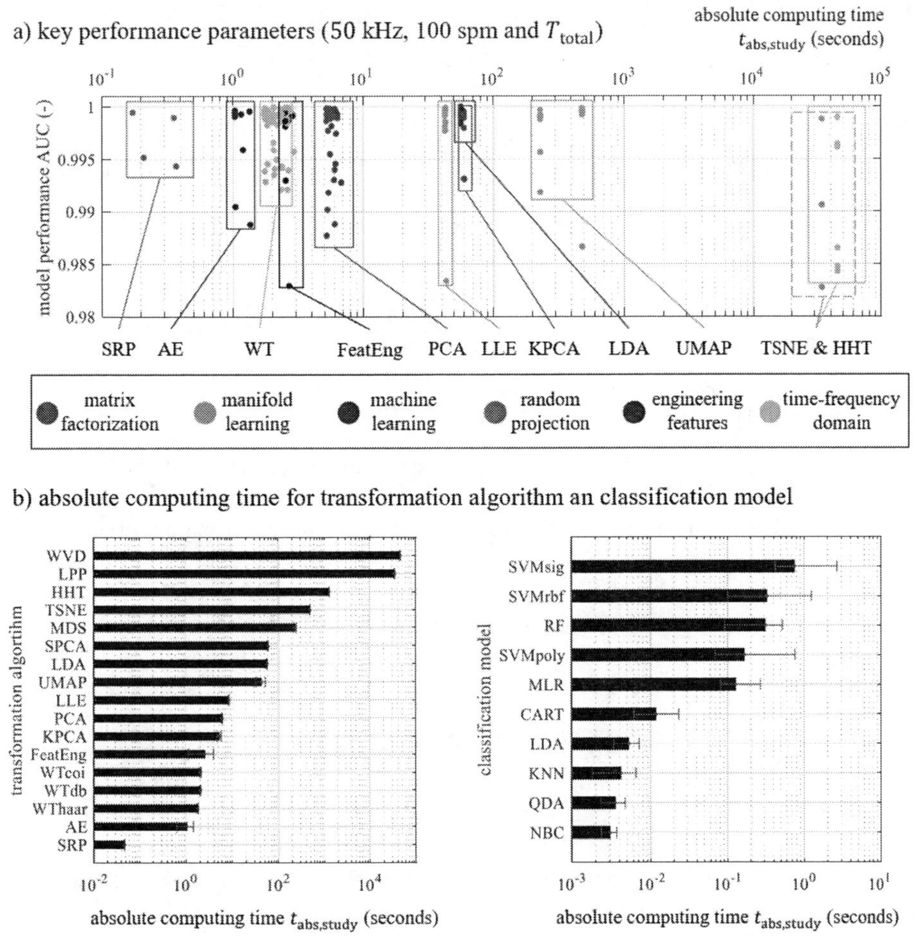

Figure 6: Key performance parameters for the T_{total} configuration (a) and the absolute computing time for different transformation algorithm as well as the time estimating the wear state by different classification models (b)

The transformation time further increases in manifold learning approaches, since they are unsupervised estimators that search for low-dimensional manifolds embedded in high-dimensional space. In doing so, they must iteratively solve a minimization problem to obtain the distances between data points in the high-dimensional matrix $\boldsymbol{F}$ and the matrix mapped to the low-dimensional space. In addition, advanced manifold learning approaches (UMPA and TSNE), attempt to minimize the divergence between two probability distributions provided by input vectors of a high-dimensional matrix compared to the low-dimensional space [47]. By iteratively solving the minimization problem, manifold learning tends to increase absolute computing time. Particularly advanced approaches exceed the absolute computing time (TSNE: 482.44 seconds and UMAP: 42.71 seconds) of other transformation algorithms.

The ML based AE turns out to be another fast algorithm for feature extraction. The structure of the AE is composed of 2 layers (input layer and compression layer) and intermediate activation functions. Here, $\boldsymbol{F}$ is compressed to 10 perceptrons by the compression layer. Since the learning procedure of the AE is based on a simple mathematical operation, multiplying the input with an activation function, the absolute computing time is low. On the other hand, LDA based on a ML classification algorithm divides $\boldsymbol{F} \in \mathbb{R}^{m \times n}$ into k classes. An attempt is made to find a linear transformation $\boldsymbol{W} \in \mathbb{R}^{m \times k}$ that maximizes the distance between the means of all classes and while minimizing the variance within a class, resulting in a higher absolute computing time.

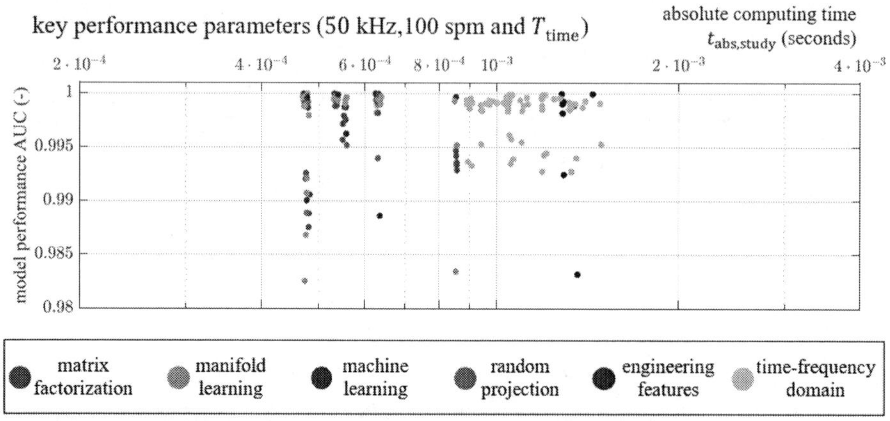

Figure 7: Key performance parameters for the T_{time} configuration

Feature extraction from the time and time-frequency domain requires less absolute computing time than model-based transformations such as matrix factorization and manifold learning methods. This is due to the fact that the absolute computing time for matrix operations increases exponentially with the number of data points, while directly extracting features from a single time series increases approximately linearly with the number of data points. Although the feature extraction in the time-frequency domain requires a pre-transformation into the frequency domain, the absolute computing time is similar to the direct extraction of engineering features. Since the engineering features are based on characteristic points determined in a time series similar to the time-frequency transformation, the FS approach is two-stage.

Implementing the DBM in an industrial application requires less absolute computing time, since the transformation procedure is limited to a single time series (T_{time}). Figure 7 shows the absolute computing time plotted against the model performance for a sampling frequency of 50 kHz for the configuration for T_{time} at 100 spm (10550 data points for each time series). Compared to the configuration T_{total}, the time for processing a stroke related time series is reduced by a factor of 10^3 while the model performance is still close to 100 %. In this context, the model-based transformation is suitable for a real-time application at the edge since the AUC is above 99.99 % and the absolute computing time is lower than 80 ms. This is because the time for the model-based transformation is reduced to a matrix multiplication, where a transformation matrix $\boldsymbol{R}$, determined during the training procedure of the model, is multiplied with the matrix $\boldsymbol{F}$. For example, in the case of PCA-based feature extraction, approximately 4.7 ms are available for a workpiece related estimation of the wear state. Assuming a stroke angle of 150 ° to 210 °, Figure 8 (a) shows the total time available for workpiece related processing of a single time series. As the stroke speed increases, the available data processing time decreases exponentially. Thus, by means of PCA features, a workpiece related estimation of the wear state is possible up to stroke speeds of 880 spm. This does not take into account that at a constant sampling frequency, the number of data points is further reduced for each time series. Not only can the sampling frequency be kept constant, but it can also be reduced further to reduce the absolute computing time. However, it must be verified that the characteristics of the sensorial acquired data still represent the physical state of the process and that the model performance remains at a high level. To this end, this study investigates the dependence of model performance and absolute computing time related to the sampling frequencies (number of samples).

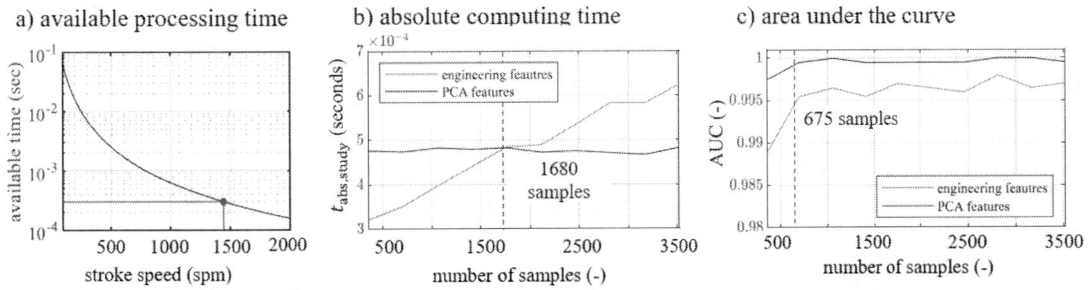

Figure 8: Available time for a workpiece related data processing depending on the stroke speed (a) and the absolute computing time (b) as well as the AUC value (c), depending on the sampling frequency (number of samples) at 300 spm comparing FS and MLS approaches.

Figure 8 (b) and (c) show the key parameters for a combination of engineering features with a QDA (FS approach) and model-based PCA features combined with a QDA (MLS approach) at a stroke speed of 300 spm for the T_{time} configuration. Since the processing time for the model-based features stay constant over the number of samples, the absolute computing time for the extraction of the engineering features decreases. At 1680 samples, which corresponds to a sampling frequency of approximately 25 kHz, the engineering features are at an advantage compared to the model-based approach. Looking at the model's performance, shows that the sampling frequency can be reduced to 675 samples, which corresponds to a sampling frequency of approximately 9.6 kHz.

6. Summary and outlook

The paper shows that DBM is feasible for real-time estimation of wear states during blanking up to strokes speeds above 880 spm. By reducing the sampling frequency to below 10 kHz, the absolute computing time can be further reduced. Even at these reduced sampling frequencies, most of the investigated FS and MLS approaches achieve model performances above 99.99 %. In particular, model-based transformation techniques such as PCA, SRP or AE allow a high estimation quality, while absolute computing time is reduced to a minimum. Furthermore, the authors show that at low sampling frequencies (< 25 kHz), engineering features become advantageous with regard to the computational effort compared to the model-based techniques. To prove the applicability of an edge device for an industrial use case depending on the available computer resources, the study provides an empirical formulation (Equation 2) to estimate the absolute computing time needed for processing data in high speed forming processes.

7. Acknowledgement

The results of this paper are achieved within the project Mittelstand-Digital Zentrum Darmstadt funded by the German Federal Ministry for Economic Affairs and Climate Action (BMWK). The authors wish to thank for funding and supporting this project. Furthermore, the authors would like to thank Bruderer AG for providing the high-speed press BSTA 810-145 and Dayton Progress GmbH for providing the blanking tools under challenging circumstances.

References

[1] Pennekamp J, Glebke R, Henze M, Meisen T, Quix C, Hai R, Gleim L, Niemietz P, Rudack M, Knape S, Epple A, Trauth D, Vroomen U, Bergs T, Brecher C, Buhrig-Polaczek A, Jarke M and Wehrle K 2019 Towards an Infrastructure Enabling the Internet of Production *Int. Conf. on Industrial Cyber Physical Systems (Taipei)* pp 31–37

[2] Lange K 1985 *Handbook of metal forming* (New York: McGraw-Hill)

[3] Kubik C, Hohmann J and Groche P 2021 Exploitation of force displacement curves in blanking-feature engineering beyond defect detection *Int. J. Adv. Manuf. Syst.* **113** pp 261–278

[4] Dilda V *et al* 2017 *Manufacturing: Analytics unleashes productivity and profitability* (McKinsey &. Company)

[5] Wuest T, Weimer D, Irgens C and Thoben K D 2016 Machine learning in manufacturing: advantages, challenges, and applications *Prod. Manuf. Res.* **4** pp 23–45

[6] Hoppe F, Hohmann J, Knoll M, Kubik C and Groche P 2019 Feature-based Supervision of Shear Cutting Processes on the Basis of Force Measurements: Evaluation of Feature Engineering and Feature Extraction *Procedia Manuf.* **34** pp 847–856

[7] Kubik C, Knauer S M and Groche P 2021 Smart sheet metal forming: importance of data acquisition, preprocessing and transformation on the performance of a multiclass support vector machine for predicting wear states during blanking *J. Intell. Manuf.* **33** p 259–282

[8] Hirsch M, Demmel P, Golle R and Hoffmann H 2011 Light Metal in High-Speed Stamping Tools *KEM* **473** pp 259–266

[9] Groche P, Hoppe F, Hesse D and Calmano S 2016 Blanking-bending process chain with disturbance feed-forward and closed-loop control *J. Manuf. Proc.***24** pp 62–70

[10] Shi W, Cao J, Zhang Q, Li Y and Xu L 2016 Edge Computing: Vision and Challenges *IEEE Internet Things J.* **3** pp 637–646

[11] Pan J and McElhannon J 2018 Future Edge Cloud and Edge Computing for Internet of Things Applications *IEEE Internet Things J.* **5** pp 439–449.

[12] Varghese B, Wang N, Barbhuiya S, Kilpatrick P and Nikolopoulos D S 2016 Challenges and Opportunities in Edge Computing *Int. Conf. on Smart Cloud (New York)* pp 20–26

[13] Al-Turjman F 2019 *Edge Computing*: *From Hype to Reality* (Springer International Publishing)

[14] Kitamura K, Makino T, Nawa M and Miyata S 2016 Tribological effects of punch with micro-dimples in blanking under high hydrostatic pressure *CIRP Annals* **65** pp 249–252

[15] Hambli R 2001 Blanking tool wear modeling using the finite element method *Int. J. Mach. Tools Manuf.* **41** pp 1815–29

[16] Archard J F and Hirst W 1956 The wear of metals under unlubricated conditions *Proc. of the R. Soc. of Lond.* **236** pp 397–410

[17] Mucha J 2010 An experimental analysis of effects of various material tool's wear on burr during generator sheets blanking *Int. J. Adv. Manuf. Technol.* **50** pp 495–507

[18] Mangalore Babu S S, Berry S, Ward M and Krzyzanowski M 2018 Numerical investigation of key stamping process parameters influencing tool life and wear *Proc. Manuf.* **15** pp 427–435

[19] Hambli R 2002 Design of Experiment Based Analysis for Sheet Metal Blanking Processes Optimisation *Int. J. Adv. Manuf. Syst.* **19** pp 403–410.

[20] Cheung C F, Lee W B and Chiu W M 2000 An investigation of tool wear in the dam-bar cutting of integrated circuit packages *Wear* **237** pp 274–282

[21] Hohmann J, Schatz T and Groche P 2017 Intelligent Wear Identification Based on Sensory Inline Information for a Stamping Process *Proc. of 5th Int. Conf. on Advanced Manufacturing Engineering and Technologies (Belgrade)* pp 285–295

[22] Cheon S and Kim H 2016 Prediction of tool wear in the blanking process using updated geometry *Wear* **352-353** pp 160–170

[23] Falconnet E, Chambert J, Makich H and Monteil G 2015 Prediction of abrasive punch wear in copper alloy thin sheet blanking *Wear* **338-339** pp 144–154

[24] Hoffmann H, Nürnberg G, Ersoy-Nürnberg K and Herrmann G 2007 A new approach to determine the wear coefficient for wear prediction of sheet metal forming tools *Prod. Eng.* **1** pp 357–363
[25] Marposs Monitoring Solutions GmbH Brankamp 2020 *X3 - All-in-one Überwachung für die Metallumformung* (http://www.brankamp.com/de/data/files/produkte/prospekt_x3_de.pdf)
[26] Anasagasti M, Godoy L, Socorro R, García R, Iturrospe A and Viguera M 2016 *MANTIS project: D1.2 Consolidated State-of-the-Art of Sensor* (ECSEL Joint Undertaking)
[27] Klingenberg W and de Boer T W 2008 Condition-based maintenance in punching/blanking of sheet metal *Int. J. Mach. Tools Manuf.* **48** pp 589–598
[28] Liewald M, Bergs T, Groche P, Behrens B-A, Briesenick D, Müller M, Niemietz P, Kubik C and Müller F 2022 Perspectives on data-driven models and its potentials in metal forming and blanking technologies *Prod. Eng. Res. Devel.* **7**
[29] Li C J Signal Processing in Manufacturing Monitoring 2006 *Condition Monitoring and Control for Intelligent Manufacturing* vol 21, ed L Wang and R X Gao (London: Springer) pp 245–265
[30] Lee W B, Cheung C F, Chiu W M and Chan L K 1997 Automatic supervision of blanking tool wear using pattern recognition analysis *Int. J. of Mach. Tools Manuf.* **37** pp 1079–95
[31] Jin J and Shi J 2000 Diagnostic Feature Extraction From Stamping Tonnage Signals Based on Design of Experiments *J. Manuf. Sci. Eng.* **122** pp 360–369
[32] Hambli R 2002 Prediction of burr height formation in blanking processes using neural network *Int. J. Mech. Sci.* **44** pp 2089–2102
[33] Ge M, Du R, Zhang G and Xu Y 2002 Fault diagnosis using support vector machine with an application in sheet metal stamping operations *MSSP* **18** pp 143–159
[34] Zhou S and Jin J 2005 Automatic feature selection for unsupervised clustering of cycle-based signals in manufacturing processes *IIE Transactions* **37** pp 569–584
[35] Asahi S, Karadogan C, Tamura S, Hayamizu S and Liewald M 2021 Process data based estimation of tool wear on punching machines using TCN-Autoencoder from raw time-series information *IOP Conf. Ser. Mater. Sci. Eng.* **1157** p 12078
[36] Niemietz P, Unterberg M, Trauth D and Bergs T 2021 Autoencoder based Wear Assessment in Sheet Metal Forming *IOP Conf. Ser. Mater. Sci. Eng.* **1157** p 12082
[37] Molitor D A, Kubik C, Hetfleisch R H and Groche P 2022 Workpiece image-based tool wear classification in blanking processes using deep convolutional neural networks *Prod. Eng. Res. Devel.*
[38] Feistle M, Koslow I, Krinninger M, Golle R and Volk W 2017 Reduction of Burr Formation for Conventional Shear Cutting of Boron-alloyed Sheets through Focused Heat Treatment *Proc. CIRP* **63** pp 493–498
[39] Caesarendra W and Tjahjowidodo T 2017 A Review of Feature Extraction Methods in Vibration-Based Condition Monitoring and Its Application for Degradation Trend Estimation of Low-Speed Slew Bearing *Machines* **5** p 21
[40] Hosameldin A and Asoke K. N 2019 *Condition Monitoring with Vibration Signals: Compressive Sampling and Learning Algorithms for Rotating Machines*, (New Jersey: Wiley-IEEE Press)
[41] Bart H *et al* 2008 *Factorization of Matrix and Operator Functions: The State Space Method*, (Basel: Birkhäuser)
[42] Bonaccorso G 2018 *Machine Learning Algorithms Second Edition*, (Birmingham: Packt Publishing)
[43] Ma Y and Fu Y *Manifold Learning Theory and Applications*, (Boca Raton: CRC Press)
[44] Li P, Hastie T J and Church K W 2006 Very sparse random projections in *Proc. of the 12th Int. Conf. on Knowledge discovery and data mining (New York)* pp 287-296
[45] Gu Q, Li Z and Han J Linear Discriminant Dimensionality Reduction in *Machine Learning and Knowledge Discovery in Databases*, vol. 6911 pp 549–564, Springer, Heidelberg 2011.
[46] Huang J and Ling C X 2005 Using AUC and accuracy in evaluating learning algorithms *IEEE Trans. Knowl. Data Eng.* **17** pp 299–310

[47] van der Maaten L 2014 Accelerating t-SNE using tree-based algorithms *J. Mach. Learning Research* pp 3221–45

Predicting Dynamic Process Limits in Progressive Die Sheet Metal Forming

D Budnick[1], A Ghannoum[1], F Steinlehner[2], A Weinschenk[3], W Volk[2], S Huhn[3], W Melek[1] and M Worswick[1]

[1] Department of Mechanical and Mechatronics Engineering, University of Waterloo, Waterloo, Canada
[2] Chair of Metal Forming and Casting, Technical University of Munich, Garching, Germany
[3] Forming Technologies, Hexagon Manufacturing Intelligence, Oakville, Canada

dbudnick@uwaterloo.ca

Abstract. Tool makers have a limited selection of tools and are afforded limited flexibility during progressive die try-outs when attempting to identify suitable process control parameters and optimize throughput. The performance of a given tooling design hinges on selecting a suitable stroke rate for the press. Cost efficiencies are realized when operating a press at higher stroke rates, but risk subjecting the sheet metal strip to larger, uncontrolled oscillations, which can lead to collisions and strip-misalignment during strip progression. Introducing active control to the strip feeder and lifters can offer increased flexibility to tool makers by allowing the strip progression to be fine-tuned to reduce strip oscillations at higher stroke rates. To alleviate uncertainties and assist in fine-tuning the process control parameters, machine learning models, such as an artificial neural network, are constructed to predict whether a given set of process parameters will lead to a collision or strip-misalignment during the strip progression. The machine learning models are trained using a dataset of FEA simulations which model the same progressive die operation using different process control inputs for the feeder, lifter and press. The machine learning models are shown to be capable of predicting the outcome of a given process permutation with a classification accuracy of about 87 % and assist in identifying the dynamic process limits in the progressive die operation.

1. Introduction

With the advancement of technologies and concepts that embody 'Industry 4.0', most industries have seen some semblance of impact from these technologies with sheet metal forming being no exception. It only takes a glance at literature review papers or conferences on sheet metal forming to notice the increase in papers that leverage advances in machine learning, sensing technology, and, advanced control. Work done by Jamli and Farid [1] reviews the use of Artificial Neural Network (ANN) to assist in predicting sheet metal spring back. They note that one benefit of using ANN is that it overcomes the complexities in constructing constitutive models that incorporate spring-back behavior and can learn directly from experimental data. Their review extends to the topic of aligning the benefits of Finite Element Analysis (FEA) and ANN to simultaneously reduce experimental costs and enhance prediction accuracy. Often these approaches focus on an inverse analysis and use the ANN trained on Finite Element (FE) simulations to more rapidly model a process with the goal of optimizing a set of process parameters. Manoochehri and Kolahan [2] generated data from FEA to train an ANN which was used to find optimal deep drawing process parameters such as blank holder force by using simulated annealing in conjunction with the ANN to minimize thinning. With respect to sensing and control technology in sheet metal forming, a review by Polyblank *et al.* [3] provides a sense for the breadth of more recent advances. They point out that while most equipment used in sheet metal forming uses closed-loop control, current implementations seen in industry rarely account for the behavior of the product and instead focus on precisely following tool-paths. More recent approaches focus on incorporating the product properties into the control loop. One such approach implemented by Endelt *et al.* [4] controlled material flow in a deep-drawing process by measuring flange draw-in with laser displacement sensors and adapting binder pressure in a closed-loop control system. However as pointed out by Allwood *et al.* [5], most of the applications for novel control implementation in sheet metal forming focus on controlling product geometry or preventing failure.

There has been limited research, which investigates control strategies that seek to minimize the probability of collisions between the work piece and tooling, particularly during progressive die operations in which a sheet metal strip is fed into a series of forming stations (tooling). Examples of such research pertain exclusively to sheet metal forming processes which seek to find an optimal collision-free tool path by using mechanical grippers to transfer the sheet metal product between operations [6,7]. In some cases, the compliance of the sheet metal is even considered in the trajectory optimization problem [8,9]. However, in the case of progressive die operations, passive spring-operated lifters raise the strip and a strip-feeder, which is in-sync with the press stroke rate, is used to progress the product between die operations. With the exception of the press stroke rate, tool operators have limited control over progressive die operations. Without additional forms of control, progressive die operations have an inherent process speed limit for a given design since increasing the stroke rate will result in the strip exhibiting a larger dynamic response which eventually may result in a collision or reduced part quality [10]. By introducing adjustable inputs to the strip feeder and lifters, excessive strip dynamics can potentially be reduced at higher stroke rates, increasing the overall maximum process speed. However, the strip layout, and in particular, the stretch-web connector selection, will have a large influence on the dynamic response of the strip. Stretch webs must be designed with the necessary compliance to permit material flow in deep-drawing operations. If the stretch web is design with too much compliance, the strip may exhibit a larger dynamic response during strip progression, increasing the chances of a collision or poor placement of the product on the subsequent die.

This paper presents a systematic approach to predicting whether a collision will occur during a progressive die operation between the tooling and sheet metal strip. This approach generates data using FE simulations, which model a progressive die operation with varying inputs to the strip feeder, lifters, and press stroke rate. Data was generated for two strip layouts, which utilize different stretch-web connectors. The generated data is subsequently used to train a machine learning model, such as an ANN, which predicts whether a given process permutation will result in a collision between the sheet metal strip and the tooling or misalignment with the locating pins due to excessive strip dynamics. The goal of this work is to create a framework for identifying the dynamical process limit of a given progressive die operation and the limitation invoked by the stretch-web selection.

IOP Conf. Series: Materials Science and Engineering 1238 (2022) 012068 doi:10.1088/1757-899X/1238/1/012068

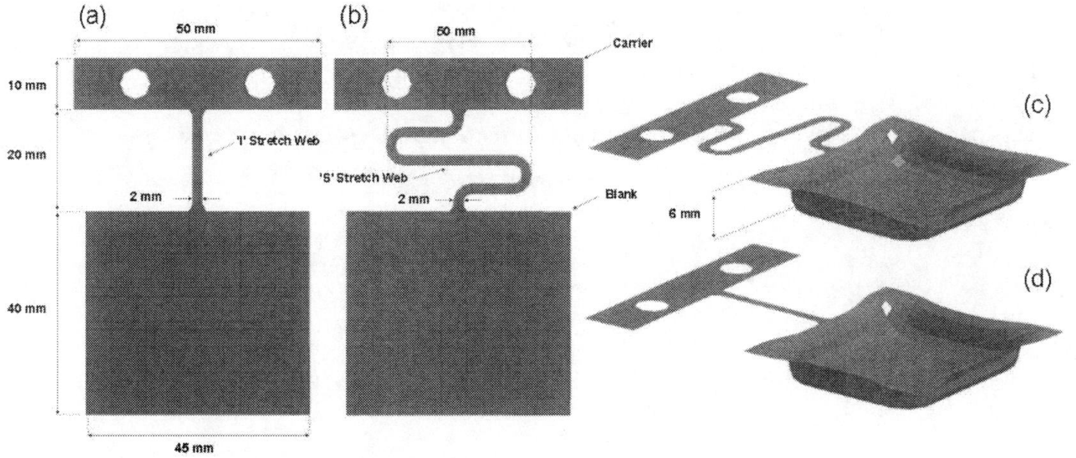

Figure 1: Mesh and dimensions for (a) 'I' stretch web strip (b) 'S' stretch web strip (c) 'S' stretch web with formed cup, (d) 'I' stretch web with formed cup.

2. Methodology

There are three distinct phases to this research, each of which are described in the following: (1) FE Modeling, (2) Data Generation, (3) Machine Learning. The physical problem being modelled comprises the transfer process and two different strip layouts are modelled using the same progressive die tooling and transfer system. The difference between these two strip layouts is limited to the stretch-web selection. As seen in Figure 1, the stretch-web in the strip layout is either selected as an 'I' or an 'S' web. A simple square-shaped cup is produced in the progressive die operation.

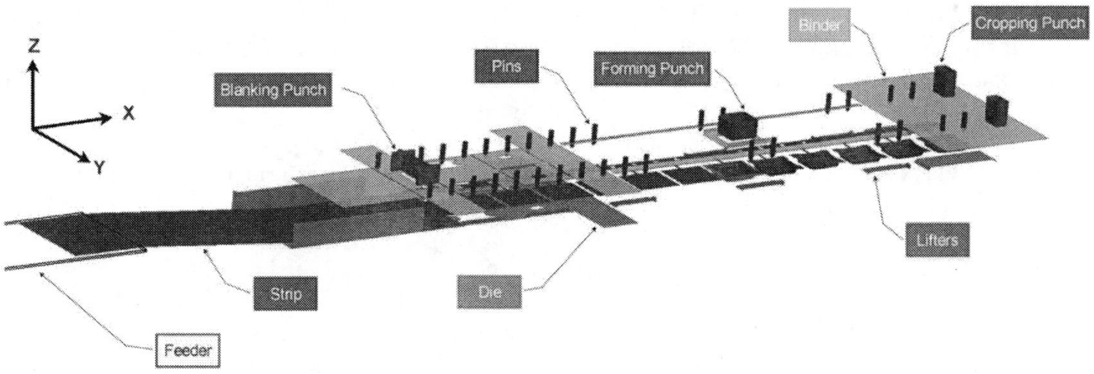

Figure 2: Simulation with labelled components. The strip is in its initial configuration, but, the binder, pins, and punches are lifted for visualization.

2.1. *Finite Element Model*

The FE simulation can be seen in Figure 2. All the simulations were completed using the LS-DYNA explicit dynamic solver R9.3 MPP with 16 cores on an Intel Xeon 8160 cluster. The progressive die operation consists of three fundamental operations: blanking, forming, and cropping. The current paper is focused on the dynamic response of the strip; consequently, the actual forming and trimming operations were not modelled and the part geometry was assigned as shown in Figure 1. The blanks ahead of the forming station were initialized using the flat blank geometry, while those after the forming station were assigned the formed part geometry. This simplification was adopted due to the long run-times of FE simulations, which stem from the requirement to model the dynamic response in real time without the use of mass scaling. The tooling elements were modelled as rigid bodies and penalty function-based contact treatment was prescribed to capture collisions with the oscillating strip during the strip lift and advance motion. The simulation begins as the upper tooling is rising and ends once the binder clamps the strip at the next die station for a total of 256° of the crank angle. The progressive die operation was designed in a way to intentionally expose the strip to more risk of increased dynamic response and, as result, may not reflect design best-practices seen in industry. The reader is referred to the paper by Budnick *et al.* [10] in which the deformation of the part being formed is considered.

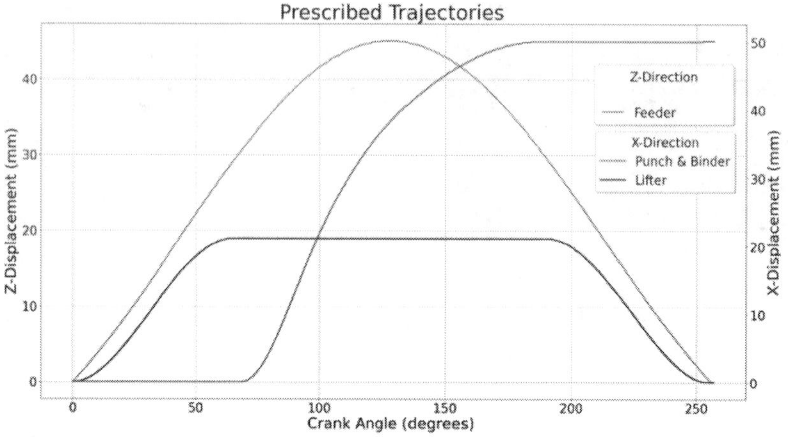

Figure 3: Prescribed trajectory for the binder, punch, lifters, and feeder.

The generated FE model is similar to that implemented in previous work by the current authors [10] with some notable differences. All motion of the punch, binder, feeder, lifters, and locating pins follow a prescribed motion. Additionally, the lifter actuation also takes place over a longer portion of the press crank angle then is possible with passive-actuation. All of the prescribed boundary conditions can be seen in Figure 3. The number of through-thickness integration points used for the stretch web was 5 and 3 for the rest of the strip. Since the focus of these simulations were to capture the dynamic response, which is largely an elastic behavior, the reduction of through-thickness integration points was justified for the advantage of shorter run-times. No mass scaling was applied in any of the simulations in order to capture the true dynamic response of the strip.

The blanks are modelled as rigid bodies to reduce the number of deformable elements and reduce the run-time. Fully-integrated shell elements are used for the rest of the strip. The strip is a 1 mm thick sheet of A5182 and modelled using a von Mises plasticity model with a yield strength of 121.4 MPa and isotropic strain hardening. Frequency independent damping was applied using 0.75 % critical damping over the first five natural frequencies. Penalty-based contact definitions were applied between the strip and the die, punch, pins, lifters, and binder for a total of five contacts.

Finally, only the first advance of the strip was modelled to simulate the initial dynamic oscillation. (In on-going work, multiple stroke simulations are being modelled to capture the steady-state response.) This initial response was used to train the machine learning models and a total of 1000 simulations were run, 500 for each strip layout.

2.2. Data Generation

The input permutations for the feeder and lifter motion used for each FE simulation were randomly sampled from a set of possible stroke rates and trajectories for the lifter and feeder. The permissible stroke rates varied between 60 and 360 strokes per minute (SPM).

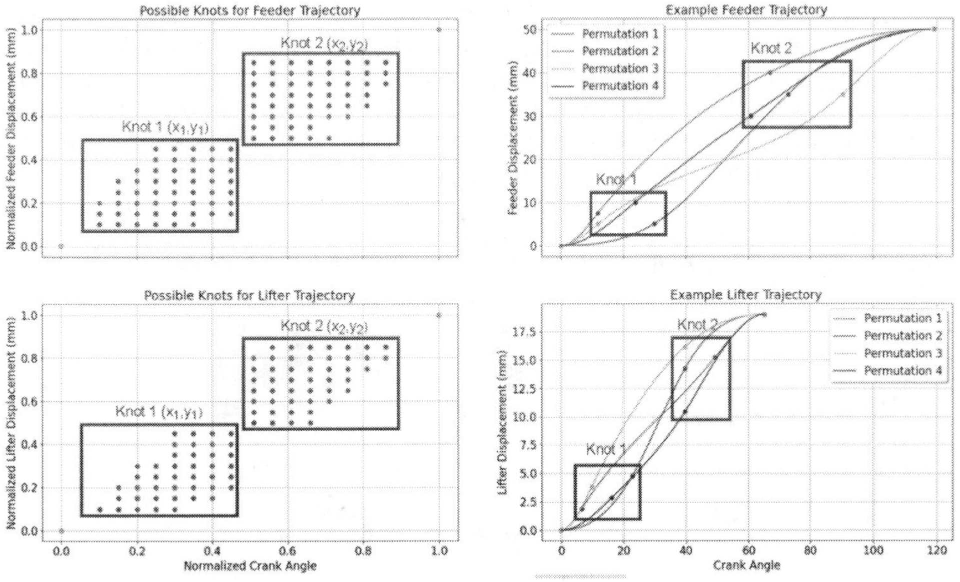

Figure 4: Possible knots for lifter and feeder trajectories (left) and examples of feeder and lifter permutations with the selected 'Knot 1' and 'Knot 2' indicated by the box (right).

2.2.1. Trajectory Selection. The set of possible trajectories for a given permutation were nearly identical for both the lifter and feeder. Cubic splines were used to define each trajectory using a total of four knots with boundary conditions that prescribe an initial and final velocity to be set to zero. Examples of possible trajectories can be seen in Figure 4. The first and last knots are fixed to the end points of the curve to define start and end points with only the inner-knots ('Knot 1' and 'Knot 2') being varied between permutations. To reduce the complexity of the solution-space to ensure that the ANN would have a reasonable chance to predict the behaviour of the system, further restrictions were applied to limit the size of the permutation space (Figure 4, left). All splines were required to be monotonically increasing to prevent the feeder or lifters from retracing their displacement and inverting their motion. As well, the velocity was restricted to avoid saddle points and intermittent periods of near zero velocity. More specifically, the normalized velocity was required to be above 0.25 between the inner two knots. Furthermore, the inner knots were restricted to a sampling area such that the X and Y normalized coordinates of 'Knot 1' are bounded by [0.1, 0.5] and 'Knot 2' is bounded by [0.5, 0.9]. Finally, the lifter trajectory was restricted to not rise faster than the upper tooling since this would cause a tooling collision.

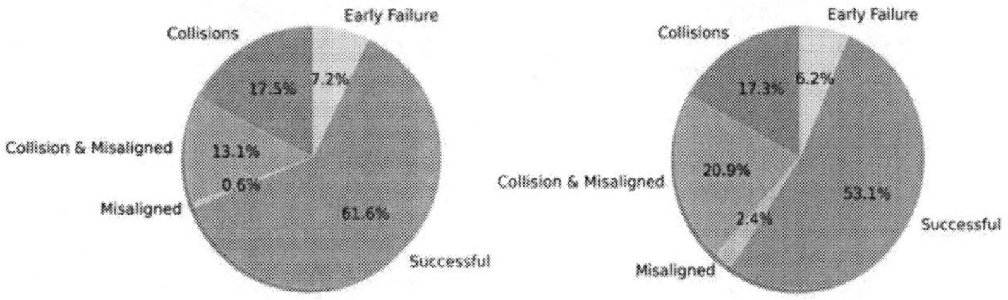

Figure 5: Simulation outcome for (left) 'I' stretch web connector and (right) 'S' stretch web connect.

2.2.2. Data Labelling. A critical component to constructing an effective machine learning model is first ensuring that the data being used is properly labelled and reflects the system that is being modelled. Referring to Figure 5, four possible outcomes were identified and labelled for each simulation:

'collision', 'strip misaligned', 'early failures', and 'successful'. (1) A simulation is labelled to have a 'collision' if there is a non-zero contact force with the tooling during the strip progression. This detection omits the initial strip lifting period since there can often be chatter between the binder and strip during this phase. (2) Referring to Figure 6, a 'misalignment' is labelled when the locating pins fail to locate the strip. Determining whether the pins accurately located the strip is done by measuring the position of a node at each locating hole and evaluating whether it is aligned with the pin within the tolerance allowed by the geometry, which is taken as 0.1 mm in this case. (3) 'Early failures' are simulations which showed erratic behaviour as a result of errors in the FE simulation setup. (4) 'Successful' simulations are any permutation that did not experience a collision, strip misalignment or early failure for which the strip was properly placed and located at the next die station.

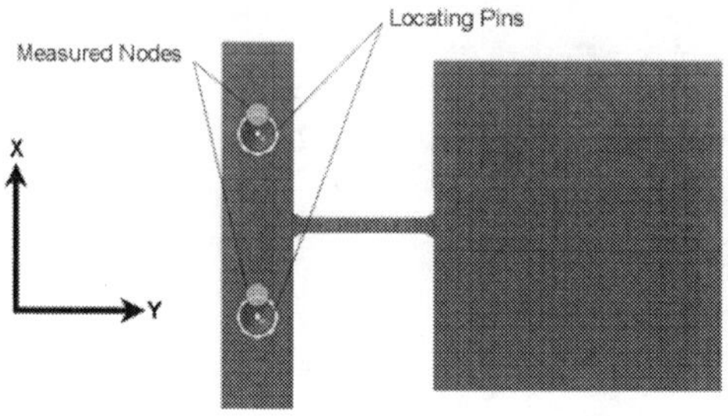

Figure 6: Single section of strip with labels indicating nodes measured for strip alignment.

With the outcome of each simulation labelled, the data is then prepared for the dataset used in the machine learning models. All of the simulations which resulted in early failures were removed from the dataset. The goal of the machine learning model is to predict whether a given set of process inputs will cause excessive strip dynamics and result in a poor outcome, or, in other words, a collision or strip misalignment. Any permutation that results in a collision or strip misalignment are classified as an unsuccessful permutation with all remaining permutations being classified as successful in this binary classification problem. Therefore, the inputs to the network are the feeder and lifter inputs and the stroke rate with the output being whether the simulation was 'successful' or 'unsuccessful'. The inputs and outputs of the network are summarized in Table 1: Labelled inputs and outputs for generated data. The data was split into a training and test set using an 80/20 split. The input variables for the training set were normalized and the same normalization scale was applied to the inputs of the test set.

Table 1: Labelled inputs and outputs for generated data.

	Inputs									Outputs
Variable	Feeder Knots				Lifter Knots				Press (SPM)	Succesful / Unsuccesful
	X_1	Y_1	X_2	Y_2	X_1	Y_1	X_2	Y_2	Stroke Rate	
Value	Normalized [0,1]									Binary {0,1}

2.3. *Machine Learning*

Two types of machine learning architectures were implemented and compared: (1) Artificial Neural Network, and (2) XGBoost. Both machine learning architecture were trained and tested using Python 3.8.3 on an Intel Core i9-9900K CPU. Separate models were implemented for each strip layout for a total of four machine learning models (two for each strip layout). After removing the simulations that resulted in 'early failures', the 'I' and 'S' stretch webs had dataset sizes of 461 and 466 samples respectively. ANN's have been shown to perform well on a wide variety of non-linear modeling problems and have been used widely in the literature. XGBoost was also chosen to be used as a comparison against the ANN as it has shown to perform well on a wide variety of problems and handle smaller datasets well, even with a larger feature size. XGBoost is a decision-tree based model, which uses an ensemble learning method based on the random forest approach to design a classifier using gradient boosting [11].

To select the hyperparameters used for each model, two rounds of hyperparameter grid-searches were performed on both models using the training data. To evaluate the performance of the hyperparameter selection for each model, K-fold cross validation (KCV) was used with 5 folds to assess the generalizability of each model by measuring the variance and average test accuracy. KCV is implemented by randomly splitting the dataset into K-folds of even size and training the model on K-1 folds and testing on the remaining fold [12]. Once hyperparameters are selected, the model is trained using the entire training data and tested on the test set to evaluate the generalizability of the selected models from the KCV hyperparameter selection procedure. The same architecture and hyperparameters were found to work well for both stretch webs and therefore the parameters were not changed between datasets. The architecture of the selected ANN has a single hidden layer with 8 neurons using a 'tanh' activation function and a single network output with a 'sigmoid' activation function and was implement using TensorFlow 2.6.0. The network was trained using small-batch gradient descent algorithm with 50 epochs, a batch-size of 32 and a learning rate of 0.005 using the Adam optimizer. The two ANN models were trained on FE simulation data for the 'I' and 'S' stretch web data respectively and used binary cross-entropy loss which can be seen in equation (1):

$$Loss = -\frac{1}{N}\sum_{i=1}^{N} y_i \cdot \log(\hat{y}_i) + (1 - y_i) \cdot \log(1 - \hat{y}_i) \quad (1)$$

where N is the number of data samples, y_i is the 'true' label of a data point, and $\hat{y}_i$ is the output of the machine learning model. The selected XGBoost algorithm was implemented with XGBoost version 1.4.2 for Python and uses a learning rate of 0.1, a max tree-depth of 2 and the regularization term, gamma, set to 0.1. The two XGBoost models were trained on the data collected for the 'I' and 'S' stretch web from the FE models and used a binary cross-entropy loss function.

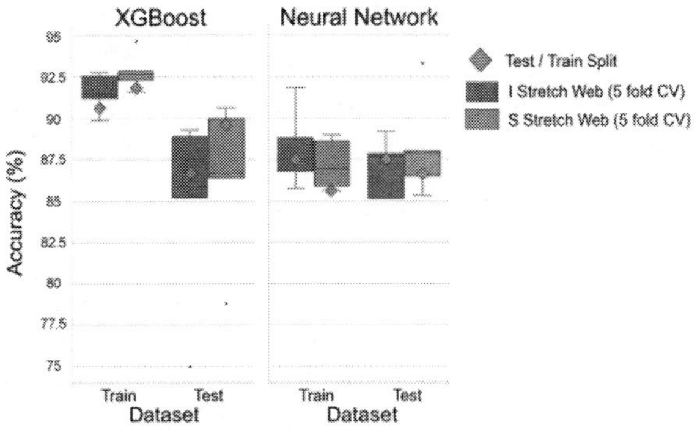

Figure 7: 5-Fold cross-validation accuracy, and test and training accuracy for each model.

3. Results & Discussion

To measure the performance of the machine learning models, prediction accuracy is assessed in terms of the number of correctly predicted simulation outcomes divided by the total number of simulations. The training and test results for the ANN and XGBoost algorithm for both stretch webs can be seen in Figure 7. The KCV results are shown using the box plots while performance of the selected models trained on the entire training data are indicated by the triangles. In general, all models performed reasonably well with an average test accuracy of about 87 %. As well, both the XGBoost and ANN models generally performed comparatively on the 'I' stretch web data and the 'S' stretch web data for both test and training sets indicating the data for both strip layouts is equally separable. Furthermore, the XGBoost models had a larger discrepancy between the test accuracy and training accuracy, as compared to the ANN, implying that the XGBoost models were subject to more overfitting. The discrepancy between prediction accuracy on the datasets reflects the hyperparameter tuning process for the XGBoost model, which was highly sensitive to overfitting the training data. Finally, both models appeared to generalize well as they achieved similar accuracy on the test set as compared to the KCV results.

To demonstrate the predictive capability of the machine learning models, the FE simulation predictions of the minimum distance measured along the vertical (press stroke) direction between the strip and binder during the strip progression are plotted as a function of stroke rate and coloured by the prediction of the XGBoost classifiers for both stretch webs overlaid (Figure 8). The predicted distance in the FE models provides a quantitative measure of how close the control inputs of the feeder and lifter were to promoting a collision. If the minimum distance is small or negative, it indicates that portions of the strip overshot the position of the upper tooling, which makes it highly susceptible to a collision. The minimum distance between the binder and strip is calculated by measuring the center-of-mass of the nearest blank with respect to the binder and therefore there will be cases for which there is a collision when there is still a positive distance. The scatter plot has large regions of correctly predicted 'successful' permutations in the upper left corner where the stroke rate tended to be low and there was a large distance between the binder and strip. Inversely, there was a large collection of conditions correctly predicted as 'failures' in the bottom right corner of the plot where the stroke rate tended to be higher, and the distance was minimal. It should be noted that when the minimum distance is negative, the strip has overshot the binder, which is likely around the forming station where there is a gap in the binder.

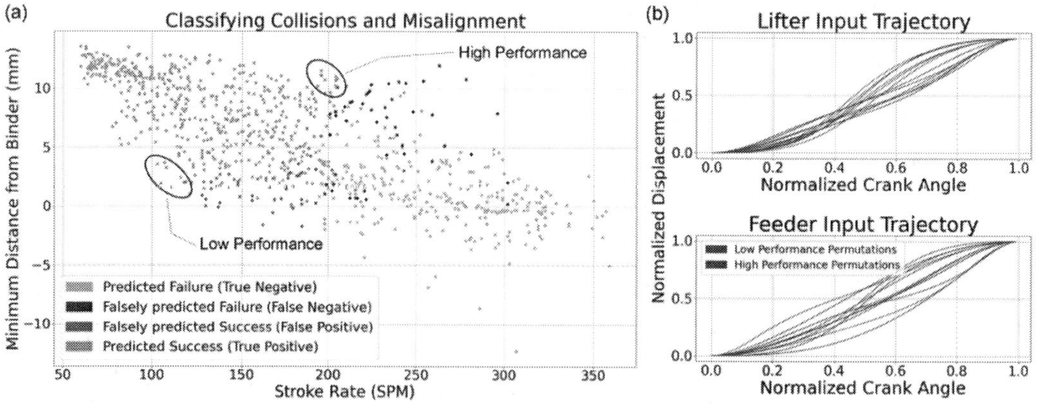

Figure 8: (a) Plot of stroke rate *vs.* the minimum vertical distance along the press stroke direction between the strip and binder as predicted by the FE models. The symbols are coloured based on the prediction of the XGBoost classifiers trained on the entire training set and evaluated on the test and training sets for each stretch web. (b) Feeder and lifter trajectories for three of each stretch web types which are 'low performance' (red) and 'high performance' (green) permutations, as indicated by the ellipses in (a).

Orange and green samples are correctly predicted 'failures' and 'successful' respectively (true negative and true positive). The pink and blue samples are permutations that were incorrectly predicted as 'successful' or as 'failures' respectively (false positive and false negatives). These misclassified permutations are the result of the given permutations outperforming the trend (false negatives) by avoiding failures at higher strokes and underperforming (false positives) by failing at relatively lower stroke rates.

Figure 8 (b) shows examples of feeder and lifter input trajectories for select permutations that were either 'low performing' and had minimal distance at lower stroke rates or were 'high performing' and achieved high stroke rates with relatively larger distance between the strip and binder. The selected permutations were comprised of three of each stretch web type for a total of six 'high performing' and six 'low performing'. The 'high performing' permutations follow a more linear path for the lifter input whereas the 'low performing' lifter inputs followed a path similar to a sigmoid function. However, in the case of feeder inputs, the trends are less evident, with a variety of inputs shapes comprising both the high and low performance permutations. This lack of trend is explained by the fact that the minimum

IOP Conf. Series: Materials Science and Engineering 1238 (2022) 012068 doi:10.1088/1757-899X/1238/1/012068

distance between the binder and strip is related to how much the blank overshoots the nominal lifting value and is directly influence by the lifter trajectory and not the feeder. The feeder trajectory has a larger impact in causing the strip to be misaligned with respect to the locating pins as well as how the blank is placed on the next tooling station.

In both cases, the excessive dynamic response of the blank is the source of the collision or strip misalignment and is initiated during the strip lifting. Once there is an excessive dynamic response of the strip, then regardless of the feeder trajectory, a collision or strip misalignment is highly probable. Therefore, since the lifter trajectory is responsible for causing excessive excitations to the blank which can result in collision or strip misalignment, there is a clear trend with respect to the lifter inputs and the minimum distance measured between the strip and binder in the FE simulations.

Ultimately, Figure 8 sheds light as to where the machine learning models struggled, and the data becomes difficult to predict. The models perform well at higher and lower stroke rates since regardless of the feeder and lifter inputs, the model can interpret based on the SPM whether or not a failure is likely. In the midrange of stroke rates where the feeder and lifter inputs tend to play a more integral role, the model begins to struggle at classifying all the results properly. The reduced classification accuracy in the midrange of stroke rates is due in part from the limited samples in both datasets and the size of the permutation space that the models are trying to learn. As well, there is a clear correlation between the stroke rate and failure and all the machine learning models will naturally assign larger weights to the input of the stroke rate and lower weights to the feeder and lifter inputs. Machine learning, and in particular, ANN provide the flexibility to employ techniques such as data augmentation, to generate additional data points, and transfer learning, to retrain the ANN models, to enhance the classification accuracy where there is opportunity for improvement, such as the midrange of SPM. Nonetheless, the current machine learning models can identify with reasonable accuracy the dynamical process limit of the progressive die operations by testing new feeder and lifter inputs and identifying the maximum stroke rate at which the model predicts a 'successful' outcome.

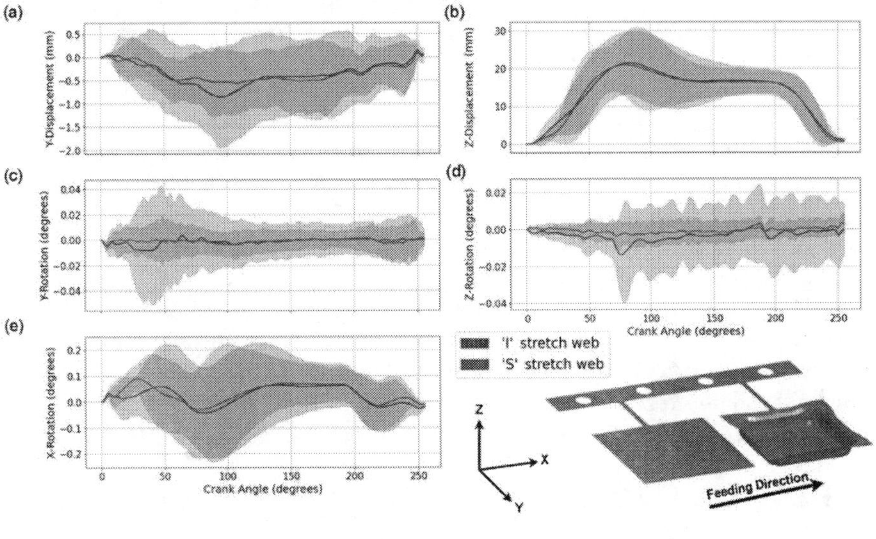

Figure 9: Dynamic response of the blank measured for 'successful' permutations in (a) y-displacement, (b) z-displacement, (c) y-rotation, (d) z-rotation, (e) x-rotation. The thin lines represent the mean and the coloured regions bound the mean by two standard deviations.

While both stretch webs appear to show a similar trend between the minimum distance from the binder and the stroke rate (Figure 8), the differences between the stretch webs are clearer when directly measuring the dynamic response of a given blank measured in the FE models (Figure 9). More specifically, by plotting two standard deviations from the mean dynamic response for each stretch web, the general behavior is more readily compared. In general, the 'S' stretch web experiences larger variability in the dynamic response, particularly when comparing the rotational response in the 'y' and 'z' orientations. The response in the z-direction is of particular importance since any rotation about this axis will influence the rotational position of the blank once it is placed on the die face (xy-plane) and will influence the final formed geometry of the blank. The large variance in the dynamic response of the 'S' stretch web can be attributed to its more compliant structure which will permit material flow during deep drawing operations but is more susceptible to an excessive dynamic response.

4. Conclusion

The machine learning models presented in this work are capable of predicting whether a collision or strip misalignment will occur after a single stroke with a classification accuracy of about 87 %. By adjusting the inputs to the feeder and lifter, a wide range of outcomes are observed with certain permutations underperforming and resulting in failures at lower stroke rates and others outperforming with no failure at higher stroke rates. By utilizing a collision prediction model in conjunction with active control to the feeder and lifters, tool makers can potentially realize higher stroke rates for a given strip layout and stretch web selection or predict the limiting stroke rate prior to die try-outs.

Acknowledgements

The authors appreciate and acknowledge the financial support from the Ontario Advanced Manufacturing Consortium, Natural Sciences and Engineering Research Council of Canada, the Industrial Research Assistance Program (IRAP), the Canadian National Research Council (NRC), the German Aerospace Center (DLR) and the German Federal Ministry of Education and Research (BMBF).

References

[1] Jamli M R and Farid N M 2019 The sustainability of neural network applications within finite element analysis in sheet metal forming: A review *J. Int. Measur. Confed.* **138** 446–60

[2] Manoochehri M and Kolahan F 2014 Integration of artificial neural network and simulated annealing algorithm to optimize deep drawing process *Int. J. Adv. Manuf. Technol.***73** 241-9

[3] Polyblank J A, Allwood J M, and Duncan S R 2014 Closed-loop control of product properties in metal forming: A review and prospectus *J. of Mater. Process. Technol.* **214** 2333–48

[4] Endelt B, Tommerup S, and Danckert J 2013 A novel feedback control system - Controlling the material flow in deep drawing using distributed blank-holder force *J. of Mater. Process. Technol.* **213** 36–50

[5] JAllwood J M *et al.* 2016 Closed-loop control of product properties in metal forming *CIRP Ann – Manuf. Technol.* **65** 573–96

[6] Han G 2016 *Transfer Die System Timing and Parameter Optimization According to an Obstacle Map* Master's theses (Vancouver, BC: University of British Columbia)

[7] Liao X and Wang G G 2003 Evolutionary path planning for robot assisted part handling in sheet metal bending *Robot. and Comp.-Integ. Manuf.* **19** 425–30

[8] Glorieux E 2017 *Multi-Robot Motion Planning Optimisation for Handling Sheet Metal Parts* Doctoral dissertation (Trollhättan, Sweden: University West)

[9] Li H and Celglarek D 2002 Optimal Trajectory Planning For Material Handling of Compliant Sheet Metal Parts *J. Mech. Design* **124** 213-22

[10] Budnick D *et al.* 2021 Simulation of Dynamic Effects in Progressive Die Operation and Control *IOP Conf. Series: Material Science and Engineering (Stuttgart)* vol. 1157, no. 1

[11] Chen T and Guestrin C 2016 XGBoost: A scalable tree boosting system *Proc. of the ACM SIGKDD Int. Conf. on Knowledge Discovery and Data Mining (San Francisco)* vol. 13-17 (New York, NY, USA: ACM) pp 785–94

[12] Hastie T, Tibshirani R, and Friedman J 2017 *The Elements of Statistical Learning Data Mining, Inference, and Prediction* vol 2 (New York: Springer) chapter 7 pp 241-7

IOP Conf. Series: Materials Science and Engineering 1238 (2022) 012069 doi:10.1088/1757-899X/1238/1/012069

Field meta modelling for process design in complex sheet metal forming

C Schwarz*[1], S Kriechenbauer[1], R Mauermann[1], W G Drossel[1]

[1] Fraunhofer Institute for Machine Tools and Forming Technology IWU, Reichenhainer Straße 88, 09126 Chemnitz, Germany

*Christian.Schwarz@iwu.fraunhofer.de

Abstract. Actually, new applications of machine learning methods like variance-based sensitivity analysis take place for developing complex sheet metal forming and joining processes. The application of this methods will help to quickly start of production when the simple numerical designing of the processes becomes too complicated. Two investigated examples show the application of the method for complex planning situations. The first one uses variable functions of force and motion during deep drawing to extend the process limits. The second one describes the complex clamping of an assembly of different sheet metal parts. Field meta modelling approaches are shown to automate tasks, visualize results and support engineers to plan and run different complex processes regarding sheet metal parts.

1. Introduction

First goal of the field meta modelling approach is the visualization of correlations and prediction of a proper process design with an interactive graphical user interface. Second goal is process optimization to achieve the greatest possible insight with the fewest possible calculations. A third goal is to stabilize critical processes with fast calculations on the field meta model in a closed-loop control system.

The limits of deep drawing processes can be extended with complex movements (e. g. superimposed pulsations) on servo presses [9]. These deep drawing technologies use a lot of process parameters (12). Therefore, it is difficult to design such processes. The presented examples show that trial-and-error is often difficult to achieve success. Also in a process chain (e.g., deep drawing, clamping, joining) many parameters have cross-correlations and strongly influence the result [13].

Field meta models calculated with the variance-based sensitivity analysis describe important parameters of complex processes. This approach helps to systematically find a suitable parameter set for which a minimum sheet thickness at the weak point is not fallen short of in order to prevent failure cracks. Another criterion is to minimize the deviation from the desired geometry at a certain point of the sheet metal part.

For calculating the field meta model the challenge will be to spatially correlate input and output of all simulations and experiments. For example the outlines of the resulting finite element mesh vary over different simulations of deep drawing processes [1]. That is why an automatic workflow for the variance based sensitivity analysis with different software tools was developed to extract results at certain locations of the sheet metal parts.

2. Application of field meta models

The objective is to achieve a three-dimensional representation of statistical measures on finite element meshes. Thereby, simulations are run with different variants, which have e. g. in deep drawing scattered material parameters, sheet thickness values or blank positions [4]. A comparable solution approach is used in the publication by Wärmefjord et al [5].

The purpose of the study is to develop a method based on principal component analysis (PCA) that can be used to determine the effects of manipulated and disturbance variables (e.g., friction or forming speed) on the part quality from the deep drawing to the assembly process. Based on this, Wolff [6] and Gerbino [7] independently perform numerical (with Finite Element Method – FEM) variance-based sensitivity analyses for sheet metal parts. Influences of material (e.g., sheet thickness) and design parameters (e.g., drawing speed, blank holder force) on the part quality are investigated. In both cases, field meta models are used for the three-dimensional representation of the sensitivities on the FE mesh surface. In a separate study [8], a PCA-based method is proposed to enable part-based visualization of variance-based sensitivities for finely discretized FE meshes in numerical parameter studies. Using a front flap inner part, it is demonstrated that by using a field meta model the influence of process parameters on specific state variables such as strain or spring back can be represented for FE meshes with over one million elements on the drawn part surface.

3. Method used for field meta modelling

Here, the method described in [8] is used for field meta modelling. A variance-based sensitivity analysis is the core of the method. The data comes from the finite element geometry and calculated results (strain, stress, plate thickness) out of the variant simulations, see 1st step in Figure 1. As described in the 2rd step, the field meta modelling uses data reduction by principal component analysis (PCA). The mathematical approach of PCA consists in a linear transformation of the original data into a new coordinate system. As a result, the variance structure of the original data can be summarized with the help of a small number of meaningful linear combinations, the so-called principal components.

After identifying the new transformation basis, the input data for dimensionality reduction is projected onto the principal components previously ordered by variance proportions, resulting in the data reduction. Based on PCA, the surrogate models are then built in the reduced feature space and by means of a variance-based sensitivity analysis, the percentage influences of the input parameters on the dimensional fluctuations or other post-size values of the FE networks are estimated (3th step). The back transformation (4th step) of the reduced data enables the field metamodeling and the 3-D visualization of the sensitivities on the entire component FE surface mesh (5th step).

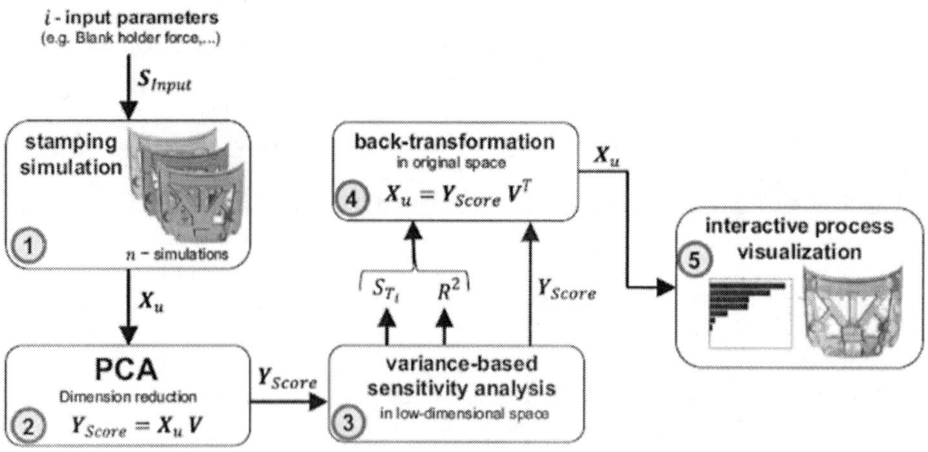

Figure 1. Overview of the method behind field meta modelling [8]

4. Toolkit for automatic field meta modelling

Overview of the developed software concept

Experimental and numerical development of deep drawing processes with variable force and motion functions are complex tasks regarding the settings of the press machine control and the parametrization of the FEM simulation, respectively.

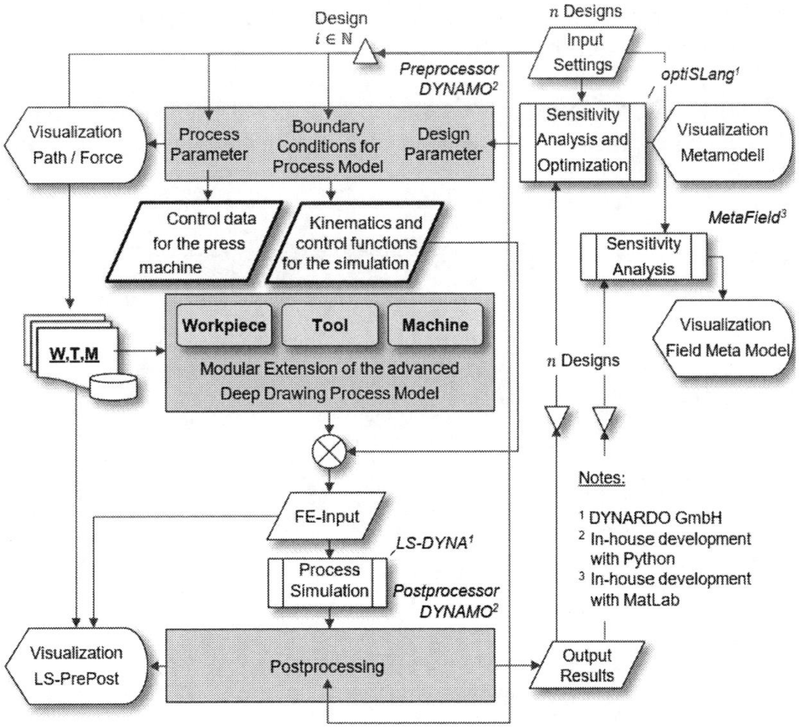

Figure 2. Developed software concept for automatic field meta modelling with advanced simulation models for deep drawing processes

Field meta modelling should help to systematically find a proper solution. That is why we setup a software concept using different tools and a data management system for field meta modelling. The developed toolkit as shown in Figure 2 is an interdisciplinary product from mathematics, computer sciences and classical mechanical engineering. The purpose of the toolkit is to automate the calculation of field meta models and simplify the pre- and post-processing for engineers.

The input is based on the process parameters (e. g. in BDD: amplitudes, forces) and the drive capabilities of the press machine. It is possible to calculate any kind of deep drawing processes with variable force and motion functions. The software concept provides an interface to an external FE solver (e. g. LS-DYNA) for the advanced forming process simulation [11]. The FE models are independent for the workpiece, the tool and the machine. This flexible concept allows to build fast and reliable simulation models for any deep drawing process. The advanced forming process model [11] will be automatically assembled, calculated and evaluated depending on the selected inputs for each design.

The developed software concept in Figure 2 is independent from any FE solver. Other solutions like e. g. AutoForm Sigma have only interfaces to the AutoForm Solver. This does not allow advanced forming simulations with variable force and motion functions to be performed. The data extraction of

MetaField is compared to other solutions (AutoForm, Deform, Simufact) much more flexible for any kind of production process because it is not restricted to numerical FE data. For example, MetaField can calculate models based on experimental data (e. g. welding micrograph structure [15]), on mixed data - numerical and experimental (see chapter 6), on process chains (consisting of several operations, e.g. in car body construction), on freely selectable formats for input and output (e.g. curves, surfaces, visual images, discrete values).

Description of the graphical user interface

The computer program DYNAMO (see (Figure 3a) calculates the input data for the simulation and the control commands for the press machine. The force and displacement functions are calculated in such a way that any deep drawing process will be realized with minimum time requirements. Thus, high productivity will be achieved.

Another important task of DYNAMO is the automation of numerical variance-based sensitivity analysis for deep drawing processes. The program supports the user at a high level in the pre- and post-processing of each process design.

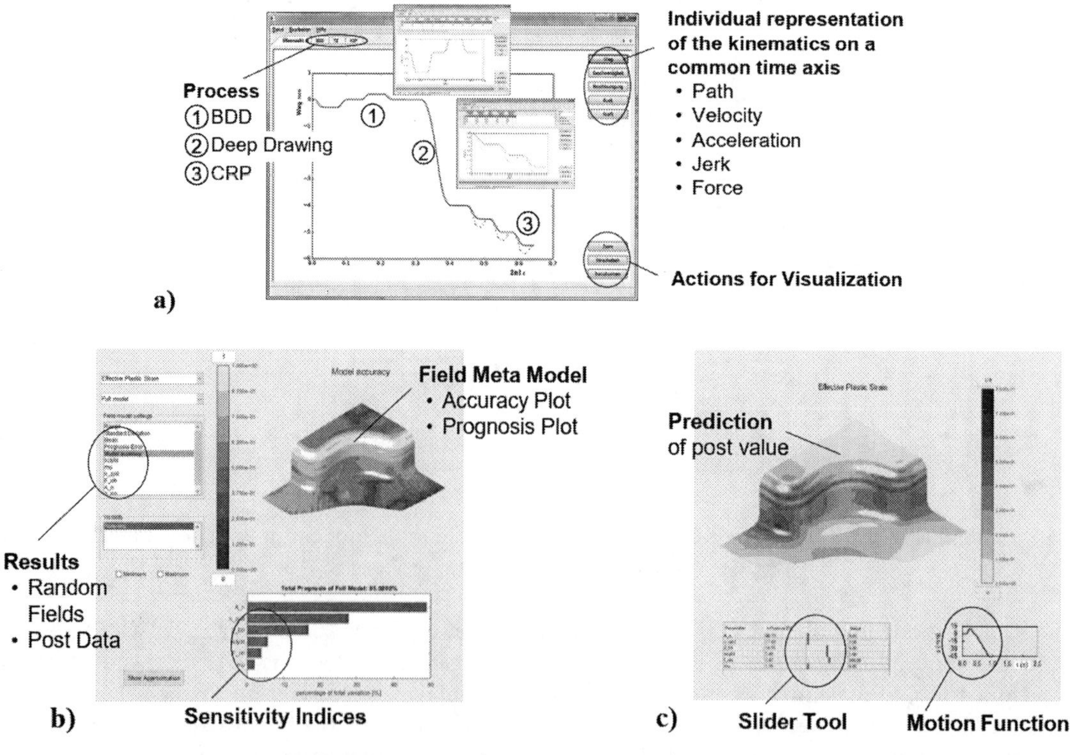

Figure 3. Graphical user interface (a) for the process design tool DYNAMO and for the field meta modelling tool MetaField with model accuray plot (b) and slider tool (c) for prediction

The development workflow should reduce the development times for deep drawing processes. Based on the flexible software concept from Figure 2, various tasks can be securely outsourced. The input and output data are given by interfaces. The graphical user interfaces (Figure 3b and c) connects all individual results and shows the field meta model to get the optimal deep drawing process.

Further this allows different departments in a company to easily work together, even though everyone is in their own area of expertise. Further developments will integrate measurements from the production process as a continuous feedback loop to improve the prediction.

5. Example for planning a complex deep drawing process

Short description of the Bidirectional-Deep-Drawing (BDD) process

As described in [10] a cylindrical cup evolves a characteristic weak point at the bottom radius during the deep drawing process. High tensile loads cause excessive thinning in the area with of least stressability. To avoid local material instabilities, one should increase the strain hardening near the bottom radius. Earlier published papers detailly shows the extension of the forming limits for a cylindrical cup [2] and a cross die [10].

BDD has a complex variable force and motion function. The sheet is clamped between blankholder and die. An upper punch fixes the bottom of the part onto the lower drawing punch. The Figure 6 in [2] shows, how the effective plastic strain increases with each alternating bending movement of the die. All details of the BDD process are described in [2] and [10]. The papers provide material properties for the simulation as well. The pre- and postprocessing of the simulation are explained in [10] for the calculation of the field meta model.

The motion function will be calculated by DYNAMO as shown in Figure 3a. The software was fully written in PYTHON at Fraunhofer IWU [3, pp. 93-104] based on the software concept from Figure 2. It is quite complicated to find a proper motion function for a specific part, which will increase the effective plastic strain in the right area. To solve this problem numerically, we use a variance-based sensitivity analysis with the Fraunhofer IWU tool MetaField (see Figure 3b and c). Based on that, we can perform a parameter study for different motion functions as described for the BDD process in [10].

Parameter study with predicted results based on the field meta model

With the bidirectional deep drawing (BDD), the work hardening can be locally adjusted by the process parameters. The Figure 4. Predicted effective plastic strain (a/b) for different motion functions (c/d) based on the field meta model [10]

illustrates the parameter study, which is based on the field meta model as described in [10]. The predicted values for the effective plastic strain will increase with larger amplitudes and more bending operations during BDD.

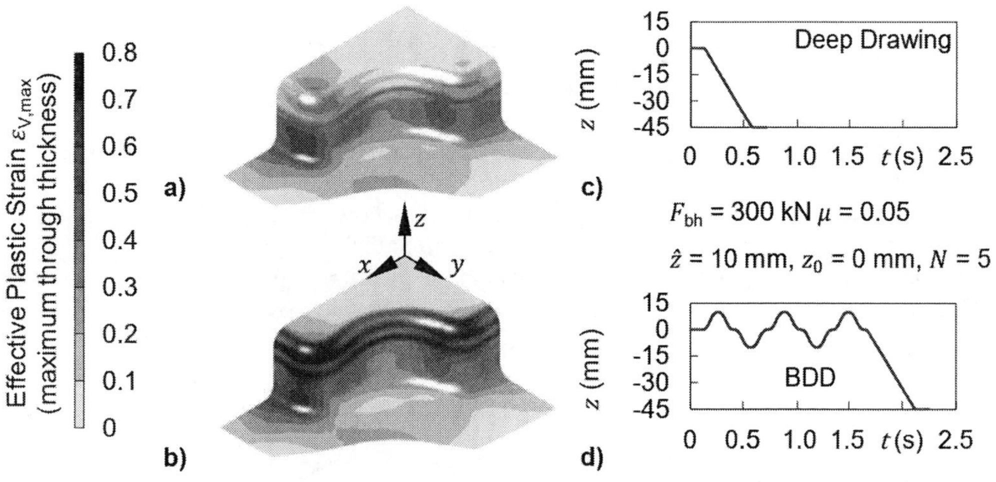

Figure 4. Predicted effective plastic strain (a/b) for different motion functions (c/d) based on the field meta model [10]

Nevertheless, the failure prediction based on the forming limit curve is currently not reliable for bidirectional deep drawing processes due to the nonlinear strain paths as shown in [10, Fig. 4]. Therefore, we cannot perform an optimization task based on this field meta model.

6. Example for clamping an assembly of sheet metal parts

The variations of the clamps in car body manufacturing have been studied for a long time [14]. The field meta modelling approach can help to find a suitable solution for this problem. The method is quite the same as explained above, only the data preparation had to be adjusted for the following investigations. The assembly, which consists out of 4 individual sheet metal parts as shown in Figure 5a, was experimentally and numerically evaluated in [13]. In the experiment each surface of the 4 single parts on 5 parts each was optically measured and digitalized. These twenty 3D-surfaces were used as parameter variations in the simulation without any known causes (e.g. property variations of the material or deep drawing parameters). In the clamping simulation the crucial clamp and two bolts were numerical varied on the assembly as shown in Figure 5a. Then the deviation between the calculated and the desired surface of the assembly was calculated.

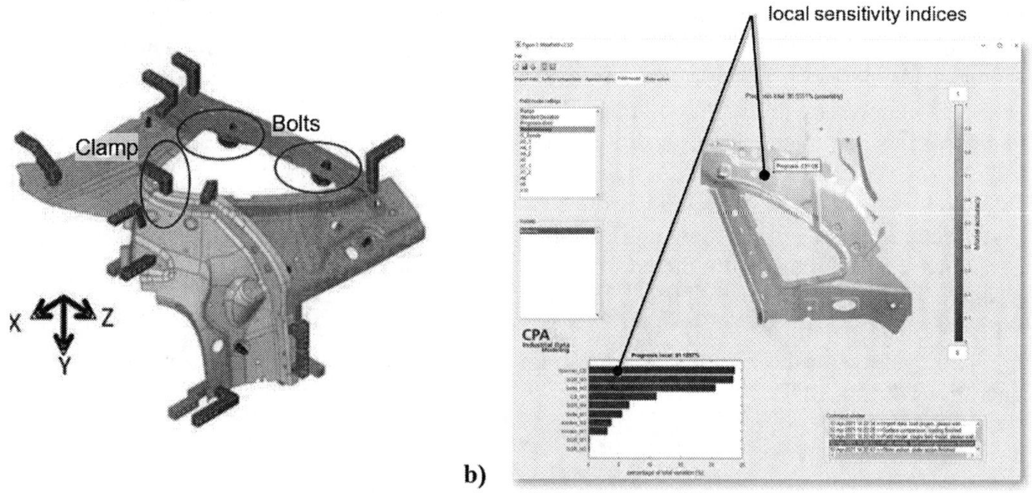

a) b)

Figure 5. Location of the varied clamp and bolts on the assembly (**a**) and model accuracy (**b**) with plot of local sensitivity indices for all input parameters in MetaField

The Figure 5b visualizes the influences of the individual parameters according to their importance. The result data are plotted onto the investigated assembly. Figure 5b shows the prediction quality with the Coefficient of Determination (COD). The value describes how much of the result variation can be explained by the input variations. In the slider tool of MetaField (Figure 6) one can play around with the field meta model to quickly find a suitable clamping situation without any remarkable delay. This variation of the input parameters, shown here manually, can be taken over by an optimization algorithm in case of a concrete target.

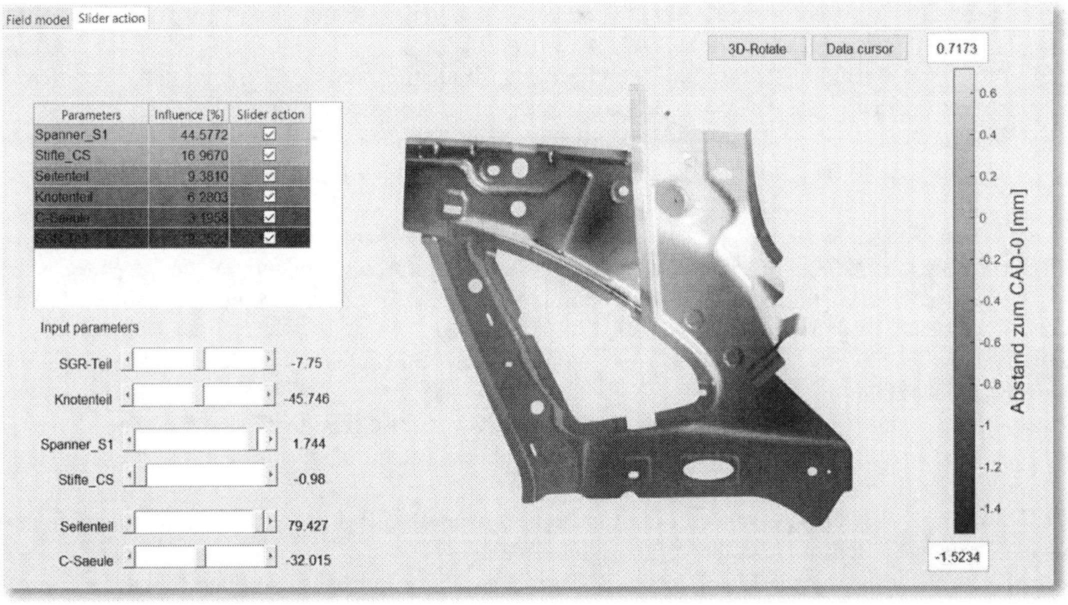

Figure 6. Predicted deviation by interactively moving sliders in MetaField

7. Conclusion

For the field meta modelling approach based on the sensitivity analysis of variances it will be necessary to automate the pre- and post-processing of experiments and simulations. The method is characterized by the design of experiments, the data generation and the result extraction.

The presented examples show how one can systematically find a suitable process design for complex or different sheet metal parts using variance-based sensitivity analysis with field meta modelling. The process design with trial-and-error has a bad cost versus effect ratio. Often the data acquisition can be the biggest cost driver. What is sought is the analytical model that achieves the best process description at the lowest cost for simulation and measurement. Therefore, the combination of numerical and experimental data is an important topic for the modelling.

Such an analytical model is then used as a basis for the automatic control of processes based on the developed software workflow via suitable actuators.

8. References

[1] Wolff S Random fields and field meta models – Correlation analysis in time and space RDO-Journal **1**, 2016. Downloaded on 02/03/2022 from https://www.dynardo.de/fileadmin/Material_Dynardo/dokumente/broschuere/JournalArtikel/RDOJournal_1_2016_Title_Random Fields_P_web.pdf
[2] Kriechenbauer S et al Deep drawing with local hardening on digital multi-axis servo press. Acta Metallurgica Sinica **12**, 2015, pp. 1490-1495.
[3] Kriechenbauer S Beitrag zur Analyse und Synthese von variablen Kraft- und Wegfunktionen zum Tiefziehen auf Servo-Spindelpressen Verlag Wissenschaftliche Scripten, Auerbach, 2022, Dissertation TU Chemnitz.
[4] Will J et al Computation and visualization of statistical measures on FE structures for forming simulations 2nd Weimar Optimization and Stochastic Days 2005

[5] Wärmefjord K et al Prediction of Geometrical Variation of Forged and Stamped Parts for Assembly Variation Simulation Proceedings of IDDRG Conference 2013: Institute of Virtual Manufacturing ETH Zurich; 2013. p. 99-104.
[6] Wolff S Robustness Analysis of Metal Forming Simulation – State of the Art in Practice Reports from the IWU **88**, Auerbach: Verlag Wissenschaftliche Scripten, pp. 319-334, 2015.
[7] Gerbino S et al Understanding Whole Shape Variability of Stamped Sheet Metal Parts for Root Cause Analysis 5th International Conference on Accuracy in Forming Technology: ICAFT 2015, 22nd SFU 2015: proceedings. Verlag Wissenschaftliche Scripten, Auerbach, 2015, pp. 423–438.
[8] Schwarz C et al Principal component analysis and singular value decomposition used for a numerical sensitivity analysis of a complex drawn part Int J Adv Manuf Technol 2018; 94: 5-8. p. 2255-2265. DOI: 10.1007/s00170-017-0980-z
[9] Nakagawa T Servo motor driven press and market trend in Japan CIRP 56th General Assembly, Kobe, 2006
[10] Kriechenbauer, S et al Extension of Process Limits with Bidirectional Deep Drawing Congress of the German Academic Association for Production Technology. Springer, Berlin, Heidelberg, 2020, S. 96-104
[11] Großmann T K et al Advanced Forming Process Model-AFPM Process Machine Interactions Springer, Berlin, Heidelberg, 2013. S. 383-401
[12] Siekirk JF Process variable effects on sheet metal quality Journal of Applied Metalworking 4.3 (1986), S. 262–269. ISSN: 0162-9700. DOI: 10.1007/BF02833934
[13] Landgrebe D et al Improving the ramp-up process of a body-construction line by numerical supported design of clamping devices and FEM based tolerance prognosis Key Engineering Materials Vols 651-653 (2015) pp 932-937; doi:10.4028/www.scientific.net/KEM.651-653.932
[14] Cai W et al Deformable sheet metal fixturing: principles, algorithms and simulation, ASME Journal of MFG Science and Engg. V118, 1996
[15] Schwarz C et al Model-Based Joining Process Design for the Body Shop Process Chain International Conference on Advanced Joining Processes (AJP 2022). Proceedings in Engineering Mechanics https://doi.org/10.1007/978-3-030-95463-5_2

Acknowledgments

The presented results are part of public research projects at Fraunhofer Institute for Machine Tools and Forming Technology IWU, which is located in Chemnitz, Germany. The Projects 17058BR, 17488BR and 19007BR of the research association Europäische Forschungsvereinigung für Blechverarbeitung (EFB) e. V. were financed by the Arbeitsgemeinschaft industrieller Forschungsvereinigungen Otto von Guericke e. V. (AiF) as part of the program to support Industrial Community Research and Development (IGF) with funds from the Federal Ministry for Economic Affairs and Energy (BMWi) following an order by the German Federal Parliament. Further development and practical work were funded by the Federal Ministry of Education and Research (BMBF) as part of the reasearch project ML@KaroProd. The financial support and cowork are gratefully acknowledged by the authors.

IOP Conf. Series: Materials Science and Engineering 1238 (2022) 012070 doi:10.1088/1757-899X/1238/1/012070

Analysis of Distributed-Ledger-Technology for the Exchange of Design, Production and Simulation Data in Roll Forming

B Kohl[1], M Krüger[2], T Dietl[1], M Lechner[3], E Trunzer[2], M Merklein[3], A Sedlmaier[1] and B Vogel-Heuser[2]

[1] data M Sheet Metal Solutions GmbH, Am Marschallfeld 17, 83626 Valley, Germany
[2] Institute of Automation and Information Systems, Technical University of Munich, Boltzmannstr.15 85748 Garching, Germany
[3] Institute of Manufacturing Technology, Friedrich-Alexander-Universität Erlangen-Nürnberg, Egerlandstr. 13, 91058 Erlangen, Germany

E-mail: b.kohl@datam.de

Abstract. Digitalization in the metal forming industry needs to be improved to achieve the goals set by Industrie 4.0. Distributed-Ledger-Technology (DLT) has been identified as a promising foundation for tackling the underlying problem of consistent information exchange. DLT-based solutions have already been developed, but none explicitly covers the roll forming use-case. This paper presents a Hyperledger-Fabric-based blockchain network to fill this research gap. This network is specifically designed for the roll forming industry, while still aiming to meet general information exchange requirements. The roll forming use-case is divided into the material supply chain and a design/simulation workflow. Participants and parameters in these two data transfer chains have been validated with the help of industry experts. Running the conceptualized network on an on-premise server has allowed for a detailed evaluation. Feedback provided by experts of the roll forming industry shows the potential of the presented network. Furthermore, the presented solution covers requirements often neglected by existing approaches like large data handling, compatibility to existing interfaces, secure communication, and access rights definition. In summary, this paper provides a pioneering implementation and evaluation of a DLT-based solution for the roll forming industry and, therefore, a foundation for the next steps towards Industrie 4.0.

1. Introduction

A system entirely in line with the concept of Industrie 4.0 requires data transfer among companies [1]. However, this transfer is rarely satisfying in the metal forming industry [2]. To tackle the challenges of information exchange in the sheet metal industry, blockchain, and with it, Distributed-Ledger-Technology (DLT), have been identified as promising technologies [2]. DLT denotes the concept of storing a blockchain decentralized on each participating node with automatic synchronization [3]. First applications of such blockchain networks for the sheet metal industry do exist, e.g., by Frey et al. [4], but the roll forming industry has been uninvestigated. In this industry, extensive supply chains with many participants are prevalent [5], providing a challenge for secure information exchange. This paper presents a pioneering approach to tackle this challenge by making use of the main properties of blockchain networks for information exchange: proof of data ownership, maintaining data integrity, system verifiability, and no need for a central authority [4]. Concluding, the contribution of this work is the unprecedented conceptualization and implementation of a blockchain network specifically for roll forming including its evaluation against the requirements for a trustworthy information exchange solution. Chapter 2 covers the definition of requirements, chapter 3 covers existing approaches in the research field and compares these solutions and requirements. Furthermore, chapter 4 presents the conceptualized network and chapter 5 the implementation. In chapter 6, the results are evaluated and discussed, followed by conclusions and future work in chapter 7.

2. Requirements for an Information Exchange Solution

An information exchange solution must follow authentication and encryption rules (R1). Having data available to the correct parties leads to the necessity of a proper access rights definition (R2). To establish trust in data, control mechanisms to prove the validity of data need to be in place. Unwanted change of data in storage needs to be prevented or at least detected (R3). Besides, maintaining integrity during transmission is essential for information exchange (R4). The high measuring rates of modern sensors must be handled without significant delay for live applications, e.g., monitoring, (R5). In addition to R5, sensors provide data over various interfaces. For integration in existing systems, these interfaces need to be supported. One of the most used protocols is the Open Platform Communications Unified Architecture (OPC UA) [6], support for which is a minimum requirement (R6). Additionally, large files are another extreme to be considered since simulation results are stored in files of up to 50 GBs, which the system needs to handle reliably (R7). Current data transfer solutions often rely on a trusted third party, like a cloud provider, but finding a third party that every participant trusts can prove complicated. To circumvent this problem, a solution that itself is trustworthy is necessary. This trustworthiness needs to be verifiable by all participants (R8). While cloud solutions are getting much attention nowadays, industrial companies might prefer to run the system on-premise (R9). Since data might contain sensitive information, using a public system would require extreme effort to prevent unauthorized access. Thus, participation in the system should be regulated (R10). Finally, the system should be applied to a real-world roll forming use-case. This does not mean realizing an industrial setup, which would require more economic considerations. Rather, actual formats and dimensions of the transferred data need to be determined, and the system evaluated accordingly (R11). Table 1 highlights the derived requirements.

Table 1. Derived requirements for data transfer in roll forming

Requirement	Formulation
R1	Entities are authenticated and communication is encrypted.
R2	The system supports CRUD (create, read, update, delete) access rights.
R3	The system prevents or detects an unwanted change of stored data.
R4	Data integrity is maintained during transmission.
R5	The system processes data points arriving at high rates.
R6	The system accepts data via the OPC UA protocol.
R7	The system processes files with a maximum size of 50 GB.
R8	The system needs to be reviewable by all participants.
R9	The system can be hosted on local hardware.
R10	Participation in the system is restricted.
R11	The system is evaluated with a real-world roll forming use-case.

3. Current State of Research and Technology

Raj et al. [7] propose an application for integrity-based authentication and secure information transfer for hospital management systems using symmetric encryption. Integrity is maintained using hash algorithms and authentication is handled by a public-key-based system. Another cloud-based solution is proposed by Huang et al. [8]. Their main goal is a secure data storage and sharing scheme handling the specific challenges of Cyber-Physical-Social-Systems. Putz et al. [9] present a blockchain-based implementation of a digital twin, including an Ethereum-based prototype. A performance-focused approach called FastFabric is presented by Gorenflo et al. [10], achieving roughly 20,000 transactions per second. Frey et al. [4] present a small-scale production example in the metal forming industry. In their work, a strong focus is placed on the accessibility of data gathered along the supply chain. In addition to the theoretical considerations, a demo factory is established. Each machine is equipped with a blockchain interface (BCI), and each BCI acts as a node for the blockchain network with its own copy of the chain. Most of the presented solutions above do not directly store data on a blockchain but mostly in undefined off-chain storage. Baumung et al. [11] present a blockchain-based order provisioning and processing system for additive manufacturing, including cloud-based off-chain storage. Public-key encryption is used to secure data stored in the cloud. A final approach concerning additive manufacturing is presented by Schmiedel [12]. This project produces a proof of concept for the use of DLT in additive manufacturing. The project bases this proof on a machine with an OPC UA-based control system as the data provider.

Table 2 puts the above-presented approaches in relation to the defined requirements of the previous section. The table shows a lack of roll forming-specific solutions (R11). In addition,

Table 2. Solutions vs. requirements

Approach	R1	R2	R3	R4	R5	R6	R7	R8	R9	R10	R11
Bindhu Raj et al. [7]	✓	✓	✓	✓	—	—	—	—	—	✓	—
Huang et al. [8]	✓	✓	✓	✓	✓	—	✓	—	—	✓	—
Putz et al. [9]	✓	✓	✓	✓	✓	—	—	✓	✓	✓	—
Gorenflo et al. [10]	—	—	✓	✓	✓	—	—	✓	✓	✓	—
Frey et al. [4]	—	✓	✓	✓	✓	—	—	✓	✓	✓	—
Baumung et al. [11]	—	—	✓	✓	—	—	✓	✓	✓	✓	—
Schmiedel [12]	—	—	✓	✓	✓	✓	—	✓	✓	✓	—

✓: fulfilled, —: not fulfilled

integration of existing interfaces is uncommon (R6), as is the explicit handling of large files (R7). Integrity (R3, R4) is generally part of the existing solutions, be it via explicit integrity checks in cloud-based approaches or via the implicit architecture of a blockchain. Only half of the presented solutions take care of authentication and encryption (R1). The situation is similar concerning access rights definition (R2). Handling data with high measuring rates is a well-handled requirement (R5). The DLT-based solutions are reviewable by all participants, which proves to be complicated for cloud-based ones (R8). Running the approach on local hardware is mainly restricted by the use of central cloud storage (R9). All of the presented systems restrict participation in some way (R10). In conclusion, the central research gap is the explicit evaluation of an information exchange system for the roll forming use-case. This gap emphasizes the need for a trustworthy data exchange solution suited for the roll forming industry, a need covered by the considerations, implementation, and evaluation presented in this paper.

4. DLT Concept for Data Transfer in Roll Forming

This chapter defines the conceptualized system for exchanging design, production, and simulation data in roll forming based on Distributed-Ledger-Technology to fill the research gap identified in the previous chapter. This paper divides data transfer in metal forming into the material supply chain and the design-simulation workflow as described in the following sections.

4.1. The Roll Forming Material Supply Chain

An up-to-date and detailed supply chain depiction is needed in order to correctly represent this area in the blockchain network. Thus, this paper presents an unprecedented overview of the cold roll forming material supply chain shown in Figure 1. The participants and the collected parameters are carefully selected and have been validated by industry experts. In addition to this overview's use for the presented work, it provides a basis for further investigations on the roll forming industry's supply chain. The categories are chosen as defined by Frey et al. [4].

	Process	Material	Organizational
Raw Material Mining	• Energy Consumption • CO2-Footpring	• Raw Material Composition	• Mining Details
Steel Manufacture	• Blast Furnace Temperature • Pouring Temperature • Flow Rate • CO2-Footprint • Vacuum Treatment • Slab Scarfing	• Width • Side Profile • Composition • Thickness • Surface Defects • Material Designation	• Storage details • Customer • Slab ID • Heat number • Heat equence • Batch seperation
Hot Rolling Post Process	• Material Temperature • Speed, Stand Forces • Cooling Time • Furnance Temperature • Coiling Temperature • CO2-Footprint • Pickling (Y/N) and Method • Input/Output in kg and m	• Width, Length • Coil Weight, Composition • Material Designation • Surface Defects • Geometrical Deviations • Mechanical Properties • Thickness over Width and Length	• Storage Details • Customer • Coil ID
Slitting	• Circular Shear Blade Configuration • Insertion Depth • Input / Output in Meters	• Width, Length, Thickness • Weight, Material Designation • Cutting Burr Position • Surface Defects and Position • Position of Cross Welds	• Strip ID • Strip Position on Mother-Coil • Processing Line
Cold Roll Forming	• Speed • Stand Forces • Setup Parameters • Input / Output in Meters • Theoretical Output-Weight	• Profile Geometry • Length • Twist • Hole Pattern Dimensions • Mechanical Properties • Material Designation • Relevant Standard	• Storage Details, Part ID • Customer • Origin (Coil, Slitting Position) • Number of Profiles/Pieces • Theoretical Weight of Bundle • Bundle-ID • Date of Production/Shipment • Date of Expiry (Corrosion Protection)
Use	• Inclusion in Assembly	• End Product Parameter	• Storage Details • Customer • Part ID

Figure 1. Participants and parameters in the roll forming material supply chain

4.2. The Roll Forming Design and Simulation Workflow

The long-standing industry experience of simulation engineers at data M Sheet Metal Solutions GmbH provides the foundation for the design and simulation workflow. Thus, the details of this workflow are based mainly on the software packages developed by data M. In general though, the handling of information in this workflow is software-independent. The participants are entities handling cold roll forming, designing, simulating, and roll tool manufacturing. The roll former acts as the initiator in this workflow. First, they send a customer inquiry to the designer in the form of a pdf file containing the necessary information on how to create the design. The designer uses COPRA® RF [13] to create a forming strategy from this information. COPRA® RF is a commonly used design software in the roll forming industry and has been used in scientific papers before, e.g. by Sedlmaier et al. [14]. To validate the created forming strategy, a specific design state is exported to a proprietary archive file format. In addition, a particular function of COPRA® RF called "COPRA2FEA" is used to create a human-readable text file containing basic simulation parameters. The archive and this file are transferred to the simulation entity. Simulation is done using COPRA® FEA RF [15]. It provides a graphical user interface to create the input file for the underlying nonlinear finite element analysis solver. The solver outputs the results in a variety of files. The most important of these are the Post-File (.t16) and the Output-File (.out). The Post-File is a binary file reaching sizes of up to 50GB. The Output-File is another human-readable text file. Finally, the simulation results are sent back to the designer, who now has a validated design. This validated design is sent to the roll tool manufacturer (RTM). The RTM uses the validated design to manufacture roll tools, which are then sent to the roll former with a corresponding spec-sheet.

4.3. Covering Information Exchange Requirements with a DLT-based Solution

The participants mentioned above are defined as independent organizations, each participating in the blockchain network (concerns R11). Their authentication is based on a public key infrastructure (PKI). Each entity runs its own certificate authority (CA) to have full control over its members (R1). In addition, this authentication scheme is used to restrict access to the system by only considering CAs of eligible organizations trustworthy (R10). Encryption is also handled by a PKI. Each organization again runs its own CA for this purpose, which provides public-/private-key pairs to entities within the organization. These key pairs are then solely used for encryption, independent of the key pairs for authentication (R1). Data in the system is stored as assets with associated public and private information. Public information contains a unique AssetID, ownership information, a timestamp, and an asset description. This information is stored directly on the blockchain and is readable by everyone with access to the blockchain. Private information covers sensitive data only the asset owner should be able to read. Of this information, only a hash is stored on the blockchain. Access rights for this hash are the same as for public data. On an organizational level, access to these different types of information is defined according to Table 3 (R2). Within an organization, more fine-grained roles are defined. These roles are readers, writers, admins, and endorsements.

Table 3. Access rights on organizational level

	Public				Private			
	C	R	U	D	C	R	U	D
Owner	✓	✓	✓	—	✓	✓	✓	✓
Non-Owner	—	✓	—	—	—	—	—	—

CRUD: create/read/update/delete, ✓: allowed, —: forbidden

To analyze the handling of sensor data (R5, R6), the presented solution covers data provided by an OPC UA server, but the base principles can be transferred to other protocols. A data processing unit is inserted between the OPC UA server and the blockchain network. This unit's primary purposes are retrieving data from the data source, identifying the asset's private information from the retrieved data, defining the asset's public information depending on the data source, and adding the asset with its private and public information to the blockchain. In addition, non-continuous data is added to the system whenever their corresponding creation occurs. Here, a retraceable path from one asset to its associated asset(s) from previous steps is established. For example, the asset created for a simulation needs to be connected to its underlying design. The retraceable path is established by including the ID(s) of the associated asset(s) in the private information of an asset. Large and binary files (R7) are handled by storing only the hash of the files in the private information of an asset. This mitigates the problem of large files overloading the system. The actual exchange can be outsourced to a solution specifically designed for such a transfer, as Baumung et al. [11] have already suggested. Having established the participants, secured communication, and creating assets, the system must allow the transfer of those assets. To achieve this, a specific smart contract for asset transfer is used, following the asset-transfer-secured-agreement example provided by HLF [16]. Smart contracts are scripts do define and automate multi-level processes. The script for this transfer logic is, like all smart contracts, stored on the blockchain. Thus, it can be reviewed by all participants (R8). Changes to the logic must be unanimously decided. Maintaining integrity during transfer (R4) is based the fact that the hash of an asset's properties is stored publicly on the blockchain during the asset's creation. Thus, before sending, the owner can verify the data against this hash, as does the receiver after getting the data. The hash on the blockchain must not to change for this concept to work. This is ensured by keeping data integrity in storage. Data integrity in storage is implicitly maintained by the blockchain architecture (R3).

5. Implementation

The developed solution is based on Hyperledger Fabric (HLF), a modular and extensible framework for running permissioned blockchain networks [17]. The fundamentals of an HLF network are organizations, channels, peers, and the ordering service. An organization is a set of members belonging together. Having the roll forming use-case in mind, every participant represents their own organization. Channels allow multiple organizations to communicate with each other. Each channel comes with its own blockchain and smart contracts. In the roll forming use-case, a separate channel is created for the material supply chain and the design and simulation workflow, as they are largely independent of each other. While communication between two participants over these channels is secure, establishing additional sub-channels would also be possible. Either way, the use of Transport Layer Security (TLS) ensures, that only the intended recipient can decrypt the transferred information. Peers (or peer nodes) host the ledgers and smart contracts. They belong to an organization and are the actual channel participants. In the designed network, each organization runs exactly one peer for each channel in which it participates. Thus, in the roll forming use-case, each organization runs one peer, except for the cold roll former running two peers, as it is part both of the material supply chain and the design and simulation workflow. Oderer nodes have two base functions. First, they order transactions and add them to new blocks. Second, they host the configurations of channels. Other entities, called clients, participate in the network as well. They are not nodes but belong to an organization, e.g. an administrator for an organization.

Setting up and running a PKI for HLF is streamlined by the Fabric CA framework [18]. The deployment guide of the Fabric CA framework is the foundation of the conceptualized networks PKI. Nodes in the network are hosted using Docker [19] images provided by Hyperledger. The (access) rights definition in Hyperledger Fabric is a multi-layered system. The configuration of channels only allows specific organizations to participate. This configuration is stored on the ordering nodes and can only be changed unanimously. Private data collections [20] provide secrecy for private information by storing only a hash on the blockchain, while the actual private data is stored in the owner's state database. If private data is transferred, it is sent directly from peer to peer without passing the orderer. Each HLF peer comes with an implicit data collection only readable by itself. The smart contracts, which allow clients to create, transfer, and verify assets, are based on the asset-transfer-secured-agreement example by HLF [21]. This example provides chaincode to transfer assets while keeping details of the asset and transaction private. In HLF, chaincode is the code implementing smart contracts. A central control hub is added to the system to act out the roll forming use-case. This control hub is a web-based user interface providing access to the blockchain network for different organizations. The design layout was provided by data M Sheet Metal Solutions GmbH, as was the basic structure of the underlying web-server and OPC UA Client. All virtual machines (VMs) and containers necessary to run the network are hosted on the same server. This server is equipped with an eight-core processor, 32 GB of RAM, and a PCIe 3.0 x4 NVMe SSD. The server runs Proxmox VE [22], an enterprise-grade, open-source server management platform for virtualization.

6. Evaluation of the Presented System

Evaluation of the system is based on hosting it on the hardware described in the previous section. Fulfillment of requirements is evaluated either with behavior/performance tests or analyzed with the help of research on the underlying technology. Furthermore, the evaluation of the roll forming use-case is based on the other requirement's fulfillment and additional expert's feedback.

The PKI handling the authentication is secure as long as the private key of an entity remains secret and as long as the underlying signing concept is not broken. Leaking of a private key, if detected, can be mitigated by revoking the certificate it belongs to and issuing a new certificate/private-key combination. The underlying signing algorithm is the Elliptic Curve Digital Signature Algorithm (ECDSA) [23]. As of January 2022, ECDSA is not considered broken, but approaches to enhance its security, especially for blockchain use-cases, do already exist [24]. HLF version 2.2 uses TLS 1.3 in its default configuration, which has been proven secure [25]. Thus, secure communication (R1) is achieved with modern algorithms and no known security flaws. However, as the situation in IT security is fluid and new weaknesses are detected regularly, the security of the used algorithms and their implementation have to be monitored.

The implementation of HLF is crucial for the system behavior, i.e., whether access rights control (R2) behaves as defined depends on the software. The following test cases are analyzed by by invoking the respective functions in the chaincode or HLF without the necessary rights:

(i) Joining a channel the organization is not allowed to join
(ii) Committing chaincode not approved by all organizations on a channel
(iii) Changing the public description of an asset not owned by the organization
(iv) Reading the private data of an asset not owned by the organization

All these calls result in an error and, thus, the system based on HLF 2.2 behaves as expected.

Ensuring the integrity of stored data (R3) is implicitly handled by the DLT/blockchain architecture. To test the behavior of the system in case of an unexpected change of one participant's ledger, the ledger is edited directly on the file system. Unfortunately, HLF does not automatically detect the corruption of a peer's ledger file. Manual detection of unwanted changes can be realized by hashing the ledger file on each peer and comparing the hashes. If an integrity breach is detected, it can be fixed by obtaining a ledger file with integrity from another organization of the same channel and replacing the faulty ledger file. Thus, data integrity in storage is achieved but needs constant monitoring. An automated solution would be preferred to reduce administration efforts and susceptibility to errors.

The integrity of transferred data (R4) can be verified by comparing the hash of the current asset's properties against the hash of the asset's properties at creation. This can be done using a specific chaincode function called VerifyAssetProperties(). It takes as arguments the assetID and private asset properties and returns whether the given properties match the ones stored on the blockchain. If sender and receiver verify the sent asset's properties before and after the transfer process, data integrity is preserved throughout the transaction.

Performance of the system (R5) is evaluated by how long it takes to create an asset, which is around 100 ms. This is far from the potential of HLF-based solutions, as performance-optimized solutions like FastFabric achieve around 20,000 transactions per second [10]. In addition, this performance is insufficient to cover the continuous gathering of sensor data. Thus, only sensor information should be stored on the blockchain, while sensor data is exchanged off-chain.

How an existing data-providing interface can be integrated into the system (R6) is investigated with OPC UA as an example. In general, retrieving data from an OPC UA server, processing it, and continuously writing to the blockchain works. This is tested by connecting to an OPC UA server, subscribing to one of its variables, and writing the processed data point as an asset to the blockchain. However, this process is limited by the performance of the system.

The handling of large files (R7) is realized by hashing them, adding the hash to the private properties of the corresponding asset, and transferring the large files off-chain. SHA256 [26] was used as the hashing algorithm, as programs for its calculation are readily available on almost every platform. Besides, it provides an adequate performance/security ratio. Even with the low spec setup used in the presented network, calculating the SHA256 hash of a 50 GB file took less than four minutes. Considering that these large files occur only after complex simulations, four minutes do not present a problem, as the duration of these simulations is significantly longer. Secure off-chain storage was not investigated in the scope of this paper and could be added in the future. One approach for off-chain storage could be a self-hosted cloud solution which, compared to cloud service providers, leaves full control of the system to the participants. Another approach, as proposed by Baumung et al. [11], could be based on the InterPlanetary File System (IPFS) [27], a peer-to-peer protocol for data exchange

To evaluate the verifiability of the system (R8), it is subdivided into channel setup and chaincode behavior. The configuration block of a channel can be fetched by any channel participant. Thus, every organization can verify the channel setup independently. In addition, the influence each organization has on the logic of the system, defined in smart contracts, is even more substantial. Smart contracts can only be added to a channel if every organization approves. Thus, each organization can verify the transaction logic independently. In addition, considering the explained setup, each VM and container runs with little resources. Each peer and ordering node is assigned two GB of RAM and four vCPUs comparable to small single-board computers. This allows the system to be hosted on-premise (R9) with low expenses. Restriction of participation (R10) is realized by explicitly listing the authorized organizations in the channels' configuration blocks. Trying to join a channel without the authorization to do so results in an error, ensuring a permissioned network.

The roll forming use-case (R11) is implicitly evaluated with the previous points, as the general requirements are also valid for the roll forming use-case. In addition, the presented network is explicitly designed to represent the material supply chain and the design and simulation workflow of roll forming. The system provides a variety of benefits:

(i) No participant needs to trust statements of a third party about setup and behavior as they can independently verify the system.

(ii) Asset ownership is unambiguously defined. The owner has full control over their assets.

(iii) An asset's properties are immutable, except for its owner and the public description. Verifying that information has not been changed is, therefore, possible at any given time.

(iv) Every asset's transaction history is retraceable. Combining this with an asset's immutable properties allows verification of an asset's origin and content, even over several transactions.

These benefits are the main reasons companies in the roll forming industry would introduce DLT. Since the system itself can evaluate ownership and integrity of data, the need to engage a third party for this evaluation is made obsolete. This leads to potential cost savings and more trust among the participants. Such incentives are needed, as adopting a new data handling system is associated with high expenses. Especially the material supply chain participants face a high effort, as production and sales departments would have to be integrated for the system to function as described. Production-ready prototypes in publicly funded research projects could be a first step for the introduction of DLT in this context. In the design and simulation workflow data M Sheet Metal Solutions GmbH provides a comprehensive software package with their COPRA® Eco System which is uniformly used by the participants. Presented with a live demonstration of the blockchain network, feedback from multiple data M experts with longstanding experience in the industry has been highly encouraging, especially concerning the integration of DLT into the existing software. This integration could reduce the effort and knowledge required from each participant while maintaining the system's benefits.

7. Summary and Outlook

Secure communication is achieved by PKIs for authentication and encryption, utilizing ECDSA and TLS 1.3. Clear access rights are defined for data within the system. Data integrity in storage is maintained by manually checking the ledger files for inconsistencies. Hashing files at creation and using this hash for comparison in the transaction process provides data integrity during transmission. System throughput is unsatisfactory, with ten transactions per second. OPC UA-provided data can be added to the system, representing existing interfaces, but the low system performance interferes with this integration. Large files are hashed, and only the hash is stored on the blockchain, while the actual data is transferred off-chain. Integrity verification and ownership proof are thus still achieved for these files. Every participant can view the necessary configurations and behavior-defining programs. The system runs on low spec hardware, making it feasible to run it on-premise. Organizations can only join the network if explicitly stated in a channel's configuration, making it non-public. Finally, the material supply chain and design and simulation workflow in roll forming have been defined and used to evaluate the system.

Future work on this topic starts with the slow performance diminishing the system's usability in a production environment. Here, more investigation of the reasons for the slow performance is needed, as the problems could be software- or hardware-related. Furthermore, the compatibility to existing interfaces needs expansion. On one side, this expansion could focus on communication protocols like Message Queuing Telemetry Transport (MQTT) [28]. On the other side, research could focus on using technology-independent data collection as described by Trunzer et al. [29] for a more general interface integration into the system. Another aspect could be a test in an industrial environment. The system's reliability needs to be improved for industrial use, as the current network does not include node redundancy. The introduction of DLT to the roll forming industry would then be another research topic, e.g., by setting up an industrial prototype in production or by doing a detailed cost-benefit analysis.

Because the presented solution tackles step three, networking and integration, of the five-step model presented by Bauernhansl et al. [1], it provides a basis for research covering the next stage towards Industrie 4.0. According to Traub et al. [30], autonomous processes are the next step in roll forming. With smart contracts, modern DLT frameworks provide an intriguing concept for this step. In conclusion, the presented DLT-based network fills the detected research gap by providing a pioneering conceptualization and implementation of a DLT-based information exchange solution for roll forming. In addition, it provides an unprecedented overview of the cold roll forming supply chain as well as detailed insight into the design and simulation workflow of the industry. Consequently, it provides a foundation for further research on DLT-based information exchange, on industrial applications, and on the next step towards Industrie 4.0 in roll forming.

References

[1] Bauernhansl T, Krüger J, Reinhart G and Schuh G 2016 *WGP-Standpunkt Industrie 4.0*
[2] Hagenah H, Schulte R, Vogel M, Hermann J, Scharrer H, Lechner M and Merklein M 2019 4.0 in metal forming – questions and challenges *Procedia CIRP* **79** 649–654
[3] Ranathunga T, Marfievici R, McGibney A and Rea S 2020 A dlt-based trust framework for iot ecosystems *2020 Int. Conf. on Cyber Security and Protection of Digital Services (Cyber Security)* pp 1–8
[4] Frey P, Lechner M, Bauer T, Shubina T, Yassin A, Wituschek S, Virkus M and Merklein M 2019 Blockchain for forming technology – tamper-proof exchange of production data *IOP Conf. Series: Materials Science and Engineering* **651** 012046
[5] English M, Castellucci M and Mynors D 2006 Eco-efficiency of the cold roll formed product supply chain *J. of Materials Processing Technology* **177** 626–629
[6] IEC TR 62541-1:2020 2020 *OPC unified architecture - part 1: overview and concepts* International Electrotechnical Commission
[7] Raj L B, Vandana R and Kumar BJ S 2020 Integrity based authentication and secure information transfer over cloud for hospital management system *2020 4th Int. Conf. on Intelligent Computing and Control Systems (ICICCS)* pp 139–144
[8] Huang L, Zhang G and Yu S 2020 A data storage and sharing scheme for cyber-physical-social systems *IEEE Access* **8** 31471–31480
[9] Putz B, Dietz M, Empl P and Pernul G 2021 Ethertwin: Blockchain-based secure digital twin information management *Information Processing & Management* **58** 102425
[10] Gorenflo C, Lee S, Golab L and Keshav S 2019 Fastfabric: Scaling hyperledger fabric to 20,000 transactions per second *2019 IEEE Int. Conf. on Blockchain and Cryptocurrency (ICBC)* pp 455–463
[11] Baumung W and Fomin V V 2018 Increasing the utilization of additive manufacturing resources through the use of blockchain technology for a production network *CEUR Workshop Proc.* **2158** 135–141
[12] Schmiedel C Vertrauen 4.0 URL `https://www.ipk.fraunhofer.de/de/publikationen/futur/futur-online-exklusiv/vertrauen40.html`
[13] data M Sheet Metal Solutions GmbH COPRA RF URL `https://www.copra.info/en/crf`
[14] Sedlmaier A, Dietl T and Harraßer J 2019 Digitizing roll forming with smart sensors *AIP Conf. Proc.* **2113** 170017
[15] data M Sheet Metal Solutions GmbH COPRA FEA RF URL `https://www.copra.info/en/fea`
[16] Hyperledger Fabric Fabric CA Deployment Guide URL `https://hyperledger-fabric-ca.readthedocs.io/en/release-1.4/deployguide/ca-deploy.html`
[17] Androulaki *et al* 2018 Hyperledger fabric: A distributed operating system for permissioned blockchains *Proc. of the 13th EuroSys Conf.* EuroSys '18 (New York, NY, USA: ACM)
[18] Hyperledger Fabric URL `https://github.com/hyperledger/fabric-ca`
[19] Docker Inc URL `https://www.docker.com/`
[20] Hyperledger Private data URL `https://hyperledger-fabric.readthedocs.io/en/release-2.2/private-data/private-data.html`
[21] Hyperledger Fabric asset-transfer-secured-agreement URL `https://github.com/hyperledger/fabric-samples/tree/master/asset-transfer-secured-agreement`
[22] Proxmox Server Solutions GmbH URL `https://www.proxmox.com/en/proxmox-ve`
[23] ANSI X9.62 2015 *Public key cryptography for the financial services industry: the elliptic curve digital signature algorithm (ecdsa)* American National Standards Institute
[24] Basha S J, Veesam V S, Ammannamma T, Navudu S and Subrahmanyam M 2021 Security enhancement of digital signatures for blockchain using eddsa algorithm *2021 3rd Int. Conf. on Intelligent Communication Technologies and Virtual Mobile Networks (ICICV)* pp 274–278
[25] Levillain O 2021 Implementation flaws in tls stacks: lessons learned and study of tls 1.3 benefits *Risks and Security of Internet and Systems* ed Garcia-Alfaro J, Leneutre J, Cuppens N and Yaich R (Springer International Publishing) pp 87–104
[26] FIPS PUB 180-4 2015 *Secure hash standard* National Institute of Standards and Technology
[27] Protocol Labs IPFS URL `https://ipfs.io/`
[28] ISO/IEC 20922:2016 2016 *Information technology - message queuing telemetry transport (mqtt) v3.1.1* International Organization for Standardization
[29] Trunzer E, Prata P, Vieira S and Vogel-Heuser B 2019 Concept and evaluation of a technology-independent data collection architecture for industrial automation *IECON 2019 - 45th Annual Conf. of the IEEE Industrial Electronics Society* pp 2830–2836
[30] Traub T, Gregório M G and Groche P 2018 A framework illustrating decision-making in operator assistance systems and its application to a roll forming process *The Int. J. of Advanced Manufacturing Technology* **97** 3701–3710

IOP Conf. Series: Materials Science and Engineering 1238 (2022) 012071 doi:10.1088/1757-899X/1238/1/012071

Control of draw-in in the deep-drawing process by regulating the force on the blank holder

M E Palmieri[1*] and L Tricarico[1]
[1]Department of Mechanics, Mathematics & Management (DMMM) - Polytechnic University of Bari, Via Orabona 4, 70125, Bari, Italy

*mariaemanuela.palmieri@poliba.it

Abstract. Sheet metal forming processes are strongly influenced by process parameters, moreover fluctuations during the process and in material properties often lead to robustness problems. Therefore, numerical simulations as a function of design and noise parameters are usually adopted to detect ahead critical regions. To ensure efficiency and quality, the new era of industry 4.0 aims to develop systems that can control critical issues online. The aim of this work is developing a methodology that allows to adjust the force on the blank holder to obtain an optimal draw-in profile. This latter assures a minimization of defects (e.g., compression areas, thickened areas, insufficient stretch areas, splits) on the final component, even when variability of friction coefficient and yield strength occur. This methodology is applied to the deep-drawing process of a component obtained from a DC05 steel sheet with a thickness of 0.75 mm. For this purpose, Finite Element (FE) simulations in AutoForm software have been performed as a function of the design parameter i.e., Blank Holder Force and of noise parameters i.e., friction coefficient and yield strength. The results explored to evaluate the quality of the stamped part are: (i) percentage of compression areas, (ii) thickened areas, (iii) insufficient stretch areas, (iv) excessive thinning areas and (v) areas with splits. The data collected from FE simulations were adopted to perform a multi-objective optimization by means of the desirability function approach. Consequently, the value of process parameters that minimize defects (high value of total desirability) were identified. In correspondence with this optimal solution, the optimal draw-in profile was obtained. Such optimal profile was compared with the draw-in profile which involves defects on the final component (non-optimal draw-in profile). This latter was obtained for a value of the force on the blank holder equal to the optimal one and for non-optimized values noise parameters. In fact, noise parameters are not controlled process parameter. Based on the point-by-point difference between the two draw-in profiles, a numerical control strategy was implemented using AutoForm and MATLAB software. This iterative strategy allows to modify step by step the force on the blank holder so that the non-optimal draw-in profile match with the optimal one. Finally, experimental tests on a 3000 kN hydraulic press were performed to verify FE results.

1. Introduction

Deep drawing is one of the most widely used processes in sheet metal forming. Generally, it is applied in the automotive industry for the manufacturing of car body parts. In the deep drawing process, the blank is formed into the die by the mechanical action of the punch. Generally, a blank holder is adopted to control the sliding of the workpiece during the process [1-2]. The force on the blank holder is one of key parameter affecting the deep-drawing process [2-4]. A proper blank holder force can prevent defects on the final component such as splits, wrinkles, thinning and thickening. For example, an increase of the BHF can led to a reduction of wrinkle. While, in the presence of zone with excessive thinning, a reduction of the BHF can avoid splits [5-6]. However, the quality of deep drawn component is also affected by material and process fluctuations, e.g., yield strength, and friction coefficient [7]. As a result, robust design is required. Academic and industrial fields are interested in studying process monitoring strategies and active control systems to minimize the effect of both type of fluctuations. The state variable most frequently chosen to detect defects is the draw-in [8]. This variable can be measured by means of laser triangulation [9], sensors based on electromagnetic field [10] or mechanical devices [11]. To adjust online the draw-in of the blank, Hardt and Fenn [12] proposed a closed-loop control technique to realize real-time control of BHF. With the aim of adopting this latter control strategy, in this work a numerical methodology, in AutoForm environment, is developed to adjust step-by-step the force on the blank holder to obtain an optimal draw-in profile that minimize defects.

2. Material and Method

The deep drawing process of a T-joint was considered as case study. Figure 1a shows the geometry of the T-joint. This component was chosen by partners of the PICO & PRO project as a formability test able to highlight typical surface defects of cold sheet metal forming due to insufficient stretch. Figure 1b shows the tools, designed by CRF (Centro Ricerche Fiat, Italy), to deep draw the T-joint during experimental tests. Tools were manufactured by Tiberina Company. For experimental tests, a 3000 kN hydraulic press machine was adopted. Press machine was produced by the Gigant Italia company.

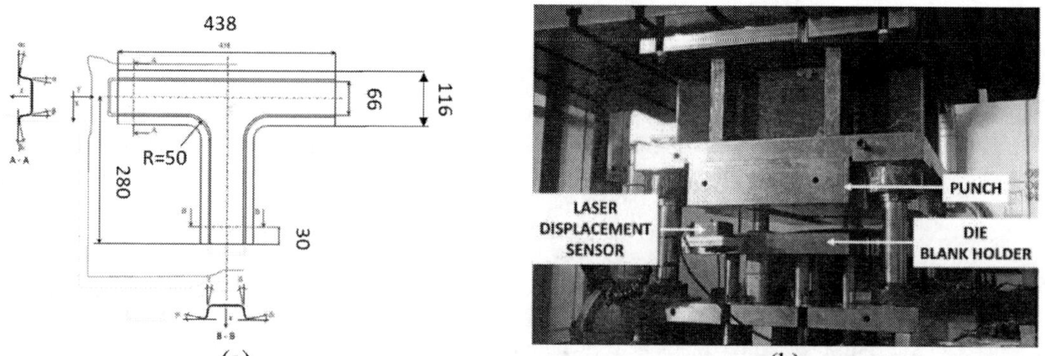

(a) (b)

Figure 1. (a) Geometry of the stamped T-joint. (b) Tools adopted for the deep-drawing of T-joint.

For this study, the methodology summarized in the graphical abstract in Figure 2 was adopted. Specifically: (i) A Finite Element (FE) model was developed in the commercial software AutoForm R10 to numerically simulate the deep-drawing process. The process was studied by varying the design parameter, i.e., the force on the blank holder (BHF) and noise parameters, i.e., the friction coefficient (μ) and yield strength (σ0). FE simulation predicted the quality of the final component in terms of percentage of compression areas, thickened areas, insufficient stretch areas, excessive thinning areas and areas with splits. Moreover, FE draw-in of the blank in critical points was evaluated. (ii) The data collected from FE simulations were processed with the MATLAB DACE toolbox to obtain kriging metamodels. Interpolated response surfaces (IRS) show how process and noise parameters variations affect the quality of the component at the end of the drawing phase. (iii) A multi-objective optimization, by means of desirability function, was performed to identify values of process parameters that minimize defects such as compression areas, thickened areas, insufficient stretch areas, excessive thinning areas and areas with splits. (iv) Once optimal solution was found, the corresponding optimal draw-in profile of the blank in critical points was derived. By setting the force on the blank holder at the optimal value and varying values of noise parameters, non-optimal solutions (deep drawn component with defects) were obtained. The draw-in in same critical points was also

measured for non-optimal solutions. Finally, optimal draw-in profile and non-optimal one was compared. Based on the point-by-point difference between the two draw-in profiles, a control strategy was implemented using AutoForm and MATLAB software. This strategy modifies step by step the force on the blank holder so that the non-optimal draw-in profile match with the optimal one. (v) At the end, experimental test of deep drawing of T-joint were carried out to verfy FE results.
Below details on each step of the adopted methodology are explained.

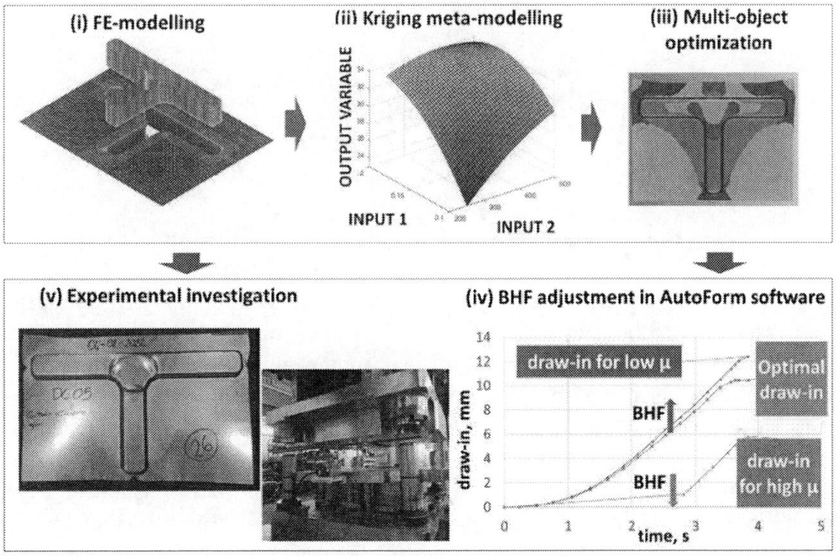

Figure 2. Graphical abstract to schematize the adopted methodology.

2.1. Material: DC05

The deep drawing of the T joint was performed on blanks obtained from DC05 steel sheets with a thickness of 0.75 mm. Chemical composition of the steel is listed in Table 1.
Figure 3 shows the mechanical characteristics of the investigated material. Specifically, Figures 3a-b-c show, respectively, hardening curve, yield surface and Formability Limit Curve (FLC). The hardening curve was defined using the Swift and Hockett-Sherby approximation. While, Hill 48 model was adopted for the yield surface.

Table 1. Chemical composition of steel (wt.%).

	C, %	*Mn, %*	P, %	*S, %*
DC05	0.06	0.35	0.025	0.025

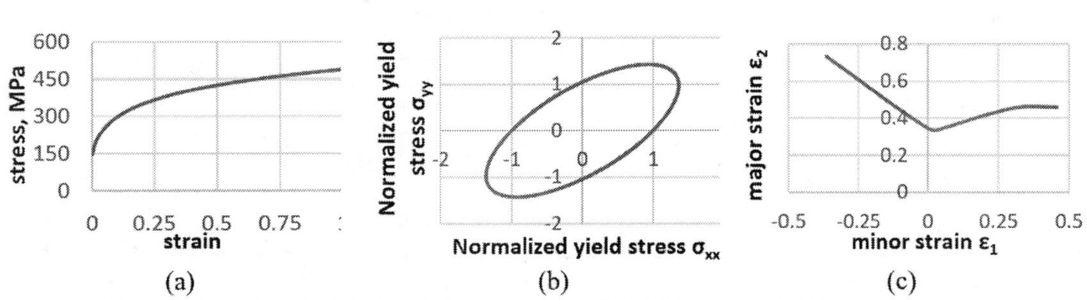

Figure 3. Properties of material. (a) Hardening curve.(b)Yield surface (c) FLC.

The mechanical characteristics shown in Figure 3 were assumed constant, however it should be noted that at the yield strength parameter a variability was assigned and consequently, the tensile strength was also varied, keeping constant the ratio between yield strength and tensile strength (0.5).

2.2. Finite Element (FE) modelling

The deep-drawing process was modelled using the commercial Finite Element (FE) software AutoFormR10. Tools geometries, i.e., die, punch and blank holder, the initial blank, their reference systems, the material and the production plan were defined. Figure 4 shows the tools modelled in AutoForm.

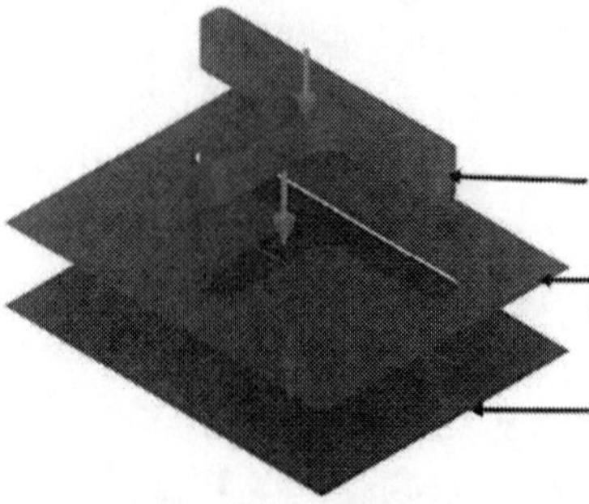

Figure 4. Punch, blank holder and die in FE-model.

The die and the punch were defined as rigid tools, while the blank holder was defined as a force-controlled tool. The initial blank is rectangular with dimensions of 450 mm x 680 mm. Numerical simulations were carried out by setting process parameters as in Table 2. The final quality of the component was evaluated as a function of the force on the blank holder, the friction coefficient and the yield strength. These two last parameters, were chosen as noise parameters since in deep drawing process, fluctuation of lubricant condition and in material properties, due to different coil or supplier, can occur. The nominal values of design and noise parameters and their variability are shown in Table 3. For FE-simulations, Coulomb lubrication model was adopted, although it is an approximation of the real friction behaviour during deep-drawing process. The value of BHF was chosen as a function of design data of the press machine. Moreover, preliminary FE simulations allowed to fix the range variability for such parameter. On the other hand, the value of yield strength was set according to the material datasheet. Finally, values of the friction coefficient were considered in the common range in deep drawing process [13].

Table 2. Settings of process parameters.

Process parameters	Value
Forming velocity	8 mm/s
Press stroke	250 mm
Forming stroke	30 mm

Table 3. Nominal value and variability range of noise and design parameters.

	Process parameters	Nominal value	Variability range
Design parameter	BHF (kN)	550	350-750
Noise parameter	μ	0.1	0.05-0.15
Noise parameter	$\sigma 0$ (MPa)	145.9	116.7-175.1

Output parameters considered for the analysis of the deep-drawing process were: (i) percentage of compression areas, (ii) thickened areas, (iii) insufficient stretch areas, (iv) excessive thinning areas and (v) areas with splits. These output parameters allowed to evaluate the quality of the stamped component as a function of the force on the blank holder under noise factors (variability of the friction coefficient and the yield strength). Since the investigated T-joint was chosen to optimize the areas with insufficient stretch, greater importance was given to this output variable in the optimization phase. Finally, FE model was also adopted to calculate the daw-in in critical points during the process in all investigated conditions.

2.3. Post-processing analysis for kriging meta-modelling and multi-object optimization

Once the process parameters were set in the FE model, a simulation plan based on a Latin Hypercube Sampling (LHS) was performed. For the case study, the plan sampling included 63 numerical simulations. For each simulation, outputs described in section 2.2 were evaluated. Data collected from FE-simulations were processed by means of MATLAB DACE toolbox (2.0, Technical University of Denmark DK-2800 Kgs, Lyngby, Denmark) to derive metamodels with kriging technique [14]. A second order polynomial was chosen as regression function and Gaussian correlation function was adopted. A grid of points on which to evaluate the new (untried) sites was generated. Specifically, a 39×39 mesh of points distributed in the area defined by the lower and upper limits of the parameters F, μ, and σ0 was chosen. Interpolated response surfaces were obtained. These IRS show the influence of the design and noise parameters on the outputs. Subsequently, with the data collected from FE simulations, a multi-objective optimization with the Desirability Function Approach (DFA) [15] was carried out in MATLAB. According to the DFA, the most desirable solution (optimal solution) is a compromise one which provides the best satisfaction in the presence of conflicting objectives. During the optimization phase, for each output parameter, the criterion "lower the better" was chosen. Desirability function (d) correlated to each output parameter (Y_i) was calculated, assigning an exponent (s) greater than one (1.5) for the calculation of the desirability function associated with the percentage of areas with insufficient stretch and an exponent equal to one for the calculation of the desirability functions associated with the other output variables.

$$d_i(Y_i) = \begin{cases} 1 & \text{if } Y_i(x) < T_i \\ \left(\frac{Y_i(x)-U_i}{T_i-U_i}\right)^s & \text{if } T_i \leq Y_i(x) \leq U_i \\ 0 & \text{if } Y_i(x) > U_i \end{cases} \quad \text{where } T_i \text{ target ith value and } U_i \text{ maximum } i\text{th value}$$

The multiplication between each desirability function gave the total desirability (D). The optimal solution is the one for which the value of total desirability is maximum. In correspondence with the optimal solution, the FE model allowed to measure the daw-in profile in critical points. These profiles were defined as optimal.

2.4. Development in AutoForm of a control strategy for BHF

The optimal draw-in profile was compared with those measured at the optimal BHF value by varying the noise parameters outside the optimal range. In this way, it is possible to simulate the industrial drawing process during which material and process fluctuations can occur. An iterative routine was developed in MATLAB; such routine compares for selected discrete instants of the deep drawing process the optimal profile with non-optimal one. If the difference between the two draw-in exceeds a minimum value, the BHF is adjusted. Such minimum value was chosen by evaluating, by means of FE simulations, the variability of the draw-in which guarantees a component free from defects. FE-simulations for the variability analysis were performed by setting BHF and σ0 to the optimal values and varying the value of μ from the minimum value (0.05) to the maximum value (0.15). BHF is increased if the non-optimal draw-in exceeds the optimal one, on the other hand, the BHF is reduced if the non-optimal daw-in is lower than the optimal one. BHF is increased or decreased proportionally to the difference between the two compared draw-in. Finally, the new BHF value obtained from MATLAB routine is imposed in AutoForm model to derive the new draw-in profile. By performing this procedure iteratively, a closed loop on-line control of the draw-in was numerically simulated.

3. Results and Discussion

3.1. Influence of process and noise parameters

FE results were collected and processed with the kriging technique. Figures 5a-b-c-d show, respectively, metamodels of the areas percentage of insufficient stretch, compression, thickened and splits, as a function of the force on the blank holder and the friction coefficient. These metamodels were obtained by setting σ0 to the nominal value. It is possible to observe that: (i) The percentage of insufficient stretch areas increases as the BHF and friction coefficient decrease (Fig.5a). (ii) For low values of the friction coefficient, the percentage of the compression areas reduces as the BHF increases. Instead, for high values of the friction coefficient, the percentage of the compression areas increases as the BHF increases (Fig.5b). (iii) The percentage of thickened areas reduces as the force on the blank holder and the friction coefficient increase (Fig. 5c). (iv) The percentage of areas with splits increases as the force on the blank holder increases for high values of the friction coefficient (Fig. 5d). Same considerations can be done observing the metamodel of the area percentage with excessive thinning, because splits and excessive thinning metamodels show the same trends.

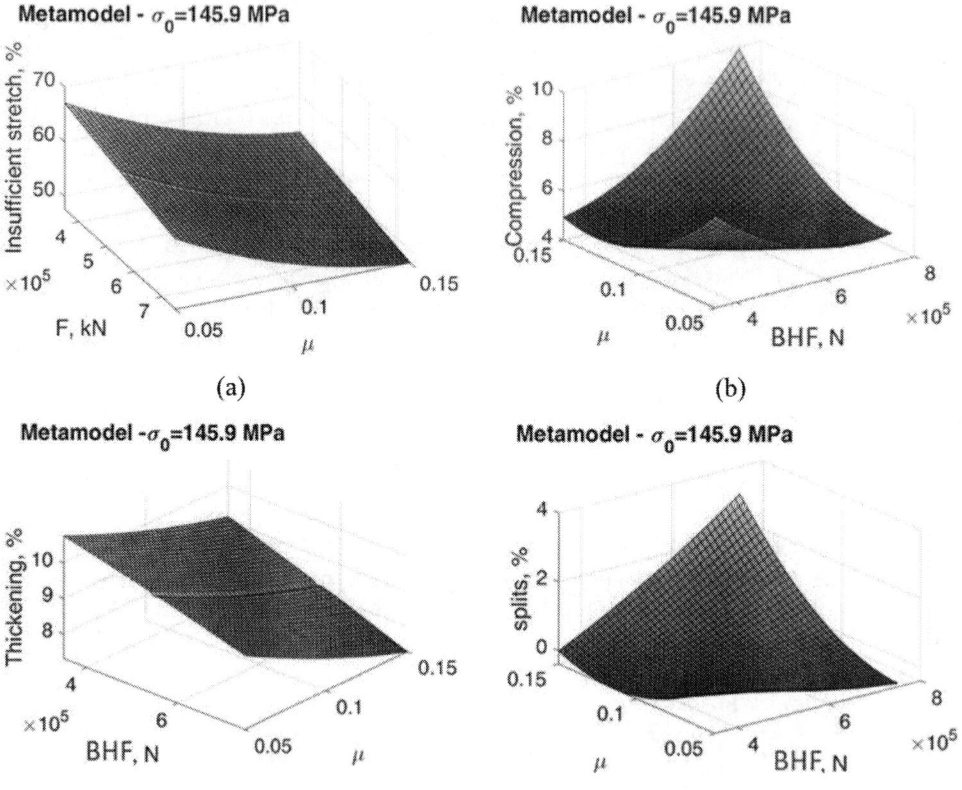

(a) (b) (c) (d)

Figure 5. Metamodel percentage of (a) insufficient stretch areas (b) compression areas (c) thickened areas (d) areas with splits as a function of F and μ.

3.2. Multi-object optimization and experimental tests

Values of the force on the blank holder, friction coefficient and yield strength corresponding to the optimized solution are equal to 550 kN, 0.1 and 145.9 MPa, respectively. The optimal solution is obtained for the maximum value of the total desirability which is equal to 0.975.

Figure 6 compares some results obtained with the FE model and with the deep drawing tests. A good correspondence between numerical and experimental results is observed. The part deep drawn with a BHF of 550 kN is similar to that obtained with the FE model, when the parameters of the optimized solution are used. Using the same friction coefficient and the same mechanical strength parameters of the optimized solution, a reduction of the blank holder force highlights an increase of the area percentage with insufficient stretching (tests with 450 kN). In the same hypothesises an increase of the blank holder force reduces the insufficient stretch areas and increases those with split and excessive thinning. Experimental result in Figure 6 shows that in correspondence of 600 kN blank holder force, the area percentage with insufficient stretching is minimal, but splits are observed in the lateral top part of the T-joint as shown in detail in Figure 7. FE result for BHF equal to 600 kN highlights a zone with excessive thinning in the same region where split occurs during deep drawing experimental test. This is more clear in Figure 7a which shows the detail of the area with the greatest thinning and the formability limit diagram (FLD).

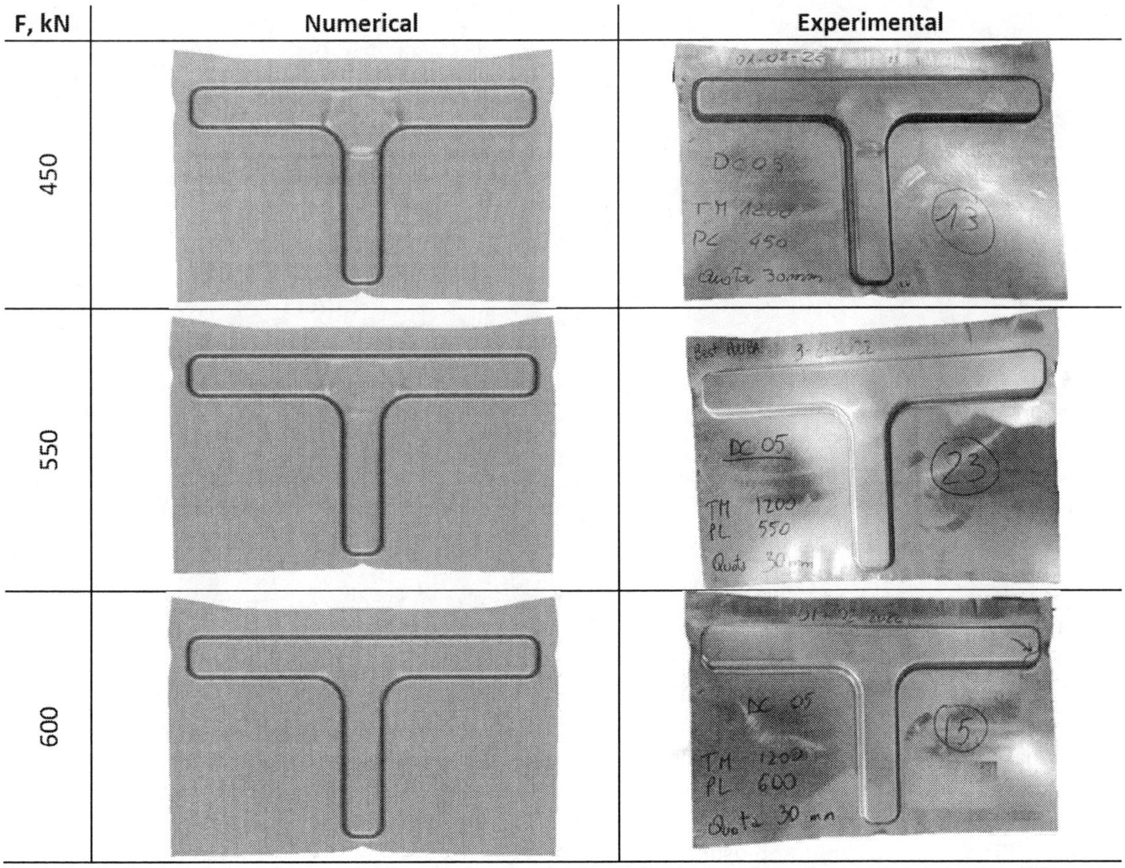

Figure 6. Comparison of numerical and experimental results in the deep drawn T-joint with different BHF.

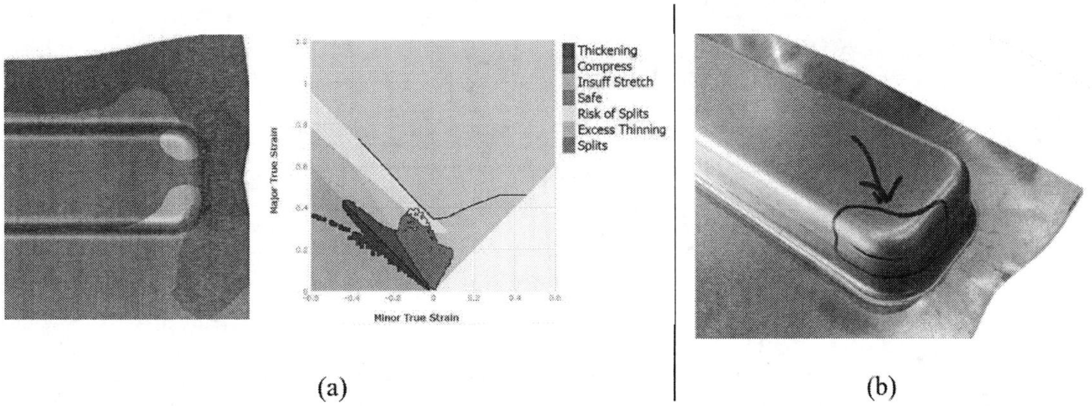

Figure 7: Detail of the region in which split occurs for BHF equal to 600 kN: (a) FE results (b) deep drawing test.

During experimental tests, draw-in was measured by means of Keyence laser displacement sensors. Moreover, numerical prediction of draw-in was derived from FE-model. The draw-in was evaluated at critical points A-B-C, highlighted in Figure 8. Such points were defined as critical since greater blank sliding occurs exactly at these points. As an example, Figure 9 compares the numerical draw-in and with the experimental one measured at point B. Point B was chosen as an example because it proved to be the point where the draw-in is most influenced by process parameters.

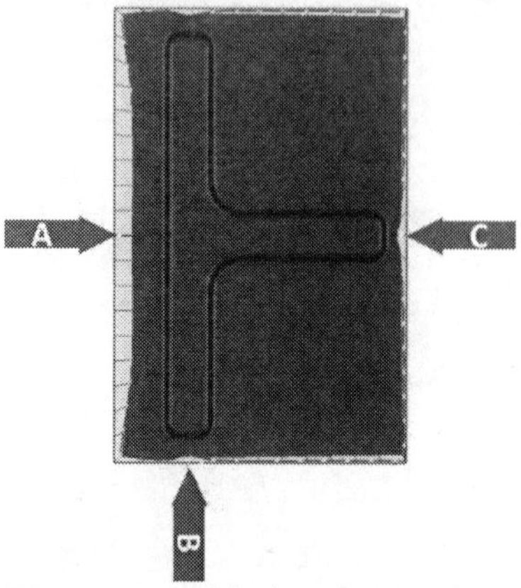

Figure 8. Critical points A-B-C- where draw-in was evaluated.

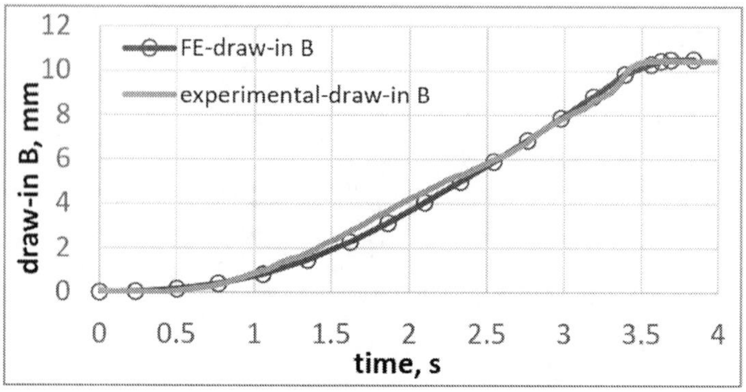

Figure 9. Comparison between numerical and experimental measurement of daw-in at point B.

From figure 9 it is possible to observe that there is a good agreement between FE prediction of the draw-in profile and the experimental one. Considering the value of the draw-in at the end of the process, the percentage error between the numerical and experimental measurement is less than 5%.

3.3. Numerical results of BHF adjustment

By varying the friction coefficient between 0.05 and 0.15, the draw-in at the end of the punch stroke varies between about 12 mm and 6 mm, respectively (Figure 10). In this work results of regulation methodology is shown in the case corresponding to a reduction of the friction coefficient up to 0.05. These results are showed in Figure 11. Specifically, the red curve with circular makers represents the optimal draw-in profile at point B (curve a). The blue curve with square markers (curve b) represents the draw-in profile obtained by setting the BHF at the optimal value, σ0 at nominal values and adopting the lowest value of the friction coefficient (0.05). The green curve with triangular markers (curve c) shows the curve obtained by adjusting the force on the blank holder using the iterative procedure.

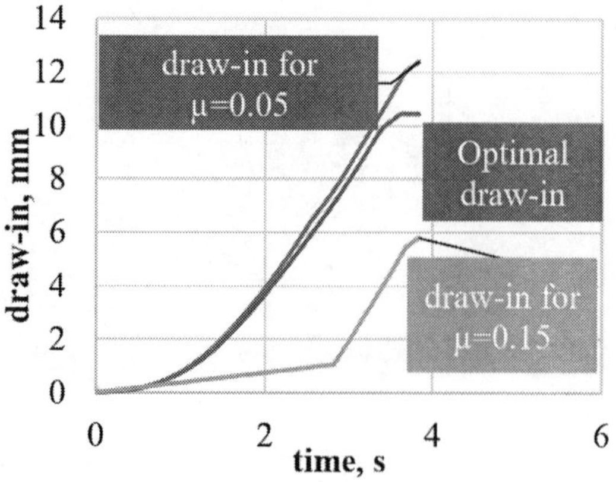

Figure 10. Trend of blank draw-in during the process as the friction coefficient varies

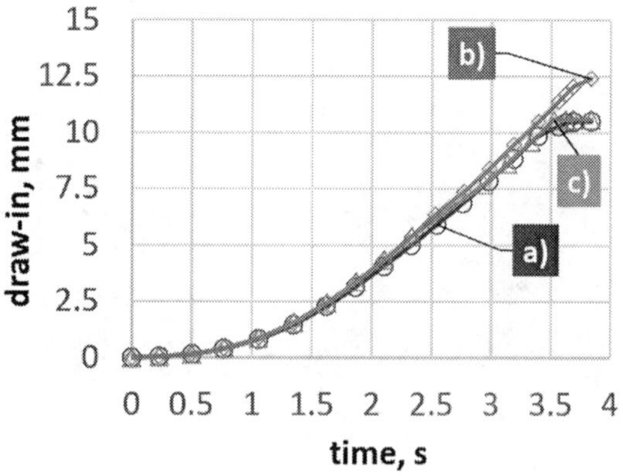

Figure 11. a) optimized draw-in profile, b) draw-in profile with lower value of friction coefficient, c) draw-in profile after BHF adjustment

The force on the blank holder was adjusted starting from a difference between the optimal and non-optimal draw-in equal to 7%. It is observed that, thanks to the effect of the BHF adjustment, the curve b tends to approach the optimal draw-in profile (curve a).
Figure 12 shows the draw-in profile obtained after the BHF adjusting phase and the corresponding law of BHF adjustment. It should be observed that the force on the blank holder starts with the optimal value of 550 kN, after which it is increased first up to 700 kN and then up to 800 kN, leading to changes in the slopes of the draw-in profile.
The effect of the friction coefficient reduction leads to a greater sliding of the blank, in fact curve b is above curve a. Moreover, a reduction of the friction coefficient induces an increase of areas percentage of insufficient stretch. This aspect is highlighted in Figure 13a which shows the FE deep drawn part obtained with lower value of the friction coefficient. Figure 13b, on the other hand, shows the FE result at the end of the BHF adjustment phase. From the comparison between Figure 12a and Figure 12b, an improvement in the quality of the final component is observed. Furthermore, the component shown in Figure 12b is similar to that obtained in the optimal solution (Figure 6 – 550 kN).

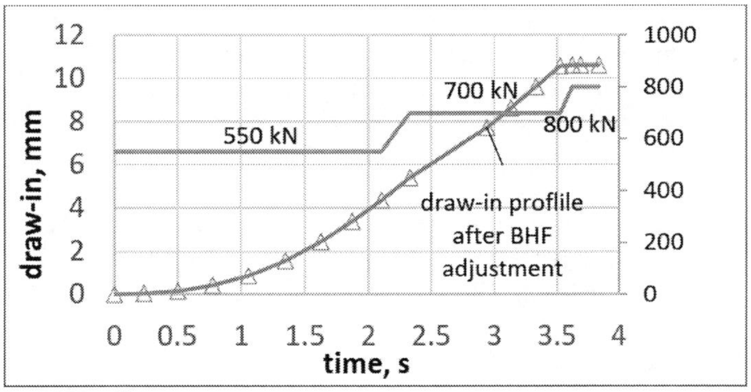

Figure 12. BHF adjustment

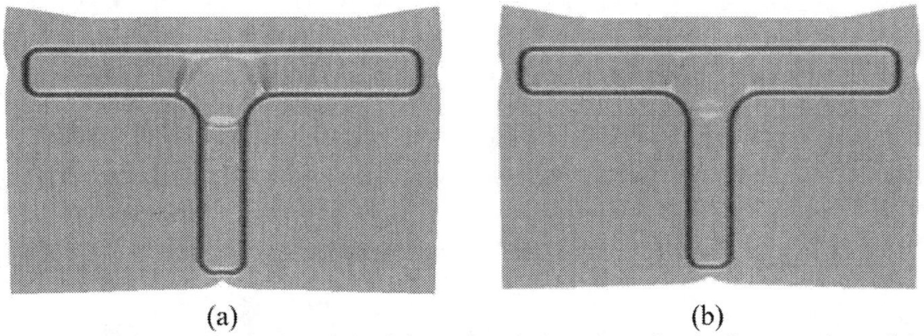

Figure 13. FE results at the end of deep drawing process for (a) lower value of the friction coefficient and (b) after BHF adjustment

4. Conclusions

The following conclusions can be drawn:

- A reduction in the BHF with respect to the optimal value leads to an increase in the areas percentage with insufficient stretch, instead, an increase in the BHF result in the reduction in areas with insufficient stretch. However, for higher value of BHF thinning and consequent risks of splits can occur. Such results were found both with numerically and experimentally investigation. A good numerical-experimental agreement was also observed in the trend of the draw-in profiles of the blank.
- FE results show that by setting the force on the blank holder to the optimal value, the variation of the noise parameters beyond the optimal values leads to draw-in profiles different from the optimal one. Consequently, deep drawn components with defects can be obtained.

- The iterative procedure, developed for adjusting the force on the blank holder, allows to improve the quality of the component at the end of the drawing phase and to obtain a draw-in profile and deep drawn component similar to the optimal ones.

5. References

[1] Ramezani, M., and Z. M. Ripin. Deep drawing of sheet metals using the friction-actuated blank-holding technique. *Rubber-Pad Forming Processes, Technology and Applications* (2012): 119-147.
[2] Zhang, Hongsheng, Siji Qin, and Liqin Cao. Investigation of the effect of blank holder force distribution on deep drawing using developed blank holder divided into double rings. *J. Brazilian Soc Mech. Sci. and Eng.* **43.6** (2021): 1-10.
[3] Krishnan, Neil, and Jian Cao. Estimation of optimal blank holder force trajectories in segmented binders using an ARMA model. *J. Manuf. Sci. Eng.* **125.4** (2003): 763-770.
[4] Ahmetoglu, Mustafa A., et al. Improving drawability by using variable blank holder force and pressure in deep drawing of round and non-symmetric parts. *SAE Technical Paper* (1993).
[5] Yagami, T., K. Manabe, and Y. Yamauchi. Effect of alternating blank holder motion of drawing and wrinkle elimination on deep-drawability. *J. Mater. Process. Tech* **187** (2007): 187-191.
[6] Wu, Peng, Youming Wang, and Peng Wan. Study on simulation of stamping process and optimization of process parameters of fender. *Adv. Mater. Sci. Eng.* (2019).
[7] Palmieri, Maria Emanuela, Vincenzo Domenico Lorusso, and Luigi Tricarico. Robust optimization and Kriging metamodeling of deep-drawing process to obtain a regulation curve of blank holder force. *Metals* **11.2** (2021): 319.
[8] Briesenick, David, Mathias Liewald, and Patrick Cyron. Potentials of an adaptive blank positioning to control material and process fluctuations in deep drawing. *IOP Conference Series: Mater. Sci. Eng.* **967**. No. 1. IOP Publishing (2020).
[9] Braunlich, H., and R. Neugebauer. Closed loop control of deep drawing processes. *SheMet* (2001): 529-538.
[10] Mahayotsanun, Numpon, Jian Cao, and Michael Peshkin. A Draw-In Sensor for Process Control and Optimization. *AIP Conference Proceedings* **778**. No. 1. American Institute of Physics (2005).
[11] Lo, Sy-Wei, and Tsu-Chang Yang. Closed-loop control of the blank holding force in sheet metal forming with a new embedded-type displacement sensor. *Int. J. Adv. Manuf. Tech.* **24.7** (2004): 553-559.
[12] Hardt, D. E., and R. C. Fenn. Real-time control of sheet stability during forming. (1993): 299-308.
[13]Fereshteh-Saniee, F., and M. H. Montazeran. A comparative estimation of the forming load in the deep drawing process. *J. Mater. Process. Technol.* **140.1-3** (2003): 555-561.
[14] Kleijnen, Jack PC. Kriging metamodeling in simulation: A review. *European journal of operational research* **192.3** (2009): 707-716.
[15] Costa, Nuno R., João Lourenço, and Zulema L. Pereira. Desirability function approach: a review and performance evaluation in adverse conditions. *Chemom. Intell. Lab. Syst.* **107.2** (2011): 234-244.

Acknowledgments

The authors wish to thank the Italian Institution MIUR (Ministry of Education, University and Research) for financing the present research activity (PICO&PRO project - COD.ID.ARS01_01061). The authors are also grateful to AutoForm Engineering GmbH for the support in FE model development and to partners of PICO & PRO project. Specifically: CRF (Centro Ricerche Fiat) for tool design and experimental test support; Gigant Industries (Bologna (BO), Italy) for press machine development and the availability to make experimental test in its plant. Tiberina industry (Sangro – Atessa (CH), Italy) for tools manufacturing, the control and telecommunications research groups of the Polytechnic of Bari (Italy) for the support on experimental tests and for the signals acquisition from the press machine.

IOP Conf. Series: Materials Science and Engineering 1238 (2022) 012072 doi:10.1088/1757-899X/1238/1/012072

Digital Twins in deep drawing for virtual tool commissioning and inline parameter optimization

Lars Klingel*[1], Lars Penter[2], Philip Mayer[1], Steffen Ihlenfeldt[2] and Alexander Verl[1]

[1] Institute for Control Engineering of Machine Tools and Manufacturing Units (ISW), University of Stuttgart, 70174 Stuttgart, Germany
[2] Institute of Mechatronic Engineering, TU Dresden, 01062 Dresden, Germany

*lars.klingel@isw.uni-stuttgart.de

Abstract. Fluctuating boundary conditions and the highly nonlinear process behavior of deep drawing operations make experience-based selection of suitable control parameters difficult. Nowadays, commissioning deep drawing tools, i.e., die spotting and identification of suitable control parameters for defect-free parts, is conducted on real world try-out presses. Transferring the tools to production machines entails adaption of these initial parameters. Once production is ramped up, any changes to material properties, lubrication and press behavior require continuous manual machine parameter tuning. Virtual tool commissioning and utilizing these simulation models in the production phase to adapt control parameters would reduce time and cost over the entire life cycle of the machine and the tool sets. The virtual representative, which provides services for a plant over several life cycle phases, is also referred to as Digital Twin. In this paper, the authors present a Digital Twin concept for deep drawing presses to predict the state of the system and optimize the control parameters during the production. The integration of all involved subsystems into one system simulation and its efficient calculation is the biggest challenge. The authors combine a virtual commissioning simulation tool with a finite element model to implement all relevant properties of the deep drawing press and the interaction of its subsystems. It is shown, how the idea of a system simulation makes predictions of system parameters specific to a production situation possible, and therefore, can help to select suitable control parameters that leads to a reduction of the error rate in deep drawing.

1. Introduction

Sheet metal forming is characterized by highly automated mass production processes with repetitive motion sequences, high process forces, expensive tool sets and a low margin per part due to high material costs. In order to save resources and costs, achieving zero-defect production is one of the major quests for manufacturing companies. Due to the complexity of the process, deep drawing operations are already extensively optimized. For further improvement, computer models and real-world equipment are combined for precise predictions of the part quality along the life-cycle of the forming tools. Such a system is often referred to as Digital Twin.

The production of large car body parts is conducted on transfer presses or on production lines with several connected presses. Deep drawing is usually the first forming operation in a multi stage process (Figure 1 right). Special deep drawing presses are equipped with die cushions to provide the blankholder force for large drawing depths. Today's multi-point die cushions generate local cylinder forces, which then enable local blankholder forces to control material flow into the die cavity. The structure of a single action deep drawing press is shown in Figure 1 on the left.

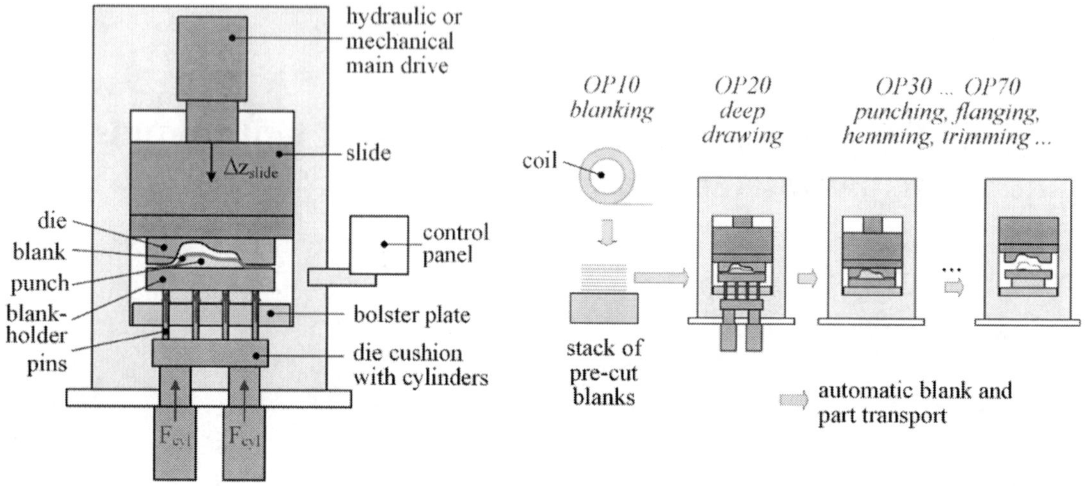

Figure 1. Left: single action deep drawing press with die cushion; right: production line of formed metal parts.

Many interacting factors define the final part quality (Figure 2). Due to large forming forces and finite press stiffness, the interactions between process and machine cause significant press and tool deformation which influence the final part properties and are the main reason for manual die spotting during tool commissioning.

Further factors that give reason for tool commissioning are deviations between real and set (simulated in the development phase) values of die cushion forces [1] and uncertainties regarding lubrication, sheet properties and final thickness distribution [2].

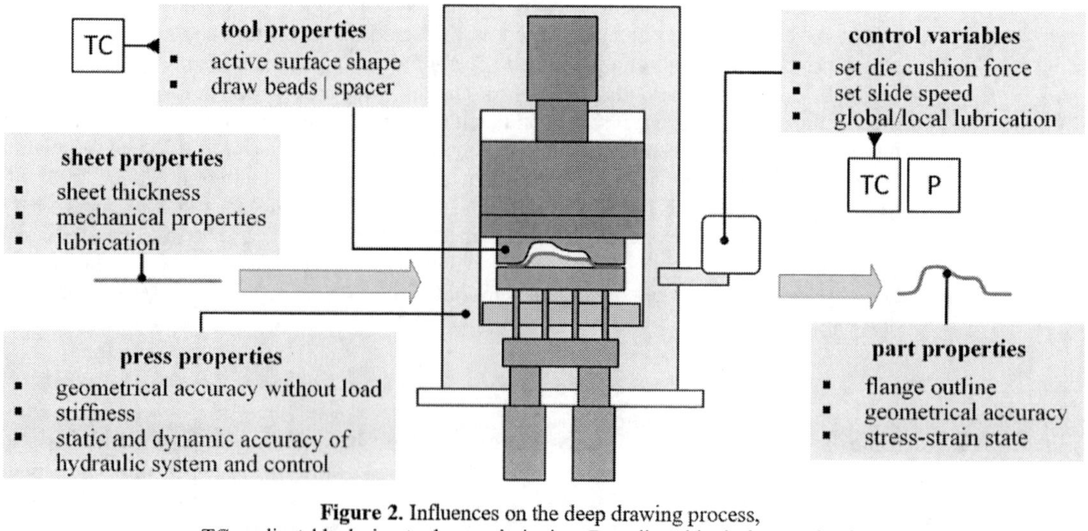

Figure 2. Influences on the deep drawing process,
TC – adjustable during tool commissioning, P – adjustable during production.

Currently, tool commissioning is conducted on extra try-out presses and comprises five major steps (Figure 3). The cost share of tool commissioning as part of the tool development process is 31% [3]. Spotted die and blankholder surface as well as identified control parameters are re-evaluated on the actual production press and, if necessary, adapted. The tool set of modern production presses changes every two hours; this additional commissioning procedure takes around five minutes [4].

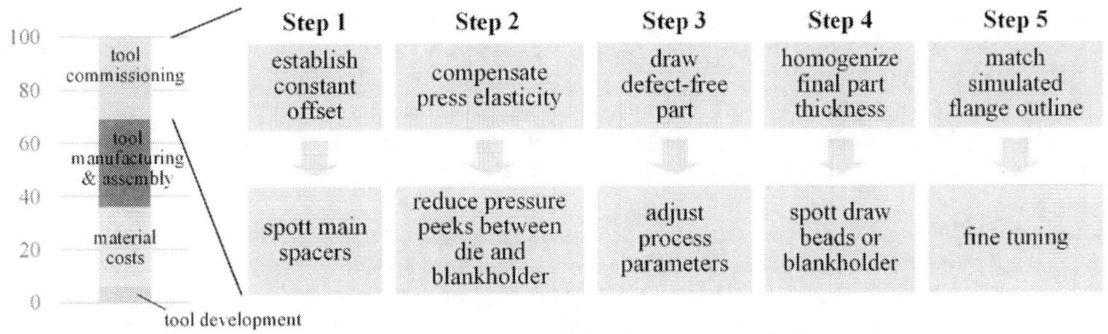

Figure 3. Cost allocation of a forming tool [3].

A complete transformation of tool commissioning into the virtual world would save time and resources. Virtual commissioning became a vital tool for many applications in the field of production engineering. It is a method to detect errors and validate control software in the engineering phase of a production system [5]. In virtual commissioning the control device is connected to a simulation model of the production system [6]. This makes software tests possible before the real machine is set up and provides a save testing environment where problems which would occur due to the control software, can be found before the machine goes into operation.

The established form of virtual commissioning is the hardware-in-the-loop simulation where a real control device (hardware and software) is connected to the model of the production system with a fieldbus [7]. In this setup, real-time requirements for the simulation arise because the simulation model must provide deterministic calculation results in the cycle time of the real control system and fieldbus. Another configuration for virtual commissioning is the software-in-the-loop simulation where the software is running on an emulated controller and the simulation and control software are synchronized on a virtual time axis [5].

Figure 4 shows virtual commissioning as a phase of a production system's lifecycle. To build the simulation models, all relevant information has to be integrated such as computer aided design (CAD) models for example. In [8] virtual commissioning simulation models are described as a path to realise the vision of a Digital Twin. This is the reason because virtual commissioning simulations are calculated in real-time, contain lots of different previous artefacts and are available anyways in a modern development process before the real production system goes into operation. In addition to simulation models, Digital Twins contain data of the system and services for the usage of the models and data.

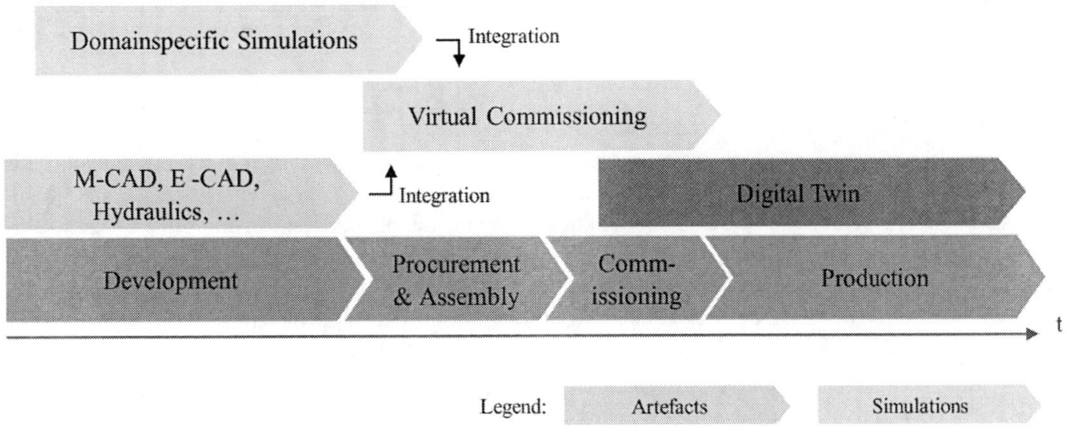

Figure 4. Simulation in the lifecycle of production systems [8].

IOP Conf. Series: Materials Science and Engineering 1238 (2022) 012072 doi:10.1088/1757-899X/1238/1/012072

Finite element (FE) forming simulations allow the prediction of the part quality [3] in advance of the actual commissioning and production process [9]. Because forming simulations neither comprise machine influences such as press and tool elasticity, dynamics of hydraulic drives, the behavior of the control device, nor do they update their parameters, the gap between numerical prediction and reality is too large to eradicate real world tryouts and manual adjustments during production [10]. During production, lubrication and sheet properties fluctuate [2], machine properties change over time due to wear and the behaviour of hydraulic system shift due to temperature changes and cause quality problems or even final part defects (Figure 2). These cannot be addressed during tool commissioning and need adjustment while the fabrication process is running, preferably without machine standstill.

A fully adaptive system is needed to achieve a completely error-free production. In literature, such a system is often referred to as Digital Twin [11]. This paper presents an approach to use a system simulation as a base for Digital Twins in deep drawing.

2. Digital Twin of forming process and equipment

In this section a concept for the usage of Digital Twins in deep drawing is presented, which can be subdivided into three steps which are listed in ascending order of complexity: Virtual tool commissioning, inline parameter optimization and online process control. These steps mainly differ in terms of the use case and the real-time requirements of the underlying simulation model, see Figure 5.

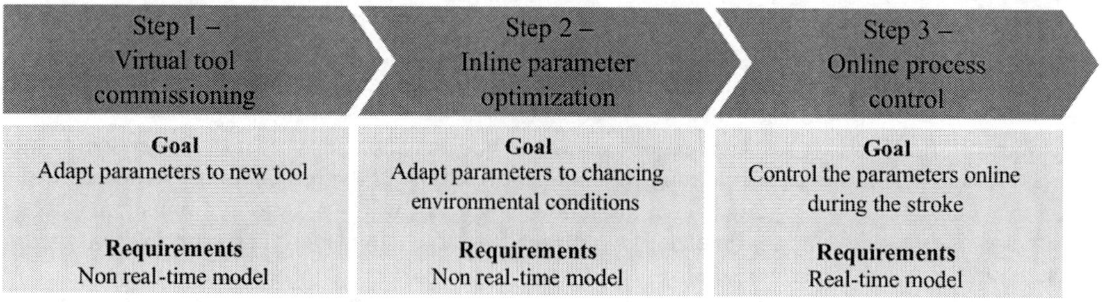

Figure 5. The steps for an integration of Digital Twins into deep drawing.

2.1. Step 1 - Virtual tool commissioning

Virtual tool commissioning can replace the necessity of try-out presses or complement them for the selection of suitable control parameters in the commissioning phase of deep drawing tools.

Necessary adaptions on the deep drawing tool and the control parameters can be evaluated in the simulation. The main reason for the current usage of try-out presses is the absence of a comprehensive system simulation and fluctuating boundary conditions. A system simulation which combines the machines behaviour and deep drawing process information would be the basis for a Digital Twin as a replacement of try-out presses.

2.2. Step 2 - Inline parameter optimization

Not everything can be tested on a try-out press or on a Digital Twin in the try-out phase. This is because there are chancing parameters in the environment, which have an influence on the deep drawing press. A prominent case is the change of material properties during the production. The variation of forming properties or thickness of the sheet often results in defective parts. To solve this problem, the authors present an approach for an inline parameter optimization in deep drawing, which is shown in Figure 6 with the described use case of changing material properties.

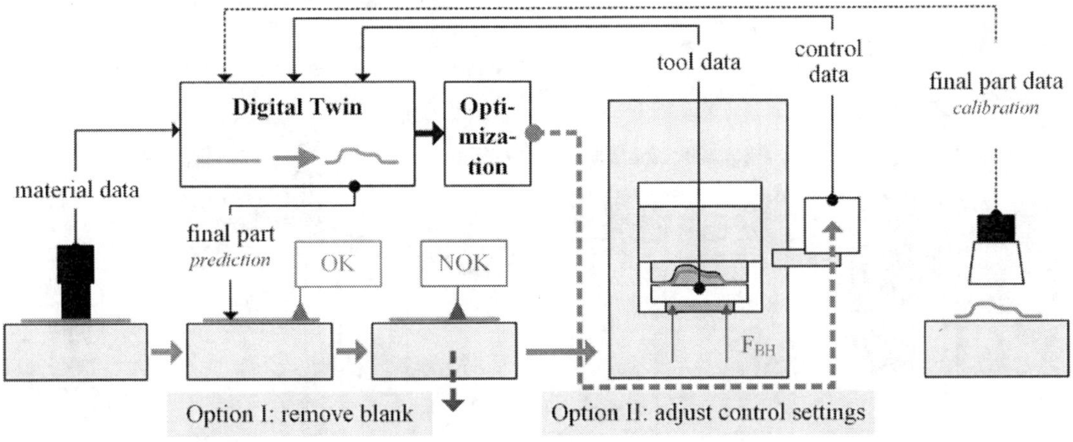

Figure 6. Digital Twin during the production for inline parameter optimization.

The concept requires measurement of the mechanical properties of the metal sheet before deep drawing. Normally the material quality changes if a new coil is used. A way to measure the material properties inline before the sheet is deep drawn are inductive sensors [12]. If the current material, tool and control data is transferred to the Digital Twin, the quality of the future part can be predicted and used by feed forward control algorithms to adapt the control parameters (Option II) or eject the blank because of poor material properties before being processed (Option I). For a continuous improvement of the Digital Twins models and a validation of the control the predicted part can be compared to a measured part in the quality control afterwards.

2.3. Step 3 – Online process control

The application of Digital Twins depends on the accuracy and calculation time of the underlying model. If the model can be calculated in real-time, it could act as an observer for closed-loop control. In [13] this is described as online closed-loop scheme. Because of the highly nonlinear behaviour and difficulty to measure the current state of the sheet during the forming process a model-based observer would allow the design of closed-loop control on parameters such as the sheet thinning. Currently, the computation of FE process models for deep drawing consumes too much time, which makes data driven models necessary for this concept.

3. Concept and implementation of the model

The concept and its implementation in the following chapter address the first two steps of chapter 2. The main focus of this work is the Digital Twins underlaying simulation model. Which is implemented in form of a co-simulation. First the concept of the co-simulation is shown which combines the machine and process behaviour. Afterwards the used application example is introduced and the used simulation tools as well as a schematic physical diagram of the model is shown.

3.1. Concept for the co-simulation environment of the Digital Twin

For the realization of the Digital Twin, a detailed model of the system which includes the machine and process is needed. In this work the system simulation is implemented in form of a co-simulation. In Figure 7 the blue parts represent the parts of the machine simulation, which is calculated in a virtual commissioning tool. The forming part on the right in the middle is calculated in an FE tool. Those simulations are coupled during the runtime with the functional mock-up interface [14]. The system boundaries and resulting coupling signals can be found in Figure 7.

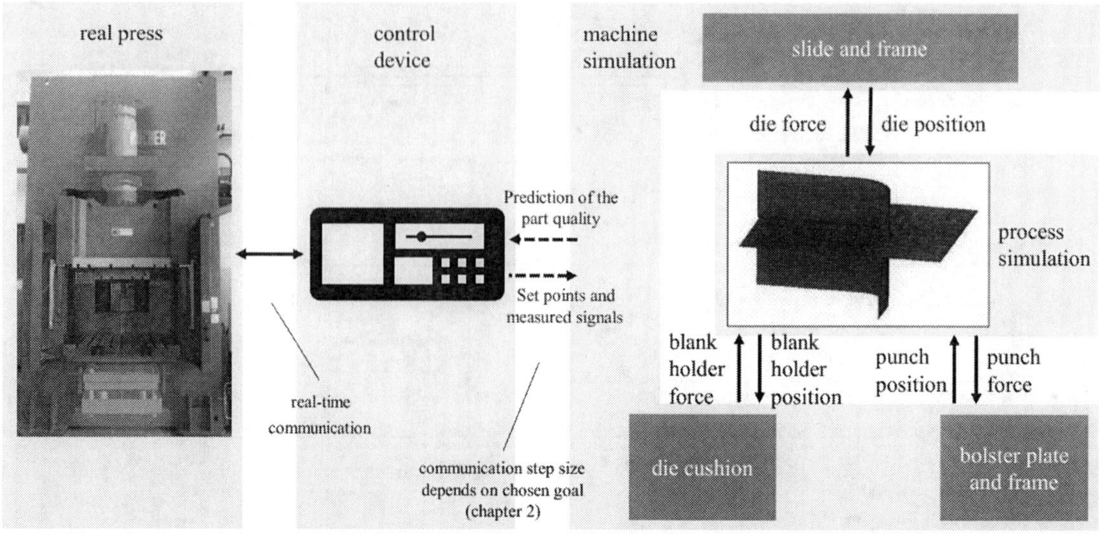

Figure 7. Structure of the control coupled co-simulation environment.

Virtual commissioning simulation tools provide a connection to control devices. This makes it possible to transfer current control parameters to the machine simulation. In case of step 2, the measured data from the real press can be transferred to the simulation model additionally. The simulation model provides a predicted part quality to the control which allows simulation-based control functionalities.

3.2. Application example and simulation setup

A hydraulic deep drawing press which is located at the Institute of Mechatronic Engineering at the TU Dresden served as example machine. This press features a hydraulic slide with a maximum force of 2500 kN and a 4-point hydraulic cushion with a maximum force of 1000 kN.

The preliminary work of TU Dresden is the basis for the implementation of the simulation model [10][15]. As an example part a rectangular pan is used.

The co-simulation employs ISG-virtuos for the simulation of the machine (see Figure 8, right top) and LS-DYNA for simulating the forming process (see Figure 8, right bottom). These simulation tools are coupled with the functional mock-up interface which comprises the interface signals shown in Figure 7. For a detailed description of the ISG-virtuos coupling with LS-SYNA refer to [16]. The left side of Figure 8 shows the physical abstraction of the system's elasticity and dynamic in ISG-virtuos and the system boundaries to LS-DYNA. A virtual CNC controller which is integrated into the system simulation generates the input trajectory, which is shown in the diagram (red arrow). From the perspective of virtual commissioning this setup is called software-in-the-loop configuration.

In a software-in-the-loop simulation the communication between the model of the plant and the control is synchronized on a virtual time axis. The simulation in ISG-virtuos is running with a step size of 1ms. LS-DYNA uses a variable step size and is not real-time capable. The co-simulation is slowed down by the FE simulation and the functional mock-up interface, which uses TCP/IP for the data transfer.

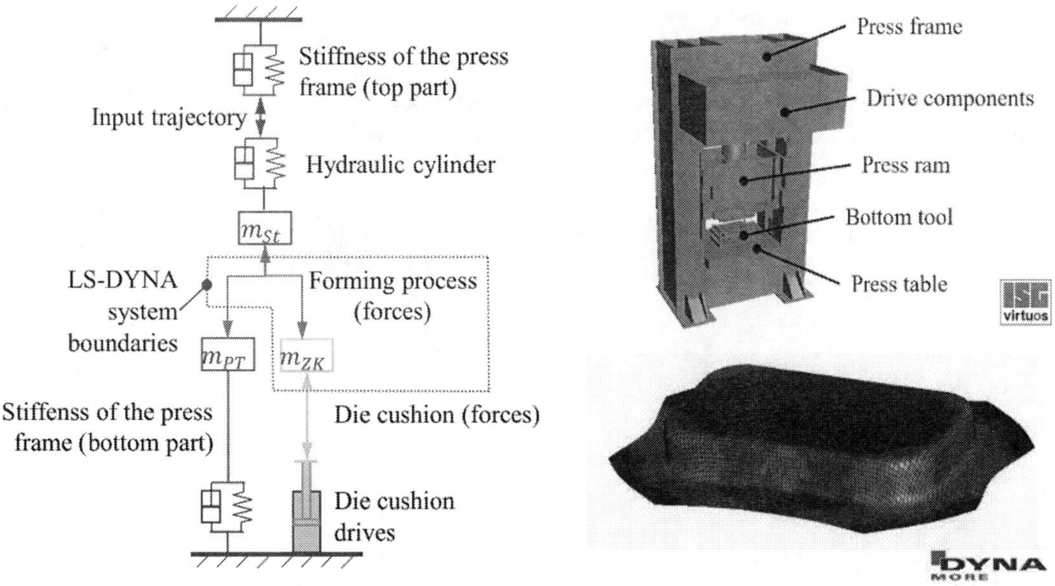

Figure 8. Physical abstraction of the press (left) and visualization in the used simulation tools (right).

4. Results

In this chapter the authors show how the simulation can be used for a prediction of the part quality, which is necessary for virtual tool commissioning and inline parameter optimization.

The part quality is measured by the maximum sheet thinning with the target to get close to 20% but not to exceed it, which leads to a hardened sheet on the one side and avoidance of cracks on the other side. In Figure 9 there is shown the distribution of the sheet thinning with the chancing parameters strength coefficient and input sheet thickness for a static blankholder force. In the bottom part of Figure 9 a parameter study is shown for the steel DC04 with the same varying material properties and with three different blankholder forces. In each graph the other material property is constant. The values and the variation of the parameters are based on DIN EN 10130 [17] and DIN EN 10140 [18]. The parameter variation which is depending on the input material can be measured before the sheet is deep drawn in the press. In the parameter study the characteristic curves for varying blankholder forces are shown. During virtual tool commissioning and inline parameter optimization those plots can be calculated after boundary parameters changed, like a new input sheet with different material properties. With this calculation the blankholder force which leads to a maximum sheet thinning close to 20% can be chosen. In Figure 9 in the left diagram there is shown how the simulation can be used to determine a suitable blankholder force. In this example the current blankholder force is 100 kN and current strength coefficient is 480 MPa (P_0 in Figure 9). With chancing material, the strength coefficient changes from 480 to 440 MPa. It shows that the combination of a strength coefficient of 440 MPa with a blankholder force of 100 kN leads to a sheet thinning over 20% (P_1 in Figure 9). To solve this problem a different blankholder force has to be chosen, which sets the operating point under the critical sheet thinning again (solid arrow).

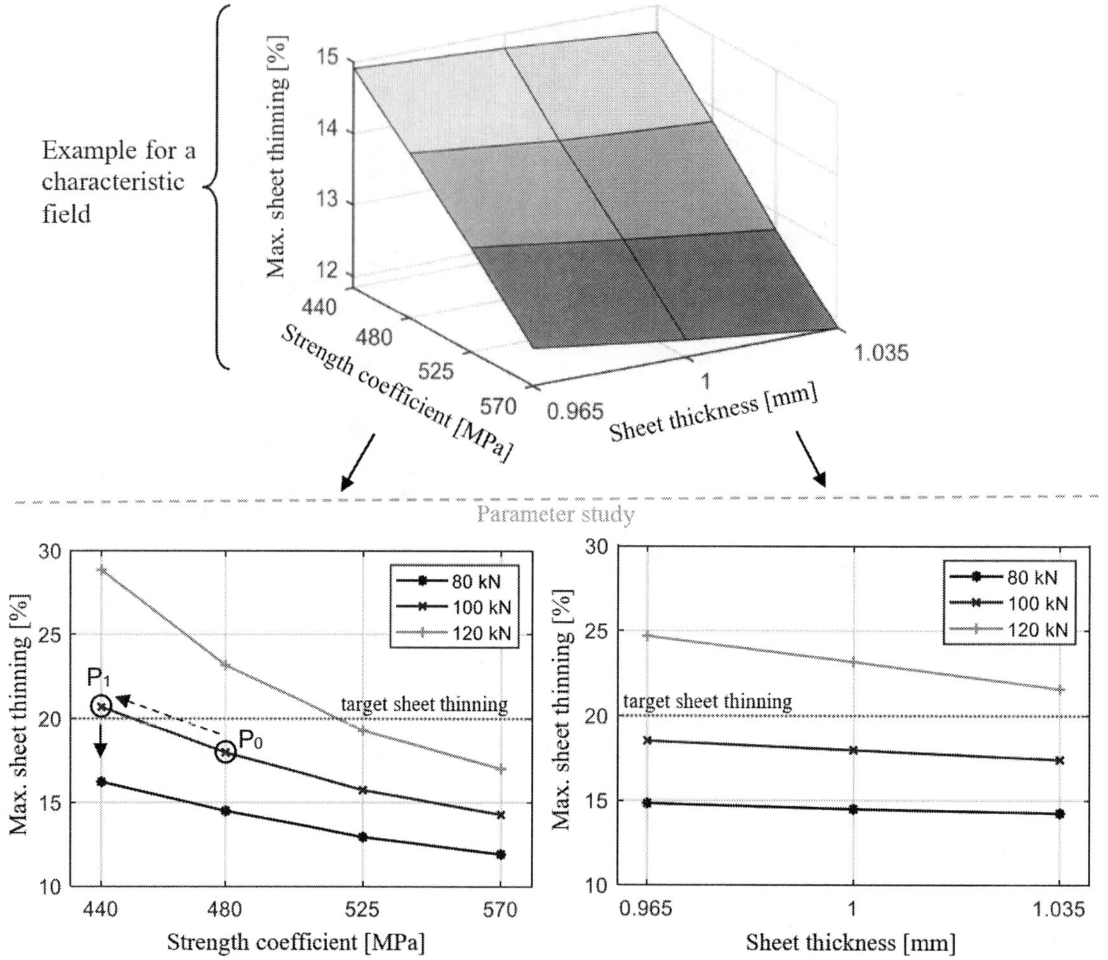

Figure 9. Identification of the sheet thinning with fluctuating material properties for one static blankholder force (top) and varying blankholder forces for the identification of a suitable blankholder force (bottom).

5. Conclusion

In this paper the vision of Digital Twins in deep drawing was introduced. A concept was presented which would allow the rollout of Digital Twins into the production process of formed metal sheets. Those three steps are virtual tool commissioning, inline parameter optimization and online process control. In this paper a co-simulation environment was designed and implemented with an example application, which could be applied on the first two steps. It was shown how the simulation could be used to detect errors in parts simulatively. In a next step this model should be validated with data from the real plant. A problem is that the process simulation is to slow, to achieve a real-time capable co-simulation which is necessary for the online process control. This is why other approaches like data driven simulations should be evaluated, this should be a focus of further research in the field of Digital Twins in deep drawing. Finally, the Digital Twin based control system should be evaluated on a deep drawing press, to validate if the usage of Digital Twins can actually increase the part quality and decrease the error rate.

Acknowledgements
This work was funded by the Deutsche Forschungsgemeinschaft (German Research Foundation DFG) within the project 438646126.

References

[1] Helmke M, Majer H, Thanassakis A 2016 *Improvement of hydraulic control quality for deep drawing presses through retrofit* 10th International Fluid Power Conference

[2] Purr S, Moelzl K, Meinhardt J, Merklein M 2016 *Das Presswerk 4.0 – Data Mining zur Vermeidung von Qualitätsproblemen bei der Herstellung von Karosseriebauteilen* 36. EFB-Kolloquiums Blechverarbeitung

[3] Birkert A, Haage S, Straub M 2013 *Umformtechnische Herstellung komplexer Karosserieteile* Springer-Verlag Berlin Heidelberg

[4] Hummelsberger R, Meinhardt, J, Grossenbacher K, Kreissl S 2016 *Strategische Ausrichtung der Presswerke der BMW Group* 36. EFB-Kolloquium Blechverarbeitung

[5] VDI 2018 *Virtuelle Inbetriebnahme* VDI 3693 Blatt2

[6] Röck S 2007 *Echtzeitsimulation von Produktionsanlagen mit realen Steuerungssystemen* Stuttgart Jost-Jetter Verlag

[7] Pritschow G, Röck S 2004 *Hardware in the Loop Simulation of Machine Tools* CIRP Annals 53 Issue 1 pages 295-298

[8] VDMA 2021 *Leitfaden Virtuelle Inbetriebnahme: Handlungsempfehlungen zum wirtschaftlichen Einstieg*

[9] Penter L 2016 *Qualifizierung von FE-Prozessmodellen zur Inbetriebnahme von Karosserieziehwerkzeugen* TU Dresden disseration.

[10] Schulze T, Weber J, Großmann K, Penter L, Schenke C 2015 *Hydraulic die cushions in deep drawing presses – analysis and optimization using coupled simulation* Proceedings of the ASME/BATH 2015 Symposium on Fluid Power & Motion Control

[11] Kritzinger W, Karner M, Traar G, Henjes J, Sihn W 2018 *Digital Twin in manufacturing: A categorial literature review and classification* IFAC PapersOnLine

[12] Bäume T 2019 *Prozessregelungen durch piezoelektrisch erweiterte Umformwerkzeuge* Technische Universität Chemnitz Dissertation

[13] Allwood J.M, Duncan S.R, Cao J, Groche P, Hiert G, Kinsey B, Kuboki T, Liewald M, Sterzing A, Tekkaya A.E 2016 *Closed-loop control of product properties in metal forming* CIRP Annals 65 Issue 2 Pages 573-596

[14] Blochwitz T, Otter M, Akesson J, Arnold M, Clauß C, Elmqvist H, Friedrich M, Junghanns A, Mauss J, Neumerkel D, Olsson H, Viel A *2012 Functional Mockup Interface 2.0: The Standard for Tool independent Exchange of Simulation Models* 9th International Modelica Conference

[15] Schulze T, Weber J 2017 *Model Based System Identification for Hydraulic Deep Drawing Presses* The 15th Scandinavian International Conference on Fluid Power

[16] Heiland S, Klingel L, Penter L, Jaensch F, Schenke C, Ihlenfeldt S, Verl A 2021 *Virtual tool commissioning using LS-DYNA functional mock-up interface* 13th European LS-DYNA Conference 2021 Ulm

[17] DIN EN 10130 2006 *Kaltgewalzte Flacherzeugnisse aus weichen Stählen zum Kaltumformen - Technische Lieferbedingungen*

[18] DIN EN 10130 2006 *Kaltband – Grenzmaße und Formtoleranzen*

On the reliability of yield functions in deep drawing simulations

H Ghiabakloo[1,*], N Manopulo[2], J Mora[3], B Carleer[4] and A Van Bael[1]

[1] Department of Materials Engineering, Katholieke Universiteit Leuven, Kasteelpark Arenberg 44/box 2450, 3001 Leuven, Belgium
[2] AutoForm Development GmbH, Technoparkstrasse 1, 8005, Zürich, Switzerland
[3] AutoForm Engineering B.V., Rhijnspoor 221, 2901 LB Capelle aan den IJssel, Netherlands
[4] AutoForm Engineering Deutschland GmbH, Joseph-von-Fraunhofer-Straße 13a, 44227 Dortmund, Germany

* hadi.ghiabakloo@kuleuven.be

Abstract. In sheet metal forming simulations, the yield functions are usually calibrated based on experimental data, and validated by comparing the modelled and measured r-values and (in more advanced models) yield stresses along various in-plane directions. The fitted yield function should ideally reproduce (or interpolate) all the experimental values used for the fitting. However, this requirement does not guarantee accurate results in a forming process simulation, and it can even lead to unexpected results. The performance of a yield function, in addition to the fitting procedure, depends on the active loading modes which the material experiences during the simulation. The active loading modes are in turn determined by the die geometry as well as the process parameters like blank holder force and lubrication. As a consequence, the performance of a yield function, which is fitted to simple experimental data, is not identical for different forming conditions. Therefore, the application of each model is usually limited to a range of materials and processes, and this applicability is often evaluated based on experience. In the present study, we examine three phenomenological yield functions and a new crystal plasticity based material model for cup drawing process simulation with an AA6016-T4 aluminium alloy. These functions are different in their input data types used for the calibration. The results surprisingly shows that the models with more experimental data (in particular, yield stresses in different directions) in their formulation not only predict unexpectedly wrong results, but also show strong sensitivity to some of those additional input data and even to modelling parameters like the friction coefficient.

1. Introduction

The accuracy of sheet metal forming simulations depends on the models used for capturing the anisotropy of the material. Traditionally, the yield functions are fitted to experimental data obtained from simple tests (e.g. tensile test) in different in-plane directions. This way the model can exactly reproduce and reflect the input data. However, the behavior of the model is not always reliable for the other loading conditions. The reliability of the yield function is basically influenced by two factors: (1) its mathematical formulation, and (2) the quantity and quality of the input data.

The mathematical formulation of the yield function should be flexible enough to fit to the unpredicted distribution of experimental data. This is obtained by including more terms in the yield function which in turn may require a larger number of tests to generate the training data set. In addition, the fitting procedure may also be a challenge in more complex yield functions. Therefore, the developers should look for a trade-off between accuracy and efficiency if the model is designed for a wide range of materials.

The training data should be designed and collected such that it covers as much loading modes as possible. This may require advanced tests which may be sometimes unfeasible or even impossible to be carried out in practice. Also, in combination with the first factor, it may end up with elevated experimental costs. Here the difference between the yield functions is in the loading modes which are thought to be more important for the considered forming process.

In addition to the reliability of the yield functions in representing material behavior under various loading modes, their robustness under experimental uncertainties is another important factor which should be tested for a given material and/or process. In other words, the results obtained from a yield function should not be too sensitive to the measured input data or to modelling parameters like the friction coefficient which are usually entered to the simulations with uncertainties. Otherwise, the risk of introducing errors in the simulations will be high unless the precision of experimental measurements and the accuracy of friction models are guaranteed.

In the present study, the performances of three phenomenological yield functions (with different levels of complexity) as well as a crystal plasticity (CP)-based model are assessed for circular cup drawing simulations. It is demonstrated how a simple phenomenological yield function requiring a minimum number of material parameters and the CP-based model can perform better in predicting the earing profile than more advanced phenomenological expressions. This observation cannot be generalized to other materials or processes, and there can be other cases that show opposite results. Moreover, it is shown that more advanced phenomenological models may lead to unrealistic results even for the current process with its simple geometry.

2. Methods and material

All the experimental data for the present study are taken from [1] and [2]. The cup drawing experiments have been carried out on an AA6016-T4 aluminum alloy, and with the following process parameters: punch diameter: 50 mm, punch fillet radius: 6 mm, die diameter: 52.82 mm, die fillet radius: 3 mm, blank diameter: 90 mm, blank thickness: 1 mm, and blank holder force: 10 kN.

The cup drawing simulations were performed with AutoForm Forming R10 with the mesh settings as follows: element type: thick shell with 11 through thickness integration points (TS-11), initial maximum element size: 1 mm, minimum element size: 0.5 mm, and number of initial elements: 18707. Four friction conditions were tested in two different ways: (1) traditional Coulomb friction with three coefficients 0.05, 0.1, and 0.15, and (2) physical modelling with TriboForm which is available in AutoForm. With TriboForm, the default settings for a contact between Al 6000 series and cast iron with drawing oil were used.

For all the simulations, an isotropic hardening behavior was assumed according to Hockett-Sherby model given by

$$\sigma = \sigma_{\text{sat}} - (\sigma_{\text{sat}} - \sigma_{\text{i}})e^{-a\varepsilon_{\text{pl}}^{p}}, \quad (1)$$

in which σ_{sat} = 438.56 MPa, σ_{i} = 115.13 MPa, a = 2.74, and p = 0.59 were found by fitting Eq. (1) to the stress-strain data obtained from the tensile test in the rolling direction.

The phenomenological yield functions were selected among those available in AutoForm R10. Hill48 [3] and Hill90 [4] were not used as they are not recommended for Al alloys due to their low r-values. The other yield functions in AutoForm R10, i.e., Barlat89 [5], BBC2005 [6], and Vegter [7] were tested. In addition, a new and reduced version of Facet yield function [8], named Facet-3D, was also implemented as an R&D Plugin material model in AutoForm R10 and tested for the simulations. This crystal plasticity based yield function is dedicated to the generalized plane stress conditions in sheet metal forming simulations. Facet-3D has the same formulation as the general 5D Facet (see[8]), except

IOP Conf. Series: Materials Science and Engineering 1238 (2022) 012073 doi:10.1088/1757-899X/1238/1/012073

that the out-of-plane shear terms are neglected in its formulation, and the virtual experiments are executed on a grid in 3D stress space created by the spiral method [9].

For calibrating the phenomenological yield functions, the material properties were measured through tensile tests in various in-plane directions. Facet-3D however is calibrated to virtual experimental data generated by the ALAMEL crystal plasticity model [10] which requires only one texture measurement as the input. The required experimental input data for all the studied yield functions are summarized in Table 1. In this table, *r*, *YS*, *PS*, and *PSH*, stand for r-value, yield stress, plane strain point, and pure shear point, respectively.

Table 1. The required experimental input data for all the studied yield functions.

model	Rolling Direction				Diagonal Direction				Transverse Direction				biaxial		Texture
	r	*YS*	*PS*	*PSH*	*r*	*YS*	*PS*	*PSH*	*r*	*YS*	*PS*	*PSH*	*r*	*YS*	
Facet-3D															*
Barlat89	*				*				*						
BBC2005	*	*			*	*			*	*			*	*	
Vegter	*	*	*	*	*	*	*	*	*	*	*	*	*	*	

Uniaxial and equi-biaxial yield stresses and r-values were experimentally measured (figure 1 and Table 2). The r-values in RD, DD, and TD were obtained at 20% plastic strain, and the equi-biaxial r-value in the point with identical plastic work to the test in RD [11]. The plane strain and pure shear points, which are required only for Vegter, were obtained by virtual experiments using Virtual Experimentation Framework (VEF) software [2] (Table 3).

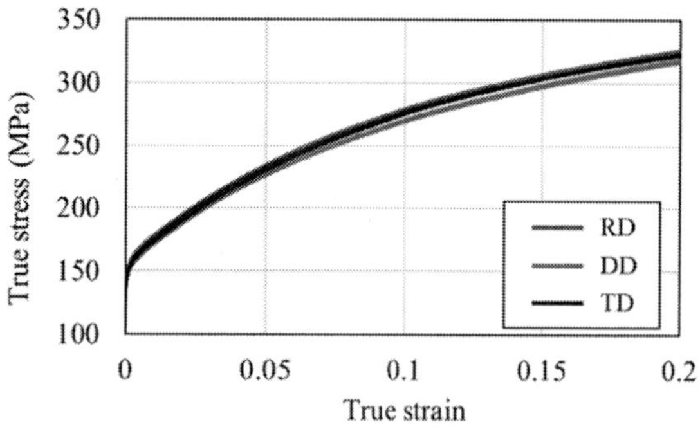

Figure 1. True stress – true strain curves obtained from uniaxial tensile tests in RD, DD, and TD (7 curves for each direction).

Table 2. Material parameters obtained by physical experiments [2].

Angle to RD (°)	r-value	Yield Stress (σ_x/σ_0)	Equi-biaxial point r-value	Equi-biaxial point stress (σ_b/σ_0)
0	0.552	1.000		
45	0.409	0.963	1.012	1.037
90	0.549	0.979		

Table 3. Material parameters obtained by virtual experiments (required for Vegter).

Angle to RD (°)	Plane strain point σ_x/σ_0	Plane strain point σ_y/σ_0	Pure shear point $\sigma_x/\sigma_0 = -\sigma_y/\sigma_0$
0	1.088	0.559	0.583
45	1.067	0.552	0.635
90	1.078	0.585	0.584

The texture of the material was measured by XRD method on plane samples taken at three different positions through the sheet thickness: sub-surface, 25% thickness, and 50% thickness. The ODF plots of the measured textures are shown in figure 2. This material exhibits a strong through-thickness texture gradient, where the cube component becomes weaker and more distributed from surface to the mid-section. The numerically merged texture (using weight factors 0.25, 0.5, and 0.25 for textures (a), (b), and (c), respectively) that was used for the virtual experiments with ALAMEL is shown in figure 2(d).

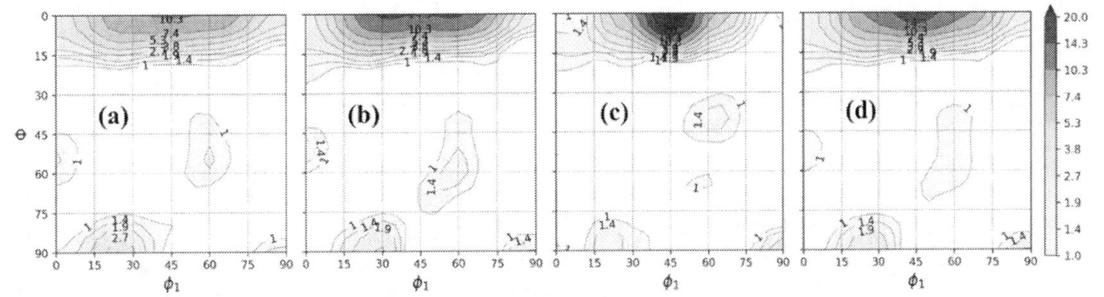

Figure 2. The ODF sections at $\phi_2 = 45°$ of the AA6016-T4 alloy, measured by XRD at (a) 50% thickness, (b) 25% thickness, (c) sub-surface, and (d) the numerically merged texture.

The measured r-values and yield stress ratios are compared with those calculated with the various yield functions in figure 3. All the three phenomenological yield functions exactly reproduce the measured r-values at 0° (RD), 45° (DD) and 90° (TD) as these are used as input parameters for their calibration. Facet-3D overestimates the planar anisotropy. Whereas its r-values in RD and TD are comparable to the experimental values, its r-value in DD is underestimated. This can be a consequence of texture gradient in the material, and a through thickness wide field EBSD measurement could be a better method than XRD for obtaining a representative texture. When the yield stress ratios are taken into account in the yield function formulation, the yield stress anisotropy is also reproduced (see BBC2005 and Vegter curves in figure 3(b)). Otherwise, the yield stress anisotropy may be different from the experiment. This is seen with both Facet-3D and Barlat89 in figure 3(b). The main reason for the inaccurate yield stress anisotropy with Facet-3D is that the ALAMEL model accounts for the crystallographic texture but neglects other microstructural elements that may be responsible for the observed yield stress anisotropy, such as particles and their distribution. The measured equi-biaxial r-value (1.037) and yield stress ratio (1.012) are exactly reproduced by BBC2005 and Vegter (as they include these parameters in their formulation); however, these values are different than the measured ones for Facet-3D ($r_b = 1.049$, $Y_b = 0.98$) and Barlat89 ($r_b = 1.060$, $Y_b = 0.97$).

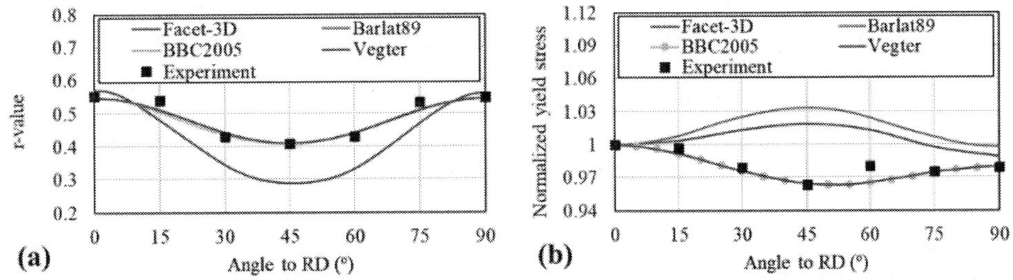

Figure 3. Measured material parameters [1] and those exploited from the yield functions: (a) r-values, and (b) yield stress ratios.

3. Results and discussion

3.1. Punch forces

The variation of punch force as a function of punch displacement for all the yield functions and friction conditions are demonstrated in figure 4. For each friction condition, the curves predicted by all the yield functions are similar to each other. With TriboForm, the peak is underestimated, and the decreasing slope after the peak is more pronounced than the measured value. These aspects both improve when Coulomb friction with a coefficient 0.05 is used. With increasing the friction coefficient to 0.1 and 0.15, the punch force peaks increase beyond the experimentally measured value, and the peaks predicted by Facet-3D and Vegter also become larger than those of Barlat89 and BBC2005. The curves here suggest that a friction coefficient around 0.1 can be a good choice for the simulations.

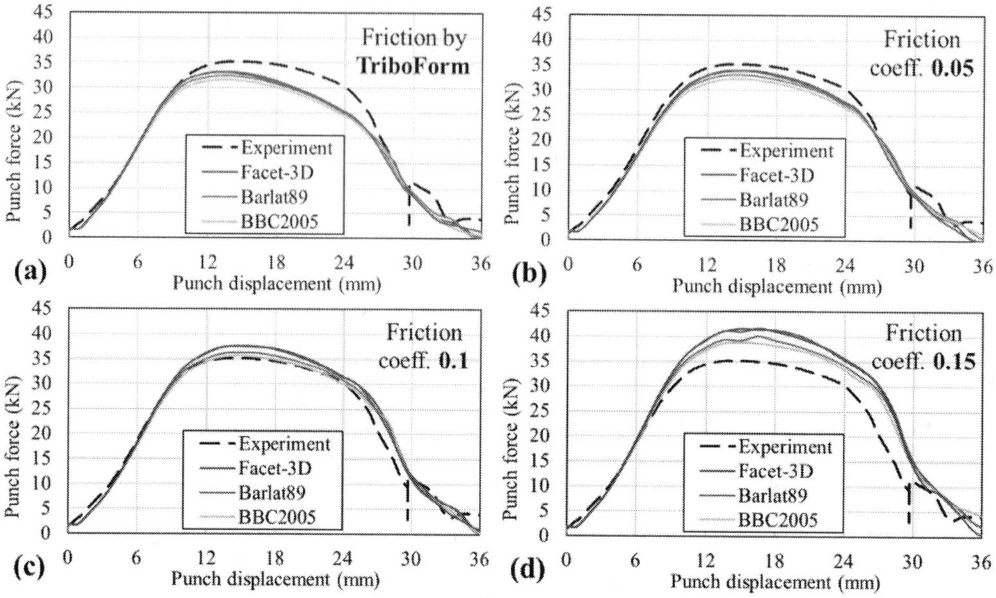

Figure 4. The measured punch force variation [1] and those predicted using: (a) TriboForm, and friction coefficients (b) 0.05, (c) 0.1, and (d) 0.15.

3.2. *Cup height profiles*

The measured and predicted cup heights were compared by calculating the relative height values using the following formula: *relative height = absolute height – average height.* Figure 5 illustrates the cup heights predicted by all the yield functions and friction conditions with the measured values. The relative heights show how the earing profile is predicted by each model-friction combination. With TriboForm, BBC2005 predicts the ear at 45°, but all the other yield functions predict acceptable earing profiles. Vegter shows the best match with the experiment, and Facet-3D and Barlat89 predict stronger earing behaviors. When Coulomb friction with a coefficient 0.05 is used, Facet-3D and Barlat89 are improved, but BBC2005 and Vegter results become worse. With increasing the friction coefficient, the gradual improvement of Facet-3D and Barlat89 continues, whereas BBC2005 and Vegter results continuously worsen and lead to erroneous positions of the ears. Vegter shows the most severe sensitivity to the friction coefficient.

As shown in figure 5(e), the average cup heights are lower than the measured value for all the studied friction conditions, but they improve with an increase in the friction coefficient. If the best friction coefficient is selected solely based on the absolute cup height values, then this figure suggests to use a higher friction coefficient even beyond 0.15. However, it was observed in the previous section that a coefficient 0.15 is too large concerning the punch force variations. If only the earing profile (i.e. the relative cup height) is considered as the reference, then a lower friction coefficient or TriboForm would be more logical as, in addition to Facet-3D and Barlat89, Vegter also predicts a better cup height profile. However, with TriboForm or a low friction coefficient like 0.05, the average cup heights become smaller and their differences with the measured value become larger.

Note that, the observed underestimation of the cup heights, as well as the inversion of the height profile in DD, is also reported in several other studies for the same material and cup drawing process [1, 2, 12] with different finite element solvers, more complex yield functions, and even higher friction coefficients. Therefore, it seems that considering the punch force variations as the reference for selecting the friction condition is safer, and this suggests that in this study a Coulomb friction with a coefficient 0.1 is the best choice among the other friction conditions.

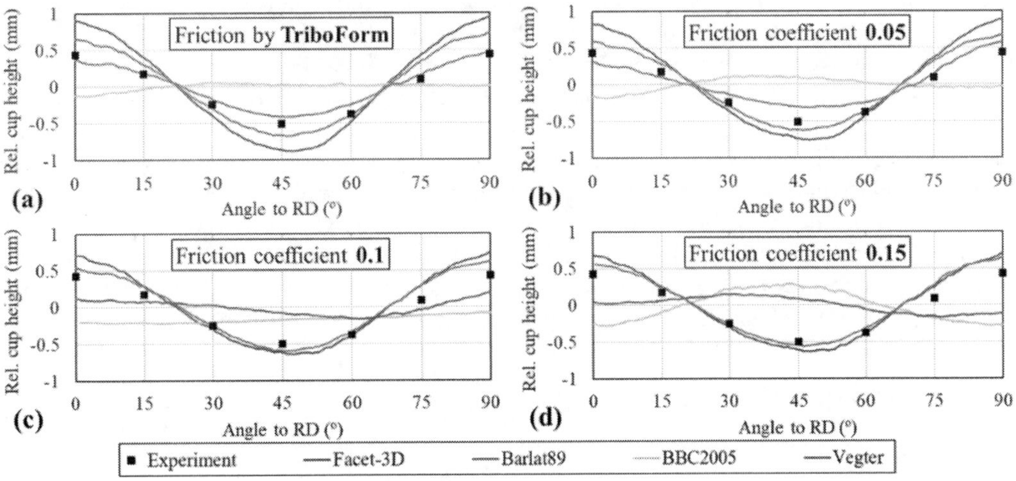

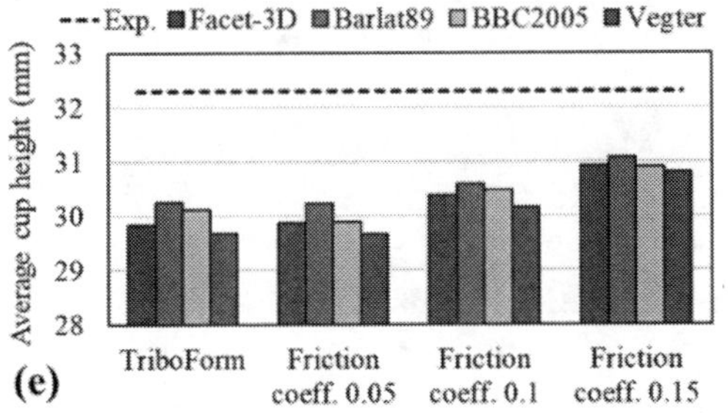

Figure 5. Measured cup heights [1] and those predicted with the yield functions for different friction conditions: (a) TriboForm, and friction coefficients (b) 0.05, (c) 0.1, and (d) 0.15, and (e) comparison of the average cup hights.

3.3. Cup thickness variations

The sheet thickness variation from center to the rim of the cup is another important factor which should be looked at. This was done in the present study in the rolling, diagonal, and transverse directions for all the yield functions and friction conditions. The sections on which the thickness variations were measured as well as the typical points on the observed curves are shown in figure 6.

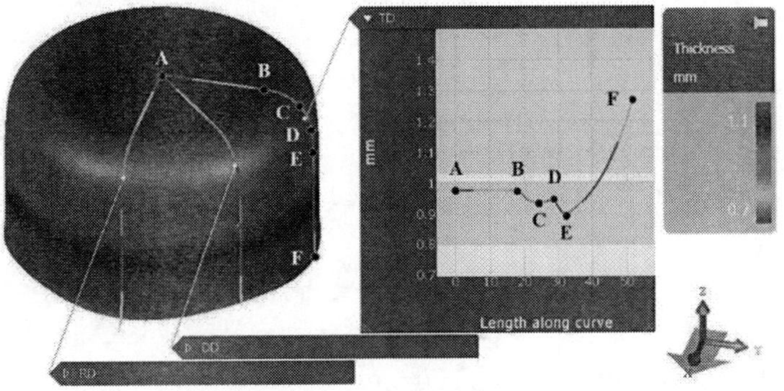

Figure 6. Thickness measurement sections in rolling (RD), diagonal (DD), and transverse (TD) directions.

The points A, B, C, D, E, and F were present in all the curves for all the yield functions and friction conditions. The curves for all the yield functions with a friction coefficient 0.1 are compared in figure 7. The studied yield functions predict roughly similar thickness variation from center of bottom to the rim of the cup. However, there are a few differences between the measured and predicted values.

For instance, the thinning on the cup bottom (AB section) is somewhat overestimated by all the yield functions. Since the loading condition on the cup bottom is close to equi-biaxial stress mode, the yield functions which take this point into account (i.e. BBC2005 and Vegter) are slightly better than the others. The predicted thickness on this section is similar for all the three directions, but the measured values are a bit larger in the diagonal direction.

In addition to the equi-biaxial stress point, the friction condition between punch and sheet can also affect the thinning on the cup bottom. Figure 8 demonstrates this only for Facet-3D, but the other yield functions also behave in the same manner. As can be seen, with a fixed friction coefficient 0.1 for the die and binder, a larger friction coefficient for the punch improves the predicted thickness on the cup bottom and corner. This variation does not influence the thickness change on the cup wall, the punch force variation, and the cup height profile. However, it affects the average cup heights as the thickness change takes place in a wide area on the cup bottom and corner, and less thinning there results in less extension and therefore a smaller contribution to the total cup height. This figure suggests that a larger friction coefficient can be assigned to the punch in the simulations to have a better prediction of thickness on the cup bottom and corner. This can be reasonable because the extremely low relative sliding speed between the punch and blank leads to more severe tribological condition in the punch-blank contact area than die-blank and binder-blank contacts [13].

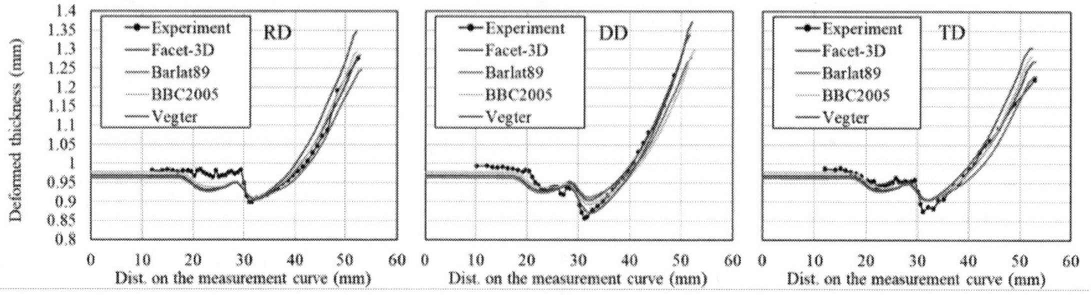

Figure 7. Comparison of the measured cup thickness variations in RD, DD, and TD [1] with those predicted by the studied yield functions.

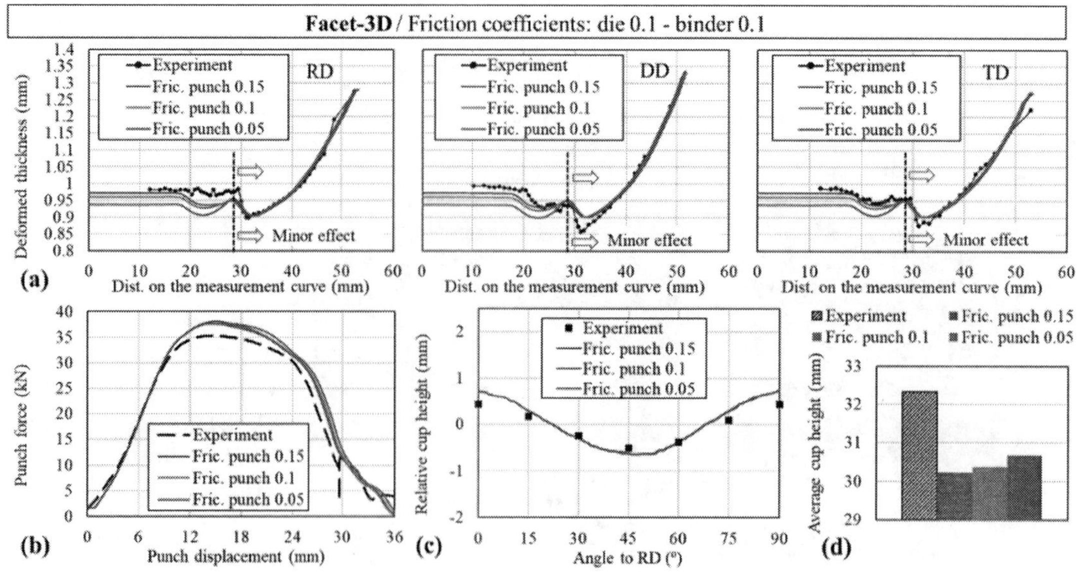

Figure 8. The effect of friction coefficient of punch on (a) cup thickness variations, (b) punch forces, (c) relative cup heights, and (d) average cup heights, with a fixed coefficient 0.1 for die and binder, and comparison to the measured values [1].

For the best friction condition, i.e., Coulomb friction with coefficients 0.1 for die-blank and binder-blank contacts, and 0.15 for the punch-blank contact, the thickness variation curves in RD, DD, and TD, punch forces, and cup height profiles are illustrated in figure 9. Except the cup profiles by BBC2005 and Vegter, the other quantities are predicted with acceptable accuracy by all the yield functions. The inverted cup profile prediction by BBC2005 and Vegter is unexpected because these models are designed to accurately reproduce the anisotropic properties of the material, especially the yield stress anisotropy (see figure 3). Note that the same phenomenon was also observed in an earlier study conducted with BBC2008 coupled to Abaqus/Explicit solver, for the same material and cup drawing process [2].

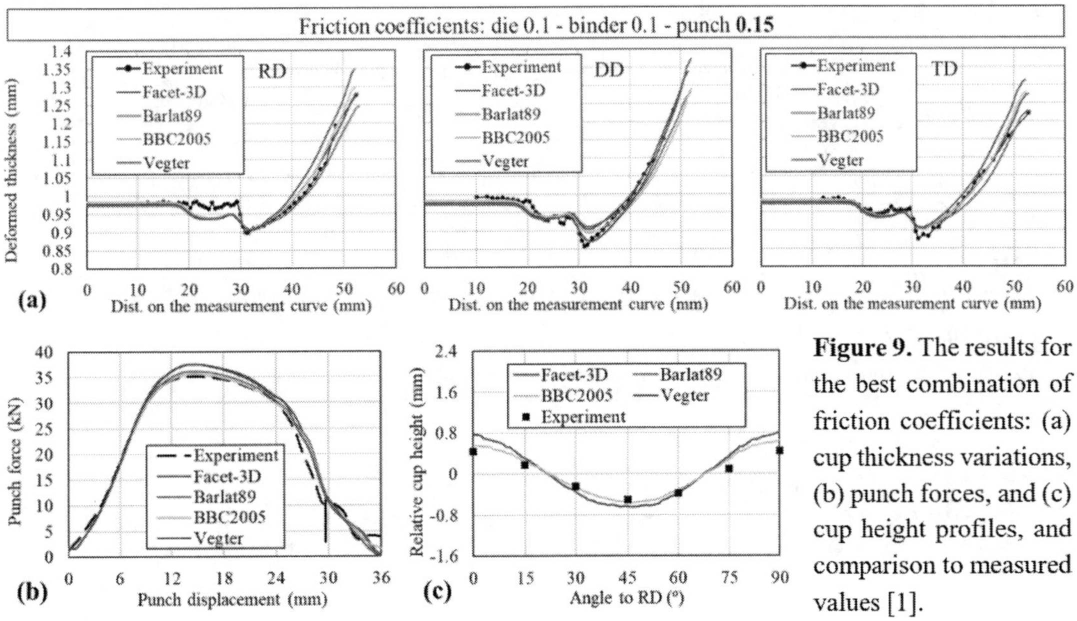

Figure 9. The results for the best combination of friction coefficients: (a) cup thickness variations, (b) punch forces, and (c) cup height profiles, and comparison to measured values [1].

3.4. *Sensitivity of cup height profiles to friction condition*

One of the most important simulation parameters including uncertainty is the friction coefficient in a sliding contact, and its value is often decided by comparing experimental and predicted forces, thickness variations, geometrical parameters, etc. The friction coefficient is not constant everywhere in the model, and using a fixed value is in fact a simplifying assumption. Therefore, the material models should be robust under small variation of tuning parameters like the friction coefficient. In other words, sensitivity of important results (e.g. the earing profile) to the friction coefficient in a simple forming process with a regular geometry like a circular cup drawing should be limited (unless such a sensitivity to friction conditions would be evidenced by experiments). It is more usual that average cup heights can significantly change by the friction coefficient, but not the position of the ears. As can be seen in figure 10, this robustness is observed with Facet-3D, Barlat89, and to some extent BBC2005 (although with inaccurate ear position). However, for Vegter, the position of ear changes from 0° and 90° (with low friction coefficient) to around 30°-45° (with higher friction coefficients) w.r.t the rolling direction.

3.5. *Sensitivity of cup height profiles to experimental data*

The experimental data can include errors, and this may influence the material models. However, the materials showing high sensitivity to the experimental input parameters should not be used in forming process simulations. In the present study, BBC2005 and Vegter take into account the yield stresses in RD, DD, and TD in their formulation, and at the same time they wrongly predict the position of ear. This stimulates investigating the sensitivity of these models to those additional input data. Figure 11 shows the variation of relative cup heights to only 2% error in the yield stress value in the diagonal direction (Y45). Both models show high sensitivity to Y45, and with Vegter, the position of the ear even changes for this small variation. This amount of uncertainty in the yield stress is not so rare, and may easily happen in 6XXX Al alloys (e.g. due to inhomogeneous distribution of precipitates and particles). Such sensitivity in these yield functions may be reduced by e.g. least-squares fitting procedure in conjunction with more material data points generated by physical or virtual experiments [2, 14, 15].

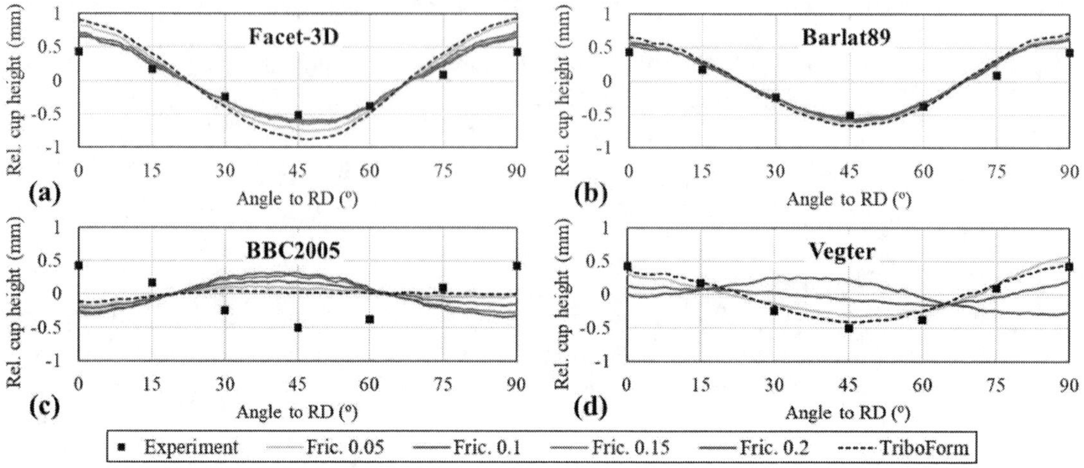

Figure 10. Sensitivity of the predicted cup height profiles by (a) Facet-3D, (b) Barlat89, (c) BBC2005, and (d) Vegter to the friction condition, and comparison to the measured values [1].

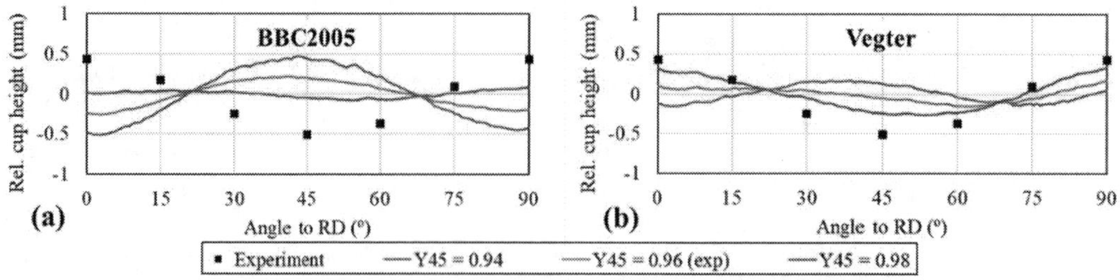

Figure 11. Sensitivity of the predicted cup height profiles by (a) BBC2005 and (b) Vegter to the yield stress ratio in the diagonal direction (Y45), and comparison to the measured values [1].

4. Conclusions

Circular cup drawing of an AA6016-T4 aluminium alloy was simulated with AutoForm R10 using three phenomenological yield functions (Barlat89, BBC2005, and Vegter) and a new crystal plasticity based yield function (Facet-3D). The performance of the yield functions was assessed by comparing the experimental and predicted punch forces, cup height profiles, and cup thickness variations under various friction conditions (TriboForm and Coulomb friction with three different coefficients). The results revealed that

- the punch force variations obtained by all the models are similar for a fixed friction condition, but their peak value varies with the friction;
- a friction coefficient 0.1 leads to the closet punch force profile to the experiment; TriboForm and lower friction coefficients underestimate the punch force peak, and a higher friction coefficient overestimates it;
- for all the studied yield functions and friction conditions, the average cup heights are underestimated; this result is in line with previous studies [1, 2, 12];

- for all the friction conditions, Facet-3D and Barlat89 predict accurate earing profiles with the correct ear position at 0° and 90°, but BBC2005 wrongly predicts the ear at 45°;
- Vegter with combination of TriboForm predicts the most accurate earing profile, but its ear position rapidly changes with an increase in the friction coefficient, and moves to around 30°;
- Vegter shows the highest sensitivity of earing profile to the friction condition;
- all the yield functions predict acceptably accurate thickness on the cup bottom, corner, and wall with close variations to the observed one;
- the punch force and cup thickness variations suggest that the best friction coefficients for the present study are 0.1 for die-blank and binder-blank contacts, and 0.15 for punch-blank contact;
- the earing profiles by BBC2005 and Vegter are very sensitive to small variations (2%) in the yield stress at 45°, where even the ear position for Vegter changes from 0° and 90° to around 30° with this variation.

The cup drawing results demonstrate the accuracy obtained with the proposed crystal plasticity based yield criterion. Thanks to its specific formulation that guarantees convexity, offers flexibility, and proves to be robust, the Facet yield locus is expected to also capture the plastic anisotropy under arbitrary loading modes. This is especially important for more complex forming geometries, and for materials with different textures and corresponding plastic anisotropy.

Acknowledgments
The authors would like to express their gratitude to Prof. Dorel Banabic for providing the data of the cup drawing experiments. A. Van Bael acknowledges financial support from the FWO (K801421N).

References

[1] Nikodim I 2013 *Creşterea eficienţei proceselor de ambutisare prin utilizarea unor sisteme de reglaj a forţei de reţinere* (Doctoral Dissertation, Technical University of Cluj-Napoca)
[2] Gawad J, Banabic D, Van Bael A, Comsa D S, Gologanu M, Eyckens P, Van Houtte P and Roose D 2015 *Int. J. Plast.* **75** pp 141–69
[3] Hill R 1948 *Proc. Roy. Soc.* A **193** pp 281–97
[4] Hill R 1990 *J. Mech. Phys. Solids* **38** pp 405–17
[5] Barlat F and Lian J 1989 *Int. J. Plast.* **5** pp 51–66
[6] Banabic D 2010 *Sheet Metal Forming, constitutive Modelling and Numerical Simulation* (Berlin Heidelberg: Springer-Verlag)
[7] Vegter H and van den Boogard A H 2006 *Int. J. Plast.* **22** pp 557–80
[8] Van Houtte P, Yerra S K and Van Bael A 2009 *Int. J. Plast.* **25** pp 332–60
[9] Staff E B and Kuijlaars A B J 1997 *Math. Intell.* **19** pp 5–11
[10] Van Houtte P, Li S, Seefeldt M and Delannay L 2005 *Int. J. Plast.* **21** pp 589–624
[11] Barlat F et al. 1997 *J. Mech. Phys. Solids* **45** pp 1727–63.
[12] Gawad J, Banabic D, Comsa D S, Gologanu M, Van Bael A, Eyckens P, Van Houtte P and Roose D 2013 *AIP Conf. Proc.* **1567** pp 350–55
[13] Sengupta A K, Fogg B and Ghosh S K 1981 *J. Mech. Working Technol.* **5** pp 181–210
[14] Hippke H, Hirsiger S, Berisha B and Hora P 2020 *Int. J. Mater. Form.* **13** pp 841–52
[15] Hippke H 2020 *Optimization methods for prediction of advanced constitutive models* (Doctoral dissertation, ETH Zurich)

Influence of the forming process on springback

U Durmaz[1*], S Heibel[1], T Schweiker[1], A. Prabhakar[2] and M Merklein[3]

[1] Mercedes-Benz AG, Umformsimulation und Beratung Stahl HPC X202, 71059 Sindelfingen, Germany
[2] Mercedes-Benz Research & Development India pvt. Ltd., Bengaluru, India
[3] Friedrich-Alexander-University Erlangen-Nürnberg, Institute of Manufacturing Technology LFT, Egerlandstr. 13, 91058 Erlangen, Germany

Abstract. The use of advanced high-strength steels (AHSS) is associated with increased springback and represents a great challenge for today's press shops with regard to the dimensional accuracy of cold-formed sheet metal parts. Springback, its appearance and characteristics are mainly dependent on the stress state introduced by the forming process and part geometry before opening the tools. With the aim of determining the influence of the forming process on springback, sheet metal forming simulation is used in this contribution to analyze a hat profile from a high-strength dual-phase steel (DH1000) by three different forming processes: crash forming, bending and drawing. For this purpose, the stresses generated by these processes and resulting springback are examined considering the respective forming history. The results show the strong influence of the material flow controlled by the process and tool geometry on the stress state before springback, which leads to significant differences in the springback of the hat profile among the forming processes: springback increases from crash forming to bending and drawing.

1. Introduction

The appearance and extent of springback are determined by the stress state, stress distribution and magnitude of elastic deformations in the direction of the sheet thickness. In this context, evenly distributed elastic stresses over the sheet thickness causes an uniform contraction or expansion in the respective cross-section upon unloading. In contrast, unloading of unevenly distributed elastic stresses over the sheet thickness leads to unfolding or curvature in the respective cross-section [1]. Compared to low alloyed steels, springback is significantly increased by the use of advanced high strength steels (AHSS) [2]. The reason for this are the mechanical properties such as higher work hardening with correspondingly higher yield stresses and consequently higher elastic strains [3]. The forming process determines the stress states and stress distribution in the part, which is why special attention must be paid to its selection with respect to springback. Frequently used demonstrators, such as u-bend or hat-shaped profiles, are formed by dominant bending stresses due to their geometry. Accordingly, springback dominates due to unevenly distributed stresses across the sheet thickness. In the press shops, hat-shaped profiles are produced by classical forming processes like drawing [4], bending [5] and crash forming [6]. For example, a hat profile produced by drawing shows a significant curvature of the part walls. This is a result of an intensive and alternate bending of the retained material as it is moved over the die radius into the part side wall [6]. In contrast, the intensity of the alternate bending as well as the curvature of the side wall can be reduced without restraining the sheet by a blank holder, as in crash forming [7]. Accordingly, the material flow associated with the process must be of great importance

IOP Conf. Series: Materials Science and Engineering 1238 (2022) 012074 doi:10.1088/1757-899X/1238/1/012074

with respect to springback. Hence, the influence of the forming process on springback of a hat profile is determined in this contribution. For this purpose, the material flow is analyzed within the forming history using sheet metal forming simulation for the forming processes drawing, bending and crash forming. Subsequently, the resulting stresses from the different processes are evaluated with regard to springback. Finally, a complex hat profile based on geometrical features of longitudinal members is produced by bending and crash forming to confirm the determined influence of the forming process on springback.

2. Scope of investigations

2.1. Parts and process tools

Figure 1 (a) shows the hat profile "geometry 1" used exclusively for the simulation in AutoForm R8.0, together with its basic dimensions. The embossing of the hat profile represents a connection surface for assembly in a serial part. In figure 1 (b), characteristic features of longitudinal members of a vehicle's body-in-white are combined in "geometry 2": inclinations in the bottom and flange areas as well as steps on the flange due to its structural position or stiffness function. "Geometry 2" is analyzed to ensure the prediction accuracy of the forming simulation with regard to the springback calculation within the respective process.

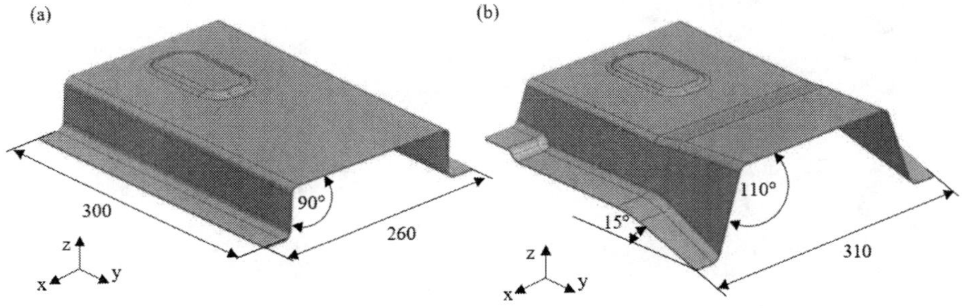

Figure 1. Hat profile "geometry 1" used for investigations in forming simulation (a), hat profile "geometry 2" used for experiments (b).

The forming processes and required tool parts are shown in figure 2. The tool set for crash forming consists of two tool parts, die and punch, between which the blank is crashed. The bending process is characterized by two operating steps, for which a four-piece and three-piece tool is used, as shown in figure 2 (b). In the first forming step, a blank holder forms the bottom of the part and simultaneously retains the sheet for the bending process. Then, the flanges are formed in a similar way to v-bending, for which the side walls initially have an angle greater than 90° to the bottom of the part, resulting in a larger radius between the bottom and the wall. In the second forming step, the walls are formed by bending while the bottom is retained by the blank holder. In the drawing process, the tool consists of four parts as shown in figure 2 (c). In contrast to the two previous processes, the target geometry is located on a stepped punch. For this reason the blank holder always acts on flat surfaces outside the target geometry. In this way, firstly, the intensity of alternate bending during forming the side walls is reduced, and secondly, flat blank holder surfaces also result for geometry 2 even with inclined flange areas. During the process, the sheet is drawn by holding the flange areas. At the same time, a pre-leading blank holder 2 reduces the strong curvature of the sheet in the bottom area of the part and is closed until the remaining drawing depth is formed. For all three processes the blank is located in the respective tools of the forming processes via two pilots.

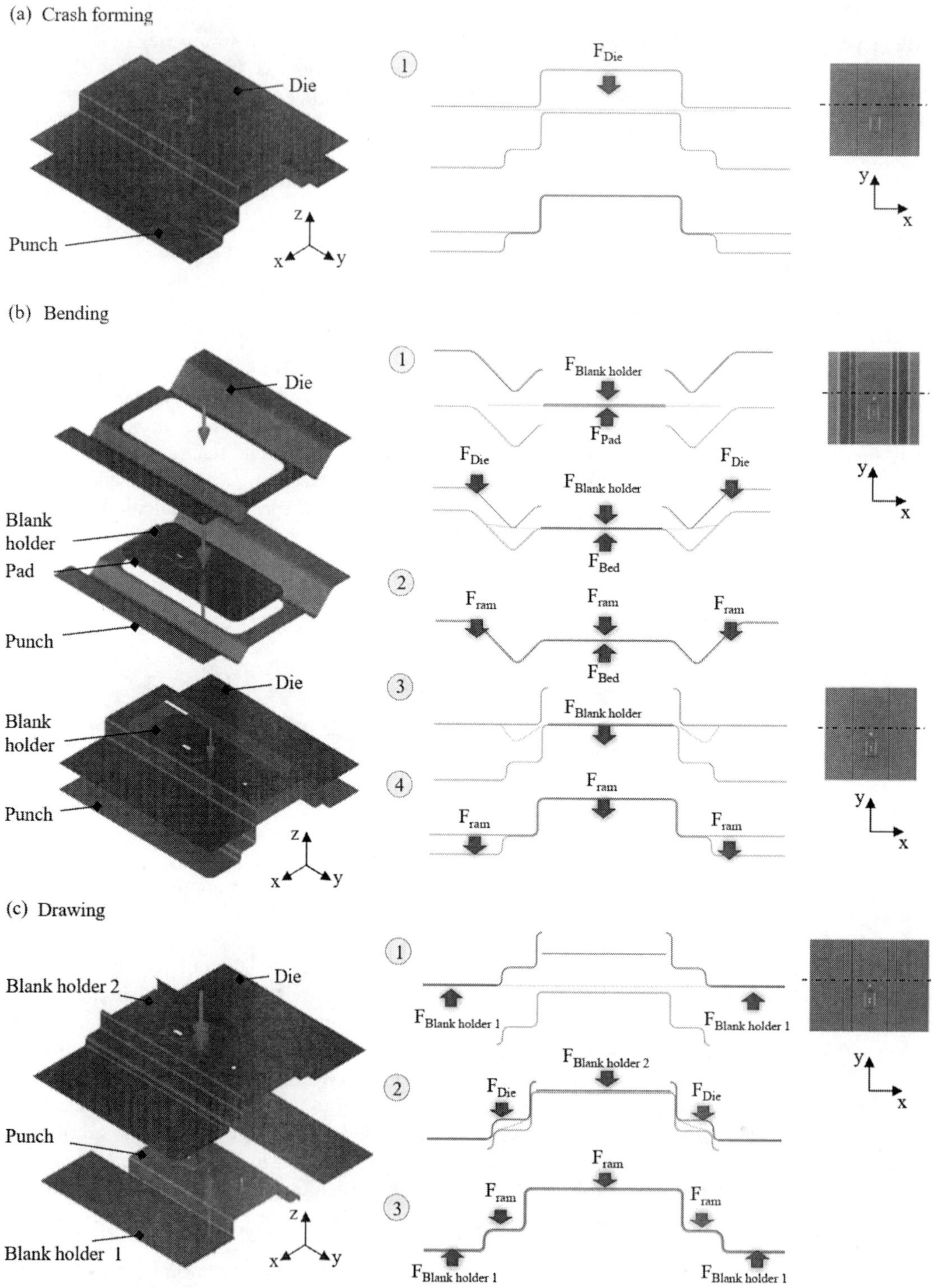

Figure 2. Tool set and process design for crash forming (a), bending (b) and drawing (c).

In comparison to previous studies, design and definition of the forming processes may differ. For this reason, Table 1 summarizes and classifies the investigated forming processes based on their tool set and process feature.

Table 1. Classification of the forming processes for based on the tool set and process features.

Process	Tool set	Process feature
Drawing	Die, Punch, Blank holder(s)	Controlled forming of sheet metal by blank holder(s)
(Wipe) Bending	Die, Punch, Counter punch	Controlled forming of sheet metal along straight axis
Crash forming	Die, Punch	Uncontrolled forming due to missing blank holder; sheet metal is "crashed" between tool parts

2.2. *Measurements*

In the simulation, springback calculation is conducted by two pilots and additional acting clamps as shown in figure 3 (a). The clamps prevent the part from tilting. In order to exclude any influence of the clamping force on springback, an internally defined force limit of the clamping force is taken into account.

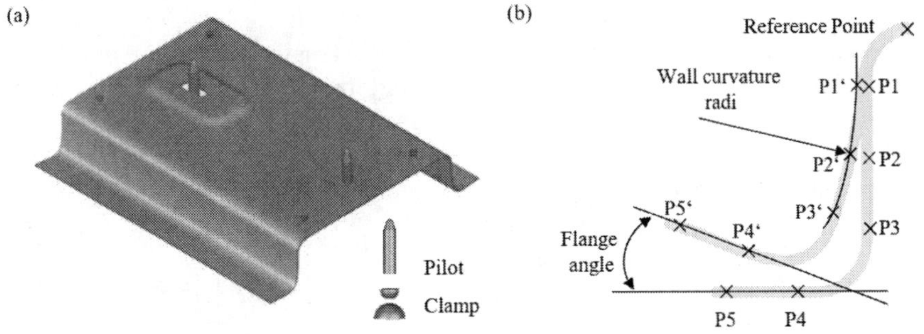

Figure 3. Measurement conditions for springback calculation (a), determination of the angular deviation and wall curvature schematically (b).

In addition, the meshes of the parts before and after springback are exported from the simulation into CAD for the determination of form deviations in a cross-section. To determine the flange angle deviation and curvature of the side wall, five points are set on the reference curve starting from a reference point. These are also transferred to the curve of the deviating component via their distances from the reference point. The points P1-P3 are used to determine the wall curvature. Since this can be different for each component, the points are adjusted for each component in such a way that the closest possible approximation of the curvature can be ensured. Points P4 and P5 as well as P4' and P5' form the lines for determining the angular deviation at the flange.

2.3. *Simulation conditions*

The material properties of the dual-phase steel used, named CR700Y980T-DH GI50/50-U (trivial name DH1000) according to VDA 239-100, are adjusted as well as supplemented in an existing material map with results from uniaxial tensile and tension-compression tests [8]. The isotropic kinematic hardening is modeled by the AutoForm model, which is based on the approach of Kubli et al [9]. In addition, the corresponding friction model from TriboForm has been imported into AutoForm R8.0, allowing the Coulomb friction coefficient to be calculated as a function of the local contact pressure, relative sliding velocity and strain of the sheet material during the process [10]. In particular, the consideration of isotropic kinematic hardening and the application of the advanced friction modeling by TriboForm have shown a significant enhancement of the springback prediction for a hat profile [11]. Therefore, sufficient accuracy of the simulation can be assumed in this study as well. The validation of the springback prediction is performed with geometry 2. More information about the simulation conditions can be found in Table 2.

Table 2. Simulation setup.

Material	DH1000	Element type	Elastic-plastic shell
Nominal sheet thickness	1.5 mm	Maximal element size	10 mm
Extrapolation hardening curve	Gosh	Refinement level	6
Yield criterion	Barlat	Forming velocity	35 mm/s

3. Results

3.1. Material flow and stress distribution

As already mentioned, the material flow during forming is determined by the tool set and process design. Therefore, the forming processes are compared with respect to the material flow and the currently acting stresses on a cross-section in several time steps. In the stress representation, the stresses in the rolling direction (r-axis) are selected due to their dominance during forming. In Figure 4 (a) it can be seen that the material initially bulges strongly during crash forming due to missing blank holders, which means that more material is stored in the bottom region. In this way, tensile stresses are introduced on the upper side and compressive stresses on the lower side in the bottom region. Subsequently, the excess material in the bottom is partially drawn into the side walls by the die, as shown in figure 4 (b).

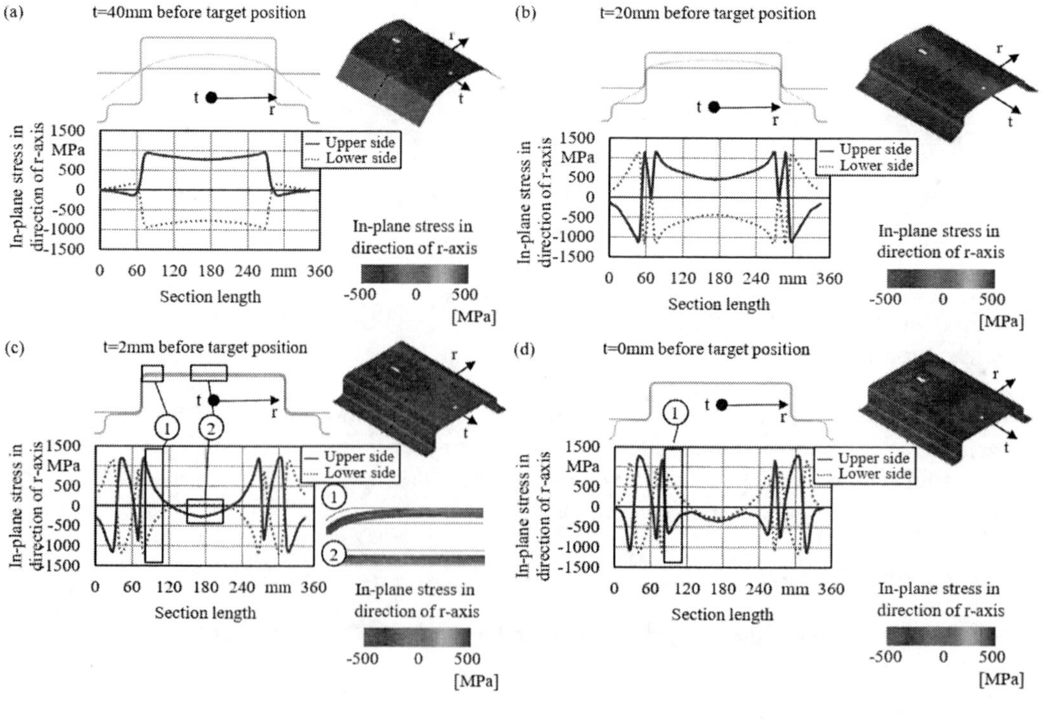

Figure 4. Images of crash forming process and stress state as well as stress curves on upper and lower side of blank for tool positions 40mm (a), 20mm (b), 2mm (c) and 0 mm (d) before target position of process tools.

At the same time, the sheet in the flange areas is moved freely over the radius into the sidewall. In the upper and lower areas of the wall, the sheet is bent alternately. This is illustrated in the stress profile by the alternating stress amplitude in the area of the wall and the radii, which is more pronounced in the further process step, see figure 4 (c). As the process progresses, the excess material in the bottom of the part is displaced towards the radii with the initial contact of the die. As shown in the detailed picture, tensile stresses continue to prevail near the radius on the upper sheet side and compressive stresses on the lower sheet side. At the end of the process, these stresses are reversed. In analogy to alternate bending at the part walls, the side of the sheet stressed by tension is superimposed with compressive stresses and the side previously stressed by compressive stresses with tensile stresses. The stresses after crash forming are concentrated in five areas of the part (high amplitudes in the stress curve), in which inverse stresses are introduced across the sheet thickness by bending and alternate bending.

In the first step of the bending process, the material is retained at the bottom of the hat profile and the embossing is freely formed. The subsequent bending is performed by the contact of the tool radii, which form the radius between the part's flange and wall. Bending stresses appear at the radii as shown in the stress profile in figure 5 (a). The compressive stresses in the center of the part are caused during the forming of the embossment. In the second forming step, the bottom of the part is reshaped by the blank holder and is held in position for the subsequent process. Afterwards the large radii between bottom and wall of the preformed geometry is formed into smaller radii by the bending die, causing the preformed component walls to collapse almost completely in the direction of the punch wall. In the stress profile on figure 5 (b), this becomes apparent through narrower amplitudes in the area of the radii.

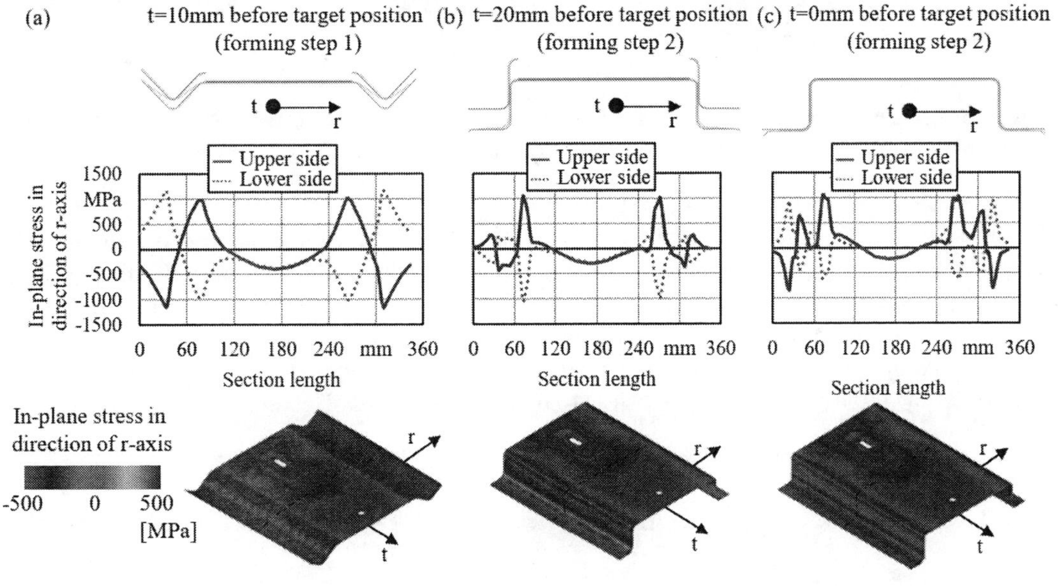

Figure 5. Images of the bending process and stress state as well as stress curves on upper and lower side of blank for tool positions 10mm (a) before target position of process tools in forming step 1, 20mm (b) and 0 mm (c) before target position of process tools in forming step 2.

When the dies are closed completely, the upper radii are completely formed and the flanges and lower component radii are calibrated as well. In this process the stresses are almost unchanged compared to the previous time step, except for the slight stress superposition near the lower radius due to the remaining shaping. After the two-step bending process, the stresses on the component are mainly concentrated in the radii.

The material flow in the drawing process is controlled by the acting blank holders 1 and 2. Firstly, the sheet is drawn over the punch while it is held outside by the blank holder 1. In the meantime, the curvature of the sheet at the bottom of the part is reduced by a pre-leading blank holder 2. Compared to the forming processes above, a subsequent trimming of the oversized blank after drawing is required to represent the target geometry. Due to the stepped punch, the material is alternately bent in the areas of the flange as can be seen in the stress changes in figure 6 (a) and (b). With the superimposed tensile stresses due to the acting blank holder 1, the following stress distribution results when the tools are closed: The bottom area is characterized by tensile stress distributed over the entire sheet thickness. Near the radii, the tensile stresses are increased successively until inverse stresses are formed on the upper and lower side of the sheet due to single or alternate bending.

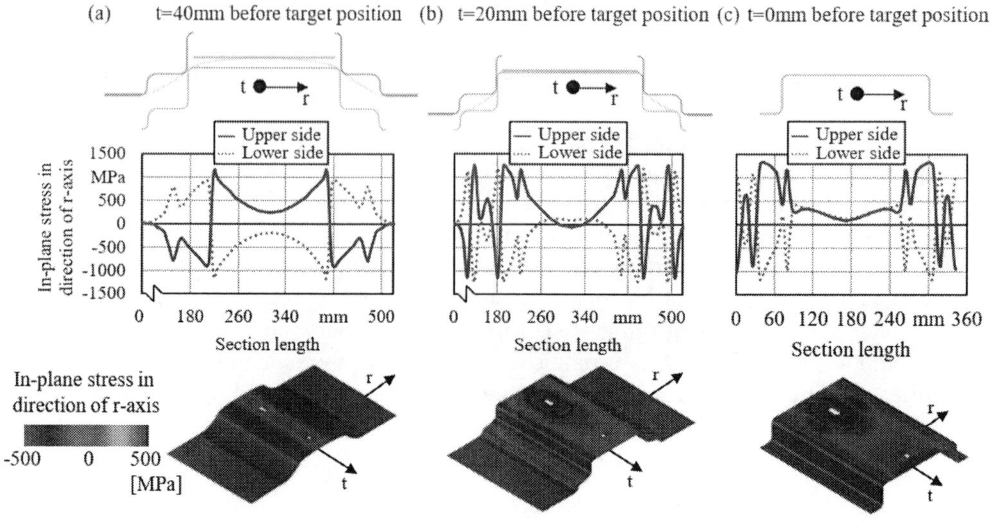

Figure 6. Images of the drawing process and stress state as well as stress curves on upper and lower side of blank for tool positions 40mm (a), 20mm (b), and 0 mm (c) before target position of process tools.

In the following, the forming processes are compared with each other. As already described in the introduction, unevenly distributed stresses over the sheet thickness leads to unfolding or curvature in the unloaded area. Rather, the stress difference in the sheet thickness direction generates a bending moment. Based on this, the bending moments acting on the component before springback are demonstrated in figure 7 (a) on almost one half of the part. In figure 7 (b) their effective direction and position for the different forming processes are listed schematically.

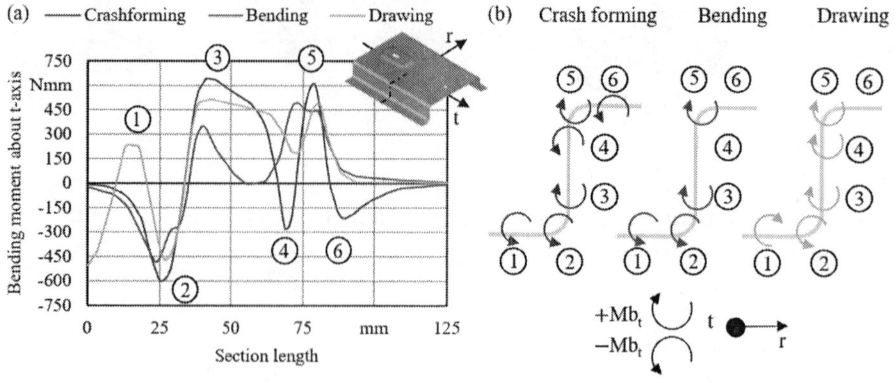

Figure 7. Bending moments along a cross section of the hat profile for the different forming processes (a), Schematic representation of the bending moments, their position and direction of rotation along a cross section before springback for the hat profile produced by crash forming, bending and drawing (b).

The comparison of the bending moments illustrates the differences of the forming processes in the numbered areas. The deviating amplitude of the drawn part in area 1 compared with the other processes can be explained by the alternate bending in the flange due to the stepped punch. Advantages of the stepped punch can be seen in area 3, where the drawn part has an even smaller bending moment than the crash forming as a result of less alternate bending. The hat profile produced by bending shows a significantly reduced bending moment due to the distribution of the bending moments and their unloading among two forming stages. In areas 4 and 6, the hat profile produced by crash forming shows negative bending moments which are generated by alternate bending during the material flow from the bottom area to the wall and by the forming of the excess material in the bottom.

3.2. Evaluation of the springback

In figure 8 (a), the springback of the parts are visualized on a limited cross-section. Starting from the reference, the crash formed hat profile shows the smallest deviations, followed by bent and drawn part. This can also be seen when comparing the flange angles in the bar chart demonstrated in figure 8 (b). In contrast, the curvature of the wall shows a different gradation: while the curvature of the bent part is negligible, the values of the crash formed and drawn hat profile are at the same level. This can be explained by the comparable bending moments in figure 7 (a) (area 3) and their interpretation in the previous chapter.

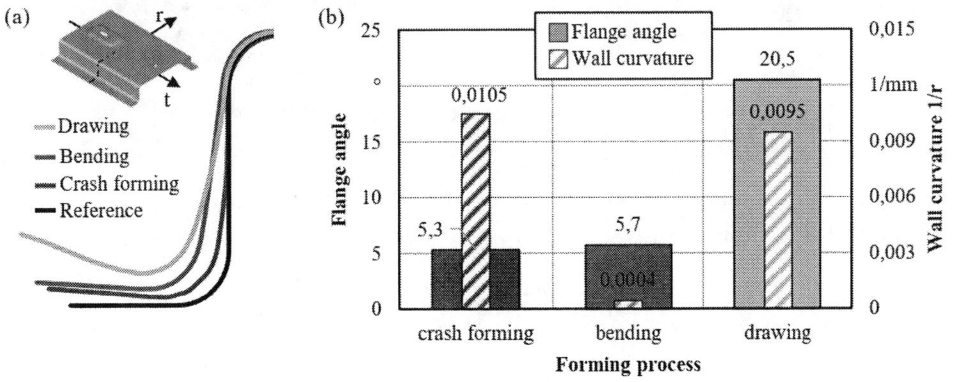

Figure 8. Comparison of the component springback for the different forming processes in a partial cross section (a), quantified angular deviation and wall curvature (b).

Beyond the extent of the bending moments, their position and direction of rotation obviously shows a strong influence on springback, which is why springback is the smallest for the crash formed hat profile. As a result of a hat shaped geometry, there is a greater influence of bending moments on springback of the flanges that are positioned closer to the center of the part. Thus, the negative bending moment of the hat profile in the upper region of the side wall (area 4 in figure 7) of the crash formed part leads to a reduction of the deviations in the wall and at the flange. A further reduction is brought about by the negative bending moment at the bottom near the radii (area 6 in figure 7). The origin of these bending moments is mainly based on the excess material in the bottom of the part, which is created when the blank is curved during eccentric tool contact. Without the excess material, there would be no material flow between the part's bottom and the part walls, which would cause the bending moment to be absent in the upper region of the side wall (area 4). On the other hand, the bending of the material at the part's bottom near the part radii is imposed by the excess material as well as its displacement to the radii. Overall, the hat profile can be graded according to its underlying forming processes with decreasing springback in the sequence drawing, bending and crash forming. Since there are several ways to design a drawing or bending process in particular, it is important to mention that the results discussed in this contribution refer to the described operations.

Based on the previous results from "geometry 1", it can be assumed that the forming processes bending and crash forming are more suitable for producing a hat profile using AHSS. For this reason, springback is only verified for these two forming processes with the help of "geometry 2". In principle, the validation of the springback prediction is legitimate, being a product of the calculated stresses and their distribution within the entire forming process. A comparison between the predicted springback from AutoForm and the optical measurements of the stamped parts by GOM Atos is visualized in figure 9. A sufficient accuracy of the forming simulation can already be determined by the visual comparison of the springback images. Furthermore, Figure 9 confirms the significantly lower springback of the hat profile produced by crash forming than by the bending process.

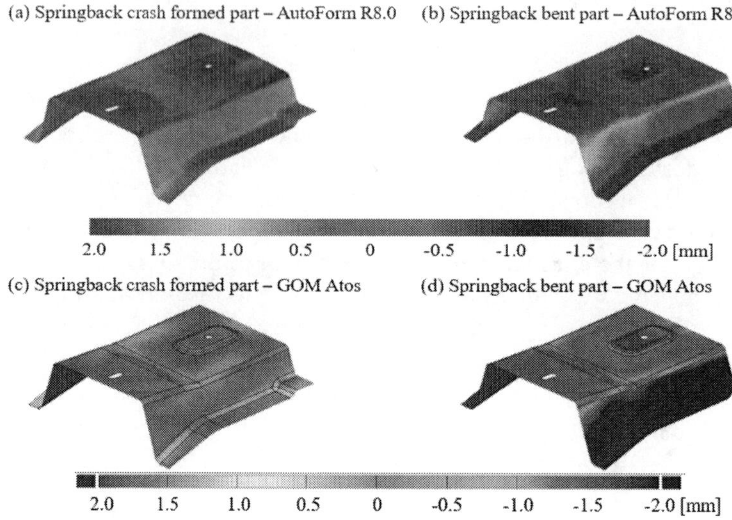

Figure 9. Optical measurement of form deviations GOM Atos for hat profile geometry 2 produced by bending (a) and crash forming (b).

Analogous to the previous investigations, figure 10 shows the bending moments before unloading along the cross section on almost one half of the hat profile. It can be seen a negative bending moment on the bottom of the crash formed part near the radii (area 6), which mainly causes the slight positive deviation near the radii and reduces the total springback.

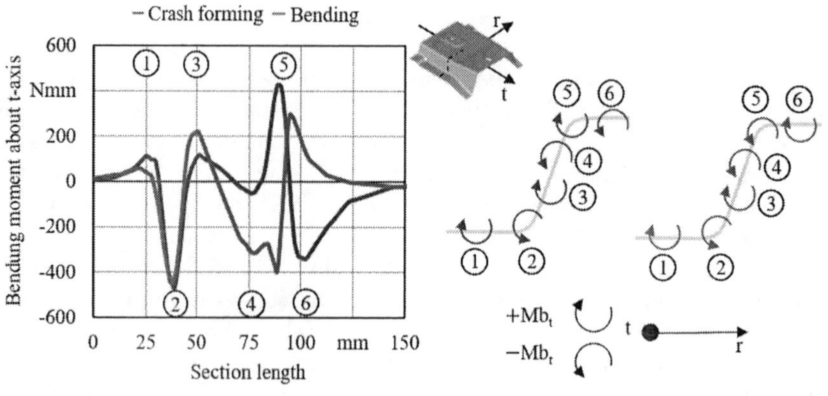

Figure 10. Bending moments before unloading along a cross section of "geometry 2" for the processes bending and crash forming as well as a schematic representation of the bending moments, their position and direction of rotation along a cross section before springback.

4. Summary

Sheet metal forming simulation is used to investigate the influence of the forming process crash forming, bending and drawing on the springback of a hat profile. A decreasing springback from drawing to bending and crash forming is found. The investigations have shown that the distribution and intensity of the bending stresses and subsequently unloaded bending moments are determined by the material flow during forming. Consequently, the material flow influences the springback significantly. In contrast to the springback of the bent part, the resulting springback from drawing and crash forming are controlled by bending stresses which are introduced during alternate bending of the material. Overall, the hat profile can be graded according to its underlying forming processes with decreasing springback in the sequence drawing, bending and crash forming. Since there are several ways to design a drawing or bending process in particular, it is important to mention that the results discussed in this contribution refer to the described operations.

5. References

[1] Ayres A 1984 *Shapeset: A process to reduce sidewall curl springback in high strength rails* Applied Metalworking **3** 127-134

[2] Mori K 2011 *Application of Servo Presses to Sheet Metal Forming* Key Engineering Materials **473** 27-36

[3] Billur E, Altan T 2013 *Three generations of advanced high-strength steels for automotive applications, Part I: The first generation*, STAMPING Journal November/December

[4] Gang L, Zhongqin L, Youxia B 2002 *Improving dimensional accuracy of a u-shaped part through an orthogonal design experiment,* Finite Elements in Analysis and Design **39** 107-118

[5] Komgrit L, Hamasaki H, Yoshida F 2016 *Elimination of springback of high-strength steel sheet by using additional bending with counter punch* Journal of Materials Processing Technology **229** 199-206

[6] Yoshida T, Hashimoto K, Katayama T, Kuriyama Y 2003 *Shape Control Techniques for High Strength Steel in Sheet Metal Forming*, Nippon steel technical report No.88

[7] Yoshida T, Isogai E, Sato K and Hashimoto K 2013 *Springback problems in forming of high-strength steel sheets and countermeasures*, Nippon steel technical report No.103

[8] Durmaz U, Heibel S, Schweiker T, Merklein M 2021 *Assessment of Springback Behaviour of 800-1200 MPa Dual-Phase Steel Grades* Key Engineering Materials **883** 151-158

[9] Kubli W, Krasovskyy A and Sester M 2008 Modeling of reverse loading effects including workhardening stagnation and early re-plastification, *Int J. Mater Form* Suppl **1** 145-148

[10] Sigvant M, Pilthammar J, Hol J, Wiebenga J H, Chezan T, Carler B and van dan Boogaard A H 2018 *Friction in Sheet Metal Forming Simulations: Modelling of New Sheet Metal Coatings and Lubricants, IOP Conf. Series: Materials Science and Engineering* **418**

[11] Durmaz U, Heibel S, Schweiker T, Merklein M, Berahmani S, Hol J and Naegele P 2021 *Enhancement of springback prediction of AHSS parts by advanced friction modelling* IOP Conf. Series: Materials Science and Engineering **1157**

Automatic generation of 3d spiral tool path for incremental sheet metal forming of mechanical parts with complex geometry

Sirine FRIKHA [1,2]*, Laurence GIRAUD-MOREAU [1], Anas BOUGUECHA [2] and Mohamed HADDAR [2]

[1] LASMIS Lab, University of Technology of Troyes, 12 Rue Marie Curie, 10004 Troyes, France; sirine.frikha@utt.fr; laurence.moreau@utt.fr ;

[2] LA2MP Lab, National Engineers School of Sfax, Route de Soukra, 3038 Sfax, Tunisie; anas.bouguecha@gmx.de; mohamed.haddar@enis.tn

E-mail: sirine.frikha@utt.fr

Abstract. Incremental sheet metal forming is a flexible manufacturing technology that allows to form of various components on the same milling machine, without expensive tools. A hemispherical tool moves along a CNC-controlled tool path and deforms the sheet into the desired shape. The tool path has a significant role in the geometric accuracy of the final part. There has been very little research on the problem of the forming of sophisticated parts with asymmetric wall angles. This paper presents a new approach to generating optimized tool paths for single-point incremental sheet forming (SPIF). At first, a strategy is proposed for the automatic generation of a 3D tool path during forming in a single-step operation. Then, a systematic technique of creating intermediate shapes is investigated by defining tool paths interpolated from the final part shape. The proposed methodology is applied to form a hip cup prosthesis. This part has a complex asymmetric geometry with important angles, a multi-step approach is used. The proposed methodology to define the different tool paths was implemented in Matlab. A numerical simulation of the incremental forming process is performed to predict the final geometry of the aluminum sheet. A comparison of desired and predicted geometries shows the reliability of the proposed method.

1. Introduction

Incremental forming at a single point SPIF is a die-less forming process that uses a moving tool head to locally deform sheet metal along a predetermined path, reaching greater forming limits [1, 2] than traditional sheet metal stamping. As a result, SPIF offers a cost-effective and efficient method of producing low-volume, bespoke functional sheet products without the need for a die. The goal of this project is to create automated 3D spiral toolpaths for SPIF.

According to published SPIF literature, toolpath has a considerable impact on forming limitations [3], geometric precision, surface quality, thickness variation, and shaping time. Although some attempts have been made to investigate various sorts of toolpaths and their impact on the aforementioned features, they are either part-specific or based on trial-and-error methods. Many academics have conducted experiments to see how different types of toolpaths, such as outline, radial, and multi-passes, affect the quality of the produced component. The majority of them utilized contour toolpaths created by commercial CAM systems designed for machining applications' surface milling modules. Forming in a contour toolpath is biaxial at the beginning and endpoints of each contour, and near to plane strain in the middle. The potential for fracture at the start and endpoints of each shape has been observed to be greater than when the rest of the shape is being created. A contour toolpath also causes stretch marks at the beginning of each contour. Wang et al. [4] recommended using a 3D spiral toolpath to eliminate equibiaxial stretching and tool marks at the end points of each contour.

However, because of the simultaneous needs of high forming angles [5, 6, 7, 8] and precision demands to retain component functioning as requested by the client, the fabrication of complicated 3D parts utilizing Single Point Incremental Forming (SPIF) is typically hard. Incremental forming [9, 10, 11] has been used on a variety of materials, including metals, which have low forming limits, which are normally specified by a maximum wall-angle at which no failure occurs. Furthermore, various attempts have been undertaken to increase part correctness. Verbert et al. [12], for example, devised a feature-based strategy that takes into consideration the behavior of specific features such as planes, rules, freeform, and ribs to produce components with high precision. The triangulated CAD model parts accessible in STL file format are utilized to detect features in this method. To build a compensated CAD model, the vertices of individual features are translated depending on actual deviations of a test component or models of projected deviations.

Verbert's feature-based technique is only shown for basic feature behavior, such as planes much below the failure wall angle. For parts on the verge of failure, this simple technique cannot be used to apply compensation strategies. Behera et al. [13] propose employing Multivariate Adaptive Regression Splines as a mathematical compensatory approach (MARS). This approach, on the other hand, is not well suited to pieces that are near the shaping limit. Similarly, pieces created in a multi-step process might help to extend the process window.

Despite these approaches, there is a need to create tool path procedures that can assist in the production of components with high shaping angles and high precision. There is also a requirement to design an automated and methodical multi-step part manufacturing technique. The multi-step technique has previously been demonstrated for basic ruled features like cones and cups, where the development of intermediate forms is visible as the final portion has a well-defined variation in wall angles. However, there is no systematic process for generating intermediate forms for sections with changing wall angles in the literature. Another challenge is that the manufacturing of complex forms with concave and convex features and forming limitations beyond failure has yet to be proved, since the traditional multi-step technique fails for such components.

As a result, an attempt was made in this study to establish a multi-step-based tool path generation approach for producing high-accuracy parts, particularly those with high forming angles. Case studies are offered to show how the suggested technique has been implemented successfully.

2. Materials and methods

2.1. Geometry and Materials

In this contribution, the test geometry is a hip cup prosthesis [14, 15] (cf. Figure 1), which has the following characteristics:

- Changing the wall angle with the depth of the part
- 120 mm in length
- Asymmetrical part
- Complex part

Aluminium (AA 1050-O), whose mechanical properties are listed in Table 1 and composition in Table 2, was employed for the appropriate investigations. The tests were carried out on

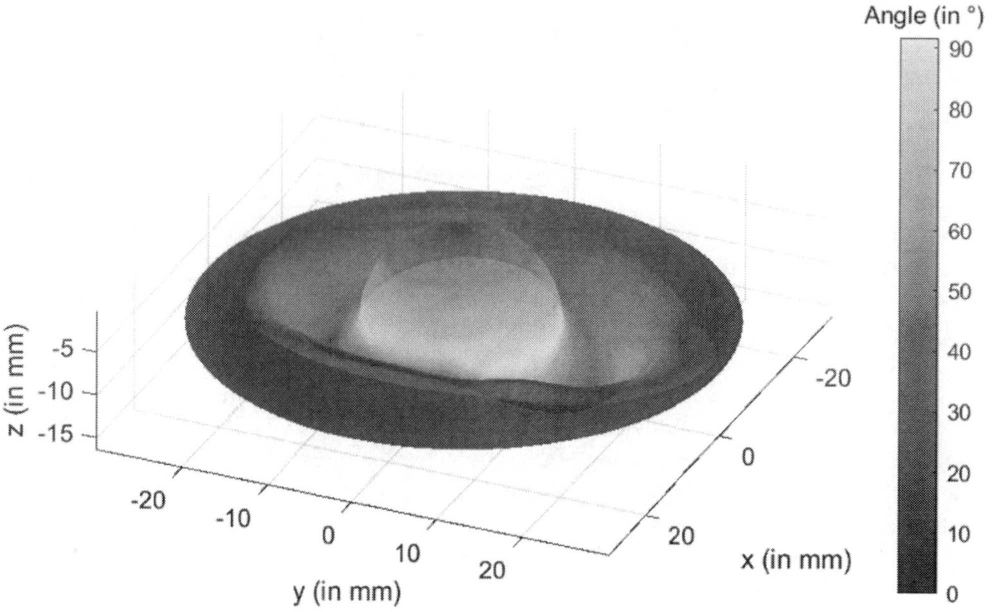

Figure 1. The analyzed part's angle variation

sheets that were 1 mm thick. Aluminium is a low-density metal ($2.6989 g/cm^3$ at 20 °C), easy to work with, and has excellent electrical and thermal conductivity as well as corrosion resistance. Aluminum crystallizes in a face-centered cubic (A1) lattice, making it a metal that may be formed cold or hot [16].

Table 1. Materials mechanical characteristics [17].

Material	E [MPa]	$Rp_{0.2}$ [MPa]	Rm [MPa]	A [%]
AA1050-O	70000	300	535	40

Table 2. Chemical composition of the material (wt.%) [18].

Si	Fe	Cu	Mn	Mg	Cr	Ni	Zn	Ti	Ga	V
0.089	0.28	0.002	0.001	0.001	0.001	0.003	0.005	0.011	0.016	0.007

2.2. Methods

The approach is taken to develop tool path techniques that can aid in the production of parts with high forming angles and high precision. The following are the main steps, which are

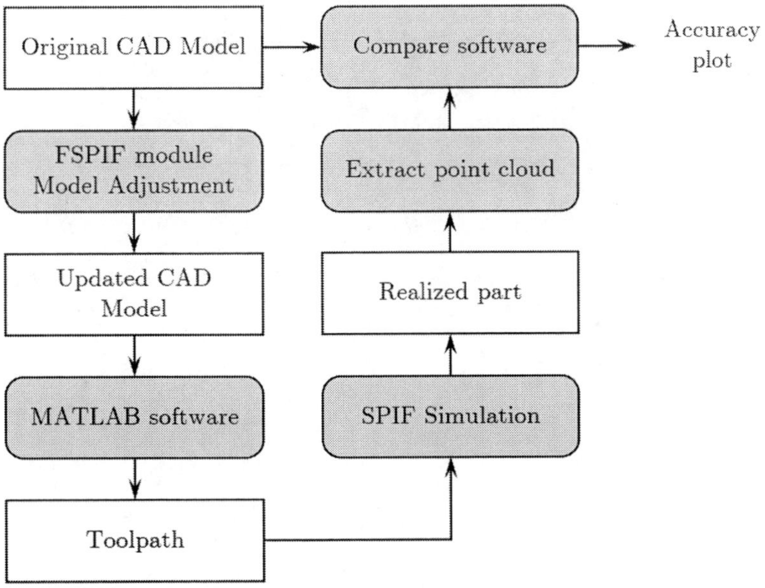

Figure 2. The developed methodology

summarized in Figure 2: A computer-based FPIF module is used to recognize features and adjust the CAD model of the part to be formed; the updated CAD model is used to generate a toolpath on the MATLAB software to simulate the forming model; and the realized part is extracted to generate a point cloud, which is compared with the original CAD to generate an accuracy plot.

Because it affects formability, surface finish, thickness variation, processing time, and dimensional accuracy, the **tool path** is critical in the design of SPIF processes. The digital control provides the tool route (CNC). The tool follows a succession of contour lines with a vertical increment step size of ($\Delta z = 0.2mm$) between them in a single forming step. The punch had a diameter of 2 mm. With a diameter of 120 mm and a thickness of 1 mm, the aluminium sheet was modeled in a round shape. The part is shaped by the tool's trajectory, as shown in Figure 3.

We start with a triangular depiction of the cup to specify the tool's trajectory in the incremental shaping phase. Indeed, the desired shape is modeled by a space mesh of triangles that define the cup's surface. We chose a spiral trajectory due to the curvature of the cup, and the trajectory is specified by multiple successive passages to lessen the stress imposed on the portion. The spiral is described by the following equation in the horizontal plane generated by the x and y axes:

$$x(t) = (R_i + (R_f - R_i)\frac{t - t_i}{t_f - t_i})cos(wt); \tag{1}$$

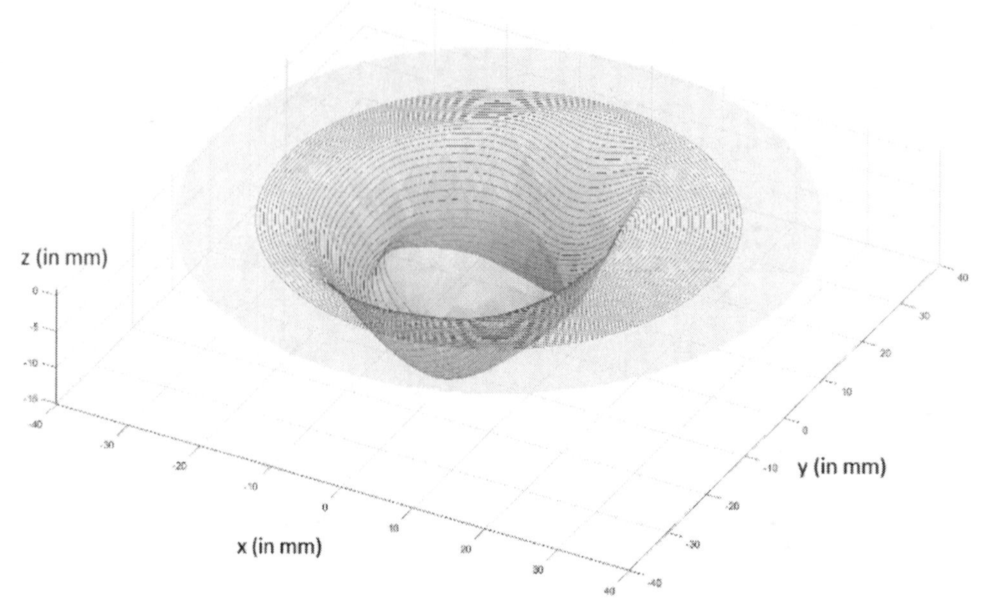

Figure 3. Tool trajectory one step

$$y(t) = (R_i + (R_f - R_i)\frac{t - t_i}{t_f - t_i})sin(wt). \tag{2}$$

with R_i and R_f the initial and final radii of the spiral, t_i and t_f the initial and final times of passage of the spiral and ω a parameter which regulates the spacing between the turns of the spiral, i.e. the angular speed of the rotation.

Then we identify the triangle that contains the point with coordinates $(x(t), y(t), z_0)$ and declare $z(t) = z_0$ for each pair $(x(t), y(t))$ on the spiral. As a result, the route $(x(t), y(t), z(t))$ is a tool path. Consecutive spirals are combined to perform many passages by alternating the values of Ri and Rf and assigning progressive values to z(t):

$$z(t) = \frac{i}{N}z_0 \tag{3}$$

with N the total number of passages and i the current passage's index $(1 \leq i \leq N)$.

While stepwise product shaping is used in many manufacturing processes, morphing is particularly useful in some manufacturing processes. The use of a source item and a target object is the fundamental notion of morphing. The input items in this study are triangulated models in STL format. The morphing problem must be handled in two phases: the correspondence problem and the interpolation step, because the STL format is a boundary representation technique. Furthermore, because various features behave differently in incremental formation, the mapping procedure must take feature behavior into account. Figure 4 depicts an illustration of this approach for morphing graphics.

2.3. Modeling

The standardized hip cup prosthesis was created using the incremental forming technique, as previously indicated. A hemispherical tool, a die with a negative model of the inner shape, and a blank holder are required for this technique. The numerical simulation was carried out

IOP Conf. Series: Materials Science and Engineering 1238 (2022) 012075 doi:10.1088/1757-899X/1238/1/012075

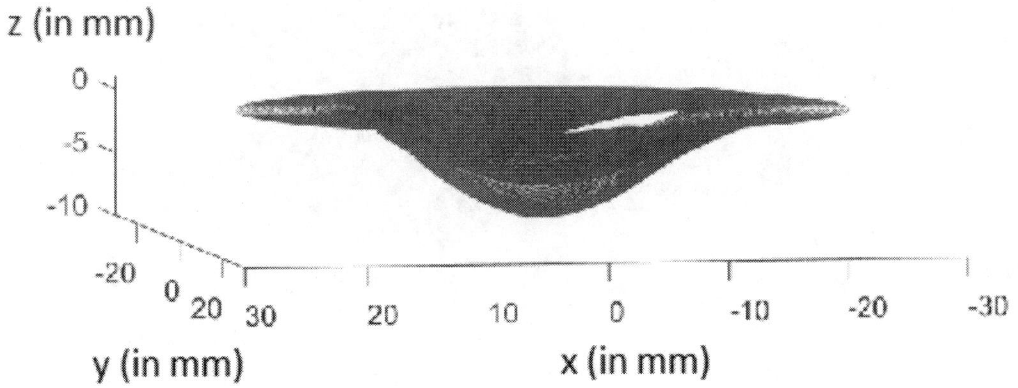

Figure 4. Multi-step Tool trajectory

using the program Abaqus 2018 (Dassault Systems, France), which was followed by a simulation of the blank springing back after the forming process. The yield curve of commercially pure aluminum, as determined by the above-mentioned tensile tests, and the anisotropic coefficients were utilized to define the material law of Hill'48, which was employed to do this numerical research. At room temperature, the material data was obtained. A substantial anisotropy of strain hardening was reported for uniaxial stress, resulting in a 30% variation in uniform tensile elongation between the extreme circumstances [18]. As a result, we'll account for anisotropy in our numerical simulation. Table 3 lists all of the discovered anisotropy coefficients and their values.

Table 3. Anisotropy coefficients for AA1050-O [18].

Orientation[°]	r
0	0.61
45	0.21
90	0.93

The blank was modeled using fully integrated rectangular shell components with 5 integration points throughout its thickness, with material yield calculations performed at each node. The rigid parts were employed for the die and the blank-holder. The friction coefficient between the blank and the tools was adjusted to 0.1.

3. Results and discussions

The standardized human hip cup's produced components were first visually inspected in Figure 5. Simulating the preforming process at room temperature produces acceptable outcomes, with a 20 percent reduction in blank thickness.

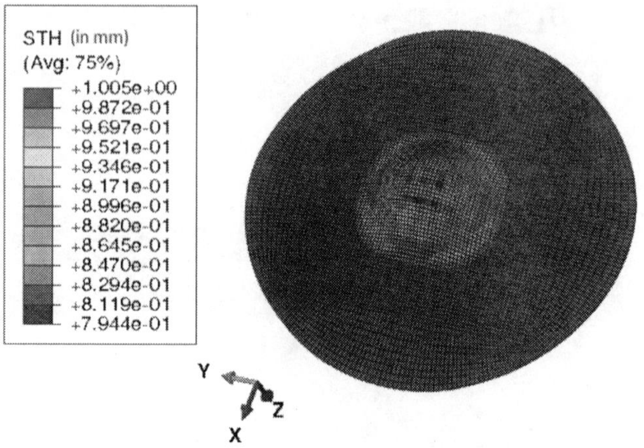

Figure 5. Thickness distribution of the first step

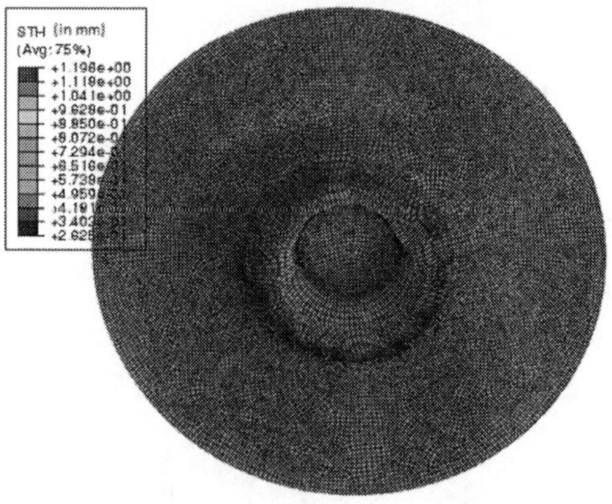

Figure 6. Thickness distribution of the incremental forming in one step

Figure 6 shows the thickness reduction of the aluminum blank as a consequence of modeling the one-step incremental forming process at room temperature. According to the modeling results, one-step incremental forming at room temperature would result in a 74 percent reduction in blank thickness, which is considered that it has exceeded tolerable limits and a rupture may occur. Because of its thinning by more than 25%, which is regarded as a limit to prevent the risk of cracking, this technique is difficult to complete without having a cracked blank at the end. Meanwhile, simulations reveal that multi-step incremental forming for aluminum blanks might lower the thickness of the blank by around 25%. Simulating multi-step incremental forming at the same temperature produces superior results in blank thickness. The aluminum blank and its thickness reduction as a consequence of modeling multi-step incremental forming at room temperature are depicted in Figure 7.

The modeling of both one-step and multi-step incremental forming procedures on an aluminum blank reveals that multi-step incremental forming of this blank at room temperature appears to be the most practical way for producing the prosthetic acetabulum.

As a result, a further step is required.

After that, a comparison is made between the simulation findings and the specified hip cup target. Catia V5 (Dassault Systems, France) was used to create the geometry of a 3D scan of

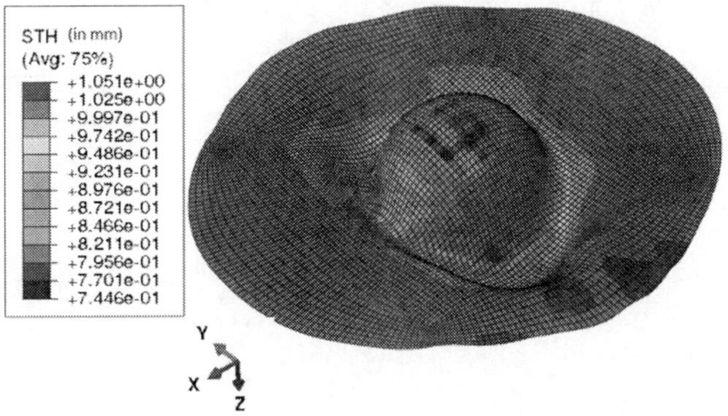

Figure 7. Thickness distribution of the incremental forming in multi-step

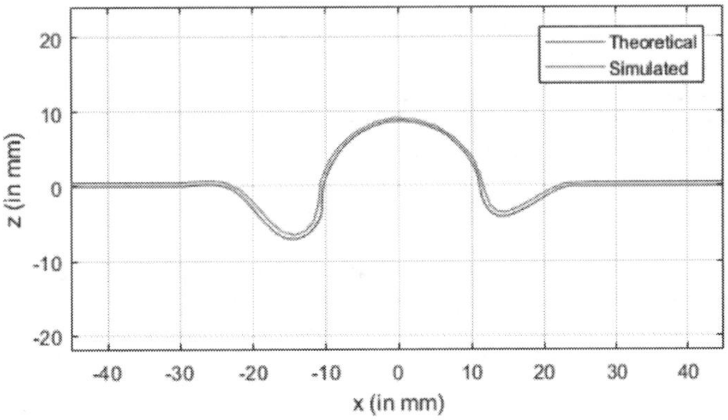

Figure 8. Comparison between theoretical and simulated profiles

the target component for this comparison. To find the best match, the digitized geometry was compared to the simulation results. Matlab was utilized in this experiment. Figure 8 depicts the results of the best fit and deviation analysis. The dimensional deviation study (Figure 8) revealed a minimum variation of 0.4633 mm in the edge region and a standard deviation of 0.2 mm for the entire dimensional deviation. As a result, there is a high level of agreement between the simulation and the theoretical component.

As a consequence of this finding, it is concluded that numerical simulation should reflect reality as it has been demonstrated.

4. Conclusion

A forming procedure has been designed based on the numerical analyses presented in this study. To begin with, a newly created standard acetabular/hip cup component is made using a preforming procedure. Individualization is achieved by an incremental shaping process using a sleeve and a backstop to adjust to the unique fit.

When the numerical findings of incremental forming the hip cup prosthesis are compared to theoretical and numerical research, it is discovered that the aluminum blank may be reduced in thickness by up to 26% without causing material damage. This criterion was used to establish the optimal forming parameters in the simulation of incremental forming the human standardized hip cup. This verification demonstrates that incremental forming may be used to manufacture the intended standardized human hip cup prosthesis.

Acknowledgments

The research is being carried out in collaboration with the University of Technology of Troyes and the National Engineers School of Sfax. The authors would like to express their gratitude to FEDER for their financial assistance.

References

[1] Ambrogio G, Palumbo G, Sgambitterra E, Guglielmi P, Piccininni A, De Napoli L, Villa T and Fragomeni G 2018 *Int. J. Adv. Manuf.* **98** 1489–1503
[2] Duflou J R, Habraken A M, Cao J, Malhotra R, Bambach M, Adams D, Vanhove H, Mohammadi A and Jeswiet J 2018 *Int. J. Mater. Form.* **11** 743–773
[3] Gupta P and Jeswiet J 2019 *Procedia Manuf.* **29** 112–119
[4] Wang J, Li L, Zhou P, Wang X and Sun S 2019 *Int. J. Adv. Manuf.* **101** 225–234
[5] Behrens B A, Nolte I, Wefstaedt P, Stukenborg-Colsman C and Bouguecha A 2009 *Biomed. Eng. Online* **8** 1–9
[6] Behrens B A, Betancur Escobar S, Almohallami A, Weigel N, Vucetic M, Stukenborg-Colsmann C, Lerch M, Nolte I, Lucas K, Wefstaedt P *et al.* 2014 *Adv Mat Res* vol 907 (Trans Tech Publ) pp 253–264
[7] Behrens B, Escobar S, Niemeier H, Vucetic M, Bouguecha A, Lucas K, Nolte I, Wefstaedt P, Lerch M, Stukenborg-Colsman C *et al.* 2016 *MatWerk* **47** 608–622
[8] Betancur Escobar S, Bouguecha A, Almohallami A, Niemeier H, Lucas K, Stukenborg-Colsman C, Nolte I, Wefstaedt P and Behrens B A 2015 *Biomedical Technology* (Springer) pp 55–68
[9] Saidi B, Giraud Moreau L, Mhemed S, Cherouat A, Adragna P A and Nasri R 2019 *Int. J. Adv. Manuf.* **101** 873–880
[10] Saidi B, Giraud Moreau L, Cherouat A and Nasri R 2020 *J BRAZ SOC MECH SCI* **42** 1–15
[11] Saidi B, Giraud Moreau L, Cherouat A and Nasri R 2021 *J. manuf.* **5** 122
[12] Verbert J, Duflou J R and Lauwers B 2007 *Key Engineering Materials* vol 344 (Trans Tech Publ) pp 527–534
[13] Behera A K, Verbert J, Lauwers B and Duflou J R 2013 *Comput. Aided Des.* **45** 575–590
[14] Bouguecha A, Elgaly I, Stukenborg-Colsman C, Lerch M, Nolte I, Wefstaedt P, Matthias T and Behrens B A 2010 *XII Mediterranean Conference on Medical and Biological Engineering and Computing 2010* (Springer) pp 562–565
[15] Behrens B A, Wirth C, Windhagen H, Nolte I, Meyer-Lindenberg A and Bouguecha A 2008 *Proc. Inst. Mech. Eng. H: J. Eng. Med.* **222** 593–600
[16] Amala Justus Selvam M, Velu R and Dheerankumar T 2017 *Innovative Design and Development Practices in Aerospace and Automotive Engineering* (Springer) pp 493–505
[17] Fratini L, Ambrogio G, Di Lorenzo R, Filice L and Micari F 2004 *CIRP Ann Manuf Technol* **53** 207–210
[18] Lopes A, Barlat F, Gracio J, Duarte J F and Rauch E 2003 *Int. J. Plast.* **19** 1–22

IOP Conf. Series: Materials Science and Engineering 1238 (2022) 012076 doi:10.1088/1757-899X/1238/1/012076

Springback inhibition of Ti-6Al-4V sheet with impact hydroforming at room temperature

H Li[1,2], Y Xu[1], S F Chen[1], H W Song[1] and S H Zhang[1,*]

1 Shi-changxu Innovation Center for Advanced Materials, Institute of Metal Research, Chinese Academy of Sciences, Shenyang 110016, People's Republic of China
2 School of Materials Science and Engineering, University of Science and Technology of China, Shenyang 110016, People's Republic of China

* shzhang@imr.ac.cn

Abstract. Springback is a tough issue in sheet forming, and always leads to dimensional inaccuracy of formed parts. Titanium alloy selected as desired light-weight alloy has been playing significant roles in aerospace industry because of its good comprehensive performance. Whereas, titanium alloy components manufactured always exhibit severe springback at room temperature, which greatly restrict their application. To explore the possibility of reducing springback, rectangular lath-shaped parts with Ti-6Al-4V alloy sheet were fabricated by impact hydroforming (IHF) with self-designed equipment. Consequently, much lower springback is obtained using IHF with high strain rate than that formed by conventional stretch-bending with low strain rate. Great efforts are paid to clarify the mechanism of springback restriction under IHF. Different from traditional forming methods, more complex and high-speed interaction behavior between liquid and Ti-6Al-4V alloy sheet occurs. It considerably increases the difficulty for analyzing the deformation process of sheet under IHF. Thus, a novel solid-liquid coupling numerical simulation technology for IHF was developed. To validate the simulation accuracy, experiments with different forming processes and parameters were performed. Given the combined analysis of experimental and simulation results, it is found that distinctive forming paths are introduced between stretch-bending and IHF. Specifically, the preferred deformation region of sheet transfers from middle region for conventional stretch-bending to end regions under IHF, which mainly attributes to the unique loading behavior of liquid at high strain rate.

1. Introduction

Springback in sheet metal forming has always been a widely concerned issue[1-3], and always cause inaccurate dimensions and assembly failure of formed light-weight alloy sheet components used in aviation industries (e.g. titanium alloys, magnesium alloys, aluminium alloys et al.). Among these light-weight alloys, titanium alloy sheet components always exhibit severe springback after forming in room temperature due to low elastic modulus and high strength of material[4]. Hence, it's necessary to develop reasonable springback reduction methods to raise the forming accuracy of titanium alloy components.

Generally, springback reduction strategies can be summarized as the optimization of loading path[5] and forming parameters[6] during forming, tool surface compensation[7] before forming, pressurization calibration[8-10] after forming et al. For instance, Feng et al. discovered non-uniform normal force and tangential adhesion force under viscous medium forming (VMF) could promote the uniformity of stress distribution on the whole sheet and reduce the bending moment of sheet, which would finally lead to the reduction of springback[5]. In addition, Odenberger et al. utilized hot forming to improve the geometric accuracy of Ti-6Al-4V double-curved aero engine components by controlling subsequent holding time and temperature[11]. Their work proved that hot forming was helpful to promote stress relaxation of deformed parts and reduce springback. Nevertheless, above forming methods will induce higher manufacturing cost and lower forming efficiency. In this regards, it is necessary to develop more reasonable forming methods to realize accurate forming of titanium alloy sheet components with high efficiency and low cost.

According to recent studies about high strain rate forming methods, high strain rate loading can promote springback reduction of light-weight alloy and high-strength steel sheet at room temperature, e.g. electromagnetic forming[12], electro-hydraulic forming[13], explosive forming[14], hydrodynamic forming[15]. Dou et al. foucused on the effect of forming speed on the springback after sheet metal bending processes. They found that springback amount of V-bending and stretch-bending would decrease as the forming speed increasing[16]. Golovashchenko et al. had a study on the springback behavior of DP600 and DP980 high-strength steel sheets. It was discovered that high strain rate loading with electromagnetic force could effectively eliminate the internal stress of deformed parts and reduce springback[17]. Although these research results can strongerly prove that high strain rate loading will promote springback inhibition of metal sheet, the potential mechanism for springback reduction under dynamic forming (e.g. IHF) on springback has not been revealed yet.

To discover the mechanism of springback reduction, the deformation history of sheet blank is always necessary to be analyzed. Since it is hard to precisely capture the instantaneous deformation behaviors of metal sheet and forming medium during high strain rate forming, it is naturally necessary to develop corresponding numerical simulation technologies. According to literatures reported in recent years, the dynamic deformation history of sheet under high strain rate could be acquired by solid-liquid coupling simulation technologies. O. Khodko quantitatively described the pressure wave during free hydrodynamic forming of tubular blanks using LS-DYNA[18]. Zohoor et al. coupled Arbitrary Lagrangian Eulerian (ALE) and Smoothed Particle Hydrodynamics (SPH) methods to simulate electrohydraulic forming process of sheet metal components[19]. Chen et al. conducted numerical simulations for impact hydroforming of AA 2B06-O sheet by fluid structure interaction (FSI) method with LS-DYNA[20]. Hajializadeh et al. simulated the impulsive hydroforming process of Al6061-T6 tube with the Coupled Eulerian Lagrangian (CEL) method by Abaqus[21]. Above simulation strategies for dynamic loading all provide positive instructions to the simulation strategy for IHF in this work.

In this research, springback behaviors of rectangular Ti-6Al-4V alloy sheets under impact hydroforming (IHF) at room temperature were investigated. Dynamic deformation histories of sheets were obtained by created IHF simulation with CEL method. Especially, the high-speed interaction behaviors between liquid and sheet blank under different strain rate were also analyzed in this work.

2. Experimental procedures

In this work, a rectangular thin-walled part is selected to test the springback behaviors of Ti-6Al-4V sheets under impact hydroforming (IHF). Experiments were conducted by self-designed IHF equipment and forming tools. The working principle of IHF can be described as that sheet blanks are deformed to the objective dimensions with provided tremendous kinetic energy of high-speed water at room temperature (Figure 1a). To contrast the deformation and springback behaviors of Ti-6Al-4V sheets under different forming velocities, experiments for IHF with different initial impact velocities (43m/s and 65m/s) and stretch-bending (0.1m/s) were performed relying on self-designed equipment and forming tools. Different from the liquid chamber with hollow structure for IHF, the dimension and geometric of the rigid punch for stretch-bending is manufactured to match that of the lower die (Figure 1b). To provide enough impact energy to make titanium alloy sheets to deform, the power source of projectile is derived from air compressor with high-power. The necessary dimensions of sheet blank and forming tools are listed in Table 1.

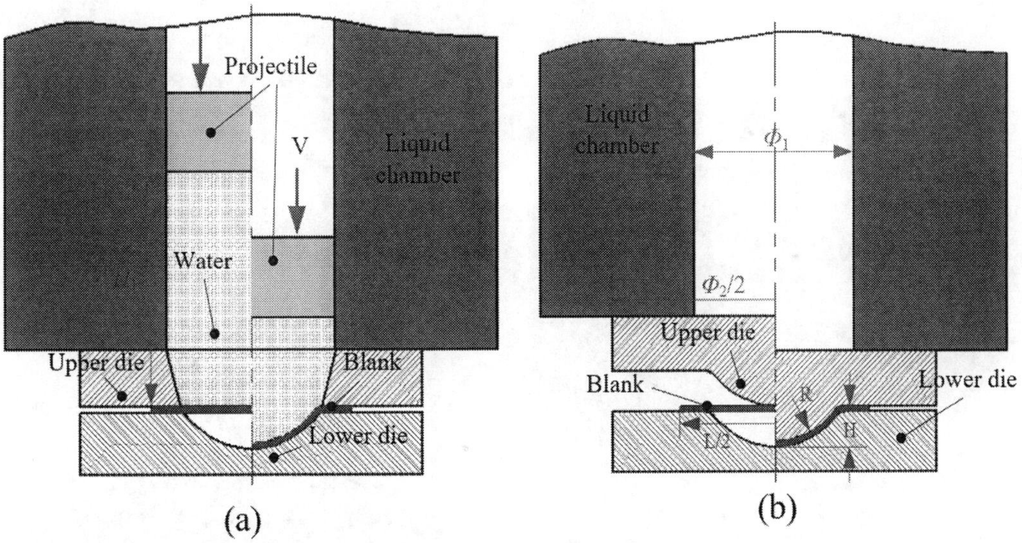

Figure 1. Forming principles and experimental configuration (a) Principle of IHF (b) Principle of stretch-bending

Table 1. Dimensions of sheet blank and critical forming tools.

Length of blank (L / mm)	Width of blank (B / mm)	Thickness of blank (t /mm)	Diameter of liquid chamber (D_1 / mm)	Diameter of the forming tools (D_2 / mm)	The maximum height of liquid (H_1 / mm)	The depth of the upper die (H_2 / mm)	Radius of the upper die (R / mm)
146	28	0.8	Φ110	Φ240	240	17	80

3. Simulation procedures for forming processes and springback behavior

Considering that it is almost impossible to precisely capture the deformation history of sheets during impact hydroforming, a solid-liquid coupling simulation method for IHF was developed by ABAQUS. To increase computation efficiency, the geometric of liquid chamber for IHF was simplified as rectangular column shape. In addition, rigid shell was selected as the element type of the forming tools (e.g. the rigid punch, the liquid chamber, the lower die and the projectile). To avoid over mesh distortion of high-speed water during IHF, the element type of water was defined as Eulerian element. Besides, the geometric of stretch-bending was modeled referring to factual dimensions, and shell element was utilized in all components in the created model. The established finite element models are shown in Figure 2. In order to accurately reflect dynamic deformation behaviors of Ti-6Al-4V alloy sheets under high strain rates, the simplified Johnson-Cook constitutive model (Equation 1) was utilized for defining material properties of Ti-6Al-4V sheet. Parameters of the constitutive model were referred to the computation value (Table 2) obtained by Hopkinson tensile tests in literature [22]. Furthermore, applied computation algorithm for forming processes relied on dynamic explicit algorithm (deformation process of sheets) and static implicit algorithm (prediction of springback), respectively. The initial velocities of the projectile and the moving speed of the rigid punch were all kept consistent with the experimental values. Other necessary simulation parameters were listed in Table 3.

$$\sigma = (A + B\varepsilon^n)(1 + C\ln\dot{\varepsilon}^*) \tag{1}$$

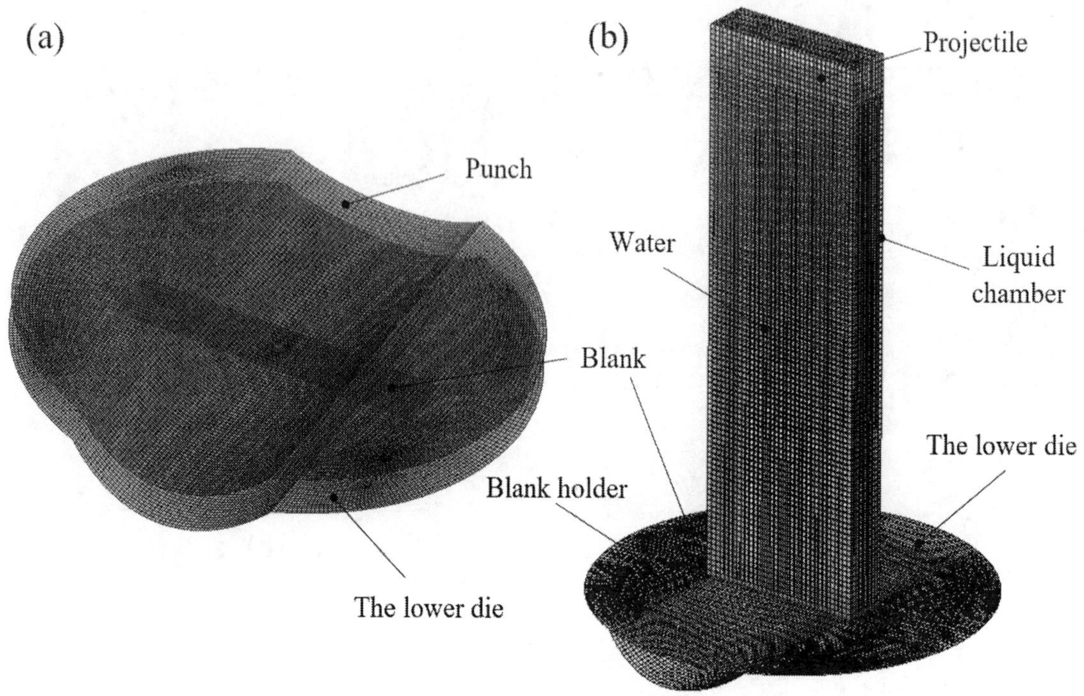

Figure 2. Geometric model for simulations (a) Stretch-bending (b) IHF

Table 2. The necessary simulation parameters in simulations.

Friction coefficience	Contact type	The maximum size of mesh (mm)	The height of water column (mm)	Blank holder force (N)	Height of projectile (mm)
0.08	Self-contact	3	290	0	150

Table 3. Parameters of the Johnson-Cook constitutive model of Ti-6Al-4V alloy[12].

A (MPa)	B (MPa)	n	C
1098	1092	0.93	0.014

4. Results and discussion

Three rectangular thin-walled Ti-6Al-4V alloy parts under different forming conditions were obtained by simulations, the results of which shew great consistence to experimental results (Figure 3). It could be learned that severe positive springback of deformed sheet occurred after stretch-bending with 0.1 m/s. Differently, it was found that severe negative springback of deformed sheet occurs when the impact velocity rises up to 65 m/s under IHF. Specifically, little springback amount was obtained under the lower impact velocity of 43 m/s. Thus, it was discovered that severe springback of Ti-6Al-4V alloy sheets could be effectively reduced under IHF.

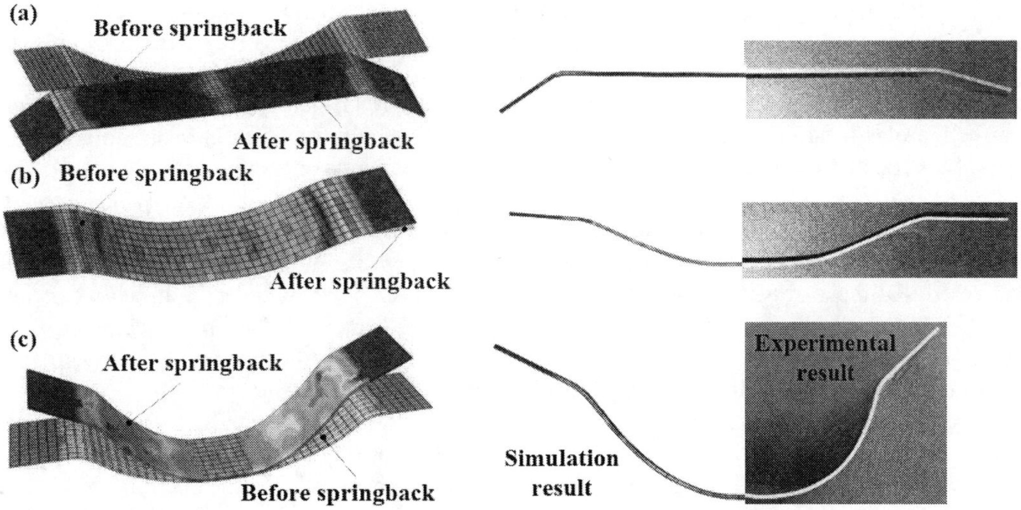

Figure 3. Simulation and experimental springback results (a) Stretch-bending with velocity of 0.1 m/s (b) IHF with velocity of 43 m/s (c) IHF with velocity of 65 m/s

To explore the mechanism of the springback inhibition under IHF at room temperature, the interaction behaviors between sheet and forming medium were discussed during stretch-bending and IHF forming processes, as well as the variation histories of strain and stress. Considering that the interaction behaviors were always related to the geometric characteristics of forming medium, the flow behaviors of the working liquid during impact hydroforming were analyzed. Several representative frames of the deformed sheets during stretch-bending and the flow liquid during IHF were displayed in Figure 4. By comparison, it could be discovered that the deformation of sheets would be initial at the middle region and then transferred to the end regions during stretch-bending, which could be explained by the first contact interaction originating from the middle region of the sheets. On the contrary, the deformation of sheets began at the end regions and ended in the middle region during IHF, which was attributed to that the high-speed liquid would firstly flow onto the edges of the sheet under impact loading.

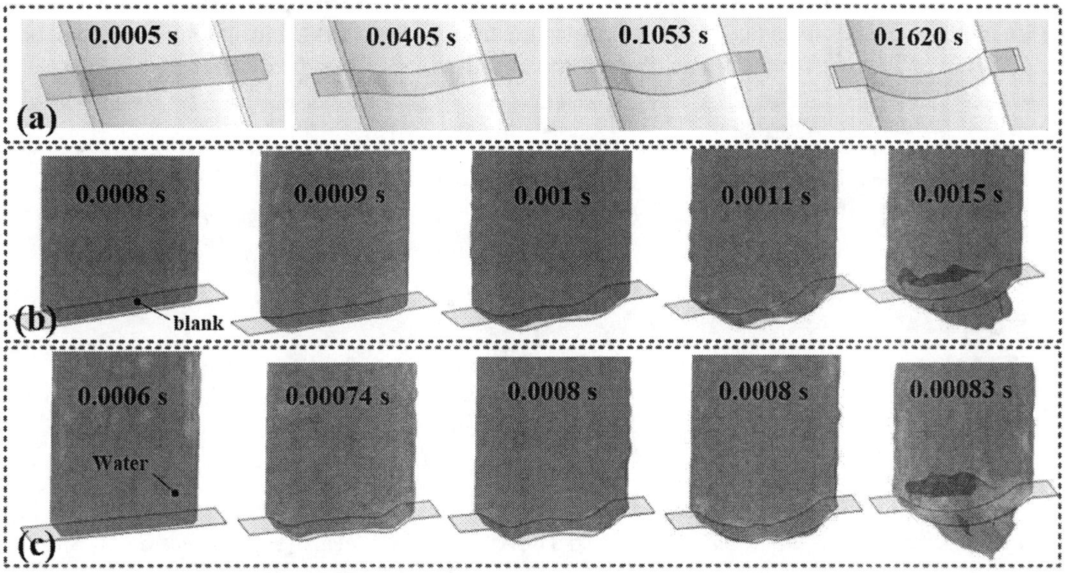

Figure 4. Deformation behaviors under different forming processes (a) Stretch-bending with velocity of 0.1 m/s (b) IHF with velocity of 43 m/s (c) IHF with velocity of 65 m/s

Change of sheet forming paths will lead to the variation of the stress and strain distributions. Obtained distributions of effective plastic strain under different impact velocities are shown in Figure 5(a). It can be seen that larger plastic strain occurs in the region of die fillet when applying over-high (65 m/s) or over-low level (0.1 m/s) forming velocities. In addition, the magnitude of plastic strain was obviously enhanced under IHF compared with that under stretch-bending (The maximum magnitude of effective plastic strain increased 0.1179 at the impact velocity of 65 m/s) in the middle region of sheets. Specifically, more uniform strain distribution was obtained at the impact velocity of 43 m/s. It illustrated that uniform distribution and increasing magnitude of plastic strain would promote the springback reduction of titanium alloy sheets. Additionally, distributions of von Mises stress along length direction of sheet obtained by simulations were shown in Figure 5(b). By computation, It was discovered that the average von Mises stress along length direction of sheet (x direction) obtained by IHF (237 MPa with the impact velocity of 43 m/s and 323 MPa with the impact velocity of 65 m/s) were much lower than that under stretch-bending (358MPa with the forming velocity of 0.1 m/s), which inferred that stress relief under IHF would lead to springback reduction.

Above all, the uniform distribution of plastic strain and the relief of residual stress under IHF will finally cause springback inhibition of Ti-6Al-4V alloy sheet at room temperature. Nevertheless, it was also discovered that the thickness reduction of obtained sheets along x direction were extremely restricted under IHF (Figure 6a). Similar variation law of thickness distributions of deformed sheets happens on the measured results (Figure 6b).

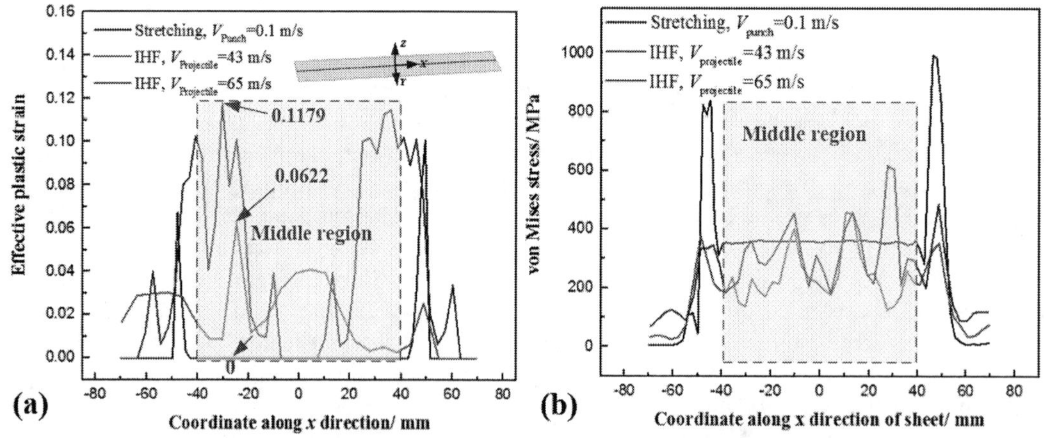

Figure 5. Stress and strain distribution along x direction of sheets (a) Distribution of effective plastic strain (b) Distribution of von Mises stress in x direction

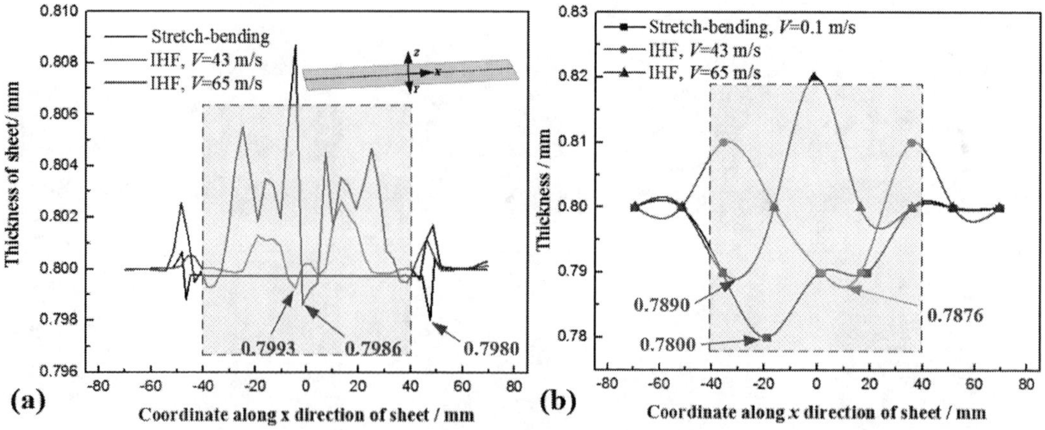

Figure 6. Thickness distributions of deformed Ti-6Al-4V alloy sheets along x direction (a) Simulation results (b) Experimental results

5. Conclusions

The springback behaviors of Ti-6Al-4V alloy sheets under IHF and stretch-bending were investigated by experiments and simulations. By comparing with the obtained results under stretch-bending and IHF, conclusions can be summarized as follows:

1. The preferred deformation region of sheet transfers from middle region for stretch-bending to end sides under IHF.
2. Stress relief and uniform distribution of plastic strain under IHF lead to the springback reduction of titanium alloy sheets at room temperature.
3. Thickness reduction of sheets will be restricted under IHF comparing with that under conventional stretch-bending.

References

[1] Badr O M, Rolfe B, Zhang P and Weiss M 2017 *Int. J. Mech. Sci.* ***28-129*** 389-400.
[2] Orallo A, Trinidad J, Galdos L, Argandoña E S and Mendiguren J 2020 Aluminum Spring-back Reduction by Post-forming Electric Pulses. *23rd Int. Conf. on Material Forming* ***47*** 1387-1391.
[3] Liu G, Lin Z Q and Xu W L 2002 *J. Mater. Process. Technol.* ***120*** 259-264.
[4] Zhao Y X, Peng L F and Lai X M 2018 Influence of the electric pulse on springback during stretch U-bending of Ti6Al4V titanium alloy sheets. *J. Mater. Process. Technol.* ***261*** 12-23.
[5] Feng Y K, Shi S G, Wang Z Y and Wang Z J 2022 *J. Mater. Process. Technol.* ***304*** 117548.
[6] Edwards W L, Grimm T J, Ragai I and Roth J T 2017 *45th SME North American Manufacturing Research Conference* ***10*** 329-338.
[7] Zhang Q F, Cai Z Y, Zhang Y and Li M Z 2013 *Mater. Design.* ***47*** 377-385.
[8] Cui X H, Du Z H, Xiao A, Yan Z Q, Qiu D Y, Yu H L and Chen B G 2021 *J. Mater. Process. Technol.* ***288*** 116889.
[9] Du Z H, Yan Z Q, Cui X H, Chen B G, Yu H L, Qiu D Y, Xia W Z and Deng Z S 2022 *J. Mater. Process. Technol.* ***299*** 117340.
[10] Xie H Y, Wang Q, Liu K, Peng F, Dong X H, Wang J F 2015 *J. Mater. Process. Technol.* ***219*** 321-327.
[11] Odenberger E L, Pederson R, Oldenburg M 2019 *Int. J. Adv. Manuf. Tech.* ***104*** 3439-3455.
[12] Lai Z P, Cao Q L, Zhang B, Han X T, Zhou Z Y, Xiong Q, Zhang X, Chen Q and Li L 2015 *J. Mater. Process. Technol.* ***222*** 13-20.
[13] Lu X and Hong J 2018 *17th Int. Conf. on Metal Forming, Metal Forming* ***15*** (Toyohashi: Japan Elsevier) p 907-914.
[14] Tong Z, Li Z, Cheng B and Zhang R 2008 *J. Mater. Process. Technol.* ***203*** 449-453.
[15] Ma Y, Chen S F, Chen D Y, Banabic D, Song H W, Xu Y, Zhang S H, Fan X S and Wang Q 2021 *Int. J. Mater. Form.* ***14*** 1221-1232.
[16] Dou L Y, Li X Q, Dong H R, Li D S and Peng X Y 2020 *Int. Deep-Drawing Research Group, Materials Science and Engineering* ***967*** (Seoul, South Korea IOP) p 012065.
[17] Golovashchenko S F, Gillard A J, Mamutov A V and Ibrahim R 2014 *J. Mater. Process. Technol.* ***214*** 2796-2810.
[18] Khodko O, Zaytsev V, Sukaylo V, Verezub N and Scicluna S 2015 *J. Manuf. Process.* ***20*** 304-313.
[19] Zohoor M and Mousavi S M 2018 *J. Manuf. Process.* ***20*** 16-28.
[20] Chen D Y, Zhang S H, Xu Y, Ma Y, Song H W and Xia L L 2020 *Int. Deep-Drawing Research Group, Materials Science and Engineering* ***967*** (Seoul, South Korea IOP) p 012028.
[21] Hajializadeh F and Mashhadi M M 2015 *J. Manuf. Process.* ***20*** 257-273.
[22] Lesuer D 2000 *Aluminium FAA Technical Report.* 1-30.

IOP Conf. Series: Materials Science and Engineering 1238 (2022) 012077 doi:10.1088/1757-899X/1238/1/012077

Understanding the influence of servo-press kinematics on a sheet metal forming process using a simulation-based approach

M Burlat[1*] and A Maillard[2]

[1] AutoForm Engineering France – 335 rue Serpentine – 13510 EGUILLES - FRANCE
[2] CETIM – Technical Centre for Mechanical Industries – 52 avenue Felix LOUAT – 60300 SENLIS - FRANCE
* matthieu.burlat@autoform.fr

Abstract. The capabilities offered by the servo-presses are not fully exploited by the sheet metal forming companies due to the lack of knowledge of their influence on the forming processes. Indeed, the control of the beneficial effects and the full use of servo-press features, such as the programmable kinematics, are increasingly becoming concerns for the companies having acquired that type of press. In this context, the finite element simulation is undoubtedly an approach to evaluate the potential gains generated by those presses and to validate the servo-press kinematics.
This study is an attempt to gain insight into the impact of servo-press kinematics on a sheet metal forming feasibility. The forming experiments presented in this study have been carried out by CETIM for a X2CrNi18-09 stainless steel. The numerical simulation has been performed with AutoForm software that enables the user to rapidly and accurately design and simulate the sheet metal forming processes, including stamping operations. All parameters influencing the stamping operation in a servo-press can be considered. They include, for example, press kinematics, sheet temperature variation during the forming process and of course material mechanical properties. The hardening curves of the studied stainless steel at different strain rates and different temperatures have been taken into account. The simulation has been carried out for two servo-press kinematics: sinusoidal and soft-touch. The feasibility criterion, i.e. the part drawing depth, has been defined to match the experimental results. The numerical simulation with AutoForm made it possible to highlight the thermal effects, in particular, the maximum temperatures reached during the drawing operation. The simulation results analysis allowed the user to optimize the motion curve of the press in order to have a minimum cycle time while respecting the lowest possible temperature the sheet might reach during a drawing operation. Thanks to this study, the sheet metal forming companies will therefore benefit from this technology to develop and optimize the servo-presses usage.

1. Introduction

Thanks to the use of servo-presses, with their programmable kinematics and – for specific configurations – a blank holder force control feature, it is now possible to consider significant progress in sheet metal forming in terms of formability [1], control and monitoring of the forming process [2] as well as tool improvement during production [3]. The increasing use of servo-presses amongst sheet metal forming companies can be explained by these potential gains which, given the higher cost of these machines, need to be accurately assessed. Depending on the type of part to be produced (sheet metal material, drawing depth, etc.), the improvements sought (prevention of fractures, reduction of springback, etc.) and in the absence of experiments, finding answers remains a difficult task. Therefore, it is important to be able to assess these gains using predictive tools, such as deep drawing simulation.

In this context, the study presented here aims to explain the gains provided by the kinematic sequences of servo presses for a stainless steel material, based on a numerical simulation of the drawing process.

2. Experimental data, reference point for the simulation

The experimental data used as reference for simulation relate to the drawing of an austenitic stainless steel bump shape using two press kinematic sequences. The experimental results of these tests are presented in [4]. The blank, part and drawing tool characteristics are provided in Table 1.

Table 1. Data of the reference experimental drawing process.

	Description	Characteristics
Blank / Part	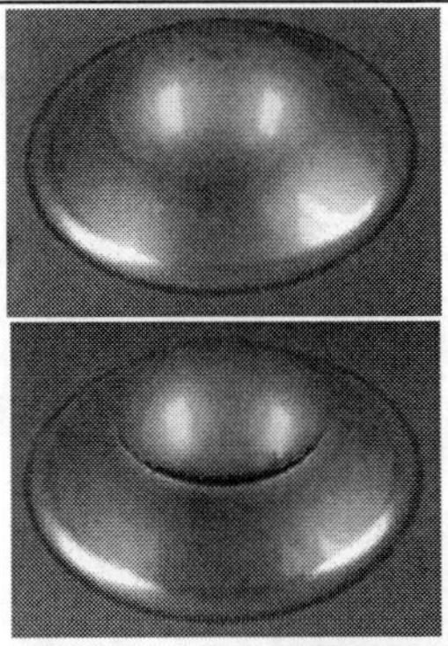	• Blank: Circular, dia.: 300 mm • Material: Stainless steel X2CrNi18-10 (type 304), thickness 1 mm • Drawing depth reached upon fracture: 23.3 mm and 25.4 mm depending on the kinematic sequences (cf. Table 2).
Tool	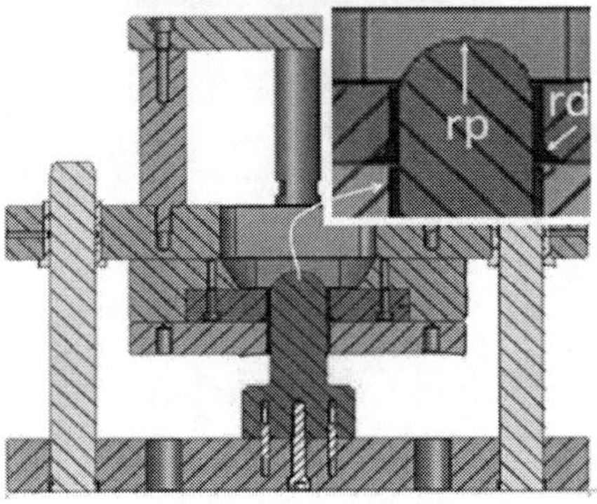	• Punch: dia. 50 mm, hemispherical end with radius rp = 25 mm • Die: inside dia.: 56 mm; die radius rd = 10 mm • Blank holder force varying from 3,600 daN to 4,000 daN • Tool / sheet lubrication with neat oil.

The two kinematic sequences taken into account for simulation are described in Table 2. The “Sinusoidal mode” kinematic sequence results in a maximum drawing depth before fracture of 23.29 mm, while the “Soft mode” kinematic sequence results in a maximum drawing depth of 25.04 mm, showing that the “Soft mode” kinematic sequence can help reach higher drawing depth [4].

Aside from the press kinematics, the sheet temperature variations during the drawing process and the material mechanical properties are parameters that can influence the forming operation in a servo-press. The X2CrNi18-09 stainless steel material used in the experimental tests was characterised by means of tensile tests carried out at different strain rates and temperatures to properly study these effects in simulation. Table 3 presents the main results obtained.

IOP Conf. Series: Materials Science and Engineering 1238 (2022) 012077 doi:10.1088/1757-899X/1238/1/012077

Table 2. Press kinematics used in the experimental tests.

Type	Diagram	Description
Sinusoidal mode	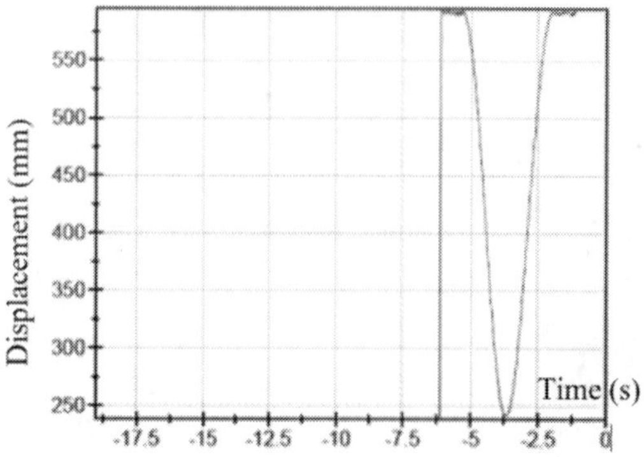	This sequence represents the movement of a conventional mechanical press. This is our reference for comparing the different simulation results. The velocity of the ram at the beginning of the drawing process (first contact with the sheet metal) is 140 mm/s.
Soft mode	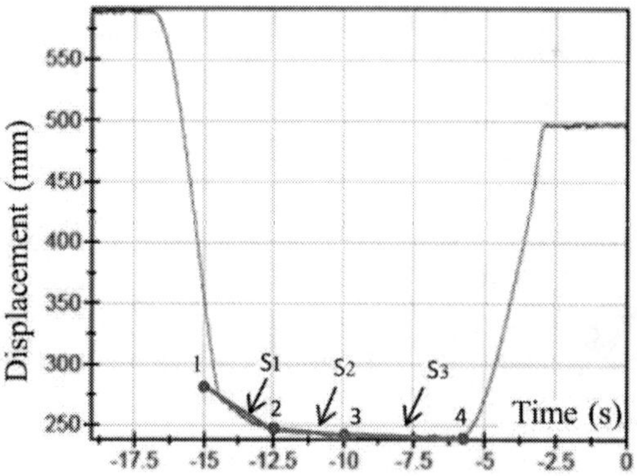	The kinematic sequence of the press ram follows 3 steps with different velocities: - velocity S1 = 14.4 mm/s, until 8 mm from the bottom dead center (BDC), - velocity S2 = 2.4 mm/s, until 2 mm from the BDC, - velocity S3 = 0.6 mm/s, until BDC.

During our tests, the X2CrNi18-09 stainless steel appeared to be highly susceptible to the temperature and almost not to the strain rate: the relationship between temperature and the work hardening of the material will be taken into account in the following forming simulations of the part.

Table 3. Results of the tensile tests on the X2CrNi18-09 stainless steel used for this study.

Title	Tensile diagram	Analysis
Strain rate influence (at room temperature) Strain rates: 0.1 s-1 0.01 s-1 0.001 s-1	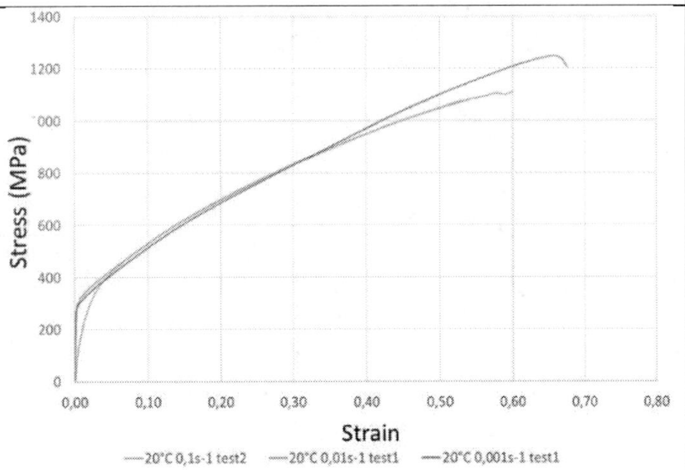	No major influence of the strain rate, within the characterized variation range, at room temperature or at a higher temperature.
Temperature influence (strain rate of 0.01s-1) Temperatures: 20°C 50°C 100°C 150°C	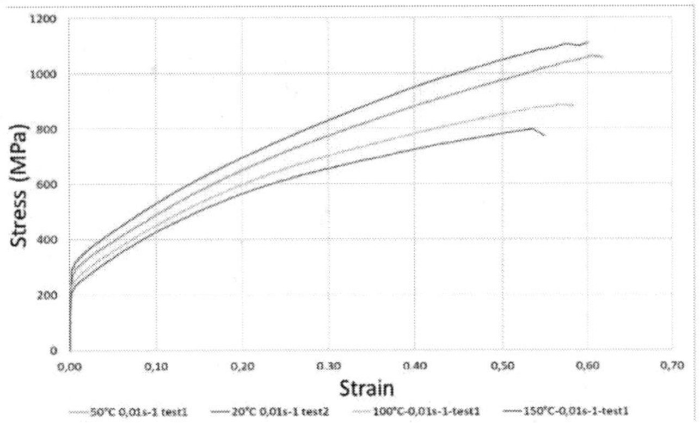	Major influence of temperature on the yield stress.

3. Simulation of the drawing processes

Numerical simulation helped us understand why the "Soft mode" kinematic sequence could reach higher drawing depths. We simulated the experimental process thanks to the AutoForm software and used its thermosolver to consider the temperature effects accurately. This solver enables the accurate prediction of the temperature increase within the sheet due to the combined effects of plastic work and friction, and of the cooling of the sheet during contact with the tools or with the ambient air.

The material description used for simulation has been adapted to include the influence of temperature as seen in Table 3 (Figure 1).

We created the kinematic curves input for both press kinematics used for the two simulations ("Sinusoidal mode" and "Soft mode" kinematic sequences), identical to the curves used in the real experiment. They are are shown on Table 4.

IOP Conf. Series: Materials Science and Engineering 1238 (2022) 012077 doi:10.1088/1757-899X/1238/1/012077

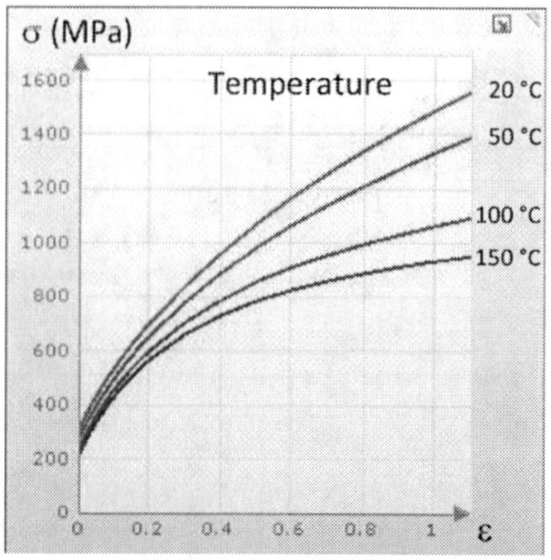

Figure 1. Temperature dependent hardening curves as input in the AutoForm simulation depending the temperature

Table 4. Press kinematics used in simulation.

Type	Diagram
Sinusoidal mode	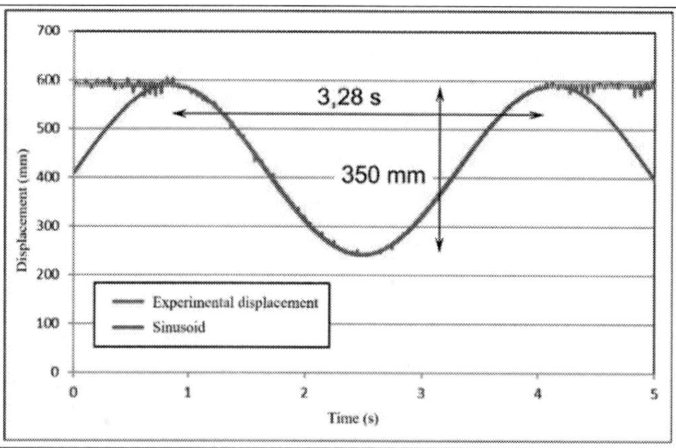
Soft mode	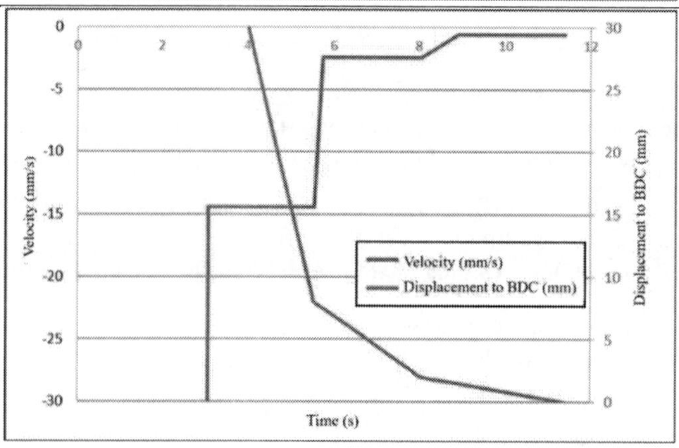

To compare both processes, we first analyzed the thinning reached on a specific point of the sheet at a specific drawing depth for the "Sinusoidal mode" simulation. In the example on the left in Figure 2, we reached a thinning of 29.8% at 2.50 mm of bottom dead center (BDC).

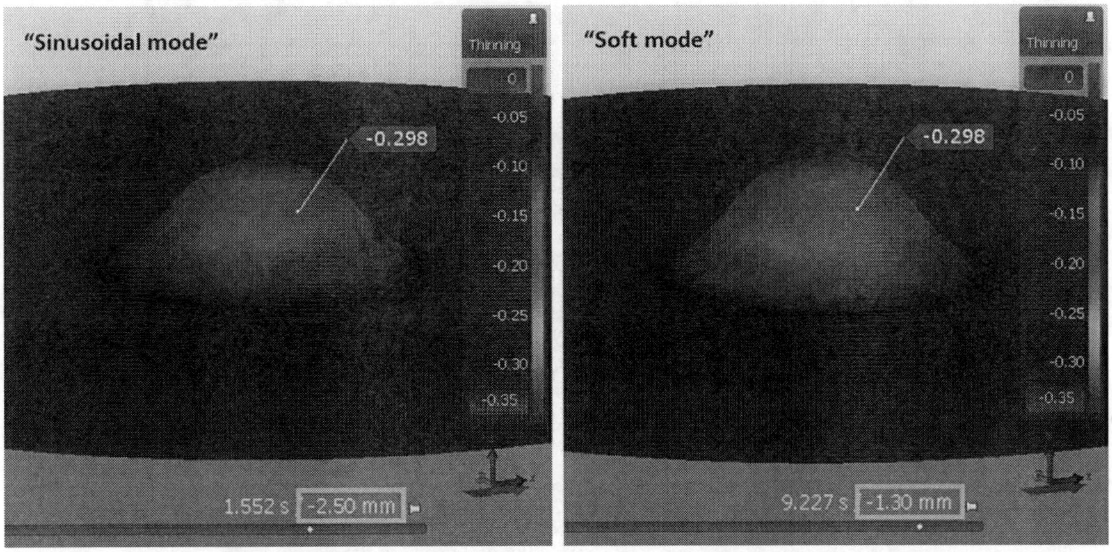

Figure 2. Comparison of the drawing depth reached for a given thinning value, for both processes. Left, the "Sinusoidal mode" simulation reached the thinning at 2.50 mm to BDC while, right, the "Soft mode" simulation reached the same thinning at 1.30 mm to BDC.

Then the same point was located on the "Soft mode" simulation and we changed the process time viewed to find the same thinning value of 29.8%. In this case, it was reached at 1.30 mm of BDC, so 1.20 mm lower, therefore corresponding to a higher drawing depth than in the "Sinusoidal mode" simulation. This confirmed the experimental result stating that the "Soft mode" enables us to draw the part deeper.
We saw that temperature has an important effect on the mechanical characteristics of the sheet between 20°C and 100°C (cf Table 3). During a classical drawing process, we can expect this range of temperatures within the sheet due to the friction and the plastic strain within the material.
If we compare the temperature reached on the same point for both simulations, we can see that the "Sinusoidal mode" simulation reached a sheet temperature of 67.5°C, with a steady increase of that temperature over process time (Figure 3 top). The temperature history for that point in the "Soft mode" simulation is not steadily increasing (Figure 3 bottom), but has a peak at 8 mm of BDC. That peak value is 39.4°C, lower than the temperature reached in the "Sinusoidal mode" simulation, and then decreases as the process velocity also decreases (cf Table 2). The temperature reached for the thinning value of 29.8% is 24.5°C. This soft touch process, offering a slower punch impact and then even smaller values of the drawing velocity, enables the strain work to dissipate more within the sheet, therefore, decreasing the temperature reached during the process and affecting the drawing depth value reached for a given thinning.

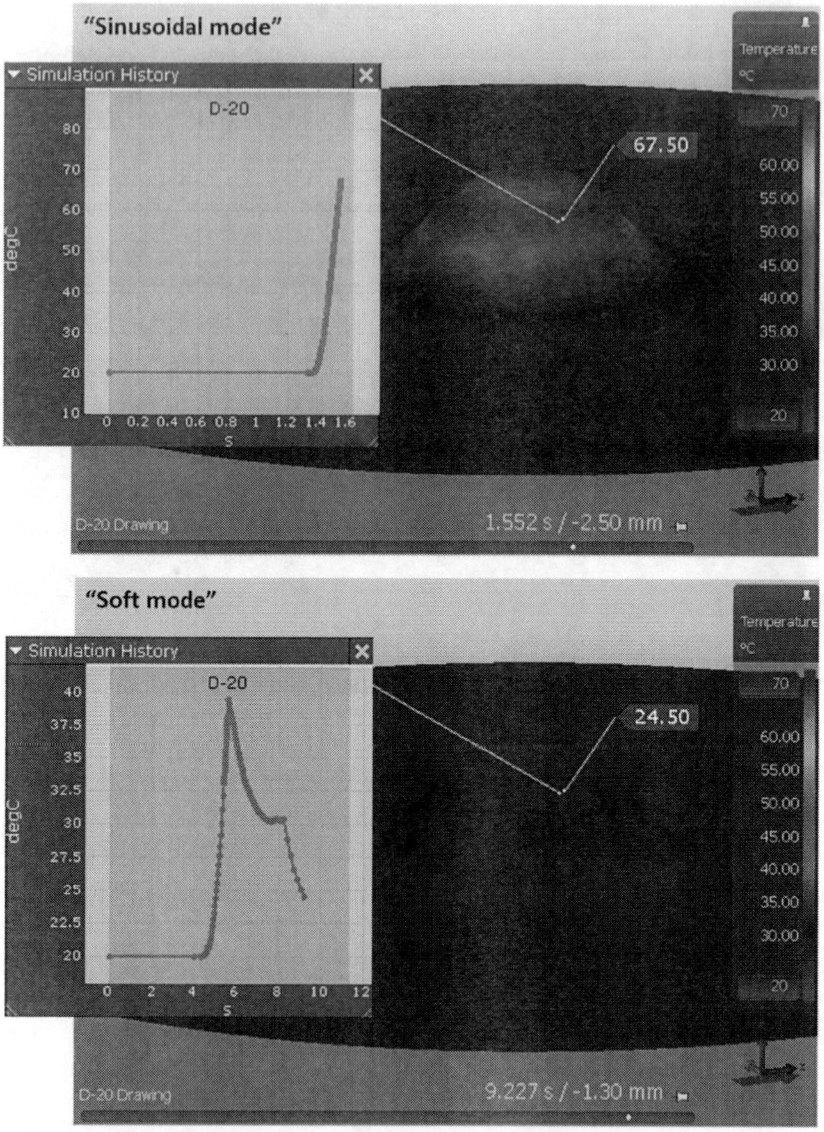

Figure 3. Temperature and temperature history within the sheet during both processes up to a thinning of 29.8%.

When we compare the hardening curves of the material depending the temperature, we could easily understand the simulated and experimental behaviour: as the temperature rises, the material softens and thus becomes less resistant. A quicker localisation of the thinning is observed as the material heats up compared to a behaviour at low temperature.

Analyzing the plastic strain rates history for that point of the sheet after both simulations (figure 4) shows us that the "Sinusoidal mode" simulation undergoes plastic strain rates up to almost 3 s-1 whereas the "Soft mode" simulation undergoes much lower values, up to almost 0.4 s-1. It would be interesting in a follow-up study to characterise the material at these higher values of plastic strain rates and simulate their influence if it proves necessary.

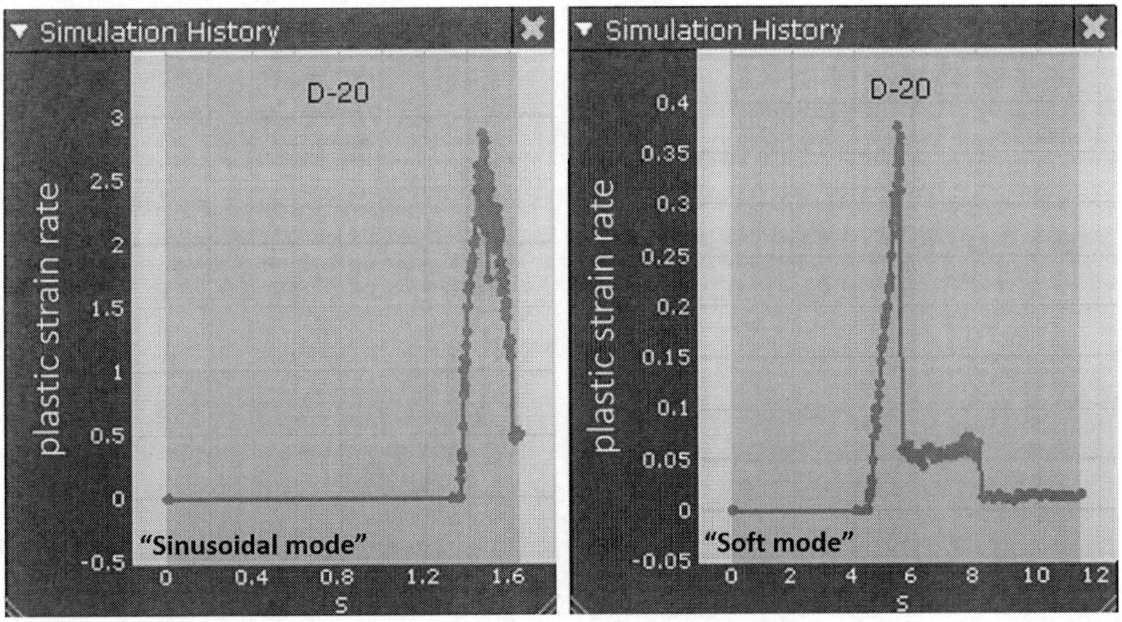

Figure 4. Plastic strain rate history within the sheet during both processes.

4. Conclusion

When we talk about servo-press, we expect the strain rate effects on the material characteristics to be the main factor influencing the forming of said material. The numerical simulation with AutoForm made it possible to highlight the thermal effects, in particular, the maximum temperatures reached during the drawing operation. For the stainless steel in this study, simulation could show that the temperatures reached within the sheet had a very prominent influence on the achievable drawing depth. The soft touch process, enabling a much lower temperature, ensures a deeper drawing. To complete this study, we could further investigate the strain rate effects on the material properties, as simulation showed that the plastic strain rates reached in this study were higher than the available data.

This study shows that simulation allows the user to optimize the motion curve of the press in order to have a minimum cycle time while respecting the lowest possible temperature the sheet might reach during a drawing operation. Thanks to this study, the sheet metal forming companies will therefore be able to confidently develop and optimize their servo-presses usage. Until now, servo-presses were not utilized to their full capacity in a production setting; now, thanks to simulation, an important gap is filled and servo-presses will be exploited to their full potential by the industry.

5. References

[1] Majidi O. and al 2015 Formability of AHSS under an Attach–Detach Forming Mode, *Steel research Int.* **86** N°2, pp98-109
[2] Hyunok Kim and al 2019 Control of the servo-press in stamping considering the variation of the incoming material properties, *International Deep Drawing Research Group 38th*
[3] Murakami T. and al 2015 Influence of blanking distance of materials on increase in blanking vibration using screw drive servo press, *Mechanical Engineering Journal*
[4] Danel A. and al 2016 Influence of servo-press kinematics on the formability of drawning parts, *IDDRG International Conference. Linz, Austria,* June 12-15

Assembly simulation development and validation project with industrial applications

Fabrizio Tinti[1], Marco Fontana[2], Leonardo Daniela Scintilla[2], Annalisa Mirabile[2], Bart Carleer[3]

[1]AutoForm Engineering Italy, Turin, Italy
[2] Fontana Group S.p.A, Calolziocorte, Italy
[3] AutoForm Engineering, Pfäffikon, Switzerland

fabrizio.tinti@autoform.it

Abstract. Full digitalization has been applied more intensively in the automotive industry. The virtual engineering of single part is established since many years. Currently, the digital process chain is expanding to the assembly process. The assembly engineering teams should be effectively supported to define the assembly process; the assembly process should make products that fulfil the required quality requirements. A detailed simulation representing the future assembly process has to be setup. In an early phase based on CAD data, in the course of the engineering process the model can be enriched with forming simulation data and even real measurement data. Based on the assembly definition, critical issues can be identified and fixed already in the virtual world. Detailed evaluation of the geometrical deviation of the assembly can be made. This methodology is validated on an industrial automotive component.

1. Introduction

The automotive industry is in a transition mode, as various mobility and people-centric trends are impacting the automotive market [1]. Moreover, various technology trends related to the development of smart factories and industry 4.0 are affecting the automotive industry. As a result of these trends, many car makers are attempting to become more agile and resilient by implementing business and organizational changes. Through the increased use of simulation technology and digitalization of stamping and assembly processes costs are further reduced. The virtual engineering of single part is established since many years. Currently, the digital process chain is expanding to the assembly process.

In this paper, we discuss a three-year project between AutoForm Engineering and Fontana Group on the simulation of the BiW assembly process. During this project three assembled subgroups were studied. AutoForm Engineering focused on the improvement of its software solutions. Fontana Group contributed with its experience and consolidated know-how of its stamping and BiW technical offices. In the first phase, it has been verified whether the software was able to represent the challenges and complexity required by the real world. In the second phase the simulation result was validated on a hood. The pillars on which the project was launched were:

- Improvement of the AutoForm-Assembly simulation software to be able to represent a variety of real assembly processes in bodywork production
- Application of AutoForm-Assembly to the assembly processes of multiple components and validation of the accuracy of the simulation in terms of deviation of the final shape (springback)
- Validation of efficient simulation techniques to model state-of-the-art assembly technologies, including hemming processes.

This assembly process involves the production of paneled sub-assemblies in which the geometric complexity of the parts is an important constraint. Just think of the variety of shapes, materials, thicknesses and artifacts (molded, extruded, die-cast, etc.) that can be found combined in a BiW. The assembly and hemming process is more than combining a few pieces of sheet metal; it can be an extremely advanced process with no more restrictions on the number of components or operations sequence. The combination according to the correct assembly method is in fact the only starting point of the process. The simulation should represent the real (future) process and should be able to define:

- how the pieces will be constrained in space
- how they will be joined
- what thermal deformations will occur (e.g. welding points and furnace firing)
- which mechanical joints are applied (clinching, riveting, hemming, etc.)
- without forgetting the final quality of the assemblies in terms of geometric deformations and aesthetic appearance.

The possibility of fully simulating and accurately representing the complete assembly processes in production is the ultimate objective. Single detailed simulations to analyze the behaviour of a specific joint or constraint had been possible for years, but it was not possible to carry out a broad spectrum of process simulations in a cycle time compatible with the car development timeline.

2. Assembly Verification

In the first phase it has been verified whether the assembly software was able to represent the challenges and complexity required by the real world. In this phase two subgroups were selected for verification, a fender group and a hood assembly. Most illustrations are shown as sketches since the real parts pictures cannot be shown due to confidentiality. First limitations identified were the representation of the extruded structural components and the flexibility in representing the sequence of assembly operations. This latter issue can be shortly explained on the fender group assembly.

The fender is produced by assembling the skin to the inner frame group and this package is inserted on the hemming bed. For proper positioning and (pre-) fixation outer and inner are not joined by hemming only, but are subjected to riveting and laser welding. So, the process definition has to be able to independently define when the individual parts are loaded into the machine, in what order and with which loading direction. In addition the method plan must be able to indicate which type of processing (assembly, hemming, curing, springback measurement) is performed in each operation. A variable number of assembly operations should be defined and alternated with assembly, hemming or curing. The scheme of the supports and the loading sequence of the pieces are also assigned manually to each operation and a series of automatic checks allow the operator to validate the hypothetical choices.

Essential process data checked are:

- the contact areas between the different parts (mating check for possible welding)
- the effectiveness of the number and position of the constraints on the stability of the loaded pieces (stability check)
- the collision areas during loading (collision check).

The possibility of having a precise and systematic control of the supports and clamps allows to speed up the engineering phase and to foresee any corrections before the equipment is built.

In the Fontana Group's assemblies there are all types of metallic and non-metallic components of different nature. We came across tubes, extrusions of different shapes and sections, die-casted elements, forged elements, composite elements, etc. which by their nature can in no way be treated like a stamped sheet. The Fontana Group required that all kind of formats be used for importing these parts. This request was fundamental for the volume of products handled by Fontana Group. It resulted in a software even more flexible and suitable for the needs of both markets: mass production and premium brands; this latter very often shows a combination of components that are different in type. We therefore made it possible to join different solid elements with stamped sheets and not just sheets with other sheets in multiple overlapping layers.

In Figure 1 we explain the assembly process sequence, showing possible complexities and occurring boundary conditions. This figure shows the assembly process of a door like it has been defined in the AutoForm graphical user interface. This overview shows the different assembly operations and which parts or sub-assemblies are joined. Operations A-10 and A-20, two small sub-assemblies, are created. Both the clamps and the weld points are indicated with little symbols in order to analyse the assembly definition. Operation A-30 shows how several sheet metal parts are joined to the door inner stamping. Additional stamped parts as well as the sub assembly from operation A-10 are joined on the door inner in operation A-40. In operation A-50 more parts including the sub-assembly from A-20 are joined. Last joining operation is A-60, the complete inner assembly is finalized. Operation G-70 is the hemming operation, the outer panel is hemmed around the inner sub-assembly. The last operation is M-80, this operation represents the measurement station. This figure illustrates a complete assembly sequence of a door and shows some of the challenges in assembly. Important boundary conditions are the above mentioned process data but also the accessibility of welding robot and the cycle time optimization.

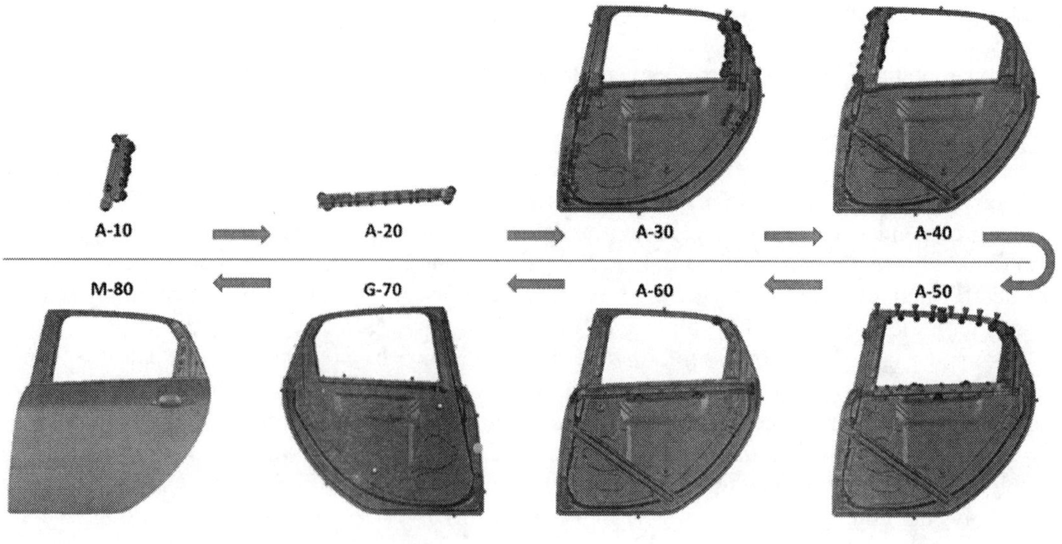

Figure 1. Schematic illustration of process sequence of a door assembly including the indication of the clamps and joints in each assembly operation.

With the second component (hood) under study, we focused on what data and what results can be generated in the different phases of process development.

2.1. *Process Engineering*

In the early stages of engineering the CAD geometries of the parts are the only inputs available for the simulation, but with these it is already possible to make a large number of analyses on the stability of the process: the verification of the loading sequence, the position of the supports, the lowest number of closures to be used and the geometric deformation of the assembled group with nominal parts (no thinning, no residual stresses). With the influence of the stress introduced by the joining technology both thermal- and mechanical-joining can be evaluated. These definitions are the foundations of the assembly process. If well-defined the final production can start efficiently with a short ramp-up times. For reasons of development time of the component, the BiW engineering cannot afford to wait for more detailed data produced by the stamping colleagues on the complete cycle of the individual component. The first analysis can start immediately with the first CAD0 mathematics just released by Product Development.

2.2. *Process Validation*

After simulating the bodywork performed with nominal CAD0 data, it is possible to replace these data with the results produced by a full cycle stamping simulation (".asm" file). Keeping the process unchanged, the tension and thinning data of each part (or of the most interesting ones) are included, reaching a further level of accuracy compared to the initial analysis. In this second calculation it is possible to evaluate the further effects induced by the residual stresses and the springback when the component is loaded and constrained on the assembly jig. The effects induced by the stamping process will then be carried over into the bodywork process and added in the final result. This analysis gives a much better representation of the real assembly process and results in more accurate results.

2.3. *Process Optimization*

However, a stamped part can slightly deviate from the simulated part, e.g. shape or geometry (springback) might slightly differ from the result generated with the stamping simulation. The causes can be manifold, manual interventions for setting up the tooling, geometric revisions not updated in simulation, variability of the material used in tool-shop and production, etc. These differences can generate such a difference that the stamping simulation data no longer reflect the physical component. To satisfy this operational need and not limit the analysis performed with CAD0 data or asm-data, a third possibility was introduced: directly import the data of a scan (".stl" file). The product development phase can therefore be simulated from start to finish, using the most up-to-date and useful data for the analysis to be carried out in succession.

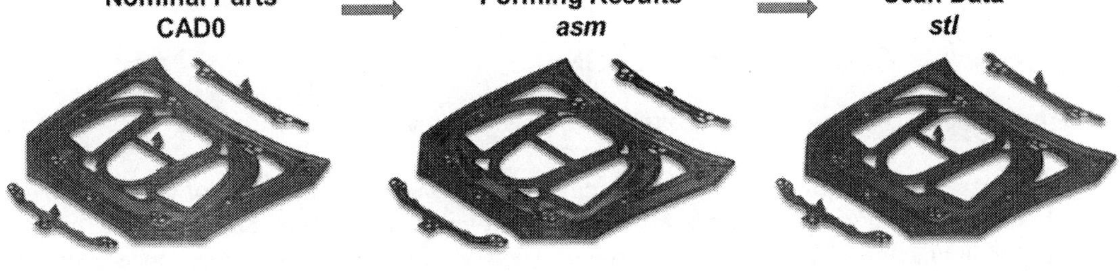

Figure 2. Maturing the assembly simulation with subsequent inputs: Nominal Parts from CAD – Forming Simulation Results, asm file – Real Measurement via geometry scan, stl data.

Starting in the early phase of the care development the assembly process is defined using CAD data. When time moves on, more mature data from the stamping simulation become available. In the last phase of the development the first physical parts fall off the press. Having the flexibility to easily exchange input data enables the engineer to use the most mature data available. This methodology allows the engineer to mature the assembly process simulation over time and a more and more reliable representation of the assembly process is available.

3. Industrial Validation

The second phase of the collaboration project focused on the analysis of actual results. The validation of the AutoForm Assembly simulation was carried out on another hood assembly. It was decided to limit the analysis to mechanical operations in order to have a direct correlation between what is actually produced and what is simulated. The thermal process step, the glue curing in the oven was omitted from the study. The validation focused on the following assembly process (see Figure 3):

- In the first assembly station A-10 four sheet metal parts are joined to an extruded component by welding and riveting. All structural components (brackets and stiffeners) have been imported as nominal CAD0 data
- In A-20 two sheet metal components are welded on the inner frame. For the inner frame it was decided to import the scan data of the real stamped part, to better replicate the real geometric deformation seen in production
- In a third assembly station, A-30, the sub-assembly extruded component is mounted on the inner frame
- The inner subgroup assembly process is followed by a hemming station G-40 in which the inner subassembly is inserted on the skin and the hemming flanges are folded by robots with three steps and two differently shaped wheels. For the hemming process, the outer panel was imported as full cycle stamping simulation result to have a better correspondence with the residual stresses existing on the real piece (important for the evaluation of possible aesthetic defects). Only three sides were hemmed, right-bottom-left, the top side hasn't been hemmed (see Figure 4). All process data (tools, forces, paths, tipping, etc.) have been set in simulation to faithfully replicate the equipment and the actions performed by the production machinery
- The result of the hemming was finally positioned on the virtual control gauge respecting the tip and the real closures, taking out a measurement map M-50 (not shown in Figure 3). For the measurement the CAD0 assembly is the reference.

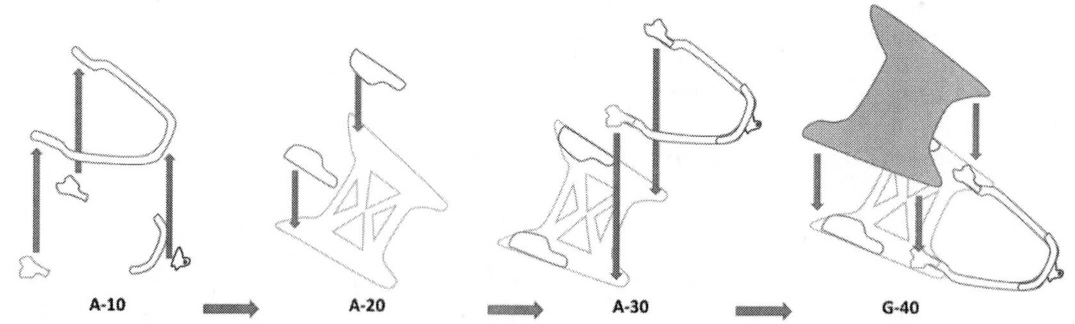

Figure 3. Schematic representation of the four assembly stations of the hood assembly process.

A hood assembly was picked out of the running production at the same stage of the process and was placed on a measuring gauge, taking out a measurement report and a scan map of the external surface. This map, combined with the measurement report, was used for the validation of the assembly process simulation results.

3.1. Assembly Measurement Results

To validate the result four sections on the outer part have been defined according to the sketch in Figure 4. On those four sections the position on 3D space was measured and compared with the ideal position according to CAD0. The measurement was done after the hemming operation and the deviation represents whether the geometry perfectly fits into the car assembly. The deviation is measured in normal direction from the hood, at the top no hemming was done. The results of the measurement (scan) and simulation (AF) are shown in Figures 5 - 8.

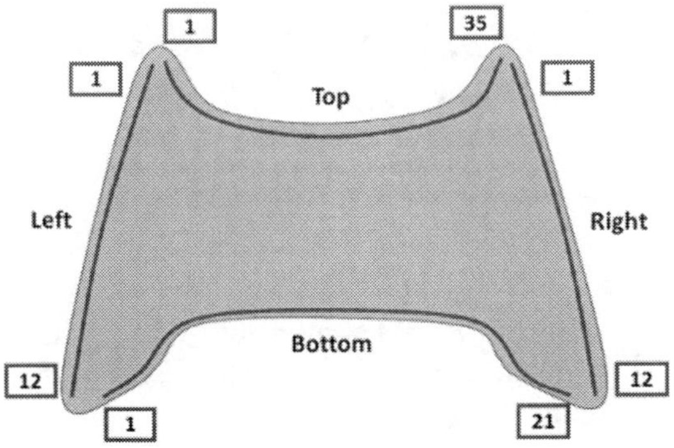

Figure 4. Sketch of hood outer with definition of measurement sections, left – right, top – bottom. Numbers indicated the start and end of the measurement points.

Figure 5 and 6 show the deviation at the left and right of the hood outer respectively. Generally the measurement and the simulation match well. On the left the deviation of the measurement points 1 – 11 is less than 0.3mm but on the right the measurement and simulation are spot on. At the corner, measurement point 12, we see a relatively high deviation of more than 1.0mm. This deviation has been introduced due to the manual modification during assembly trials. These modifications haven't been re-simulated. What also catches the eye when comparing left and right is that although the hood is symmetrical the deviation isn't symmetrical. The unsymmetrical results, both of the measurement and the simulation, are caused by the unsymmetrical stamping result of the inner frame. Keep in mind that we imported for the simulation the inner frame stl-data of the real stamped part to better replicate the unsymmetrical deformation seen in stamping.

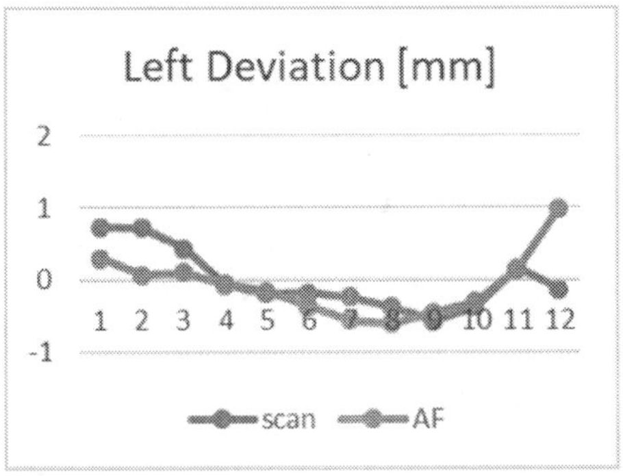

Figure 5. Measurement (scan) and simulation (AF) at the left side of the hood

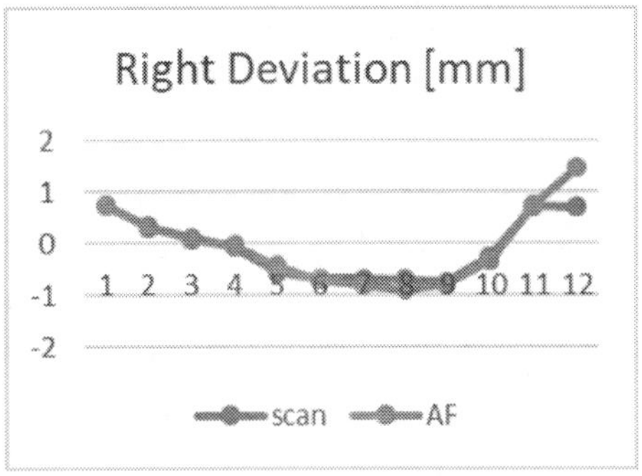

Figure 6. Measurement (scan) and simulation (AF) at the right side of the hood

Figure 7 and 8 show the deviation at the top- and bottom-section of the hood. Since the four sections go around the hood, point 1 from the top corresponds with point 1 from left, point 35 from the top corresponds with point 1 from right. The relative large deviation in the corner at point 12 of the left section corresponds with point 1 of the bottom point 12 of the right corresponds with point 21 of the bottom section. Also the sections are not symmetric, neither the measurement nor the simulation. The deviation in the top- and bottom-sections show that the unsymmetrical inner frame was very dominant for the overall geometrical deviation. The methodology to replace simulated data by measurement data turned out to be crucial in representing the real assembly result. The simulation result matches very well with the measurements.

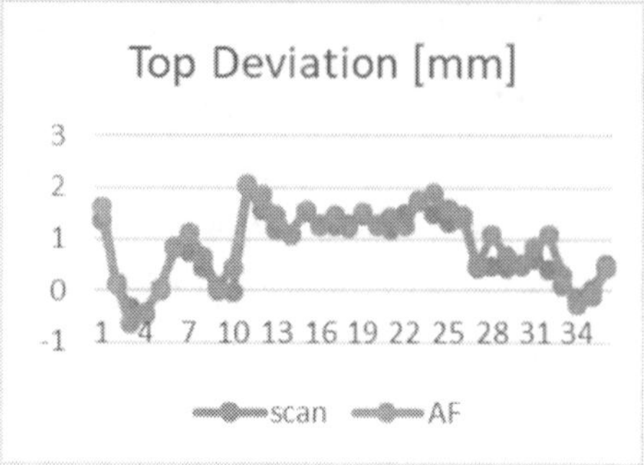

Figure 7. Measurement (scan) and simulation (AF) at the top side of the hood

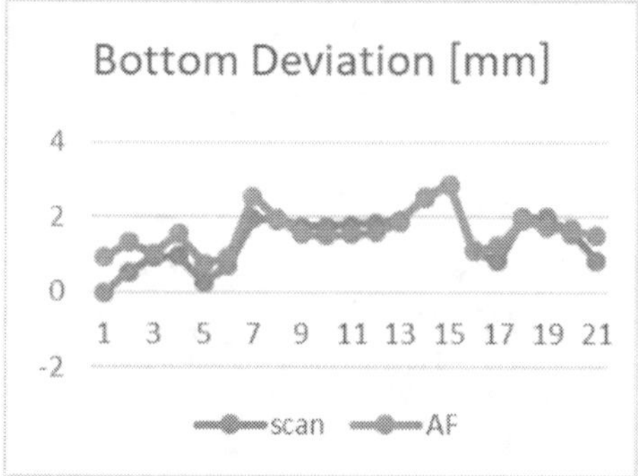

Figure 8. Measurement (scan) and simulation (AF) at the bottom side of the hood

This virtual assembly validation showed that the assembly process can be very well simulated. The geometrical deviation can be predicted within a few tenths of a mm. Knowing the deviation before the first physical trials and having identified the most critical area enables to take countermeasures immediately. The countermeasures effectiveness can be validated with help of the simulation on beforehand. With this simulation methodology the number of physical assembly trials can be reduced and cumbersome trial and error can be eliminated. After this validation activity, Fontana Group started to use AutoForm Assembly to analyse the production process of different parts (fenders, hoods, doors, etc.) to anticipate and prevent possible critical scenarios in the production start-up phase.

4. Conclusions

The first phase of the project confirmed that it is possible to properly represent productive assembly processes. All critical and essential process steps cab be modelled in the available software. In the second phase, we were able to validate currently available simulation software on a complex industrial assembly process. Several critical aspects of the real process have been highlighted during the project thanks to the transparency carried out by the digital process twin modelled in simulation. The validation of the results allowed to certify the accuracy of the calculated result, demonstrating that the precise description of the assembly process allows to obtain the real geometric deformation of the assembled component. The simulation methodology enables engineers to mature the assembly process in the developments phase by defining a digital twin. This reduces the cumbersome trial and error loops and it contributes to a more effective car development.

Acknowledgements

The authors would like to thank the management of Fontana Group and AutoForm Engineering for enabling the fruitful and great collaboration.

References

[1] Automotive Industry Trends and Challenges: Process Management Today; Dr. Ir. Jan Harmen Wiebenga (TriboForm), Jochen Hambrecht, Ph.D. (AutoForm) – 2021/02/15; www.formingworld.com

Evaluating the influence of the deformation of the forming tools in the thickness distribution along the wall of a cylindrical cup

M C Oliveira[1,*], D M Neto[1], A.F.G. Pereira[1], J L Alves[2] and LF Menezes[1]

[1]CEMMPRE, Department of Mechanical Engineering, University of Coimbra, Polo II, Rua Luís Reis Santos, Pinhal de Marrocos, 3030-788 Coimbra, Portugal
[2]CMEMS, Microelectromechanical Systems Research Unit, University of Minho, Campus de Azurém, Guimarães, 4800-058, Portugal

*E-mail: marta.oliveira@dem.uc.pt

Abstract. The Swift cup drawing test has been adopted for evaluating sheet metal deformation properties, namely the material formability assessed through the limiting drawing ratio. Since the strain path of the points located on the flange is between uniaxial compression and pure shear, this region is subjected to thickening. If the thickness of the drawn flange is larger than the gap between the punch and the die, cup wall ironing will occur. Therefore, the ironing forces can lead to significant elastic deformation of the forming tools (punch and die). The main objective of this study is to evaluate numerically the deformation of the forming tools during the deep drawing-ironing process of a cylindrical cup. Then, the effect of that deformation on both the earing profile and the evolution of thickness along the circumferential direction, at different heights, is analysed. The material studied is the AA 6016-T4. Since both the thickness strain and the earing profile are strongly influenced by the adopted yield criterion, the study considers the classical quadratic one proposed by Hill and a non-quadratic proposed by Cazacu and Barlat. The process conditions considered are the ones from EXACT, the ESAFORM Benchmark 2021, enabling the comparison with experimental results.

1. Introduction

The tools used in sheet metal forming processes are commonly made from carbon and alloy steels with distinct characteristics, such as hardness, wear resistance, toughness, and resistance to softening at elevated temperatures. Besides, comparing the geometry of the forming tools with the blank, the tools are significatively stiffer. This enables adopting the assumption that they can be treated like rigid bodies, in the numerical simulation of sheet metal forming processes. This simplifies the contact algorithms reducing the computational effort. Nevertheless, in some cases this assumption can lead to inaccurate predictions of the material flow and, consequently, of the thickness distribution [1]. The elastic deformation of the tools depends on their geometry, mechanical properties and magnitude of the contact forces. Therefore, the accurate prediction of the tools deformation requires their modelling and treatment as a solid deformable body, leading to a considerable increase of the computational time.

This study presents the finite element analysis of the deep drawing of a cylindrical cup, considering that the forming tools are modelled either as rigid or as deformable bodies. The numerical results are compared with the experimental ones, in order to assess the impact of the tools deformation resulting from the ironing stage.

2. Swift cup drawing test

The forming conditions adopted in this study correspond to the ones defined in the ESAFORM 2021 Benchmark: EXACT - Experiment and Analysis of Aluminum Cup Drawing Test [2]. The tools consist of a cylindrical punch with a flat bottom, a blank holder and a die, as schematically shown in Figure 1. The drawing operation is performed considering a constant blank holder force of 40 kN. Moreover, a stopper with the same thickness of the blank was used to avoid the pinching of the ears. The AA 6016-T4 blank sheet is circular in shape with a diameter of 107.5 mm and a thickness of 0.98 mm. The ESAFORM 2021 benchmark committee did not recommended a value for the friction coefficient. The results for the punch force evolution with its displacement and the earing profile were supplied to enable the participants the adjustment of its value. All details about the experimental procedure for the characterization of the material behaviour and forming of the cup are described in [3].

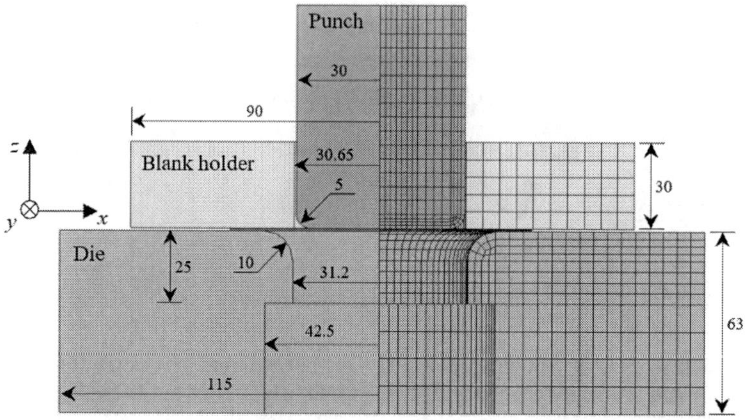

Figure 1. Schematic representation of the forming tools including main dimensions [mm] and the mesh adopted when considering deformable tools.

2.1. Mechanical behaviour

The mechanical behaviour of the AA 6016-T4 aluminium alloy is assumed to be isotropic in the elastic regime, being described by the Young's modulus, E, and the Poisson ratio, υ. Regarding the hardening behaviour, the Swift law suggested by the benchmark committee was adopted. The orthotropic behaviour was mainly described by the Hill'48 yield criterion, described by the equation

$$\bar{\sigma}^2 = F(\sigma_{22}-\sigma_{33})^2 + G(\sigma_{33}-\sigma_{11})^2 + H(\sigma_{11}-\sigma_{22})^2 + 2L\tau_{23}^2 + 2M\tau_{13}^2 + 2N\tau_{12}^2 \qquad (1)$$

where σ_{ij}, $i,j=1,2,3$ are the stress components of the Cauchy stress tensor $\boldsymbol{\sigma}$, defined in the principal axes of orthotropy, and F, G, H, L, M and N are the material parameters describing the anisotropy of the metal sheet. $\bar{\sigma}$ is the equivalent stress, which is equal to the flow stress, Y, under loading conditions. The parameters were determined based on the Lankford values for the rolling, diagonal and transverse directions and assuming that the flow stress corresponds to the stress-strain curve along the rolling direction ($G+H=1$). The L and M are the anisotropy parameters corresponding to the off-plane properties that cannot be evaluated for metallic sheets. Thus, they are assumed as equal to the isotropic values, i.e. 1.5.

Since an associated flow rule is adopted, in order to improve the understanding about the influence of the normal to the yield surface on the thickness distribution predicted, the yield criterion proposed by Cazacu and Barlat in 2001 (CB2001) [4] was also considered. The CB2001 is a generalization of the Drucker's isotropic criterion to orthotropy, such that:

$$\bar{\sigma} = \left\{27\left[\left(J_2^0\right)^3 - c\left(J_3^0\right)^2\right]\right\}^{\frac{1}{6}} \qquad (2)$$

IOP Conf. Series: Materials Science and Engineering 1238 (2022) 012079 doi:10.1088/1757-899X/1238/1/012079

where J_2^0 and J_3^0 are the second and third generalized invariants of the deviatoric Cauchy stress tensor, defined as:

$$J_2^0 = \frac{a_1}{6}(\sigma_{11} - \sigma_{22})^2 + \frac{a_2}{6}(\sigma_{11} - \sigma_{33})^2 + \frac{a_3}{6}(\sigma_{11} - \sigma_{33})^2 + a_4\sigma_{12}^2 + a_5\sigma_{13}^2 + a_6\sigma_{23}^2 \tag{3}$$

$$\begin{aligned} J_3^0 = {} & (1/27)(b_1 + b_2)\sigma_{11}^3 + (1/27)(b_3 + b_4)\sigma_{22}^3 + (1/27)[2(b_1 + b_4) - b_2 - b_3]\sigma_{33}^3 \\ & -(1/9)(b_1\sigma_{22} + b_2\sigma_{33})\sigma_{11}^2 - (1/9)(b_3\sigma_{33} + b_4\sigma_{11})\sigma_{22}^2 \\ & -(1/9)[(b_1 - b_2 + b_4)\sigma_{11} + (b_1 - b_3 + b_4)\sigma_{22}]\sigma_{33}^2 \\ & +(2/9)(b_1 + b_4)\sigma_{11}\sigma_{22}\sigma_{33} - (\sigma_{13}^2/3)[2b_9\sigma_{22} - b_8\sigma_{33} - (2b_9 - b_8)\sigma_{11}] \\ & -(\sigma_{12}^2/3)[2b_{10}\sigma_{33} - b_5\sigma_{22} - (2b_{10} - b_5)\sigma_{11}] - (\sigma_{23}^2/3)[(b_6 - b_7)\sigma_{11} - b_6\sigma_{22} - b_7\sigma_{33}] \\ & +2b_{11}\sigma_{12}\sigma_{23}\sigma_{13} \end{aligned} \tag{4}$$

where $a_1,...,a_6$ and $b_1,...,b_{11}$ are the anisotropy parameters and c is a weighting parameter. The conditions that guarantee the convexity of CB2001 are unknown, except when assuming in-plane isotropic behaviour, for which $c \in [-3.75, 2.25]$. In this case, the a_5, a_6 and b_k $(k = 6, 7, 8, 9, 11)$ are the anisotropy parameters corresponding to the off-plane properties, for which the isotropic values, i.e. 1.0 are assumed. The anisotropy parameters were identified considering the normalized tensile yield stresses and r-ratios extracted from uniaxial tensile tests performed at every 15°, between the rolling and the transverse directions, and the direction of the plastic strain-rate, extracted from biaxial tests performed with cruciform specimens. The material parameters adopted are all available in [3].

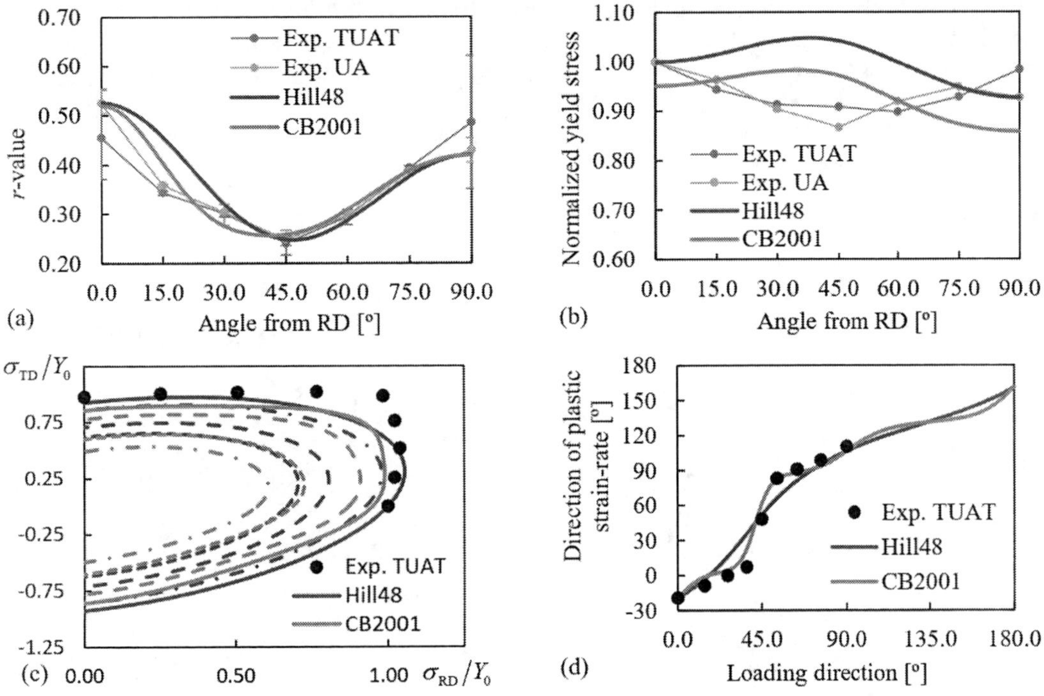

Figure 2. Comparison between experimental and predicted: (a) r-values; (b) yield stresses; (c) projection of the yield surface in the biaxial plane $(\sigma_{RD};\sigma_{TD})$ (with $\tau_{RD}/Y_0 = 0.0$; 0.25; 0.43; 0.50) and (d) direction of the plastic strain-rate. All stresses are normalized by the yield stress along the rolling direction Y_0.

Figure 2 presents the comparison between the experimental results (obtained by the Tokyo University of Agriculture and Technology (TUAT) and by the University of Aveiro (UA)) [3]. For both yield criteria, the uniaxial tensile test results considered in the identification were the ones obtained by UA. Figure 2 shows that both yield criteria lead to a similar in-plane evolution of the r-values and an identical trend for the normalized yield stress; both yield criteria describe well the experimental r-values but are unable to capture the yield stress distribution. Nevertheless, the projection of the yield surface in the biaxial plane shows that both yield surfaces present quite different descriptions of the shear stresses. Moreover, the direction of plastic strain-rate is also quite different for both yield criteria, as shown in Figure 2 (d), which highlights the flexibility of the CB2001.

2.2. Finite Element model: rigid tools

Due to geometrical and material symmetries, only a quarter of the global structure is modelled. In a first approach, all forming tools are considered rigid and are modelled using Nagata patches [5], which allows the use of a coarse finite element mesh. The contact with friction conditions are described by the Coulomb's law. As previously mentioned, there was no recommendation for a value for the friction coefficient. Therefore, a constant value for the friction coefficient was considered (μ=0.07) in order to best describe the drawing force.

The blank sheet is discretized with 3D 8-node hexahedral finite elements, combined with a selective reduced integration technique. The mesh comprises a non-structured mesh for a radius lower than 22 mm (bottom of the cup) and a structured mesh for the remaining region. The mesh is composed of two layers through the thickness, which allows an accurate evaluation of the contact forces and the through-thickness stress gradients. The total number of elements is 15408 and the total number of nodes is 16006.

2.2.1. Results analysis and discussion. Figure 3 (a) presents the punch force evolution with its displacement, comparing the experimental result with the one obtained with rigid tools. The results show that, although the constant value of μ=0.07 enables an accurate description of the drawing force, the ironing force is clearly overestimated. Moreover, the analysis of the thickness distribution along the cup circumference at different heights, shown in Figure 3 (b), indicates that its trend is clearly underestimated. This is particularly evident for the top of the cup (height of 30 mm (H30)), for which the numerical simulation predicts a constant value, equal to the gap between the punch and the die (see Figure 1).

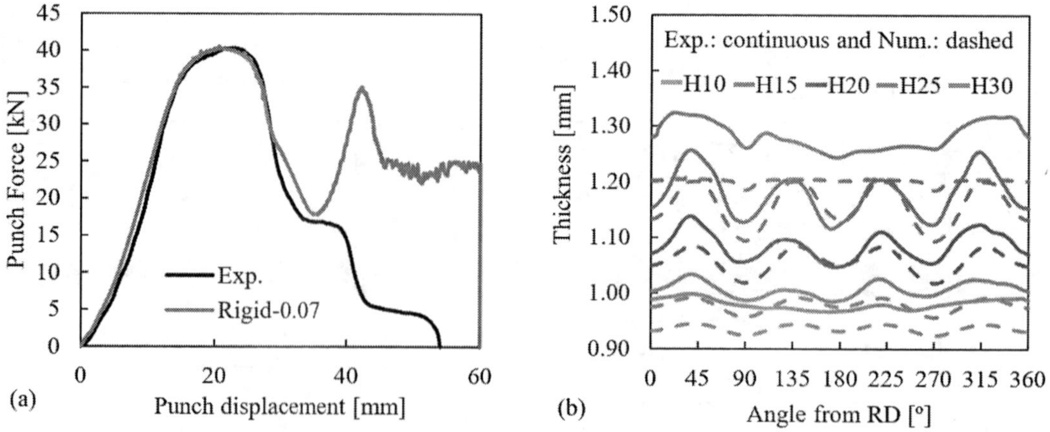

Figure 3. Comparison between experimental and predicted results using rigid tools, Hill'48 and μ=0.07: (a) punch force-displacement and (b) thickness distribution along the cup circumference at different heights H = 10, 15, 20, 25 and 30 mm from the cup bottom (Experimental: continuous line and Numerical: dashed line).

The thickness was experimentally measured using different techniques [3], all of which confirmed the occurrence of values higher than the gap between the punch and the die at the top of the cup. The tools were also measured using different techniques, which revealed deviations on dimensions related with the tools clearance in the order of ± 0.001 mm [3]. Therefore, unlike in [6], it is not possible to associate the thickness results with slight wear of the tools.

3. Finite Element Model: deformable tools

The forming tools were discretized with solid finite elements to allow considering their elastic deformation. As shown in Figure 1, the discretization adopted is quite coarse because a surface smoothing method is applied. The surface smoothing method is based on Nagata patches, following the procedure proposed in [7] for frictional contact problems between deformable bodies. This way, the discretization of the deformable tools in the curved surfaces is identical to the one used for the rigid tools, providing a similar shape error, resulting from the surface interpolation with Nagata patches.

The blank holder is controlled with an imposed displacement on its top face that increases linearly with the punch displacement, until a maximum value of 0.17 mm, when the blank loses contact with the blank holder. This allows to mimic the behaviour observed when using rigid tools. The blank discretization adopted is the same used in the simulation performed with rigid tools.

3.1. Tools stiffness

The mechanical behaviour of the forming tools is assumed elastic and isotropic (von Mises). Since the tools are made of steel, the elastic properties initially considered were a Young's modulus of 210 GPa (E=210) and a Poisson ratio of 0.30. Nevertheless, in order to evaluate the influence of the tools stiffness, a 3 time lower value was considered for the tools Young's modulus, i.e. a value of 70 GPa (E= 70) was also considered. The constant value for the friction coefficient was adjusted to fit the maximum experimental drawing force. In this case, a slightly smaller value of μ =0.06 was selected for both values of stiffness of the deformable tools.

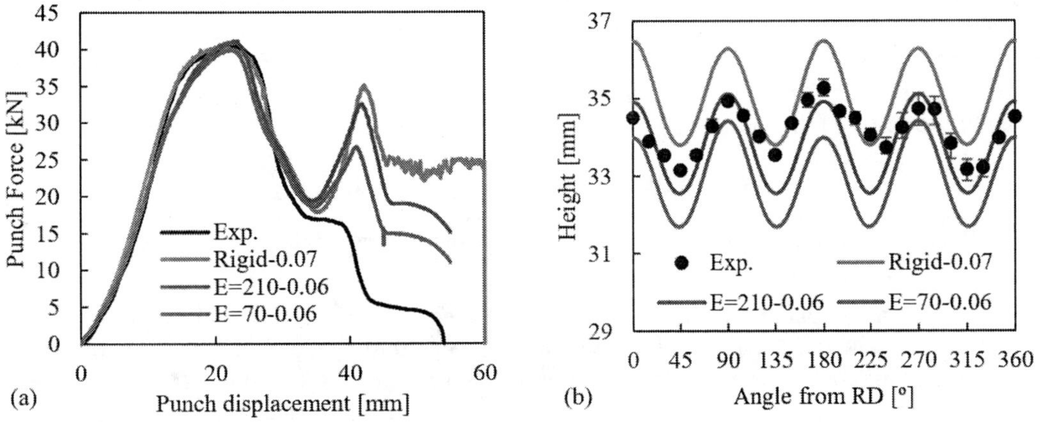

Figure 4. Comparison between experimental and predicted results using Hill'48 with rigid and deformable tools: (a) punch force-displacement and (b) earing profile.

Figure 4 (a) presents the comparison between the experimental and the numerical punch force evolution, confirming the small influence of the tools stiffness in the drawing stage. However, in the ironing stage, the punch force reduces with the decrease of the tools stiffness. The tools stiffness also has an impact on the cup height, as shown in Figure 4 (b). Note that the earing profile predicted with Hill'48 is similar to the experimental one, i.e. four ears with a minima at 45°. The average cup height decreases with the decrease of the tools stiffness. This is related with the fact that the magnitude of the friction force is proportional to the contact pressure. Thus, a lower tool stiffness generates lower friction forces, contributing to a smaller cup with higher thickness values for the same height of the cup, as shown in Figure 5. The maximum values of thickness for each height are always predicted at 45°, which correlate with the minima height (see Figure 4 (b)). The increase of the tools stiffness leads to lower values of thickness, along the vertical wall of the cup and, consequently, to a higher average height. Note that the impact of the tools stiffness when considering deformable tools is mainly observed on the top of the vertical wall of the cup. Nevertheless, as shown in Figure 4 (a) the ironing force is still overestimated. Therefore, in the following section the influence of the friction coefficient is analysed.

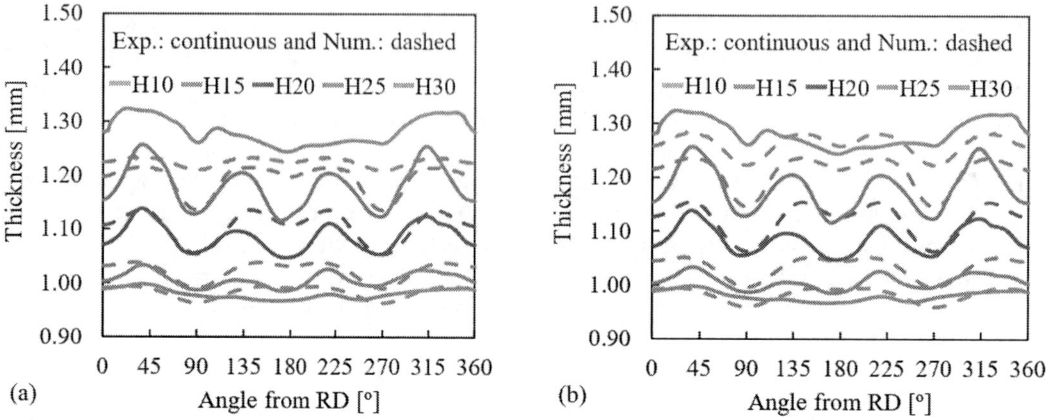

Figure 5. Comparison between experimental and predicted results using Hill'48 and deformable tools with μ=0.06, for the distribution of the thickness along the cup circumference at different heights (Experimental: continuous line and Numerical: dashed line): (a) E=210 GPa and (b) E=70 GPa.

3.2. Friction law

Numerical simulations were also performed with the deformable tools considering a null value for the friction coefficient, in order to evaluate the minimum value predicted for the punch force in the ironing stage. The analysis of the results, including the evolution of the contact pressure between the blank and the tools, enabled the definition of a pressure sensitive law for the friction coefficient. The friction law adopted is defined as follows:

$$\mu = B-(B-A)\exp\left(-mP^{n}\right)=0.1\exp\left(-0.01P\right) \tag{5}$$

where P is the contact pressure (see e.g. [8]). When there is no contact $\mu = A = 0.1$ while for high contact pressure values μ will tend to the null value.

Figure 6 (a) shows that the evolutional friction law enables a more accurate global description of the punch force for both the drawing and the ironing stage, due to the reduction of the friction value with the increase of the contact pressure. It is also possible to observe that the ironing force is also sensitive to the tools stiffness, presenting lower values for the less stiff tools. The analysis of the cup height, presented in Figure 6 (b) indicates that this variable is more sensitive to changes in the friction coefficient value for the stiffer tools. It is also interesting to note that a similar earing profile is predicted for the stiffer tools, with a variable friction coefficient, and the softer tools, with the higher value for the friction coefficient. Nevertheless, in the last case the ironing force is clearly overestimated. Figure 7 presents the distribution of the thickness along the cup circumference at different heights, when considering the deformable tools with a null friction coefficient. The comparison with Figure 5 shows that the average thickness increases with the decrease of the friction coefficient. This is particularly evident for the intermediate heights of 15 and 20 mm, which justifies the lower average height (see Figure 6 (b)). Moreover, the change in thickness is not uniform along the cup circumferential coordinate.

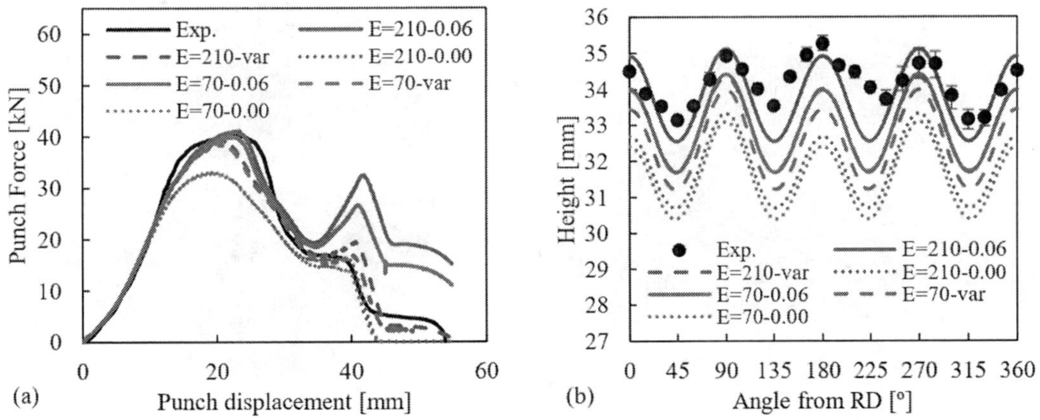

Figure 6. Comparison between experimental and predicted results using Hill'48, deformable tools and different values for the friction coefficient: (a) punch force-displacement and (b) earing profile.

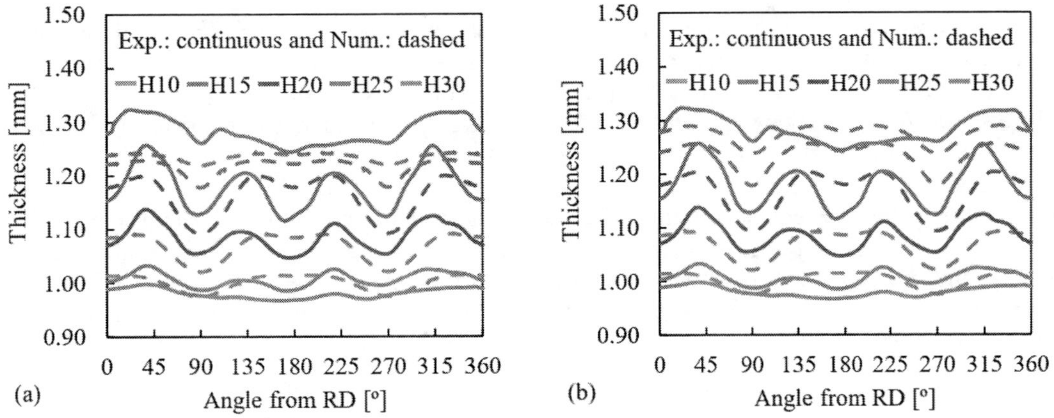

Figure 7. Comparison between experimental and predicted results using Hill'48, deformable tools and a null friction coefficient, for the evolution of the thickness along the cup circumference at different heights (Experimental: continuous line and Numerical: dashed line):
(a) E=210 GPa and (b) E=70 GPa.

Figure 8 (a) presents the contour plot of the norm of the radial nodal displacements in the less stiff tools, for a punch displacement of 40mm. Note that, according with the punch force evolution, this punch displacement corresponds to the ironing stage. Therefore, it is possible to see that the maximum deflection of the tools occurs in the plane (marked in red) corresponding to the transition between the die curvature radius and the vertical wall. The die presents a positive radial displacement while the punch presents a negative one, meaning that the gap between both tools increases. Figure 8 (b) compares the norm of the radial nodal displacements obtained in that plane, for both deformable tools and the two constant values of the friction coefficient. This confirms that the tools deflection decreases with the increase of the friction coefficient. Moreover, the tools deflection is higher between 0° and 45° to RD, i.e. where there is a greater thickening of the blank (see Figure 7), which is related with the lower *r*-value observed between 45° and 90° (see Figure 2 (a)). This also justifies the higher sensitivity of the thickness distribution along the circumferential direction to the friction value, for angles between 0° and 45° to RD.

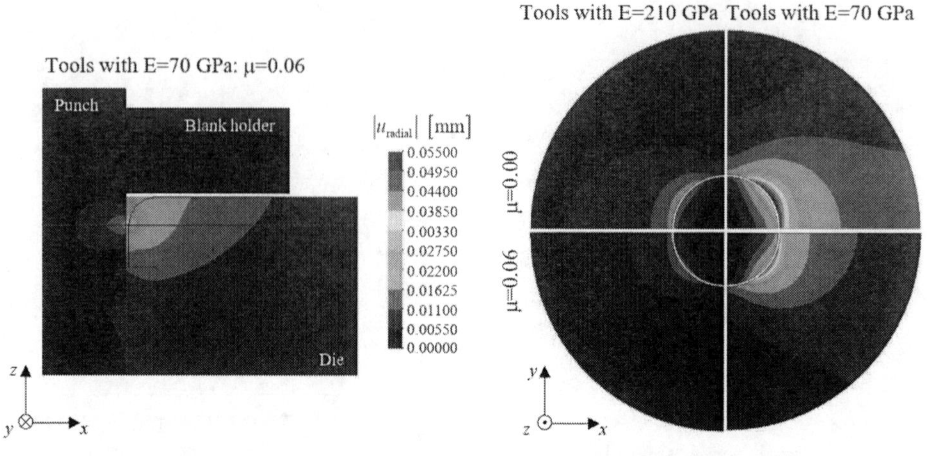

Figure 8. Contour plot of the norm of the radial nodal displacements [mm] in the tools for a punch displacement of 40 mm: (a) for the less stiff tools and (b) for both deformable tools and constant friction coefficient values, in the plane marked with the red line in (a).

Figure 9 presents the contour plot of the thickness strain, for a punch displacement of 35 mm, i.e. at the end of the drawing stage. Whatever the tools stiffness, for the higher friction coefficient value the thinning in the bottom of the cup is slightly higher while the thickening of the flange is slightly smaller. However, the thickness along the vertical wall of the cup is quite similar. This confirms the small impact of the tools stiffness in the drawing stage (see also Figure 4 (a)). The ironing of the vertical wall changes the thickness along the entire cup wall, even for relatively small ironing strains. The same was previously observed regarding the reduction in the residual stresses after ironing [9]. Although not shown here, the results obtained for the thickness distribution in the cup wall with the evolutional friction law are in between the ones reported for the two constant values of friction coefficient considered. This means that the average height of the cup is underestimated while the thickness values are overestimated, except for the highest height of 30 mm.

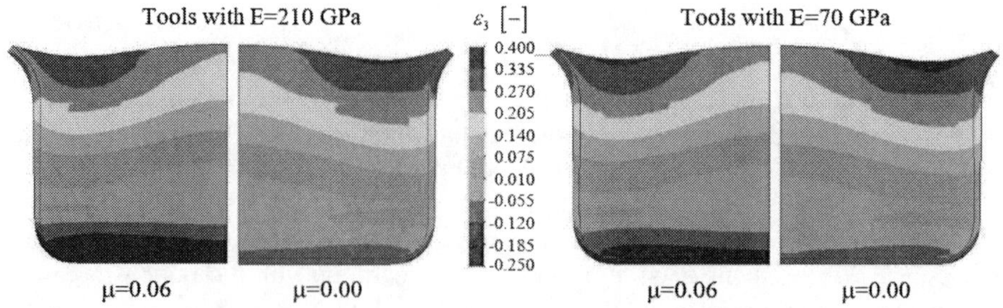

Figure 9. Contour plot of the thickness strain for a punch displacement of 35 mm.

3.3. *Yield criteria*

The thickness distribution is dictated by the normal to the yield surface since an associated flow rule is adopted. A numerical simulation was performed considering the stiffer deformable tools, the evolutional law for the friction coefficient and the CB2001 yield criterion. Figure 10 shows the evolution predicted for the punch force. The punch force predicted with the CB2001 is slightly lower than the one observed with the Hill'48, since the CB2001 predicts lower in-plane yield stress values (see Figure 2 (b)). In fact, the ironing force resembles more the one predicted with the less stiff tools. The same happens with the cup height at 0°, although at 90° the result is closer to the one obtained with the stiffer tools, as shown in Figure 10 (b). This means that the CB2001 predicts an earing amplitude (3.86 mm) slightly higher than the Hill'48 (2.61 mm); both overestimate the experimental value (2.29 mm).

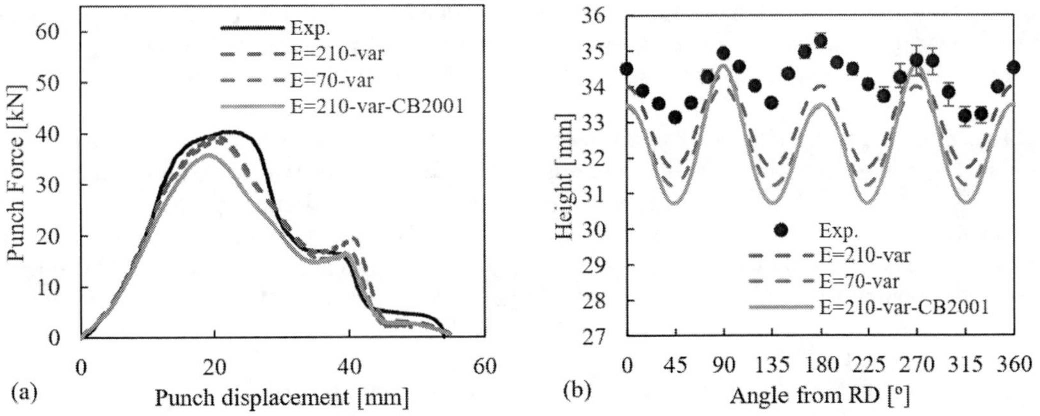

Figure 10. Comparison between experimental and predicted results considering deformable tools and an evolutional friction coefficient with the Hill'48 and the CB2001 yield criteria: (a) punch force-displacement and (b) earing profile.

Figure 11 (a) presents the distribution of the thickness along the cup circumference at different heights, predicted with the CB2001 yield criterion. The major differences with the values predicted with Hill'48 occur for angles between 45° and 90°. This results from the thickness distribution obtained at the end of the drawing stage, which presents lower thickening values for those directions, as shown in Figure 11 (b) (see also Figure 9). Note that the material initially located in the flange region is submitted to compression in the circumferential direction and tension in the radial one. Only for the material points in the free edge at RD and TD there will be a null component of the shear stress, meaning that the stress state is located in the compression-tension quadrant and the thickness strain is dictated by the normal to the yield surface in the region between uniaxial compression and pure shear stress states [10]. These corresponds to loading directions between 90° and 180°, which were not covered by the experimental tests performed, as shown in Figure 2 (d).

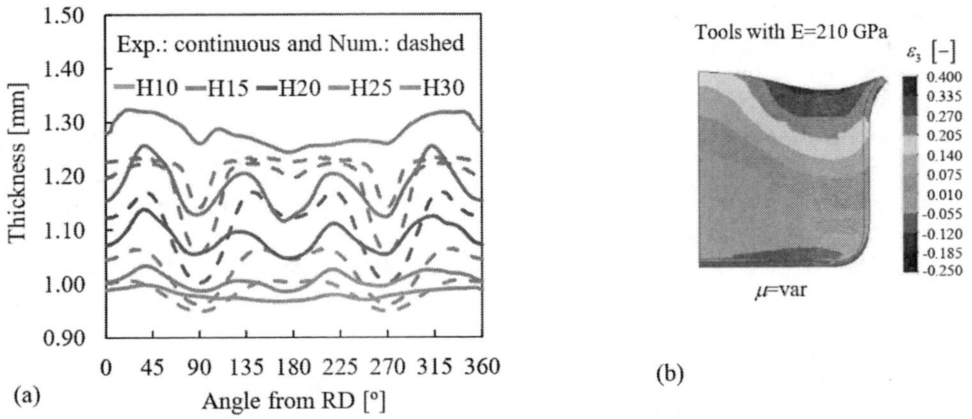

Figure 11. Comparison between experimental and predicted results considering deformable tools and an evolutional friction coefficient: (a) distribution of the thickness along the cup circumference at different heights with the Hill'48 and the CB2001 yield criteria (Experimental: continuous line and Numerical: dashed line) and (b) contour plot of the thickness strain for a punch displacement of 35 mm as predicted with the CB2001 yield criterion.

4. Conclusions

The analysis of the influence of the tools stiffness on the thickness distribution in a cylindrical cup wall was performed, showing that it affects not only the average value, at each height, but also the trend along the circumferential direction. This results from the orthotropic behaviour of the blank, which imposes a different thickening of the flange. Higher thickening values, associated with lower *r*-values, result in higher contact pressure values and, consequently, higher radial displacement of the tools. Nevertheless, the tool deflection is not enough to reduce the value of the punch force during the ironing stage. This force is quite sensitive to the friction coefficient, meaning that an evolutional friction law is required to enable capturing the drawing and ironing forces. The interaction between the tools stiffness and the friction law is highlighted by the similar earing profile predicted for the stiffer tools, with a variable friction coefficient, and the softer tools, with the higher value for the friction coefficient. This is related with similar thickness values excepted for the top of the cup. The thickness distribution at the end of the ironing stage is dictated by the one predicted at the end of the drawing stage, which requires improved knowledge about the direction of the plastic strain-rate in the compression-tension quadrant.

Acknowledgments

The authors gratefully acknowledge the financial support of the Portuguese Foundation for Science and Technology (FCT) under projects with reference PTDC/EME-EME/30592/2017 and PTDC/EME-EME/31657/2017 and by European Regional Development Fund (ERDF) through the Portugal 2020 program and the Centro 2020 Regional Operational Programme (CENTRO-01-0145-FEDER-031657) under the project UIDB/00285/2020.

References

[1] Neto D M, Coër J, Oliveira M C, Alves J L, Manach P Y and Menezes L F 2016 Numerical analysis on the elastic deformation of the tools in sheet metal forming processes *Int. J. Solids Struct.* **100** 270–85

[2] Vincze G and Habraken A 2021 EXACT - the ESAFORM Benchmark 2021 *ESAFORM 2021, 24th International Conference on Material Forming*

[3] Habraken A M, Toros A A, Alves J L, Amaral R L, Betaieb E, Chandola N, Corallo L, Cruz D J, Duchêne L, Engel B, Esener E, Firat M, Frohn-Sörensen P, Galán-López J, Ghiabakloo H, Kestens L, Lian J, Lingam R, Liu W, Ma J, Menezes L F, Nguyen-Minh T, Miranda S S, Neto D M, Pereira A F G, Prates P A, Reuter J, Revil-Baudard B, Rojas-Ulloa C, Sener B, Shen F, Van Bael A, Verleysen P, Barlat F, Cazacu O, Kuwabara T, Lopes A, Oliveira M C, Santos A D and Vincze G Analysis of ESAFORM 2021 cup drawing benchmark of an Al alloy , critical factors for accuracy and efficiency of FE simulations *Submitted to Int. J. Mater. Form.*

[4] Cazacu O and Barlat F 2001 Generalization of Drucker's yield criterion to orthotropy *Math. Mech. Solids* **6** 613–30

[5] Neto D M, Oliveira M C, Alves J L and Menezes L F 2015 Comparing faceted and smoothed tool surface descriptions in sheet metal forming simulation *Int. J. Mater. Form.* **8** 549–65

[6] Coër J, Laurent H, Oliveira M C, Manach P-Y and Menezes L F 2018 Detailed experimental and numerical analysis of a cylindrical cup deep drawing: Pros and cons of using solid-shell elements *Int. J. Mater. Form.* **11** 357–73

[7] Neto D M, Oliveira M C, Menezes L F and Alves J L 2016 A contact smoothing method for arbitrary surface meshes using Nagata patches *Comput. Methods Appl. Mech. Eng.* **299**

[8] Magny C 2002 Lois de frottement évolutives destinées à la simulation numérique de l'emboutissage *La Rev. Métallurgie* **99** 145–56

[9] Ragab M S and Orban H Z 2000 Effect of ironing on the residual stresses in deep drawn cups *J. Mater. Process. Technol.* **99** 54–61

[10] Oliveira M C, Cazacu O, Chandola N, Alves J L and Menezes L F 2021 On the effect of the ratio between the yield stresses in shear and in uniaxial tension on forming of isotropic materials *Mech. Res. Commun.* **114** 103693

IOP Conf. Series: Materials Science and Engineering 1238 (2022) 012080 doi:10.1088/1757-899X/1238/1/012080

A novel incremental sheet metal forming process for long and open section profiles

A Essa[1], B Abeyrathna[1], B Rolfe[2], and M Weiss[1]

[1] Institute for Frontier Materials, Deakin University, Waurn Ponds, Pigdons Rd., Geelong, VIC 3216, Australia
[2] School of Mechanical Engineering, Deakin University, Waurn Ponds, Pigdons Rd., Geelong, VIC 3216, Australia

E-mail: aessa@deakin.edu.au

Abstract. Complex components from High Strength Steel (HSS) can be formed with Flexible Roll Forming (FRF). However, flange wrinkling limits the part shape complexity that can be achieved and hence limits its application in the automotive industry. A novel forming technology is introduced in this study, Incremental Shape Rolling (ISR), where a pre-cut blank is clamped between a top and bottom die and then a single forming tool incrementally forms the material according to the bottom die's geometry. It is believed that ISR can significantly reduce the wrinkling issue compared to FRF. In the contrast to the incremental sheet forming process, ISR allows the manufacture of long and open sections from HSS sheets. In this work, the experimental prototyping trials of the ISR process are performed using a 5-axis milling machine and then are used to validate the FEA model which is developed to simulate and analyse the deformation behaviour of the ISR process. The results show a clear development of transverse tensile strain along the flange which can overcome the wrinkling issue when the ISR process is applied to complex profiles.

1. Introduction

High Strength Steel HSS has attracted a large amount of attention for use in the automotive industry for structural and crash components due to the high strength and low weight of parts made from such materials [1, 2]. However, forming of HSS with the conventional Flexible Roll Forming (FRF) comes up with flange wrinkling which occurs when the required longitudinal compressive strain in the flange edge exceeds the buckling limit [3] and this restricts the part shape complexity achievable with FRF [4, 5].

Recently, a prototype flexible roll forming facility has been developed, where the pre-cut blank is fully clamped between two dies. Then, a single forming roll stand incrementally bends up the pre-cut blank around the top die to obtain the required shape. The roll stand is placed on two hexapods on either side of the clamping dies, each hexapod has six degrees of freedom so that the forming tool can follow the part shape contour [6]. However, similar to the conventional FRF process, the results showed high potential for the occurrence of excessive wrinkling which limits the part shape complexity that can be achieved.

In the Incremental Sheet Forming (ISF) process, such as sheet metal spinning, the developed axial tensile stress along the flange balances the compressive hoop stress, hence, wrinkle-free components can be formed [7, 8]. However, warm forming techniques need to be implemented when ISF is used to form HSS [9]. In addition, due to the limited working range of the milling machines, the forming of long and open sections is not feasible with ISF [10].

In this paper, the ISR process will be introduced and applied to a U-profile to study potential forming issues and material deformation. In ISR, the pre-cut blank is clamped between two dies, and a single forming tool then incrementally forms the sheet to the desired shape. Because the flange is being wrapped around the forming tool in ISR, a tensile transverse strain is introduced along the flange which balances the compressive longitudinal strain similar to ISF, and hence allows the forming of wrinkle-free components with complex cross-sections. In this way, this novel study develops the platform for the understanding and development of ISR to form complex profiles from high strength materials in future work.

2. Material properties and profile geometry

An Instron 5967 with a 30 kN load cell was used to carry out the standard tensile tests. Three tensile specimens were tested at a test speed of 0.025 mm s^{-1} and the averaged true stress–true strain curve of the stainless steel specimens along the rolling direction is shown in Figure 1-a. The mechanical properties of the tested 0.45 mm ferritic stainless steel strip shown in Table 1 are obtained by fitting the Hollomon's power law of the tested specimens. The U-profile illustrated in Figure 1-b is selected for this study. The profile has a uniform cross-section along its length (Figure 1-c). Due to the shape symmetry, only one-half of the U-profile was addressed in this investigation.

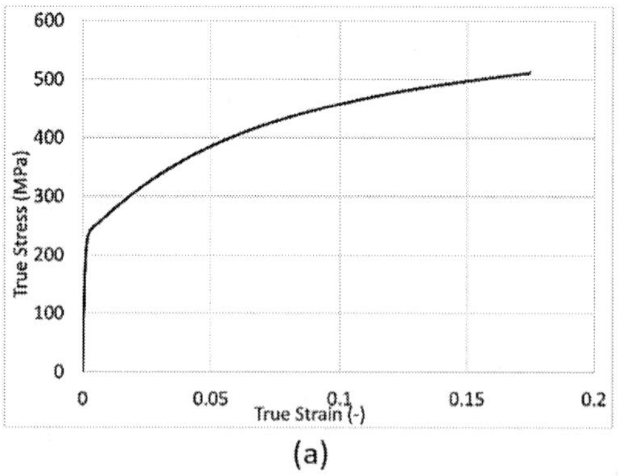

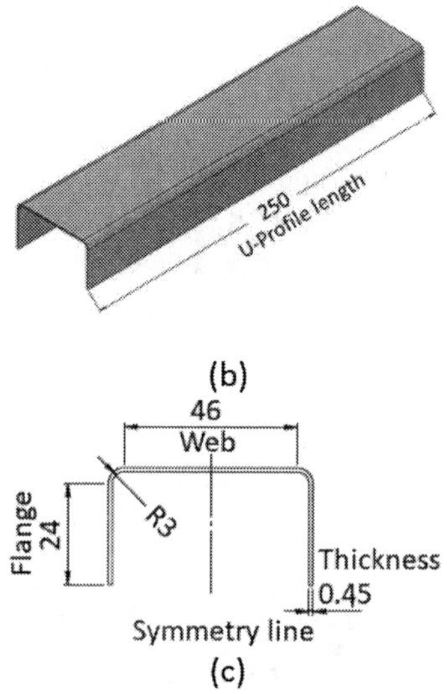

Figure 1 Averaged true stress-true strain curve of the stainless steel, (b) 3D view of the required profile and (c) the cross-section of the U-profile – dimensions in mm

Table 1 Tensile parameters

Material	Ferritic stainless steel		
Young's modulus (GPa)	202.02		
Poisson's ratio v	0.3 [11]		
The angle from the rolling direction	0°	45°	90°
Yield Strength (MPa)	243.03	242.68	244.33
K (MPa)	699.56	697.67	712.92
Strain hardening n	0.1868	0.1877	0.1856
Lankford parameters $r=\varepsilon_w/\varepsilon_t$	1.042	1.025	1.047
Planer anisotropy $\Delta R=(r_0-2r_{45}+r_{90})/2$	0.0195		

3. The concept of the ISR process

In the ISR process, the pre-cut blank is fully clamped between the top clamp and the bottom die, and one forming roll is initially placed at a distance from the bottom die equal to the blank thickness. The tool is then incrementally fed down in Y-direction (d_y) before it moved linearly in the longitudinal (Z) direction. The incremental vertical feed (d_y) and the linear longitudinal movement in Z-direction are repeated until the blank is formed to the required shape according to the geometry of the bottom die, see Figure 2. The experimental trials were done on a 5-axis milling machine where the die-set is placed on a milling table that provides the vertical (d_y) and linear longitudinal (Z) movements. The blank holding force is provided by five equally distributed bolts. The roll is mounted on a tool holder which is attached to the tilted spindle of the milling machine. Due to the symmetry of the shape, only one transversal half of the U-profile was formed, as shown in Figure 2.

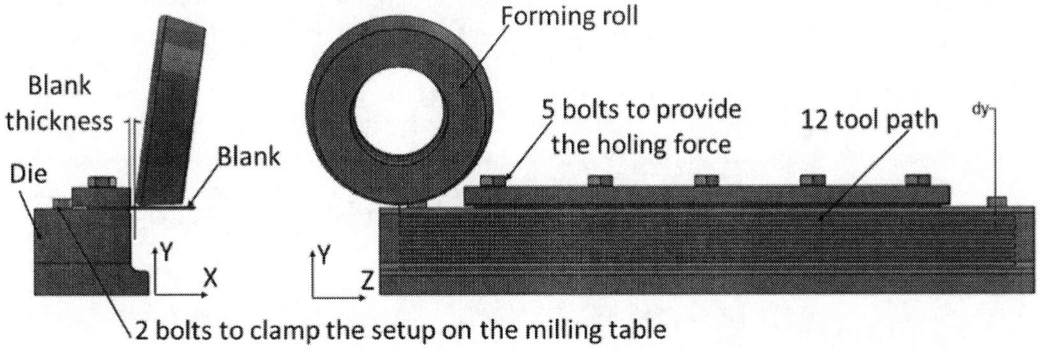

Figure 2 Schematic drawing of the experimental setup of the incremental shape rolling process

Figure 3 shows a schematic drawing for the forming sequence of the ISR process of the U-channel profile, where 12 forming steps are implemented with increment step size d_y equal to 2.35 mm. In this forming sequence, the flange is wrapped around the forming tool in each forming pass and incrementally rolled onto a forming die until the blank is formed to the desired shape according to the die's geometry. Thus, bending and stretching in the transverse direction are the main deformation mechanisms in the ISR process.

A strain gauge was glued to the centre length of the sheet and near to the blank edge to measure the transverse strain during the 1st forming pass.

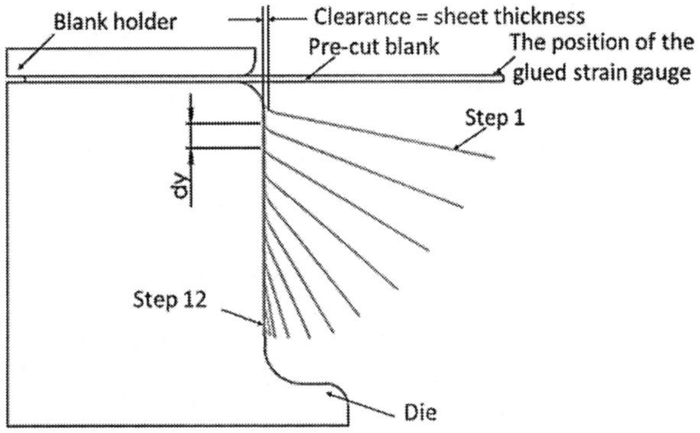

Figure 3 Schematic of the forming sequence of ISR

4. The finite element model

The finite element simulation of the ISR process was performed with Abaqus Implicit. The dies and the forming roll were modelled as rigid bodies, while the pre-cut blank has been discretized with reduced integration, hexahedral, linear brick elements. Four elements through the blank thickness were used with an aspect ratio of less than 10. Only the transversal half of the profile was modelled due to the symmetry of the shape about the Y-Z plane. A "frictionless contact" was assumed between the forming roll and the blank surfaces to avoid convergence issues [12]. To minimize the penetration of the rigid bodies into the blank surfaces at the constraint locations, the "hard contact condition" was applied [13]. As shown in Table 1, the stainless steel sheet showed isotropic material behaviour, hence, isotropic hardening and the von Mises yield criteria were used to define the plastic material behaviour of the stainless steel sheet [14, 15].

5. Shape analysis

The results illustrated that the formed flange deviates from the desired shape. The flange edge is not fully formed to 90° degrees, and this shape error starts close to the middle length of the flange, see Figure 4-b. Above the starting point of that shape error, no spring-back was observed after the forming tool lost contact with the sheet. Thus, instead, the maximum deviation from the desired shape in the X-direction (max. X deviation) that is generated at the flange edge is used to determine the severity of that shape error. In addition to that f intact defines the length of the well-formed flange length measured to the starting point of X deviation, see Figure 4-b. The ideal shape has the maximum value of f intact which is equal to the flange length, this means that the full flange length follows the required shape.

The formed profiles were scanned with a 3D laser scanner after they were removed from the tooling to be compared with the desired shape and the FEA results. The 2D shape analysis of the formed profiles has been done at three X-Y sections along the part' length, i.e., the section where the forming started (lead section), the tail section, and the centre of the formed part (middle section), see Figure 4-a.

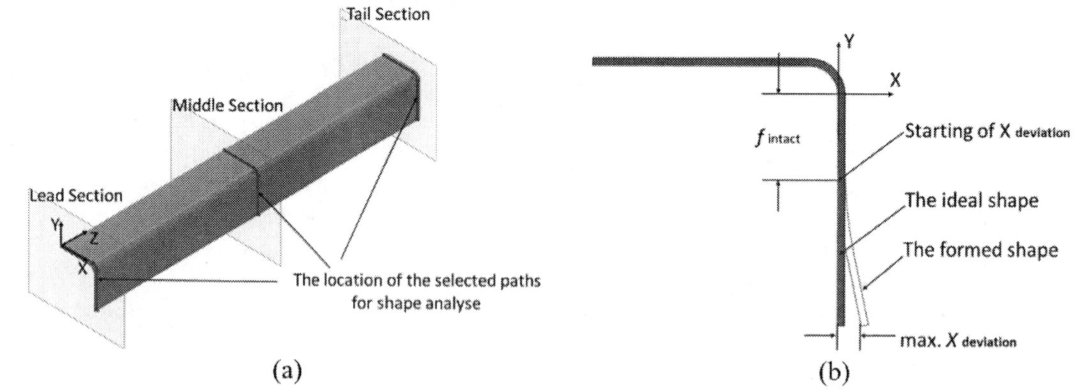

Figure 4 Schematic drawings show (a) the three X-Y sections along the part's length (b) the transverse shape deviation from the desired shape

6. Results and discussion

6.1. The shape quality in the longitudinal and transverse directions

All the ISR trials in this study were done with increment size d_y=2.35 mm i.e., 12 forming steps. Figure 5 shows the top view of the FEA and the experimental results of the formed profile which prove that the ISR process is capable to produce straight U-profiles with a high flange quality. Nevertheless, the experimental result shows a shape error in the longitudinal direction that the FEA cannot capture. Moreover, both the experimental and the FEA results demonstrate that there is a deviation between the formed flange and the ideal shape in the X-direction (the flange edge is not fully formed to 90° degree) and that shape error starts from different values of f_{intact}. The experimentally measured values and FEA results of the f_{intact} and the max. $X_{deviation}$ are shown in Figure 6. FEA appears to accurately represent

IOP Conf. Series: Materials Science and Engineering 1238 (2022) 012080 doi:10.1088/1757-899X/1238/1/012080

trends observed in the experimental data, however, it underestimates the magnitude of the generated shape deviations.

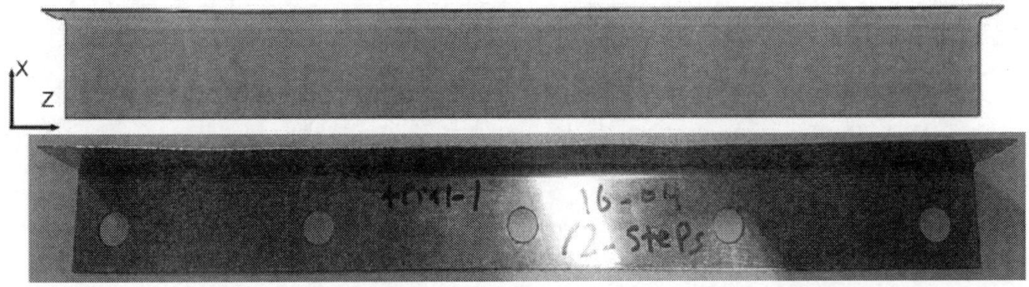

Figure 5 Top view of the FEA and experimental results show the formed profile after the final pass

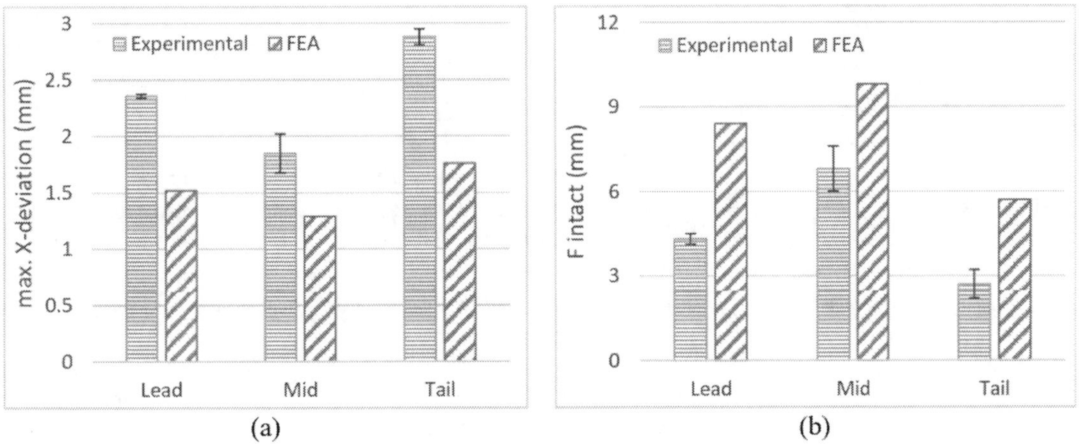

Figure 6 The values of (a) max. X deviation and (b) f intact measured after the last forming pass

6.2. *State of strain in the sheet edge*

Figure 7 shows the transverse strain (PE11) measured versus time during the 1st forming step. Note that the transverse strain component, PE11, is measured with respect to the local coordinate system which follows the deformed flange as shown in Figure 8-a, while X, Y, Z represent the global coordinate system. The experimental strain measurements are obtained using a strain gauge. Figure 7 shows that there is a compression transverse strain in front of the forming roll and this decreases as the forming roll gets closer to the attached strain gauge (at the middle of the strip). When the forming tool passes the location of the attached strain gauge (approximately 150 seconds) transverse strain reverses to transversal tension. It therefore can be concluded that the flange is first compressed in front of the roll and stretched behind the forming roll. However, the result shows near-zero permanent transverse strain, given that this strain reading was measured at the sheet edge (where the strain gauge is located).

In conclusion, even though the FEA model underestimates the cross-sectional shape error measurements, it reproduces the experimental trends in regard to the effect of material and process parameters on the final shape. This suggests that the FEA model is valid to be used for further investigation of the ISR process in this study.

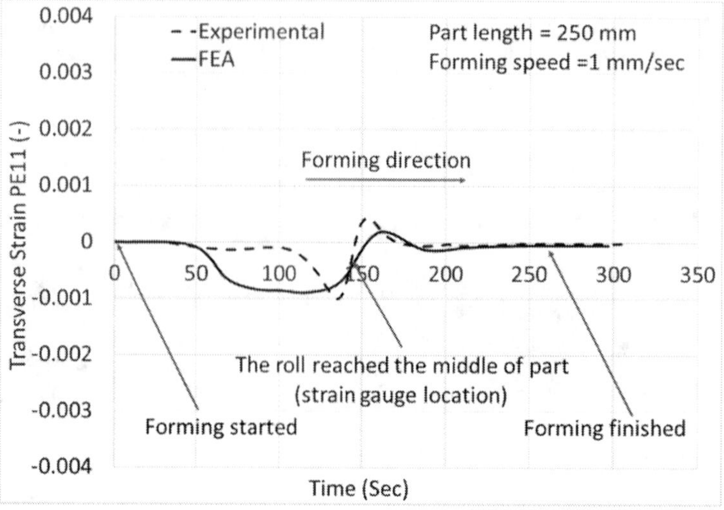

Figure 7 The history of the transverse strain during the 1st step

6.3. The reason for the shape error in the transverse direction

The results demonstrate that the flange edge is not fully formed to 90° and that there is a deviation from the ideal shape in the X-direction. This may be related to the level of the flange wrapping over the tool radius (see Figure 8-a). To prove this hypothesis, the transverse plastic strain is numerically analysed along the three cross-sections after the final forming pass (Figure 8-b). This shows that the maximum transverse strain occurred just under the profile corner which is formed in the earlier forming passes where a higher level of sheet wrapping over the tool is developed. The transverse strain reduces towards the flange edge and reached near-zero transverse strain at the flange edge; this is due to the low level of sheet deformation by the tool in the later forming steps. This suggests that the plastic transverse strain that is formed into the flange decreases in the later stages of the forming passes, and that this results in $X_{deviation}$. The small level of transverse plastic strain at the flange edge shown in Figure 8-b is confirmed with the experimental transverse strain result shown in Figure 7 which shows near-zero permanent transverse deformation, given that the strain gauge was located at the strip edge.

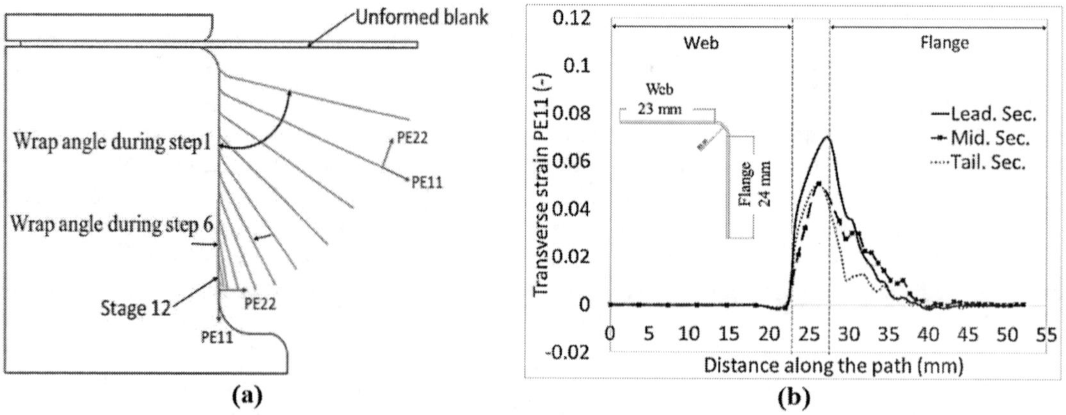

Figure 8 (a) Schematic shows the level of sheet wrapping over the tool during different forming passes (b) FEA results of the transverse plastic strain along the three cross-sections after the 12th forming step

It is believed that the permanent stretching in the transverse direction can overcome the wrinkling issues when ISR is applied to complex profiles. This argument is based on previous studies that have shown that flange wrinkling is significantly reduced when a transverse stretch is introduced into the flange [8]. Therefore, we can say that ISR may provide an alternative forming process that can form wrinkle-free complex components that would be wrinkled if produced with the FRF process [6].

7. Conclusion

- A simplified straight profile was successfully formed with the incremental shape rolling process.
- The result shows a clear development of permanent transverse stretching along the flange.
- It is believed that the transverse stretching will overcome the wrinkling issue when the ISR process is applied to complex profiles with variable cross-sections over the part length.
- The smaller level of flange wrapping over the tool in the later forming passes leads to a lower permanent transverse deformation, hence, the $X_{deviation}$ increases towards the flange edge with a decrease in the transverse deformation, and reaches its maximum value "max. $X_{deviation}$" at the flange edge where the transverse deformation is almost zero.

Acknowledgements
The authors would like to thank Deakin University Postgraduate Research Scholarships (DUPRS) for the provided scholarship and financial support and acknowledge data M Sheet Metal Solutions GmbH for the developed 3D Rollforming Center.

References

[1] Baluch N, Udin Z M and Abdullah C S 2014 "Advanced high strength steel in auto industry: an overview" *Eng. Tech. Appl. Sci. Res.* **4** 686-90.
[2] Sun G, Deng M, Zheng G and Li Q 2019 "Design for cost performance of crashworthy structures made of high strength steel" *Thin-Walled Struct.* **138** 458-72.
[3] Kasaei M M, Naeini H M, Abbaszadeh B, Mohammadi M, Ghodsi M, Kiuchi M, Zolghadr R, Liaghat G, Tafti R A and Tehrani M S 2014 "Flange wrinkling in flexible roll forming Process" *Procedia Eng.* **81** 245-50.
[4] Kasaei M M, Naeini H M, Liaghat G H, Silva C M A, Silva M B and Martins P A F 2015 "Revisiting the wrinkling limits in flexible roll forming" *J. Strain Anal. Eng. Des.* **50** 528-41.
[5] Groche P, Zettler A, Berner S and Schneider G 2010 "Development and verification of a one-step-model for the design of flexible roll formed parts" *Int. J. Mater. Form.* **4** 371-77.
[6] Abeyrathna B, Ghanei S, Rolfe B, Taube R and Weiss M 2021 "Optimising part quality in the flexible roll forming of an automotive component" *Int. J. Adv. Manuf. Technol.* **118** 3361-73.
[7] Music O, Allwood J M and Kawai K 2010 "A review of the mechanics of metal spinning" *J. Mater. Process. Technol.* **210** 3-23.
[8] Jia Z, Li L, Han Z R, Fan Z J and Liu B M 2018 "Experimental study on wrinkle suppressing in multi-pass drawing spinning of 304 stainless steel cylinder" *Int. J. Adv. Manuf. Technol.* **100** 111-16.
[9] Kumar A, Gulati V, Kumar P and Singh H 2019 "Forming force in incremental sheet forming: a comparative analysis of the state of the art" *J. Braz. Soc. Mech. Sci. Eng.* **41**.
[10] Behera A K, de Sousa R A, Ingarao G and Oleksik V 2017 "Single point incremental forming: An assessment of the progress and technology trends from 2005 to 2015" *J. Manuf. Process.* **27** 37-62.
[11] Ellobody E and Young B 2005 "Structural performance of cold-formed high strength stainless steel columns" *J. Constr. Steel Res.* **61** 1631-49.
[12] Weiss M, Abeyrathna B, Gangoda D S, Mendiguren J and Wolfkamp H 2017 "Bending behaviour and oil canning in roll forming a steel channel" *Int. J. Adv. Manuf. Technol.* **91** 2875-84.
[13] Woo Y Y, Han S W, Oh I Y and Moon Y H 2019 "Shape defects in the flexible roll forming of automotive parts" *Int. J. Automot. Technol.* **20** 227-36.

[14] Abeyrathna B, Rolfe B, Pan L, Ge R and Weiss M 2016 "Flexible roll forming of an automotive component with variable depth" *Adv. Mater. Process. Technol.* **2** 527-38.
[15] Abeyrathna B, Rolfe B, Harrasser J, Sedlmaier A, Rui G, Pan L and Weiss M 2017 "Prototyping of automotive components with variable width and depth" *36th IDDRG Conf.* Munich, Germany.

IOP Conf. Series: Materials Science and Engineering 1238 (2022) 012081 doi:10.1088/1757-899X/1238/1/012081

Quantitative Surface Quality Assessment of Car Outer Panels with a Virtual Light Room

Ing. Frédéric Guillon, Ing. Arthur Camanho, (ESI Group, France & USA)
fgu@esi-group.com

aca@esi-group.com

Ing. Renan Padovani (AUTAZA, Brazil)

Ing. Jorge Gripp (AUTAZA, Brazil)

Dr. Carlos Sakuramoto (AEA, Brazil)

Abstract. Perception is reality. A customers' perception about the quality of a car is closely linked to the aspect of its outer panels and that reality translates into what motivates customers to choose one car over another. The smoothness of the panels, how much gap is visible and how that looks; these visual cues are considerations within a customer's decision-making process. The process of choosing a new car is certainly subjective and varies from customer to customer, but Original Equipment Manufacturers (OEMs) agree that the visual appeal of the body of a car is a key factor. Building a car only to find surface imperfections results in huge costs in postproduction rectifying those perceived defects. The further into production these defects are found the more expensive it is to fix them. Some OEMs release cars into the buying cycle knowing that their engineering teams are still struggling to fix surface defects. There is a solution to this though. Numerical simulation. This action can and should be used to find surface defects very early in the design process. Surface quality assessment via numerical simulation is achieved by analysing different contours that relate stress or strain values. Other subjective criteria exist, but it is often very difficult to ascertain if a surface defect is severe enough to be detected by human eye. This paper describes a simulation methodology that, when used along with a software that reads surface light distortions and rates them, will give reliable results allowing engineers to fix problems before a car is built. This technology was developed using actual data supplied by a partnering OEM. Current examples will be presented, comparing parts "as designed" with parts "as manufactured". After manufacturing parts will show variations caused by the stamping process such as skid marks, bumps, hollows, etc. All that can be accurately predicted by simulation.

1. Introduction

Over the past 20 years stamping simulation has proven to be accurate enough to well predict the outcome of almost all existing sheet metal forming processes. Initially we were able to check formability, looking mainly at thinning, wrinkling and Forming Limit Diagrams. Few years after that the bar was set higher by the industry who was looking for more complex results, such as springback. Materials got more complex requiring new and more advanced models. One of the more complex requirements is the ability to well predict surface defects, especially on car outer panels in loose and mounted condition, and doors & closures as presented in 2015 by G. Prajescu, and D. Schlatter at Forming in car body engineering in Bad Nauheim [6].

Traditionally, stamping simulations are analysed via some colour mapping on the part related to certain "contours" that allow to identify the potential issues on the parts. This methodology, which has been proved to be very efficient for the most common issues (thinning, plastic strain level, springback magnitude, distance to the targeted geometry…) has been evaluated also for checking the severity of the cosmetic defects such as dips and bumps. But, as presented by G. Prajescu, D. Schlatter and B. Changeux at NAFEMS France in Paris in 2016 [1], this methodology isn't robust enough for cosmetic defects evaluation for two main reasons:

- No single contour (or formula combining several contours) that would be extracted from simulation and that would correlate with the quality assessments done in real life has been identified so far.
- The evaluation of the severity of the defect is highly subjective because is driven by human perception, which takes into consideration not only the defect itself, but also its position on the car body (example: a defect that would be in a zone where no light reflexion would happen wouldn't be seen by the final customer). Therefore, evaluating a colour mapping may underestimate, or oppositely overestimate, the defect severity, which would lead to respectively the late discovery of some defects, or to over engineering parts and/or process to fight defects that would never be perceived by the final customer.

Under these circumstances, some alternative solutions have been searched and digital stoning solutions have been developed and are available in industrial software (figure 1). These new ways of evaluating the cosmetic defects, aiming at reproducing on simulated parts one of the tests (stoning) done in real life, have been industrially proven to reflect the reality (figure 2) and to allow identifying some zones of "risk of defects". This methodology was still not allowing users to know accurately if the final customer would see those defects, and even less if they would damage the "feeling of high quality" of the entire vehicle.

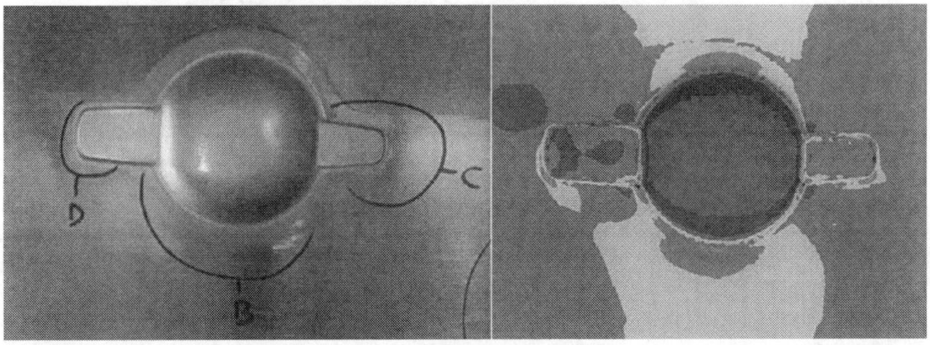

Figure 1: comparison between real and digital stoning based on a numerical stamping simulation.

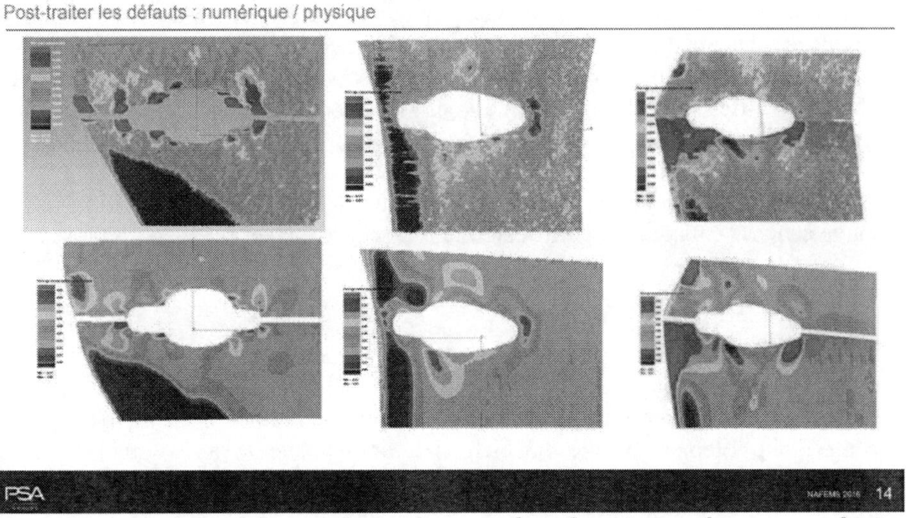

Figure 2: comparison of three real parts between digital stoning made on scanned parts (top row) and simulated parts (bottom row) – PSA Groupe, NAFEMS 2016.

Based on that status, the complete "cosmetic defects evaluation process" used at several OEMs has been analysed and it has been observed that the first step of this process was not the stoning, but rather a visual inspection performed by a trained individual that, after positioning the part on a device (in a position that would correspond to its real position on the car, such as a door is placed vertical and a roof horizontal…) in front of a light room, looks at the part and rates the potential defects according to an established severity criterion. Only then, a stoning is performed at the location of each identified defect to understand the root causes so that some action plans can be engaged to correct those defects.

Based on this refined understanding of the real cosmetic defects' evaluation process, some trials have been engaged to reproduce this visual inspection inside a virtual lightroom, in which a simulated part was positioned (figure 3).

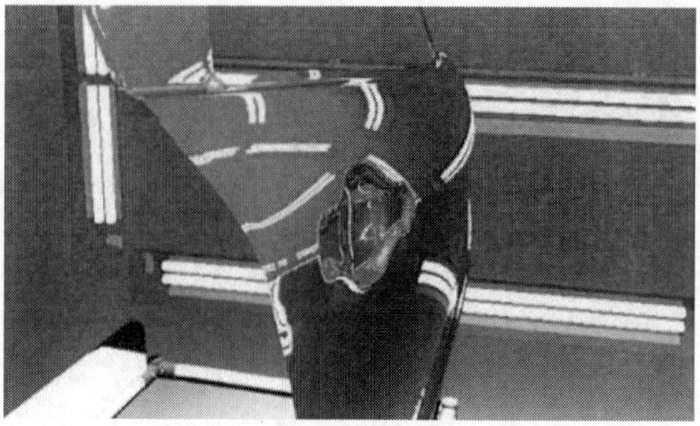

Figure 3: Example of a simulated door (zoom on the door handle area) inside a virtual light room.

This new approach allowed to quickly prove that if a proper simulation methodology was used (chapter 2), the worker was able to better acknowledge the severity of the defects, but then, the issue of subjectivity became more present, exactly like in real life where two different workers are sometimes giving quite different defects estimation, in the real light room like in the virtual one.

To solve this situation, some tests of automatic defect severity evaluation based on image analysis coupled with artificial intelligence have been engaged.

2. Finite Element Method Simulation Methodology

Several work steps are required to allow an accurate prediction of the part geometry using Finite Element Method, which is a prerequisite to any virtual light room inspection:

- Tool meshing

A detailed tool mesh is required to minimize the impact of the discretization of the tool geometry on the blank. Advanced meshing parameters are used in a dedicated template to ensure that the mesh is at the right size and with the right density. A larger mesh, even localized, could result in fake marks in the blank that will be visible in the virtual light room.

- Blank initial mesh and refinement during the calculation

Since the size of the defect that must be captured during the simulation is rather small (figure 4), a fine blank mesh must be obtained at the end of the forming simulation (before springback during where no further refinement will occur). For this purpose, and to save computation time, an initial mesh size of 12mm is chosen, coupled with 5 levels of refinement. These settings will result in a minimum element size at the end of the complete simulation of 0.75mm which is still compatible with the shell element formulation used during the computation.

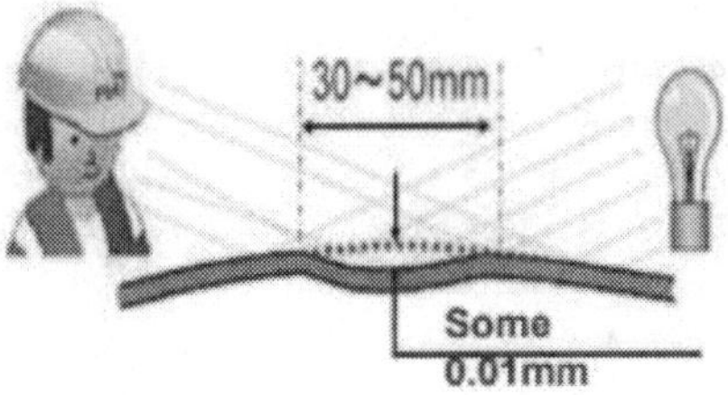

Figure 4: Typical size of a cosmetic defect around a door handle [2] justifying the requirement of a very fine mesh at the end of the stamping simulation

- Forming speed reduction at the end of the forming stage

To avoid numerical artefacts associated to a too high forming velocity, where the blank would excessively oscillate at the very last instants of the forming stroke, and as in reality where the forming speed is progressively reducing to zero near the bottom dead point, the velocity of the forming tools is reduced to a low value few millimetres before the reach of the Bottom Dead Point (BDP):

- Until 20mm to BDP, velocity is constant and set to 5000mm/sec.
- Between 20mm and 10mm to BDP, a linear ramp from 5000mm/sec to 1000mm/sec is applied.
- From 10mm to 0mm to BDP, velocity is constant again and set to 1000mm/sec.

- Springback

At the end of the forming, the part is in good agreement with the tool's geometry. To reveal the dips and bumps that will appear when the tools will be reopened, a springback stage is added after the forming stage, which will correspond to the tool's reopening happening at the end of the forming stroke (figure 5).

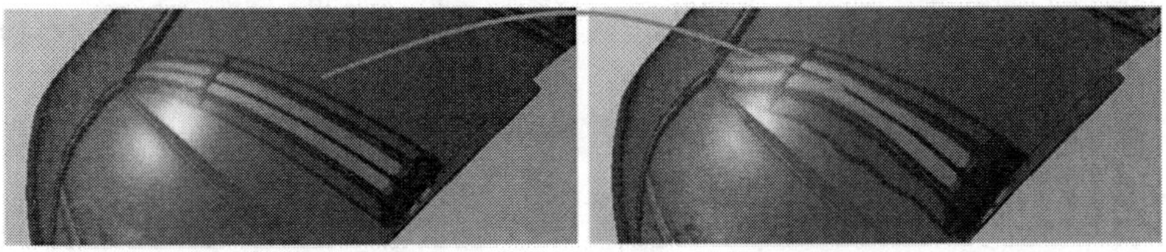

Figure 5: Reflection in the light room, before and after springback [3]

An accurate springback prediction is very important as most surface defects will show up or get worse in a sprung part. A wrong springback prediction may lead to defects that do not exist or defects that have a different magnitude than the actual ones.

To ensure a good springback calculation some advanced material models are being used. Since the deformation path in some zones of the part may correspond to a succession of traction and compression, a YOSHIDA-UEMORI kinematic hardening model is used.

In case where some additional forming stages (corresponding to other trimming and/or forming operations) would be required, some intermediate springback between each operation must be added, coupled with gravity stage at the beginning of the next forming operation to properly represent the real and full forming process.

To ensure a fast-enough turnaround time for refined models, the computation can be performed on several cores in parallel (typically 16 or 32 cores).

- Realistic Rendering

Along with a good springback result, the quality of results rendering is of the same importance. A bad rendering may change the perception of existing defects.

Most stamping simulation software work with a finite element mesh, which essentially do not produce a smooth curved surface. A mesh with big elements of varying sizes has a wavy reflection when viewed in a virtual light room. The virtual light room is a 3D environment with computer graphics functions able to simulate the reflection on the mesh surface. It considers that the angle between the incident ray and the mesh element normal is equal to the angle of the reflected ray and the mesh element normal.

Having a fine enough mesh to allow for a smoother surface and a better reflection would result in too many elements and unrealistic calculation times using today's available hardware. Because of that, a specific methodology was developed and fine-tuned to allow best fringe reflection results with a reasonable mesh size.

To create the reflection image, the computer graphics software traces back the rays arriving at the camera, reflecting on the surface, and exiting the lighting. This surface mesh is obtained by simulating a homogeneous mesh with final element size around blank thickness. The size of the final polygonal mesh elements is homogeneous and in a small size, such that the image does not present waviness in the reflections.

- Assembly Simulation

During assembly processes, panel shapes may change and that can affect the surface quality. As demonstrated by H. PORZNER at IABC in Dearborn in 2018 [5], the assembly process, considering positioning, closing, hemming, spot-welding, gluing, etc. can make the surface defect look less visible or even more important.

The assembly simulation has then two main advantages: The first one is to understand, beforehand, if after assembly the part will be within the expected tolerances. Maybe after assembly many of the component level deviations will get fixed, so all the time and cost used to fix each component was lost. Secondly, in terms of surface defects, the same may occur as defects may get better or worse after assembly (figure 6).

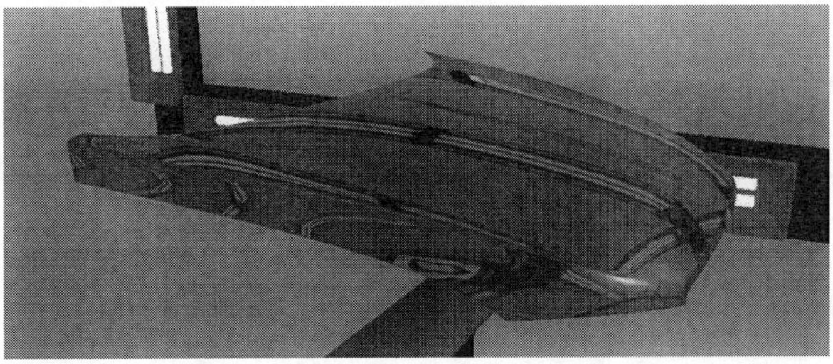

Figure 6: Assembled hood in the virtual light room (Courtesy of Ford) [4]

3. Light Fringes and Defects Classification

When a software can correctly predict the shape of a part or assembly after springback, the light reflections seen in a light room is again subjective : some might say that one part looks good and agrees with the expected quality standards, some might say that part is not acceptable [7]. The challenge is now to classify the defects automatically, using an algorithm that could represent what most of the auditors would say about parts in an actual light room.

To resolve these problems, a software solution was developed using actual data supplied by a partnering OEM. The software can read light fringe reflections (Figure 7) and rate them based on a standard criterium (Figure 8), and it can be customized to fit different criteria used by any OEM. The software is based on artificial intelligence techniques, so the more parts are analysed the more precise and smart the software becomes.

Figure 7: Fringe light reflections

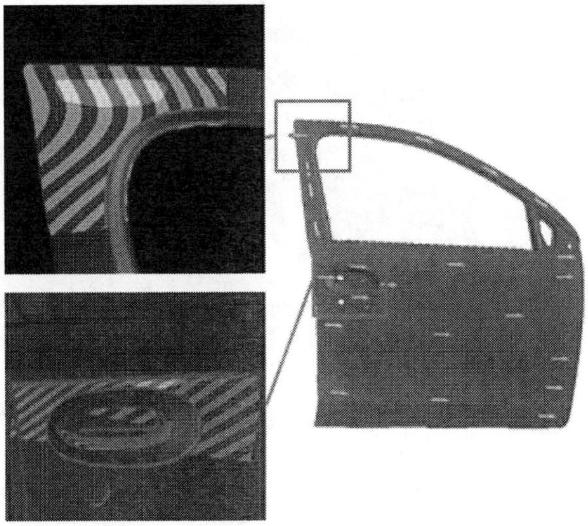

Figure 8: Defects classifier example

IOP Conf. Series: Materials Science and Engineering 1238 (2022) 012081 doi:10.1088/1757-899X/1238/1/012081

The camera and lighting system is always positioned at a standard size and distance from the surface (Figure 9), allowing for standardized image capture. Thus, it is possible to obtain a smooth and realistic reflection on the surface with: i) a realistic simulation of the physical phenomena in the mesh, ii) a mesh with a uniform size of the elements, iii) a mesh with an element size smaller than the spatial resolution of the camera rendering, iv) a set of lighting patterns with high contrast, v) a lighting with standardized size, distance, and angle, and vi) a virtual camera that captures images with standardized resolution, size, distance, and angle.

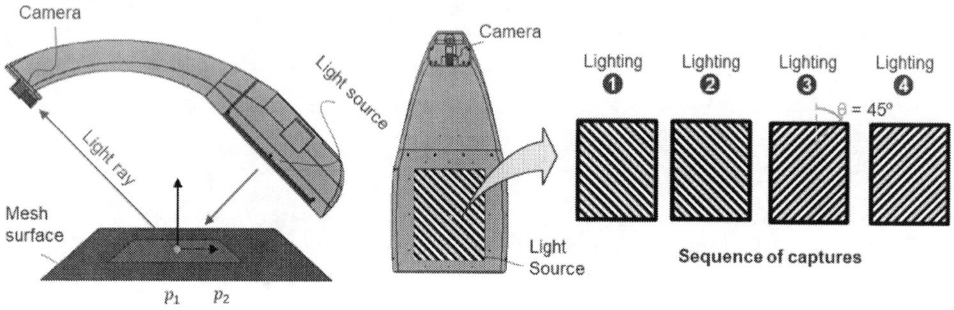

Figure 9: Standard camera and lighting position

In terms of accuracy and validation against actual measurements, the software shows a very good agreement as shown hereunder (Figure 10 and 11).

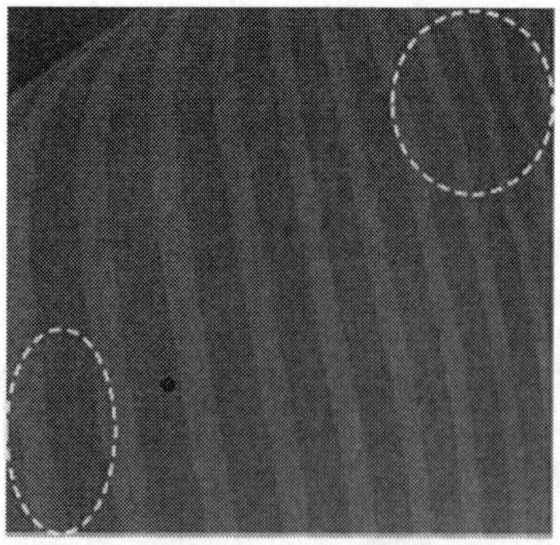

Figure 10: Defects spotted by inspector

Figure 11: Defects spotted by software

4. Conclusion

By using a stamping simulation methodology that can produce reliable results with good turnaround time, one is able to provide surface defects assessment virtually in a light room, in the same manner OEMs check their cars. The rendering quality from stamping simulation results is good enough to enable light fringes reflection analysis, from which one can classify the defects automatically using artificial intelligence. The system used to classify the defects can learn by itself and better detect defects the more parts are analysed. The defects classification is automatic and does not require a calibration phase.

A potential next step to further increase accuracy of the shape predicted by the simulation software could be to ease the consideration of the impact of tools deformation during the forming stage, which can induce variation on the contact pressure between the tool and blank, leading to slight modifications of the panel shape, and thus on the perceived quality level, as presented by M. Vrolijk, T. Ogawa, A. Camanho, M. Biasutti and D. Lorenz at ESAFORM 2018 in Palermo [8].

5. References

[1] Simulation et cotation de défauts d'aspect sur pièces embouties automobiles, G. Prajescu, D. Schlatter, B. Changeux, NAFEMS France, Paris, 2016

[2] Prediction and compensation of surface deflection using PAM-STAMP, T. Ohtsuka, ESI Global Forum, 2010

[3] IABC 2017 Dearborn, MI "PAM-STAMP Assembly Technologies for Process Design in Sheet Metal Forming", Biasutti, Manfredi (ESI GmbH)

[4] ACC 2018 Livonia, MI "Single Component Fabrication, Assembly and Light Room Inspection for Doors and Closures", Liasi, Evangelos (Ford)

[5] IABC 2018 Dearborn, MI "Remove uncertainty from try-out and enable virtual light room inspection for large flexible outer panels", Porzner, Harald (ESI North America)

[6] Cosmetic defect inspection through digital simulation, G. Prajescu, D. Schlatter, Forming in car body engineering, Bad Nauheim, 2015

[7] IABC 2019 Livonia, MI "Quantitative Surface Quality Assessment of Car Outer Panels with a Virtual Light Room", Camanho, Porzner (ESI Group), Padovani (Autaza), Sakuramoto (AEA).

[8] ESAFORM 2018 Palermo, Italy. "A study with ESI PAM-STAMP on the influence of tool deformation on final part quality during a forming process.", Vrolijk, Ogawa, Camanho, Biasutti, Lorenz (ESI Group)

Advanced part design method for springback minimization of stamped sheet metal car body components

A Birkert[1*], B Hartmann[2], M Nowack[1], A Petker[1], M Scholle[1], P Zimmermann[1] and T. Kraft[2]

[1] Centre for Metal Forming and Car Body Manufacturing, Faculty of Mechanics and Electronics, Heilbronn University of Applied Sciences, Heilbronn, Germany

[2] inigence gmbh, Bretzfeld, Germany

*E-mail: arndt.birkert@hs-heilbronn.de

Abstract. The requirements for high-strength and low-weight car body structures prove challenging for the stamping process and cause growing problems with the elastic springback. Compensation approaches such as the geometrical adjustment of the stamping dies by the amount of and in opposite direction to the springback can be an effective solution in many cases. However, this approach can only lead to satisfactory results if the scatter of the springback is considered, controlled and in the best case reduced. The reduction of the scatter can be provoked by adaption of the part design. Therefore, different methods such as the topology optimization method, the so-called line-of-force method and shape optimization can be applied. In the present work an advanced topology optimization method is suggested. Based on the decomposition of the springback deformation into membrane and bending components, the most efficient optimization of the part design is found relying on unconventional topology optimization. For the proposed adapted topology optimization approach the excellent results are also confirmed when applied to the industry-relevant case of a truck rear panel.

1 Introduction

In automotive industry, the stamping process of complex lightweight car body components of high stiffness is considerably complicated by the occurrence of springback effects compromising the process robustness and the dimensional accuracy of the resulting parts. In order to address this problem several compensation approaches have been developed and are used today. The first and most widely known compensation approach can be described as tool geometry-based and relies on the idea to adjust the stamping dies in the opposite direction of the springback; thus, causing the sprung-back part to approximately take on the desired geometry [1]. The second approach has the goal to minimize the springback by optimizing the forming process itself, i.e. to modify the draw-in behavior of the draw shell by suitable selection of the stamping parameters [2]. The third approach is part-geometry-based and focuses on slight modifications of the part design, realized by introduction of stiffening beads, in order to locally manipulate the part stiffness in a desired way. The optimal part geometry with regard to a minimum amount of springback can depend on the shape of the stiffening elements, their orientation and position [3]. However, in industry it is still common practice to derive such modifications of the part design according to experience or on a trial-and-error basis leading to time-consuming iteration processes and suboptimal results. In [4] the authors have shown, that the springback of a round hat profile could be reduced significantly by inserting an additional bead into this simple geometry.

Therefore, three novel simulation-based approaches for a suitable modification of structural component design have been suggested in [5]. In the same paper, for demonstration purposes, the part design of a truck rear panel has been optimized by placing predefined stiffening beads in those areas of the part indicated by the method of unconventional topology optimization. It should be mentioned that the beads have not been placed directly within the detected areas – as described in the paper – but slightly outside of them in order to avoid deforming areas of small radius. Nevertheless, a forming simulation based on the optimized geometry revealed the springback of the truck rear panel to be reduced by 86% from 10.5 mm to 1.5 mm. Even though the unconventional topology optimization method indicates where on the part stiffening elements should be introduced the question of the precise shape and orientation of the stiffening elements remains.

In the present paper an advanced part design method for springback minimization of stamped sheet metal car body components is explained in detail and applied to and evaluated for a truck rear panel. The method involves a split of the springback into its membrane and bending components, Sec. 2, the advanced usage of the topology optimization method based on these bending and membrane components in Sec. 3 and the introduction of stiffening bead elements based on the calculated stress distribution in Sec. 4. For the advanced topology optimization approach excellent results have been achieved in the industry-relevant case of a truck rear panel.

2 Calculation of the membrane and bending components of springback

To exemplify the advanced topology optimization method a truck rear panel is employed using a bake hardening steel HX180BDbb with 0.8 mm sheet thickness. In a preliminary forming simulation, hereafter referred to as the base simulation, the springback of the non-optimized part is determined which for the truck rear panel is up to 10.5 mm measured in normal direction, see Fig. 1.

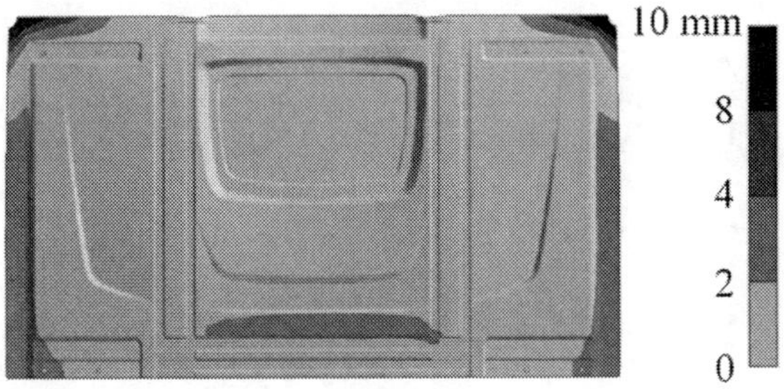

Figure 1. Truck rear panel with the calculated springback of the base simulation [5].

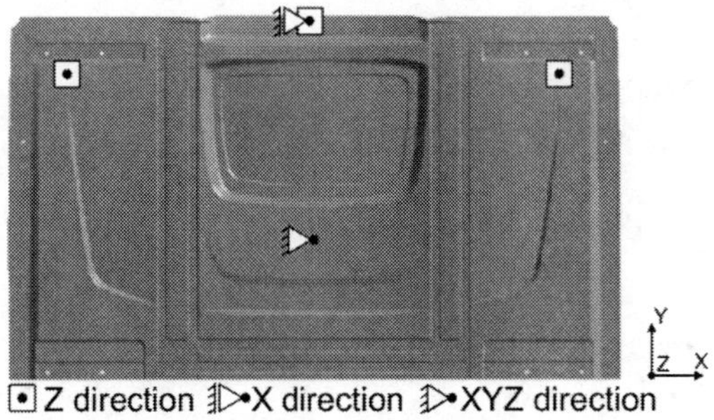

Figure 2. Boundary conditions to fix translational and rotational degrees of freedom during simulation.

The *advanced topology optimization method (ATO)* applies the classical *topology optimization method (TO)* where the simulated springback is taken as predefined load case. However, the simulated springback is not taken as a whole but rather the bending and membrane components of springback are used separately; this is the reason why the calculation of separate bending and membrane springback is considered in the following.

The bending springback is calculated using combined application of the commercial tools *AutoForm* and *Ansys Workbench*. First, the forming simulation and the calculation of the springback is performed in *AutoForm* using a multilayer shell element [6]. The calculated stress state is subsequently, for reasons of calculation flexibility, transformed to and approximated by a simple SHELL181 element used in *Ansys Workbench* [7]. This allows the element-wise separation of the stress tensor into bending and membrane stresses. The results are calculated in the element coordinate system. The shell bending stress tensor (σ_{11}^b, σ_{22}^b, σ_{12}^b) represents the linear variation portion of the in-plane stress tensor (σ_{11}, σ_{22}, σ_{12}) along the shell thickness direction, equation (1) [8]:

$$\sigma_{11}^{b} = \frac{6}{t^2}\int_0^t \sigma_{11}\left(\frac{t}{2} - z\right) dz, \qquad \sigma_{22}^{b} = \frac{6}{t^2}\int_0^t \sigma_{22}\left(\frac{t}{2} - z\right) dz, \qquad \sigma_{12}^{b} = \frac{6}{t^2}\int_0^t \sigma_{12}\left(\frac{t}{2} - z\right) dz, \quad (1)$$

where:

t is the total shell thickness, and

z is the position in thickness direction where the in-plane stress is evaluated.

The bending springback is obtained from linear elastic FEM by using the calculated bending stresses as a predefined load case for the nominal part geometry (Fig. 3), while translational and rotational degrees of freedom are fixed as shown in Fig. 2.

Based on the same procedure, the membrane springback is calculated from the membrane stresses. The shell membrane stress tensor (σ_{11}^{m}, σ_{22}^{m}, σ_{12}^{m}) is defined as the average of the in-plane stress tensor (σ_{11}, σ_{22}, σ_{12}) along the shell thickness direction, equation (2) [8]:

$$\sigma_{11}^{m} = \frac{1}{t}\int_0^t \sigma_{11}(z)dz, \qquad \sigma_{22}^{m} = \frac{1}{t}\int_0^t \sigma_{22}(z)dz, \qquad \sigma_{12}^{m} = \frac{1}{t}\int_0^t \sigma_{12}(z)dz. \quad (2)$$

where:

t is the total shell thickness, and

z is the position in thickness direction where the in-plane stress is evaluated.

Now, the calculated membrane stresses are used as a predefined load case and, for this purpose, are mapped on the nominal part geometry. Figure 4 shows the absolute values of the calculated membrane springback.

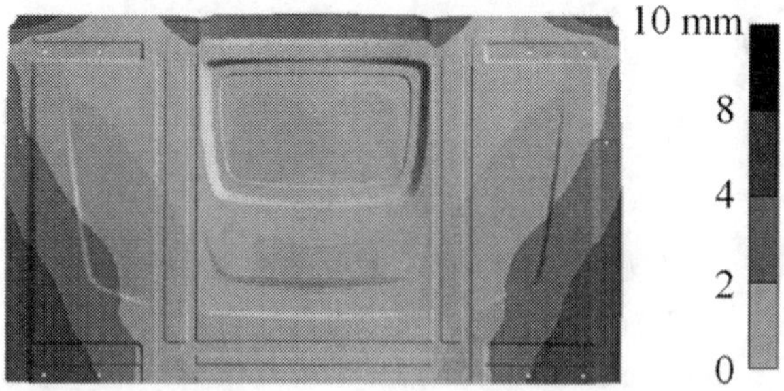

Figure 3. Truck rear panel with the calculated bending springback.

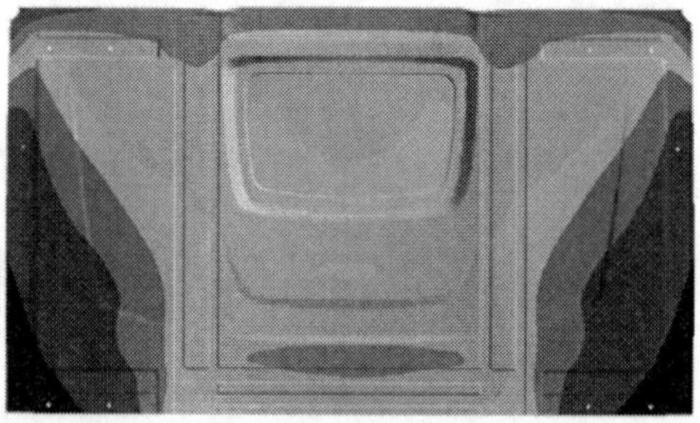

Figure 4. Truck rear panel with the calculated membrane springback.

3 Advanced usage of topology optimization method

The common application of the *TO* is the stiffness optimization of the structural component layout for a specified set of boundary conditions, loads and constraints. In practice, this means that for a given initial part the mass density is locally reduced in regions that do not contribute to the part stiffness significantly until a user defined threshold for the total mass is reached.

The aim of the present paper is to finalize the part geometry based on the simulated bending and membrane springback. The results are used to identify areas a 'stiffening' of which potentially reduces the respective component of the springback notably. Stiffening is achieved by the introduction of stiffening beads. The split of the springback into bending and membrane components reveals the part areas where to place the stiffening beads and enables the definition of the direction of the stiffening beads as well.

IOP Conf. Series: Materials Science and Engineering 1238 (2022) 012082 doi:10.1088/1757-899X/1238/1/012082

3.1 Identifying areas for stiffening beads to reduce bending springback

In a first step, stiffening areas for a potential reduction of bending springback are identified. For this purpose, a mechanical load has been defined based on the nodal displacements of the finite element simulation using *Ansys Workbench.* The respective nodal displacements refer to the calculated bending springback of the part. The target of the optimization has been the minimization of the compliance of the part. Hereby two mass thresholds have been defined, one being 5% and the other 20% of the original value. These percentages have been chosen, first, to identify the areas for the most efficient reduction of bending springback as precisely as possible (5% threshold, small areas), and second, to have an additional design proposal if the areas obtained with 5% are too small to allow for a feasible positioning of stiffening beads (20% threshold, larger areas). The *TO* has been realized in 19 (5% threshold) and 12 (20% threshold) iterations. Fig. 5 shows the resulting density distribution of a possible design proposal by applying the nodal displacements of the bending springback and with this results the location of potential stiffening bead areas to reduce bending springback are visualized.

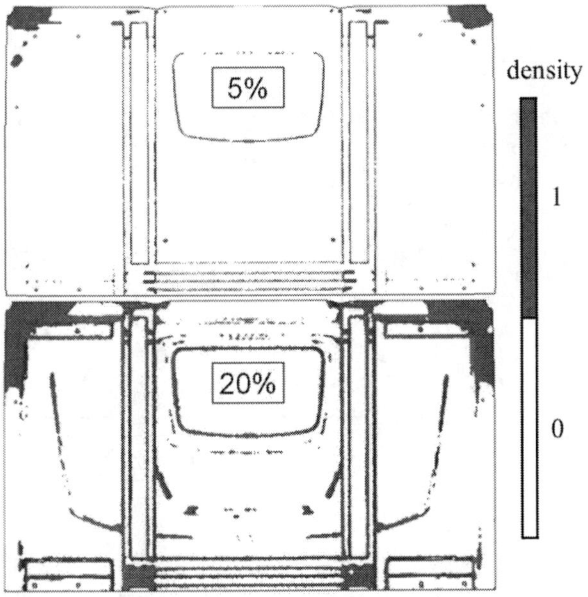

Figure 5. Design proposal for the truck rear panel for reduction of bending springback based on topology optimization.

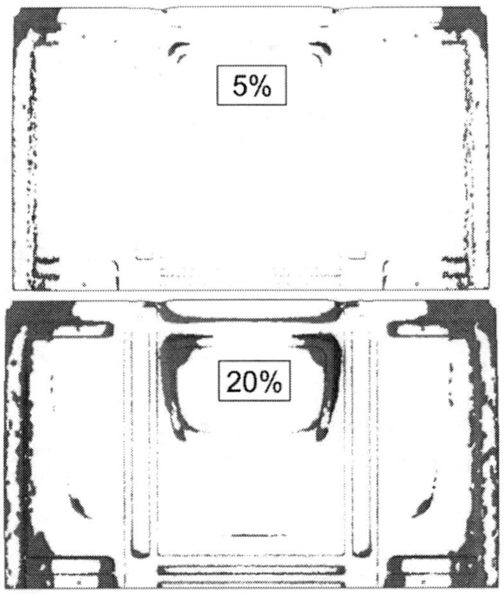

Figure 6. Design proposal for the truck rear panel for reduction of membrane springback based on topology optimization.

3.2 Identifying areas for stiffening beads to reduce membrane springback

In a second step the design proposal for the truck rear panel for the reduction of membrane springback has been calculated. Now, the nodal displacements referring to the membrane springback of the part are used to define the mechanical load. As before, the target of the optimization has been the minimization of the compliance and the mass has been defined to be 5% and 20% of the original value. The *TO* has been realized in 29 (5% threshold) or rather 9 (20% threshold) iterations. Fig. 6 shows the resulting density distribution of a possible design proposal by applying the nodal displacements of the membrane springback and with this results the location of potential stiffening bead areas to reduce membrane springback are visualized.

4 Introduction of stiffening bead elements

In the previous chapter the identification of areas for the potential reduction of membrane and bending springback has been presented. Although areas for the introduction of stiffening beads have been detected, the optimal orientation of the beads remains unclear. In the following the workflow for the definition of the orientation of stiffening beads is described.

4.1 *Stiffening beads for reduction of the bending springback*

The orientation of stiffening beads within the calculated design proposal based on *TO* can have a huge impact on the resulting springback but has not been discussed in [6]. The split of the springback in the present paper allows the clear definition of the orientation of the beads. To define the orientation with the goal to reduce bending springback the direction is defined in accordance with the direction of the *Principal Bending Moment*, which has been calculated using *AutoForm*. This variable gives the element-wise direction of the maximum and the minimum bending moment resulting from bending stresses. The calculation of the *related bending moment* (related by length) in a given direction is calculated based on equation (3) in *AutoForm* [9]:

$$M_a = \int_h \sigma_a n dn, \tag{3}$$

where:

a is a given direction,
h is the element thickness, and
n is the sheet normal direction.

The method of calculation of the direction of the maximum *related bending moment* is not specified in the *Autoform* manual [9]. However, it is assumed to be based on the eigenvectors of the major bending stresses in accordance with *Krauss* [10]. Fig. 6 shows the directions of the maximum and minimum bending moments from *AutoForm* within the resulting design proposal from *TO* in a selected area of the part.

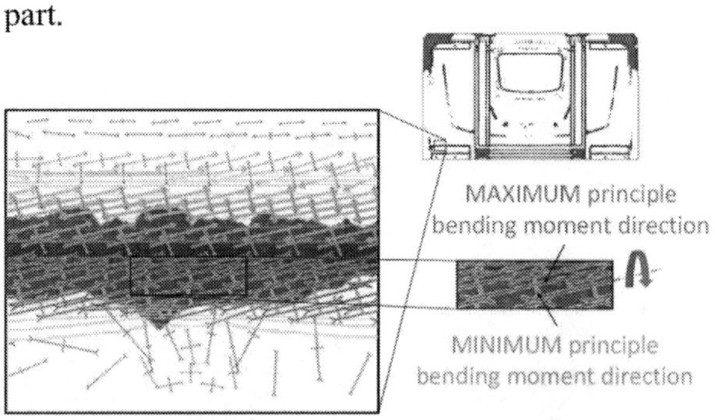

Figure 7. *Principal bending moment* within the resulting design proposal in a selected area of the part.

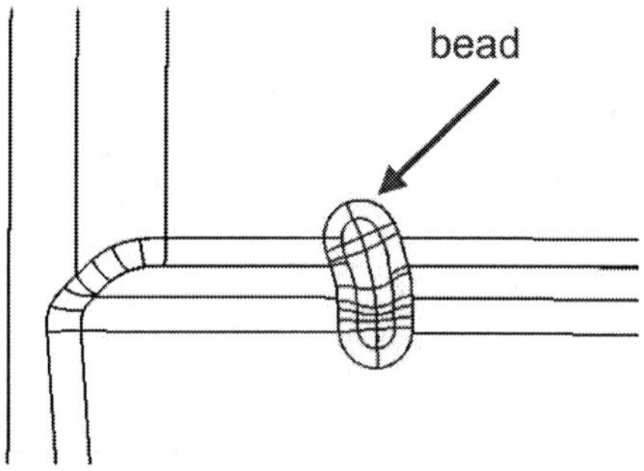

Figure 8. Adjusted CAD-model of the truck rear panel for reduction of bending springback.

Based on the resulting design proposal and the direction of the *Principal Bending Moment* stiffening beads have been defined. Figure 8 shows the adjusted CAD model optimized for a reduced bending springback.

4.2 *Stiffening beads for reduction of the membrane springback*

To define the orientation of the beads with the goal to reduce the membrane springback the direction of the *Principle In-Plane Force* has been calculated using *AutoForm*. This variable gives the direction of the maximum and minimum force in plane direction as a result of membrane stresses. The calculation of the *related in-plane forces* (related by length) in a given direction is calculated based on equation (4) in *AutoForm* [9]:

$$N_a = \int_h \sigma_a dn, \tag{4}$$

where:

a is a given direction,

h is the element thickness, and

n is the sheet normal direction.

The method of calculation of the direction of the maximum *related in-plane forces* is not defined in [9]. It is assumed to be based on the eigenvectors of the major membrane stresses in accordance with *Krauss* [10]. Fig. 9 shows the variable *Principle In-Plane Force Dir* in *AutoForm* within the resulting design proposal of the *TO*.

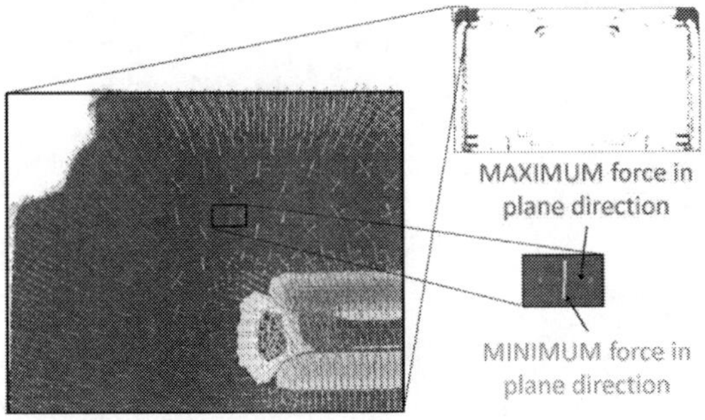

Figure 9. *Principle In-Plane Force Dir* from *AutoForm* within the resulting design proposal.

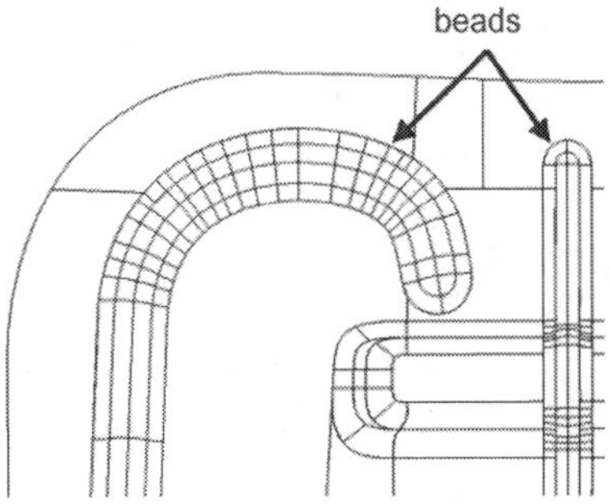

Figure 10. Adjusted CAD-model of the truck rear panel for reduction of membrane springback.

Based on the design proposal of the *TO* and the *Principle In-Plane Force Dir* the CAD model has been adjusted by implementing predefined beads within the detected area in the calculated direction. The form of the stiffening beads and the positioning is shown in Fig. 10. It has to be mentioned that for reasons of feasibility some stiffening beads are placed slightly outside of the area indicated by the design proposal.

A forming simulation based on the optimized geometry reveals that the springback of the truck rear panel could be reduced by 65.7% from 10.5 mm to 3.6 mm. Fig. 11 shows the final adjusted CAD-model of the truck rear panel and Fig. 12 shows the resulting springback.

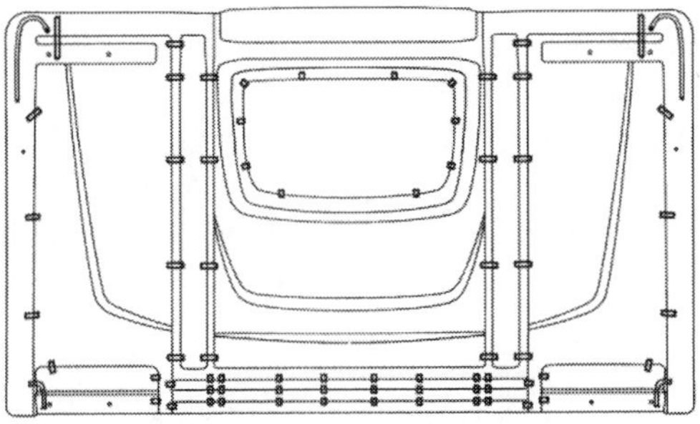

Figure 11. Adjusted CAD model of the truck rear panel optimized for a reduced bending and membrane springback.

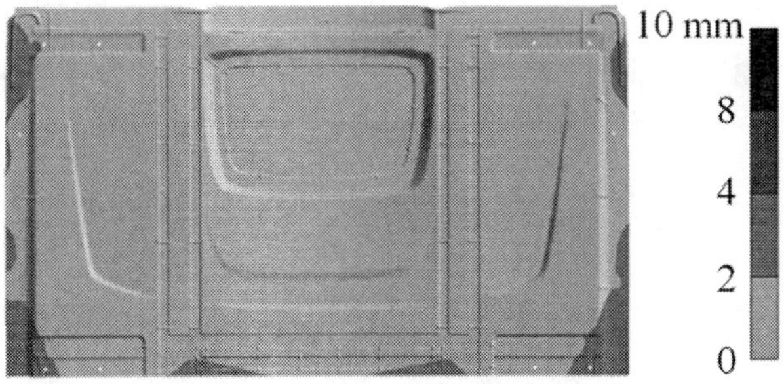

Figure 12. Truck rear panel with the calculated springback based on the optimized geometry.

5 Discussion and Conclusion

In the present paper a simulation-based approach for the optimization of car body components with the goal to minimize elastic springback after the forming process has been suggested. The method is based on classical topology optimization. However, to define the required set of boundary conditions, loads and constraints, the elastic springback has been split into its bending and membrane components. Subsequently, each of the calculated components have been defined as predefined load cases for the *TO*. This resulted in two different mass density distributions regarding the bending and membrane springback. In contrast to the classical topology optimization the mass density distributions have not been used to receive an optimized design proposal but rather to identify areas of the car body component in which an increase of local stiffness by adding stiffening beads reduces the springback most likely.

The key idea of the advanced *TO* is to use the membrane and bending springback independently of each other. This enables not only to identify areas of the car body component in which a local increase of stiffness is the most effective but also provides information on how the respective stiffening beads have to be oriented. The orientation is defined based on the maximum bending moments and the maximum in-plane forces resulting from the forming simulation in *AutoForm* while the tools are closed. For the proposed adapted topology optimization approach excellent results could be achieved when applied to the industry-relevant case of a truck rear panel. The springback could be reduced by 65.7% from 10.5 mm to 3.6 mm.

Acknowledgements

The authors would like to thank the German Federal Ministry for Economic Affairs and Energy for the financial support within the framework of this ZIM project.

References

[1] Birkert A, Hartmann B and Straub M 2017 New method for springback compensation for the stamping of sheet metal components *J. Phys.: Conf. Ser.* **896** 012067
[2] Radonjic R and Liewald M 2016 Approaches for springback reduction when forming ultra high-strength sheet metals *IOP Conf. Ser.: Mater. Sci. Eng.* **159** 012028
[3] Oehler G and Garbers F 1968 Untersuchung der Steifigkeit und Tragfähigkeit von Sicken (Investigation of the stiffness and load capacity of beads) *Forschungsberichte des Landes Nordrhein-Westfalen* Nr. 1917 (Köln and Obladen: Westdeutscher Verlag) DOI 10.1007/978-3-663-07144-0
[4] Gösling M, Kracker H, Brosius A, Kuhnt S and Tekkaya A E 2011 Strategies for springback compensation regarding process robustness *Prod. Eng. Res. Devel.* **5** 49–57
[5] Birkert A, Dreiseitel F, Hartmann B, Held T, Hetterle O, Markin M and Scholle M 2021 Adapted part design methods for springback minimization of stamped sheet metal car body components *IOP Conf. Ser.: Mater. Sci. Eng.* ***1157*** *012038*
[6] AutoForm R8 Software Manual 2018 Elements
[7] Ansys 2022 Element Reference page 673, Release 2022 R1
[8] Ansys 2022 Mechanical User's Guide page 1955, Release 2022 R1
[9] AutoForm R8 Software Manual 2018 Directional Variables
[10] Krauß A 2017 *Optimierte Sickengestaltung im Konstruktionsprozess dünnwandiger Bauteile (Optimized bead creation within the construction process of thin-walled parts)* (Chemnitz: University of Technology Chemnitz, Dissertation)

Advanced Springback Compensation Strategy through elimination of avoidable elastic strain energy

A Birkert[1*], P Zimmermann[1], B Hartmann[2] and M Nowack[1]

[1] Center for Metal Forming and Car Body Manufacturing, Faculty of Mechanics and Electronics, Heilbronn University of Applied Sciences, Heilbronn, Germany

[2] inigence gmbh, Bretzfeld, Germany

*E-mail: arndt.birkert@hs-heilbronn.de

Abstract. The stamping of dimensionally accurate sheet metal body components still represents a huge challenge in the automotive industry. This is partly caused by the multi-stage production process which complicates the design of an appropriate compensation strategy dramatically. A key prerequisite of an appropriate compensation strategy is to eliminate any undesired influence of elastic energy being potentially induced when closing the blankholders in the several operations throughout the production process. Here, a universally applicable compensation strategy is presented which fulfils this requirement thoroughly; by applying this process, the involved strain energy is reduced to a minimum due to proper part position in all operations. Technically this compensation strategy is achieved by, first, simulating all operations of the stamping process individually, second, by individually calculating the springback after each operation and, third, by accumulating the calculated deviation vector fields for each operation appropriately, which are then used for springback compensation. The process is time and cost effective and the required efforts are moderate even for complicated multistage stamping operations.

1. Introduction

During the opening of the tools in the manufacturing process of stamped parts the part geometry is changing due to the release of elastic energy. This phenomenon is called springback and leads to undesired dimensional deviations between the produced part and the target geometry. If the respective parts are out of dimensional tolerance, subsequent quality issues and difficulties in the assembly process of the parts result. To reduce the dimensional deviations caused by springback, a common approach is the so-called *geometrical springback compensation*. Here, the geometries of the stamping dies are modified in the opposite direction of the springback. The geometries modified in this way will be called *compensated tools* in the following.

In order to determine the compensated tools based on the springback result and the dimensional deviations, respectively, there are different approaches referred to as *compensation methods*. Well known compensation methods are, for example, the *Displacement Adjustment Method* [1] or the *Force Descriptor Method* [2], mainly differing in the way how a displacement rule for the compensated tools is generated from the springback results.

Most body parts are manufactured in several consecutive forming operations and, consequently, after each operation different springback results are obtained. The *compensation strategy* defines which operation of a multistage manufacturing process has to be compensated and which springback result is used. The decision for a compensation strategy is quite often based on practical experience and often requires several trial-and-error loops. The main goal of the compensation process is to obtain produced parts within the required dimensional tolerances which can, however, be achieved with very different compensation strategies. In addition, also the part position in each operation should be stable in order to guarantee a robust stamping process; an important condition being neglected in many compensation strategies. Besides the mentioned aims, also the required time and cost efforts have to be considered when assessing a compensation strategy.

In this paper an *Advanced Springback Compensation Strategy* is presented which delivers produced parts within given dimensional tolerances while a stable and reproducible part position is guaranteed. Nevertheless, the effort to apply the new method is small compared to other strategies.

2. Review of currently used compensation strategies

Due to the multistage character of common stamping processes in car body manufacturing and due to the impact of every individual stamping operation on the final springback result, *Roll et al.* [3] recommend the independent compensation of each stamping operation based on their respective springback. Just in cases where this approach fails, a different strategy should be used.

Birkert et al. [4] describe different such compensation strategies and compare their advantages and drawbacks. Subsequently, the most common strategies for practical application are summarized. One popular compensation strategy uses the deviations at the end of the process chain to compensate the drawing operation only. This strategy is very popular because only one operation has to be modified and thus the effort of the overall compensation process is very small in comparison to other strategies. However, it is very unlikely that this strategy works for components with complex manufacturing processes.

Another popular compensation strategy contains the compensation of all stamping and trimming operations, each based on the deviations at the end of the process chain. This means that the amount of compensation is the same for every single operation. The advantage of this strategy is the relatively small compensation effort due to the fact that (more or less) the same compensation vector field can be used for all operations. The third worth mentioning strategy is the *Drawshell-Compensation.* This strategy is mainly used for the compensation of trimming operations and is applied to ensure a proper positioning of the incoming part and to avoid unwanted plastic deformations. To adapt the tools of the trimming operation to the incoming part, the calculated or measured springback before the respective trimming operation is used.

Birkert et al. [5] recommend an *Improved Springback Compensation Strategy* (*Improved Strategy*) through optimized part position in the dies. This strategy can be generally applied to all multistage car body component stamping processes. It leads to an improved dimensional accuracy compared to other common strategies, caused by a reduction of the involved elastic energy when the blankholders are closed. This reduction is achieved by ensuring a proper part position in all trimming and forming dies of the stamping process. The disadvantage of the strategy is a very complex and time-consuming work process compared to other compensation strategies.

3. Deficits of today's compensation strategies

As previously discussed in Sec. 2, different compensation strategies have different pros and cons with regard to general applicability, compensation amount, effort and complexity. The *Improved Strategy* described in [5] is the only strategy, which can fulfil the requirement of a general applicability, i.e. a reliable reduction of dimensional deviations of any car body component after compensation. However, the main disadvantage is the very complex and time-consuming application, mainly caused by two reasons. Firstly, the amount of compensation is different for each forming operation in a multistage car body stamping process; this means each operation must be compensated differently in each compensation loop. Secondly, the compensation of the operations must be executed consecutively with each compensation depending on the respective springback of the previous operation. This requires an alternating process of adaption and simulation (Figure 1), which prevents a continuous simulation of the entire process chain and thus causes a high time effort, especially when a large number of operations is involved.

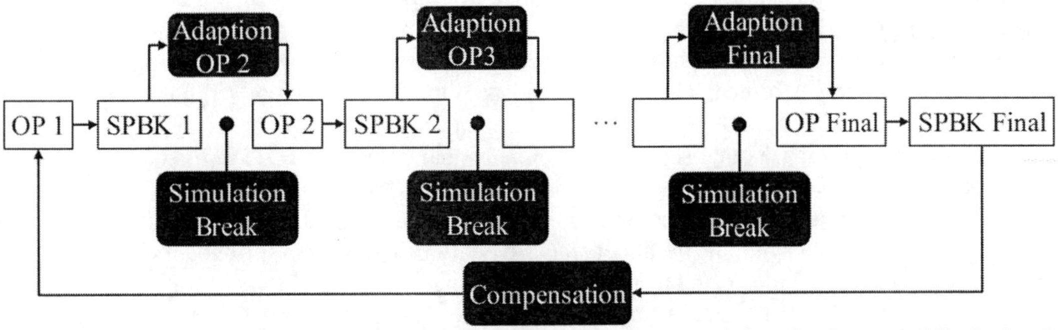

Figure 1. Alternating process of adaption and simulation when applying the *Improved Springback Compensation Strategy*

Figure 2 shows the qualitative representation of the dimensional deviations in the initial and last compensation loop of the compensation process when the *Improved Strategy* is applied. In the initial compensation loop the focus is not the improvement of the dimensional deviation of the respective part at the end of the process chain but rather on a proper part position in all operations to eliminate any negative influence of elastic energy while closing the blankholders. Indeed, the additional effort in the initial loop can lead to even higher dimensional deviations than those observed in an uncompensated process. In the last compensation loop, however, the desired state should be reached; the dimensional deviations are then expected to be within tolerance while, additionally, in every operation a proper part position without any elastic energy is provided when closing the blankholders.

To maintain the advantages of the *Improved Strategy* and, at the same time, to reduce the computational effort, it would be desirable to obtain the initial deviations (free of the negative influence of the elastic energy while closing the blankholders) without the possibly higher dimensional deviations. Also, it would be preferable if the tool-compensation of the respective compensation loop could be completed before starting the simulation of the next compensation loop, so that the simulation can be performed without the breaks indicated in Figure 1.

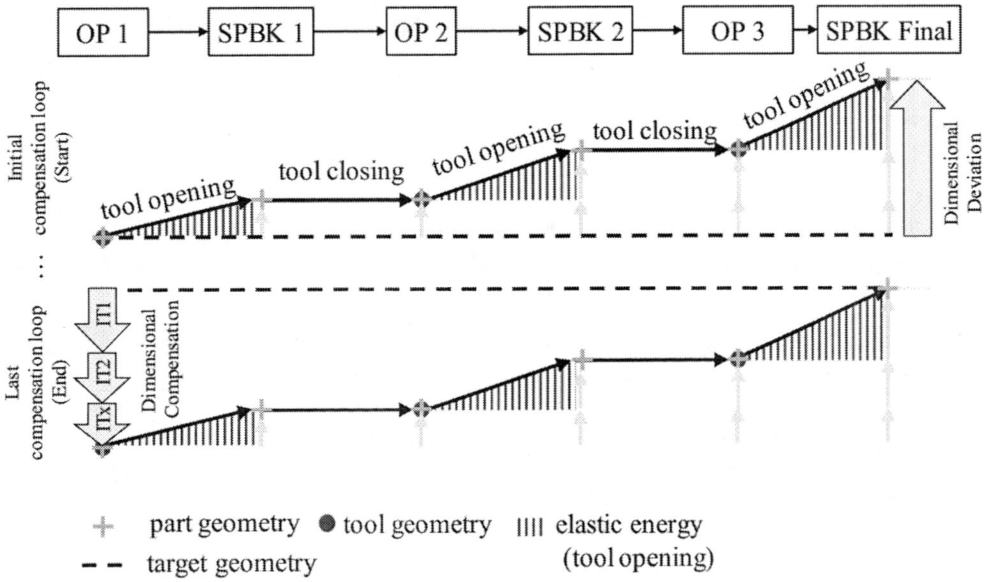

Figure 2. Qualitative representation of the dimensional deviations when applying the *Improved Springback Compensation Strategy*

4. New advanced springback compensation strategy

The *Advanced Springback Compensation Strategy* (*Advanced Strategy*) differs from the *Improved Springback Compensation Strategy* insofar as it does not need an adaption of the forming tools to calculate the initial dimensional deviations (without any influence of elastic energy while closing the blankholders) caused by springback. As can be seen in Figure 3, the entire process chain will be split up. For each operation (OP) a separate forming simulation with a subsequent springback calculation step (SPBK) is being performed. Thereby the process in the first simulation (Simulation 1) is identical to an uncompensated process. After the forming simulation and subsequent springback calculation the part is in its first springback-state (P1) and the dimensional deviations (D1) to the target geometry can be measured.

In the second simulation (Simulation 2) a part is needed exhibiting all geometric characteristics resulting from OP 1 but without the dimensional deviations caused by springback (P1*). To achieve this, the part from Simulation 1 is taken *after* the forming operation but *before* opening the tools (i.e. before calculation of the springback), and in this state all stresses are deleted. The same geometry could be obtained by other methods, e.g. a CAD geometry of the part state could be meshed and mechanical parameters – for instance the sheet thickness – mapped on this mesh. The most important thing is to have the same geometry for the part and the receiving tool surface so that no stresses are induced into the part when closing the blankholders. After closing the blankholders and forming the part in OP 2, the subsequent springback (SPBK 2) and the resulting dimensional deviations (D2) to the target geometry can be calculated. It is important that this deviation is only caused by the desired forming in this operation, without any other (negative) influences.

The procedure in Simulation 3 is the same as in Simulation 2 but taking the part without any dimensional deviations after OP 2 (P2*). Subsequently, the corresponding springback (SPBK Final) and the resulting dimensional deviation (D3) are calculated. Now, all the isolated dimensional deviations have been calculated and can be used to compensate the forming tool surfaces.

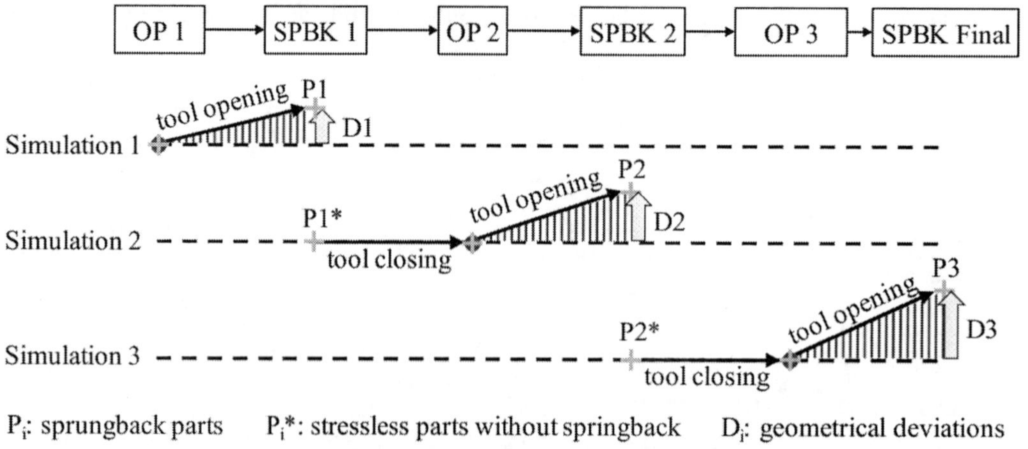

Figure 3. Simulation setups and theoretical results by applying the first iteration of the *Advanced Springback Compensation Strategy*

In order to achieve a comparable situation as with the *Improved Strategy* (Figure 2, bottom), the following steps have to be carried out. For the first operation (OP 1) all deviations (D1+D2+D3) have to be accumulated whereas for the second operation (OP 2) only D2 and D3 have to be used, as demonstrated in Figure 4. Finally, the last operation (OP 3) is compensated based only on the last deviation (D3). The general rule is, that for a distinct operation the deviations of the corresponding operation and those of all subsequent operations have to be accumulated to get the basis for the compensation.

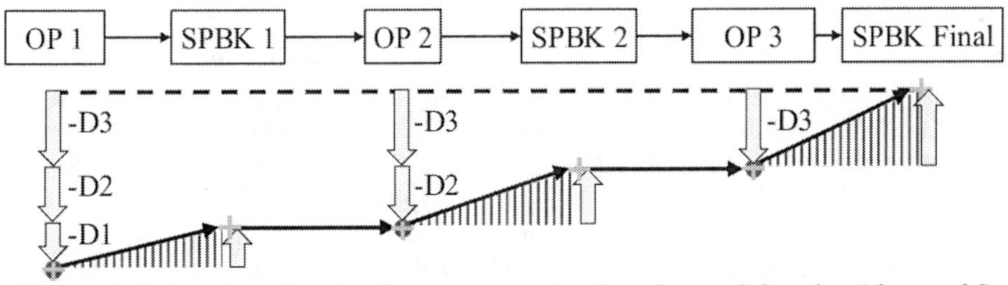

Figure 4. Theoretical situation after the first compensation loop by applying the *Advanced Springback Compensation Strategy* (if springback is not affected by compensation)

Assuming that the amount of springback is not affected by the compensation, the part will now be in tolerance and the compensation process will be finished. Normally, however, the springback is affected when modifying the tool surfaces with the effect that the part will still be out of tolerance, as shown in Figure 5. On the one hand, there are dimensional deviations between the part and the target geometry at the end of the process chain (D3-1); on the other hand, the incoming part does not properly fit to the respective receiving tool surface geometry in the other operations (D1*-1, D2*-1) causing elastic energy in the part and thus further dimensional deviations after the last operation. Only in case that the dimensional deviations between the incoming part and the respective tool geometries (D1*-1, D2*-1) are negligible, the resulting dimensional deviations between the part and the target geometry at the end of the process chain (D3-1) can be used to compensate the tools again. In case that the deviations are not negligible, the tools have to be adapted to the respective incoming part with the consequence that the advantage in computational effort with respect to the *Improved Strategy* is lost.

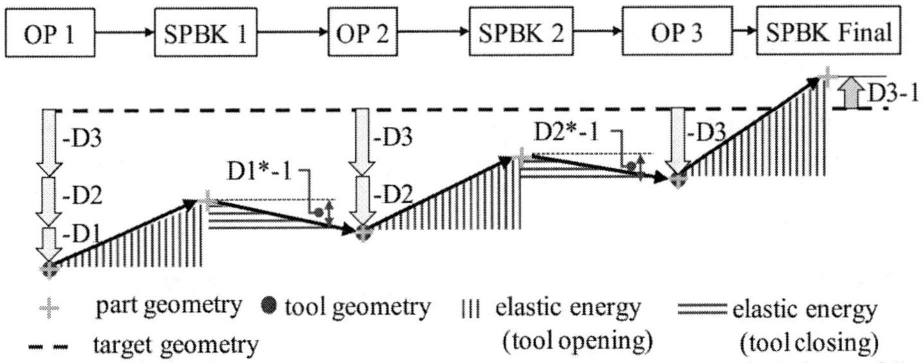

Figure 5. Theoretical situation after the first compensation loop by applying the *Advanced Springback Compensation Strategy* (if springback is affected by compensation)

In the second compensation loop the deviations (D3-1) are used to compensate all tools by the same amount (-D3-1), as displayed in Figure 6. Thus, the deviations at the end of the process are being reduced while the part position in the distinct operations stays more or less the same (D1*-1≈D1*-2, D2*-1≈D2*-2). This process is being repeated until the part is in tolerance.

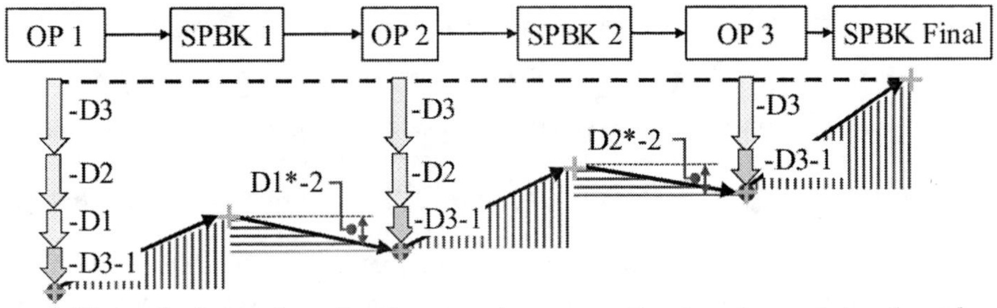

Figure 6. Theoretical situation after the second compensation loop by applying the *Advanced Springback Compensation Strategy* (dimensional deviations after OP 3 in tolerance)

5. Results of the new *Advanced Strategy* and comparison with those of the older *Improved Strategy*

In this chapter the results of the *Improved Springback Compensation Strategy* and those of the *Advanced Springback Compensation Strategy* are compared to each other. The present studies have been carried out using the example of an aluminium fender (modified *AC120*) (Figure 7). The manufacturing process starts with a drawing operation (*D20*); then the part is being separated from the addendum by a virtual laser cut (*T30*); finally, different flange areas are formed in two flanging operations (*F40, F50*). The production process of the fender has been simulated with the FE-Software *AutoForm R8*. To compensate the tool geometries, the *Physical Compensation Method* [6] has been used.

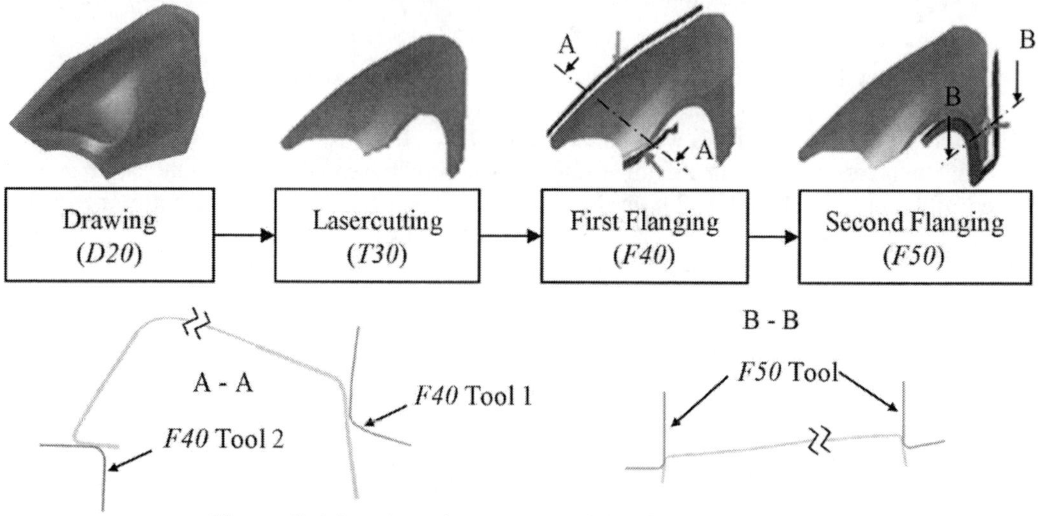

Figure 7. Manufacturing process of the aluminium fender

At first the part position in the first flanging operation (*F40*) is considered, see Figure 8. The distance between the incoming part and the receiving die surface is evaluated after three compensation loops and is split up into three distance classes, marked with different colours. When the *Improved Strategy* is applied, this distance is lower than 0.5 mm nearly on the whole surface; only two small areas exhibit higher distances but still below 1.0 mm. The maximum distance is 0.7 mm. If the *Advanced Strategy* is used, there are two larger areas with distances of more than 1.5 mm being again surrounded by areas with distances of more than 0.5 mm. The maximum distance amounts to 2.0 mm.

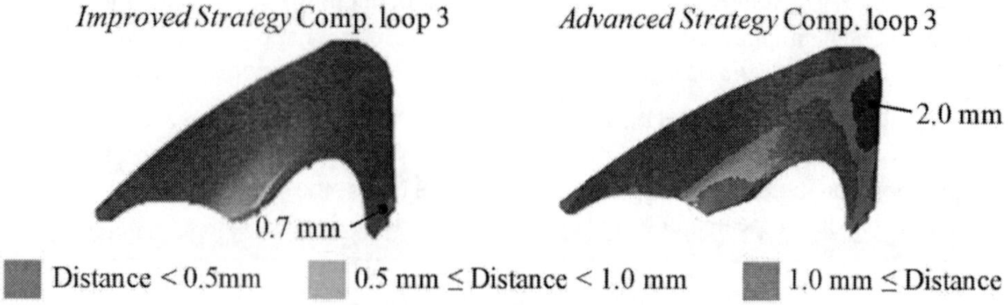

Figure 8. Part position in the first flanging operation (*F40*) after three compensation loops

In the second flanging operation (*F50*) the results are similar to those of the first flanging operation (*F40*), see Figure 9. With the *Improved Strategy* nearly the complete surface exhibits distances below 0.5 mm and a maximum distance of only 0.6 mm. By using the *Advanced Strategy,* more than 50% of the part has got distances larger than 0.5 mm with a maximum of 2.3 mm.

With the *Improved Strategy* the part position is nearly perfect; this being a result of the continuous adaption of the receiving tools to the respective incoming parts in every compensation loop. The

Advanced Strategy leads to higher distances between the part and the receiving tools, but not so much that it would necessarily cause a so-called instable part position. Thus it needs to be assessed, whether the additional time for the *Improved Strategy* is worth spending or whether the slightly worse part position caused by the *Advanced Strategy* is acceptable.

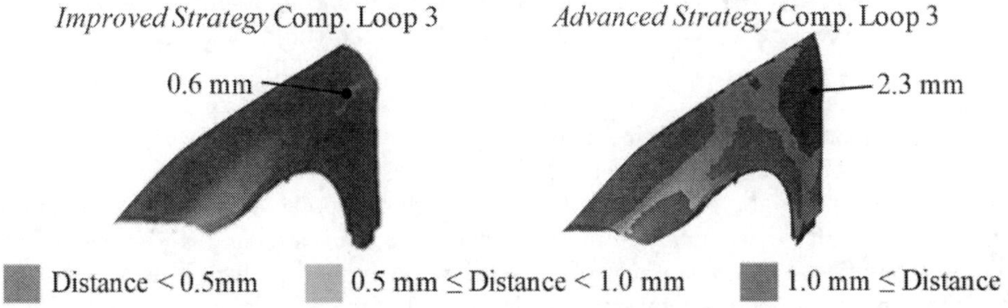

Figure 9. Part position in the second flanging operation (*F50*) after three compensation loops

In Figure 10 the average dimensional deviations are plotted for consecutive compensation loops and for both strategies. Due to the adaption of the flanging operations without compensating the springback, the average deviation with the *Improved Strategy* (A) is even higher than in the simulation without compensation (B). The areas that exceed a deviation of 1.0 mm increase from 34.1% to 51.3%. After the first compensation loop with the *Advanced Strategy* the average deviation is lower (0.44 mm) than with the *Improved Strategy* (0.53 mm). Through all following compensation loops this advantage remains preserved. Both strategies show a good convergence behaviour, which can be seen by the continuously decreasing average deviation from one compensation loop to the next. At the end, after three compensation loops, the *Advanced Strategy* benefits from its initial lower deviations so that 99.5% (!) of the part area show a dimensional deviation below 0.5 mm.

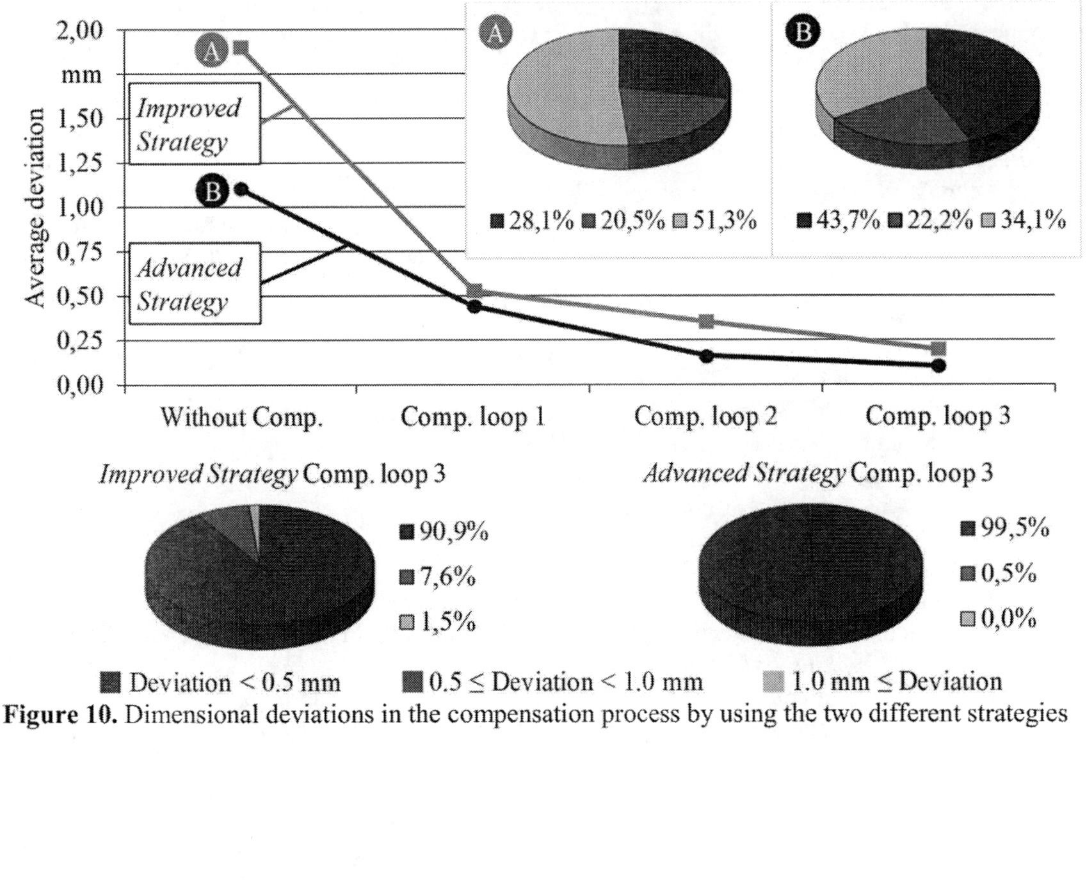

Figure 10. Dimensional deviations in the compensation process by using the two different strategies

6. Conclusion

A good springback compensation process is characterised by continuously decreasing dimensional deviations over several compensation loops until the part is within tolerance. Furthermore, it is very important that in every additional operation after drawing a proper part position is guaranteed. A proper part position means, that in the respective die the incoming part lies in a stable and reproducible position and that any deformations by the blankholders are reduced to a minimum. This can be achieved by minimizing the distance between the incoming part and the receiving tool surface.

The so called *Improved Springback Compensation Strategy* [5] fulfils these requirements in a nearly perfect way. However, by applying this strategy, the time effort is huge, especially if the manufacturing process contains more than one additional operation after drawing.

In the present paper a compensation strategy has been described which can be conducted much faster. In a first step all the operations of a multistage manufacturing process and the subsequent springback calculations are simulated separately from each other. Here, it is essential that the respective incoming parts do perfectly fit to the tool surfaces of the receiving operations, which is achieved by eliminating the springback of the previous operation through deletion of the corresponding stresses. By doing this, only those deviations are used for compensation which are caused by the generic manufacturing process itself (drawing, trimming, flanging…) and not by any elastic energy resulting from elastic deformations when the blankholders are closed. Subsequently, for the compensation of any distinct operation the deviations of the corresponding operation and those of all subsequent operations have to be accumulated. If the dimensions are still out of tolerance after the first loop, a second compensation loop has to be performed. In this loop all tools are compensated with the (same) final deviation obtained after the first loop. This process is being repeated until the part is in tolerance.

Based on the example of an aluminium fender, the *Advanced Strategy* has been tested and the results have been compared to those of the *Improved Strategy*. The part position in the flanging operations is slightly worse when using the new *Advanced Strategy* but with a maximum distance of 2.3 mm between the part and the receiving tool surface this is valued as still acceptable. With regard to the dimensional accuracy the *Advanced Strategy* delivers better results than the *Improved Strategy*. After three compensation loops the average dimensional deviation has been reduced to 0.1 mm (*Improved Strategy*: 0.2 mm) and 99.5% of the part surface shows dimensional deviations below 0.5 mm. This advantage results from the lower initial deviations. The main benefit of the new *Advanced Strategy* is the significantly lower time effort which, in the present study, amounts to less than 50% of the *Improved Strategy*.

In future work, the described strategy will be tested on different body components, particularly on structural components of high-strength steel. Here, the impact of a slightly worse part position in the tools in connection with a significantly higher part stiffness needs to be investigated.

References

[1] Gan W and Wagoner RH 2004 Die design method for sheet springback *International Journal of Mechanical Sciences* 46(7) 1097-113

[2] Karafillis A and Boyce M 1992 Tooling design in sheet metal forming using springback calculations *Int. J.Mach. Tools. Manuf.* 34113

[3] Roll K, Lemke T and Wiegand K 2004 Simulationsgestützte Kompensation der Rückfederung, LS-Dyna Anwenderforum, Bamberg

[4] Birkert A, Straub M and Haage S 2013 Umformtechnische Herstellung komplexer Karosserieteile, Springer-Verlag Berlin Heidelberg pp 304-307

[5] Birkert A, Hartmann B, Lepple F, Straub M and Zimmermann P 2019 Improved Spring-Back Compensation Strategy through Location Optimized Part Position in the Dies *IOP Conf. Ser.: Mater. Sci. Eng. 651 012036*

[6] Birkert A, Hartmann B, and Straub M 2017 New method for springback compensation for the stamping of sheet metal components *IOP Conf. Series: Journal of Physics: Conf. Series 896 012067*

IOP Conf. Series: Materials Science and Engineering 1238 (2022) 012084 doi:10.1088/1757-899X/1238/1/012084

Evaluation of testing methods for the characterization of material properties under plane strain

M Lenzen[1]* and M Merklein[1]

[1]Friedrich-Alexander-Universität Erlangen-Nürnberg, Institute of Manufacturing Technology, Egerlandstr. 13, Erlangen, 91058, Germany

*matthias.lenzen@fau.de

Abstract. With the first numerically calculated forming simulations of a complex component by Hora in 1987, a renewed need emerged for a large number of stress-based material parameters for the yield locus modelling. Material characterization and modelling under plane strain is an important step towards an improved mapping accuracy in numerical calculated sheet metal forming processes. There are numerous of different testing methods, which can be used for this purpose. It is crucial to choose the right experiment depending on which material properties or characteristic values are needed. The aim of this investigation is to analyze the suitability of the currently mostly used testing methods that induce a plane strain in the material, the notched tensile test, the biaxial tensile test and the hydraulic bulge test with an elliptical die. For this purpose, the stress-based material properties of DC06 in all three tests are determined and compared. Subsequently, recommendations for material characterization are derived depending on the application.

1. Introduction

Nowadays, efficient and versatile manufacturing processes are based on a precise numerical design. Thus, material models with a high mapping accuracy are needed. Due to the improved possibilities in material characterization, models with an increasing number of degrees of freedom can be identified. The use of optical measuring systems has further increased the possibilities for determining characteristic values. Since the first numerical calculation of a sheet metal forming process by Hora in 1987 [1], there has been a further demand for improved mathematical models to describe the plastic material behavior as accurately as possible. These mathematical descriptions of the observed material behavior enabled finite element (FE) analyses of the forming processes and can significantly reduce development time and costs in the tooling design. A mathematical formulation that has the degree of freedom to correctly represent the plasticity of deep drawing steel grades requires significantly more input parameters than available in conventional models. These parameters are provided by experiments in order to be able to characterize the material behavior under different types of loading in laboratory tests. Thus, the experimental effort goes hand in hand with a better mapping accuracy and is contrary to a time- and cost-efficient process design. A stronger interaction between experiments and modelling offers the possibility of an increase in efficiency in numerically calculated sheet metal forming processes.

Due to the minimal forming limit under plane strain and a relatively high rate of failure of about 80 % in the plane strain area [2] there is currently a focus on the material characterization and modelling under plane strain. Among others, three testing methodologies are mostly used, when it comes to the material characterization under plane strain loading, the notched tensile test, the biaxial tensile test or the hydraulic bulge test with an elliptical die, that induces a near plane strain load in the material. All three test have their special limitations and advantages. In the notched tensile test, only the first principle stress can be characterized. Wagoner [WAG80] proposed an analytical formulation with reference to the Lankford coefficient for a calculation of the second principle stress component. However, for accurate modelling of yield locus curves, the first and second principal stress components are required. This could be overcome by Kuwabara [4] with his biaxial tensile test. The limitation in these investigations was the maximum true strain of 0.09 for an IF steel grade. Investigations of Hippke et al. [5] used the biaxial tensile test for testing of many sampling points in the first quadrant of the yield locus. In this case, the maximum elongations were in the true strain range between 0.01 and 0.02, depending on the load case. Another test setup that is suitable for material characterization in plane strain is the hydraulic bulge test with an elliptical die. As valid for all die geometries in the bulge test, the initial yield stress cannot be calculated with sufficient accuracy due to the evaluation methodology. Possible applications of the material properties under plane strain are the identification of the yield criterion of Veger [6] or an implementation in other yield criteria, for example Yld2000-2d via variation of the yield locus exponent m [7] with the additional experimental data [8]. Hippke et al. [5] proved the better mapping accuracy of the optimized yield locus model through a comparison of the prognosed stress values in the first quadrant to those characterized in biaxial tensile tests.

It can be concluded, that there is a need for experimental parameters for the numerical simulation but there is a lack of comprehensive understanding of which experiment can or should be used for which material information due to the different limitations of each test setup. In this investigation the suitability of the three mentioned testing methods are analyzed. Therefore, a material characterization is done with the notched tensile test, the biaxial tensile test and the hydraulic bulge test with an elliptical die. Further, the stress-based material properties of DC06 in all three tests are analyzed and compared. Subsequently, recommendations for material characterization are derived depending on the application.

2. Methodology and testing setups

For a comprehensive understanding of the material behavior under plane strain, tests with the different testing setups are performed. The used material should have a load path dependent anisotropic hardening. This is an important attribute because the load paths are slightly different in the elliptic hydraulic bulge test and in the notched tensile test. In the notched tensile test, there is no second principle strain component while in the bulge test there is a ratio of 1:5 for the second to the first principle strain [8]. For this reason these two strain paths are adapted in the biaxial tensile test for this investigation by a special specimen design and different loading conditions. As a result, there are four different yield curves out of the three tests and the two load paths which are further qualitatively and quantitatively analyzed with respect to the maximum elongation and the resulting flow stress values. Based on these findings, recommendations for the material characterization under plane strain are derived.

Material

The used material is the DC06 with a sheet thickness of 1.0 mm. Due to its good formability and the associated excellent suitability for deep drawing, the material is also referred to as deep drawing steel. It is characterized by a low initial yield stress, a distinct work hardening and a high ductility. However, the anisotropic hardening behavior [9] is a challenge for this steel grade, which makes precise numerical mapping difficult.

Notched tensile test

The notched tensile tests are performed on the universal testing machine Z100 (Zwick, Germany). The strain is measured with a 3D DIC device ARAMIS (GOM, Germany). The specimen has a width of 60 mm with notches that result in a transverse section reduction to 30 mm. In the notch there is a parallel length of 2 mm and two radii of 2 mm. Due to the localization of plastic strain in the notch, a constant testing velocity of 3.0 mm/min for the steel material is used. The measuring frequency for force and strain is 2 Hz. Specimens are tested in the rolling direction (RD). The determination of yield curves under plane strain is based on the methodology of Flores et al. [10].

Elliptic hydraulic bulge test (HBT)

The hydraulic bulge tests with an elliptical die are performed with the tool presented in [8]. The tool is adapted in a hydraulic press Hydrap HDPZb630 (Hydrap, Germany). Again the strain measurement is done by 3D DIC. The strain velocity of 4.0 %/s is online controlled by the DIC system through a regulation of the volumetric flow of the hydraulic fluid. Thus, a constant deformation rate can be guaranteed within the whole test. Analogous to the notched tensile test, the material characterization is done in rolling direction.

Biaxial tensile test

The biaxial tensile tests are performed with the testing machine published by Merklein and Biasutti [11]. The test rig is based on a base plate that can be moved in the z-axis, with four clamping jaws mounted on it, which are fixed to a lower immovable table plate by means of a joint kinematics. The lifting movement of the base plate generates a tensile force on all four arms due to the fixation of the arms. By setting different angles in the joint kinematics, a different increase in strain in each tensile direction can be generated. This allows strain paths to be realized under equi-biaxial or near plane strain. The test forces occurring during deformation are characterized using four load cells, one in each arm. The strain is recorded with a 3D DIC system. Different strain paths can be realized not only by the kinematics of the test rig. Numerous publications on specimen geometries, summarized for example by Bruschi et al. [12], show the relevance of the specimen shape for material characterization in biaxial tensile testing. The investigations are carried out on the basis of two strain path based target variables. On the one hand, the strain path obtained in the elliptical hydraulic bulge tests is reproduced via different loading angels of the joints as mentioned above. On the other hand the strain path in notched tensile tests has to be replicated. Thus, the specimen shape and the test rig are adapted in such a way that an ideal plane strain state results.

3. Results and discussion

An analysis of the plastic material behavior is carried out in a first step by experimental data provided from notched tensile tests and hydraulic bulge tests with an elliptical die. The stress calculation in the notched tensile test is done according to Flores et al. [10] with a maximum allowable perpendicular strain of 2.0 %. The parameter α, for the homogeneous length, was set to $\alpha = 0.98$. The yield stress calculation in the hydraulic bulge tests is done according to [8]. The resulting flow curves of both tests and the corresponding strain paths are shown in Fig. 1.

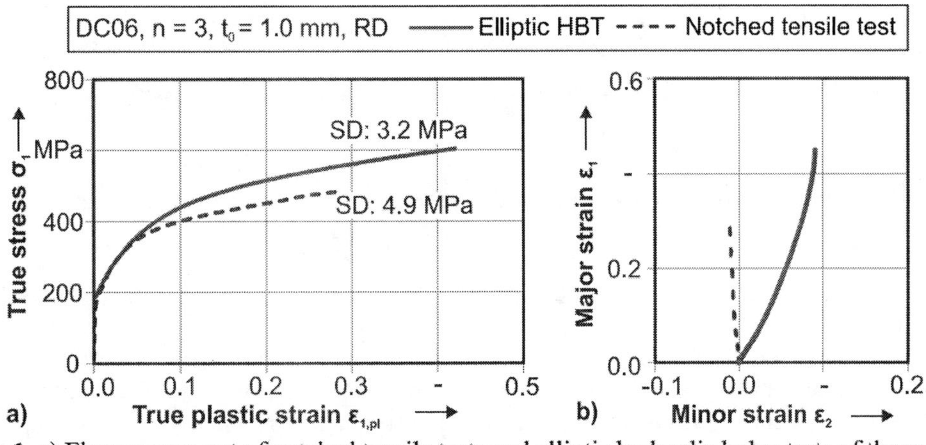

Figure 1. a) Flow curves out of notched tensile tests and elliptic hydraulic bulge tests of the materials DC06 and b) resulting strain paths

A qualitative comparison of the flow curves shows a very good comparability of the curves between the elliptical hydraulic bulge test and the notched tensile test up to a first principle strain of 0.05. In the beginning of yielding, however, a comparison is not possible, since the stress evaluation in the hydraulic bulge test is not possible in this strain level and starts at approximately 0.01 true strain. Above 0.05 true strain, the strain hardening behavior in the notched tensile test is lower than in the hydraulic bulge test. The reason for this is, on the one hand, the localization of the strain and the associated locally different sheet thickness along the forming zone due to the tensile stress superposition at the specimen edge [10]. For clarification, optically measured strain fields at different pre-strains are shown in Fig. 2. While at 0.05 true strain a relatively homogeneous straining is observable in the whole forming zone, except a small region at the radii. In contrast to this, significant inhomogeneities can be identified above 0.05 true strain for both major and minor strain with a clearly tendency to tensile loading conditions with a negative minor strain and positive major strain. In addition, the strain components are evaluated on the sheet plane as a basis for the thickness calculation along a measuring field in the center of the specimen. In this forming zone, locally different strain values in this area are averaged and leads to a mean strain value that is too low [13]. The third influencing factor is the anisotropic work hardening of this material, that leads to a different yield stress under tensile loading and under plane strain, whereby the biaxial strain of this material class leads to a higher stress values [9]. For this reason, in combination with an insufficiently precise measurement of the local plate thickness, the yield stress determined in the notched tensile test is output with a value that is too low.

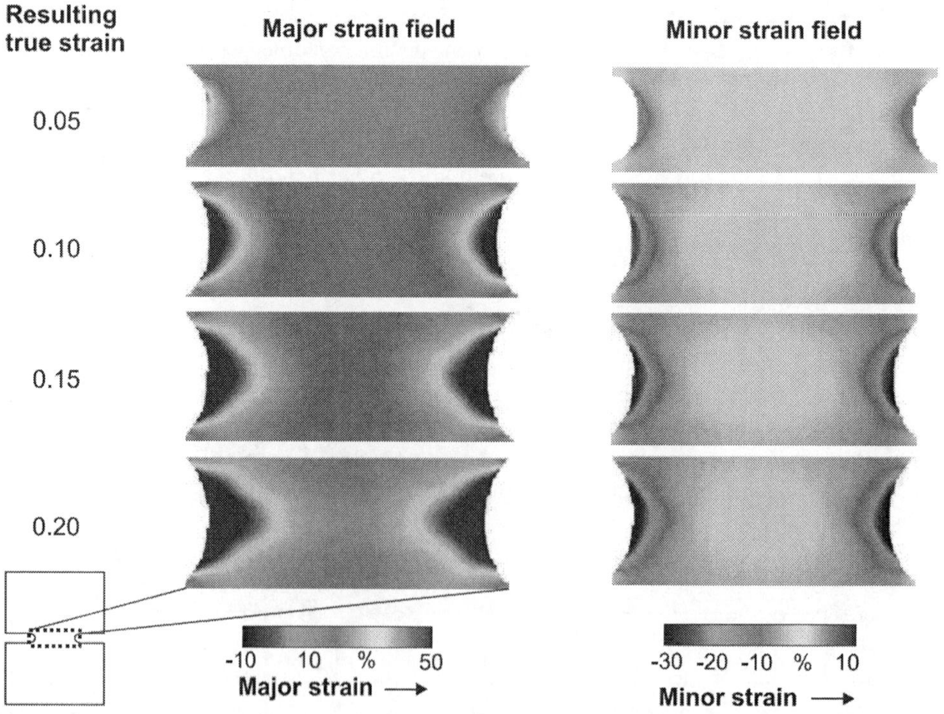

Figure 2. Distribution of major and minor strain in the notched tensile test at defined stages of deformation for DC06

Both tests for the characterization of the plane strain lead to a similar result at low degrees of deformation. Limitations in the determination of the onset of yielding in the HBT and the characterization of strain hardening in the notched tensile test make it necessary to conduct a further test in order to be able to characterize and analyze the plastic material behavior holistically. On the basis of the different strain paths of notched tensile test and elliptical hydraulic bulge test, a testing procedure is therefore developed which can characterize the onset of yielding for both stress components at ideal plane strain as well as at near plane strain. The analysis for both load cases is necessary because the investigated material exhibits a stress-state-dependent strain hardening [9]. Therefore, it is essential to analyze the difference of the flow curves for the different load paths. For this reason, specimen geometries and loading conditions are being developed which correspond to the tests and can be

IOP Conf. Series: Materials Science and Engineering 1238 (2022) 012084 doi:10.1088/1757-899X/1238/1/012084

reproduced in one test rig. With the two resulting strain paths in Fig. 1, the loading conditions and specimen geometries in the biaxial tensile test are improved. The resulting geometries and loading conditions are shown in Fig. 3.

To achieve the objective of a strain path with ideal plane strain (variant on the left), concave contours of slits in the specimen in the tensile loading direction are investigated. The shape of the concave slot structure and load distribution is adapted so that quasi-ideal plane strain ($\varepsilon_2 = 0$) dominates in the middle region of the specimen. For this purpose, elongation of the specimen in the transverse direction is prevented by a solid fixing of the tensile arms. Examination of the remaining width in the transverse direction results in a width in the center of the specimen of 2 mm. A higher width leads to a major strain below 3 %. The slits in the arms in the tensile direction are applied to homogenize the strain in the loading direction [11]. A strain path can be achieved with the specimen shape shown in Fig. 3 (variant on the right), that is comparable to elliptical hydraulic bulge tests. Here, the main loading direction is along the slitted arms. The minor strain is in the direction of the unslitted arms. In this direction, a lower testing velocity $v_2 < v_1$ is applied. The resulting strain paths based on both geometries and loading conditions are shown in Fig. 3 on the right.

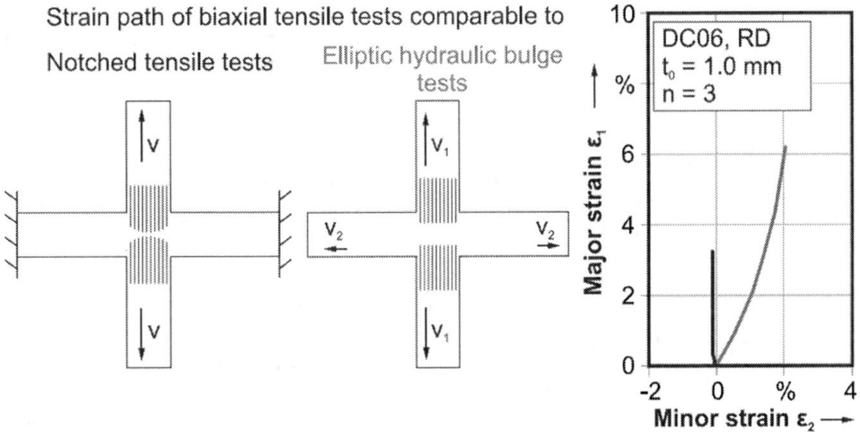

Figure 3. Resulting specimen geometries and load paths for the biaxial tensile test

Testing of the biaxial tensile test specimen, which corresponds to the strain path in the notched tensile test, leads to a vertically load path with no significant minor strain (black line). A limiting factor for this geometry is the maximum achievable strain in the specimen center of 3 %. In the second variant, shown in grey, a major strain of 6 % is reachable before failure of the tensile arms. The early failure in the arms can be explained by the reduced cross section in this area due to the slits. A milling of the sheet metal in the area of the specimen center which is relevant for material characterization to a lower sheet thickness, would on the one hand increase the achievable maximum strain. On the other hand due to the mechanical processing, an influence on the material properties cannot be completely excluded.

For the biaxial tensile test setup with ideal plane strain, the curves of the first and second principal stress are shown in Fig. 4. The stress curves resulting from biaxial tensile tests correlate with the stress curves from elliptical hydraulic bulge tests. The mean difference of 4.8 MPa between the two tests is within the standard deviation of the respective tests. A stagnating course of the second principal stress at the beginning of plastic deformation is visible. This can be explained by the existing preload, which is necessary for fixing the specimen in the tool. Also the flow curves at near plane strain of the biaxial tensile tests in Fig. 5 are in good agreement with the curves from elliptical hydraulic bulge tests. In this configuration, the mean difference between the principal stress curves is 2.8 MPa. Due to this significant correlation, this testing setup is suitable for determining the onset of plastic deformation under plane strain. The resulting yield stresses for both principal stresses is $\sigma_1 = 199.4$ MPa and $\sigma_2 = 104.7$ MPa with a maximum standard deviation of 3.0 MPa.

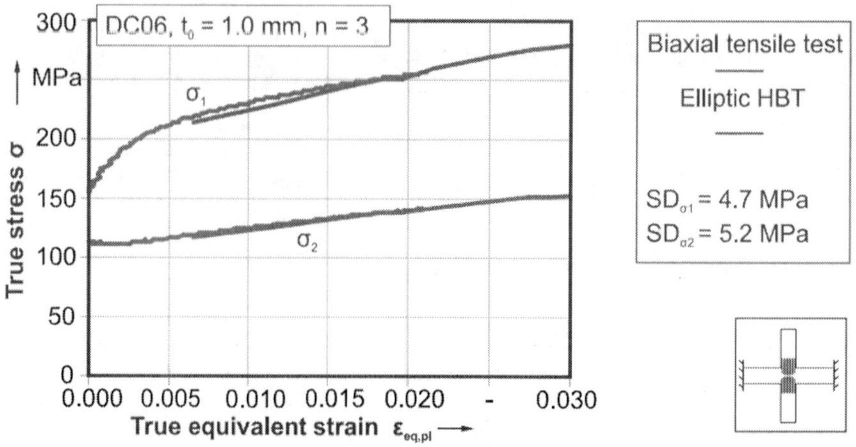

Figure 4. Resulting flow curves under ideal plane strain for biaxial tensile tests and elliptic hydraulic bulge tests (HBT)

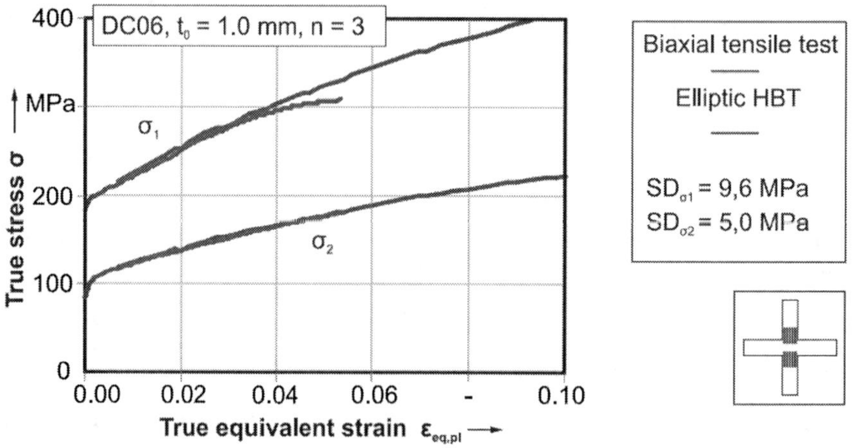

Figure 5. Resulting flow curves under near plane strain for biaxial tensile tests and elliptic hydraulic bulge tests (HBT)

4. Discussion

A comparison of all three curves at the onset of plastic deformation shows a very good agreement between the stress curves up to a plastic strain of 0.05, see Fig. 6. Based on this, the notched tensile test is also suitable for characterizing the first principal stress component under plane strain. For notched tensile tests, the onset of yielding is 199.9 MPa. In comparison, the biaxial tensile specimen yields at 201.1 MPa. As a limitation, it must be taken into account for the notched tensile test that the second principal stress component, which is required for yield locus modeling, cannot be determined. Based on the analysis of the first principal stress, both test methods - biaxial tensile test and elliptical hydraulic bulge test - are suitable to perform a material characterization under plane strain, since the deviation from the established procedure with notch tensile tests [10] does not show significant differences in the yield stress. The use of these test methods for material characterization in the plane strain condition is thus possible. Based on the results, the individual tests are evaluated on the basis of the characteristic values that can be determined. Furthermore, the flexibility and suitability for material modeling of the three tests are categorized in Table 1. This means that, depending on the application or the required material characteristic value, a different test method is appropriate.

In both the biaxial tensile test and the elliptical hydraulic bulge test, the characterization of both principal stress components is possible. In the bulge test, the material hardening can be characterized up to failure (ε large). In the biaxial tensile test the characterization of the onset of flow is possible (ε small). The notched tensile test is basically suitable for characterizing the first principal stress up to an equivalent plastic strain of 0.08 for DC06. This allows initial conclusions to be drawn about the work hardening behavior of the used material. Furthermore, the notched tensile test requires a complex preparation of the measured values, since the material thinning is unevenly due to an inhomogeneous stress distribution in the notch [10]. A positive aspect of the notched tensile test, however, is that an ideally plane strain state can be achieved for the material characterization. This is also possible for the biaxial tensile test. In this test, the flexibility for strain path variations with a positive or negative second principle strain can be easily adjusted. In the elliptical hydraulic bulge test, only a near plane strain state with slightly positive second principle strain can be adjusted. Due to the test concept, the ideal plane strain state cannot be achieved. In order to generate a further support point in the yield locus curve, the bulge test is well suited due to the determination of first and second principal stress, since the determined point can be displayed directly in the two-dimensional stress space and therefore be used for an optimization of the yield locus geometry. This is also possible for the biaxial tensile test. The notched tensile test does not offer this possibility. It is only possible to define a boundary for the first principal stress component, which must not exceed the yield locus curve at the point of the largest first principal stress. This can be represented as a vertical line in the diagram of the yield locus curve in the system of principal stresses. A direct application for a yield locus curve modeling results from the first principal stress component only to a limited extent in that the yield locus curve must tangentially touch the determined yield stress in notched tensile tests. The value of the second principal stress results only from the contour of the yield surface and does not necessarily have to correspond to the real material behavior. This test is therefore not suitable for the model parameter identification of the stress space.

Table 1. Evaluation of the test methods for plane strain with regard to determinable parameters, flexibility in strain path design and suitability for yield locus modeling

Testing setup	Determinable parameters	Flexibility in strain path changes	Suitability for yield locus modelling
Notched tensile test	σ_1, ε medium	-	-
Elliptic hydraulic bulge test	σ_1, σ_2, ε high	-	+
Biaxial tensile test	σ_1, σ_2, ε low	++	+

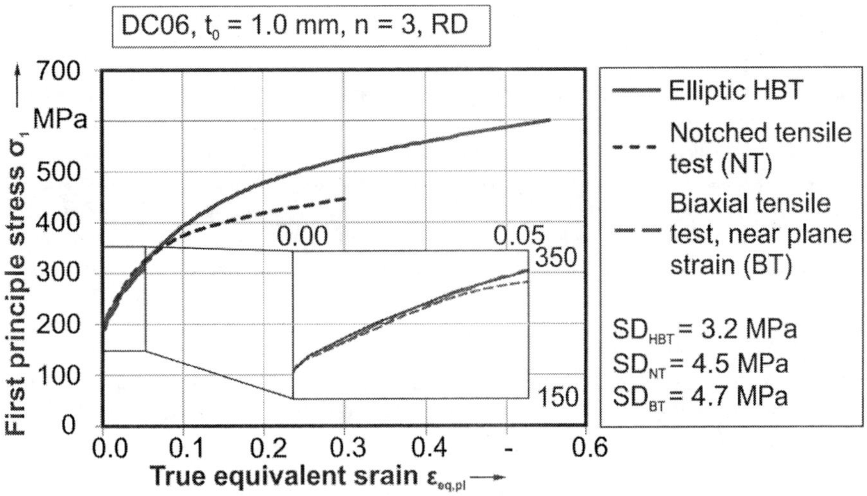

Figure 6. Comparison of the flow curves characterized in all three tests under plane stain loading for DC06

5. Conclusion

A precise determination of the material behavior under plane strain conditions is important to obtain correct results in the numerical process design of sheet metal forming processes. In 1987, at the time of the first numerical analysis of complex components, the demand for better material data was already formulated by Prof. Hora. This demand could only be met later with improved measurement capabilities, especially DIC. Thus, more than 30 years ago, Prof. Hora's demands for better material data influenced the rethinking of material characterization, especially since the end of the 1990s, and the adoption of new methods with DIC. Based on this history, tests and testing setups for improved material characterization are available today in order to meet the requirements of that time. In this context, too, are the results of this investigation and the following conclusions can be drawn:

- For a complete understanding of the plane strain material behavior it is necessary to conduct two different tests
- Depending on the necessary parameters, it is possible to perform the notched tensile test and the elliptic hydraulic bulge test or the biaxial tensile test and the hydraulic bulge test
- Also for materials with anisotropic hardening the slightly different load paths lead to comparable flow curves and therefore can equally be used for the material characterization

Further investigations should focus the transferability of the findings to other material classes with and without a stress state dependent hardening behavior.

Acknowledgment

The authors would like to thank the German Research Foundation (DFG) for supporting the present investigations which were performed within the scope of the research project "Improvement of the mapping accuracy in the material modelling with the integration of the yield locus at plane strain (DFG 412137480)".

References

[1] Hora P, Vu T C, Wollrab P M and Reissner J 1987 Simulation of the Forming Process for Irregularyly Shaped Autobody Panels *Advanced Technology of Plasticity* **1** 633-640
[2] Xavier M D, Plaut R L and Schön C G 2014 Uniaxial Near Plane Strain Tensile Tests Applied to the Determination of the FLC Formabillity Parameter *Materials Research* pp. 982-986
[3] Kuwabara T 2005 Advances of Plasticity Experiments on Metal Sheets and Tubes and Their Applications to Constitutive Modeling *AIP Conference Proceedings* **778** 20-39
[4] Wagoner R H 1980 Measurement and Analysis of Plane-Strain Work Hardening *Metallurgical Transactions A* **11** 165-175
[5] Hippke H, Hirsiger S, Berisha B and Hora P 2020 Optimized and validated prediction of plastic yielding supported by cruciform experiments and crystal plasticity *Int J Mater Form* **13** 841-852
[6] Vegter H and van den Boogaard A H 2006 A plane stress yield function for anisotropic sheet material by interpolation of biaxial stress states *International Journal of Plasticity* **22** 557-580
[7] Manopulo N, List J, Hippke H and Hora P 2016 A non-associated flow rule based Yld2000-2d model *Challenges in Forming High-Strength Sheets Proc. of the IDDRG Conference*
[8] Lenzen M and Merklein M 2018 Improvement of Numerical Modelling Considering Plane Strain Material Characterization with an Elliptic Hydraulic Bulge Test *J. Manuf. Mater. Process* **2(1)** 6
[9] Dorn J E 1949 Stress-Strain Rate Relations for Anisotropic Plastic Flow *J. Appl. Phys.* **20** 15-20
[10] Flores P, Tuninetti V, Gilles G, Gonry P, Duchêne L and Habraken A 2010 Accurate stress computation in plane strain tensile tests for sheet metal using experimental data *J. Mater. Process. Technol.* **210** 1772-1779
[11] Merklein M and Biasutti M 2013 Development of a biaxial tensile machine for characterization of sheet metals *Journal of Materials Processing Technology* **213** 939-946
[12] Bruschi S, Altan T, Banabic D, Bariani P F, Brosius A, Cao J, Ghiotti A, Khraisheh M, Merklein M and Tekkaya A E 2014 Testing and modelling of material behaviour and formability in sheet metal forming *CIRP Annals* **63** 727-749
[13] Pan B, Xie H, Wang Z, Qian K and Wang Z 2008 Study on subset size selection in digital image correlation for speckle patterns *Optics express* **16** 7037-7048

Superposing tensile stresses into single point incremental forming to affect martensitic transformation of SS304

E M Mamros[1], F Maaß[2], M Hahn[2], A E Tekkaya[2], J Ha[1], and B L Kinsey[1]

[1] Department of Mechanical Engineering, University of New Hampshire, Durham, NH USA
[2] Institute of Forming Technology and Lightweight Components, TU Dortmund University, Dortmund, Germany

elizabeth.mamros@unh.edu

Abstract. Superposing pre-stress on a SS304 sheet metal blank in biaxial tension and performing a single point incremental forming operation on the stretched blank is investigated experimentally. By applying a pre-stress to the sheet metal blank prior to incremental forming, the resulting microstructural change can be affected to obtain functionally graded materials according to the intended application. In austenitic stainless steels, this variation of the stress states alters the phase transformation, specifically the martensitic transformation kinetics, by influencing key process parameters, such as process force, temperature, and equivalent plastic strain. The phase transformation in truncated square pyramids is measured using magnetic induction. These measurements validate the effectiveness of the stress superposition method for achieving the desired mechanical properties based on altering the final microstructure of a simple geometry.

1. Introduction

For rapid-prototyping purposes and applications that require customizable manufacturing, e.g., trauma fixation hardware to hold fractured bones together during healing, which necessitate patient-specific geometries, incremental forming is an excellent candidate among sheet metal forming processes. Incremental forming is a manufacturing process that focuses on localized deformation as one or more tools move across a clamped sheet at varying depths according to the programmed toolpath [1]. This inherent flexibility is essential for low quantity jobs and reducing rapid-prototyping costs. However, incremental forming is not yet commonly used in industry due to existing gaps in knowledge related to how the process parameters affect the final part and the adaptability of the process for specific products.

The most common type of incremental forming is single point incremental forming (SPIF), which uses one tool to deform a clamped blank. This can be performed in a computer numerical control (CNC) milling machine, i.e., it does not require specialized equipment, and is easily integrated into existing manufacturing environments [2]. Alternatively, a robotic arm can be used to perform this process in an industrial setting [3]. For axisymmetric geometries, a SPIF process, commonly referred to as metal spinning, can be used [4]. These processes are suitable for a variety of materials ranging from steels to composites to polymers.

Austenitic stainless steels, which undergo strain-induced transformation into a stronger, more brittle phase, i.e., martensite [5,6], are of interest for many applications, including the automotive, aerospace, and biomedical industries. Several studies have investigated the extent to which the phase transformation from γ-austenite to α'-martensite can be influenced by adjusting the process parameters e.g., in SPIF [7,8] and deep drawing [9]. This also affects the material formability [10] and residual stress development [11]. Increased phase transformation is desired in areas of the part requiring increased strength, e.g., mounting locations in trauma fixation hardware.

A novel incremental forming process variant introduced in [12] is tensile stress-superposed incremental forming (TSSIF), which uses a custom frame to pre-stress the sheet in-plane to affect the final part properties, e.g., residual stress development. TSSIF is an example of stress superposition, which is generally defined as the incorporation of additional stresses into an existing manufacturing process during a single operation [13,14]. By adjusting the intensity of the stress superposition, i.e., changing the amount of pre-stress imposed on the clamped blank, the final part properties can be tailored for the intended application of the final product.

In this paper, the effect of stress superposition on the phase transformation in SS304 truncated square pyramids formed by SPIF and TSSIF is presented. Force and temperature data are recorded during experiments. After forming, the geometrical accuracy and final strain state are assessed using a photogrammetry camera. Magnetic induction is used to measure the phase transformation from γ-austenite to α'-martensite at four locations on each wall of the formed pyramids. Increased martensite transformation is observed at the base and middle of parts formed by TSSIF, indicating that stress superposition can be used in incremental forming to create functionally graded materials.

2. Incremental Forming Experiments

2.1. Material, Geometry, and Experimental Setup

Experiments are conducted at the Institute of Forming Technology and Lightweight Components (IUL). Blanks for incremental forming, with the geometry shown in **Figure 1**a, are laser cut from sheets of 0.8 mm thick SS304 (EN 1.4307). A dot pattern with 1 mm diameter dots and 2 mm grid spacing is electrochemically etched onto one side of the blank prior to forming for 3D optical measurements of the formed part. The investigated geometry is a truncated square pyramid with a 45° wall angle (β), 30 mm height (h), and 85 mm base (b) as shown in **Figure 1**b. A 5-axis DMU 50 milling machine (DMG Mori) performs the incremental forming operation by using its computer numerical control to follow the user-specified toolpath. A custom tensile frame with hydraulic cylinders is placed inside of the milling machine to serve as the blank holder during SPIF and the means to induce tensile stress superposition during TSSIF. During SPIF, all four sides of the blank are clamped and fixed in the tensile frame with the cylinders pressurized just enough to prevent displacement. During TSSIF, the blank is fixed on two sides while the remaining two sides are pulled in tension along the positive x- and y-axes by the pressurized cylinders in the frame. In this work, the application of the pre-stress for TSSIF results in an average of 1% von Mises strain in the forming area, which was determined using ARGUS (GOM Inc.). The material's rolling direction (RD) is aligned with the x-axis. The complete experimental setup is shown in **Figure 2**.

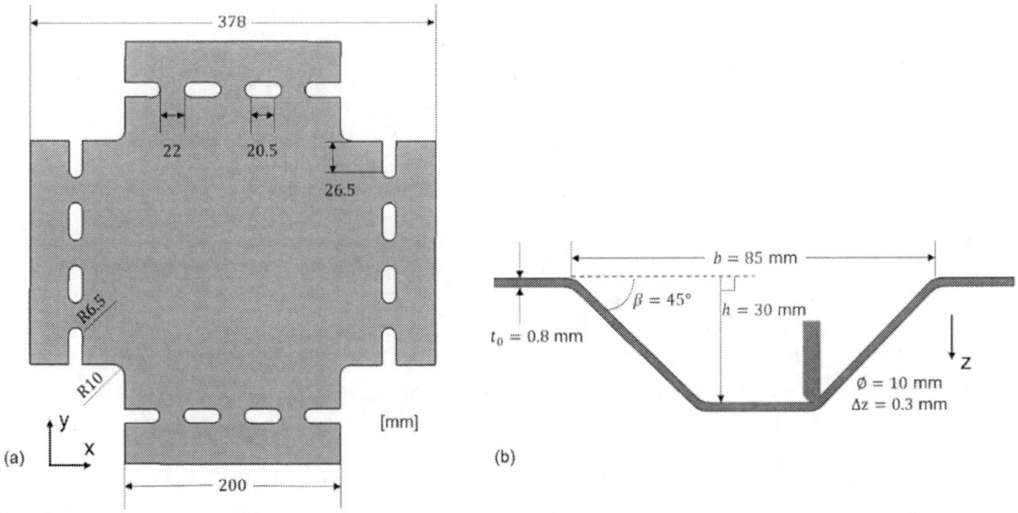

Figure 1. Incremental forming a) blank geometry and b) target part geometry: truncated square pyramid.

For both incremental forming processes, i.e., SPIF and TSSIF, the following parameters are used. The hemispherical tool has a diameter ($\emptyset$) of 10 mm, and the step down (Δz) is 0.3 mm (Figure 1b). The tool rotational speed is set to 0 rpm, i.e., fixed, and the feed rate is 1500 mm/min. The tool follows a bidirectional toolpath with the start point located in the same corner of the geometry for each layer and forms the pyramid from the base to the top, i.e., outside-in toolpath. The surface of the blank is lubricated with a thin layer of deep-drawing oil (Castrol) prior to forming. The forming process time is approximately 840 s.

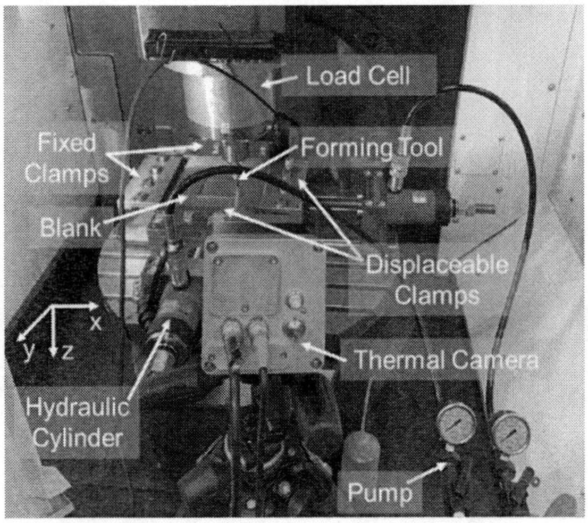

Figure 2. Incremental forming experimental setup in DMU 50 milling machine.

2.2. Data Collection

A three-directional load cell (Kistler) measures the forces from the tool, and a thermal camera (VarioCAM HD head) measures the temperature field during the process. These data are analyzed using Catman Easy (HBM) and IRBIS 3 (InfraTec) software, respectively. The geometrical accuracy of the unclamped part is compared to the computer-aided design (CAD) target geometry using ARGUS. Using the electrochemically etched dot pattern, the strain state after forming on the non-tool-side is also analysed with ARGUS. An FMP30 Feritscope (Fischer Technology Inc.) is used to determine the α'-martensite volume fraction at four locations along each wall of the truncated square pyramid through magnetic induction with the conversion factor from the ferrite number to the α'-martensite content proposed by Talonen et al. [15]. Feritscope measurements, performed on the tool-side of the workpiece, are repeated three times at each location and then averaged.

3. Results and Discussion

3.1. Force and Temperature Results

In the first half of the forming process (~420 s), the axial force is noticeably higher for TSSIF than SPIF (**Figure 3**a). This is due to the differing stress states of the blank and boundary conditions between the two processes. The biaxial tensile stress superposition in-plane increases the axial, i.e., z- or normal, force initially during TSSIF since the sheet is less susceptible to bending than during SPIF, and the geometry is still relatively planar. However, near the end of the forming process (~>600 s), TSSIF results in axial forces that are approximately equal to those in SPIF.

Correspondingly, the maximum observable temperature in the forming area is higher for TSSIF than SPIF for the entire forming process until the end, where the temperature is similar for the two processes (**Figure 3**b). The insets in **Figure 3**b are thermal images captured approximately when Location 2 was being formed by TSSIF and SPIF. Note that the maximum temperature value in the forming area is located beneath the tool and not visible to the thermal camera, i.e., the total temperature increase is > 50°C. The similar temperatures near the end of the forming process, i.e., at the top of the truncated pyramid geometry, can be attributed to similar axial forces, heat dissipation, and plastic strain in this location for SPIF and TSSIF. Note that the four approximate locations for strain and α'-martensite measurements are indicated in Figure 3.

From **Figure 3**, the superposed stress is effective as long as the part is still nearly planer (up to ~500 s and including Locations 1 and 2). For the remainder of the TSSIF forming process, the effect of the planar superposed stress diminishes. This explains the similar axial force and temperature levels for SPIF and TSSIF near the end of the forming processes.

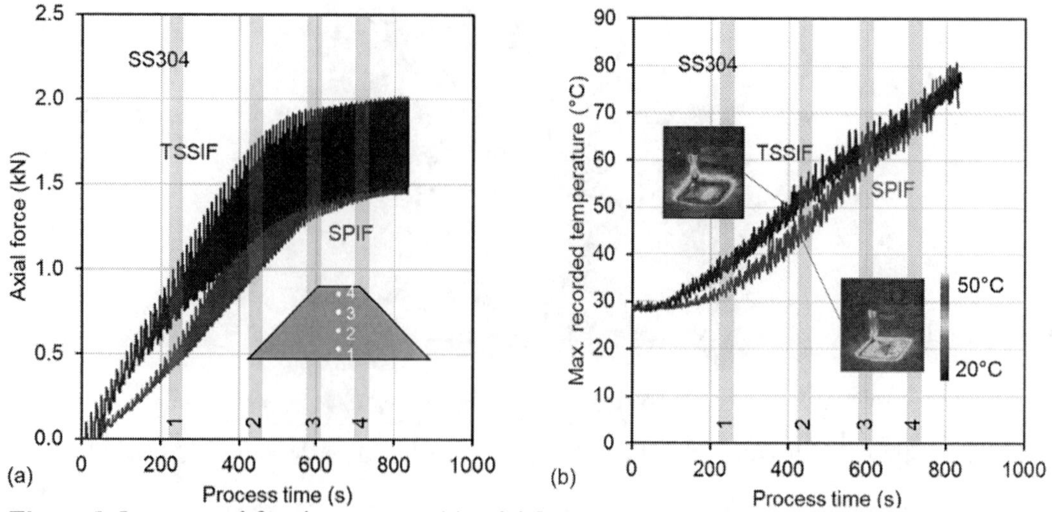

Figure 3. Incremental forming process: (a) axial force (z-force) and (b) maximum temperature recorded by thermal camera and thermal images captured at Location 2.

3.2. Geometry Comparison

Springback, resulting from the elastic recovery of the material, occurs when the formed part is removed from the clamping mechanisms [16]. In the case of TSSIF, springback also occurs in-plane when the pressure is released from the cylinders in the frame. To compare the geometrical accuracy of the SPIF and TSSIF processes, a 100 mm subsection of the formed parts, which includes the truncated square pyramid geometry and ~7.5 mm of the flange, is considered. A section cut is made along the x- and y-axes through the geometrical center of the truncated square pyramid (see inset figure in **Figure 4**). The resulting profiles for SPIF, TSSIF, and the target geometry along the x-direction (solid) and y-direction (dashed) are shown in **Figure 4**.

For SPIF and TSSIF, the profiles along the x- and y- directions are nearly symmetrical, and the deviation from the target heights are ~7 and 5.6 mm, respectively, at the corner near the base of the truncated pyramid. For TSSIF, all four sides of the truncated pyramid formed by TSSIF have increased geometrical accuracy in the z-direction near the base of the truncated pyramid compared to SPIF. When comparing the wall angles in the formed parts, the TSSIF profiles tend to follow the target profile more closely than SPIF. Overall, the truncated square pyramid manufactured by TSSIF achieves greater geometrical accuracy than the part manufactured by SPIF with respect to the target geometry.

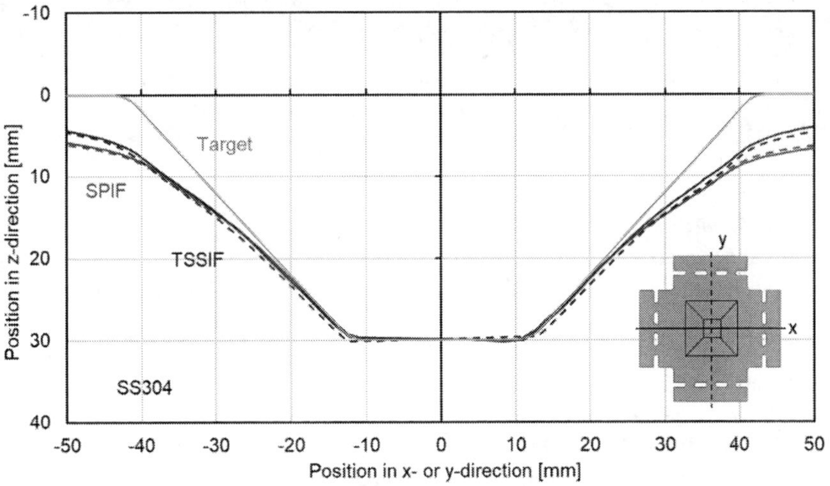

Figure 4. Profiles along the center of truncated square pyramids in the x- (solid) and y- (dashed) directions for target geometry and parts manufactured by SPIF and TSSIF.

3.3. Final Strain State

To compare the final strain state of the parts manufactured by TSSIF and SPIF after removal from the frame, a 90x90 mm^2 section is analyzed. The von Mises strain contours are shown in **Figure 5**. The white circles represent the four measurement locations along the pyramid walls (see inset figure in **Figure 5**b) plotted in **Figure 6**a. The von Mises strain near the outer edges of the analyzed area, i.e., at the base, and at the top of the truncated pyramid is nearly zero as expected due to minimal deformation in these areas of the geometry. In TSSIF, a larger uniform strain region is visible on the truncated pyramid walls compared to SPIF. For both forming processes, the maximum strain is located along the diagonals, where the tool changes directions, with a slightly increased strain along the top right diagonal, which corresponds to the starting point for each layer of the toolpath.

In Figure 6a, discrete von Mises strain values are shown for SPIF (squares) and TSSIF (triangles) at the four locations, i.e., 1 – 4, for all four faces of the truncated square pyramid geometry (see inset figure in **Figure 6**a for color map) on the non-tool-side. As is similarly presented in **Figure 5**, TSSIF produces a larger plastically deformed area along the wall (see the orange to red zone in Figure 5), with higher von Mises strain values near the base and middle of the truncated square pyramid, i.e., Locations 1 – 3 in Figure 6a. This leads to reduced springback when the stress superposition is imposed in the experiments as is shown in Figure 4.

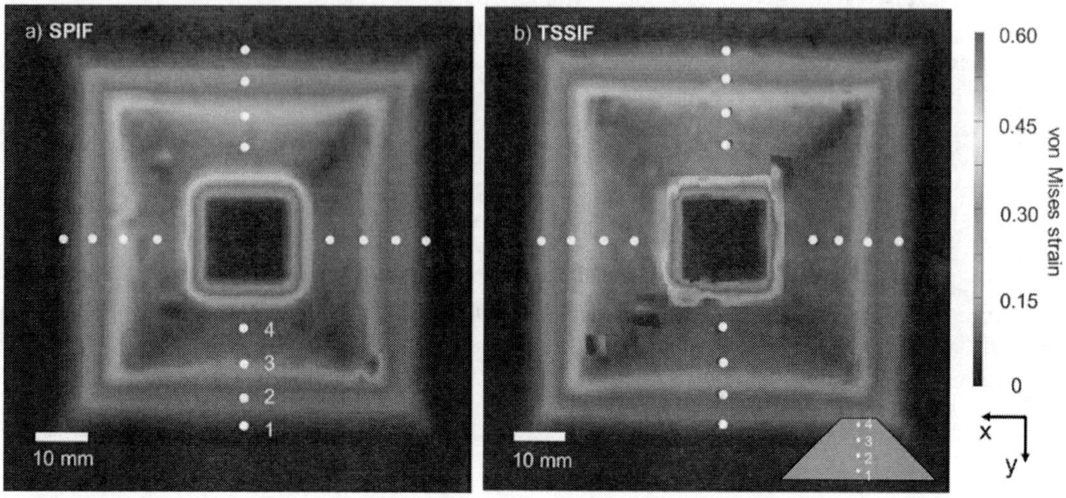

Figure 5. Final strain state of unclamped parts manufactured by a) SPIF and b) TSSIF. Circles represent Locations 1 – 4 on pyramid wall.

3.4. Phase Transformation

The as-received sheet material is fully austenitic and is subjected to strain-induced transformation into α'-martensite as a result of incremental forming. In Figure 6b, the α'-martensite volume fraction is shown for SPIF (squares) and TSSIF (triangles) at the four locations on the tool-side of the truncated square pyramid geometry. The average of the four sides for each location is also included as solid (TSSIF) and dashed (SPIF) lines to indicate the phase transformation trends in **Figure 6**b. Near the base at Locations 1 and 2, the largest differences of α'-martensite volume fraction are observed between TSSIF and SPIF, i.e., 0.051 and 0.049, respectively. Assuming that the average von Mises strain of ~0.01 in the forming area after pre-stressing the blank is negligible, this indicates that the effect of the tensile stress superposition on the phase transformation is strongest in the first few layers of the toolpath. The increased axial force at these locations in **Figure 3**a is due in part to the presence of the stronger α'-martensite material phase. In the middle at Location 3, TSSIF still experiences greater phase transformation than SPIF, i.e., 0.012, and the largest phase transformation overall for the four locations measured in the TSSIF part. At the top of the geometry, i.e., Location 4, the effect of the tensile stress superposition is no longer apparent, and other parameters affecting the phase transformation, e.g., temperature, may become dominant since the strain levels at Locations 3 and 4 are comparable for TSSIF (**Figure 6**a). The decrease in volume percent of α'-martensite observed for TSSIF when comparing Locations 3 and 4 may be caused by the increased temperature in the forming area (see **Figure 3**b), which is known to inhibit austenite to martensite phase transformation [8,17]. Note that correction factors based on the material thickness are provided by the manufacturer (but not included in these results) and may result in martensitic volume fractions up to 1.15 times those reported in this work [18].

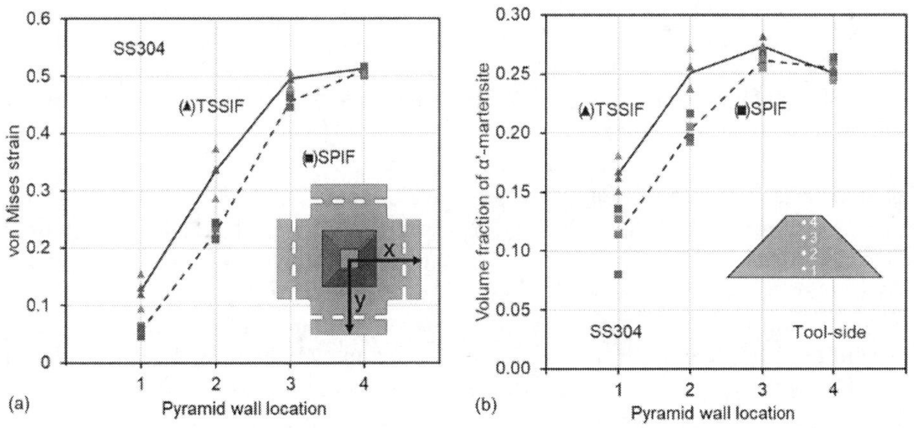

Figure 6. Truncated square pyramid parts manufactured by TSSIF (triangles) and SPIF (squares): (a) von Mises Strain and (b) volume fraction of α'-martensite measured on tool-side. Average values shown as solid (TSSIF) and dashed (SPIF) lines.

4. Conclusions and Future Work

In this paper, biaxial tensile stress is superposed into the single point incremental forming process, i.e., TSSIF, to affect the martensitic transformation of SS304 for a truncated square pyramid geometry. Initially, increased and then, ultimately, similar axial forces and temperatures are observed for TSSIF when compared to the conventional process, i.e., SPIF. The overall geometrical accuracy of the part manufactured by TSSIF is greater than that of the part manufactured by SPIF, particularly for the part height. The stress superposition in TSSIF aids the reduction of the springback in the final part. A larger uniform strain area along the pyramid walls and increased phase transformation is observed at Locations 1, 2, and 3 of the pyramids manufactured by TSSIF compared to SPIF. Therefore, the methodology of using stress superposition in incremental forming is successful in influencing the phase transformation and altering the microstructure in the final part.

Several parameters, e.g., temperature and stress and strain states, contribute to the phase transformations in this austenitic stainless steel. Additional modelling efforts are needed to better understand which effects are dominating at varying points throughout the toolpath. Future work will aim to establish a fundamental understanding of this phenomenon and to more effectively influence the transformation kinetics with the stress superposition effect, e.g., through higher pre-stress or lowering the temperature in the process. An additional material characterization method, e.g., electron backscatter diffraction, will be used to validate the martensite volume fractions.

5. Acknowledgements

Support for the New Hampshire Center for Multiscale Modeling and Manufacturing of Biomaterials (NH BioMade) project is provided by the US National Science Foundation (NSF) EPSCoR award (#1757371). This research is also supported by the Institute of International Education Graduate International Research Experiences program (NSF #1829436), the German-American Fulbright Commission, and the Germanistic Society of America.

References

[1] Cao, J., Huang, Y., Reddy, N. V., Malhotra, R., and Wang, Y., 2008, "Incremental Sheet Metal Forming: Advances and Challenges," pp. 1967–1982.
[2] Martins, P. A. F., Bay, N., Skjoedt, M., and Silva, M. B., 2008, "Theory of Single Point Incremental Forming," CIRP Ann., **57**(1), pp. 247–252.
[3] Schafer, T., and Dieter Schraft, R., 2005, "Incremental Sheet Metal Forming by Industrial Robots," Rapid Prototyp. J., **11**(5), pp. 278–286.
[4] Rentsch, B., Manopulo, N., and Hora, P., 2017, "Numerical Modelling, Validation and Analysis of Multi-Pass Sheet Metal Spinning Processes," Int. J. Mater. Form., **10**(4), pp. 641–651.
[5] Feng, Z., Mamros, E. M., Ha, J., Kinsey, B. L., and Knezevic, M., 2021, "Modeling of Plasticity-Induced Martensitic Transformation to Achieve Hierarchical, Heterogeneous, and Tailored Microstructures in Stainless Steels," CIRP J. Manuf. Sci. Technol., **33**, pp. 389–397.
[6] Mamros, E. M., Bram Kuijer, M., Davarpanah, M. A., Baker, I., and Kinsey, B. L., 2021, "The Effect of Temperature on the Strain-Induced Austenite to Martensite Transformation in SS 316L During Uniaxial Tension," *Forming the Future*, G. Daehn, J. Cao, B. Kinsey, E. Tekkaya, A. Vivek, and Y. Yoshida, eds., Springer International Publishing, Cham, pp. 1853–1862.
[7] Katajarinne, T., and Kivivuori, S., 2013, "Strain Induced Martensite in Incremental Forming - Formation, Effect and Control," Mater. Sci. Forum, **773–774**, pp. 119–129.
[8] Katajarinne, T., Louhenkilpi, S., and Kivivuori, S., 2014, "A Novel Approach to Control the Properties of Austenitic Stainless Steels in Incremental Forming," Materials Science and Engineering: A, **604**, pp. 23–26.
[9] Krauer, J., and Hora, P., 2012, "Enhanced Material Models for the Process Design of the Temperature Dependent Forming Behavior of Metastable Steels," Int. J. Mater. Form., **5**(4), pp. 361–370.
[10] Lonardelli, I., Bosetti, P., Bruschi, S., and Molinari, A., 2011, "On the Formability and Microstructural Characteristics of AISI 301 Parts Formed by Single-Point Incremental Forming," Key Eng. Mater., **473**, pp. 869–874.
[11] Afzal, M. J., Maqbool, F., Hajavifard, R., Buhl, J., Walther, F., and Bambach, M., 2020, "Modeling the Residual Stresses Induced in the Metastable Austenitic Stainless Steel Disc Springs Manufactured by Incremental Sheet Forming by a Combined Hardening Model with Phase Transformation," Procedia Manuf., **47**, pp. 1410–1415.
[12] Maaß, F., Hahn, M., and Tekkaya, A. E., 2022, "Setting Residual Stresses in Tensile Stress-Superposed Incremental Sheet Forming," *Proceedings of the 22nd International Conference on Material Forming.*
[13] Mamros, E., Ha, J., Korkolis, Y., and Kinsey, B., 2020, "Experimental Investigation and Plasticity Modeling of SS316L Microtubes Under Varying Deformation Paths," J. Micro Nano-Manuf.
[14] Maaß, F., Hahn, M., and Tekkaya, A. E., 2021, "Adjusting Residual Stresses by Flexible Stress Superposition in Incremental Sheet Metal Forming," Arch. Appl. Mech., **91**(8), pp. 3489–3499.
[15] Talonen, J., Aspegren, P., and Hänninen, H., 2004, "Comparison of Different Methods for Measuring Strain Induced Martensite Content in Austenitic Steels," Mater. Sci. Technol., **20**, pp. 1506–1512.
[16] Hosford, W. F., and Caddell, R. M., 2011, *Metal Forming: Mechanics and Metallurgy*, Cambridge University Press.
[17] Olson, G., and Cohen, M., 1975, "Kinetics of Strain-Induced Martensitic Nucleation," Met. Trans A, **6**, pp. 791–795.
[18] Helmut Fischer GmbH, 2008, "Operators Manual Feritscope FMP30."

Design of a simple shear test for large strains with sequential re-machining of the specimen edges

Xavier Colon[1,2], Morwan Adlafi[1,3], Bertrand Galpin[1,3], Laurent Maheo[1,3], Vincent Grolleau[1,2]

[1] Univ. Bretagne Sud, CNRS UMR 6027, IRDL, F-56100 Lorient, France
[2] Department of Mechanical and Process Engineering, ETH Zurich, Zurich, Switzerland
[3] Ecoles Saint-Cyr Coëtquidan, CREC, F-56380 Guer, France

E-mail: vincent.grolleau@univ-ubs.fr

Abstract. The simple shear test is a standard experiment used for the calibration of plasticity models. For thin sheets, the specimen can be as simple as a rectangular strip, or feature optimized geometries of the free edges. These enhancements are mainly motivated by the undesirable initiation of fracture from the free edges, which limit the usable strain range of the test. Previous studies have shown that fracture from the edges occurs due to a local stress state close to uniaxial tension in these areas. In an attempt to increase the maximum strain, a sequential simple shear test is proposed. The specimen is a sheet metal strip with two opposing cut-outs with rounded concave edges. The specimen is mounted in a shear testing device, composed of two jaws with a prismatic joint connection. The shear test protocol includes multiple two-steps sequences. First, an interrupted shear test is performed up to a given value of displacement. Second, the shear testing device, along with the clamped specimen, is positioned in a milling machine and the rounded free edges of the specimen are re-machined. The application of the proposed testing protocol is presented for three engineering materials. It is found that the valid range of this experiment (i.e. the maximum strains attained before specimen failure) can be substantially extended through repeated re-machining of the specimen boundaries.

1. Introduction

The accuracy of sheet metal forming numerical simulations is highly dependent upon the material behaviour model employed. This need is further emphasised by the recent development of advanced steels and aluminum alloys that can have high local plastic strains. Simple shear test, for example, has proved to be an efficient technique to evaluate the mechanical behaviour of sheet metal. However, the occurrence of fracture along the free edges of the specimen, as well as the tearing of the specimen from the clamping jaws limit the achievable maximum shear strain. As a result, the usable strain range is not large enough to enable an accurate choice of the hardening model and its proper parameter identification.

The existing shear techniques can be divided into two categories whether the shear is induced by a translation displacement, see figure 1 a and b, or a rotation displacement, see figure 1 c and d. A second classification can be performed from the geometry of the sheared area, with (see figure 1 a, b and c) or without free edges (without slits in the case of in-plane torsion, see figure 1 d). Note that for the latter, a circular groove is usually machined on the specimen surface in order to localise the strain on a narrow circular band. In the presence of free edges (see figure 1 b, c), the shorter the gage section (for a given width), the more heterogeneous the stress state; thereby the determination of the plastic behaviour generally requires a hybrid numerical and experimental analysis. It should be noted that heterogeneity of the stress state may also arise during in-plane torsion experiments of slit free specimens due to material anisotropy.

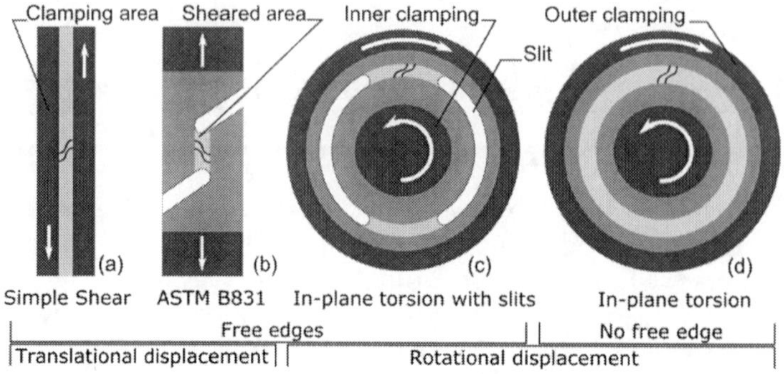

Figure 1. Different categories of shear tests, inspired from Yin et al. [1] and Reyne et al. [2].

The simplicity of the simple shear test geometry and set-up and ability of fully experimental post-processing led us to favour it. Previous studies by Bouvier et al. [3] and An et al. [4] have shown that fracture from the edges is due to a stress state close to uniaxial tension in these areas. In an attempt to double the maximum strain, a sequential simple shear test is proposed. Each sequence is then composed of a simple interrupted shear loading step followed by a machining step, in order to reshape the free edges and remove the damaged material. In the present study, sequential simple shear experiments are performed on three engineering materials and discussed.

2. Experimental set-up

The specimen, see figure 2, is a sheet metal strip with two opposing cut-outs with rounded concave edges which diameter $d = 3.4$ mm is very close to the shear gage section height $h = 4$ mm. Shear cross-section refers to the minimal cross-section $S = L \cdot t$. Materials thicknesses are larger than $t \geq 0.8$ mm. Hereby, a shear gage height to sheet thickness ratio $h/t < 5$ is ensured, and the shear gage length $L > 37$ mm leads to a gage length to gage height ratio larger than $L/h > 9$, in line with previous recommendations by Bouvier at al. [3]. In vue of sequential machining and further specimen shortening, the initial length is chosen to be larger than 65 mm.

The shear test protocol includes multiple sequences. Prior to testing, the specimen is mounted in a shear testing device, composed of two frames (label 1 and 2 in figure 3), with a prismatic joint connection between each-other. Each sequence is composed of two steps. Step 1, figure 2, an interrupted shear test is performed up to a given value of displacement. This displacement corresponds to a strain between one third and half of the strain at load maximum of a monotonic shear test of the same material. During the second step, figure 2, the shear testing device along with the clamped specimen are positioned in a universal milling machine and the rounded free edges of the specimen are re-machined.

Shear tests are performed in a universal tensile machine (label 5 in figure 3), using a compression loading in order to avoid any clamping and un-clamping of the shear device. The shear device is made up of a fixed frame (label 1 in figure 3), and a movable frame (label 2). Both frames include a clamping jaw (label 6) with 4 clamping screws (label 7) in order to clamp the specimen (label 8). The shear load F is measured using the 50 kN load cell of the machine. The test is performed at a constant speed of the movable jaw of 0.5 mm/min ensuring a strain rate about 10^{-3} /s.

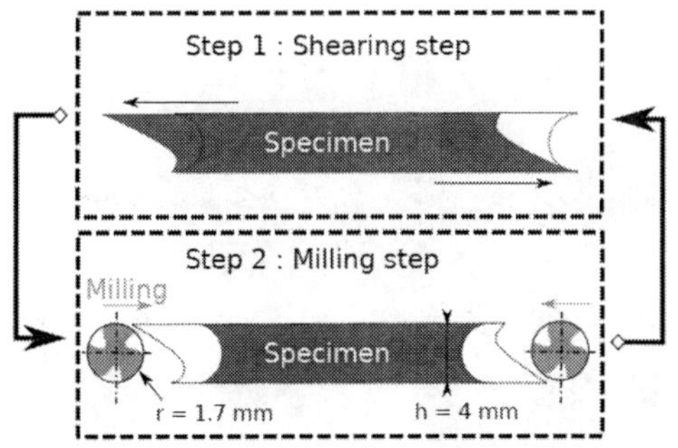

Figure 2. Scheme of the sequential shear. Step 1, simple shear test ; step 2, machining of the free edges.

The samples for the experiments are waterjet cut as shown in figure 3 with milled free edges. To allow for Digital Image Correlation (DIC), a random speckle pattern with a white background and black speckles is applied. A 4 Mpx camera (label 3 figure 3 ; 2000x2000 px) equipped with a 50 mm f2.0 lens is used to record the images of the test at a frequency of 1 Hz with a spatial resolution of 50μ m/px. Two LED spots (label 4 figure 3) are used for illumination. Images from the experiments are post processed with the commercial DIC software Aramis (GOM). The numerical and the experimental campaign led us to select a square subset size of 14 px and a step size of 8 px for the strain field calculations using a linear interpolation on a 3x3 points grid. These subset parameters enable to calculate the strain at the center on an area as small as 1.5x1.5 mm. The effective strain is computed from the principal strains by the equation:

$$\bar{\varepsilon} = \frac{2}{\sqrt{3}}\sqrt{\varepsilon_I^2 + \varepsilon_{II}^2 + \varepsilon_I \varepsilon_{II}} \tag{1}$$

After the simple shear step, the device is clamped on the table of a milling machine (figure 4). Free edges are machined using a 3.4 mm diameter cylindrical mill. A two teeth micro-grain tungsten carbide end milling tool with sharp edges has been selected to minimise the surface hardening of the edge. The removed machined length is about 3 mm on each side.

3. Numerical simulation of the test, stress and strain evaluations

Finite element simulations are performed for the design of the test and validation of the measurement technique using the commercial FE solver LS-dyna (R10) and implicit time integration. Mesh is made of 0.1 mm length edge single integration point brick elements resulting in 40 elements through the gage section height, see figure 5. One side of the gage section is fully clamped while a rigid body translation displacement u is applied to the opposite side. In order to gain detailed insight into the mechanical fields, a large range of engineering materials from soft aluminum alloys up to high strength steels and gage section length $L \in [30; 76]$ mm is involved. For the sake of brevity, only the main conclusions of the numerical study are presented.

Ideally, the shear stress and strain would be homogeneous along the specimen length. As already stated by An et al. [4], our calculations confirm that stress and strain homogeneity along the shear gage length evolve in opposite ways for rectangular and notched specimens. Nevertheless, the central element is in an almost perfect shear stress state and simple shear deformation. The goal of the shear test experimental measurement protocol is then to provide an accurate approximation of the shear stress and the shear strain at the central position from the load, the geometry of the specimen and the DIC displacement field. The shear stress is almost constant along the shear gage length of a notched specimen, while the shear strain is not homogeneous. For rectangular specimens, the shear strain is almost constant along the shear gage length, while the shear stress is not homogeneous. Hereby, for a notched specimen with gage length $L > 30$ mm, the shear stress of the central element can be approximated using the

IOP Conf. Series: Materials Science and Engineering 1238 (2022) 012086 doi:10.1088/1757-899X/1238/1/012086

Figure 3. Simple shear device

Figure 4. Machining set up

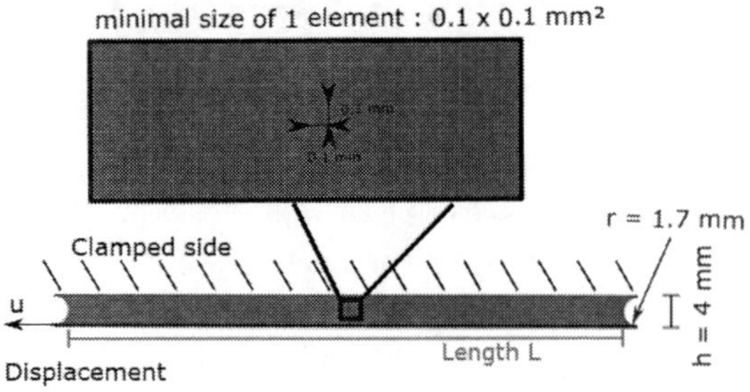

Figure 5. View of the mesh used for the simulation

average shear stress $\tau_{ave} = F/S$, and the error is found to be lower than 1.2% for effective strains up to 0.5 in the center of the specimen.

Thanks to the DIC parameters, the shear strain at the center can be calculated on a minimum area of 1.5x1.5 mm. The systematic use of the smallest ROI could lead to increase the noise of the strain value and the risk of mistake in the selection of the ROI on specimen surface. Hereby, the use of a notched specimen requires the definition of a proper Region Of Interest (ROI) around the center of the shear gage section for a robust and accurate calculation of an averaged strain which is an approximation of the shear strain of the element at the center. Considering a maximum discrepancy of $\pm 2\%$ of the effective strain, a capable $ROI_{2\%}$ is found to be 5 mm long and 2 mm high for the proposed geometries. Note that during our tensile test of the DC01 steel, the maximum discrepancy in stress was about $\pm 0.5\%$ for an engineering strain up to 0.5. Considering the behaviour law of the DC01 steel, a $\pm 0.5\%$ error in stress corresponds to a $\pm 2\%$ error in strain. In the following, the shear stress will be approximated using the averaged shear stress $\tau_{ave} = F/S$ value, the strain values will be approximated using the average strain in the $ROI_{2\%}$=5x2 mm central area.

4. Experimental results

Sequential and monotonic shear test were carried out on DP450 steel ($t = 0.8$ mm), AA2024-T3 aluminum alloy ($t = 1.2$ mm) and DC01 steel ($t = 1$ mm). For the latter, an additional in-plane torsion test was performed and already presented in Grolleau and al. [5].

Table 1. Specimen length at the beginning of each sequence

	Sequence 1	Sequence 2	Sequence 3	Sequence 4	Sequence 5
AA2024-T3	L = 76 mm	L = 68 mm	L = 63mm		
DP450	L = 77 mm	L = 72 mm	L = 66 mm	L = 60 mm	
DC01	L = 65.5 mm	L = 59 mm	L = 50 mm	L = 43 mm	L = 37 mm

An example of the shear stress as function of the effective strain evolution is presented for a monotonic simple shear test and a sequential shear test in the case of the DP450 steel in figure 6. A 0.2 shear step increment value was used for the sequential test. For both tests the initial specimen length is 77 mm, see Tab. 1. In the case of the monotonic test, a maximum stress is reached at an effective strain value about 0.4, and the hardening rate is monotonically decreasing during the test. It is clear from figure 6 that even with the interruption of the test and the re-machining of the free edges the sequential testing enables to recover the same behaviour as the monotonic test. Moreover, the sequential test result doesn't exhibit any saturation of the shear stress, and the hardening rate is almost constant up to an effective strain of 0.7. Similar results are obtained for the three materials.

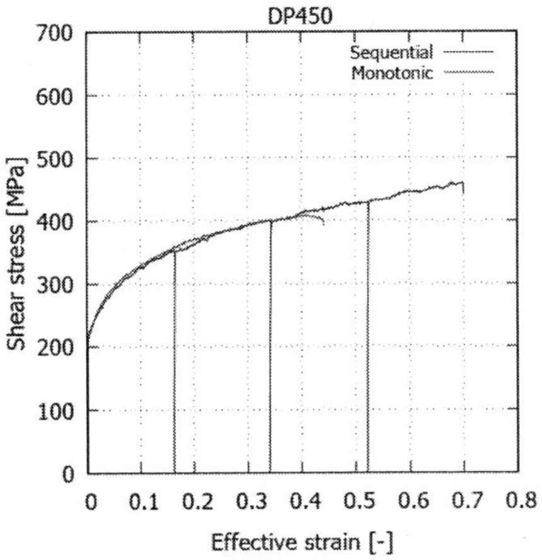

Figure 6. Comparison of monotonic (red solid line) and sequential (blue solid line) shear test of DP450 steel, shear stress vs effective strain.

The homogeneity of the mechanical fields is presented in the case of the sequential test of a DC01 steel in figure 7 and 8 showing the evolution of the effective strain as a function of the position along and across the shear gage section respectively. The 5 curves, in blue solid lines, correspond to the final stage of each sequence, and the corresponding shear gage lengths are given in Tab 1.

In the case of the sequences number 2 and 4, the absolute values of the major and the minor strains are plotted as function of the position along the gage section in figure 7 using green and red solid lines. All along the test, and along the gage section, major and minor strains have almost the same absolute value. The discrepancy is negligible in the central area compared to

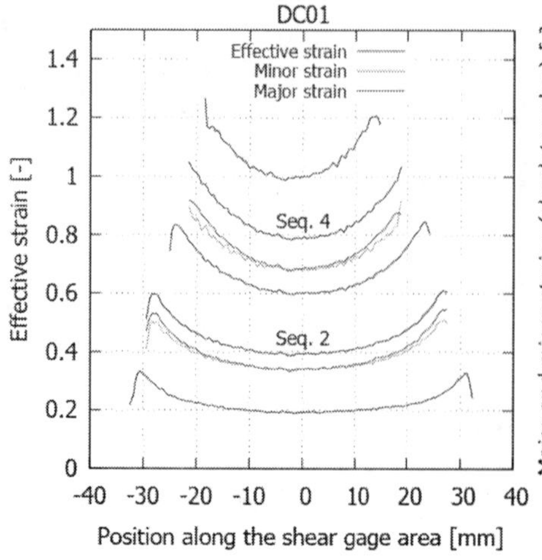

Figure 7. Effective strain evolution along the shear gage section.

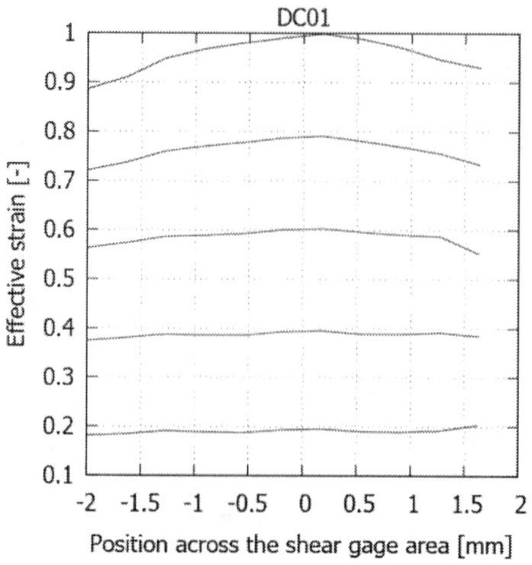

Figure 8. Effective strain evolution across the shear gage section.

the noise in strain, and increases towards the edges up to a strain discrepancy about 0.03, which confirms the simple shear kinematics state hypothesis in the center. As discussed in the section 3, the effective strain field is not homogeneous, with a pronounced convex shape along the gage length and a slightly concave shape across the gage section. These experimental results confirm that a reduced ROI is required for the approximation of the strain value at the center of the specimen.

In figure 9, the relative error from the strain at the specimen center is plotted for various ROI sizes. The reference value of the strain at the center is calculated on a 2x2 mm ROI at specimen center. When the ROI corresponds to the full specimen surface (see blue solid line), the error is as high as 20% at the beginning of the test. Each machining step leads to a drop in the error, see for example between sequences 1 and 2. The smaller the ROI, the smaller the error, and the error is lower than 2% for a ROI size of 5x2 mm. Note that for smaller sizes, the signal to noise ratio can become prohibitive.

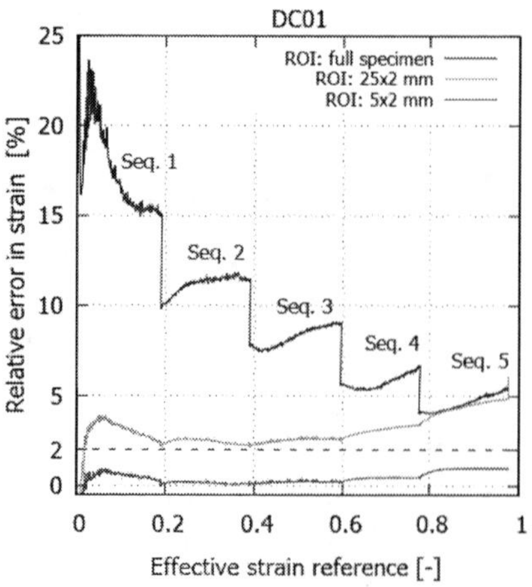

Figure 9. Evolution of the relative error in strain for three ROI: full specimen ROI (blue solid line), 25x2 mm ROI (green solid line) and 5x2 mm ROI (red solid line). The reference strain is calculated using 2x2 mm ROI at specimen center

The figures 10 anf 11 show the results obtained in the case of monotonic, sequential and in-plane torsion of the DC01 steel. The figure 10 (left) focuses on the repeatability obtained using monotonic shear test. The same conclusions as for the DP450 steel can be drawn, with a maximum load occurring at a strain about 0.6 and a decreasing hardening rate in the case of the monotonic shear. The stress vs. strain curve is again characterised by a concave shape and the occurence of a maxima in stress. The error bars indicates the min and max values obtained over three repeats, and show a large discrepancy for effective strain above 0.45. This value corresponds to the strain at which monotonic and sequential results diverge. The sequential result is plotted in green solid lines up to a strain value about unity. A shear strain increment of 0.2 is used, highlighted with solid circles.

The repeatability of the sequential shear test results can be appreciated in figure 10 (right) for three repeats. The dispersion is slightly increasing at large strain. For example at an effective strain of 0.8, see figure 10 (e), the maximum dispersion is 15 MPa. This corresponds to the dispersion observed in the monotonic tests at an effective strain of 0.4.

The repeats have been performed using various shear strain increments values from 0.2 up to 0.35. The effect of the shear strain increment value can be appreciated comparing the test number 2, blue curve (shear strain increment about 0.2), and test 1, red curve (shear strain increment about 0.35). At a strain about 0.35, see figure 10 (d), the test 1 is at the beginning of its second sequence and a drop in the shear stress value is clearly visible. At the same strain, the test number 3 is at the beginning of the third sequence, and the use of smaller shear step increment lead to negligible drop in stress. From our experiments, it seems that an optimal shear step increment value is about one third and half the strain at force maximum of the corresponding monotonic shear test.

Nevertheless, whatever the length of the shear strain increment, a stress maximum always occurs during the last sequences. This is attributed to the initiation of fracture along the free edges. From sequence to sequence, this phenomenon occurs closer to reloading. The test is stopped when the fracture from the side starts almost immediately after reloading.

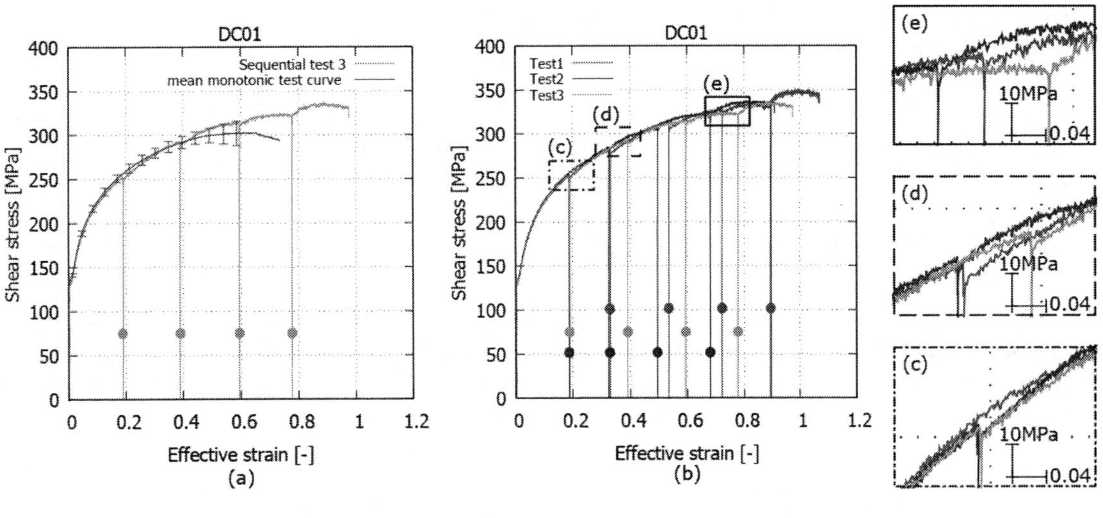

Figure 10. Comparison of monotonic shear tests (red solid line) with sequential (green solid line) shear test on DC01 steel (left), repeatability of the sequential shear test on three tests (right). Each end of shear step is highlighted with a dot at the corresponding strain value, vertical position is meaningless.

The figure 11 shows the comparison of the monotonic, sequential and in-plane torsion tests results in terms of stress versus strain. All results agree well up to an effective strain of 0.4.

Above that strain, the sequential and the torsion test results only agree up to an effective strain of 0.85.

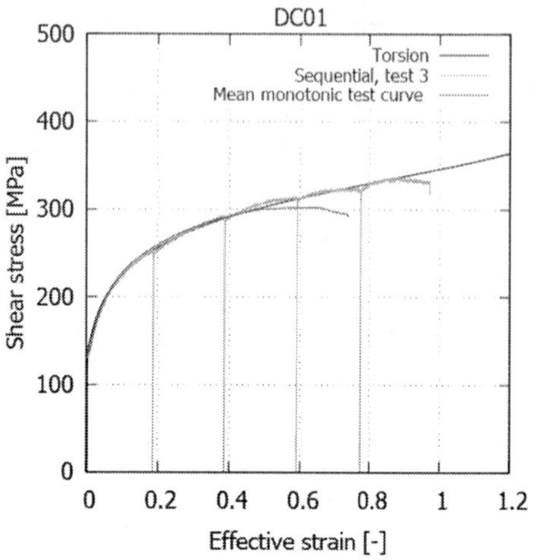

Figure 11. Comparison of simple shear (monotonic, red solid line ; sequential, green solid line) and in-plane torsion (black solid line) results on DC01 steel, shear stress vs effective strain.

5. Conclusion

In order to extend the shear strain range achievable with the simple shear test, we propose a sequential shear test protocol. The specimen is a sheet metal strip with two opposing cut-outs with rounded concave edges. The shear test protocol includes multiple two-steps sequences. First, an interrupted shear test is performed up to a given value of displacement. Second, the rounded free edges of the specimen are re-machined. Special care is devoted to the position of the specimen by transferring the entire testing frame from the testing to the milling machines. Stress versus strain curves obtained from monotonic tests are characterised by the occurence of a stress maxima, a monotonically decreasing hardening rate and an increasing dispersion of the results close to the maximum stress. On the contrary, sequential tests results lead to an almost constant hardening rate and a higher repeatability of the results up to a maximum strain value which is twice higher than for the monotonic test. After a numerical study of the experimental set-up, the sequential test has been applied to three engineering materials. In the case of a DC01 steel, the results have been validated from against a newly-developed in plane torsion test.

Aknowledgments

The authors greatly appreciated financial support from the Région Bretagne (ARED PhD grant). This research was also supported by the chair SAFRAN - Fondation Saint-Cyr (PhD grant). Mr. Arnaud Penin, CEO of the MATandSIM company, is gratefully acknowledged for his helpful and instructive discussions about the experimental part of the project.

Reference

[1] Yin Q, Soyarslan C, Isik K and Tekkaya A 2015 *International Journal of Solids and Structures* **66** 121–132
[2] Reyne B, Hérault D, Thuillier S and Manach P Y 2021 *International Journal of Mechanical Sciences* **208**
[3] Bouvier S, Haddadi H, Levée P and Teodosiu C 2006 *Journal of Materials Processing Technology* **172** 96–103
[4] An Y, Vegter H and Heijne J 2009 *Journal of Materials Processing Technology* **209** 4248–4254
[5] Grolleau V, Roth C and Mohr D 2022 *International Journal of Solids and Structures* **236-237** 111341

Effect of Initial Temper on the Warm Forming Characteristics of a High Strength 7000-series Al-Zn-Mg-Cu Alloy

S DiCecco, M Di Ciano, C Butcher, M Worswick

University of Waterloo, 200 University Ave W, Waterloo, ON N2L 3G1, Canada

sdicecco@uwaterloo.ca

Keywords: 7000-series, Warm Formability, Aging, Curvature, Aluminium

Abstract. In this work, the formability of a developmental 7000-series copper containing aluminium alloy was assessed at room temperature (RT), 150°C, 175°C and 200°C in pre-aged (PA), peak-aged (T6) and overaged (T76) tempers using Nakazima tests with stereoscopic digital image correlation (DIC) strain measurement. The limit strains were identified using a novel curvature-based approach to detect the formation of an acute neck. The tensile mechanical properties in these warm forming processing routes were characterized with and without a paint bake cycle. Finally, a thermo-mechanical tensile simulator was used to evaluate the constitutive response of the PA and T76 tempers as a function of strain-rate and time at 175°C. Formability results found the selected PA temper to have a good room temperature formability and a mild positive response to the selected warm-forming cycles. The T6 and T76 tempers both exhibited increases in formability in response to warm forming. The PA temper had a significant positive response to short-duration warm forming and subsequent paint baking, with the yield strength increasing from 420 MPa to 512 MPa following this thermal cycle. For the T6 temper, the warm-forming cycle showed a trend characteristic of retrogression and re-aging, with the warm-forming cycle dropping the yield strength from 566 MPa to 534 MPa and the subsequent paint-bake re-aging to 554 MPa. The effect of aging during pre-heating prior to warm forming on the warm constitutive response of the PA and T76 tempers was also investigated. Both tempers exhibited rather different aging responses to short-duration thermal cycles. In the PA temper, this manifested as an increase in at-temperature yield strength and loss of hardening rate. In contrast, the T76 temper exhibited a drop in strength since this temper is already over-aged prior to warm forming. Both the PA and T76 tempers showed comparable at-temperature strain-rate sensitivity.

1. Introduction

While the strength –to-weight ratio of 7000-series aluminium alloys is attractive for automotive light weighting applications, the manufacturing of structural automotive components is challenging due to their limited room temperature formability. The 7000-series alloys are also susceptible to stress corrosion cracking (SCC), which must be accounted for via process route design prior to use in automotive applications, *e.g.* by over-aging as a means to reduce SCC [1], [2], [3]. To address formability limitations, there has been a keen interest in warm forming (WF), [4], [5], [6], [7], [8], [9], as well as hot stamping [10], [11], [12], [13] of high strength 7000-series alloys in recent years. In hot stamping, the sheets are first solutionized by heating above approximately 470°C to dissolve the precipitates, after which they are formed non-isothermally in a cooled die system. In warm forming, sheets are formed isothermally or non-isothermally at temperatures above approximately 150°C, but below the alloy recrystallization temperature. An advantage of hot forming is the enhanced formability and reduced springback but the parts require a secondary ageing heat treatment to obtain a peak strength T6 condition. An advantage of warm forming is that secondary ageing treatments can be avoided since the material does not require solutionization prior to forming. The starting temper can be selected such that the secondary ageing that will occur during warm forming and the subsequent paint bake cycle can produce a final temper near the peak strength or intentionally over-aged to increase resistance to SCC.

Sotirov *et al.* [14] have identified formability limits for peak-aged AA7075-T6 sheet and considerable benefit was found for warm forming in the 200°C - 230°C temperature range. Kumar *et al.* [4] demonstrate that the limiting draw ratio of AW7020-T6 benefits from warm forming up to 250°C, but with significant degradation in the as-formed yield and ultimate tensile strengths after 200°C. In the study of warm formability of AA7075-T6 by Wang *et al.* [7], a decrease in mechanical properties was observed when warm forming above 230°C which was attributed to precipitate coarsening (over-aging); however, they note some strength loss is recovered during a paint-bake cycle as compared to the as-received material that saw a mild decrease in strength after paint-baking.

At present, there is not a clear indication of which initial temper should be used for 7000-series warm forming, for automotive applications. This research investigates the (i) warm formability and (ii) final property characteristics, such as strength and ductility, of a developmental 7000-series alloy, designated herein as AA7xxx, for three different initial tempers comprising (i) a pre-aged (PA) temper, a peak-aged (T6) temper and (iii) an over-aged (T76) temper. The isothermal formability of each temper was characterized at room temperature (~23°C), 150°C, 175°C and 200°C under near plane-strain loading. Tensile samples were extracted from the flanges of the PA and T6 samples after warm forming to understand the effect of the forming process on the subsequent mechanical properties with and without a 30 minute, 177°C simulated paint-bake. In parallel with the formability experiments, the warm constitutive response of the PA and T76 tempers is characterized at 175°C, as a function of time at temperature and strain-rates relevant to the forming experiments to provide insight into the observed formability trends. Based on the culmination of these tests, final recommendations are made for the warm forming of this developmental AA7xxx alloy.

2. Testing Methodology and Analysis Techniques

2.1. Material

The starting sheet material used throughout this study was a 2 mm thick, developmental 7000-series aluminium alloy, designated AA7xxx, by Arconic in the T76 condition. The T76 is an overaged temper which imparts improved stress corrosion cracking (SCC) resistance [2], [3]. The nominal composition for the developmental AA7xxx alloy is given in Table 1.

Table 1 - Nominal composition of AA7xxx, in wt %.

Alloy	Cr	Cu	Fe	Mg	Mn	Si	Ti	Zn	Zr	Other	Al
AA7xxx	<0.04	1.3-2.0	<0.08	1.2-1.8	<=0.4	<=0.06	<=0.06	7.0-8.0	0.08-0.15	0.15	Bal.

To produce the pre-aged (PA) and peak-aged (T6) tempers, a batch of the AA7xxx alloy was solutionized at approximately 473 °C for 10 minutes in a fluidized sand furnace (FB-08C Fluidized Bath Furnace, Techne®), water quenched, and subsequently aged. The aging process for the PA temper [10], [15] was a two-step process of 48 hours of natural aging at room temperature followed by 4 hours at 100°C of artificial aging in the fluidized sand furnace. The T6 (peak aged) treatment was taken as a single step 24 hour aging process at 120°C. Between any lulls in heat-treating, samples were held in dry ice (-78.5°C) to prevent unintentional aging.

Representative stress-strain curves for the PA, T6 and as-received T76 tempers are included in Figure 1. The PA temper shows significant work hardening, with a yield strength (YS) of 419 MPa ultimate tensile strength (UTS) of 524 MPa at a uniform elongation (UE) of 18.2%. In the T6 peak-aged condition the YS and UTS increase to 507 MPa and 566 MPa, respectively, but the UE decreases to 14.1% as the overall work-hardenability decreases and the stress-strain curve becomes flatter relative to the PA condition. In the overaged T76 condition, the work-hardenability further decreases and the strength (494 MPa/524 MPa YS/UTS) and UE (8.8%) also decrease relative to the T6 condition. In the T76 temper, there is a distinctive flat region after yielding not observed in the PA and T6 tempers. Fribourg *et al.* [16] have explored this behaviour in detail for AA7449, an Al-Zn-Mg-Cu class alloy, and showed that this flat region after yielding emerges and becomes more prominent as the extent of overaging increases. This trait is thought to be associated with the transition from shearable to non-shearable precipitates with overaging.

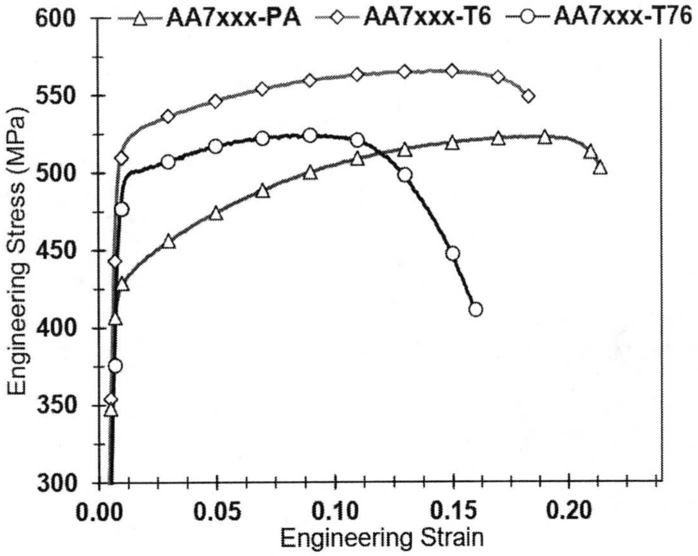

Figure 1 – Representative stress-strain curves of AA7xxx in the PA, T6 and T76 tempers, tested along the rolling direction.

2.2. *Limiting Dome Height Testing and Analysis*

Formability characterization of the PA, T6 and T76 tempers was undertaken at room temperature (23°C), 150°C, 175°C and 200°C. Testing was accomplished using a double-acting hydraulic press with tooling heated by closed-loop controlled embedded heater cartridges, as described in [9] and [17]. The specimens were heated *in-situ*, with typical heating curves measured at the specimen flanges plotted in Figure 2c). The initial heating rate corresponds to the placement of the specimen within the tooling. At approximately 60 seconds, the tooling is closed, and the clamping load is applied, giving rise to an increase in heating rate, followed by asymptotic heating to the target test temperature. A minimum of three repeated LDH tests were completed for each test condition.

The warm-forming cycle entailed a 3 minute heating time of the blank to the target temperature after clamping, with the blank within 20°C of the targeted temperature during approximately the final 60 seconds of heating. The forming duration at elevated temperature varied as a function of dome height. To minimize variation, all samples were kept at elevated temperatures for 60 seconds from the time testing began prior to water quenching after testing was completed.

A 101.6 mm diameter Nakazima punch was used in the formability testing per the ISO12004-2:2008 standard [18], as shown in Figure 2a), with a corresponding annular Nakazima die set with an inner diameter of 106 mm and die entry radius of 6.35 mm. The formability was evaluated using a near plane-strain *dog-bone* type geometry with a gauge width of 76.2 mm and entry radius of 20 mm [18] under limiting dome height (LDH) testing. Samples were tested in the transverse sheet direction (TD), *i.e.* the sheet rolling direction was perpendicular to the sample major axis and major strain direction. A clamping force of 330 kN was applied to prevent material draw-in. The punch speed during testing was

1.0 mm/s and lubrication between the punch and blank interface consisted of 2-3 layers of 0.127 mm Teflon™ film to promote centre fracture.

In the formability testing, the Nakazima samples were tested to failure and stereoscopic digital image correlation (DIC) techniques were used to capture the deformation and corresponding strain fields on the sample surface not in contact with the punch over the duration of each test. A sample strain profile prior to cracking is included in Figure 2b). Deformation of speckled samples was tracked at 140 frames per second using two 4 megapixel FLIR Gazelle Camera Link cameras. The camera, lens and test setup produced a pixel density of approximately 14 pixels/mm. Image analysis was completed using the Correlated Solutions Vic3d© version 8 software package. The following key analysis parameters were used: a subset of 25 – 35 pixels, a step size of 3 pixels, and a Gaussian strain filter of 5 pixels. These analysis parameters influence the minimum strain resolution and control the range over which averaging occurs [11], [19] and thus their effect is more significant in the presence of large strain gradients.

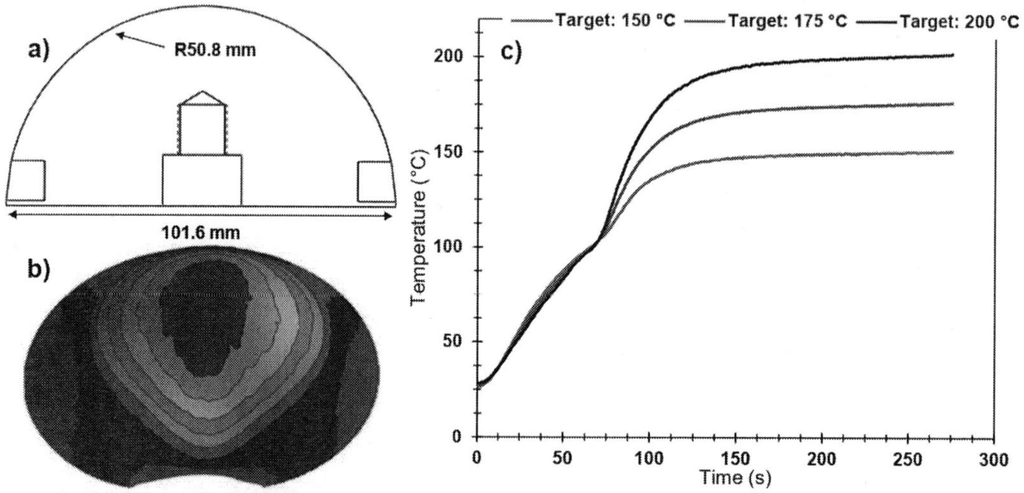

Figure 2 – a) Schematic of Nakazima punched used in limiting dome height testing with b) a 3D strain contour of a dome sample prior to failure. In c), sample time-temperature profiles are included for the heating period of LDH testing, as measured at the specimen flange.

Necking onset, corresponding to reaching the material forming limit, was determined using the acquired DIC data in conjunction with a necking detection scheme referred to herein as the “curvature approach”. Necking is assumed to have initiated once changes in the surface curvature occur due to initiation of an acute neck. The curvature approach applied in this work is termed the Enhanced Curvature Method (ECM) and was built upon the method developed by DiCecco *et al.* [9], [17]. For comparison purposes, limit strains are also computed using the ISO12004:2-2008 methodology.

2.3. Aging Response of PA and T6 Tempers to Warm Forming Process

Tensile testing was performed on as-thermally processed samples to characterize their response to the various aging treatments. ASTM-E8 sub-sized tensile coupons [20] were machined from the flanges of the warm formed LDH samples in the PA and T6 tempers. The gauge length and width of the sub-sized specimens were 25 mm and 6 mm, respectively. The samples were tested under two conditions: (i) as-extracted from the flanges following warm forming, and (ii) as-extracted with a subsequent simulated paint-bake of 177°C for 30 minutes in the fluidized sand bath furnace. Three repeated tensile tests were completed for each condition.

Tensile testing was completed in the sheet rolling direction on a 100 kN load frame (MTS Criterion 45) and DIC techniques were used to extract nominal strain data, using 25 mm virtual extensometers, from each tensile sample. The crosshead speed for all tests was 0.25 mm/s.

2.4. *Warm Constitutive Response of PA and T76 Tempers*

Constitutive characterization of the AA7xxx–PA and –T76 tempers was completed at 175°C for nominal strain rates of 0.01/s and 0.1/s. The PA and T76 tempers were chosen since they were expected to exhibit the strongest and lowest response to elevated temperature deformation, respectively.

Testing was performed using a closed-loop thermo-mechanical simulator (Gleeble-3500, Dynamic Systems Incorporated) with the ability to resistively heat samples up to 10,000°C/s. A modified ASTM-E8 sub-sized tensile sample was used to minimize thermal gradients along the gauge length, in which the grip length was increased to 60 mm and the gauge length was reduced to 20 mm. The sample width was nominally 6 mm. Further sample details are given in [21]. Virtual extensometers using DIC techniques were used to measure strain histories over each test.

The heating profile entailed a linear ramp from room temperature to 175°C over 10 seconds, followed by a *hold time* at the prescribed temperature, after which isothermal tensile testing was performed. In this work, testing was done with hold times of 5 s and 180 s. The intent of the two hold times was to assess any changes in the elevated temperature constitutive response that may arise from aging prior to forming during heating to the forming temperature 175°C.

3. Results and Discussion

3.1. *Formability Results*

Major true limit strains as a function of temper and forming temperature are plotted in Figure 3, as well as dome heights measured at the time of failure using DIC analysis, with the PA (red) results in a) T6 (blue) results in b), and T76 (black) results in c). The minor limit strains ranged from 0.016 to 0.030 across all conditions.

The limit strains calculated using the enhanced curvature method (ECM) are represented by the solid bars and the limit strains determined using the ISO12004-2:2008 method (hereafter referred to as the ISO standard) by the dashed bars. For each temper, some disagreement is observed between the quantitative values using the ECM and ISO methods; however, both approaches generally showed the same trends. As such, the following discussion focuses primarily on the ECM limit strains.

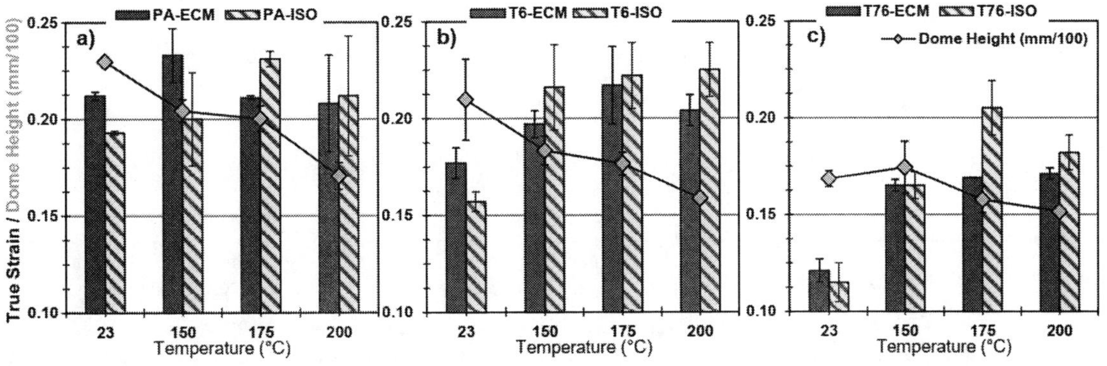

Figure 3 – Major true limit strain as a function of temperature for the different AA7xxx tempers: a) PA, b) T6, and c) T76. Solid bars indicate limit strains based on ECM, while dashed bars indicate values from ISO method. Dome heights corresponding to failure are plotted on the same axis in units of mm/100.

In Figure 3, the RT limit strain of the PA temper is approximately 0.21, while the T6 and T76 tempers are lower at 0.18 and 0.12, respectively. These rankings are consistent with the RT work-hardenability trends in Figure 1, in that the tempers with the higher degree of work hardening exhibit the highest RT formability. For the PA temper, the maximum limit strains are achieved between 150°C and 175°C, depending on the limit detection scheme applied. Nevertheless, at 200°C, both limit strain measures show a reduction of formability of the PA temper relative to RT as well as a marked increase in measurement scatter.

The trends in the T6 formability (Figure 3b) roughly mirror those of the PA temper, with the T6 formability improving up to 0.22 major true strain at 175°C before plateauing or degrading at 200°C based on the ECM data. The T76 temper exhibits an improvement in formability from RT to 150°C, from 0.12 to 0.17 major true strain; however, the formability of the T76 temper lies below that of the PA and T6 tempers for all temperatures.

Although the forming limit strains were improved with warm forming for all tempers, the measured dome heights at failure decreased with increases in temperature for all tempers. Comparison of the data in Figure 3 reveals that the dome heights for the PA temper are slightly higher than those for the T6 temper, while the T76 samples had inferior dome heights. The PA temper also offers excellent RT formability which increased by 10% with warm forming at 150°C.

3.2. Secondary Aged Tensile Properties of PA and T6 Tempers Following Warm Forming

Engineering stress-strain curves from tensile samples extracted from the warm-formed (**WF**) PA and T6 LDH specimens are plotted in Figure 4 for each forming temperature (open symbols). Corresponding data from samples that were paint-baked after forming are also plotted and are distinguished by solid symbols. The room-temperature curves from Figure 1, corresponding to the material condition prior to forming, are replotted for comparison purposes (triangular symbols).

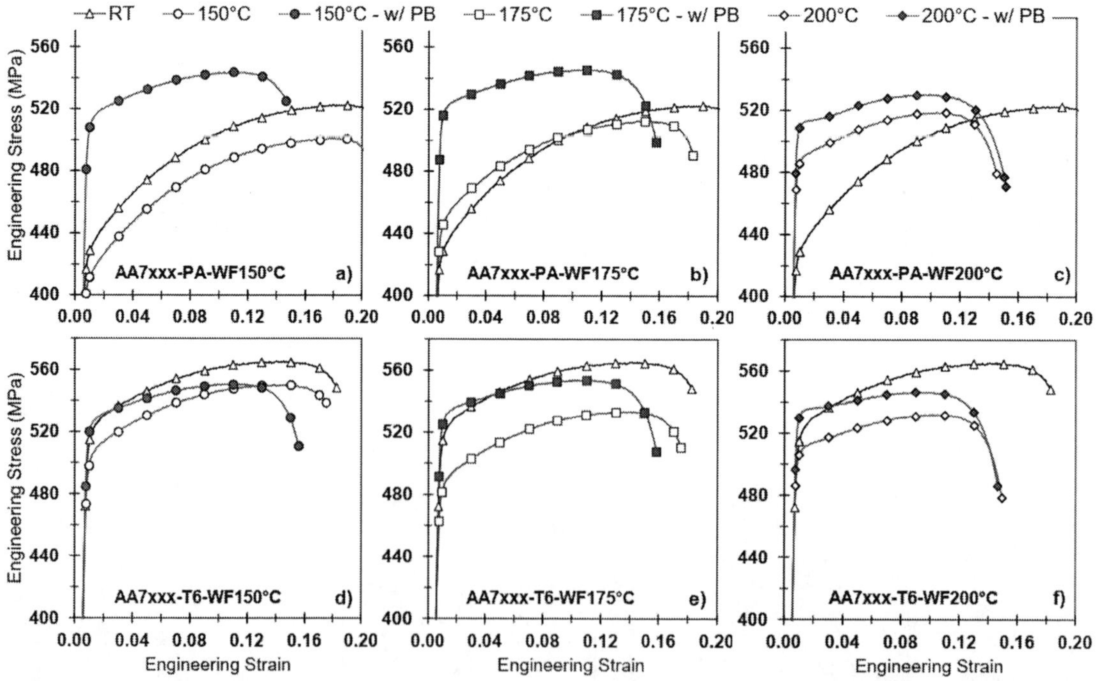

Figure 4 – Tensile response of a) – c) the PA temper and d) – f) the T6 temper prior to forming (RT - triangular symbols), after warm forming at the temperatures indicated (open symbols), and after a subsequent paint-bake of 177°C for 30 minutes solid symbols. The curves plotted are the average of three tests.

For the PA condition, a decrease of 16 MPa in YS relative to RT is observed in the stress-strain response after exposure to the lowest WF temperature of 150°C (Figure 4a)), for which the flanges were at the target temperature for under 120 seconds. After the paint-bake, a significant aging response is observed and the strength of the warm formed PA samples approach that of the RT T6 condition (seen in Figure 4 d-f). The observed drop and subsequent recovery in strength may be indicative of retrogression and re-aging [22] for the 150°C WF+PB cycle.

After WF at the higher temperatures of 175°C and 200°C, the PA temper shows a positive aging response to both WF and WF+PB, in that the strength increases relative to the RT PA condition, while the ductility decreases. The highest tensile strength of 546 MPa ± 2 MPa is achieved after the 175°C WF+PB cycle, while the strength following the 200°C WF+PB is lower which may indicate overaging during the PA+200°C+PB cycle.

The trends in YS, UTS, UE and TE (total elongation) are summarized in Figure 5 a) – d). After a 200°C forming cycle and paint bake, the ductility in UE and TE of the PA and T6 tempers approximate that of the as-received T76. The yield and tensile strengths also appear to converge towards the overaged T76 properties, although are higher after paint-bake.

From a SCC susceptibility standpoint, the trends towards overaging in the PA temper following the 200°C+PB route may be optimal [2], [3], while offering improved strength, albeit with no formability improvements relative to the lower warm-forming temperatures. The RT and 150°C forming routes, by contrast, offer excellent formability for a 7000-series alloy; however, corrosion studies are required to assess the SCC resistance following this processing route.

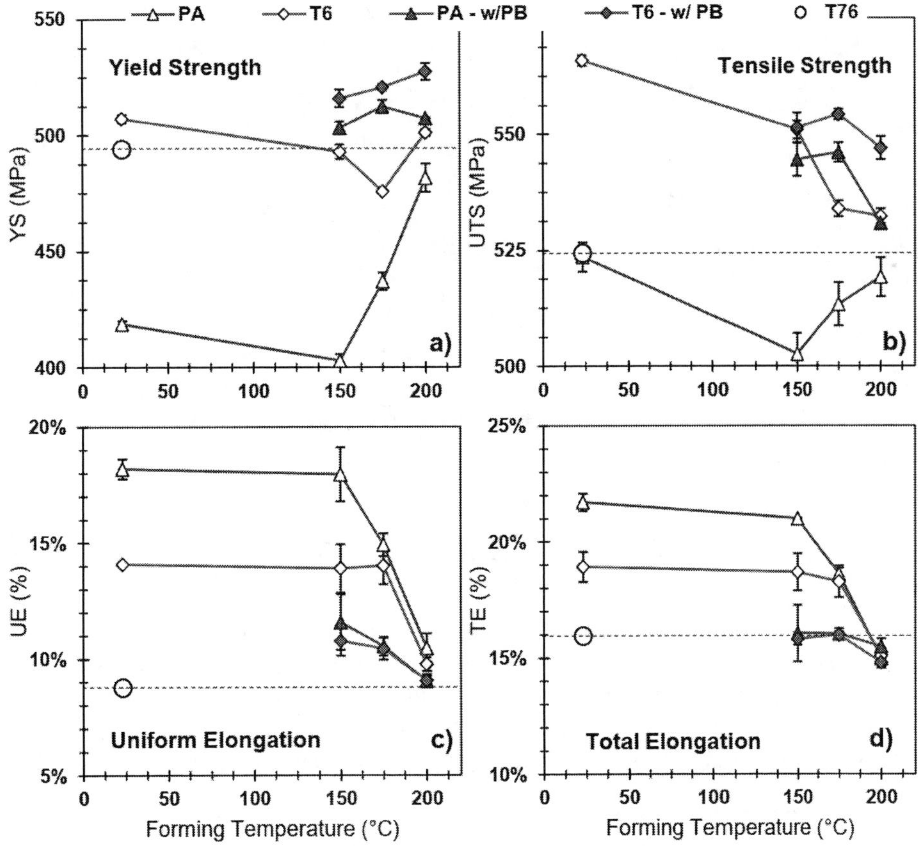

Figure 5 – Effect of the warm forming cycle temperature on the tensile properties of the PA and T6 tempers with (solid symbols) and without (open symbols) a paint bake. a) and b) show trends in yield and tensile strength, respectively. Uniform (c)) and total (d)) elongation capture trends in ductility.

A softening in the stress-strain response of the T6 temper relative to the RT baseline is observed after warm forming (Figure 4 d-f) which is attributed to a retrogression response [16]. This effect is most significant at 175°C, with a drop in YS and UTS of 31 MPa (6%) and 32 MPa (6%), respectively. A positive re-aging response is noted following PB of the T6 samples for all WF cycles, with the most significant re-aging occurring again at 175°C.

With consideration to formability and final strength characteristics, the T6 temper formed at 175°C offers improved formability without a significant degradation in strength from over-aging and is expected to show improved corrosion performance [3], while offering improved strength relative to the as-received AA7xxx-T76 alloy. At 200°C, the formability benefits are more muted, but signs of over-aging are more prominent.

3.3. Warm Constitutive Behaviour of the PA and T76 Tempers

To further understand the effects of the various starting tempers during warm forming, the engineering stress-strain curves for the PA and T76 tempers from thermo-mechanical testing at 175°C are plotted in Figure 6 a) and b), respectively. The tensile properties, comprising YS, UTS, and UE, are summarized in Table 2. Material rate sensitivity was examined by performing these warm stress-strain tests at strain rates of 0.01 and 0.1/s. In addition, the effect of aging during heating prior to forming was examined by introducing a *hold time* at elevated temperature prior to isothermal tensile testing. The warm constitutive behaviour of the two starting tempers is dramatically different. The T76 temper begins to diffusely neck immediately after yielding. By contrast, the PA temper exhibits conventional work-hardening at 175°C for all hold times and strain rates.

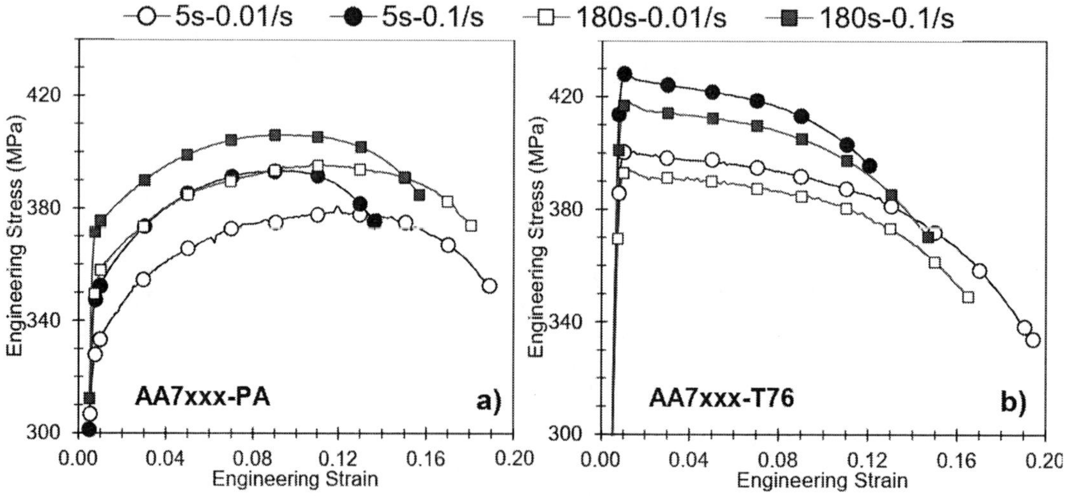

Figure 6 – Engineering stress-strain response of the PA (a)) and T76 (b)) tempers at 175°C. Note the T76 temper begins to diffusely neck immediately after yielding in figure b). Conversely, the PA temper shows conventional positive work hardening.

Given the lack of work hardening in the elevated temperature T76 data, there is effectively no plastic uniform elongation, and yet the T76 temper positively responded to warm forming. In Figure 6 b), a positive rate sensitivity is observed for each hold time, whereby there is an approximately 7% increase in yield strength from 0.01/s to 0.1/s. This positive rate sensitivity is the driving force for promoting stability during diffuse necking, leading to the formability improvements in the T76 temper with warm forming.

Table 2 – Summary of tensile properties for the PA and T76 tempers under warm tensile testing at 175°C. Both tempers exhibit similar degrees of positive strain-rate sensitivity. The underaged temper shows at-temperature strengthening as a function of time at temperature and the overaged alloy, by contrast, decreases in strength with time at elevated temperature.

Hold	Nominal	AA7xxx-PA			AA7xxx-T76		
Time (s)	Rate (s^{-1})	YS (MPa)	UTS (MPa)	UE (%)	YS (MPa)	UTS (MPa)	UE (%)
5	0.01	328.3 [1.6]	379.4 [2]	11.8 [0]	399.3 [3.3]	400.6 [2.7]	1.3 [0.7]
5	0.10	347.8 [2.8]	394.1 [3.2]	9.1 [0]	426.4 [6.3]	427.4 [6.4]	0.9 [0.2]
180	0.01	352.8 [3.2]	396 [3.9]	10.7 [0.9]	390.6 [4.9]	392.4 [5.4]	2.2 [1]
180	0.10	373.2 [2.5]	406.8 [1.8]	9.2 [0.1]	414.6 [2.8]	416.5 [3.8]	1.4 [0.4]

For the PA condition, a positive rate sensitivity (increase in strength) of approximately 6% between the 0.01/s and 0.1/s conditions was observed, independent of hold time and the shape of all tensile curves were relatively consistent, per Figure 6a). At temperature, the warm formability (ECM-based data) saw a maximum improvement of only 10% at 150°C and almost no difference to RT at 175°C. In contrast to the T76 temper, the PA temper had strong RT work-hardenability and showed correspondingly good formability. In addition, the PA temper exhibited consistently higher formability than the T76 temper for all temperatures, which is largely attributed to the prominent work-hardening in the PA condition.

For the PA temper, the effect the longer (180s) hold time at 175°C prior to tensile testing is an increase in yield strength of about 24 MPa, compared the shorter (5s) data. This increase is attributed to precipitation strengthening (aging) prior to commencing the tensile test. In contrast, the overaged T76 temper experiences additional over-aging with increases in hold time, and exhibits an at-temperature decrease in yield strength of 9 MPa from a 5 s hold time to a 180 s hold time at a nominal rate of 0.01/s.

4. Conclusions

The warm formability and final strength characteristics of a developmental AA7xxx alloy were assessed in the PA and T6 tempers, as well as a baseline T76 temper as a function of temperature. The PA temper offered the highest room temperature limit strain of 0.21, compared to 0.18 for the T6 temper at RT. The PA temper saw a 10% improvement in formability up to 150°C to reach a limit strain of 0.23, while the highest formability of the T6 temper was 0.22 at 175°C. The overaged temper saw some benefit from warm forming; but was in general the least formable temper for all conditions.

Tensile testing of samples exposed to the warm forming time-temperature cycle was completed with and without a paint bake. It was observed that the PA temper showed a strong positive response in strength to the combination of WF and PB at 150°C and 175°C. With the 200°C WF+PB cycle, the PA temper had tensile properties approaching the baseline T76 properties, indicating this cycle may lead to overaging of the initial PA temper. The T6 temper showed a softening response with warm forming, notably at 175°C, consistent with retrogression and exhibited recovery in strength following the simulated paint-bake.

Warm tensile testing of the PA and T76 tempers at 175°C and corresponding formability results suggest the positive strain-rate sensitivity of the alloy at elevated temperatures may promote stability during diffuse necking, hence the positive WF response of the T76 temper, despite diffusely necking immediately after yielding in uniaxial tension testing at 175°C. Time at 175°C had little influence on the strain-rate sensitivity, but did lead to in-situ aging (strengthening) for the PA temper and coarsening of the T76 temper.

Acknowledgments
The authors greatly appreciate financial support from the Honda R&D Americas, Promatek Research Centre (Cosma International), Arconic Technical Center, the Natural Sciences and Engineering Research Council of Canada, the Canada Research Chairs Secretariat, the Ontario Advanced Manufacturing Consortium, the Ontario Research Fund and Ontario Centres of Excellence.

References

[1] Davis J, Corrosion of aluminum and aluminum alloys, 1999 *ASM International*

[2] Sun X, Zhang B, Lin H, Zhou Y, Sun L, Wang J, Han E-H and Ke W, Correlations between stress corrosion cracking susceptibility and grain boundary microstructures for an Al-Zn-Mg alloy, 2013 *Corros. Sci.* **77** 103-12

[3] Chen S-y, Chen K-h, Dong P-x, Ye S-p and Huang L-p, Effect of heat treatment on stress corrosion cracking, fracture toughness and strength of 7085 aluiminum alloy, 2014 *Trans. Nonferrous Met. Soc. China* **24** 2320-25

[4] Kumar M, Sotirov N and Chimani C M, Investigations on warm forming of AW-7020-T6 alloy sheet, 2014 *J. Mater. Process. Technol.* **214** 8 1769-76

[5] Kumar M and Ross N, Warm forming and paint bake response of AW-7xxx sheet in various tempers, 2015 *Graz, Austria Key Eng. Mater.*

[6] Kumar M and Ross N, Influence of temper on the performance of a high-strength Al-Zn-Mg alloy sheet in the warm forming processing chain, 2016 *J. Mater. Process. Technol.* **231** 189-98

[7] Wang H, Luo Y B, Friedman M H and Gao L, Warm forming behavior of high strength aluminum alloy AA7075, 2012 *Trans. Nonferrous Met. Soc. China* **22** 1 1-7

[8] Lee M, Sohn S, Kang C, Suh D and Lee S, Effects of pre-treatment conditions on warm hydroformability of 7075 aluminum tubes, 2004 *J. Mater. Process. Technol.* **155** 1337-43

[9] DiCecco S, Butcher C, Worswick M, Chu C, Boettcher E and Shi C, Application of Time and Position Dependent Localized Necking Criteria to High Strength Aluminum Alloys at Elevated Temperatures, 2016 *Ohlstadt, Germany Forming Technology Forum*

[10] Omer K, Abolhasani A, Kim S, Nikdejad T, Butcher C, Wells M, Esmaeili S and Worswick M, Process parameters for hot stamping of AA7075 and D-7xxx to achieve high performance aged products, 2018 *J. Mater. Process. Technol.* **257** 170-179

[11] DiCecco S, Di Ciano M, Butcher C and Worswick M, Limit Strain Characterization in an Aluminum Die-Quenching Process, 2018 *IOP Conf. Ser. Mater. Sci. Eng.* **418** 1 012035

[12] Kumar M and Ross N, Investigations on the Hot Stamping of AW-7921-T4 Alloy Sheet, 2017 *Adv. Mater. Sci. Eng.*

[13] Garrett R, Lin J and Dean T, Solution Heat Treatment and Cold Die Quenching in Forming AA 6xxx Sheet Components: Feasibility Study, 2005 *Adv. Mater. Res.* **6-8** 673-80

[14] Sotirov N, Simon P, Chimani C, Uffelmann D and Melzer C, Warm Deep Drawability of Peak-Aged 7075 Aluminium Sheet Alloy, 2012 *Key Eng. Mater.* **2**, 504-6

[15] Abolhasani A, Nikdejad T, Omer K, Esmaeili S, Wells M and Worswick M, Effect of the thermal processing history on the age hardening behaviour of 7000-series aluminum alloys, 2017 *San Diego, Ca TSM2017*

[16] Fribourg G, Brechet Y, Deschamps A and Simar A, Microstructure-based modelling of isotropic and kinematic strain hardening in a precipitation-hardened aluminum alloy, 2011 *Acta Mater.* **59** 3621-35

[17] DiCecco S, Di Ciano M, Butcher C and Worswick M, Numerical and experimental investigation of the formability of AA6013-T6, 2017 *J. Phys. Conf. Ser.* **896** 1 012114

[18] International Organization for Standardization, ISO/DIS 12004-2:2008. Metallic Materials - Sheet and Strip - Determination of Forming Limit Curves - Part 2: Deformination of Forming limit Curves in the Laboratory, 2008

[19] Jones E M C and Iadicola M A, A Good Practices Guide for Digital Image Correlation, 2018 *International Digital Image Correlation Society*

[20] ASTM Standard E8/E8M-11, Standard Test Methods for Tension Testing of Metallic Materials, 2011 *ASTM International*

[21] Di Ciano M, DiCecco S, Esmaeili S, Wells M and Worswick M, Coarsening of AA6013-T6 Precipitates During Sheet Warm Forming Applications, 2018 *J. Mater. Eng. Perform.* **27** 939-47

[22] Ward N, Tran A, Abad A, Lee E, Hahn M, Fordan E and Es-Said O, The Effects of Retrogression and Reaging on Aluminum Alloy 2099 (C458), 2011 *J. Mater. Eng. Perform.* **20** 6 989-96

Simulations of plastic deformation by anisotropic hardening yield functions for QP1180

Zhe Chen and Yanshan Lou*

Xi'an Jiaotong University, Xi'an, 710049 Shaanxi, China

*ys.lou@xjtu.edu.cn

Abstract. This research compares the accuracy, computation efficiency and user-friendliness of the Hill48 and three anisotropic hardening models (S-Y 2009, CQN and newly proposed one) for QP1180 steel sheet. Experiments are conducted with dogbone and notched specimens along RD, DD and TD and bulging specimens for strain hardening behavior under equibiaxial tension. These models are calibrated by the experimental data above. The calibrated models are applied to simulate the uniaxial tension and plane strain tension tests for the metal. The simulation time and accuracy of the models are compared for both uniaxial tension and plane strain tension tests to evaluate their accuracy and computation efficiency. The results suggest that the CQN and proposed models are the most accurate among the four models compared, but the computation efficiency of the proposed model can dramatically reduce the numerical simulation time by 20%~40% compared to the CQN model. The comparison shows that the proposed model is the most accurate and is higher in numerical computation efficiency for engineers. Therefore, the proposed model is recommended to be utilized to describe anisotropic hardening behaviors during sheet metal forming processes.

1. Introduction

Accurate prediction of the plastic behaviour of textured sheet metals is of great importance for reliable numerical analysis of sheet metal forming. Numerous yield criteria are proposed to describe the plasticity of sheet metals [1-9]. These functions can provide accurate prediction of the yield surface of different materials. However, these yield functions are based on the assumption of isotropic hardening which assume that the yield surface shape does not change during plastic deformation. The function coefficients in these models are calibrated and determined in advance and will not change during numerical application. Experiments [10] show that hardening behaviors are different for different loading conditions and directions under proportional loading, which is refered to as anisotropic hardening. Various anisotropic hardening models [10-22] are proposed to more accurately describe the anisotropic hardening behaviour of metals by assuming that the yield surface changes during plastic deformation. The newly developed anisotropic hardening functions are more friendly to engineers since the inputs to numerical simulation are strain hardening curves at different loading conditions and directions instead of anisotropic parameters which are generally calibrated by an optimization code.

In this research, an anisotropic hardening model [21] is applied to describe the anisotropic hardening under uniaxial and plane strain tension for QP1180. Tensile tests are conducted with dogbone and notched specimens to characterize strain hardening behaviour under uniaxial and plane strain tension. Bulge tests are also conducted to acquire the equibiaxial tension data. Then the anisotropic hardening model is applied to simulate the experiments. The simulation result is compared with other isotropic hardening models and anisotropic hardening models to evaluate their performance.

2. Experiments

As illustrated in figure 1, three kinds of specimens with different shapes were tested to characterize plastic behaviour of QP1180 steel sheet including (a) dogbone specimen, (b) notched specimen and (c) bulge specimen. The material is 1.2 mm in thickness. Uniaxial tension tests were conducted with dogbone and notched specimens with the universal mechanical testing system. Deformation during experiments was measured by digital image correlation (DIC) as shown in figure 2. The bulge specimens are tested on the hydraulic bulge device with DIC recording as shown in figure 3.

To make sure that the experimental results are reliable, at least four of the dobgone and notched specimen were tested along the rolling direction (RD), diagonal direction (DD) and transverse direction (TD) for good repeatability. The yield stress is determined at 0.2% plastic strain and the hardening curves are described with the Swift-Voce hardening function as below:

$$\sigma(\varepsilon_U) = \alpha K(\varepsilon_U + r_0)^n + (1-\alpha)(A-(A-B)\exp(-C\varepsilon_U)) \tag{1}$$

The material properties of QP1180 are then given in table 1. The force-stroke curves of dogbone and notched specimens are compared along different directions in figure 4, while the flow curves along different loading directions are compared in figure 5 (a). Figure 5 (b) shows the relative anisotropic strength with respect to plastic strain along different loading directions. Experimental results show that anisotropic hardening cannot be ignored for QP1180.

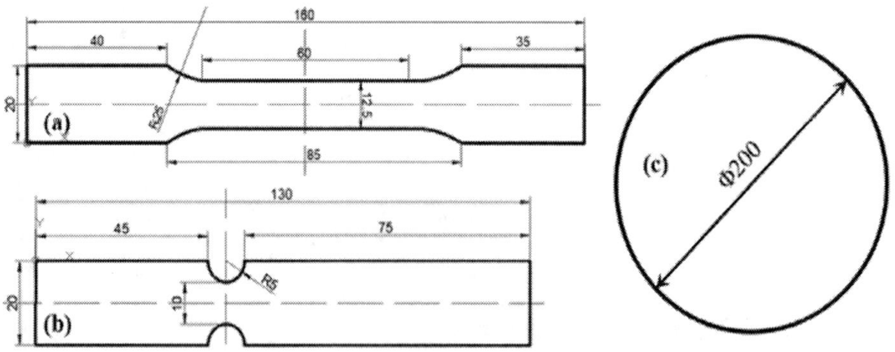

Figure 1. Adopted Specimens: (a) dogbone; (b) notched; and (c) bulging specimen.

Figure 2. Universal mechanical testing system and the XTOP digital image correlation system

Table 1: Mechanical properties of QP1180 (σ_y: yield stress)

stress state	σ_y [MPa]	Strength [MPa]	Elongation [%]	r-value	K [MPa]	e0	n	A [MPa]	B [MPa]	C	α
RD	969	1215	20.4	0.750	2093.0	0.028	0.215	1531.8	973.7	11.04	0.5
DD	956	1192	20.8	0.775	2057.7	0.028	0.216	1498.6	954.0	11.18	0.5
TD	971	1209	17.8	0.698	2084.6	0.029	0.2156	1511.3	980.1	11.48	0.5
EB	1211	NA	NA	1	1769.0	0.0072	0.1122	1603.5	1077.9	8.71	0.5

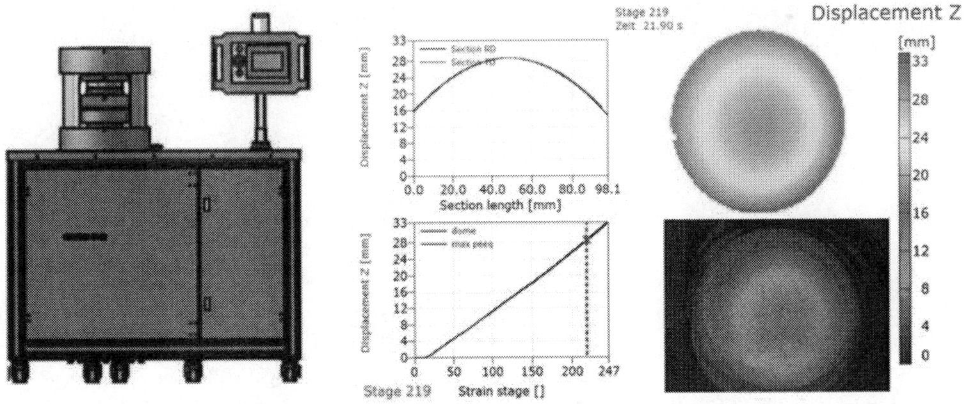

Figure 3. Hydraulic bulge device and bulge Strain nephogram

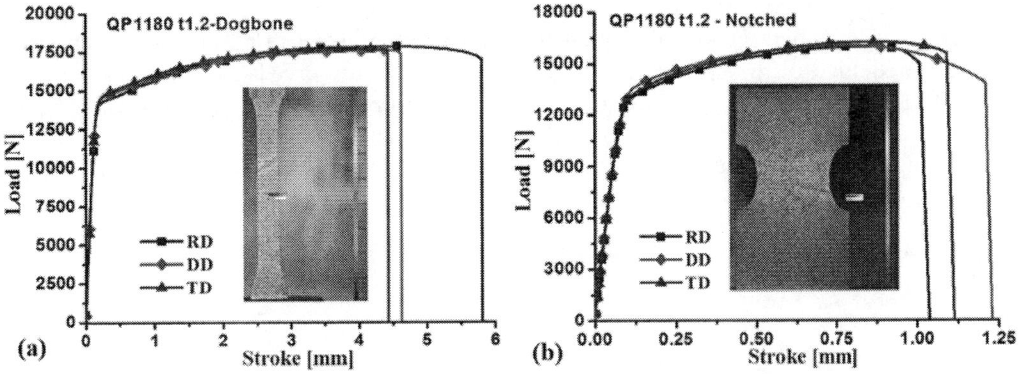

Figure 4. Force-stroke curves of QP1180 for: (a) dogbone specimens; and (b) notched specimens

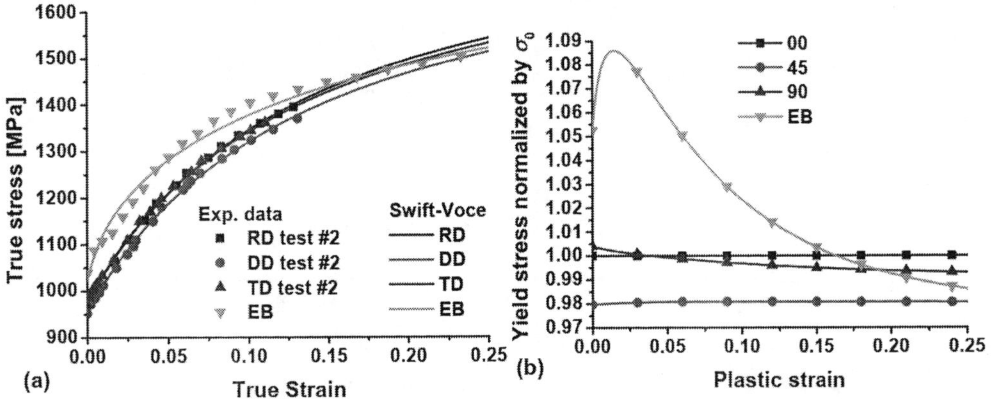

Figure 5. (a) Flow curves of QP1180 along different loading directions; (b) anisotropic evolution

3. Newly proposed anisotropic hardening function

The newly proposed anisotropic hardening model [21] is based on the S-Y 2009 [10] anisotropic hardening model, CQN model [12] and Cazacu'2018 function [7]. It is achieved by coupling the anisotropic hardening model S-Y 2009 and the Cazacu'2018 model. The S-Y 2009 part in this model can describe anisotropic hardening accurately while the Cazacu'2018 part can describe the yield surface for BCC and FCC materials. Besides, the new model is convenient to be applied in numerical simulation since it is in a form of stress invariants. Compared with the S-Y'2009 and CQN models, the new model can differentiate the plastic behaviour between BCC and FCC metals and is high in numerical computation efficiency. Its form is given under plane stress condition as:

$$f_c\left(\boldsymbol{\sigma},\bar{\lambda}\right)=b\left\{\left[\left(\frac{\sigma_{11}}{\sigma_0^8(\bar{\lambda})}-\frac{\sigma_{22}}{\sigma_{90}^8(\bar{\lambda})}\right)(\sigma_{11}-\sigma_{22})+\frac{\sigma_{11}\sigma_{22}-\sigma_{12}\sigma_{21}}{\sigma_b^8(\bar{\lambda})}+\frac{4\sigma_{12}\sigma_{21}}{\sigma_{45}^8(\bar{\lambda})}\right]\left(J_2^3-cJ_3^2\right)\right\}^{\frac{1}{8}} \tag{2}$$

with

$$b=\left(\frac{729}{27-4c}\right)^{\frac{1}{8}} \tag{3}$$

$\bar{\lambda}$ is the plastic compliance. The parameter c is 1.5776 for BCC metals and 2.5116 for FCC metals. Besides, by restricting the parameter c within the range of [-27/5, 3], the yield surface is guaranteed to be convex. It also should be mentioned that the newly proposed model can be extended to spatial stress condition as described by Chen et al. [21].

4. Application of the new model to QP1180

The material properties of QP1180 are characterized by the newly proposed model in Eq. (2) to describe the anisotropic hardening. Figure 6 illustrates the comparison of experimental hardening curves and the theoretical prediction of the new model of QP1180. It can be observed that the predicted hardening curves along RD, DD, TD and EB are identical with experimental results. It is quite natural because these four hardening curves are the input parameters of the new model. Therefore, it can be concluded that the new model can accurately describe the anisotropic hardening along RD, DD, TD and EB directions.

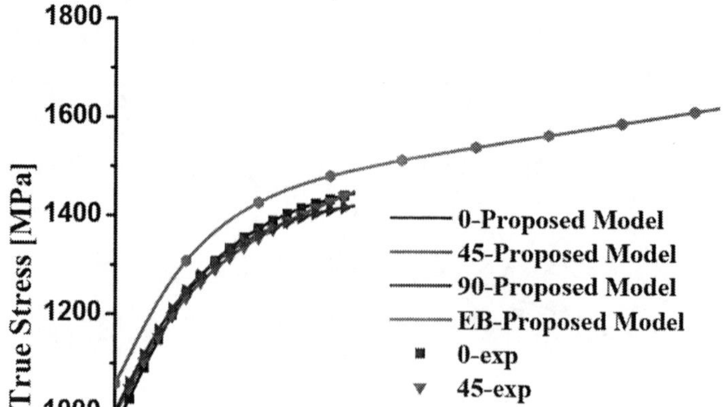

Figure 6. Comparison of true stress-true strain curves under uniaxial tension (RD, DD and TD) and equibiaxial tension between experiments and prediction for QP1180

To provide a more intuitive description of the anisotropic hardening of the material and the performance of the anisotropic hardening models, the evolution of the yield surface predicted by Hill48, S-Y 2009, CQN and the new model and the experimental results are given in figure 7. Obviously, the isotropic hardening Hill48 model can only describe initial yield surface. The prediction error of the Hill48 function increases as the plastic strain increases. For the three anisotropic hardening models, their description remains accurate as the plastic strain increases. It is also noted that the shape of the S-Y 2009 yield surface is larger than the other two yield surfaces around plane strain. Theoretically, it is the advantage of the CQN and proposed models since the Hill48-based S-Y 2009 model describes higher strength under plane strain than experimental result of BCC and FCC metals.

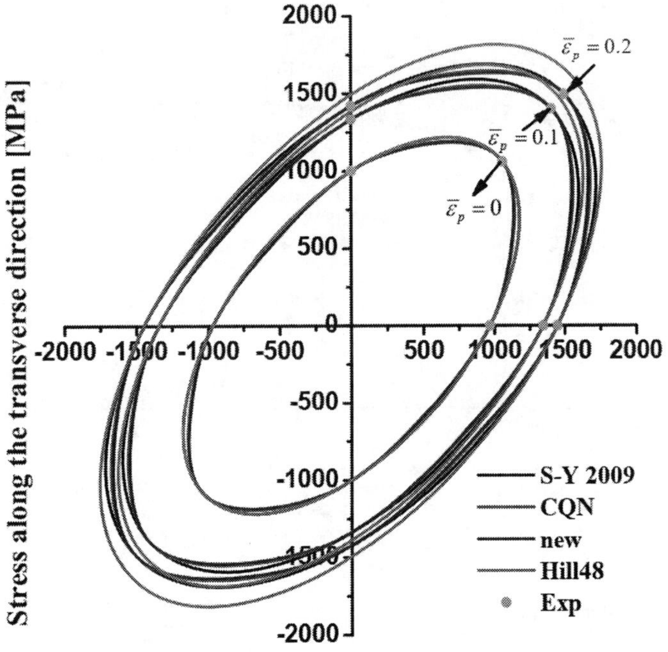

Figure 7. Yield surface evolution prediction compared with experimental results

To evaluate the performance of the proposed model, numerical simulations were carried out with ABAQUS/Explicit with solid elements (C3D8R) for uniaxial tension test of dogbone and notched specimen along RD, DD and TD. The force-stroke curves are extracted and calculated from simulation and are compared with experimental results. Figure 8 illustrates the force-stroke curves of dogbone specimens along RD, DD and TD between experiments and numerical simulation. For the load-stroke prediction, it is obvious that all the yield functions can precisely predict the experimental results along RD. All the three anisotropic hardening models can predict the reaction forces along DD and TD with very high accuracy, but the Hill48 function predicts the force-stroke curves along DD and TD with poor accuracy. It is quite natural because the flow curves along these three directions are the input parameters for the three anisotropic hardening models, but the isotropic hardening Hill48 function only takes the flow curve along RD into account during simulation.

The prediction error of the load-stroke curves is compared in figure 9 for the dogbone specimens. It is apparent that the maximum prediction error is less than 2% for the load-stroke curves of dogbone specimens along three directions for all the three anisotropic hardening models. The comparison shows that all the three anisotropic hardening models are capable of modelling different strain hardening behaviors along three loading directions with high accuracy. The isotropic hardening Hill48 function predicts satisfactory experimental results along RD, but the error exceeds 5% for the dogbone specimens along DD and TD, which indicates the limitation of Hill48 for anisotropic hardening characterization.

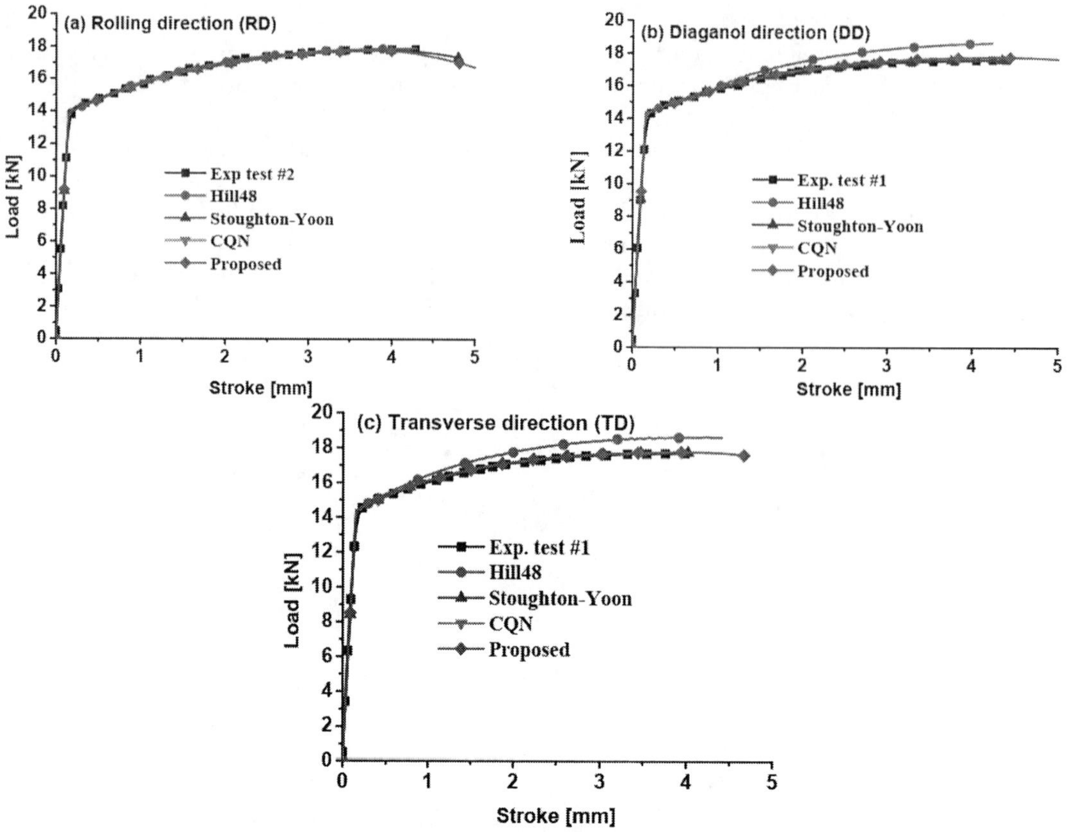

Figure 8. Comparison of force-stroke curves of dogbone specimens between experiments and simulation for QP1180: (a) RD; (b) DD; and (c) TD

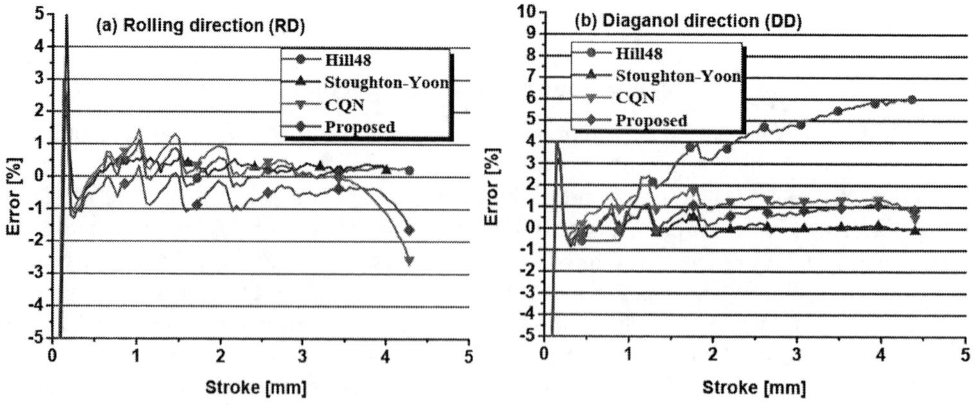

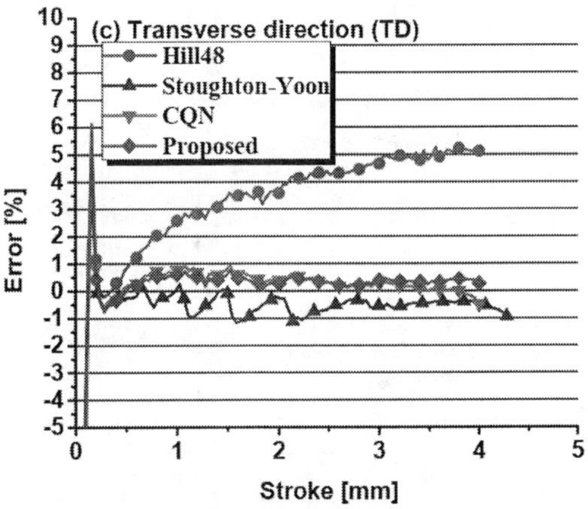

Figure 9. Prediction error of flow curves of the dogbone specimens: (a) RD; (b) DD; and (c) TD

Figure 10 illustrates the force-stroke curves of the notched specimens along RD, DD and TD between experiments and numerical simulation. It is obvious that the Hill48 function over-predicts the reaction forces along all the three directions, especially along DD and TD. The prediction accuracy is highly improved by the three anisotropic hardening models. Among the three anisotropic hardening functions, the S-Y 2009 model slightly over-estimates the reaction force along the three directions and the trend is very similar with the experimental results. The CQN and newly proposed models predict the results with higher accuracy than the Hill48 and S-Y 2009 functions.

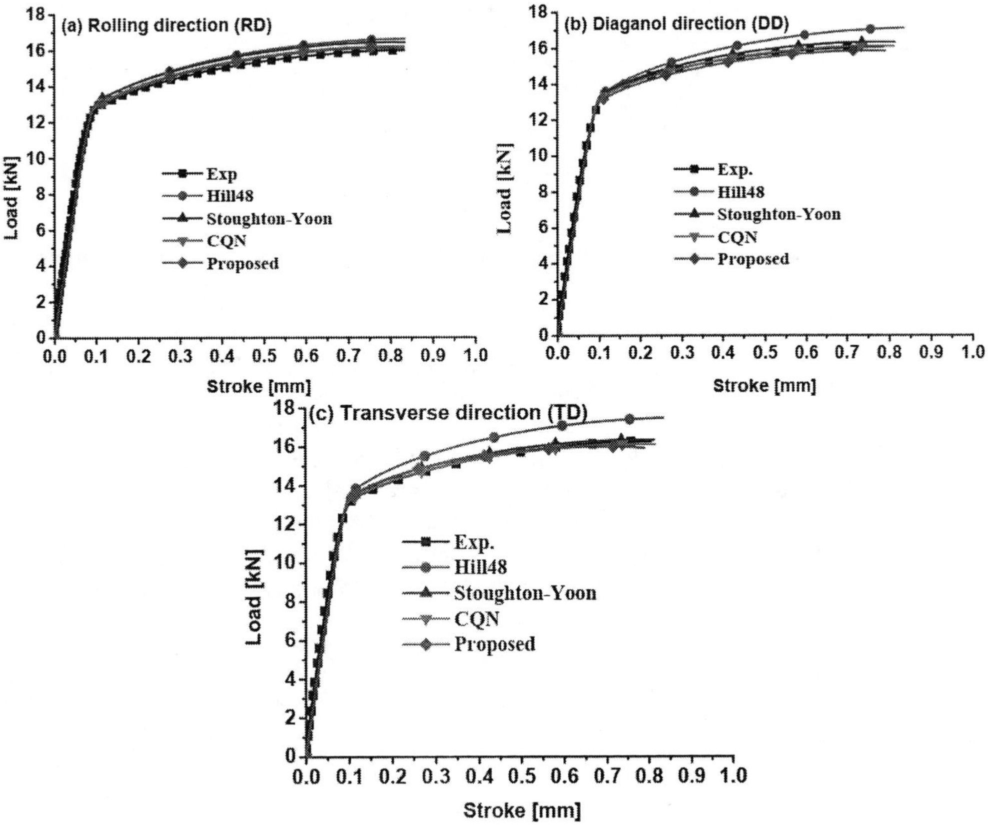

Figure 10. Comparison of force-stroke curves of notched specimens between experiments and simulation: (a) RD; (b) DD; and (c) TD

The prediction error of the load-stroke curves is compared in figure 11 for the notched specimens. It is apparent that the error of the Hill48 function is higher than 5%, which is not acceptable. For the S-Y 2009 model, the maximum error exceeds 3% along RD. For the CQN and proposed models, the error is within 2%. Therefore, the CQN and proposed models provide the best prediction accuracy for the notched specimens along the three directions.

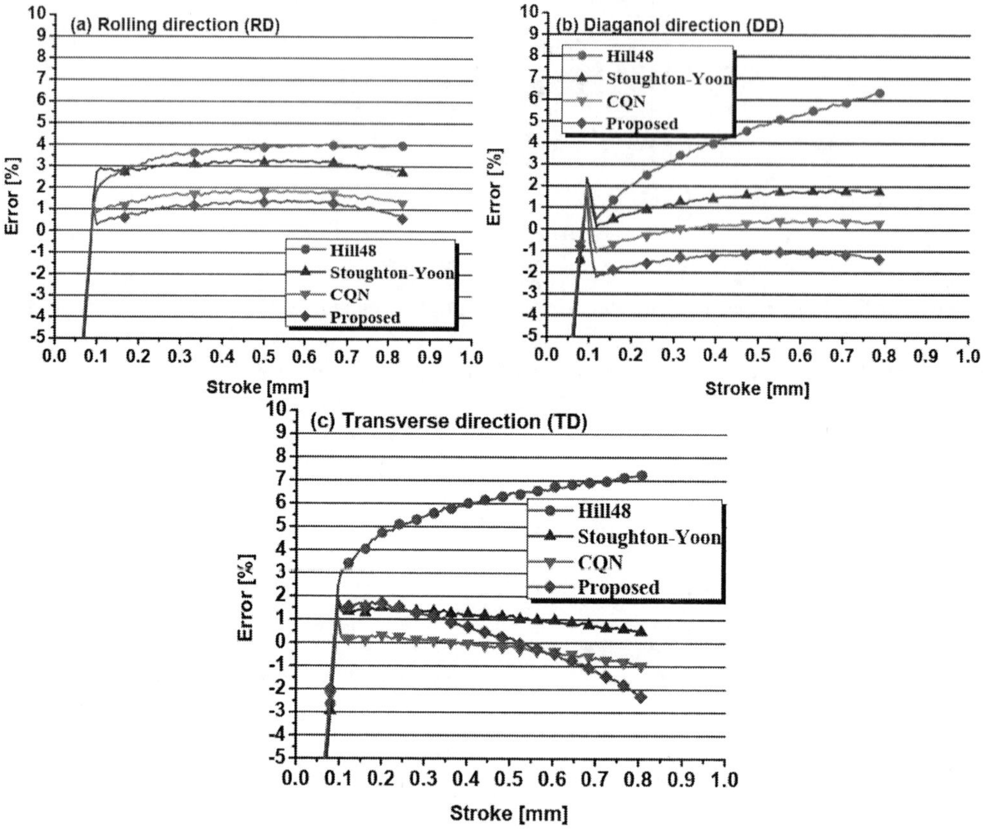

Figure 11 Prediction error of flow curves the notched specimens: (a) RD; (b) DD; and (c) TD

From the accuracy evaluation above, the Hill48 function is the worst in accuracy, the CQN and the proposed models are the best and the S-Y 2009 function is inbetween. Therefore, these functions are further evaluated by comparison of the numerical computation cost. The computation time are listed in table 2 for the notched specimen along TD of these four constitutive models. It is observed that the Hill48 function is the most efficient in numerical computation because of its simplicity. It takes about 40% longer for the Hill48-based S-Y 2009 function compared with the Hill48 function. The CQN model needs the longest time to run the simulation, which is about 3.0 times longer for the notched specimen compared with the Hill48 function. For the proposed model, it takes shorter time to finish the simulation for notched specimens compared to the CQN model, with 48% increase in numerical computation efficiency

Table 2: Simulation time of notched specimens for QP1180 (unit: s)

specimen	Hill48	S-Y 2009	CQN	proposed
Notched TD	201	284	595	307

5. Conclusions

This research characterizes the anisotropic hardening behaviour of QP1180 steel sheet by a newly proposed anisotropic hardening model as well as the Hill48, S-Y 2009 and CQN models. The comparison with experimental results shows that the CQN and new models are the best in predicting accuracy. Further evaluation of computation costs shows that the new model can reduce about 48% of computation time compared with the CQN model. Therefore, the proposed anisotropic hardening model is suggested to characterize the anisotropic behaviour of sheet metal due to its advantages in high accuracy, low computation cost and user-friendliness.

Acknowledgments

The authors would like to acknowledge the financial supports by the National Natural Science Foundation of China (52075423 and U2141214)

References

[1] Hill R 1948 *Proc. Roy. Soc. London Ser. A* **193** 281
[2] Barlat F and Lian K 1989 *Int. J. Plasticity* **5** 51
[3] Barlat F, Lege DJ and Brem JC 1991 *Int. J. Plasticity* **7** 693
[4] Barlat F, Brem JC, Yoon JW, Chung K, Dick RE, Lege DJ, Pourboghrat F, Choi S-H and Chu E 2003 *Int. J. Plasticity* **19** 1297
[5] Lou Y, Huh H and Yoon JW 2013 *Int. J. Mech. Sci.* **66** 214
[6] Yoon JW, Lou Y, Yoon J and Glazoff MV 2014 *Int. J. Plasticity* **56** 184-202
[7] Cazacu O 2018 *Int. J. Solids Struct.* **139** 200
[8] Lou Y and Yoon JW 2018 *Int. J. Plasticity* **101** 125
[9] Lou Y, Zhang S and Yoon JW 2019 *Int. J. Mech. Sci.* **161** 105027
[10] Stoughton TB and Yoon JW 2009 *Int. J. Plasticity* **25** 1777
[11] Cai Z, Diao K, Wu X and Wan M 2016 *Int. J. Mech. Sci.* **107** 43
[12] Lee EH, Stoughton TB and Yoon JW 2017 *Int. J. Plasticity* **99** 120
[13] Li H, Hu X, Yang H and Li L 2016 *Int. J. Plasticity* **82** 127
[14] Li H, Zhang HQ, Yang H, Fu MW and Yang H 2017 *Int. J. Plasticity* **90**, 177
[15] Park N, Stoughton TB and Yoon JW, 2019 *Int. J. Plasticity* **121** 76
[16] Hou Y, Min J, Stoughton TB, Lin J, Carsley JE and Carlson BE 2020 *Int. J. Plasticity* **135**, 102808
[17] Hu Q, Yoon JW, Manopulo N and Hora P 2021 *Int. J. Plasticity* **136**, 102882
[18] Hu Q and Yoon JW 2021 *Int. J. Plasticity* **140**, 102978
[19] Hu Q, Yoon JW and Stoughton TB 2021 *Int. J. Mech. Sci.* **201**, 106467
[20] Hu Q and Yoon JW 2022 *Int. J. Plasticity* **151**, 103214
[21] Chen Z, Wang Y and Yoon JW 2022 *Mech. Mater.* **165**, 104190
[22] Hu Q, Chen J and Yoon JW 2022 *Mech. Mater.* **167**, 104245

AUTHOR INDEX